D0216879

Plant Design and Economics for Chemical Engineers

McGraw-Hill Chemical Engineering Series

Bailey and Ollis
Biochemical Engineering Fundamentals

Bennett and Myers
Momentum, Heat and Mass Transfer

Coughanowr and LeBlanc
Process Systems Analysis and Control

Davis and Davis
Fundamentals of Chemical Reaction Engineering

de Nevers
Air Pollution Control Engineering

de Nevers
Fluid Mechanics for Chemical Engineers

Douglas
Conceptual Design of Chemical Processes

Edgar, Himmelblau, and Lasdon
Optimization of Chemical Processes

Gates, Katzer, and Schuit
Chemistry of Catalytic Processes

King
Separation Processes

Luyben
Process Modeling, Simulation, and Control
for Chemical Engineers

Marlin
Process Control: Designing Processes and Control
Systems for Dynamic Performance

McCabe, Smith, and Harriott
Unit Operations of Chemical Engineering

Middleman and Hochberg
Process Engineering Analysis in Semiconductor Device
Fabrication

Perry and Green
Perry's Chemical Engineers' Handbook

Peters, Timmerhaus, and West
Plant Design and Economics for Chemical Engineers

Reid, Prausnitz, and Poling
Properties of Gases and Liquids

Smith, Van Ness, and Abbott
Introduction to Chemical Engineering Thermodynamics

Treybal
Mass Transfer Operations

Plant Design and Economics for Chemical Engineers

Fifth Edition

Max S. Peters
Klaus D. Timmerhaus
Ronald E. West
University of Colorado

Boston Burr Ridge, IL Dubuque, IA Madison, WI New York San Francisco St. Louis
Bangkok Bogotá Caracas Kuala Lumpur Lisbon London Madrid Mexico City
Milan Montreal New Delhi Santiago Seoul Singapore Sydney Taipei Toronto

McGraw-Hill Higher Education

A Division of The McGraw-Hill Companies

PLANT DESIGN AND ECONOMICS FOR CHEMICAL ENGINEERS
FIFTH EDITION

Published by McGraw-Hill, a business unit of The McGraw-Hill Companies, Inc., 1221 Avenue of the Americas, New York, NY 10020.

This book is printed on acid-free paper.

International 2 3 4 5 6 7 8 9 0 QFR/QFR 1 5 4 3 2 1
Domestic 8 9 0 QFR/QFR 1 5 4 3 2 1

ISBN-13: 978-0-07-239266-1
ISBN-10: 0-07-239266-5
ISBN-13: 978-0-07-119872-1 (ISE)
ISBN-10: 0-07-119872-5 (ISE)

Publisher: *Elizabeth A. Jones*
Sponsoring editor: *Suzanne Jeans*
Developmental editor: *Kate Scheinman*
Marketing manager: *Sarah Martin*
Senior project manager: *Kay J. Brimeyer*
Production supervisor: *Kara Kudronowicz*
Senior media project manager: *Tammy Juran*
Coordinator of freelance design: *Rick D. Noel*
Cover designer: *John Rokusek/Rokusek Design*
Lead photo research coordinator: *Carrie K. Burger*
Compositor: *Interactive Composition Corporation*
Typeface: *10.5/12 Times Roman*
Printer: *Quad/Graphics Fairfield, PA*

Library of Congress Cataloging-in-Publication Data

Peters, Max Stone, 1920–.
Plant design and economics for chemical engineers. — 5th ed. / Max S. Peters, Klaus D. Timmerhaus, Ronald E. West.
p. cm. — (McGraw-Hill chemical engineering series)
Includes bibliographical references and index.
ISBN 0–07–239266–5 (acid-free paper)
1. Chemical plants—Design and construction. I. Timmerhaus, Klaus D. II. West, Ronald E. (Ronald Emmett), 1933–. III. Title. IV. Series.

TP155.5 .P4 2003
660'.28—dc21 2002032568
CIP

INTERNATIONAL EDITION ISBN 0–07–119872–5

www.mhhe.com

ABOUT THE AUTHORS

Max S. Peters is currently professor emeritus of chemical engineering and dean emeritus of engineering at the University of Colorado at Boulder. He received his B.S. and M.S. degrees in chemical engineering from Pennsylvania State University, worked for Hercules Power Company and Treyz Chemical Company, and returned to Penn State for his Ph.D. Subsequently he joined the faculty of the University of Illinois, and later he came to the University of Colorado as dean of the College of Engineering and Applied Science and professor of chemical engineering. He relinquished the position of dean in 1978 and became emeritus in 1987.

Dr. Peters has served as president of the American Institute of Chemical Engineers, as a member of the board of directors for the Commission on Engineering Education, as chairman of the President's Committee on the National Medal of Science, and as chairman of the Colorado Environmental Commission. A fellow of the American Institute of Chemical Engineering, Dr. Peters is the recipient of the George Westinghouse and Lamme Award of the American Society of Engineering Education, the Award of Merit of the American Association of Cost Engineers, and the Founders and W. K. Lewis Award of the American Institute of Chemical Engineers. He is a member of the National Academy of Engineering.

Klaus D. Timmerhaus is currently President's Teaching Scholar at the University of Colorado at Boulder. He received his B.S., M.S., and Ph.D. degrees in chemical engineering from the University of Illinois. After serving as a process design engineer for Standard Oil of California Research Corporation, Dr. Timmerhaus joined the chemical engineering department of the University of Colorado, College of Engineering. He was subsequently appointed associate dean of the College of Engineering and director of the Engineering Research Center. This was followed by a term as chairman of the chemical engineering department. The author's extensive research publications have been primarily concerned with cryogenics, energy, and heat and mass transfer. He has edited 25 volumes of *Advances in Cryogenic Engineering* and coedited more than 30 volumes in the *International Cryogenics Monograph Series.*

He is past president of the American Institute of Chemical Engineers, past president of the Society of Sigma Xi, and past president of the International Institute of Refrigeration; and he has held offices in the American Institute of Chemical Engineers (AIChE), the Cryogenic Engineering Conference, the International Cryocooler Conference, Society of Sigma Xi, American Astronautical Society, American Association for the Advancement of Science (AAAS), American Society for Engineering Education (ASEE)—Engineering Research Council, Accreditation Board for Engineering and Technology, and National Academy of Engineering.

A fellow of AIChE and AAAS, Dr. Timmerhaus has received the ASEE George Westinghouse and Fred Merryfield Design Award, the AIChE Alpha Chi Sigma Award, the AIChE W. K. Lewis Award, the AIChE Founders Award, the SAE Ralph Teeter Award, the USNC/IIR W. T. Pentzer Award, NSF Distinguished Service Award, University of Colorado Stearns Award, Colorado Professor of the Year Award, and Samuel C. Collins Award of the Cryogenic Engineering Conference. He is a member of the National Academy of Engineering and the Austrian Academy of Science.

Ronald E. West is currently professor emeritus of chemical engineering after serving for 38 years as a member of the faculty in the department of chemical engineering at the University of Colorado at Boulder. In that capacity, he directed both the undergraduate laboratory program and the senior plant design course for more than 25 years. He received his B.S., M.S., and Ph.D. degrees in chemical engineering from the University of Michigan. His professional work has been mainly in the areas of water pollution control and renewable energy technology. Dr. West is a member of the American Institute of Chemical Engineers and the American Solar Energy Society.

CONTENTS

PREFACE

Process design is probably one of the most creative activities enjoyed by chemical engineers. Throughout the activity, there are many opportunities to develop imaginative new chemical or biochemical processes or to introduce changes in existing processes that could alter the environmental or economic aspects of the process. This activity clearly involves creative problem-solving abilities on the part of the chemical engineer, often in a team effort, in which basic knowledge of chemical engineering and economic principles is applied.

Chemical engineering design of new processes and the expansion and revision of existing processes require the use of engineering principles and theories combined with a practical understanding of the limits imposed by environmental, safety, and health concerns. However, the development of a new process from concept evaluation to profitable reality can become rather complex since process design problems are open-ended. Thus, there may be many solutions that are profitable even when not optimal.

Advances over the past two decades in the level of understanding of chemical principles, combined with the availability of new techniques and computer-based tools, have led to an increased degree of sophistication that now can be applied to the design of chemical and biochemical processes. This fifth edition takes advantage of this widened spectrum of chemical engineering knowledge with special emphasis on the engineering and economic principles involved in the design of processes that meet a societal or industrial need.

The purpose of this textbook is to present economic and design principles as applied in chemical engineering processes and operations. No attempt is made to train the reader as a skilled economist, and, obviously, it would be impossible to present all the possible ramifications of the multitude of different plant designs. Instead, the goal has been to give a clear concept of the important principles and general methods. The subject matter and manner of presentation are such that the book should be of value to advanced chemical engineering undergraduates, graduate students, and practicing engineers. The information should also be of interest to administrators, operation supervisors, and research or development workers in the process industries.

Chapters 1 through 3 provide an overall analysis of the major factors involved in process design. The use of computer software in process design is described early in the text as a separate chapter to introduce the reader to this important topic, with the understanding that this tool will be useful throughout the text. The various costs involved in industrial processes, capital investments and investment returns, cost estimation, cost accounting, optimum economic design methods, and other subjects dealing with economics are covered both qualitatively and quantitatively in Chaps. 6 through 9. A guide for selecting materials of construction is presented in Chap. 10, followed by a discussion on report writing in Chap. 11. The last four chapters provide extensive design procedures for many of the equipment items utilized in process design. Generalized subjects, such as waste disposal, structural design, and equipment fabrication, are included along with design methods for different types of process equipment. Basic cost data and cost correlations are also presented for use in making cost estimates.

Illustrative examples and sample problems are used extensively in the text to illustrate the applications of the principles to practical situations. Problems are included at the end of the most chapters to give readers a chance to test their understanding of the material. Practice session problems, as well as longer design problems of varying degrees of complexity, are included in App. C. Suggested recent references are presented as footnotes to show the reader where additional information can be obtained. Earlier references are listed in prior editions of this book.

A large amount of cost data is presented in tabular and graphical form. The table of contents for the book lists chapters where equipment cost data are presented, and additional cost information on specific items of equipment or operating factors can be located by reference to the subject index. To simplify use of the extensive cost data given in this book, all cost figures are referenced to the *Chemical Engineering* plant cost index of 390.4 applicable for January 1, 2002. The McGraw-Hill website http://www.mhhe.com/peters-timmerhaus provides the mathematical cost relations for all the graphical cost data provided in the text integrated with economic evaluation routines. Because exact prices can be obtained only by direct quotations from manufacturers, caution should be exercised in the use of the data for other than approximate cost estimation purposes.

This textbook is suitable for either a one- or two-semester course for advanced undergraduate chemical engineers. A one-semester course covering plant and process design would utilize Chaps. 1 through 11. A second-semester course involving equipment design would utilize Chaps. 12 through 15. For either approach it is assumed that the student has a background in stoichiometry, thermodynamics, and chemical engineering principles as taught in normal undergraduate degree programs in chemical engineering. Explanations of the development of various design equations and methods are presented. The book provides a background of design and economic information with a large amount of quantitative interpretation so that it can serve as a basis for further study to develop a complete understanding of the general strategy of process engineering design.

Although nomographs, simplified equations, and shortcut methods are included, every effort has been made to indicate the theoretical background and assumptions for these relationships. The true value of plant design and economics for the chemical engineer is not found merely in the ability to put numbers into an equation and solve for a final answer. The true value is found in obtaining an understanding of the reasons *why* a given calculation method gives a satisfactory result. This understanding gives the engineer the confidence and ability necessary to proceed when new problems are encountered for which there are no predetermined methods of solution. Thus, throughout the study of plant design and economics, the engineer should always attempt to understand the assumptions and theoretical factors involved in the various calculation procedures, particularly when computer software is being used for the first time.

Because applied economics and plant design deal with practical applications of chemical engineering principles, a study of these subjects offers an ideal way for tying together the entire field of chemical engineering. The final result of a plant design may be expressed in dollars and cents, but this result can only be achieved through the application of various theoretical principles combined with industrial and practical knowledge. Both theory and practice are emphasized in this book, and aspects of all phases of chemical engineering are included.

The authors are indebted to the many industrial firms and individuals who have supplied information and comments on the material presented in this edition. The authors also express their appreciation to the following reviewers who have supplied constructive criticism and helpful suggestions on the presentation for this edition: Luke Achenie, University of Connecticut; Charles H. Barron, Clemson University; James R. Beckman, Arizona State University; David C. Drown, University of Idaho; Steinar Hauan, Carnegie Mellon University; Marianthi Ierapetritou, Rutgers University; Jan A. Puszynski, South Dakota School of Mines and Technology; Johannes Schwank, University of Michigan; Thomas L. Sweeney, Ohio State University; Eric J. Thorgerson, Northeastern University; and Bruce Vrana, Dupont Engineering Technology.

Acknowledgment is made to Barr Halevi, HBarr, Inc., for his many contributions to the text and his preparation of Chap. 5 and much of Chap. 13. Special thanks are also expressed to L. T. Fan, Kansas State University, and F. Friedler, Veszprem University, for their preparation of the "Algorithmic Flowsheet Generation" section in Chap. 4. The assistance in typing of the manuscript by Cynthia Ocken, CSLR University of Colorado, and Ellen Romig, chemical engineering department, University of Colorado, is greatly appreciated.

Max S. Peters
Klaus D. Timmerhaus
Ronald E. West

PROLOGUE

As the United States moves toward acceptance of the International System of units (SI), it is particularly important for the design engineer to be able to think in both SI units and U.S. Customary System (USCS) units. For this reason, a mixture of SI and USCS units will be found in this text. For those readers who are not familiar with all the rules and conversions for SI units, App. A of this text presents the necessary information. This appendix gives descriptive and background information for the SI units alongwith lists of conversion factors presented in various forms which should be of special value for chemical engineering usage.

The use of computer software is an important part of process design and economic evaluation. The McGraw-Hill website for this textbook, http://www.mhhe.com/peters-timmerhaus, provides a number of useful aids to both the instructor and students using this textbook. These include a restricted-access instructor solutions manual, graphical files useful to the instructor, and a set of website links useful to instructors and students. Also available on the textbook website is an electronic equipment costing and process economic evaluation spreadsheet file. This downloadable spreadsheet contains a complete range of process equipment costing relations, corresponding to the complete set of costing charts presented in the textbook covering a wide selection of process equipment costs, as well as the economic evaluation procedure tables corresponding to those presented in the text. This downloadable spreadsheet can therefore be used to facilitate process economic evaluation throughout the course and beyond.

CHAPTER

1

Introduction

A successful chemical engineer in this modern age of national and international competition needs more than a knowledge and full understanding of the fundamental science and the related engineering concepts of material and energy balances, thermodynamics, reaction kinetics, heat transfer, mass transfer, and computer technology. The engineer must also have the ability to apply this knowledge to practical situations to initiate and develop new or improved processes and products that will be beneficial to society. However, in achieving this goal, the chemical engineer must recognize the economic, environmental, and ethical implications that are involved in such developments and proceed accordingly.

Chemical engineering design of new chemical or biochemical processes and the expansion or revision of existing processes require the use of engineering principles and theories combined with a practical realization of the limits imposed by environmental, safety, and health concerns. Development of a new process or plant from concept evaluation to profitable reality often is a very complex operation. It is important to keep in mind that process design problems are open-ended and thus may have many solutions that are profitable even when not entirely optimal, yet meet the design constraints noted above.

CHEMICAL ENGINEERING PLANT DESIGN

As used in this text, the general term *plant design* includes all engineering aspects involved in the development of a new, modified, or expanded commercial process in a chemical or biochemical plant. In this development, the chemical engineer will be making economic evaluations of new processes, designing individual pieces of equipment for the proposed new venture, or developing a plant layout for coordination of the overall operation. Because of these many design duties, often the chemical engineer is referred to here as a *design engineer*. On the other hand, a chemical engineer specializing in the economic aspects of the design is often referred to as a *cost engineer*. In many instances, the term *process engineering* is used in connection with economic evaluation and general economic analyses of commercial processes, while *process design* refers to the actual design of the equipment and facilities necessary for providing the desired products and services. Similarly, the meaning of plant design is limited by some engineers to items related directly to the complete plant, such as plant layout, general service facilities, and plant location.

The purpose of this text is to present the major aspects of plant design as related to the overall design project. Although one individual cannot be an expert in *all* the phases involved in plant design, it is necessary to be acquainted with the general problems and approaches in each phase. The process engineer may not be connected directly with the final detailed design of the equipment, and the designer of the equipment may have little influence on a decision by management as to whether a given return on an investment is adequate to justify construction of the proposed process or plant. Nevertheless, if the overall design project is to be successful, close teamwork is necessary among the various groups of engineers working on various phases of the project. The most effective teamwork and coordination of efforts are obtained when each of the engineers in the specialized groups is aware of the many functions in the *overall* design project.

GENERAL OVERALL DESIGN CONSIDERATIONS

The development of the overall design project involves many different design considerations. Failure to include these considerations in the overall design project may, in many instances, alter the entire economic situation so significantly as to make the venture unprofitable.

Some of the factors that require particular attention in the development of a process or complete plant are those associated with various aspects of environmental protection, as well as the safety and health needs of plant personnel and the public. Other factors that affect the profitability of a process design include plant location, plant layout, plant operation and control, utility requirements, structural design, storage and buildings, materials handling, and patent considerations. Because of their importance, these general overall design considerations are individually reviewed and detailed in Chap. 2.

Process Design Development

The development of a process design, as presented in Chap. 3, involves many different steps. The first step, of course, must be the inception of the basic idea. This idea may originate in the sales department, as a result of a customer request, or to meet a competing product. It may occur spontaneously to someone who is acquainted with the aims and needs of a particular company, or it may be the result of an orderly research program or an offshoot of such a program. The operating division of the company may develop a new or modified chemical product, generally as an intermediate in the final product. The engineering department of the company may originate a new process or modify an existing process to create new products. In all these possibilities, if the initial analysis indicates that the idea may have possibilities of developing into a worthwhile project, a preliminary research or investigation program is initiated. Here, a general survey of the possibilities for a successful process is made by considering the physical and chemical operations involved as well as the economic aspects. Next comes the process research phase, including preliminary market surveys, laboratory-scale experiments, and production of research samples of the final product. When the potentialities of the process are fairly well established to meet the economic goals of the company, the project is ready for the development phase. At this point, a pilot plant or a commercial development plant may be constructed. A pilot plant is a small-scale replica of the full-scale final plant, while a commercial development plant is usually constructed from odd pieces of equipment that are already available and is not meant to duplicate the exact setup to be used in the full-scale plant.

Design data and other process information are obtained during the development stage. This information is used as the basis for carrying out the next phase of the design project. A complete market analysis is made, and samples of the final product are sent to prospective customers to determine if the product is satisfactory and if there is a reasonable sales potential. Capital cost estimates for the proposed process or plant are made. Probable returns on the required investment are determined, and a complete cost-and-profit analysis of the process is developed.

Before the final process design starts, company management normally becomes involved to decide whether significant capital funds will be committed to the project. It is at this point that the engineers' preliminary design work along with the oral and written reports which are presented becomes particularly important because these documents will provide the primary basis on which management will decide whether further funds should be provided for the project. When management has made a firm decision to proceed and provide sufficient capital funds for the project, the engineering then involved in further work on the project is known as *capitalized engineering* while that which has gone on before while the consideration of the project was in the development stage is often referred to as *expensed engineering*. This distinction is used for tax purposes to allow capitalized engineering costs to be amortized over several years.

If the economic picture is still satisfactory, the final process design phase is ready to begin. All the design details are worked out in this phase including controls, services, piping layouts, firm price quotations, specifications and designs for individual pieces of equipment, and all the other design information necessary for the construction of the final plant. A complete construction design is then made with elevation drawings, plant layout arrangements, and other information required for the actual construction of the plant. The final stage consists of procurement of the equipment, construction of the plant, start-up of the plant, overall improvements in the operation, and development of standard operating procedures to provide the best possible results.

The development of a design project proceeds in a logical, organized sequence requiring more and more time, effort, and expenditure as one phase leads into the next. It is extremely important, therefore, to stop and analyze the situation carefully before proceeding with each subsequent phase. Many projects are discarded as soon as the preliminary investigation or research on the original idea is completed. The engineer working on the project must maintain a realistic and practical attitude while advancing through the various stages of a design project and must not be swayed by personal interest and desires when deciding if further work on a particular project is justifiable. Remember, if the engineer's work is continued through the various phases of a design project, it will eventually end up in a proposal that money be invested in the process. If no tangible return can be realized from the investment, the proposal will be turned down. Therefore, the engineer should have the ability to eliminate unprofitable ventures before the design project approaches a final-proposal stage.

The design steps outlined above are summarized in Table 1-1. Note that not all the design steps may be necessary, particularly in small or simple design projects. The order of the design steps may also be altered to better meet the design approach

Table 1-1 Typical design steps for chemical and biochemical processes

1. Recognize a societal or engineering need.
   a. Make a market analysis if a new product will result.
2. Create one or more potential solutions to meet this need.
   a. Make a literature survey and patent search.
   b. Identify the preliminary data required.
3. Undertake preliminary process synthesis of these solutions.
   a. Determine reactions, separations, and possible operating conditions.
   b. Recognize environmental, safety, and health concerns.
4. Assess profitability of preliminary process or processes (if negative, reject process and create new alternatives).
5. Refine required design data.
   a. Establish property data with appropriate software.
   b. Verify experimentally, if necessary, key unknowns in the process.
6. Prepare detailed engineering design.
   a. Develop base case (if economic comparison is required).
   b. Prepare process flowsheet.
   c. Integrate and optimize process.
   d. Check process controllability.
   e. Size equipment.
   f. Estimate capital cost.
7. Reassess the economic viability of process (if negative, either modify process or investigate other process alternatives).
8. Review the process again for environmental, safety, and health effects.
9. Provide a written process design report.
10. Complete the final engineering design.
    a. Determine equipment layout and specifications.
    b. Develop piping and instrumentation diagrams.
    c. Prepare bids for the equipment or the process plant.
11. Procure equipment (if work is done in-house).
12. Provide assistance (if requested) in the construction phase.
13. Assist with start-up and shakedown runs.
14. Initiate production.

followed by a design group in a company. For example, the retrofitting of a process to meet more stringent environmental regulations may require the replacement of a compressor utilizing seals that provide significantly lower vapor losses with proper maintenance than those in the present unit. In this case, the design may also be concerned with establishing a monitoring system that can automatically identify any increases in emissions from the compressor and supporting auxiliaries.

Development of a complex process that incorporates a new, but commercially unproven technology will probably require additional design steps besides those listed in Table 1-1. Besides preparing a base case for economic comparison, it may be necessary to perform some pilot plant testing to investigate the viability of using a new feedstock, a new catalyst, or a different type of membrane. Dynamic simulation may be required to assess the plantwide controllability or the development of a safe start-up procedure. Depending on the situation, if additional design steps are necessary, they can be inserted at the appropriate location in the table.

Regardless of the simplicity or complexity of a creative chemical or biochemical design, Table 1-1 indicates that the design process involves a wide variety of skills. Among these are research, market analyses, computer simulation, software programming, equipment design, cost estimation, profitability analysis, and technical communications. In fact, the services of a chemical engineer as a design engineer are needed in each step of the design procedure, either in a central creative role or as a key adviser.

Flowsheet Development

Once a need has been identified, the chemical engineer creates one or more solutions to meet this need, possibly entailing different feeds and different intermediates to obtain the same products. Since these potential solutions generally require quite different process steps and unique operating conditions, the chemical engineer must establish separate flowsheets or road maps for each solution. Preliminary process synthesis for each solution begins with the designer creating flowsheets involving just the reaction, separation, and temperature change and pressure change operations, and selecting process equipment in a task integration step. Only those simplified flowsheets that provide a favorable gross profit are retained; the others are rejected. Thus, detailed work on any potential process is avoided when the projected cost of the feed materials exceeds those of the products and by-products. These various steps are described in detail in Chap. 4, where both the traditional flowsheet development, utilizing heuristic guidelines, and the more modern flowsheet development, based on algorithmic approaches, are addressed.

To evaluate the most promising flowsheet alternatives, the design engineer generally develops *base-case designs* for each of these alternatives. This is accomplished by creating a detailed process flowsheet with a listing of steady-state material and energy balances and a designation of major equipment items. The material and energy balances are generally performed with the use of a computer-aided process simulator. Attempts are made to improve the design of the process units and to achieve more efficient process integrations through the use of separation train synthesis, heat and power integration, and second-law analysis. The results from using these algorithmic methods permit the design engineer to compare the base case with other promising alternatives, and often identify flowsheets that should be developed with, or in place of, the base-case design.

After the detailed process flowsheet has been completed, an assessment of the controllability of the process is generally made, beginning with the qualitative synthesis of control structures for the entire flowsheet. Measures are available that can be used, before final sizing of equipment, to assess the ease of controlling the process and the degree to which the design is resilient to possible process disturbances. Once control systems are added to the process, dynamic simulations can be carried out to confirm the earlier projections.

Computer-Aided Design

Computing hardware and software have become indispensable tools in process and plant design. The capabilities provided by computers for rapid calculations, large storage, and logical decisions plus the available technical and mathematical software permit design engineers to examine the effect that various design variables will have on the process or plant design and to be able to do this more rapidly than was required in the past to complete a single design by hand calculation. The emphasis of the design engineer has therefore shifted from problem solving to one of conceiving, examining, and implementing alternative solutions that meet a specified need. Chapter 5 provides a brief introduction to computer software useful for process and plant design and to approaches in design that take advantage of the capabilities of computers and appropriate software.

Process simulators are often useful in generating databases because of their extensive data banks of pure-component properties and physical property correlations for ideal and nonideal mixtures. When such data are unavailable, simulation programs can regress experimental data obtained in the laboratory or pilot plant for empirical or theoretical curve fitting. The computer-aided process simulator is also a useful tool in preparing material and energy balances.

Spreadsheet software is another tool used by all chemical engineers because of the availability of personal computers, ease of use, and adaptability to many types of design problems. This software is especially useful for mass and energy balances, approximate sizing of equipment, cost estimating, and economic analysis steps of process design. It is less useful for more detailed equipment design since the complex algorithms usually necessary for this step are somewhat more difficult to incorporate into a spreadsheet.

Cost Estimation

As soon as the final process design stage is completed, it becomes possible to make accurate cost estimations because detailed equipment specifications and definite plant facility information are available. Direct price quotations based on detailed specifications can then be obtained from various equipment vendors. However, as mentioned earlier, no design project should proceed to the final stages before costs are considered, and cost estimates should be made throughout all the early stages of the design, even when complete specifications are not available. Evaluation of costs in the preliminary design phases is referred to as *predesign cost estimation*. Such estimates should be capable of providing a basis for company management to decide whether further capital should be invested in the project.

The chemical engineer (or cost engineer) must be certain to consider all possible factors when making a cost analysis. Fixed costs, direct production costs for raw materials, labor, maintenance, power, and utilities must all be included along with costs for plant and administrative overhead, distribution of the final products, and other miscellaneous items.

Chapter 6 presents many of the special techniques that have been developed for making predesign cost estimations. Labor and material indices, standard cost ratios, and special multiplication factors are examples of information used in making design estimates of costs. The final test as to the validity of any cost estimation can come only when the completed plant has been put into operation. However, if the design engineer is well acquainted with the various estimation methods and their accuracy, it is possible to make remarkably close cost estimations even before the final process design gives detailed specifications.

Profitability Analysis of Investments

A major function of the directors of a chemical or biochemical company is to maximize the long-term profit to the owners or the stockholders. A decision to invest in fixed facilities carries with it the burden of continuing interest, insurance, taxes, depreciation, manufacturing costs, etc., and reduces the fluidity of the company's future actions. Capital investment decisions, therefore, must be made with great care. Chaps. 6 and 7 present guidelines for making these capital investment decisions.

Money, or any other negotiable type of capital, has a time value. When a manufacturing enterprise invests money, it expects to receive a return during the time the money is being used. The amount of return demanded usually depends on the degree of risk that is assumed. Risks differ between projects which might otherwise seem equal on the basis of the best estimates of an overall plant design. The risk may depend upon the process used, whether it is well established or a complete innovation; on the product to be made, whether it is a staple item or a completely new product; on the sales forecast, whether all sales will be outside the company or whether a significant fraction is internal; etc. Since means for incorporating different levels of risk into *profitability* forecast are not too well established, the most common methods are to increase the minimum acceptable *rate of return* for the riskier projects.

The time value of money has been integrated into investment evaluation systems by means of *compound interest* relationships. Dollars, at different times, are given different degrees of importance by means of compounding or discounting at some preselected compound interest rate. For any assumed interest value of money, a known amount at any one time can be converted to an equivalent but different amount at a different time. As time passes, money can be invested to increase at the interest rate. If money is needed for investment in the future, the present value of that investment can be calculated by discounting from the time of investment back to the present at the assumed interest rate.

Expenses for various types of taxes and insurance can materially affect the economic situation for any industrial process. Because various taxes can account for a major portion of a company's net earnings, it is essential that the chemical engineer be conversant with the fundamentals of taxation. For example, income taxes apply differently to projects with different proportions of fixed and working capital. Profitability, therefore, should be based on income after taxes. Insurance costs, on the other hand, are normally only a small part of the total operational expenditure of an industrial enterprise; however, before any operation can be carried out on a sound economic basis, it is necessary to determine the insurance requirements to provide adequate coverage against unpredictable emergencies or developments.

Since all physical assets of a commercial facility decrease in value with age, it is normal practice to make periodic charges against earnings so as to distribute the first cost of the facility over its expected service life. This *depreciation* expense as detailed in Chap. 7, unlike most other expenses, entails no current outlay of cash. Thus, in a given accounting period, a firm has available, in addition to the net profit, additional funds corresponding to the depreciation expense. This cash is *capital recovery,* a partial regeneration of the first cost of the physical assets.

Income tax laws permit recovery of funds by one accelerated depreciation schedule as well as by straight-line methods. Since cash flow timing is affected, the choice of depreciation method affects profitability significantly. Depending on the ratio of depreciable to nondepreciable assets involved, two projects which look equivalent before taxes, or rank in one order, may rank entirely differently when considered after taxes. Although cash costs and sales values may be equal on two projects, their reported net incomes for tax purposes may be different, and one will show a greater net profit than the other.

Optimum Design

In almost every case encountered by a chemical engineer, several alternative methods can be used for any given process or operation. For example, formaldehyde can be produced by catalytic dehydrogenation of methanol, by controlled oxidation of natural gas, or by direct reaction between CO and H_2 under special conditions of catalyst, temperature, and pressure. Each of these processes contains many possible alternatives involving variables such as the gas-mixture composition, temperature, pressure, and choice of catalyst. It is the responsibility of the chemical engineer, in this case, to choose the best process and to incorporate into the design the equipment and methods that will give the best results. To meet this need, various aspects of chemical engineering plant design optimization are described in Chap. 9, including presentation of design strategies that can be used to establish the desired results in the most efficient manner.

Optimum Economic Design If there are two or more methods for obtaining exactly equivalent final results, the preferred method is the one involving the least total cost. This is the basis of an *optimum economic design*. One typical example of an optimum economic design lies in determining the pipe diameter to use when pumping a given amount of fluid from one point to another. Here the same final result (i.e., a set amount of fluid pumped between two given points) can be accomplished by using an infinite number of different pipe diameters. However, an economic balance will show that one particular pipe diameter gives the least total cost. The total cost includes the cost for pumping the liquid and the cost (i.e., fixed charges) for the installed piping system.

A graphical representation showing the meaning of an optimum economic pipe diameter is presented in Fig. 1-1. As shown in this figure, the pumping cost increases with decreased size of pipe diameter because of frictional effects, while the fixed charges for the pipeline become lower when smaller pipe diameters are used because of the reduced capital investment. The optimum economic diameter is located where the sum of the pumping costs and fixed costs for the pipeline becomes a minimum, since this represents the point of least total cost. In Fig. 1-1, this point is represented by *E*.

The chemical engineer often selects a final design on the basis of conditions giving the least total cost. In many cases, however, alternative designs do not give final products or results that are exactly equivalent. It then becomes necessary to consider the quality of the product or the operation as well as the total cost. When the engineer speaks of an optimum economic design, it ordinarily means the least expensive one selected from a number of equivalent designs. Cost data, to assist in making these decisions, are presented in Chaps. 12 through 15.

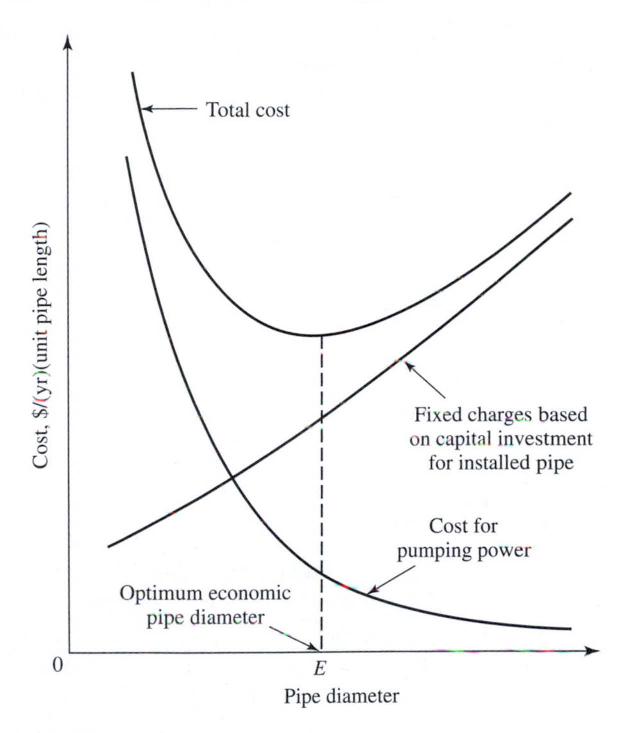

Figure 1-1
Optimum economic pipe diameter for a constant mass-throughput rate

Various types of optimum economic requirements may be encountered in design work. For example, it may be desirable to choose a design that gives the maximum profit per unit of time or the minimum total cost per unit of production.

Optimum Operation Design Many processes require specific conditions of temperature, pressure, contact time, or other variables if the best results are to be obtained. It is often possible to make a partial separation of these optimum conditions from direct economic considerations. In cases of this type, the best design is designated as the *optimum operation design*. The chemical engineer should remember, however, that economic considerations ultimately determine most quantitative decisions. Thus, the optimum operation design is usually merely the first step in the development of an optimum economic design.

An excellent example of an optimum operating design is the determination of operating conditions for the catalytic oxidation of sulfur dioxide to sulfur trioxide. Suppose that all the variables, such as converter size, gas rate, catalyst activity, and entering-gas concentration, are fixed, and the only possible variable is the temperature at which the oxidation occurs. If the temperature is too high, the yield of SO_3 will be low because the equilibrium between SO_3, SO_2, and O_2 is shifted in the direction of SO_2 and O_2. On the other hand, if the temperature is too low, the yield will be poor because the reaction rate between SO_2 and O_2 will be low. Thus, there must be one temperature at which the amount of sulfur trioxide formed will be a maximum. This particular temperature would give the optimum operation design. Figure 1-2 presents a graphical method for determining the optimum operation temperature for the sulfur

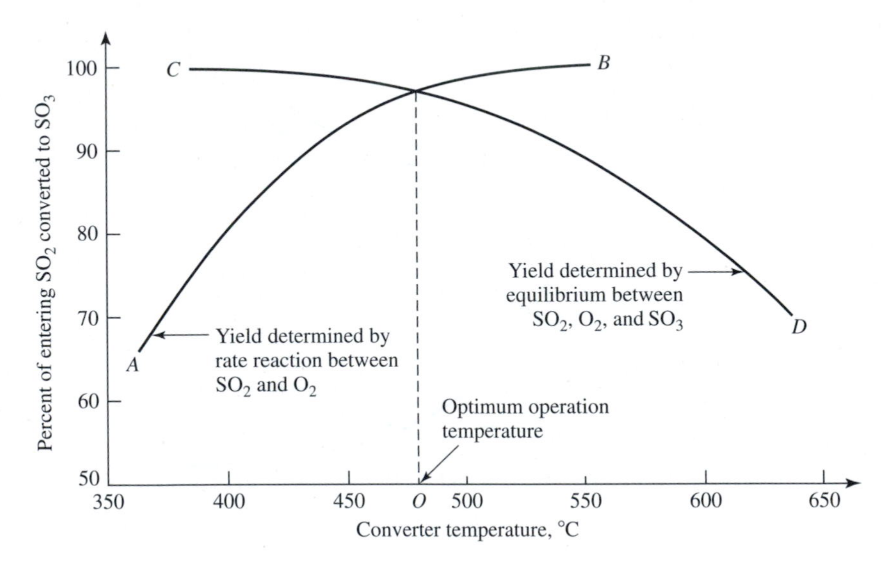

Figure 1-2
Determination of optimum operation temperature in a sulfur dioxide converter

dioxide converter in this example. Line *AB* represents the maximum yields obtainable when the reaction rate is controlling, while line *CD* indicates the maximum yields on the basis of equilibrium conditions controlling. Point *O* represents the optimum operation temperature where the maximum yield is obtained.

The preceding example is a simplified case of what an engineer might encounter in a design. In reality, it would usually be necessary to consider various converter sizes and operations with a series of different temperatures to arrive at the optimum operation design. Under these conditions, several equivalent designs would apply, and the final decisions would be based on the optimum economic conditions for the equivalent designs.

PRACTICAL CONSIDERATIONS IN DESIGN

The chemical engineer must never lose sight of the practical limitations involved in a design. It may be possible to determine an exact pipe diameter for an optimum economic design, but this does not imply that this exact size must be used in the final design. Suppose the optimum diameter were 0.087 m. It would be impractical to have a special pipe fabricated with this inside diameter. Instead, the engineer would choose a standard pipe size which could be purchased at regular market prices. In this case, the recommended pipe size would probably be a U.S. *standard* $3\frac{1}{2}$-in.-diameter pipe having an inside diameter of 0.090 m.

If the engineer happened to be very conscientious about getting an adequate return on all investments, he or she might say, "A U.S. *standard* 3-in.-diameter pipe would require less investment and would probably only increase the total cost slightly; therefore, I think we should compare the costs between the two pipes before making a final decision." Theoretically, the conscientious engineer in this case is correct. Suppose the total cost of the installed 0.090-m-diameter pipe is $5000 and the total cost of the installed 0.078-m-diameter pipe is $4500. If the total yearly savings in power and fixed charges of using the larger-diameter pipe instead of the smaller-diameter pipe were $25, the yearly percent return on the extra $500 investment would only be 5 percent. Since it should be possible to invest the extra $500 elsewhere to obtain more than a 5 percent return, it would appear that the smaller-diameter pipe would be preferred over the larger-diameter pipe.

The logic presented in the preceding example is perfectly sound. It is a typical example of investment comparison and should be understood by all chemical engineers. Even though the optimum economic diameter was 0.087 m, the practical engineer knows that this diameter is only an exact mathematical number and may vary from month to month as prices or operating conditions change. Therefore, all one expects to obtain from this particular optimum economic calculation is a good estimation as to the best diameter, and investment comparisons may not be necessary.

The practical engineer understands the physical problems involved in the final operation and maintenance of the designed equipment. In developing the plant layout, key control valves must be placed where they are easily accessible to the operators. Sufficient space must be available for maintenance personnel to check, take apart, and repair equipment. The engineer should realize that cleaning operations are simplified if a scale-forming fluid is passed through the inside of the tubes rather than on the shell side of a tube-and-shell heat exchanger. Obviously, sufficient plant layout space should be made available that the maintenance workers can remove the head of the installed exchanger and force cleaning drills or brushes through the inside of the tubes or remove the entire tube bundle when necessary.

The theoretical design of a distillation unit may indicate that the feed should be introduced on one particular tray in the tower. Instead of specifying a tower with only one feed inlet on the calculated tray, the practical engineer will include inlets on several trays above and below the calculated feed point, since the actual operating conditions for the tower can vary and the assumptions included in the calculations make it impossible to guarantee absolute accuracy.

The preceding examples typify the type of practical problems the chemical engineer encounters. In design work, theoretical and economic principles must be combined with an understanding of the common practical problems that will arise when the process finally comes to life in the form of a complete plant or a complete unit.

The Design Approach

The chemical engineer has many tools to choose from in the development of a profitable plant design. None, when properly utilized, will probably contribute as much to the optimization of the design as the use of available computers and accompanying software. Many problems encountered in the process development and design can be solved rapidly with a higher degree of completeness with computers and at lower cost than with ordinary hand or desk calculators. Generally overdesign and safety factors can be reduced with a substantial savings in capital investment.

At no time, however, should the engineer be led to believe that plants are designed around computers. They are used to determine design data and are used as models for optimization once a design is established. They are also used to maintain operating plants on the desired operating conditions. The latter function is a part of design and supplements and follows process design.

The general approach in any plant design involves a carefully balanced combination of theory, practice, originality, and plain common sense. In original design work, the engineer must deal with many different types of experimental and empirical data. The engineer may be able to obtain accurate values of heat capacity, density, vapor-liquid equilibrium data, or other information on physical properties from the literature. In many cases, however, exact values for necessary physical properties are not available, and the engineer is forced to make approximate estimates of these values. Many approximations also must be made in carrying out theoretical design calculations. For example, even though the engineer knows that the ideal gas law applies exactly only to simple gases at very low pressures, this law is used in many of the calculations when the gas pressure is as high as 500 kPa. With common gases, such as air or simple hydrocarbons, the error introduced by using the ideal gas law at ordinary pressure and temperature is usually negligible in comparison with other uncertainties involved in the design calculations. The engineer prefers to accept this error rather than spend additional time determining virial coefficients or other factors to correct for ideal gas deviations.

In the engineer's approach to any design problem, it is necessary to be prepared to make many assumptions. Sometimes these assumptions are made because no absolutely accurate values or methods of calculation are available. At other times, methods involving close approximations are used because exact treatment would require long and laborious calculations giving little gain in accuracy. The good chemical engineer recognizes the need for making certain assumptions but also knows that this type of approach introduces some uncertainties into the final results. Therefore, assumptions are made only when they are necessary and reasonably correct and will not adversely affect the overall design and its economic conclusions.

Another important factor in the approach to any design problem involves economic conditions and limitations. The engineer must consider costs and probable profits constantly throughout all the work. It is almost always better to sell many units of a product at a low profit per unit than a few units at a high profit per unit. Consequently, the engineer must take into account the unit production rate when determining costs and total profits for various types of designs. This obviously leads to considerations of customer needs and demands. These factors may appear to be distantly removed from the development of a plant design, but they are extremely important in determining its ultimate success.

ENGINEERING ETHICS IN DESIGN

In any professional activity, engineers are obligated to pursue their profession with the highest level of ethical behavior.[†] Specifically, engineering ethics is directly related with the personal conduct of engineers as these uphold and advance the integrity, honor, and dignity of the engineering profession. This conduct of behavior has obligations to the individual engineer, employer and/or client, colleagues and coworkers, the public, and the environment. No process design should be initiated unless the engineer has this level of ethical behavior.

Specific examples of these obligations are detailed in codes of ethics adopted by the various engineering societies. The Code of Ethics adopted by the American Institute of Chemical Engineers is given below.[‡]

> Members of the American Institute of Chemical Engineers shall uphold and advance the integrity, honor, and dignity of the engineering profession by: being honest and impartial and serving with fidelity their employers, their clients, and the public; striving to increase the competence and prestige of the engineering profession; and using their knowledge and skill for the enhancement of human welfare. To achieve these goals, members shall:
>
> **1.** Hold paramount the safety, health, and welfare of the public in performance of their professional duties.

[†]M. W. Martin and R. Schinzinger, *Ethics in Engineering,* 2d ed., McGraw-Hill, New York, 1989; C. F. Mascone, A. G. Santaquilana, and C. Butcher, *Chem. Eng. Prog.*, **87:** 61 (1991).

[‡]Reprinted with permission of the American Institute of Chemical Engineers.

2. Formally advise their employers or clients (and consider further disclosure, if warranted) if they perceive that a consequence of their duties will adversely affect the present or future health or safety of their colleagues or the public.
3. Accept responsibility for their actions and recognize the contributions of others; seek critical review of their work and offer objective criticism of the work of others.
4. Issue statements or present information only in an objective and truthful manner.
5. Act in professional matters for each employer or client as faithful agents or trustees, and avoid conflicts of interest.
6. Treat fairly all colleagues and co-workers, recognizing their unique contributions and capabilities.
7. Perform professional services only in areas of their competence.
8. Build their professional reputations on the merits of their services.
9. Continue their professional development throughout their careers, and provide opportunities for the professional development of those under their supervision.

The National Society of Professional Engineers (NSPE) adopted a more detailed code of ethics in 1947 and updated this in 1996. This code can be retrieved from the NSPE through its website: http://www.nspe.org.

CHAPTER

2

General Design Considerations

The development of a complete plant design involves consideration of many different topics. Quite understandably, the overall economic picture generally dictates whether the proposed facility will receive management approval. However, before proceeding with the development of a process design and its associated economics, it is useful to provide an overall view of other key aspects that are involved in a complete process or plant design. In this discussion, particular emphasis will be placed on important health, safety, loss prevention, and environmental considerations. Other aspects that will be discussed briefly include plant location, plant layout, plant operation and control, utility use, structural design, materials handling and storage, and patent considerations.

With respect to safety, health, and environmental concerns, it is rather difficult to provide detailed descriptions of the applicable regulations and codes since these are periodically revised or updated. Consequently, this chapter addresses the types of regulations that are relevant to chemical process and plant design and provides guidance where to obtain information on current and proposed regulations. The major emphasis, therefore, will be on general concepts and strategies of risk assessment and reduction that are incorporated in those regulations.

HEALTH AND SAFETY HAZARDS

The potential health hazard to an individual by a material used in any chemical or biochemical process is a function of the inherent toxicity of the material and the frequency and duration of exposure. It is common practice to distinguish between the short-term and long-term effects of a material. A highly toxic material that causes immediate injury is classified as a safety hazard while a material whose effect is apparent only after long exposure at low concentrations is considered as an industrial health and hygiene hazard. Both the permissible limits and the precautions to be taken to ensure that such limits will not be exceeded are quite different for these two classes of toxic materials. Information on the effects of many chemicals and physical agents is accessible through a number of computer databases, such as TOXLINE and TOXNET.

The inherent toxicity of a material is measured by tests on animals. The short-term effect is expressed as LD_{50}—the lethal dose at which 50 percent of the test animals do not survive. Estimates of the LD_{50} value for humans are extrapolated from the animal tests. On the other hand, the *permissible exposure limit* (PEL) of concentration for the long-term exposure of humans to toxic materials is set by the *threshold limit value* (TLV). The latter is defined as the upper permissible concentration limit of the material believed to be safe for humans even with an exposure of 8 h/day, 5 days/week over a period of many years. The handbook prepared by Sax and Lewis[†] provides a comprehensive source of data as well as guidance on the interpretation and use of the data. Recommended TLVs are published in bulletins by the Occupational Safety and Health Administration (OSHA), the American Conference of Governmental Industrial Hygienists (ACGIH), the American Industrial Hygiene Association (AIHA), the National Institute for Occupational Safety and Health (NIOSH), and the United Kingdom Health and Safety Executive (HSE).

To search for published data on chemical and physical hazards, a good place to begin is the *Registry of Toxic Effects of Chemical Substances* (see http://www.cdc.gov/niosh/rtecs.html) and the NIOSHIC bibliographic database (see http://www.cd.gov/nioshtic.html).

Sources of Exposure

The main objective of health hazard control is to limit the chemical dosage of a chemical by minimizing or preventing exposure. It is not practical to measure or control the chemical dosage directly; rather, exposure is measured, and limits are set for the control of such exposure.

The most common and most significant source of workplace exposure to chemicals—and the most difficult to control—is inhalation. Workers become exposed when the contaminant is picked up by the air they breathe. Thus, an understanding of the sources of contaminants to which workers are exposed is important for the recognition, evaluation, and control of occupational health hazards. For example, mechanical abrasions of solid materials by cutting, grinding, or drilling can produce small particles that can form an airborne dust cloud or solid aerosol. Liquid aerosols, on the other hand, may be produced by any process that supplies sufficient energy to overcome the surface tension of the liquid. Contaminant vapors are normally formed by allowing the liquid to evaporate into the air.

Gases are usually stored or processed in closed systems. Contamination of air with such gas occurs from fugitive emissions (leaks) or from venting. Essentially all closed systems leak to some degree. [The Environmental Protection Agency (EPA) through various studies has determined that emissions from just the synthetic organic chemical

[†]N. I. Sax and R. J. Lewis, Sr., *Dangerous Properties of Industrial Materials,* 10th ed., J. Wiley, New York, 1999.

manufacturing industry in the United States were greater than 80,000 Mg/yr before emission controls were instituted.] Obviously, the tightness of a system is directly related to the engineering and leak monitoring effort expended. This, in turn, depends on the consequences resulting from these emissions. High-value and very toxic materials are usually very tightly controlled. Contaminants that are neither valuable nor toxic but that create an undesirable atmosphere in neighboring communities are also controlled to maintain good public relations. Flammable materials likewise are carefully controlled because a leak may lead to a fire and a possible major loss in life and facility. Lipton and Lynch[†] provide a listing of more than 30 potential sources of air contamination that may exist in a typical chemical process plant, many of which can be minimized or reduced to acceptable contamination levels by the use of appropriate design measures.

In typical well-maintained plants, pumps and valves are probably the major source of fugitive emissions. Monitoring and maintenance efforts are therefore generally focused on these sources. Taken as a whole, fugitive emissions, even without major seal failure, are the origin of the continuous background exposure of workers. This source of exposure may not, by itself, result in overexposure; but its presence reduces the margin within which other emissions may vary while still remaining under the acceptable TLV.

The continuous movement of materials through a process unit generally does not involve any situations for emission release and consequent exposure. However, some material-handling steps are difficult to accomplish with total containment. For example, liquids entering fixed tankage generally displace air that must be vented to avoid overpressurizing the tank. Control of such liquid-transfer operations can be achieved by using variable-volume tanks, particularly those with floating roofs, or by scrubbing, flaring, or recovering the vented gas stream.

Solids handling can provide considerable exposure to contaminants whenever the operation is performed in an open atmosphere. Where possible, such operations should be retrofitted with a closed system. Even then, potential release problems exist, particularly during maintenance and repair of the system.

It should be recognized that the maintenance of any closed system can pose a hazardous exposure problem since most maintenance is performed while the plant is in operation and requires that workers be in close proximity to the operating equipment for long periods of time. Under such conditions, it is necessary to consider not only local contaminant releases but also physical hazards that may be present, such as noise and thermal radiation.

In a closed system, equipment that must be repaired should first be cleaned to reduce exposure before the system is opened. Where highly toxic process materials are present, it may be necessary to flush equipment with a low-toxicity stream, strip with steam, and then purge with nitrogen. In such situations, the equipment design should include special fittings necessary for the flushing and purging procedures.

Turnarounds, or major periodic overhauls of chemical plant units, are a special case of plant maintenance. Since the units are shut down, some exposure risks are

[†]S. Lipton and J. Lynch, *Health Hazard Control in the Chemical Process Industry,* J. Wiley, New York, 1987.

avoided. However, since the unit is not in production, there is a time pressure to complete the turnaround and resume production. In such an environment, there is the potential for disorganization and misunderstanding on the part of workers with the unanticipated release of contaminants. To conduct a safe turnaround requires careful planning. Contingencies need to be anticipated to the greatest extent possible and plans made to deal with them.

Note that the materials and operations used in a plant maintenance effort may involve a new set of hazards quite separate from the exposure hazards encountered with feedstocks, intermediates, and products for the process plant. For example, proper maintenance often involves such operations as welding, sandblasting, painting, chemical cleaning, catalyst handling, and insulation replacement. The maintenance of safe conditions requires extensive worker training in each one of these operations.

In the same vein, certain waste-handling procedures, even those performed intermittently, can result in very serious contaminant exposure without proper precautions. Workers need to be instructed in the proper procedures for cleaning up spills and accumulated debris. Spilled materials can become airborne and pose an inhalation hazard. Spills and chemical process wastes may end up in the wastewater treatment facilities where they again can be volatilized into the air and result in unexpected worker exposure.

Exposure Evaluation

If health hazards are to be controlled, they must be recognized and evaluated. A logical place to initiate the process of health hazard recognition is with a total inventory of all materials present in the various stages of the process. Even when materials are present in only trace amounts, there can be more than the acceptable concentration to produce a potentially hazardous situation in a localized work area. Generally, feedstocks and products of a process are well known. Intermediates, by-products, and waste materials may be less conspicuous and may not even have been identified. Other materials such as catalysts, additives, cleaning agents, and maintenance materials need to be identified to complete the inventory.

An estimate of the toxicity or intrinsic hazard is required by the OSHA Hazard Communication Standard in the form of a Material Supply Data Sheet (MSDS). A summary of this form is given in Table 2-1. (Other countries besides the United States have similar requirements.)

Standard sources of hazard data may need to be consulted for those chemical compounds for which no MSDSs are presently available. Adequate hazard data may be lacking for various mixtures that are unique to the plant. For such mixtures, it may be necessary to analyze the contents and then estimate the overall hazard based on the individual components.

To perform a risk assessment and then prioritize the exposure measurement effort requires an approximate assessment of initial exposure potential. For each chemical present and for each source of exposure for that chemical, an estimate of exposure can be made. These exposure estimates combined with a toxicity estimate from the hazard data can then be combined to yield a risk estimate which can be used as a basis for prioritization of the measurement and monitoring effort.

Table 2-1 Summary of a material safety data sheet

1. Product identification
   a. Precautionary labeling
   b. Precautionary label statements
   c. Laboratory protective equipment required
2. Hazardous components
3. Physical data
4. Fire and explosion hazard data
   a. Fire extinguishing media
   b. Special fire-fighting procedures
   c. Unusual fire and explosion hazards
   d. Toxic gases produced
5. Health and hazard data
   a. Effects of overexposure
   b. Target organs
   c. Medical conditions generally aggravated by exposure
   d. Emergency and first aid procedures
6. Reactivity data
7. Spill and disposal procedures
   a. Steps to be followed in the event of a spill or discharge
   b. Disposal procedure
8. Protective equipment
9. Storage and handling precautions
   a. Special precautions
10. Transportation data and other information
    a. Domestic
    b. International

It is generally not necessary to make an exposure estimate for every chemical/exposure source combination since many will be of such low significance that they can be neglected. For those chemical/exposure source combinations that could be near the top of the priority list, the exposure estimate is probably not needed beyond an order of magnitude. Methods for making this type of estimate have been developed by the EPA for the purpose of evaluating Premanufacturing Notifications (PMNs).

Contaminant concentrations in a typical plant environment are highly variable. The background level of exposure in a chemical plant is generally the result of a large number of small fugitive emissions, each varying with time. These variability aspects in the contaminant concentrations and the exposure of workers require that a sufficient number of samples be taken to permit both characterization of the statistical distribution and estimation of exposure over the appropriate averaging time. In mathematical terms, the averaging time should be no longer than the biological half-life of a substance acting in the body. Although the range of biological half-lives is continuous, for simplicity only a few discrete averaging times are commonly used. For fast-acting substances 15 and 30 min are used, while 8 h is most often used for substances with biological half-lives longer than 8 h. The latter is generally labeled as the 8-h time-weighted average (TWA).

Table 2-2 Air analysis methods

Method	Substance analyzed
Atomic absorption spectroscopy	Metals
Gas chromatography	Volatile organic compounds
Gravimetric	Nuisance dust, coal dust
Particle count	Asbestos
Ion-specific electrode	Halogens, HCN, NH_3
X-ray diffraction	Silica
Colorimetry	Miscellaneous

The most commonly used methods for the analysis of airborne contaminants are listed in Table 2-2. Any method used for a particular contaminant must be appropriate for the sampling media, have sufficient sensitivity, and be reasonably free from interference. The ultimate confidence that can be placed on an analytical result depends in part on the accuracy of the method, but to a greater extent on how well the method has been validated for the particular purpose and on the reliability of the laboratory performing the test.

As noted earlier, the EPA has determined that fugitive emissions from process equipment are a large source of volatile organic compounds (VOCs). The latter are defined by the EPA as organic compounds that participate in photochemical reactions. These reactions are of significance since the ozone level in the atmosphere is affected by the concentration of volatile organic compounds. Standards for ozone concentration in nonplant areas were originally one of the major concentration targets in the U.S. Clean Air Act.

To meet requirements of the U.S. Clean Air Act, the EPA in 1997 established two types of national air quality standards. Primary standards for six principal pollutants set limits to protect public health, including the health of asthmatics, children, and the elderly. Secondary standards for the same pollutants set limits to protect public welfare, including protection against decreased visibility and against damage to animals, crops, vegetation, and buildings. These air quality standards are specified in Table 2-3.

In addition to these types of regulations, the EPA has added other regulations that control a number of compounds which neither are carcinogenic agents nor cause serious health problems to the public. These hazardous pollutants, controlled under the National Exposure Standards for Hazardous Pollutants (NESHAP), include benzene, vinyl chloride, mercury, asbestos, arsenic, beryllium, and radionuclides. The NESHAP regulations in combination with VOC emission control regulations reduce exposures in the plant environment through equipment emission control systems. This is in contrast to the specific objective of the Occupational Safety and Health Act (OSHA), which is the control of occupational exposures in the workplace. This is considered in the next section.

Control of Exposure Hazards

When it is concluded that an exposure problem exists, decisions need to be made regarding the implementation of hazard control measures for the purpose of reducing exposure and correspondingly reducing the risks. However, a given set of exposure conditions does not lead to a fixed set of control strategies. There are many options.

Table 2-3 National ambient air quality standards†

Pollutant	Standard value‡	Standard type
Carbon monoxide		
8-h average	9 ppm (10 mg/m^3)	Primary
1-h average	35 ppm (40 mg/m^3)	Primary
Nitrogen dioxide		
Annual arithmetic mean	0.053 ppm (100 $\mu g/m^3$)	Primary and secondary
Ozone		
1-h average	0.12 ppm (235 $\mu g/m^3$)	Primary and secondary
Lead		
Quarterly average	1.5 $\mu g/m^3$	Primary and secondary
Sulfur dioxide		
Annual arithmetic mean	0.03 ppm (80 $\mu g/m^3$)	Primary
24-h average	0.14 ppm (365 $\mu g/m^3$)	Primary
3-h average	0.50 ppm (1300 $\mu g/m^3$)	Secondary
Particulates (< 10 μm)		
Annual arithmetic mean	50 $\mu g/m^3$	Primary and secondary
24-h average	150 $\mu g/m^3$	Primary and secondary

†For additional information see http://www.epa.gov/airs/criteria.html.
‡Parenthetical value is an approximately equivalent concentration.

Since zero risk is not attainable, a decision must be made relative to the degree of risk reduction that is to be attained. Then a series of choices must be made from a wide range of options available to achieve the desired risk reduction. This choice of options is a judgment decision since the precise degree of risk assessment achievable by a specific strategy is usually not known in advance. Furthermore, the strategy selected must meet company safety standards, comply with regulatory requirements, receive worker acceptability, and not adversely impact production and operability.

There are three general control principles utilized in reducing the exposure of workers to occupational health hazards. These involve source controls, transmission barriers, and personal protection. In the first strategy, measures are taken to prevent the release of the toxic contaminant to the air. The second strategy provides means for capturing or blocking the contaminant before it reaches the worker. The final strategy assumes the first two were unsuccessful and requires workers to wear some protective device to prevent contact with the toxic contaminant.

Containment eliminates most opportunities for exposure and is the preferred method of control in chemical production. Actually, containment in many chemical plants is dictated by pressure, temperature, fire, or product loss requirements and is really not a health-hazard control option. However, it must be recognized that containment is never perfect, that releases and exposure opportunities will still occur, and that additional control will probably be required.

Basic or detailed changes in the way the process is permitted to operate can eliminate or reduce exposure. For example, rather than handling a material as a dry powder, it might be handled as a slurry or in solution. A special case of process change involves the substitution of a less hazardous material in the process for a more hazardous one. If such a substitution is not possible, then it may be necessary to completely isolate the process from the worker, as has been done in the manufacture of HCN (prussic acid).

The primary purpose of local exhaust ventilation is to control contaminant exposure by establishing a control surface or barrier between the emission source and the worker so that the contaminant is captured and does not reach the workplace. Local exhaust ventilation is cumbersome and inconvenient and requires considerable maintenance. It is an effective form of control that can be retrofitted to an existing plant, thereby minimizing a problem that was not anticipated in the original design. However, local exhaust ventilation is rarely completely effective since capture is not complete and not all release points are adequately covered.

Dilution ventilation, on the other hand, adds a dilutent to a contaminant-filled space. The objective of dilution ventilation is not to prevent any exposure, but to keep the exposure to acceptable levels by dilution. This strategy should only be used in low-release-rate, low-toxicity (low-hazard) situations.

There are also procedures and precautions that can be taken by workers themselves to minimize exposure while on the job. Such practices do not generally eliminate a hazard by themselves but are necessary to prevent overexposure by emission sources not controlled by engineering design. Personal protection against exposure by inhalation can be accomplished by respirators. Such devices are capable of providing considerable protection when selected and used properly.

Various control options or combinations of options need to be selected to reduce the evaluated exposure level to an acceptable one. The best option or combination of options is then selected by means of a cost analysis. The latter is most useful when comparing two or more options that have approximately an equal probability of reducing the exposure below an appropriate occupational exposure limit. Costs, including capital and expense, of the various options may then be compared by using such economic parameters as net present worth or annualized cost.

An engineering system or work procedure that is utilized to eliminate a health effect should be evaluated to determine the degree to which it reduces the occurrence of the health effect. Measurements of exposure, for use in comparison with occupational-exposure limits, need to be made over the averaging time appropriate to the standard.

Fire and Explosion Hazards

Besides toxic emissions, fire and explosion are the two most dangerous events likely to occur in a chemical plant. Considerable resources are expended to prevent both of these hazards or to control them when they do occur because of an accident. These two hazards account for the major loss of life and property in the chemical and petroleum industry.

For a fire to occur, there must be a fuel, an oxidizer, and an ignition source. In addition, the combustion reaction must be self-sustaining. If air is the oxidizer, a certain minimum concentration of fuel is necessary for the flame to be ignited. While the minimum concentration required depends on the temperature of the mixture and, to a lesser extent, on the pressure, greatest interest generally is focused on the ignition conditions necessary at ambient temperature. The minimum concentration of fuel in air required for ignition at ambient temperature is known as the *lower flammable limit* (LFL). Any mixture of fuel and air below the LFL is too lean to burn. Conversely, the concentration above which ignition will not occur is labeled the *upper flammable limit* (UFL). Both limits of flammability are published in various literature sources[†] for many hydrocarbons and chemicals. Note that there is also a concentration of oxidizer that must be present for ignition, called the *limiting oxygen index* (LOI), with a meaning analogous to the LFL.

The flammability limits of mixtures can be estimated from the data for individual fuels by using le Chatelier's principle

$$\sum \frac{y_i'}{\text{LFL}_i} = 1.0 \tag{2-1}$$

where y_i' is the mol fraction of each component of the fuel in the air and LFL_i is the corresponding LFL value for each component. A similar relationship can be used to estimate the UFL for a gas mixture. If the concentration of a mixture of fuel gases is known, the LFL and UFL for the mixture can be approximated from

$$(\text{LFL})_{\text{mix}} = \frac{1}{\sum_{i=1}^{n} (y_i/\text{LFL}_i)} \tag{2-2a}$$

$$(\text{UFL})_{\text{mix}} = \frac{1}{\sum_{i=1}^{n} (y_i/\text{UFL}_i)} \tag{2-2b}$$

where LFL_i and UFL_i are the flammability limits of component i, y_i the mol fraction of component i in the vapor, and n the number of components in the mixture, excluding air. To extend the flammability limits to elevated temperatures and pressures, the following approximations have been developed:

$$(\text{LFL})_T = \text{LFL}_{25}\left[1 - \frac{0.75(T - 25)}{\Delta H_c}\right] \tag{2-3a}$$

$$(\text{UFL})_T = \text{UFL}_{25}\left[1 - \frac{0.75(T - 25)}{\Delta H_c}\right] \tag{2-3b}$$

and

$$(\text{UFL})_p = \text{UFL} + 20.6(\log p + 1) \tag{2-4}$$

where T is the temperature in °C, ΔH_c the net heat of combustion in kcal/mol at 25°C, p the pressure in megapascals (absolute), and UFL the upper flammability limit at

[†]N. I. Sax and R. J. Lewis, Sr., *Hazardous Chemicals Desk Reference,* Van Nostrand Reinhold, New York, 1987; N. I. Sax and R. J. Lewis, Sr., *Dangerous Properties of Industrial Materials,* 10th ed., J. Wiley, New York, 1999; D. B. Crowl and J. Louvar, *Chemical Process Safety: Fundamentals and Applications,* Prentice-Hall, Englewood Cliffs, NJ, 1990; Center for Chemical Process Safety, *Safety, Health and Loss Prevention in Chemical Processes,* American Institute of Chemical Engineers, New York, 1990.

101.3 kPa. The LFL does not change significantly with elevation in pressure. All these equations provide reasonably good LFL and UFL values for mixtures of hydrocarbon gases and mixtures of hydrogen, carbon monoxide, and methane. These relationships provide poorer results for other gas mixtures. Flammability limits for some widely used chemicals and gases are listed in various handbooks.

If the concentration of fuel is within the flammability limits and the temperature of the mixture is high enough, the mixture will ignite. The temperature at which ignition will occur without the presence of a spark or flame is designated the *autoignition temperature* (AIT). If the temperature is less than the AIT, a minimum amount of energy (as low as a few millijoules for hydrocarbons) is required for ignition of flammable mixtures.

When the fuel is a gas, the concentration required for flammability is reached by allowing more fuel to mix with a given quantity of air. However, if the fuel is a liquid, first it must be vaporized before it will burn. When the vapor concentration reaches the LFL, the vapor will ignite if an ignition source is present. The liquid temperature at which the concentration of the fuel in the air becomes large enough to ignite is labeled the *flash point.* The latter is a measure of the ease of ignition of a liquid fuel.

Prevention of fires is best accomplished by keeping all flammable materials under close control. In most industrial operations, once the confined materials are released, it becomes very difficult to keep air from mixing with the materials to form a flammable mixture. It is thus essential to eliminate as many ignition sources as possible. In fact, a number of codes, such as the National Electrical Code promulgated by the National Fire Protection Association (NFPA), specify in NFPA Standard 70(1) the elimination of all ignition sources or the use of protective devices to prevent potential ignition in areas where flammable mixtures are apt to occur. However, damage from the release may make this difficult. Thus, most designers of fire protection systems assume that ignition generally will occur when a flammable material is released.

The heat-transfer rate in a fire depends on two mechanisms: convection and radiation. Calculation of the heat-transfer rate must be made by considering each of the mechanisms separately and then combining the result. If the fire is large, it will radiate at a constant flux; for most hydrocarbons and combustible chemicals, the radiant flux averages close to 95 kW/m^2.

Fires are classified into four groups: Class A fires are those burning ordinary solids; class B fires are those burning liquids or gases; class C fires are those that burn either class A or class B fuels in the presence of live electric circuits; and class D fires consume metals. Fire protection systems can be divided into two large categories: passive and active. Active systems include such agents as water sprays, foam, and dry chemicals; these require that some action be taken, either by plant personnel or as a response by an automatic fire protection system. Passive fire protection systems do not require any action at the time of the fire. They are designed and installed at the time the plant is built and remain passively in place until needed.

One example of passive fire protection is insulating material (called *fireproofing*) that is applied to steel structural members and equipment supports in the plant. The time required for unprotected steel supports to fail during a fire is rather short. Fireproofing can significantly extend the failure time and provide additional time for firefighters to reach the scene, apply cooling water to the supports, and bring the fire under control.

An explosion is a sudden and generally catastrophic release of energy, causing a pressure wave. An explosion can occur without a fire, such as the failure through overpressure of a steam boiler. It is necessary to distinguish between detonation and deflagration when describing the explosion of a flammable mixture. In a *detonation,* the chemical reaction propagates at supersonic velocity, and the principal heating mechanism is shock compression. In a *deflagration,* the combustion process is the same as in the normal burning of a flammable mixture with the reaction propagating at subsonic velocity and experiencing a slow pressure buildup. Whether detonation or deflagration occurs in a flammable mixture depends on such factors as the concentration of the mixture and the source of ignition. Unless confined or ignited by a high-intensity source, most materials will not detonate. However, the pressure wave caused by a deflagration can still cause considerable damage.

An explosion can result from a purely physical reaction, from a chemical reaction, or from a nuclear reaction. A physical explosion is one in which a container fails, releasing its contents to the surroundings. The damage to the surroundings from the sudden expansion of the confined gas can be approximated by determining the maximum energy released from an isentropic expansion of the gas and converting this energy quantity to a TNT equivalent. (The energy released by an explosion of TNT is 4.52 MJ/kg.) A useful relation for this estimation is given by

$$E = \frac{p_b V_G[1 - (p_a/p_b)^{(k-1)/k}]}{k - 1} \tag{2-5}$$

where E is the maximum energy release, V_G the volume of the gas in the container, p_b the burst pressure of the container, p_a the pressure of the surrounding air, and k the ratio of the specific heats.

The amount of energy that is released from a chemical reaction involving a flammable fuel and oxidizer can be estimated from the heat of combustion of the fuel. The damage expected from the resulting explosion may be approximated by comparison with a similar energy release from a known charge of TNT.

There are two special kinds of explosions of particular importance to the chemical industry, namely, the boiling-liquid expanding-vapor explosion (BLEVE) and the unconfined vapor cloud explosion (UVCE). In the former, heat leakage into a container filled with a boiling liquid results in an excessive vaporization accompanied by a steady pressure buildup that ruptures the tank. The sudden depressurization causes very rapid vaporization with a substantial explosive force. An unconfined vapor cloud explosion, on the other hand, can result when a large cloud of gas or vapor forms following release of a flammable material. If ignition occurs, either the cloud may deflagrate, burning with a relatively low burning speed, or the burning speed may accelerate until the flame front reaches detonation velocities. Substantial destruction will occur if the flame front reaches high velocities. A method for approximating the potential for probable loss caused by a vapor-cloud explosion consists of estimating the quantity of combustible that can be released during an accident and then estimating the fraction of the material that is vaporized immediately after the spill. The explosive load is then considered to be 2 percent of the heat of combustion of the material vaporized.[†]

It is important to recognize that dusts and mists may also explode when ignited. A large number of solids can form explosive mixtures in air if they are sufficiently pulverized to remain well dispersed and suspended over a period of time. Some dusts are more sensitive than others to ignition whereas some dusts cause more severe explosions than others when ignited. The ignition sensitivity depends on the ignition temperature, the minimum ignition energy, and the minimum explosion concentration. The explosion severity, on the other hand, is a function of the maximum pressure measured during a test explosion and the maximum rate of pressure rise during the test. Since small dust particles are usually easier to ignite and burn more rapidly than larger particles, both the ignition sensitivity and the explosion severity appear to be a function of particle size. Extensive data on the explosion characteristics of dusts can be found in the *Fire Protection Handbook.*[‡]

If an explosion occurs, whether it is from a physical reaction or a chemical reaction, an overpressure will be generated. Data are available to estimate the effects of overpressure on personnel and equipment. To use the available information, it is necessary to equate the energy of the explosion in terms of equivalent quantities of TNT, as discussed earlier. The explosive yield data are usually scaled in terms of $L/M_{\mathrm{TNT}}^{1/3}$, where L is the distance from the blast center and M is the equivalent yield in terms of mass of TNT. Even though present attempts at using this scaling parameter are rather crude, they do provide reasonable guidelines for locating process equipment and control facilities.[§]

It becomes clear that the chance that a single fire or explosion will spread to adjoining units can be reduced by careful plant layout and judicious choice of construction materials. Hazardous operations should be isolated by location in separate buildings or by the use of brick fire walls. Brick or reinforced concrete walls can serve to limit the effects of an explosion, particularly if the roof is designed to lift easily under an explosive force.

Equipment should be designed to meet the specifications and codes of recognized authorities, such as the American Standards Association, American Petroleum Institute (API), American Society for Testing and Materials (ASTM), Factory Mutual Laboratories (FML), National Fire Protection Association (NFPA), and Underwriters' Laboratories (UL). The design and construction of pressure vessels and storage tanks should follow API and American Society of Mechanical Engineers (ASME) codes, and the vessel should be tested at 1.5 to 2 or more times the design pressure. Adequate venting is necessary, and it is advisable to provide protection by using both spring-loaded valves and rupture disks.

Possible sources of fire are reduced by eliminating all unnecessary ignition sources, such as flames, sparks, or heated materials. Matches, smoking, welding and cutting, static electricity, spontaneous combustion, and non-explosion-proof electrical

[†]Center for Chemical Process Safety, *Guidelines for Hazard Evaluation Procedures,* American Institute of Chemical Engineers, New York, 1992.

[‡]G. P. McKinnon and K. Tower, *Fire Protection Handbook,* National Fire Protection Association, Boston, 1986.

[§]K. Gugan, *Unconfined Vapor Cloud Explosions,* Gulf Publishing, Houston, TX, 1979.

equipment are all potential ignition sources. The installation of sufficient fire alarms, temperature alarms, fire-fighting equipment, and sprinkler systems must be specified in the design.

Personnel Safety

Every attempt should be made to incorporate facilities for health and safety protection of plant personnel in the original design. This includes, but is not limited to, protected walkways, platforms, stairs, and work areas. Physical hazards, if unavoidable, must be clearly defined. In such areas, means for egress must be unmistakable. All machinery must be guarded with protective devices.† In all cases, medical services and first aid must be readily available for all workers.

Safety Regulations

The expressed intent of the Occupational Safety and Health Act originally enacted in 1970 is "to assure so far as possible every working man and woman in the Nation safe and healthful working conditions and to preserve our human resources. . . ." The act presently affects approximately 6 million workplaces and 70 million employees. Over 500 amendments to the act have been introduced since the original legislation. A recent printing of the OSHA standards can be found in Title 29, Chapter XVII, Part 1910 of the *Code of Federal Regulations.*

Two of the standards directly related to worker health and important in design work are *Toxic Hazardous Substances* and *Occupational Noise Exposure.* The first concerns the normal release of toxic and carcinogenic substances, carried via vapors, fumes, dust fibers, or other media. Compliance with the act requires the designer to make calculations of concentrations and exposure time of plant personnel to toxic substances during normal operation of a process or plant. These releases could emanate from various types of seals and from control-valve packings or other similar sources. Normally, the designer can meet the limits set for exposure to toxic substances by specifying special valves, seals, vapor recovery systems, and appropriate ventilation systems.

The list of materials declared hazardous is being updated at a rapid rate. Acceptable material exposure times and concentrations are likewise undergoing continuous revision. Thus, it is important that one examine closely the *Federal Register* before beginning the detailed design of a project. A useful publication entitled *Chemical Regulation Reporter*‡ detailing these proposed and new regulations is available to the design engineer. This weekly information service includes information concerning the Toxic Substances Control Act (a law administered by EPA rather than OSHA).

The *Occupational Noise Exposure* standard requires a well-planned, timely execution of steps to conform to the 90-dBA rule in the design stages of a project. Since

---

†A general requirement for safeguarding all machinery is provided in Section 212 of the Occupational Safety Standard for General Industry (OSHA Standards, 29 *CFR* 1910).

‡*Chemical Regulation Reporter,* Bureau of National Affairs, Inc., 1231 25th Street NW, Washington, DC 20037.

many cities have adopted EPA's recommended noise-level criteria, or have stringent regulations of their own, design-stage noise control must also consider noise leaving the plant.

Other standards in the safety area that are most often cited by OSHA and that must be considered in detailed designs are the *National Electric Code* and *Machinery and Machinery Guarding*. A cursory investigation by a designer of these and other OSHA standards should be made to comply with the most recent regulations. When there is uncertainty in the interpretation of a regulation, the design engineer should contact the regional or area OSHA office to request a clarification of the regulation. Since many states also have approved plans comparable to that of the federal government, the designer must be aware of these regulations.

Note that maintaining an awareness of federal regulations is not an end in itself, but a necessary component for legally acceptable plant design. To aid the design engineer, Table 2-4 presents a listing of federal repositories for environmental and safety regulations.

Table 2-4 Federal repositories of federal regulations

1. *Federal Register* (*FR*) is published daily, Monday through Friday, excepting federal holidays. It provides regulations and legal notices issued by federal agencies. The *Federal Register* is arranged in the same manner as the *CFR* (see below), as follows:
   a. *Title.* Each title represents a broad area that is subject to federal regulations. There are a total of 50 titles. For example, Title 29 involves labor, and Title 40 is about protection of the environment.
   b. *Chapter.* Each chapter is usually assigned to a single issuing agency. For example, Title 29, Chapter XVII, covers the Occupational Safety and Health Administration; Title 40, Chapter I, covers the Environmental Protection Agency.
   c. *Part.* Chapters or subchapters are divided into parts, each consisting of a unified body of regulations devoted to a specific subject. For example, Title 40, Chapter I, Subchapter C, Part 50, is the National Primary and Secondary Ambient Air Quality Standards. Title 29, Chapter XVII, Part 1910, is Occupational Safety and Health Standards. Parts can be further divided into subparts, relating sections within a part.
   d. *Section.* The section is the basic unit of the *CFR* (see below), and ideally it consists of a short, simple presentation of one proposition.
   e. *Paragraph.* When internal division of a section is necessary, sections are divided into paragraphs (which may be further subdivided).
2. *FR Index* is published monthly, quarterly, and annually. The index is based on a consolidation of content entries appearing in the month's issues of the *Federal Register* together with broad subject references. The quarterly index and annual index consolidate the previous 3 months' and 12 months' issues, respectively.
3. *Code of Federal Regulations* (*CFR*) is published quarterly and revised annually. A codification in book form of the general and permanent rules is published in the *Federal Register* by the executive departments and agencies of the federal government.
4. *CFR General Index* is revised annually on July 1. It contains broad subject and title references.
5. *Cumulative List of CFR Sections Affected* is published monthly and revised annually according to the following schedule: Titles 1 to 16 as of January 1; 17 to 27 as of April 1; 28 to 41 as of July 1; 42 to 50 as of October 1. The *CFR* is also revised according to these dates. This list provides users of the *CFR* with amendatory actions published in the *Federal Register.*

LOSS PREVENTION

The phrase *loss prevention* in the chemical industry is an insurance term where the loss represents the financial loss associated with an accident. This loss not only represents the cost of repairing or replacing the damaged facility and taking care of all damage claims, but also includes the loss of earnings from lost production during the repair period and any associated lost sales opportunities.

As noted in the previous section, there are numerous hazards associated with chemical processing. The process designer must be aware of these hazards and ensure that the risks involved with these hazards are reduced to acceptable levels through the application of engineering principles and proven engineering practice. In its simplest terms, loss prevention in process design can be summarized under the following broad headings:

1. Identification and assessment of the major hazards.
2. Control of the hazards by the most appropriate means, for example, containment, substitution, improved maintenance, etc.
3. Control of the process, i.e., prevention of hazardous conditions in process operating variables by utilizing automatic control and relief systems, interlocks, alarms, etc.
4. Limitation of the loss when an incident occurs.

Identification, as required by the Process Safety Management of Highly Hazardous Chemicals regulation effective in mid-1997, can be as simple as asking "what if" questions at design reviews. It can also involve the use of a checklist outlining the normal process hazards associated with a specific piece of equipment. The major weakness of the latter approach is that items not on the checklist can easily be overlooked. The more formalized hazard assessment techniques include, but are not limited to, hazard and operability (HAZOP) study, fault-tree analysis (FTA), failure mode-and-effect analysis (FMEA), safety indexes, and safety audits.

HAZOP Study

The hazard and operability study, commonly referred to as the HAZOP study, is a systematic technique for identifying all plant or equipment hazards and operability problems.† In this technique, each segment (pipeline, piece of equipment, instrument, etc.) is carefully examined, and all possible deviations from normal operating conditions are identified. This is accomplished by fully defining the intent of each segment and then applying guide words to each segment as follows:

No or *not*—No part of the intent is achieved, and nothing else occurs (e.g., no flow).

More—quantitative increase (e.g., higher temperature).

Less—quantitative decrease (e.g., lower pressure).

As well as—qualitative increase (e.g., an impurity).

†T. Kletz, *Plant Design to Safety—A User Friendly Approach,* Hemisphere, Washington, DC, 1991.

Part of—qualitative decrease (e.g., only one of two components in a mixture).

Reverse—opposite (e.g., backflow).

Other than—No part of the intent is achieved, and something completely different occurs (e.g., flow of wrong material).

These guide words are applied to flow, temperature, pressure, liquid level, composition, and any other variable affecting the process. The consequences of these deviations on the process are then assessed, and the measures needed to detect and correct the deviations are established. Since a majority of the chemical process industry now uses some version of HAZOP for all new facilities and selectively uses it on existing ones, an example of this technique, as originally described by Ozog,† is given in the following paragraphs.

Assume that a HAZOP study is to be conducted on a new flammable reagent storage tank and feed pump, as presented by the piping and instrument diagram shown in Fig. 2-1. In this scheme, the reagent is unloaded from tank trucks into a storage tank maintained under a slight positive pressure until the reagent is transferred to the reactor in the process. For simplification, the system is divided into two elements: the tank T-1 and the pump P-1 plus the feed line. Application of the guide words to these two elements is shown in Table 2-5 along with a listing of the consequences that result from the process deviations. Note that not all guide words are applicable to the process deviations listed. Also, some of the consequences identified with these process deviations have raised additional questions that need resolution to determine whether a hazard exists. This will require either more detailed process information or an estimation of release rates. For example, similar release rates could be the consequence of either event 3 (V-3 open or broken) or event 4 (V-1 open or broken); however, the total quantity released through V-3 could be substantially reduced over that with V-1 open or broken by closing V-2. Of the 41 events listed in Table 2-5, event 5 (tank rupture) and event 24 (external fire) would provide the most unsafe consequences since both would result in instantaneous spills of the entire tank contents.

Hazard assessment is a vital tool in loss prevention throughout the life of the facility. Ideally, the assessment should be conducted during the conceptual design phase, final design stage, and prestart-up period as well as when the plant is in full operation. In the conceptual-design phase, many potential hazards can be identified and significant changes or corrections made at minimal cost. Results of these assessments are key inputs to both site selection and plant layout decisions. The major hazards usually include toxicity, fire, and explosions; however, thermal radiation, noise, asphyxiation, and various environmental concerns also need to be considered.

A thorough hazard and risk assessment of a new facility is essential during the final design stage. At this stage, the piping and instrument diagrams, equipment details, and maintenance procedures are finalized. However, since equipment often has not been ordered, it is still possible to make changes without incurring major penalties or delays.

---

†H. Ozog, *Chem. Eng.*, **92**(4): 161 (1985).

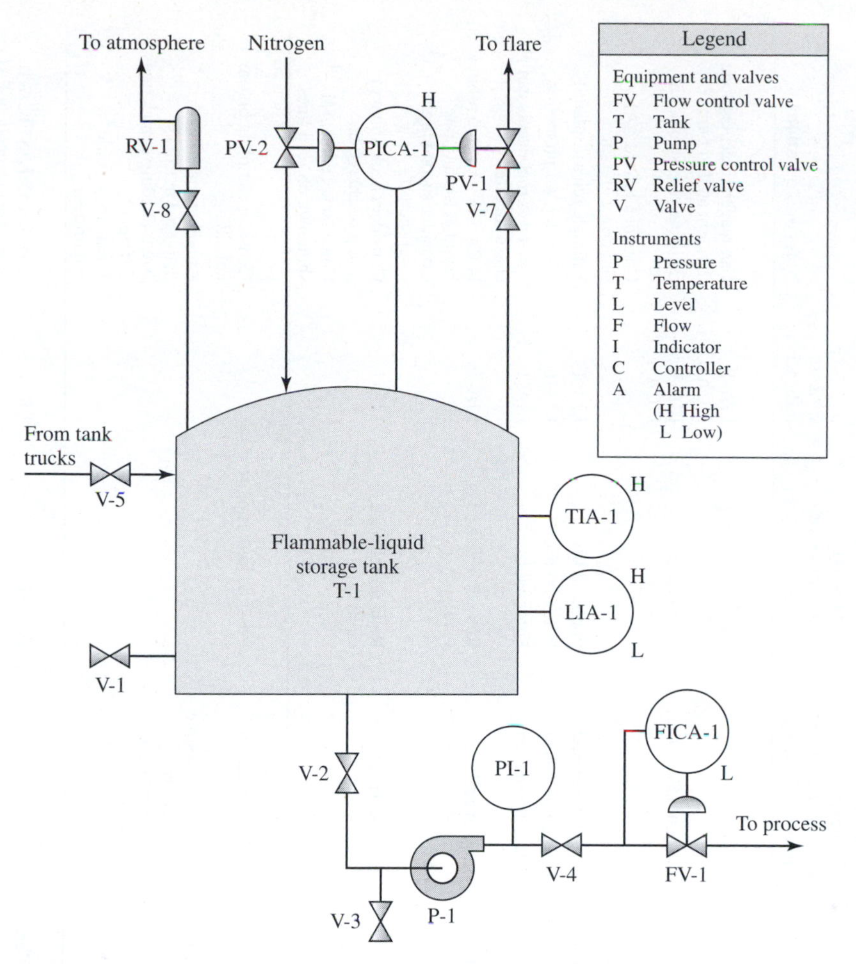

Figure 2-1
Piping and instrumentation diagram used in HAZOP example

A hazard assessment during the prestart-up period should be a final check rather than an initial assessment. This review should include the status of recommended changes from previous hazard studies and any significant design changes made after the final design. If serious hazards are identified at this time, it is unlikely that they can be eliminated without significant cost or start-up delay.

Since process and operating procedure changes are often made during or shortly after plant start-up, it is strongly advised that hazard assessment not stop after start-up. Rather, periodic hazard assessment studies should be used to define the hazard potential of such changes throughout the life of the facility. The average time between reviews is about 3 years; more-hazardous facilities are reviewed more frequently.

Table 2-5 HAZOP evaluation of the process shown

Equipment reference and operating conditions	Deviations from operating conditions	What event could cause this deviation?	Consequences of this deviation on item of equipment under consideration	Additional implications of this consequence	Process indications	Notes and questions
Storage tank	Level					
T-1	Less	1. Tank runs dry	Pump cavitates	Damage to pump	LIA-1, FICA-1	Can reagent react/explode if overheated in pump?
		2. Rupture discharge line	Reagent released	Potential fire	LIA-1, FICA-1	Estimate release quantity.
						Consider second LAL shutdown on pump.
		3. V-3 open or broken	Reagent released	Potential fire	LIA-1	Estimate release quantity.
		4. V-1 open or broken	Reagent released	Potential fire	LIA-1	Consider V-1 protection.
		5. Tank rupture	Reagent released	Potential fire	LIA-1	What external events can cause rupture?
	More	6. Unload too much from tank truck	Tank overfills	Reagent released via RV-1	LIA-1	Is RV-1 designed to relieve liquid at loading rate? Consider second high-level shutoff.
		7. Reverse flow from process	Tank overfills	Reagent released via RV-1	LIA-1	Consider check valve in pump discharge line.
	No	Same as less				Consider second LAH shutdown on feed lines.
	Composition					
	Other than	8. Wrong reagent	Possible reaction	Possible tank rupture		Consider sampling before unloading.
	As well as	9. Impurity in reagent	If volatile, possible overpressure Possible problem in reactor			Are other materials delivered in trucks? Are unloading connections different? What are possible impurities?
	Pressure					
	Less	10. Break in line to flare or to nitrogen line	Reagent released	Potential fire	PICA-1	Consider PAL to PICA-1. Consider independent PAL.
		11. Lose nitrogen	Tank implodes	Reagent released	PICA-1	Consider vacuum-break valve.

	12. PV-2 fails closed	Tank implodes	Reagent released	PICA-1	Consider PAL on PICA-1.
	13. PICA-1 fails, closing PV-2	Tank implodes	Reagent released	PICA-1	Tank not designed for vacuum.
More	14. PICA-1 fails closing PV-1	Reagent released via RV-1	Tank rupture if RV-1 fails	PICA-1	What is capacity of PV-1? RV-1? Consider independent PAH.
	15. PV-1 fails closed	Reagent released via RV-1	Tank rupture if RV-1 fails	PICA-1	Consider independent PAH.
	16. V-7 closed	Reagent released via RV-1	Tank rupture if RV-1 fails	PICA-1	Is V-7 locked open? Is V-8 locked open? Consider independent PAH.
	17. Overfill tank	See Event 6	Tank rupture if RV-1 fails	PICA-1	Consider second high-level shutoff.
	18. Temperature of inlet is hotter than normal	Reagent released via RV-1	Tank rupture if RV-1 fails	PICA-1	What prevents high temperatures of inlet? Consider independent PAH.
More	19. High pressure in flare header	Reagent released via RV-1	Tank rupture if RV-1 fails	PICA-1	Can pressure in flare header exceed tank design? Consider alternative venting.
	20. Volatile impurity in feed	Reagent released via RV-1	Tank rupture if RV-1 fails	PICA-1	Consider independent PAH. Consider sampling before unloading.
No	Same as less				
Temperature					
Less	21. Temperature of inlet is colder than normal	Possible vacuum (see less pressure)	Thermal stress on tank		What are temperature limits of tank?
	22. Low tank pressure	See events 10–13	Thermal stress on tank		What are pressure limits of tank?
More	23. Temperature of inlet is hotter than normal	See event 18	Thermal stress on tank		What are temperature limits of tank?
	24. External fire	Tank fails	Reagent released		What could cause an external fire? What are fire protection capabilities? Is fire protection adequate?

(Continued)

Table 2-5 *Continued*

Equipment reference and operating conditions	Deviations from operating conditions	What event could cause this deviation?	Consequences of this deviation on item of equipment under consideration	Additional implications of this consequence	Process indications	Notes and questions
Feed pump P-1	Flow Less	25. V-2 closed	Pump cavitates	Damage to pump	FICA-1	See event 1.
		26. V-4 closed	Deadhead pump	Damage to pump	FICA-1	Any other problem with deadhead?
		27. Line plugs	Pump cavitates	Damage to pump	FICA-1	See event 1.
		28. FV-1 fails closed	Deadhead pump	Damage to pump	FICA-1	See event 26.
		29. FICA-1 fails, closing FV-1	Deadhead pump	Damage to pump	None	See event 26.
		30. V-3 open	Reagent released		FICA-1	Estimate release quantity.
	More	31. FV-1 fails open	Upset in reactor	Reagent released	FICA-1	Possible problem in reactor.
		32. FICA-1 fails, opening FV-1	Upset in reactor	Reagent released	None	See event 31.
	Pressure More	33. V-4 closed	Deadhead pump	Damage to pump	PI-1, FICA-1	See event 26.
		34. FV-1 fails closed	Deadhead pump	Damage to pump	PI-1, FICA-1	See event 26.
		35. FICA-1 fails, closing FV-1	Deadhead pump	Damage to pump	PI-1	See event 26.
		36. V-2 and V-4 closed	Deadhead pump	Overpressure in pump or line	PI-1, FICA-1	Evaluate need for hydraulic relief.
	Less	37. V-2 closed	Pump cavitates	Damage to pump	PI-1, FICA-1	See event 1.
		38. V-3 open	Reagent released		PI-1	See event 3.
	Temperature More	39. V-4 closed	Deadhead pump	Damage to pump	None	See event 26.
		40. FV-1 fails closed	Deadhead pump	Damage to pump	None	See event 26.
		41. FICA-1 fails, closing FV-1	Deadhead pump	Damage to pump	None	See event 26.

Fault-tree Analysis

The *fault-tree analysis* (FTA) is primarily a means of analyzing hazardous events after they have been identified by other techniques such as HAZOP. The FTA is used to estimate the likelihood of an accident by breaking it down into its contributing sequences, each of which is separated into all its necessary events. The use of a logic diagram or fault tree then provides a graphical representation between certain possible events and an undesired consequence. The sequence of events forms pathways on the fault tree, provided with logical AND and OR gates. The AND symbol is used where coincident lower-order events are necessary before a more serious higher-order event occurs. By multiplying the probabilities of each event in this set, the probability of the next higher-order event is obtained. Correspondingly, when the occurrence of any one of a set of lower-order events is sufficient to cause a more serious higher-order event, the events in the set are joined by an OR gate and the probabilities are added to obtain the probability of the higher-order event. Probabilities of the various events are expressed as a yearly rate. For example, a 1×10^{-3} chance occurrence per year would represent an event on average would occur only once every 1000 years. Estimation of failure rates with any precision is generally difficult because of the limited prior data. In such cases, information from various sources is used and then revised to incorporate information that is site-specific.

Once a fault-tree analysis has been completed, it becomes rather easy to investigate the impact of alternative preventive measures. For example, in the development of an FTA for Fig. 2-1 and its associated HAZOP study presented in Table 2-5, Ozog[†] has determined that the most probable event is a liquid release from the storage tank (event 6) due to overfilling. However, by adding an independent high-level shutoff to the tank truck unloading pump, the probability of a liquid release by this event is significantly reduced and event 12 or 13 (PV-2 closed) becomes the most probable event. The probability of these events, in turn, could be reduced by the installation of an independent low-pressure alarm to the tank. This process of reducing the probability of the most probable event could be continued until an overall acceptable risk level is eventually achieved.

Magnitudes of events are typically expressed in terms of the amount of flammable or toxic material released during an event. Release rates are estimated by using appropriate single-phase and two-phase flow models. Since release duration is directly related to the cause and context of the release, its estimation is generally quite subjective.

The severity of the hazard usually cannot be related directly to the magnitude of a release since this is often a function of both the proximity and the number of on- and off-site ignition sources. To determine the hazard severity requires quantifying, with the aid of state-of-the-art hazard models, the likely extent of toxic or flammable vapor cloud travel under different atmospheric conditions, the thermal radiation fields around vapor and liquid pool fires, the overpressure from any anticipated explosions, and any missile or fragmentation activity that may result from a confined explosion. These hazard events can then be translated into hazard zone estimates by incorporating criteria

[†]H. Ozog, *Chem. Eng.*, **92**(4): 161 (1985).

for human injury and property damage. Finally, the results of various loss scenarios can be combined and presented in risk profiles listing injuries, fatalities, and/or property damage. These results can be compared with data for other risks to the public and to workers in various related areas, and these serve as the basis for an assessment of whether the risks of the facility as designed are acceptable.

Failure Mode and Effect Analysis

The failure mode and effect analysis (FMEA) is generally applied to a specific piece of equipment in a process or a particularly hazardous part of a larger process. Its primary purpose is to evaluate the frequency and consequences of component failures on the process and surroundings. Its major shortcoming is that it focuses only on component failure and does not consider errors in operating procedures or those committed by operators. As a result, it has limited use in the chemical process industry.

Safety Indexes

The safety and loss prevention guide developed by Dow Chemical Company[†] provides a method for evaluating the potential hazards of a process and assessing the safety and loss prevention measures needed. In this procedure, a numerical fire and explosion index is calculated, based on the nature of the process and the properties of the materials. The index can be used in two different ways. In the preliminary design, the Dow index will indicate whether alternative, less hazardous processes should be considered in the manufacture of a specific chemical product. In the final design, after the piping and instrumentation diagrams and equipment layout have been prepared, the calculated index is used as a guide to the selection and design of the preventive and protective equipment needed for safe plant operation.

Safety Audits

The principal function of most safety audits in the past has been to verify the adequacy of safety equipment and safety rules. The former includes equipment for fire protection, personnel protection, and on-site emergency responses. In addition to reviewing the general safety rules, the audit has provided explicit safety rules for new process areas and associated emergency response procedures. However, with the greatly increased concerns for environmental health, safety, community relations, and loss prevention, safety audits have become significant as well as continuous activities for all chemical process companies. Detailed checklists have been developed that cover every aspect of health, safety, and loss prevention. An example of such a checklist has been prepared by Whitehead[‡] and is shown in greatly condensed form in Table 2-6. (For complete details, the original table should be consulted.) A critical analysis of all the items on this

---

[†]*Dow's Fire and Explosion Index Hazard Classification Guide,* 5th ed., American Institute of Chemical Engineers, New York, 1981.

[‡]L. W. Whitehead, *Appl. Ind. Hyg.,* **2:** 79 (1987); the unabridged table is also reproduced by S. Lipton and J. Lynch, *Health Hazard Control in the Chemical Process Industry,* J. Wiley, New York, 1987, pp. 85–96.

Table 2-6 General safety checklist for identifying process hazards*

I. External-plant considerations: location, site qualities and use, availability of services

A. Site

1. Location variables—worker, materials and product access, emergency services, population exposure, security, rights-of-way, size
2. Isolation, separation of on-site and adjacent hazard areas; considerations of quantities of materials, separation, and barriers
3. Protection of materials, storage, plant access, and other critical areas against earthquakes, hurricanes, tornadoes, vandalism, air crashes, sabotage, etc., to practical limits
4. Ice, snow, and water removal and control or storage, grounds management, site drainage
5. Air quality aspects of site

B. People, materials, and energy flow into and out of site and within site—optimization of patterns and reduction of interferences

1. Raw materials—inflow, handling, storage, distribution
2. Products—outflow, handling, storage, packaging
3. Workers
4. Power services—electric, gas, cogeneration, backup power
5. Water—supply, quality, treatment, storage
6. Liquid wastes—wastewater, chemicals
7. Solid wastes—collection, management, and disposition
8. Material and energy losses to the surroundings

II. Internal plant, structure, and services considerations to provide maximum flexibility in a fully serviced area of controlled space

A. Flexible layout of space

1. Power available throughout, without circuit overloads; local power takeoffs equipped with lockout provisions
2. Water available throughout with adequate drainage; for cleaning
3. Adequate roof support for weather extremes including snow, water, and wind
4. Foundation load capacity
5. Vertical space use and access
6. Horizontal spatial layout
7. Steam available as needed
8. Compressed air and/or vacuum available as needed, including different breathing air system if needed (breathing air intake point not contaminated); also inerting gas systems, as needed
9. Central coolant systems provided, as needed
10. Drainage in plant—adequate for cleaning, does not leave standing water; size to include sprinkler flow
11. Adequate preplanning for expansion

B. Materials maintenance and selection

1. Fewest feasible dust-collecting surfaces and spaces; access for cleaning of surfaces, around and under machines
2. Durable, noncorroding, cleanable structure and surfaces, resistant to chemicals used
3. Fewest possible materials or fixtures requiring routine hand maintenance or cleaning
4. Provision for cleaning outside and inside of structure
5. Central and dispersed space for maintenance equipment and supplies, and repair of maintenance equipment

C. General environmental control system

1. Thermal environment control
2. General lighting levels—intensity, color, glare, contrast, etc.

(Continued)

Table 2-6 *Continued*

3. General ventilation for odor, humidity, temperature control
4. Adequate makeup air to prevent indoor air-quality problems
5. Where feasible, general ventilation design for low-level control of toxic air contaminants
6. Buildings, processes, utility lines, process connections, and controls all have appropriate degree of environmental protection (e.g., against freezing or corrosion)

D. Minimum construction requirements (usually building codes)
1. Construction types
2. Allowable areas—those in codes are minima, usually not optimum
3. Allowable heights are minima, usually not optimum
4. Fire separations and susceptibility to ignition from adjacent fires
5. Exterior wall protection—meets adequate over- or underpressure requirements, explosion-relieving sash, walls, roof, as appropriate
6. Fire-fighting requirements—water
7. Fire-resistant materials or insulation, absence of openings to transmit flames
8. Interior finishes—cleanable with reasonable methods
9. Means of egress
10. Fire protection systems—sprinklers, adequate water, extinguishers (halocarbons, CO_2, or special fire-fighting procedures for processes used), fire trucks, carts, etc.
11. Vertical openings—controlled or minimized to reduce fire spread
12. Hazardous areas
13. Light and ventilation—minimum and optimum
14. Sanitation, including drainage, water, sewerage
15. Electrical wiring and grounding; lightning arrestors
16. Provisions for handicapped
17. Energy conservation
18. High-hazard provisions

E. Rodent control

III. In-plant physical and organizational considerations

A. Design of work—implications for workplace design
1. Teams versus assembly lines
2. Individual team facilities
3. Underlying work organization versus space organization
4. Materials-handling system to coincide with designed work patterns
5. Worker participation in work and workplace design; worker participation in planning phase and ongoing hazard anticipation and recognition

B. Process health and safety review: preconstruction written occupational health assessment of each work site or process or situation, prepared prior to final design
1. Consideration of material or process substitutions possible to reduce hazard and risk
2. Review of toxicity of feedstocks and products and their typical impurities, by-products, and intermediates, and effluents, catalysts, and solvents, additives of all types, unexpected products generated under abnormal process conditions
3. Normal exposure potential
4. Start-up, shutdown, turnaround emergency, exposure potential
5. Immediate control of occasional peak exposure
6. Occupational health of maintenance staff considered, as well as operators
7. Health and safety issues reviewed in major modification or automation planning
8. Avoidance of process overcrowding; maintain adequate headroom under equipment
9. Failure modes and repairs considered in industrial hygiene evaluations
10. Positive or negative pressure areas to control flow of contaminated air

Table 2-6 ***Continued***

11. Whether toxicity or flammability, stability, etc., justify extreme engineering controls, e.g., much higher standards, special facilities for dump, blowdown, quench, or deluge
12. Microscale dispersion processes versus concentration
13. Inclusion of needed automatic monitoring/sampling systems
14. Hierarchical emission control strategy
15. Selection of design and maintenance standards adequate for risk category; equipment designed to appropriate recognized standards
16. Detailed considerations of specific process conditions and safety (do not treat as all-inclusive)
17. Ensure written operating procedures include safety checks, actions, maintenance
18. Ensure operator training, formal and informal, includes safety and health training, including emergency procedures, protective equipment use, and hazard communication
19. Adequacy of normal staff personnel and distribution for fire fighting and emergencies, particularly in small plants

C. Isolation and separation of risks to minimize exposure
1. Identification of toxic or radiation areas; minimization of number of people exposed, degree of exposure
2. Noise protection
3. Radiation protection
4. Fire explosion risks separated from ignition hazards, other such risk areas—fire walls, curbs, dikes, barriers, etc.
5. Boilers, other major pressure vessels
6. Storage areas for hazardous materials, also compressed gas storage and restraint systems
7. Carcinogen or biohazard areas
8. Emergency chemical or other exposure refuge points
9. Stored amounts of materials less than acceptable hazard amounts and areas affected if explosion or fire occurs

D. Ergonomics considerations
1. Job analysis needed for repetitive motion, biochemical stresses, etc. (Can machines take over repetitive tasks?)
2. Acceptable information flow and control design at human-machine interfaces
3. Workstations and materials handling evaluated for above, redesigned as needed

E. Material flow and handling systems and organization
1. Horizontal transport systems—mechanical and pneumatic conveyors, carts, robot tugs, piping, augers
2. Vertical transport—elevators, piping, etc.; gravity flow where possible
3. Review of transfer, measurement, and packaging points for exposure potential
4. Storage—long-term and temporary, ready-to-use
5. Proper location of controls, valves, etc., including access during emergencies
6. Review of materials packaging with respect to high-risk quantities and risk reduction
7. Temperature and moisture control systems for dry bulk materials

F. People flow
1. Worker access to worksites and facilities (carts, bicycles, scooters, foot)
2. Visitors
3. Conflicts between worker flows/locations and material flows/storage

(Continued)

Table 2-6 *Concluded*

G. Occupational health input into automation and mechanization
   1. Programming of machines to include safe movements; software interlocks, etc., as well as hardware interlocks
   2. Robots require two to three times more space

H. Industrial sanitation and services
   1. Water—quantity and quality as needed
   2. Food handling, eating, lounge/rest space away from worksite; vending machine sanitation
   3. Solid waste collection and in-plant handling
   4. Bacterial and insect control
   5. Air cleaning where required
   6. Sanitary facilities: toilets, washrooms—adequate number, size, and distribution; internal circular traffic patterns, not in busy areas
   7. Personnel services, where required or useful
   8. Space and access for first aid and medical services
   9. Facilities for industrial hygiene staff, laboratories, information handling
   10. Space, facilities, fixtures planned throughout plant for health and safety equipment

I. Hazard communication within plant—consistent system of signage, placarding, content labeling, etc.; planned facilities for MSDS access

*From American Conference of Governmental Industrial Hygienists, *Applied Occupational and Environmental Hygiene,* Volume 2, page 79, 1987. Reprinted with permission.

checklist will generally identify the major hazards in a proposed or existing facility and assist in prescribing preventive actions. Because of their importance, several of the items on the checklist are amplified in later sections of this chapter.

A typical example of the steps involved in the development of a process plant was presented earlier in Table 1-1. The enormity of the task confronting the design engineer is illustrated by the fact that each of the items in Table 2-6 must be considered for most of the steps in Table 1-1. It becomes apparent that considering these items only at the end of the design is unwise because decisions have been made that foreclose what might have been the optimum control option for occupational health reasons. Experience has shown that continuous integration of environmental, safety, and occupational health issues into all design stages leads to the most cost-effective design. Examples of the kinds of interactions and hazard control choices that need to be made at the various design stages are highlighted in the text by Lipton and Lynch.[†]

ENVIRONMENTAL PROTECTION

Because of the greater concern for the continued degradation of the environment, the Environmental Protection Agency has systematically been rewriting and tightening many policies and regulations. Early effects of the agency designed to protect the environment focused on the removal of pollutants from gas, liquid, and solid waste streams.

[†]S. Lipton and J. Lynch, *Health Hazard Control in the Chemical Process Industry,* J. Wiley, New York, 1987.

The effort through the passage of the Pollution Prevention Act of 1990 has now been shifted to waste minimization through waste reduction and pollution prevention.†

It is very clear that chemical engineers must be versed in the latest federal and state regulations involving environmental protection, worker safety, and health. This need is especially great for engineers in design-related functions, such as capital cost estimating, process and equipment design, and plant layout. It is particularly important to determine what is legally required by EPA, OSHA, and corresponding regulatory groups at the state and local levels. As a minimum, every design engineer should understand how the federal regulatory system issues and updates its standards.

Every design engineer must be certain that a standard being used has not been revised or deleted. To be sure that a regulation is up to date, first it must be located in the most recent edition of the *Code of Federal Regulations* (*CFR*). Next, the *Cumulative List of CFR Sections Affected* must be checked to see if actions have been taken since the *CFR* was published. If action has been taken, the *Cumulative List* will indicate where the changes can be found in the *Federal Register.* The latter provides the latest regulations and legal notices issued by various federal agencies.

Environmental Regulations

Several key aspects of the U.S. federal environmental regulation are spelled out in legislation entitled *Protection of the Environment* (Title 40, Chapter 1 of the *CFR*). Part 6 of Title 40, Chapter 1, requires the preparation of an Environmental Impact Statement (EIS). The National Environment Policy Act (NEPA) requires that federal agencies prepare such a statement in advance of any major "action" that may significantly alter the quality of the environment. To prepare the EIS, the federal agencies require the preparation of an Environmental Impact Assessment (EIA). The latter is required to be a full-disclosure statement. This includes project parameters that will have a positive environmental effect, negative impact, or no impact whatsoever. Generally, design engineers will only be involved with a small portion of the EIA preparation, in accordance with their expertise. However, each individual should be aware of the total scope of work necessary to prepare the EIA as well as the division of work. This will minimize costly duplication as well as provide the opportunity for developing feasible design alternatives.

The preparation of an EIA requires determining what environmental standards require compliance by the project, obtaining baseline data, examining existing data to determine environmental safety of the project, preparing an effluent and emission summary with possible alternatives to meet acceptable standards, and finally preparing the environmental statement or report. Since it may require a full year to obtain baseline data such as air quality, water quality, ambient noise levels, ecological studies and social surveys, emissions, and effluents, studies should take place concurrently to avoid delay in preparing the EIA.

---

†B. W. Piasicka, *In Search of Environmental Excellence,* J. Wiley, New York, 1990; J. Eisenhauer and S. McQueen, *Environmental Considerations in Process Design and Simulation,* Energetics, Inc., Columbia, MD, 1993; T. E. Graedel and B. R. Allenby, *Industrial Ecology,* Prentice-Hall, Upper Saddle River, NJ, 1995.

Development of a Pollution Control System

Developing a pollution control system involves an engineering evaluation of several factors which encompass a complete system. These include investigation of the pollution source, determining the properties of the pollution emissions, design of the collection and transfer systems, selection of the control device, and dispersion of the exhaust to meet applicable regulations.[†]

A key responsibility of the design engineer is to investigate the pollutants and the total volume dispersed. It is axiomatic that the size of equipment is directly related to the volume being treated, and thus equipment costs can be reduced by decreasing the exhaust volume. Similarly, stages of treatment are related to the quantity of pollutants that must be removed. Any process change that favorably alters the concentrations will result in savings. Additionally, consideration should be given to changing raw materials used and even process operations if a significant reduction in pollution source can be attained. The extent to which source correction is justified depends on the cost of the proposed treatment plant.

For example, the characteristics of equipment for air pollution control, as specified in Table 2-7, often limit the temperature and humidity of inlet streams to these devices. Three methods generally considered for cooling gases below 260°C are dilution with cool air, quenching with a water spray, and use of cooling columns. Each approach has advantages and disadvantages. The method selected will be dependent on cost and limitations imposed by the control device.

Selection of the most appropriate control device requires consideration of the pollutant being handled and the features of the control device. Often, poor system performance can be attributed to the selection of a control device that is not suited to the pollutant characteristics. An understanding of the equipment operating principles will enable the design engineer to avoid this problem.

Air Pollution Abatement

The most recent changes in the U.S. Clean Air Act Amendments have altered the regulatory ground rules so that almost any air-pollutant-emitting new facility or modification is subject to the provisions of the law. For most situations, a New Source Review (NSR) application will have to be approved before construction is allowed. Source categories covered at this time include petroleum refineries, sulfur recovery plants, carbon-black plants, fuel conversion plants, chemical process plants, fossil-fuel boilers, and petroleum storage and transfer facilities.

To obtain a construction permit, a new or modified source governed by the Clean Air Act must meet certain requirements. These include a demonstration that *best-available control technology* (BACT) is to be used at the source. In addition, an air quality review must demonstrate that the source will not cause or contribute to a violation of the Ambient Air Quality Standard (AAQS), particularly in terms of sulfur dioxide and particulate concentrations in any area.

[†]D. T. Allen and K. S. Rosselot, *Pollution Prevention for Chemical Process,* J. Wiley, New York, 1997; A. Ghassemi, ed., *Handbook of Pollution Control and Waste Minimization,* Marcel Dekker, New York, 2001.

Table 2-7 Air pollution control equipment characteristics

Control equipment	Minimum[†] size particle, μm	Optimum concentration, g/m^3	Temperature limitations, °C	Pressure drop, Pa	Efficiency	Space requirements[‡]	Collected pollutant	Remarks
				Particulate pollutant				
Mechanical collectors								
Settling chamber	>50	>175	370	<25	<50	L	Dry dust	Good as precleaner, low initial cost
Cyclone	5–25	>35	370	250–1250	50–90	M	Dry dust	
Dynamic precipitator	>10	>35	370	—	<80	M	Dry dust	
Impingement separator	>10	>35	370	<1000	<80	S	Dry dust	
Bag filter	<1	>3.5	260	>1000	>99	L	Dry dust	Bags sensitive to humidity, filter velocity, and temperature
Wet collector								
1. Spray tower	25	>35	5–370	125	<80	L	Liquid	Waste treatment required
2. Cyclone	>5	>35	5–370	>500	<80	L	Liquid	Visible plume possible
3. Impingement	>5	>35	5–370	>500	<80	L	Liquid	Corrosion
4. Venturi	<1	>3.5	5–370	250–15,000	<99	S	Liquid	High-temperature operation possible
5. Electrostatic precipitator	<1	>3.5	450	<250	95–99	L	Dry or wet dust	Sensitive to varying condition and particle properties
				Gaseous pollutant				
Gas scrubber		>1%	5–370	<2500	>90	M-L	Liquid	Same as wet collector
Gas adsorber		§	5–370	<2500	>90	L	Solid or liquid	Adsorbent life critical, high initial and operating costs
Direct incinerator		Combustible vapors	1100	<250	<95	M	None	High operating costs
Catalytic combustion		Combustible vapors	540	<250	<95	L	None	Possible catalyst poisoning by contaminants

[†]Minimum particle size (collected at approximately 90 percent efficiency under usual operating conditions).

[‡]Space requirements: S = small, M = moderate, L = large.

[§]Adsorber (concentrations less than 2 ppm, nonregenerative system; greater than 2 ppm, regenerative system).

Air pollution control equipment can essentially be classified into two major categories: those suitable for removing particulates and those associated with removing gaseous pollutants. Particulates are generally removed by mechanical forces, while gaseous pollutants are removed by chemical and physical means.[†]

Particulate Removal The separating forces in a cyclone are the centrifugal and impact forces imparted on the particulate matter. Similar forces account for the particulate capture in mechanical collectors such as impingement and dynamic separators. In settling chambers, the separation is primarily the result of gravitational forces on the particulates. The mechanism in a wet collector involves contact between a water spray and the gaseous pollutant stream. Separation results primarily from a collision between the particulates and the water droplets. Separation also occurs because of gravitational forces on the large particles, or electrostatic and thermal forces on the small particles. The main separating forces in a bag filter are similar to those described in the wet collector, that is, collision or attraction between the particle and the filter of the bag. Finally, the principal components in an electrostatic precipitator are a discharge plate and a collecting surface. The separation is effected by charging the particles with a high voltage and allowing the charged particles to be attracted to the oppositely charged collection plates.

To obtain the greatest efficiency in particulate removal, additional attention must be given to particle diameter and air velocity. The particle size determines the separating force required, while the effectiveness of the control equipment is related to the stream velocity. Generally, the greater the relative velocity between the airstream and the collision obstacle for the particulates, the more effective the separating mechanism. The electrostatic precipitator is an exception to this generality, since here the particle diameter influences the migratory velocity and the power required to maintain the electrical field that influences the equipment performance. Figure 2-2 illustrates the characteristics of various pollution particulates and the range of application for several control devices as related to particle size.

A review of Table 2-7 and Fig. 2-2 indicates that large-diameter particles can be removed with low-energy devices such as settling chambers, cyclones, and spray chambers. Submicrometer particles must be removed with high-energy units such as bag filters, electrostatic precipitators, and venturi scrubbers. Intermediate particles can be removed with impingement separators or low-energy wet collectors. Obviously, other equipment performance characteristics, as noted in Table 2-7, will also influence the final equipment selection. Costs for much of the equipment considered in this section are given in Chap. 12.

Noxious Gas Removal Gaseous pollutants can be removed from airstreams by absorption, adsorption, condensation, or incineration. A list of typical gaseous pollutants that can be treated with these four methods is given in Table 2-8. Generally, condensation is not utilized as a method for removing a solvent vapor from air or other carrier gas unless the concentration of the solvent in the gas is high and the solvent is worth

[†]N. deNevers, *Air Pollution Control Engineering,* McGraw-Hill, New York, 1995.

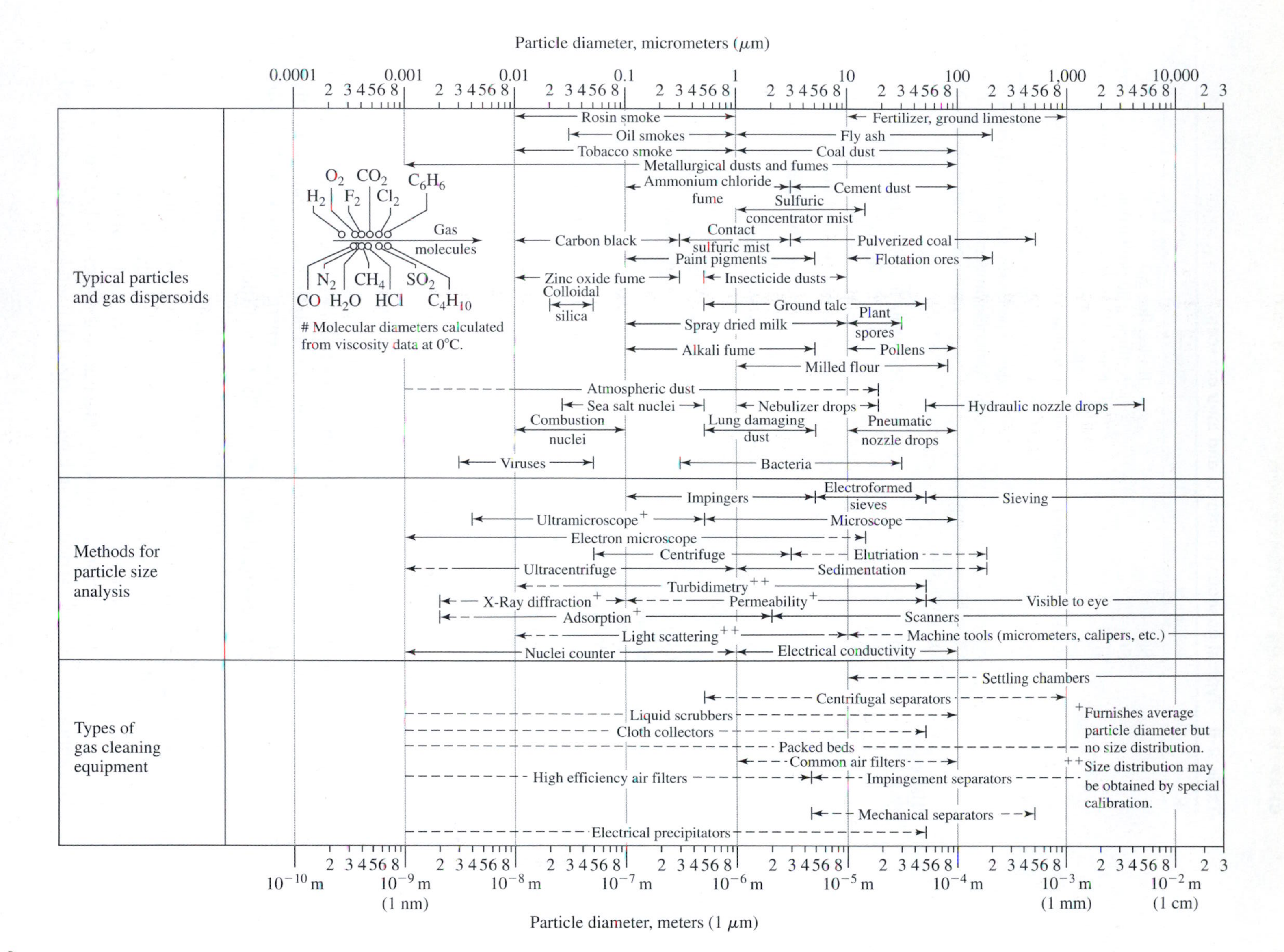

Figure 2-2
Characteristics of pollution particulates and control equipment for removal

Table 2-8 Typical gaseous pollutants and their sources

Key element	Pollutant	Source
S	SO_2	Boiler, flue gas
	SO_3	Sulfuric acid production
	H_2S	Natural gas processing, sewage treatment, paper and pulp industry
	R-SH (mercaptans)	Petroleum refining, pulp and paper
N	NO, NO_2	Nitric acid production, high-temperature oxidation processes, nitration processes
	NH_3	Ammonia production
	Other basic N compounds, pyridines, amines	Sewage, rendering, pyridine-base solvent processes
Halogen:		
F	HF	Phosphate fertilizer, aluminum
	SiF_4	Ceramics, fertilizers
	CFC	Cleaning operations, refrigeration and air conditioning systems, insulation foams
Cl	HCl	HCl production, PVC combustion, organic chlorination processes
	Cl_2	Chlorine production
C	*Inorganic*	
	CO	Incomplete combustion processes
	CO_2	Combustion processes
	Organic	
	Hydrocarbons—paraffins, olefins, and aromatics	Solvent operations, gasoline, petrochemical operations, solvents
	Oxygenated hydrocarbons—aldehydes, ketones, alcohols, phenols, and oxides	Partial oxidation processes, surface coating operations, petroleum processing, plastics, ethylene oxide
	Chlorinated solvents	Dry-cleaning, degreasing operations

recovery. Since condensation cannot remove all the solvent, it can only be used to reduce the solvent concentration in the carrier gas.

Gas-liquid absorption processes are normally carried out in vertical, countercurrent flow through packed, plate, or spray towers. For absorption of gaseous streams, good liquid-gas contact is essential and is partly a function of proper equipment selection. Optimization of absorbers or scrubbers (as applied to noxious gas removal) is also important. The power consumption of a modern, high-energy scrubber can be significant because of the high pressure drop involved. The latter difficulty has been alleviated in the spray scrubber with its low pressure drop even when handling large volumes of flue gases. In addition, sealing and plugging, which can be a problem in certain scrubbing processes (e.g., the use of a limestone-slurry removal of sulfur dioxide from a flue gas), do not present difficulties when a spray is used in a chemically balanced system.

The use of dry adsorbents such as activated carbon and molecular sieves has received considerable attention in removing final traces of objectional gaseous pollutants. Adsorption is generally carried out in large, horizontal fixed beds often equipped with blowers, condensers, separators, and controls. A typical installation usually consists of two beds; one is on stream while the other is being regenerated.

For those processes producing contaminated gas streams that have no recovery value, incineration may be the most acceptable route when the gas streams are combustible. There are presently two methods in common use: direct flame and catalytic oxidation. The former usually has lower capital cost requirements, but higher operating costs, particularly if an auxiliary fuel is required. Either method provides a clean, odorless effluent if the exit-gas temperature is sufficiently high.

Each technique for removing pollutants from process gas streams is economically feasible under certain conditions. Each specific instance must be carefully analyzed before a commitment is made to any type of approach.

Water Pollution Abatement

Better removal of pollutants from wastewater effluents was originally mandated by the federal government in the Water Pollution Control Act of 1970 (P.L. 92-500). Since then, the performance requirements for the various treatment technologies have been raised to new and higher standards with additional legislation aimed at regulating the amounts of toxic and hazardous substances discharged as effluents. Increased legal and enforcement efforts by various government agencies to define toxic and hazardous substances give evidence of the demands that will be placed on pollution technology in the future. The trend in effluent standards is definitely away from the broad, nonspecific parameters (such as chemical oxygen demand or biochemical oxygen demand) and toward limits on specific chemical compounds.

The problems of handling a liquid waste effluent are considerably more complex than those of handling a waste gas effluent. The waste liquid may contain dissolved gases or dissolved solids, or it may be a slurry in either concentrated or dilute form. Because of this complexity, priority should be given first to the possibility of recovering part of or all the waste products for reuse or sale. Frequently, money can be saved by installing recovery facilities rather than more expensive waste treatment equipment. If product recovery is not capable of solving a given waste disposal problem, waste treatment must be used. One of the functions of the design engineer, then, is to decide which treatment process, or combination of processes, will best perform the necessary task of cleaning up the wastewater effluent involved. This treatment can be physical, chemical, or biological in nature, depending upon the type of waste involved and the amount of removal required.

Physical Treatment The first step in any wastewater treatment process is to remove large floating or suspended particles by sources. This is usually followed by sedimentation or gravity settling using circular clarifiers with rotating sludge scrapers or rectangular clarifiers with continuous chain sludge scrapers. These units permit removal of settled sludge from the floor of the clarifiers and scum removal from the surface. Numerous options for improving the operation of clarifiers are presently available.

Sludge from primary or secondary treatment that has been initially concentrated in a clarifier or thickener can be further concentrated by vacuum filtration or centrifugation. The importance of first concentrating a thin slurry by clarifier or thickener action needs to be recognized. For example, concentrating the sludge from 5 to 10 percent solids before centrifuging can result in a 250 percent increase in solids recovery for the same power input to the centrifuge.

Solid-liquid separation by flotation may be achieved by gravity alone or induced by dissolved-air or vacuum techniques. The mechanisms and driving forces are similar to those found in sedimentation, but the separation rate and solids concentration can be greater in some cases.

Adsorption processes, and in particular those using activated carbon, are used in wastewater treatment for removal of refractory organics, toxic substances, and color. The primary driving forces for adsorption are a combination of the hydrophobic nature of the dissolved organics in the wastewater and the affinity of the organics to the solid adsorbent. The latter is due to a combination of electrostatic attraction, physical adsorption, and chemical adsorption. Operational arrangements of the adsorption beds are similar to those described for gaseous adsorption.

Three different membrane processes—ultrafiltration, reverse osmosis, and electrodialysis—are receiving increased interest in pollution control applications as end-of-pipe treatment and for in-plant recovery systems. There is no sharp distinction between ultrafiltration and reverse osmosis. In the former, the separation is based primarily on the size of the solute molecule which, depending upon the particular membrane porosity, can range from about 2 to 10,000 nm. In the reverse-osmosis process, the size of the solute molecule is not the sole basis for the degree of removal, since other characteristics of the solute such as hydrogen bonding and valency affect the membrane selectivity. In contrast to these two membrane processes, electrodialysis employs the removal of the solute (with some small amount of accompanying water) from solution rather than the removal of the solvent. The other major distinction is that only ionic species are removed. The advantages of using electrodialysis are primarily due to these distinctions.

Chemical Treatment In wastewater treatment, chemical methods are generally used to remove colloidal matter, color, odors, acids, alkalies, heavy metals, and oil. Such treatment is considered as a means of stream upgrading by coagulation, emulsion breaking, precipitation, or neutralization.

Coagulation is a process that removes colloids from water by the addition of chemicals. The chemicals upset the stability of the system by neutralizing the colloid charge. Additives commonly used introduce a large multivalent cation such as Al^{3+} (from alum) or Fe^{3+} (from ferric chloride). Emulsion breaking is similar to coagulation. The emulsions are generally broken with a combination of acidic reagents and polyelectrolytes. The common ion effect can also be useful in wastewater treatment. In this case an unwanted salt is removed from solution by adding a second soluble salt to increase one of the ion concentrations. Coagulant aids may then be needed to remove the precipitate.

One method for treating acid and alkaline waste products is by neutralization with lime or sulfuric acid (other available materials may also be suitable). Even though this treatment method may change the pH of the waste stream to the desired level, it does not remove the sulfate, chloride, or other ions. Therefore, the possibility of recovering the acid or alkali by distillation, concentration, or in the form of a useful salt should always be considered before neutralization or dilution methods are adopted.

Chemical oxidation is frequently another tool in wastewater treatment. Chemical oxidants in wide use today are chlorine, ozone, and hydrogen peroxide. The historical use of chlorine and ozone has been in the disinfection of water and wastewater. All three oxidizers are, however, receiving increased attention for removing organic materials from wastewaters that are resistant to biological or other treatment processes. The destruction of cyanide and phenols by chlorine oxidation is well known in waste treatment technology. However, the use of chlorine for such applications has come under intense scrutiny because of the uncertainty in establishing and predicting the products of the chlorine oxidation reactions and their relative toxicity. Ozone, on the other hand, with only a short half-life, is found to be effective in many applications for color removal, disinfection, taste and odor removal, iron and manganese removal, as well as in the oxidation of many complex inorganics, including lindane, aldrin, surfactants, cyanides, phenols, and organometal complexes. With the latter, the metal ion is released and can be removed by precipitation.

The need for chemical reduction of wastewaters occurs less often. The most common reducing agents are ferrous chloride or sulfate which may be obtained from a variety of sources.

Biological Treatment In the presence of the normal bacteria found in water, many organic materials will oxidize to form carbon dioxide, water, sulfates, and similar materials. This process consumes the oxygen dissolved in the water and may cause a depletion of dissolved oxygen. A measure of the ability of a waste component to consume the oxygen dissolved in water is known as the *biochemical oxygen demand* (BOD). This oxygen demand of a waste stream is often the primary factor that determines its importance as a pollutant. The BOD of sewage, sewage effluents, polluted waters, or industrial wastes is the oxygen, reported as parts per million, consumed during a set time by bacterial action on the decomposable organic matter.

One of the more common biological wastewater treatment procedures today involves the use of concentrated masses of microorganisms to break down organic matter, resulting in the stabilization of organic wastes. These organisms are broadly classified as aerobic, anaerobic, and aerobic-anaerobic facultative. Aerobic organisms require molecular oxygen for metabolism, anaerobic organisms derive energy from organic compounds and function in the absence of oxygen, while facultative organisms may function in either an aerobic or an anaerobic environment.

Basically, the aerobic biological processes involve either the activated sludge process or the fixed-film process. The activated sludge process is a continuous system in which aerobic biological growths are mixed with wastewater and the resulting excess flocculated suspension separated by gravity clarification or air flotation. The predominant fixed-film process has, in the past, been the conventional trickling filter. In this process, wastewater trickles over a biological film fixed to an inert medium. Bacterial action in the presence of oxygen breaks down the organic pollutants in the wastewater. (An attempt to improve on the biological efficiency of the fixed-film process has resulted in the development of a useful rotating disk biological contactor.)

Table 2-9 Comparative requirements for processing 775 kg BOD/day by various aerobic systems

System	Area, m^2	Biological loading, kg BOD/100 m^3	BOD removal, %
Stabilization pond	230,000†	1.5–3.7	70–90
Aerated lagoon	23,300‡	18.5–25.5	80–90
Activated sludge			
Extended	930	175–480	95‡
Conventional	325	530–6400	90
High rate	185	915–2400	70
Trickling filter			
Rock	800–2000	11.2–800	40–70
Plastic disks	80–200	320–3200	50–70

†1.5-m depth
‡3.0-m depth

A comparison of biological loading and area requirements for various aerobic biological processes is shown in Table 2-9.

Many organic industrial wastes (including those from food processing, meat packing, pulp and paper, refining, leather tanning, textiles, organic chemicals, and petrochemicals) are amenable to biological treatment; however, a fair number may prove to be refractory, that is, nonbiodegradable. Thus, although the BOD removal may be excellent, the removal of the *chemical oxygen demand* (COD) may be low. Process evaluation prior to system design should center on characterization of the waste stream, particularly to determine the presence of inhibitory or toxic components relative to biological treatment, and the establishment of pollutant removal rates, oxygen requirements, nutrient requirements (nitrogen and phosphorus), sludge production, and solids settleability.

Anaerobic treatment is important in the disposal of municipal wastes but has not been widely used on industrial wastes. It has found some use in reducing highly concentrated BOD wastes, particularly in the food and beverage industries as a pretreatment or roughing technique. However, compared to aerobic systems, anaerobic treatment is more sensitive to toxic materials and more difficult to control.

Solid Waste Disposal

Solid wastes differ from air and water pollutants since these wastes remain at the point of origin until a decision is made to collect and dispose of them. There are several means of disposal available, including recycling, chemical conversion, incineration, pyrolysis, and landfill. Federal regulations, local conditions, and overall economics generally determine which method is the most acceptable.

Recycling and Chemical Conversion Resource recovery is a factor often overlooked in waste disposal. For example, specific chemicals may often be recovered by stripping, distillation, leaching, or extraction. Valuable solids such as metals and plastics can be recovered by magnets, electrical conductivity, jigging, flotation, or hand picking. Process wastes at times may also be converted to salable products or innocuous materials that can be disposed of safely. The former would include hydrogenation of organics to produce fuels, acetylation of waste cellulose to form cellulose acetate, or nitrogen and phosphorus enrichment of wastes to produce fertilizer.

Incineration The controlled oxidation of solid, liquid, or gaseous combustible wastes to final products of carbon dioxide, water, and ash is known as incineration. Since sulfur and nitrogen-containing waste materials will produce their corresponding oxides, they should not be incinerated without considering their effect on air quality. A variety of incinerator designs are available. Multiple-chamber incinerators, rotary kilns, and multiple-hearth furnaces are most widely used in industrial waste disposal.

In the past, incineration has provided certain advantages, particularly where land disposal sites were not available or were too remote for economic hauling. A properly designed and carefully operated incinerator can be located adjacent to a process plant and can be adjusted to handle a variety and quantity of wastes. Not only can heat recovery through steam generation reduce operating costs, but also it can save on pollution control equipment. Additionally, the residue is a small fraction of the original weight and volume of the waste and may be acceptable for landfill.

Pyrolysis The most acceptable route to recycling wastes in the future may be through pyrolytic techniques in which wastes are heated in an air-free chamber, at temperatures as high as 1650°C. Pyrolysis seems to provide several advantages over incineration. These systems encounter far fewer air pollution problems, handle larger throughputs resulting in lower capital costs, provide their own fuel, degrade marginally burnable materials, and have the added potential for recovering chemicals or synthesis gas.

Landfill Sanitary landfill is basically a simple technique that involves spreading and compacting solid wastes into cells that are covered each day with soil. Care needs to be exercised that wastes disposed of in this fashion are either inert to begin with or are capable of being degraded by microbial attack to harmless compounds. The principal problems encountered in landfill operation are the production of leachates that may contaminate the surrounding groundwater and the potential hazards associated with the accumulation of flammable gases produced during the degradation of the waste material.

Thermal Pollution Control

Temperature affects nearly every physical property of concern in water quality management, including density, viscosity, vapor pressure, surface tension, gas solubility, and gas diffusion. The solubility of oxygen is probably the most important of these parameters, inasmuch as dissolved oxygen in water is necessary to maintain many forms of aquatic life. This potential damage to the aquatic environment by changes in temperature, the reduction in the assimilative capacity of organic wastes due to increased temperature, and the federal enactment of more stringent water temperature standards has led design engineers to investigate various offstream cooling systems to handle thermal discharges from processes and plants. Cooling towers are most often considered for this service, followed by cooling ponds and spray ponds in that order.

Cooling towers may be classified on the basis of the fluid used for heat transfer and on the basis of the power supplied to the unit. In wet cooling towers, the condenser cooling water and ambient air are intimately mixed. Cooling results from the evaporation of a portion of the water and to a lesser extent from the loss of sensible heat to the air. In dry cooling towers, the temperature reduction of the condenser water depends upon conduction and convection for the transfer of heat from the water to the air.

Mechanical draft cooling towers either force or induce the air which serves as the heat-transfer medium through the tower. For their driving force, natural-draft cooling towers depend upon the density difference between the air leaving the tower and the air entering the tower. Cooling ponds are generally considered for heat removal only when suitable land is available at a reasonable price, since such systems are simple, cheap, and frequently less water-intensive. It is normally assumed that all heat discharged to a cooling pond is lost through the air-water interface. With low heat-transfer rates, large surface areas are required.

When land costs are too high, spray ponds often provide a viable alternative to cooling ponds. It is estimated that a spray pond requires only about 5 to 10 percent of the area of a cooling pond due to the more intimate air-water contact. In addition, drift losses and corrosion problems are less severe than in cooling towers.

Noise Control

The design engineer should include noise studies in the design stage of any commercial facility. Generally, acoustical problems left for field resolution cost roughly twice as much. Unnecessary costs incurred in postconstruction noise work may include the replacement of insulation, redesign of piping configuration to accommodate silencers, modification of equipment, additional labor costs, and possible downtime to make necessary changes. Considerable judgment, therefore, must be exercised by the designer to establish final design-stage noise recommendations. These not only should consider the results of the equipment data analysis procedure, but also should recognize additional factors such as administrative controls, feasibility of redesign, economic alternatives, intrusion of noise into the community, and the basic limitations of the equations employed in the applicable computer programs.

To attain efficient, effective, and practical noise control, it is necessary to have an understanding of the individual equipment or process noise sources, their acoustic properties and characteristics, and how they interact to create the overall noise situation. Table 2-10 presents typical process design equipment providing high noise levels and potential solutions to this problem.

PLANT LOCATION

The geographic location of a final plant can have a strong influence on the success of an industrial venture. Considerable care must be exercised in selecting the plant site, and many different factors must be considered. Primarily, the plant should be located where the minimum cost of production and distribution can be obtained; but other factors, such as room for expansion and safe living conditions for plant operation as well as the surrounding community, are also important.

Table 2-10 Equipment noise sources, levels, and potential control solutions

Equipment	Sound level, dBA,† at a distance of 1 m	Possible noise control treatments
Air coolers	87–94	Aerodynamic fan blades; decrease in rpm and increase in pitch; tip and hub seals; decrease in pressure drop
Compressors	90–120	Installed mufflers on intake and exhaust, enclosed machine casings, vibration isolation, and lagging of piping systems
Electric motors	90–110	Acoustically lined fan covers, enclosures, and motor mutes
Heaters and furnaces	95–110	Acoustic plenums, intake mufflers, ducts lined and damped
Valves	<80–108	Avoidance of sonic velocities, limited pressure drop and mass flow, replacement with special low-noise valves, vibration isolation, and lagging
Piping	90–105	In-line silencers, vibration isolation and lagging

†Defined as the sound intensity measured in units equal to 10 times the logarithm of the square of the relative pressure associated with the sound wave.

Factors Involved

A general consensus as to the plant location should be obtained before a design project reaches the detailed-estimate stage, and a firm location should be established upon completion of the detailed-estimate design. The choice of the final site should be based first on a complete survey of the advantages and disadvantages of various geographic areas and, ultimately, on the advantages and disadvantages of available real estate. The following factors should be considered in selecting a plant site:

1. Raw materials availability
2. Markets
3. Energy availability
4. Climate
5. Transportation facilities
6. Water supply
7. Waste disposal
8. Labor supply
9. Taxation and legal restrictions
10. Site characteristics
11. Flood and fire protection
12. Community factors

The factors that must be evaluated in a plant location study indicate the need for a vast amount of information, both quantitative (statistical) and qualitative. Fortunately, a large number of agencies, public and private, publish useful information of this type, greatly reducing the actual gathering of the data.

Raw Materials Availability The source of raw materials is one of the most important factors influencing the selection of a plant site. This is particularly true if large volumes of raw materials are consumed, because location near the source of raw materials permits considerable reduction in transportation and storage charges. Attention should be given to the purchased price of the raw materials, distance from the source of supply, freight or transportation expenses, availability and reliability of supply, purity of the raw materials, and storage requirements.

Markets The location of markets or intermediate distribution centers affects the cost of product distribution and the time required for shipping. Proximity to the major markets is an important consideration in the selection of a plant site, because the buyer usually finds it advantageous to purchase from nearby sources. Note that markets are needed for by-products as well as for major final products.

Energy Availability Power and steam requirements are high in most industrial plants, and fuel is ordinarily required to supply these utilities. Consequently, power and fuel can be combined as one major factor in the choice of a plant site. Electrolytic processes require a cheap source of electricity, and plants using electrolytic processes are often located near large hydroelectric installations. If the plant requires large quantities of coal or oil, location near a source of fuel supply may be essential for economic operation. The local cost of power can help determine whether power should be purchased or self-generated.

Climate If the plant is located in a cold climate, costs may be increased by the necessity for construction of protective shelters around the process equipment, and special cooling towers or air conditioning equipment may be required if the prevailing temperatures are high. Excessive humidity or extremes of hot or cold weather can have a serious effect on the economic operation of a plant, and these factors should be examined when a plant site is selected.

Transportation Facilities Water, railroads, and highways are the common means of transportation used by major industrial concerns. The kind and amount of products and raw materials determine the most suitable type of transportation facilities. In any case, careful attention should be given to local freight rates and existing railroad lines. The proximity to railroad centers and the possibility of canal, river, lake, or ocean transport must be considered. Motor trucking facilities are widely used and can serve as a useful supplement to rail and water facilities. If possible, the plant site should have access to all three types of transportation; certainly, at least two types should be available. There is usually need for convenient air and rail transportation facilities between the plant and the main company headquarters, and effective transportation facilities for the plant personnel are necessary.

Water Supply The process industries use large quantities of water for cooling, washing, steam generation, and as a raw material. The plant, therefore, must be located where a dependable supply of water is available. A large river or lake is preferable, although deep wells or artesian wells may be satisfactory if the amount of water required is not too great. The level of the existing water table can be checked by consulting the state geological survey, and information on the constancy of the water table and the year-round capacity of local rivers or lakes should be obtained. If the water supply shows seasonal fluctuations, it may be desirable to construct a reservoir or to drill several standby wells. The temperature, mineral content, silt or sand content, bacteriological content, and cost for supply and purification treatment must also be considered when choosing a water supply.

Waste Disposal In recent years, many legal restrictions have been placed on the methods for disposing of waste materials from the process industries. The site selected for a plant should have adequate capacity and facilities for correct waste disposal. Even though a given area has minimal restrictions on pollution, it should not be assumed that this condition will continue to exist. In choosing a plant site, the permissible tolerance levels for various methods of waste disposal should be considered carefully, and attention should be given to potential requirements for additional waste treatment facilities.

Labor Supply The type and supply of labor available in the vicinity of a proposed plant site must be examined. Consideration should be given to prevailing pay scales, restrictions on number of hours worked per week, competing industries that can cause dissatisfaction or high turnover rates among the workers, and variations in the skill and productivity of the workers.

Taxation and Legal Restrictions State and local tax rates on property income, unemployment insurance, and similar items vary from one location to another. Similarly, local regulations on zoning, building codes, nuisance aspects, and transportation facilities can have a major influence on the final choice of a plant site. In fact, zoning difficulties and obtaining the many required permits can often be much more important in terms of cost and time delays than many of the factors discussed in the preceding sections.

Site Characteristics The characteristics of the land at a proposed plant site should be examined carefully. The topography of the tract of land and the soil structure must be considered, since either or both may have a pronounced effect on construction costs. The cost of the land is important as well as local building costs and living conditions. Future changes may make it desirable or necessary to expand the plant facilities. Therefore, even though no immediate expansion is planned, a new plant should be constructed at a location where additional space is available.

Flood and Fire Protection Many industrial plants are located along rivers or near large bodies of water, and there are risks of flood or hurricane damage. Before a plant site is chosen, the regional history of natural events of this type should be examined and the consequences of such occurrences considered. Protection from losses by fire is another important factor in selecting a plant location. In case of a major fire, assistance from outside fire departments should be available. Fire hazards in the immediate area surrounding the plant site must not be overlooked.

Community Factors The character and facilities of a community can have quite an effect on the location of the plant. If a certain minimum number of facilities for satisfactory living of plant personnel do not exist, it often becomes a burden for the plant to subsidize such facilities. Cultural facilities of the community are important to sound growth. Churches, libraries, schools, civic theaters, concert associations, and other similar groups, if active and dynamic, do much to make a community progressive. Recreation opportunities deserve special consideration. The efficiency, character, and history of both state and local government should be evaluated. The existence of low taxes is not in itself a favorable situation unless the community is already well developed and relatively free of debt.

Selection of the Plant Site

The major factors in the selection of most plant sites are (1) raw materials, (2) markets, (3) energy supply, (4) climate, (5) transportation facilities, and (6) water supply. For a preliminary survey, the first four factors should be considered. Thus, on the basis of raw materials, markets, energy supply, and climate, acceptable locations can usually be reduced to one or two general geographic regions. For example, a preliminary survey might indicate that the best location for a particular plant would be in the south central or southeastern part of the United States.

In the next step, the effects of transportation facilities and water supply are taken into account. This permits reduction of the possible plant location to several general target areas. These areas can then be reduced further by considering all the factors that have an influence on plant location.

As a final step, a detailed analysis of the remaining sites can be made. Exact data on items such as freight rates, labor conditions, tax rates, price of land, and general local conditions can be obtained. The various sites can be inspected and appraised on the basis of all the factors influencing the final decision. Many times, the advantages of locating a new plant on land or near other facilities already owned by the company outweigh the disadvantages of the particular location. In any case, however, the final decision on selecting the plant site should take into consideration all the factors that can affect the ultimate success of the overall operation.

PLANT LAYOUT

After the process flow diagrams are completed and before detailed piping, structural, and electrical design can begin, the layout of process units in a plant and the equipment within these process units must be planned. This layout can play an important part in determining construction and manufacturing costs and thus must be planned carefully, with attention being given to future problems that may arise. Since each plant differs in many ways and no two plant sites are exactly alike, there is no one ideal plant layout. However, proper layout in each case will include arrangement of processing areas, storage areas, and handling areas in efficient coordination and with regard to such factors as

1. New site development or addition to previously developed site
2. Type and quantity of products to be produced
3. Type of process and product control
4. Operational convenience and accessibility
5. Economic distribution of utilities and services

6. Type of buildings required and building code requirements
7. Health and safety considerations
8. Waste disposal requirements
9. Auxiliary equipment
10. Space available and space required
11. Roads and railroads
12. Possible future expansion

Preparation of the Layout

Scale drawings complete with elevation indications can be used for determining the best location for equipment and facilities. Elementary layouts are developed first. These show the fundamental relationships between storage space and operating equipment. The next step requires consideration of the safe operational sequence and gives a primary layout based on the flow of materials, unit operations, storage, and future expansion. By analyzing all the factors that are involved in plant layout, a detailed recommendation can be presented, and drawings and elevations, including isometric drawings of the piping systems, can be prepared, with appropriate software.

Errors in a plant layout are easily located when three-dimensional models prepared with today's software are used, since the operations and construction engineers can immediately see errors which might have escaped notice on two-dimensional templates or blueprints. In addition to increasing the efficiency of a plant layout, these three-dimensional models are very useful during plant construction and for instruction and orientation purposes after the plant is completed.

PLANT OPERATION AND CONTROL

In the design of an industrial plant, the methods which will be used for plant operation and control help determine many of the design variables. For example, the extent of instrumentation can be a factor in choosing the type of process and setting the labor requirements. Remember that maintenance work will be necessary to keep the installed equipment and facilities in good operating condition. The engineer must recognize the importance of such factors which are directly related to plant operation and control and must take them into proper account during the development of a design project.

Instrumentation

Instruments are used in the chemical industry to measure process variables, such as temperature, pressure, density, viscosity, specific heat, conductivity, pH, humidity, dew point, liquid level, flow rate, chemical composition, and moisture content. By use of instruments having varying degrees of complexity, the values of these variables can be recorded continuously and controlled within narrow limits.

Automatic control is the norm throughout the chemical industry, and the resultant savings in labor combined with improved ease and efficiency of operations have more than offset the added expense for instrumentation. (In most cases, control is achieved through the use of high-speed computers. In this capacity, the computer serves as a

Figure 2-3
Example of a graphic panel for a modern industrial plant with a computer-controlled system
(*Courtesy of C.F. Braun and Company.*)

vital tool in the operation of the plant.) Effective utilization of the many instruments employed in a chemical process is achieved through centralized control, whereby one centrally located control room is used for the indication, recording, and regulation of the process variables. Panel boards have been developed that present a graphical representation of the process and have the instrument controls and indicators mounted at the appropriate locations in the overall process. This helps a new operator to quickly become familiar with the significance of the instrument readings, and rapid location of any operational variance is possible. An example of a graphic panel in a modern industrial plant is shown in Fig. 2-3.

Maintenance

Many of the problems involved in maintenance are often caused by not thoroughly evaluating the original design and layout of plant and equipment. Sufficient space for maintenance work on equipment and facilities must be provided in the plant layout, and the engineer needs to consider maintenance and its safety requirements when making decisions on equipment.

Too often, the design engineer is conscious only of first costs and fails to recognize that maintenance costs can easily nullify the advantages of selecting a less expensive installation. For example, a close-coupled motor pump utilizing a high-speed motor may require less space and have a lower initial cost than a standard motor combined with a coupled pump. However, if replacement of the impeller and shaft becomes necessary, the repair cost with a close-coupled motor pump is much greater than with a regular coupled pump. The use of a high-speed motor reduces the life of the impeller and shaft, particularly if corrosive liquids are involved. If the engineer fails to consider the excessive maintenance costs that may result, an error can be made in recommending the cheaper and smaller unit. Similarly, a compact system of piping, valves, and equipment may have a lower initial cost and be more convenient for the operators' use, but maintenance of the system may require costly and time-consuming dismantling operations.

Utilities

The primary sources of raw energy for the supply of power are found in the heat of combustion of fuels and in elevated water supplies. Fuel-burning plants are of greater industrial significance than hydroelectric installations because the physical location of fuel-burning plants is not restricted. At present, the most common sources of energy are oil, gas, coal, and nuclear energy. The decreasing availability of the first two sources of energy will necessitate the use of alternate forms of energy in the not-too-distant future.

In the chemical industries, power is supplied primarily in the form of electrical energy. Agitators, pumps, hoists, blowers, compressors, and similar equipment are usually operated by electric motors, although other prime movers such as steam engines, internal combustion engines, and hydraulic turbines are sometimes employed.

When a design engineer is setting up the specifications for a new plant, a decision must be made on whether to use purchased power or have the plant set up its own power unit. It may be possible to obtain steam for processing and heating as a by-product from the self-generation of electricity, and this factor may influence the final decision. In some cases, it may be justified by means of a HAZOP study to provide power to the plant from two independent sources to permit continued operation of the plant facilities if one of the power sources fails.

Power can be transmitted in various forms, such as mechanical energy, electrical energy, heat energy, and pressure energy. The engineer should recognize the different methods for transmitting power and must choose the ones best suited to the particular process under development.

Steam is generated from whatever fuel is the cheapest, usually at pressures of 3100 kPa or more, and expanded through turbines or other prime movers to generate the necessary plant power; and the exhaust steam is used in the process as heat. The quantity of steam used in a process depends upon the thermal requirements, plus the mechanical power needs, if such a power is generated in the plant.

Water for industrial purposes can be obtained from one of two general sources: the plant's own source or a municipal supply. If the demands for water are large, it is more economical for the plant to provide its own water source. Such a supply may be obtained from drilled wells, rivers, lakes, dammed streams, or other impounded supplies. Before a company agrees to go ahead with any new project, it must ensure a sufficient supply of water for all industrial, sanitary, and safety demands, both present and future.

The value of an abundance of good water supplies is reflected in the selling price of plant locations that have such supplies. Any engineering techniques that are required to procure, conserve, and treat water can significantly increase the operational cost for a plant or process. Increased costs of water processing have made maximum use of processed water essential. In fact, the high costs of constructing and operating a waste treatment plant have led to concentration of industrial wastes with the smallest amount of water, except where treatment processes require dilution.

Structural Design

One of the most important aspects in structural design for the process industries is a correct foundation design with allowances for the heavy equipment and vibrating machinery used. The purpose of the foundation is to distribute the load so that excessive or damaging settling will not occur. The type of foundation depends on the load involved and the material on which the foundation acts. It is necessary, therefore, to know the characteristics of the soil at a given plant site before the structural design can be started.

The allowable bearing pressure varies for different types of soils, and the soil should be checked at the surface and at various depths to determine the bearing characteristics. The allowable bearing pressure for rock is at least 30×10^5 kg/m^2, while that for soft clay may be as low as 10^4 kg/m^2. Intermediate values of 4×10^4 to 1×10^5 kg/m^2 apply for mixtures of gravel with sand, hard clay, and hardpan.

A foundation may simply be a wall supported on rock or hardpan, or it may be necessary to increase the bearing area by the addition of a footing. Plain concrete is usually employed for footings, while reinforced concrete, containing steel rods or bars, is commonly used for foundation walls. If possible, a foundation should extend below the frost line and should always be designed to handle the maximum load. Pilings are commonly used for supporting heavy equipment or for other special loads.

Maintenance difficulties encountered with floors and roofs should be given particular attention in a structural design. Concrete floors are used extensively in the process industries, and special cements and coatings are available which make the floors resistant to heat or chemical attack. Flat roofs are often specified for industrial structures. Felt saturated with coal-tar pitch combined with a coal-tar pitch-gravel finish is satisfactory for roofs of this type. Asphalt-saturated felt may be used if the roof has a slope of more than 4 percent.

Corrosive effects of the process, cost of construction, and climatic effects must be considered when choosing structural materials. Steel and concrete are the materials of construction most commonly used, although wood, aluminum, glass blocks, cinder blocks, glazed tile, bricks, and other materials are also of importance. Allowances must be made for the type of lighting and drainage, and sufficient structural strength must be provided to resist normal loads as well as extreme loads due to high winds or other natural causes.

In any type of structural design for the process industries, the function of the structure is more important than the form. The style of architecture should be subordinated to the need for supplying a structure that is adapted to the proposed process and has sufficient flexibility to permit changes in the future. Although cost is certainly important, the engineer preparing the design should never forget the fact that the quality of a structure remains apparent long after the initial cost is forgotten.

Storage

Adequate storage facilities for raw materials, intermediate products, final products, recycle materials, off-grade materials, and fuels are essential to the operation of a process plant. A supply of raw materials permits operation of the process plant regardless of temporary procurement or delivery difficulties. Storage of intermediate products may be necessary during plant shutdown for emergency repairs while storage of final products makes it possible to supply the customer even during a plant difficulty or unforeseen shutdown. An additional need for adequate storage is often encountered when it is necessary to meet seasonal demands from steady production.

Bulk storage of liquids is generally handled by closed spherical or cylindrical tanks to prevent the escape of volatiles and minimize contamination. Since safety is an important consideration in storage tank design, the American Petroleum Institute[†] and the National Fire Protection Association[‡] publish rules for safe design and operation. Floating roof tanks are used to conserve valuable products with vapor pressures that are below atmospheric pressure at the storage temperature. Liquids with vapor pressures above atmospheric must be stored in vapor-tight tanks capable of withstanding internal pressure. If flammable liquids are stored in vented tanks, flame arresters must be installed in all openings except connections made below the liquid level.

Gases are stored at atmospheric pressure in water dry-seal gas holders. The wet-gas holder maintains a liquid seal of water or oil between the top movable inside tank and the stationary outside tank. In the dry-seal holder, the seal between the two tanks is obtained through the use of a flexible rubber or plastic curtain. Recent developments in bulk natural gas or gas-product storage show that pumping the gas into underground strata is the cheapest method available. High-pressure gas is stored in spherical or horizontal cylindrical pressure vessels.

Solid products and raw materials are stored either in weather-tight tanks with sloping floors or in outdoor bins and mounds. Solid products are often packed directly in bags, sacks, or drums.

Materials Handling

Materials-handling equipment is logically divided into continuous and batch types, and into classes for the handling of liquids, solids, and gases. Liquids and gases are handled by means of pumps and blowers; in pipes, flumes, and ducts; and in containers such as drums, cylinders, and tank cars. Solids may be handled by conveyors, bucket elevators, chutes, lift trucks, and pneumatic systems. The selection of materials-handling equipment depends upon the cost and the work to be done. Factors that must be considered in selecting such equipment include

1. Chemical and physical nature of material being handled
2. Type and distance of movement of material
3. Quantity of material moved per unit time
4. Nature of feed and discharge from materials-handling equipment
5. Continuous or intermittent nature of materials handling

[†]American Petroleum Institute, 50 W. 50th St., New York, NY.

[‡]National Fire Protection Association, 60 Batterymarch St., Boston, MA.

The major movement of liquid and gaseous raw materials and products within a plant to and from the point of shipment is done by pipeline. Many petroleum plants also transport raw materials and products by pipeline. When this is done, local and federal regulations must be strictly followed in the design and specification of the pipeline.

Movement of raw materials and products outside of the plant is usually handled by rail, ship, truck, or air transportation. Some type of receiving or shipping facilities, depending on the nature of the raw materials and products, must be provided in the design of the plant. Information for the preparation of such specifications usually can be obtained from the transportation companies serving the area.

In general, the materials-handling problems in the chemical engineering industries do not differ widely from those in other industries except that the existence of special hazards, including corrosion, fire, heat damage, explosion, pollution, and toxicity, together with special service requirements, will frequently influence the design. The most difficult of these hazards often is corrosion. This is generally overcome by the use of a high-first-cost, corrosion-resistant material in the best type of handling equipment or by the use of containers which adequately protect the equipment.

PATENT CONSIDERATIONS

A patent is essentially a contract between an inventor and the public. In consideration of full disclosure of the invention to the public, the patentee is given exclusive rights to control the use and practice of the invention. A patent gives to the holder the power to prevent others from using or practicing the invention for a period of 17 years from the date of granting. In contrast, trade secrets and certain types of confidential disclosures can receive protection under common-law rights only as long as the secret information is not public knowledge.

A new design should be examined to make certain no patent infringements are involved. If the investigation can uncover even one legally expired patent covering the details of the proposed process, the method can be used with no fear of patent difficulties. Although most large corporations have patent attorneys to handle investigations of this type, the design engineer can be of considerable assistance in determining if infringements are involved. An engineer, therefore, should have a working knowledge of the basic practices and principles of patent law.[†]

PROBLEMS[‡]

2-1 Cleanup of a mercury spill in an unventilated stockroom fails to remove the mercury trapped in small cracks in the floor. Is the maximum concentration of mercury in the air at 23°C

[†]Statutes and general rules applying to U.S. patents are presented in the two pamphlets "Patent Laws" and "Rules of Practice of the United States Patent Office in Patent Cases," U.S. Government Printing Office.

[‡]A number of the problems are adapted by permission from the Center for Chemical Process Safety, *Safety, Health, and Loss Prevention,* American Institute of Chemical Engineers, New York, 1990.

acceptable if the federal standard permissible exposure limit for mercury in the air is $0.1\ mg/m^3$?

2-2 Corrosion inside a closed container can result in an oxygen-deficient atmosphere within the container. The typical corrosion rate for carbon steel in very moist air is about 0.127 mm/yr and is approximately first-order with respect to the oxygen concentration. Assuming that the corrosion reaction can be approximated by the reaction

$$2Fe + \frac{3}{2}O_2 \rightarrow Fe_2O_3$$

determine the exposure time required to reduce the oxygen content in a closed 7.6-m-diameter spherical container from 21 to 19.5 mol %.

2-3 Vinyl chloride, with a *permissible exposure limit* (PEL) of 1 ppm, evaporates into the air at a rate of 7.5 g/min. Assuming that a safety factor of at least 5 is used to ensure proper dilution, determine the flow rate of air that will be required to achieve the PEL, utilizing dilution ventilation.

2-4 Benzene has a PEL of 1 ppm for an 8-h exposure. If liquid benzene is evaporating at a rate of 2 ml/min in air whose temperature and pressure are 22°C and 88 kPa, respectively, what must the ventilation rate of the air be to keep the benzene concentration below the PEL value?

2-5 Chemical cartridge respirators are personal protective devices used to adsorb harmful vapors and gases. A performance analysis of such respirators is similar to that used in a fixed-bed adsorber. The generalized correlation of adsorption potential shows that the logarithm of the amount adsorbed is linear with the function $(T/V)[\log(f_s/f)]$, where T is the temperature in kelvins, V is the molar volume of liquid at the normal boiling point in cubic centimeters per gram mol, f_s is the saturated fugacity (approximate as the vapor pressure), and f is the fugacity of the vapor (approximate as the partial pressure). Data for the adsorption of dichloropropane on a selected charcoal are as follows:

Amount adsorbed, cm^3_{liq}/100 g charcoal	$(T/V)[\log(f_s/f)]$ (units as given)
1	21
10	11

If a respirator contains 100 g of this charcoal and breakthrough occurs when 82 percent of the adsorbent is saturated, how long can the respirator be used in a dichloropropane concentration of 750 ppm when the temperature is 26.9°C? Assume that the flow rate of contaminated air through the adsorbent is 45 l/min. The molar volume of the dichloropropane may be assumed to be 100 cm^3/g mol.

2-6 The flash point of a liquid mixture can be estimated by determining the temperature at which the equilibrium concentration of the flammable vapors in the air reaches a concentration such that $\sum(y_i/\text{LFL}_i) = 1.0$, where y_i is the vapor-phase mol percent of component i and LFL_i is the lower flammability limit concentration of component i, is expressed in mol percent. Estimate the flash point of a liquid mixture containing 50 mol percent n-octane and 50 mol percent n-nonane. The LFL values for n-octane and n-nonane are 1.0 and 0.8 percent by volume, respectively.

2-7 Estimate the flash point of acetone and compare it with the experimental value given in the literature. *Hint:* Start with the basic principle that the fugacity in the vapor phase must equal that in the liquid phase. The lower flammable limit for acetone is 2.55 percent by volume.

2-8 Which of the following liquids commonly used in the laboratory could form flammable air-vapor mixtures if spilled in a storage cabinet and allowed to reach equilibrium?

Acetone	Carbon disulfide	Methyl alcohol
Benzene	Ethyl ether	*n*-Pentane

What type of fire extinguishers would be appropriate if a small spill were ignited? Note that if equilibrium is not attained, the vapor concentrations will be lower than calculated and a flammable air-vapor mixture could still be present.

2-9 When a flammable material is burned, there will be an increase in either the volume of the gas produced (provided the pressure is constant) or the pressure in the container (provided the volume is constant). Calculate the volume of gas formed during the adiabatic combustion of 100 kg mol of gaseous propane at a constant pressure of 1 atm. Assume that the 200 percent theoretical air and the propane involved in the combustion are at 25°C and that the combustion goes to completion.

2-10 A number of accidents have occurred as a result of the discharge of static electricity generated by the flow of fluids. To provide additional protection against explosion or fire due to the discharge of static electricity, determine at what temperature an acetone drum filling operation would have to be maintained so that flammable mixtures would not be produced by vapor-air mixtures in equilibrium with the liquids. The lower flammable limit for acetone vapor in air is 2.55 percent by volume.

2-11 A number of chemicals decompose under appropriate conditions and liberate a fair quantity of energy that could initiate considerable property damage. Determine the temperature and pressure attained in a closed spherical container when acetylene, initially at 25°C and 10 atm, rapidly decomposes to carbon and hydrogen after accidentally being subjected to an electric spark. What thickness of carbon steel would have been required to contain the decomposition reaction if no relief valve had been available to relieve the pressure buildup? (Assume a safety factor of 4 in the thickness calculation.)

2-12 If the heat generated in an exothermic reaction is not appropriately removed, the reaction rate can get out of control and the reactor will be damaged if it is not adequately vented or protected. Consider a reaction in a continuously stirred reactor that has an activation energy of 28,000 cal/g mol. The water-cooled jacket surrounding the reactor utilizes cooling water with an inlet temperature of 15°C. What is the maximum temperature at which the reaction can operate without having the reaction run away? What actions might be taken to maintain safe operation of the reactor?

2-13 Liquid chlorine may be transferred from a chlorine storage container by pressurizing with dry chlorine gas. Develop two systems (*a*) using recompressed chlorine vapor and (*b*) using vaporized chlorine; provide the appropriate control and alarm system for each system.

2-14 A distillation unit has been designed to handle a very hazardous material. The unit utilizes a reflux drum and buffer storage. List several ways in which the inventory of the hazardous material can be reduced or eliminated. Sketch and instrument the system that is recommended.

2-15 Review the benzene storage system shown in Fig. P2-15 in which benzene enters tank T-101 through line L-101 and is subsequently pumped out through pumps P-101A and P-101B at

35 atm into line L-102. The accompanying table shows part of a HAZOP study of line L-102 that has been completed. Provide the missing information to complete the study.

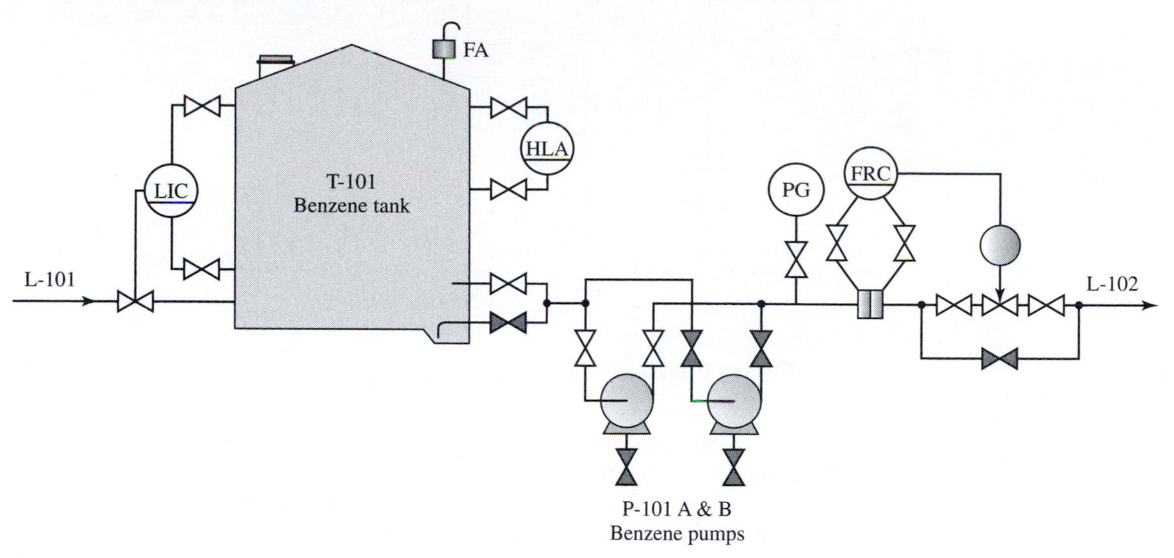

Figure P2-15

HAZOP for line L-102

Guide word	Property	Possible cause	Possible consequence	Action required
More	Temperature	—	Pump seal failure, vapor lock	Install a feedback line
Less	Temperature	Low ambient temperature	—	Steam tracing
More	Flow	Line fracture	Spillage (possible large explosion)	—
		Pump seal failure	Spillage (possible small explosion)	—
		Control fault		Consider bypass
No	Flow	—	Shutdown	Low-level alarm
			Shutdown	Automatic start-up of standby pump
Reverse	Flow	Pump failure	Backpressure on storage vessel	—
As well as	Impurities	—	Possible small detonation	Priming line

2-16 Produce a fault tree for the system shown in Fig. P2.16, on page 66, where the highest-order event is overpressure of the vessel. The possible causes of various failures should be suggested.

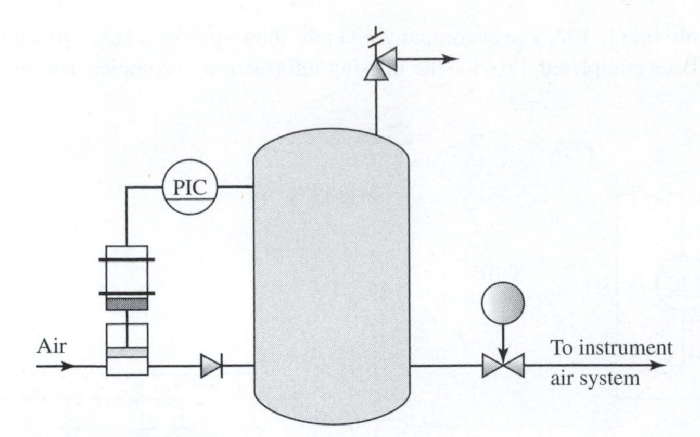

Figure P2-16

2-17 The trend in the fertilizer industry during the past few years has been toward larger and larger fertilizer plants. In terms of plant location, what are the more important factors that should be considered and which factors become even more important as the size of the plant is increased? Are these factors of equal importance regardless of the type of fertilizer produced? Analyze this situation for ammonia, urea, and phosphate fertilizer process plants.

2-18 The following information has been obtained during a test for the BOD of a given industrial waste: 15 ml of the waste sample was added to a 500-ml BOD bottle, and the bottle was then filled with standard dilution water seeded with waste organisms. A second 500-ml BOD bottle was filled with standard dilution water. After 5 days at 20°C, the blank and the diluted waste samples were analyzed for dissolved-oxygen content by the sodium azide modification of the Winkler method. The blank results indicated a dissolved-oxygen content of 9.0 ppm. Results for the diluted sample showed a dissolved-oxygen content of 4.0 ppm. On the basis of the following assumptions, determine the BOD for the waste. The specific gravities of the liquids are 1.0; the waste sample contains no dissolved oxygen.

2-19 Liquid chlorine is transferred from a storage tank, vaporized, and sent to a reactor. The vaporization may be carried out by a vaporizer using hot water, steam, or closed-circuit heater using a heat-transfer fluid that is inert to chlorine. The vaporizer may be a coil immersed in a heating bath, a vertical tube bundle, a concentric tube unit, or an evaporator. Indicate the relative merits of these different systems, and select the one most suitable for the vaporization process. Develop a piping and instrument diagram for the system selected, making certain that all temperature, pressure, liquid level and flow control systems, and alarms are included for safe operation.

2-20 A cylindrical tank with a diameter of 15 m is used to store 1600 m^3 of liquid benzene. To provide adequate protection during an external fire, determine the venting rate and the vent area if the maximum tank pressure is to be maintained at 250 Pa gauge. Is the calculated vent area realistic, or should the tank specifications call for a weak-seam roof? Heat transfer to the tank is by radiation from the hot soot particles and gases in the flame and convection from the hot gases. The radiant flux emitted from many burning hydrocarbon fuels is approximately 95 kW/m^2. This average accounts for fluctuations in temperature, emissivity, composition, and other variables in the flame. Note that the tank when heated from the fire will reradiate some of the energy it receives from the fire.

CHAPTER

3

Process Design Development

Process design can provide chemical engineers with probably the most creative activity enjoyed by the engineering profession. There are many opportunities to come up with imaginative new processes or the introduction of changes that can positively alter existing processes. Chemical engineers in process design face challenges in creating ingenious and often complex flowsheets for new or revised processes. In this task they will make many design decisions that can affect the success or failure of a process design. Without a doubt, process design is rarely straightforward or routine. Rather, such developments involve innovative approaches to entirely new processes or revisions to existing processes that are more profitable, better controlled, operationally safe, as well as environmentally sound.

This chapter will review in greater detail some of the principal steps, as outlined earlier in Table 1-1, that are followed in designing and retrofitting chemical and biochemical processes. A number of these steps will be highlighted in an abbreviated form with the development of a preliminary process design. The design examines a practical problem of the type frequently encountered within the chemical industry, one which involves both process design and economic considerations.

DEVELOPMENT OF DESIGN DATABASE

The development of a design project is always initiated with the creation of one or more potential solutions to meet a recognized societal or engineering need. A careful analysis of each proposed solution must be made so that the approach recommended can be stated as clearly and concisely as possible to define the scope of the project, recognize potential design problems, and identify data that will be required in the solution.

Literature Survey

In the development of solutions to a design need, it is important to make a thorough search of the literature to obtain the latest data, flowsheets, equipment, and simulation models that may lead to a more profitable design. Several literature indexes are extremely helpful in searching the current literature. These provide electronic access to many of the technical articles published since the late 1970s, including kinetics data, thermophysical property data, and related information for many chemicals. Of primary interest to a design engineer are the *Chemical Abstracts* (one of the most comprehensive scientific indexing and abstracting services in biochemistry, organic chemistry, physical and analytical chemistry, macromolecular chemistry, applied chemistry, and chemical engineering, available electronically with entries since 1907), the *Engineering Index* (with access to 4500 journals, technical reports, and books, electronically since 1985), the *Applied Science and Technology Index* (with electronic access to 350 journals since 1985), and the *Science Citation Index* (with access to 3300 journals since 1955 and electronically since 1985, with searches that indicate the work for which an author has been cited).

A primary source of information on all aspects of chemical engineering principles, design, costs, and applications is *Perry's Chemical Engineers' Handbook* (1997). Other handbooks and reference texts well known to chemical engineers include the *Handbook of Chemistry and Physics* (published annually by CRC Press), *Chemical Processing Handbook* (1993), *Unit Operations Handbook* (1993), *Riegel's Handbook of Industrial Chemistry* (1992), *JANAF Thermochemical Tables* (1985), *Handbook of Reactive Chemical Hazards* (1990), *Standard Handbook of Hazardous Waste Treatment and Disposal* (1989), and *Data for Process Design and Engineering Practice* (1995).

Several literature resources are widely used by design engineers. These include the SRI (Stanford Research Institute) Design Reports, which provide detailed documentation of many chemical processes. Three of the more comprehensive, multivolume encyclopedias that describe the uses, history, typical process flowsheets, operating conditions, and related information for the production of most chemicals produced today are the *Kirk-Othmer Encyclopedia of Chemical Technology* (1994), *Ullman's Encyclopedia of Industrial Chemistry* (1988), and *Encyclopedia of Chemical Processing and Design* (1976). Other encyclopedias that may provide additional information in this creative task include the *McGraw-Hill Encyclopedia of Science and Technology* (1987), *Van Nostrand's Scientific Encyclopedia* (1988), and *Encyclopedia of Materials Science and Engineering* (1986).

In addition to the data available from the literature, extensive databases for some 2000 compounds are provided by process simulators such as ASPEN PLUS, HYSYS, CHEM CAD, and PRO-II. These are very useful since they are accessed by large libraries of programs that carry out material and energy balances as well as make estimates of equipment sizes and costs.

The extensive phase equilibria data available electronically from the DECHEMA data bank can often eliminate the need to survey the literature for this important property. Each set of data in this compilation has been regressed to determine interaction coefficients for the binary pairs to be used to estimate liquid-phase activity coefficients for the NRTL, UNIQUAC, Wilson, etc., equations. This database is also accessible by process simulators.

As noted earlier in Chap. 2, a number of common chemicals have toxic effects on humans and should not be used in a process. One source of information on these chemicals is *Toxic Chemical Release Inventory* (TCRI), maintained by the EPA and the NFPA. Another source is *Sittig's Handbook of Toxic and Hazardous Chemicals* (2002). Flammability data are tabulated in *Perry's Chemical Engineers' Handbook* (1997). For those compounds not included in the tabulation, methods are available to estimate the data.

To determine the gross profitability of a specific chemical reaction, a source for chemical prices available is the *Chemical Market Reporter,* a biweekly newspaper that provides up-to-date prices on many chemicals. Since these prices may not reflect the market situation at a particular location, it may be necessary to contact the manufacturers of those specific chemicals.

Appropriate prices for utilities such as steam, cooling water, and electricity can generally be obtained from local utility companies. For design purposes, a range of costs for many of the utilities is tabulated in App. B.

New information is constantly becoming available in periodicals, books, trade bulletins, government reports, university bulletins, and many other sources. Many of the publications are devoted to shortcut methods for estimating physical properties or making design calculations, while others present compilations of essential data in the form of monographs or tables.

The effective design engineer must make every attempt to keep an up-to-date knowledge of the advances in the field. Personal experience and contacts, attendance at meetings of technical societies and industrial expositions, and reference to the published literature are very helpful in giving the engineer the background information necessary for a successful design.

Patent Search

Design engineers must be aware of patents to avoid duplication of designs protected by these patents. However, expired patents can provide helpful information in the design of second-generation processes to produce these chemicals or chemicals that have similar properties or chemical reactions. Patents from the United States, Germany, Japan, and other countries are available in major libraries from which copies may be obtained. Since the early 1990s, patents have also become available on the Internet.

PROCESS CREATION

After careful review of the societal or engineering need and surveying the literature for possible leads to one or more practical solutions, the design engineer frequently encounters a major challenge in creating a new process or significantly improving an existing process to satisfy this need. This part of process design involves the synthesis of various configurations of processing operations that will produce a product in a reliable, safe, and economical manner with a high yield and minimum by-product or waste. In the past this creative activity was normally performed from experience gained in similar processing situations and the use of *heuristics,* or rules of thumb. However, this approach has changed rather dramatically over the past two decades as synthesis strategies have become more quantitative and scientific with the use of decision-tree analysis and mathematical programming aided by the use of modern computers.

In the process of synthesizing a flowsheet of process operations to convert raw materials to desired products, the design engineer first must select the processing mode: either batch or continuous. Then a decision must be made on the raw materials, products, and by-products involved in the process. This is followed by a review of the principal operations required to achieve the desired product and then assembling different configurations of these process operations. It is quite apparent that a process with several raw material choices, each with a number of process operations, can very quickly become a large combinatorial problem. However, the quantitative methods of synthesis discussed in Chap. 4 and the optimization procedures reviewed in Chap. 9 now provide tools for the design engineer to minimize the number of options that need to be investigated.

Batch Versus Continuous Operation

In general, continuous processing is the preferred mode of operation used in the production of commodity chemicals, petroleum products, plastics, paper, solvents, etc., because of reduced labor costs, improved process control, and more uniform product quality. However, batch and semicontinuous processes are often utilized when production rates are small, such as in the manufacture of specialty chemicals, pharmaceuticals, and electronic materials, or when the product demand is intermittent. This is particularly true when the demand for the product is interspersed with the demand for one or more other products that can be manufactured using essentially the same processing equipment. In such cases, the processing equipment is cleaned between batches and used alternately to carry out the steps involved in the production of different products.

The choice between continuous or batch as well as semicontinuous processing is commonly made very early in the process synthesis step. Normally, continuous processing is assumed unless a qualitative analysis indicates that batch or semicontinuous processing provides a better mode of operation. Besides in the case of low or intermittent demand, the design engineer may opt for batch and semicontinuous operation when the process involves hazardous or toxic chemicals or when safety of the process is a major concern.

Raw Materials and Product Specifications

Once the operational mode is determined, the design engineer should establish a number of specifications that can define the state condition of the raw material and product.

Generally, the flow rate required for the product, as established by a market analysis, is of primary interest. This will have a direct effect on the flow rate of the raw material(s). After the flow rates have been established, the composition (mol or mass fraction of each chemical species), phase (liquid, gas, or solid), form (particle size distribution and particle size), temperatures, and pressures of each raw material and product stream are established as well. Although other specifications may be required, the specifications listed are usually sufficient to establish the condition of both the raw material and the product streams.

Process Synthesis Steps

The next step in process synthesis involves the selection of processing operations to convert the raw material to products. The basic processing operations listed below are used to eliminate property differences between inlet and outlet streams.†

Synthesis step	*Basic processing operation*
Molecular change	Chemical reaction
Composition change	Separation of mixture
Phase elimination	Phase separation
Temperature difference	Change in temperature
Pressure difference	Change in pressure
Phase difference	Change in phase
Distribution change	Mixing of streams

Most chemical engineering equipment involves one or more of these basic programming operations. For example, reactors are used in a chemical process to eliminate differences in the molecular structure between the raw material and product streams. The positioning of the reaction operation in the arrangement of process operations proceeding from the raw material to product involves the degree of conversion, reaction rates, competing side reactions, and the existence of reversible reactions. All these reaction concerns, in turn, are closely related to the temperature and pressure at which the reactions are carried out, the methods for adding or removing energy during the reaction, and the choice of catalysts that provide both the reaction rate and the selectivity required in the process.

It is apparent that the use of any one process operation in a process stream will have the effect of reducing or eliminating one or more of the property differences between the raw materials(s) and the product. As each process operation is inserted into the process flowsheet, the resulting streams from that operation move closer to those of the required product. In essence, the process operations are inserted with the goal of reducing the differences in molecular structure, composition, phase, temperature, pressure, etc., until the process stream leaving the last process operation is identical to that required by the desired product.

†D. F. Rudd, G. J. Powers, and J. J. Siirola, *Process Synthesis,* J. Wiley, New York, 1973.

Note that as the process synthesis proceeds, many alternatives are generated that should be considered in the application of each step. Many of these alternatives cannot be rejected before proceeding to the next step. As a result, each step generates additional process flowsheets. In a complex process this can become a huge combinatorial problem. Obviously, approaches are required to eliminate the less promising alternatives as soon as possible. Because of their importance, several of these approaches are covered in some detail in Chap. 4.

To achieve the objective of creating the most promising process flowsheets, care must be taken to include sufficient analysis for each synthesis step to make certain that each new step does not lead to a less profitable process or exclude the most profitable process prematurely. For this reason, it has become common practice to make a quick economic analysis of various process arrangements in the synthesis step with the aid of a suitable simulator.

PROCESS DESIGN

Quite often the process creation step, after elimination of the less profitable processes, still provides several entirely different processing flowsheets for the manufacture of the same product. These processes must be compared in order to select the one that is best suited for the existing conditions. This comparison certainly can be accomplished through the development of complete designs for each process. In many cases, however, all but one or two of the potential processes can be eliminated by a weighted comparison of the essential variable items, and detailed design calculations for each process will not be required. The following items should be considered in a comparison of this type:

1. Technical factors
   - *a.* Process flexibility
   - *b.* Continuous, semicontinuous, or batch operation
   - *c.* Special controls involved
   - *d.* Commercial yields
   - *e.* Technical difficulties involved
   - *f.* Energy requirements
   - *g.* Special auxiliaries required
   - *h.* Possibility of future developments
   - *i.* Health and safety hazards involved
2. Raw materials
   - *a.* Present and future availability
   - *b.* Processing required
   - *c.* Storage requirements
   - *d.* Materials
3. Waste products and by-products
   - *a.* Amount produced
   - *b.* Value
   - *c.* Potential markets and uses

   d. Manner of discard
   e. Environmental aspects

4. Equipment
   a. Availability
   b. Materials of construction
   c. Initial costs
   d. Maintenance and installation costs
   e. Replacement requirements
   f. Special designs

5. Plant location
   a. Amount of land required
   b. Transportation facilities
   c. Proximity to markets and raw material sources
   d. Availability of service and power facilities
   e. Availability of labor
   f. Climate
   g. Legal restrictions and taxes

6. Costs
   a. Raw materials
   b. Energy
   c. Depreciation
   d. Other fixed charges
   e. Processing and overhead
   f. Special labor requirements
   g. Real estate
   h. Patent rights
   i. Environmental controls

7. Time factor
   a. Project completion deadline
   b. Process development required
   c. Market timeliness
   d. Value of money

8. Process considerations
   a. Technology availability
   b. Raw materials common with other processes
   c. Consistency of product within company
   d. General company objectives

In many cases, this comparison may indicate that additional research, laboratory, or pilot plant data are necessary, and a program to obtain this information may be initiated. Process development on a pilot-plant or semiworks scale is usually desirable in order to obtain accurate design data. Valuable information on material and energy balances can be obtained, and process conditions can be examined to supply data on temperature and pressure variation, yields, rates, grades of raw materials and products, materials of construction, operating characteristics, and other pertinent design variables. In the case of a new product, a pilot plant or semiworks plant may be needed in order to provide sufficient product for market evaluation.

Types of Process Designs

Depending on the accuracy and detail required, design engineers generally classify process designs in the following manner:

1. Order-of-magnitude designs
2. Study or factored designs
3. Preliminary designs
4. Detailed-estimate designs
5. Final process designs

The first two types really are not process designs themselves, but are quick estimating procedures that sometimes are used to determine the level of investment required for a proposed design project. These types are discussed in Chap. 6.

Preliminary designs are ordinarily used as a basis for determining whether further work should be done on a proposed process. The design is based on approximate process methods, and approximate cost estimates are prepared. Few details are included, and the time spent on calculations is kept to a minimum.

If the results of the preliminary design show that further work is justified, a *detailed-estimate design* may be developed. In this type of design, the cost-and-profit potential of an established process is determined by detailed analyses and calculations. However, exact specifications are not given for the equipment; piping and layout work is minimized.

When the detailed-estimate design indicates that the proposed project should be a commercial success, the final step before developing construction plans for the plant is the preparation of a *final process design*. Complete specifications are presented for all components of the plant, and accurate costs based on quoted prices are obtained. The final process design includes detailed printouts and sufficient information to permit immediate development of the final plans used during the construction phase of the project.

Preliminary Design In preparing the preliminary design, the chemical engineer utilizes the flowsheet that was developed in the process creation step. If the solution to the societal or engineering need did not involve the process creation step, the chemical engineer first must establish a workable manufacturing process for producing the desired product. Obviously, if a number of alternative processes or methods are developed, a comparison between these processes needs to be made as outlined above, or each process should receive equal consideration.

The first step in preparing the preliminary design is to establish the *bases for design*. In addition to the known specifications for the product and availability of raw materials, the design can be controlled by such items as the expected annual operating factor (fraction of the year that the plant will be in operation), temperature of the cooling water, available steam pressures, fuel used, value of by-products, etc. The next step consists of preparing a simplified flow diagram showing the processes that are involved and deciding upon the unit operations that will be required. A preliminary material balance at this point may very quickly eliminate some of the alternative cases. Flow rates and stream conditions for the remaining cases are now evaluated by complete material balances, energy balances, and a knowledge of raw material and product specifications, yields, reaction rates, and time cycles. The temperature, pressure, and composition of every process stream are determined. Stream enthalpies; percent vapor, liquid, and solid; heat duties; and so on are included where pertinent to the process.

Unit process operations are used in the design of the specific pieces of equipment. (Assistance with the design and selection of various types of process equipment is given in Chaps. 12 through 15.) Equipment specifications are generally summarized in the form of tables and included with the final design report. These tables usually include the following:

1. *Columns (distillation).* In addition to the number of plates and operating conditions it is necessary to specify the column diameter, materials of construction, plate layout, etc.
2. *Vessels.* In addition to size, which is often dictated by the holdup time desired, materials of construction and any packing or baffling should be specified.
3. *Reactors.* Catalyst type and size, bed diameter and thickness, heat-transfer configuration, cycle and regeneration arrangements, materials of constructions, etc., must be specified.
4. *Heat exchangers and furnaces.* Manufacturers are usually supplied with the duty, corrected log mean-temperature difference, percent vaporized, pressure drop desired, and materials of construction.
5. *Pumps and compressors.* Specify type, power requirement, pressure difference, gravities, viscosities, and working pressures.
6. *Instruments.* Designate the function and any particular requirements.
7. *Special equipment.* Provide specifications for mechanical separators, mixers, dryers, etc.

The foregoing is not intended as a complete checklist, but rather as an illustration of the type of summary that is required. (The headings used are particularly suited for the petrochemical industry; others may be desirable for other industries.) As noted in the summary, the selection of materials is intimately connected with the design and selection of the proper equipment.

As soon as the equipment needs have been firmed up, the utilities and labor requirements can be determined and tabulated. Estimates of the capital investment and the total product cost (as outlined in Chap. 6) complete the preliminary-design calculations. Economic evaluation plays an important part in any process design. This is particularly true not only in the selection of a specific process, choice of raw materials used, and operating conditions chosen, but also in the specification of equipment. No design of a piece of equipment or a process is complete without an economic evaluation. In fact, as mentioned in Chap. 1, no design project should ever proceed beyond the preliminary stages without a consideration of costs. Evaluation of costs in the preliminary-design phases greatly assists the engineer in further eliminating many of the alternative cases.

The final step, and an important one in preparing a typical process design, involves preparing a report which will present the results of the design work. Unfortunately, this phase of the design work quite often receives very little attention by the chemical engineer. As a consequence, untold quantities of excellent engineering calculations and ideas are sometimes discarded because of poor communications between the engineer and management.† Finally, it is important that the preliminary design be carried out as soon as sufficient data are available from the process development step. In this way, the preliminary design can serve its main function of eliminating an undesirable project before large amounts of money and time are expended.

Detailed-Estimate Design The preliminary design and the associated process development work provide sufficient information necessary for a detailed-estimate design. The following factors should be established within narrow limits before a detailed-estimate design is developed:

1. Manufacturing process
2. Material and energy balances
3. Temperature and pressure ranges
4. Raw material and product specifications
5. Yields, reaction rates, and time cycles
6. Materials of construction
7. Utilities' requirements
8. Plant site

When the preceding information is included in the design, the result permits accurate estimation of required capital investment, manufacturing costs, and potential profits. Consideration should be given to the types of buildings, heating, ventilating, lighting, power, drainage, waste disposal, safety facilities, instrumentation, etc.

Final Process Design The final process design (or detailed design) is prepared for purchasing and construction from a detailed-estimate design. Detailed drawings are made for the fabrication of special equipment, and specifications are prepared for purchasing standard types of equipment and materials. A complete plant layout is prepared, and printouts and instructions for construction are developed. Piping diagrams

†See Chap. 11 for assistance in preparing more concise and clearer design reports.

and other construction details are included. Specifications are given for warehouses, laboratories, guardhouses, fencing, change houses, transportation facilities, and similar items. The final process design must be developed with the assistance of individuals skilled in various engineering fields, such as architectural, ventilating, electrical, and civil. Safety conditions and environmental impact factors must always be included in the design.

PROCESS FLOW DIAGRAMS

For many years the chemical engineer has used flow diagrams to show the sequence of equipment and unit operations involved in the overall process, principally to simplify visualization of the manufacturing procedures and indicate the quantities of materials and energy that were transferred. These diagrams were generally identified as qualitative, quantitative, or combined-detail.

A qualitative flow diagram indicated the flow of material, unit operations involved, equipment necessary, and special information on operating temperatures and pressures. A quantitative flow diagram showed the quantities of materials required for the process operation. Preliminary flow diagrams were made during the early stages of the design project. As the design moved toward completion, detailed information on flow quantities and equipment specifications became available, and combined-detail flow diagrams were prepared. This type of diagram showed the qualitative flow pattern and served as a base reference for providing equipment specifications, quantitative data, and sample calculations.

With the aid of computer software, a more complete modeling of the chemical process being designed is now possible, once the overall flowsheet and required equipment have been established. This is accomplished with *process flow diagrams* (PFDs). The latter are approximate models to the actual chemical process that include all major processing operation units, all necessary auxiliary equipment such as pumps and compressors, and all material and energy streams. This permits PFDs to present a more complete picture of the process design requirements and allows for improved accuracy in design when compared with the flowsheets developed in the past.

Initially, the detail available in PFDs was not very useful to the design engineer; however, with modern simulation software, PFDs have achieved a high degree of reliability that permits some processes to go directly from the PFD to actual construction. Present process flow diagrams can also assist in other aspects of design, including equipment sizing, piping network design, and control schemes for the process. Through simulation, the operational behavior of the design can be investigated, and appropriate control methods can be examined for applicability to the specific design. This in turn allows for a more accurate visualization of the overall process performance and perhaps can either eliminate or resurrect design options that were considered in the initial flowsheet process. An example of a modern PFD illustrating the degree of detail possible is shown in Fig. 3-1. Additional detail possible in terms of control schemes is demonstrated in Fig. 3-2, which highlights the main column shown

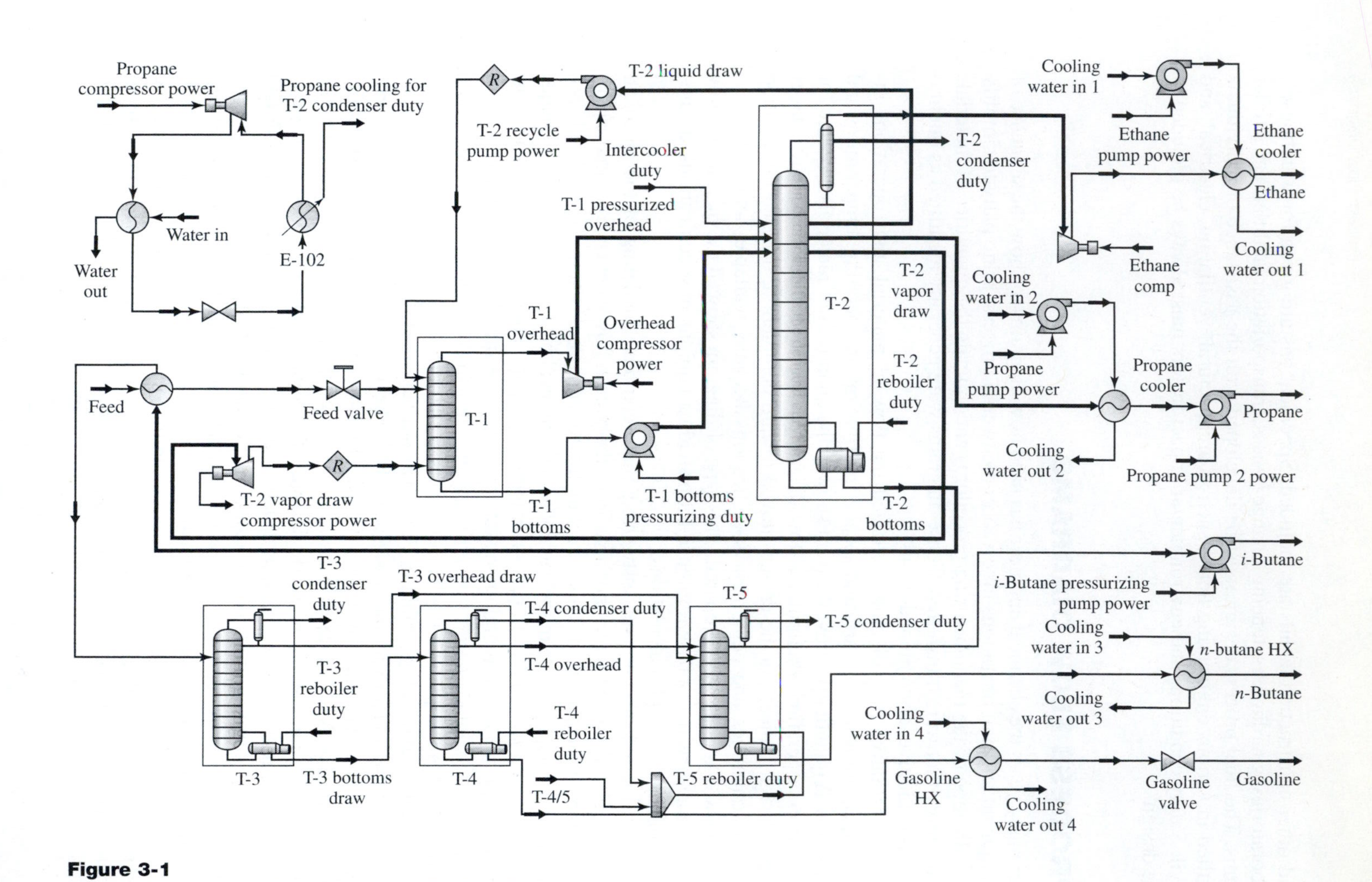

Figure 3-1
Process flow diagram for a natural gas liquid refinery

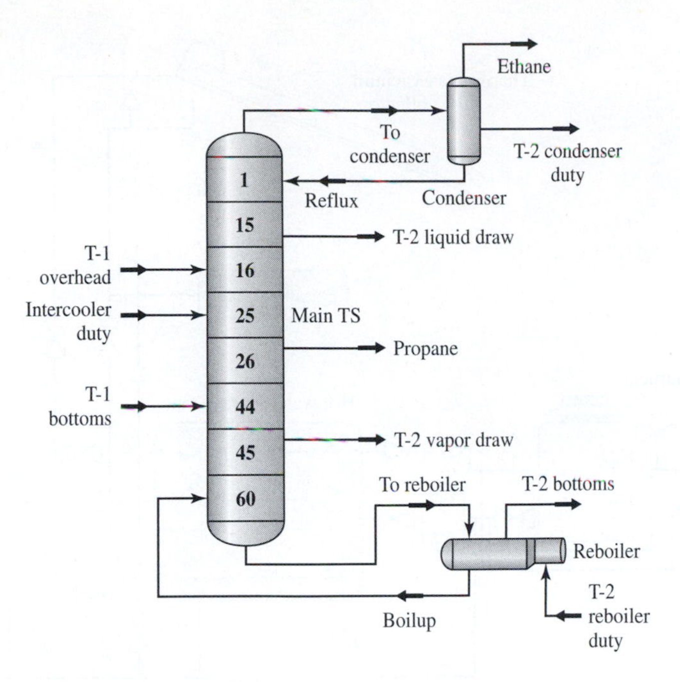

Figure 3-2
Process flow diagram for the main column of the natural gas liquid refinery presented in Fig. 3-1

in the previous PFD. The availability, use, and selection of PFD software are reviewed in Chap. 5.

PIPING AND INSTRUMENTATION DIAGRAMS

Another important level in process design is the generation of piping and instrument diagrams (P&IDs). The latter provide an additional level of detail for the overall design process that is not necessarily critical for the design aspect of the process, but essential for simulation and later construction and operation of the process. Piping and instrument diagrams contain schematics for all piping, associated fixtures such as valves, various components of instrumentation such as pneumatic air lines, and control mechanisms such as control valves. (These can be manually generated or produced from PFDs using available software that is also described in Chap. 5.) Such diagrams allow the design engineer to simulate various operating conditions and investigate the effect that these changes will have on the operability and economics of the process. The diagrams are therefore required for reference while operating a process to serve as guides for operators. This has led to legal requirements for maintaining the currency of P&IDs and making them available to facilitate operational safety in chemical plants. An example of a P&ID is given in Fig. 3-3.

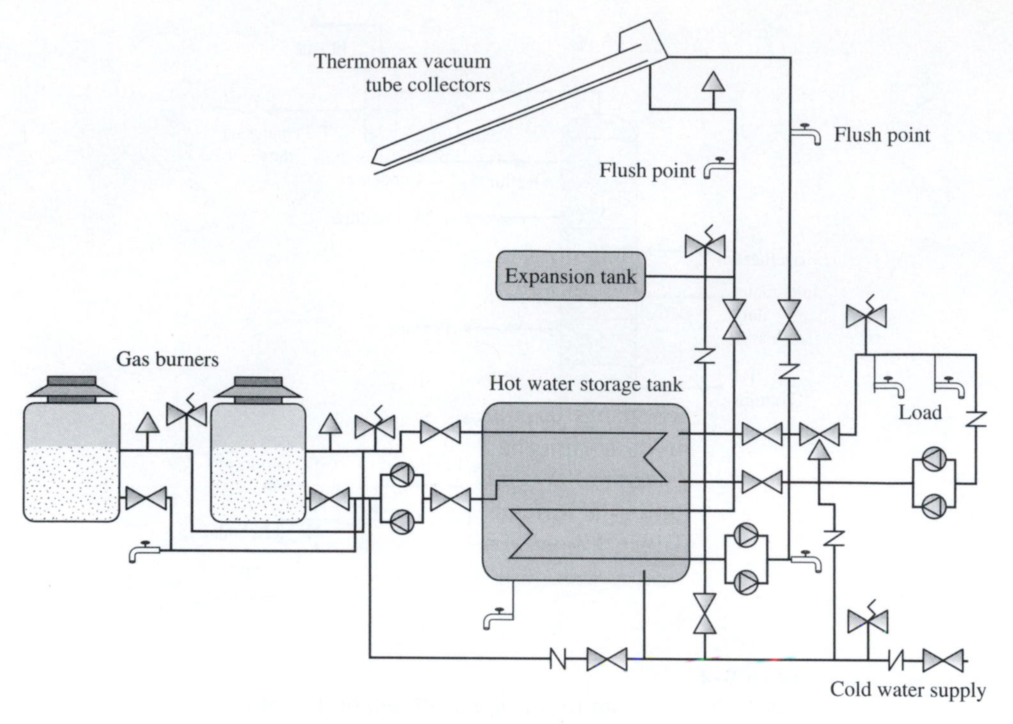

Figure 3-3
Piping and instrumentation diagram for a commercial integrated solar water heating system

VESSEL AND PIPING LAYOUT ISOMETRICS

The last general design task is to establish correct vessel and piping layouts. This requires detailed specification of equipment and piping locations and orientations in three dimensions, including orientation of vessels, piping, and piping components. Considerations influencing this process besides those involved in safety aspects include connection, and access requirements for both operation and maintenance, heat and materials transfer, climate sensitivity, and economic piping and auxiliary equipment. The goal of the vessel and piping layout isometrics is to produce a layout that fulfills the layout considerations mentioned in Chap. 1 while meeting process flow diagram requirements, at a minimum investment and operating cost.

The production of vessel and piping isometrics has been greatly facilitated by software specifically designed to establish vessel and piping requirements, determine potential layouts, and allow designers to evaluate isometrics graphically in three dimensions.† These software packages have progressed to the point where producing vessel and piping layout isometrics is easily accomplished.

†See Chap. 5 for more details regarding software use and availability.

EQUIPMENT DESIGN AND SPECIFICATIONS

The goal of the chemical engineer in process or plant design is to develop and present a complete chemical or biochemical process that can operate on an effective industrial basis. To achieve this goal, the chemical engineer must be able to combine many separate units or pieces of equipment into one smoothly operating plant. If the final process or plant is to be successful, each piece of equipment must be capable of performing its necessary function. The design of equipment, therefore, is an essential part of a design.

The engineer developing a process design must accept the responsibility of preparing the specifications for individual pieces of equipment and should be acquainted with methods for fabricating different types of equipment. The importance of appropriate materials of construction in this fabrication must be recognized. Design data must be developed, giving sizes, operating conditions, number and location of openings, types of flanges and heads, codes, variation allowances, and other information. Many of the machine design details are handled by the fabricators, but the chemical engineer must supply the basic design information.

Scale-up of Equipment in Design

When accurate data are not available in the literature or when past experience does not give an adequate design basis, pilot-plant tests may be necessary to design effective plant equipment. The results of these tests must be scaled up to the plant capacity. A chemical engineer, therefore, should be acquainted with the limitations of scale-up methods and should know how to select the essential design variables.

Pilot-plant data are almost always required for the design of filters unless specific information is already available for the type of materials and conditions involved. Heat exchangers, distillation columns, pumps, and many other types of conventional equipment can usually be designed adequately without using pilot-plant data. Table 3-1 presents an analysis of important factors in the design of different types of equipment. This table shows the major variables that characterize the size or capacity of the equipment and the maximum scale-up ratios of these variables. Information on the need for pilot-plant data, safety factors, and essential operational data for the design is included in this table.

Safety Factors

Some examples of recommended safety factors for equipment design are also shown in Table 3-1. These factors represent the amount of overdesign that would be used to account for not only the changes in the operating performance with time, but also the uncertainties in the design process.

The indiscriminate application of safety factors can be very detrimental to a design. Each piece of equipment should be designed to carry out its necessary function. Then if uncertainties are involved, a reasonable safety factor can be applied. The role of the particular piece of equipment in the overall operation must be considered along

Table 3-1 Factors in equipment scale-up and design†

Type of equipment	Is pilot plant usually necessary?	Major variables for operational design (other than flow rate)	Major variables characterizing size or capacity	Maximum scale-up ratio based on indicated charac-terizing variable	Approximate recommended safety or over-design factor, %
Agitated batch crystallizers	Yes	Solubility–temperature relationship	Flow rate Heat-transfer area	>100:1	20
Batch reactors	Yes	Reaction rate Equilibrium state	Volume Residence time	>100:1	20
Centrifugal pumps	No	Discharge head	Flow rate Power input Impeller diameter	>100:1 >100:1 10:1	10
Continuous reactors	Yes	Reaction rate Equilibrium state	Flow rate Residence time	>100:1	20
Cooling towers	No	Air humidity Temperature decrease	Flow rate Volume	>100:1 10:1	15
Cyclones	No	Particle size	Flow rate Diameter of body	10:1 3:1	10
Evaporators	No	Latent heat of vaporization Temperatures	Flow rate Heat-transfer area	>100:1 >100:1	15
Hammer mills	Yes	Size reduction	Flow rate Power input	60:1 60:1	20
Mixers	No	Mechanism of operation System geometry	Flow rate Power input	>100:1 20:1	20
Nozzle-discharge centrifuges	Yes	Discharge method	Flow rate Power input	10:1 10:1	20 20
Packed columns	No	Equilibrium data Superficial vapor velocity	Flow rate Diameter Height/diameter ratio	>100:1 10:1	15
Plate columns	No	Equilibrium data Superficial vapor velocity	Flow rate Diameter	>100:1 10:1	15

Plate-and-frame filters	Yes	Cake resistance or permeability	Flow rate Heat-transfer area	>100 : 1 >100 : 1	20
Reboilers	No	Temperatures Viscosities	Flow rate Heat-transfer area	>100 : 1 >100 : 1	15
Reciprocating compressors	No	Compression ratio	Flow rate Power input Piston displacement	>100 : 1 >100 : 1 >100 : 1	10
Rotary filters	Yes	Cake resistance or permeability	Flow rate Filtration area	>100 : 1 25 : 1	20
Screw conveyors	No	Bulk density	Flow rate Diameter Drive horsepower	90 : 1 8 : 1	20
Screw extruders	No	Shear rate	Flow rate Power input	100 : 1 100 : 1	20 10
Sedimentation centrifuges	No	Discharge method	Flow rate Power input	10 : 1 10 : 1	20 20
Settlers	No	Settling velocity	Volume Residence time	>100 : 1	15
Spray columns	No	Gas solubilities	Flow rate Power input	10 : 1	20
Spray condensers	No	Latent heat of vaporization Temperatures	Flow rate Height/diameter ratio	70 : 1 12 : 1	20
Tube-and-shell heat exchangers	No	Temperatures Viscosities Thermal conductivities	Flow rate Heat-transfer area	>100 : 1 >100 : 1	15

[†]Modified from data presented by A. Bisio and R. L. Kabel, *Scaleup of Chemical Processes,* J. Wiley, New York, 1985; H. Z. Kister, *Distillation Design,* McGraw-Hill, New York, 1992; S. M. Walas, *Chemical Process Equipment—Selection and Design,* Butterworth-Heinemann, Newton, MA, 1985; and D. R. Woods, *Data for Process Design and Engineering Practice,* Prentice-Hall, Englewood Cliffs, NJ, 1995.

with the consequences of underdesign. Fouling, which may occur during operation, should never be overlooked when a design safety factor is determined. Potential increases in capacity requirements are sometimes used as an excuse for applying large safety factors. This practice, however, can result in so much overdesign that the process or equipment never has an opportunity to prove its economic value.

In general design work, the magnitudes of safety factors are dictated by economic or market considerations, the accuracy of the design data and calculations, potential changes in the operating performance, background information available on the overall process, and the amount of conservatism used in developing the individual components of the design. Each safety factor must be chosen on the basis of the existing conditions, and the chemical engineer should not hesitate to use a safety factor of zero if the situation warrants it.

Equipment Specifications

A generalization for equipment design is that standard equipment should be selected whenever possible. If the equipment is standard, the manufacturer may have the desired size in stock. In any case, the manufacturer can usually quote a lower price and give better guarantees for standard equipment than for special equipment.

Before a manufacturer is contacted, the engineer should evaluate the design needs and prepare a preliminary specification sheet for the equipment. This preliminary specification sheet can be used by the engineer as a basis for the preparation of the final specifications, or it can be sent to a manufacturer with a request for suggestions and fabrication information. Preliminary specifications for equipment should show the following:

1. Identification
2. Function
3. Operation
4. Materials handled
5. Basic design data
6. Essential controls
7. Insulation requirements
8. Allowable tolerances
9. Special information and details pertinent to the particular equipment, such as materials of construction including gaskets, installation, necessary delivery date, supports, and special design details or comments

Final specifications can be prepared by the engineer; however, care must be exercised to avoid unnecessary restrictions. The engineer should allow the potential manufacturers or fabricators to make suggestions before preparing detailed specifications. In this way, the final design can include small changes that reduce the first cost with no decrease in the effectiveness of the equipment. For example, the tubes in standard heat exchangers are usually 2.44, 3.66, 4.88, or 6.10 m in length, and these lengths are ordinarily kept in stock by manufacturers and maintenance departments. If a design

<table>
<tr><th colspan="2">HEAT EXCHANGER</th></tr>
<tr><td>Identification: Item: Condenser
Item No. H-5
No. required 1</td><td>Date 1-1-02
By PTW</td></tr>
<tr><td colspan="2">Function: Condense overhead vapors from methanol fractionation column</td></tr>
<tr><td colspan="2">Operation: Continuous</td></tr>
<tr><td colspan="2">Type: Horizontal
Fixed tube sheet
Expansion ring in shell
Duty 1000 kW Outside area 44 m^2</td></tr>
<tr><td>Tube side:
Fluid handled Cooling water
Flow rate 0.025 m^3/s
Pressure 240 kPa
Temperature 15 to 25°C
Head material Carbon steel</td><td>Tubes: 0.0254 m diam. 14 BWG
0.03175 m Centers Δ Pattern
225 Tubes each Length 2.44 m
2 Passes
Tube material Carbon steel</td></tr>
<tr><td>Shell side:
Fluid handled Methanol vapor
Flow rate 0.9 kg/s
Pressure 101 kPa
Temperature 65°C to (constant temp.)</td><td>Shell: 0.56 m diam. 1 Passes
(Transverse baffles Tube support Req'd)

(Longitudinal baffles 0 Req'd)
Shell material Carbon steel</td></tr>
<tr><td colspan="2">Utilities: Untreated cooling water
Controls: Cooling-water rate controlled by vapor temperature in vent line
Insulation: 0.051 m rock cork or equivalent; weatherproofed
Tolerances: Tubular Exchangers Manufacturers Association (TEMA) standards
Comments and drawings: Location and sizes of inlets and outlets are shown on drawing</td></tr>
</table>

Figure 3-4
Specification sheet for a heat exchanger

specification called for tubes 4.57 m in length, the manufacturer would probably use 4.88-m tubes and cut them off to the specified length. Thus, an increase from 4.57 to 4.88 m for the specified tube length could cause a reduction in the total cost for the unit, because the labor charge for cutting the standard-length tubes would be eliminated. In addition, if replacement of the tubes became necessary after the heat exchanger had been in use, the replacement costs with 4.88-m tubes would probably be less than with the 4.57-m tubes.

Figures 3-4 and 3-5 show typical types of specification sheets for heat exchangers and distillation columns. These sheets apply for the normal types of equipment encountered by a chemical engineer in design work. The details of mechanical design, such as shell or head thicknesses, are not included, since they do not have a direct effect on the performance of the equipment. However, for certain types of equipment involving unusual or extreme operating conditions, the engineer may need to extend the specifications to include additional details of the mechanical design. Locations and sizes of outlets, supports, and other essential fabrication information can be presented with the specifications in the form of comments or drawings.

SIEVE-TRAY COLUMN				
Identification:	Item: Item No. No. required		**Date** **By**	
Function:				
Operation:				
Materials handled:	*Feed*	*Overhead*	*Reflux*	*Bottoms*
Quantity				
Composition				
Temperature				
Design data:	No. of trays			Reflux ratio
	Pressure			Tray spacing
	Functional height			Skirt height
	Material of construction			
	Diameter	Liquid density	kg/m^3	(lb/ft^3)
		Vapor density	kg/m^3	(lb/ft^3)
	Maximum allowable vapor velocity (superficial)			m/s (ft/s)
	Maximum vapor flow rate			m^3/s (ft^3/s)
	Recommended inside diameter			
	Hole size and arrangement			
	Tray thickness			
Utilities:				
Controls:				
Insulation:				
Tolerances:				
Comments and drawings:				

Figure 3-5
Specification sheet for a sieve-tray distillation column

Materials of Construction

The effects of corrosion and erosion must be considered in the design of chemical plants and equipment. Chemical resistance and physical properties of construction materials, therefore, are important factors in the choice and design of equipment. The materials of construction must be resistant to the corrosive action of any chemicals that may contact the exposed surfaces. Possible erosion caused by flowing fluids or other types of moving substances must be considered, even though the materials of construction may have adequate chemical resistance. Structural strength, resistance to physical or thermal shock, cost, ease of fabrication, necessary maintenance, and general type of service required, including operating temperatures and pressures, are additional factors that influence the final choice of construction materials.

If there is any doubt concerning suitable materials for construction of equipment, reference should be made to the literature,† or laboratory tests should be carried out under conditions similar to the final operating conditions. The results from the laboratory

†Detailed information on materials of construction is presented in Chap. 10.

tests indicate the corrosion resistance of the material and also the effects on the product caused by contact with the particular material. Further tests on a pilot-plant scale may be desirable to determine the amount of erosion resistance or the effects of other operational factors.

THE PRELIMINARY DESIGN—A SPECIFIC EXAMPLE

To amplify the remarks made earlier in this chapter concerning the design project procedure, it is appropriate to look more closely at a specific preliminary design. Because of space limitations, only a brief summary can be presented at this point. However, sufficient detail will be given to outline the steps that are necessary to prepare a preliminary design.

Problem Statement

A natural gas company is considering diversifying its operation into the petrochemical field to increase its profitability. From its natural gas separation process, the company produces large quantities of ethane and propane gas that are used by customer companies in the manufacture of higher-priced ethylene and propylene products. To diversify its operations, the company is investigating the possibilities of producing these commodity chemicals itself. Process engineers have informed management that because of the competitive nature between producers it would be difficult to enter the market unless there were some significant improvement in the process that could provide an economic advantage for the company. The process engineers have also reported that the most promising area for achieving such an improvement in the ethylene process is in the separation section of the process since this section is very energy-intensive, utilizing nearly 70 percent of the energy required for the entire process.

Based on this information, management has requested a review of the separation section of the ethylene process to determine whether any new separation technologies could provide sufficient incentives to justify a major investment in a grass-roots ethylene plant with a capacity of 500,000 t (metric)/yr. To assist management in this decision, the latter has requested that an economic comparison be made between a conventional base-case ethylene plant and an ethylene plant utilizing new or improved technology in the separation section.

Literature Survey

Since the ethylene/ethane and propylene/propane separations are the more difficult separations in terms of energy and refrigeration requirements, they also provide greater opportunities for altering the economics of the ethylene process. A literature search indicates that separation of these two paraffin/olefin mixtures can be accomplished through the use of selective absorption, selective adsorption, extractive distillation, and membrane separation. The first three unit operations are not new and have been investigated rather thoroughly over the years. The selective absorption process does not provide the high purity of ethylene required for the production of polyethylene and therefore is not an acceptable replacement for the present distillation technology.[†] In the selective adsorption process, several types of zeolite molecular sieves have been developed to achieve the desired separation goals. However, the molecular sieves encounter problems over long-time operation which deter their use in commercial plants.[‡] Likewise, extractive distillation systems have been developed for these two specific separations. In this case, the separation is more expensive than the distillation process, particularly when polymer grade propylene is produced because of the additional costs of handling the extractive solvents.[§]

The use of simple diffusion membrane systems for the separation of a low relative volatility mixture such as ethylene/ethane is impractical because of the small transfer rates and low selectivity in the separation process. However, utilizing the facilitated transport concept, membranes can make this process feasible.[¶] In such transport, the simple diffusional process is coupled with a chemical reactant that reversibly binds with one of the species to be separated and thereby increases the net transport rate. Facilitated transport membrane technology (FTMT) has been successfully demonstrated both in bench-scale laboratories and in pilot-plant studies for the selective separation of ethylene from ethane and propylene from propane.

Process Creation

The desired objectives for any ethylene separation process are to obtain a high-purity ethylene product combined with a high-percentage recovery of the ethylene. Conventional distillation technology accomplishes both of these objectives, but at considerable energy consumption. On the other hand, the FTMT is capable of producing a high-purity product, but with a lower percentage of ethylene recovery. The lower percentage of recovery dictates the use of a multistage membrane system incorporating the use of additional compressors.

The combination of distillation and membrane technologies to form a hybrid system is another design alternative for replacing the current distillation technology. The hybrid system incorporates the high-purity product aspect of both technologies as well as the high-percentage recovery of the distillation process. The combination of these two technologies can be represented with numerous different schemes. For example, hybrid configurations may differ with regard to the location of the membrane relative to the distillation column. Operating conditions may be varied for each configuration in order to achieve optimum performance.

Figure 3-6 suggests some possible hybrid system configurations.* If the membrane is located at the top of the distillation column, the membrane performs the final

---

[†]S. B. Zdonik et al., *Oil and Gas J.*, **64**(49): 62 (1966).

[‡]T. C. Kumar et al., *Sep. Sci. & Tech.*, **27**(15): 2157 (1992).

[§]R. Kumar, J. M. Prausnitz, and C. J. King, *Advances in Chemistry Series*, vol. 115, American Chemical Society, Washington, DC, 1972.

[¶]M. Teramoto et al., *J. Membrane Sci.*, **35:** 115 (1989); R. J. Valus et al., U.S. Patent 5,057,641, Oct. 15, 1991.

*A. A. Al-Rabiah, Ph.D. thesis, Department of Chemical Engineering, University of Colorado, Boulder, CO, 2001.

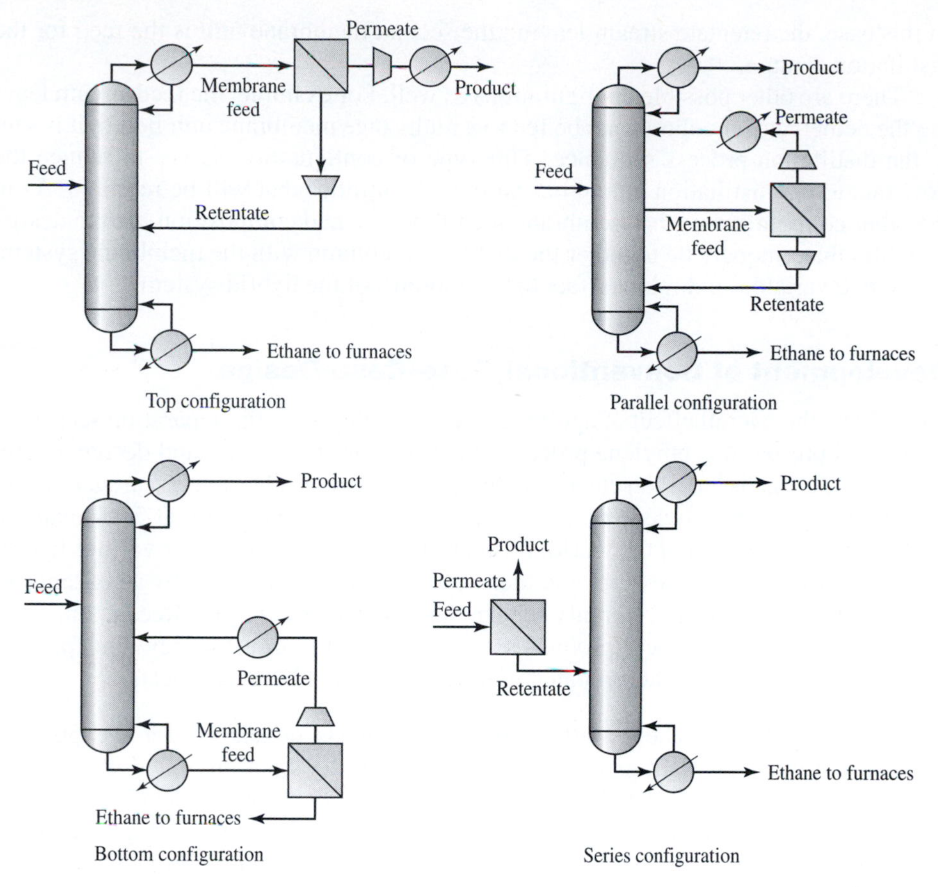

Figure 3-6
Four possible configurations for the hybrid distillation membrane separation system

purification of the product. The retentate in this scheme is recycled back to the distillation column at a location where the composition of the retentate matches that of the stage to which it is recycled.

The membrane may also be combined with the distillation column in a parallel configuration. In this scheme, the membrane feed stream is withdrawn from one of the intermediate stages of the distillation column. The permeate and retentate streams leaving the membrane unit are recompressed and returned to the column.

In the bottom configuration, the membrane unit performs the final purification of the ethane recycle stream from the distillation column. The permeate rather than the retentate, as in the top configuration, is recycled and recompressed to match the pressure of the entry point into the distillation column.

In a special case of the parallel configuration, the feed stream can first undergo a membrane separation before entering the distillation column to complete the separation.

In this case, the retentate stream leaving the second membrane unit is the feed for the distillation column.

There are other possible configurations as well. For example, the feed stream leaving the demethanizer column can be fed to a multistage membrane unit before it is sent to the distillation process sequence. This type of configuration serves to reduce the flow rate to the distillation column as well as the utilities that will be required. As in the other configurations, the membrane feed flow rate and composition are the design variables that uniquely tie together the distillation column with the membrane system. These two variables indirectly affect the economics of the hybrid system.

Development of Conventional Base-Case Design

To evaluate the overall effect of applying new technologies in the separation section of the typical present-day ethylene process, it is necessary to develop and design an ethylene process that is based on current technology. Based on a literature search, the conceptual design of the ethylene process is similar to that shown in Fig. 3-7 for many of the ethylene plants around the world. It is based on thermal cracking of various hydrocarbon feedstocks, compression, and then separation of the products by a sequence of distillation units using refrigerants for condensation of the reflux. Recent improvements in thermal efficiency, product selectivity, separation efficiency, and process modifications need to be incorporated in the conventional base-case design.

Process Conditions The detailed process flowsheets developed for the 500,000 t (metric)/yr ethylene plant using propane as the feedstock are presented in Figs. 3-8 through 3-13. The flowsheets in sequence schematically show the details associated with the thermal cracking process, quenching, compression and acid gas removal, drying, deethanization and acetylene hydrogenation, cracked gas chilling and demethanization, MAPD hydrogenation, and product separation. (Flowsheets for the propylene refrigeration, ethylene refrigeration, and steam process systems are also required in the preliminary design, but are omitted for brevity reasons.) The flowsheets include the major equipment used in this base-case ethylene plant. The numbering technique used in these flowsheets is a standard one, although techniques may vary from one organization to another.

As shown in Fig. 3-8, the fresh feed and recycle are fed to a bank of parallel pyrolysis furnaces. In the convection zone of the furnace, the feed is preheated to about 600°C and then diluted with steam. The steam reduces coking and improves product selectivity. In the radiation zone, the feedstock-steam mixture flows through vertical coils where pyrolysis occurs at temperatures above 630°C. The coil outlet temperature for the pyrolysis products is about 835°C. The product from the thermal cracking is immediately quenched to about 340°C in transfer line exchangers to stop the cracking reactions and recover waste heat for steam generation. The cracked gas is cooled in a water quench tower to about 40°C to condense the fuel oil and most of the dilution steam. The cooled gas is compressed to about 3590 kPa generally with four or five compression stages. During the compression stages, the gas stream is scrubbed with caustic to remove H_2S and CO_2. Before the last stage of compression, the gas is cooled

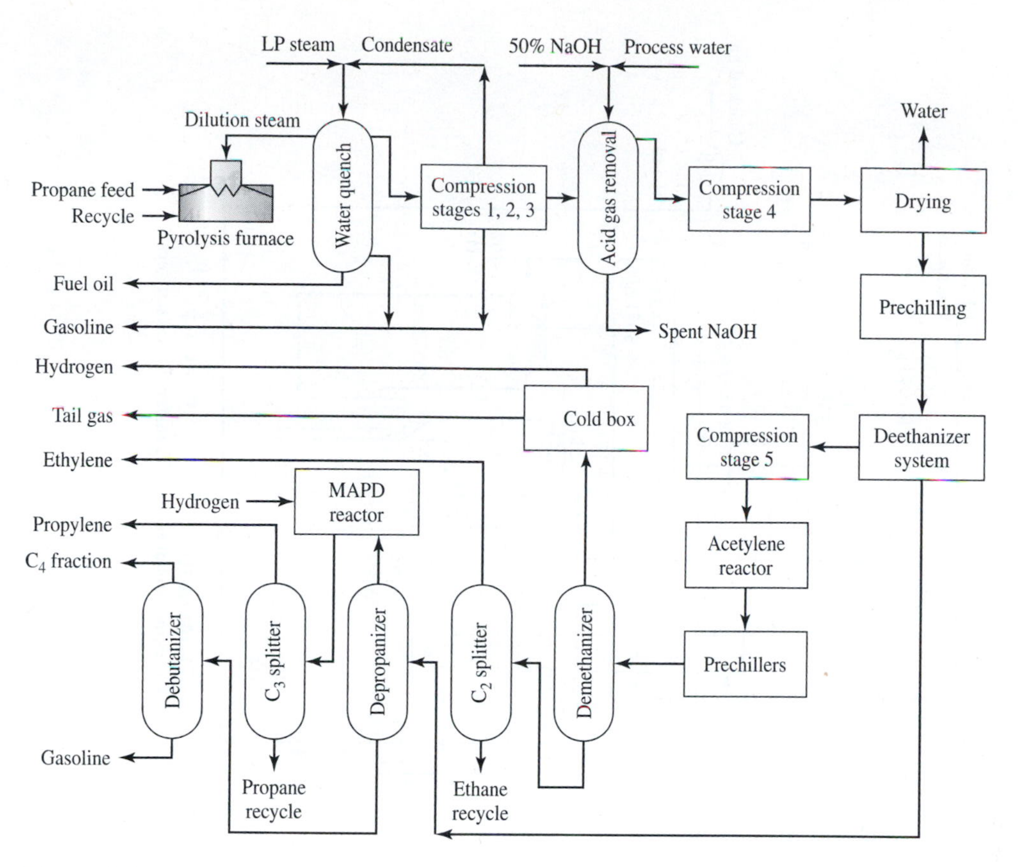

Figure 3-7
Flow diagram for a conventional ethylene process using the front-end deethanizer process scheme

to about 15°C with propylene refrigerant and dehydrated with molecular sieves. The dried gas is further cooled in a series of heat exchangers before it enters the deethanizer, in which the C_2's and lighter components in the gas stream are separated from the C_3's and heavier components.

The overhead from the deethanizer, after further compression, is first sent to a catalytic reactor to convert any acetylene in the gas to ethylene and ethane. The gas and hydrogenated products then proceed to the demethanizer which separates the C_2's from the tail gas (methane) and hydrogen. The latter are further separated in a cold box. The recovered C_2's are separated into ethylene and ethane in the C_2 splitter. The ethylene product is of polymer grade (99.9 mol %) with methane and ethane as the minor (0.1 mol %) impurities.

The bottoms from the deethanizer pass to a depropanizer which separates the C_3 components from the heavier components. The latter are separated in a debutanizer into a C_4 product and a heavier pyrolysis gasoline. The depropanizer overhead is

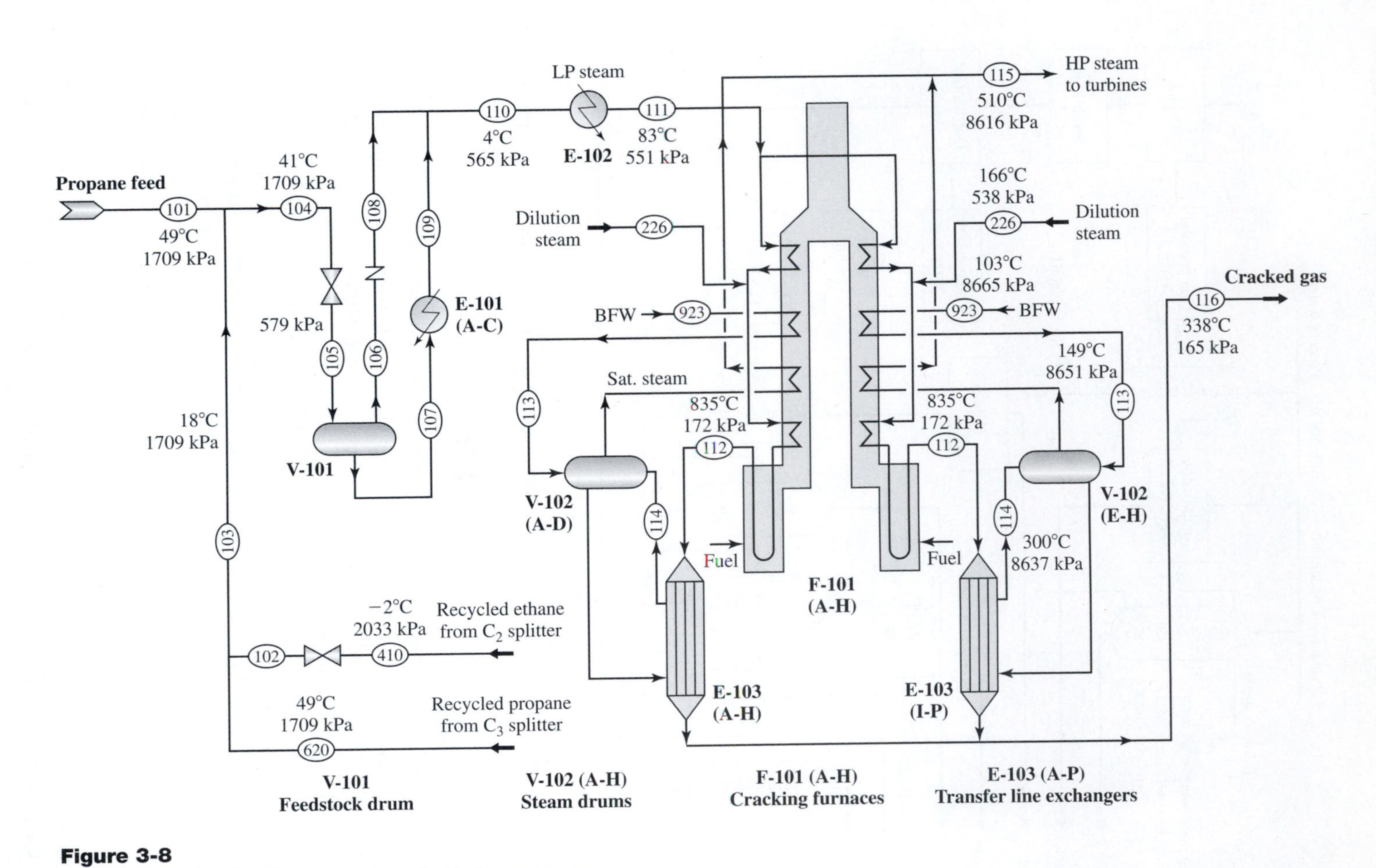

Figure 3-8
Cracking section for the conventional ethylene process

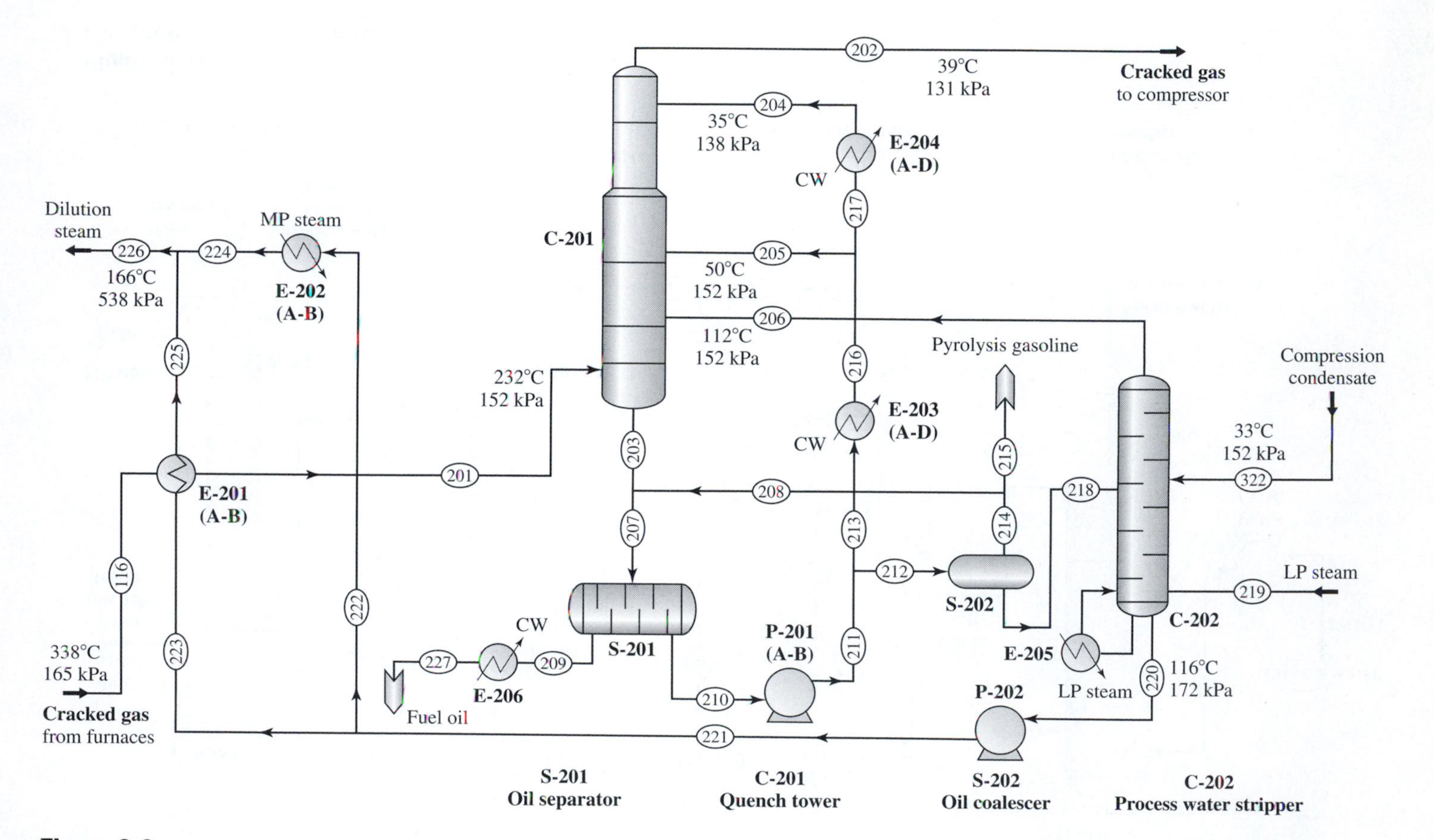

Figure 3-9
Quenching section

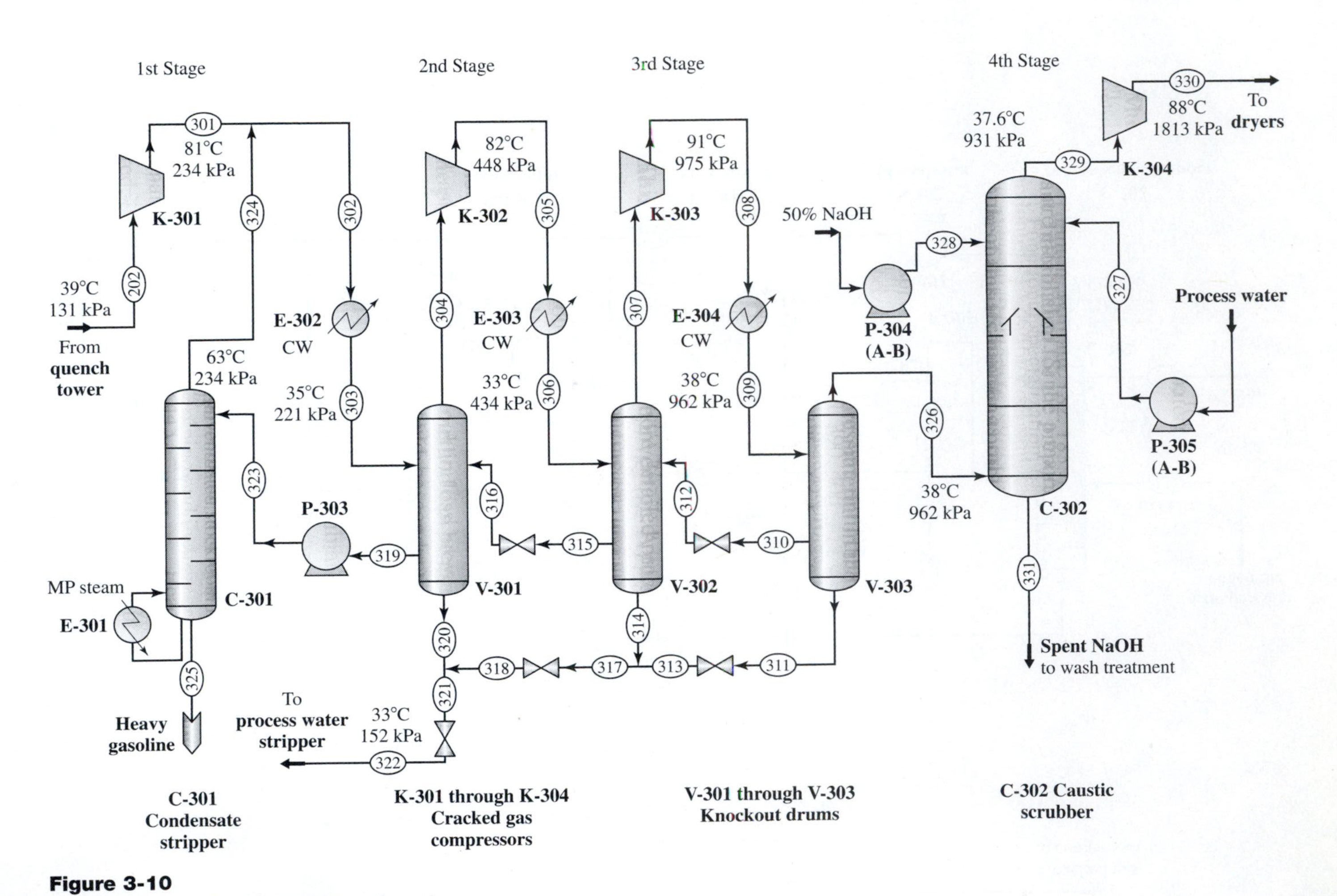

Figure 3-10
Compression and acid gas removal section

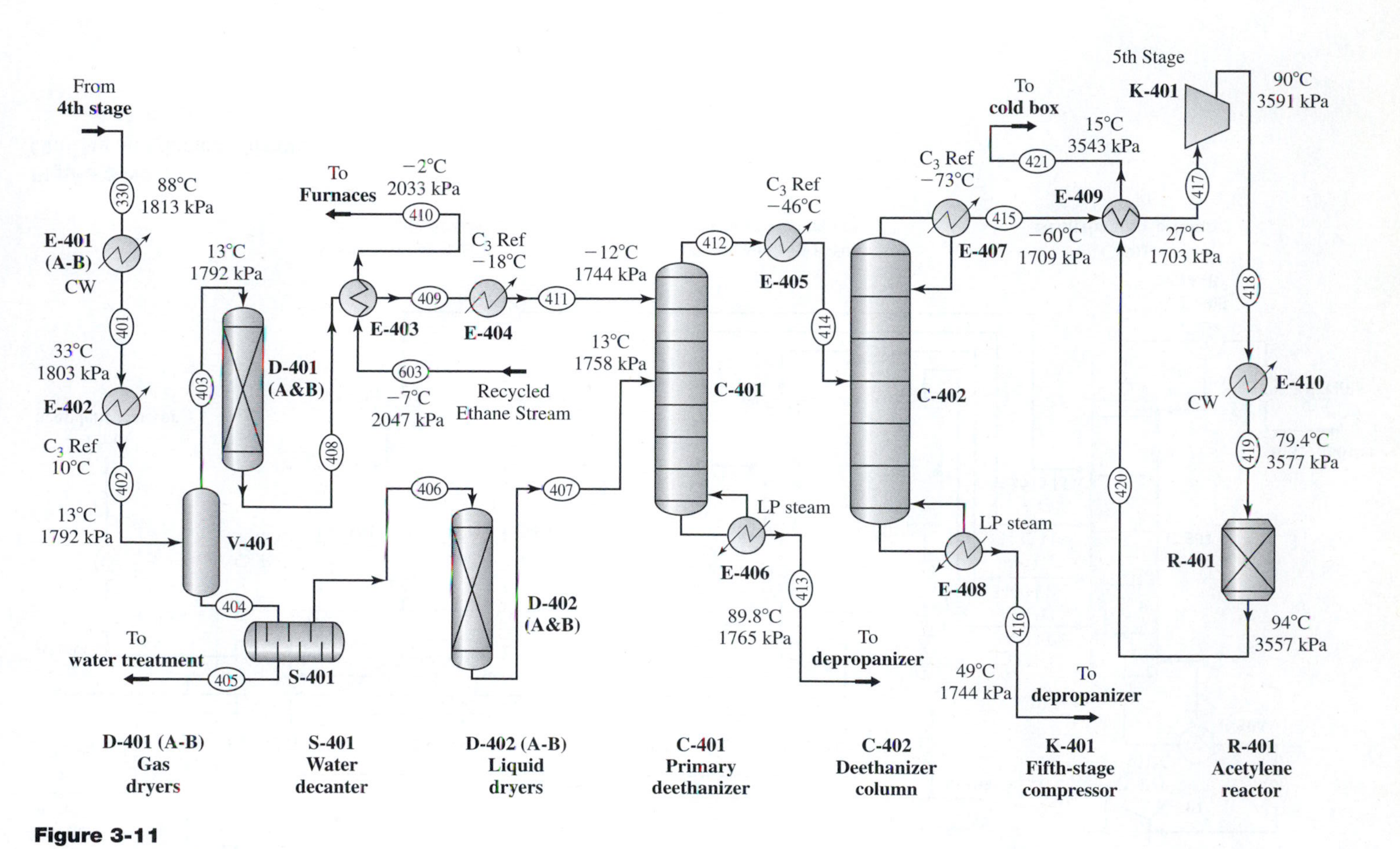

Figure 3-11
Drying, deethanization and acetylene hydrogenation section

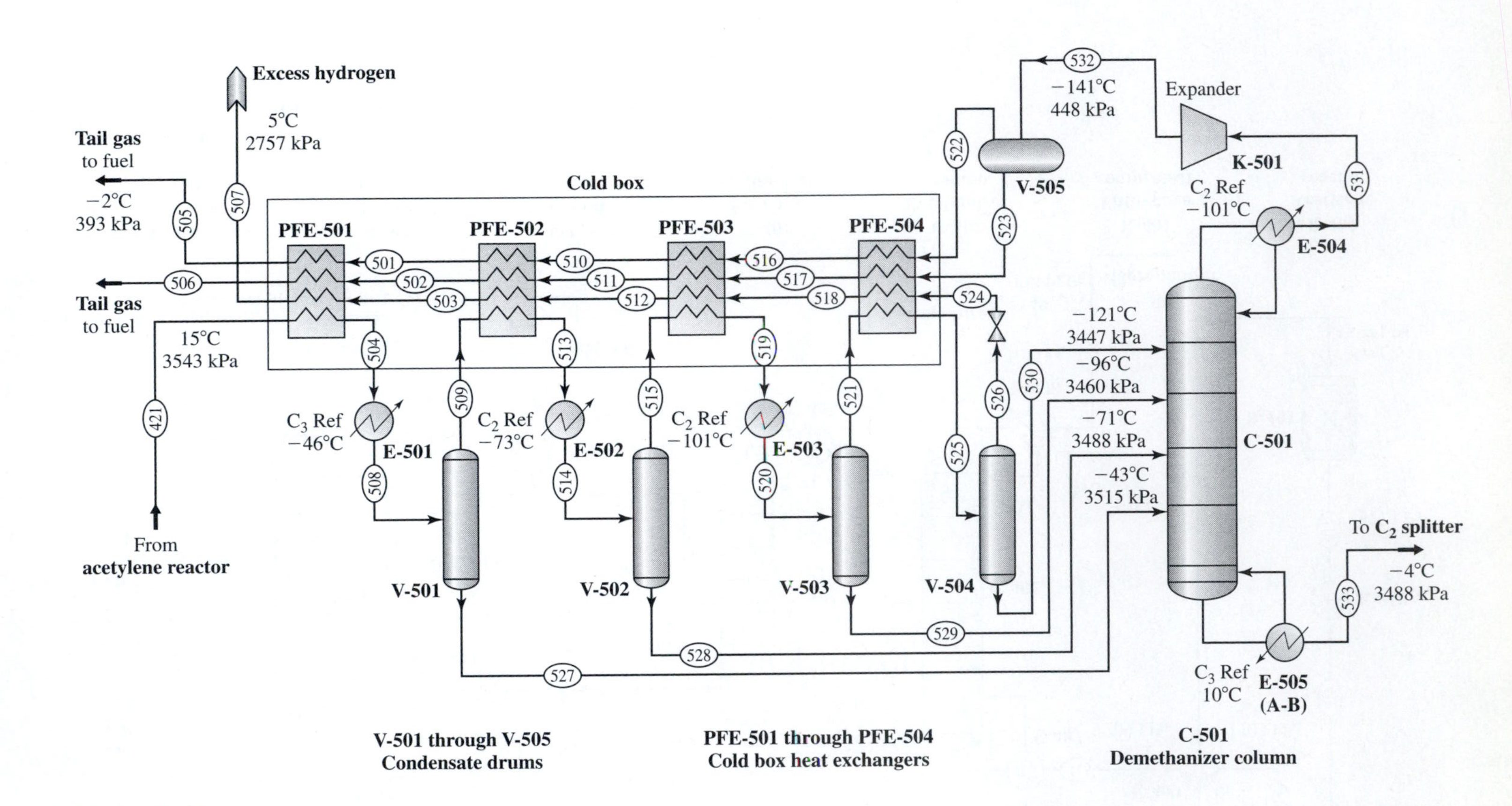

Figure 3-12
Chilling and demethanization section

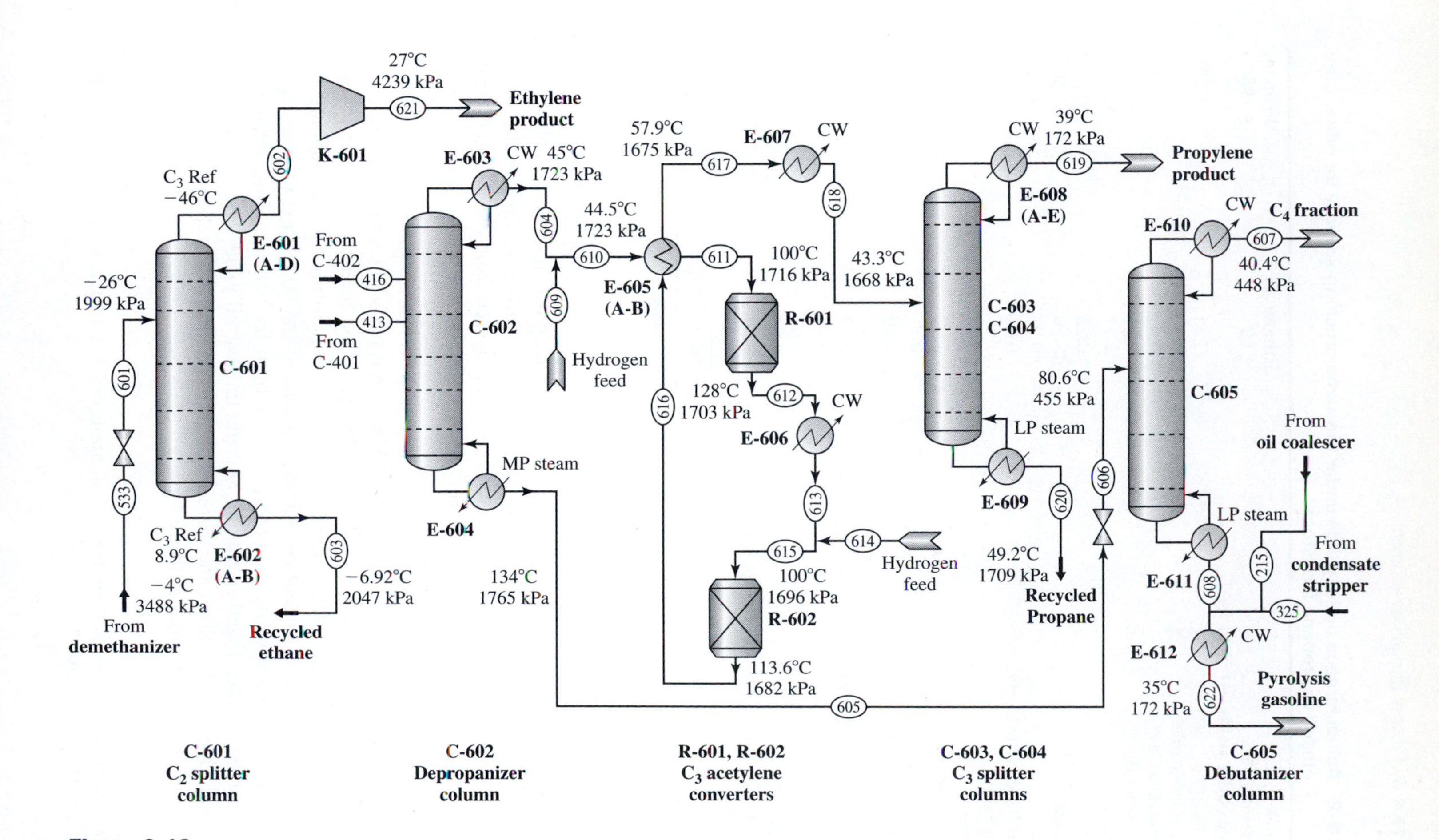

Figure 3-13
Product separation section

Table 3-2 Actual furnace outlet yields under the same cracking severity for either pure ethane or propane feedstock†

Feedstock	Ethane	Propane
Conversion (wt %)	60	90
Furnace outlet yields (wt % of feed)		
Hydrogen	3.55	1.29
Carbon monoxide	0.01	0.01
Carbon dioxide	0.01	0.01
Hydrogen sulfide	0.01	0.01
Methane	4.17	24.67
Acetylene	0.25	0.33
Ethylene	48.20	34.50
Ethane	40.00	4.40
Propadiene/methylacetylene	0.02	0.34
Propylene	1.11	13.96
Propane	0.17	10.00
1,3-Butadiene	1.07	2.65
Isobutene	0.11	0.52
Butene-1	0.10	0.48
n-Butane	0.27	0.05
C_5s PON*(paraffin, olefin, napthene)	0.27	1.81
Benzene	0.48	2.20
Toluene	0.06	0.48
Xylene	0.00	0.00
Gasoline C_6-C_8 (paraffin, olefin, napthene)	0.14	1.44
Fuel oil	0.00	0.85
Total	100.0	100.0

†From L. Kniel, O. Winter and K. Storch, *Ethylene Keystone to the Petrochemical Industry,* Marcel Dekker, NY, 1978. Reprinted by courtesy of Marcel Dekker, Inc.

hydrogenated to convert any methylacetylene and propadiene (MAPD) to propylene and propane. The gas and hydrogenated products are then fractionated in a C_3 splitter to produce a polymer-grade (99.6 mol %) propylene with paraffins as the minor (0.4 mol %) impurity. Both the ethane and propane recovered from the C_2 and C_3 splitters, respectively, are recycled to the pyrolysis furnace.

Design Data The cracking yields for feedstocks of either pure ethane or pure propane are listed in Table 3-2. The yield distribution is based on actual operating data of present-day ethylene steam crackers. The nature of the feedstock and the level of pyrolysis severity largely determine the operating conditions in the cracking and quenching section. At 60 percent conversion for ethane and 90 percent conversion for propane, the cracking severity is about equal for both feedstocks. Thus, ethane and propane can be cracked together without altering the overall yield. The value in using these pyrolysis yields to initiate the material balance process is that the ultimate yields derived from them are similar to those reported in the literature.†

†J. J. L. Ma, *Ethylene Report No. 29 E,* SRI International, Menlo Park, CA, 1991; L. K. Ng, C. N. Eng, and R. S. Zack, *Hydrocarbon Process.*, **62**(12): 99 (1983).

Table 3-3 Design bases and assumptions for the base-case ethylene plant

Plant location	U.S. Gulf Coast
Plant capacity	500,000 t (metric)/yr
Design on-stream factor	0.92 (336 days/yr)
Feedstock	Propane at 49°C and 1710 kPa
Main products	Polymer-grade ethylene 99.9 mol %
	Polymer-grade propylene 99.6 mol %
Cracking conditions:	
Coil outlet temperature/pressure	835°C and 170 kPa for ethane-propane
Residence time	0.4 s for ethane-propane
Steam dilution ratios	0.5 mol steam/mol ethane
	0.856 mol steam/mol propane
Conversions	90% per pass for propane; 60% per pass for ethane
Recycled streams	Ethane and propane
Furnace thermal efficiency	95% (based on LHV)
Transfer line exchangers (TLXs)	
Outlet temperature and pressure	338°C and 165 kPa
Steam temperature and pressure	510°C and 8615 kPa
Separation scheme	Front-end deethanization and acetylene removal
Cracked gas compression	
Number of stages	5
Maximum discharge temperature	105°C
Last-stage discharge pressure	3590 kPa
Acid gas removal	
System	Caustic scrubber
Scrubber location	Between 3rd and 4th stages of compression
Scrubber temperature and pressure	38°C and 930 kPa
Caustic solution concentration	10 wt% in water
Caustic utilization	70%
Cracked gas drying	
Dryer location	After 4th stage of compression
Type of desiccant	Type 3A molecular sieves
Moisture loading	8 wt %
Desiccant regeneration cycle	24 h
Dew point of dried gas	−173°C
Process efficiencies	
Distillation column	0.85
Compressor	0.85 (polytropic)
Expander	0.85 (polytropic)
Motor	0.90
Pump	0.75

The design bases and assumptions for the conventional base-case ethylene plant are listed in Table 3-3. All assumptions listed are those currently used in ethylene plants as verified by a careful search of the literature. Propylene and ethylene refrigerants are used as the cooling fluids. The propylene refrigeration system has three levels of temperature (10, −18, and −46°C) while the ethylene refrigeration system has only two levels (−73 and −101°C). Methane-rich gas recovered in the low-temperature section of the plant is utilized as fuel. The high-pressure steam, generated in the thermal cracking section, provides the energy required by the plant turbines. The medium-pressure steam (291°C and 1650 kPa) and the low-pressure steam (162°C and 450 kPa) are utilized in the process heat exchangers. Cooling, process, and boiler-feed water are all available at 29°C and 450 kPa. The maximum temperature rise for the water has been set at 11°C.

The separation process requires maximum temperature limits to prevent polymerization fouling and minimum temperature limits to prevent hydrocarbon freezing or hydrate formation. Additionally, temperature limits are observed for the design of the cracked gas compressors to avoid polymerization fouling. This is achieved by employing a five-stage compressor with interstage cooling to reduce the discharge temperature.

Equilibrium Conversion of Ethane and Propane For ethane pyrolysis, the primary reaction is

$$C_2H_6 \rightleftarrows C_2H_4 + H_2 \tag{3-1}$$

At 800°C the equilibrium constant for this reaction is essentially 1.0.[†] With 50 mol percent steam, the observed equilibrium conversion at 800°C and 103 kPa is 0.781.[‡] Pyrolysis of propane is similar to ethane, but involves two reactions

$$C_3H_8 \rightleftarrows C_2H_4 + CH_4 \tag{3-2}$$

$$C_3H_8 \rightleftarrows C_3H_6 + H_2 \tag{3-3}$$

The reported equilibrium conversion of propane at 800°C and 50 mol percent steam is greater than 0.90.[‡]

To obtain the equilibrium conversion of ethane at the design conditions of the base-case design of 835°C and 170 kPa, assume the same steam ratio of 0.5 mol steam/mol ethane. Take a basis of 1 mol of ethane. From chemical reaction equilibria

$$-RT \ln K = \sum n_i G_i^o = \Delta G^o \tag{3-4}$$

and

$$\frac{d \ln K}{dT} = \frac{\Delta H^o}{RT^2} \tag{3-5}$$

The general expression for ΔH^o is given by

$$\Delta H^o = J + \int \Delta C_p^o \, dT \tag{3-6}$$

where

$$\frac{\Delta C_p^o}{R} = \Delta A + (\Delta B)T + (\Delta C)T^2 + \frac{\Delta D}{T^2} \tag{3-7}$$

[†]L. Kramer and J. Happel in *The Chemistry of Petroleum Hydrocarbons,* vol. 21, B. T. Brooks et al., eds., Reinhold, New York, 1951, p. 126.

[‡]C. McConnell and B. Head in *Pyrolysis: Theory and Industrial Practice,* L. F. Albright et al., eds., Academic, New York, 1983, p. 25.

Eliminating ΔC_p^o from Eq. (3-6) and integrating result in

$$\frac{\Delta H^o}{R} = \frac{J}{R} + (\Delta A)T + \frac{\Delta B}{2}T^2 + \frac{\Delta C}{3}T^3 - \frac{\Delta D}{T} \tag{3-8}$$

Substitution of Eq. (3-6) in Eq. (3-5) and integrating give

$$\ln K = \frac{-J}{RT} + \Delta A \ln T + \frac{\Delta B}{2}T + \frac{\Delta C}{6}T^2 + \frac{\Delta D}{2T^2} + I \tag{3-9}$$

where ΔG^o is the standard Gibbs energy change for the reaction, K the equilibrium constant, R the universal gas constant, n_i the stoichiometric coefficient of component i, ΔH^o the standard enthalpy change for the reaction, T the absolute temperature of the reaction, J and I constants of integration, ΔC_p^o the standard heat capacity change for the reaction, and ΔA, ΔB, ΔC, ΔD constants in the heat capacity relation. For the reaction shown in Eq. (3-1), the following data are available for a temperature of 298 K:

$$\Delta G^o = 100{,}315 \text{ J/mol} \qquad \Delta H^o = 136{,}330 \text{ J/mol}$$

$$\Delta A = 3.542 \qquad \Delta B = -4.409 \times 10^{-3}$$

$$\Delta C = 1.169 \times 10^{-6} \qquad \text{and} \qquad \Delta D = 0.083 \times 10^5$$

Solving Eqs. (3-8) and (3-9) at a temperature of 298 K provides values of 15,555.4 for J/R and -7.856 for I. The general expression for $\ln K$ is given by

$$\ln K = \frac{-15{,}555.4}{T} + \Delta A \ln T + \frac{\Delta B}{2}T + \frac{\Delta C}{6}T^2 + \frac{\Delta D}{2T^2} - 7.856 \tag{3-10}$$

Solving the above equation at the design temperature of 1108 K (835°C) establishes an equilibrium constant value of 2.086. At equilibrium, this constant based on the partial pressures is given by

$$K = \frac{\bar{p}_{C_2H_4}\bar{p}_{H_2}}{\bar{p}_{C_2H_6}} = \frac{y_{C_2H_4}y_{H_2}}{y_{C_2H_6}}p \tag{3-11}$$

where $\bar{p}_i$ is the partial pressure of component i, y_i is the mol fraction of component i, and p is the total pressure. If ε is defined as the equilibrium number of mols of ethane reacted, then the number of mols of the component in the reacting mixture is a function of ε. The mol fractions of the components specified in Eq. (3-11) can be defined in terms of ε as

$$y_{C_2H_6} = \frac{1-\varepsilon}{1.5+\varepsilon} \qquad y_{C_2H_4} = \frac{\varepsilon}{1.5+\varepsilon} \qquad y_{H_2} = \frac{\varepsilon}{1.5+\varepsilon} \tag{3-12}$$

Substitution of these quantities in Eq. (3-11) for a K value of 2.086 and an operating temperature and pressure of 1108 K and 170 kPa, respectively, provides a value for ε of 0.782. Thus, the theoretical equilibrium conversion of ethane at the operating conditions of the thermal furnace is 78.2 percent. This indicates that a 60 percent conversion of ethane is feasible.

The equilibrium conversion of propane based on Eqs. (3-2) and (3-3) is obtained in the same manner that the equilibrium conversion for ethane was calculated. This calculation will involve two equilibrium constants which will be used to determine ε_1, the number of mols of propane reacted in Eq. (3-2), and ε_2, the number of mols of propane reacted in Eq. (3-3). The equilibrium conversion is the summation of ε_1 and ε_2, or 0.9406 and 0.0591, respectively. Since this summation is 0.9997, a 90 percent conversion of propane can be performed in the cracking furnace.

Material Balance for the Ethylene Process The base-case ethylene process was simulated using the CHEMCAD-III™ and ASPEN PLUS™ commercial simulation packages. These computer programs can perform rigorous mass and energy balances based on any equation of state selected to provide the properties of the fluid mixtures encountered in the ethylene process. Both computer programs provide similar results when the SRK (Soave-Redlich-Kwong) equation of state is selected. This equation of state provided a good match between simulated properties and actual properties reported in the literature.

To meet the desired annual production capacity of the plant, the feed flow rate to the plant must be determined. The initial estimate was obtained by conducting an overall material balance based on the actual outlet yields from an industrial thermal cracking unit, as presented in Table 3-2. The estimated component mass balance was then fed to a rigorous simulator that recognized both major and minor reactions occurring in the ethylene plant to determine the precise flow rates of each component for the detailed process flowsheet.

Figure 3-14 illustrates a simple flow diagram for the ethylene process. For a feed F of 100 kg/h as a basis, utilizing the nomenclature provided on the figure and assuming theoretical separation in the purification processes, the overall material balance can be represented by the relation

$$F = E - (R_1 + R_2) = P \tag{3-13}$$

The ethane material balance is given by

$$R_1 = 0.4R_1 + 0.044(100 + R_2) \tag{3-14}$$

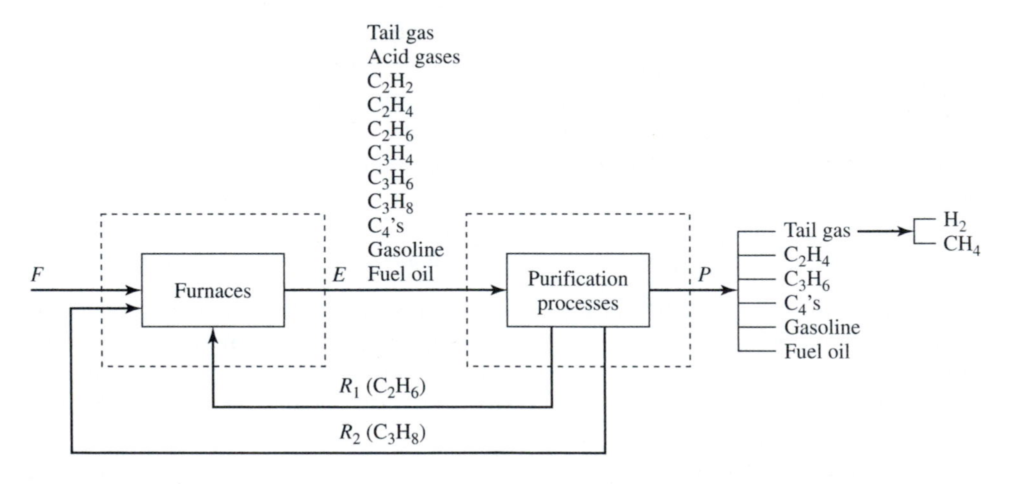

Figure 3-14
Simple flow diagram used to calculate the overall material balances for the conventional ethylene process

According to Table 3-2, 40 percent of the ethane is unreacted and theoretically can be returned in R_1 while 4.4 percent of the ethane obtained from the propane reaction theoretically can be returned in R_1.

The propane material balance is

$$R_2 = 0.0017R_1 + 0.1(100 + R_2) \tag{3-15}$$

where the propane in R_2 is obtained from a 0.17 percent conversion from ethane and a nonreaction of 10 percent of the propane fed to the thermal furnace and therefore could theoretically be returned in R_1. Solving Eqs. (3-14) and (3-15) simultaneously indicates that R_1 is 8.15 kg/h and R_2 is 11.13 kg/h.

An ethylene material balance can now be made to determine the mass of ethylene in the product based on a feed of 100 kg/h from

$$P_{C_2H_4} = 0.482R_1 + 0.345(100 + R_2) \tag{3-16}$$

Note that the ethylene yield is 48.2 percent from the ethane conversion and 34.5 percent from the propane conversion (see Table 3-2). From Eq. (3-16), $P_{C_2H_4}$ is equal to 42.26 kg/h. This indicates that each unit of ethylene produced in the ethylene process will require about 2.36 units of fresh propane feed.

A propylene material balance is given by

$$P_{C_3H_6} = 0.0111R_1 + 0.1396(100 + R_2) \tag{3-17}$$

Again, note that the propylene yield is 1.11 percent from the ethane conversion and 13.96 percent from the propane conversion. The sum of these conversions indicates a yield of 15.60 kg/h of propylene product.

A similar procedure, as outlined above, can be followed to arrive at material balances for the other by-products. Table 3-4 shows the calculated product distribution that is obtained with this procedure. For a production rate of 62,010 kg/h (500,000 t (metric)/yr) this would provide an estimated feed flow rate of about 146,730 kg/h. A more accurate product can only be determined by a simulation program that includes

Table 3-4 Calculated and simulated product distribution for the base-case ethylene process

	Calculated[†]		Simulated[‡]	
Product	wt %	Flow rate, kg/h	wt %	Flow rate, kg/h
Ethylene	42.26	62,010	42.58	62,008
Propylene	15.60	22,890	15.86	23,100
Tail gas + H_2	29.49	43,270	24.69	43,241
C_2H_2[§]	0.39	570	—	—
C_3H_4[§]	0.38	560	—	—
C_4	4.25	6,240	4.24	6,184
Gasoline	6.68	9,800	6.68	9,727
Fuel oil	0.95	1,390	0.95	1,378
	100.00	146,730	100.00	145,638

[†]Does not include the hydrogenation steps and the ethylene loss in by-products.
[‡]Includes effect of all unit operations affecting ethylene production.
[§]Generated in pyrolysis reaction, but removed by the hydrogenation step.

all the unit operations. The program includes the two hydrogenation steps and the loss of a small amount of ethylene with other by-products. For the base-case ethylene process, sensitivity calculations using the simulator are performed until convergence matches the desired production capacity.

The material balance summary for the base-case ethylene process, described in Figs. 3-8 through 3-13, is the one provided in Fig. 3-15 and specifies the hourly flow

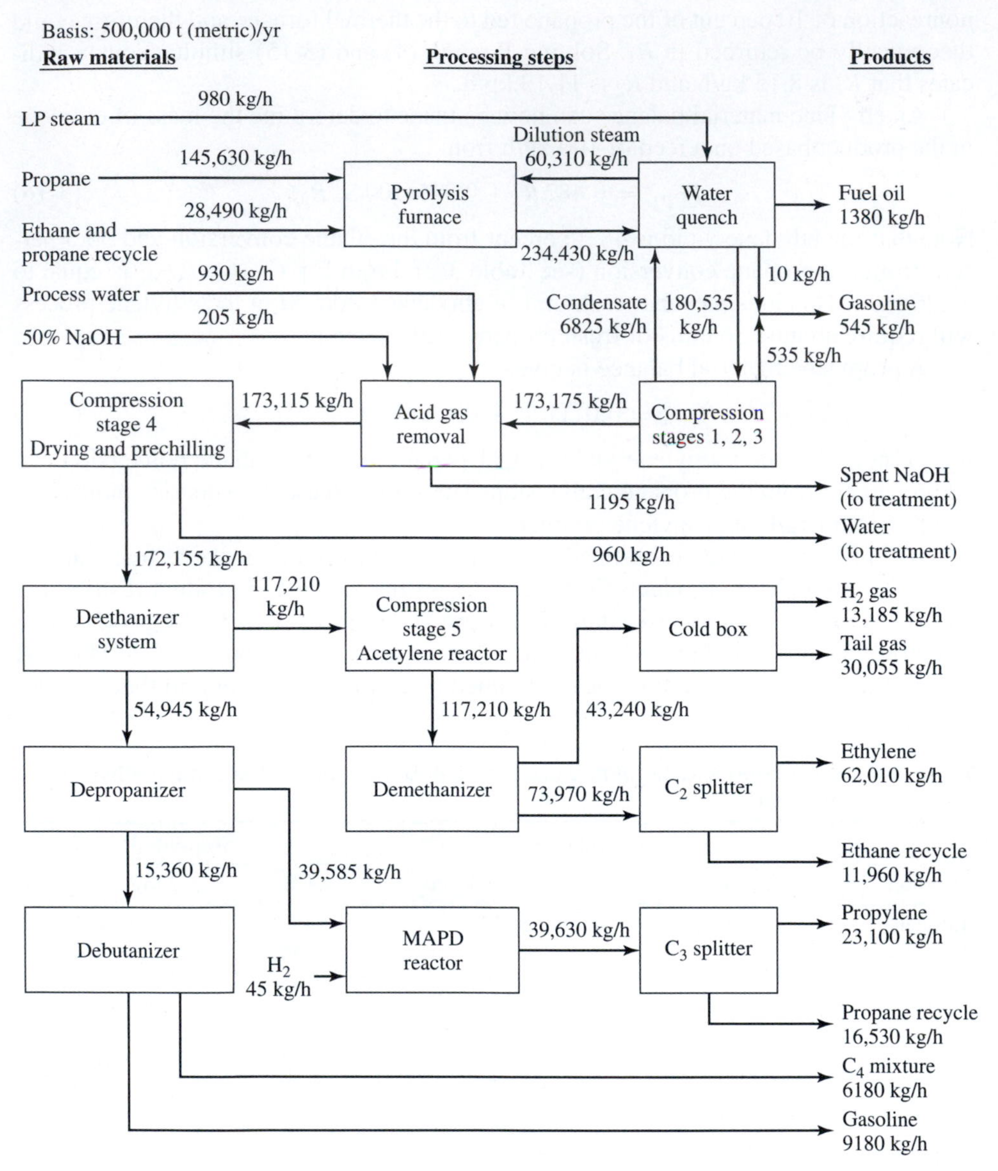

Figure 3-15
Material balance summary around the various sections of the ethylene process

rates of the streams to the various sections of the ethylene process. Note that heat exchangers, condensers, drums, pumps, etc., are not included in Fig. 3-15 since these equipment items do not affect the material balance.

Once the complete material balance is made, the mass quantities are used to prepare energy balances around each piece of equipment. Temperature and pressure levels at various key points in the process, particularly at the reactors and compressors, serve as guides in making these balances. The complete calculations for the material and energy balances for each piece of equipment, because of their length, are not presented in this discussion.

Equipment Design For the preliminary design of the base-case ethylene process, this involves determining the size of the equipment normally selected in terms of volume, surface area, or flow per unit time. However, because of the large number of different pieces of equipment involved in the ethylene process, the design procedures even for the key equipment items will be brief. Further details can be obtained by referring to Chaps. 12 through 15.

The *thermal furnace* includes both a convection and a radiation zone, as shown in Fig. 3-8. In the design of this unit, the energy requirements of the cracking and high-pressure steam must first be obtained so that the fuel requirement can be determined. A furnace model was developed with the design of the thermal cracking process. Figure 3-16 illustrates the different elements of the furnace model. The feed to the

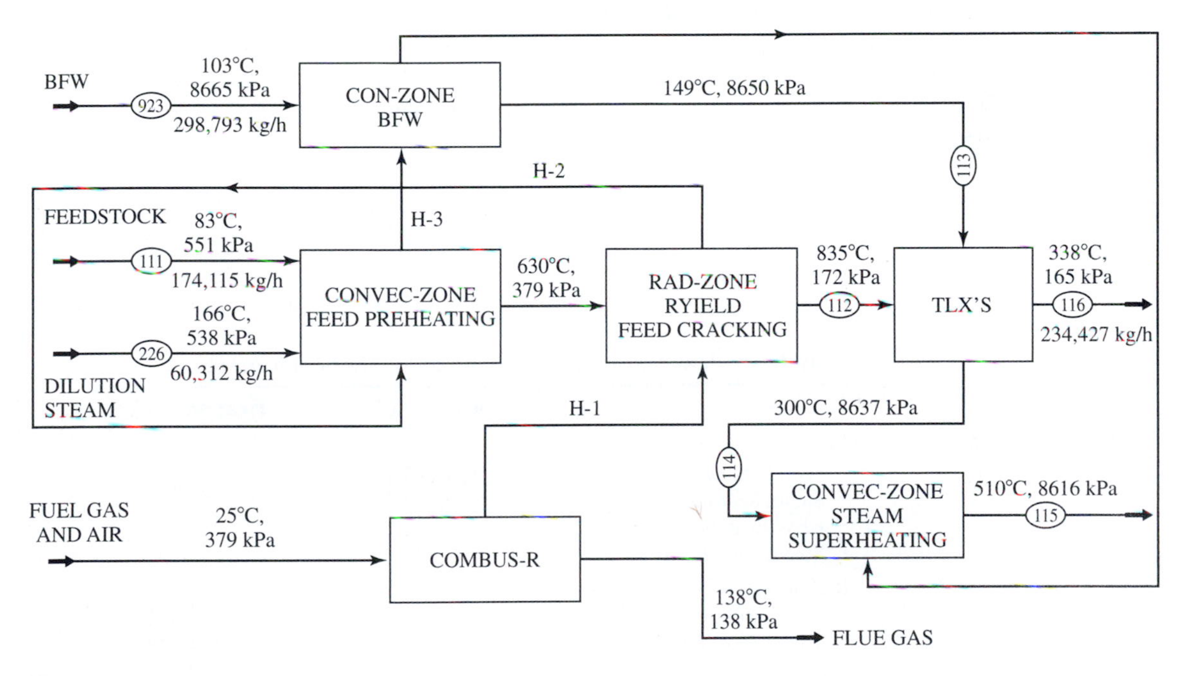

Figure 3-16
Diagram of the thermal cracking furnace model

furnace is preheated externally to about 83°C before entering the uppermost coil of the convection zone for further heating. The dilution steam is added to the feed, and the mixture is further heated in the bottom coil of the convection zone to about 630°C. The steam-hydrocarbon mixture then enters the radiation zone where it is further heated and cracked into various products, as shown in Table 3-2. The fuel utilized for the heating is the tail gas component of the cracking process which is combined with 10 percent excess air and admitted to the lowest end of the radiation zone where it is combusted. About 34 percent of the combustion heat is absorbed in the radiation zone. The rest of the heat is recovered in the convection zone where the flue gas is used for preheating the feed and the boiler feed water (BFW) as well as superheating the steam.

For modeling the thermal process, the RYIELD package of the ASPEN PLUS™ simulator was used to simulate the radiation zone of the furnace. Since the yield distribution of this cracking process is experimentally known, the heat of reaction can be determined by using the enthalpies of formation of the reactants and products. The combustion process of the tail gas was simulated using the RSTOIC package of the ASPEN PLUS™ simulator. For purposes of accuracy, trace contaminants of carbon monoxide, ethylene, and ethane in the tail gas were included in the combustion calculations.

On the basis of 1 mol of tail gas consumed, the mol flow rate of oxygen in the air (with 10 percent excess) required is 2.127 mol. The lower heating value (LHV) of the tail gas is about 216 kW/kg mol for an outlet temperature of 25°C. Based on a thermal efficiency of 95 percent, the energy required from the combustion of the tail gas in the conventional base-case ethylene process is about 3.75×10^5 kW. Table 3-5 shows the heat duties for the two sections of the furnace and the tail gas for the base-case design.

Since the cracking reactions involving propane and ethane are first-order, the residence time θ_R can be obtained from the relation

$$k = (1/\theta_R) \ln \left(\frac{1}{1 - x} \right) \tag{3-18}$$

Table 3-5 Heat duties for each zone of the thermal furnace

Heat zone	Flow rate, kg/h	Heat duty, kW
Convection zone		
Feed heating	174,115	95,579
Steam heating	60,312	
BFW heating	298,793	18,562
Steam superheating	298,793	132,702
Radiation zone		
Feed cracking	234,427	127,635
Total process demand		374,478
Combustion zone		
Total heat supplied	560,980	−374,478

Table 3-6 Design details of the thermal furnace

Number of furnaces	7 furnaces, one spare
Number of coils per furnace	10
Heat load per furnace	53,500 kW
Residence time	0.4 s
Tube ID	0.102 m
Tube OD	0.119 m
Coil length	73.7 m
Mass velocity	143.3 kg/(m^2·s)
Heat flux	85.8 kW/m^2
Material of construction	HP Mod (25% Cr, 35% Ni, 0.4% C, 1.25% Nb)

where x is the conversion of propane and ethane to cracked products and the reaction rate constant is obtained from the Arrhenius equation

$$k = A \exp(-E/RT) \tag{3-19}$$

where A is the frequency factor and E the activation energy. These factors covering the process conditions shown in Fig. 3-8 are available from the literature.[†] With such data, the residence time for a 60 percent conversion of ethane and a 90 percent conversion of propane is about 0.4 s. This indicates that both hydrocarbons can be mixed and cracked in the same furnace coils.

Once the residence time and the allowable volumetric gas flow rates for a given tube diameter have been established, the coil length of the tube can be determined. In this calculation, the average physical properties of the cracked gas inside the tubes must be considered. Table 3-6 presents the design geometry and design aspects of the thermal furnace recommended for the base-case ethylene process.

Acetylene hydrogenation over Pd/Al_2O_3 (palladium/alumina) catalyst in a fixed-bed reactor is the favored industrial process since the acetylene concentration can be reduced to less than 5 ppm while undesirable hydrogenation of ethylene to ethane is minimized.[‡] For the hydrogenation of acetylene, the selectivity of the reaction is usually in the range of 90 percent. The selectivity for this reaction is defined as

$$S = \frac{R_{C_2H_4}}{R_{C_2H_4} + R_{C_2H_6}} \tag{3-20}$$

where $R_{C_2H_4}$ is the net rate of ethylene produced from the acetylene hydrogenation and $R_{C_2H_6}$ is the net rate of ethane produced from the ethylene hydrogenation.

The reaction rate constants for the acetylene and ethylene hydrogenation reactions can be determined by using the Arrhenius equation. Parameters for both hydrogenation equations are available in the literature for various operating conditions.[§] The two

[†]K. M. Sundaram and G. F. Froment, *Chem. Eng. Sci.*, **32:** 601 (1977); H. C. Schutt, *Chem. Eng. Prog.*, **55**(1): 68 (1959).

[‡]J. M. Moses et al., *J. Catalysis,* **86:** 417 (1984).

[§]F. E. Hancock and J. K. Smith, in *Proc. 4th Ethylene Producers Conference,* vol. 1, American Institute of Chemical Engineers, New York, 1992, p. 177.

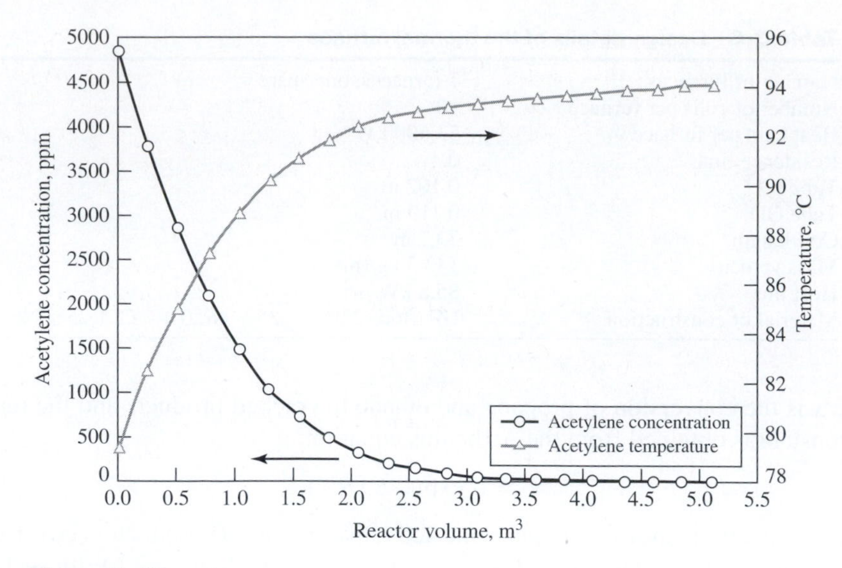

Figure 3-17
Temperature and acetylene concentration as a function of reactor volume in the acetylene reactor

reactions are highly exothermic and are usually carried out in an adiabatic tubular reactor. The Ergun equation is used to calculate the pressure drop through the packed porous bed of Pd/Al_2O_3 catalyst.† The reactor volume and catalyst weight W are related through the relation

$$W = [(1 - \phi)A_c L]\,\rho_c \tag{3-21}$$

where ϕ is the porosity of the packed bed, A_c is the cross-sectional area of the bed, L is the length of the tubular reactor, and ρ_c is the density of the solid catalyst.

An adiabatic plug flow reactor model is normally used to represent the operation of the acetylene reactor. Based on this model, Fig. 3-17 shows the decrease in acetylene concentration as a function of reactor volume. Note for this adiabatic exothermic reaction in which the temperature increases along the length of the reactor, the reaction virtually ceases when the reactor volume is increased beyond 2.5 m^3 because of the decrease in acetylene concentration. However, a volume of at least 5 m^3 is required to lower the desired acetylene concentration to about 2 ppm. This low acetylene concentration is only attained at the endpoint of the reactor. The selectivity for the acetylene reaction is calculated with the aid of Eq. (3-20) and is 89.1 percent.

The pressure drop through the reactor depends on the catalyst particle diameter selected as well as the reactor dimensions. Calculations show that for a catalyst particle diameter of 7.5 mm, a pressure drop of less than 20 kPa can be achieved in the base-case design for the acetylene reactor with a diameter of 1.83 m and a length of 1.94 m. Approximately 2500 kg of catalyst will be required.

†H. S. Fogler, *Elements of Chemical Reaction Engineering,* 2d ed., Prentice-Hall, Englewood Cliffs, NJ, 1992.

Rigorous design models are used to design the *distillation columns* and *quench tower*. For example, the distillation columns in the separation section of the base-case design were simulated with the CC-SCDS model using the SRK equation of state. This model is a multistage vapor-liquid equilibrium module that can handle five feed streams, four side products, and side heaters or coolers. The model uses a Newton-Raphson convergence method and rigorously calculates the derivatives for the material, energy, and equilibria relationships involved. The results of such a simulation are compared with those obtained from a hand calculation for the base-case C_3 splitter in two examples outlined in Chap. 15. Mechanical design equations for the columns are presented in Chap. 12.

Shell-and-tube exchangers are used at different locations within the ethylene process. Each heat exchanger has its unique design which depends on the nature of the streams involved in the thermal process. Generally, the design of the shell-and-tube exchanger is an interactive procedure since heat-transfer coefficients and pressure drop depend on different geometric factors such as tube length and diameter, shell length and diameter, number of shell passes, and baffle type. Numerous software programs are available to perform this iterative task. For the base-case design, CC-THERM software was used to perform the thermal and mechanical design, rating, and simulation of the TEMA shell-and-tube heat exchangers. An example of its use is provided in Chap. 14.

A *caustic scrubber* is used to remove the acid gases (H_2S, CO_2) from the cracked gas stream prior to entering the fourth stage of the gas compressor. Both contaminants must be reduced to concentration levels less than 1 ppm before entering the separation section. Both acid gas components react with caustic to form soluble inorganic salts.

The design procedure developed by Raab was used to design the caustic scrubber. The latter is designed to prevent any flooding within the column and maintains the flood percent at less than 80 percent of total flood. The scrubber can be simulated by using the Sour Water Model of CHEMCAD.

Gas and liquid dehydrators are used to dry the cracked gas after the fourth compression stage to a water content of less than 1 ppm to avoid problems with freezing and hydrate formation in the low-temperature equipment located downstream. To achieve this design goal, the cracked gas stream is first cooled to about 13°C, condensing some of the water which is removed in a flash drum. About 13 weight percent of the water is separated as a vapor and removed from the cracked gas in a gas dryer to less than 1 ppm. The remaining water leaves with the hydrocarbon condensate stream, whereupon 90 percent of the water in the stream is removed in a water decanter. The remaining hydrocarbon stream is sent to a liquid dryer to reduce the water content to less than 1 ppm.

Type 3A molecular sieve is the commercial desiccant used for the drying unit operation in the ethylene process. Even though its equilibrium water content is high, the drying process is designed for a dry gas capacity of 8 kg of water per 100 kg of desiccant. The gas and liquid dryers are designed for an adsorbent regeneration cycle of 24 h. Under these conditions, the desiccant required and the corresponding dryer volume are 37,900 kg and 436 m^3 for the gas dryer, respectively, and 24,950 kg and 289 m^3 for the liquid dryer, respectively.

Economic Assessment of Base-Case Design

An economic analysis of the base-case ethylene process is necessary for comparing the profitability of this process with the proposed design alternatives that utilize various distillation/membrane hybrid configurations in the separation section of the process. The economic parameters and conditions that are utilized in the base-case design and which are also used for the modified process include plant location, raw material costs, product costs, utility costs, depreciation, and tax rates.

Capital Investment The purchased cost of each piece of equipment is determined by using the cost data provided in Chaps. 12 through 15, current ethylene literature, and vendor information. Regardless of the source, the purchased-cost data must be corrected to the current cost index, as described in some detail in Chap. 6. An example of such a tabulation is provided in Table 3-7, which lists design details and purchased cost of individual pieces of equipment associated with the separation section of the ethylene process. Purchased cost data for pumps, pressure vessels, and storage tanks have not been included in this tabulation only because they were not shown in Fig. 3-13. To complete the cost tabulation for this section, such items must also be included.

Rather than tabulating the purchased cost of each piece of equipment associated with the other sections of the ethylene process, it is simpler to summarize the purchased-cost data in terms of the types of equipment that are required in the overall process, as shown in Table 3-8.

The required fixed-capital investment for the base-case ethylene process may be estimated from the total purchased-equipment cost using the equipment/cost ratio method outlined in Chap. 6. The estimated fixed-capital cost tabulation is given in Table 3-9. The probable error in this method of estimating the fixed-capital investment can be as much as ± 20 percent. However, since the probable error in estimating the fixed-capital investment associated with the proposed modification in the separation section of the ethylene process will be very similar, this estimation procedure is acceptable for making the economic comparison.

Production Costs Such costs include direct costs, indirect costs, and general expenses. Table 3-10 provides the cost in cents per kilogram for the raw materials, the utilities, and the credit for the sale of by-products in the base-case ethylene process. Direct costs for labor and maintenance in the conventional ethylene plant are generally between 3 and 5 percent of the fixed-capital investment, with approximately one-half of it for material replacements and one-half for maintenance labor.[†] For this preliminary design, 4 percent of the fixed-capital investment was designated for maintenance charges.

The indirect costs include depreciation, taxes, insurance, and plant overhead charges. For ethylene plants, the allowable depreciation rate typically ranges from 7 to 12 percent.[‡] For the cost analysis, the depreciation rate was assumed to be 10 percent

---

[†]L. Kniel, O. Winter, and K. Stork, *Ethylene: Keystone to the Petrochemical Industry,* Marcel Dekker, New York, 1978.

[‡]L. Kniel et al., loc. cit.

Table 3-7 Base-case separation section equipment specification and purchased cost†

Equipment code	Equipment function	Equipment specification		Materials of construction		Purchased cost,‡ $
Fractionators		**Diam., m**	**# Trays**			
C-601	C_2 splitter	4.57	111[a]	304 s.s		4,232,000[c]
C-602	Depropanizer	2.60	51[a]	c.s		418,000[c]
C-603	C_3 splitter—I	5.33	106[b]	c.s		3,575,000[c]
C-604	C_3 splitter—II	5.33	106[b]	c.s		3,575,000[c]
C-605	Debutanizer	0.92	59[a]	c.s		36,900[c]
Reactors		**Diam., m**	**Volume, m**			
R-601	1st-stage MAPD reactor	1.83	10	s.s. clad		131,900
R-602	2nd-stage MAPD reactor	2.3	40	s.s. clad		302,500
Compressor		**Power reqm't., kW**				
K-601	Ethylene prod. compressor	910		c.s		408,000
Heat exchangers or condensers		**# Shells**	**Area/shell, m²**	**Shell**	**Tube**	
E-601	C_2 splitter condenser	4	303	A-240-316	A-240-316	189,500
E-602	C_2 splitter reboiler	2	730	A-515-55	A-214	107,700
E-603	Depropanizer condenser	1	576	A-285-C	A-214	43,400
E-604	Depropanizer reboiler	1	60	A-285-C	A-214	8,900
E-605	Process exchanger	2	537	A-285-C	A-214	82,500
E-606	1st-stage C_3 cooler	1	14	A-285-C	A-214	4,100
E-607	Water-cooled exchanger	1	72	A-285-C	A-214	9,900
E-608	C_3 splitter condenser	5	1109	A-285-C	A-214	549,700
E-609	C_3 splitter reboiler	1	211	A-285-C	A-214	19,200
E-610	Debutanizer condenser	1	222	A-285-C	A-214	20,900
E-611	Debutanizer reboiler	1	17	A-285-C	A-214	4,200
E-612	Pyrolysis gasoline cooler	1	67	A-285-C	A-214	9,600

†See Fig. 3-13 for separation section flowsheet and equipment location.
‡January 1, 2000, purchased-cost data
[a]0.61-m tray spacing
[b]0.76-m tray spacing
[c]Cost includes columns, trays, and auxiliary accessories.

Table 3-8 Summary of purchased equipment cost for the ethylene process†

Equipment type	Purchased cost
Heat exchangers and condensers	$ 5,015,000
Pressure vessels	3,069,000
Columns and trays	17,235,000
Compressors	34,835,000
Furnaces	19,944,000
Reactors	551,000
Dryers	364,000
Pumps	1,457,000
Separators	96,000
Storage vessels	940,000
Total	$83,506,000

†January 1, 2000, purchased-cost data.

Table 3-9 Fixed-capital investment estimate†

Investment items	Cost
Purchased equipment, E	$ 83,506,000
Purchased-equipment installation, $0.47E$	39,248,000
Instrumentation and control, $0.36E$	30,062,000
Piping (installed), $0.68E$	56,784,000
Electrical (installed), $0.11E$	9,186,000
Buildings (including services), $0.18E$	15,031,000
Yard improvements, $0.1E$	8,351,000
Service facilities (installed), $1.05E$‡	87,681,000
Total direct plant cost D	329,849,000
Engineering and supervision, $0.33E$	27,557,000
Construction expenses, $0.41E$	34,237,000
Legal expenses, $0.04E$	3,340,000
Total direct and indirect cost, $D+I$	393,983,000
Contractor's fee, $0.05(D+I)$	19,699,000
Contingency, $0.1\ (D+I)$	39,398,000
Fixed-capital investment (FCI)	$453,080,000

†Equipment/cost ratio percentages are factors applicable to a fluid-processing plant similar to that outlined in Table 6-9. No land purchase was assumed.
‡The estimated cost for service facilities was increased by 50 percent to meet the need for additional environmental and refrigeration requirements in the process.

of the fixed-capital investment. Insurance costs are about 0.5 percent annually, and taxes are on the order of 1.5 percent annually of the fixed-capital investment. The plant overhead charges are 80 percent of the total labor costs. The general expenses include administration costs, distribution and selling costs, research and development costs, and financing.

Table 3-10 Material, utility, and by-product costs for the base-case ethylene process

Variable costs	Quantity consumed per kilogram of ethylene produced	Market price†	Cost contribution, ¢/kg, of ethylene produced
Raw materials			
Propane (feedstock)	2.3486 kg	26.66 ¢/kg	62.633
Caustic (100% basis)	0.0017 kg	11.24 ¢/kg	0.019
Catalyst and chemicals	—	—	0.740
Total			63.392
Utilities			
Cooling water	0.3796 m^3	2.25 ¢/m^3	0.854
Process water	1.67×10^{-4} m^3	31.17 ¢/m^3	0.005
Fuel gas	25,416 kJ	$2.79/$10^6$ kJ	7.091
Electricity	0.1337 kWh	4.3 ¢/kWh	0.552
Total			8.502
By-products (produced)	Quantity produced per kilogram of ethylene produced		
Hydrogen-rich gas	0.2127 kg	73.7 ¢/kg	−15.675
Methane-rich fuel gas	27,200 kJ	$2.79/$10^6$ kJ	−7.589
Propylene (polymer grade)	0.373 kg	57.3 ¢/kg	−21.374
C_4 fraction	0.0979 kg	42.5 ¢/kg	−4.237
Pyrolysis gasoline	0.1569 kg	31.0 ¢/kg	−4.864
Fuel oil	0.0222 kg	13.7 ¢/kg	−0.304
Total			−54.043

†January 1, 2000, cost data

A production cost estimate for the base-case ethylene design is shown in Table 3-11. The cost of raw materials is 63.4 ¢/kg of ethylene produced while the total credit for the by-products is 54.0 ¢/kg of ethylene produced. All methane-rich gas has been credited at $2.79/$10^6$ kJ (HHV), and part of this fuel has been charged as utility fuel for the thermal furnaces. The *net production cost* (NPC) of ethylene as estimated from the base-case design is 47.8 ¢/kg.

Profitability Analysis One measure of profitability often used is the *return on investment* (ROI). The latter is defined as the annual profit before taxes, divided by the fixed-capital investment. The annual profit is the difference between the annual revenue and expenses. Based on a price of 70.5 ¢/kg for the ethylene product, the ROI for the base-case ethylene process before taxes is 25.1 percent.

Assessment of Proposed Base-Case Design Modification

As noted in the process creation step, the use of a hybrid distillation/membrane system to replace the conventional C_2 splitter in the ethylene/ethane separation provides the potential to reduce the energy requirements for this separation. Several different distillation membrane configurations were noted in Fig. 3-6, and each of these needs to be

Table 3-11 Total product cost estimate

Cost items	Cost, ¢/kg, of ethylene
Direct production costs	
Raw materials (see Table 3-10)	63.39
Operating labor (9 workers/shift)	0.49
Operating supervision (15% operating labor)	0.07
Maintenance (labor and materials, 4% FCI)	3.62
Utilities (see Table 3-10)	8.50
Operating supplies (15% maintenance)	0.54
Laboratory charges (20% operating labor)	0.10
Indirect production costs	
Depreciation (10% FCI)	9.07
Insurance and taxes (2% FCI)	1.81
Plant overhead costs (80% total labor costs)	1.90
General expenses	
Administrative costs (25% overhead)	0.48
Distribution and selling costs (6% NPC)	2.87
Research and development costs (4% NPC)	1.91
Financing (interest, 7% TCI)†	7.05
Credit for sale of by-products (see Table 3-10)	−54.04
Annual net product cost (NPC), ¢/kg	47.76

†TCI = \$503,400,000 = \$453,040,000 + 10% (TCI) for working capital.

investigated to determine which configuration would provide the best economic alternative. Because of the extensive simulation and optimization effort that is involved with each configuration, only the general procedures involved in this effort will be outlined with one of the configurations. However, the results of this economic comparison will be summarized to complete the preliminary design example.

To obtain a meaningful economic comparison for the ethylene/ethane separation, the material balance around the hybrid distillation/membrane system is the same as that established earlier for the C_2 splitter in the base-case design. Likewise, the operating conditions associated with the entering and leaving streams are the same for each separation system. Any adjustments in temperature or pressure requiring additional capital investment, utilities, labor, maintenance, etc., become part of the operating cost for the hybrid system.

Technical Aspects Figure 3-18 illustrates the ethylene/ethane facilitated membrane system in which the membrane consist of an aqueous solution of silver nitrate held by capillarity in the pores of a microporous membrane. Ethylene in the gas mixture, fed at high pressure, reacts with the silver ion to form a complex, which then diffuses freely across the membrane. When this complex reaches the opposite low-pressure side of the membrane, the complex decomposes, releasing the ethylene into the permeate and thereby permitting the aqueous solution to return to the membrane inlet. The diffusion of the uncomplexed ethylene and ethane is small because the uncomplexed forms are considerably less soluble in solution.

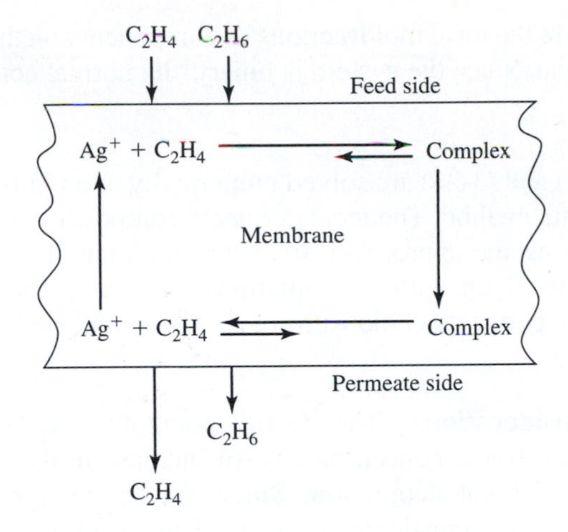

Figure 3-18
Mechanism for facilitated transport of ethylene by silver ion

Table 3-12 Assumptions used in modeling of cross-flow model with FT membrane

1. Cross-flow along the permeate side and plug flow along the feed side of the membrane.
2. Negligible pressure drop along either side of the membrane.
3. Isothermal operation.
4. Instantaneous ethylene/silver reaction, forming a one-to-one complex.
5. Equilibrium between product and reactant species at the membrane boundaries.
6. Steady-state process.
7. The diffusivity and solubility coefficients of each species are independent of pressure.

The assumptions used in developing the membrane model are summarized in Table 3-12. The membrane properties used in the simulations are based on the experimental data of Teramoto et al.† The cross-flow model is based on the model developed by Shindo et al.‡ In this model, the local permeate-side composition along the membrane length is only a function of the local feed-side composition.

The overall mass balance across a differential area of membrane dS can be described by

$$\frac{dN}{dS} = -\sum_{i=1}^{n} J_i \tag{3-22}$$

Here N is the molar flow rate of gas on the feed side of the membrane, and J_i is the steady-state flux of component i. The component mass balance is given by

$$\frac{d\bar{y}_{F,i}}{dS} = \frac{J}{N}(\bar{y}_{F,i} - \bar{y}_{P,i}) \tag{3-23}$$

†M. Teramoto et al., *J. Chem. Eng. Japan*, **19**(5): 419 (1986).

‡Y. Shindo et al., *Sep. Sci. Tech.*, **20**(2): 445 (1985).

where $\bar{y}_{F,i}$ and $\bar{y}_{P,i}$ are the local mol fractions of component i on the feed and permeate sides of the membrane. Since the system is binary, the normal conservation equations apply, that is, $\sum_{i=1}^{n} y_i = 1$.

Equations (3-22) and (3-23) are solved numerically as an initial-value problem by using the Runge-Kutta method. The local permeate composition $\bar{y}_{P,i}$ and the local flux are calculated by using the appropriate transport equations for ethylene and ethane across the membrane along with the equations describing the feed and permeate boundaries. The local permeate composition $\bar{y}_{P,i}$ can be obtained from the local fluxes (that is, $\bar{y}_{P,i} = J_{P,i}/J$).

Hybrid Process Considerations The freezing point of the aqueous solution of silver nitrate depends on the solute concentration. An increase in the solute concentration increases the freezing point depression. Since the temperature profile within the conventional distillation column decreases from −7 to −29°C for an operating pressure of 2000 kPa, an innovative design is required to effectively combine the membrane and distillation technology. To avoid freezing, 6 mol/l $AgNO_3$ is used in the ethylene/ethane separation scheme.

The FT membrane performance also is sensitive to impurities such as acetylene and sulfur. Use of the front-end deethanization scheme removes the acetylene impurity in the hydrogenation catalytic reactors. Any sulfur contaminants can be removed by caustic scrubbers as in the conventional ethylene process.

Simulation and Optimization Strategy Simulation studies are required to identify the optimal configuration which will provide the maximum economic advantage. Such simulations are also useful for determining the operational limits for each system.

For the facilitated transport membrane unit, the most important parameters that can be varied are the permeate pressure and the carrier concentration. These two parameters directly affect the requirements for membrane area and compression duty. Generally, there is a tradeoff between these two parameters. An increase in the carrier concentration results in an increase in the driving force across the membrane, which minimizes the area required by the membrane. On the other hand, an increase in the permeate pressure provides a decrease in the driving force across the membrane and thus requires a larger membrane area. However, this increase in permeate pressure results in a decrease in compression duty.

For the simulation and optimization process, two limiting pressures of 103 and 345 kPa have been selected. The lower value provides higher selectivity, which minimizes the membrane area required by the hybrid system but increases the required compression duty. The upper limit of 345 kPa, on the other hand, decreases the compressor requirement, but the selectivity decreases with an increase in membrane area. Permeate pressures greater than 345 kPa provide too small a driving force across the membrane for a feed pressure of 2000 kPa. Thus, both selectivity and permeation rate would be decreased.

For the distillation process, there is a tradeoff between the number of stages required for the separation and the condenser duty. The key parameter that relates these two important factors in an inverse manner is the reflux ratio. For example, when the reflux ratio is reduced, the process utilities are reduced, resulting in a decrease in operating costs. However, this decrease in reflux ratio necessitates an increase in the number of stages in the distillation column which, in turn, increases the capital cost of the column. For low-temperature distillation systems, the optimum reflux ratio is generally maintained between 1.05 and 1.1 times the minimum reflux ratio to minimize the refrigeration costs associated with such units. To maintain economic comparability, the low-temperature columns in the base-case design and the various hybrid configurations have used the 1.05 value to determine the actual reflux.

For the distillation/membrane hybrid system, the variables that affect each technology are related. Combination of these variables makes the optimization process more complex. Since the simulation and optimization of the hybrid system are not local but global, the process of optimization must be carefully analyzed with at least two simulation steps to set limits so that a simple and suitable approach can be developed.

In the first simulation step, an intensive local optimization of the hybrid system is pursued by conducting a series of parametric studies. For each configuration, the effects of key parameters on the hybrid system are established. To determine whether the reduction of the refrigerant or the reduction of the hybrid system capital cost has the greater effect on the entire ethylene process requires analysis over a wide range of operating conditions. For that purpose, the reflux ratio of the distillation column is established initially so that the condenser duty of the hybrid system is maintained at the same value as the condenser duty of the base-case distillation system. This is done so that the addition of the hybrid system to the capital costs of the distillation column can be established.

In the second simulation, the number of theoretical stages in the distillation column of the hybrid system is maintained at the same value as that required for the base-case C_2 splitter. This results in a hybrid system that achieves a minimum condenser duty. The effect of this case on other parts of the ethylene process is then determined. These two limits provide guidelines for determining which variables have the greater effect on the overall ethylene process.

Once the hybrid system that provides the minimum condenser duty or the minimum capital costs is established, a major redesign of the process is necessary to observe the effects of the new configurations on the entire process. Finally, an economic criterion is used to compare the optimized hybrid designs with the base-case design.

The membrane model used in this design is closely integrated with a rigorous distillation model using SRK as the equation of state. This distillation model (CHEMCAD-SCDS) is a multistage vapor-liquid equilibrium module. Equipment dimensions and operating conditions are evaluated by computer routines that are included with the CHEMCAD simulator. CC-THERM is utilized to design all heat exchangers and condensers that are part of each hybrid system.

The simulation sequences pursued in evaluating the hybrid distillation/membrane systems depend on the location of the membrane unit relative to the distillation column. For example, for the parallel configuration shown in Fig. 3-6, the membrane feed composition and flow rate are the design variables that need to be optimized to determine the number of stages or the optimal reflux ratio in the column. This is initiated by using the cross-flow model for the membrane with an appropriate value for the membrane area. To solve for the minimum number of stages, the reflux ratio is set at the same value as that originally used for the base-case C_2 splitter. Then the two design variables are varied until optimal values are attained which provide the minimum number of stages in the column. The initial value for the membrane area is then changed and the optimization process is repeated. To solve for the optimal reflux ratio which establishes the minimum condenser duty, the number of stages is first set at the same value as that used in the base-case C_2 splitter. Figure 3-19 illustrates the performance of the parallel configuration when this simulation is conducted for a permeate pressure of 345 kPa and a carrier concentration of 6 mol/l $AgNO_3$. Note that with a decrease in reflux ratio, there is both an increase in the membrane feed flow rate and an increase in the membrane feed composition. As the condenser duty is reduced by decreasing the reflux rates in the column, the capacity and the required utilities for the auxiliary equipment are increased.

Economic Evaluations Processing cost is the economic measure that was employed to evaluate the economic performance of the different hybrid systems. The processing cost includes all capital and operating expenses that are involved with the separation of ethylene from ethane. Capital charges include the depreciated costs of each piece of equipment involved in this separation.

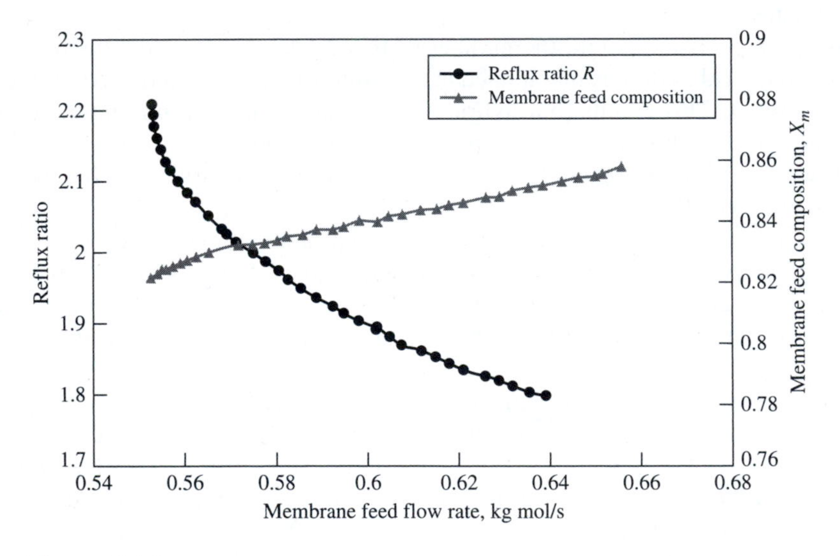

Figure 3-19
Simulated performance of the parallel hybrid configuration with a permeate pressure of 345 kPa and a carrier concentration of 6 mol/l $AgNO_3$

Savings in processing cost are determined from the differences in operating costs of the base-case design shown in Table 3-10 and the design incorporating the different hybrid systems. As noted earlier, the only sections in the ethylene process that are affected by the substitution of the hybrid system for the C_2 splitter are the ethylene/ethane separation system, the propylene refrigeration system, and the steam system. The ethylene refrigeration system is only affected when the series-configuration hybrid system is considered since additional ethylene refrigerant is utilized in the low-pressure column condenser. The operating costs of other sections of the ethylene process are not affected since the feed, product, and recycle streams of the ethylene/ethane separation system are maintained at the same level as the base-case design.

Table 3-13 shows additional assumptions and cost data besides those shown in Tables 3-3 and 3-10 used for the economic evaluation.

The annual savings in capital, utilities, and processing costs for the parallel configuration hybrid system are shown in Fig. 3-20. The optimum hybrid system with this configuration and a permeate pressure of 103 kPa occurs when a product recovery of about 0.70 is achieved. Under these conditions, the maximum annual savings are about $2.38 million. The annual savings decrease when the product recovery exceeds 0.7, indicating that the membrane performance is best for bulk separation where the composition is close to the feed composition. If the permeate pressure is increased to 345 kPa, the maximum annual savings decrease to about $1.84 million.

The optimum hybrid configuration, however, occurs when the membrane is coupled with the distillation column in a series configuration. The series hybrid system provides an annual savings of about $5.7 million. Table 3-14 summarizes the savings obtained for the four hybrid configurations shown earlier in Fig. 3-6.

Table 3-13 Hybrid system economic parameters and assumptions†

Assumptions	
Life of membrane	5 years
Life of other equipment	10 years
Cost data	
Membrane module	$260/m^2 (membrane material, housing, piping, instrumentation and installation labor)
Membrane replacement	$150/m^2 (membrane material and installation labor)
$AgNO_3$	$488/kg

†January 1, 2000, cost data.

Table 3-14 Annual and unit processing savings with hybrid systems

Hybrid configuration	Annual process savings, $	Unit process savings, ¢/kg
Top	505,000	0.10
Parallel	2,383,000	0.47
Bottom	333,000	0.066
Series	5,700,000	1.14

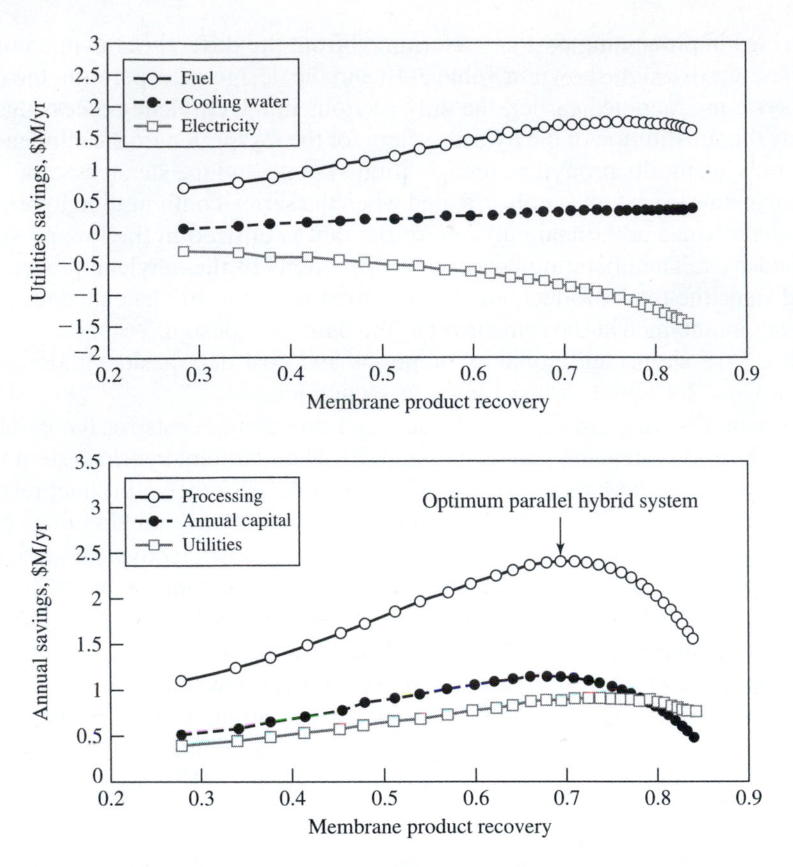

Figure 3-20
Economic performance of the parallel hybrid configuration with a permeate pressure of 103 kPa and a carrier concentration of 6 mol/l $AgNO_3$

SUMMARY

The preliminary design presented in this section was developed to show the step-by-step approach which is quite often followed for a new process design. The exact procedure may vary from company to company and from one design engineer to another. Likewise, the assumptions and rule-of-thumb factors used may vary from one company to the next depending to a large extent on design experience and company policy. Nevertheless, the basic steps for a process design are those outlined in this preliminary design covering the manufacture of a major petrochemical commodity.

No attempt has been made to present a complete design. In fact, to minimize the length, many assumptions were made which would have been verified or justified in a normal process design. Neither were other potential alternatives presented that could have provided additional savings in the ethylene process. These would include using a hybrid distillation/membrane system as a replacement for the C_3 splitter or the use of membranes in the hydrogen separation section of the ethylene process. The investigation of these alternatives is left to the reader.

PROBLEMS

3-1 Using the various literature sources identified in the chapter, list the original source, title, author, and brief abstract of three published articles presenting three different processes for the production of formaldehyde.

3-2 Prepare, in the form of a flowsheet, an outline showing the sequence of steps in the complete development of a plant for producing formaldehyde. A detailed analysis of the points to be considered at each step should be included. The outline should take the project from the initial idea to the stage where the plant is in efficient operation.

3-3 A process for making a single product involves reacting two liquids in a continuously agitated reactor and distilling the resulting mixture. Unused reactants are recovered as overhead and are recycled. The product is obtained in sufficiently pure form as bottoms from the distillation tower.

a. Prepare a qualitative flowsheet for the process, showing all pieces of equipment.

b. With cross reference to the qualitative flowsheet, list each piece of equipment and tabulate for each the information needed concerning chemicals and the process, in order to design the equipment.

3-4 A search of the literature reveals many different processes for the production of acetylene. Select four different processes, prepare qualitative flowsheets for each, and discuss the essential differences between each process. When would one process be more desirable than the others? What are the main design problems which would require additional information? What approximations would be necessary if data were not available to resolve these questions?

3-5 Ethylene is produced commercially in a variety of different processes. Feedstocks for these various processes range from refinery gas, ethane, propane, butane, natural gasoline, light and heavy naphthas to gas and oil and heavier fractions. Prepare three different qualitative flowsheets to handle a majority of these feedstocks. What are the advantages and disadvantages of each selected process?

3-6 One of the ethylene processes using propane as a feedstock with front-end deethanization and front-end acetylene hydrogenation was presented as the preliminary design example in this chapter. For the product separation section for the base-case design, shown in Fig. 3-13, complete the material balance around each piece of equipment shown. The feed in kg/h for the C_2 splitter and the depropanizer column are as follows:

Component	Stream 533	Stream 413	Stream 416
Methane	2.92	—	—
Acetylene	0.24	—	—
Ethylene	62.008	—	0.005
Ethane	11,953	—	1.18
MAPD	—	97.5	458
Propylene	2.10	2,220	20,572
Propane	0.005	1,912	14,321
C_4s—C_6	—	8,734	4,776
Heavy gasoline	—	1,858	0.55

3-7 With the information developed in Prob. 3-6, prepare a quantitative energy balance for the separation section of the base-case ethylene process and size the equipment in sufficient detail for a preliminary cost estimate.

3-8 Figure P3-8 shows the series configuration hybrid system that was used to replace the C_2 splitter in the preliminary design outlined in this chapter. If the feed is 84.75 mol percent ethylene, the product stream is 99.9 mol percent ethylene, and both recycle streams are 99.45 mol percent ethane, establish a material and energy balance around each piece of equipment of the hybrid system.

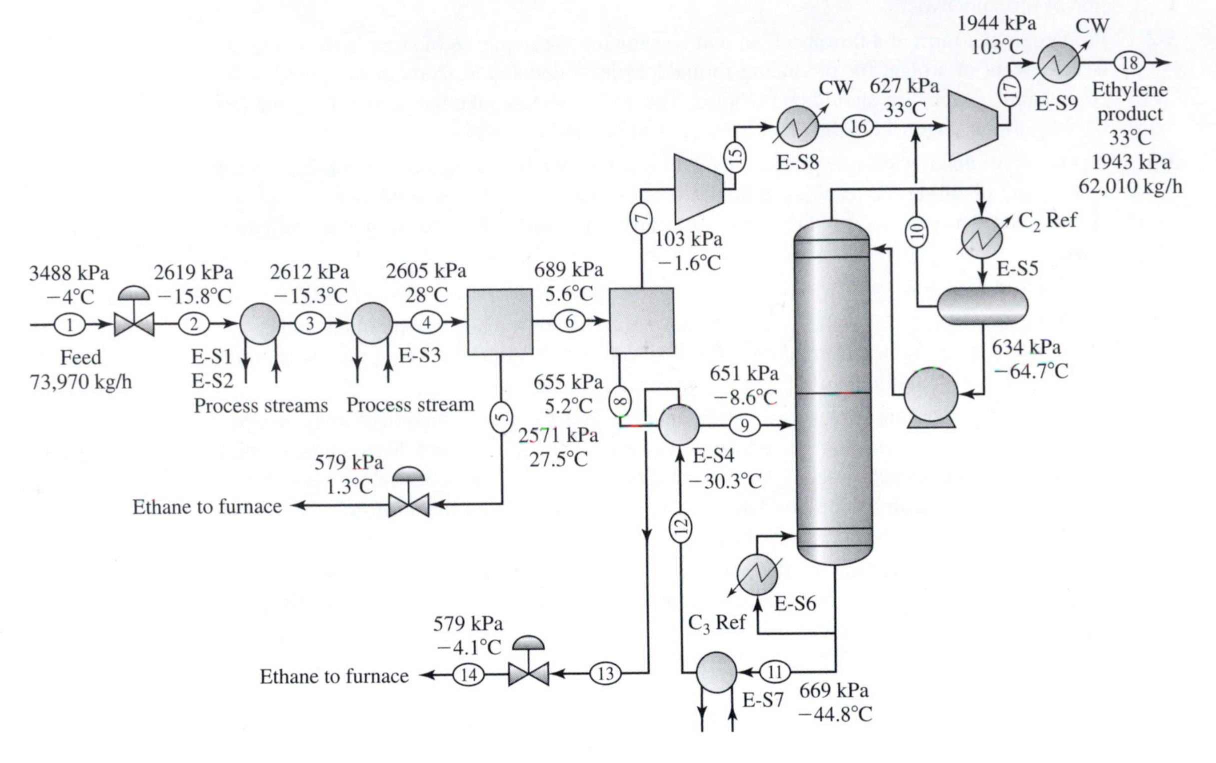

Figure P3-8
Series configuration of the ethane/ethylene hybrid separation system

3-9 Size and cost the equipment in the series configuration hybrid system described in Prob. 3-8. Facilitated transport membrane properties for the ethylene/ethane separation are as follows:

Component	Diffusivity, m^2/s	Solubility, $mol/(m^3 \cdot Pa)$
Ethylene	1.87×10^{-9}	4.72×10^{-5}
Ethane	1.8265×10^{-9}	1.78×10^{-5}
Ethylene-silver complex	1.66×10^{-9}	
$K_{\text{ethylene-silver complex}} = 0.125\text{–}0.13\ m^3/mol$		
Silver nitrate concentration = 6 mol/l		
Membrane porosity/tortuosity ratio = 0.258		
Membrane thickness = 1.7×10^{-6} m		

3-10 Develop a membrane cascade system that could replace the C_2 splitter without the use of a distillation column. Show whether a recycle of the retentate would be necessary to achieve the desired separation specified for the ethylene process. List the advantages and disadvantages of using such a system relative to the other distillation/membrane hybrid configurations examined in the preliminary design of the base-case ethylene process.

3-11 The utilization of facilitated transport membrane technology to accomplish the separation of propylene from propane is another processing alternative in the ethylene process. Develop several hybrid systems for the separation similar to those identified for the ethylene/ethane separation. Examine advantages and disadvantages of these different systems. Which of these systems would provide the greater annual savings and warrant further evaluation?

3-12 Synthesis gas may be prepared by a continuous, noncatalytic conversion of any hydrocarbon by means of controlled partial combustion in a firebrick-lined reactor. In the basic form of this process, the hydrocarbon and oxidant (oxygen or air) are separately preheated and charged to the reactor. Before entering the reaction zone, the two feedstocks are intimately mixed in a combustion chamber. The heat produced by combustion of part of the hydrocarbon pyrolyzes the remaining hydrocarbons into gas and a small amount of carbon in the reaction zone. The reactor effluent then passes through a waste-heat boiler, a water-wash carbon-removal unit, and a water cooler-scrubber. Carbon is recovered in equipment of simple design in a form which can be used as fuel or in ordinary carbon products.

Prepare a suitable flowsheet for the process, with temperature and pressure conditions at each piece of equipment. Make a material balance and a qualitative flowsheet for the process. Assume an operating factor of 95 percent and a feedstock with an analysis of 84.6 percent C, 11.3 percent H_2, 3.5 percent S, 0.13 percent O_2, 0.4 percent N_2, and 0.07 percent ash (all on a weight basis). The oxidant in this process will be oxygen having a purity of 95 percent. Production is to be 8.2 m^3/s.

3-13 One method of preparing acetaldehyde is by the direct oxidation of ethylene. The process employs a catalytic solution of copper chloride containing small quantities of palladium chloride. The reactions may be summarized as follows:

$$C_2H_4 + 2CuCl_2 + H_2O \xrightarrow{PdCl_2} CH_3CHO + 2HCl + 2CuCl$$

$$2CuCl + 2HCl + \frac{1}{2}O_2 \longrightarrow 2CuCl_2 + H_2O$$

In the reaction $PdCl_2$ is reduced to elemental palladium and HCl, and is reoxidized by $CuCl_2$. During catalyst regeneration the CuCl is reoxidized with oxygen. The reaction and regeneration steps can be conducted separately or together.

In the process, 99.8 percent ethylene, 99.5 percent oxygen, and recycle gas are directed to a vertical reactor and are contacted with the catalyst solution under a slight pressure. The water evaporated during the reaction absorbs the exothermic heat evolved, and makeup water is fed as necessary to maintain the catalytic solution concentration. The reacted gases are water-scrubbed, and the resulting acetaldehyde solution is fed to a distillation column. The tail gas from the scrubber is recycled to the reactor. Inerts are eliminated from the recycle gas in a bleed stream which flows to an auxiliary reactor for additional ethylene conversion.

Prepare, in the form of an equipment flowsheet, the sequence of steps in the development of a plant to produce acetaldehyde by this process. An analysis of the points to be considered at each step should be included. List the additional information that will be needed to complete the preliminary design evaluation. Identify temperature, pressure, and composition, wherever possible, at each piece of equipment.

3-14 Prepare a material balance for the production of 7800 kg/h of acetaldehyde, using the process described in Prob. 3-13. However, because 99.5 percent oxygen is unavailable, it will be necessary to use 830-kPa air as one of the raw materials. What steps of the process will be affected by this substitution in feedstocks? Assume an operating factor of 90 percent and a 95 percent yield on the ethylene feed.

3-15 In the face of world food shortages accompanying an exploding world population, many futurists have suggested that the world look to crude oil as a new source of food. Explore this possibility and prepare a flowsheet that utilizes the conversion of petroleum to food by organic microorganisms. What are the problems that must be overcome to make this possibility an economic reality?

3-16 A chemical engineering consultant for a large refinery complex has been asked to investigate the feasibility of manufacturing 1.44×10^{-2} kg/s of thiophane, an odorant made from a combination of tetrahydrofuran (THF) and hydrogen sulfide. The essential reaction is

$$\begin{array}{c} H_2C-CH_2 \\ |\quad\quad | \\ H_2C\quad CH_2 \\ \diagdown\ \diagup \\ O \end{array} + H_2S \rightleftharpoons \begin{array}{c} H_2C-CH_2 \\ |\quad\quad | \\ H_2C\quad CH_2 \\ \diagdown\ \diagup \\ S \end{array} + H_2O$$

The process consists essentially of the following steps:

a. THF is vaporized and mixed with H_2S in a ratio of 1.5 mol H_2S to 1 mol THF and reacted over an alumina catalyst at an average temperature of 672 K and pressure of 207 kPa.

b. Reaction vapors are cooled to 300 K and phase-separated.

c. The noncondensable gases are removed and burned in a fume furnace while the crude thiophane is caustic-washed in a batch operation.

d. The caustic-treated thiophane is then batch-distilled in a packed tower and sent to storage before eventual shipment to commercial use.

e. Recoverable THF is recycled to the reactors from the batch column.

f. The aqueous bottoms stream is stored for further processing in the plant.

g. Carbon deposition on the catalyst is heavy (4 percent of the THF feed), and therefore provision for regeneration of the catalyst must be made.

Assist the consultant in analyzing this process with a complete flowsheet and material balance, assuming 85 percent operating factor, 80 percent conversion in the reactor, and 90 percent recovery after the reactor. Outline the types of equipment necessary for the process. Determine the approximate duties of heat exchangers, and list overall heat balances on the plant. The heat of formation is −59.4 kcal/g mol for THF, −4.77 kcal/g mol for H_2S, and −17.1 kcal/g mol for thiophane. What additional information would be required to complete the project analysis?

Physical properties: THF: MW = 72, sp gr = 0.887, boiling point = 65°C, vapor pressure at 25°C = 176 mmHg. Thiophane: MW = 88, boiling point = 121°C.

CHAPTER

4

Flowsheet Synthesis and Development

As noted earlier, a schematic representation of a chemical process is made by the use of a flowsheet. The preparation of such a flowsheet encompasses the steps of synthesis, development, evaluation, and selection of the most appropriate processing arrangement of a chemical process. The development of a flowsheet involves a branching nature resulting from the multiple possible types and arrangements of equipment and from the selection of process and equipment conditions. This cascading nature in the flowsheet development creates complexities and intricacies in any process-design investigation. However, by constructing increasingly complex and complete flowsheets in an orderly fashion and then rationally evaluating them, it is possible to obtain optimal or near optimal results. Selection of the flowsheet is one of the most important steps in the design of a chemical plant, because only from the most optimal flowsheet can the most profitable, safe, and environmentally sound final design be obtained.

This chapter discusses two design procedures for flowsheet synthesis and design, namely, the hierarchical and algorithmic methods. The former is based on heuristic rules derived from past experience, and the latter utilizes mathematical programming procedures that include optimization techniques. The hierarchical approach has been used with reasonable success in flowsheet development of more routine chemical processes but with considerably smaller success in the more complex chemical processes. The algorithmic approach has had similar experiences in flowsheet synthesis since most conventional algorithmic methods require the manual construction of the complete network containing all possible flowsheets, many of which are infeasible. However, some recent work by Friedler and Fan and their associates is providing some valuable insights as to how the approach through computerization can become a valuable tool in the synthesis of more complex chemical processes.[†] To demonstrate the procedures involved, each approach is used to synthesize and develop an acceptable flowsheet for the production of the same chemical product, vinyl chloride.

[†]M. H. Brendel, F. Friedler, and L. T. Fan, *Comp. & Chem. Eng.*, **24:** 1859 (2000).

The intuitive approach for methodically developing a set of flowsheets is through a multilevel analysis of process requirements and a hierarchical synthesis of potential processes. The essence of this approach is to generate from processing requirements a number of more concrete functions, which are then expanded into better-defined operations. These operations are then integrated into a final flowsheet.

FLOWSHEET SYNTHESIS AND DEVELOPMENT

General Procedure

A generalized hierarchical scheme for flowsheet synthesis and development is illustrated as a six-step procedure in Fig. 4-1.[†] These steps given below are not unique, but suggest an organized procedure to follow.

Process Information Given a product that is to be manufactured, conduct a search of the technical and patent literature for information about the product.[‡] Study market conditions and the pricing of the product. Key property data are obtained from the literature or must be estimated. For a new product, there presumably may be laboratory and perhaps pilot-plant data plus a market evaluation.

Input/Output Diagram Examine all potential chemical reaction paths, and do a preliminary economic analysis of each path. Eliminate those for which the value of the raw materials exceeds that of the products and those which appear infeasible for other reasons. For the rest, construct an *input/output* (I/O) diagram, showing all major material input and output streams with a stoichiometric balance. Figure 4-2a illustrates an *I/O diagram* for manufacturing sodium dodecylbenzene sulfonate, a detergent ingredient known as ABS that is no longer used because of environmental reasons.

Functions Diagram For each particular chemical reaction path, indicate all the major functions of the process and the material flows to and from these functions. There is a *reaction box* for each reaction that must be carried out separately from the rest of the process. For each raw material input, determine, if possible, whether preprocessing is needed to meet process requirements. When it is needed, or if it is uncertain, show a *preprocessing box*. Each reactor output stream enters a *separation box* that produces a *recycle* to the reactor, intermediate streams to the next reaction step or steps, and possibly products to the finishing steps. A *finishing box* represents converting each final product to its final form. Heat flows are suggested by the symbol Q into or out of each reactor, and into and out of each preprocessing, separation, and finishing step, to serve as reminders to examine the heating, cooling, and other energy requirements of each processing step. The molar flow rate of each main component is indicated based on 1 mol of a key raw material, 100 percent selectivity for the principal products, and perfect separations.

[†]A much more detailed presentation of hierarchical flowsheets is presented by J. M. Douglas, *Conceptual Design of Chemical Processes,* McGraw-Hill, New York, 1988.

[‡]A former colleague has remarked that "A week spent in the laboratory may save an hour or two in the library." To which we now add, "or online."

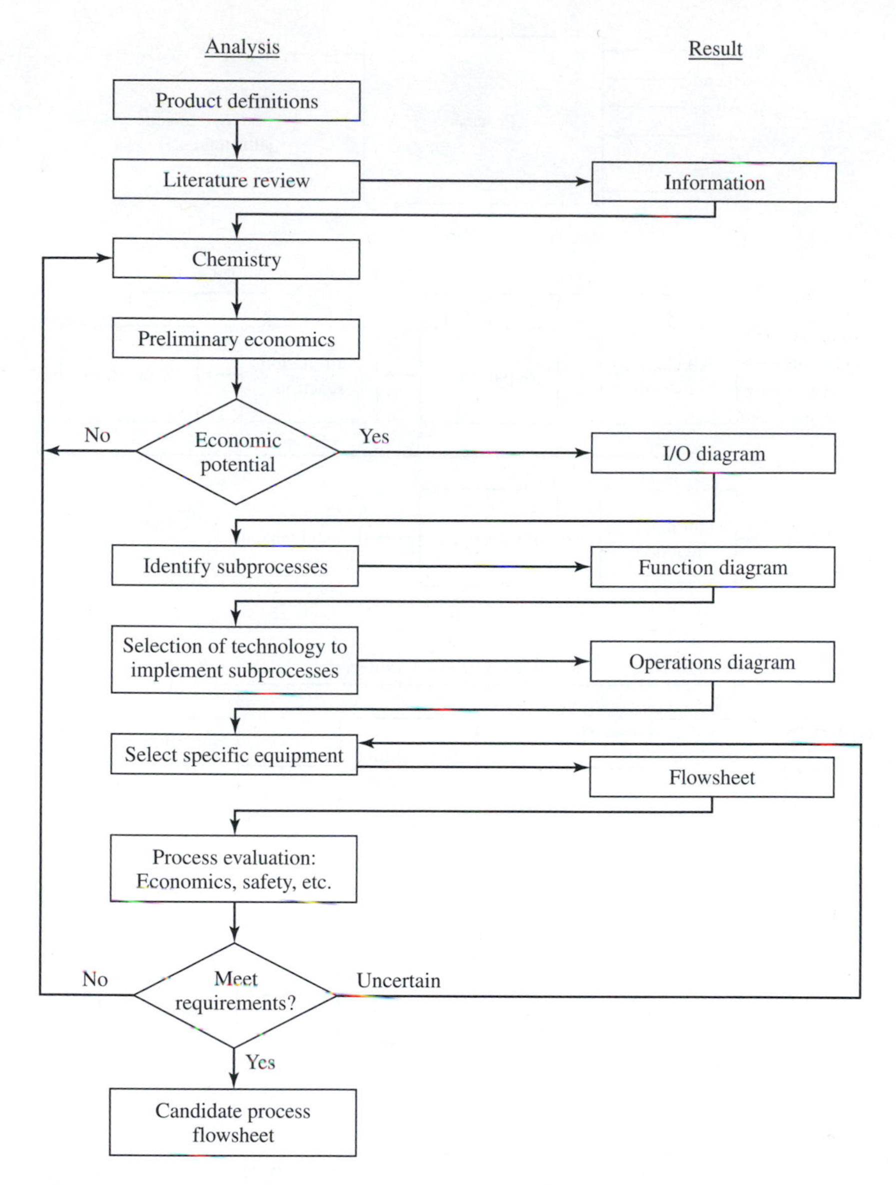

Figure 4-1
Flow diagram illustrating hierarchical process—flowsheet synthesis, development, evaluation, and selection

The result of this step is a *functions diagram,* as illustrated in Fig. 4-2b for ABS.

Operations Diagram In this step the technology to be used to accomplish each of the operations is selected. Preprocessing technologies are chosen to accomplish the temperature change, phase change, or purification needed. A choice of plug flow or

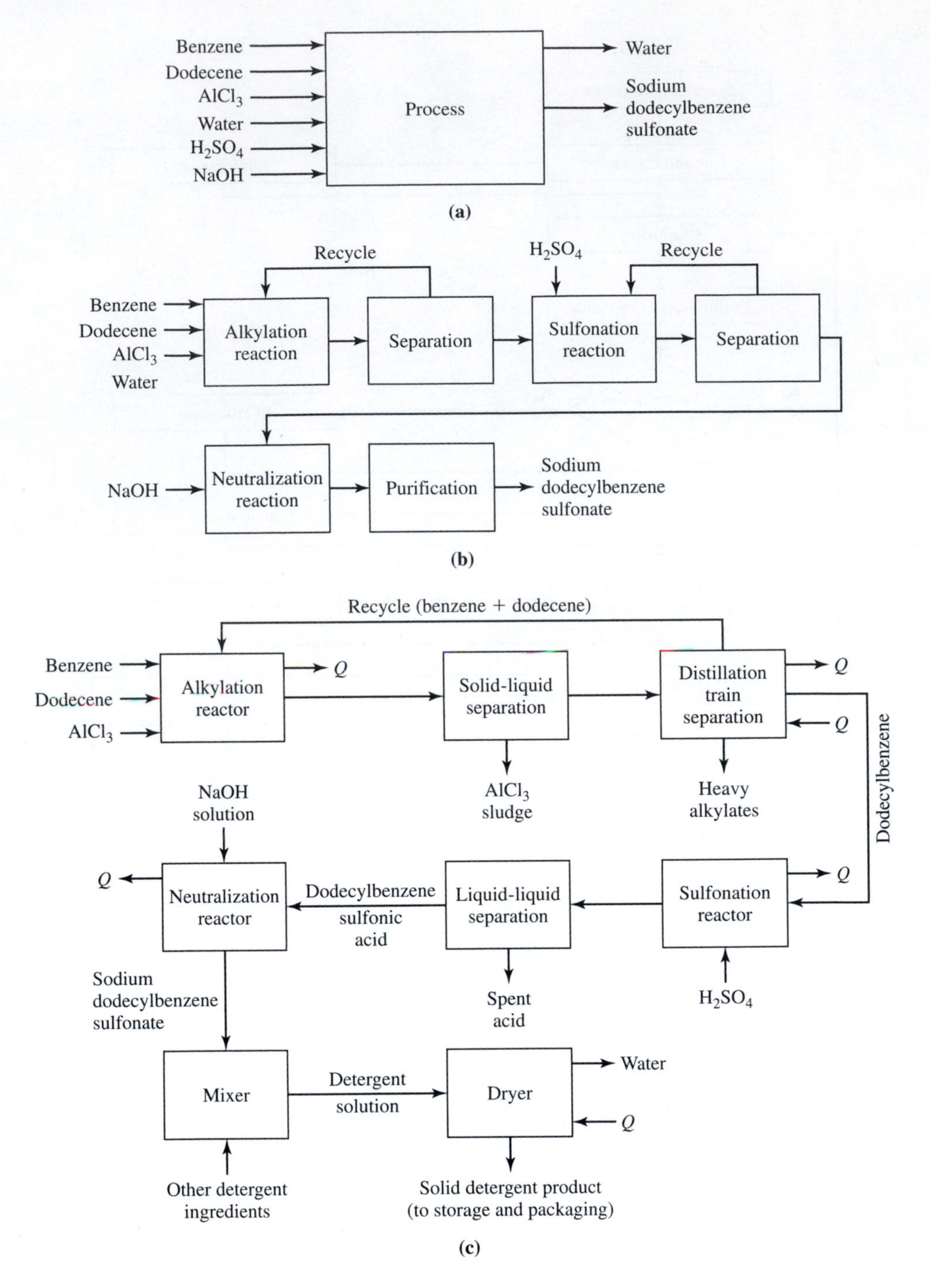

Figure 4-2

Continued on next page

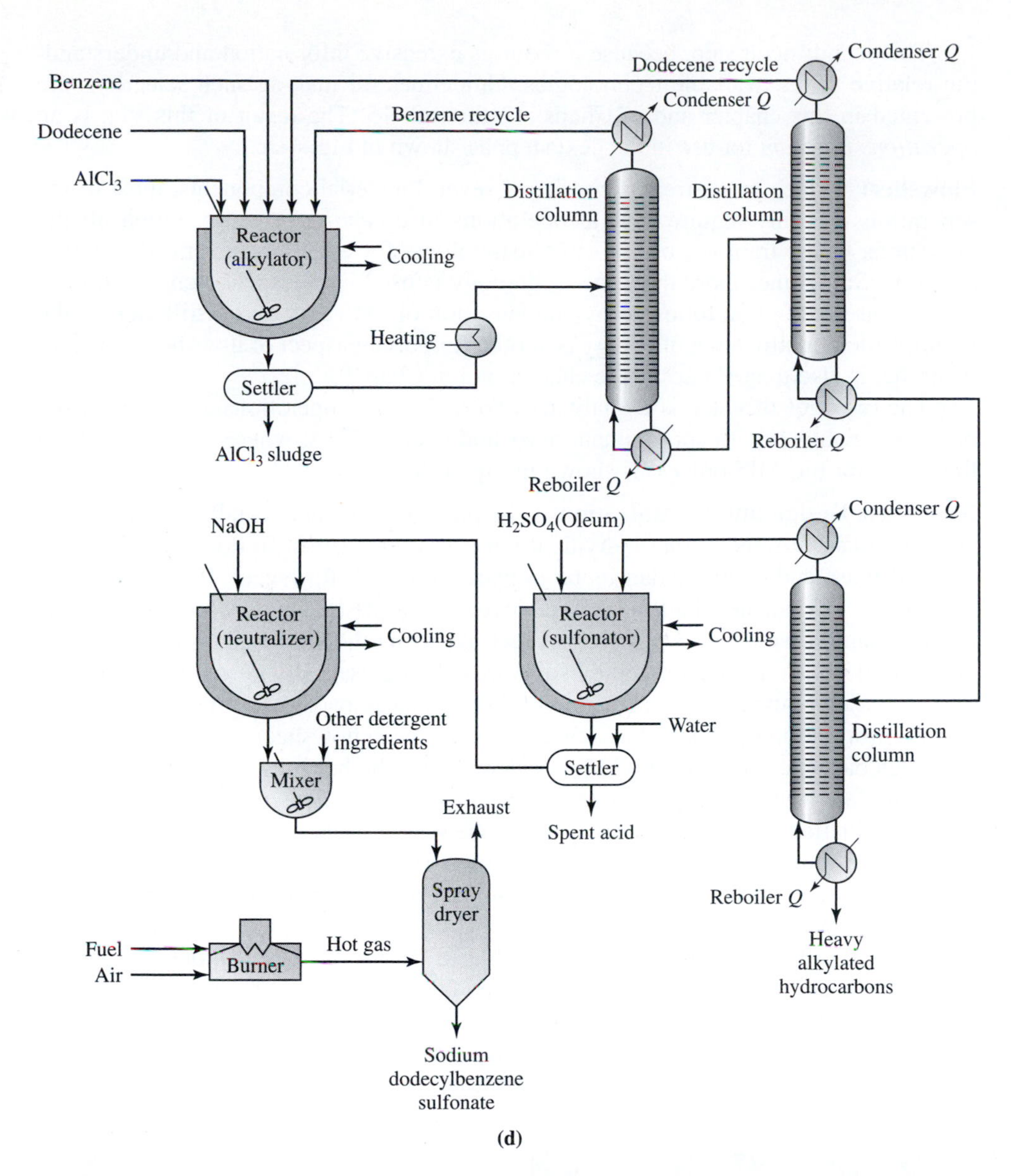

Figure 4-2
Development stages for sodium dodecylbenzene sulfonate production; (a) input/output diagram, (b) functions diagram, (c) operations diagram, and (d) final flowsheet

continuous stirred tank reactors is made. Ranges of temperature, pressure, reactant conversion, and product selectivity are estimated. Each purification and separation is examined to determine whether phase separation, stripping, distillation, membrane separation, extraction, or some other technology should be used. Similarly, the technologies needed for finishing are selected.

This is a difficult step, because it requires extensive information and understanding relative to the available technologies. Guidelines for making such selections are presented in this chapter and in Chaps. 12 through 15. The result of this step is an *operations diagram* for use in the next step, as shown in Fig. 4-2c.

Flowsheet Since most streams consist of several material components, most of the separations actually require several operations in a *separation train*. Often all the operations in the train are of the same technology, for example, several distillation columns. Sometimes more than one technology is used in the train, such as a liquid-vapor phase separation followed by condensation of the vapor and distillation of the liquid. Efficient utilization of energy is another important aspect of flowsheet development that is also treated later in the chapter and in Chap. 9.

The result of this step is a qualitative *flowsheet*. The operations and steps of the process are shown with approximate mass and energy flows, where possible. A final flowsheet for the ABS process is shown in Fig. 4-2d.

Base-Case Design and Optimization The preceding method usually results in several candidate flowsheets that individually must be evaluated. To do so requires that detailed mass and energy balances be prepared for each flowsheet. If the preceding steps are well done, these balances should reveal no surprises, but they provide the basis for the quantitative analysis of the flowsheet. By using the mass and energy balances, or in parallel with them if simulation software is being used, key design parameters of the process equipment, such as reactor volumes and heat loads, can be calculated. Such design information provides a base-case design for each flowsheet.

Each base-case design is reviewed to determine whether it can meet product specifications, environmental requirements, and health and safety needs. Also, an economic evaluation is conducted. With these reviews and evaluations, it may be possible to eliminate some of the flowsheets because process or economic requirements are not met. However, before a process is eliminated, it should be determined whether, by changing conditions and design parameters, the process can be brought within requirements. It may or may not be possible to determine this without performing some optimization studies. All remaining candidate processes should be optimized, and the best alternative should be selected.

The first five steps in developing a flowsheet are covered in this chapter. The design and optimization step is discussed in Chap. 9.

PROCESS INFORMATION

Background Information

There are three main approaches to establishing the basis for a potential chemical process: use of complete existing designs, modification of or use of part of an existing design, and generation of a completely new design. All three of these require extensive research at the beginning of the design process, to determine the availability of existing designs or basic chemical processes that will enable production of the desired product. The importance of this initial step cannot be overstated since an incomplete search at this early stage will cause, at the very least, difficulties later in the design and potentially can overlook attractive process alternatives.

The search for a process begins with general sources such as handbooks, encyclopedias, textbooks, and professional books. These will often provide a reasonably clear picture of current technologies and processes that can be used as is or modified. Next, more specific sources such as journals, patents, databases, and expert consultants can be utilized to both expand on the more general information and to suggest alternate or new processing techniques. Specific information such as reaction kinetic data, thermodynamic data, and raw materials and product prices must be obtained for all the identified potential processes. This step may require laboratory measurements to determine or confirm values, especially when a new product or processing condition is being investigated.

Molecular Path Synthesis

If a new desired molecule has been identified, a chemical pathway for its synthesis must be identified. The bonds that attach atoms of a molecule together are the basis for many of the properties of a molecular species that often make the molecule valuable. These same bonds are also a major source of difficulty in assembling a desired molecule, because methodical assembly of atoms and bonds on a large scale is generally not feasible. Instead, a series of reactions, designated as a *reaction train,* must be used to achieve the desired product. The reaction train may require a series of steps that lead in a direction seemingly independent of the desired molecule, because this may only be achieved with the final few steps of the train. The reactions and their sequence leading to the desired molecule are rarely unique.

Factors such as yield, by-products, and additional processing steps can make the assessment of the viability of molecular synthesis schemes quite difficult. Further, it is not always evident what feed materials should be used. In some cases it is possible to start with simple species, such as carbon monoxide and hydrogen. In other cases, such as cellulose esters and ethers, synthesis begins with complex molecules from plant or animal sources. In any case, starting with the most basic potential reactants may not be worthwhile since the cost associated with the required additional synthesis may exceed the cost of using feed materials from higher up the synthesis train.

Selection of appropriate reactions and reaction combinations usually proceeds by starting with the desired product and working backward to the required feed materials. The major chemically active functional groups in the target molecule are identified for this procedure. Functional groups are relatively stable groupings of atoms with potential external bonding sites. The composition, structure, and bonding of many functional groups are fully identified and available in ordered databases. Synthesis of new functional groups is demanding, although general guidelines are available.[†]

Some rules of thumb for molecular assembly are as follows:

1. Divide molecules, especially large organic molecules, into subsections.
2. Divide molecules into repeating subsections.

[†]For more details regarding molecular synthesis, see advanced chemical synthesis sources, including the many available organic molecule synthesis texts.

3. Divide molecular subsections into functional groups.
4. "Cap" sensitive areas with active groups, and remove the cap once the molecule is complete and stable.
5. Limit bond formation through steric hindrance, either permanent or temporary.
6. Build more complex molecules, then cut or cleave them.
7. Rings create steric hindrance and orientation issues; use and then cut the ring if necessary to make a linear molecule.
8. Look for readily available, inexpensive raw materials.

Once the functional groups and their subassembly possibilities are known, the process of investigating functional-group assembly reactions produces a diagram of the molecular synthesis train.

Selecting a Process Pathway

If a commercial chemical process exists for the desired product, it may be utilized. If patent rights cover the process and are available for licensing, then the cost of using this process must be compared with the cost of a different process not covered by patents or the development of a new process. To develop a new process usually is expensive and time-consuming, but potentially it may provide a competitive advantage. Using nonprotected aspects of the known technology may reduce the cost and time. However, for a new product, research and development generally must be performed. The ultimate decision is based largely on economic comparisons.

Production Mode

As noted in Chap. 3, a key criterion in designing a chemical process is the production mode. The choices are batch, continuous, or combinations of the two modes. Continuous processes are common in the chemical process industries. Continuous processes result in relatively high production rates per unit of equipment size, consistent product quality, and relatively constant operating conditions and requirements. Moreover, labor requirements are usually much less than for batch operation. Batch operation may result in less consistent product quality, lower average production rates per unit of equipment size, and variable operating conditions and requirements. Batch operation does provide for greater flexibility of operating schedule, opportunity to use the same equipment for different products, and product identification and characterization by batch. Combination modes can offer many of the advantages and disadvantages of both continuous and batch modes.

Some rules of thumb for the selection of batch processing are as follows:[†]

1. Usually when the production rate is less than 5×10^5 kg/yr.
2. If heavy fouling is expected.

[†]J. M. Douglas, *Conceptual Design of Chemical Processes,* McGraw-Hill, New York, 1988.

3. For biological processes.
4. For pharmaceutical processes.
5. Short, 1- to 2-yr, product life spans.
6. When product value far exceeds total product cost.

Recording Decisions

The branching nature that occurs in flowsheet development requires that at each branch point a decision be made to determine the path to be followed in the process. Some of these decisions are based upon clear and unambiguous criteria and need not be revisited. Other decisions, however, may be more or less arbitrary at the time they are made and eventually need to be revisited to investigate the implications of pursuing the alternate decision. One way of doing this is to keep a list of decision points that need to be reexamined. Examples 4-4 and 4-5 use this technique and provide each decision point in italics.

Information on Vinyl Chloride Production — EXAMPLE 4-1

Develop information on potential reaction pathways and physical and operational properties involved in the production of vinyl chloride, C_2H_3Cl.

■ Solution

Vinyl chloride is a major chemical, which essentially serves as the monomer for making polyvinyl chloride (PVC). Production processes are well established, and there is considerable information on the chemical product available in the open literature.[†] There are several patented processes for the manufacture of vinyl chloride in the United States and other countries. Some 23,000,000 t (metric) was produced worldwide in 1997, about 26 percent of which was produced in the United States. The average U.S. plant capacity is about 500,000 t (metric)/yr (5×10^8 kg/yr). Because of this, it is reasonable to make a firm decision to use a continuous process.

Vinyl chloride monomer and nearly all the materials used and coproduced in its production are flammable and toxic. This requires strict limitations on emissions from such plants and on the concentrations within the plant working areas. Plant design must allow for control of these substances.

The chlorinated species, hydrogen chloride in particular, are corrosive at higher temperatures, especially in the presence of water. Feed and hydrogen chloride streams must be very dry, less than 10 parts per million (ppm) of water. Equipment at elevated temperatures must use corrosion-resistant alloys, such as Inconel, Hastelloy B, or nickel cladding, which greatly increases equipment costs, and although selection of materials of construction comes later in the design, it should be borne in mind when making decisions about process conditions.

[†]Much of the information on vinyl chloride in this example and Examples 4-2 to 4-4 is attributed to that reported in the *Kirk-Othmer Encyclopedia of Chemical Technology,* 4th ed., M. Howe-Grant, ed., J. Wiley, New York, 1994.

A few key physical and thermodynamic properties of the most important materials in the various vinyl chloride processes are listed in the table below.

Properties of vinyl chloride reaction path components[†]

Material	Molecular weight, kg/kg mol	Vapor pressure, MPa					ΔH_R	ΔH_V	C_p at 20°C kJ/(kg mol·K)	
		0.1	0.2	0.5	1.0	2.0	MJ/kg mol			
		Temp., °C					25°C	NBP	Liquid	Vapor
O_2	32.0	−183	−176	−165	−153	−140	—	7	—	29
C_2H_4	28.1	−104	−91	−71	−53	−29	—	14	—	59
HCl	36.5	−85	−71	−50	−32	−9	—	16	—	31
Cl_2	70.9	−34	−17	10	36	65	—	20	—	36
C_2H_3Cl	62.5	−14	5	33	58	92	71[‡]	20	74	33
$C_2H_4Cl_2$	99.0	84	108	148	183	226	−180[§] −239[¶]	32	125	52

[†]*Kirk-Othmer Encyclopedia of Chemical Technology,* 4th ed., M. Howe-Grant, ed., J. Wiley, New York, 1994.
[‡]Pyrolysis of EDC to vinyl chloride.
[§]Direct chlorination.
[¶]Oxychlorination.

Five potential reaction pathways to vinyl chloride from two possible organic raw materials were identified in the literature:

1. Acetylene hydrochlorination:

$$C_2H_2 + HCl \rightarrow C_2H_3Cl$$

2. Ethylene direct chlorination:

$$C_2H_4 + Cl_2 \rightarrow C_2H_3Cl + HCl$$

3. Ethylene chlorination plus 1,2-dichloroethane pyrolysis:

$$C_2H_4 + Cl_2 \rightarrow C_2H_4Cl_2$$
$$C_2H_4Cl_2 \rightarrow C_2H_3Cl + HCl$$
$$C_2H_4 + Cl_2 \rightarrow C_2H_3Cl + HCl \text{ (overall)}$$

4. Oxychlorination plus 1,2-dichloroethane pyrolysis:

$$C_2H_4 + 2HCl + \tfrac{1}{2}O_2 \rightarrow C_2H_4Cl_2 + H_2O$$
$$C_2H_4Cl_2 \rightarrow C_2H_3Cl + HCl$$
$$C_2H_4 + HCl + \tfrac{1}{2}O_2 \rightarrow C_2H_3Cl + H_2O \text{ (overall)}$$

5. Chlorination plus oxychlorination plus pyrolysis:

$$C_2H_4 + Cl_2 \rightarrow C_2H_4Cl_2$$
$$C_2H_4 + 2HCl + \tfrac{1}{2}O_2 \rightarrow C_2H_4Cl_2 + H_2O$$
$$2C_2H_4Cl_2 \rightarrow 2C_2H_3Cl + 2HCl$$
$$2C_2H_4 + Cl_2 + \tfrac{1}{2}O_2 \rightarrow 2C_2H_3Cl + H_2O \text{ (overall)}$$

These reaction paths are compared in greater detail in the next section. ■

INPUT/OUTPUT STRUCTURE

The purpose of an investment in a chemical process is to produce a product and derive a profit. The raw materials typically represent a substantial fraction of the total cost of producing a product. Clearly, if the value of the products does not exceed the value of the raw materials, the process cannot be economical. However, if the reverse is true, only further analysis can find whether the process is truly economical. Any reaction path for which this is not true can be eliminated from further consideration. Examination of the value of inputs relative to outputs is therefore an important screening tool. For processes showing favorable economic potential, an input/output diagram can be constructed that shows all the major material inputs and outputs as the basis for further analysis. The vinyl chloride reactions, presented in Example 4-1, are examined, and the procedure for making the preliminary economic analysis is described in Example 4-2.

Product and Raw Material Values for the Vinyl Chloride Paths — EXAMPLE 4-2

Compare the product and raw material values for the five vinyl chloride reaction paths identified earlier based on a production of 1 kg of vinyl chloride.

■ Solution

Prices for all the chemicals involved are shown in the following table.†

Vinyl chloride reaction path comparisons using Jan. 2002 prices†

		Reaction path, kg/kg vinyl chloride				
Species	**Price, $/kg**	**1**	**2**	**3**	**4**	**5**
Cl_2	0.03	0.00	−1.13	−1.13	0.00	−0.57
HCl	0.22	−0.58	0.58	0.58	−0.58	0.00
C_2H_2	1.39	−0.42	0.00	0.00	0.00	0.00
C_2H_4	0.45	0.00	−0.45	−0.45	−0.45	−0.45
C_2H_3Cl	0.45	1.00	1.00	1.00	1.00	1.00
O_2	0.04	0.00	0.00	0.00	−0.26	−0.13
Excess value of products over raw materials, $/kg vinyl chloride		−0.26	0.34	0.34	0.11	0.23

It is clear from the results in the table that reaction path 1 is not economical because of the high price of acetylene and can therefore be eliminated. Reaction paths 2 and 3 give the same result because the net reaction is the same in both cases. The literature makes it clear, however, that when chlorine and ethylene react, the product is almost entirely EDC and not vinyl chloride. Thus, reaction path 3 is chemically feasible, but reaction path 2 is not and can also be eliminated.

†Prices for all species except acetylene and oxygen are from *Chemical Market Reporter,* Jan. 7, 2002. The acetylene price is from the *Chemical Market Reporter,* Dec. 3, 2001, and the oxygen price is an 8-yr average from *Kirk-Othmer Encyclopedia of Chemical Technology,* 4th ed., M. Howe-Grant, ed., J. Wiley, New York, 1994. The price for HCl is that of 22 degrees Be′ muriatic acid, divided by 0.36, the fraction HCl in solution.

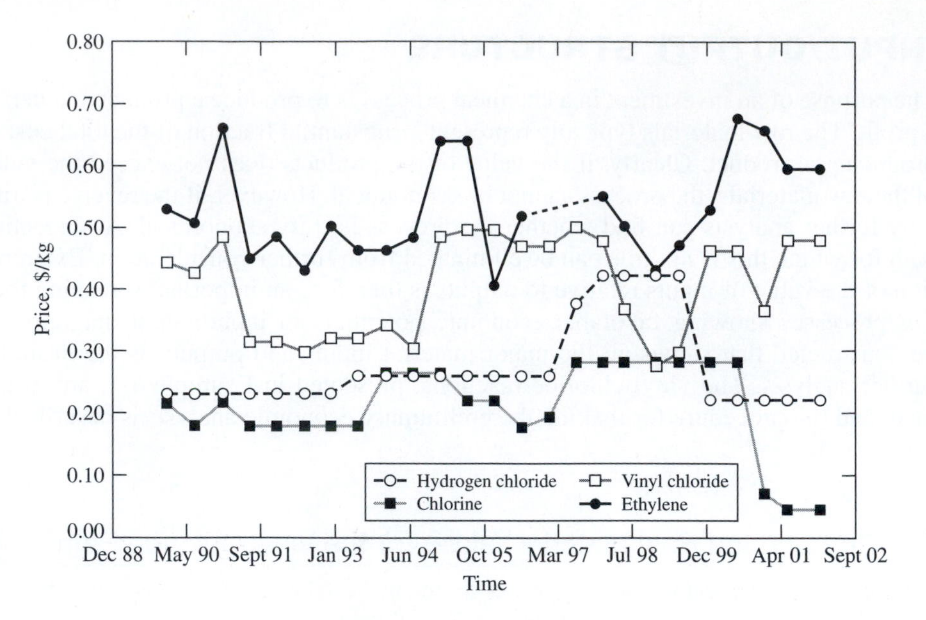

Figure 4-3
Chemical cost trends for vinyl chloride production process raw materials and products

Reaction paths 4 and 5 show economic potential although apparently less than that indicated for reaction path 3. A major factor that makes reaction path 3 appear attractive is the high price of the by-product HCl coupled with the low price of Cl_2. Because HCl is a raw material in reaction path 4, its high price greatly affects the overall attractiveness of this path. Reaction path 5, which neither uses nor produces HCl, falls between reaction paths 3 and 4 in apparent economical potential. Since the relative prices of HCl and Cl_2 profoundly affect the economics of reaction paths 3 and 4, their relationship needs to be examined more closely. In fact, all the prices used should not be just one-time values, but should be examined over an extended period of time when possible. This has been done for the vinyl chloride synthesis with the results presented in Fig. 4-3.†

The prices of the four species included have fluctuated over the 12-yr period shown. Whereas the early-2002 prices of ethylene, vinyl chloride, and hydrogen chloride are all close to their 12-year averages, the January 2002 price of $0.03/kg for chlorine is dramatically lower than the long-term average of $0.21/kg. It would be unwise to base the economics of a process on this unusually low price. Therefore, the preliminary economic calculations for reaction paths 3, 4, and 5 have been recalculated using the average price between 1990 and 2002 for ethylene, chlorine, vinyl chloride, and hydrogen chloride and the 8-year average for oxygen as mentioned earlier. The results are shown in the following table.

†Prices are from the *Chemical Market Reporter* over the period from 1990 to 2002.

Vinyl chloride reaction path comparisons using long-term average prices†

		Reaction path, kg/kg vinyl chloride		
Species	**Price, $/kg**	**3**	**4**	**5**
Cl_2	0.21	−1.13	0.00	−0.57
HCl	0.27	0.58	−0.58	0.00
C_2H_4	0.53	−0.45	−0.45	−0.45
C_2H_3Cl	0.41	1.00	1.00	1.00
O_2	0.04	0.00	−0.26	−0.13
Excess value of products over raw materials, $/kg vinyl chloride		0.09	0.00	0.04

†See Fig. 4-3 for long-term prices.

The results show that reaction path 4 shows essentially zero excess value of products over reactants and can therefore be eliminated from further consideration. Path 3 shows a larger excess than does path 5, but without the sale of the by-product HCl, path 3 shows a negative excess. The choice between paths 3 and 5, then, depends on the market potential for the by-product HCl. About 95 percent of all HCl produced in the United States is a by-product of chlorination processes. The producers use about two-thirds of the HCl so produced, and 60 percent of the total HCl produced is used in the manufacture of vinyl chloride. The actual market for HCl sales is only about one-third of that produced. Since the market for HCl is only just over one-half the amount generated by vinyl chloride producers alone, the possibility of selling a significant new amount of HCl at the market price is quite small. Therefore, it seems advisable for the purpose of the preliminary economic analysis to assign a zero value to HCl as a salable by-product. When this is done, the result for reaction path 3 is negative. Only reaction path 5 remains with a positive excess value of products over reactants, so it becomes the choice for further analysis.

An input/output diagram is therefore constructed for reaction path 5. A single box represents the process; the chemical reactions are shown inside the box. Arrows into and out of the box show the raw material inputs and product outputs on the basis of 2 mols of vinyl chloride product. The result is shown here:

Inputs: $2C_2H_4 \rightarrow$, $Cl_2 \rightarrow$, $\frac{1}{2}O_2 \rightarrow$

$$C_2H_4 + Cl_2 \rightarrow C_2H_4Cl_2$$
Liquid-phase chlorination
$$2C_2H_4Cl_2 \rightarrow 2C_2H_3Cl + 2HCl$$
Gas-phase pyrolysis
$$C_2H_4 + 2HCl + \tfrac{1}{2}O_2 \rightarrow C_2H_4Cl_2 + H_2O$$
Gas-phase oxychlorination
$$2C_2H_4 + Cl_2 + \tfrac{1}{2}O_2 \rightarrow 2C_2H_3Cl + H_2O$$

Outputs: $\rightarrow 2C_2H_3Cl$, $\rightarrow H_2O$

FUNCTIONS DIAGRAM

This step involves the identification of the major functions or subprocesses that must be achieved by the process to produce the desired product. The result is a diagram showing those functions with material flows for major products and raw materials.

A *reactor* box is shown for each of the reactions in the path that must take place separately from the others because the reaction conditions are distinctly different. Arrows into the reactor box show raw materials fed to the reactor as well as any recycles that are necessary. One arrow out of the box shows the product stream. A *preprocessing* box is added for each entering raw material stream, unless it is apparent from specifications and associated chemistry that the material is available at the appropriate conditions and purity required for the reaction. Following each reactor is a *separations* box. Nearly every chemical reaction requires separation of the desired products from unreacted reactants, by-products of the reaction, and other materials that may be used in the reaction. Any of the separated materials that have a useful function in the process, such as raw materials and solvents, are shown as *recycle* streams from the separation to their point of use in the process. This internal recycle of materials within a process is one of the key concepts of chemical engineering and is vital to the economic viability of many, if not most, commercial chemical processes.

Purges of undesired species are shown from the separation system, as a reminder that such substances are inevitably produced and must be removed, to prevent their buildup within the process. Such streams usually must be treated and properly disposed of for legal and environmental reasons. Additional processing of products and by-products, such as drying, blending, further purification, or packaging and storage, is indicated by a *finishing* box.

Preprocessing

The state and composition of feed streams may not meet the needs of the chemical reactions in a process. When this is so, the feed streams must be processed so as to meet those requirements. Here are some guidelines (sometimes referred to as rules of thumb, or heuristics) for preprocessing:[†]

1. Raw material feeds may need to be heated or cooled or may undergo a change in phase before being fed to a process.
2. Unless the effects of feed impurities on process operations are known to be negligible, they should be removed to an acceptable level.
3. Impurities that are known to be catalyst poisons should be removed as completely as possible.
4. If a raw material stream contains final products or by-products, it may be useful to feed that stream to the separation box where these products or by-products are recovered.
5. If a feed impurity is present in a quantity large enough to affect the economics of the process, remove it from the feed.
6. Inert components of feeds that are necessary for the process, or are difficult to remove, should not be removed.
7. There are exceptions to every heuristic.

[†]Adapted from J. M. Douglas, *Conceptual Design of Chemical Processes,* McGraw-Hill, New York, 1988.

Reactions

Data are needed for each major reaction in the reaction path. The information is necessary to determine the need, if any, of a catalyst and the type required; operating conditions of temperature and pressure; kinetic information for the primary reactions; and the phase in which the reaction takes place. Kinetic data for series and parallel reactions that consume products or raw materials and produce undesired products are also necessary. It is important to evaluate, by means of experiment or calculation, the conversion of the raw materials, that is, the fraction of each raw material that reacts while in the reactor, since reactions are rarely complete. Equally important is the fraction of the reacted raw material that is converted to the desired product, usually defined as the *yield*.

Recycle

Whenever a material used in a process appears in the output stream of a unit, the rule is to recycle. Most chemical reactions do not completely react one or more of the raw material feeds due to equilibrium-, kinetic-, or transport-limited conversion. Thus, unreacted raw materials occur in the reactor output streams. Since these materials typically represent a high percentage of the cost of production, there is a large economic incentive to recover and recycle them to the reactor. It is usually desirable to separate such materials from the desired reaction products before recycling since recycling desired products to a reactor is rarely advantageous. Other materials such as solvents and catalysts are frequently recycled as well. These are some guidelines for recycling:[†]

1. Always consider separation and recycle of materials useful in the process when they occur in output streams.
2. Mixtures of components that are to be recycled to the same unit do not need to be separated.
3. There is generally an optimum recycle/purge ratio.
4. If possible, recycle materials as liquids rather than gases, because it is much less expensive to repressurize liquids than gases.

By-products, Intermediates, and Wastes

When a chemical reaction is carried out, other products in addition to the desired product may be obtained in the output stream. These may be products of the same reaction that produces the desired product, for example, the HCl produced along with vinyl chloride in the pyrolysis of EDC, shown earlier. This HCl is a *coproduct* of the main reaction; it is also an *intermediate* in reaction path 5, as is the EDC, because they are synthesized in the reaction path but are subsequently consumed in other reactions. In reaction path 4, HCl is a product of the process and is then known as a *by-product* of the vinyl chloride process. Another type of by-product from a chemical process is a material produced from reactions other than the intended reaction, such as the trichloroethane and the 1,1-dichloroethane produced in small percentages in the

[†]J. M. Douglas, *Conceptual Design of Chemical Processes,* McGraw-Hill, New York, 1988.

oxychlorination of ethylene. By-products may themselves have a commercial value and contribute to the sales revenue of a process; the sodium hydroxide by-product in the manufacture of chlorine is an example. If the by-product has insignificant value, it must be removed from the process and disposed of in a responsible manner. Water from the oxychlorination of ethylene, which is contaminated with chlorinated hydrocarbons, is such a by-product. In such cases, the by-product is a waste and a liability because of the expense of its treatment and disposal. The chemistry of the process should always be examined for ways to minimize the amount of waste generated.

Separations

The mixtures inherent in chemical processing must be separated into nearly pure components or less complex mixtures. Mixture separation steps are therefore present in nearly every chemical process. Separation technologies are based on a variety of physical and physicochemical properties, a large percentage of which depend upon phase differences and phase changes. More information on selection of separation technologies is provided in Chap. 15.

In this early stage of flowsheet development, the purpose is only to identify where mixture separations are needed. General guidelines are that

1. Reactor product streams need to be separated into intermediate products, final products, by-products, and reactants for recycle.
2. Separations involve energy in many forms, often as inputs and outputs, indicated by Q arrows both into and out of each separation box.

EXAMPLE 4-3 Functions Diagram for Vinyl Chloride Process

Based upon reaction path 5 from Example 4-2, develop a function diagram for the vinyl chloride process.

■ Solution

Information on the conditions for the reactions in the vinyl chloride pathways is shown below.[†]

Reaction	Data from literature sources
Chlorination of acetylene	
$C_2H_2 + HCl \rightarrow C_2H_3Cl$	This reaction was once the main pathway to vinyl chloride, but now acetylene is too expensive to use.
Direct chlorination of ethylene to vinyl chloride	
$C_2H_4 + Cl_2 \rightarrow C_2H_3Cl + HCl$	This occurs to a negligible extent at conditions where the following reaction occurs at high conversion; so this reaction is not feasible.

(Continued)

[†]M. Howe-Grant, ed., *Kirk-Othmer Encyclopedia of Chemical Technology,* J. Wiley, New York, 1994.

Reaction	Data from literature sources
Direct chlorination of ethylene to 1,2-dichloroethane (EDC)	
$C_2H_4 + Cl_2 \rightarrow C_2H_4Cl_2$	It occurs rapidly at temperatures greater than 65°C up to about 200°C in liquid EDC with a solid $AlCl_3$ catalyst. Pressure must be equal to or greater than the vapor pressure of EDC at the reaction temperature. Conversion of reactants is essentially 100 percent and conversion to EDC is >99.5 percent.
Oxychlorination of ethylene to 1,2-dichloroethane	
$C_2H_4 + 2HCl + \frac{1}{2}O_2 \rightarrow C_2H_4Cl_2 + H_2O$	This occurs rapidly at temperatures in the range of 220 to 300°C over a $CuCl_2$ catalyst on a solid. Pressures of 250 to 1500 kPa absolute are typical. HCl must contain <10 ppm H_2O. Air or oxygen can be used as the O_2 source. With air, conversion of ethylene is 94 to 99 percent, HCl is 98 to 99+ percent, and conversion to EDC is 94 to 99 percent. With O_2 results are similar, and the conversion to EDC from ethylene is about 90 to 92 percent. The EDC from this reaction is not as pure as that from the direct chlorination, and purification is required before its use in pyrolysis. It must be cooled rapidly to avoid product loss.
Pyrolysis of 1,2-dichloroethane to vinyl chloride	
$C_2H_4Cl_2 \rightarrow C_2H_3Cl + HCl$	This noncatalytic thermal reaction takes place in the range of 475 to 525°C and 1.5 to 3.1 MPa absolute. Conversion of reactants is about 53 to 63 percent per pass, with 2- to 30-s residence time. Conversion of EDC to vinyl chloride is >99 percent. Increasing conversion decreases selectivity. It must be cooled rapidly to avoid product loss.

A completed functions diagram is shown in Fig. 4-4. The three principal reactions of the path—direct chlorination, oxychlorination, and pyrolysis—require distinctly different conditions, so each must be conducted in a separate reactor.

The raw material feeds for the ethylene chlorination reactor R-1 are ethylene and chlorine; the product is 1,2-dichloroethane (EDC). The stoichiometric amounts of each feed are shown. This reaction is catalyzed by solid $AlCl_3$ and takes place in liquid EDC. Consider preprocessing for each of the feeds. According to the literature references, the commercial form of each of the feeds is sufficiently pure for introduction into the reactor. Introducing the feeds at their storage conditions (chlorine is usually stored as a liquid) absorbs some of the heat of reaction and reduces the need for preheating of the feeds. Therefore, no preprocessing is necessary for these two feed streams.

The reaction is carried out over a temperature range of 65 to 200°C and a pressure range from 150 kPa to 1.2 MPa (absolute); the pressure is equal to the vapor pressure of EDC at the reaction temperature. Chlorine and ethylene are gases at the reaction condition, so they are bubbled through the liquid EDC into which they dissolve and react. The conversion rate is controlled by the rate of mass transfer of ethylene and chlorine from the gas to the liquid phase. The liquid-phase reaction itself is very rapid. If the reactor is designed so as to allow sufficient mixing and contact time for mass transfer, the conversion of reactants is essentially complete. The conversion of both reactants to EDC is

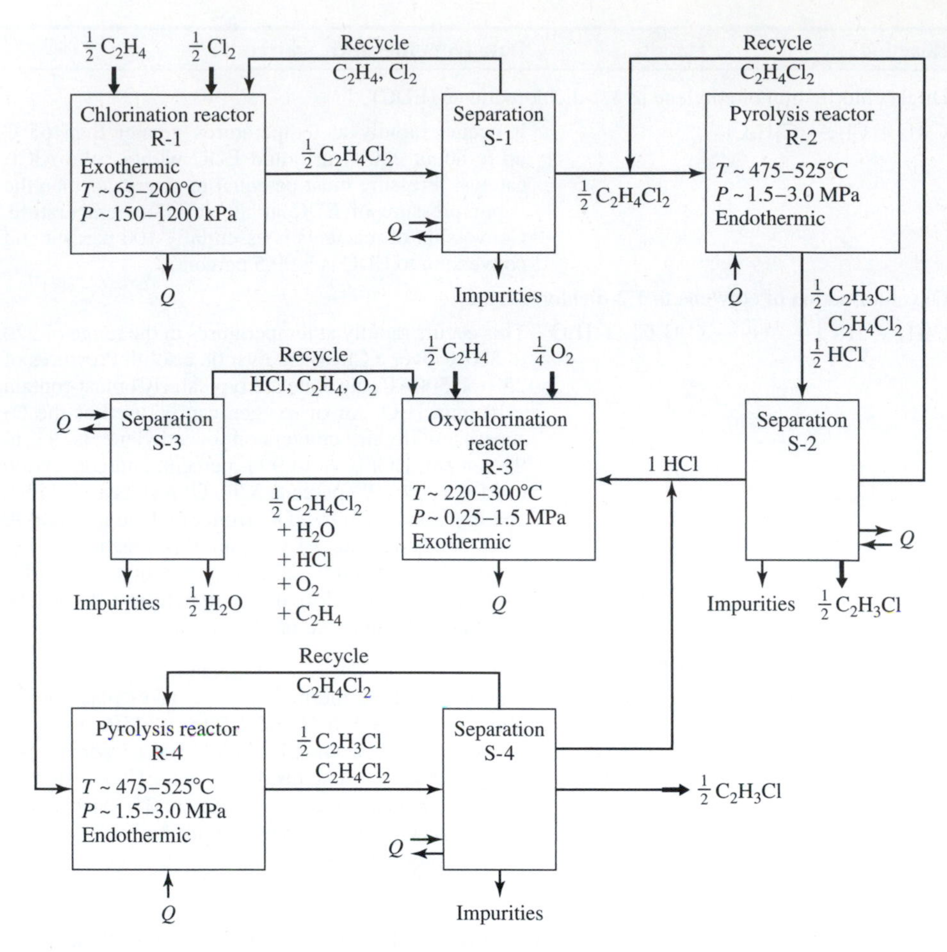

Figure 4-4
Functions diagram for vinyl chloride production

reported to be greater than 99.5 percent. Therefore, no recycle of reactants to R-1 is necessary. Also no preprocessing of the EDC feed to the pyrolysis reactor R-2 is required because of the high purity of the product from reactor R-1. Since the direct chlorination reaction is exothermic, heat must be removed from the reactor, as shown by the Q arrow leaving the reactor.

The EDC, a process intermediate, is the only feed to reactor R-2; vinyl chloride (VC) and HCl are the primary products. In the reactor, EDC is heated to a temperature between 475 to 525°C where it pyrolizes to form vinyl chloride and HCl. This reaction is endothermic shown by the Q arrow into the reactor box. The reaction can be conducted at atmospheric pressure, but usually is carried out at a pressure in the range of 1.5 to 3.0 MPa (absolute). The elevated pressure makes the separation of HCl from vinyl chloride easier, plus the reactor and downstream equipment volumes can be smaller.

More than 99 percent of the EDC that reacts is converted to VC, but the conversion per pass is only 53 to 63 percent of the EDC at typical design conditions. The reaction can be driven to higher fractions of EDC conversion, but the conditions required to do so tend to cause the decomposition of vinyl chloride. In fact, vinyl chloride will decompose at typical reaction conditions if the time of residence at high temperatures is too long. Therefore the residence time purposely is kept short, in the range of 2 to 30 s. The reactor effluent must be cooled quickly (quenched) to avoid vinyl chloride loss. Following the quench the effluent is separated into HCl, an intermediate product used in the oxychlorination reaction, vinyl chloride product, and unreacted EDC which is recycled to the pyrolysis reactor. A separation box S-2, showing these output streams plus an impurity waste stream, follows reactor R-2. The VC stream is a final product. It can be sent directly to storage, because it can be purified sufficiently in S-2 to meet polymerization specifications.

Next is the oxychlorination reactor, shown as box R-3 in the diagram. The chemistry of pyrolysis dictates that HCl is a by-product of the reaction, while the realities of marketing, as discussed in Example 4-2, indicate that the HCl cannot be expected to be a very salable product. Fortunately, the oxychlorination reaction makes it possible to use the HCl to produce more EDC. The feed streams to R-3 are HCl, ethylene, and oxygen (as either air or purified oxygen); EDC and water are the primary products. This reaction takes place on a $CuCl_2$ catalyst, sometimes including additives, impregnated into a solid support such as alumina or silica-alumina. A temperature in the range of 20 to 300°C and a pressure of 0.25 to 1.5 MPa (absolute) are typical. Since the reaction is highly exothermic, a high rate of heat removal is important to avoid reaching reactor temperatures that are too high. A Q removal arrow is shown on the reactor box. Conversions typically are 94 to 99 percent for ethylene, 98 to 99.5 percent for HCl, and essentially 100 percent for oxygen. The EDC yield from ethylene is in the range of 94 to 97 percent. A separation box S-3 follows the reactor. The EDC produced must again be purified to at least 99.5 percent before being sent to the pyrolysis reactor R-4; in particular, the water content must be reduced to less than 10 ppm to avoid serious corrosion problems.

The functions diagram is now complete. There are several streams of by-product impurities, consisting mainly of chlorinated carbon compounds other than EDC and VC. A strategy for recovery for treatment and disposal, sale, or use within the process must be developed for these impurities. ■

OPERATIONS DIAGRAM

Selection of the types of processing equipment, or unit operations, necessary to accomplish the functions previously identified, along with specification of some of the key conditions such as temperatures and pressures, is the next step. An *operations diagram* is constructed indicating these choices.

Preprocessing

The need for preprocessing is determined by raw materials purity, temperature, pressure, and state compared to what is required in the process. The considerations in selecting preprocessing operations are similar to those described later for separations and heat transfer.

Reactors

The type of vessels, operating temperatures and pressures, residence time, catalyst needs, means of promoting contact between reactants and catalysts, space velocity over solid catalysts, and provision for heat addition or removal are based upon the requirements of each reaction. Chapter 13 contains further information on the selection of reactor types.

The flow rates of major components in the operations diagram are based on the simple reaction stoichiometry, with little consideration for reaction yields or degrees of separations.

If ranges of operating temperatures and pressures are available, select values near the midpoint of each range. Consistently doing so will give a final set of base-case conditions that are all near midrange. This may well not be optimal, but it provides a convenient, average starting point for optimization studies.

Separations Methods

Selection of the physicochemical properties upon which to base the separations processes is a key decision in flowsheet development. It depends upon establishing which properties of the materials in question can be exploited for separation purposes. Table 4-1 lists some of these exploitable property differences that can form the basis for the separations process with the corresponding separations methodology. Extensive information on the selection of the appropriate separation process is given in Chap. 15.

Table 4-1 Property bases of separation methods

Property	Separation method
Size	Screening Molecular sieves
Charge	Ion exchange Electrostatic Electrophoretic
Volatility	Flash vaporization Distillation Drying Evaporation
Solubility/phase affinity	Gas adsorption Liquid absorption Solid adsorption Liquid extraction Solid leaching
Different phases	Settling Coalescence Filtration Centrifugation Flotation
Permeability	Membrane
Phase change	Crystallization Melting Solidification

One way of providing a basis for selecting the appropriate separations technology is to find and list measures of potential separation properties for the major species in the process. For example, the solubility of species in a variety of solvents would provide guidance as to the possibility of using one of the solvents as an extractive agent. Another example is to list a measure of the volatility of species, either the normal boiling point or the vapor pressures at particular temperatures, as indicators of the prospects of separating the components by flash vaporization or multistage distillation. Permeability data for components through various membranes would be indicative of the possibilities of separation via membrane permeation.

Separation Guidelines Some guidelines for selecting separations technologies are to

1. Use distillation if the relative volatility of key components is greater than 1.2.
2. Favor known techniques, such as distillation, extraction, and filtration.
3. Avoid introducing foreign species; but if they are used, remove them immediately after use.
4. Favor energy over mass separation technologies.

Heating and Cooling

The operations diagram also requires specifying the sources of heating and cooling to be used. Utilities often can be used for meeting those requirements. For example, all cooling to temperatures of about 20 to 30°C could be done with cooling water; refrigeration is used for lower temperatures. Heat at temperatures above about 400°C can be supplied by combustion, and at lower temperatures by hot liquid utilities or steam. However, since there are process equipment and streams that require heating as well as ones that require cooling, opportunities exist for exchanging heat and reducing utility requirements, resulting in energy savings. Of course, such savings come at a cost of investment in heat exchangers. Experience shows that such heat exchange is often economically favorable. A simple approach to identifying heat exchange opportunities is illustrated later in this chapter.

Minimization of Processing

Equipment and operating cost savings are potentially achievable by combining processing steps, or integration. For example, sometimes reaction and distillation can both be accomplished in a distillation column, as can extraction and distillation. The keys to this kind of integration are the characteristics of the system and the ability to control the operating conditions of the process. Another type of integration occurs when similar operations are being conducted in different parts of the same process, and possibly in different but adjacent processes. A good example of this is seen in the vinyl chloride functions diagram, Fig. 4-4. Two different reactors are producing dichloroethane which subsequently undergoes similar but slightly different processing. It may be possible to combine the EDC streams for some of the processing steps, as shown in Example 4-4. Opportunities to combine processing steps should always be investigated when developing a flowsheet.

EXAMPLE 4-4 Operations Diagram for Vinyl Chloride

Select reactor types, operating conditions, and separation technologies for reaction path 5 of the vinyl chloride process presented in Example 4-2.

■ Solution

The simple mass balance to produce $\frac{1}{2}$ kg mol of EDC requires $\frac{1}{2}$ kg mol each of ethylene and chlorine as feeds to reactor R-1. Because mass transfer of the gaseous reactants is rate-limiting, it is important to obtain active dispersion of the gases into the liquid EDC. This is usually performed with a sparger, utilizing a multihole pipe to obtain widespread gas distribution in a vertical cylindrical vessel. The high exothermic heat of reaction requires that heat be removed rapidly. Typical methods of cooling vessels with either a water-cooled jacket or an internal coil are probably inadequate; so another scheme is needed. Since the reaction takes place in a liquid EDC phase, one option is to circulate the EDC through an external heat exchanger. Another option is to allow part of the EDC to vaporize and then remove the heat in a condenser. Because of the high heat-transfer rates generally associated with phase changes, an external condenser is selected and shown as E-1 in Fig. 4-5. This decision can be revisited later.

Decision: *Use a vapor condenser for cooling.*

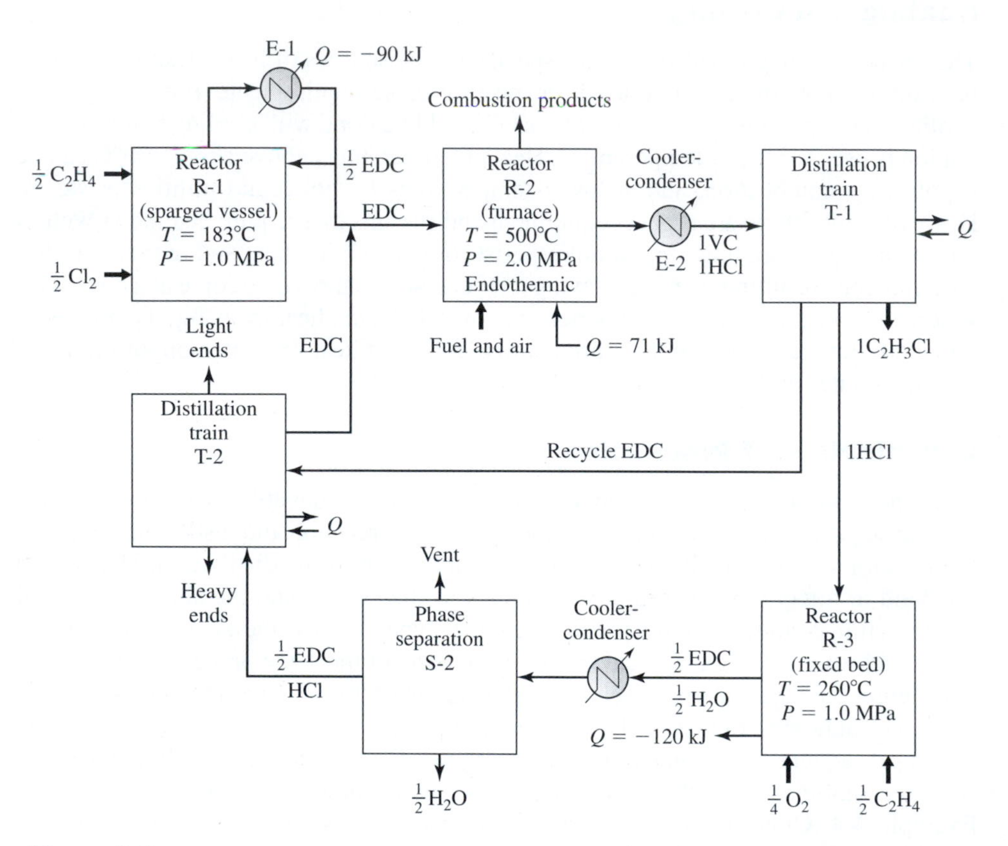

Figure 4-5
Operations diagram for vinyl chloride production

The temperature and pressure of the reactor can be selected within the ranges shown in the reaction conditions table established in Example 4-3. The higher the temperature, the less heat that must be added in R-2 and the more likely it is that the heat of condensation can be used for other purposes in the process. In this case, the approximate midpoint of the pressure range is selected, namely, a pressure of 1 MPa, giving a corresponding temperature of 183°C.

Decision: *Operate R-1 at 1 MPa and 183°C.*

Figure 4-4 shows a separation step following reactor R-1, but the EDC produced by direct chlorination has a purity of 99.5+ percent, indicating that a purification step is not required; thus, the S-1 step is deleted.

The EDC from reactor R-1 must be heated to 475 to 525°C, and the endothermic heat of pyrolysis must be added to reactor R-2. A pressure in the range of 1.5 to 3.0 MPa is recommended. For the operating conditions, select the approximate midpoint temperature and pressure.

Decision: *Operate R-2 at 500°C and 2 MPa.*

The pyrolysis reaction is thermally driven, and no catalyst is required. Because of the high temperature, a heat source such as combustion is called for. This in turn suggests that a process heat furnace be used in which the EDC within the tubes flows essentially countercurrent to the combustion gases of the furnace. The product gases leaving the furnace must be cooled very quickly to stop the thermal decomposition of the vinyl chloride. This type of rapid quenching is often accomplished by spraying a cool, volatile liquid into the vapor that is to be cooled. A liquid process component, either EDC or VC, should be used so that a contaminant is not added to the system. The liquid must be cooled in a separate heat exchanger and recirculated to the quench spray. The disadvantage of the liquid quench is that much of the liquid vaporizes into the product stream and subsequently must be separated. Another alternative is to use a transfer-line heat exchanger located in the effluent line immediately downstream from the reactor. Although the liquid spray provides a faster quench, the in-line exchanger allows greater opportunity for heat recovery. For these reasons, make this choice:

Decision: *Use an in-line heat exchanger to quench the R-2 effluent.*

The in-line heat exchanger E-2 to accomplish the rapid quench is shown in Fig. 4-5. Once the effluent has been cooled to a temperature of about 250°C to quench the reactions, it may or may not need to be condensed before further processing. E-2 is designated as a cooler-condenser, but further analysis is needed to make this decision.

The product stream from the pyrolyzer includes VC, HCl, EDC, and impurities that mostly boil at higher temperatures, designated as heavy ends. Because of their substantially different volatilities, as shown in the properties table assembled in Example 4-1, distillation is the logical candidate for separating the HCl, VC, and EDC. The distillation train T-1 in Fig. 4-5 accomplishes this task. The vinyl chloride is a final product and can be produced with sufficient purity for polymerization purposes by proper design of the separation train. The EDC is not of sufficient purity to be returned directly to the pyrolysis reactor and must receive further treatment to remove the heavy ends before being recycled to R-2.

Ethylene and oxygen along with the HCl from distillation train T-1 are fed to R-3, the oxychlorination reactor. The oxygen may be in the form of air or enriched (~99.5 percent) oxygen. Air is relatively inexpensive while enriched oxygen has a moderate cost relative to the cost of the other raw materials. The advantage of using enriched oxygen is that the quantity of the vent gas streams is significantly reduced. This results in a corresponding decrease in the amount of chlorinated hydrocarbons and ethylene in the vent gas that must be treated and thereby provides a substantial reduction in treatment difficulty and cost. This decision may require additional review.

Decision: *Use enriched oxygen in reactor R-3.*

The oxychlorination reaction is a partial oxidation that occurs on a solid-supported $CuCl_2$ catalyst. The reactor therefore must contain the catalyst and allow the reactant gases to flow through the catalyst structure and directly contact the catalyst. The reaction is highly exothermic; thus, the reactor must be capable of high rates of heat removal. Since the reaction should be approximately isothermal, a midrange temperature is selected. The pressure will be approximately 20 to 40 kPa lower than the pressure in the distillation train, or approximately 1 MPa.

Decision: *Select temperature of 260°C and ~1-MPa pressure for R-3.*

The typical reactor arrangement is either a fluidized bed or a packed bed of the catalyst. The fluidized bed has an internal cooling coil or exchanger and remains essentially isothermal due to the excellent mixing during fluidization. The fluidized-bed reactor operates much as a completely mixed reactor with relatively low concentration of reactants and low rates of reaction and heat generation per unit volume. The packed-bed reactor resembles a shell-and-tube heat exchanger with the catalyst particles and reactant gas in the tubes and a cooling medium on the shell side. Because the flow in the tubes of the packed-bed reactor is essentially plug flow, there is a tendency for temperature gradients and hot spots to occur, which can adversely affect the conversion to VC. One way to avoid this problem is to have a variable catalyst loading on the support, with less catalyst at the inlet end of the tubes where the reactant concentrations are highest, followed by a gradually increasing catalyst loading in the direction of flow. Another way is to have multiple reactors in series with the oxygen feed split among the reactors. The shell-and-tube packed bed with distributed catalyst loading is selected, primarily because of the division of this reaction between two reactors, one of which must be in tubes to supply the heat required for a distillation column reboiler. In fact, it would be possible to have one fluidized-bed and one packed-bed reactor, or just one fluidized-bed reactor.

Decision: *Use packed-bed oxychlorination reactors.*

The product stream from the oxychlorination reactor(s) is a gas containing EDC and water, plus small amounts of ethylene and HCl. The gas is cooled, condensing the EDC and water while the bulk of the ethylene and HCl remains in the gas phase, as does some EDC (far less if enriched oxygen rather than air is used). It is especially important to reduce the water concentration in the EDC to <10 ppm, to reduce corrosion of downstream equipment. This product stream goes to a separation tank S-2, which provides sufficient holdup time that the stream may separate into a gas phase, a water phase, and an EDC phase. The gas phase is vented for further treatment, the water phase is drawn off and sent to treatment, and the EDC goes to the distillation train T-2. The latter removes the light ends (including more water) and heavy ends which go to treatment.

The functions diagram, represented by Fig. 4-4, showed the EDC from S-3 being sent to the pyrolysis reactor R-4 and to a separation unit S-4. Examination shows, however, that R-4 and S-4 perform the same functions as R-2 and S-2. Here is a clear opportunity for process integration. The EDC from S-4 can, in fact, be sent to R-2, followed by S-2. The capacity of R-2 and S-2 will have to be larger to accommodate both sources of EDC, but a useful rule of thumb is that

> One process unit for a given throughput is nearly always less expensive than two separate units, each handling a fraction of the same total throughput (an economy-of-scale rule).

Therefore, an unambiguous decision is made to use one pyrolysis reactor R-2 followed by one effluent separation train T-2 for the effluents from both R-1 and distillation train T-1. The operations diagram is now adequately defined.

■

PROCESS FLOWSHEET

The final steps in defining a flowsheet are to estimate key equipment performance parameters, improve the mass balances, estimate approximate energy balances, define the separation trains, and develop heat integration opportunities.

Reactors

The conversion of raw materials and the yield of products form the design basis for chemical reactors. At this stage of flowsheet evolution, it is common to select a conversion to the desired product. This information is necessary to improve the mass balances. The expectation is that appropriate equilibrium, kinetic, or mass-transfer models will later be used to size the reactor to achieve the selected conversion. The decision variables of temperature and pressure, or conversion, then can become optimization variables in the final reactor design.

Mass and Energy Balances

Before separation trains and heat exchange opportunities can be examined, at least approximate mass and energy balances must be conducted. Mass balances are initiated by selecting a convenient basis. A simple basis that can be easily scaled to any production rate is 1 or 100 mol (or kg mol) of the desired product. Mass balances are then calculated using the conditions developed in the preparation of the operations diagram; energy balances follow, using thermodynamic properties from the literature. Separation operations must be sufficiently defined that their energy requirements can be estimated if they are to be included in the heat integration. For example, as a minimum, the operating pressure and reflux ratio must be specified for distillation columns. The pressure usually is established either by the pressure of neighboring process equipment or by iterative calculations to determine the appropriate dew point and bubble point temperatures. A procedure for estimating an operating reflux ratio is described in Example 4-5.

If a chemical process simulation computer code is available, it can be used for the mass and energy balances; such codes are discussed in Chap. 5. Even in that case, however, it is useful to have approximate balances and identification of potential heat exchanges before undertaking a computer-based simulation, because this provides a more complete flowsheet for simulation and provides a basis for realistically checking the simulation results.

Separation Trains

Most separation processes result in two product streams. Therefore, to separate a mixture containing c components into c pure products[†] requires $c - 1$ distinct separation operation steps. Separating a multicomponent mixture into its various pure components thus requires a *train* of separations. These separations need not all be of the same technology, although they often are. Assuming that the separation technology has been

[†]It is common that the nearly pure product stream is 98+ percent pure. For convenience, that purity or higher will be referred to as a pure product.

selected, the sequence in the separation train then needs to be established. Some general rules of thumb for separation train sequencing are:†

1. When possible, reduce the separation load by stream splitting and blending.
2. Difficult separations are best saved for last.
3. Remove high-concentration components early in the sequence.
4. Remove components not normally present in the process soon after use.
5. Avoid wide excursions in temperature and pressure, particularly since cooling at temperatures above ambient is relatively inexpensive and heating is more expensive while cooling at temperatures below ambient is very expensive.

In addition to the just-mentioned heuristics, others that are more specific to distillation trains include:‡

1. Remove thermally unstable, corrosive, or reactive components early in the sequence.
2. Remove the final products one by one, starting with the most volatile and continuing in order of decreasing volatility.
3. Remove the components present in the largest amounts early.
4. Sequence in order of decreasing relative volatility; that is, do the more difficult separations later.
5. Sequence in order of increasing product purity requirements.
6. Sequence so that distillate and bottom product flow rates are as equal as possible.

It is not possible to satisfy all these heuristics simultaneously, but it is recommended that they be applied in the order listed.

Heat Exchange

Since there are process equipment and streams that require heating as well as ones that require cooling, opportunities nearly always exist for exchanging heat and reducing utility requirements. Therefore, some initial searching for heat exchange opportunities needs to be done. The heating and cooling requirements of separation operations should also be included in the search. Effective algorithms are available for such searches once a base-case flowsheet has been defined. The purpose here is to illustrate a simple approach that can be used to identify some of the opportunities for heat exchange that may be investigated much more thoroughly with commercial heat integration software.§

EXAMPLE 4-5 Final Flowsheet for Vinyl Chloride Process

Develop a flowsheet showing the major process units, approximate mass and energy balances, and apply heat integration for the vinyl chloride process selected in Example 4-2 and the operations diagram of Example 4-4.

---

†Adapted from D. F. Rudd, G. J. Powers, and J. J. Siirola, *Process Synthesis,* Prentice-Hall, Englewood Cliffs, NJ, 1973; and W. D. Seider, J. D. Seader, and D. R. Lewin, *Process Design Principles,* J. Wiley, New York, 1999.

‡Adapted from D. F. Rudd, G. J. Powers, and J. J. Siirola, ibid.

§See Chap. 9 for more information.

■ Solution

Reactor types and operating conditions were established in Example 4-4. Reactant conversions and product yields can now be estimated from the data provided in the table summarizing the reaction information for vinyl chloride production pathways in Example 4-3. The two distillation trains need to be developed in greater detail. Additional operating conditions of the process units will be specified to allow the preparation of more detailed, though still approximate, mass balances and energy balances. Consideration is given to disposal of the waste streams. The results of the following analysis are shown in Fig. 4-6 as a flowsheet schematically representing reaction path 5 for the vinyl chloride process.

In reactor R-1, the reaction between chlorine and ethylene is essentially complete, and the conversion to EDC is 99.5+ percent. No separation step and no recycle are necessary, and the EDC is sufficiently pure to feed it directly to the pyrolysis reactor. In the pyrolysis reactor R-2, the conversion per pass of reactant EDC is 53 to 63 percent. The midpoint of 58 percent is selected as a design target for the reactor. The conversion of EDC to VC is expected to be 99 percent.

Decision: *Design reactor R-2 to obtain 58 percent conversion of EDC per pass with 99 percent of the reacted EDC converted to VC.*

The oxychlorination reactor R-3 is expected to provide 96 percent conversion of ethylene and 98 percent conversion of HCl per pass, with 92 percent of the ethylene being converted to EDC.

Decision: *Design R-3 to meet these criteria.*

Each distillation train is now expanded. Train T-1 of Fig. 4-5 must separate a three-component mixture into three essentially pure components. Therefore two distillation columns are required. The columns operate between 1 and 2 MPa. The first column should be at the higher pressure, just below 2 MPa, and the HCl should be removed as a product because of its high vapor pressure and corrosive nature. Even at this pressure, refrigeration is required to condense the overhead vapor from this column. Only a partial condenser is needed, however, because the HCl product can be fed as a vapor to R-3 without the need of a compressor since R-3 is at 1 MPa. Any light ends in the feed stream will tend to leave with the HCl.

Decision: *Remove HCl in column C-1, operate below 2 MPa, and use a partial condenser.*

The second column designated C-2 separates the vinyl chloride as overhead product and EDC as bottom product. The EDC will contain heavy ends from the feed stream. This stream needs further purification before being recycled to the pyrolysis reactor R-2. Therefore it is sent to distillation train T-2, as shown in Fig. 4-5. A pressure somewhat above 1 MPa, probably 1.2 MPa, is satisfactory.

Decision: *Operate column C-2 at 1.2 MPa.*

Distillation train T-2 receives one feed of EDC containing light and heavy ends from separator V-1 and a second feed of EDC containing heavy ends from C-2. This constitutes essentially three components, so two columns are required. A reasonable combination is to remove the light ends from the first feed in column C-3 and then feed both the C-3 bottoms and the C-2 bottoms into a fourth column to remove the heavy ends and recycle the EDC overhead product to the pyrolysis reactor R-2 via HX-2 and R-3. It is unnecessary and undesirable to mix these two feeds since the mixture would have to go through both columns. Because the light ends and heavy ends presumably have volatilities substantially different from that of EDC and each constitutes less than 1 percent of the feed streams, these separations are simple and require few plates and low reflux ratios. These columns can operate near 1 MPa. Column C-3 should operate at a sufficiently higher pressure than column C-4 so that no pump is required to obtain flow between them.

Decision: *Operate C-3 at 1.1 MPa and C-4 at 1 MPa.*

A recovery of 99.5 percent is assigned for each product (EDC, VC, and HCl) in these separation steps.

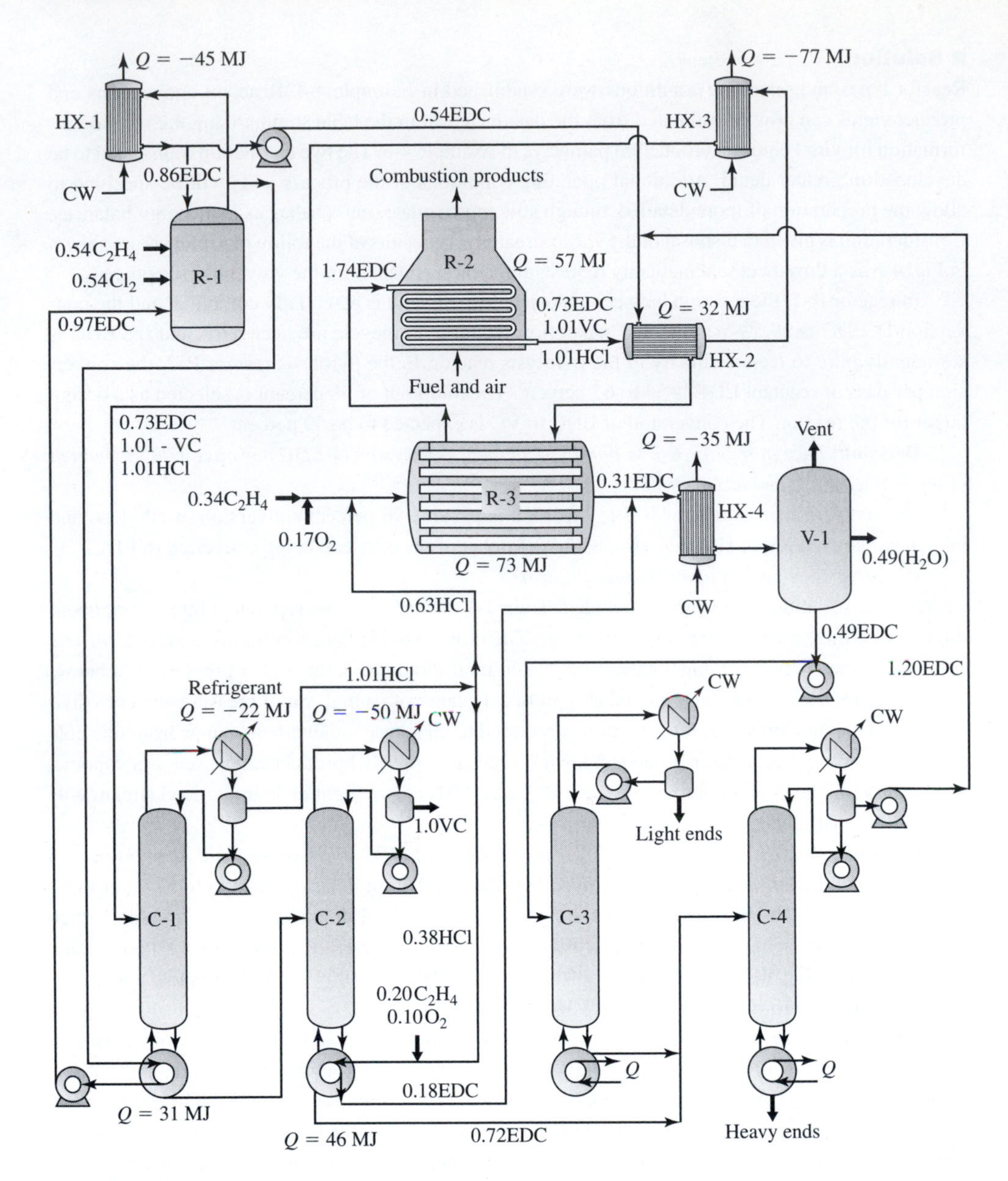

Figure 4-6
Flowsheet for vinyl chloride production. Stream quantities are in kg mol and heat quantities are in MJ, on a basis of 1 kg mol vinyl chloride produced. CW = cooling water

Decision: *Design separations to accomplish at least 99.5 percent recovery of each reaction product in each reaction step.*

The phase separation vessel V-1 must have sufficient residence time that separation of the three phases (gas, water, and EDC) is at least 99.5 percent complete.

Decision: *Design adequate residence time in the phase separation vessel.*

Approximate mass and energy balances are required before heat exchange opportunities can be investigated. The basis is to produce 1 kg mol of vinyl chloride product. The primary feeds to the process are ethylene and chlorine to R-1 and ethylene and oxygen to R-3. The balanced chemical equations indicate that the same amount of ethylene should be provided for each reactor. However, since the ethylene conversions are not the same in the two reactors, it is not obvious that this should be true. Nonetheless, more detailed mass balances show that, with the conversions assumed in this solution, the feeds should remain equal. The amount of the various feeds required is obtained by solving a vinyl chloride balance, based upon the conversions of ethylene and EDC, and recoveries of EDC, namely,

$$1 \text{ kg mol vinyl chloride} = \frac{FE}{2}(CR1)(CR2)(0.995)^{\text{ss1}} + \frac{FE}{2}(CR3)(CR2)(0.995)^{\text{ss2}}$$

where FE is the total feed of ethylene in kg mol split equally between the two reactors R-1 and R-2, $CR1$ the conversion of ethylene to EDC in R-1 (0.995), $CR2$ the conversion of EDC to vinyl chloride in R-2 (0.99), $CR3$ the conversion of ethylene to vinyl chloride in R-3 (0.92), and ss1 and ss2 the number of separation steps encountered in going from ethylene feed to vinyl chloride product for R-1 (2) and R-2 (5), respectively (these are shown on Fig. 4-6). Therefore,

$$1 = \frac{FE}{2}[(0.995)(0.99)(0.995)^2 + (0.92)(0.99)(0.995)^5]$$

Solving the relation gives a value of 1.073 kg mol for FE. Each of the two ethylene feeds is one-half of this, or 0.537 kg mol. In fact, this gives a slight excess of ethylene over HCl in the oxychlorination reactor, but that is how this reaction is typically run. The ethylene feed rate is rounded to 0.54 kg mol. All the mass flow values are rounded off as shown on the flowsheet based for the production of 1 kg mol of vinyl chloride.

The rest of the major component quantities shown on the flowsheet are calculated from these feed rates and the specified conversions and recoveries. The conversion of EDC in R-2 has been set at 58 percent. Therefore 1.01 kg mol of VC must exit from R-2 in order to still have 1.0 kg mol after two separation steps. The amount of EDC in the reactor feed must be

$$(F_{\text{EDC}})(0.58) = 1.01 \qquad \text{or} \qquad F_{\text{EDC}} = 1.74 \text{ kg mol}$$

where F_{EDC} is the EDC in the reactor R-2 feed in kg mol. The unreacted EDC in the R-2 effluent is $1.74 - 1.01 = 0.73$ kg mol.

It is now possible to perform heat balances for the units and streams. These are calculated using the mass balance results and the specified temperatures, heats of reaction, and heat capacities. For each kg mol of VC produced in the process, 0.54 kg mol of EDC is produced in reactor R-1. The energy balance for R-1 is given by

$$Q_{\text{R-1}} = n_{\text{EDC}}\,\Delta H_{\text{R}} + \sum_{\text{reactants}} [nC_p(25 - T_{\text{in}}) + n\Delta H_{\text{V}}] + \sum_{\text{products}} [nC_p(T_{\text{out}} - 25) + n\Delta H_{\text{V}}]$$

where $Q_{\text{R-1}}$ is the heat to be added (removed if negative) in MJ, ΔH_{R} the heat of reaction in MJ/kg mol, n_{EDC} the number of kg mol of EDC, n the number of kg mols, C_p the heat capacity in MJ/(kg mol·K), T the temperature in °C, and ΔH_{V} the heat of vaporization in MJ/kg mol. For R-1, the reactants are chlorine which enters as a liquid at 35°C and ethylene which enters as a gas at 25°C. The product is EDC which leaves as a vapor at 183°C. Substituting appropriate values in this relation results in

$$\begin{aligned} Q_{\text{R-1}} &= (0.54)(-180) + (0.54)(0.036)(25 - 35) + (0.54)(20) + (0.54)(0.125)(183 - 25) \\ &= -76 \text{ MJ} \end{aligned}$$

This amount of heat must be removed in the condensation of 2.37 kg mol of EDC in HX-1 and the reboiler of C-1. Since only 0.54 kg mol of EDC is produced and fed to R-2, the rest of the EDC is refluxed back to reactor R-1.

The heat duties of the distillation column condensers and reboilers need to be estimated to be used in the heat integration. For columns C-1 and C-2 this is done by first estimating the minimum reflux ratio for each column, using the Underwood binary equation for key components.[†] Vapor pressure data from the properties table presented in Example 4-1 are used to estimate relative volatilities. The actual reflux ratio is then estimated as 1.1 times the minimum for C-1 and 1.25 times the minimum for C-2, based on a heuristic that the ratio of actual to minimum reflux is in the range of 1.1 to 1.5 and at the low end for temperatures below ambient.[‡]

Decision: *Use 1.1 times the minimum reflux ratio for C-1 and 1.25 times the minimum reflux ratio for C-2.*

To complete the material balance, the composition and concentration of the light-end and heavy-end constituents need to be established and quantitatively defined. However, since the amounts of these impurities are very low, their effect on the flowsheet development and heat integration will be negligible. Accordingly, establishment of these quantities has not been included in this analysis.

Energy balances for each stream or unit in the flowsheet can now be made, using the results of the material balances. The heat exchange duties obtained from these energy balances, for the vinyl chloride process developed to this point, indicate that a total of 239 MJ of heat needs to be added to the process, 97 MJ from fuel combustion and 142 MJ from steam or hot oil per kg mol of vinyl chloride produced. Also, 411 MJ of heat must be removed, 22 MJ by refrigerants and 389 MJ by coolants. In this energy balance, 11 heat exchanges take place, 6 in process units (reactors, column reboilers, and condensers) and 5 in separate heat exchangers (exclusive of the heat exchangers associated with columns C-3 and C-4).

Summary of the heat balance before process heat integration[†]

Unit	Heat exchange	Heat source or sink, MJ/kg mol VC	T or ΔT, °C	Utility required
R-1	Heat of reaction	−76	183	Cooling water
R-2 feed	Heat and vaporize	+65	183–226	Steam or oil
	Heat gas	+25	226–500	Fuel
	Heat of reaction	+72	500	Fuel
R-2 effluent	Quench and cool	−49	500–140	Cooling water
	Condense	−60	140–40	Cooling water
C-1	Reboiler	+31	150	Steam or oil
	Condenser	−22	−10	Refrigerant
C-2	Reboiler	+46	180	Steam or oil
	Condenser	−50	58	Cooling water
R-3	Heat of reaction	−119	260	Cooling water
R-3 effluent	Cool and condense	−35	260–50	Cooling water

[†]Basis: 1 kg mol of vinyl chloride produced.

[†]J. M. Douglas, *Conceptual Design of Chemical Processes,* McGraw-Hill, New York, 1988.

[‡]R. H. Perry and D. W. Green, eds., *Perry's Chemical Engineers' Handbook,* 7th ed., McGraw-Hill, New York, 1997.

Obviously, the 389 MJ that need to be removed by coolants could be removed by cooling water. However, considerable energy is available from the R-2 effluent quench and reactor R-3 to generate high-temperature (~245°C) steam that can supply heat to other parts of the process. This represents an opportunity for heat integration for the process. From the heat exchange duties summary table shown above, it can be seen that steam could be generated in reactors R-3 and R-2 to supply all the duty of the C-2 reboiler and nearly all that required to heat and evaporate the R-2 feed. The remainder of the heat required for the R-2 feed is supplied in reactor R-2, while the 31-MJ duty of the C-1 reboiler is still provided by utility steam or hot oil. This integration reduces the heat input required via utilities by 60 percent, from 239 to 96 MJ, by using the 143 MJ of steam generated within the process. The cooling water duty is reduced 37 percent, from 389 to 246 MJ, by the integrated use of the 143 MJ of steam. However, this heat integration still requires 11 total heat exchanges and 5 separate heat exchanges, as was required without integration.

Despite the substantial utility savings achieved by integrating the steam generation and condensation in the process, there are still further opportunities for heat integration. The use of steam as an intermediary requires an extra heat-transfer step that entails a loss of 25 to 35°C in temperature difference when compared to a direct heat transfer. Since there are other opportunities for heat exchange between process streams, a more useful approach to heat integration needs to be applied.

Each of the heat duties listed in the heat exchange duties summary table is graphed on a plot of temperature versus heat duty in Fig. 4-7. The horizontal difference between two endpoints on any line is the heat duty Q, in MJ per kg mol of vinyl chloride produced, required to accomplish the corresponding phase or temperature change. Phase changes are represented by horizontal lines for pure

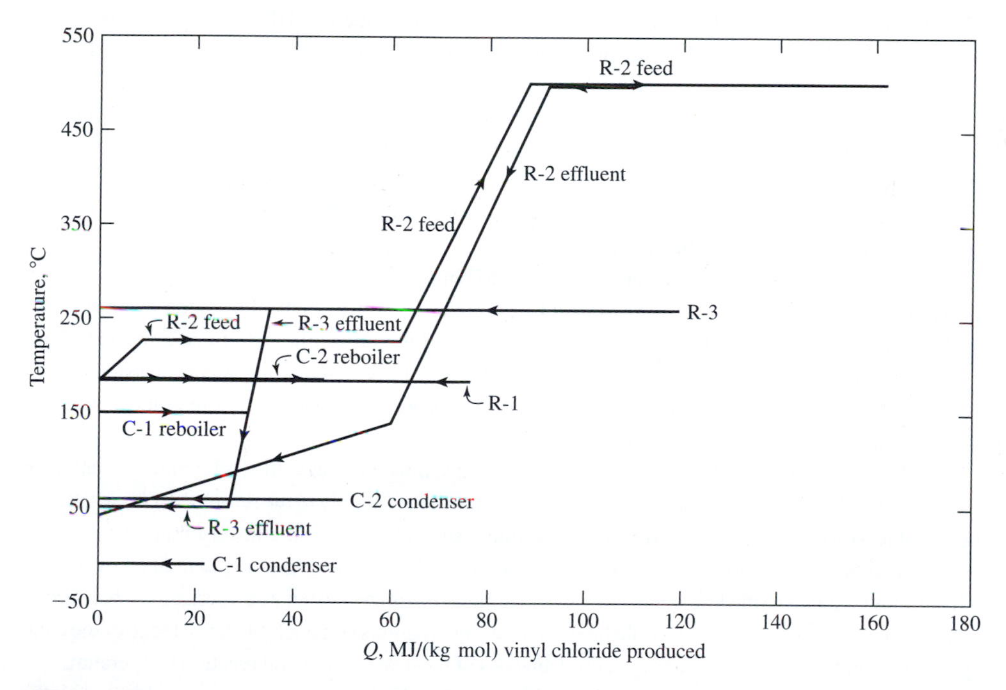

Figure 4-7
Temperatures versus heat duty for the vinyl chloride process

components and for mixtures in reboilers and condensers. Nonhorizontal lines represent sensible-heat temperature changes, except for the lower part of the curve for the R-2 effluent which is a condensation of a mixture showing a temperature change from the dew point to the bubble point as condensation takes place. Arrows on the lines point to increasing values of enthalpy when a stream is being heated or vaporized, and decreasing values of enthalpy when a stream is being cooled or condensed. All lines are shown as straight. This assumes constant heat capacities for sensible heats and equal and constant heats of vaporization for vaporizing or condensing mixtures. These are reasonably accurate, but inexact assumptions adequate for the purpose at hand, since more exact values will be incorporated into calculations for the preliminary design.

The advantage of this type of graph is that the relationship between stream temperature and heating and cooling requirements becomes evident, making it possible to locate heat exchange opportunities. Of course, as the number of heat exchanges increases, the graph becomes more difficult to use. Initially, notice that all streams at temperatures below 150°C require cooling, and there are no streams below 150°C that require heating. Therefore, all streams below 150°C must be cooled with utilities; the C-1 condenser (at −10°C) must be cooled by a refrigerant, and the C-2 condenser and the R-2 and R-3 effluents by cooling water. In fact, the R-2 effluent below 180°C could only be used for heating the C-1 reboiler, but the quantity of heat available is too small to make this economically worthwhile; so it, too, should be cooled with cooling water.

Reactor R-1 has 76 MJ per kg mol of vinyl chloride product that needs to be removed at 183°C. Reboiler C-1, requiring 31 MJ at 150°C, provides the only heating load at a low enough temperature to utilize heat from reactor R-1. This 31-MJ duty requires 0.97 kg mol of EDC vapor from R-1 to be fed to the C-1 reboiler. The condensed EDC from C-1 must be pumped back to R-1. The remaining 45-MJ duty at 183°C must be removed by cooling water in condenser HX-1, which will condense 1.4 kg mol of EDC per kg mol of VC produced. Only 0.54 kg mol of EDC is fed to reactor R-2, so the remaining 0.86 kg mol is refluxed back to reactor R-1. It may appear unnecessary to condense the 0.54 kg mol of EDC that is being fed to reactor R-2 where, because of the higher pressure, it must be heated and revaporized. Note that if it is not condensed, the condensation duty of HX-1 does not change, but the heating load for reactor R-2 decreases. Based upon this result plus the heuristic that it is more economical to increase the pressure on a liquid than on the same amount of gas or vapor, condensation and liquid pumping are a firm choice.

The oxychlorination reaction can provide 119 MJ of heat at 260°C. This heat source can be used to supply all the heat required at 183°C in the C-2 reboiler as well as the heat needed to vaporize and heat the R-2 feed. This requires two oxychlorination reactors; one is the tube bundle in the C-2 reboiler, and the other is the R-3 reactor. Even though this requires using alloy tubes, heads, and tube sheets in the C-2 reboiler, the equipment should be no more expensive since a reboiler is required anyway and reactor R-3 can be smaller.

A disadvantage of having a solid-catalyzed reaction supply the heat to the C-2 reboiler is that, with the inevitable decay of catalyst activity over time, the rate of heat supply to the reboiler decreases. It may be possible to compensate for this decay by installing extra catalyst or using lower initial feed rates, or doing a combination of both, so that nearly constant boil-up and reflux can be maintained over time. Changes in feed rate to the C-2 boiler can be accommodated by the design and operation of the second oxychlorination reactor R-3. Accomplishing such a design requires considerable knowledge of the catalyst decay conditions as well as a very careful analysis of the heat transfer and reaction temperature.

Reactor R-2 effluent must be cooled quickly from 500°C. An effective way to do this is with the in-line heat exchanger HX-2, cooled by a boiling liquid. The R-2 feed at 183°C is heated to 226°C and partially vaporized in exchanger HX-2 by the removal of 32 MJ from the reactor effluent and cooling the latter to 200°C. The reactor R-2 effluent is further cooled and condensed in HX-3 with cooling water, transferring 77 MJ. The reactor R-2 effluent also is used to remove 73 MJ from reactor R-3 to further vaporize the feed to reactor R-2.

Additional sensible heat and heat of reaction amounting to 57 MJ must be added in reactor R-2 at 500°C. With the lack of a heat source greater than 500°C an external utility is required. Since only fuel combustion can provide such a high temperature, a process heat furnace is used to supply this heat.

There are no streams remaining to be heated at temperatures below 260°C, so reactor R-3 effluent is cooled and condensed from 260 to 50°C by cooling water, transferring 35 MJ in HX-4.

Decision: *Select the integrated flowsheet shown in Fig. 4-6.*

The results from this process integration are summarized in the following table showing the heat exchange requirements for the integrated flowsheet provided in Fig 4-6. Note that now the heat exchange requires 10 heat exchangers, 6 are with process units and 4 are with heat exchangers. The heat input has been reduced by 76 percent, and the heat removed has been reduced by 48 percent with the process heat integration performed with the aid of Fig. 4-7.

Consideration should be given to the disposal of the vent gas, light and heavy ends, and wastewater streams shown on the flowsheet. Each of these streams contains chlorinated hydrocarbons, and as such their disposal is highly regulated. Gaseous chlorinated hydrocarbons can be burned with air,

Summary of the heat balance after process heat integration†

Unit	Heat source	Source Q, MJ/kg mol	T or ΔT, °C	Heat sink	Sink Q, MJ/kg mol	T or ΔT, °C
HX-1	R-1 heat of reaction	−45	183	Cooling water	+45	25–40
C-1 reboiler	R-1 heat of reaction	−31	180	Column boil-up	+31	150
C-1 condenser	C-1 vapor	−22	−10	Refrigerant	+22	∼−15
HX-2	R-2 effluent quench and cool	−32	500–40	R-2 feed, heat and vaporize	+32	183–226
R-3	R-3 heat of reaction	−73	260	R-2 feed, vaporize	+73	226
R-2	Fuel	−57	500	R-2 feed, heat of reaction	+57	226–500 500
HX-3	R-2 effluent cool and condense	−77	240–40	Cooling water	+77	25–40
C-2 reboiler	R-3 heat of reaction	−46	260	Column boil-up	+46	180
C-2 condenser	C-2 vapor	−50	58	Cooling water	+50	25–40
HX-4	R-3 and C-2 effluents quench and condense	−35	260–50	Cooling water	+35	25–40

Sum of heat inputs = +57 MJ (fuel)
Sum of heat outputs = −229 MJ (22 to refrigerant, 207 to cooling water)
Sum of heat exchanged = 182 MJ (process streams)
Requires 10 heat exchanges; 6 are process units and 4 are exchangers (excluding columns C-3 and C-4).
†Basis: 1 kg mol of vinyl chloride produced.

or preferably enriched oxygen, to produce carbon dioxide and HCl. The HCl can be recovered by absorption into water or condensation and then dried and returned to the process. The chlorinated compounds may be air-stripped from the wastewater and burned with the other waste streams, and the wastewater is then treated by more conventional methods.

Decision: *For effluent control, employ wastewater stripping followed by conventional wastewater treatment with combustion of all gaseous effluents followed by HCl recovery and recycle.*

Movement of materials throughout the process is necessary. Movement is caused in this process by pumps or pressure differences. Pumps are shown in Fig. 4-6 where they are needed to overcome pressure increases or flow energy losses. All other movement results from pressure decreases in the direction of flow.

The basic process flowsheet, shown in Fig. 4-6, is now complete. This flowsheet can be used to provide a basis for further refinement and optimization of the process design. However, some 16 reviewable decisions (several apply to more than one unit) were recorded in the course of obtaining this flowsheet. These decisions should be revisited in order to generate alternative flowsheets for evaluation and comparison. ■

ALGORITHMIC FLOWSHEET GENERATION[†]

The traditional approach to flowsheet synthesis employs a series of hierarchical steps to obtain a result. This is not a firm, concrete method, but rather the application of common sense, heuristics, and analysis. The problem with this approach is that identifying all alternatives, let alone analyzing all of them, is virtually impossible. Therefore, flowsheets generated in this way are not necessarily even near optimal.

An alternative method that involves an ordered, mathematical analysis is the algorithmic approach. Here mathematical tools are used to generate and evaluate all possible flowsheet arrangements to possibly eliminate many of them as nonoptimal. However, the overall effort involved in the development of the total possible flowsheet structures as well as the computational effort in optimization can be enormous when conventional algorithmic methods are applied. For example, only 3465 structures may be feasible flowsheets among 3.4×10^9 possible flowsheets from the 35 operating units of a real industrial process.[‡] Since the feasible flowsheets only represent 0.00001 percent of the possible flowsheets, it becomes clear that the task of locating the optimal or near optimal flowsheet can be a very difficult one.

The need mentioned above to circumvent the profound complexity and difficulty of flowsheet synthesis has given rise to the development of an algorithmic method by resorting to graph theory, based on a unique class of graphs identified as *process graphs*, or P-graphs.[§] With these graphs, this method generates the rigorous structural

[†]Acknowledgment for the preparation of this section on algorithmic flowsheet generation is gratefully extended to L. T. Fan, University Distinguished Professor, and the Mark H. and Margaret H. Hulings, Chair in Engineering at Kansas State University, and F. Friedler, Professor and Head, Department of Computer Science, University of Veszprem, Veszprem, Hungary.

[‡]F. Friedler, J. B. Varga, and L. T. Fan, *Chem. Eng. Sci.*, **50:** 1755 (1995).

[§]F. Friedler, J. B. Varga, E. Feher, and L. T. Fan, *State of the Art in Global Optimization, Computational Methods and Applications*, C. A. Floudas and P. M. Pardalos, eds., Kluwer, Norwell, MA, 1996, pp. 609–626.

model of the process network which is to be synthesized through the implementation of algorithms derived from a set of axioms portraying the unique combinatorial features involved in material processing systems. The method can significantly reduce the search area for finding the optimal flowsheet or process network.

Fundamentals of Algorithmic Process-Network Synthesis

The framework of the algorithmic method for process-network synthesis (PNS) based on P-graphs is depicted in Fig. 4-8.

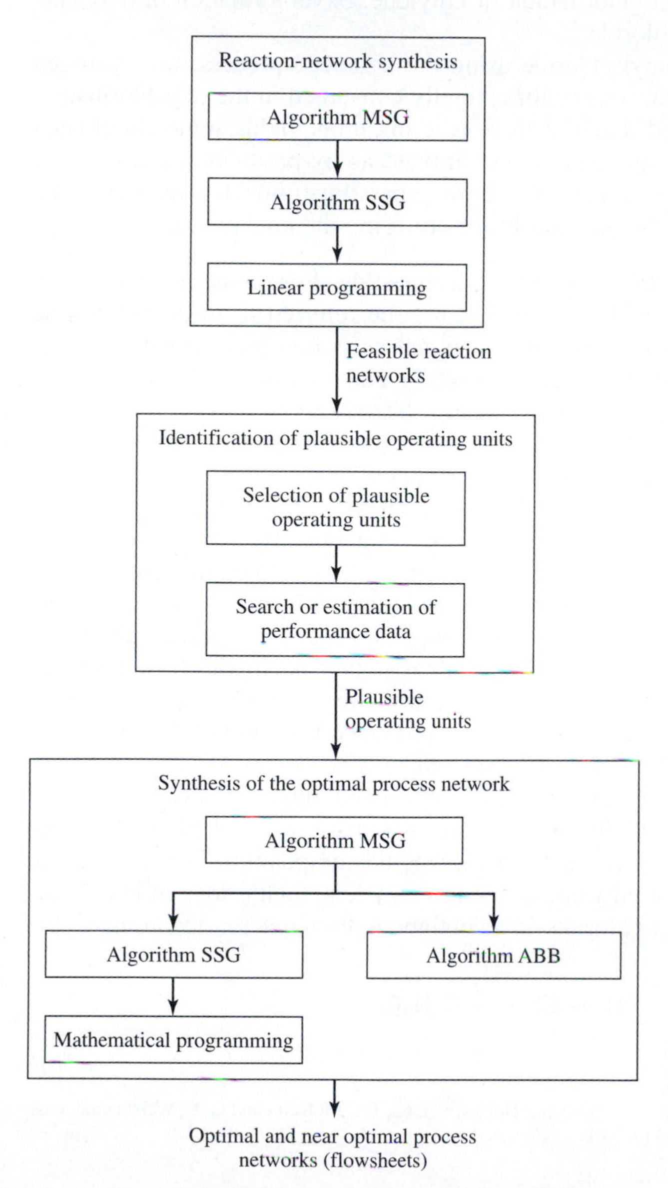

Figure 4-8
Major steps for the algorithmic process-network synthesis based on graph theory

For clarity, the basic terminologies as well as the mathematical concepts and operations involved in this method can be illustrated with simple examples from the synthesis of the same chemical product, vinyl chloride, in the flowsheet development based on the hierarchical approach.

Vinyl Chloride Synthesis This specific chemical can be synthesized by a number of reactions that have been identified in Example 4-1. Vinyl chloride currently is produced almost exclusively by the balanced process from the conversion of ethylene, chlorine, and oxygen. The synthetic steps of this process, as noted in Example 4-3, involve three reactions: direct chlorination of ethylene, oxychlorination of ethylene, and pyrolysis of ethylene dichloride.[†]

In the manufacture of vinyl chloride using the balanced process, the hydrogen chloride generated in the pyrolysis reactor is totally consumed in the oxychlorination reactor to which it is recycled. Each of these reacting units yields some chlorinated hydrocarbons, carbon monoxide, and carbon dioxide as by-products. Each reacting unit is accompanied by one or possibly more separating units for removing the by-products, recycling the hydrogen chloride or purifying the final product.

Materials and Operating Units Before we discuss this algorithmic method, it will be beneficial to define some of the terms that will be referred to in describing the method. For example, operating units are functional units in a process network performing various operations, such as mixing, reacting, and separating. These operating units correspond to the blocks in the flowsheet of the process to be synthesized. The various units of processing equipment, such as reactors, separators, and mixers, in each operating unit transform the physical and/or chemical states of the materials being processed. The chemical process represented by the flowsheet converts the raw materials into the desired products by utilizing these operating units. By-products generated in the process can be gainfully recovered or regarded as wastes to be treated.

In process-network synthesis, a material is uniquely defined by its components and concentrations, that is, by its composition. Suppose that M is a given finite set of materials involved in the process-network synthesis, that is, in the flowsheet. An example for M can be found in the separation of a feed consisting of three components A, B, and C, denoted as ABC, into the three pure products A, B, and C, by employing only high-purity separators. For example, feed ABC can be separated into AB, AC, or BC; thus, by considering all the materials in the inputs and outputs of every separator, we have $M = \{A, B, C, AB, AC, BC, ABC\}$. Since the quantities of A, B, and C in the feed are fixed, every element in M represents a specific composition. A more specific example is the chlorination of ethylene to produce ethylene dichloride, an intermediary in the production of vinyl chloride, in a continuous-flow reactor according to the following reaction:

$$C_2H_4 + Cl_2 \longrightarrow C_2H_4Cl_2$$

[†]F. Borelli, *Encyclopedia of Chemical Processing and Design,* vol. 62, J. J. McKetta and G. E. Weismantel, eds., Marcel Dekker, New York, 1998, pp. 313–340.

Even when one of the two feed streams to the reactor is pure C_2H_4 and the other is pure Cl_2, the single product stream from the reactor is a homogeneous gaseous mixture of $C_2H_4Cl_2$, trace amounts of unreacted C_2H_4 and Cl_2, and by-product $C_2H_2Cl_2$, all with a specific composition; in other words, the product stream is not pure $C_2H_4Cl_2$. If this homogeneous mixture is designated as m_p, then

$$M = \{C_2H_4, Cl_2, m_p\}$$

The process-network synthesis involves not only the materials represented by set M, including all final products in set P, all raw materials in set R, all by-products and intermediate products but also all operating units represented by set O. Each of the operating units, in general, comprises one or more items of processing equipment or unit operations and corresponds to a block in the flowsheet. As a result, the following basic relationships exist among M, P, R, and O:

$$P \subseteq M \qquad R \subseteq M \qquad \text{and} \qquad M \cap O = \emptyset \tag{4-1}$$

The above relationships are self-evident: By definition, the final products in set P and the raw materials in set R must be part of the materials in set M, which also contains all other concomitant materials, including intermediates. Moreover, the materials in set M and the operating units in set O are mutually exclusive. In the manufacture of vinyl chloride utilizing the balanced process, $P = \{C_2H_3Cl\}$ and $R = \{C_2H_4, Cl_2, O_2\}$. In addition to C_2H_3Cl in set P, and C_2H_4, Cl_2, and O_2 in set R, set M includes by-product water and the materials in the streams exiting from all the reacting and separating units that contain various intermediates and trace by-products; hence, P and R are subsets of M. Since the materials and operating units are entirely different entities, sets M and O do not share any common elements; in other words, these two sets do not overlap, and thus their intersection is empty, thereby resulting in $\emptyset$. These relationships appear to be very trivial from the standpoint of chemical engineering. Nevertheless, it is essential in ensuring mathematical rigor and error-free execution of the computer algorithms for process-network synthesis.

Two classes of materials or material streams are associated with any operating unit. The first class includes the input materials, and the second includes the output materials. Note that a material can be composed of one or more components. Operating units are defined when their input and output materials are specified. Moreover, the output materials from one operating unit can be the input materials for all other operating units.

Process Graph (P-graph) This type of graph is a directed bipartite graph exhibiting two types of vertices or nodes. As depicted in Fig. 4-9, one type, designated by circles, is of the M-type, representing materials, while the other type, designated by horizontal bars, is of the O-type, representing operating units. An arc, with an arrow indicating the direction of flow of a material stream, is either from a vertex (node) representing a material to an operating unit, or vice versa. For convenience, four classes of circles are defined in Fig. 4-9 as symbols for the M-type vertices. Once again, consider the ethylene chlorination reactor mentioned earlier for which $M = \{C_2H_4, Cl_2, m_p\}$; note that this relation is represented in Fig. 4-9a by designating C_2H_4 as A, Cl_2 as B, and m_p as C.

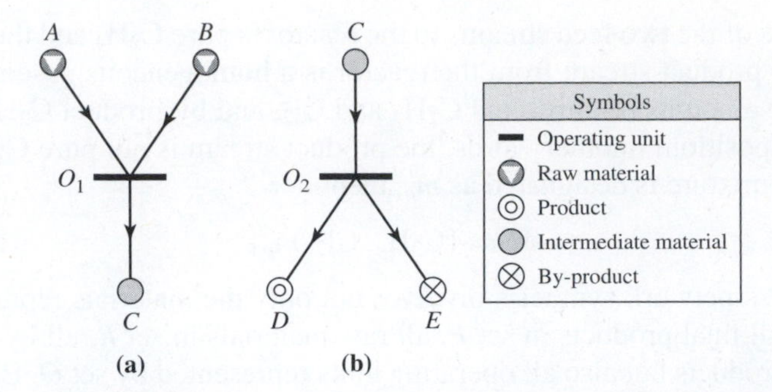

Figure 4-9
P-graphs of some operating units and their concomitant material streams: (a) Materials A, B, and C and operating unit $(\{A, B\}, \{C\}) = O_1$. (b) Materials C, D, and E and operating unit $(\{C\}, \{D, E\}) = O_2$

Problem Definition At the outset of the flowsheeting process, the raw materials as well as the desired final products manufactured from the raw materials are known. The first step of the process-network synthesis (PNS) is to identify all plausible operating units and concomitant intermediate materials that are to be involved in the transformation of the raw materials to the final products. Thus, any PNS problem is symbolically expressed as a synthesis problem (P, R, O), signifying that it is defined upon the specifications of the set of final products P, the set of raw materials R, and the set of all plausible candidate operating units O. Note that the set of all materials M is automatically defined through P, R, and O; that is, all the by-products and intermediate products are specified in conjunction with all the operating units in set O.

The P-graph (M, O) signifies the P-graph representation of a network containing all the materials and operating units that are contained within the synthesis. Similarly, P-graph (m, o) represents the P-graph representation of a network containing part of the materials in M, designated as subset m, and part of the operating units in O, designated as subset o.

Axioms The concept of process graphs (P-graphs) constitutes a cornerstone for the current graph-theoretic or combinatorial approach to process synthesis. The P-graphs facilitate the representation of a process-network structure so that it can be combinatorially manipulated with ease. The basis for the current approach to process synthesis involves a set of five axioms. (An axiom generally is defined as a principle that is accepted without question.) During the development of this algorithmic method, it will become abundantly clear that these five axioms will satisfy this definition. Accordingly, the P-graph (m, o) is defined to be combinatorially feasible or to be a solution-structure of synthesis problem (P, R, O) if it satisfies the following axioms.[†]

S1. Every final product is represented in the graph.

S2. A vertex of the M-type has no input if and only if it represents a raw material.

[†]F. Friedler, K. Tarjan, Y. W. Huang, and L. T. Fan, *Chem. Eng. Sci.,* **47:** 1973 (1992).

S3. Every vertex of the O-type represents an operating unit defined in the synthesis problem.

S4. Every vertex of the O-type has at least one path leading to a vertex of the M-type representing a final product.

S5. If a vertex of the M-type is to be part of the graph, it must be an input to or output from at least one vertex of the O-type in the graph.

The implications of these axioms can be defined in the following statement: axiom S1 implies that each product is produced by at least one of the operating units of the system; axiom S2 says that a material is not produced by any operating unit of the system if and only if this material is a raw material; axiom S3 means that only the plausible operating units of the problem are considered in the synthesis; axiom S4 says that any operating unit of the system has a series of connections eventually leading to the operating unit producing at least one of the products; and axiom S5 states that each material appearing in the system is an input to or an output from at least one operating unit of the system.

The statement of each axiom evidently satisfies the meaning of "axiom" since it is simply a restatement a terminology, for example, raw material or product, the statement of a totally logical relationship, and/or a consequence of the law of conservation of mass. Each axiom by itself may sound innocent and insignificant in practice. Nonetheless, the five axioms collectively act as a filter to eliminate all combinatorially infeasible or invalid networks that are invariably included in the network superstructure. The five axioms together with the P-graph representation of the network structure of a process, that is, the flowsheet, give rise to three algorithms (computing procedures) that make it possible to carry out efficient, automated generation of optimal and all other feasible flowsheets on a computer. These flowsheets can be ranked according to the resultant values of the objective function. The three algorithms are algorithm MSG for the *maximal structure generation,* algorithm SSG for *solution-structure generation,* and algorithm ABB for *accelerated branch and bound.* As denoted in Fig. 4-8, however, algorithm SSG in the network development must be supplemented by a mathematical optimization algorithm.

Algorithm MSG This algorithm for generating the maximal structure is established on the basis of the five axioms presented above. The computational steps required by the algorithm do not exceed $n_m(n_m + 1)n_o$, where n_m and n_o are the numbers of materials and operating units involved in the problem definition, respectively; thus, algorithm MSG is a polynomial algorithm.[†]

The maximal structure of synthesis problem (P, R, O) contains all the combinatorially feasible structures capable of yielding the desired final products from the given raw materials. The maximal structure of synthesis problem (P, R, O) comprises all the combinatorially feasible structures capable of yielding the desired final products from the given raw materials. The implementation of algorithm MSG involves two major phases; for convenience, the initial, or input, structure of the network is constructed at the outset by merging all the common material nodes, each represented by •.

[†]F. Friedler, K. Tarjan, Y.W. Huang, and L. T. Fan, *Comp. Chem. Eng.,* **17:** 929(1993).

In the first phase, the materials and operating units that should not belong to the maximal structure are eliminated, stepwise and layer by layer. The procedure is started from the deepest layer, that is, the raw material end, of the input structure by assessing alternatively the nodes (vertices) in a material layer with those in the succeeding operating-unit layer to ascertain that none of the nodes violates one or more of the five axioms.

In the second phase, the nodes (vertices) are linked, again stepwise and layer by layer, starting from the shallowest end, that is, the final product end, of the remaining input structure by assessing if any of the linked nodes violates one or more of the axioms.

The generation of the maximal structure by algorithm MSG will be illustrated with the synthesis of vinyl chloride.

Algorithm SSG This algorithm is also established on the basis of the five axioms presented earlier. The maximal structure generated by algorithm MSG contains exclusively all the solution structures (combinatorially feasible flowsheets) capable of yielding the desired final products from the given raw materials. These flowsheets range from the simplest to the most complex or complete, as represented by the maximal structure itself. Obviously, the optimal structure in terms of a specific objective function, such as the cost, is included within the maximal structure; nevertheless, the simplest is not necessarily the optimal.

Algorithm SSG makes it possible to generate all the solution-structures (combinatorially feasible flowsheets) of the process to be synthesized. To construct all these combinatorially feasible flowsheets, alternative decisions or decision sequences must be performed for including or excluding the operating units.[†] At least one or a combination of two or more operating units must be included in the structure to generate each product. Similarly, at least one or a combination of two or more additional operating units must be included to generate each material consumed in the structure unless it is a raw material.

A sequence of decisions is consistent unless it contains inconsistent decisions. Inconsistency may result when a certain operating unit is included by a decision and excluded by another decision. To avoid inconsistency, algorithm SSG classifies operating units into three sets: the set of those included in the structure, the set of those excluded from the structure, and the set of free operating units that have not been included in or excluded from the structure. Initially, every operating unit is included in the set of free operating units.

The set of materials whose production is to be decided on is called the *active set*. Initially, the final products are included in the active set. The decision to produce a material in the active set is consummated by including at least one of the candidate operating units capable of producing this material and excluding the others. Any one of the input materials from the operating units included in the structure is contained in the active set unless it is a raw material or a material whose production has already been decided; such a material is removed from the active set.

[†]F. Friedler, J. B. Varga, and L. T. Fan, *Chem. Eng. Sci.*, **50:** 1755 (1995).

If the active set becomes empty as a consequence of consistent decisions, the set of operating units included in the structure represents a solution-structure. A sequence of decisions may end in a contradiction, resulting in the exclusion of all the operating units capable of producing a certain material in the active set. The enumeration of all the alternative consistent decision sequences leads to the complete set of solution-structures (combinatorially feasible flowsheets).

For further elaboration, the generation of solution-structures (combinatorially feasible flowsheets) will be demonstrated again with the synthesis of vinyl chloride.

Optimization and Feasible Flowsheets Each resultant solution-structure (combinatorially feasible flowsheet) can be individually optimized by a linear or nonlinear programming method depending on the linearity or nonlinearity of the objective function, and the mathematical models of the operating units (see Fig. 4-8). In the course of optimization, one or more of the combinatorially feasible flowsheets may fail to yield solutions because, for example, they violate the mass balance constraints that are always linear. In other words, the feasibility of the combinatorially feasible flowsheets is automatically determined in the course of optimization. Obviously, the feasible flowsheets can be ranked in terms of the values of the objective function or profit function simply by comparison. The optimization of combinatorially feasible flowsheets to determine the optimal and other feasible flowsheets will also be illustrated with the synthesis of vinyl chloride.

Algorithm ABB When the number of solution-structures generated by algorithm SSG is exceedingly large, their optimization can indeed be time-consuming. In practice, only a limited number of the optimal and near optimal flowsheets are of interest to the designer. The elimination of the remaining flowsheets is accomplished with the use of algorithm ABB.[†] This algorithm is developed by modifying conventional branch-and-bound algorithms. However, algorithm ABB is implemented quite differently from the conventional branch-and-bound methods.

The conventional branch-and-bound methods encounter deficiencies in process synthesis because they often give rise to an unduly large number of networks, the majority of which is redundant for any sizable process. The reason is that the methods do not exploit the structural features of the process system. In contrast, algorithm ABB judiciously exploits the structural features of the process to be synthesized; consequently, no redundant network is generated.

Application of Algorithmic Process-Network Synthesis to Vinyl Chloride Production

The algorithmic method for process-network synthesis presented up to this point is illustrated by applying it to the structural flowsheet optimization of the balanced process for producing vinyl chloride. It is worth recalling that the method consists mainly of the

[†]F. Friedler, J. B. Varga, E. Feher, and L. T. Fan, *State of the Art in Global Optimization, Computational Methods and Applications,* C. A. Floudas and P. M. Pardalos, eds., Kluwer, Norwell, MA, 1996, pp. 609–626.

P-graph representation of process networks (flowsheets), a set of the five axioms pertaining to the unique structural features of such networks, and the algorithms based on the P-graphs and axioms. The major steps for implementing the method are shown in Fig. 4-8, which indicates that the first main step involves reaction-network synthesis.

Earlier work has established that reaction-network synthesis must precede process-network synthesis for a chemical process involving multiple chemical reactions.[†] These multiple reactions invariably give rise to multiple reaction networks, representing alternative synthetic routes (reaction paths), even for a specific set of reactants and final products. Naturally, it is anticipated that different process networks (flowsheets) are required to implement such alternative synthesis routes, none of which should be prematurely eliminated in the structural flowsheet optimization process.

Qualitatively speaking, the outline of a flowsheet without ancillary facilities, such as pumps and heat exchangers, can be generated from a reaction network. This is carried out by replacing each reaction required with a reacting unit, with one or more reactors, each followed by a separating unit with one or more separators to separate desired products from intermediates, by-products, and trace impurities. The intermediates are fed to the succeeding reacting unit.[‡] This implies that the generation of reaction networks corresponding to the alternative synthetic routes (reaction paths) facilitates the selection of operating units for inclusion in the flowsheets.

The present algorithmic method for process-network synthesis is applied straightforwardly and rigorously to reaction-network synthesis as indicated in Fig. 4-8. The notion of hardware does not come into play in reaction-network synthesis. Nevertheless, a reaction can be regarded as an operating unit identified as a functional unit transforming reacting materials expressed in integer molar units. The reaction products leave the operating unit, again expressed in integer molar units according to the stoichiometry of the reaction. Since the mathematical framework for the reaction-network synthesis is identical to that for the process-network synthesis, the present algorithmic method can be adopted as well for the former synthesis. In other words, the reaction networks can be represented by P-graphs. Thus, the feasible reaction networks can be synthesized by means of algorithms MSG and SSG followed by the linear programming optimization of combinatorially feasible reaction networks to determine their feasibility, as indicated in the first main block of Fig. 4-8.

Reaction-Network Synthesis The three reactions involved in the balanced process for manufacturing vinyl chloride, previously noted in Example 4-3, are repeated here for convenience (but the coefficients are doubled to eliminate coefficients less than 1):

$$\begin{array}{ll} r_1: & 2C_2H_4 + 2Cl_2 \longrightarrow 2C_2H_4Cl_2 \\ r_2: & 2C_2H_4 + 4HCl + O_2 \longrightarrow 2C_2H_4Cl_2 + 2H_2O \\ r_3: & 4C_2H_4Cl_2 \longrightarrow 4C_2H_3Cl + 4HCl \\ \hline \text{Overall:} & 4C_2H_4 + 2Cl_2 + O_2 \longrightarrow 4C_2H_3Cl + 2H_2O \end{array}$$

[†]D. F. Rudd, G. J. Powers, and J. J. Siirola, *Process Synthesis,* Prentice-Hall, Englewood Cliffs, NJ, 1973, pp. 40–45; and A. Lakshmanan and L. T. Biegler, *Computer & Chem. Eng.,* **21:** S785 (1997).

[‡]J. M. Douglas, *Conceptual Design of Chemical Processes,* McGraw-Hill, New York, 1988.

Note that the overall reaction contains only the reactants and final reaction products.

Problem Definition The problem of the reaction-network synthesis, denoted as synthesis problem (P, R, O), is defined by identifying the set of products P comprising the desired final reaction products, the set of reactants R comprising the starting reactants (precursors), and the set of operating units O comprising the set of reactions leading to the overall reaction. For the current example,

$$P = \{C_2H_3Cl\}$$
$$R = \{C_2H_4, Cl_2, O_2\}$$
$$O = \{r_1, r_2, r_3\}$$
$$= \{(\{C_2H_4, Cl_2\}, \{C_2H_4Cl_2\}), (\{C_2H_4, HCl, O_2\}, \{C_2H_4Cl_2, H_2O\}), (\{C_2H_4Cl_2\}, \{C_2H_3Cl, HCl\})\}$$

Note that

$$(P, R, O) = (\{C_2H_3Cl\}, \{C_2H_4, Cl_2, O_2\}, \{r_1, r_2, r_3\})$$

and

$$M = \Big\{ \overset{\text{starting reactants}}{\underset{\text{raw materials}}{C_2H_4, Cl_2, O_2}} \quad \overset{\text{final products}}{\underset{\text{desired product and by-product}}{C_2H_3Cl, H_2O}} \quad \overset{\text{intermediates}}{\underset{\text{intermediate materials}}{C_2H_4Cl_2, HCl}} \Big\}$$

Graph Representation Figure 4-10 depicts P-graphs of the three reactions r_1, r_2, and r_3 that are involved. The maximal structure representing the maximal reaction network is generated from these P-graphs.

Generation of the Maximal Structure Corresponding to the Maximal Reaction Network with Algorithm MSG As noted previously, two phases are involved in generating the maximal reaction network as the maximal structure with algorithm MSG. This structure contains the three reactions as the operating units. The two phases are the

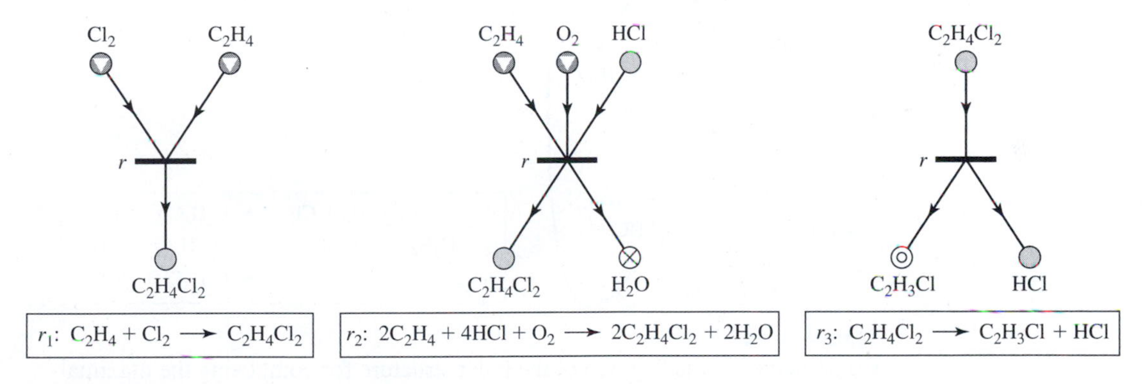

Figure 4-10
P-graphs of reactions r_1, r_2, and r_3 regarded as operating units

elimination of the invalid vertices (nodes) from the input structure and the composition of the maximal structure.

The input structure is generated by merging all the common vertices for each material, that is, the components participating in the reactions, each represented by • in the P-graphs of reactions r_1, r_2, and r_3. Referring to Fig. 4-10, the procedure is as follows:

Step 1. Merge the vertices for C_2H_4 in r_1 and r_2.
Step 2. Merge the vertices for $C_2H_4Cl_2$ in r_1, r_2, and r_3.
Step 3. Merge the vertices for HCl in r_2 and r_3.

Since no additional common material vertices exist, the resultant input structure appears as in Fig. 4-11.

a. *Elimination of the invalid vertices from the input structure.* This task is accomplished by assessing each vertex (node) in light of the five axioms. Axiom S1 pertains only to the vertices for the desired final products, axioms S2 and S5 apply to the vertices of the *M*-type for the materials, and axioms S3 and S4 apply to the vertices of the *O*-type for the operating units. The elimination is executed stepwise, starting from the raw material (starting reactant) end of the P-graph of the input structure.

Step 1. The vertices for the starting reactants Cl_2, C_2H_4, and O_2 are not eliminated: These vertices do not violate axioms S2 and S5; note that each of the vertices has no input, but is an input to the vertices for reactions r_1 and/or r_2.
Step 2. The vertex for HCl is not eliminated: This vertex does not violate axioms S2 and S5; observe that this vertex has an input and also is an input to reaction r_2.

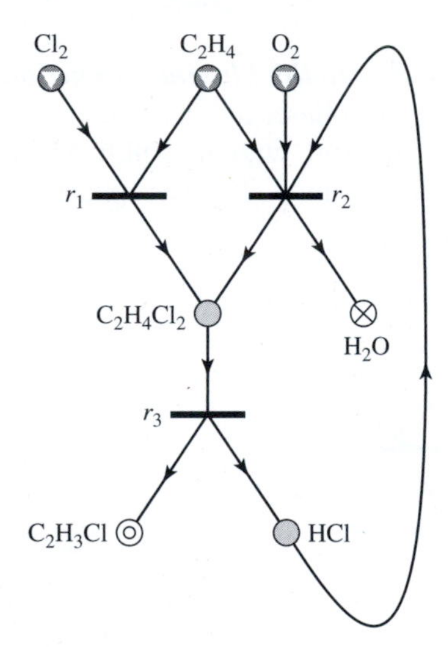

r_1:	$C_2H_4 + Cl_2 \longrightarrow C_2H_4Cl_2$
r_2:	$2C_2H_4 + 4HCl + O_2 \longrightarrow 2C_2H_4Cl_2 + 2H_2O$
r_3:	$C_2H_4Cl_2 \longrightarrow C_2H_3Cl + HCl$

Figure 4-11
Construction of the P-graph of the input structure for composing the maximal structure. This is also the final maximal structure, that is, the maximal reaction network.

Step 3. The vertices for reactions r_1 and r_2 are not eliminated: These vertices do not violate axioms S3 and S4; each of these vertices has one path leading to the desired final product C_2H_3Cl, and both are defined in the synthesis problem.

Step 4. The vertices for $C_2H_4Cl_2$ and H_2O are not eliminated: These vertices do not violate axioms S2 and S5; each of the vertices has an input and also is an output from the vertices for reactions r_1 and/or r_2.

Step 5. The vertex for reaction r_3 is not eliminated: This vertex does not violate axioms S3 and S4; observe that the vertex has one path leading to the desired product C_2H_3Cl.

Step 6. The vertex for C_2H_3Cl is not eliminated: This vertex does not violate axioms S2 and S5; in this case the vertex has an input and is also an output from the vertex for reaction r_3.

For the relatively simple application under consideration, the structure resulting from the elimination phase is identical to the input structure. This is not always the case.

b. *Composition of the maximal structure* In general, one or more of the feasible paths and valid vertices in the input structure may disappear if some of the invalid vertices are eliminated. Thus, the final maximal structure is composed or reconstructed from the remaining parts of the input structure after the elimination has been completed. This is accomplished stepwise by alternately linking the vertices of the M-type for materials to the vertices of the O-type for operating units, and vice versa. Similar to the elimination phase, the vertices that are linked are assessed in view of the five axioms at each step. The vertices of the M-type must satisfy axioms S1, S2, and S5; those of the O-type must satisfy axioms S3 and S4. The execution is initiated from the desired final product end since this is the shallowest layer of the structure. The stepwise procedure for the composition is illustrated in Fig. 4-12.

Again, because of the simplicity of the vinyl chloride problem, the maximal structure resulting from the composition is identical to the structure resulting from the elimination, which in turn is identical to the input structure. This maximal structure is shown in Fig. 4-11, which contains all the solution structures or the combinatorially feasible reaction networks.

Generation of the Solution-Structures Corresponding to the Combinatorially Feasible Reaction Networks as the Solution-Structures with Algorithm SSG This is executed by a series of decisions on the production of the materials included in the active sets identified in the P-graph of the maximal structure representing the maximal reaction network. Recall that algorithm SSG in general and the decisions on the production have been described in a preceding subsection.

The matrix given in Tables 4-2a through 4-2c is constructed by referring to Fig. 4-11 to demonstrate the identification of the active set in each step and the implementation of algorithm SSG to generate the solution-structure (combinatorially feasible reaction networks) on the P-graph for the maximal structure (maximal reaction network). Figure 4-13 illustrates the search tree for implementing algorithm SSG.

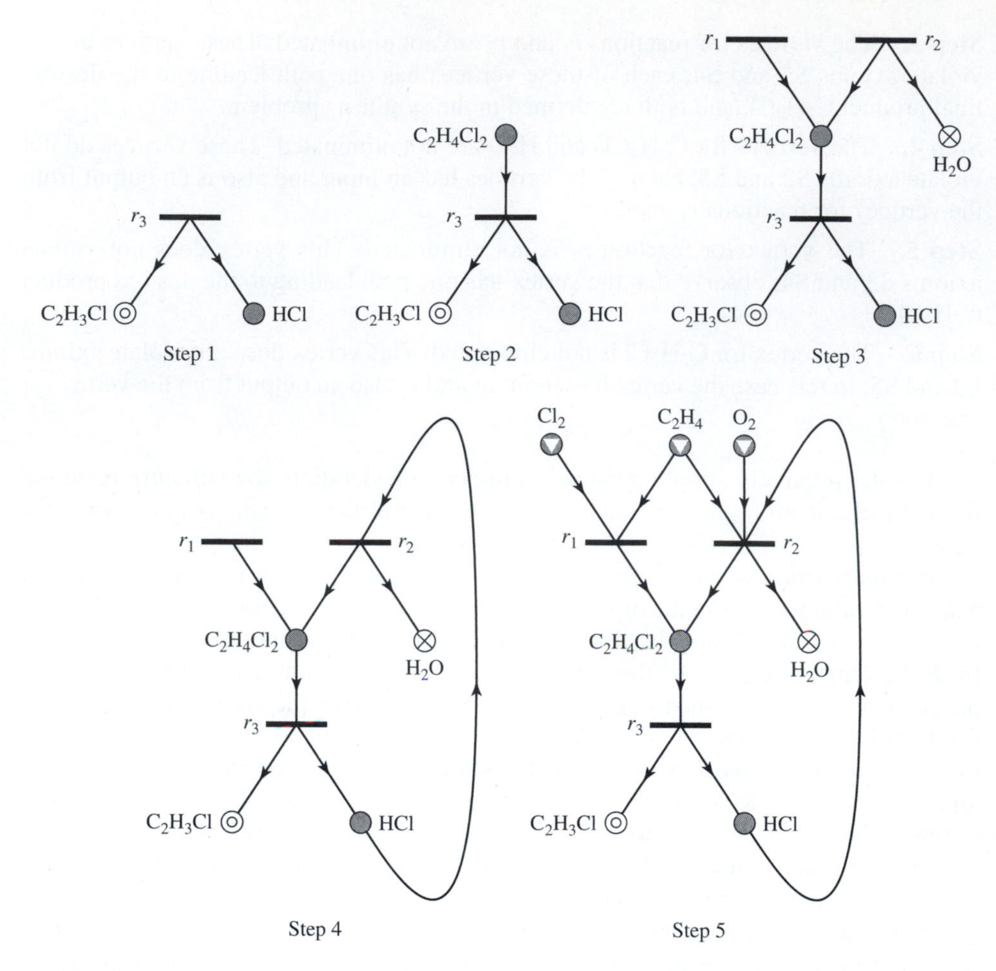

Figure 4-12
Steps for the composition of the maximal structure representing the maximal reaction network

Initially, only the final product C_2H_3Cl is included in the set of active materials whose production is to be decided. Moreover, no reaction is included or excluded; that is, all reactions are considered free. At this juncture, the active set has the single element C_2H_3Cl; it can only be produced by reaction r_3, thereby giving rise to no alternative decision on its production. In the first step, reaction r_3 ($C_2H_4Cl_2 \rightarrow C_2H_3Cl + HCl$) is included to produce C_2H_3Cl, and its input material $C_2H_4Cl_2$ becomes active; that is, it becomes a new element in the active set. This material can be produced by reaction r_1 only, by reaction r_2 only, or by both reactions r_1 and r_2. These three decision alternatives are enumerated in the second, third, and fifth steps. Both of the input materials of reaction r_1 ($C_2H_4 + Cl_2 \rightarrow C_2H_4Cl_2$) are raw materials. After including r_1 to produce $C_2H_4Cl_2$ and excluding r_2, the active set becomes empty, thereby indicating

Table 4-2a Steps of algorithm SSG for generating solution-structure S1

Step	Operating units			Materials	
	Included	Free	Excluded	Active	Decided
1	—	r_1, r_2, r_3	—	C_2H_3Cl	—
2	r_3	r_1, r_2	—	$C_2H_4Cl_2$	C_2H_3Cl
	r_1, r_3	—	r_2	—	C_2H_3Cl, $C_2H_4Cl_2$

Table 4-2b Steps of algorithm SSG for generating solution-structure S2

Step	Operating units			Materials	
	Included	Free	Excluded	Active	Decided
1	—	r_1, r_2, r_3	—	C_2H_3Cl	—
3	r_3	r_1, r_2	—	$C_2H_4Cl_2$	C_2H_3Cl
	r_2, r_3	—	r_1	HCl	C_2H_3Cl, $C_2H_4Cl_2$
4	r_2, r_3	—	r_1	—	C_2H_3Cl, $C_2H_4Cl_2$, HCl

Table 4-2c Steps of algorithm SSG for generating solution-structure S3

Step	Operating units			Materials	
	Included	Free	Excluded	Active	Decided
1	—	r_1, r_2, r_3	—	C_2H_3Cl	—
5	r_3	r_1, r_2	—	$C_2H_4Cl_2$	C_2H_3Cl
	r_3, r_2, r_3	—	—	HCl	C_2H_3Cl, $C_2H_4Cl_2$
6	r_3, r_2, r_3	—	—	—	C_2H_3Cl, $C_2H_4Cl_2$, HCl

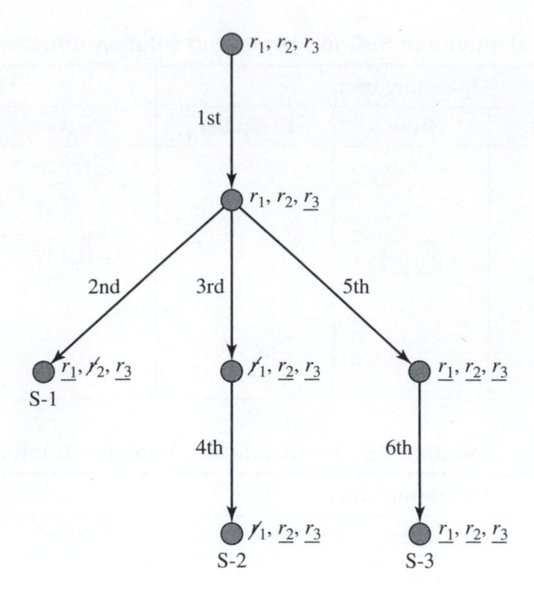

Figure 4-13
Search tree for implementing algorithm SSG to generate the solution-structures, i.e., combinatorially feasible reaction networks: S1, S2, and S3 represent the three solution-structures

that a solution-structure is generated in the second step. After including reaction r_2 ($2C_2H_4 + 4HCl + O_2 \rightarrow 2C_2H_4Cl_2 + 2H_2O$) in both the third and fifth steps, material HCl becomes active since it is not a raw material. In the fourth and sixth steps, reaction r_3 is included to produce HCl consistent with the decision in the first step; therefore, each of these steps yields a solution-structure. Figure 4-14 shows the three resultant solution-structures (combinatorially feasible reaction networks).

Identification of the Feasible Reaction Networks Among the Combinatorially Feasible Ones by Linear Programming A combinatorially feasible reaction network is not necessarily feasible since it may violate mass balance in terms of molar balance. Its feasibility can be conveniently tested by linear programming or, more specifically, integer linear programming (ILP), yielding only the integer solution.

Since no objective function can be subjectively defined in adapting the algorithmic method of process-network synthesis for the reaction-network synthesis, the feasibility of any combinatorially feasible reaction network generated by algorithm SSG must be individually assessed in view of its mass balances, specifically in terms of molar balances. (The mass of a molar species in the stoichiometric expression of any chemical reaction is given in the molar unit.) The molar balances are obtained by multiplying the reactions (reaction steps) in the network with unknown positive integer multipliers

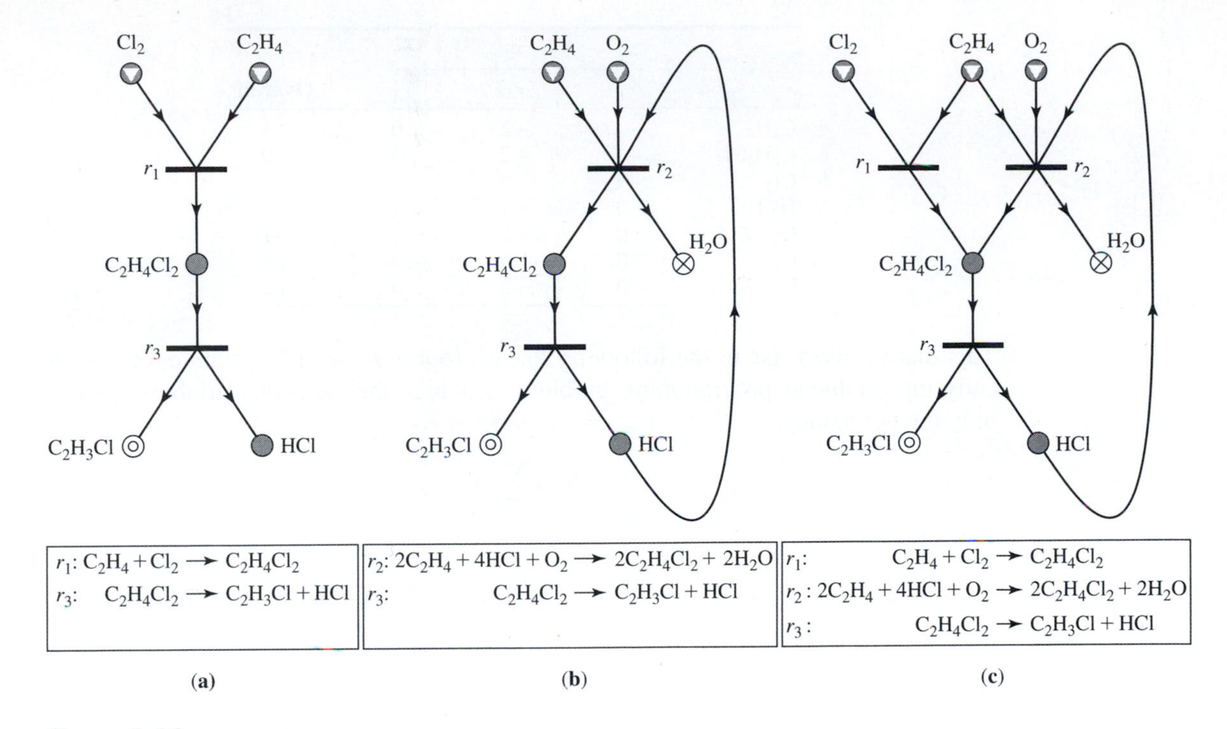

Figure 4-14
Solution structures, that is, combinatorially feasible reaction networks: only (c) is feasible, and thus it is a feasible reaction network

that are manipulated so that the sum of these reactions leads to the overall reaction. Moreover, the sum should be a minimum. (This becomes apparent by recognizing that $2A + 4B \rightarrow 2C$ is equivalent to $A + 2B \rightarrow C$.) Thus, such multipliers can be efficiently determined by minimizing their sum by integer linear programming (ILP) subject to the molar balance constraints. A network failing to yield a solution is infeasible. For illustration, consider the combinatorially feasible reaction network in Fig. 4-14c.

By denoting the positive integer by which reaction r_i $(i = 1, 2, 3)$ is to be multiplied as x_i, we can formulate the problem as

$$
\begin{array}{ll}
x_1 \times r_1 & C_2H_4 + Cl_2 \rightarrow C_2H_4Cl_2 \\
x_2 \times r_2 & 2C_2H_4 + 4HCl + O_2 \rightarrow 2C_2H_4Cl_2 + 2H_2O \\
x_3 \times r_3 & C_2H_4Cl_2 \rightarrow C_2H_3Cl + HCl \\
\hline
\text{Overall} & 4C_2H_4 + 2Cl_2 + O_2 \rightarrow 4C_2H_3Cl + 2H_2O
\end{array}
$$

For convenience, the above expressions are transformed into the following matrix in which the minus sign represents consumption and the plus sign represents generation.

	x_1	x_2	x_3	
	r_1	r_2	r_3	**Overall**
C_2H_4	−1	−2	0	−4
$C_2H_4Cl_2$	+1	+2	−1	0
Cl_2	−1	0	0	−2
HCl	0	−4	+1	0
O_2	0	−1	0	−1
C_2H_3Cl	0	0	+1	+4
H_2O	0	+2	0	+2

This matrix gives rise to the following linear programming (LP) problem or specifically integer linear programming problem, in which the decision variables x_i's take only integer values.

$$\text{Min} \sum_{i=l}^{3} x_i$$

subject to

$$x_1(-1, 1, -1, 0, 0, 0, 0) + x_2(-2, 2, 0, -4, -1, 0, 2) + x_3(0, -1, 0, 1, 0, 1, 0)$$
$$= (-4, 0, -2, 0, -1, 4, 2)$$

or

$$\begin{aligned} -x_1 - 2x_2 \quad &= -4 \\ x_1 + 2x_2 - x_3 &= \ \ 0 \\ -x_1 \quad &= -2 \\ -4x_2 + x_3 &= \ \ 0 \\ -x_2 \quad &= -1 \\ x_3 &= \ \ 4 \\ 2x_2 \quad &= \ \ 2 \end{aligned}$$

The solution is $x_1 = 2$, $x_2 = 1$, and $x_3 = 4$. Thus,

r_1: $2C_2H_4 + 2Cl_2 \rightarrow 2C_2H_4Cl_2$

r_2: $2C_2H_4 + 4HCl + O_2 \rightarrow 2C_2H_4Cl_2 + 2H_2O$

r_3: $4C_2H_4Cl_2 \rightarrow 4C_2H_3Cl + 4HCl$

This set of reactions yields the overall reaction for the balanced process

$$\text{Overall:} \quad 4C_2H_4 + 2Cl_2 + O_2 \rightarrow 4C_2H_3Cl + 2H_2O$$

It can be easily shown that the combinatorially feasible reaction networks in Figs. 4-14a and 4-14b do not yield positive integer solutions for x_i's, thus indicating that these networks do not satisfy the molar balance constraints and therefore are infeasible.

Process-Network Synthesis As described earlier, a feasible reaction network guides the initial selection of major operating units that are reacting units, each of which is followed by a separating unit. With such a network in hand, we are in a position to proceed to the next phase of the algorithmic method corresponding to the second main block in Fig. 4-8. In this phase, the operating units that are possible, or plausible, for inclusion in the flowsheet are identified to the maximum extent possible; each operating unit consists of one or more processing units or chemical reactors. The identification of the operating units is accomplished through an extensive online and/or offline search of available technical information pertaining to the process being synthesized. Such information is obtainable from various sources, as outlined in Chap. 3.

Often, the necessary information or data need to be generated through computer-aided simulation based on first principles, semiempirical and empirical correlations of the existing data, and/or heuristics. This is especially the case when synthesizing a novel process. In contrast, much of the necessary information and data is already available when an existing process is examined to assess its structural optimality, as in the current application to the production of vinyl chloride.

Identification of Plausible Operating Units This phase of the present algorithmic method is indicated in the first subblock of the second main block in Fig. 4-8. According to the feasible reaction network for the balanced production of vinyl chloride, as shown in Fig. 4-11 or 4-14c, it is highly likely that any of the plausible flowsheets would incorporate three reacting units, each followed by a separating unit. The three reacting units are designated as R-1 for carrying out reaction r_1, R-2 for reaction r_2, and R-3 for reaction r_3. The corresponding three separating units are designated as S-1, S-2, and S-3, respectively.

The output from reacting unit R-3 for pyrolyzing diethylene chloride ($C_2H_4Cl_2$) to yield vinyl chloride (C_2H_3Cl) contains a substantial amount of unreacted $C_2H_4Cl_2$ besides trace amounts of by-products. The equilibrium of the pyrolysis reaction does not favor its completion. Hence, it is logical to purify $C_2H_4Cl_2$ to be recycled to the inlet of R-3 from S-3 together with $C_2H_4Cl_2$ from R-1 and R-2. This is accomplished by locating S-1 behind S-2. The latter removes H_2O, HCl, and trace amounts of excess O_2 from the main stream of $C_2H_4Cl_2$ and becomes suitable as the feed to S-1.

In addition to $C_2H_4Cl_2$, S-3 separates HCl for recycling to R-2 and C_2H_3Cl as the desired product. For convenience, it is visualized that a mixing unit identified as M-1 precedes R-3; this mixing unit functionally represents inevitable in-line mixing of the three streams of low-purity $C_2H_4Cl_2$ from R-1, S-2, and S-3 prior to providing the feed to R-3.

The plausible operating units mentioned above are shown in the forms of P-graphs in Fig. 4-15. Table 4-3 presents the compositions of the input streams and output streams from each operating unit with flow rates of components in these streams similar to those observed in industrial production.[†]

[†]F. Borelli, *Encyclopedia of Chemical Processing and Design,* vol. 62, J. J. McKetta and G. E. Weismantel, eds., Marcel Dekker, New York, 1998, p. 313; M. Turkay and I. E. Grossmann, *Computer & Chem. Eng.,* **22:** 673 (1998); A. Lakshmanan, W. C. Rooney, and L. T. Biegler, *A Case Study for Reaction Network Synthesis, An Aspen Plus Simulation Case Study,* Aspen Technology, Cambridge, MA, 1999.

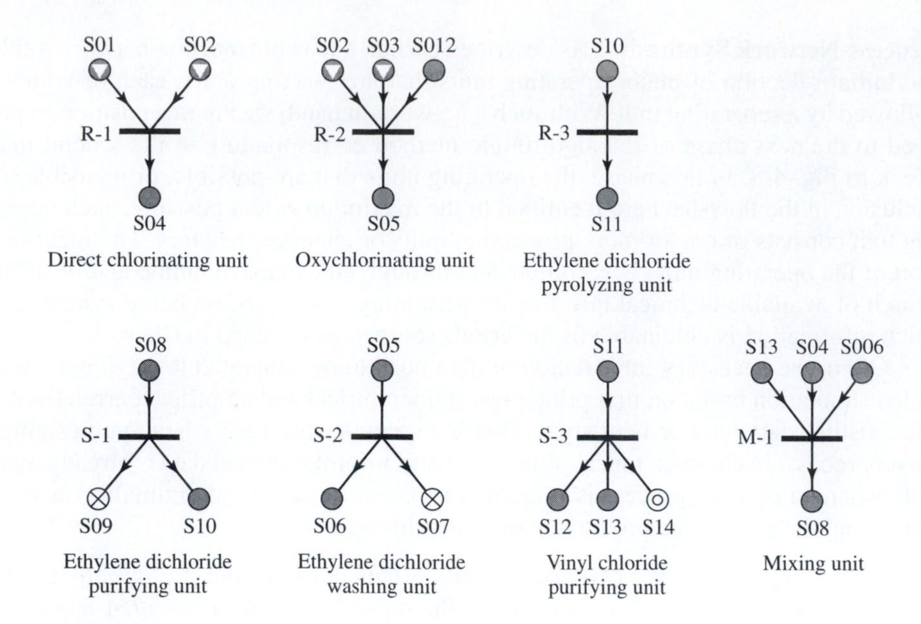

Figure 4-15
P-graph representations of plausible operating units identified for the balanced process to manufacture vinyl chloride

The mass balances around each operating unit are based on a production rate of 4535 kg/h of vinyl chloride. The mass balances have been computed from stream composition data also reported in the literature based on the production of 1 kg mol of vinyl chloride. The computations for the mass balances have been carried out to four to six decimal places to ensure the inclusion of all trace components. Such knowledge is essential from the standpoint of emission control and pollution prevention. However, for estimating the capacities or sizes of processing equipment and reactors in the operating units, such computational specificity generally is not necessary. For completeness the included mass balance solution tables meet these computational precision levels.

Generation of the Maximal Structure with Algorithm MSG Figure 4-8 indicates that once the plausible operating units have been identified, the maximal structure of the process is generated with algorithm MSG. The general procedure involved in this task was outlined in a prior section and also illustrated with the reaction-network synthesis for vinyl chloride that resulted in the development of Fig. 4-11. At the outset, the input structure is constructed from the P-graphs of plausible operating units, as illustrated in Fig. 4-15. This is executed by merely connecting all the common vertices (nodes) of the M-type (nodes denoted by symbols Sij, where $i = 0, 1$ and $j = 1, 2, \ldots, 5$), as demonstrated in the generation of the reaction network. The resultant input structure is illustrated in Fig. 4-16. The procedure then goes forward in two phases.

a. *Elimination of the invalid vertices from the input structure* As demonstrated earlier with the reaction network generation in conjunction with the development of

Table 4-3 Compositions of the input streams to and the output streams from plausible operating units of the balanced process to manufacture vinyl chloride in terms of the flow rates†

Material stream		Material species			
				Flow rate	
Designation	Type	Component		kg mol/h	kg/h
R-1 (direct chlorinating unit)					
S01	Input	Cl_2		38.32224	2,720.87904
			Total	38.32224	2,720.87904
S02	Input	C_2H_4		38.68514	1,083.18392
			Total	38.68514	1,083.18392
S04	Output/(input)	C_2H_4		0.3629	10.1612
		$C_2H_4Cl_2$		37.81418	3,743.60382
		$C_2H_3Cl_3/Cl_2$		Trace	
			Total	38.18912	3,754.6930
R-2 (oxychlorinating unit)					
S02	Input	C_2H_4		34.91098	977.5074
			Total	34.91098	977.5074
S03	Input	O_2		17.49178	559.737
			Total	17.49178	559.737
S12	(Output)/input	HCl		72.58	2,649.17
			Total	72.58	2,649.17
S05	(Output)/input	$C_2H_4Cl_2$		34.8384	3,449.002
		HCl		2.75804	100.6685
		H_2O		34.8384	627.0912
		$CO_2/C_2H_4/O_2$		Trace	
			Total	72.54371	4,180.094
S-2 (ethylene dichloride washing unit)					
S05	(Output)/input	$C_2H_4Cl_2$		34.8384	3,449.002
		HCl		2.75804	100.6685
		H_2O		34.8384	627.0912
		$C_2H_4/CO_2/O_2$		Trace	
			Total	72.5437	4,180.094
S06	Output/(input)	C_2H_4		0.07258	2.03224
		$C_2H_4Cl_2$		34.8384	3,449.002
		O_2/CO_2		Trace	
			Total	34.9328	3,451.87
S07	Output	HCl		2.75804	100.6685
		H_2O		34.8384	627.0912
		O_2		Trace	
			Total	37.611	728.2242
M-1 (mixing unit)					
S04	(Output/input)	C_2H_4		0.3629	10.1612
		Cl_2		0.01089	0.77297
		$C_2H_4Cl_2$		37.8142	3,743.604
		$C_2H_3Cl_3$		0.00116	0.155031
			Total	38.1891	3,754.693

(Continued)

Table 4-3 *Continued*

Material stream		Material species		
			Flow rate	
Designation	Type	Component	kg mol/h	kg/h
S06	(Output)/input	C_2H_4	0.07258	2.03224
		$C_2H_4Cl_2$	34.8384	3,449.002
		O_2	0.01016	0.325158
		CO_2	0.01016	0.510963
		Total	34.9328	3,451.87
S13	(Output)/input	C_2H_4	0.00051	0.014226
		$C_2H_4Cl_2$	48.6286	4,814.231
		$C_2H_2/C_4H_6/C_2H_2/Cl_2$	Trace	
		Total	48.6312	4,814.37
S08	(Output)/input	C_2H_4	0.50806	14.22568
		Cl_2	0.01089	0.772977
		$C_2H_4Cl_2$	121.281	12,006.84
		O_2	0.01016	0.325158
		CO_2	0.01016	0.510963
		$C_2H_3Cl_3$	0.00116	0.155031
		$C_2H_2/C_4H_6/C_2H_2/Cl_2$	Trace	
		Total	121.825	12,022.95
S-1 (ethylene dichloride purifying unit)				
S08	(Output)/input	C_2H_4	0.50806	14.22568
		$C_2H_4Cl_2$	121.28118	12,006.836
		$Cl_2/O_2/CO_2/ C_2H_3Cl_3/ C_2H_2,C_4H_6,C_2H_2Cl_2$	Trace	
		Total	121.82517	12,022.951
S09	Output	C_2H_4	0.50806	14.22568
		$C_2H_4Cl_2$	0.07258	7.18542
		O_2, CO_2, $C_2H_2,C_4H_6,C_2H_2Cl_2$	Trace	
		Total	0.6045189	22.371406
S10	Output/(input)	$C_2H_4Cl_2$	121.2086	11,999.651
		C_2H_4	Trace	
		Total	121.20912	11,999.666
R-3 (ethylene dichloride pyrolyzing unit)				
S10	(Output)/input	$C_2H_4Cl_2$	121.2086	11,999.651
		C_2H_4	Trace	
		Total	121.2091	11,999.666
S11	Output/(input)	$C_2H_4Cl_2$	48.6286	4,814.2314
		HCl	72.58	2,649.17
		C_2H_3Cl	72.58	4,536.25
		$C_2H_4/C_2H_2,C_4H_6, C_2H_2Cl_2$	Trace	
		Total	193.79121	11,999.79

Table 4-3 *Continued*

Material stream		Material species		
			Flow rate	
Designation	Type	Component	kg mol/h	kg/h
S-3 (vinyl chloride purifying unit)				
S11	(Output)/input	$C_2H_4Cl_2$	48.6286	4,814.2314
		HCl	72.58	2,649.17
		C_2H_3Cl	72.58	4,536.25
		$C_2H_4/C_2H_2, C_4H_6, C_2H_2Cl_2$	Trace	
		Total	193.79121	11,999.790
S12	Output/(input)	HCl	72.58	2,649.17
		Total	72.58	2,649.17
S13	Output/(input)	$C_2H_4Cl_2$	48.6286	4,814.2314
		$C_2H_4/C_2H_2, C_4H_6, C_2H_2Cl_2$	Trace	
		Total	48.631213	4,814.3698
S14	Output	C_2H_3Cl	72.58	4,536.25
		Total	72.58	4,536.25

[†]Flow rates are based on a production capacity of 4,535 kg/h of vinyl chloride using the balanced process. The stream type (output)/input signifies that the stream is the output from another operating unit but is the input to this operating unit; conversely, output/(input) indicates that the stream is the output of this operating unit but is the input to another operating unit. The same statement applies to all other operating units.

Fig. 4-11, the vertices (nodes) for materials and operating units that should not be in the maximal structure are removed from the input structure. This is accomplished, step by step, starting at the raw material end of the structure while applying the five axioms to the feasibility of the combinatorial structures. As expected, no vertices are eliminated from the P-graph of the well-established process to produce vinyl chloride, whose structure is relatively simple. In fact, it is highly likely that the structure of the process, as represented by the flowsheet, has been optimized during its long existence. However, this may not always be the case even for a process which has been in operation for many years, if it is complex. In the application of the present method to such a process, some redundant operating units have been identified.[†]

b. *Composition of the maximal structure* Since none of the operating units are eliminated from the input structure in Fig. 4-16, the maximal structure is developed from these operating units. The elimination is carried out stepwise according to the procedure outlined in Fig. 4-12 for constructing the maximal reaction network. The nodes (vertices) of the M-type for materials are alternately connected to the vertices of the O-type for operating units, and vice versa. Starting from the vertex for the desired

[†]F. Friedler, K. Tarjan, Y. W. Huang, and L. T. Fan, *Computers & Chem. Eng.*, **16:** S313 (1992); and F. Friedler, K. Tarjan, Y. W. Huang, and L. T. Fan, *Chem. Eng. Sci.*, **47:** 1973 (1992).

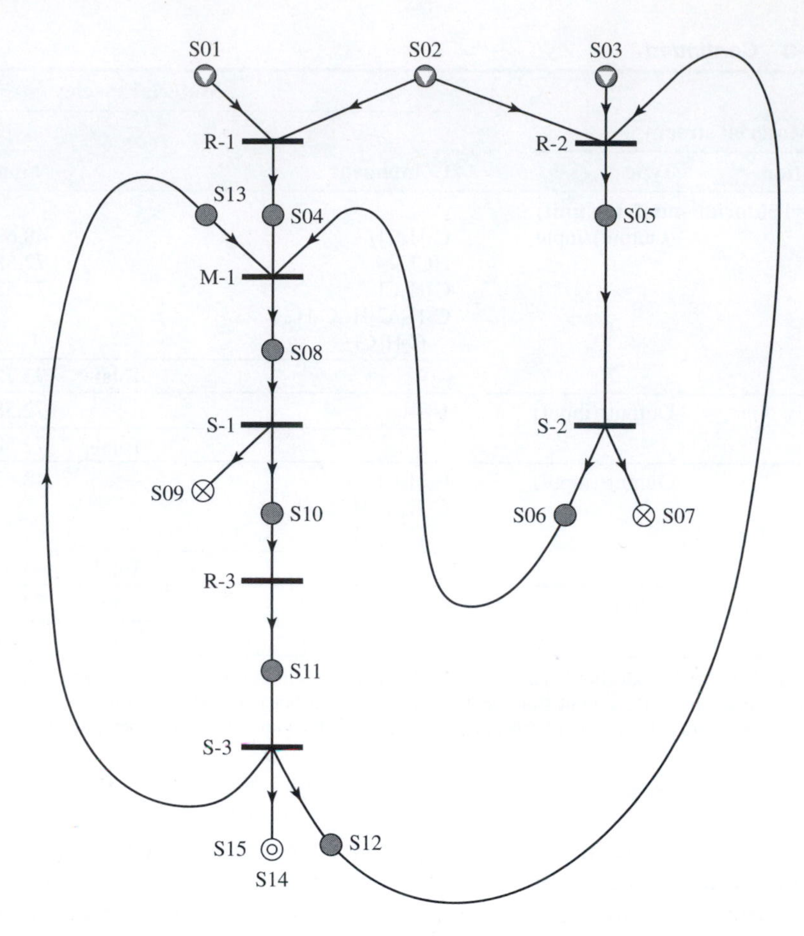

Figure 4-16
Input structures for the maximum-structure generation for the balanced process to manufacture vinyl chloride. This is also the maximal structure as well as the only solution-structure (combinatorially feasible flowsheet): each operating unit consists of one or more types of processing equipment jointly performing a specific task

product in stream S14, and proceeding upward toward the vertices for the raw materials (Cl_2 in stream S02, C_2H_4 in stream S02, and O_2 in stream S03). At every step, each vertex that is connected is examined if it violates any of the five axioms of a combinatorially feasible process network. As expected and also indicated on Fig. 4-16, the input structure turns out to be the maximal structure for this structurally simple and mature process.

Generation of the Solution-Structures Corresponding to the Combinatorially Feasible Process Networks with Algorithm SSG According to Fig. 4-8, the solution-structures, that is, the combinatorially feasible process networks (flowsheets), are generated by means of algorithm SSG. The procedure for implementing algorithm SSG is described in an earlier subsection on fundamentals and then described in greater detail in the reaction-network synthesis for vinyl chloride synthesis by resorting to Table 4-2a through 4-2c and Fig. 4-13. The procedure gives rise to only one solution structure (combinatorially feasible flowsheet) identical to the input or maximal structure, as indicated in Fig. 4-16. The matrix given in Table 4-4 is constructed by referring to Fig. 4-16 to demonstrate the identification of the active set in each step and the implementation of algorithm SSG to generate this single solution-structure. This flowsheet is also given in Fig. 4-17 as a conventional block flow diagram.

Table 4-4 Steps of algorithm SSG for generating the single combinatorially feasible process network identical to the maximal structure

Step	Operating units			Materials	
	Included	Free	Excluded	Active	Decided
	—	R-1, R-2, R-3, M-1 S-1, S-2, S-3	—	S14	—
1	S-3	R-1, R-2, R-3, M-1, S-1, S-2	—	S11	S14
2	R-3, S-3	R-1, R-2, M-1, S-1, S-2	—	S10	S11, S14
3	R-3, S-1, S-3,	R-1, R-2, M-1, S-2	—	S08	S10, S11, S14
4	R-3, M-1, S-1, S-3	R-1, R-2, S-2	—	S04, S06, S13	S08, S10, S11, S14
5	R-3, M-1, S-1, S-3	R-1, R-2, S-2	—	S04, S06	S08, S10, S11, S13, S14
6	R-3, M-1, S-1, S-2, S-3	R-1, R-2	—	S04, S05	S06, S08, S10, S11, S13, S14
7	R-2, R-3, M-1, S-1, S-2, S-3	R-1	—	S04, S12	S05, S06, S08, S10, S11, S13, S14
8	R-2, R-3, M-1, S-1, S-2, S-3	R-1	—	S04	S05, S06, S08, S10, S11, S12, S13, S14
9	R-1, R-2, R-3, M-1, S-1, S-2, S-3	—	—	—	S04, S05, S06, S08, S10, S11, S12, S13, S14

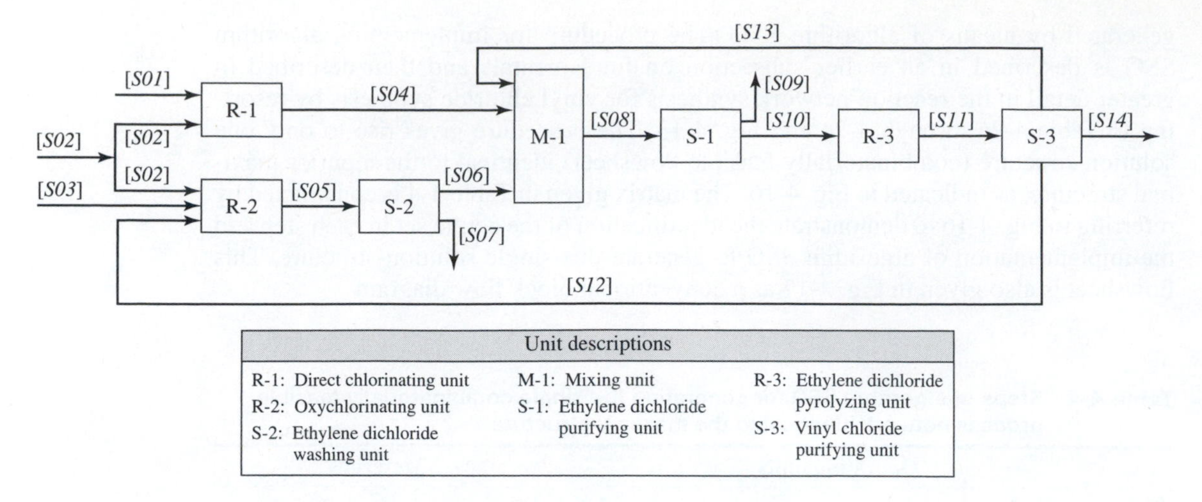

Figure 4-17
Combinatorially feasible flowsheet in which each operating unit may consist of one or more types of processing equipment jointly performing a specific task

Network Structures Within Plausible Operating Units In general, each plausible operating unit, whose incorporation into the flowsheet is determined at the initial stage of process synthesis, may consist of two or more units of processing equipment and/or reactors, as discussed at the outset. This naturally leads to the second stage, where the network structures of the operating units are determined separately and independently with their input and output conditions already established through earlier literature searches. The available information is given below.

Separating unit S-1 The separating unit for purifying ethylene dichloride shown in Fig. 4-17 requires two types of distillation columns: S-11 for removing the light residues or impurities and S-12 for removing the heavy residues or impurities.

Separating unit S-3 The separating unit for purifying the desired product also consists of two distillation columns: S-31 for removing the HCl that is to be recycled and S-32 for separating the vinyl chloride from the ethylene dichloride that is to be recycled. The mass balances around the four separators S-11, S-12, S-31, and S-32 are listed in Table 4-5. Figure 4-18 is the P-graph representation of the solution-structure (combinatorially feasible flowsheet), which explicitly shows these four separators. Figure 4-19 is the conventional counterpart of Fig. 4-18. The next section will show that this flowsheet is feasible as well as structurally optimal.

Optimization and Feasible Flowsheets Table 4-6 lists the types and operating conditions of the processing equipment in the operating units in terms of the temperature, pressure, and mean residence times. With such information in hand, the combinatorially feasible flowsheet in Fig. 4-18 or 4-19 is optimized and its feasibility simultaneously verified in the final phase of the algorithmic process-network synthesis based on the P-graph, as noted in the third main block of Fig. 4-8.

Table 4-5 Flow rate of the input streams and output streams for the separation units†

Material stream		Material species		
			Flow rate	
Designation	Type	Component	kg mol/h	kg/h
S-11 (lights column)				
S08	(Output)/ input	C_2H_4	0.50806	14.22568
		$C_2H_4Cl_2$	121.28118	12,006.8368
		$Cl_2/O_2/CO_2/C_2H_3Cl_3$, $C_2H_2,C_4H_6,C_2H_2Cl_2$	Trace	
		Total	121.825167	12,022.95081
S091	Output/ (input)	O_2	0.0101612	0.3251584
		CO_2	0.0116128	0.5109632
		$C_2H_2,C_4H_6,C_2H_2Cl_2$	0.00210482	0.1241844
		Total	0.02387882	0.9603060
S15	Output/ (input)	C_2H_4	0.50806	14.22568
		$C_2H_4Cl_2$	121.28118	12,006.83682
		$Cl_2/C_2H_3Cl_3$	Trace	
		Total	121.801288	12,021.990504
S-12 (heavies column)				
S15	(Output)/ input	C_2H_4	0.50806	14.22568
		$C_2H_4Cl_2$	121.28118	12,006.83682
		$Cl_2/C_2H_3Cl_3$	Trace	
		Total	121.801288	12,021.990504
S092	Output	C_2H_4	0.50806	14.22568
		Cl_2, $C_2H_4Cl_2$, $C_2H_3Cl_3$	Trace	
		Total	0.5926883	22.3391079
S10	Output/ (input)	$C_2H_4Cl_2$	121.2086	11,999.6514
		Total	121.2086	11,999.6514
S-31 (HCl column)				
S11	(Output)/ input	$C_2H_4Cl_2$	48.6286	4,814.2314
		HCl	72.58	2,649.17
		C_2H_3Cl	72.58	4,536.25
		$C_2H_4/C_2H_2,C_4H_6,$ $C_2H_2Cl_2$	Trace	
		Total	193.791213	11,999.78981
S16	Output/ (input)	$C_2H_4Cl_2$	48.6286	4,814.2314
		C_2H_3Cl	72.58	4,536.25
		$C_2H_4/C_2H_2,C_4H_6,$ $C_2H_2Cl_2$	Trace	
		Total	121.2112129	9,350.6198101
S12	Output/ (input)	HCl	72.58	2,649.17
		Total	72.58	2,649.17

(Continued)

Table 4-5 ***Continued***

Material stream		Material species		
			Flow rate	
Designation	Type	Component	kg mol/h	kg/h
S-32 (Vinyl chloride monomer column)				
S16	(Output)/ input	$C_2H_4Cl_2$	48.6286	4,814.2314
		C_2H_3Cl	72.58	4,536.25
		$C_2H_4/C_2H_2,C_4H_6,$ $C_2H_2Cl_2$	Trace	
		Total	121.211212	9,350.619810
S13	Output/ (input)	$C_2H_4Cl_2$	48.6286	4,814.2314
		$C_2H_4/C_2H_2,C_4H_6,$	0.0021048	0.1241844
		$C_2H_2Cl_2$	Trace	
		Total	48.6312129	4,814.36981
S14	Output	C_2H_3Cl	72.58	4,536.25
		Total	72.58	4,536.25

†Flow rates are based on a production capacity of 4,535 kg/h of vinyl chloride using the balanced process. The stream type (output)/input signifies that the stream is the output from another operating unit but is the input to this operating unit; conversely; output/(input) indicates that the stream is the output of this operating unit but is the input to another operating unit. The same statement applies to all other operating units.

Integer variables, each signifying the exclusion (1) or inclusion (0) of an operating unit, no longer come into play in optimizing any combinatorially feasible flowsheet. The inclusion of all the operating units is unequivocally and rigorously determined through algorithm MSG followed by algorithm SSG. In general, therefore, the objective function in terms of the total annualized capital cost that is to be minimized in optimizing a combinatorially feasible flowsheet has the form of

$$S = \sum_{i=1}^{n} \left(a_i + b_i x_i^{\eta}\right) \qquad x_i \leq M \tag{4-2}$$

where M is a large positive value denoting the upper bound of x_i. In this expression, a_i is the fixed part of the capital cost, b_i the proportionality constant representing the size dependent part of the capital cost of unit O_i with a unit size, x_i the size of the operating unit O_i involving one piece of processing equipment, and n the number of operating units in the combinatorially feasible flowsheet under consideration. The evaluation of a_i and b_i is treated in Chap. 6. The value of η in Eq. (4-2) is unity for the linear cost function, thus requiring linear programming for its minimization. Generally, η takes a value less than 1; in practice, it is often 0.6 and requires a nonlinear optimization method. Some of the methods for linear and nonlinear optimization are covered in Chap. 9.

The minimization of the annualized total cost S, as given in Eq. (4-2), is carried out subject to a set of constraints comprising the mass balances for all the materials involved, which are linear in x_i's. The flow rate, or amount, of any material for the operating unit of size x_i is just the product of x_i and that of the material for the corresponding operating unit

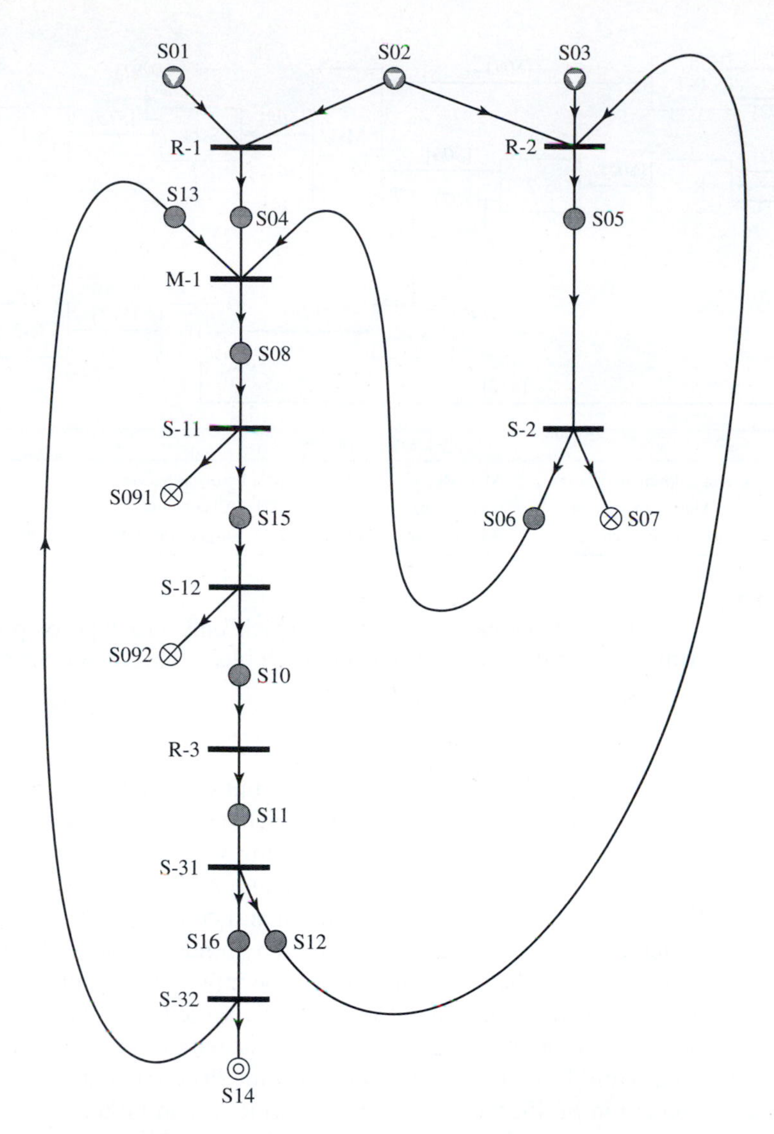

Figure 4-18
P-graph of the combinatorially feasible flowsheet with all the network structures of plausible operating units identified for the balanced process to manufacture vinyl chloride. This is also the feasible as well as optimal flowsheet in which each operating unit consists of only one type of processing equipment performing a specific task

of a unit size. In carrying out the structural flowsheet optimization of a novel process or for extensively renovating an existing process to alter its capacity, these linear constraints usually appear as inequalities. These inequalities indicate that the amount of any material produced must be equal to or more than the amount consumed, the amount of any desired

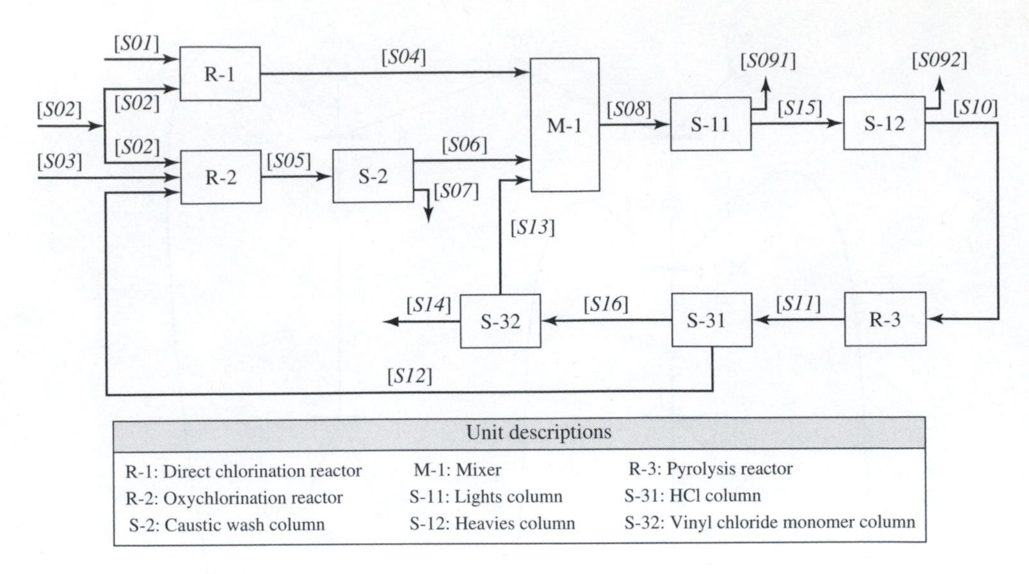

Figure 4-19
Combinatorially feasible and optimal flowsheet in which only one type of processing equipment performing a specific task is assigned to each block representing an operating unit

product produced must be equal to or more than the amount specified, and the amount of any raw material consumed must be equal to or less than the amount available.

In exploring the structural optimality of the existing vinyl chloride process flowsheet, the flow rates are known for the materials into and out of each operating unit in a combinatorially feasible flowsheet. The production and consumption rates of every material can be calculated from these flow rates, thereby rendering possible the verification of the balance between the two. If necessary, the input and output flow rates of the materials for some operating units are proportionally varied to achieve the balancing. The data and information pertaining to the performance of operating units might have been obtained from different sources. Based on the flow-rate data in Tables 4-3 and 4-5 for a production of 4535 kg/h of vinyl chloride, it can be readily confirmed computationally that the mass balances are satisfied by all the materials involved, thus indicating that the combinatorially feasible flowsheet is indeed feasible. It is also structurally optimal since no other feasible flowsheets exist.

The sizes of the operating units in Eq. (4-2) are estimated either from their optimal performance data that may be available or by optimizing their performances as functions of the operating conditions that are continuous variables, on the basis of mathematical models developed for the units. This can be facilitated with the aid of existing software for simulation, such as ASPEN PLUS, PRO III, or HYSYS. For the process under consideration and any other highly mature processes, there exists a wealth of information pertaining to the performance of the operating units, much of which has already been optimized. A substantial portion of such information as listed in Table 4-6,

Table 4-6 Summary of the plausible operating units of the balanced process to manufacture vinyl chloride in terms of individual processing equipment, operating conditions, sizes, and costs[†]

Operating units		Operating conditions			Equipment cost estimation parameters and cost				
Name	Type	Temperature, °C	Pressure, kPa	Holding time, s	Number of trays	Diameter, m	Height, m	Volume, m^3	Cost, million US$
R-1 Direct chlorination reactor	CSTR followed by PFR	59.9	200	9293 (CSTR) 0.01 (PFR)				14.9	0.21
R-2 Oxychlorination reactor	PFR (fixed bed) followed by CSTR	Inlet: 193.1 Outlet: 192.5	640	324 (PFR) 5.6 (CSTR)				14.5	0.31
R-3 Pyrolysis reactor	PFR	500	2130	0.35					0.62
S-11 Lights column	Distillation column	Overhead: 68.9 Bottom: 85.6	200		20	0.5	7.3	14.5	0.07
S-12 Heavies column	Distillation column	Overhead: 86.7 Bottom: 88.9	200		40	1.8	15.2		0.32
S-2 Caustic wash column	Flash tank	Inlet: 90.6 Outlet: 90.6	210			1.0	3.0		0.05
M-1 Mixer	(Online)								
S-31 HCl column	Distillation column	Overhead: −1.1 Bottom: 140.8	2630		30	1.1	11.3		0.58
S-32 Vinyl chloride monomer column	Distillation column	Overhead: 32.8 Bottom: 140.0	610		26	0.6	11.3		0.09
								Total	2.25

[†]Based on a production rate of 4535 kg/h of vinyl chloride using the balanced process.

especially for the three reactors, has been optimized in their practical implementation and thoroughly analyzed for verification.

In concluding this section on structural flowsheet optimization, note that the current illustration is seemingly rather trivial. It is partly due to the fact that the balanced process to produce vinyl chloride is highly mature and has been thoroughly optimized, and the process involves only three chemical reactions. It is also due to the fact that the algorithmic method for process-network synthesis based on the P-graph is profoundly efficient in significantly and rigorously removing all the redundant networks from consideration. This is accomplished by using a set of combinatorial algorithms that are implemented twice in the procedure, as shown in Fig. 4-8. In the first implementation of the algorithms, the reaction network is rapidly determined, serves as the starting point for the flowsheet, and provides the initial guidance for selecting and locating the operating units. In the second implementation, the optimal structure of the flowsheet itself is actually identified among a limited number of combinatorially feasible flowsheets. This is diametrically opposite to any conventional algorithmic method based on the superstructure. For example, with just 9 operating units in a process, such a superstructure contains 511 networks. To identify the optimal flowsheet requires that all 511 networks be taken into account. Such an undertaking is indeed difficult and time-consuming. All efforts to date to rigorously circumvent this serious difficulty have been unsuccessful.

It is highly advantageous to implement algorithm ABB, which is a modified form of the conventional branch-and-bound method, if the number of combinatorially feasible flowsheets becomes excessive. Such a situation can easily arise, as can be illustrated by considering an "unbalanced" process to manufacture vinyl chloride, where reaction r_4 given below is added to a set of the three reactions r_1, r_2 and r_3 of the balanced process.

$$r_4: \qquad C_2H_2 + HCl \rightarrow C_2H_3Cl$$

The addition of merely a single reaction would increase the number of combinatorially feasible reaction networks from 2 to 12, of which 10 are feasible. This would result in an "exponential" magnification. Naturally, the number of combinatorially feasible flowsheets obtained from the feasible reaction networks would increase accordingly depending on the types and numbers of reactors and separators that will need to be included to accommodate reaction r_4. This simple example is indicative of the usefulness and need of algorithm ABB to identify in a given order a specified number of optimal and near optimal flowsheets. The profound acceleration of algorithm ABB is achieved through the judicious exploitation of the unique structural features of process flowsheets.[†]

The procedures for implementing algorithms MSG, SSG, and ABB are available[‡] at http://www.p-graph.com. for demonstration. Moreover, highly automated and integrated implementation of all three algorithms after introducing synthesis problem (P, R, O) is possible through software SYNPHONY.

[†]F. Friedler, J. B. Varga, E. Feher, and L. T. Fan, *State of the Art in Global Optimization, Computational Methods and Applications,* C. A. Floudas and P. M. Pardalos, eds., loc. cit.

[‡]Please see the textbook website for a expanded demonstration of the P-graph approach to process flowsheeting.

COMPARISON OF HIERARCHICAL AND ALGORITHMIC RESULTS

Flowsheets for the vinyl chloride process have been obtained by both a hierarchical and an algorithmic approach. Comparing Fig. 4-7 with Fig. 4-19 shows that these two diagrams are remarkably similar, differing only as to the type of separation between the dichloroethane (EDC), water, and gaseous products from the oxychlorination reaction. Because the vinyl chloride process is well developed and described in the literature, such agreement is to be expected. For a less well-developed process or a more complex process, the algorithmic approach would yield much more definitive results than the hierarchical approach.

There are some differences in details between the two. The mass balances in the algorithmic method are much more detailed and dependable. For instance, while the approximate mass balances used in the hierarchical method indicate that 1.08 kg mol of ethylene is required to produce 1 mol of vinyl chloride and that this quantity is equally divided between the direct and oxychlorination reactors, the algorithmic balances show 1.06 kg mol of ethylene per kg mol of vinyl chloride, with 53 percent to direct chlorination and 47 percent to oxychlorination. This is due to differences in the values used for the conversion of ethylene to EDC in the two chlorination reactors, as well as a different conversion of EDC to vinyl chloride in the pyrolysis reactor. The more detailed balances include components at very low concentration, which makes it possible to deal more effectively with the light- and heavy-component distillation columns.

The algorithmic method did not include the heat integration, because excellent algorithms are already available for that purpose. The hierarchical approach illustrates a simple approach to heat integration that results in a minimization of the operating units for the process. The algorithmic method includes an optimization based on equipment size. Such an optimization is feasible only with a computer-based algorithm. Clearly the algorithmic approach is more defined and thorough, and its use is recommended with the appropriate software, particularly when more complex chemical reactions than the vinyl chloride process are considered.

GENETIC ALGORITHMS

Genetic algorithms (GAs) are procedures based on the mimesis of mechanics of natural selection and genetics. Such algorithms emulate biological evolutionary theory to solve optimization problems.

Applying GAs to the chemical process flowsheet synthesis process is complex. Setting up the structural component analogies to process design is not difficult. Traits correspond to tasks. Genes correspond to subtasks, etc. The generation analogy is also simple, corresponding to algorithmic iterations. The main difficulty lies in establishing natural-selection-analogous criteria.[†] A number of approaches, including absolute

[†]C. Laquerbe, J. C. Laborde, S. Soares, P. Floquet, L. Pibouleau, and S. Domenech, *Comp. & Chem. Eng.*, **25:** 1169 (2001).

energy requirements beginning with primary elements, have been proposed but have yet to gain wide support or prove themselves. Also, the important concept of mutation is difficult to emulate as forces driving mutation are not certain. However, initial investigations do show some success by simple introduction of a degree of randomness into decisions that seem to correspond to mutation. Despite these difficulties, GAs are pursued because they demonstrate a large-set efficiency, yet to be matched by human-generated algorithms.

FUTURE APPROACHES TO FLOWSHEET SYNTHESIS AND DEVELOPMENT

The advent of algorithmic approaches to flowsheet synthesis, evaluation, and selection offers a potential for great improvement in accuracy, thoroughness, and scale. Algorithmic flowsheet procedures give structure to the flowsheet process. This is done by combining hierarchical approaches not dissimilar to those used in traditional flowsheets with nonlinear mathematical principles and advanced mathematical models of various chemical unit process operations. As of this date, algorithmic approaches have not yet reached the sophistication, maturity, and applicability level where they are widely employed in the process industry. However, this is certain to change as algorithms mature. Also, the availability of abundant and exponentially growing computational power guarantees that even the most calculation-intensive algorithms, those that currently are showing the greatest promise, will be tractable.

SOFTWARE USE IN FLOWSHEET SYNTHESIS

There are currently a number of software applications which aid and facilitate the generation, selection, and development of process flowsheets. These include both custom algorithmic programs and commercial software packages. The custom programs aimed at studying and implementing algorithms have demonstrated certain usefulness in a limited number of studies. They do, however, require human interaction at almost all stages of synthesis and are by definition customized to specific study cases. There are a number of commercial packages that are aimed at flowsheet synthesis. The commercial packages are more widely applicable, but again usually require intensive human interaction and constant supervision. Current software is therefore not sufficiently developed to allow completely autonomous flowsheet synthesis and evaluation, although it does prove useful in analyzing very complex defined process systems and in the generation of process arrangements once the processes have been defined.

ANALYSIS AND EVALUATION OF FLOWSHEETS

Criteria for Evaluating Designs

The complex, varied, and evolutionary nature of flowsheets seldom allows for easy selection of the more desirable possibilities from among the great variety of candidate flowsheets. There are, however, a number of factors that must always be considered carefully when selecting flowsheets, with varying importance depending on the particular design criteria. These criteria include economic viability, safety, environmental, and legal issues. Evaluating prospective flowsheets with regard to these factors helps determine which flowsheet most closely meets all the design criteria.

Economic Viability The economic analysis of a process flowsheet is the most important factor in selecting flowsheets since the goal of any design is to produce an economically viable process. The economic analysis must take into account equipment and construction costs and operating costs, such as raw materials, utilities, and operator wages. Further, a complete economic analysis must investigate future process viability by assessing possible future shifts in operating costs and product price, including foreseeable changes in legislation and regulations. The aspects of conducting a complete economic evaluation are discussed further in Chaps. 6, 7, and 8.

Safety Issues Increasing social and thus regulatory and economic pressures have catapulted the safety aspects of design to a great importance in the initial phases of the process design. The safety aspects that are scrutinized include chemical-exposure effects on operators and materials compatibility, as well as shutdown and start-up issues. They also entail detailed studies of equipment failures and operational changes, such as the effect of an undetected shift in raw material purity or contaminants. While many of these safety considerations and evaluation procedures are straightforward, well documented, and standardized, many are not. Furthermore, planning to meet increasingly strict regulations and possible future regulations can be worthwhile in the long run, although they may require greater initial time and expenditures.

Environmental Issues These are generally legally imposed regulations, established by various levels of government. These regulations can vary with even small shifts in locale, due to nothing more than a change in available waste disposal services. Further, environmental regulations are generally increasing in number, coverage, and strictness even while being applied to an ever-growing number of substances and conditions. It is imperative to fully understand the complete environmental regulatory regime of all proposed plant locations to establish if a process can meet those requirements.

The types of limits generally imposed by environmental laws can be grouped into three areas: substances that can be used, utilities that can be used, and allowed waste. Of these, waste is generally most heavily regulated. Wastes are controlled by composition, temperature, rate, and even phase. Monitoring, meeting, and adhering to the prescribed waste regulations often require several specialists with the support of a good amount of specialized equipment or employing companies that specialize in environmental monitoring and compliance. Further, in many cases environmental regulations require additional processing of waste before disposal, often adding greatly to operating costs. Finally, increasingly environmentally aware employees can sometimes place even stricter environmental demands on employers. Again, foresight at the planning stage can often assist in meeting future environmental regulations, or at least help prepare for possible tightening of regulations.

Legal Issues The complexity of chemical processes and the resulting large expenses for research and development lead to issues of proprietary, patented, and otherwise legally constrained process information. Laws governing intellectual property, under which chemical process information falls, change by location and time, requiring expert knowledge and appraisal in order to meet their requirements. Further, while some processes are widely known to be legally constrained, others are not, so that careful research of any possible legal aspects of a process is absolutely necessary before a process can be implemented or used. At the same time, patent data are often a good source of information for determining potential processes at the initial stage of research.

The sensitivity of legal matters and the major economic repercussions of legal entanglements require the involvement of legal specialists, in-house, or more often companies specializing in chemical process law.

SUMMARY

The criteria for evaluation of a design may be summarized by stating that any flowsheet should be evaluated by how well it can produce a desired product most profitably while conforming to regulations and keeping future requirements in mind.

NOMENCLATURE

a_i = annualized fixed part of the capital cost for operating unit O_i, \$/yr

b_i = coefficient of the annualized, size-dependent part of capital cost for operating unit O_i, (\$/yr)/[(equipment size units)^n]

c = a constant in expression for the number of computational steps required, dimensionless

C_p = heat capacity at constant pressure, kJ/kg mol·K

CR_i = conversion of principal reactant in reaction *i*, dimensionless

F_{EDC} = reactor R-2 ethylene dichloride feed, kg mol/(kg mol vinyl chloride product)

FE = total ethylene feed, kg mol/(kg mol vinyl chloride product)

ΔH_R = heat of reaction, kJ/kg mol

ΔH_V = heat of vaporization, kJ/kg mol

m = subset of M including all materials in feasible flowsheets

m_p = homogeneous mixture of components

M = finite set of materials involved in process network synthesis; upper bound on the size of operating units, x_i

n = number of operating units in maximal structure; number of mols of reactant or product; number of operating units in a particular feasible flowsheet; number of components

n_{EDC} = number of kg mol ethylene dichloride/(kg mol vinyl chloride product)

o = subset of O including all operations in feasible flowsheets

O = set of all operating units; subscript *i* denotes operating unit *i*

P = set of all final products

Q = amount of heat transferred, MJ/(kg mol vinyl chloride product)

r_i = reaction number *i*

R = set of all raw materials

S = total annual cost, \$/yr

ssi = number of separation steps between ethylene feed and vinyl chloride product via reactor i, dimensionless

T = temperature, K; subscripts i and o denote in and out

ΔT = temperature change, K

x_i = integer reaction equation multiplier, dimensionless; the size of the single equipment in operating unit O_i, units depend on the type of equipment

Greek Symbols

η = exponent of equipment size term in cost equation

PROBLEMS[†]

4-1 A natural gas processing plant is to separate natural gas into an ethane stream, a propane/n-butane stream, and a sales gas stream that is mainly methane. The natural gas feed rate is 2.5×10^3 kg mol/h, and the composition is 83, 12, 4, and 1 mol percent methane, ethane, propane, and n-butane, respectively. The ethane product stream contains 1 mol percent methane, 2 mol percent propane, and negligible butane. The propane product stream contains all the butane, 2 mol percent ethane, and negligible methane. The sales gas contains 0.5 mol percent ethane, and the rest is methane.

a. Draw an I/O diagram for this process.

b. Calculate the flow rate of each product stream.

c. Find the monetary values of all streams using the following costs:

\$4.00/GJ of lower heating value (LHV) for natural gas

\$0.40/kg for the ethane product stream

\$0.44/kg for the propane product stream

\$3.25/GJ of lower heating value (LHV) for the sales gas product

Component	LHV, GJ/kg mol
Methane	0.802
Ethane	1.428
Propane	2.044
n-Butane	2.686

d. Calculate the cost of separation, in dollars per hour, at which the sales revenue less the sum of the raw material cost and separation cost equals zero for this process.

4-2 Produce a separation train for Y-grade NGL (natural gas liquids) separation into nearly pure species up to C_4 and gasoline.

4-3 Research and establish two of the main chemical processes used to produce the styrene monomer. Show an I/O diagram for these processes. Analyze the basic economics of the styrene monomer processes, and establish which process is more promising.

[†]The problems for this chapter are based on well-established processes, thus reducing some of the potential complexities of novel process design.

4-4 Create a functions diagram for one of the two styrene-producing processes analyzed in Prob. 4-3.

4-5 Create an operations diagram for one of the styrene-producing processes analyzed in Prob. 4-3. Additional literature research may be necessary for this problem.

4-6 Develop a final flowsheet, without heat integration, from the operations diagram developed for the styrene process. Then develop the final flowsheet with heat integration. What savings can be generated through the heat integration step? What type of major considerations do these flowsheets evoke? Do any of these issues call for the redesign of the flowsheet?

4-7 Evaluate the economic potential of the final flowsheet created for the styrene monomer production, using the following assumptions: The solid catalyst costs \$5/kg, operates for 800 h before requiring regeneration via a process costing \$0.15/kg and requiring 50 h. Further, assume the ratio of the feed per hour to catalyst weight to be 100.

4-8 Generate a computerized process flow diagram (PFD) for the styrene monomer process, assuming 30 percent ethylbenzene conversion per pass. The volumetric flow of the fluid that is returned with a recycle to the reactor inlet is 2.6 times that of the product leaving the reactor system. All separations are by single pure product per column distillation.

4-9 Research and establish potential chemical processes that may be used to produce the major reactants for the processes used to make styrene.

4-10 Are any of the reactant-producing processes established in Prob. 4-9 worth integrating into the styrene process? Justify.

4-11 Introduce and integrate the potential raw material processes found in Prob. 4-9 into the styrene monomer process.

4-12 Evaluate the new combined final styrene flowsheet, using the assumptions in Prob. 4-10.

4-13 Analyze the basic economics and show an I/O diagram for producing hydrogen from water, coal, and natural gas. What production mode should be utilized to obtain production rates of 3×10^7 and 1×10^8 kg/yr?

4-14 For the v^ess profitable styrene production process found in Problem 4-3, assume

Catalyst costs = \$6.6/kg

Ratio of feed per hour to catalyst weight of 100

Catalyst life of 1100 h

Catalyst regeneration cost = \$0.25/kg

Catalyst regeneration time = 40 h

Catalyst replacement time = 8 h

100 percent separation of all ethylbenzene from the reactor exit stream, and recycle of all remaining material

100 percent purity, single species separation

Ideal distillation with a 20 percent adjustment to the distillation heat requirements used for all separations

a. Establish a functions diagram for this process.

b. Establish an operations diagram for this process.

c. Establish a final flowsheet, with heat integration, for this process.

d. Evaluate the operating economics of this process, neglecting equipment costs.

4-15 Acrylic acid can be produced from propylene.

a. Research this process and determine sufficient details required for process flowsheet synthesis.

b. Determine the gross economics of this proposed process.
c. Generate a functions diagram, an operations diagram, and a final diagram, without heat integration, for this process.
d. What heat integration can be carried out in this process?
e. Create a computerized PFD for this process.

4-16 Vinyl acetate is a raw material for a number of major chemical processes.

a. Find possible reaction paths for vinyl acetate production.
b. Determine the gross economics for these potential synthesis paths.
c. Most current methods for vinyl acetate production rely on the oxidation of ethylene and acetic acid. Generate a progression of flowsheets for this process, taking into account the side reaction combustion of ethylene. Also, make sure to include packaging in your flowsheets.
d. Evaluate the most promising process.
e. Create a computerized PFD for this process.

4-17 Industrial ethanol can be produced by reacting feedstocks or by fermentation of organic matter. Generate I/O and functions diagrams for both the feedstock reaction and the organic matter fermentation processes for producing ethanol. Compare these two conceptual processes.

4-18 The average automobile catalytic converter contains approximately 0.5 kg of platinum-group metal powder contained in a 2-kg ceramic honeycomb block. Design a flowsheet for a process to separate and recover the platinum-group metals in the spent catalytic converters.

CHAPTER

5

Software Use in Process Design[†]

As with all other aspects of engineering as well as those in everyday living, computer software use and utilization are now an ingrained and indispensable part of process design and economic evaluation. Computers ease and enhance the ability of the design engineer to carry out preexisting tasks, allowing for expansion on previously manual tasks, as well as facilitate previously impossible tasks that have now become part of the process design and evaluation process.

In the very first steps of the development process, computerized databases, research results, and electronic versions of traditional publications aid in the selection of appropriate reactions and raw materials to determine potential chemical processes that provide the desired product. Computers are also useful whenever experimentation is required, to plan, run, and analyze the experiments. Software can then be used to generate, evaluate, and select from the best process flow diagrams.[‡] Process simulation software is then utilized in generating process flow diagrams based on the selected process flowsheets. These simulations, perhaps the most emphasized aspect of software use in the design process, allow for very accurate simulation of the process over a wide range of operating conditions and scenarios. The process simulations for the candidate chemical processes can then be examined under a wide range of conditions to ensure safety and operability. The simulations, so accurate that some processes bypass the traditional pilot-plant stage of process design, can then be used to evaluate the economics of the process using economic evaluation software. Computerized optimization of the process is then also possible using either dedicated optimization software or combinations of non-optimization-specific software. Once the process components and operating conditions are established, the physical arrangement of the process can also be evaluated and optimized. Finally, the schematics, drawings, and all

[†]This chapter was prepared by B. Halevi, Computer Consultant, HBarr Inc., Boulder, CO.

[‡]The term process flow diagram is used interchangeably with process simulation when used in reference to software use.

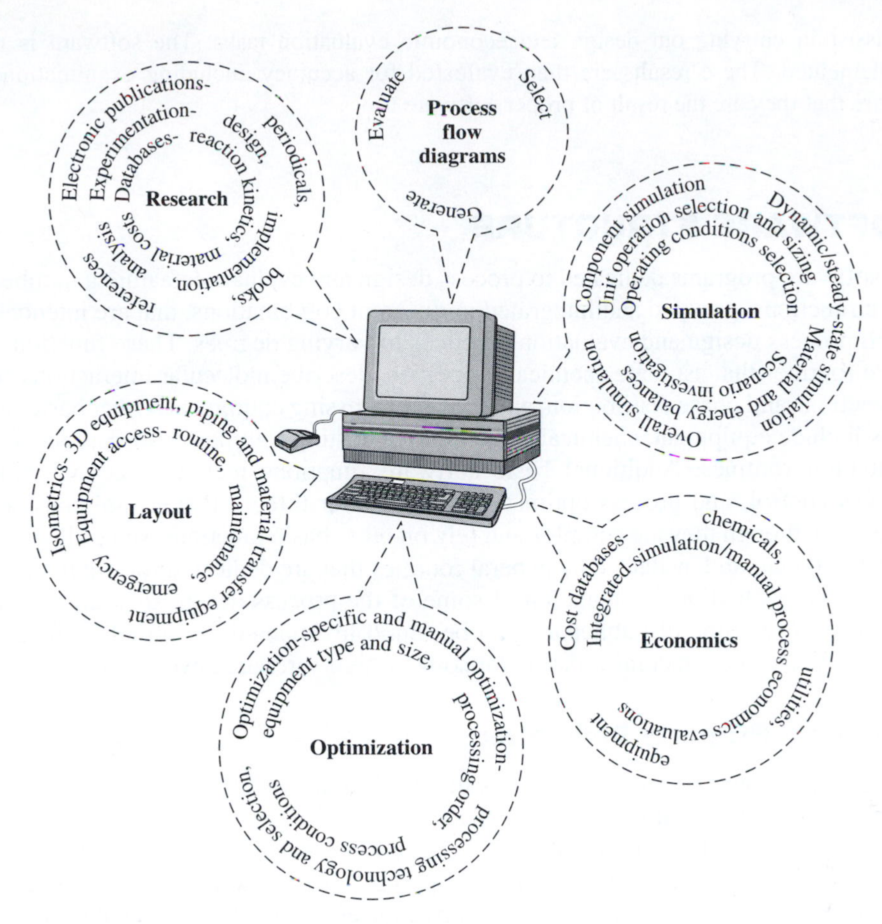

Figure 5-1
The role of software in process design and evaluation

the necessary planning aspects for constructing the process can be generated using computers. All of these aspects of software in process design and evaluation are displayed in Fig. 5-1.

The use of software also extends to the less obvious parts of process design and evaluation. These include word processors that greatly ease and enhance all aspects of writing, communication software that facilitates quick and effective communications, and any number of mathematical evaluation software programs that aid calculation and numerical analysis. These types of generic software do not play critical roles in the process design and evaluation, yet they are in many ways just as integral as the software specifically intended for process design and development.

Correct understanding, selection, use, and evaluation of software in process and economic evaluation begins with an examination of the structure of software. This reveals the major software functions that form the basis for the software, and forms the basis for the subsequent selection, use, and evaluation of the software. Next, software is selected to assist in carrying out design and economic evaluation tasks. The software is then implemented. These results are then evaluated for accuracy, including examinations to ensure that they are the result of proper software use.

SOFTWARE STRUCTURE

All software programs dedicated to process design and evaluation feature a number of basic functions, grouped and integrated in different combinations, that are intended to fulfill process design and evaluation practices to varying degrees. These functions include routines that estimate chemical properties, describe molecular interactions such as reactions and phase distribution, and model processing equipment. Other basic functions include equipment, chemicals, and utilities costing routines as well as economic evaluation routines. Additional basic software functions include heat integration, process control, and process optimization. The latter three functions are considered basic even though they are complex and rely on other basic functions, since all are controlled, guided, and evaluated by general routines that are dedicated for use in process design and evaluation. A tabulation of some of the process design software that was available at the time of publication can be found in Appendix F, sorted by function. Also provided is contact information for some design software developers.

Chemical Property Estimation

A key requirement of process design is the need to accurately reproduce the various physical properties that describe chemical species. The physical properties required for accurate modeling and simulation often include molecular reaction and kinetic data, thermodynamic properties, and transport properties.[†] In the case of reaction kinetic data, the critical properties are rate equations, activation energies, and reaction mechanisms. Necessary thermodynamic properties include enthalpy, entropy, fugacity coefficients, Gibbs free energy, etc. The transport properties of greatest interest are diffusion coefficients and models, thermal conductivities, and viscosities.

Estimation of these key chemical properties is done through three methods. The first is that of interpolation, extrapolation, and empirical curve fitting of tabulated, experimentally determined properties. These empirical property estimators can be very accurate, but must be used only under the same conditions as the experimentally determined data used as the basis for the empirical relations. The second approach to property estimation is through the use of equations of state. The latter are generalized equations derived from fundamental molecular principles of molecular repulsion and attraction.[‡] Such relationships directly describe pressure, temperature, and volume behavior through the use of species-specific coefficients and subsequently yield other thermodynamic properties, such as enthalpies and fugacity coefficients. The coeffi-

[†]S. Bumble. *Computer Generated Physical Properties,* Lewis Publishers, Washington, 1999.

[‡]For more about equations of state and activity coefficient models, consult general thermodynamics texts.

cients used in the equations of state can be either experimentally determined or derived from molecular behavior principles. Equations of state are useful and convenient as long as the assumptions used to derive them are adhered to and as long as the species-specific coefficients are known for the chemicals of interest. This means that equations of state are useful for ideal or nearly ideal fluid systems over wide ranges of pressure and temperature, including sub- and supercritical regions, as well as for highly non-ideal systems at high pressures.[†] Further, many software programs allow users to augment the accuracy of the physical property estimation based on equations of state by entering their own coefficients. The third method, termed the *activity coefficient method,* describes an approach to accounting for nonideal behavior of chemicals through the modification of basic molecular interaction principles and the resultant thermodynamic phenomena. Nonideality compensation is accomplished by adding molecular size disparity and stronger molecular interaction terms to equations that describe the basic molecular thermodynamic behavior of the chemical species. This adds complexity to molecular behavior equations but does improve the accuracy of chemical property estimation, particularly for highly nonideal fluids at low pressure and specifically for liquid mixtures. The nature of the nonideality compensation techniques used in activity coefficient methods does lead to a number of requirements and limitations. The coefficients used in the activity coefficient models must be experimentally determined and are only valid for the conditions for which they were established. Also, the accuracy of activity coefficient models is limited to the conditions for which the models are derived. Therefore, while useful and accurate for specific cases, especially for liquid-liquid equilibria, the activity coefficient method must be used with care.

The nature of the physical property estimation methods is such that not all properties can be obtained from a single property estimation method and/or equation. Some methods and property relations are restricted to very specific pressure, temperature, molecular weight, and molecular size ranges that further limit property estimation abilities. Such restrictions and the availability of data for the three property estimation approaches cause property estimation routines to combine different approaches to estimate different properties for different chemicals. This, more or less, allows complete and appropriate physical property estimation for chemicals, with varying accuracy. These physical property estimation methods and relation assemblies are varyingly termed thermodynamic, physical, or fluid property packages.[‡] A summary of the principal physical property estimation models, their limitation, and applicability is listed in Table 5-1.

Electrolyte solutions have a larger variety of interactions and phenomena than found in nonelectrolyte solutions. Such solutions are evaluated in a variety of empirical models that account for short, long, and multiphase interactions. These models are

---

[†]J. M. Smith, H. C. Van Ness, and M. Abbott. *Introduction to Chemical Engineering Thermodynamics,* 6th ed., McGraw-Hill, New York, 2001; and R. Span, *Multiparameter Equations of State: An Accurate Source of Thermodynamic Property Data,* Springer-Verlag, Berlin, Germany, 2000.

[‡]J. P. O'Connell and M. Neurock, *Fifth International Conference on Foundations of Computer-Aided Process Design,* M. F. Malone, J. A. Trainham, and B. Carnahan, eds., **96,** series 323, pp. 5–19, American Institute of Chemical Engineers-CACHE, New York, 2000.

Table 5-1 Physical property estimation model summary†

	Description	Typical uses	Range of application
Equation of state			
Lee-Kesler-Plocker	Most accurate general method for nonpolar substances and mixtures	Refinery and gas processing	All (least accurate near critical property values)
Peng-Robinson	Model appropriate for vapor-liquid equilibrium calculations. Many modified models based on basic PR model are available with extended nonideal applicability ranges, including sour systems.	Refinery, petrochemical, and gas processing	All (least accurate near critical property values)
Soave-Redlich-Kwong	Similar to PR, but its range of application is more limited and is less reliable for nonideal systems.	Refinery, petrochemical, and gas processing	All (least accurate near critical property values)
Activity model			
Braun K10	Based on K10 charts for 70 hydrocarbons and light gases	Petrochemicals	Hydrocarbons under low pressures and low to moderately high temperatures (100–800 K)
NRTL *(nonrandom two-liquid)*	Extension of the Wilson equation. Uses statistical mechanics and the liquid cell theory to represent the liquid structure. Many modifications to model available for modeling of dissimilar fluid mixtures	Strongly nonideal VLE, LLE, and VLLE phase behavior	Fluids and fluid mixtures with similar phase behavior characteristics. Modifications allow modeling of dissimilar phase behavior characteristics
UNIFAC	Modification of UNIQUAC. Applies UNIQUAC principles to functional groups instead of molecules	Strongly nonideal VLE, LLE, and VLLE phase behavior	Similar to UNIQUAC
UNIQUAC *(universal quasi chemical)*	Uses statistical mechanics and quasi-chemical theory to represent the liquid structures. Many modifications of model available to allow modeling of dissimilar fluid mixtures	Strongly nonideal VLE, LLE, and VLLE phase behavior	Similar to NRTL
Wilson	First activity coefficient equation. Uses local composition model to derive the Gibbs excess-energy expression. Offers a thermodynamically consistent approach to predicting multicomponent behavior from regressed binary equilibrium data	Highly nonideal systems, especially alcohol-aqueous systems	Limited to single-liquid-phase system and nonelectrolytes
Semiempirical			
Chao-Seader	A semiempirical method developed for systems containing hydrocarbons and light gases, such as CO_2 and H_2S, but with the exception of H_2	Nonpolar systems, especially hydrocarbons	Valid for high pressures (<10 MPa) and moderate temperatures (200–550 K)

Table 5-1 *Continued*

	Description	Typical uses	Range of application
Grayson-Streed	An extension of Chao-Seader, with an emphasis on H_2	Nonpolar systems, especially hydrocarbons and heavy hydrocarbons	High pressures (<20 MPa) and moderate temperatures (300–700 K). Caution should be observed for temperatures below 300 K, pressures above 4 MPa, and separation of mixtures with similar VLE behavior

[†]S. Bumble, *Computer Generated Physical Properties,* Lewis Publishers, Washington, 1999; L. T. Biegler, I. E. Grossman and W. Westerberg, *Systematic Methods of Chemical Process Design,* Prentice-Hall, Upper Saddle River, NJ, 1997; and various software manuals including products by Aspen Tech, Chemstations, Epcon, and Hyprotech.

often combined with activity, equation of state, and semiempirical models to extend model applicability to electrolyte systems.

Process Equipment Models

Process equipment models are collections of rigorous material and energy balances that are derived from basic process equipment operating principles and are solved using numerical solution techniques.[†] These collections are combined with logical routines that select, group, and assemble a number of equations appropriate to the specific operating and physical arrangement that the model is intended to reflect. The resulting model can then be used to reproduce a range of processing equipment setup specifications for any type, quality, and quantity of feeds and withdrawals. For instance, a multistage distillation column is simply a cascade of thermodynamic equilibrium stages. The material and energy balances are established from cascade flows combined with thermodynamic equilibrium calculations, and the actual flow rates and equilibria compositions and states are established using chemical property data.[‡]

In addition to the general material and energy balance equations that are derived from basic process equipment operating principles, models must account for all operation characteristics of the process equipment that is being modeled. It is this latter incorporation of more specific characteristics and features that is the key to accurate and useful modeling. The basic operating principles of most process equipment are well understood, simple, and easy to represent mathematically, although in many cases it is the required repetitive use of simple principles that results in overall model complexity. However, actual equipment is seldom, if ever, accurately described by basic operating principles because of the mechanical design and construction considerations

[†]More detailed descriptions of mathematical modeling of process equipment can be found in general chemical engineering numerical analysis or modeling texts.

[‡]The principles underlying various process operations are described in greater detail in later chapters.

necessary to transform process principles to useful equipment. The more specific characteristics and features of real process equipment are not easy or simple to quantify, represent mathematically, or incorporate into the general-principles models. Such equipment characteristics and features as efficiencies, pressure drops, and friction losses are, however, necessary to describe the actual process equipment setup and operation.[†] Models account for these features and characteristics through three main approaches and combinations of those approaches. One approach is to create several explicit models based on a general model where the explicit models incorporate assumed values and features. The second approach requires users to customize and specify possible or reasonable equipment features, characteristics, and values. The third modeling approach incorporates reasonable average assumed features and characteristics while allowing users to change or specify values. The actual features and characteristics and their corresponding quantified operation performance values are derived from concessions to mechanical operation and construction and are generally available from equipment manufacturers and process equipment texts.

Process Equipment Cost Estimation

In the past, process equipment costing was based on general equations or charts that related process equipment costs to an attribute of the process equipment, usually the key processing factor. The design engineer had to obtain a cost relation for the selected equipment type, cost an appropriate-sized unit, and adjust the unit cost for pressure, temperature, materials of construction and auxiliary equipment considerations. The engineer then had to estimate installation costs for the equipment, often more than the actual equipment cost. Current process equipment cost estimation software routines carry out all these tasks automatically, with the increased accuracy inherent in computerized procedures and with a few improvements to the process equipment costing practice.

The first major improvement on traditional costing routines is the use of computerized equipment cost databases such as the one available on a website for users of this text.[‡] Computerized databases can contain more types and arrangements of equipment costs as well as allow for noncontinuous, nonlinear cost relations. The accuracy of the equipment availability and cost data available in the equipment cost databases can be improved through regular updates, often many times per year and reflective of real-time equipment manufacturing markets. Additional database accuracy is realized by some software companies through inducements to the equipment manufacturers supplying the equipment availability and cost data that establish the basis for the equipment cost databases. These inducements include reimbursing equipment manufacturers for regular and accurate equipment manufacturing abilities and cost reporting, as well as facilitating engineer-manufacturer direct communication lines to the equipment manufacturers as a form of advertisement.

[†]A. Rutherford. *Mathematical Modeling: A Chemical Engineer's Perspective,* Academic Press, San Diego, CA, 1999.

[‡]The computerized Jan. 2002 cost data for process equipment are obtainable from .

A second major improvement in computerized equipment cost estimation, and perhaps the most significant, is in the estimation of equipment installation costs.[†] As with equipment availability and cost databases, equipment cost routines may now contain extensive databases and relations of equipment installation factors, requirements, and costs for most equipment types.[‡] This allows for much more accurate, and realistic, installation cost estimations.

Other improvements to traditional manual equipment costing are the increased accuracy and reduction of calculation errors inherent in computerized operations, incorporation of equipment sizing utilities into some equipment costing software routines, and the ability to integrate equipment costing routines with other software components for use in optimization and similar tasks.

Process Economic Evaluation

Computerized economic evaluation routines are essentially a compilation of the economic evaluation relations, equations, and routines used in manual economic evaluation.[§] The advantage of computerized economic evaluation routines is again automation, increased speed, and increased accuracy of process economic evaluation. In addition, computerization of economic evaluation routines serves to facilitate integration with other software routines, permitting the computerization of other tasks inherent to process design and evaluation. Of these, the critical process flowsheet generation, evaluation, and analysis and process optimization are dependent on computerized economic evaluation routines.

Heat Integration

Process heat integration is traditionally a complex and problematic part of process design, requiring massive and repetitive calculations. At the same time, heat integration is also a mature and relatively well-understood process with well-established practices and algorithms.[¶] These well-understood practices coupled with a need for substantial and repetitive calculations have been conveniently incorporated into computerized form. Current process heat integration routines are therefore largely a computerized expression of standard heat integration practices, with the added benefit of the ability to incorporate heat integration with other process design and economic evaluation basic functions for increased usefulness, accuracy, and effectiveness.

Process Control

Process control software is divided into a number of areas. First is the analysis of experimental data, simulated or real, for the purpose of establishing, altering, or testing

[†]See Chap. 6 for the various factors that influence installation costs.

[‡]Aspen Tech, *Icarus Process Evaluator 6.0 Documentations,* Aspen Tech®, Cambridge, MA, 2002.

[§]See Chaps. 6, 7, and 8 for a more detailed description of process economic evaluation.

[¶]Process heat integration is discussed in further detail in Chap. 9.

process control schemes and parameters.[†] This is a straightforward task using established control practices that will often include designing and analyzing the experiments under a range of expected process conditions. These tasks are possible using manually driven procedures, although computerization serves to greatly ease implementation and analysis of traditional methods. A second task that process control software handles is the construction of virtual control schemes through the control of computerized process models. This is accomplished through the analysis of process models, establishment of control schemes for the model, tuning of the design parameters for the model control scheme, and finally testing of the control scheme and parameters established earlier. The final use of process control software is in the control of operating processes. Here actual operating processes are monitored and controlled, allowing the implementation of modern real-time, plantwide control and control schemes. The latter two abilities have allowed significant improvements in process control procedures over the traditional control methods since they were impossible, or nearly so, without computerization of process modeling and control.[‡]

Process Optimization

As with process control, the computerization of process optimization takes the form of transferring established optimization practices, as described in Chap. 9, into software in combination with the use of process modelling and analysis.[§] Computerization of established optimization techniques and practices provides significant improvements in both accuracy and speed of optimization, allowing for the usual benefits associated with computerization of calculation-intensive procedures. The ability of process optimization function routines to integrate and manipulate other basic software functions is an additional improvement over manual optimization through gains in speed and accuracy from process models. These gains have led to some modifications of traditional process optimization through increased scope and accuracy, making previously prohibitive procedures possible. In addition to improvements in speed and accuracy to traditional manual optimization and making hitherto impractical optimization practices feasible, computerization has allowed for a number of new approaches and techniques in process optimization that are promising significant improvements.

SOFTWARE CAPABILITIES

The types of software used in process design can be divided essentially into two categories, nonspecific and dedicated. Nonspecific software originally was not developed for use in process design or evaluation. Rather, these tools were intended for more

[†]More details regarding accepted process control practices are widely available in process control texts and general references.

[‡]E. F. Camacho and C. Bordons, *Model Predictive Control,* Springer-Verlag, Berlin, Germany, 1999.

[§]Process optimization in general, the role of software on process optimization, and computerized process optimization are discussed in Chap. 9.

generalized use such as word processing, spreadsheets, databases, numerical calculation, and computer-aided design and manufacture. These software tools, however, serve to ease and facilitate many aspects of the design and evaluation process and in some cases are integral parts of process design and evaluation practice. Further, many aspects of process design and evaluation that may be made with dedicated design and evaluation software can also be done using nonspecific software, but with a cost of greatly increased manual labor and reduced accuracy. Dedicated software, in turn, is intended specifically for use in process design and evaluation, or in procedures inherent to the design and evaluation practice. The main general types of software that are used for process design and economic evaluation include process simulation, process equipment cost estimation, process economics evaluation, heat integration and savings management, flowsheet generation, and process optimization. These are often integrated with other general process design and economic evaluation software or can, at the very least, interface with other software to facilitate data transfer.

General-Type Software

There are a number of aspects of process design and economic evaluation that can be greatly facilitated, enhanced, or even made possible through the use of software that is not specifically intended for use in process design and evaluation practices. These aspects include the initial literature survey, patent search, all necessary numerical calculations, and the preparation and presentation of the design and economic evaluation results.

Research The initial stages of process design almost always require a certain amount of research.[†] This research generally begins with a survey of existing literature and may then require the design, implementation, and analysis of experiments. All these tasks generally require correct use of available software if they are to be carried out efficiently.

Literature research today, and more so in the foreseeable future, requires the use of electronic databases, of both actual data and references. Since pertinent publications are increasingly being computerized and many offer electronic search options with a level of detail not offered by reference databases, it has become necessary to correctly select and search appropriate databases. The actual selection and search of reference databases should take into account the limitations of the databases, particularly for the references included in the databases and the information that the databases contain. Therefore, it is often wise to search a number of reference databases and then search the indicated references electronically. This procedure helps ensure that nearly all potential reference sources are established and that appropriate articles within those databases are retrieved.

The importance of correct design, implementation, and analysis of experimental results cannot be overstated and should be given appropriate consideration. Proper design and analysis of experiments require great care, attention to detail, understanding

[†]See Chaps. 3 and 4 for greater detail regarding research requirements.

of experimental design and procedures, and time for both design and analysis. These tasks can be done manually, although the high quality, ease of use, low cost, and ability to reduce calculation errors so often associated with experimental design and analysis are such that experimental design and analysis software should be used for even the simpler tasks. This is particularly true for scenarios where large numbers of variables must be analyzed, a situation often encountered in process design-related experiments. In addition, it is often convenient to not only control the experiments but also collect the experimental data electronically. While this is not strictly a software issue, as sensors and control equipment are required, the software and software settings used to collect and control the experiment are important, if not strictly required. Here again, great care with consideration of specific software applicability and use of guidelines is required in the selection and use of software utilized to both control the experiments and collect the data. Special attention should be given to the interface between the software and the electric or electronic signal from and to the sensors and control mechanisms. The interface generally includes a set of procedures, often programmable and almost always tunable, that converts electronic and electric signals to values recognized and accepted by the software.

Calculations, Numerical and Analytical Both process design and economic evaluations require significant calculations, mostly numerical but also some analytical. Calculations are required to describe the various design aspects of the process, to evaluate process performance, to generate alternate designs, and to evaluate the economics of a process. These calculations can be done manually, but manual calculations become prohibitive in any design with some complexity because of the intensive calculations required to solve material and energy balances, chemical property estimations, and process equipment operation performances. Instead, it is much more efficient and useful to employ one or more computerized calculation techniques.

First consideration should be given to the various program languages. Programming software with varying emphasis on calculations is widely available, as are many preprogrammed routines that greatly ease calculations.[†] These are fairly standardized, and most have extensive documentation and usage guides that ease programming, although they do not greatly reduce programming time requirements. Further, currently there are a number of available programming software packages that are specifically intended for computation and are therefore more efficient and may be easier to program for calculations. The nature of software programming is such that programmed computations do allow for complete simulations of designs and automatic evaluation as well, although at the cost of prohibitive programming time requirements. However, this is generally not a good choice since this replicates commercial software packages intended for process design and economic evaluation, without the expertise that goes into the development of the commercial software. Instead it is often more efficient to

---

[†]Two programming software packages particularly well suited to the calculation requirements of process design and economic evaluation are Fortran in its various incarnations and Matlab. Both have extensive preexisting functions as well as extensive manuals and guides.

use one of the many software packages intended for process design and economic evaluation and only program the selected section of the process when necessary.

Another useful type of software is the software dedicated to calculation.[†] These are software programs designed to facilitate calculation, removing much of the time-consuming mechanical aspects of programming. Use of these does indeed simplify process design and evaluation calculations, but they are not sufficiently structured or tailored to process design to make the calculation software useful for any but the simpler tasks. Calculation software is therefore similar to programming calculations in that it is quite time-intensive and therefore appropriate only for use in small portions of the design and evaluation process.[‡] In addition, calculation software does lose some flexibility and autonomy when compared to programmed calculations because of the structure imposed by the software. This also limits the usefulness of calculation software for all but simple procedures.

Perhaps the most widely used type of calculation software is the spreadsheet. The latter, with its combination of built-in functions, customizability, visual representation format, and ability to do massive and repetitious calculations effectively, is ideally suited to the types of calculations encountered in process design and economic evaluation. Spreadsheets allow users to make large calculations repetitively, can be set so that intermediate calculation values are available for inspection, and can operate customized programmed routines. In addition, spreadsheets contain many built-in mathematical tools that can be augmented with marketed additions that contain tools customized for tasks encountered in the process design and evaluation. This is of particular interest for experimental design, implementation, and analysis because of the nature of the calculations that occur in these tasks, as well as the proliferation of add-ins specifically designed for experiment purposes. Spreadsheets have the additional advantage of being compatible with many types of software, including software specifically intended for process design and evaluation. This allows for easy transfer of data between software, as well as the outright control and manipulation of other software. This ability is very useful for both optimization and economic evaluation routines, as well as experimental design and analysis. Spreadsheets are powerful enough that simple and small complete process simulation can be made using spreadsheets. However, they are best suited for simpler calculations and the tasks specifically mentioned such as experimental design, implementation, and analysis.[§] They are also well suited to optimization, economic evaluation, and tasks requiring integration of multiple software packages as long as the designs are not too large or are simplified to accommodate spreadsheet limitations.[¶]

[†]Some dedicated calculation software packages are Mathcad, Polymath, and TK Solver. Also, very useful is the calculation-friendly programming language Matlab.

[‡]A. Rutherford, *Mathematical Modeling: A Chemical Engineer's Perspective,* Academic, San Diego, CA, 1999.

[§]P. J. Pritchard, *Spreadsheet Tools for Engineers: Excel 2000,* McGraw-Hill, Boston, 2000.

[¶]More information regarding spreadsheet use can be found in the many spreadsheet engineering calculation guides and manuals, in both traditional printed and electronic, and often free, form.

Reports and Presentations An important part of the process design and economic evaluation practice is preparing written reports and making oral presentations.[†] Software programs to assist with these two important tasks are available in many forms and levels of sophistication. Even though the mainstream software form is sufficient for most writing and the presentation of most design projects, there are specific adaptations and issues unique to process design and evaluation practices. The main deficiencies in the word processing, presentation, desktop publishing, and drawing programs software used for report compositions and presentations are their limited equation-writing and annotation abilities. However, these abilities can be augmented by software specifically designed to supplement the abilities of mainstream writing and presentation software.

SOFTWARE FOR PROCESS DESIGN

A major consideration in the use of software specifically intended for use in process design and economic evaluation is the type of software that is available. Appendix F contains a tabulation of some currently available software directed to process design and economic evaluation. Generally speaking, software today is increasingly intended for use in packages that accommodate all aspects of process design and economic evaluation. Also, many of these tasks can be done with an increasingly great degree of accuracy and with varying degrees of facility, although some tasks, such as process optimization and conceptual design, are not sufficiently supported by current software. It is therefore important to know what types of software are available for use in process design and economic evaluation practices.

Molecular Reaction Databases and Simulators

Most chemical processes involve molecular reactions, often requiring multiple reactions operating in series. Determining the potential reactions that can produce a desired set of products and the reaction mechanisms with their inherent operating condition requirements and limitations is therefore of great interest. This task is facilitated by a number of software tools. First among these are electronic chemical reaction databases that list the chemical reactions and associated kinetic and mechanism aspects of the reactions. Some of these databases are products of federal research and are freely available;[‡] others are commercially available. Some of these commercial electronic chemical reaction databases have the additional advantage of being designed for integration with flowsheet and reaction path design software. A second software tool is in the form of software that seeks to find potentially desirable reactions through molecular simulations. This software tool would be useful in determining potential reactions, although experimental confirmation of results is required and some experimentation

[†]Guides for the use of software for writing and presenting engineering reports are available, as are regularly published articles in trade journals.

[‡]The U.S. National Institute of Standards and Technology (NIST) has a free online chemical kinetics database, listing 38,000 reactions as well as reaction mechanisms and associated kinetic constants.

for establishing molecular characteristics may also be required. While present versions of this tool at the time of publication of this text have been limited, there are indications that molecular simulations may become more useful and prevalent in reaction selection and determination as inexpensive computing power increases.

Chemical Cost Databases

The main purpose of any proposed chemical process is to make a profit. Two key factors in profitability are the selling price of the product and the cost of the raw materials. Chemical costs are generally available as market averages, direct supplier-published costs, or direct manufacturer-quoted costs. Determining one, and often all three, of these types of costs is critical to the process design. While chemical cost determination is possible without the use of computers, the use of computerized databases, electronic publications, and software specifically intended to aid in the task of determining chemical costs does serve to ease and greatly reduce the time required to establish such costs.

Market-average chemical costs are generally most often estimated by trade publications or chemical-specific associations.† These are often available electronically in real time and online to paid subscribers and members or are available at nominal costs. In addition, many chemical suppliers and manufacturers make their product costs available online to the public, again greatly facilitating chemical cost acquisition. Chemical costs are also available using proprietary chemical cost database software. These are databases of chemical costs and available delivery schedules that are updated continuously, often integrated with manufacturer databases and generally designed for integration with other process design and economic evaluation software. This is the most accurate method for establishing current chemical costs, yielding both real-time and time- and market-averaged costs, although it is more expensive than other methods. The ability to integrate these databases with design and evaluation software is also very useful, particularly for chemical reaction path synthesis and separation train design.

Flowsheeting Software

The requirements and principles of flowsheet practices do not translate easily to software. However, as discussed in Chap. 4, there are a number software tools that assist in implementing flowsheet subdisciplines, or even the entire flowsheet process, albeit without the automation usually associated with other software application areas. The difficulty of translating flowsheet principles and practices to software has not allowed the development of completely autonomous computerized flowsheets, thereby necessitating intense user involvement, supervision, evaluation, and guidance for best results. Further, users must often use nearly customized software or spend significant efforts to customize and integrate available software, flowsheet routines, and

†The publication *Chemical Market Reporter* is particularly noted for publishing weekly market-average costs of a wide range of chemicals.

algorithms to the point where the software can be used for the specific project of flowsheet development. An advantage that does mitigate the difficulties encountered in flowsheets that utilizes software when compared to manual flowsheets is the ability to integrate computerized unit operation models into the flowsheet software, thereby allowing for increased quantitative accuracy.

Flowsheet Subdiscipline Software This type of software deals with reaction, heat-transfer, and separation aspects of the flowsheet process. Chemical reaction flowsheet software aims to select, order, and assemble chemical reactions, reaction synthesis paths, and reaction sequences intended to produce desired process products. This is accomplished by using routines that select appropriate reactions from reaction databases, generate potential reaction paths and sequences, and then evaluate and select the lowest-cost paths using a combination of predicted reaction performance and chemical costs. The chemical costs can be either supplied by users or obtained from integrated chemical cost databases.

Another major flowsheet subdiscipline deals with meeting process heat exchange requirements. Heating and cooling are required for many processing operations, aside from actual cooling and heating of materials. While most heating and cooling utility costs are generally low on a per-unit basis, the rates of utilities required for most chemical processes are such that heating and cooling utility requirements account for a significant portion of the operating costs. The significant role that heating and cooling play in operating costs makes proper heating and cooling system design a worthwhile design activity and has led to the development of a number of successful software types aimed at reducing process heating and cooling costs. These software packages include specific unit operation simulators for heat exchangers and other heat-transfer equipment that help reduce costs through proper equipment selection and sizing. Also included in these packages are software for heating train design and software for heat integration, which are at times combined. These two tasks are carried out by software programs that replicate established manual practices. The advantage of software in these practices is in the ability of the software to efficiently carry out the repetitive practices and routines inherent in heat exchange network design and heat integration. Heat exchange train design and heat integration software also have the added benefit over manual practices in the ability to better quantify heat transfer and heat transfer effects. This allows a much tighter and more accurate estimation of heat transfer with a more accurate and worthwhile heat exchange network design and heat integration.

The last major flowsheet subdiscipline is separation train design. The basic principles and operations of separation train design software are fundamentally those that are employed in manual separation train synthesis, but software does exhibit some advantages over manual practices. The first major difference is in the source of physical property data. Separation train design software packages determine these data, using integrated physical property estimation routines and databases. This allows a more complete determination of property differences that are potentially exploitable for separation purposes than is generally possible in manual analysis. Relative properties of species are examined to determine maximum property differences between the species to be separated. Again, software does this more extensively and accurately than is possible manually. These properties are then correlated to separation techniques that exploit the differences in physical properties; and these, in turn, are used to select processing equipment derived from the correlated separation techniques. This step is also done more quickly and completely with software. All the various separation techniques and equipment are then combined into every possible combination and examined to see if they meet the process separation requirements. This last step is where software really demonstrates supremacy over manual separation train design since this is a repetitive and calculation-intensive step. The last major advantage of separation train design software is the ability to use process equipment models in the evaluations of potential processing techniques and equipment combinations. Manual execution of this step is not easily combined with accurate process equipment models, so that the overall separation train performance as determined by separation train design software is more accurate than the performance obtained manually.

Complete Flowsheet Software These types of flowsheet programs are used to complete the entire flowsheet process for a given product. These programs are essentially packages that combine all the capabilities of the previously described flowsheet subdiscipline software, integrated and governed by the results of the evaluation routines.[†] Proven and tested complete flowsheet software packages were not available at the time of this review of flowsheet software. However, there are a number of indications that such software may become available in the next decade. These indications from recent publications show progress in developing the processing technique selection, result evaluation, and integration routines necessary for computerized flowsheets.

One type of currently available flowsheet software does offer a truncated form of complete flowsheet ability. This software type requires users to specify reactions and reaction equipment, while requiring some user guidance in the selection of some major processing steps, especially in the separation technique selection. This is a useful software type, and current implementations have been positively received and tested, serving to further recommend the use of software in flowsheeting.

Unit Operations Simulators

One of the most common types of software used in process design, and eventually in the economic evaluation of a process, is process simulation. Process simulators use a combination of process operation models, physical property estimation routines, and numerical solution methodologies to create process models. These process models generally evaluate and determine material and energy balances for streams and process equipment as well as determine performance estimates for specified equipment. Differentiation of the various available simulation software depends on a number of factors including equipment behavior, chemical-physical property estimation capabilities, and the number of operations the software can model. There is also a differentiation among simulation software in the mathematical model type used, either discrete or continuous, and the time nature of the simulation, either steady-state or dynamic. The

[†]See Chap. 4 for descriptions of flowsheet procedures, practices, and requirements.

effect and importance of the latter factors are more difficult to estimate, but evaluating these with regard to simulation requirements is critical for proper process simulation software.

Mathematical Model Type Models of this type are classified according to their treatment of model dimensional dependence. The most general mathematical models for chemical processes require consideration of gradients in both time and spatial coordinates. The gradients that are involved, such as composition or temperature, can therefore be visualized with up to four dimensions and require the use of partial differential equations for proper mathematical representation. These mathematical models are discrete or distributed[†] because the model properties depend on more than one dimension; accordingly, their properties are said to be discrete or dimensionally distributed. Solving the partial differential equations necessary to depict discrete mathematical models can be very difficult. It is therefore often useful to assume that gradients manifest themselves in only a single dimension or to assume that gradients are much larger in one dimension than they are in others, and thus essentially assume only one-dimensional gradients. These models incorporating assumptions of dimensional continuity or lumping of dimensional effects are identified as continuous or lumped models. The reduction in dimensional order dependence also allows the differential equations necessary to mathematically model a process to be reduced from partial to ordinary form. This reduction to single-dimension representation and use of ordinary differential equations greatly reduces the effort and time requirements of solving the mathematical equations representing the chemical process being modeled. It is therefore often expedient to make continuous models of a process or to simplify discrete models.

In general, most unit operation simulation software programs are based on continuous rather than discrete models. This allows significant savings in computational requirements and for most cases is sufficiently descriptive of actual operating equipment. However, some unit operations, such as an extractive distillation column, do require some level of discrete behavior modeling because of their greater multidimensional dependence. These types of models are available and supported by various software developers, with varying levels of discreditization. Many can be integrated with other software and software components. It is therefore important to consider the type of mathematical model required to fulfill the process simulation requirements based on the sensitivity of a process to multidimensional effects, the difficulties inherent in discrete-model use, and the loss of accuracy when continuous models are used to simplify models that are really discrete.

Steady-State and Dynamic Simulators Simulation of processes can be either dynamic or steady-state.[‡] Steady-state simulation software endeavors to establish the constant process behavior over a long time period. This type of simulation is especially

[†]A. Rutherford. *Mathematical Modeling: A Chemical Engineer's Perspective,* Academic, San Diego, CA, 1999.

[‡]Software manuals, including software by Aspen Tech, AEA technologies, Chemstations, Scientific Computing, and others, amplify these simulation types.

useful for establishing material and energy balances and compositions, as well as investigating long-term equipment performance, and thus provides a reasonably good estimate for process equipment, raw material needs, and utility requirements. Steady-state simulations are also, in general, easier to set up and solve than dynamic simulations. However, because of their disregard for time, steady-state simulators do not account for process performance variations over time including such aspects as equipment holdup and cyclical requirements. Further, steady-state simulations can only simulate continuous processes and may not describe nonlinear behavior with sufficient accuracy and therefore are not able to fulfill simulation requirements for some processes.

Dynamic simulators evaluate and model processes at a series of sequential time steps, where the entire process model is reevaluated at the end of a time step based on the model obtained at the beginning of the time step. Generally, it is possible to set the time step size, the number of steps to be modeled sequentially, or an overall modeling time period. This allows simulations of time-sensitive processes, cyclical process requirements such as start-up and shutdowns, nonlinear systems, and noncontinuous processes. The ability to model time-dependent behavior is especially valuable for nonlinear systems and for use in creating, establishing, setting, and testing process control schemes. Also, the types of variables that can be included may be more reflective of real-world situations than steady-state simulations. Last, dynamic simulations can be set to model the behavior of processes over entire lifetimes of cyclical behavior and fouling. The lifetime simulation can then be used to specify equipment requirements and sizes, and thus costs, more accurately. The increase in capability of dynamic simulations over steady-state software does, however, come at the cost of much greater computing capability and is generally more difficult and time-consuming to develop and operate. A good compromise is therefore to create steady-state simulations and only dynamically simulate particularly sensitive and nonlinear portions of the design.

Specific Unit Operation Simulators These types of simulations are prepared for a specific piece of equipment. They are developed using a general process equipment model, with all but the operating condition variables preprogrammed. Prespecification and programming of unit operation simulators can yield very accurate, if inflexible, modeling. This is especially true when the simulators are developed by equipment manufacturers since this allows incorporation of the empirical design considerations and experimental operation data for the operating equipment into the simulation models.

Type-Specific Unit Operations Simulators These are simulators for a single type of chemical process unit operations. They are based on general modeling equations and principles, incorporate a certain degree of flexibility in configuration specifications, such as tube arrangement in heat exchangers, and allow variable performance specifications. Type-specific unit operation simulators can therefore model a range of configurations for a single type of unit operation as long as the performance variables necessary to describe specific performance aspects of the equipment are known. These types of models are available, as are generalized performance variables for the models, from general software designers as well as companies and associations that specialize in certain types of equipment. Additionally, some manufacturers provide performance variables and efficiencies for their equipment that can be used in the more widely used type-specific simulators. This allows for good modeling accuracy while allowing some customization and flexible use of the unit operation simulator software.

Overall Process Simulators These types of software programs are generalized models intended to simulate complete processes. They are perhaps the most useful types of simulators for process design and economic evaluation because simulation of both the individual unit operation and the effect of unit interactions is possible. These types are generally structured as multiple interconnected type-specific unit operation models which allow complete process unit operations and stream simulation. As such, they have the same configuration and operation variable requirements as do the type-specific unit operation simulators. This permits more accurate modeling of complete processes, even though overall process simulators do not model single unit operations with the same level of accuracy as do prespecified unit operation models. A further feature of many overall process simulators is their ability to incorporate user-programmed equipment models and some prespecified or type-specific simulators. This feature yields software that combines ease of use and reasonable accuracy for entire processes, while allowing highly accurate and flexible modeling of unusual, customized, or key equipment.

Piping System Design

One of the later stages of process design requires that piping systems be prepared to meet process fluid-transfer requirements, as described in Chap. 3. Piping system design is a well-understood, calculation-intensive, and repetitive process that can be greatly simplified through use of one of the many piping system design software packages currently available. These software packages will design piping systems for given sets and arrangements of processing equipment, including in many cases all the necessary valves, pumps, compressors, and pressure change requirements. Further, some of these software packages will generate piping system isometrics and test their spatial position to ensure that they do not interfere with other piping, ducts, or processing equipment. An added advantage to many such piping system design software packages is their ability to take process simulations as an input source from which the piping system can be derived.

Plant Layout

Another of the later design process stages aided by software use is the determination of appropriate and optimal plant layout. This type of software carries out one or more of a number of basic tasks including determination of equipment spatial requirements, equipment spatial orientation, minimization of building size, and ensuring equipment spatial placement. One of the advantages of using such software is that the programs often have built-in routines to determine the actual, access, and safety spatial requirements for most equipment. Further, these tasks are often integrated, both with each other and with piping design and process simulation software. The plant layout can be displayed and examined using three-dimensional rendering to help ensure that equipment, piping, and ducting do not interfere with spatial flow or with each other.

Economic Evaluation

Process economics software has two basic functions: process equipment cost estimation and process economics evaluation. Specification of process equipment for cost evaluation purposes can be by manual user selection of equipment from types available in the software, automatic integration with process simulators, or reading process simulations. Software that determines equipment automatically will often use simulation information such as material stream conditions, material stream compositions, and actual equipment operation requirements to select and size the specific process equipment to fit the simulated equipment type.† They also determine the required equipment construction materials using the simulation data.‡ As mentioned previously, equipment costs can often be reduced if standard commercial equipment types, sizes, and construction materials are selected. Some software will implement this automatically or suggest the user do so; others do not.

The second function of process economics software, the economic evaluation of processes, depends on a number of economic factors. Some of these factors are relatively constant across geographic locations and over a period of a year or more. These factors, such as tax rates and cost of capital, are therefore generally prespecified in more sophisticated economic evaluation software. Other more project-specific factors, such as local wage rates and utilities, must be entered by software operators. In general, software will prompt users for necessary economic evaluation factor values. The other requirements for economic evaluation are equipment costs, process utility requirements, and chemical costs. The first of these can be either acquired directly from separate or integrated process equipment cost evaluation software or entered manually. Process utility requirements can be acquired similarly, or they can be read from simulator duty calculations and entered manually. Chemical costs are established similarly from integrated chemical cost databases, or manually from separate databases or other chemical cost sources. Once all necessary economic factors, equipment costs, utility costs, and chemical costs are established, the software evaluates the process, producing the standard set of economic indicators used to determine process economic viability.

There is an additional advantage to the speed, accuracy, and ease-of-use benefits inherent with economic evaluation software, and that is the ability to be integrated with process simulators under the control of optimization software. This integration and control allows highly accurate, automated optimization of processes.

---

†Equipment type selection is described in Chaps. 12 through 15.

‡Equipment materials of construction selection is described in Chap. 10.

SOFTWARE SELECTION

Selection of software for use in process design and economic evaluation is often a delicate balance between software capabilities and costs. The integration of software supporting various design and evaluation tasks does serve to ease and improve software use, at an often greater financial cost than that of less integrated packages. It is therefore important to assess software capabilities and the advantages that software use affords over manual practices as well as the cost of software.

Appendix F contains a tabulation of some available design and economic evaluation software and various software developer contacts. Also, the American Institute of Chemical Engineers (AIChE) maintains a database of available software, software types and applications, and software manufacturers and contact information.† These are good places to locate product manufacturers. Various trade publications will also contain both software reviews and product advertisements that can be used to at least establish the availability of desired software. Another source of software availability is companies that specialize in areas similar to the focus of the design and evaluation project of interest. Once manufacturers for the type of software of interest are established, their actual capabilities and intended use can be found in product literature, usually available on company product websites. In addition to providing product literature, most software manufacturers provide demonstration software upon request. These demonstrations range from examples showcasing software capability to fully operational installations that are limited to a month or so of operation. Demonstration software acquisition sometimes requires direct contact with manufacturer sales representatives, and its availability is not publicized. It is therefore good practice to inquire about demonstration software when contacting sales representatives for software cost quotations. When contacting sales departments, it is also helpful to inquire about other users of the software of interest and to request contact information for them. Consulting with these established software users can be most enlightening.

These practices, especially the examination of demonstration software and software costs, should result in sufficiently accurate and detailed information for a balanced evaluation of software costs and abilities. At this point it is again a good idea to contact companies that already use the candidate software packages, if this has not already been done.

SOFTWARE USE

Software is a tool, and how it is used is as important as the tool itself. This is as true for the simplest pump simulators as it is for the integrated simulation and economic evaluation packages. Fortunately, ease of use is a primary consideration for software manufacturers. Software is steadily becoming easier to use, often including user-guiding routing or multiple levels of complexity. Despite this trend it is still critical to carefully read product manuals and often wise to enroll in software use courses and seminars.

---

†Available on the AIChE website and published annually in *Chem. Eng. Prog.*

With this in mind there are a number of general procedures and practices common to most, if not all, software.

It is vital to remember and consider the basic principles behind software at all times. Software is an aid, *not* a replacement for users.

Physical Property Estimation Guidelines

Physical properties are critical to process design. Selection of appropriate physical property estimation methods is therefore a key part of the design practice. When selecting physical property estimation methods, users must first consider the capabilities, intended uses, and limitations of the methods. Of these, availability of method data for specific chemical species is paramount. The limitations and inaccuracy of models can also be somewhat reduced by employing multiple-property estimation methods. That is, create a simulation using one property estimation method, then change the method. This practice is generally easily done, and will provide a sense of the sensitivity of the simulated process to property estimation method limitations.

A general guide to physical property estimation model selection is presented in Table 5-2 and Fig. 5-2. Users should also consult the manuals for the specific software that is used since the definitions, names, and physical property estimation methods do vary from software to software.

Process Simulation Guidelines

Process simulations are essentially a series of material and energy balances coupled with process equipment models and physical property estimation software. The

Table 5-2 Recommended physical property estimation models for various applications

Application, specific use, or specific chemical	Recommended physical property estimation model†
Air	CS, GS
Ammonia	PR, SRK
Azeotropic separation alcohol systems	Wilson, NRTL, UNIQUAC, and variants
Carboxylic acids	Wilson, NRTL, UNIQUAC, and variants
Esters	Wilson, NRTL, UNIQUAC, and variants
Phenols	Wilson, NRTL, UNIQUAC, and variants
Liquid-phase reactions	Wilson, NRTL, UNIQUAC, and variants
Cryogenic gas processing	PR and variants
Aromatics	Wilson, NRTL, UNIQUAC, and variants
Ethylene tower	Lee-Kesler Plocker
Ethers	Wilson, NRTL, UNIQUAC, and variants
Solvent recovery hydrocarbon stripping	Wilson, NRTL, UNIQUAC, and variants
Substituted aromatics	PR, SRK
Substituted hydrocarbons	PR, SRK
Vacuum towers	PR and variants, GS, Braun K10
Steam	CS, GS

†Consult Table 5-1 for further details on the estimation model listed.

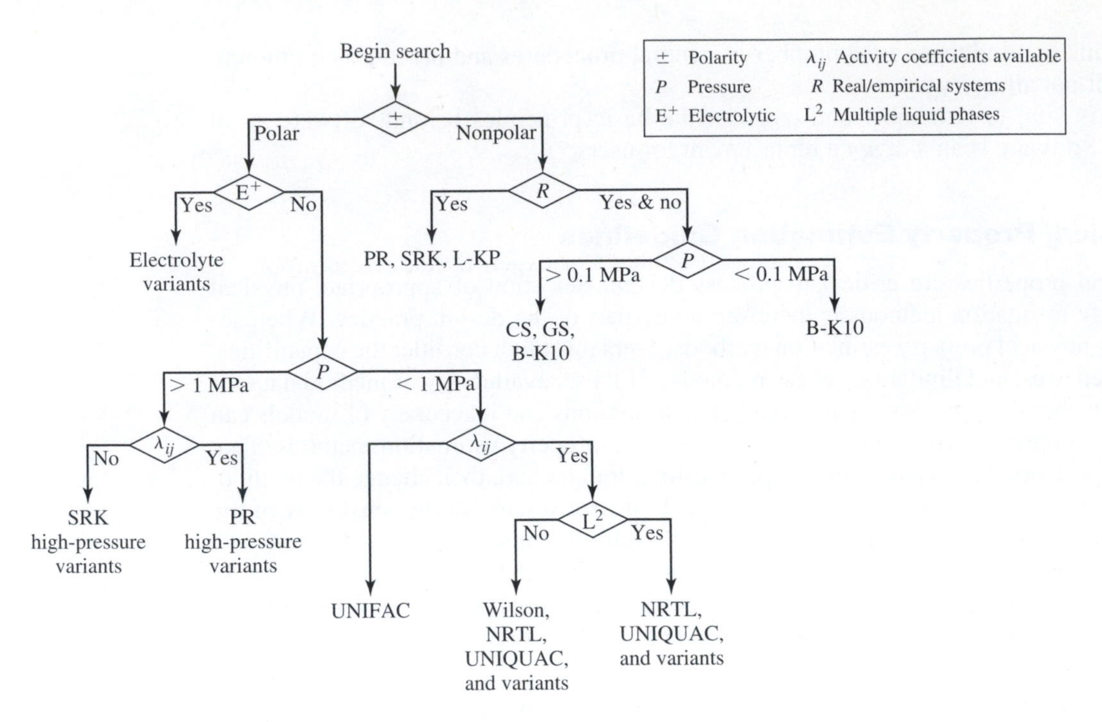

Figure 5-2
Graphical guide to physical property estimation model selection

following are suggested practice guidelines:

1. Start at the beginning and proceed to the end. Do not proceed with simulation until all the connected equipment and streams are evaluated and specified. For cases where streams are recycled, make informed estimates of stream flow rates and compositions based on material and component balances. This is critical to ensure that recycle stream effects are propagated through the system before the recycled stream values are determined.
2. It is often useful to isolate particularly difficult sections or unit operations. These can be studied and optimized separately from the overall process and then reintegrated once they are properly arranged. Failure to do this may cause the overall simulation to crash and will generally take longer to solve due to the increased computational requirements.
3. Use experimental design and analysis techniques to improve designs. Treat the simulation as an experimental setup; then run experiments with various equipment and flow settings. These results can be analyzed for qualitative and quantitative trends in the simulation.
4. Use realistic parameters. The simulation may allow reflux ratios in excess of 100; real equipment may not.
5. Make sure numerical solution parameters are appropriately set, as are all equipment tolerances.

6. Remember to use process simulation software logical tools such as goal seekers, spreadsheets, and recycle controllers.
7. Make sure correct phase equilibria type of interaction coefficients are selected in the physical property estimation model. Also, the user should test the thermodynamic data and routines to ensure that they produce the known, prominent features of the modeled mixture.
8. It may be wise to consult software consultants or expert users, before problems arise.
9. Do not overlook the underlying principles of unit operations and how they operate. Several examples of these operation principles with some general use hints are provided in Table 5-3.

Table 5-3 Selected unit operation underlying principles and useful hints

Operation	Principle	Useful hints
Distillation	Separation by differing volatilities	Uses energy inputs, energy removal, and multiple real or virtual equilibrium stages to separate mixtures by differing volatilities. Energy in the system is controlled through coolers, condensers, and reboilers.
Extraction	Phase affinity	Correct species selection and their ratios are key. Phase equilibrium is often sensitive to temperature change
Heat exchange	Fluid-fluid heat transfer	Do not disregard pressure losses
Reaction	Chemical reaction	Select correct type of reactor, and ensure reactions are correctly specified for type, phase, and reaction mechanism and constants. Ensure proper temperature and pressure operation conditions

Steady-State Simulations

1. Steady-state simulators require the use of logical operations, such as recycling, equipment residence times, and utility requirements.
2. Steady-state simulators are often not sufficiently sensitive to unreasonable operating parameters; extra caution must be taken to avoid these.
3. Determine if equipment heat losses are important to the simulation, and select appropriate heat loss models to calculate these losses.
4. Consider which parts of the simulation are appropriately simulated in steady state and which require dynamic simulation.

Dynamic Simulations

1. Remember to make correct specification types for simulation, such as pressure rather than mass flow to set material stream flow rates.
2. Take extra care in specifying numerical calculation methods and parameters. Be wary of preset values.

3. Consider decreasing time step size if simulation crashes, and increasing time step size when establishing long-term process behavior.
4. Consider which parts of the simulation are appropriately simulated in steady state and which require dynamic simulation. Dynamic simulation is generally more time-consuming.
5. Consider the time scale required to achieve steady state.

Heat Exchange Design and Integration

1. Pay attention to setup parameters, especially pinch and convergence criteria.
2. Examine overall results for consistency.

Separation Networks

1. Pay attention to property estimation packages.
2. Evaluate the need for distributed versus continuous models.
3. Pay attention to solution mixing parameters.

Piping

1. Pay attention to heat loss model.
2. Verify friction loss parameters for piping material and fittings.

Economics

1. Ensure correct local economic factors, such as wage rates.
2. Ensure correct utility selection. Utility availability and costs change locally, especially for waste disposal standards and costs.
3. Ensure correct evaluation method and parameters, for instance, the depreciation method.
4. Investigate the benefits of using standardized equipment types and sizes over those selected and established by the software.

Optimization

1. Define the objective function.
2. Evaluate more general, less defined representations of the process first, then fill in gaps. (Go down the hierarchical decomposition method as described in Chap. 4.)
3. Use multiple solution methods to ensure appropriate numerical solutions.

AVOIDING PITFALLS IN SOFTWARE USE

Design-related software is an invaluable tool, allowing for design accuracy that could not be imagined 20 or 30 years or even a decade ago. In addition to exponential increases in realism, flexibility, and applicability, design-related software has become much easier to use and understand. The benefits of the ever-increasing trend supporting increasingly easy-to-use software with steep learning curves are many. There are, however, a number of potential design and economic evaluation pitfalls that result from the trend supporting the creation of software that is easy to use and learn. Avoiding potential software use-related errors in the design process therefore becomes as important as the actual software use.

Graphic Interface and Ease of Use

There are four main sources of errors in design and evaluations that stem from attempts to ease software use. The first is due to the practice of embedment, or showing multiple operations as one object on the screen. Many chemical process software packages embed complex operations composed of several pieces of equipment within the larger overall simulation. This reduces the clutter that occurs in compound simulations, thereby providing a clearer view of the process simulation. Embedment can also reduce computational tasks by splitting up large calculation tasks into smaller, more manageable simulation routines. Embedment can, however, be a source of simulation error as it is easy to overlook design and simulation issues in the embedded operations, functions, or procedures. Further, the interface between the embedded and overall simulations in the past has shown errors in transferring information back and forth. The resultant transfer error can then give an impression that the simulation is complete, when in fact only the overall simulation is correct. The obvious example of embedment-related errors is the treatment of distillation columns by many of the off-the-shelf simulation software packages. The simulation scheme displays a single unit, whereas the embedded simulation of the distillation column contains the reboiler and condenser as well as pumps and any number of potential subcomponents. The embedded simulations may also contain entire multiple separation columns even though the simulation flow diagram only graphically displays a single column. Thus, embedment can lead to misleading equipment specifications and resulting incorrect economic evaluations. Further, the overall and embedded simulation interface may not function properly, resulting in false indications of embedded simulation completion or solution.

A second source of design and evaluation error resulting from attempts to make software easier to use is attributed to numerical calculation tolerance levels.[†] The numerical evaluation techniques used in software are very sensitive to convergence criteria specified by the software. These tolerance levels can generally be changed manually and have default values specified by the software programmers. The specified default tolerances are often sufficient for simple calculations. However, default tolerance levels may be inappropriate for delicate, nonlinear, or complex systems. In fact, tolerance levels should be seriously considered and evaluated in any situation where nondestructive error propagation is possible. Automatic use of default tolerances without due consideration can lead to faulty simulations, estimates, and thus incorrect designs. For instance, 0.1 percent mass balance tolerances are quite reasonable for single-unit simulations such as those for heat exchangers. Use of this seemingly

[†]See general numerical evaluation method texts for more information about the effect of tolerance levels on numerical solutions.

reasonable tolerance in an average refinery simulation containing mass-transfer operations consisting of 6 distillation columns, 15 pumps, and 14 heat exchangers can lead to an unacceptable propagated mass error of 3 percent or more. A good approach to avoid these problems is to begin with larger convergence criteria and reduce them after convergence, until convergence limits are reached.

The third common source of software use error resulting from aesthetic and user-friendliness considerations is the utilization of nonautomatic procedures. In an attempt to ease software use, many packages allow users to employ various degrees of accuracy as well as the complexity associated with this use. This trend can cause users to overlook steps required for accuracy, but not for software operation. For example, many software packages utilize external processing for the completion of such items as separation tray sizing, heat exchanger area sizing, and various equipment pressure drops. Failure to employ and use external software utility functions such as these can be a source of considerable faults.

The final, and perhaps leading, source of software ease-of-use related error is due to built-in default software assumptions. All software packages have default values for the various factors required to operate the software. Most of these are reasonable and appropriate. However, the broad-spectrum use of software does not allow for very accurate defaults. It is therefore up to the software user to determine the correct default values appropriate for the particular circumstance. Ignoring the need to determine and correctly change default values leads to major problems, ranging from simulation error, to economic evaluation error, to reporting grammatical errors. Some of the most common default value types that must be set correctly are provided in Table 5-4.

Thermodynamic Property Packages

As discussed in greater detail earlier in the chapter, correct selection of thermodynamic property packages is critical to accurate, worthwhile simulations. However, the validity of a simulation that only works under one set of thermodynamic property estimation methods is questionable as thermodynamic property packages are estimates. Therefore, use of multiple property packages to simulate a process is recommended. The benefits of evaluating designs under various assumptions upon which a number of thermodynamic packages are based yields a more robust design. Furthermore, it is easy to perform in almost all simulators available today. Any associated design changes required for accommodating multiple thermodynamic packages are reflective of the need for process design robustness required for real-world situations, and equipment specifications that meet these requirements yield a correspondingly more realistic process. Thus, utilization of multiple thermodynamic packages in simulating a design gives an added degree of realism, robustness, and acceptability while requiring relatively little effort.

Simulation Realism

Perhaps the most serious source of software use errors arises from unrealistic use of simulation software. Many, if not all, current simulation software programs can

Table 5-4 Commonly defaulted software parameters

Application	Commonly defaulted parameters
Simulators	
Thermodynamic property package	Binary coefficient types Possible phases
Equipment	
Pumps and compressors	Efficiencies Values and types
Liquid storage vessels	Liquid level
Vapor storage vessels	Pressure
Separation columns	Tray/packing efficiencies Subcomponent equipment types Specification types Reflux ratios Draw types Tray performance
Heat exchangers	Type Pressure drop Heat-transfer efficiency
Economic evaluators	Utility availability Utility costs Tax rates Local economic factors Salaries Construction costs

function in an unrealistic manner. It is therefore critical that the software user employ sound engineering judgment when employing software capable of unrealistic behavior, as the point of simulation is to approximate reality. Simulator unreal behavior extends into a number of areas. First, simulators allow unrealistic parameters to be used, generally without warning of the improbable scale of parameters used. These generally apply to such factors as efficiencies, sizing of units, recycle ratios, and other such criteria. Table 5-5 displays some of the most commonly encountered unreal simulation parameters and an explanation of their effect.

Use of inappropriate parameter scales can be prevented to a certain extent by understanding the various levels of detail provided by the software, especially simulations. The utilization of dynamic simulations, which emulate material and energy flow more accurately, gives a much more realistic behavior picture for the design. Because dynamic simulations are more realistic in their treatment of material and energy flow, they will generally not work very well for unrealistic parameters and will often indicate that unrealistic parameters are specified.

Another major potential source of error is due to the utilization of logical operations in simulators. Many simulators allow the use of logical operations. These commonly include recycle, splitters, mixers, and theoretical heat-transfer equipment. While these logical operations are useful, care must be taken when they are employed since

Table 5-5 Examples of commonly encountered unreal simulation parameters

Operation	Parameter	Error	Explanation
Distillation columns	Reflux ratio/ reboil ratio	High reflux/reboil ratios are attractive to use since they allow easy control and good performance	High reflux and reboil ratios, compared to their minimum, results in poor economics.
	Pressure drop	No pressure gradient across the tray/packed section	Lack of a pressure gradient affects vapor flow
Extraction columns	Pressure drop	No pressure gradient across the tray/packed section	Lack of a pressure gradient can result in anomalous numerical solutions and similar simulation results
All operations	Pressure drop	No frictional pressure losses	Avoiding frictional losses does not give an accurate picture of necessary pressurizing equipment
Pressurization units	Efficiency	Unrealistically high efficiencies	Does not portray real equipment requirements and possibly allows physically impossible processing

enacting similar processing with real equipment can be difficult, if not impossible. For instance, a two-vapor-stream mixing operation can be as simple as connecting two pipes to a third through two one-way valves. On the other hand, a liquid stream splitter requiring one logical operation may require a series of distillation and extraction columns, with all their associated auxiliaries. Of course, some processes carried out by logical operations are impossible.

EVALUATION OF SOFTWARE RESULTS

Software is a sophisticated, often complex tool that can seemingly operate independently and with near consciousness. The sophistication and power of software unfortunately result in a tendency of users to assume that software results are correct with little or no discrimination. This is a tendency that must be avoided. Software is a tool and only responds to the command of the user. In fact, the very complexity of software creates an even greater potential for inaccuracy in its use. But how can software results be evaluated when they often arise from calculations so complex that manual reconstruction is not feasible? There are three ways of ensuring the quality of software results. The first is the proper use of the software. A majority of erroneous software results are due to human error in employing the correct software, entering the correct design parameters, or following the correct software procedures. Paying attention to all the steps of software use, use of real and sensible parameters, and use of judgment at all times greatly reduce the possibility of human error in setting up the software.

Next, manual approximation of the sensibility for each individual operation is generally possible without too much difficulty and can again reduce human error. Finally, examining results with a critical mind set can help determine if software errors are present. An example of critical evaluation of results is determination that it is not reasonable for a carbon dioxide stream at 350 kPa and 700 K to be in the liquid state. Investigation of the simulation reveals no errors, thus indicating a potential software error. The key to correct software use is for the user to be active, critical, and thoughtful at all times.

CHAPTER

6

Analysis of Cost Estimation

An acceptable plant design must represent a plant that can produce a product which will sell at a profit. Initially, sufficient capital must be committed to construct all aspects of the facility necessary for the plant. Since net profit equals total income minus *all* expenses, it is essential that the chemical engineer be aware of the various types of costs associated with each manufacturing step. Funds must be available for direct plant expenses, such as those for raw materials, labor, and utilities, and for indirect expenses, such as administrative salaries, product sales, and distribution costs. In this chapter, investment and plant operation costs are reviewed as well as cash flow and gross and net profits.

CASH FLOW FOR INDUSTRIAL OPERATIONS

Cash Flow

Figure 6-1 shows a simplified representation of the flow of funds for an overall industrial operation based on a corporate treasury serving as a reservoir and source of capital. Inputs to the capital reservoir normally are in the form of loans, stock issues, bond sales, and other capital sources, and the cash flow from project operations. Outputs from the capital reservoir are in the form of capital investments in projects, dividends to stockholders, repayment of debts, and other investments.

Figure 6-1 illustrates capital inputs and outputs for an industrial operation using a tree growth analogy, depicting as the trunk the total capital investment, excluding land cost, necessary to initiate the particular operation. The total capital investment comprises the fixed-capital investment in the plant and equipment, including the necessary investment for auxiliaries, and nonmanufacturing facilities, plus the working-capital investment. Some of the capital investments can usually be considered to occur as a lump sum, such as the provision of working capital required at the start of operation

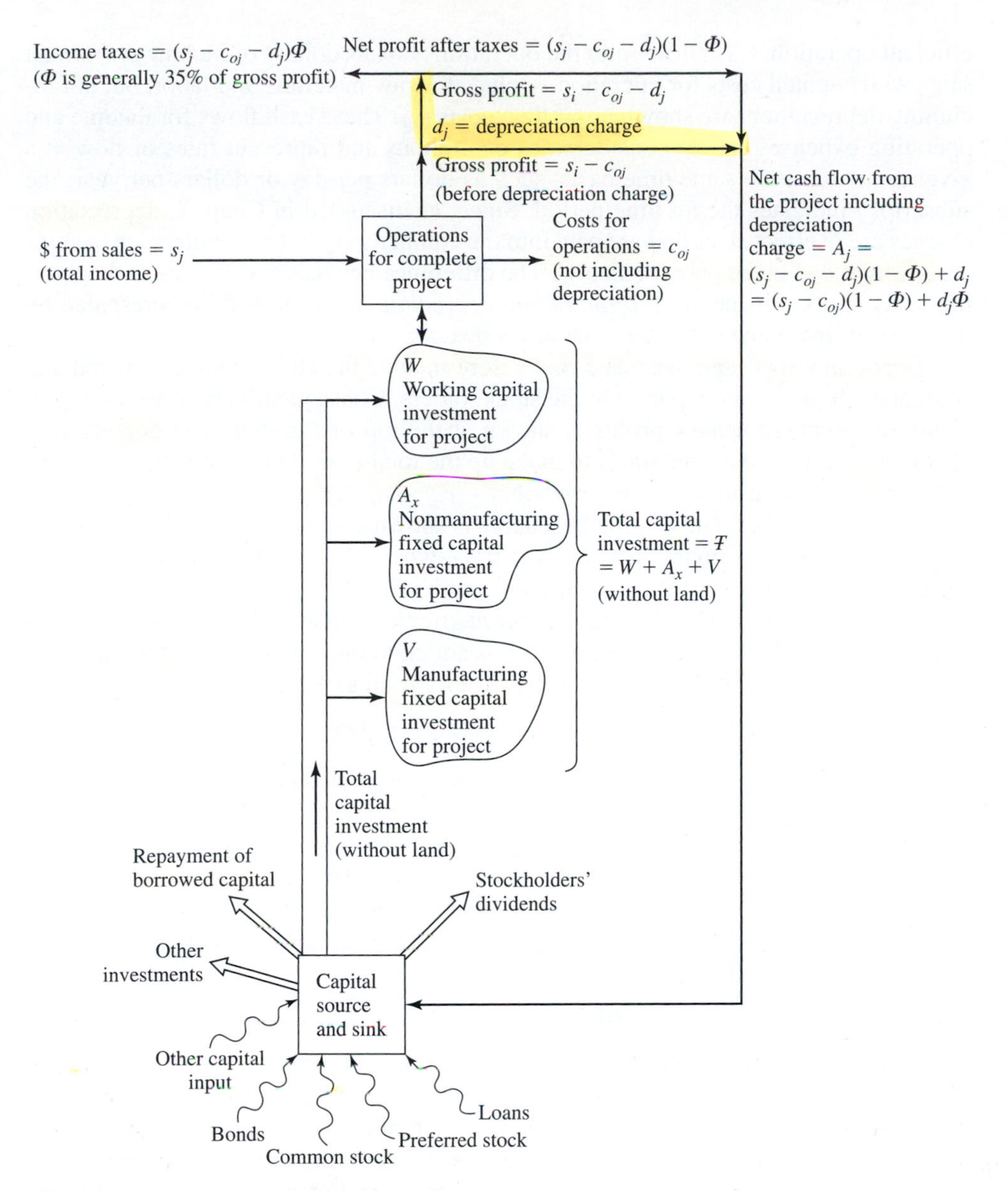

Figure 6-1
Tree diagram showing cash flow for industrial operations

of the completed plant. The flow of cash for the fixed-capital investment is usually spread over the entire construction period. Because income from sales and the costs of operation may occur on an irregular time basis, a reservoir of working capital must be available to meet these requirements.

The rectangular box near the top of Fig. 6-1 represents the operating phase for the complete project with working-capital funds maintained at a level acceptable for efficient operation. Cash flows into the operations box as dollars of income s_j from all sales while annual costs for operation, such as for raw materials and labor, but not including depreciation, are shown as outflow costs c_{oj}. These cash flows for income and operating expenses can be considered as continuous and represent rates of flow at a given time using the same time basis, such as dollars per day or dollars per year; the subscript j indicates the jth time period. Since, as discussed in Chap. 7, depreciation charges are in effect costs that are paid into the company capital reservoir, such charges are not included in the operation costs. The difference between the income and operating costs $s_j - c_{oj}$ is the *gross profit before depreciation charge* and is represented by the vertical line rising out of the operations box.

Depreciation is subtracted as a cost before income tax charges are calculated and paid, and net profits are reported to the stockholders. Consequently, removal of depreciation as a charge against profits is shown at the top of Fig. 6-1. The depreciation charge d_j is added to the net profit to make up the total *cash flow* for return to the capital reservoir. The resulting *gross profit* of $s_j - c_{oj} - d_j$ that accounts for the depreciation charge is taxable. The income tax charge is shown at the top of the diagram where it is removed in the amount $(s_j - c_{oj} - d_j)(\Phi)$, where Φ is the fixed income tax rate designated as a fraction of the annual gross profits. The remainder after income taxes are paid $(s_j - c_{oj} - d_j)(1-\Phi)$ is the *net profit* after taxes that is returned to the capital reservoir. When the depreciation charge d_j is added to the net profit, the total project-generated cash flow returned to the capital reservoir on an annual basis is

$$A_j = (s_j - c_{oj})(1-\Phi) + d_j\Phi \tag{6-1}$$

where A_j is the *cash flow* from the project to the corporate capital reservoir resulting from the operation in year j in dollars, s_j the sales rate in year j in dollars, c_{oj} the cost of operation (depreciation not included) in year j in dollars, d_j is the depreciation charge in year j in dollars, and Φ the fractional income tax rate. This cash flow is used for new investments, dividends, and repayment of loans, as indicated by the various branches emanating from the capital source in Fig. 6-1, as well as for retained earnings.

Cumulative Cash Position

The cash flow diagram in Fig. 6-1 represents the rates of cash flow with s_j, c_{oj}, and d_j all based on the same time increment. Figure 6-2 is for the same type of cash flow for an industrial operation except that it depicts the situation as the *cumulative cash position* over the life cycle of a project. The numerical values are only for illustration.

In the situation depicted in Fig. 6-2, land value is included as part of the total capital investment to show clearly the complete sequence of steps in the full life cycle for an industrial project. The zero point on the time coordinate represents the point at which the plant has been completely constructed and begins start-up of operation. The total capital investment at the zero time point includes land cost, manufacturing and nonmanufacturing fixed-capital investment, and working capital. The cash position is negative by the amount of the total capital investment at zero time. In the ideal situation, revenues come in from the operation as soon as time is positive. Cash flow to the company treasury, in the form of net profits after taxes plus depreciation, starts to

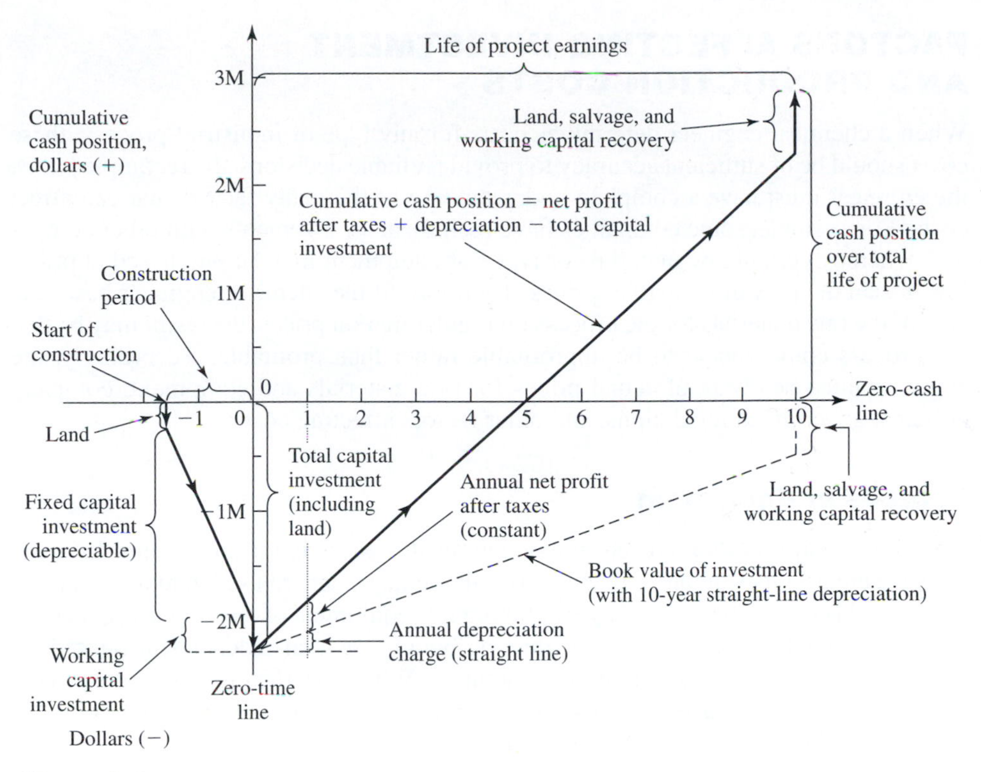

Figure 6-2
Graph of cumulative cash position showing effects of cash flow over the full life cycle for an industrial operation, neglecting the time value of money

accumulate and gradually repays the total capital investment. In the figure, a constant cash flow rate has been assumed from time zero until the end of operation, although in reality a constant cash flow would not be expected. For the conditions shown in Fig. 6-2, the total capital investment is repaid in 5 years, and the cumulative cash position is zero. After that time, profits accumulate on the positive side of the cumulative cash position until the end of the project life, when the plant is shut down and project operation ceases. At shutdown, the working capital and land value are recovered. The working capital is recovered by the sale of materials, supplies, and equipment. Land can be either sold or transferred to another company use. For evaluation purposes it is generally assumed that the dollar amount recovered for working capital and land is the same as that spent originally. Thus, the final cumulative cash position over the 10-year life of the project is shown in the upper right-hand bracket in Fig. 6-2.

The relationships presented in Fig. 6-2 are very important for the understanding of the factors to be considered in cost estimation. To put emphasis on the basic nature of the role of cash flow, Fig. 6-2 has been simplified considerably by neglecting the time value of money and using constant annual profit and constant annual depreciation. In the chapters to follow, more complex cases will be considered in detail.

FACTORS AFFECTING INVESTMENT AND PRODUCTION COSTS

When a chemical engineer determines costs for any type of industrial process, these costs should be of sufficient accuracy to provide reliable decisions. To accomplish this, the engineer must have a complete understanding of the many factors that can affect costs. For example, some companies have reciprocal arrangements with other companies whereby certain raw materials or types of equipment may be purchased at prices lower than the prevailing market prices. Therefore, if the chemical engineer bases the cost of the raw materials for the process on regular market prices, the result may be that the process could appear to be unprofitable rather than profitable. Accordingly, the engineer must be aware of actual prices for raw materials and equipment, company policies, government regulations, and other factors affecting costs.

Sources of Equipment

One of the major costs involved in any chemical process is for equipment. In many cases, standard types of tanks, reactors, or other equipment are used, and a substantial reduction in cost can be realized by employing idle equipment or by purchasing secondhand equipment. If new equipment must be bought, several independent quotations should be obtained from different manufacturers. When specifications are given to the manufacturers, the chances for a low-cost estimate are increased if overly strict limitations on the design are kept to a minimum.

Price Fluctuations

In today's economic market, prices may vary widely from one period to another. For example, plant operators or supervisors cannot be hired today at the same wage rate as in 1985. The same statement applies to comparing prices of equipment purchased at different times. The chemical engineer, therefore, must keep up to date on price and wage fluctuations. One of the most complete sources of information on existing price conditions is the *Monthly Labor Review,* published by the U.S. Bureau of Labor Statistics. This publication gives up-to-date information on present prices and wages for different types of industries.

Company Policies

Policies of individual companies have a direct effect on costs. For example, some companies have particularly strict safety regulations, and these must be met in every detail. Accounting procedures and methods for allocating corporate costs vary among companies. Company policies with reference to labor unions must be considered, because these can affect overtime labor charges and the type of work that operators or other employees can perform. Labor union policies may, for example, even dictate the amount of wiring and piping that can be done on a piece of equipment before it is brought into the plant and thus have a direct effect on the total cost of installed equipment.

Operating Time and Rate of Production

One of the factors that has a major effect on the profits is the fraction of time a process is in operation. If equipment stands idle for an extended period, raw materials and labor costs are usually low; however, many other costs, designated as fixed costs, for example, maintenance, protection, and depreciation, continue even though the equipment is not in active use. More importantly, anytime that a plant is not producing a product, it is also not producing revenue. Some time must be allowed periodically to perform scheduled routine maintenance; however, downtime should be kept to a necessary minimum, as it is one of the chief sources of poor profitability in process plants.

Sales demand, rate of production, and operating time are closely interrelated. The ideal plant should operate under a time schedule that gives the maximum production rate consistent with market demand, safety, maintainability, and economic operating conditions. In this way, the total cost per unit of production is minimized because the variable costs averaged over time are low. If the production capacity of the process is greater than the sales demand, the operation can be operated continuously at reduced capacity or periodically at full capacity.

Figure 6-3 shows the effect on costs and profits based on the rate of production. As indicated in this figure, the fixed costs remain constant, and the total product cost

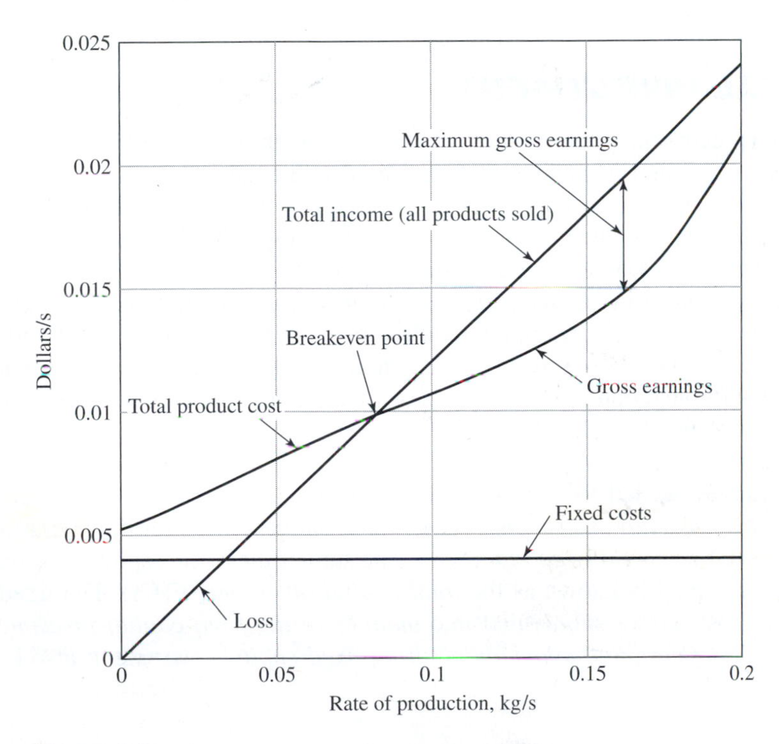

Figure 6-3
Breakeven chart for chemical processing plant

increases as the rate of production increases. The point where the total product cost equals the total income is designated as the *breakeven point.* Under the conditions shown in Fig. 6-3, a desirable production rate for this chemical processing plant would be approximately 5×10^6 kg/yr, because this represents the point of maximum gross and net profit. By considering sales demand along with the capacity and operating characteristics of the equipment, the engineer can recommend the production rate and operating schedules that will give optimal economic results.

Government Policies

The national government has many laws and regulations that have a direct effect on industrial costs. Some examples of these are import and export tariff regulations, depreciation rates, income tax rules, and environmental and safety regulations. Of these, income tax regulations and depreciation have the largest impact on most businesses.

As of the writing of this text, modifications of federal corporate tax laws were under consideration in the U.S. Congress. However, the last major change in federal corporate income tax rates was in 1993 and in depreciation was in 1988. The important point to remember is that tax law is subject to change at any time, and the design engineer must consult with tax experts to be sure that the most current tax codes are used in economic analyses. More details on tax policies may be found in Chap. 7.

CAPITAL INVESTMENT

A traditional economic definition of *capital* is "a stock of accumulated wealth." In an applied sense, capital is savings that may be used as the owner decides. One use of the savings is *investment;* that is, to use the savings ". . . to promote the production of other goods, instead of being available solely for purposes of immediate enjoyment" with ". . . the view of obtaining an income or profit."[†]

Before an industrial plant can be put into operation, a large sum of money must be available to purchase and install the required machinery and equipment. Land must be obtained, service facilities must be made available, and the plant must be erected complete with all piping, controls, and services. In addition, funds are required with which to pay the expenses involved in the plant operation before sales revenue becomes available.

The capital needed to supply the required manufacturing and plant facilities is called the *fixed-capital investment* (FCI), while that necessary for the operation of the plant is termed the *working capital* (WC). The sum of the fixed-capital investment and the working capital is known as the *total capital investment* (TCI). The fixed-capital portion may be further subdivided into *manufacturing fixed-capital investment,* also known as *direct cost,* and *nonmanufacturing fixed-capital investment*, also known as *indirect cost.*

[†]W. A. Neilson, ed., *Webster's New International Dictionary,* 2d ed., G. & C. Merriam Company, Springfield, MA, 1957.

Fixed-Capital Investment

Manufacturing fixed-capital investment represents the capital necessary for the installed process equipment with all components that are needed for complete process operation. Expenses for site preparation, piping, instruments, insulation, foundations, and auxiliary facilities are typical examples of costs included in the manufacturing fixed-capital investment.

The capital required for construction overhead and for all plant components that are not directly related to the process operation is designated the *nonmanufacturing fixed-capital investment.* These plant components include the land; processing buildings, administrative and other offices, warehouses, laboratories, transportation, shipping, and receiving facilities, utility and waste disposal facilities, shops, and other permanent parts of the plant. The *construction overhead cost* includes field office and supervision expenses, home office expenses, engineering expenses, miscellaneous construction costs, contractors' fees, and contingencies. In some cases, construction overhead is proportioned between manufacturing and nonmanufacturing fixed-capital investment.

Working Capital

The working capital for an industrial plant consists of the total amount of money invested in (1) raw materials and supplies carried in stock; (2) finished products in stock and semifinished products in the process of being manufactured; (3) accounts receivable; (4) cash kept on hand for monthly payment of operating expenses, such as salaries, wages, and raw material purchases; (5) accounts payable; and (6) taxes payable.

The raw material inventory included in working capital usually amounts to a 1-month supply of the raw materials valued at delivered prices. Finished products in stock and semifinished products have a value approximately equal to the total manufacturing cost for 1 month's production. Because credit terms extended to customers are usually based on an allowable 30-day payment period, the working capital required because of accounts receivable ordinarily amounts to the production cost for 1 month of operation.

The ratio of working capital to total capital investment varies with different companies, but most chemical plants use an initial working capital amounting to 10 to 20 percent of the total capital investment. This percentage may increase to as much as 50 percent or more for companies producing products of seasonal demand, because of the large inventories which must be maintained for appreciable periods.

ESTIMATION OF CAPITAL INVESTMENT

Most estimates of capital investment are based on the cost of the equipment required. The most significant errors in capital investment estimation are generally due to omissions of equipment, services, or auxiliary facilities rather than to gross errors in costing. Table 6-1 provides a checklist of items for a new facility and is an invaluable aid in making a complete estimation of the fixed-capital investment.

Table 6-1 Breakdown of fixed-capital investment items for a chemical process

Direct costs

1. *Purchased equipment*
   - All equipment listed on a complete flowsheet
   - Spare parts and noninstalled equipment spares
   - Surplus equipment, supplies, and equipment allowance
   - Inflation cost allowance
   - Freight charges
   - Taxes, insurance, duties
   - Allowance for modifications during start-up
2. *Purchased-equipment installation*
   - Installation of all equipment listed on complete flowsheet
   - Structural supports
   - Equipment insulation and painting
3. *Instrumentation and controls*
   - Purchase, installation, calibration, computer control with supportive software
4. *Piping*
   - Process piping utilizing suitable structural materials
   - Pipe hangers, fittings, valves
   - Insulation
5. *Electrical systems*
   - Electrical equipment switches, motors, conduit, wire, fittings, feeders, grounding, instrument and control wiring, lighting, panels
   - Electrical materials and labor
6. *Buildings (including services)*
   - Process buildings—substructures, superstructures, platforms, supports, stairways, ladders, access ways, cranes, monorails, hoists, elevators
   - Auxiliary buildings—administration and office, medical or dispensary, cafeteria, garage, product warehouse, parts warehouse, guard and safety, fire station, change house, personnel building, shipping office and platform, research laboratory, control laboratory
   - Maintenance shops—electric, piping, sheet metal, machine, welding, carpentry, instrument
   - Building services—plumbing, heating, ventilation, dust collection, air conditioning, building lighting, elevators, escalators, telephones, intercommunication systems, painting, sprinkler systems, fire alarm
7. *Yard improvements*
   - Site development—site clearing, grading, roads, walkways, railroads, fences, parking areas, wharves and piers, recreational facilities, landscaping
8. *Service facilities*
   - Utilities—steam, water, power, refrigeration, compressed air, fuel, waste disposal
   - Facilities—boiler plant incinerator, wells, river intake, water treatment, cooling towers, water storage, electric substation, refrigeration plant, air plant, fuel storage, waste disposal plant, environmental controls, fire protection
   - Nonprocess equipment—office furniture and equipment, cafeteria equipment, safety and medical equipment, shop equipment, automotive equipment, yard material-handling equipment, laboratory equipment, locker-room equipment, garage equipment, shelves, bins, pallets, hand trucks, housekeeping equipment, fire extinguishers, hoses, fire engines, loading stations
   - Distribution and packaging—raw material and product storage and handling equipment, product packaging equipment, blending facilities, loading stations
9. *Land*
   - Surveys and fees
   - Property cost

Table 6-1 *Continued*

Indirect costs

1. *Engineering and supervision*
   Engineering costs—administrative, process, design and general engineering, computer graphics, cost engineering, procuring, expediting, reproduction, communications, scale models, consultant fees, travel
   Engineering supervision and inspection
2. *Legal expenses*
   Identification of applicable federal, state, and local regulations
   Preparation and submission of forms required by regulatory agencies
   Acquisition of regulatory approval
   Contract negotiations
3. *Construction expenses*
   Construction, operation, and maintenance of temporary facilities, offices, roads, parking lots, railroads, electrical, piping, communications, fencing
   Construction tools and equipment
   Construction supervision, accounting, timekeeping, purchasing, expediting
   Warehouse personnel and expense, guards
   Safety, medical, fringe benefits
   Permits, field tests, special licenses
   Taxes, insurance, interest
4. *Contractor's fee*
5. *Contingency*

Types of Capital Cost Estimates

An estimate of the capital investment for a process may vary from a predesign estimate based on little information except the magnitude of the proposed project to a detailed estimate prepared from complete drawings and specifications. Between these two extremes of capital investment estimates, there can be numerous other estimates that vary in accuracy depending upon the stage of development of the project. These estimates are called by a variety of names, but the following five categories represent the accuracy range and designation normally used for design purposes:

1. *Order-of-magnitude estimate* (*ratio estimate*) based on similar previous cost data; probable accuracy of estimate over ±30 percent.
2. *Study estimate* (*factored estimate*) based on knowledge of major items of equipment; probable accuracy of estimate up to ±30 percent.
3. *Preliminary estimate* (*budget authorization estimate* or *scope estimate*) based on sufficient data to permit the estimate to be budgeted; probable accuracy of estimate within ±20 percent.
4. *Definitive estimate* (*project control estimate*) based on almost complete data but before completion of drawings and specifications; probable accuracy of estimate within ±10 percent.
5. *Detailed estimate* (*contractor's estimate*) based on complete engineering drawings, specifications, and site surveys; probable accuracy of estimate within ±5 percent.

Figure 6-4 shows the information required for the preparation of these five levels of estimates and the approximate limits of error in these methods. There is a large probability that the actual cost will be more than the estimated cost when information is incomplete or during periods of rising costs. For such estimates, the positive spread is likely to be wider than the negative, say, +40 and −20 percent for a study estimate.

Predesign cost estimates (defined here as order-of-magnitude, study, and preliminary estimates) require much less detail than firm estimates such as the definitive or detailed estimates. However, the predesign estimates are extremely important for determining whether a proposed project should be given further consideration or comparing alternative designs. For this reason, much of the information presented in this chapter is devoted to predesign estimates, although it should be understood that the distinction between predesign and firm estimates gradually disappears as more and more details are included.

Predesign estimates may be used to provide a basis for requesting and obtaining a capital appropriation from company management. Later estimates, made during the progress of the design, may indicate that the project will cost more or less than the amount appropriated. Management is then asked to approve a *variance,* which may be positive or negative.

COST INDEXES

Most cost data that are available for making a preliminary or predesign estimate are only valid at the time they were developed. Because prices may have changed considerably with time due to changes in economic conditions, some method must be used for updating cost data applicable at a past date to costs that are representative of conditions at a later time.[†] This can be done by the use of cost indexes.

A cost index is an index value for a given time showing the cost at that time relative to a certain base time. If the cost at some time in the past is known, the equivalent cost at present can be determined by multiplying the original cost by the ratio of the present index value to the index value applicable when the original cost was obtained, namely,

$$\text{Present cost} = \text{original cost}\left(\frac{\text{index value at present}}{\text{index value at time original cost was obtained}}\right)$$

Cost indexes can be used to give a general estimate, but no index can take into account all factors, such as special technological advancements or local conditions. The common indexes permit fairly accurate estimates if the period involved is less than 10 years. Indexes are frequently used to extrapolate costs into the near future. For example, the cost estimator may project costs forward from the time a study is being done until the expected start-up time of a plant. Such projections are done by using extrapolated values of an index, or an expected inflation rate.

Many different types of cost indexes are published regularly. Some can be used for estimating equipment costs; others apply specifically to labor, construction, materials, or other specialized fields. The most common of these indexes are the *Marshall and*

[†]See Chap. 8 for a discussion of the effects of inflation or deflation on costs and revenues in the future.

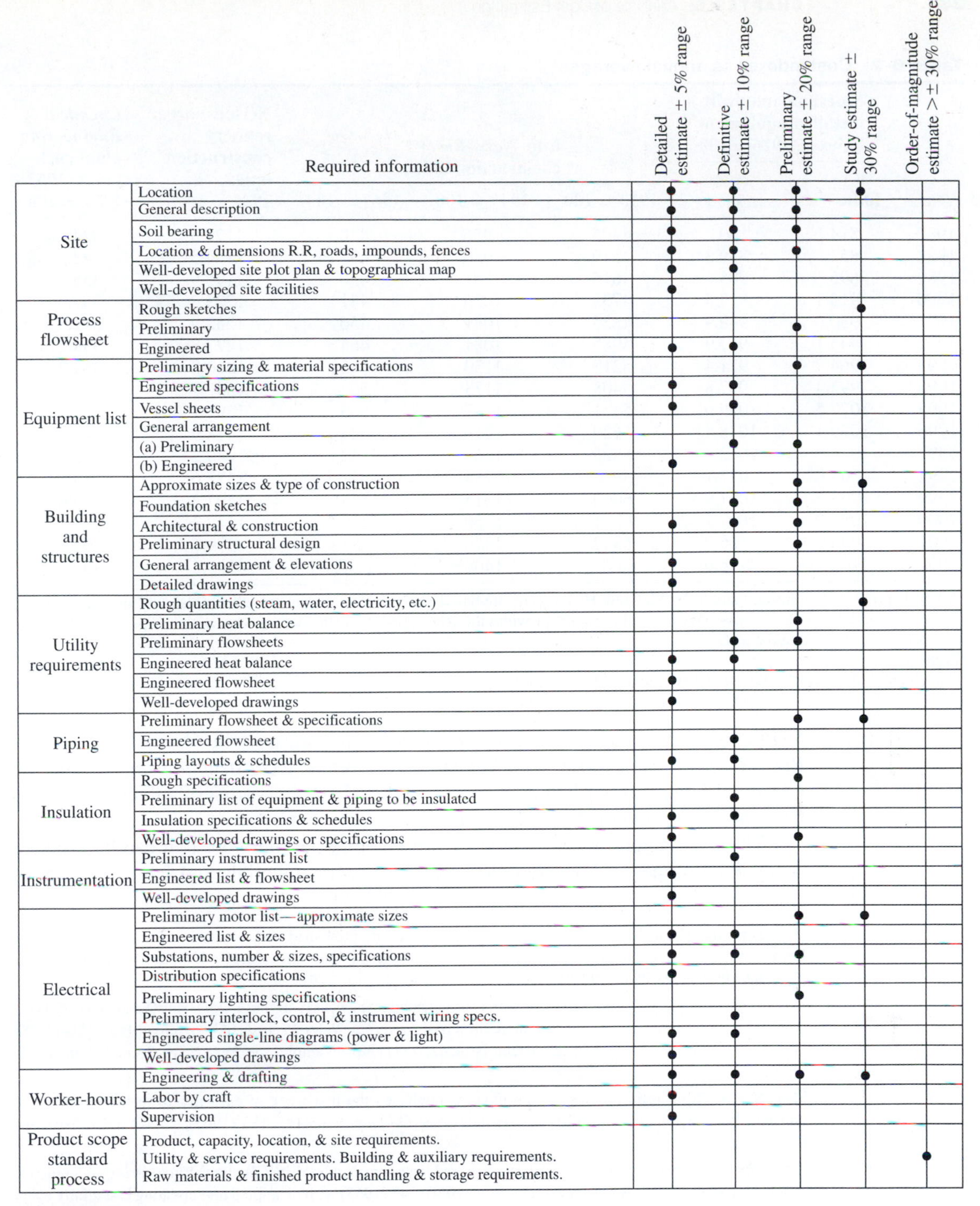

Required information		Detailed estimate ± 5% range	Definitive estimate ± 10% range	Preliminary estimate ± 20% range	Study estimate ± 30% range	Order-of-magnitude estimate >± 30% range
Site	Location	●	●	●	●	
	General description	●	●	●		
	Soil bearing	●	●	●		
	Location & dimensions R.R, roads, impounds, fences	●	●	●		
	Well-developed site plot plan & topographical map	●	●			
	Well-developed site facilities	●				
Process flowsheet	Rough sketches				●	
	Preliminary			●		
	Engineered	●	●			
Equipment list	Preliminary sizing & material specifications			●	●	
	Engineered specifications	●	●			
	Vessel sheets	●	●			
	General arrangement					
	(a) Preliminary		●	●		
	(b) Engineered	●				
Building and structures	Approximate sizes & type of construction			●	●	
	Foundation sketches		●	●		
	Architectural & construction	●	●	●		
	Preliminary structural design			●		
	General arrangement & elevations	●	●			
	Detailed drawings	●				
Utility requirements	Rough quantities (steam, water, electricity, etc.)				●	
	Preliminary heat balance			●		
	Preliminary flowsheets		●	●		
	Engineered heat balance	●	●			
	Engineered flowsheet	●				
	Well-developed drawings	●				
Piping	Preliminary flowsheet & specifications			●	●	
	Engineered flowsheet		●			
	Piping layouts & schedules	●	●			
Insulation	Rough specifications			●		
	Preliminary list of equipment & piping to be insulated		●			
	Insulation specifications & schedules	●	●			
	Well-developed drawings or specifications	●		●		
Instrumentation	Preliminary instrument list		●			
	Engineered list & flowsheet	●				
	Well-developed drawings	●				
Electrical	Preliminary motor list—approximate sizes			●	●	
	Engineered list & sizes	●	●			
	Substations, number & sizes, specifications	●	●	●		
	Distribution specifications	●				
	Preliminary lighting specifications			●		
	Preliminary interlock, control, & instrument wiring specs.		●			
	Engineered single-line diagrams (power & light)	●	●			
	Well-developed drawings	●				
Worker-hours	Engineering & drafting	●	●	●	●	
	Labor by craft	●				
	Supervision	●				
Product scope standard process	Product, capacity, location, & site requirements. Utility & service requirements. Building & auxiliary requirements. Raw materials & finished product handling & storage requirements.					●

Figure 6-4
Cost-estimating information guide

Table 6-2 Cost indexes as annual averages

Year	Marshall and Swift installed-equipment indexes, 1926 = 100		*Eng. News-Record* construction index			Nelson-Farrar refinery construction index, 1946 = 100	*Chemical Engineering* plant cost index, 1957–1959 = 100
	All industries	Process industry	1913 = 100	1949 = 100	1967 = 100		
1987	814	830	4406	956	410	1121.5	324
1988	852	859.3	4519	980	421	1164.5	343
1989	895	905.6	4615	1001	430	1195.9	355
1990	915.1	929.3	4732	1026	441	1225.7	357.6
1991	930.6	949.9	4835	1049	450	1252.9	361.3
1992	943.1	957.9	4985	1081	464	1277.3	358.2
1993	964.2	971.4	5210	1130	485	1310.8	359.2
1994	993.4	992.8	5408	1173	504	1349.7	368.1
1995	1027.5	1029.0	5471	1187	509	1392.1	381.1
1996	1039.1	1048.5	5620	1219	523	1418.9	381.7
1997	1056.8	1063.7	5825	1264	542	1449.2	386.5
1998	1061.9	1077.1	5920	1284	551	1477.6	389.5
1999	1068.3	1081.9	6060	1315	564	1497.2	390.6
2000	1089.0	1097.7	6221	1350	579	1542.7	394.1
2001	1093.9	1106.9	6342	1376	591	1579.7	394.3
2002	1102.5[‡]	1116.9[‡]	6490[‡]	1408[‡]	604[‡]	1599.2[‡]	390.4[†,§]

[†]All costs presented in this text and in the McGraw-Hill website are based on this value for January 2002, obtained from the *Chemical Engineering* index unless otherwise indicated. The website provides the corresponding mathematical cost relationships for all the graphical cost data presented in the text.
[‡]Projected.
[§]Calculated with revised index; see *Chem. Eng.*, **109:** 62 (2002).

Swift all-industry and *process-industry equipment indexes,*[†] the *Engineering News-Record construction index,*[‡] the *Nelson-Farrar refinery construction index,*[§] and the *Chemical Engineering plant cost index.*[¶] Table 6-2 presents a list of values for various types of indexes over the past 15 years.

There are numerous other indexes presented in the literature that can be used for specialized purposes. For example, cost indexes for materials and labor for various

[†]Values for the Marshall and Swift equipment cost indexes are published each month in *Chemical Engineering*. For a complete description of these indexes, see R. W. Stevens, *Chem. Eng.*, **54**(11): 124 (1947). See also *Chem. Eng.*, **85**(11): 189 (1978) and **92**(9): 75 (1985).

[‡]The *Engineering News-Record* construction cost index appears weekly in the *Engineering News-Record*. For a complete description of this index and the revised basis, see *Eng. News-Record,* **143**(9): 398 (1949), **178**(11): 87 (1967). A history is presented in the March issue each year; for example, see *Eng. News-Record,* **220**(11): 54 (1988).

[§]The Nelson-Farrar refinery construction index is published the first week of each month in the *Oil and Gas Journal*. For a complete description of this index, see *Oil Gas J.,* **63**(14): 185 (1965), **74**(48): 68 (1976), and **83**(52): 145 (1985).

[¶]The *Chemical Engineering* plant cost index is published each month in *Chemical Engineering*. A complete description of this index is found in *Chem. Eng.,* **70**(4): 143 (1963) with recapping and updating essentially every 3 years. The index is being revised in 2002 to provide a better relationship between the various cost factors involved in the index; see W. M. Vavatuk, *Chem. Eng.,* **109**(1): 62 (2002) for details.

types of industries are published monthly by the U.S. Bureau of Labor Statistics in the *Monthly Labor Review*. These indexes can be useful for special kinds of estimates involving particular materials or unusual labor conditions. Another example of a cost index which is useful for worldwide comparison of cost charges with time is published periodically in the *International Journal of Production Economics* (formerly *Engineering Costs and Production Economics*). This presents cost indexes for plant costs for various countries in the world including Australia, Belgium, Canada, Denmark, France, Germany, Italy, Netherlands, Norway, Japan, Sweden, the United Kingdom, and the United States.[†]

All cost indexes are based on limited sampling of the goods and services in question; therefore, two indexes covering the same types of projects may give results that differ considerably. The most that any index can hope to do is to reflect general trends. These trends may at times have little meaning when applied to a specific case. For example, a contractor may, during a slack period, accept a construction job with little profit just to keep the construction crew together. On the other hand, if there are current local labor shortages, a project may cost considerably more than a similar project in another geographic location.

The Marshall and Swift equipment cost indexes and the *Chemical Engineering* plant cost indexes are recommended for process equipment and chemical-plant investment estimates. These two cost indexes give very similar results, while the *Engineering News-Record* construction cost index has increased with time much more rapidly than the other two because it does not include a productivity improvement factor. Similarly, the Nelson-Farrar refinery construction index has shown a very large increase with time and should be used with caution and only for refinery construction.

COST COMPONENTS IN CAPITAL INVESTMENT

Capital investment is the total amount of money needed to supply the necessary plant and manufacturing facilities plus the amount of money required as working capital for operation of the facilities. Let us now consider the proportional costs of each major component of fixed-capital investment, as outlined previously in Table 6-1. The cost factors presented here are based on a careful interpretation of recent sources[‡] with input based on industrial experience.

Table 6-3 summarizes these typical variations in component costs as percentages of fixed-capital investment (FCI) for multiprocess *grass-roots* plants or large *battery-limit* additions. A *grass-roots* plant is defined as a complete plant erected on a new site.

[†]For methods used, see *Eng. Costs Prod. Econ.*, **6**(1): 267 (1982).

[‡]K. M. Guthrie, *Process Plant Estimating, Evaluation, and Control,* Craftsman Book Company of America, Solana Beach, CA, 1974; G. D. Ulrich, *A Guide to Chemical Engineering Process Design and Economics,* J. Wiley, New York, 1984; R. K. Sinnott, *An Introduction to Chemical Engineering Design,* Pergamon Press, Oxford, United Kingdom, 1983; P. F. Ostwald, *AM Cost Estimator,* McGraw-Hill, New York, 1988; D. R. Woods, *Process Design and Engineering Practice,* Prentice-Hall, Upper Saddle River, NJ, 1995; R. H. Perry and D. W. Green, eds., *Perry's Chemical Engineers' Handbook,* 7th ed., McGraw-Hill, New York, 1997.

Table 6-3 Typical percentages of fixed-capital investment values for direct and indirect cost segments for multipurpose plants or large additions to existing facilities

Component	Range of FCI, %
Direct costs	
Purchased equipment	15–40
Purchased-equipment installation	6–14
Instrumentation and controls (installed)	2–12
Piping (installed)	4–17
Electrical systems (installed)	2–10
Buildings (including services)	2–18
Yard improvements	2–5
Service facilities (installed)	8–30
Land	1–2
Indirect costs	
Engineering and supervision	4–20
Construction expenses	4–17
Legal expenses	1–3
Contractor's fee	2–6
Contingency	5–15

Investment includes all costs of land, site development, battery-limit facilities, and auxiliary facilities. A geographic boundary defining the coverage of a specific project is a *battery limit*. Usually this encompasses the manufacturing area of a proposed plant or addition, including all process equipment but excluding provision of storage, utilities, administrative buildings, or auxiliary facilities unless so specified. Normally this excludes site preparation and therefore may be applied to the extension of an existing plant.

EXAMPLE 6-1 Estimation of Fixed-Capital Investment Using Ranges of Process-Plant Component Costs

Make a study estimate of the fixed-capital investment for a process plant if the purchased-equipment cost is $100,000. Use the ranges of process-plant component cost outlined in Table 6-3 for a process plant handling both solids and fluids with a high degree of automatic controls and essentially outdoor operation. Do not include land.

■ Solution

A percentage is selected within the range in Table 6-3 for each of the components of fixed-capital investment; this selection is somewhat arbitrary, with selection made of average values unless process-plant characteristics suggest lower or upper values. Generally, when all these percentages are added, they will not total 100 percent. Therefore, all the percentages must be normalized to a total of 100 by dividing each percentage by the total sum over 100. The estimated cost for a component cost is then calculated as $100,000 multiplied by the normalized percentage for that component, and then divided by the normalized percentage for the purchased equipment. All values are rounded to the nearest $1000.

These computations are summarized in the following table.

Components	Selected percentage of FCI	Normalized percentage of FCI	Estimated cost
Purchased equipment	25	22.9	$100,000
Purchased-equipment installation	9	8.3	36,000
Instrumentation (installed)	10	9.2	40,000
Piping (installed)	8	7.3	32,000
Electrical (installed)	5	4.6	20,000
Buildings (including services)	5	4.6	20,000
Yard improvements	2	1.8	8,000
Service facilities (installed)	15	13.8	60,000
Engineering and supervision	8	7.3	32,000
Construction expense	10	9.2	40,000
Legal expense	2	1.8	8,000
Contractor's fee	2	1.8	8,000
Contingency	8	7.3	32,000
Total	109	99.9	$436,000

It should be recognized that the $436,000 has a large uncertainty, on the order of ±30 percent.

Purchased Equipment

The cost of purchased equipment is the basis of several predesign methods for estimating capital investment. Sources of equipment prices, methods of adjusting equipment prices for capacity, and methods of estimating auxiliary process equipment are therefore essential to the estimator in making reliable cost estimates. The various types of equipment can often be divided conveniently into (1) processing equipment, (2) raw materials handling and storage equipment, and (3) finished-products handling and storage equipment.

The sizes and specifications of the equipment needed for a chemical process are determined from the equipment parameters fixed or calculated along with the material and energy balances. In a process simulation to obtain the material and energy balances for a distillation column, for example, the engineer must specify the number of equilibrium stages, reflux ratio, total or partial condensation of the overhead stream, and operating pressure at a particular point, such as at the top of the column. With these parameters and feed conditions, a distillation algorithm calculates the product compositions, temperatures, and pressures as well as the condenser and reboiler duties. The number of actual plates needed can be obtained by specifying the plate efficiency. This information plus the materials of construction is sufficient to make an estimate of the purchased cost of the column, condenser and reboiler, and associated piping. Similarly, for other types of process equipment the specifications required to complete the material and energy balances are usually sufficient to make a cost estimate.

The most accurate method for determining process equipment costs is to obtain firm bids from fabricators or suppliers. Often, fabricators can supply quick estimates that will be close to the bid price but will not take too much time. Second-best in reliability are cost values from the file of past purchase orders. When used for pricing new equipment, purchase-order prices must be corrected with the appropriate cost index ratio. Limited information on process-equipment costs has also been published in various engineering journals. Costs estimates for a large number of different types and capacities of equipment are presented in Chaps. 12 through 15.

Estimating Equipment Costs by Scaling

It is often necessary to estimate the cost of a piece of equipment when cost data are not available for the particular size or capacity involved. Predictions can be made by using the power relationship known as the *six-tenths factor rule,* if the new piece of equipment is similar to one of another capacity for which cost data are available. According to this rule, if the cost of a given unit b at one capacity is known, the cost of a similar unit a with X times the capacity of the first is $X^{0.6}$ times the cost of the initial unit.

$$\text{Cost of equipment } a = (\text{cost of equipment } b)X^{0.6} \tag{6-2}$$

The preceding equation indicates that a log-log plot of capacity versus cost for a given type of equipment should be a straight line with a slope equal to 0.6. Figure 6-5 presents a plot of this sort for shell-and-tube heat exchangers. The application of the 0.6 rule of thumb for most purchased equipment is, however, an oversimplification, since the actual values of the cost capacity exponent vary from less than 0.3 to greater than 1.0, as shown in Table 6-4. Because of this, the 0.6 power should be used only in

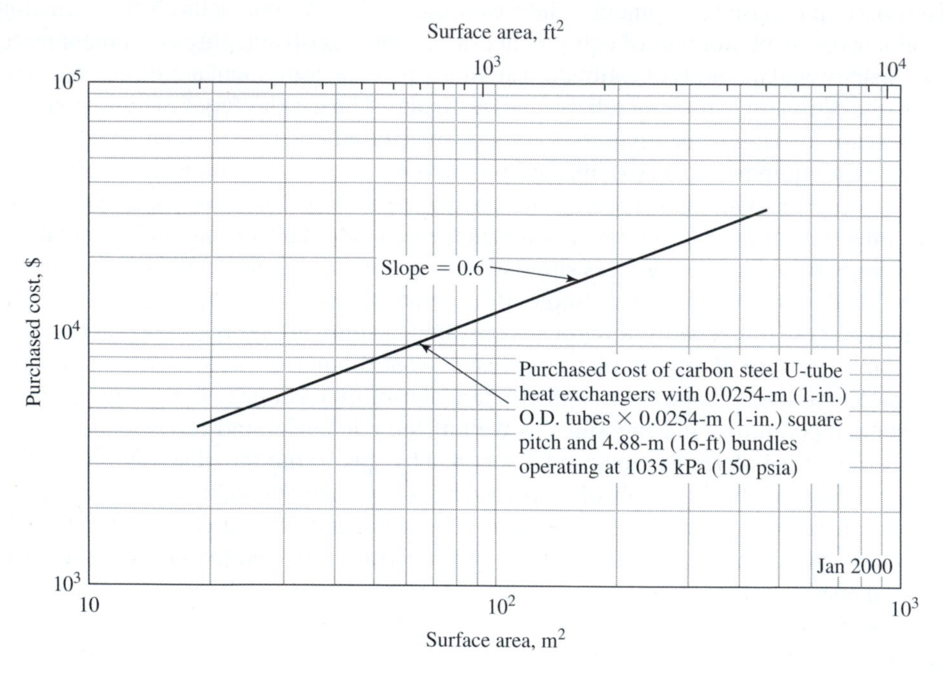

Figure 6-5
Application of "six-tenth factor" rule to costs for U-tube heat exchangers

Table 6-4 Typical exponents for equipment cost as a function of capacity

Equipment	Size range	Exponent
Blender, double cone rotary, carbon steel (c.s.)	1.4–7.1 m^3 (50–250 ft^3)	0.49
Blower, centrifugal	0.5–4.7 m^3/s (10^3–10^4 ft^3/min)	0.59
Centrifuge, solid bowl, c.s.	7.5–75 kW (10–10^2 hp) drive	0.67
Crystallizer, vacuum batch, c.s.	15–200 m^3 (500–7000 ft^3)	0.37
Compressor, reciprocating, air-cooled, two-stage, 1035-kPa discharge	0.005–0.19 m^3 (10–400 ft^3/min)	0.69
Compressor, rotary, single-stage, sliding vane, 1035-kPa discharge	0.05–0.5 m^3/s (10^2–10^3 ft^3/min)	0.79
Dryer, drum, single vacuum	1–10 m^2 (10–10^2 ft^2)	0.76
Dryer, drum, single atmospheric	1–10 m^2 (10–10^2 ft^2)	0.40
Evaporator (installed), horizontal tank	10–1000 m^2 (10^2–10^4 ft^2)	0.54
Fan, centrifugal	0.5–5 m^3/s (10^3–10^4 ft^3/min)	0.44
Fan, centrifugal	10–35 m^3/s (2×10^4–7×10^4 ft^3/min)	1.17
Heat exchanger, shell-and-tube, floating head, c.s.	10–40 m^2 (100–400 ft^2)	0.60
Heat exchanger, shell-and-tube, fixed sheet, c.s.	10–40 m^2 (100–400 ft^2)	0.44
Kettle, cast-iron, jacketed	1–3 m^3 (250–800 gal)	0.27
Kettle, glass-lined, jacketed	0.8–3 m^3 (200–800 gal)	0.31
Motor, squirrel cage, induction, 440-V, explosion-proof	4–15 kW (5–20 hp)	0.69
Motor, squirrel cage, induction, 440-V, explosion-proof	15–150 kW (20–200 hp)	0.99
Pump, reciprocating, horizontal cast-iron (includes motor)	1×10^{-4}–6×10^{-3} m^3/s (2–100 gpm)	0.34
Pump, centrifugal, horizontal, cast steel (includes motor)	4–40 $m^3/s\cdot kPa$ (10^4–10^5 gpm·psi)	0.33
Reactor, glass-lined, jacketed (without drive)	0.2–2.2 m^3 (50–600 gal)	0.54
Reactor, stainless steel, 2070-kPa	0.4–4.0 m^3 (10^2–10^3 gal)	0.56
Separator, centrifugal, c.s.	1.5–7 m^3 (50–250 ft^3)	0.49
Tank, flat head, c.s.	0.4–40 m^3 (10^2–10^4 gal)	0.57
Tank, c.s., glass-lined	0.4–4.0 m^3 (10^2–10^3 gal)	0.49
Tower, c.s.	5×10^2–10^6 kg (10^3–2×10^6 lb)	0.62
Tray, bubble cap, c.s.	1–3 m (3–10 ft) diameter	1.20
Tray, sieve, c.s.	1–3 m (3–10 ft) diameter	0.86

the absence of other information. In general, the cost capacity concept should not be used beyond a 10-fold range of capacity, and care must be taken to make certain the two pieces of equipment are similar with regard to type of construction, materials of construction, temperature and pressure operating range, and other pertinent variables. Nonetheless, this six-tenths rule is widely used in approximations of equipment and even total process costs.

Estimating Cost of Equipment Using Scaling Factors and Cost Index — EXAMPLE 6-2

The purchased cost of a 0.2-m^3, glass-lined, jacketed reactor (without drive) was \$10,000 in 1991. Estimate the purchased cost of a similar 1.2-m^3, glass-lined, jacketed reactor (without drive) in 1996. Use the annual average *Chemical Engineering* plant cost index to update the purchase cost of the reactor.

■ Solution

The *Chemical Engineering* plant cost index in 1991 was 361 and in 1996 was 382 (Table 6-2). From Table 6-4 the equipment-cost versus capacity exponent is 0.54:

$$\text{Cost of reactor in 1996} = (\$10{,}000)\left(\frac{382}{361}\right)\left(\frac{1.2}{0.2}\right)^{0.54} = \$27{,}850$$

Purchased-equipment costs for vessels, tanks, and process and materials-handling equipment can often be estimated on the basis of weight. The fact that a wide variety of types of equipment have about the same cost per unit weight is quite useful, particularly when other cost data are not available. Generally, the cost data generated by this method are sufficiently reliable to permit study-estimates. ■

Purchased-Equipment Delivery

Purchased-equipment prices are usually quoted as f.o.b. (free on board, meaning that the purchaser pays the freight). Clearly freight costs depend upon many factors, such as the weight and size of the equipment, distance from source to plant, and method of transport. For predesign estimates, a delivery allowance of 10 percent of the purchased-equipment cost is recommended.

Purchased-Equipment Installation

Installation of process equipment involves costs for labor, foundations, supports, platforms, construction expenses, and other factors directly related to the erection of purchased equipment. Table 6-5 presents the general range of installation costs as a percentage of the purchased-equipment costs for various types of equipment. Installation labor cost as a function of equipment size shows wide variations and is difficult to predict.

Table 6-5 Installation cost for process equipment as a percentage of purchased-equipment cost†

Type of equipment	Installation cost, %
Centrifugal separators	20–60
Compressors	30–60
Dryers	25–60
Evaporators	25–90
Filters	65–80
Heat exchangers	30–60
Mechanical crystallizers	30–60
Metal tanks	30–60
Mixers	20–40
Pumps	25–60
Towers	60–90
Vacuum crystallizers	40–70
Wood tanks	30–60

†Modified from K. M. Guthrie, *Process Plant Estimating, Evaluation, and Control,* Craftsman Book Company of America, Solana Beach, CA, 1974.

Analyses of a number of typical chemical plants indicates that the cost of the purchased equipment varies from 65 to 80 percent of the total installed cost depending upon the complexity of the equipment and the type of plant in which the equipment is installed. Installation costs for equipment, therefore, are estimated to vary from 25 to 55 percent of the delivered purchased-equipment cost. Expenses for equipment insulation and piping insulation are often included under the respective headings of equipment installation costs and piping costs. The total cost for the labor and materials required for insulating equipment and piping in ordinary chemical plants is approximately 8 to 9 percent of the delivered purchased-equipment cost. This is equivalent to approximately 2 percent of the total capital investment.

Instrumentation and Controls

Instrument costs, installation labor costs, and expenses for auxiliary equipment and materials constitute the major portion of the capital investment required for instrumentation. Total instrumentation and control cost depends on the amount of control required and may amount to 8 to 50 percent of the total delivered equipment cost.

For the normal solid-fluid chemical processing plant, a value of 26 percent of the delivered purchased-equipment cost is recommended as an estimate for the total instrumentation and control cost. This cost represents approximately 5 percent of the total capital investment.

Piping

The cost for piping covers labor, valves, fittings, pipe, supports, and other items involved in the complete erection of all piping used directly in the process. This includes raw material, intermediate-product, finished-product, steam, water, air, sewer, and other process piping. Since process-plant piping can run as high as 80 percent of delivered purchased-equipment cost or 20 percent of the fixed-capital investment, the importance of this item in capital cost estimation is clear.

Piping estimation methods involve either some degree of piping takeoff from detailed flowsheets or use of a factor technique when neither detailed drawings nor flowsheets are available. Factoring by percentage of purchased-equipment cost and percentage of fixed-capital investment is based strictly on experience gained from piping costs for similar previously installed chemical process plants. Table 6-6 presents a

Table 6-6 Estimated cost of piping

	Percent of purchased equipment			
Types of process plant	Material	Labor	Total	Percent of fixed-capital equipment
Solid[†]	9	7	16	4
Solid-fluid[‡]	17	14	31	7
Fluid[§]	38	30	68	13

[†]A coal briquetting plant would be a typical solid-processing plant.
[‡]A shale oil plant with crushing, grinding, retorting, and extraction would be a typical solid-fluid processing plant.
[§]A distillation separation system would be a typical fluid-processing plant.

rough estimate of the piping costs for various types of chemical processes. Additional information for estimating piping costs is presented in Chap. 12. Labor for installation is estimated as approximately 40 to 50 percent of the total installed cost of piping. Material and labor for pipe insulation are estimated to vary from 15 to 25 percent of the total installed cost of the piping and are influenced greatly by the extremes in temperature, which are encountered by the process streams.

Electrical Systems

The electrical systems consist of four major components, namely, power wiring, lighting, transformation and service, and instrument and control wiring. In most chemical plants the installed cost of electrical systems is estimated to be 15 to 30 percent of the delivered purchased-equipment cost or between 4 and 8 percent of the fixed-capital investment.

Buildings

The cost of buildings, including services, consists of expenses for labor, materials, and supplies involved in the erection of all buildings connected with the plant. Costs for plumbing, heating, lighting, ventilation, and similar building services are included. The cost of buildings, including services, for different types of process plants is shown in Table 6-7 as a percentage of purchased-equipment cost and fixed-capital investment.

Yard Improvements

Costs for fencing, grading, roads, sidewalks, railroad sidings, landscaping, and similar items are all considered part of yard improvements. The cost for these items in most chemical plants approximates 10 to 20 percent of the purchased-equipment cost. This is equivalent to approximately 2 to 5 percent of the fixed-capital investment.

Service Facilities

Utilities for supplying steam, water, power, compressed air, and fuel are part of the service facilities of a chemical process plant. Waste disposal, fire protection, and

Table 6-7 Cost of buildings including services based on purchased-equipment cost or on fixed-capital investment

	Percentage of purchased-equipment cost			Percentage of fixed-capital investment		
Type of process plant†	New plant at new site‡	New unit at existing site§	Expansion at existing site	New plant at new site‡	New plant at existing site§	Expansion at existing site
Solid	68	25	15	18	7	4
Solid-fluid	47	29	7	12	7	2
Fluid	45	5–18¶	6	10	2–4¶	2

†See Table 6-6 for description of types of process plants.
‡Generally referred to as a grass-roots plant.
§Designated as a battery-limit plant.
¶Smaller figure is applicable to petroleum refining and related industries.

miscellaneous service items, such as shop, first aid, and cafeteria equipment and facilities, require capital investments that are included under the general heading of service facilities cost.

The total cost for service facilities in chemical plants generally ranges from 30 to 80 percent of the purchased-equipment cost with 55 percent representing an average for a normal solid-fluid processing plant. For a single-product, small, continuous process plant, the cost is likely to be in the lower part of the range. For a large, new multiprocess plant at a new location, the costs are apt to be near the upper limit of the range. The cost of service facilities, in terms of fixed-capital investment, generally ranges from 8 to 20 percent with 14 percent considered an average value. Table 6-8 lists the typical ranges in percentages of fixed-capital investment that can be encountered for various components of service facilities. Except for entirely new facilities, it is unlikely that all service facilities will be required in every process plant. This accounts to a large degree for the wide variation range assigned to each component in Table 6-8. The range also reflects the degree to which utilities requirements depend on energy balances for the process. Service facilities largely are functions of plant physical size and will be present to some degree in most plants. The omission of unneeded utilities tends to increase the relative percentages of the necessary service facilities for the plant. Recognition of this fact, coupled with a careful appraisal of the extent to which service facilities are used in the plant, should result in selecting from Table 6-8 a reasonable cost percentage applicable to a specific process design.

Health, Safety, and Environmental Functions

Over time, the requirements for occupational health and safety and environmental functions in plants have increased substantially. Table 6-8 includes modest allowances

Table 6-8 Typical variation in percent of fixed-capital investment for service facilities

Service facilities	Range, %	Typical value, %
Steam generation	2.6–6.0	3.0
Steam distribution	0.2–2.0	1.0
Water supply, cooling, and pumping	0.4–3.7	1.8
Water treatment	0.5–2.1	1.3
Water distribution	0.1–2.0	0.8
Electric substation	0.9–2.6	1.3
Electric distribution	0.4–2.1	1.0
Gas supply and distribution	0.2–0.4	0.3
Air compression and distribution	0.2–3.0	1.0
Refrigeration including distribution	0.5–2.0	1.0
Process waste disposal	0.6–2.4	1.5
Sanitary waste disposal	0.2–0.6	0.4
Communications	0.1–0.3	0.2
Raw material storage	0.3–3.2	0.5
Finished-product storage	0.7–2.4	1.5
Fire protection system	0.3–1.0	0.5
Safety installations	0.2–0.6	0.4

for these functions, but in reality, many plants require much higher expenditures than suggested here. There do not seem to be general guidelines for estimating these expenditures at this time. It is highly recommended that they all be considered in the design of a plant. These functions should not be mere add-ons, but should be integrated into the process design itself. Pollution prevention and pollutant minimization techniques should be part of the design strategy. Pollution minimization is sometimes the driving force for new process development, design, and construction.

Land

The cost for land and the accompanying surveys and fees depends on the location of the property and may vary by a cost factor per acre as high as 30 to 50 between a rural district and a highly industrialized area. As a rough average, land costs for industrial plants amount to 4 to 8 percent of the purchased-equipment cost or 1 to 2 percent of the total capital investment. By law, the cost of land cannot be depreciated; therefore it is usually not included in the fixed-capital investment. Rather, it is shown as a one-time investment at the beginning of plant construction.

Engineering and Supervision

The costs for construction design and engineering, including internal or licensed software, computer-based drawings, purchasing, accounting, construction and cost engineering, travel, communications, and home office expense plus overhead, constitute the capital investment for engineering and supervision. This cost, since it cannot be directly charged to equipment, materials, or labor, is normally considered an indirect cost in fixed-capital investment and is approximately 30 percent of the delivered-equipment cost or 8 percent of the fixed-capital investment for the process plant.

Legal Expenses

Legal costs result largely from land purchases, equipment purchase, and construction contracts. Understanding and proving compliance with government, environmental, and safety requirements also constitute major sources of legal costs. These usually total on the order of 1 to 3 percent of fixed-capital investment.

Construction Expenses

Another indirect plant cost is the item of construction or field expense and includes temporary construction and operation, construction tools and rentals, home office personnel located at the construction site, construction payroll, travel and living, taxes and insurance, and other construction overhead. This expense item is occasionally included under equipment installation, or more often under engineering, supervision, and construction. For ordinary chemical process plants, the construction expenses average roughly 8 to 10 percent of the fixed-capital investment for the plant.

Contractor's Fee

The contractor's fee varies for different situations, but it can be estimated to be about 2 to 8 percent of the direct plant cost or 1.5 to 6 percent of the fixed-capital investment.

Contingencies

A contingency amount is included in all but the smallest estimates of capital investment in recognition of the fact that experience shows there will be unexpected events and changes that inevitably increase the cost of the project. Events, such as storms, floods, transportation accidents, strikes, price changes, small design changes, errors in estimation, and other unforeseen expenses, will occur even though they cannot be predicted. Contingency factors ranging from 5 to 15 percent of the fixed-capital investment are commonly used, with 8 percent being considered a reasonable average value.

METHODS FOR ESTIMATING CAPITAL INVESTMENT

Various methods can be employed for estimating capital investment. The choice of any one method depends upon the amount of detailed information available and the accuracy desired. Seven methods are outlined in this chapter, with each method requiring progressively less detailed information and less preparation time. Consequently, the degree of accuracy decreases with each succeeding method.

Method A: Detailed-Item Estimate A detailed-item estimate requires careful determination of each individual item shown in Table 6-1. Equipment and material needs are determined from completed drawings and specifications and are priced either from current cost data or preferably from firm delivered quotations. Estimates of installation costs are determined from accurate labor rates, efficiencies, and employee-hour calculations. Accurate estimates of engineering, field supervision employee-hours, and field expenses must be detailed in the same manner. Complete site surveys and soil data must be available to minimize errors in site development and construction cost estimates. In fact, in this type of estimate, an attempt is made to firm up as much of the estimate as possible by obtaining quotations from vendors and suppliers. Because of the extensive data needed and the large amounts of engineering time required to prepare such a detailed-item estimate, this type of estimate is almost exclusively prepared by contractors bidding on lump-sum work from finished drawings and specifications. An accuracy in the ±5 percent range is expected from a detailed estimate.

Method B: Unit Cost Estimate The unit cost method results in good estimating accuracies for fixed-capital investment provided accurate records have been kept of previous cost experiences. This method, which is frequently used for preparing definitive and preliminary estimates, also requires detailed estimates of purchased price obtained either from quotations or index-corrected cost records and published data. Equipment installation labor is evaluated as a fraction of the delivered-equipment cost. Costs for concrete, steel, pipe, electrical systems, instrumentation, insulation, etc., are obtained by takeoffs from the drawings and applying unit costs to the material and labor needs. A unit cost is also applied to engineering employee-hours, number of drawings, and specifications. A factor for construction expense, contractor's fee, and contingency is estimated from previously completed projects and is used to complete this type of estimate. Equation (6-3) summarizes this method as†

$$C_n = \left[\sum(E + E_L) + \sum(f_x M_x + f_y M'_L) + \sum f_e H_e + \sum f_d d_n\right] f_F \tag{6-3}$$

where C_n is the new capital investment, E the delivered purchased-equipment cost, E_L the delivered-equipment labor cost, f_x the specific material unit cost, M_x the specific material quantity in compatible units, f_y the specific material labor unit cost per employee-hour, M'_L the labor employee-hours for the specific material, f_e the unit cost for engineering, H_e the engineering employee-hours, f_d the unit cost per drawing or specification, d_n the number of drawings or specifications, and f_F the construction or field expense factor (always greater than 1). Depending on the detail included, a unit cost estimate should give ±10 to 20 percent accuracy.

Method C: Percentage of Delivered-Equipment Cost This method for estimating the fixed-capital and total capital investment requires determination of the delivered-equipment cost. The other items included in the total direct plant cost are then estimated as percentages of the delivered-equipment cost. The additional components of the capital investment are based on average percentages of the total direct plant cost, total direct and indirect plant costs, or total capital investment. This is summarized in the following cost equation:

$$C_n = \sum(E + f_1 E + f_2 E + f_3 E + \cdots + f_n E) = E \sum(1 + f_1 + f_2 + \cdots + f_n) \tag{6-4}$$

where $f_1, f_2, f_3, \ldots, f_n$ are multiplying factors for piping, electrical, indirect costs, etc. The factors used in making an estimation of this type should be determined on the basis of the type of process involved, design complexity, required materials of construction, location of the plant, past experience, and other items dependent on the particular unit under consideration. Average values of the various percentages have been determined for typical chemical plants, and these values are presented in Table 6-9.

Estimating by percentage of delivered-equipment cost is commonly used for preliminary and study estimates. The expected accuracy is in the ±20 to 30 percent range. It yields more accurate results when applied to projects similar in configuration to recently constructed plants. For comparable plants of different capacity, this method sometimes has been reported to yield definitive estimate accuracies, that is, close to ±10 percent.

---

†H. C. Bauman, *Fundamentals of Cost Engineering in the Chemical Industry,* Reinhold, New York, 1964.

Table 6-9 Ratio factors for estimating capital investment items based on delivered-equipment cost

Values presented are applicable for major process plant additions to an existing site where the necessary land is available through purchase or present ownership.[†] *The values are based on fixed-capital investments ranging from under $1 million to over $100 million.*

	Percent of delivered-equipment cost for		
	Solid processing plant[‡]	Solid-fluid processing plant[‡]	Fluid processing plant[‡]
Direct costs			
Purchased equipment delivered (including fabricated equipment, process machinery, pumps, and compressors)	100	100	100
Purchased-equipment installation	45	39	47
Instrumentation and controls (installed)	18	26	36
Piping (installed)	16	31	68
Electrical systems (installed)	10	10	11
Buildings (including services)	25	29	18
Yard improvements	15	12	10
Service facilities (installed)	40	55	70
Total direct plant cost	269	302	360
Indirect costs			
Engineering and supervision	33	32	33
Construction expenses	39	34	41
Legal expenses	4	4	4
Contractor's fee	17	19	22
Contingency	35	37	44
Total indirect plant cost	128	126	144
Fixed-capital investment	397	428	504
Working capital (15% of total capital investment)	70	75	89
Total capital investment	467	503	593

[†]Because of the extra expense involved in supplying service facilities, storage facilities, loading terminals, transportation facilities, and other necessary utilities at a completely undeveloped site, the fixed-capital investment for a new plant located at an undeveloped site may be as much as 100 percent greater than that for an equivalent plant constructed as an addition to the existing plant.

[‡]See Table 6-6 for descriptions of types of process plants.

Estimation of Fixed-Capital Investment by Percentage of Delivered-Equipment Cost — EXAMPLE 6-3

Prepare a study estimate of the fixed-capital investment for the process plant described in Example 6-1 if the delivered-equipment cost is $100,000.

■ Solution

Use the ratio factors outlined in Table 6-9 with modifications for instrumentation and outdoor operation. Take instrumentation as 10 percent of fixed-capital investment, that is, 0.1(428/100), or 43 percent, of the purchased equipment delivered. Take buildings as 15 percent of purchased equipment.

Component	Cost
Purchased equipment (delivered), E	\$100,000
Purchased equipment installation, 39%E	39,000
Instrumentation (installed), 43%E	43,000
Piping (installed), 31%E	31,000
Electrical (installed), 10%E	10,000
Buildings (including services), 15%E	15,000
Yard improvements, 12%E	12,000
Service facilities (installed), 55%E	55,000
Total direct plant cost, D	305,000
Engineering and supervision, 32%E	32,000
Construction expenses, 34%E	34,000
Legal expenses, 4%E	4,000
Contractor's fee, 19%E	19,000
Contingency, 37%E	37,000
Total indirect plant cost, I	126,000
Fixed-capital investment, $D + I$	\$431,000

■

Figure 6-6 shows a spreadsheet for estimating the fixed and total capital investment by the delivered-equipment ratio factor method, as detailed in Table 6-9. Default factors for the three general process types—solid, solid-liquid, and liquid processing—are included. A set of these factors, or individual values, can be copied to the corresponding user input location. Alternatively, the user can supply values for as many of these factors as desired.

The user must supply the total purchased-equipment cost for the major equipment items, as determined from material and energy balances and equipment operating characteristics. These costs can be estimated from information supplied in Chaps. 12 through 15.

Method D: Lang Factors for Approximation of Capital Investment This technique, proposed originally by Lang[†] and used quite frequently to obtain order-of-magnitude cost estimates, recognizes that the cost of a process plant may be obtained by multiplying the equipment cost by some factor to approximate the fixed or total capital investment. These factors vary depending upon the type of process plant being considered. The percentages given in Table 6-10 are rough approximations that hold for the types of process plants indicated. These values may be used as Lang factors for estimating the fixed-capital investment or the total capital investment.

Greater accuracy of capital investment estimates can be achieved in this method by using not one but a number of factors. One approach is to use different factors for different types of equipment. Another approach is to use separate factors for installation of equipment, foundations, utilities, piping, etc., or even to divide each item of cost into material and labor factors.[‡] With this approach, each factor has a range of

[†]H. J. Lang, *Chem. Eng.*, **54**(10): 117 (1947); H. J. Lang, *Chem. Eng.*, **55**(6): 112 (1948).

[‡]Further discussions on these methods may be found in W. D. Baasel, *Preliminary Chemical Engineering Plant Design*, 2d ed., Van Nostrand Reinhold, New York, 1990; S. G. Kirkham, *AACE Bull.*, **15**(5): 137 (1972); C. A. Miller, *Cost Engineers' Notebook*, ASCE A-1000, 1978.

See Table 6-9 for details.					
Project Identifier:	Fraction of delivered equipment				
	Solid-processing plant	Solid-fluid processing plant	Fluid-processing plant	User values	Calculated values, $M
Direct costs					
Purchased equipment					
Delivery, percent of purchased equipment				0.10	
Subtotal: delivered equipment					
Purchased equipment installation	0.45	0.39	0.47		
Instrumentation and controls (installed)	0.18	0.26	0.36		
Piping (installed)	0.16	0.31	0.68		
Electrical systems (installed)	0.10	0.10	0.11		
Buildings (including services)	0.25	0.29	0.18		
Yard improvements	0.15	0.12	0.10		
Service facilities (installed)	0.40	0.55	0.70		
Total direct cost					
Indirect costs					
Engineering and supervision	0.33	0.32	0.33		
Construction expenses	0.39	0.34	0.41		
Legal expenses	0.04	0.04	0.04		
Contractor's fee	0.17	0.19	0.22		
Contingency	0.35	0.37	0.44		
Total indirect cost					
Fixed capital investment					
Working capital	0.70	0.75	0.89		
Total capital investment					

Figure 6-6
Estimation of capital-investment items based on delivered-equipment cost

values, and the chemical engineer must rely on past experience to decide, in each case, whether to use a high, average, or low figure.

Since tables are not convenient for computer calculations, it is better to combine the separate factors into an equation similar to the one proposed by Hirsch and Glazier†

$$C_n = f_I[E'(1 + f_F + f_p + f_m) + E_i + A] \tag{6-5}$$

†H. Hirsch and E. M. Glazier, *Chem. Eng. Prog.*, **56**(12): 37 (1960).

Table 6-10 Revised Lang factors for estimation of fixed-capital investment or total capital investment

Factor × delivered-equipment cost = fixed-capital investment or total capital investment for major additions to an existing plant.

	Lang factors	
Type of plant†	**Fixed-capital investment**	**Total capital investment**
Solid	4.0	4.7
Solid-fluid	4.3	5.0
Fluid	5.0	6.0

†See Table 6-6 for description of types of process plants.

where the three installation-cost factors are, in turn, defined by the following three equations:

$$\log f_F = 0.635 - 0.154 \log(0.001\,E') - 0.992\left(\frac{e}{E'}\right) + 0.506\left(\frac{f_v}{E'}\right) \tag{6-6}$$

$$\log f_p = -0.266 - 0.014 \log(0.001\,E') - 0.156\left(\frac{e}{E'}\right) + 0.556\left(\frac{p}{E'}\right) \tag{6-7}$$

$$\log f_m = 0.344 + 0.033 \log(0.001\,E') + 1.194\left(\frac{t}{E'}\right) \tag{6-8}$$

where E' is the purchased equipment on an f.o.b. basis, f_I the indirect cost factor that is always greater than 1 (normally taken as 1.4), f_F the cost factor for field labor, f_p the cost factor for piping materials, f_m the cost factor for miscellaneous items, including the materials cost for insulation, instruments, foundations, structural steel, building, wiring, painting, and the cost of freight and field supervision, E_i the cost of equipment already installed, A the incremental cost of corrosion-resistant alloy materials, e the total heat exchanger cost (less incremental cost of alloy), f_v the total cost of field-fabricated vessels (less incremental cost of alloy), p the total pump plus driver cost (less incremental cost of alloy), and t the total cost of tower shells (less incremental cost of alloy). Note that Eq. (6-5) is designed to handle both purchased equipment on an f.o.b. basis and completely installed equipment.

Method E: Power Factor Applied to Plant/Capacity Ratio This method for study or order-of-magnitude estimates relates the fixed-capital investment of a new process plant to the fixed-capital investment of similar previously constructed plants by an exponential power ratio. That is, for certain similar process plant configurations, the fixed-capital investment of the new facility is equal to the fixed-capital investment of the constructed facility C (adjusted by a cost index ratio), multiplied by the ratio R, defined as the capacity of the new facility divided by the capacity of the old facility, raised to a power x. This power has been found to average between 0.6 and 0.7 for many process facilities. Table 6-11 gives the capacity power factor x for various kinds of processing plants

$$C_n = C f_e R^x \tag{6-9}$$

where f_e is the cost index ratio at the time of cost C_n to that at the time of C.

Table 6-11 Capital cost data for chemical and petroleum processing plants (2000)[†]

Product or process	Process	Typical plant size	Fixed-capital investment, million $	Power factor x[‡] for specified process plant
		10^3 kg/yr (10^3 ton/yr)		
Acetic acid	CH_3OH and CO—catalytic	9×10^3 (10)	8	0.68
Acetone	Propylene-copper chloride catalyst	9×10^4 (100)	33	0.45
Ammonia	Steam reforming	9×10^4 (100)	29	0.53
Ammonium nitrate	Ammonia and nitric acid	9×10^4 (100)	6	0.65
Butanol	Propylene, CO, and H_2O—catalytic	4.5×10^4 (50)	48	0.40
Chlorine	Electrolysis of NaCl	4.5×10^4 (50)	33	0.45
Ethylene	Refinery gases	4.5×10^4 (50)	16	0.83
Ethylene oxide	Ethylene—catalytic	4.5×10^4 (50)	59	0.78
Formaldehyde (37%)	Methanol—catalytic	9×10^3 (10)	19	0.55
Glycol	Ethylene and chlorine	4.5×10^3 (5)	18	0.75
Hydrofluoric acid	Hydrogen fluoride and H_2O	9×10^3 (10)	10	0.68
Methanol	CO_2, natural gas, and steam	5.5×10^4 (60)	15	0.60
Nitric acid (high-strength)	Ammonia—catalytic	9×10^4 (100)	8	0.60
Phosphoric acid	Calcium phosphate and H_2SO_4	4.5×10^3 (5)	4	0.60
Polyethylene (high-density)	Ethylene—catalytic	4.5×10^3 (5)	19	0.65
Propylene	Refinery gases	9×10^3 (10)	4	0.70
Sulfuric acid	Sulfur—contact catalytic	9×10^4 (100)	4	0.65
Urea	Ammonia and CO_2	5.5×10^4 (60)	10	0.70
		10^3 m^3/day (10^3 bbl/day)		
Alkylation (H_2SO_4)	Catalytic	1.6 (10)	23	0.60
Coking (delayed)	Thermal	1.6 (10)	31	0.38
Coking (fluid)	Thermal	1.6 (10)	19	0.42
Cracking (fluid)	Catalytic	1.6 (10)	19	0.70
Cracking	Thermal	1.6 (10)	6	0.70
Distillation (atm.)	65% vaporized	16 (100)	38	0.90
Distillation (vac.)	65% vaporized	16 (100)	23	0.70
Hydrotreating	Catalytic desulfurization	1.6 (10)	3.5	0.65
Reforming	Catalytic	1.6 (10)	34	0.60
Polymerization	Catalytic	1.6 (10)	6	0.58

[†]Adapted from K. M. Guthrie, *Chem. Eng.*, **77**(13): 140 (1970); and K. M. Guthrie, *Process Plant Estimating, Evaluation, and Control*, Craftsman Book Company of America, Solana Beach, CA, 1974. See also J. E. Haselbarth, *Chem. Eng.*, **74**(25): 214 (1967), and D. E. Drayer, *Petro. Chem. Eng.*, **42**(5): 10 (1970).
[‡]These power factors apply within roughly a 3-fold ratio extending either way from the plant size as given.

Table 6-12 Relative labor rate and productivity indexes in chemical and allied products industries for the United States (1999)†

Geographic area	Relative labor rate	Relative productivity factor
New England	1.14	0.95
Middle Atlantic	1.06	0.96
South Atlantic	0.84	0.91
Midwest	1.03	1.06
Gulf	0.95	1.22
Southwest	0.88	1.04
Mountain	0.88	0.97
Pacific Coast	1.22	0.89

†Adapted from J. M. Winton, *Chem. Week,* **121**(24): 49 (1977), and updated with data from M. Kiley, ed., *National Construction Estimator,* 37th ed., Craftsman Book Company of America, Carlsbad, CA, 1989. Productivity, as considered here, is an economic term that gives the value added (products minus raw materials) per dollar of total payroll cost. Relative values were determined by taking the average of Kiley's weighted state values in each region divided by the weighted-average value of all the regions. See also Tables 6-14 and 6-15.

A closer approximation for this relationship which involves the direct and indirect plant costs has been proposed as

$$C_n = f(DR^x + I) \tag{6-10}$$

where f is a lumped cost index factor relative to the original facility cost, D the direct cost, and I the total indirect cost for the previously installed facility of a similar unit on an equivalent site. The value of the power factor x approaches unity when the capacity of a process facility is increased by adding identical process units instead of increasing the size of the process equipment. The lumped cost index factor f is the product of a geographic labor cost index, the corresponding area labor productivity index, and a material and equipment cost index. Table 6-12 presents the relative median labor rate and productivity factors for various geographic areas in the United States.

EXAMPLE 6-4 Estimating Relative Costs of Construction Labor as a Function of Geographic Area

If a given chemical process plant is erected near Dallas, Texas (Southwest area), with a construction labor cost of $100,000, what would be the construction labor cost of an identical plant if it were erected at the same time near Los Angeles (Pacific Coast area) for the time when the factors given in Table 6-12 apply?

■ Solution

Relative median labor rates from Table 6-12 are 0.88 and 1.22 for the Southwest and Pacific Coast areas, respectively. The relative productivity factors for these same geographic areas are 1.04 and 0.89, respectively.

Relative labor rate ratio $= \frac{1.22}{0.88} = 1.3864$

Relative productivity factor ratio $= \frac{0.89}{1.04} = 0.8558$

Labor cost for Pacific Coast relative to Southwest $= \frac{1.3864}{0.8558} = 1.620$

Labor cost at Los Angeles = (1.62)($100,000) = $162,000 ■

Estimation of Fixed-Capital Investment with Power Factor Applied to Plant/Capacity Ratio — EXAMPLE 6-5

If the process plant described in Example 6-1 was erected in the Dallas area for a fixed-capital investment of $436,000 in 1990, estimate the fixed-capital investment in 1998 for a similar process plant located near Los Angeles with twice the process capacity but with an equal number of process units. Use the power factor method to evaluate the new fixed-capital investment, and assume the factors given in Table 6-12 apply.

■ Solution

If Eq. (6-9) is used with a 0.6 power factor and the Marshall and Swift all-industry index (Table 6-2), the fixed-capital investment is

$$C_n = Cf_e R^x$$

$$= 436{,}000\left(\frac{1062}{915}\right)(2)^{0.6} = \$767{,}000$$

If Eq. (6-9) is used with a 0.7 power factor and the Marshall and Swift all-industry index, the fixed-capital investment is

$$C_n = 436{,}000\left(\frac{1062}{915}\right)(2)^{0.7} = \$822{,}000$$

If Eq. (6-10) is used, the fixed-capital investment is given by

$$C_n = f(DR^x + I)$$

where $f = f_E f_L e_L$ and D and I are obtained from Example 6-1. With a 0.6 power factor, the Marshall and Swift all-industry index (Table 6-2), and the relative labor and productivity indexes (Table 6-12), the result is

$$C_n = \left(\frac{1062}{915}\right)\left(\frac{1.22}{0.88}\right)\left(\frac{1.04}{0.89}\right)[(316{,}000)(2)^{0.6} + 120{,}000] = \$1{,}126{,}000$$

Changing the power factor to 0.7 gives

$$C_n = (1.161)(1.620)[(316{,}000)(2)^{0.7} + 120{,}000] = \$1{,}191{,}000$$

Results obtained by using this latter procedure have shown high correlation with fixed-capital investment estimates that have been obtained with more detailed techniques. Properly used, these power factor methods can yield quick fixed-capital investment requirements with accuracies sufficient for preliminary estimates. ■

Method F: Investment Cost per Unit of Capacity Many data have been published giving the fixed-capital investment required for various processes per unit of annual production capacity. Such values may be obtained from Table 6-11 by dividing investment values by the corresponding capacity in the preceding column. Although these values depend to some extent on the capacity of the individual plants, it is possible to determine the unit investment costs which apply for average conditions. An order-of-magnitude estimate of the fixed-capital investment for a given process can then be obtained by multiplying the appropriate investment cost per unit of capacity by the annual production capacity of the proposed plant. The necessary correction for change of costs with time can be made with the use of cost indexes.

Method G: Turnover Ratio A rapid evaluation method suitable for order-of-magnitude estimates is known as the *turnover ratio* method. The turnover ratio is defined as the ratio of gross annual sales to fixed-capital investment

$$\text{Turnover ratio} = \frac{\text{gross annual sales}}{\text{fixed-capital investment}} \tag{6-11}$$

where the product of the annual production rate and the average selling price of the commodities is the gross annual sales figure. The reciprocal of the turnover ratio is sometimes called the *capital ratio* or the *investment ratio*.† Turnover ratios of up to 4 are obtained for some business establishments while some are as low as 0.2. For the chemical industry, as a very rough rule of thumb, the ratio can be approximated as 0.5.‡

ESTIMATION OF REVENUE

Thus far in this chapter methods for estimating the total capital investment required for a given plant have been presented. Determination of the necessary capital investment is only one part of a complete cost estimate. The revenue generated by plant operation clearly is very important. Revenue comes from sale of the product or products produced by the plant. The total annual revenue from product sales is the sum of the unit price of each product multiplied by its rate of sales.

$$\text{Annual sales revenue, \$/yr} = \sum (\text{sales of product, kg/yr})(\text{product sales price, \$/kg}) \tag{6-12}$$

A plant is designed for a specific rate of production of the major product. Rates of production of other products (by-products) are determined in turn by the chemistry and operating characteristics of the process. Mass balances for the process establish the by-product flow rates.

In conducting an economic analysis of a process, the engineer must establish production rates, as a fraction or percentage of the design capacity, for each year of process operation. It is common in preliminary economic studies to use 50 percent for the first year of operation because, during the start-up period, production rates are very low, the length of the start-up period is uncertain, and the time of the year for the beginning of start-up is unknown. After the first year, it is common to use the design annual capacity of the plant as the production and sales rate for each subsequent year. This is based on the usual practice of rating the annual capacity of chemical plants as the actual annual production, with an allowance for downtime. This downtime

†When the term *investment ratio* is used, the investment is usually considered to be the total capital investment which includes working capital as well as other capitalized costs.

‡R. H. Perry and D. W. Green, eds., *Perry's Chemical Engineers' Handbook,* 7th ed., McGraw-Hill, New York, 1997.

allowance is typically 10 to 20 percent, based on a 24 h/day, 7 days/week, 52 weeks/year production for continuous processes. So the actual operating time is from approximately 300 to 330 days per year. This number is usually established as part of the plant design basis, based on experience. For 90 percent operating time, the hourly production rate during operation must be the annual output rate divided by 0.90 times the 8760 h/yr.

Product prices are best established by a market study. For established products, price information is available in sources such as the *Chemical Market Reporter.* More information on chemical prices is given in the next section.

Other sources of revenue may include sale of obsolete equipment, recovery of working capital, and sale of other capital items. Revenue from such one-time events is included at the time it is expected to occur. But revenues due to product sales occur regularly, and more or less continuously, throughout the operating life of the product.

ESTIMATION OF TOTAL PRODUCT COST

The third major component of an economic analysis is the total of all costs of operating the plant, selling the products, recovering the capital investment, and contributing to corporate functions such as management and research and development. These costs usually are combined under the general heading of *total product cost.* The latter, in turn, is generally divided into two categories: *manufacturing costs* and *general expenses.* Manufacturing costs are also referred to as *operating* or *production costs.* Further subdivision of the manufacturing costs is somewhat dependent upon the interpretation of variable, fixed, and overhead costs.

Accuracy is as important in estimating total product cost as it is in estimating capital investment costs. The most important contribution to accuracy is to include all the costs associated with making and selling the product. The largest sources of error in total product cost estimation often are those of overlooking one or more elements of cost. A tabular form is very useful for estimating total product cost and provides a valuable checklist to preclude omissions. Figure 6-7 displays a suggested checklist of all the costs involved in chemical processing operations and is presented in the form of a spreadsheet in Fig. 6-8. The entries in Figs. 6-7 and 6-8 are discussed in detail below.

Figure 6-8 shows the user inputs that are required and provides default values for the other quantities needed to calculate the total product cost. The required user inputs include fixed-capital investment, annual amounts and unit prices of raw materials, operating labor, utilities, catalysts, and solvents. The remaining items are supplied with default values that can be changed. Depreciation is calculated separately because it changes from year to year under the most used method. However, if desired, the depreciation can be charged at a constant rate for a fixed number of years, such as 20 percent per year for 5 years.[†]

Total product costs are commonly calculated on one of three bases: namely, daily basis, unit of product basis, or annual basis. Annual cost is probably the best choice for

[†]Constant annual depreciation for a fixed number of years is appropriate if the time value of money is not to be considered. See Chap. 7 for a discussion of depreciation.

Raw materials
Operating labor
Operating supervision
Utilities
- Electricity
- Fuel
- Refrigeration
- Steam
- Waste treatment and disposal
- Water, process
- Water, cooling

Maintenance and repairs
Operating supplies
Laboratory charges
Royalties (if not on lump-sum basis)
Catalysts and solvents

Subtotal: Variable production costs

Depreciation
Taxes (property)
Financing (interest)
Insurance
Rent

Subtotal: Fixed charges

Medical
Safety and protection
General plant overhead
Payroll overhead
Packaging
Restaurant
Recreation
Salvage
Control laboratories
Plant superintendence
Storage facilities

Subtotal: Plant overhead costs
Total of above = Manufacturing costs

Executive salaries
Clerical wages
Engineering
Legal costs
Office maintenance
Communications

Subtotal: Administrative expenses

Sales offices
Sales personnel expenses
Shipping
Advertising
Technical sales service

Subtotal: Distribution and marketing expenses

Research and development

Total of administrative, distribution and marketing, R&D = General expenses
Total of all above = Total product cost

Figure 6-7
Costs involved in total product cost for a typical chemical process plant

Title:		Date:		
Product:		Capacity, kg/h:		
Operating time, h/yr:		Capacity, kg/s:		
Capacity, kg/yr:		Fixed Capital Investment (FCI)		

	Suggested factor	User variables: Rate or quantity per year	User variables: Cost per rate or quantity unit	Calculated values, $M
Raw materials				
1		______	______	______
2		______	______	______
3		______	______	______
4		______	______	**Total** ______
Operating labor†		______	______	______
Operating supervision	0.15	of operating labor		______
Utilities†				
Water				
Cooling		______	______	______
Process		______	______	______
Electricity		______	______	______
Fuel		______	______	______
Refrigeration		______	______	______
Steam		______	______	______
Waste treatment and disposal		______	______	______
Maintenance and repairs	0.07	of FCI		______
Operating supplies	0.15	of maintenance and repairs		______
Laboratory charges	0.15	of operating labor		______
Royalties (if not on lump-sum basis)	0.04	of TPC without depreciation		______
Catalysts and solvents				______
			Total variable production costs	______
Depreciation—calculated separately below				
Taxes (property)	0.02	of FCI		______
Financing (interest)	0.00	of FCI		______
Insurance	0.01	of FCI		______
Rent	0.00	of FCI		______
Depreciation‡				______
			Fixed charges (without depreciation)	______
			Plant overhead costs	______
			Administrative costs	______
			Distribution + marketing costs	______
			Research and development	______
			General expenses	______
			Total product cost (without depreciation)	______

‡Calculated according to MACRS 5-year schedule. (See Chap. 7 for details.)

Year	Depreciation: % of FCI	Depreciation: d_j, $/yr	TPC = $(d_j + c_o)$, $/yr
1	20		
2	32		
3	19.2		
4	11.52		
5	11.52		
6	5.76		
7	0		
⋮	⋮		
20	0		

†See Table 6-14 for suggested general utility and labor costs.

Figure 6-8
Spreadsheet for first-year, annual total product cost for 100 percent capacity

the purpose of economic analyses. Moreover, annual estimates (1) smooth out the effect of seasonal variations, (2) include plant on-stream time or equipment operation, (3) permit more rapid calculation of operating costs at less than full capacity, and (4) provide a convenient way of considering large expenses that occur infrequently such as annual planned maintenance shutdowns.

The best source of information for total product cost estimates is data from similar or identical projects. Most companies have extensive records of their operations, so that quick, reliable estimates of manufacturing costs and general expenses can be obtained from existing records. Adjustments for increased costs due to inflation must be made, and differences in plant site and geographic location must be considered. Methods for estimating total product cost in the absence of specific information are discussed in the following paragraphs.

Manufacturing Costs

All expenses directly connected with the manufacturing operation or the physical equipment of a process plant itself are included in the manufacturing costs. These expenses, as considered here, are divided into three classifications: (1) variable production costs, (2) fixed charges, and (3) plant overhead costs.

Variable production costs include expenses directly associated with the manufacturing operation. This type of cost involves expenditures for raw materials (including transportation, unloading, etc.), direct operating labor, supervisory and clerical labor directly applied to the manufacturing operation, utilities, plant maintenance and repairs, operating supplies, laboratory supplies, royalties, catalysts, and solvents. These costs are incurred for the most part only when the plant operates, hence the term *variable costs*. It should be recognized that some of the variable costs listed here as part of the direct production costs have an element of fixed cost in them. For instance, maintenance and repair costs decrease with reduced production level, but some maintenance and repair still occurs when the process plant is shut down.

Fixed charges are expenses which are practically independent of production rate. Expenditures for depreciation, property taxes, insurance, financing (loan interest), and rent are usually classified as fixed charges. These charges, except for depreciation, tend to change due to inflation. Because depreciation is on a schedule established by tax regulations, it may differ from year to year, but it is not affected by inflation.

Plant overhead costs are for hospital and medical services; general plant maintenance and overhead, safety services, payroll overhead including social security and other retirement plans, medical and life insurance, and vacation allowances, packaging, restaurant and recreation facilities, salvage services, control laboratories, property protection, plant superintendence, warehouse and storage facilities, and special employee benefits. These costs are similar to the basic fixed charges since they do not vary widely with changes in production rate.

Variable Production Costs

Raw Materials In the chemical industry, one of the major costs in a production operation is for the raw materials used in the process. The category *raw materials* refers in general to those materials that are directly consumed in making the final products; this includes chemical reactants and constituents and additives included in the product. Materials necessary to carry out process operations but which do not become part of the final product, such as catalysts and solvents, are listed separately.

Direct price quotations from prospective suppliers are desirable for the raw materials. When these are not available, published prices are used. For preliminary cost analyses, market prices are often used for estimating raw material costs. Prices for many commercial chemicals are published weekly in the *Chemical Market Reporter*. Prices for some commodity chemicals, such gases as hydrogen, oxygen, and nitrogen, are published occasionally in *Chemical and Engineering News* and *Chemical Week*. Other price sources are industry organizations and their publications. Chemical prices are usually quoted on an f.o.b. (free-on-board) basis. Any transportation charges should be included in the raw material costs when available; they may be estimated as 10 percent of the raw material cost, but are highly variable.

The amounts of raw materials that must be supplied per unit of time or per unit of product are determined from process material balances. One of the most important steps of the design process is to calculate accurate material balances for the process; these are essential to establishing the process raw materials requirements. Usually the basis for a process design is the production rate of a key product, that is, an output. Mass balances are nearly always calculated by starting with inputs. Thus, mass balance calculations begin by setting the flow rate of a key feed stream as a basis. This basis often is set as the amount of the feed required to yield the key product rate for ideal conversion of that feed to product. Alternatively, a conversion efficiency may be assumed, especially if there is past experience on which to make this assumption. The feed rates of other raw materials are set in relation to the key feed rate, using the process chemistry and conversion assumptions. The process mass balances are conducted with this set basis, and the rates of production of the products are calculated. The calculated production rate of the key product will usually not agree with that specified as the plant design basis. The feed rates of the raw materials are then adjusted so as to meet the output target, and the mass balances are repeated until convergence is obtained.

The ratio of the cost of raw materials to total product cost varies considerably for different types of plants. In chemical plants, raw material costs are usually in the range of 10 to 60 percent of the total product cost.

Operating Labor In general, operating labor may be divided into skilled and unskilled labor. Hourly wage rates for operating labor in different industries at various locations can be obtained from the U.S. Bureau of Labor publication entitled *Monthly Labor Review*. For chemical processes, operating labor usually amounts to about 10 to 20 percent of the total product cost.

In preliminary cost analyses, the quantity of operating labor can often be estimated either from company experience with similar processes or from published information on similar processes. The relationship between labor requirements and production rate is not a linear one; a 0.2 to 0.25 power of the capacity ratio when plant capacities are scaled up or down is often used.

Table 6-13 Typical labor requirements for process equipment†

Type of equipment	Workers/unit/shift
Blowers and compressors	0.1–0.2
Centrifugal separator	0.25–0.50
Crystallizer, mechanical	0.16
Dryer, rotary	0.5
Dryer, spray	1.0
Dryer, tray	0.5
Evaporator	0.25
Filter, vacuum	0.125–0.25
Filter, plate and frame	1.0
Filter, rotary and belt	0.1
Heat exchangers	0.1
Process vessels, towers (including auxiliary pumps and exchangers)	0.2–0.5
Reactor, batch	1.0
Reactor, continuous	0.5

†For expanded process equipment labor requirements see G. D. Ulrich, *A Guide to Chemical Engineering Process Design and Economics,* J. Wiley, New York, 1984.

If a flowsheet and drawings of the process are available, the operating labor may be estimated from an analysis of the work to be performed. Consideration must be given to such items as the type and arrangement of equipment, multiplicity of units, amount of instrumentation and control for the process, and company policy in establishing labor requirements. Table 6-13 indicates some typical labor requirements for various types of process equipment.

Another method of estimating labor requirements as a function of plant capacity is based on adding the various principal processing steps on the flowsheet. In this method, a process step is defined as any unit operation, unit process, or combination thereof that takes place in one or more units of distillation, evaporation, drying, filtration, etc. Once the plant capacity is fixed, the number of employee-hours per day per step is obtained from Fig. 6-9 and multiplied by the number of process steps to give the total employee-hours per day. Variations in labor requirements from highly automated processing steps to batch operations are provided by selection of the appropriate curve on Fig. 6-9.

EXAMPLE 6-6 Estimation of Labor Requirements

Consider a highly automated processing plant having an output rate of 1.0 kg/s of product and requiring principal processing steps of heat transfer, reaction, and distillation. What are the operating labor requirements for an annual operation of 300 days?

■ Solution

The process plant is considered to require three process steps. From Fig. 6-9, for a capacity of 1.0 kg/s (8.6×10^4 kg/day) the highly automated process plant requires approximately 33 employee-hours/(day)(processing step). Thus, for 300 days of annual operation, operating labor required = (3)(33)(300) = 29,700 employee-hours/year.

■

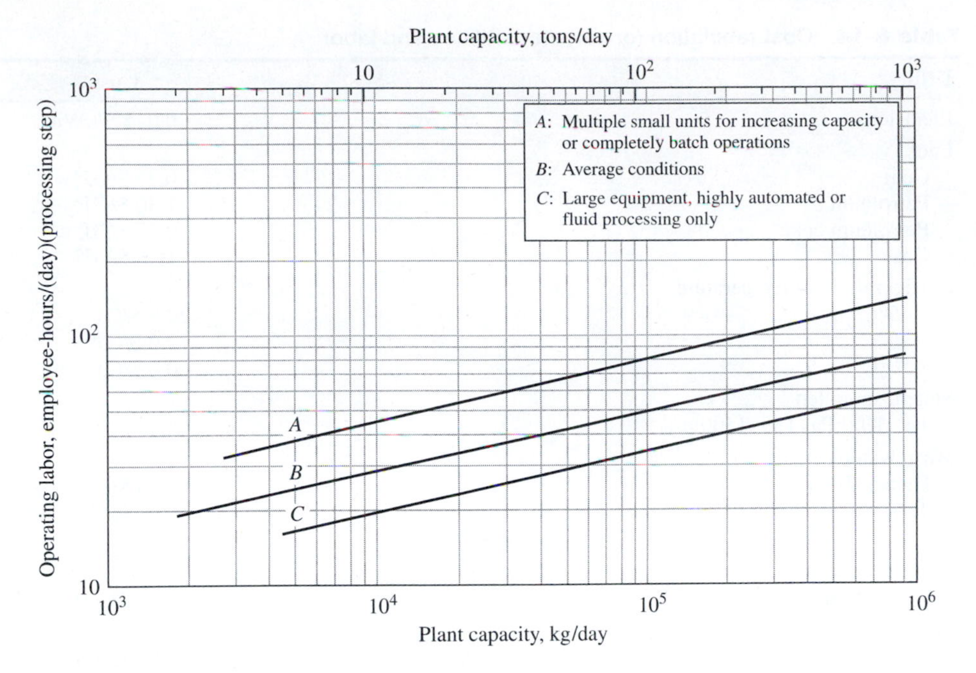

Figure 6-9
Operating labor requirements in the chemical process industry

Because of new technological developments including computerized controls and long-distance control arrangements, the practice of relating employee-hour requirements directly to production quantities for a given product can give inaccurate results unless very recent data are used. As a rule of thumb, the labor requirements for a fluids processing plant, such as an ethylene oxide plant, would be in the low range of 0.33 to 2 employee-hours per 1000 kg of product; for a solid-fluids plant, such as a shale-oil plant, the labor requirement would be in the intermediate range of 2 to 4 employee-hours per 1000 kg of product; for plants primarily engaged in solids processing, such as a coal briquetting plant, a range of 4 to 8 employee-hours per 1000 kg of product would be reasonable.

Certainly better estimates for the labor requirements than those obtained from the preceding rule of thumb can be made based on experience with similar processes. In determining costs for labor, account must be taken of the type of worker required, geographic location of the plant, prevailing wage rates, and worker productivity. Table 6-12 presents data that can be used as a guide for relative median labor rates and productivity factors for workers in various geographic areas of the United States. Table 6-14 gives some average labor rates. The *Engineering News-Record* provides data on prevailing labor rates in many U.S. cities, as shown in Table 6-15.

Operating Supervision and Clerical Assistance A certain amount of direct supervisory and clerical assistance is always required for a manufacturing operation. The necessary amount of this type of labor is closely related to the total amount of operating labor, complexity of the operation, and product quality standards. The cost for direct

Table 6-14 Cost tabulation for selected utilities and labor

Utility	Cost
Electricity	0.045 $/kWh[a]
Fuel	
Coal	0.35 $/GJ[b]
Petroleum	1.30 $/GJ[b]
Petroleum coke	0.17 $/GJ[b]
Gas	1.26 $/GJ[b]
Refrigeration, to temperature	
5°C	20.0 $/GJ[c]
−20°C	32.0 $/GJ[c]
−50°C	60.0 $/GJ[c]
Steam, saturated	
10^3–10^4 kPa (150–1500 psi)	4.40 $/1000 kg[e,d]
Wastewater	
Disposal	0.53 $/1000 kg[e]
Treatment	0.53 $/1000 kg[e]
Waste	
Hazardous	145.00 $/1000 kg[c]
Nonhazardous	36.00 $/1000 kg[c]
Water	
Cooling	0.08 $/1000 kg[e,f]
Process	0.53 $/1000 kg[e]
Labor	
Skilled	33.67 $/h[g]
Common	25.58 $/h[g]

[a]Based on U.S. Department of Energy, Energy Information Administration form EIA-861, 2001. U.S. average for year 2000.
[b]Based on U.S. Department of Energy, Energy Information Administration form EIA-0348, 2001. U.S. average for year 2000.
[c]R. Turton, R. C. Bailie, W. B. Whiting, and J. A. Shaeiwitz, *Analysis, Synthesis, and Design of Chemical Processes,* Prentice-Hall, Upper Saddle River, NJ, 1998.
[d]U.S. Department of Energy, Office of Industrial Technologies, DOE/GO-102000-1115, December 2000.
[e]U.S. Department of Energy, Office of Industrial Technologies, DOE/GO-10099-953, June 2001.
[f]M. S. Peters and K. D. Timmerhaus, *Plant Design and Economics for Chemical Engineers,* 4th ed., McGraw-Hill, New York, 1991.
[g]*Engineering News-Record* indexes, December 2001.

supervisory and clerical labor averages about 15 percent of the cost for operating labor. For reduced capacities, supervision usually remains fixed at the 100 percent capacity rate.

Utilities The cost for utilities, such as steam, electricity, process and cooling water, compressed air, natural gas, fuel oil, refrigeration, and waste treatment and disposal, varies widely depending on the amount needed, plant location, and source. A detailed list of ranges of rates for various utilities is presented in App. B. Some typical costs for utilities are given in Table 6-14.

The required types of utilities are established by the flowsheet conditions; their amount can sometimes be estimated in preliminary cost analyses from available information about similar operations. More often the utility requirements are determined

Table 6-15 ***Engineering News-Record*** **labor indexes†‡**

Location	Common labor 1997	1998	1999	2000	2001	Skilled labor 1997	1998	1999	2000	2001
Atlanta	6,305	6,305	6,563	7,195	7,326	3,623	3,698	3,850	4,152	4,393
Baltimore	8,442	8,442	8,592	8,745	8,745	4,587	4,671	4,704	4,917	4,917
Birmingham	7,853	7,853	8,432	8,747	9,300	3,589	3,659	3,914	4,084	4,205
Boston	14,605	15,132	15,526	15,526	15,526	6,560	6,820	6,951	7,360	7,360
Chicago	13,974	15,168	15,842	16,553	16,552	6,074	6,553	6,810	7,139	7,235
Cincinnati	11,184	11,316	11,789	12,211	12,211	4,659	4,716	4,839	5,198	5,198
Cleveland	12,771	13,156	13,350	14,232	14,753	5,544	5,701	5,846	6,002	6,201
Dallas	6,637	6,742	6,742	6,742	6,911	3,289	3,383	3,386	3,474	3,819
Denver	7,739	8,255	8,387	8,926	8,926	3,892	4,033	4,189	4,442	4,517
Detroit	13,668	14,216	14,742	15,374	16,032	6,123	6,410	6,631	6,886	7,177
Kansas City	11,621	12,053	12,053	12,834	13,437	4,875	5,015	5,260	5,503	5,653
Los Angeles	14,011	14,458	14,458	15,018	15,574	5,852	5,953	5,953	6,111	6,285
Minneapolis	13,368	13,979	14,532	15,084	15,953	5,429	5,672	5,937	6,222	6,580
New Orleans	6,842	6,842	6,842	6,995	7,274	3,358	3,359	3,587	3,669	3,848
New York	19,284	19,955	20,597	21,368	23,176	9,132	9,416	9,535	9,906	10,634
Philadelphia	14,605	15,211	15,737	16,000	16,974	6,450	6,705	6,978	7,158	7,476
Pittsburgh	11,884	12,234	12,497	12,839	13,195	5,438	5,584	5,733	5,911	6,099
St. Louis	13,726	14,121	14,553	15,000	15,474	5,447	5,696	5,874	6,094	6,252
San Francisco	14,157	14,411	14,411	16,005	16,011	6,477	6,740	6,740	7,057	7,142
Seattle	14,026	14,811	15,300	15,879	15,879	5,425	5,671	5,975	6,173	6,343
Montreal	12,379	12,379	12,742	13,074	13,674	5,381	5,387	5,520	5,685	5,988
Toronto	16,579	16,611	16,913	16,897	16,897	6,320	6,350	6,522	6,651	6,738
National labor index	11,835	12,233	12,547	13,063	14,461	5,294	5,473	5,635	5,873	6,067
Wages, $/h	22.48	23.24	23.84	24.82	25.58	29.38	30.37	31.27	32.6	33.67

†Published in December issues of *Engineering News-Record* (with permission from *Engineering News Record,* McGraw-Hill, New York).
‡Indexes = 100 in 1913.

from material and energy balances calculated for the process. A utility may be purchased at a predetermined rate from an outside source, or the service may be available within the company. If the company supplies its own service and this is utilized for just one process, the entire cost of the service installation is usually charged to the manufacturing process. If the service is utilized for the production of several different products, the service cost is apportioned among the different products at a rate based on the amount of individual consumption.

Steam requirements include the amount consumed in the manufacturing process plus that necessary for auxiliary needs. An allowance for radiation and line losses must also be made.

Electric power must be supplied for lighting, motors, and various process-equipment demands. These direct power requirements should be increased by a factor of 1.1 to 1.25 to allow for line losses and contingencies. As a rough approximation, utility costs for ordinary chemical processes amount to 10 to 20 percent of the total product cost.

The cost for pollution control and waste disposal is best estimated from pollutant quantities calculated from the process material balances. These quantities may require

Table 6-16 Estimation of costs for maintenance and repairs

	Maintenance cost as percentage of fixed-capital investment (on annual basis)		
Type of operation	**Wages**	**Materials**	**Total**
Simple chemical processes	1–3	1–3	2–6
Average processes with normal operating conditions	2–4	3–5	5–9
Complicated processes, severe corrosion operating conditions, or extensive instrumentation	3–5	4–6	7–11

special attention in simulation programs. Some estimates for the unit costs of treatment of various wastes are included in Table 6-14.†

Maintenance and Repairs Annual costs for equipment maintenance and repairs may range from 2 to 20 percent of the equipment cost. Charges for plant buildings average 3 to 4 percent of the building cost. In the process industries, the total plant cost per year for maintenance and repairs ranges from 2 to 10 percent of the fixed-capital investment, with 7 percent being a reasonable value. Table 6-16 provides a guide for estimation of maintenance and repair costs as a function of process conditions.

For operating rates less than plant capacity, the maintenance and repair cost is generally estimated as 85 percent of that at 100 percent capacity for a 75 percent operating rate, and 75 percent of that at 100 percent capacity for a 50 percent operating rate.

Operating Supplies Consumable items such as charts, lubricants, test chemicals, custodial supplies, and similar supplies cannot be considered as raw materials or maintenance and repair materials, and these are classified as operating supplies. The annual cost for these types of supplies is about 15 percent of the total cost for maintenance and repairs.

Laboratory Charges The cost of laboratory tests for control of operations and for product quality control is covered in this manufacturing cost. This expense is generally calculated by estimating the employee-hours involved and multiplying this by the appropriate rate. For quick estimates, this cost may be taken as 10 to 20 percent of the operating labor.

Patents and Royalties Patents cover many products and manufacturing processes. To use patents owned by others, it is necessary to pay for patent rights or a royalty based on the amount of material produced. Even when the company involved in the operation obtained the original patent, a certain amount of the total expense involved in the development and procurement of the patent rights should be borne by the plant as an operating expense. In cases of this type, these costs are usually amortized over the legally protected life of the patent. Although a rough approximation of patent and

†R. Turton, R. C. Bailie, W. B. Whiting, and J. A. Shaeiwitz, *Analysis, Synthesis and Design of Chemical Processes,* Prentice-Hall, Upper Saddle River, NJ, 1998.

royalty costs for patented processes is 0 to 6 percent of the total product cost, costs specific to the patent position in question are always preferred.

Catalysts and Solvents Costs for catalysts and solvents can be significant and should be estimated based on the catalyst and solvent requirements and prices for the particular process.

Fixed Charges Costs that change little or not at all with the amount of production are designated as *fixed costs* or *fixed charges*. These include costs for depreciation, local property taxes, insurance, and loan interest. Expenses of this type are a direct function of the capital investment and financing arrangement. They should be estimated from the fixed-capital investment. Rent is usually taken as zero in preliminary estimates. As a rough approximation, these charges amount to about 10 to 20 percent of the total product cost.

Depreciation The equipment, buildings, and other material objects comprising a manufacturing plant require an initial investment that must be paid back, and this is done by charging depreciation as a manufacturing expense. Since depreciation rates are very important in determining the amount of income tax, the Internal Revenue Service, under U.S. tax law, determines the rate at which depreciation may be charged for various types of industrial facilities.

In the most widely used method of depreciation calculation (MACRS), the amount of depreciation changes year by year. Therefore, depreciation is calculated separately in the table in the bottom of the spreadsheet in Fig. 6-8. In economic studies in which the time value of money is not to be considered, it is acceptable to use a constant yearly depreciation rate for a fixed period.

Financing *Interest* is considered to be the compensation paid for the use of borrowed capital. A fixed rate of interest is established at the time the capital is borrowed; therefore, interest is a definite cost if it is necessary to borrow the capital used to make the investment for a plant. Although the interest on borrowed capital is a fixed charge, there are many persons who claim that interest should not be considered as a manufacturing cost, but that it should be listed as a separate expense under the general heading of management or financing cost. Annual interest rates amount to 5 to 10 percent of the total value of the borrowed capital. For income tax calculations, interest on money supplied by the corporation cannot be charged as a cost. In design calculations, however, interest can be included as a cost if the required funds need to be borrowed from external sources.

Local Taxes The magnitude of local property taxes depends on the particular locality of the plant and the regional laws. Annual property taxes for plants in highly populated areas are ordinarily in the range of 2 to 4 percent of the fixed-capital investment. In less populated areas, local property taxes are about 1 to 2 percent of the fixed-capital investment.

Property Insurance Insurance rates depend on the type of process being carried out in the manufacturing operation and on the extent of available protection facilities. These rates amount to about 1 percent of the fixed-capital investment per year.

Rent Annual costs for rented land and buildings amount to about 8 to 12 percent of the value of the rented property. In preliminary estimates, rent is usually not included.

Plant Overhead Costs

The costs considered in the preceding sections are directly related to the production operation. In addition, however, many other expenses are always involved if the complete plant is to function as an efficient unit. The expenditures required for routine plant services are included in *plant overhead costs*. Nonmanufacturing machinery, equipment, and buildings are necessary for many of the general plant services, and the fixed charges and direct costs for these items are part of the plant overhead costs. Other components of the overhead are listed in Fig. 6-7. These charges are closely related to the costs for all labor directly connected with the production operation. The plant overhead cost for chemical plants is about 50 to 70 percent of the total expenses for operating labor, supervision, and maintenance.

General Expenses

In addition to the manufacturing costs, other general expenses are involved in the operations of a company. These general expenses may be classified as (1) administrative expenses, (2) distribution and marketing expenses, and (3) research and development expenses.

Administrative Costs The expenses connected with executive and administrative activities cannot be charged directly to manufacturing costs; however, it is necessary to include the administrative costs if the economic analysis is to be complete. Salaries and wages for administrators, secretaries, accountants, computer support staff, engineering, and legal personnel are part of the administrative expenses, along with costs for office supplies and equipment, outside communications, administrative buildings, and other overhead items related to administrative activities. These costs may vary markedly from plant to plant and depend somewhat on whether the plant under consideration is a new one or an addition to an existing plant. In the absence of more accurate cost figures from company records, or for a preliminary estimate, the administrative costs may be approximated as 15 to 25 percent of operating labor.

Distribution and Marketing Costs These types of general expenses are incurred in the process of selling and distributing the various products. From a practical viewpoint, no manufacturing operation can be considered a success until the products have been sold or put to some profitable use. Included in this category are salaries, wages, supplies, and other expenses for sales offices, salaries, commissions, and traveling expenses for sales representatives, shipping expenses, cost of containers, advertising expenses, and technical sales service.

Distribution and marketing costs vary widely for different types of plants depending on the particular material being produced, other products sold by the company, plant location, and company policies. These costs for most chemical plants are in the range of 2 to 20 percent of the total product cost. The higher figure usually applies to a new product or to one sold in small quantities to a large number of customers. The lower figure applies to large-volume products, such as bulk chemicals.

Research and Development Costs New methods and products are constantly being developed in the chemical industries as a result of research and development. Any progressive company that wishes to remain in a competitive industrial position incurs research and development expenses. Research and development costs include salaries and wages for all personnel directly connected with this type of work, fixed and operating expenses for all machinery and equipment involved, costs for materials and supplies, and consultants' fees. In some industries, such as pharmaceuticals, research may be the largest component of the total product cost. In the chemical industry, these costs amount to about 2 to 5 percent of every sales dollar, or about 5 percent of total product cost.

GROSS PROFIT, NET PROFIT, AND CASH FLOW

The product sales revenue minus the total product cost gives the *gross profit,* also called *gross earnings*. Gross profit is expressed both with and without depreciation included as follows:

$$g_j = s_j - c_{oj} \tag{6-13}$$

where g_j is gross profit, depreciation not included, in year j, and

$$G_j = s_j - c_{oj} - d_j \tag{6-14}$$

where G_j is gross profit, depreciation included, in year j.

Net profit, also called *net earnings,* is the amount retained of the profit after income taxes have been paid

$$N_{pj} = G_j(1-\Phi) \tag{6-15}$$

where N_{pj} is the net profit in year j.

The cash flow resulting from process operations is given by Eq. (6-1) and also by

$$A_j = N_{pj} + d_j \tag{6-16}$$

Breakeven Point, Gross and Net Profit for a Process Plant — EXAMPLE 6-7

The annual variable production costs for a plant operating at 70 percent capacity are \$280,000. The sum of the annual fixed charges, overhead costs, and general expenses is \$200,000, and may be considered not to change with production rate. The total annual sales are \$560,000, and the product sells for \$4/kg. What is the breakeven point in kilograms of product per year? What are the gross annual profit G_j (depreciation included) and net annual profit for this plant at 100 percent capacity if the income tax rate is 35 percent of gross profit?

■ Solution

The breakeven point (Fig. 6-3) occurs when the total annual product cost equals the total annual sales. The total annual product cost is the sum of the fixed charges (depreciation included), overhead, and general expenses, and the variable production costs. Total annual sales are the product of the number of kilograms of each product and corresponding selling price per kilogram. Thus,

$$\text{Direct production cost/kg} = \frac{\$280{,}000}{\$560{,}000/(\$4/\text{kg})} = \$2/\text{kg}$$

and the kg/yr needed for a breakeven point are given by

$$\$200{,}000 + (\$2)(\text{kg/yr}) = (\$4)(\text{kg/yr})$$

$$\text{kg/yr required} = 100{,}000$$

Since the annual capacity is

$$\frac{\$560{,}000}{(\$4/\text{kg})0.70} = 200{,}000 \text{ kg}$$

the breakeven point is

$$\frac{100{,}000}{200{,}000}(100) = 50\% \text{ of capacity}$$

The gross annual profit = total annual sales − total annual costs. So at 100 percent capacity

$$G_j = (\$4/\text{kg})(200{,}000) - [\$200{,}000 + (200{,}000 \text{ kg})(\$2/\text{kg})]$$
$$= \$200{,}000$$

and the annual net profit $= \$200{,}000 - (0.35)(\$200{,}000)$, so

$$N_{pj} = \$130{,}000$$

CONTINGENCIES

Unforeseen events, such as strikes, storms, floods, price variations, and other *contingencies,* may have an effect on the costs for a manufacturing operation. Contingencies are usually taken into account for product costs in estimating the number of operating days per year.

SUMMARY

This chapter has outlined the economic considerations necessary when a chemical engineer prepares estimates of capital investment cost or total product cost for a project. Predesign cost estimates purposely have been emphasized because of their importance in determining the feasibility of a proposed investment in comparison to alternative designs. It should be remembered, however, that predesign estimates are approximations with ±20 percent or greater uncertainty. Tables 6-17 and 6-18 summarize the predesign estimates for capital investment costs and total product costs, respectively. The percentages indicated in both tables give the ranges encountered in typical chemical plants. Because of the wide variations in different types of plants, the factors presented should be used *only* when more accurate data are not available.

Table 6-17 Estimation of capital investment cost (showing individual components)

The percentages indicated in the following summary of the various costs constituting the capital investment are approximations applicable to ordinary chemical processing plants. It should be realized that the values given vary depending on many factors, such as plant location, type of process, and complexity of instrumentation.

I. **Direct costs** = material and labor involved in actual installation of complete facility (65–85% of fixed-capital investment)
   A. Equipment + installation + instrumentation + piping + electrical + insulation + painting (50–60% of fixed-capital investment)
      1. Purchased equipment (15–40% of fixed-capital investment)
      2. Installation, including insulation and painting (25–55% of purchased-equipment cost)
      3. Instrumentation and controls, installed (8–50% of purchased-equipment cost)
      4. Piping, installed (10–80% of purchased-equipment cost)
      5. Electrical, installed (10–40% of purchased-equipment cost)
   B. Buildings, process, and auxiliary (10–70% of purchased-equipment cost)
   C. Service facilities and yard improvements (40–100% of purchased-equipment cost)
   D. Land (1–2% of fixed-capital investment or 4–8% of purchased-equipment cost)

II. **Indirect costs** = expenses which are not directly involved with material and labor of actual installation of complete facility (15–35% of fixed-capital investment)
   A. Engineering and supervision (5–30% of direct costs)
   B. Legal expenses (1–3% of fixed-capital investment)
   C. Construction expense and contractor's fee (10–20% of fixed-capital investment)
   D. Contingency (5–15% of fixed-capital investment)

III. **Fixed-capital investment** = direct costs + indirect costs

IV. **Working capital** (10–20% of total capital investment)

V. **Total capital investment** = fixed-capital investment + working capital

Table 6-18 Estimation of total product cost (showing individual components)

The percentages indicated in the following summary of the various costs involved in the complete operation of manufacturing plants are approximations applicable to ordinary chemical processing plants. It should be realized that the values given vary depending on many factors, such as plant location, type of process, and company policies.

I. **Manufacturing cost** = direct production costs + fixed charges + plant overhead costs
   A. Direct production costs (about 66% of total product cost)
      1. Raw materials (10–80% of total product cost)
      2. Operating labor (10–20% of total product cost)
      3. Direct supervisory and clerical labor (10–20% of operating labor)
      4. Utilities (10–20% of total product cost)
      5. Maintenance and repairs (2–10% of fixed-capital investment)
      6. Operating supplies (10–20% of maintenance and repair costs, or 0.5–1% of fixed-capital investment)
      7. Laboratory charges (10–20% of operating labor)
      8. Patents and royalties (0–6% of total product cost)
   B. Fixed charges (10–20% of total product cost)
      1. Depreciation (depends on method of calculation—see Chap. 7)
      2. Local taxes (1–4% of fixed-capital investment)
      3. Insurance (0.4–1% of fixed-capital investment)

(Continued)

Table 6-18 ***Continued***

4. Rent (8–12% of value of rented land and buildings)
5. Financing (interest) (0–10% of total capital investment)

C. Plant overhead costs (50–70% of cost for operating labor, supervision, and maintenance; or 5–15% of total product cost) include costs for the following: general plant upkeep and overhead, payroll overhead, packaging, medical services, safety and protection, restaurants, recreation, salvage, laboratories, and storage facilities

II. **General expenses** = administrative costs + distribution and selling costs + research and development costs (15–25% of the total product cost)

A. Administrative costs (about 20% of costs of operating labor, supervision, and maintenance; or 2–5% of total product cost) include costs for executive salaries, clerical wages, computer support, legal fees, office supplies, and communications

B. Distribution and marketing costs (2–20% of total product cost) include costs for sales offices, salespeople, shipping, and advertising

C. Research and development costs (2–5% of every sales dollar, or about 5% of total product cost)

III. **Total product cost**[†] = manufacturing cost + general expenses

IV. **Gross earnings cost** (gross earnings = total income − total product cost; amount of gross earnings cost depends on amount of gross earnings for entire company and income tax regulations; a general range for gross earnings cost is 15–40% of gross earnings)

[†]If desired, a contingency factor can be included by increasing the total product cost by 1–5%.

NOMENCLATURE

A = incremental cost of corrosion-resistant alloy materials, dollars

A_j = annual cash flow in year j, dollars

A_x = nonmanufacturing fixed-capital investment, dollars

c_o = first-year cost of operation, all total product costs except depreciation, at 100% of capacity, dollars

c_{oj} = total product costs except depreciation in year j, dollars

C = original capital investment, dollars

C_n = new capital investment, dollars

d_j = depreciation charge in year j, dollars

d_n = number of drawings and specifications

D = total direct cost of plant, dollars

e = total heat exchanger cost (less incremental cost of alloy), dollars

e_L = labor efficiency index in new location relative to cost of E_L and M'_L

E = purchased-equipment cost (delivered), dollars

E' = purchased-equipment cost on f.o.b. basis, dollars

E_i = installed-equipment cost (delivered and installed), dollars

E_L = purchased-equipment labor cost (base), dollars

f = lumped cost index relative to original installation cost

f_1, f_2, f_3 = multiplying factors for piping, electrical, instrumentation, etc., dimensionless

f_d = unit cost per drawing and specification, dollars per drawing or specification

f_e = unit cost for engineering, dollars per engineering employee-hour
f_E = current equipment cost index relative to cost index at time of original cost
f_F = construction or field labor expense factor (always greater than 1)
f_I = indirect cost factor (normally taken as 1.4)
f_L = current labor cost index in new location relative to cost of E_L and M'_L
f_M = current material cost index relative to cost of M, dimensionless
f_m = cost factor for miscellaneous items, dimensionless
f_p = cost factor for piping materials, dimensionless
f_v = total cost of field-fabricated vessels (less incremental cost of alloy), dollars
f_x = specific material unit cost, dollars
f_y = specific material labor unit cost, dollars per employee-hour
g_j = gross profit, depreciation not included, in year j, dollars
G_j = gross profit, depreciation included, in year j, dollars
H_e = engineering employee-hours
I = total indirect cost of plant, dollars
M = material cost, dollars
M'_L = labor employee-hours for specific material
M_L = direct labor cost for equipment installation and material handling, dollars
M_x = specific material quantity in compatible units
N_{pj} = net profit in year j, dollars
p = total pump plus driver cost (less incremental cost of alloy), dollars
R = ratio of new to original capacity
s_j = total income from sales in year j, dollars
t = total cost of tower shells (less incremental cost of alloy), dollars
$\mathcal{T}$ = total capital investment, dollars
V = manufacturing fixed-capital investment, dollars
W = working-capital investment, dollars
x = power value for cost-capacity relationships, dimensionless

Greek Symbol

Φ = fractional income tax rate

PROBLEMS

6-1 The purchased cost of a shell-and-tube heat exchanger (floating-head and carbon-steel tubes) with 100 m^2 of heating surface was \$4200 in 1990. What will be the 1990 purchased cost of a similar heat exchanger with 20 m^2 of heating surface if the purchased cost capacity exponent is 0.60 for surface areas ranging from 10 to 40 m^2? If the purchased cost capacity exponent for this type of exchanger is 0.81 for surface areas ranging from 40 to 200 m^2, what will be the purchased cost of a heat exchanger with 100 m^2 of heating surface in 2000?

6-2 Plot the 2000 purchased cost of the shell-and-tube heat exchanger outlined in Prob. 6-1 as a function of the surface area from 10 to 200 m^2. Note that the purchased cost capacity exponent is not constant over the range of surface area requested.

6-3 The purchased and installation costs of some pieces of equipment are given as a function of weight rather than capacity. An example of this is the installed costs of large tanks. The 1990 cost for an installed aluminum tank weighing 45,000 kg was $640,000. For a size range from 90,000 to 450,000 kg, the installed cost weight exponent for aluminum tanks is 0.93. If an aluminum tank weighing 300,000 kg is required, what capital investment is needed in the year 2000?

6-4 The 1990 cost for an installed 304 stainless steel tank weighing 135,000 kg was $1,100,000. The installed cost weight exponent for stainless steel tanks is 0.88 for a size range from 100,000 to 300,000 kg. What weight of installed stainless steel tank could have been obtained for the same capital investment as in Prob. 6-3?

6-5 The purchased cost of a 5-m^3 stainless steel tank in 1995 was $10,900. The 2-m-diameter tank is cylindrical with a flat top and bottom. If the entire outer surface of the tank is to be covered with 0.05-m-thickness of magnesia block, estimate the current total cost for the installed and insulated tank. The 1995 cost for the 0.05-m-thick magnesia block was $40 per square meter while the labor for installing the insulation was $95 per square meter.

6-6 A one-story warehouse 36 m by 18 m is to be added to an existing plant. An asphalt pavement service area 18 m by 9 m will be added adjacent to the warehouse. It will also be necessary to put in 150 linear m of railroad siding to service the warehouse. Utility service lines are already available at the warehouse site. The proposed warehouse has a concrete floor and steel frame, walls, and roof. No heat is necessary, but lighting and sprinklers must be installed. Estimate the total cost of the proposed addition. Consult App. B of the 4th ed. for necessary cost data.

6-7 The purchased cost of equipment for a solid processing plant is $500,000. The plant is to be constructed as an addition to an existing plant. Estimate the total capital investment and the fixed-capital investment for the plant. What percentage and amount of the fixed-capital investment are due to cost for engineering and supervision, and what percentage and amount for the contractor's fee?

6-8 The purchased-equipment cost for a plant which produces pentaerythritol (solid-fluid processing plant) is $300,000. The plant is to be an addition to an existing formaldehyde plant. The major part of the building cost will be for indoor construction. The contractor's fee will be 7 percent of the direct plant cost. All other costs are close to the average values found for typical chemical plants. On the basis of this information, estimate the total direct plant cost, the fixed-capital investment, and the total capital investment.

6-9 Estimate by the turnover ratio method the fixed-capital investment required in 2000 for a proposed sulfuric acid plant (battery-limit) which has an annual capacity of 1.3×10^8 kg/yr of 100 percent sulfuric acid (contact-catalytic process), using the data from Table 6-11, when the selling price for the sulfuric acid is $86 per metric ton. The plant will operate 325 days/year. Repeat the calculation, using the cost capacity exponent method with data from Table 6-11.

6-10 The total capital investment for a chemical plant is $1 million, and the working capital is $100,000. If the plant can produce an average of 8000 kg of final product per day during a 365-day year, what selling price in dollars per kilogram of product would be necessary to give a turnover ratio of 1.0?

6-11 A process plant was constructed in the Philadelphia area (Mid-Atlantic) at a labor cost of $425,000 in 1990. What would be the labor cost for the same plant located in the Miami, Florida, area (south Atlantic) if it were constructed in late 1998? Assume, for simplicity, that the relative labor rate and relative productivity factor have remained constant.

6-12 A company has been selling a soap containing 30 percent by weight water at a price of $20 per 50 kg f.o.b. (i.e., the customer pays the freight charges). The company offers an equally effective soap containing only 5 percent water. The water content is of no importance to the laundry, and it is willing to accept the soap containing 5 percent water if the delivered costs are equivalent. If the freight rate is $1.50 per 50 kg, how much should the company charge the laundry per 50 kg f.o.b. for the soap containing 5 percent water?

6-13 The total capital investment for a conventional chemical plant is $1,500,000, and the plant produces 3 million kg of product annually. The selling price of the product is $0.82/kg. Working capital amounts to 15 percent of the total capital investment. The investment is from company funds, and no interest is charged. Delivered raw materials costs for the product are $0.09/kg; labor, $0.08/kg; utilities, $0.05/kg; and packaging, $0.008/kg. Distribution costs are 5 percent of the total product cost. Estimate the following:

a. Manufacturing cost per kilogram of product

b. Total product cost per year

c. Profit per kilogram of product before taxes

d. Profit per kilogram of product after income taxes at 35 percent of gross profit

6-14 Estimate the manufacturing cost per 100 kg of product under the following conditions:

Fixed-capital investment = $4 million.
Annual production output = 9 million kg of product.
Raw materials cost = $0.25/kg of product.
Utilities
  800-kPa steam = 50 kg/kg of product.
  Purchased electric power = 0.9 kWh/kg of product.
  Filtered and softened water = 0.083 m^3/kg of product.
Operating labor = 12 persons per shift at $25.00 per employee-hour.
Plant operates three hundred 24-h days per year.
Corrosive liquids are involved.
Shipments are in bulk carload lots.
A large amount of direct supervision is required.
There are no patent, royalty, interest, or rent charges.
Plant overhead costs amount to 50 percent of the cost for operating labor, supervision, and maintenance.

6-15 A company has direct production costs equal to 50 percent of total annual sales, and fixed charges, overhead, and general expenses equal to $200,000. If management proposes to increase present annual sales of $800,000 by 30 percent with a 20 percent increase in fixed charges, overhead, and general expenses, what annual sales amount is required to provide the same gross earnings as the present plant operation? What would be the net profit if the expanded plant were operated at full capacity with an income tax on gross earnings of 35 percent? What would be the net profit for the enlarged plant if total annual sales remained at $800,000? What would be the net profit for the enlarged plant if the total annual sales actually decreased to $700,000?

6-16 A process plant making 5000 kg/day of a product selling for $1.75/kg has annual variable production costs of $2 million at 100 percent capacity and fixed costs of $700,000. What is the fixed cost per kilogram at the breakeven point? If the selling price of the product is increased by 10 percent, what is the dollar increase in net profit at full capacity if the income tax rate is 35 percent of gross earnings?

6-17 A rough rule of thumb for the chemical industry is that $1 of annual sales requires $2 of fixed-capital investment. In a chemical processing plant where this rule applies, the total capital investment is $2,500,000, and the working capital is 20 percent of the total capital investment.

The annual total product cost amounts to \$1,500,000. If the income tax rates on gross earnings total 35 percent, determine the following:

a. Percent of total capital investment returned annually as gross earnings

b. Percent of total capital investment returned annually as net profit

6-18 The total capital investment for a proposed chemical plant, which will produce \$1,500,000 worth of goods per year, is estimated to be \$1 million. It will be necessary to do a considerable amount of research and development (R&D) work on the project before the final plant can be constructed, and management wishes to estimate the permissible research and development costs. It has been decided that the after tax return from the plant should be sufficient to pay off the total capital investment plus all research and development costs in 7 years. A return after taxes of at least 12 percent of sales must be obtained. Because R&D is an expense and the company's income tax rate is 35 percent of gross earnings, only 65 percent of the funds spent on R&D must be recovered after taxes are paid. Under these conditions, what is the total amount the company can afford to pay for research and development?

6-19 A chemical processing unit has a capacity for producing 1 million kg of a product per year. After the unit has been put into operation, it is found that only 500,000 kg of the product can be sold per year. An analysis of the existing situation shows that all fixed and other invariant charges, which must be paid whether or not the unit is operating, amount to 35 percent of the total product cost when operating at full capacity. Raw material costs and other production costs that are directly proportional to the quantity of production (i.e., constant per kilogram of product at any production rate) amount to 40 percent of the total product cost at full capacity. The remaining 25 percent of the total product cost is for variable overhead and miscellaneous expenses, and the analysis indicates that these costs are directly proportional to the production rate during operation raised to the 1.5 power. What will be the percent change in total cost per kilogram of product if the unit is switched from the original design rate of 10^6kg/yr of product to a time and rate schedule which will produce 0.5×10^6 kg or "half that amount" of product per year at the least total cost?

CHAPTER

7

Interest, Time Value of Money, Taxes, and Fixed Charges

As noted earlier, there are many economic parameters besides capital investment and operating expenses that can have an effect on the decision of whether to appropriate funds for a proposed project. This is particularly true when the required funds for the project may have to be borrowed. If such is the case, there will be an interest charge for the use of the required funds. If internal funds are available, a decision must be made between the use of the funds for the proposed project or for some other, more profitable project. In the decision-making process, a careful analysis of the time value of money will help establish the worth of earnings and investments.

Consideration must also be given to the effect of taxes on the net profit of the proposed project. For example, income taxes presently range from 15 to 39 percent of the taxable income, and such a tax may have a major impact on the net, after-taxes, earnings of a project. In addition to income taxes, there are other fixed charges such as property taxes, depreciation, and insurance that, once the proposed assets have been acquired, continue no matter what level of business is maintained.

This chapter examines the various forms of interest available to the borrower or the lender. It identifies various means of assessing the worth of earnings and investments that can provide meaningful economic comparisons between various investment opportunities. Finally, the chapter provides a summary of the taxes and fixed charges that need to be considered in the preparation of the economic assessment of the proposed project.

INTEREST

Interest is the cost of borrowed money, or the earnings on money loaned. *Principal* refers to the original amount or the remaining unpaid amount of a loan. *Interest rate* is defined as the amount of money earned by, or paid on, a unit of principal in a unit of time, expressed as a fraction or percentage per year. Interest paid is an expense of business operation that must be included in the analysis of business profitability.

Simple Interest

The simplest form of interest requires compensation payment at a constant interest rate based only on the original principal. Thus, if $1000 were loaned for a total time for 4 years at a constant interest rate of 10 percent per year, the simple interest earned would be

$$(\$1000)(0.1)(4) = \$400$$

If P represents the principal, N the number of time units or interest periods, and i the interest rate based on the length of one interest period, the amount of simple interest I accumulated during N interest periods is

$$I = PiN \tag{7-1}$$

The principal must be repaid eventually; therefore, the entire amount of principal plus simple interest due after N interest periods is

$$F = P + I = P(1 + iN) \tag{7-2}$$

where F is the total amount of principal and accumulated interest at time N.

Compound Interest

Compound interest is almost universally used in business transactions. *Compound interest* is the interest earned on accumulated, reinvested interest as well as the principal amount. This applies to payments on loans or interest on investments. Thus, an investment of $1000 at an interest rate of 10 percent per year payable annually would earn $100 in the first year. If no principal were removed and the interest were left in the investment to earn at the same rate, then at the end of the second year $(\$1000 + \$100)(0.10) = \$110$ in interest would be earned and the total *compound amount* would be

$$\$1000 + \$100 + \$110 = \$1210$$

The compound amount earned after any discrete number of interest periods can be determined as follows:

Period	Principal at start of period	Interest earned during period (i = interest rate based on length of one period)	Compound amount F at end of period
1	P	Pi	$P + Pi = P(1+i)$
2	$P(1+i)$	$P(1+i)(i)$	$P(1+i) + P(1+i)(i) = P(1+i)^2$
3	$P(1+i)^2$	$P(1+i)^2(i)$	$P(1+i)^2 + P(1+i)^2(i) = P(1+i)^3$
N	$P(1+i)^{N-1}$	$P(1+i)^{N-1}(i)$	$P(1+i)^N$

Therefore, the total amount of principal plus interest earned after N interest periods is

$$F = P(1+i)^N \tag{7-3}$$

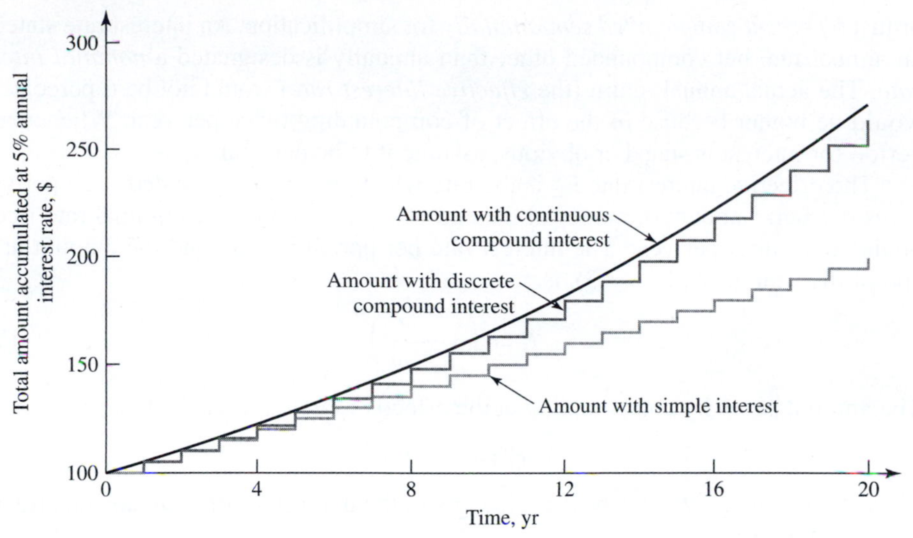

Figure 7-1
Comparison among total amounts accumulated with simple annual interest, discrete annually compounded interest, and continuously compounded interest

The term $(1 + i)^N$ is commonly referred to as the *discrete single-payment compound amount* factor, or the *discrete single-payment future-worth* factor. It is represented by the functional form $(F/P, i, N)$,† and

$$(F/P, i, N) = (1 + i)^N \tag{7-4}$$

Equation (7-3) can also be written as

$$F = P(F/P, i, N) \tag{7-5}$$

The factor value is easily calculated from Eq. (7-4).

Figure 7-1 shows a comparison among the total amounts earned at different times for cases using simple interest, discrete annually compounded interest, and continuously compounded interest, as described later in this chapter.

Nominal and Effective Interest Rates

In common industrial practice, the length of the discrete interest period is taken to be 1 year, and the fixed interest rate i is based on 1 year. However, there are cases where other time periods are employed. Even though the actual interest period is not 1 year, the interest rate is often expressed on an annual basis. Consider an example in which the interest rate is 3 percent per period and the interest is compounded at half-year periods. A rate of this type is referred to as *6 percent per year compounded semiannually,*

†This notation is from the American National Standards Institute publication ANSI Z94.7-2000. It can be read as: multiply P by the factor value to obtain F, for interest rate i and time period N. It is, as the symbol implies, the ratio of F to P for a given i and N.

or just *6 percent compounded semiannually* for simplification. An interest rate stated as an annual rate but compounded other than annually is designated a *nominal interest rate*. The actual annual return (the *effective interest rate*) would not be 6 percent, but would be higher because of the effect of compounding twice per year. Whenever no period for interest is stated or obvious, assume it to be per year.

The effective interest rate i_{eff} is the rate which, when compounded once per year, gives the same amount of money at the end of 1 year, as does the nominal rate r compounded m times per year. The interest rate per period is r/m, and the amount at the end of the year, using Eq. (7-3), is

$$F = \left(1 + \frac{r}{m}\right)^m \tag{7-6}$$

The amount given by compounding at the effective rate i_{eff} at the end of 1 year is

$$F = (1 + i_{\text{eff}})^1 \tag{7-7}$$

These two values of F must be the same, given the definitions of nominal and effective interest; therefore,

$$1 + i_{\text{eff}} = \left(1 + \frac{r}{m}\right)^m \tag{7-8}$$

The effective annual interest rate can be determined from this relation as

$$i_{\text{eff}} = \left(1 + \frac{r}{m}\right)^m - 1 \tag{7-9}$$

Thus, if $r = 0.06$ and $m = 2$ (twice on an annual basis), then

$$i_{\text{eff}} = \left(1 + \frac{0.06}{2}\right)^2 - 1 = 1.0609 - 1 = 0.0609$$

Nominal interest rates should always include a qualifying statement indicating the compounding period. For example, using the common annual basis, \$100 invested at a nominal interest rate of 20 percent compounded annually would amount to \$120.00 after 1 year; if compounded semiannually, the amount would be \$121.00; and, if compounded continuously, the amount would be \$122.14. The corresponding effective interest rates are 20.00, 21.00, and 22.14 percent, respectively.

EXAMPLE 7-1 Applications of Different Types of Interest

It is desired to borrow \$1000 to meet a financial obligation. This money can be borrowed from a loan agency at a monthly interest rate of 2 percent. Determine the following:

a. The total amount of principal plus simple interest due after 2 years if no intermediate payments are made.

b. The total amount of principal plus compounded interest due after 2 years if no intermediate payments are made.

c. The nominal interest rate when the interest is compounded monthly.

d. The effective interest rate when the interest is compounded monthly.

■ **Solution**

a. The length of one interest period is 1 month, and the number of interest periods in 2 years is 24. For simple interest, the total amount due after n periods at a periodic (in this case monthly) interest rate of i is given by Eq. (7-2):

$$F = P(1 + iN)$$
$$= 1000[1 + (0.02)(24)] = \$1480$$

b. For compound interest, the total amount due after N periods at a periodic interest rate i is obtained from Eq. (7-3):

$$F = P(1 + i)^N$$
$$= \$1000(1 + 0.02)^{24} = \$1608$$

c. Nominal interest rate is (2)(12), or 24 percent, per year compounded monthly.

d. Use Eq. (7-9) where $m = 12$ and $r = 0.24$.

$$i_{\text{eff}} = \left(1 + \frac{r}{m}\right)^m - 1$$

$$= \left(1 + \frac{0.24}{12}\right)^{12} - 1 = 0.268 = 26.8\% \text{ per year}$$

■

Continuous Interest

The preceding discussion of simple and compound interest considered only the form of interest in which the payments are charged at periodic and discrete intervals, where the intervals represent a finite length of time with interest accumulating in a discrete amount at the end of each interest period. Although in practice the basic time interval for interest accumulation is usually taken as 1 year, shorter periods can be used, such as 1 month, 1 day, 1 h, or 1 s. The extreme case, of course, occurs when the time interval becomes infinitesimally small so that the *interest is compounded continuously*. The concept of continuous interest is that the cost or income due to interest flows regularly, and this is just as reasonable an assumption for most cases as the concept of interest accumulating only at discrete intervals.

Equations (7-6) and (7-9) form the basis for developing continuous interest relationships. The symbol r represents the nominal interest rate with m interest periods per year. If the interest is compounded continuously, m approaches infinity. For N years Eq. (7-6) can be written as

$$F = P \lim_{m\to\infty}\left(1 + \frac{r}{m}\right)^{mN} = P \lim_{m\to\infty}\left(1 + \frac{r}{m}\right)^{(m/r)(rN)} \tag{7-10}$$

The fundamental definition for the base of the natural system of logarithms ($e = 2.71828$) is[†]

$$\lim_{m\to\infty}\left(1 + \frac{r}{m}\right)^{m/r} = e = 2.71828\cdots \tag{7-11}$$

[†]See, for example, W. Fulks, *Advanced Calculus,* 3d ed., J. Wiley, New York, 1978, pp. 55–56.

Thus†

$$F = Pe^{rN} \tag{7-12}$$

The term e^{rN} is known as the *continuous single-payment compound amount* factor, or the *continuous single-payment future-worth* factor. In functional form‡

$$e^{rN} = (F/P, r, N) \tag{7-13}$$

thus,

$$F = P(F/P, r, N) \tag{7-14}$$

and the factor $(F/P, r, N)$ is the ratio of future worth F to present worth P, when compounding is continuous at a nominal rate of r per year for N years.

Based on Eq. (7-9), the effective interest rate can be expressed in terms of the continuous nominal interest rate as

$$i_{\text{eff}} = e^{r} - 1 \tag{7-15}$$

from which

$$r = \ln(1 + i_{\text{eff}}) \tag{7-16}$$

and therefore,

$$e^{rN} = (1 + i_{\text{eff}})^{N} \tag{7-17}$$

and

$$F = Pe^{rN} = P(1 + i_{\text{eff}})^{N} \tag{7-18}$$

EXAMPLE 7-2 Calculations with Continuous Interest Compounding

For the case of a nominal annual interest rate of 20 percent per year, determine

a. The total amount to which \$1 of initial principal would accumulate after 1 year with annual compounding and the effective annual interest rate.

b. The total amount to which \$1 of initial principal would accumulate after 1 year with monthly compounding and the effective annual interest rate.

c. The total amount to which \$1 of initial principal would accumulate after 1 year with daily compounding and the effective annual interest rate.

---

†The same result can be obtained by noting that, for the case of continuous compounding, the differential change of F with time equals the nominal, fractional continuous interest rate r times F, or $dF/dN = rF$. This expression may be separated and integrated as follows to give Eq. (7-12):

$$\int_{P}^{F} (dF/F) = r \int_{0}^{N} dN$$

$$\ln\left(\frac{F}{P}\right) = rN \qquad \text{or} \qquad F = Pe^{rN}$$

‡ANSI Z94.7-2000.

d. The total amount to which \$1 of initial principal would accumulate after 1 year with continuous compounding and the effective annual interest rate.

■ **Solution**

a. Using Eq. (7-3), $P = \$1.0$, $r = 0.20$, $N = 1$, and

$$r = i = i_{\text{eff}} = 0.20 \text{ or } 20\% \text{ because compounding is annual}$$
$$F = P(1 + i_{\text{eff}})^N = (1 + 0.20)^1 = \$1.20$$

b. With Eq. (7-6), $P = \$1.0$, $r = 0.20$, $m = 12$, and

$$F = P\left(1 + \frac{r}{m}\right)^m = (1.0)\left(1 + \frac{0.20}{12}\right)^{12} = \$1.2194$$

Using Eq. (7-9),

$$i_{\text{eff}} = \left(1 + \frac{r}{m}\right)^m - 1 = \left(1 + \frac{0.20}{12}\right)^{12} - 1 = 0.2194, \text{ or } 21.94\%$$

c. With Eq. (7-6), $P = \$1.0$, $r = 0.20$, $m = 365$, and

$$F = P\left(1 + \frac{r}{m}\right)^m = (1.0)\left(1 + \frac{0.20}{365}\right)^{365} = \$1.2213$$

Using Eq. (7-9),

$$i_{\text{eff}} = \left(1 + \frac{r}{m}\right)^m - 1 = \left(1 + \frac{0.20}{365}\right)^{365} - 1 = 0.2213, \text{ or } 22.13\%$$

d. With Eq. (7-12), $P = (\$1.0)$, $r = 0.20$, $n = 1$, and

$$F = Pe^{rN} = (1.0)e^{(0.20)(1)} = \$1.2214$$

With Eq. (7-15),

$$i_{\text{eff}} = e^r - 1 = e^{0.20} - 1 = 0.2214, \text{ or } 22.14\%$$

■

Example 7-2 illustrates that continuously compounded interest gives a result which agrees much better with the results from monthly and daily compounding than does the result given by annual compounding when the same interest rate is used for each case. On the other hand, if the effective and nominal rates agree according to Eq. (7-9) or (7-15), then each case will give the same annual result. These results emphasize the importance of determining the appropriate interest rate to be used in economic calculations.

COST OF CAPITAL

There are several possible sources of capital for business ventures, including loans, bonds, stocks, and corporate funds. Corporate funds, primarily from undistributed profits and depreciation accumulations, are usually a major source of capital for established businesses. Borrowed funds are often used to supply all or part of corporate investments. The interest paid on the portion of an investment that comes from loans is one of the costs of making a product. Interest paid on bonds is also a cost of doing business. The question sometimes arises as to whether interest on investor-owned funds can be charged as a cost of doing business. The answer, based on court decisions and income tax regulations, is definitely no. Therefore, interest on corporate funds and earnings paid on preferred or common stocks are not a manufacturing cost, because they are a return to owners of equity in the corporation. Furthermore, the borrowed principal, which is first a gain, but must be repaid, is not taxable as a gain; nor is repaying it deductible as a business expense.

In the preliminary design of a project, unless more specific information is available, one of the following two methods is usually employed to account for interest costs:

1. No interest costs are included. This assumes that all the necessary capital comes from internal corporate funds, and comparisons to alternative investments must be on the same basis.
2. Interest is charged on the total capital investment, or a predetermined fraction thereof, at a set interest rate, usually equivalent to rates charged for bank loans.

As the design proceeds to the final stages, the actual sources of new capital should be considered in detail, and more appropriate interest costs can then be used.

Because of the different methods used for treating interest costs, a definite statement should be made concerning the particular method employed in any given economic analysis. Interest costs become especially important when comparisons are made among alternative investments. These comparisons as well as the overall cost picture are simplified if the role of interest in the analysis is clearly defined.

Income Tax Effects

The effect of income taxes on the cost of capital is very important. In determining income taxes, interest on loans and bonds can be considered as a cost, while the return of both preferred and common stock cannot be included as a cost. If the incremental annual income tax rate is 35 percent, every \$1 spent for interest on loans or bonds has a true cost after taxes of only \$0.65. Thus, after income taxes are taken into consideration, a bond issued at an annual interest rate of 5 percent actually has an interest rate of only $(5)(65/100) = 3.25$ percent. On the other hand, dividends on preferred stock must be paid from net profits after taxes. If a preferred stock has an annual dividend rate of 8 percent, the equivalent rate before income taxes is $(8)(100/65) = 12.3$ percent, and after income taxes it is $(12.3)(65/100) = 8$ percent.

A comparison of representative interest and dividend rates for different types of externally financed capital is presented in Table 7-1.

The choice of capital sources to be used to fund a project is usually made at the highest corporate level based upon corporate circumstances and policies. It would be typical for a new business to rely primarily on external financing, while a successful, established business would use a higher proportion of internal funds.

Table 7-1 Representative costs for externally financed capital†

Source of capital	Indicated interest or dividend rate, %/yr	Interest or dividend rate before taxes, %/yr	Interest or dividend rate after taxes, %
Bonds	5	5	3.25
Bank or other loans	8	8	5.2
Preferred stock	8	12.3	8
Common stock	n/a	13.8	9

†Income tax rate of 35 percent of taxable income.

Loan Payments

There are many types of loan repayment terms. A very common type, used for nearly all home mortgages and many business loans, calls for constant periodic payments for a fixed period. Each payment covers the current interest due and repays some of the remaining principal balance. The total payment is constant, but the principal balance decreases, so that the interest portion of each payment is smaller than the previous one and the principal portion of each payment is larger than the previous one. Because this loan type is so common, the equations used to calculate the payment will be developed and illustrated.

The constant amount that must be paid each period so that the required interest is paid and the original amount borrowed is returned over the total life of the loan is given by

$$L = I_j + p_j \tag{7-19}$$

where L is the constant payment each period, I_j the jth period interest payment, and p_j the jth period principal payment. The index j begins at 1, because payment is made at the end of the period. [The symbol i in the equations in this section represents the effective annual interest rate for annual payments or the nominal rate per period (r/m) for other periods.] The interest payment is

$$I_j = iP_{j-1} \tag{7-20}$$

where i is the interest rate and P_{j-1} the principal balance after payment $j-1$. The remaining principal balance after $j-1$ periods is

$$P_{j-1} = P_0 - \sum_{m=1}^{j-1} p_m \tag{7-21}$$

where p_m is the mth principal payment and P_0 the initial amount of the loan. From Eqs. (7-19), (7-20), and (7-21),

$$p_j = L - I_j = L - i\left(P_0 - \sum_{m=1}^{j-1} p_m\right) \tag{7-22}$$

From Eq. (7-22),

$$p_1 = L - iP_0 \tag{7-23}$$

similarly,

$$p_2 = L - i(P_0 - p_1) = L - i[P_0 - (L - iP_0)] = L(1+i) - iP_0(1+i) \tag{7-24}$$

and

$$\begin{aligned} p_3 &= L - i(P_0 - p_i - p_2) = L - i\{P_0 - (L - iP_0) - [L(1+i) - iP_0(1+i)]\} \\ &= L[(1+i) + i(1+i)] - iP_0[1 + i + i(1+i)] \end{aligned} \tag{7-25}$$

from which

$$p_3 = L(1+i)^2 - iP_0(1+i)^2 \tag{7-26}$$

In general,

$$p_j = L(1+i)^{j-1} - iP_0(1+i)^{j-1} \tag{7-27}$$

The sum of all the principal payments must equal the original loan principal (i.e., pay off the original loan). Accordingly,

$$\begin{aligned} \sum_{j=1}^{N} p_j = P_0 &= \sum_{1}^{N} L(1+i)^{j-1} - \sum_{1}^{N} iP_0(1+i)^{j-1} \\ &= L\sum_{1}^{N}(1+i)^{j-1} - iP_0\sum_{1}^{N}(1+i)^{j-1} \end{aligned} \tag{7-28}$$

Solving for L gives

$$L = \frac{P_0\left[1 + i\sum_{1}^{N}(1+i)^{j-1}\right]}{\sum_{1}^{N}(1+i)^{j-1}} \tag{7-29}$$

After first calculating the summation term in Eq. (7-29), we can use this equation to calculate the constant payment amount for any loan amount, interest rate, and total payment period. Equations (7-20) and (7-22) can then be used, in that order, to calculate the interest and principal portions of each payment. Equation (7-21) can be used to calculate the remaining principal after each payment. These calculations are repeated for each payment until the end of the period. Example 7-3 illustrates these calculations.

EXAMPLE 7-3 Calculation of Loan Payments with Annually Compounded Interest

Present a spreadsheet for the following calculations: A loan of $100,000 at a nominal interest rate of 10 percent per year is made for a repayment period of 10 years. Determine the constant payment per period, the interest and principal paid each period, and the remaining unpaid principal at the end of each period by using constant end-of-month payments (assume 12 equal-length months per year).[†] Repeat for annual, end-of-the-year payments.

■ Solution

In both cases the procedure is as follows:

Calculate the constant payment L from Eq. (7-29), after first calculating the summation of the $(1+i)^{j-1}$ term. Then calculate the interest payment for the first period, using Eq. (7-20). Determine the principal payment for the first period by subtracting the interest payment from the total constant payment, using Eq. (7-22). Establish the remaining unpaid balance by subtracting the first principal payment from the original principal amount, Eq. (7-21). Repeat this procedure period by period until the full term of the loan is reached. The results are provided in the following tables.

Check: If the calculations have been done correctly, the final remaining principal will equal zero.

Monthly loan payments

P_0 = \$100,000.00; *nominal interest* $r = 0.1$ *fraction/yr; payments/yr* = 12, *for 10 yr*

Month j	**$(1+0.1/12)^{j-1}$ Column sum = 204.84498**	**Constant payment L, \$/month, Eq. (7-29)**	**Interest payment, \$/month, Eq. (7-20)**	**Principal payment, \$/month, Eq. (7-22)**	**Remaining principal, \$, Eq. (7-21)**
0					100,000.00
1	1.0000000	1,321.51	833.33	488.17	99,511.83
2	1.0083333	1,321.51	829.27	492.24	99,019.58
3	1.0167361	1,321.51	825.16	496.34	98,523.24
4	1.0252089	1,321.51	821.03	500.48	98,022.76
5	1.0337523	1,321.51	816.86	504.65	97,518.11
6	1.0423669	1,321.51	812.65	508.86	97,009.25
7	1.0510533	1,321.51	808.41	513.10	96,496.15
8	1.0598121	1,321.51	804.13	517.37	95,978.78
9	1.0686439	1,321.51	799.82	521.68	95,457.10
10	1.0775492	1,321.51	795.48	526.03	94,931.07
11	1.0865288	1,321.51	791.09	530.42	94,400.65
12	1.0955832	1,321.51	786.67	534.84	93,865.82
From here, most results are deleted; only those for every 12th month are shown.					
24	1.2103051	1,321.51	730.67	590.84	87,089.30
36	1.3370398	1,321.51	668.80	652.71	79,603.20
48	1.4770454	1,321.51	600.45	721.06	71,333.20
60	1.6317113	1,321.51	524.95	796.56	62,197.23
72	1.8025728	1,321.51	441.54	879.97	52,104.60
84	1.9913258	1,321.51	349.39	972.11	40,955.15
96	2.1998436	1,321.51	247.60	1,073.91	28,638.19
108	2.4301960	1,321.51	135.15	1,186.36	15,031.50
120	2.6846692	1,321.51	10.92	1,310.59	0.00
		Total payments \$158,580.88	Total interest paid \$58,580.88	Total principal paid \$100,000.00	

[†]*Note:* At a 50 percent per year interest compounded monthly, the error caused by this assumption is less than 8×10^{-4} percent in 1 year and 8×10^{-3} percent in 10 years. The error increases with increasing interest rate and length of period, but it takes extraordinarily large values of interest or time to make this error significant.

The only changes from the monthly payment are that i_{eff} is 0.10 and N is 10.

P_0 = \$100,000.00; *interest* i = 0.1 *fraction/yr; payments/yr* = 1, *for 10 yr*

Year j	$(1 + 0.1)^{j-1}$ Column sum = 15.937425	Constant payment, \$/yr, Eq. (7-29)	Interest payment, \$/yr, Eq. (7-20)	Principal payment, \$/yr, Eq. (7-22)	Remaining principal, \$, Eq. (7-21)
0					100,000.00
1	1.000000	16,274.54	10,000.00	6,274.54	93,725.46
2	1.100000	16,274.54	9,372.55	6,901.99	86,823.47
3	1.210000	16,274.54	8,682.35	7,592.19	79,231.27
4	1.331000	16,274.54	7,923.13	8,351.41	70,879.86
5	1.464100	16,274.54	7,087.99	9,186.55	61,693.31
6	1.610510	16,274.54	6,169.33	10,105.21	51,588.10
7	1.771561	16,274.54	5,158.81	11,115.73	40,472.37
8	1.948717	16,274.54	4,047.24	12,227.30	28,245.07
9	2.143589	16,274.54	2,824.51	13,450.03	14,795.04
10	2.357948	16,274.54	1,479.50	14,795.04	0.00
		Total payments \$162,745.39	Total interest paid \$62,745.39	Total principal paid \$100,000.00	

In this example, 7.1 percent less total interest is required by paying monthly compared to annually. In general, the more frequently loan payments are made, the less total interest is paid. ■

TIME VALUE OF MONEY

Money can be used to earn money by investment, for example, in savings accounts, bonds, stocks, or projects. Therefore, an initial amount of money that is invested increases in value with time. This effect is known as the *time value of money*. By virtue of this earning capacity, an amount of money available now is worth, or equivalent to, a greater amount in the future. For example, \$1000 invested at 10 percent per year compounded annually is worth \$1100 one year later and \$2594 ten years later. The value at a future time is the *future worth* (also known as the *future value*) of the money. This earning power of money can be included in the analysis of business profitability. Methods for calculating the worth of money at different times are similar to those for calculating interest.

An amount of money available at some time in the future is worth, or equivalent to, a smaller amount at present. For example, \$1100 available at the end of 1 year after being compounded at 10 percent over that 1-year period is actually only worth \$1000 at present. The \$1000 is the *present worth* (also known as *present value*) of the future \$1100. This comparison again exemplifies the concept of the time value of money. These amounts—\$1000 now, \$1100 one year from now, and \$2594 ten years from now—are all said to be *equivalent* at a 10 percent rate of interest, and only at that rate, compounded annually. Amounts of money are equivalent if they are equal when calculated at the same time with a specified interest rate.

The time value of money often is a very important consideration when investments are compared that require or generate different amounts of funds at different times. In fact, the timing of expenses and income may significantly impact the present

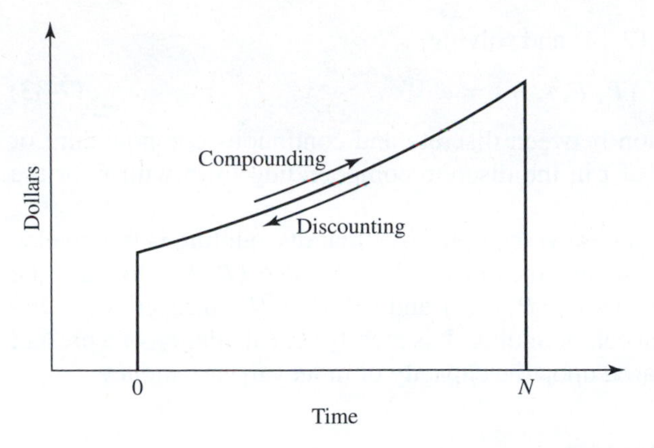

Figure 7-2
Schematic of compounding and discounting

worth of such funds. Put another way, the most appropriate way to make economic comparisons is to make all cash flows equivalent. Thus, it is necessary to be able to determine the value of investments at any selected time.

The time value of money is related solely to the capacity of money to earn money. It has nothing to do with inflation. Inflation, or deflation, refers to the change in the prices of goods and services, not to the amount of money available. For example, if the rate of inflation is 4 percent per year and if the cost of an item is \$100 now, it will cost \$104 one year from now—regardless of the time value of money.[†]

The calculation of the future worth of a present amount of money, known as compounding, was illustrated earlier in this chapter. The inverse of compounding that is, obtaining the present worth of a future amount, is designated as *discounting*. The expression for discrete compounding of a single present amount to obtain its future worth was presented in Eq. (7-3) or (7-5). The equation for discrete discounting of a single future amount to obtain its present worth is obtained by rearranging Eq. (7-3) to give

$$P = F(1 + i)^{-N} \tag{7-30}$$

This can also be expressed in terms of the ANSI functional form of the discount factor as

$$P = F(P/F, i, N) \tag{7-30a}$$

hence the *discrete single-payment discount* factor or the *discrete single-payment present-worth* factor is

$$(P/F, i, N) = (1 + i)^{-N} \tag{7-31}$$

The operations of compounding and discounting are inverses of each other. The present-worth factor of Eq. (7-31) is the inverse of the future-worth factor of Eq. (7-4). Figure 7-2 illustrates that compounding and discounting are inverse operations.

The future worth of a present amount of money, with continuous compounding, was given by Eq. (7-12) or (7-14). The equation for discounting future values to the present, obtained by rearranging Eq. (7-12), is therefore

$$P = Fe^{-rN} \tag{7-32}$$

[†]The effect of inflation on economic analyses is discussed in Chap. 8.

or, by equating Eqs. (7-12) and (7-14) and solving,

$$(P/F, r, N) = e^{-rN} \tag{7-33}$$

In the ANSI form, the distinction between discrete and continuous compounding or discounting is the replacement of i in the discrete compounding form with r for the continuous compounding.

Comparison of the equations involved emphasizes that discounting is the inverse of compounding and that the discount factors $(P/F, i, N)$ and $(P/F, r, N)$ are the inverse of the compounding factors $(F/P, i, N)$ and $(F/P, r, N)$, respectively. Discounting is no more difficult than compounding. It is merely the calculation of a present amount from a future amount based upon the capacity of money to earn money.

CASH FLOW PATTERNS

Cash flow was introduced in Chap. 6 and is the amount of funds that enter the corporate treasury as a result of the activities of the project. The annual cash flow is equal to the net (after-tax) profit plus the allowed depreciation charges for the year. Since cash flows of a project occur over the lifetime of a project, it is necessary to convert them to equivalent values. This is done either by discounting future cash flows or by compounding earlier cash flows to a particular point in time. While it is essential that all cash flows be converted to the same time, the time selected is not critical, because different present-worth amounts at any one time will be in the same rank order and in the same ratio as at any other time. Often the selected time is not the present, but rather sometime in the future. It is common, while in the design phase of a project, to select the projected start-up time as the time at which all "present" values are calculated.

Discrete Cash Flows

Cash flow patterns are often best perceived graphically. Figure 7-3 shows equal discrete cash flows occurring once per month at the end of the month for a period of 1 year; each cash flow is represented by a bar.

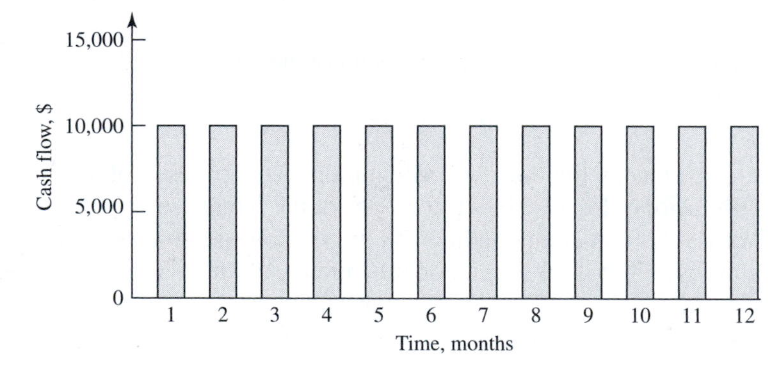

Figure 7-3
Constant, end-of-month cash flows for 1 yr

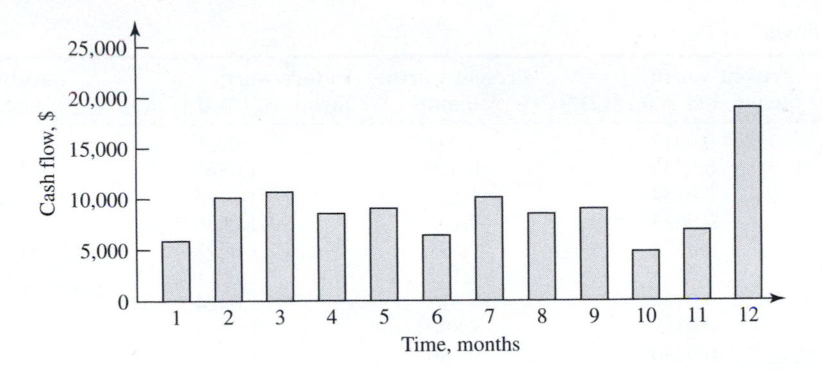

Figure 7-4
Series of unequal end-of-month cash flows

Figure 7-4 shows unequal cash flows occurring at the end of each month for 1 year. The cash flows shown in Figs. 7-3 and 7-4 are equivalent at a discount rate of 10 percent per year. That is, they have the same worth at a particular time when calculated at that discount rate, whether that worth is calculated at time zero, at 12 months, or at any other time. It follows that if two different cash flows are equivalent at one interest rate, they are not equivalent at any other interest rate.

Present Worth of Two Different Cash Flow Patterns — EXAMPLE 7-4

The cash flow patterns in Figs. 7-3 and 7-4 are shown numerically in the second column of the two spreadsheets shown below. With interest compounded monthly at a rate of 10 percent per year, calculate the total amount of the cash flow, the present worth at zero time, and the future worth at 12 months for both series of cash flows.

■ Solution

Equal monthly cash flows

End of month j	Cash flow, $	Present-worth factor $= (1 + 0.1/12)^{-j}$	Present worth, $, @ 0 mo	Future-worth factor $= (1 + 0.1/12)^{(12-j)}$	Future worth, $, @ 12 mo
1	10,000	0.9917	9,917	1.0956	10,956
2	10,000	0.9835	9,835	1.0865	10,865
3	10,000	0.9754	9,754	1.0775	10,775
4	10,000	0.9673	9,673	1.0686	10,686
5	10,000	0.9594	9,594	1.0598	10,598
6	10,000	0.9514	9,514	1.0511	10,511
7	10,000	0.9436	9,436	1.0424	10,424
8	10,000	0.9358	9,358	1.0338	10,338
9	10,000	0.9280	9,280	1.0252	10,252
10	10,000	0.9204	9,204	1.0167	10,167
11	10,000	0.9128	9,128	1.0083	10,083
12	10,000	0.9052	9,052	1.0000	10,000
Total	$120,000	Total	$113,745	Total	$125,656

Unequal monthly cash flows

End of month j	Cash flow, \$	Present-worth factor = $(1+0.1/12)^{-j}$	Present worth, \$, @ 0 mo	Future-worth factor = $(1+0.1/12)^{(12-j)}$	Future worth, \$, @ 12 mo
1	7,000	0.9917	6,942	1.0956	7,669
2	11,000	0.9835	10,819	1.0865	11,952
3	11,500	0.9754	11,217	1.0775	12,392
4	9,500	0.9673	9,190	1.0686	10,152
5	10,000	0.9594	9,594	1.0598	10,598
6	7,500	0.9514	7,136	1.0511	7,883
7	11,000	0.9436	10,379	1.0424	11,466
8	9,500	0.9358	8,890	1.0338	9,821
9	10,000	0.9280	9,280	1.0252	10,252
10	6,000	0.9204	5,522	1.0167	6,100
11	8,000	0.9128	7,302	1.0083	8,067
12	19,304	0.9052	17,474	1.0000	19,304
Total	\$120,304	Total	\$113,745	Total	\$125,656

The results show that the present and future worths are the same for both cash flows; therefore, the two cash flows are equivalent at the specified interest rate. This is true even though the patterns are very different and the actual totals of the two cash flows are somewhat different.

Annual compounding and discounting are often used in project evaluation. It is also common to represent total annual cash flows as a single (discrete) amount received at the end of the year (year-end convention). For periods other than 1 year, 1 month for example, the end of the period is usually used as the time for the discrete cash flow. In present-worth calculations, these are approximations for cash flows that occur frequently over the year.

Continuous Cash Flows

A continuous cash flow is one in which receipts and expenditures occur continuously over time. This is an idealized approximation of cash flows that occur frequently compared to once per year. Good business practice dictates investments of available funds as soon as they are received. Therefore, in the case of a continuous cash flow, the cash flow is invested continuously as it is received. If interest is compounded continuously, the rate of earning at any instant consists of two terms: (1) a continuous, constant rate of cash flow $\bar{P}$ in dollars per period (usually 1 year) and (2) the compound rate of earning on the accumulated amount M that has been invested at the rate r. This can be expressed as

$$\frac{dM}{d\theta} = \bar{P} + rM \tag{7-34}$$

Rearranging and forming the integrals give

$$\int_0^F \frac{dM}{\bar{P} + rM} = \int_{j-1}^{j} d\theta \tag{7-35}$$

where F is the future worth of the cash flow plus the earnings at the end of a 1-year period. Performing the integration gives

$$\frac{1}{r} \ln \frac{\bar{P} + rF}{\bar{P}} = j - (j - 1) = 1 \tag{7-36}$$

In exponential form this can be simplified to

$$\frac{\bar{P} + rF}{\bar{P}} = e^r \tag{7-37}$$

and solving for F results in

$$F = \bar{P}\left(\frac{e^r - 1}{r}\right) \tag{7-38}$$

The term $(e^r - 1)/r$ multiplied by the cash flow rate $\bar{P}$ equals the amount of funds accumulated at the end of the year from a 1-year, continuous, constant cash flow invested at a rate r. It can be represented as a compounding factor in ANSI functional form as

$$\frac{e^r - 1}{r} = (F/\bar{P}, r, j) \tag{7-39a}$$

The end-of-year N future-worth factor for such a cash flow is

$$(F/\bar{P}, r, N) = \left(\frac{e^r - 1}{r}\right) e^{(N-j)} \tag{7-39b}$$

The present worth of a 1-year, continuous, constant cash flow starting at the end of year $j - 1$ and ending at the end of year j, with continuous discounting, is determined by multiplying Eq. (7-38) by the discount factor e^{-rj} of Eq. (7-32) to give

$$P = \bar{P}\left(\frac{e^r - 1}{r}\right) e^{-rj} \tag{7-40}$$

The present-worth factor for this case is

$$(P/\bar{P}, r, j) = \left(\frac{e^r - 1}{r}\right) e^{-rj} \tag{7-41}$$

The term $(e^r - 1)/r$ in Eq. (7-41) accounts for earnings from the cash flow that was invested as it was received over the year. It is this term which distinguishes the continuous interest, continuous cash flow discounting factor from the continuous interest, discrete, year-end cash flow factor given by Eq. (7-33). The term e^{-rj} in Eq. (7-41) discounts the worth of the accumulated funds at the end of the year to time zero, just as it does in Eq. (7-33). It is possible to use different interest rates for the r in the two components of the discount factor in Eq. (7-41). However, the same interest rate is generally used in both terms.

For a cash flow occurring over a period of N years, starting at time zero, the future-worth factor is

$$(F/\bar{A}, r, N) = \frac{e^{rN} - 1}{r} \tag{7-42}$$

where $\bar{A}$ is a continuous, equal cash flow occurring in each year (or period) over N years (or periods). The present-worth factor of such a cash flow is

$$(P/\bar{A}, r, N) = \left(\frac{e^{rN} - 1}{r}\right) e^{-rN} \tag{7-43}$$

EXAMPLE 7-5 Continuous Cash Flow Calculations

Find the future worth at the end of the cash flow period and the present worth at the beginning of the cash flow period of a constant, continuous cash flow of $450,000 per year for 6 years. Use a continuously compounded interest rate of 8 percent per year. Determine the future worth and present worth, using the simplest approach. Also tabulate the cash flow a year at a time, using Eqs. (7-39*b*) and (7-41).

■ Solution

Calculate the future worth at 6 years with Eq. (7-42):

$$F = \bar{A}(F/\bar{P}, r, N) = \bar{A}\left(\frac{e^{rN} - 1}{r}\right)$$

$$= \$450{,}000 \frac{e^{(0.08)(6)} - 1}{0.08} = \$3.465 \times 10^6$$

Notice that the total of the cash flow in 6 years of $450,000 annually is $2.7 million. The additional $0.765 million shown in the future-worth calculation is interest earned on the cash flow itself and by compounding the earned interest.

The present worth at time zero of the cash flow is given by Eq. (7-43):

$$P = \bar{A}\left(\frac{e^{rN} - 1}{r}\right) e^{-rN}$$

Notice that the same result can be obtained by multiplying the future worth obtained with Eq. (7-42) by the discount factor e^{-rN}; thus

$$P = Fe^{-rN}$$

$$= (\$3.465 \times 10^6)(e^{-(0.08)(6)}) = \$2.144 \times 10^6$$

Tabulation of the cash flow a year at a time with a continuous cash flow of $450,000 per year is presented here.

Year	$\bar{P}$, \$1000/yr	$F/\bar{P}$, yr, Eq. (7-39*b*)	Future worth F at 6 yr, \$1000	P/F, yr at 6 yr, Eq. (7-33)	Present worth P at 0 yr, \$1000
1	450	1.5531	699	0.6188	432
2	450	1.4337	645	0.6188	399
3	450	1.3235	596	0.6188	369
4	450	1.2217	550	0.6188	340
5	450	1.1278	508	0.6188	314
6	450	1.0411	468	0.6188	290
Totals	\$2700		\$3465		\$2144

Notice that these values agree with those obtained directly by using Eqs. (7-42) and (7-43). The use of these equations is much simpler for this example. However, the year-by-year method is useful for the more common situation, where the cash flow changes from year to year. The present worth also can be found directly one year at a time using Eq. (7-40).

COMPOUNDING AND DISCOUNTING FACTORS

Several compounding and discounting factors have already been introduced. Cash flows and interest compounding can be either discrete or continuous. This gives four possible combinations of cash flow and interest types, as shown in Table 7-2.

The period of discrete compounding can be any time interval, and a discrete cash flow can occur at any time throughout the period. The most common choices are a compounding period of 1 year and the year-end convention for cash flow. These two choices will be used for discrete times and cash flows throughout the rest of this chapter, unless otherwise indicated.

For simple cash flow patterns, such as a single amount as shown in Fig. 7-5 or equal flows for a period of years as presented in Fig. 7-6, it is possible to derive compounding and discount factors. Some of these were derived earlier in this chapter. The rest of these factors will be presented without derivation and are available from many

Table 7-2 Combinations of cash flow and interest types

	Interest compounding	
Cash flow pattern	**Annual**	**Continuous**
Annual, end of year	Case 1	Case 2
Continuous	Case 3	Case 4

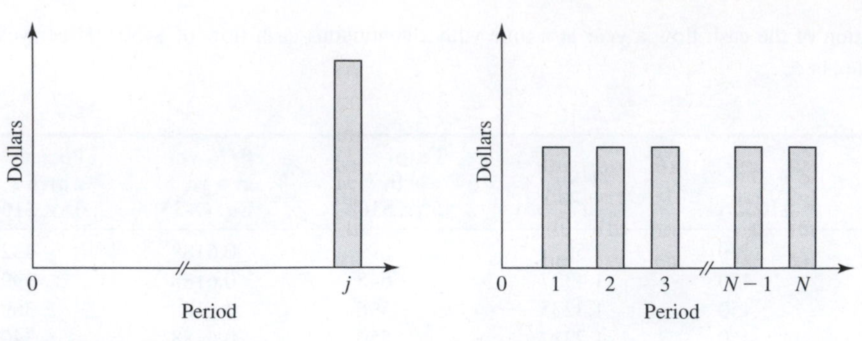

Figure 7-5
Single-amount cash flow

Figure 7-6
Uniform series cash flow

Table 7-3 Compounding and discounting factors for discrete interest compounding and discrete or continuous cash flows

	Compounding			Discounting		
Cash flow pattern	**Factor name; description; units**	**Symbol**	**Formula**	**Factor name; description; units**	**Symbol**	**Formula**
Single amount at end of year *j*. See Fig. 7-5.	Future worth; multiplies a single amount at time *j* to give a single amount at *N*; dimensionless	$(F/P, i, N-j)$	$(1+i)^{N-j}$	Present worth; multiplies a single amount at time *j* to give a single amount at time zero; dimensionless	$(P/F, i, j)$	$(1+i)^{-j}$
Series of equal amounts from time 1 to *N*; uniform series. See Fig. 7-6.	Future worth; multiplies the annual rate of a series of *N* equal amounts to give a single amount at *N*; years	$(F/A, i, N)$	$\frac{(1+i)^N - 1}{i}$	Sinking fund; multiplies a single amount at *N* to give the annual rate of a series of *N* equal amounts; $(\text{years})^{-1}$	$(A/F, i, N)$	$\frac{i}{(1+i)^N - 1}$
Series of equal amounts from time 1 to *N*; uniform series. See Fig. 7-6.	Capital recovery; multiplies a single amount at time zero to give the annual rate of a series of *N* equal amounts; $(\text{years})^{-1}$	$(A/P, i, N)$	$\frac{i(1+i)^N}{(1+i)^N - 1}$	Present worth; multiplies the annual rate of a series of *N* equal amounts to give a single amount at time zero; years	$(P/A, i, N)$	$\frac{(1+i)^N - 1}{i(1+i)^N}$

sources.[†] Table 7-3 shows the common compounding and discounting factors for case 1, discrete interest and discrete cash flow. These same factors are also applicable to case 3 since the final result is independent of the pattern of cash flow over the year when interest for a year is calculated based on the amount of principal at the beginning of the year.

Present Worth of Annual Cash Flows with Annual Compounding — EXAMPLE 7-6

A cash flow consisting of $10,000 per year is received in one discrete amount at the end of each year for 10 years. Interest will be at 10 percent per year compounded annually. Determine the present worth at time zero and the future worth at the end of 10 years of this cash flow.

■ Solution

Identify the cash flow pattern as a series of discrete, equal amounts received at the end of years 1 to N, with annual interest compounding. In Table 7-3, the factor to use to obtain the present worth of this series is $(P/A, i, N)$. Notice that the factor has units of years.

With $A = \$10{,}000/\text{yr}$, $i = 0.1$, and $N = 10$, the present worth is

$$P = A(P/A, i, N) = \frac{A[(1+i)^N - 1]}{i(1+i)^N} = \frac{10{,}000[(1.1)^{10} - 1]}{0.1(1.1)^{10}}$$

$$= 10{,}000(6.144567) = \$61{,}445.67, \text{ or } \$61{,}446$$

The total cash flow over the 10 years is 10($10,000), or $100,000. This present worth reflects the fact that $61,446 invested at time zero would grow to the same amount—the future worth calculated below—in 10 years, as does the actual cash flow.

From Table 7-3 the factor used to get the future worth from the annual amount is $(F/A, i, N)$. Alternatively, the future worth may be obtained by multiplying the present worth P by the factor $(F/P, i, N)$. The same result is obtained either way,

$$F = P(F/P, i, N) = P(1+i)^N = \$61{,}446(1.1)^{10} = \$159{,}375$$

Notice that the $100,000 received and invested over the 10 years is equivalent to $61,446 invested at time zero, and is also equivalent to $159,375 received 10 years later.

The results would be the same even if the cash flow were continuous over each year, provided that interest was compounded on the principal at the beginning of the year (case 3). ■

Compounding and discounting factors for case 2, continuous interest, and annual cash flows are given in Table 7-4.

[†]For example, see ANSI Z94.7-2000; J. A. White, M. H. Agee, and K. E. Case, *Principles of Engineering Economic Analysis,* 2d ed., J. Wiley, New York, 1977; and M. S. Peters and K. D. Timmerhaus, *Plant Design and Economics for Chemical Engineers,* 4th ed., McGraw-Hill, New York, 1991.

Table 7-4 Compounding and discounting factors for continuous interest compounding and discrete cash flows

	Compounding			Discounting		
Cash flow pattern	**Factor name; description; units**	**Symbol**	**Formula**	**Factor name; description; units**	**Symbol**	**Formula**
Single amount at end of year j. See Fig. 7-5.	Future worth; multiplies a single amount at j to give a single amount at N; dimensionless	$(F/P, r, N-j)$	$e^{r(N-j)}$	Present worth; multiplies a single amount at j to give a single amount at zero; dimensionless	$(P/F, r, j)$	e^{-rj}
Series of equal amounts from time 1 to N; uniform series. See Fig. 7-6.	Future worth, annuity; multiplies the annual rate of a series of N equal amounts to give a single amount at N; years	$(F/A, r, N)$	$\frac{e^{rN}-1}{e^r-1}$	Sinking fund; multiplies a single amount at N to give the annual rate of a series of equal amounts; $(\text{years})^{-1}$	$(A/F, r, j)$	$\frac{e^r-1}{e^{rN}-1}$
Series of equal amounts from time 1 to N; uniform series. See Fig. 7-6.	Capital recovery; multiplies a single amount at time zero to give the annual rate of a series of N equal amounts; $(\text{years})^{-1}$	$(A/P, r, N)$	$\frac{e^{rN}(e^r-1)}{e^{rN}-1}$	Present worth; multiplies the annual rate of a series of equal amounts to give a single amount at time zero; years	$(P/A, r, j)$	$\frac{e^{rN}-1}{e^{rN}(e^r-1)}$

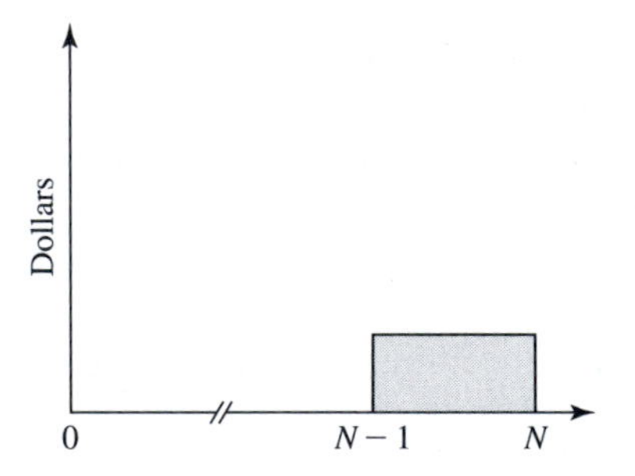

Figure 7-7
One-year continuous constant-rate cash flow

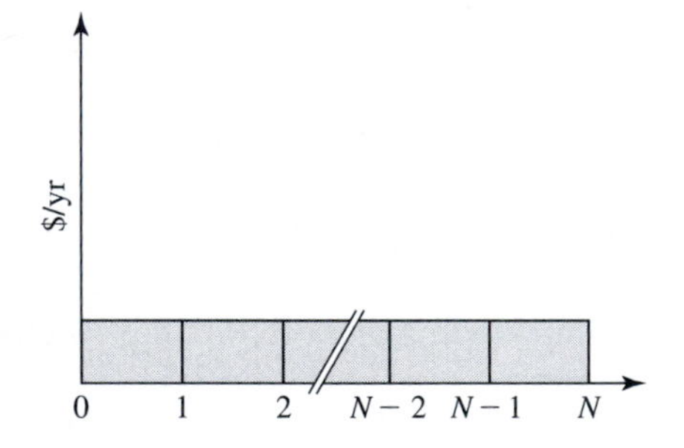

Figure 7-8
Series of continuous, constant-rate cash flows

Future and Present Worths of Discrete Annual Cash Flows Compounded Continuously — EXAMPLE 7-7

Repeat Example 7-6 but require that the compounding be performed continuously. Do the calculations two ways: first using the nominal interest rate $r = 0.1$ and then using $i_{\text{eff}} = 0.1$.

■ Solution

The cash flow pattern is the same as in Example 7-6, but the appropriate equations are from Table 7-4. The future worth is given by

$$F = \bar{A}\left(\frac{e^{rN} - 1}{e^{r} - 1}\right)$$

$$= \$10{,}000\left(\frac{e^{1.0} - 1}{e^{0.1} - 1}\right) = \$163{,}380$$

The present worth is found by using

$$P = \bar{A}\left(\frac{e^{rN} - 1}{e^{r} - 1}\right)e^{-rN}$$

$$= Fe^{-rN}$$

$$= \$163{,}380(e^{-1.0}) = \$60{,}104$$

Notice that these values are somewhat different from those calculated in Example 7-6. If $i_{\text{eff}} = 0.1$, then $r = \ln(1 + i_{\text{eff}})$

$$r = \ln(1 + 0.1) = 0.09531$$

which gives

$$F = \$10{,}000\left(\frac{e^{0.9531} - 1}{e^{0.09531} - 1}\right) = \$159{,}375$$

and

$$P = (\$159{,}375)(e^{-0.9531}) = \$61{,}446$$

Both of these values agree with those obtained in Example 7-6. ■

The compounding and discounting factor for case 4, continuous interest and continuous cash flows, are given in Table 7-5. Figure 7-7 illustrates a 1-year, continuous, constant rate cash flow and Fig. 7-8 shows a series of continuous, constant cash flow rates for N years.

Table 7-5 Compounding and discounting factors for continuous interest compounding and continuous cash flows

	Compounding			Discounting		
Cash flow pattern	Factor name; description; units	Symbol	Formula	Factor name; description; units	Symbol	Formula
Continuous, constant rate for 1 year only, ending at the end of year j. See Fig. 7-7.	Future worth; multiplies continuous annual rate to give a single amount at N; dimensionless	$(F/\bar{P}, r, N)$	$\frac{e^r - 1}{r} e^{r(N-j)}$	Present worth; multiplies continuous annual rate to give a single amount at time zero; dimensionless	$(P/\bar{P}, r, j)$	$\left(\frac{e^r - 1}{r}\right) e^{-rj}$
Series of continuous, constant annual amounts from time zero to N. See Fig. 7-8.	Future worth, multiplies the annual rate of a series of continuous, equal amounts to give a single amount at N; years	$(F/\bar{A}, r, N)$	$\frac{e^{rN} - 1}{r}$	Sinking fund; multiplies a single amount at N to give the annual rate of a series of continuous, equal amounts, $(\text{years})^{-1}$	$(\bar{A}/F, r, N)$	$\frac{r}{e^{rN} - 1}$
Series of continuous, constant annual amounts from time zero to N. See Fig. 7-8.	Capital recovery; multiplies a single amount at time zero to give the annual rate of a series of continuous, equal amounts; $(\text{years})^{-1}$	$(\bar{A}/P, r, N)$	$\frac{re^{rN}}{e^{rN} - 1}$	Present worth; multiplies the annual rate of a series of continuous, equal amounts to give a single amount at time zero; years	$(P/\bar{A}, r, N)$	$\frac{e^{rN} - 1}{r} e^{-rN}$

EXAMPLE 7-8 Worth of a Capital Investment Converted to a Continuous Constant Annual Series

Use a continuous nominal earning rate of 12 percent per year.

a. Find the future worth at start-up (time zero) of an investment made according to this schedule: \$3 million is invested continuously at a constant rate for 1 year, beginning 2 years before start-up; \$7 million is invested continuously at a constant rate for 1 year, beginning 1 year before start-up.

b. Convert the worth of the investment at start-up to a continuous, constant annual series for a 10-year period.

■ Solution

The investment and annual cash flow patterns can schematically be shown below in Fig. 7-9:

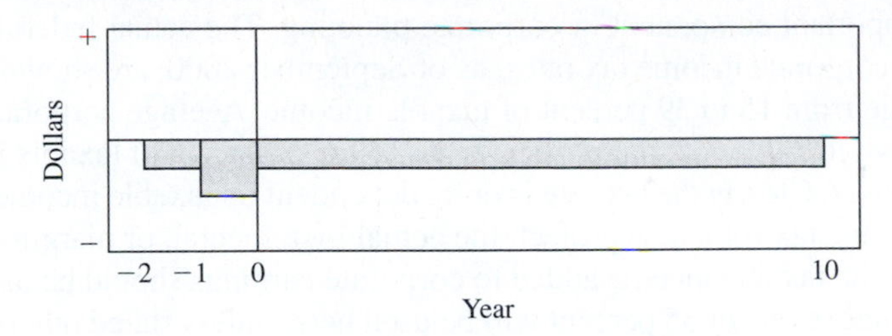

Figure 7-9

a. The two investments are converted to their worth at time zero by the future worth of a continuous cash flow factor from Table 7-5.

$$F = \bar{P}\left(\frac{e^r - 1}{r}\right)e^{r(N-j)}$$

So

$$F_1 = (3 \times 10^6)\left(\frac{e^{0.12} - 1}{0.12}\right)e^{0.12[0-(-2)]} = \$4.052 \times 10^6$$

and

$$F_2 = (7 \times 10^6)\left(\frac{e^{0.12} - 1}{0.12}\right)e^{0.12[0-(-1)]} = \$8.386 \times 10^6$$

The worth of the investment at start-up is the sum of these two values and is $\$12.44 \times 10^6$. Because the \$10 million invested could have been earning 12 percent per year, its worth at start-up is greater than the amount actually spent.

b. To convert the worth at zero time to an annual series, use the capital recovery factor from Table 7-5.

$$\bar{A} = P\left(\frac{re^{rN}}{e^{rN} - 1}\right)$$

$$\bar{A} = (\$12.44 \times 10^6)\left(\frac{0.12e^{(0.12)(10)}}{e^{(0.12)(10)} - 1}\right) = \$2.14 \times 10^6 \text{ per year}$$

Notice that \$21.4 million must be earned over 10 years to be equivalent to the original \$10 million invested (i.e., to be equal to the capital plus earnings at 12 percent per year). ■

INCOME TAXES

Corporate income taxes are levied by the U.S. government and by some of the states as well. The federal income taxes are by far the largest of these. State income taxes are often calculated by closely following federal policies. Therefore, federal corporate taxes are emphasized here. The federal government levies some other taxes as well.

Federal Income Taxes

Because the federal corporate income tax rate is as high as 39 percent of net profit, it is an extremely important component in corporate planning. The actual federal marginal, or incremental, corporate income tax rates, as of September 2000, are shown in Table 7-6, and they range from 15 to 39 percent of taxable income. Average corporate income tax rates are shown in Table 7-7; these range from 15 to 35 percent of taxable income as the income increases. Clearly the tax rate is quite dependent on taxable income. To estimate corporate income tax for a new project, the actual incremental, or marginal, tax rate associated with the taxable income added to corporate earnings should be used, if known. For convenience, a rate of 35 percent will be used here, unless stated otherwise.

Taxable Income

Income taxes are paid on a corporatewide basis. The taxable income of a corporation is the total gross profit. Gross profit, or gross earnings, equals total revenue minus

Table 7-6 U.S. corporate incremental income tax rates[†]

Taxable income		Incremental tax rate, %
Over	But not over	
$ 0	$ 50,000	15
50,000	75,000	25
75,000	100,000	34
100,000	335,000	39
335,000	10,000,000	34
10,000,000	15,000,000	35
15,000,000	18,333,333	38
18,333,333	—	35

[†]Source: © 2002 CCH Incorporated. All Rights Reserved. Reprinted with permission from *2000 U.S. Master Tax Guide.*

Table 7-7 U.S. corporate average income tax rates[†]

Taxable income		Average tax rate, % (increases are linear with income over each range)	
Over	But not over	From	To
$ 0	$ 50,000	15	15
50,000	75,000	15	18.333
75,000	100,000	18.333	22.25
100,000	335,000	22.25	34
335,000	10,000,000	34	34
10,000,000	15,000,000	34	34.333
15,000,000	18,333,333	34.333	35
18,333,333	—	35	35

[†]Source: © 2002 CCH Incorporated. All Rights Reserved. Reprinted with permission from *2000 U.S. Master Tax Guide.*

total product cost (TPC).[†] Total revenue is income from all sources, primarily product sales, but including sales of assets and supplies, royalties, and other revenues. Dividends and interest paid to the corporation and its shareholders are not considered as allowable costs of doing business for income tax purposes, nor are repayments of loan principal, and therefore cannot be subtracted from revenues in the calculation of gross profit.

In the assessment of the performance of a particular unit or process within a corporation, the revenues and costs associated with that process are determined and used in the evaluation. In addition, there are expenses incurred at levels above that of the individual operating units, at the plant, division, or corporate level. These include such items as safety, payroll, restaurant, recreation, control laboratories, waste disposal, administrative costs, donations, advertising, and research and development. Such costs are allocated to particular processes, usually as a percentage of the capital investment in each process. In evaluating a new investment, the income tax attributable to that investment is the gross income it is expected to generate, multiplied by the marginal income tax rate associated with the addition of that income to corporate income. It is common in evaluations to use a fixed income tax rate, such as the 35 percent noted above.

Corporate income tax payments are due in installments of 25 percent of the projected annual total in mid-April, June, September, and December. In cash flow calculations, they are usually treated in the same manner as other cash flows, that is, as occurring either once a year at year end or continuously.

Tax law is subject to legislative actions, Internal Revenue Service rulings, and court interpretations; so there are frequent changes. Methods for calculating and including taxes in economic evaluations are often established by corporate policy. The engineer charged with conducting evaluations should consult with company tax accounting and legal departments for current policy interpretations and for corporate procedures.[‡]

Capital Gains Tax

A *capital gains tax* is levied on profits made from the sale of capital assets, such as land, buildings, and equipment. The profit on the sale of land equals the selling price less the acquisition price, costs of selling, and costs of improvements. Land is not depreciable. For depreciable assets, such as buildings and equipment, profit is the selling price less the costs of selling, and less the cost of acquisition reduced by the amount of depreciation that has already been charged. Thus, a piece of equipment originally purchased for $80,000 which has already had $50,000 of depreciation charged as an expense, and is sold for $45,000 with $2000 in selling expense—advertising and removal from service—would show a capital gain of $45,000 − (80,000 − 50,000) − 2000 = $13,000.

[†]Total product cost is summarized in detail in Fig. 6-7 and Table 6-18.

[‡]Complete details are available in *Income Tax Regulations* and periodic *Income Tax Bulletins* issued by the U.S. Treasury, Superintendent of Documents, and Internal Revenue Service. Helpful information may also be found in annual private publications, such as *Prentice-Hall Federal Tax Advisor,* Prentice-Hall Information Services, Paramus, NJ, and *U.S. Master Tax Guide,* CCH, Inc., Chicago.

The capital gain on an item held for 1 year or more before sale is known as a long-term capital gain. The capital gain on items held for less than 1 year is designated as a short-term capital gain. Under current income tax laws, the tax rate on a long-term capital gain is 20 percent, while that on a short-term capital gain is equal to the incremental income tax rate. Normally, a 35 percent rate can be used for estimates.

Losses

The tax laws also make provisions for losses as well. Losses within a company may be used to offset gains within the company in the same year, thereby reducing the taxable income. If total corporate operations show a loss within a given year, that loss can be carried back for up to 3 years to offset past profits; or they may be carried forward for up to 5 years to offset future profits.

A particular project may be profitable overall, but show losses in one or more years. In such cases, the question arises as to whether the losses should be used to reduce overall corporate income taxes or whether these reductions should be used as a "negative income tax" in the evaluation of the process. Even though it may be realistic to do so, it is unwise to use this argument for justifying the economic viability of a project. Here, again, company policy determines the decision.

Other Federal Taxes

Employers pay a percentage of employees' wages as a contribution to Social Security and Medicare taxes. These are considered part of the total employee benefits package which includes medical insurance and retirement plans that are usually combined with wages to obtain labor costs. Such benefits depend upon individual corporate benefits plans, but a typical value is 40 percent of wages.

State Taxes

Some states in the United States levy a corporate income tax, while many do not. Moreover, the tax rate differs among states that do have an income tax. If state income tax is to be included in an economic analysis, the state in which the project is to take place and the corporate taxation policies of that state must be determined and factored into the analysis. Some analysts choose to add a number to the federal income tax rate to reflect the possibility of state taxes; for example, the average federal rate of 35 percent might be increased to 40 percent for preliminary economic analyses. This is another case where specific information or corporate policy should be used to guide the choice.

There are other state taxes as well, for example, workers' compensation taxes. Here again, specific knowledge of the taxation policies at the site of the project is needed to determine the allowances to be made.

Nonincome Taxes

In addition to income taxes, property and excise taxes may be levied by federal, state, or local governments. Local governments usually have jurisdiction over property taxes, which are commonly charged on a county and city basis. Taxes of this type are referred to as *direct* since they must be paid directly by the particular business and cannot be passed on as such to the consumer.

Property taxes may vary widely from one locality to another, but the average annual amount of these charges is 1 to 4 percent of the assessed valuation. The method and the amount of assessment of property value differ considerably with locale as well. For economic evaluation purposes, 2 percent of the fixed-capital investment in a project is a typical property tax rate charge.

Excise taxes are levied by federal and state governments. Federal excise taxes include charges for import customs duties, transfer of stocks and bonds, and a large number of similar items. Manufacturers' and retailers' excise taxes are levied on the sale of many products such as gasoline and alcoholic beverages. Taxes of this type are often referred to as *indirect* since they can be passed on to the consumer. Many businesses must also pay excise taxes for the privilege of carrying on a business or manufacturing enterprise in their particular localities. Excise taxes are included in economic evaluations only when they are known to apply specifically to the project under consideration.

FIXED CHARGES

Among the many costs included in the total product cost, as listed in Fig. 6-7, are a group of *fixed charges*. The first three of these charges—depreciation, property taxes (discussed above), and insurance—are costs related to the capital investment in a project. Once the facility for the project has been built, these charges are fixed. The other fixed charge, rent, is also fixed if there has been a contractual agreement for the rental of space. These charges are fixed in that they are not related to the activity level of the plant and continue even if the plant does not operate. Interest is sometimes included as a fixed charge, but because of its importance has been treated separately in this chapter.

Depreciation

Depreciation is an unusual charge in that it is paid into the corporate treasury. There are other kinds of intracorporate transfers, such as material and utility purchases from one division by another. Such transfers generally have little or no overall impact on the corporate finances. Depreciation, however, has a significant effect on corporate cash flow.

The concept of depreciation is based upon the fact that physical facilities deteriorate and decline in usefulness with time; thus, the value of a facility decreases. *Physical depreciation* is the term given to the measure of the decrease in value of a facility due to changes in the physical aspects of a property. Wear and tear, corrosion, accidents, and deterioration due to age or the elements are all causes of physical depreciation. With this type of depreciation, the serviceability of a property is reduced because of the physical changes.

Depreciation due to all other causes is known as *functional depreciation*. One common type of functional depreciation is obsolescence. This is caused by technological advances which make an existing property obsolete. Other causes of functional depreciation could be (1) decrease in demand for the service rendered by the property, (2) shifts in population, (3) changes in requirements of public authority, (4) inadequacy or insufficient capacity, and (5) abandonment of the enterprise.

Depreciation and Income Tax

Because depreciation impacts the federal income tax liability of a corporation, the federal government, as well as the corporation, has an interest in depreciation. Indeed, federal tax laws closely control the manner in which depreciation is charged.

Depreciation is a charge to the revenue resulting from an investment in real property. It is entirely reasonable that invested principal should be recovered by the investor and that project revenues be charged to pay that principal. In the case of other investments, such as savings accounts, the original investment is available in addition to any return that has been earned, and thus a recovery of invested capital is to be expected in plant investments as well. Depreciation is charged as an expense and then paid to the corporation. It is added and subtracted on the corporate books, and because of this, it is sometimes referred to as an accounting artifact. Depreciation is more than an artifact, however, because of the effect it has on the amount of income tax that a corporation must pay. One definition of *depreciation* is as follows: "A deduction for depreciation may be claimed each year for property with a limited useful life that's used in a trade or business or held for the production of income. This deduction allows taxpayers to recover their costs for the property over a period of years."[†]

Amortization, a word sometimes used interchangeably with depreciation, has a more restricted meaning in tax policy: "You may claim an *amortization* deduction for intangibles with limited useful lives that can be estimated with reasonable accuracy. For example, patents and copyrights are amortizable."[‡]

It has been shown in Chap. 6 that depreciation adds to the corporate treasury as indicated by the following relationship:

$$A_j = (s_j - c_{oj})(1 - \Phi) + \Phi d_j \tag{6-1}$$

where the subscript j indicates an annual value in year j, A the annual cash flow, s the annual sales revenue, c_o the operating cost (all of total product cost except depreciation), Φ the fractional income tax rate, and d the annual depreciation. The positive term Φd_j in Eq. (6-1) results from subtracting depreciation as an expense before income taxes are calculated. This reduces the taxable income by d_j, the taxes owed by Φd_j, and the net income by $(1 - \Phi)d_j$. The recovery of the depreciation increases the net cash flow from the project to the capital source of the company (see Fig. 6-1 for further details). This reflects recognition in the tax code that depreciation is the recovery of the original investment.

[†] *1993 Prentice-Hall Federal Tax Advisor,* Prentice-Hall Tax and Professional Practice, Englewood Cliffs, NJ, 1993, chap. 11.

[‡] *1993 Prentice-Hall Federal Tax Advisor,* loc. cit.

Depreciable Investments

In general, all property with a limited useful life of more than 1 year that is used in a trade or business, or held for the production of income, is depreciable. Physical facilities, including such costs as design and engineering, shipping, and field erection, are depreciable. Land is not depreciable, but improvements to the land, such as grading and adding utility services, are depreciable. Working capital and start-up costs are not depreciable. Inventories held for sale are not depreciable. In project terminology, the fixed-capital investment, not including land, is depreciable.

Maintenance is necessary for keeping a property in good condition; repairs connote the mending or replacing of broken or worn parts of a property. The costs of maintenance and repairs are direct operating expenses and thus are not depreciable.

The total amount of depreciation that may be charged is equal to the amount of the original investment in a property—no more and no less. Depreciation does not inflate or deflate.

Current Value

The *current value* of an asset is the value of the asset in its condition at the time of valuation. *Book value* is the difference between the original cost of a property and all the depreciation charged up to a time. It is important, because it is included in the values of all assets of a corporation. The method of determining depreciation may be different for purposes of obtaining the book value than that which is used for income tax purposes, depending on corporate policy. The price that could be obtained for an asset if it were sold on the open market is designated the *market value*. It may be quite different from the book value and clearly is important for determining the true asset value of the company.

Salvage Value

Salvage value is the net amount of money obtainable from the sale of used property over and above any charges involved in removal and sale. The term *salvage value* implies that the property can be of further service. If the property is not useful, it can often be sold for material recovery. Income obtainable from this type of disposal is known as *scrap value*. As of 2002, tax laws do not permit considering salvage or scrap value in the calculation of depreciation. Income from the sale of used property, to the extent that it exceeds the undepreciated value of the property, is therefore taxed as a capital gain. If the net sale price is less than the undepreciated value, it is not taxable but it is included as an income to the project at the time of the sale.

Recovery Period

The period over which the use of a property is economically feasible is known as the *service life* of the property. The period over which depreciation is charged is the *recovery period,* and this is established by tax codes. While originally the recovery period was at least approximately related to the service life, the reality now is that there is little relationship between the two. Recovery periods for some chemical- and process-industries-related depreciation are shown in Table 7-8.

Table 7-8 Recovery periods for selected chemical-industry-related asset classes†

Type of assets	Recovery period, years	
	MACRS	Straight line
Heavy general-purpose trucks	5	5
Industrial steam and electric generation and/or distribution systems	15	22
Information systems (e.g., computers)	5	5
Manufacture of chemicals and allied products (including petrochemicals)	5	9.5
Manufacture of electronic components, products, and systems	5	5
Manufacture of finished plastic products	7	11
Manufacture of other (than grain, sugar, and vegetable oils) food and kindred products	7	12
Manufacture of pulp and paper	7	13
Manufacture of rubber products	7	14
Manufacture of semiconductors	5	5
Petroleum refining	10	16
Pipeline transportation	15	22
Gas utility synthetic natural gas (SNG)	7	14
SNG—coal gasification	10	18
Liquefied natural gas plant	15	22
Waste reduction and resource recovery plant	7	10
Alternative energy property	5	12

†Source: © 2002 CCH Incorporated. All Rights Reserved. Reprinted with permission from *1997 Depreciation Guide Featuring MACRS.*

Methods for Calculating Depreciation

There are several methods for calculating depreciation;† however, because of the federal income tax rules in effect since 1987, we shall consider here only three of the methods: straight-line, double-declining balance, and the modified accelerated cost recovery system (MACRS).

Depreciation results in a reduction in income tax payable in the years in which it is charged. The total amount of depreciation that can be charged is fixed and equal to the investment in depreciable property. Thus, over any recovery period, the same total amount is depreciated; hence, the same total amount of tax is paid—assuming that the incremental tax rate is the same in all those years. However, because money has a time value, it is economically preferable to receive benefits, including tax savings, sooner rather than later. Therefore, it is usually in the taxpayer's interest to depreciate property as rapidly as possible. From the federal government's perspective, however, for

†M. S. Peters and K. D. Timmerhaus, *Plant Design and Economics for Chemical Engineers,* 4th ed., McGraw-Hill, New York, 1991.

the same reason, it is preferable to receive tax revenues sooner rather than later. Counterbalancing this, from the government's point of view, is the desire to encourage business activity and thus the overall economy. For these reasons, the rate and length of time during which depreciation can be charged are a matter of government policy.

Straight-Line Method This method may be elected under the federal tax code as an *alternative depreciation system*. It depreciates property less rapidly than does MACRS, and therefore, it would only be chosen for use in tax computations under special circumstances. For example, a new company might wish to conserve depreciation deductions for use in the future when its incremental tax rate is expected to be higher. For purposes of economic evaluation of projects, straight-line depreciation is often used when employing a profitability measure that does not consider the time value of money, because under these circumstances the rate of depreciation is not important.

In the straight-line method, the property value is assumed to decrease linearly with time over the recovery period. No salvage or scrap value may be taken. Thus the amount of depreciation in each year of the recovery period is

$$d = \frac{V}{n} \tag{7-44}$$

where d is the annual depreciation in dollars per year, V the original investment in the property at the start of the recovery period, and n the length of the straight-line recovery period. For tax purposes, the recovery period for straight-line depreciation is 9.5 years for chemical plants, as shown in Table 7-8. For purposes other than tax calculations, the corporation may select a value for n. If straight-line depreciation is being used in conjunction with a profitability measure that does not take into account the time value of money, then one reasonable recovery period to use is the length of the evaluation period. Another reasonable choice is 6 years, because this gives the same average rate of depreciation as does MACRS.

Modified Accelerated Cost Recovery System MACRS is the depreciation method used for most income tax purposes and therefore also for most economic evaluations. The MACRS method is based upon the classical double-declining-balance method,[†] but with no salvage or scrap value allowed, a switch to straight-line at a point, and use of the half-year convention. There are also mid-month and mid-quarter conventions, but they rarely occur in corporate project tax situations.

The double-declining-balance method allows a depreciation charge in each year of the recovery period that is twice the average rate of recovery on the remaining undepreciated balance for the full recovery period. Thus, in the first year of a 5-year recovery period, the fraction of the original depreciable investment that can be taken as depreciation is $(2)(\frac{1}{5})$, or 40 percent. The undepreciated portion is now 60 percent of the original investment; thus, in the second year, the allowable amount is $(2)(\frac{0.6}{5})$, or

[†]M. S. Peters and K. D. Timmerhaus, *Plant Design and Economics for Chemical Engineers,* 4th ed., McGraw-Hill, New York, 1991, chap. 9.

24 percent; and so on. Because this method always takes a fraction of the remaining balance, the asset is never fully depreciated. The MACRS method overcomes this by employing a shift to the straight-line method in the first year that the straight-line depreciation provides a higher depreciation rate than the declining-balance method.

The *half-year convention* indicates that in the first year only one-half of the double-declining-balance method is allowed and the balance remaining after the end of the recovery period is depreciated in the next year. This leads to the strange result that the MACRS depreciation always requires an additional year over the length of the recovery period.

EXAMPLE 7-9 Calculation of the MACRS Yearly Depreciation Percentage

Calculate the annual percentage rate of depreciation for a 5-year recovery period asset, such as a chemical plant, using the double-declining-balance method and the half-year convention, and switching to the straight-line method on the remaining balance when it gives a higher annual depreciation than that obtained with the double-declining-balance method. This is the MACRS method.

■ Solution

First year: The double-declining-balance (DDB) method allows a deprecation of $(2)(\frac{1}{5}) = 0.4$, but the half-year convention reduces this by one-half to 0.20, or 20 percent.

The straight-line (SL) method for the first year permits a depreciation of $\frac{1}{5} = 0.20$, or 20 percent, the same as the DDB method with the half-year convention.

Second year: The undepreciated balance is now $1 - 0.2 = 0.8$ of the original. The DDB method allows a depreciation of $(2)(0.8/5) = 0.32$, or 32 percent. The SL method, with 4.5 years remaining, allows $0.8/4.5 = 0.177$, so the DDB method should be used.

Third year: The undepreciated balance is now $1 - 0.52 = 0.48$ of the original. The DDB method allows a depreciation of $(2)(0.48/5) = 0.192$, or 19.2 percent. The SL method, with 3.5 years remaining, allows $0.48/3.5 = 0.137$; so again, the DDB method should be used.

Fourth year: The undepreciated balance is now $1 - 0.712 = 0.288$. The DDB method allows a depreciation of $(2)(0.288/5) = 0.1152$, or 11.52 percent. The SL method, with 2.5 years remaining, shows a depreciation of $0.288/2.5 = 0.1152$. Both methods give the same value.

Fifth year: The undepreciated balance is now $1 - 0.8272 = 0.1728$. The DDB method allows a depreciation of $(2)(0.1728/5) = 0.06912$, or 6.192 percent. The SL method, with 1.5 years remaining, allows a depreciation of $0.1728/1.5 = 0.1152$, or 11.52 percent. This time the SL method provides the higher rate, and that value is used.

Sixth year: Because of the half-year convention, there is a depreciation charge left for this year. It is the remaining undepreciated balance, amounting to $1 - 0.9424 = 0.0576$, or 5.76 percent. ■

Table 7-9 shows the MACRS annual rates for the half-year convention for all the allowed recovery periods up to 20 years, the half-year convention, and switching to straight line. The 3-, 5-, 7- and 10-year recovery periods employ the double-, or 200 percent, declining-balance method, whereas the 15- and 20-year recovery periods use the 150 percent declining-balance method. U.S. tax law prescribes all these conditions.

Table 7-9 MACRS depreciation rates†

General depreciation system
Applicable depreciation method: 200 or 150 percent
Declining balance switching to straight-line method
Applicable recovery periods: 3, 5, 7, 10, 15, 20 years
Applicable convention: half-year

Recovery year	Recovery period					
	3-year	5-year	7-year	10-year	15-year	20-year
	Depreciation rate, %					
1	33.33	20.00	14.29	10.00	5.00	3.750
2	44.45	32.00	24.49	18.00	9.50	7.219
3	14.81	19.20	17.49	14.40	8.55	6.677
4	7.41	11.52	12.49	11.52	7.70	6.177
5		11.52	8.93	9.22	6.93	5.713
6		5.76	8.92	7.37	6.23	5.285
7			8.93	6.55	5.90	4.888
8			4.46	6.55	5.90	4.522
9				6.56	5.91	4.462
10				6.55	5.90	4.461
11				3.28	5.91	4.462
12					5.90	4.461
13					5.91	4.462
14					5.90	4.461
15					5.91	4.462
16					2.95	4.461
17						4.462
18						4.461
19						4.462
20						4.461
21						2.231

†Source: © 2002 CCH Incorporated. All Rights Reserved. Reprinted with permission from *2000 U.S. Master Tax Guide.*

Insurance

The annual insurance costs for ordinary industrial projects are approximately 1 percent of the fixed-capital investment. Despite the fact that insurance costs may represent only a small fraction of total costs, it is necessary to consider insurance requirements carefully to make certain the economic operation of a plant is protected against emergencies or unforeseen developments.

The design engineer can aid in reducing insurance requirements if all the factors involved in obtaining adequate insurance are understood. In particular, the engineer should be aware of the different types of insurance available and the legal responsibilities of a corporation with regard to accidents and other unpredictable emergencies.

Legal Responsibility A corporation can obtain insurance to protect itself against loss of property due to any of a number of different causes. Protection against unforeseen emergencies other than direct property loss can also be obtained through insurance.

For example, injuries to employees or others due to a fire or explosion can be covered. It is impossible to insure against every possible incidence, but it is necessary to consider the results of a potential emergency and understand the legal responsibility for various types of events. The payments required for settling a case in which legal responsibility has been proved can be much greater than any costs due to direct property damage.

The design engineer should be familiar with laws and regulations governing the type of plant or process involved in a design. In case of an accident, failure to comply with the laws involved is a major factor in establishing legal responsibility. Compliance with all existing laws, however, is not a sufficient basis for disallowance of legal liability. Every known safety feature should be included, and extraordinary care in the complete operation must be proved before a good case can be presented for disallowing legal liability.

Liability for product safety has become a major concern for manufacturers in recent years, due to heightened public awareness of producer's liability. Product testing and hazard warnings are minimal activities to undertake before releasing a product for distribution.

Types of Insurance Many different types of insurance are available for protection against property loss or charges based on legal liability. Despite every precaution, there is always the possibility of an unforeseen event causing a sudden drain on a corporation's finances, and an efficient management protects itself against such potential emergencies by taking out insurance to cover such risks.

The major insurance requirements for manufacturing concerns can be classified as follows:

1. Fire insurance and similar emergency coverage on buildings, equipment, and all other owned, used or stored property. Included in this category would be losses caused by lightning, wind, hailstorms, floods, automobile accidents, explosions, earthquakes, and similar occurrences.
2. Public-liability insurance, including bodily injury and property loss or damage, on all operations such as those involving automobiles, elevators, attractive nuisances, aviation products, or any corporate function carried out at a location away from the plant premises.
3. Business-interruption insurance. The loss of income due to a business interruption caused by fire or other emergency may far exceed any loss in property. Consequently, insurance against a business interruption of this type is extremely important.
4. Power plant, machinery, and special-operations hazards.
5. Workers' compensation insurance.
6. Marine and transportation insurance for all property in transit.
7. Comprehensive crime coverage.
8. Employee-benefit insurance, including life, hospitalization, accident, health, personal property, and pension plans.
9. Product liability.

Self-Insurance

Because the payout of claims by insurance companies is perhaps only 55 to 60 percent for each dollar of premium they receive, self-insurance is sometimes used to minimize the cost of insurance. The decision whether to purchase or self-insure requires balancing the possible savings versus the chances of large losses. The tax implications must be considered as well, because insurance premiums for standard insurance are tax-deductible while funds paid into a self-insurance reserve ordinarily are not. Overall corporate policies dictate the type and amount of insurance that will be held. It should be realized, however, that a well-designed insurance plan needs input from persons who understand all the aspects of insurance as well as the problems involved in the manufacturing operation.

NOMENCLATURE

A = uniform series; series of constant, year-end cash flows, dollars
A_j = cash flow in year j, dollars
$\bar{A}$ = continuous cash flow for N years at a constant rate, dollars
c_{oj} = outflow cost of operation in year j, including all expenses except depreciation, dollars
d = amount of straight-line depreciation, dollars
d_j = depreciation in year j, dollars
F = future amount or future worth of money, dollars
i = interest rate based on length of one interest period, percent/100
i_{eff} = effective interest rate defined as the interest rate with annual compounding which provides the same amount of interest in 1 year as that earned in 1 year with another compounding period, percent/100
I = amount of interest earned in a period, dollars per period
I_j = amount of interest paid in period j, dollars per period
j = specified interest period, dimension of time
L = constant amount of loan repayment per period, dollars per period
m = number of interest periods per year, dimensionless
N = total number of interest periods, dimensionless
n = recovery period, straight-line depreciation, years
p_j = amount of principal repaid in period j, dollars per period
p_m = amount of principal repaid in period m, dollars per period
P = principal or present worth, dollars
P_{j-1} = amount of unpaid principal remaining after $j - 1$ loan payments, dollars
P_0 = initial amount of a loan, dollars
$\bar{P}$ = continuous, constant cash flow rate for 1 year, dollars/yr
r = nominal interest rate—for compounding other than annual, percent/100
s_j = amount of sales in year j, dollars
V = amount of depreciable capital investment, dollars

Greek Symbol

Φ = fractional income tax rate, percent/100

PROBLEMS

7-1 What funds will be available 10 years from now if $10,000 is deposited at present at a nominal interest rate of 6 percent compounded semiannually?

7-2 The original cost for a distillation tower is $50,000, and the useful life of the tower is estimated to be 10 years. How much must be placed annually in an annuity at an interest rate of 6 percent to obtain sufficient funds to replace the tower at the end of 10 years? If the scrap value of the distillation tower is $5000, determine the asset value (i.e., the total book value of the tower) at the end of 5 years based on straight line depreciation.

7-3 With total yearly payments of $10,000 for 10 years, compare the compound amount accumulated at the end of the 10 years if the payments are (1) at the end of the year, (2) weekly, and (3) continuous. The effective (annual) interest rate is 8 percent, and the payments are uniform. Also determine the present worth at time zero for each of the three types of payments.

7-4 Derive two expressions for capitalized cost based on (1) annual discrete interest compounding and (2) continuous interest compounding. *Capitalized cost* is defined as the sum of the original cost C_v of the equipment or asset plus the amount P that must be invested when the original equipment or asset is purchased so that when the original equipment or asset is replaced in N years at a cost C_R, the value of the investment equals P plus C_R.

7-5 A heat exchanger is to be used in a heating process. A standard type of heat exchanger with a negligible scrap value costs $20,000 and will have a useful life of 6 years. Another type of heat exchanger with equivalent design capacity is priced at $34,000 but with a useful life of 10 years and a scrap value of $4000. Assume an effective compound interest rate of 6 percent per year and that the replacement cost of each exchanger is the same as that of the original exchanger. Determine which heat exchanger is cheaper by comparing the capitalized cost of each. See Prob. 7-4 for a definition of capitalized cost.

7-6 A new storage tank can be purchased and installed for $10,000. The estimated service life of this tank is 10 years. It has been proposed that an available tank with the capacity equivalent to the new tank be used instead of buying the new tank. If the latter tank were repaired, it would have a service life of 3 years before similar repairs would be needed again. Neither tank has any scrap value. Money is worth 6 percent compounded annually. On the basis of equal capitalized costs for the two tanks, how much can be spent for repairing the existing tank? See Prob. 7-4 for a definition of capitalized cost.

7-7 The total investment required for a new chemical plant is estimated at $20 million. Fifty percent will be supplied from the company's noncapital resources. Of the remaining investment, one-half will come from a loan at an effective interest rate of 8 percent, and one-half will come from an issue of preferred stock paying dividends at a stated effective rate of 8 percent. The income tax rate for the company is 35 percent of pretax earnings. Under these conditions, how many dollars per year does the company lose (i.e., after taxes) by issuing preferred stock at 8 percent dividend instead of bonds at an effective interest rate of 6 percent?

7-8 It has been proposed that a company invest $1 million of its own funds in a venture which will yield a gross income of $1 million per year. The total annual costs will be $800,000 per year. In an alternate proposal, the company can invest a total of $600,000 and receive annual net earnings of $220,000 from the project. Depreciation and income tax effects are not to be considered. The remaining $400,000 can be loaned at an effective 6 percent annual interest rate. What alternative would be more profitable for the company with regard to investing its available funds?

7-9 The fixed-capital investment for an existing chemical plant is $20 million. Annual property taxes amount to 1 percent of the fixed-capital investment, and state income taxes are 5 percent of the gross earnings. The net income after all taxes is $2 million, and the federal income taxes amount to 35 percent of gross earnings. If the same plant had been constructed for the same fixed-capital investment but at a location where property taxes were 4 percent of the fixed-capital investment and state income taxes were 2 percent of the gross earnings, what would be the net income per year after taxes, assuming all other cost factors were unchanged?

7-10 Self-insurance is being considered for one area of a chemical company. The fixed-capital investment involved is $50,000, and insurance costs for complete protection amount to $500 per year. If self-insurance is used, a reserve fund will be set up under the company's jurisdiction, and annual premiums of $400 will be deposited in this fund under an ordinary annuity plan. All money in the fund can be assumed to earn interest at a compound interest rate of 6 percent. Neglecting any charges connected with administration of the fund, how much money must be deposited in the fund at the beginning of the program in order to have sufficient funds accumulated to replace a $50,000 loss after 10 years?

7-11 The initial installed cost for a new piece of equipment is $10,000. After the equipment has been in use for 4 years, it is sold for $7000. The company that originally owned the equipment employs a straight-line method for determining depreciation costs. If the company had used the MACRS 5-year method for determining depreciation costs, the asset or book value for the piece of equipment at the end of 4 years would have been $1728. The total income tax rate for the company is 35 percent of all gross earnings. Capital gains taxes amount to 20 percent of the gain. How much net savings would the company have achieved by using the MACRS method instead of the straight-line depreciation method?

7-12 Solve Prob. 7-8 on an after tax basis employing depreciation at 20 percent per year over the project investment and an income tax rate of 35 percent per year of gross earnings.

7-13 A piece of equipment having a negligible salvage and scrap value is estimated to have a MACRS and straight line recovery period of 5 years. The original cost of the equipment was $50,000. Determine (1) the depreciation charge for the second year if straight-line depreciation is used and the percent of the original investment paid off in the first 2 years, and (2) the depreciation charge for the fifth year if the modified acceleration cost recovery system (MACRS) is used, and the percent of the original investment paid off in the first 2 years.

7-14 A chemical company has a total income of $1 million per year and total expenses of $600,000 not including depreciation. At the start of the first year of operation, a composite account of all depreciable assets shows a value of $850,000 with a MACRS recovery period of 5 years, and a straight-line recovery period of 9.5 years. Thirty-five percent of all profits before taxes must be paid out for income taxes. What would be the reduction in income tax charges for the first year of operation if the MACRS method were used for the depreciation accounting instead of the straight-line method?

7-15 The total value of a new liquefied natural gas plant is $10 million. A certificate of necessity to meet a national need has been obtained, permitting a write-off of 60 percent of the initial cost of the plant in the first 5 years. The balance of the plant investment requires a write-off period of 10 years. Using the straight-line depreciation method and assuming negligible salvage and scrap value, determine the total depreciation cost during the first year. If the MACRS method of depreciation (15 yr recovery period for this type of plant) had been selected, what would have been the total depreciation cost during the first year?

7-16 A small company is using the unit-of-production method for determining depreciation costs. The original value of the property is $110,000. It is estimated that the company can produce 11,000 units before the equipment will have a salvage or scrap value of zero; that is, the depreciation cost per unit produced is $10. The equipment produces 200 units during the first year, and the production rate is doubled each year for the first 4 years. The production rate obtained in the fourth year is then held constant until the value of the equipment is paid off. What would have been the annual depreciation cost if the straight-line method based on this same time period had been used?

7-17 A laboratory piece of equipment was purchased for $35,000 and is estimated to be used for 5 years with a salvage value of $5000. Tabulate the annual depreciation allowances and year-end book values for the 5 years by using (1) the straight-line depreciation method, (2) the MACRS 5-yr recovery period depreciation method, and (3) the sum-of-the-digits depreciation method.†

7-18 A piece of equipment with an original cost of $10,000 and no salvage value has a depreciation allowance of $2381 during its second year of service when depreciated by the sum-of-the-digits method.† What recovery period has been used?

†The sum-of-the-digits depreciation method is outlined in M. S. Peters and K. D. Timmerhaus, *Plant Design and Economics for Chemical Engineers,* 4th ed., McGraw-Hill, New York, 1991, p. 283.

CHAPTER

8

Profitability, Alternative Investments, and Replacements

A new project, such as constructing and operating a new chemical plant, requires a commitment of capital funds. The decision to make such a commitment is based upon many factors. In the private sector of a capitalistic economic system, earning a rewarding profit is perhaps the most important of these factors. Other factors weighing into such a decision include availability of capital; market position; health, safety, and environmental concerns; experience of personnel; and need to diversify. Such factors, however, are not part of the economic evaluations considered in this chapter.

The resources required to undertake a project are always limited. Therefore, it follows that these resources should be used in an appropriate and efficient manner. The wise investor selects investments that are expected to maximize the return from the capital that is available.

A proposed investment must be evaluated for its economic feasibility. When a new technical project is proposed, a design study must be carried out. The design study produces specifications from which cost estimates can be made. These cost estimates, in turn, become the data for evaluating the economic consequences of the project. This chapter introduces principles and methods for carrying out economic evaluations. Such economic evaluations provide information that is essential, although not necessarily sufficient, for making decisions about how to use these limited resources.

A number of methods for calculating profitability are employed in economic analyses. The basis for the most commonly used methods is presented, and the measures are defined. Considerations involved in using each of the methods are also reviewed.

When decisions are made about committing resources to a project, they are based upon anticipated future performance of the project as well as on future economic and other related conditions. Since predicting the future is far from an exact science, there is always some uncertainty associated with investing. There is, however, a wide range in this uncertainty depending on the type of investment. Government bonds are less risky than stocks; investing in an established chemical process with a large, unfilled market is less risky than investing in an unproven process producing a new chemical. On the other hand, the more risky investments often provide the promise of greater rewards. Making allowance for risk in economic evaluations is considered in this chapter.

Even in the research and development stages of a new chemical processing concept, plant designs and economic evaluations can be useful. Such studies will necessarily be approximate, but can yield useful guidance as to the economic potential of a project. In addition, these studies can identify those parts of the process where major cost savings can be expected, and thus where research and development efforts might be focused.

In the economic evaluation of a particular project, there are always alternative uses for the resources. For example, the available capital can be invested in another project or projects, or invested in various financial instruments such as stocks, bonds, and savings accounts. Since economic evaluation is always performed in the context of these alternatives, evaluation methods must permit rational comparisons of these alternatives. Several such methods are discussed in this chapter. Equipment replacement strategy also is reviewed in the light of these economic evaluation criteria.

PROFITABILITY STANDARDS

In the process of making an investment decision, the profit anticipated from an investment must be judged relative to some profitability standard. A *profitability standard* is a quantitative measure of profit with respect to the investment required to generate that profit.

Profit is the goal of any investment, but maximizing profit is an inadequate profitability standard. The profit must be judged relative to the investment. For example, suppose two equally sound investment opportunities are available. One of these requires a $100,000 capital investment and will yield a profit of $10,000 per year, while the second requires $1 million of capital investment and will yield a profit of $25,000 per year. While the second investment provides a greater yearly profit than the first, the annual rate of return on that investment is only ($25,000/$1,000,000)(100), or 2.5 percent, while it is 10 percent for the first investment. If one had $1 million to invest and there were no alternative uses for these funds, then one might select the second investment. However, because there are numerous reliable alternatives, such as bonds, that will yield annual returns greater than 2.5 percent, the second investment is not attractive. In this case, it would be advisable to invest $100,000 in the first alternative, assuming that it is not significantly more risky than the second alternative, and invest the remaining $900,000 in other reliable alternatives that provide higher annual returns.

Cost of Capital

The cost of capital based upon corporate experience is often used as a basic profitability standard. *Cost of capital* is the amount paid for the use of capital from such sources as bonds, common and preferred stock, and loans. The cost of capital after income taxes is found by weighting the cost of each of these outside sources according to its fraction of the total capital from these sources. Bond dividends and loan interest are contractual amounts and as such are charged as an expense of doing business. The argument for using the cost of capital as a basic profitability standard is that any project must earn at least that rate just to repay these external capital sources.

Minimum Acceptable Rate of Return

A commonly used profitability standard is the minimum acceptable rate of return (also known as the minimum attractive rate of return, or MARR). The *minimum acceptable rate of return* is a rate of earning that must be achieved by an investment in order for it to be acceptable to the investor. The symbol m_{ar} will be used for the minimum acceptable annual rate of return, and it is used as a fraction per year but often expressed as a percentage per year. The m_{ar} generally is based on the highest rate of earning on safe investments that is available to the investor, such as corporate bonds, government bonds, and loans. The argument for this choice is that any investment in a project must show earnings at a rate that is at least equal to the highest safe alternative opportunity available to a company or corporation. The cost of capital can be used as an alternative basis from which to establish a minimum acceptable rate of return. The basic, safe value selected for m_{ar} is then normally adjusted to account for the uncertainties associated with a new project. From the unknown future behavior of the overall economy to the uncertain future of price and demand for a particular product, there are risks associated with investing in a project. The risk is further increased whenever most of the capital is invested in equipment and plant construction, because that capital is not liquid; that is, it is not easily recoverable on demand. Therefore, the practice is to adjust the safe cost of capital or earning rate in order to make the minimum acceptable rate of return commensurate with the risk. There is no given formula for doing this. The compensation for risk is largely a judgment call established for each corporation by upper management. Some guidance for establishing this level of risk compensation is suggested below.

Suppose that a corporation has a weighted cost of capital, based on Table 7-1, of 8 percent per year after taxes. This rate of return on safe investments available to the company likewise could be used as the value for m_{ar}. If 8 percent per year is used as the after-tax rate of earning on all the investment in the project, including that from corporate funds as well as from outside sources, then just enough funds are available to pay the dividends and interest on the outside capital sources and provide earnings at that same rate on the corporate investment. Such a rate of earnings on the corporate investment is not likely to be considered satisfactory to either management or shareholders. Rather, the basic rate is increased sufficiently to make it attractive in the face of risk, thereby arriving at a m_{ar} value that can be used in project evaluations.

The cost of capital and the rate of earnings on alternative investments are generally expressed as a single, end-of-the-year amount after income taxes, calculated as a percentage per year of the capital obtained or invested. This is equivalent, in project terms, to the net profit divided by the total capital investment.

Table 8-1 Suggested values for risk and minimum acceptable return on investment

Investment description	Level of risk	Minimum acceptable return m_{ar} (after income taxes), percent/year
Basis: Safe corporate investment opportunities or cost of capital	Safe	4–8
New capacity with established corporate market position	Low	8–16
New product entering into established market, or new process technology	Medium	16–24
New product or process in a new application	High	24–32
Everything new, high R&D and marketing effort	Very high	32–48+

Various degrees of risk and corresponding m_{ar} values are presented in Table 8-1. These are suggestions, for use when no other guidance is supplied, inferred from rates of return used in numerous sources.† Both risk and m_{ar} should be considered to be on a continuum.

For new products that require large investments in research, development, and testing before their future is determined, such as biotechnology products, the minimum acceptable return may be even higher than those listed in the table, especially for the earlier stages of the project study.

METHODS FOR CALCULATING PROFITABILITY

The calculation of profitability is generally performed with one of the methods listed below. The methods that do not consider the time value of money include *rate of return on investment, payback period*, and *net return*. The methods that consider the time value of money involve the *discounted cash flow rate of return* and *net present worth*. The listing does not include the discounted cash flow payback period method because the latter is rarely used. On the other hand, the net return method also is not widely used, but has some useful properties that warrant its inclusion with the other methods. Table 8-2 shows the results of a recent survey on the use of these methods in economic analyses.

Table 8-2 notes that there are methodology differences between small and large companies and also indicates that large companies often use more than one method.

Methods That Do Not Consider the Time Value of Money

For those methods that do not consider the time value of money, it is not important what depreciation schedule is used in the evaluation. Therefore, straight-line depreciation is

†M. S. Peters and K. D. Timmerhaus, *Plant Design and Economics for Chemical Engineers,* 4th ed., McGraw-Hill, New York, 1991; R. H. Perry and D. W. Green, eds., *Perry's Chemical Engineers' Handbook,* 7th ed., McGraw-Hill, New York, 1997; and W. D. Baasel, *Preliminary Chemical Engineering Plant Design,* 2d ed., Van Nostrand and Reinhold, New York, 1990.

Table 8-2 Use of profitability measures[†]

Evaluation method	Percentage use: Small companies	Percentage use: Large companies
Payback period	43	52
Return on investment	22	34
Net present worth	16	80
Discounted cash flow rate of return	11	78

[†]E. J. Farragher, R. T. Kleiman, and A. P. Sahu, *Eng. Econ.*, **44**(2): 137 (1999).

often used for convenience. Any depreciation period that is less than or equal to the evaluation period is usable for evaluations, although the recovery period specified by the Internal Revenue Service (IRS) must be used for income tax purposes.

Return on Investment (ROI) This profitability measure is defined as the ratio of profit to investment. Although any of several measures of profit and investment can be used, the most common are net profit and total capital investment. This can be expressed as

$$\text{ROI} = \frac{N_p}{T} \qquad (8\text{-}1a)$$

where ROI is the annual return on investment expressed as a fraction or percentage per year, N_p the annual net profit, and T the total capital investment. This definition agrees with the way in which the m_{ar} values in Table 8-1 are defined. Gross profit, before income taxes, or cash flow is sometimes used in place of net profit. Fixed-capital investment can be used rather than total investment. Corporate policy or the preference of the resource analyst determines the choice.

Net profit usually is not constant from year to year for a project; total investment also changes if additional investments are made during project operation. In such a case, no one year is necessarily likely to be representative of the entire project life; therefore, it becomes a question of what value to use for the net profit in Eq. (8-1*a*). The recommended procedure is to take the average ROI over the entire project life as given by

$$\text{ROI} = \frac{(1/N)\sum_{j=1}^{N}(N_{p,j})}{\sum_{j=-b}^{N}(T_j)} \qquad (8\text{-}1b)$$

where N is the evaluation period, $N_{p,j}$ the net profit in year j, $-b$ the year in which the first investment is made in the project with respect to zero as the startup time, and T_j the total capital investment in year j. Note that for $j > 0$, that is, anytime after the original investment, T_j may often be zero or at most small compared to the original investment, and the denominator can be replaced by the initial total capital investment to simplify Eq. (8-1*b*) to

$$\text{ROI} = \frac{(1/N)\sum_{j=1}^{N}(N_{p,j})}{T} = \frac{N_{p,\text{ave}}}{T} \qquad (8\text{-}1c)$$

where $N_{p,\text{ave}}$ is the average value of net profit per year over the evaluation period.

An ROI calculated from any of these three prior equations can be compared directly with a m_{ar} value supplied or selected from Table 8-1. If the ROI equals or exceeds the minimum acceptable rate of return m_{ar}, then the project offers an acceptable rate of return. If it does not, then the conclusion is that the project is not desirable for the investment of either borrowed or corporate funds.

Payback Period The profitability measure of payback period, or payout period, is the length of time necessary for the total return to equal the capital investment. The initial fixed-capital investment and annual cash flow are usually used in this calculation, so the equation is

$$\text{PBP} = \frac{V + A_x}{A_j} \tag{8-2a}$$

where PBP is the payback period in years, V the manufacturing fixed-capital investment, A_x the nonmanufacturing fixed-capital investment, $V + A_x$ the fixed-capital investment, and A_j the annual cash flow. This PBP represents the time required for the cash flow to equal the original fixed-capital investment. It is subject to the fact that the cash flow usually changes from year to year, thereby raising the question of which annual values to use. A particular year can be selected, or the average of the A_j values may be used to give

$$\text{PBP} = \frac{V + A_x}{(1/N)\sum_{j=1}^{N}(A_j)} = \frac{V + A_x}{(A_j)_{\text{ave}}} \tag{8-2b}$$

A PBP calculated from either Eq. (8-2*a*) or (8-2*b*) should be compared to a PBP obtained from the minimum acceptable rate of return. Because $V + A_x$ is approximately equal to $0.85\mathcal{F}$[†] and $(A_j)_{\text{ave}}$ is equal to $N_{p,\text{ave}} + d_{j,\text{ave}} = m_{ar}\mathcal{F} + 0.85\mathcal{F}/N$,[†]

$$\text{PBP} = \frac{0.85\mathcal{F}}{m_{ar}\mathcal{F} + 0.85\mathcal{F}/N} = \frac{0.85}{m_{ar} + 0.85/N} \tag{8-2c}$$

To be acceptable, a project payback period should be less than or equal to the reference value given by Eq. (8-2*c*).

Net Return Another profitability measure is the amount of cash flow over and above that required to meet the minimum acceptable rate of return and recover the total capital investment. This quantity is calculated by subtracting the total amount earned at the minimum acceptable rate of return, as well as the total capital investment, from the total cash flow. Each of these quantities represents the total amount obtained over the length of the evaluation period. The net return is given by

$$R_n = \sum_{j=1}^{N}(N_{p,j} + d_j + rec_j) - \sum_{j=-b}^{N}\mathcal{F}_j - m_{ar}N\sum_{j=-b}^{N}\mathcal{F}_j \tag{8-3a}$$

where R_n is the net return in dollars and rec_j the dollars recovered from the working capital and the sale of physical assets (equipment, buildings, land, etc.) in year j. By noting that the sum of d_j plus the sum of the recovered amounts is equal to the total

[†]The 0.85 value is obtained by assuming that the working capital is $0.15\mathcal{F}$.

capital investment or the sum of $\mathcal{T}_j$, the equation simplifies to

$$R_n = \sum_{j=1}^{N} N_{p,j} - m_{ar}\left(N \sum_{j=-b}^{N} \mathcal{T}_j\right) \qquad (8\text{-}3b)$$

and when divided by N, the equation becomes

$$R_{n,\text{ave}} = N_{p,\text{ave}} - m_{ar}\mathcal{T} \qquad (8\text{-}3c)$$

where $R_{n,\text{ave}}$ is the average net return in dollars per year.

The right-hand side of Eq. (8-3*a*), (8-3*b*), or (8-3*c*) is the total cash flow less the amount of money required to repay the total capital investment and also provide the earnings that are anticipated at the minimum acceptable rate. Any positive value for R_n indicates that the cash flow to the project is actually greater than the amount necessary to repay the investment and obtain a return that meets the minimum acceptable rate. Therefore, it is earning at a rate greater than the minimum acceptable rate. If R_n happens to equal zero, then the project is repaying the investment and matching the required m_{ar}. Either result indicates a favorable rating for the project. A negative value for R_n, however, indicates that the project obtains a return that is less than the m_{ar}, and therefore the project should be unfavorably rated. While not widely used, this method is effective as an optimization criterion.

Calculation of Profitability Measures Not Considering Time Value of Money — EXAMPLE 8-1

The research department of a large specialty monomer and polymer company has developed and formulated a new product. Early tests have been encouraging regarding the use of this product as a high-performance adhesive and sealant for cracks and joints in new and old cured concrete. The company foresees a substantial, virtually competition-free market in construction and repair if detailed product development and marketing studies are successful and the company enters the market early.

A preliminary design study has just been completed. The estimated economic parameters relevant to the project include the following items:

Production at 100 percent of capacity = 2×10^6 kg/yr
Batch process, total capital investment = \$28 million
Fixed-capital investment = \$24 million
Working capital = \$4 million
Sum of the variable product costs at full capacity = \$5 million/yr
Sum of the fixed costs (except for depreciation) = \$1 million/yr

Company evaluation policies are as follows:

Use 5-year recovery period, half-year convention, MACRS depreciation with all evaluation methods. Neglect working capital and salvage-value recovery. Use a 35 percent per year income tax rate.

Use a 10-year evaluation period; base the calculations on 50 percent of rated output the first year, 90 percent the second year, and 100 percent each year thereafter.

Assume that all the capital investment occurs at zero time.

Because of the high risk factor, a minimum acceptable return of 30 percent per year is the profitability standard for this preliminary economic evaluation.

Calculate the product sales price that is required to achieve the m_{ar} obtained by using the methods of return on investment, payback period, and net return.

■ Solution

First calculate the quantities to be used in the calculation of the evaluation criteria.

Summary of economic data used in the profitability analysis

	Year										
	1	2	3	4	5	6	7	8	9	10	Sum
A. Percent of operating time	50	90	100	100	100	100	100	100	100	100	—
B. Product rate, $10^6 kg/yr	1	1.8	2	2	2	2	2	2	2	2	18.8
C. All variable costs, $10^6/yr	2.5	4.5	5	5	5	5	5	5	5	5	47
D. All fixed costs (except depreciation), $10^6/yr	1	1	1	1	1	1	1	1	1	1	10
E. Depreciation,† $10^6/yr	4.8	7.7	4.6	2.8	2.7	1.4	0	0	0	0	24
F. Total product cost (C + D + E), $10^6/yr	8.3	13.2	10.6	8.8	8.7	7.4	6	6	6	6	81

†Fixed-capital investment of $24 million multiplied by MACRS factors from Table 7-9.

Return on Investment

For this method Eq. (8-1*c*) is used; so first calculate the average net profit

$$N_{p,\text{ave}} = \frac{1}{10}[p(18.8) - 81](1 - 0.35)(10^6) = (1.222p - 5.265)(10^6)$$

where p is the unit price of the product in \$/kg. Since $\mathcal{F} = \$28 \times 10^6$,

$$\text{ROI} = 0.30 = \frac{(1.222p - 5.265)(10^6)}{28 \times 10^6}$$

Solving for p gives

$$p = \frac{0.30(28) + 5.265}{1.222} = \$11.18/\text{kg}$$

Payback Period

The PBP corresponding to the given ROI of 30 percent is given by Eq. (8-2*c*)

$$\text{PBP} = \frac{0.85}{0.3 + 0.85/10} = 2.21 \text{ years}$$

Now, using Eq. (8-2*b*) gives

$$2.21 = \frac{(24)(10^6)}{p(1.222)(10^6) - (5.265)(10^6) + (2.4)(10^6)}$$

Solving for p gives

$$p = \frac{24/2.21 + 2.865}{1.222} = \$11.23/\text{kg}$$

Net Return

The net return is given by Eq. (8-3a). Note that when the rate of return is fixed, the product price must be such that the net return equals zero. Thus,

$$0 = p(12.22)(10^6) - (52.65)(10^6) + (24)(10^6) - (28)(10^6) - (0.30)(10)(28)(10^6)$$

Solving for p gives

$$p = \frac{52.65 - 24 + 28 + 84}{12.22} = \$11.51$$

In summary, these three methods give results that are not identical, but they agree well within the accuracy of the estimates. On the basis of these results, a product price in the range of \$11.20/kg to \$11.50/kg would be a reasonable recommendation, assuming no time value for money.

Methods That Consider the Time Value of Money

The methods that do consider the time value of money include net present worth and discounted cash flow rate of return. These methods account for the earning power of invested money by the discounting techniques described in Chap. 7. As shown in Table 8-2, they are the methods of economic analysis most often used by large companies. *Market value added,* another popular evaluation method, is equivalent to the net present worth method,[†] so it is not discussed here.

Net Present Worth The net present worth (NPW) is the total of the present worth of all cash flows minus the present worth of all capital investments, as defined by

$$\text{NPW} = \sum_{j=1}^{N} \text{PWF}_{cf,j}[(s_j - c_{oj} - d_j)(1 - \Phi) + rec_j + d_j] - \sum_{j=-b}^{N} \text{PWF}_{v,j}T_j \qquad (8\text{-}4)$$

where NPW is the net present worth, $\text{PWF}_{cf,j}$ the selected present worth factor (from Table 7-3, 7-4, or 7-5) for the cash flows in year j, s_j the value of sales in year j, c_{oj} the total product cost not including depreciation in year j, $\text{PWF}_{v,j}$ the appropriate present worth factor for investments occurring in year j, and T_j the total investment in year j. An earning rate is incorporated into the present worth factors by the discount rate used. Thus, the net present worth is the amount of money earned over and above the repayment of all the investments and the earnings on the investments at the discount (earning) rate used in the present worth factor calculations.

[†]J. C. Hartman, *Eng. Econ.*, **45**(2): 158 (2000).

The appropriate discount rate to use for discrete compounding is the minimum acceptable rate of return or m_{ar} originally selected as the evaluation standard. For continuous compounding the nominal interest rate is used, as given by

$$r_{ma} = \ln(1 + m_{ar}) \tag{8-5}$$

where r_{ma} is the minimum acceptable nominal rate for continuous compounding.

The net present worth is the time value of money equivalent of the net return defined in Eq. (8-3). If the net present worth is positive, then the project provides a return at a rate greater than the discount (earning) rate used in the calculations. In making comparisons of investments, the larger the net present worth, the more favorable is the investment. If the net present worth is equal to zero, then the project provides a return that matches the discount rate. In either of these cases, the project is judged as favorable compared to the m_{ar} selected. If the net present worth is less than zero, then the project rates unfavorably with respect to the m_{ar} standard.

Discounted Cash Flow Rate of Return[†] The discounted cash flow rate of return, or DCFR, is the return obtained from an investment in which all investments and cash flows are discounted. It is determined by setting the NPW given by Eq. (8-4) equal to zero and solving for the discount rate that satisfies the resulting relation. Thus,

$$0 = \sum_{j=1}^{N} \mathrm{PWF}_{cf,j}[(s_j - c_{oj} - d_j)(1 - \Phi) + rec_j + d_j] - \sum_{j=-b}^{N} \mathrm{PWF}_{v,j} T_j \tag{8-6}$$

The DCFR is only of concern when the project rates favorably compared to the value of m_{ar} used in calculating the net present worth. Clearly, if the NPW that is calculated equals zero, then the m_{ar} or r_{am} used is the DCFR. However, if the NPW is greater than zero, then the DCFR must be calculated from Eq. (8-6). Since the discount rate appears in numerous exponents, it is generally impossible to solve for the discount rate analytically and an iterative solution is required. As guidance, the discounted cash flow rate of return will be greater than the m_{ar} or r_{am} used. Thus, the m_{ar} or r_{am} value used is a good starting point. When the NPW is favorable, the DCFR will necessarily be favorable and will be the actual earning rate of the investment. The two methods—-net present worth and DCFR—are nearly always used together.

EXAMPLE 8-2 Calculation of Measures of Profitability Including Time Value of Money

With the same economic information and company policies as supplied in Example 8-1, use discrete, year-end cash flows and discrete compounding to determine the product price that will be required to provide a discretely compounded earning (discount) rate of 30 percent per year. If a price of \$11.50 is established for the product, determine the discounted cash flow rate of return.

[†]Other common names for this method are *internal rate of return, profitability index, interest rate of return, true rate of return,* and *investor's rate of return.*

■ Solution

Since the cash flows differ from year to year, the single-year discount factor, or the present worth factor, must be applied to each year. The discount factor $(P/F, i, j) = (1 + i)^{-j}$ in Table 7-3 must be used. The $\text{PWF}_{v,j}$ is equal to 1 because all the investment is made at time zero. Since the discount rate is fixed, the net present worth must be zero, and Eq. (8-6) is applicable. Substitution of the production rates into this equation results in

$$0 = \sum_{j=1}^{N} (1 + i)^{-j}\{[p(P_{r,j}) - c_{oj} - d_j](1 - \Phi) + d_j\} - \mathcal{T}$$

where P_r is the product rate in kg/yr.

Note that $(c_{oj} + d_j)$ is the total product cost in year j. Since p is a constant, we see that this equation may be rewritten in a more solvable form as

$$0 = p \sum_{j=1}^{N} P_{r,j}(1 + i)^{-j}(1 - \Phi) - \sum_{j=1}^{N}(c_{oj} + d_j)(1 + i)^{-j}(1 - \Phi) + \sum_{j=1}^{N} d_j(1 + i)^{-j} - \mathcal{T}$$

Solving for p gives

$$p = \frac{\mathcal{T} + \sum_{j=1}^{N}(c_{oj} + d_j)(1 + i)^{-j}(1 - \Phi) - \sum_{j=1}^{N}[d_j(1 + i)^{-j}]}{\sum_{j=1}^{N}[P_{r,j}(1 + i)^{-j}(1 - \Phi)]}$$

Each of these summations is calculated in the table below; rows A, B, and C are from the table in Example 8-1.

Summary of economic data used in the NPW profitability analysis

	Year										Sum of row
	1	2	3	4	5	6	7	8	9	10	
A. Product rate, 10^6 kg/yr	1	1.8	2	2	2	2	2	2	2	2	18.8
B. Total product cost, $\$10^6$/yr	8.31	13.18	10.61	8.77	8.76	7.38	6	6	6	6	81
C. Depreciation, $\$10^6$/yr	4.8	7.68	4.61	2.77	2.76	1.38	0	0	0	0	24
D. $(1 + 0.3)^{-j}$	0.769	0.592	0.455	0.350	0.269	0.207	0.159	0.123	0.094	0.073	3.09
E. (A)(0.65)(D), 10^6 kg/yr	0.500	0.692	0.592	0.455	0.350	0.269	0.206	0.159	0.122	0.095	3.44
F. (B)(0.65)(D), $\$10^6$/yr	4.15	5.08	3.13	2.00	1.53	0.99	0.62	0.48	0.37	0.28	18.63
G. (C)(D), $\$10^6$/yr	3.69	4.55	2.10	0.97	0.74	0.29	0	0	0	0	12.34

Substituting from the table into the preceding equation permits evaluation of the sales price.

$$p = \frac{(28)(10^6) + (18.63)(10^6) - (12.34)(10^6)}{(3.44)(10^6)} = \$9.97/\text{kg}$$

This value is considerably lower than the \$11.20/kg to \$11.50/kg sales price obtained in Example 8-1 by using methods that do not consider the time value of money. If this calculation is repeated with straight-line depreciation for 10 years at $\$2.4 \times 10^6$ per year, the resulting price is \$11.40 per kilogram, in excellent agreement with the values calculated in Example 8-1. Therefore, this lower value is essentially due to the recovery of most of the depreciation in the early years so that it is discounted less, on average, than either the sales revenue or total product cost. Its contribution to cash flow is therefore relatively greater.

To determine the discounted cash flow rate of return when the product price is established at \$11.50/kg requires an iteration of Eq. (8-6). Again, the data presented in Example 8-1 are used to develop the required values as shown in the following table:

Summary of economic data used in the DCFR profitability analysis

	Year										Sum of row
	1	2	3	4	5	6	7	8	9	10	
A. Product rate, 10^6 kg/yr	1	1.8	2	2	2	2	2	2	2	2	18.8
B. Sales revenue A (\$11.50/kg), $\$10^6$/yr	11.5	20.7	23	23	23	23	23	23	23	23	216.2
C. Total product cost, $\$10^6$/yr	8.31	13.18	10.61	8.77	8.76	7.38	6	6	6	6	81
D. B − C, $\$10^6$/yr	3.19	6.92	12.39	14.23	14.24	15.62	17	17	17	17	134.6
E. Depreciation, $\$10^6$/yr	4.8	7.68	4.61	2.77	2.76	1.38	0	0	0	0	24

Substituting into Eq. (8-6) gives

$$0 = -(28 \times 10^6) + 0.65 \sum_{j=1}^{N} (s_j - c_{oj} - d_j)(1+i)^{-j} + \sum_{j=1}^{N} d_j (1+i)^{-j}$$

For convenience, divide the equation by (28×10^6) so that all terms are of order 1. It is known that $i = 0.30$ causes the sum to be zero when the price of the product is \$9.97/kg, so in the present case i must be greater than 0.30. Starting with $i = 0.35$, several iterations give a value for the normalized right-hand side of less than 0.001 for $i = 0.3603$. So, at a price of \$11.50/kg, the DCFR is 36.0 percent. ■

Selecting a Profitability Method

The net present worth method, combined with the discounted cash flow rate of return method, is strongly recommended for making economic decisions. These methods not only include all the pertinent information of the other methods, but also take into account the time value of money. In that way they give a more realistic picture of the value of the earnings in relationship to the investment than do those methods that do not include the time value of money.

Income tax effects can be included in all the evaluation methods, and there is no good reason not to include them. Income taxes affect the net profit and cash flow into the corporate treasury, but not the capital investment. Thus, taxes impact all measures

Table 8-3 Definitions to clarify income tax situation for profitability evaluation

Revenue = total income
Net profits = revenue − all expenses − income tax
All expenses = cash expenses + depreciation
Income tax = (revenue − all expenses)(tax rate)
Cash flow = net profits + depreciation
Cash flow = (revenue)(1 − tax rate) − (cash expenses)(1 − tax rate) + (depreciation)(tax rate)
Cash flow = (revenue)(1 − tax rate) − (all expenses)(1 − tax rate) + depreciation

For the case of a 35 *percent tax rate*:
$1.00 of revenue (as either sales income or savings) yields a cash flow of $0.65.
$1.00 of cash expenses (as raw materials, labor, etc.) yields a cash outflow of $0.65.
$1.00 of depreciation yields a cash inflow of $0.35.

of return with respect to investment, and that includes all five profitability measures considered here. Table 8-3 reviews the quantities involved in order to emphasize and clarify many of these income tax effects.

It is recommended that all investment opportunities that are being analyzed be compared using the same profitability methods. Since each of the methods discussed has advantages and disadvantages, it is reasonable to apply and examine the results obtained from each one in the course of any economic analysis. However, it is quite possible to compare a series of alternative investments by any one of the profitability methods discussed and find that different alternatives could be recommended depending on the evaluation technique used.[†] If there is any question as to which method should be used for a final determination, the net present worth method should be chosen, since this specific method best accounts for the factors affecting the profitability of the investment.

Comparison of Alternative Investments by Different Profitability Methods — EXAMPLE 8-3

A company has three alternative investments that are being considered. Because all three investments are for the same type of unit and provide the same service, only one of the investments can be accepted. The risk factors are the same for all three cases. Company policies, based on the current economic situation, dictate that a minimum annual return on the original investment of 15 percent after taxes be predicted for any unnecessary investment. (This may be assumed to mean that other, equally sound investments yielding a 15 percent return after taxes are available.) Company policies also dictate that straight-line depreciation be used over the service life of the investment with accounting for the salvage value. For time-value-of-money calculations, the continuous cash flow and continuous compounding relationships are to be used. Land value and prestart-up costs can be ignored.

Use the following data to determine the following profitability evaluation criteria:

a. Rate of return on investment
b. Payback period

[†]This situation is illustrated in Example 8-3.

c. Net return
d. Net present worth
e. Discounted cash flow rate of return

Investment number	Total initial fixed-capital investment	Working-capital investment	Salvage value at end of service life	Service life, years	Annual earnings after taxes[†]
1	\$100,000	\$10,000	\$10,000	5	See yearly tabulation[‡]
2	170,000	10,000	15,000	7	\$42,000 (constant)
3	210,000	15,000	20,000	8	48,000 (constant)

[†]This is total annual income minus all costs except depreciation.
[‡]For investment 1, annual earnings flow to project are: year 1 = \$30,000, year 2 = \$31,000, year 3 = \$36,000, year 4 = \$40,000, year 5 = \$43,000.

■ Solution

Dollar values in this solution are generally rounded to three digits.

a. *Return on investment.* This is given by Eq. (8-1c), ROI $= N_{p,\text{ave}}/T$.

Investment 1

Now $N_{p,j}$ is the after-tax earnings (depreciation not included) $-0.65d_j$.

Year	Annual earnings – $(0.65)d_j$, \$	$N_{p,j}$
1	30,000 – (100,000 – 10,000)(0.65)/5	\$18,300
2	31,000 – 11,700	19,300
3	36,000 – 11,700	24,300
4	40,000 – 11,700	28,300
5	43,000 – 11,700	31,300
		\$121,500

$$N_{p,\text{ave}} = \frac{121{,}500}{5} = \$24{,}300/\text{yr}$$

For investment 1, the ROI $= (24{,}300)(100)/110{,}000 = 22.2$ percent per year. This ROI exceeds the m_{ar}, so investment 1 is acceptable.

Investment 2

For investment 2,

$$N_{p,\text{ave}} = \$42{,}000 - \frac{(170{,}000 - 15{,}000)(0.65)}{7} = \$27{,}600$$

$$\text{ROI} = \frac{(27{,}600)(100)}{180{,}000} = 15.3\%/\text{yr}$$

Investment 2 is acceptable because the ROI exceeds m_{ar}.

Investment 3

For investment 3,

$$N_{p,\text{ave}} = \$48{,}000 - \frac{(210{,}000 - 20{,}000)(0.65)}{8} = \$32{,}560$$

$$\text{ROI} = \frac{(32{,}560)(100)}{225{,}000} = 14.5\%/\text{yr}$$

This return is less than the m_{ar} of 15 percent per year, so investment 3 is not acceptable.

b. *Payback period.* The payback period is defined by Eq. (8-2*b*) as PBP = fixed-capital investment/$(N_{p,\text{ave}} + d_{j,\text{ave}})$. The reference value for PBP was given as PBP $= 0.85/(m_{ar} + 0.85/N)$ by Eq. (8-2*c*), based on a typical ratio of fixed to total capital investment; however, specific ratios can be estimated for this situation, and they are used for each case.

Investment 1

Reference value for PBP: $0.91/(0.15 + 0.91/5) = 2.74$ years.

$$\text{PBP} = \frac{100{,}000}{24{,}300 + 18{,}000} = 2.36 \text{ years}$$

The payback period is less than the reference, so investment 1 is acceptable.

Investment 2

For investment 2, the reference PBP $= 0.94/(0.15 + 0.94/7) = 3.31$ years.

$$\text{PBP} = \frac{170{,}000}{27{,}600 + 22{,}140} = 3.41 \text{ years}$$

where \$27,600 is the average annual cash flow and \$22,140 the annual depreciation recovered in investment 2. This PBP is greater than the reference value, and therefore investment 2 is not acceptable.

Investment 3

The reference value for investment 3 is $0.93/(0.15 + 0.93) = 3.49$ years.

$$\text{PBP} = \frac{210{,}000}{32{,}560 + 23{,}750} = 3.72 \text{ years}$$

This payback period is more than the reference value, investment 3 is not acceptable.

c. *Net return.* The net return from Eq. (8-3*b*) is $R_n = \sum_{j=1}^{N}(N_{p,j}) - m_{ar} N \mathcal{T}$.

Investment 1

$$R_n = 24{,}300(5) - (0.15)(5)(110{,}000) = \$39{,}000$$

This value is positive, and therefore investment 1 is acceptable.

Investment 2

$$R_n = 27{,}600(7) - (0.15)(7)(180{,}000) = \$4{,}200$$

Since this value is positive, investment 2 is acceptable.

Investment 3

$$R_n = 32{,}560(8) - (0.15)(8)(225{,}000) = -\$9{,}500$$

This value is negative; consequently, investment 3 is not acceptable.

d. *Net present worth.*

Investment 1

Because the income varies from year to year, the net present worth, as given by Eq. (8-4), can be simplified for investment 1 to

$$\text{NPW} = -\bar{T} + \sum_{n=1}^{N} \text{PWF}_j (N_{p,j} + d_j + rec_j)$$

The PWF_j to be used in this case is for 1-year constant, continuous cash flows from Table 7-5; $(P/\bar{P}, r, j) = [(e^r - 1)/r]e^{-rj}$. The PWF_j for recovery is $(P/F, r, j) = e^{-rj}$. From Eq. (8-5), $r = r_{am} = \ln 1.15 = 0.1398$; d_j is \$18,000/yr as established earlier, and rec_j is the recovery of working capital plus salvage value of \$20,000 at the end of year 5. The present worth of the cash flow is calculated in the following table:

Year	$N_{p,j}$	d_j	rec_j	$[(e^r - 1)/r]e^{-rj}$	Present worth
1	18,300	18,000	0	0.933	\$33,900
2	19,300	18,000	0	0.811	30,300
3	24,300	18,000	0	0.705	29,900
4	28,300	18,000	0	0.613	28,400
5	31,300	18,000	0	0.533	26,300
				e^{-rj}†	
5 end	0	0	20,000	0.497	9,900
				Total	\$158,700

†Since the recovery of the working capital and the salvage occurs at the end of the fifth year, the discount factor is e^{-rj} rather than $[(e^r - 1)/r]e^{-rj}$.

The net present worth is therefore \$158,700 − 110,000, or \$48,700. Since the value is positive, investment 1 is acceptable.

Investment 2

The present worth can be calculated with the constant, continuous series form from Table 7-5; $\text{PWF} = (P/A, r, N) = [(e^{rN} - 1)/r](e^{rN})$. The recovery amount is \$25,000, and the discount factor is e^{-rN}. For investment 2, the annual cash flow is

$$\text{Cash flow} = 42{,}000 + \frac{(155{,}000)(0.35)}{7} = \$49{,}750$$

Accordingly,

$$\text{Present worth} = 49{,}750(4.465) + 25{,}000(0.376) = \$231{,}500$$

and the net present worth is

$$\text{NPW} = 231{,}500 - 180{,}000 = \$51{,}500$$

The value is positive, so investment 2 is acceptable.

Investment 3

The present worth of investment 3 is

$$\text{Cash flow} = 49{,}000 + \frac{(190{,}000)(0.35)}{8} = \$57{,}300/\text{yr}$$

$$\text{Present worth} = (57{,}300)(4.815) + (35{,}000)(0.326) = \$287{,}400$$

and the net present worth is 287,400 − 225,000 = $62,400. Since this result is positive, investment 3 is acceptable.

e. *Discounted cash flow rate of return.* The DCFR is the value of r that makes the net present worth equal to zero. It is calculated by iterating on r.

Investment 1

The DCFR for investment 1 is the value of r that will make the sum of present worths in the preceding table equal to $110,000, the total investment. By assuming various values of r greater than 0.14 and calculating the sum of the present worths, it is established that a value of $r = 0.33$ is satisfactory. Thus, investment 1 earns at a discounted cash flow rate of return of 33 percent per year.

Investment 2

It is simpler to find the DCFR for this case because a series compounding factor can be used, but it is still iterative. A value of 22.6 percent per year is determined for this investment; it also is acceptable.

Investment 3

This investment earns at a DCFR of 21.5 percent per year, and this also makes it acceptable compared to the 14 percent per year standard.

In summary, investment 1 is found to be acceptable by all five profitability measures, investment 2 is found to be acceptable by all of the measures except the payback period, and investment 3 is found to be acceptable based on the net present worth and discounted cash flow rate of return. Investment 3 is found to be unacceptable by the return on investment, payback period, and net return, that is, by all of the measures that do not consider the time value of money. These different findings occur even though a consistent acceptance standard has been used with each measure. It is not surprising that the methods do not always provide the same conclusion, because different quantities are utilized in the definitions of the profitability measure.

■

The net present worth (including the discounted cash flow rate of return) is recommended as the most suitable measure for assessing profitability, because it includes the largest number of factors affecting profitability.

Effect of Inflation on Profitability Analysis

Inflation is an increase in prices of goods and services over time, and deflation is a decrease in prices. Inflation affects the amount of money required to purchase goods and services. It has nothing to do with the time value of money. Time value of money is a consequence of the earning power of money and relates to the amount of money available, but not the amount required.

Inflation is measured as the percentage or fractional change in price of a quantity with time and is almost universally expressed as percent per year compounded annually; that is, it is an effective annual rate. The effect of inflation on the price of a product or service is shown by the relation

$$\text{Price at } j \text{ years} = (\text{price at 0 years})(1 + i')^j \tag{8-7a}$$

for annual compounding, or for continuous compounding by

$$\text{Price at } j \text{ years} = (\text{price at 0 years})(e^{r'j}) \tag{8-7b}$$

where i' is the annual rate of inflation, j the number of years, and r' the nominal value for continuous compounding that corresponds to i'. As discussed in Chap. 7, $r' = \ln(1 + i')$.

The effects of inflation on project evaluation can best be explained by examining such effects before and after project time zero.

Inflation Before Time Zero Inflation that occurs between the time of an estimate and the time of the actual expenditure for equipment and construction impacts the number of dollars of capital investment required in a project. It also affects the total product cost at time zero as well as the sales price of the product. Since these three quantities are fundamental values for any economic analysis, inflation needs to be included in the profitability analysis to provide the best cost estimates possible. The inflation rates that need to be used must be carefully selected. A capital cost index, such as the CE index described in Chap. 6, can provide a reasonable value to account for inflation of capital investment. However, this inflation rate must be extrapolated to the time at which the capital expenditures are expected to occur. The total product cost includes many items that can be expected to inflate at different rates. In the absence of better information, a single inflation rate, such as the gross national product implicit price deflator index also known as the consumer price index,† can be extrapolated to zero time. Other indexes developed for chemicals and for labor can also be used. Chemical prices typically do not inflate at the general rate of inflation. In fact, they tend to be relatively stable except for significant changes in supply and demand. The product and raw materials prices would be inflated based on the best material-specific information available to estimators. Another choice is not to inflate the sales price of the product at all; this is almost certainly the best choice for an entirely new product. This same strategy may be adopted for raw materials.

EXAMPLE 8-4 Effects of Inflation Prior to Time Zero

Cost estimates have just been completed using current prices for a new project. The projected time schedule for the project assumes that actual construction will be initiated 6 months from now; the first fixed-capital expenditure of 30 percent of the estimated total, inflated at 5 percent per year, would occur 1 year from now; and the remaining 70 percent, also inflated at 5 percent per year, would be expended 2 years from now. The working capital investment is expected to inflate at 4 percent per year

†See a current *Survey of Current Business,* by the Bureau of Economic Analysis, U.S. Department of Commerce.

and will be required in a lump sum 2 years from now. No inflation of the sales revenue is anticipated. The total product cost, excluding depreciation, is expected to inflate at 4 percent per year. Assume that all inflation rates given are for annual compounding. Zero time is to be taken as 2 years from now.

Established current costs:

Fixed-capital investment	\$6.45 million
Working capital investment	\$1.14 million
Annual sales revenue	\$3.88 million
Annual cost of operation (depreciation not included)	\$2.05 million

At the time of plant start-up (2 years from now), estimate the inflated fixed-capital investment required and the working capital investment. Determine the annual cost of operation at plant start-up.

■ Solution

The fixed-capital investment presently is estimated at \$6.45 million. The 30 percent of the investment that is to be expended 1 year from now will have inflated to

$$(6.45 \times 10^6)(0.30)(1.05) = \$2.03 \times 10^6$$

The 70 percent of the investment that is to be used 2 years from now will inflate to

$$(6.45 \times 10^6)(0.7)(1.05)^2 = \$4.98 \times 10^6$$

Thus, the total fixed-capital investment 2 years from now will inflate to

$$(2.03 + 4.98)(10^6) = \$7.01 \times 10^6$$

The working capital investment in 2 years will inflate to

$$(1.14 \times 10^6)(1.04)^2 = \$1.23 \times 10^6$$

The annual cost of operation inflates at 4 percent per year. In 2 years the estimated cost will be

$$(2.05 \times 10^6)(1.04)^2 = \$2.22 \times 10^6$$

Inflation After Time Zero Product prices and all costs of operation except depreciation are subject to changes due to inflation over the period of plant operation. Such changes will impact the cash flow resulting from operation and will directly affect the profitability of the project. However, once a plant has been constructed, the capital investment is a firm value that does not inflate. New investments may be added over time; these will be made at the then-prevailing costs, but they, too, are firm once they have been made. Depreciation charges, which are fractions of the fixed-capital investment, therefore do not inflate or deflate over time. However, the charges may change with time, due to the depreciation schedule that is selected. In fact, tax law does not permit any inflation of depreciation; only the initial actual dollar amount of fixed-capital investment can be depreciated.

To understand the effects of inflation that occur after plant operation begins, the equation for cash flow must be examined. From Eq. (6-1) the cash flow in year j is

$$A_j = (s_j - c_{oj})(1 - \Phi) + \Phi d_j \tag{8-8}$$

The effect of inflation on dollar amounts is found by applying the inflation rate to year 0 values to obtain year j values, as demonstrated by Eq. (8-7a), to give

$$A_j = [s_o(1+i')^j - c_{o,o}(1+i')^j](1-\Phi) + \Phi d_j \tag{8-9}$$

where s_o is the sales revenue at time zero and $c_{o,o}$ is the total product cost not including depreciation at time zero. Depreciation is not inflated for the reasons stated above. As Eq. (8-9) illustrates, the effect of inflation is to increase the sales and total product cost (without depreciation) by the same factor, if the same inflation rate is used for both. Thus, a process that has a positive cash flow at zero time has even more positive annual and total cash flows in future years, if the same inflation rate is used for all terms. This approach is known as using *current* dollars, that is, the amount of dollars expressed in the year in question.

Usually of greater concern, though, is the purchasing power of the cash flow at time zero of the analysis. The purchasing power at time zero is the number of dollars it would take at time zero to purchase the same amount of goods and services as A_j dollars would purchase at time j. This is obtained by dividing Eq. (8-9) by $(1+i')^j$, which results in

$$\frac{A_j}{(1+i')^j} = A_{j,o} = (s_o - c_{o,o})(1-\Phi) + \frac{\Phi d_j}{(1+i')^j} \tag{8-10}$$

where $A_{j,o}$ is the purchasing power of the jth-year cash flow in time zero dollars. The only quantity on the right-hand side of Eq. (8-10) that is affected by the time zero computation is the depreciation, because the inflation factor has been canceled out of the other terms. Notice that the effect of inflation is to *decrease* the dollar amount of the zero-time equivalent of the cash flow by an amount $\Phi d_j[1-(1+i')^{-j}]$ which, for an inflation rate of 5 percent, varies from 0 to 3 percent of the total capital investment. Over a 10-year project life, this decrease averages around 1 percent of $\mathcal{T}$.

Because of this rather small effect due to inflation and the uncertainty in predicting the inflation rate, a common approach in project economic evaluations is to entirely ignore inflation after time zero, or, put another way, to perform the entire analysis in *constant,* zero-time dollars. In periods of double-digit inflation, this would not be advisable. Another approach is to perform a sensitivity analysis using different inflation rates for various components of revenue and total product cost. In particular, it is common to use one inflation rate for the product and a different inflation rate for the raw materials.

EXAMPLE 8-5 Inflation After Time Zero

Use the sales and cost values determined in Example 8-4 at time zero. Assume straight-line depreciation for 6 years with no salvage value. Use the same inflation rates as in Example 8-4. The income tax rate is 35 percent.

a. Determine the inflated annual and total net profits and cash flows for the project over an 8-year period.

b. Find the annual and total net profits and cash flows in zero-time dollars.

c. Obtain the annual and total net profits and cash flows assuming no inflation after zero time.

d. Find the return on investment for the three previous cases.

■ **Solution**

For all three cases the annual depreciation for a 6-yr service life is $(7.01 \times 10^6)/6 = \$1.17 \times 10^6$. The zero-time cost values are

$$s_o = \$3.88 \times 10^6 \qquad c_{o,o} = \$2.22 \times 10^6 \qquad d_j = \$1.17 \times 10^6$$

Assume $\Phi = 0.35$.

a. The net profit and cash flow for these inflated conditions are given by

$$N_{p,j} = (s_j - c_{oj} - d_j)(1 - \Phi) = [s_o - c_{o,o}(1 + i')^j - d_j](1 - \Phi)$$

$$A_j = N_{p,j} + d_j$$

b. The net profit and cash flow for part (a) in zero-year dollars are obtained from

$$N_{p,j} = \left[\frac{s_j}{(1 + i')^j} - c_{o,o} - \frac{d_j}{(1 + i')^j}\right](1 - \Phi)$$

$$A_j = N_{p,j} + \frac{d_j}{(1 + i')^j}$$

c. The net profit and cash flow assuming no inflation after time zero are

$$N_{p,j} = (s_o - c_{o,o} - d_j)(1 - \Phi)$$

$$A_j = N_{p,j} + d_j$$

The results for all three cases are tabulated below.

Summary of net profit and cash flow calculations in 10^6 dollars

				Case (a)		Case (b)		Case (c)	
Year	s_j	c_{oj}	d_j	$N_{p,j}$	A_j	$N_{p,j}$	A_j	$N_{p,j}$	A_j
0	3.88	2.22	1.17	0.32	1.49	0.32	1.49	0.32	1.49
1	3.88	2.31	1.17	0.26	1.43	0.25	1.38	0.32	1.49
2	3.88	2.40	1.17	0.20	1.37	0.19	1.27	0.32	1.49
3	3.88	2.50	1.17	0.14	1.31	0.12	1.16	0.32	1.49
4	3.88	2.60	1.17	0.08	1.25	0.06	1.06	0.32	1.49
5	3.88	2.70	1.17	0.01	1.18	0.01	0.97	0.32	1.49
6	3.88	2.81	1.16	−0.06	1.10	−0.04	0.87	0.33	1.49
7	3.88	2.92	0	0.63	0.62	0.48	0.47	1.08	1.08
8	3.88	3.04	0	0.55	0.55	0.40	0.40	1.08	1.08
Total (1–8)	31.04	21.28	7.01	1.81	8.81	1.47	7.58	4.09	11.10

d. For case (a): $\text{ROI} = \dfrac{100(1.81/8)}{7.01} = 3.2\%$

For case (b): $\text{ROI} = \dfrac{100(1.47/8)}{7.01} = 2.6\%$

For case (c): $\text{ROI} = \dfrac{100(4.09/8)}{7.01} = 7.3\%$

The effect of inflation on ROI is readily apparent. This is because all costs *except* the product cost were inflated. This shows the importance of doing sensitivity studies on the effect of inflation rates.

■

Start-up Costs

After plant construction has been completed, there is a start-up period during which all units are tested for operational characteristics and safety. The start-up period process is continued until the desired operating conditions with proper control have been achieved. Depending on the complexity of the plant and the experience of the engineering and operating personnel, this period may last from a few weeks to as long as a year with little acceptable product being produced. In the overall cost analysis, start-up expenses over and above normal operating expenses are represented as a one-time-only expenditure in the first year of the plant operation. These start-up expenses can vary from 5 to 20 percent of the fixed-capital investment. A typical value is 10 percent. The production rate for the first year of operation of a new plant is usually estimated as a fraction of the annual design capacity because of unproductive time at start-up, and because start-up may begin at any time of year. In preliminary economic estimates, it is common to use a fraction of one-half of annual capacity for the first year. This fraction, of course, has a profound impact on the sales revenue and variable expenses for that period, and it may be changed as a start-up date becomes more accurately known.

During start-up, it sometimes becomes apparent that equipment changes have to be made before the plant can operate at maximum design conditions. These changes involve expenditures for materials and equipment that become part of the capital investment. It is usually expected that such investments will be relatively small and are to be covered from the contingency allowance; but a separate investment entry for these items may be included in the estimate if it is desired to monitor their expense. Losses in product revenue due to such changes are usually included in the capacity allowance mentioned in the previous paragraph.

Spreadsheet for Economic Evaluation Calculations

Figure 8-1 shows the structure of a model spreadsheet for performing the economic evaluation calculations described above. The user of the spreadsheet inputs basic data from, for example, the spreadsheets in Figs. 6-7 and 6-8. Minimal required user inputs are at the top of Fig. 8-1 and in items 1, 9, and 11. The user may modify default values and relations as needed for a particular project. The five profitability measures considered here are calculated and tabulated near the bottom of the spreadsheet.

ALTERNATIVE INVESTMENTS

In industrial operations, it is often possible to produce equivalent products in different ways. Although the physical results may be approximately the same, the capital required and the expenses involved can vary considerably depending on the particular method chosen. Similarly, alternative methods that require various levels of capital and operating systems can often be used to carry out other types of business ventures. It may be necessary, therefore, to decide not only if a given business venture

All data through items 11 are user-supplied, except as indicated by a default value.

Inputs: Design capacity, kg/yr (produced during the projected number of operated days per year)
Minimum acceptable rate of return (as an after-tax, discrete annual fraction per year)
Number of years in evaluation period, N
Discount rate, fraction/yr

Enter expenditures as negative numbers and incomes as positive numbers.
Numbers in parentheses refer to line numbers.

	Default value	Year j from −3 to 20	Row sum −3 to 20
1. Fixed capital investment, $			
2. Working capital, $	0.176 (1)		
3. Salvage value, $	0		
4. Total capital investment, $	(1 + 2)		
5. Annual investment, $	0		
6. Start-up cost, $	0.1(1)		
7. Operating rate, % of capacity per year	$0, j \leqq 0$ $50, j = 1$ $90, j = 2$ $100, j = 3, \ldots, N$		
8. Annual sales, $	(sales @ 100% capacity)(7)		
9. Annual manufacturing cost, $			
a. Raw materials			
b. Labor			
c. Utilities			
d. Maintenance and repair			
e. Operating supplies			
f. Royalties			
g. Catalysts and solvents			
h. Depreciation	MACRS		
i. Property taxes and insurance			
j. Interest			
k. Plant overhead			
l. Other	0		
10. Total of (9)	calculated		
11. Annual general expenses			
a. Administration			
b. Distribution and marketing			
c. Research and development			
d. Other	0		
12. Total of (11)	calculated		
13. Total product cost			
14. Annual gross profit			
15. Annual net profit			
16. Annual operating cash flow			
17. Total annual cash flow			
18. Average return on investment, %/yr			
19. Payback period			
20. Net return			
21. Discount factors			
22. Annual present worth, $/yr			
23. Net present worth			
24. Discounted cash flow rate of return			

Figure 8-1
Spreadsheet for economic evaluation calculations

would be profitable, but also which of several possible methods would be the most desirable.

The final decision as to the best among alternative investments is simplified if it is recognized that each dollar of additional investment should yield an adequate rate of return. In practical situations, there are usually a limited number of choices, and the alternatives must be compared on the basis of incremental increases in the necessary capital investment.

The following simple example illustrates the principle of investment comparison. A chemical company is considering adding a new production unit that will require a total investment of $1,200,000 and yield an annual after-tax profit of $240,000. An alternative addition has been proposed requiring an investment of $2 million that will yield an annual after-tax profit of $300,000. Although both proposals are based on reliable estimates, the company executives feel that other equally sound investments can be made with at least a 14 percent annual after-tax rate of return. Therefore, the minimum acceptable rate of return for the new investment is 14 percent after taxes.

The rate of return on the $1,200,000 unit is 20 percent, and that for the alternative addition is 15 percent. Both of these returns exceed the minimum required value, and it might appear that the $2 million investment should be recommended because it yields the greater amount of profit per year. However, a comparison of the incremental investment between the two proposals shows that the extra investment of $800,000 gives a profit of only $60,000, or an incremental return of 7.5 percent. Therefore, if the company had $2 million to invest, it would be more profitable to accept the $1,200,000 proposal and put the other $800,000 in another investment at the indicated 14 percent return.

A general rule for making comparisons of alternative investments can be stated as follows: *The minimum investment which will give the necessary functional results and the required rate of return should always be accepted unless there is a specific reason for accepting an alternative investment requiring more initial capital.* When alternatives are available, therefore, the base plan would be that one requiring the minimum acceptable investment. The alternatives should be compared with the base plan, and additional capital should not be invested unless an acceptable incremental return or some other distinct advantage can be shown.

Alternatives When an Investment Must Be Made

The design engineer often encounters situations where it is absolutely necessary to make an investment and the only choice available is among various alternatives. An example might be found in the design of a plant that requires the use of an evaporation system. The latter must have a given capacity based on the plant requirements, but there are several alternative methods for carrying out the operation. A single-effect evaporator would be satisfactory. However, the operating expenses would be less if a multiple-effect evaporator were used principally because of the reduction in steam consumption. Under these conditions, the optimum number of effects can be determined

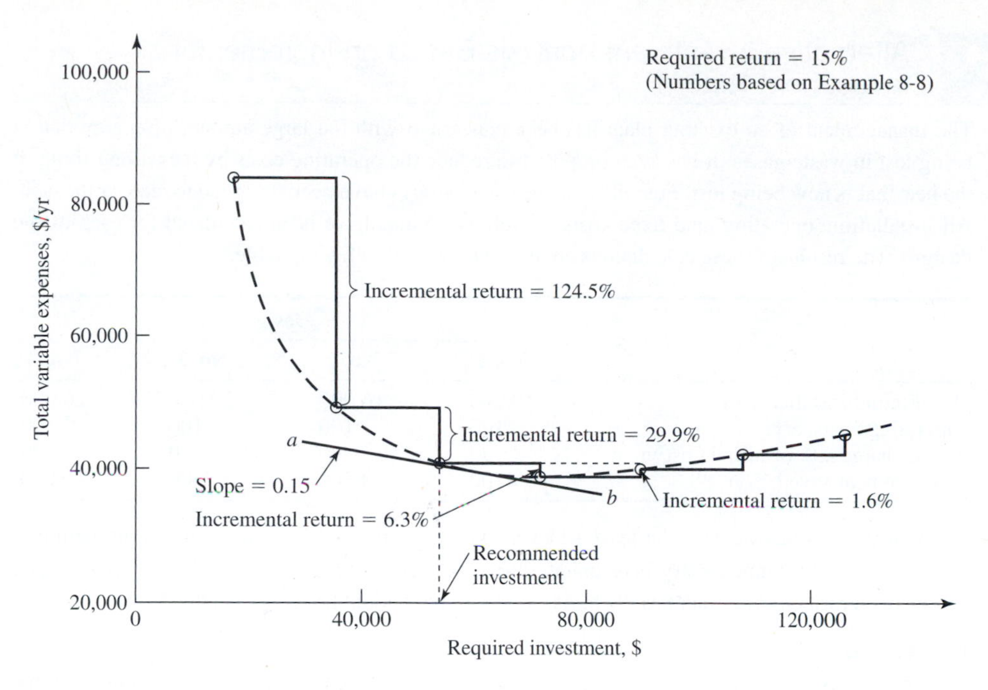

Figure 8-2
Comparison of alternative investments when one investment must be made for a given service and there are a limited number of choices

by comparing the increased savings associated with the added investment required for each additional effect. A graphical representation showing this kind of investment comparison is presented in Fig. 8-2.

The base plan for an alternative comparison of the type discussed in the preceding paragraph would be the minimum investment that gives the necessary functional results. The alternatives should then be compared with the base plan, and an additional investment would be recommended only if it gave a definite advantage.

When investment comparisons are made, alternatives requiring more initial capital are compared only with lower investments *that have been found to be acceptable.* Consider an example in which an investment of $50,000 will give a desired result, while alternative investments of $60,000 and $70,000 will give the same result with less annual expense. Suppose that comparison between the $60,000 and the $50,000 cases shows the $60,000 investment to be unacceptable. Certainly, there would be no reason to give further consideration to the $60,000 investment, and the next comparison should be between the $70,000 and the $50,000 cases. In making a choice among various alternative investments, it is necessary to recognize the need to compare one investment to another on a mutually exclusive basis in such a manner that the *return on the incremental investment* is satisfactory. Example 8-6 illustrates this principle.

EXAMPLE 8-6 Alternative Investment Analysis Based on Incremental Investment Return

The management of an existing plant has been concerned with the large amount of energy that is being lost in waste gases. It has been proposed to reduce the operating costs by recovering some of the heat that is now being lost. Four different heat exchangers have been designed to recover the heat. All installation, operating, and fixed costs as well as savings have been calculated for each of the designs. The results of these calculations are presented in the following table.

	Design			
	No. 1	**No. 2**	**No. 3**	**No. 4**
Total initial installed cost, \$	10,000	16,000	20,000	26,000
Operating costs, \$/yr	100	100	100	100
Fixed charges, % of initial cost/yr	20	20	20	20
Value of heat saved, \$/yr	4100	6300	7300	8850

Company policy demands at least a 15 percent annual return before taxes based on the initial investment for any unnecessary investment. Only one of the four designs can be accepted. Using before-tax return on investment as the basis, which design should be recommended?

■ Solution

The first step is to determine the amount of money saved per year for each design, from which the annual percent return on the initial investment can be determined. The net annual savings equals the value of heat saved minus the sum of the operating costs and fixed charges.

Design 1

$$\text{Annual savings} = 4100 - (0.2)(10{,}000) - 100 = \$2000$$

$$\text{Annual percent return} = \frac{2000}{10{,}000}(100) = 20\%$$

Design 2

$$\text{Annual savings} = 6300 - (0.2)(16{,}000) - 100 = \$3000$$

$$\text{Annual percent return} = \frac{3000}{16{,}000}(100) = 18.8\%$$

Design 3

$$\text{Annual savings} = 7300 - (0.2)(20{,}000) - 100 = \$3200$$

$$\text{Annual percent return} = \frac{3200}{20{,}000}(100) = 16\%$$

Design 4

$$\text{Annual savings} = 8850 - (0.2)(26{,}000) - 100 = \$3550$$

$$\text{Annual percent return} = \frac{3550}{26{,}000}(100) = 13.7\%$$

Because the indicated percent return for each of the first three designs is above the minimum of 15 percent required by the company, any one of these three designs would be acceptable. Now it becomes necessary to choose one of the three alternatives. One way to do this is by using the method of return on incremental investment.

Analysis by means of return on incremental investment is accomplished by a logical step-by-step comparison of an acceptable investment to another investment that might be better. If design 1 is taken as the starting basis, comparison of design 2 to design 1 shows that an annual saving of $\$3000 - \$2000 = \$1000$ is obtained with an additional investment of \$6000. Thus, the percent return on the incremental investment is $(1000/6000)(100) = 16.7$ percent. Design 2 is acceptable by company policy in preference to design 1.

Comparing design 2 to design 3, the annual percent return is $(200/4000)(100) = 5$ percent. Thus, design 3 compared to design 2 shows that the incremental return is unacceptable, and design 2 is the preferred alternative.

Identical results to the preceding can be obtained by selecting the alternative that provides the greatest annual profit or saving if the required return is included as a cost for each case, as in the net return method. This is demonstrated in Example 8-7.

Alternative Investment Analysis Incorporating Minimum Return as a Cost — EXAMPLE 8-7

For the alternative investment analysis of the heat exchangers in Example 8-6, the annual cost for the required return would be 15 percent of the total initial investment. Thus,

Design 1 The annual savings above the required return is

$$2000 - (0.15)(10{,}000) = \$500$$

Design 2 The annual savings above the required return is

$$3000 - (0.15)(16{,}000) = \$600$$

Design 3 The annual savings above the required return is

$$3200 - (0.15)(20{,}000) = \$200$$

Design 4 The annual savings above the required return is

$$3550 - (0.15)(26{,}000) = -\$350$$

Because annual saving is greatest for design 2, this would be the recommended alternative and is the same result as was obtained by the analysis based on return on incremental investment.

These last two simplified examples, 8-6 and 8-7, have been used to illustrate the basic concepts involved in making comparisons of alternative investments. The approach was based on using the simple return on initial investment in which the time value of money is neglected. Although this method may be satisfactory for preliminary and approximate estimations, for final evaluations a more sophisticated approach is needed in which the time value of money is considered along with other practical factors, to ensure the best possible chance for future success. More advanced approaches of this type are presented in the following sections.

Analysis with Small Investment Increments

The design engineer often encounters the situation in which the addition of small investment increments is possible. For example, in the design of a heat exchanger for recovering waste heat, each square meter of additional heat-transfer area can cause a reduction in the amount of heat lost; but the amount of heat recovered per square meter of heat-transfer area decreases as the area is increased. Since the investment for the heat exchanger is a function of the heat-transfer area, a plot of net savings (or net profit due to the heat exchanger) versus total investment can be developed. This results in a smoothed curve of the type shown in Fig. 8-3.

The point of maximum net savings, indicated by *O* in Fig. 8-3, represents a classical optimum condition. However, the last incremental investment before this maximum point is attained is at essentially a 0 percent return. On the basis of alternative investment comparisons, therefore, some investment less than that for maximum net savings should be recommended.

The exact investment where the incremental return is a given value occurs at the point where the slope of the curve of net savings versus investment equals the required return. Thus, a straight line with a slope equal to the necessary return is tangent to the net savings versus investment curve at the point representing the recommended investment. Such a line for an annual return on incremental investment of 20 percent is shown in Fig. 8-3, and the recommended investment for this case is indicated by *RI*. If an analytical expression relating net savings and investment is available, it is obvious that the

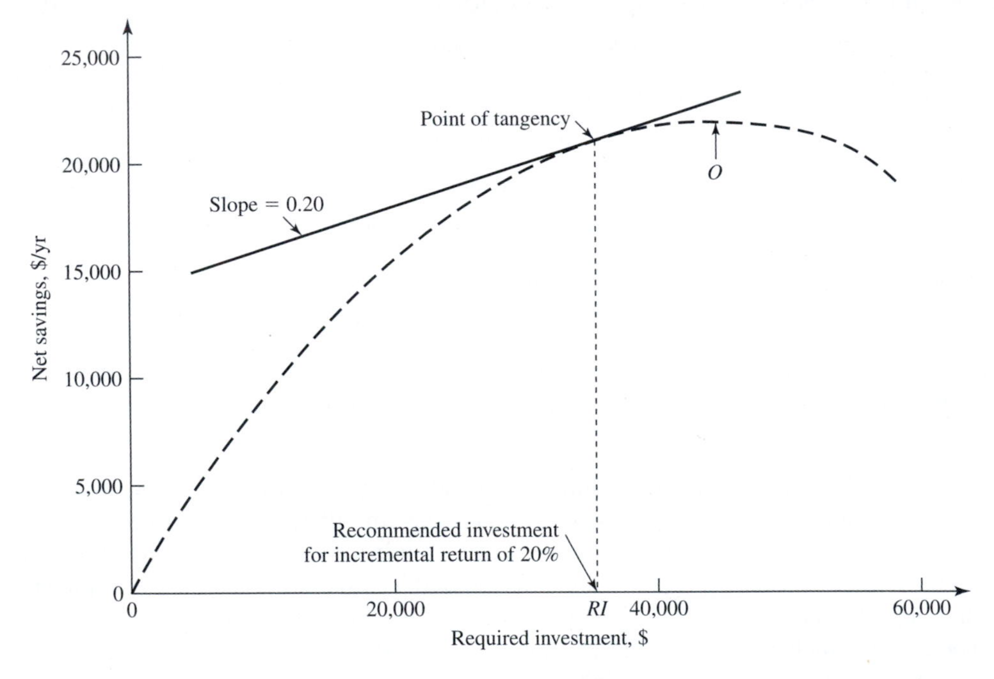

Figure 8-3
Graphical method for determining investment for a given incremental return when investment increments can approach zero

recommended investment can be determined directly by merely setting the derivative of the net savings with respect to the investment equal to the required incremental return.

The method described in the preceding paragraph can also be used for continuous curves of the type represented by the dashed curve in Fig. 8-2. Thus, the line *ab* in Fig. 8-2 is tangent to the dashed curve at the point representing the recommended investment for the case of a 15 percent incremental return.

Investment Comparison for Required Operation with Limited Number of Choices — EXAMPLE 8-8

A plant is being designed in which 204,000 kg per 24-h day of a water–caustic soda solution containing 5 percent by weight caustic soda must be concentrated to 40 percent by weight. A single-effect or multiple-effect evaporator will be used, and a single-effect evaporator of the required capacity requires an initial investment of \$18,000. This same investment is required for each additional effect. The service life is estimated to be 10 years, and the salvage value of each effect at the end of the service life is estimated to be \$6000. Fixed charges minus depreciation amount to 20 percent yearly, based on the initial investment. Steam costs \$1.32/10^3 kg, and administration, labor, and miscellaneous costs are \$40 per day, no matter how many evaporator effects are used.

If X is the number of evaporator effects, $0.9X$ equals the number of kilograms of water evaporated per kilogram of steam for this type of evaporator. Assume that there are 300 operating days per year and an income tax rate of 35 percent. If the minimum acceptable return after taxes on any investment is 15 percent, how many effects should be used?

■ Solution

Basis: 1 operating day. Let X equal the total number of evaporator effects. Depreciation per operating day (straight-line method) is

$$\frac{X(18{,}000 - 6{,}000)}{10(300)} = \$4.00X/\text{day}$$

$$\text{Fixed charges} - \text{depreciation} = \frac{X(18{,}000)(0.20)}{300} = \$12.00X/\text{day}$$

The amount of water evaporated per day is given by

$$(204{,}000)(0.05)\left(\frac{95}{5}\right) - (204{,}000)(0.05)\left(\frac{60}{40}\right) = 178{,}500\ \text{kg/day}$$

$$\text{Steam costs} = \frac{(178{,}500)(1.32)}{X(0.9)(1000)} = \$261.80/(X/\text{day})$$

Summary of operating costs associated with number of effects used

No. of effects X	Steam costs per day	Fixed charges minus depreciation per day	Depreciation per day	Labor, etc., per day	Total cost per day
1	\$261.8	\$12	\$ 4	\$40	\$317.8
2	130.9	24	8	40	202.9
3	87.3	36	12	40	175.3
4	65.5	48	16	40	169.5
5	52.4	60	20	40	172.4
6	43.6	72	24	40	179.6

Comparing two effects with one effect gives

$$\text{Percent return} = \frac{(317.8 - 202.9)(300)(100)(0.65)}{36{,}000 - 18{,}000} = 124.5\%$$

Therefore, two effects are better than one effect.

Comparing three effects with two effects gives

$$\text{Percent return} = \frac{(202.9 - 175.3)(300)(100)(0.65)}{54{,}000 - 36{,}000} = 29.9\%$$

Therefore, three effects are better than two effects.

Comparing four effects with three effects yields

$$\text{Percent return} = \frac{(175.3 - 169.5)(300)(100)(0.65)}{72{,}000 - 54{,}000} = 6.3\%$$

Since a return of at least 15 percent is required on any investment, three effects are better than four effects, and the four-effect evaporator should receive no further consideration.

Comparing five effects with three effects, we get

$$\text{Percent return} = \frac{(175.3 - 172.4)(300)(100)(0.65)}{90{,}000 - 54{,}000} = 1.6\%$$

Therefore, three effects are better than five effects.

Since the total daily costs continue to increase as the number of effects is increased above five, no further comparisons need to be made.

A three-effect evaporator should be used.

■

REPLACEMENTS

The term *replacement,* as used in this chapter, refers to a special type of alternative in which facilities are currently in existence and it may be desirable to replace these facilities with different ones. Although intangible factors may have a strong influence on decisions relative to replacements, the design engineer must understand the tangible economic implications when a recommendation is made as to whether existing equipment or facilities should be replaced. The reasons for making replacements can be divided into two general classes, as follows:

1. An existing property *must* be replaced or changed in order to continue operation and meet the required demands for service or production. Some examples of reasons for this type of necessary replacement are that
   a. The property is worn out and can give no further useful service.
   b. The property does not have sufficient capacity to meet the demand placed upon it.
   c. Operation of the property is no longer economically feasible because changes in design or product requirements have caused the property to become obsolete.

2. The existing property is capable of yielding the necessary product or service, but more efficient equipment or property is available which can operate with lower expenses.

When the reason for a replacement is of the first general type, the only alternatives are to make the necessary changes or else go out of business. Under these conditions, the final economic analysis is usually reduced to a comparison of alternative investments.

The correct decision as to the desirability of replacing an existing property that is capable of yielding the necessary product or service depends on a clear understanding of theoretical replacement policies plus a careful consideration of many practical factors. To determine whether a change is advisable, the operating expenses with the present equipment must be compared with those that would exist if the change were made. Practical considerations, such as amount of capital available or benefits to be gained by meeting a competitor's standards, may also have an important effect on the final decision.

Methods of Profitability Evaluation for Replacements

The same methods that were presented and applied earlier in this chapter are applicable for replacement analyses. Net present worth and discounted cash flow methods give the most reliable results for maximizing the overall future worth of an investment. However, for the purpose of explaining the *basic principles* of replacement economic analyses, the simple rate of return on investment method of analysis is just as effective as those methods involving the time value of money. Thus, to permit the use of direct illustrations which will not detract from the principles under consideration, the following analysis of methods for economic decisions on replacements uses the annual rate of return on the initial investment as the profitability measure. The identical principles can be treated by more complex rate of return and net present worth solutions by merely applying the methods described earlier in this chapter.

Typical Example of Replacement Policy

The following example illustrates the type of economic analysis involved in determining if a replacement should be made. A company is using a piece of equipment that originally cost $30,000. The equipment has been in use for 5 years and presently has a net realizable value of $6000. At the time of installation, the service life was estimated to be 10 years and the salvage value at the end of the service life was estimated to be zero. Operating costs amount to $22,000 per year. At present, the remaining service life of the equipment is estimated to be 3 years.

A proposal has been made to replace the present piece of property by one of more advanced design. The proposed equipment would cost $40,000, and the operating costs would be $15,000 per year. The service life is estimated to be 10 years with a zero salvage value. Each piece of equipment will perform the same service, and all costs other than those for operation and depreciation will remain constant. Depreciation cost is determined by the straight-line method. The company will not make any unnecessary investments in equipment unless it can obtain an annual return on the necessary capital of at least 10 percent.

The two alternatives in this example are to continue the use of the present equipment or to make the suggested replacement. To choose the better alternative, it is necessary to consider both the reduction in expenses obtainable by making the change and the amount of new capital necessary. The only variable expenses are those for operation and depreciation. Therefore, the annual variable expenses for the proposed equipment would be $15,000 + $40,000/10, or $19,000.

The annual cost of continuing to operate the old equipment is $22,000 + $30,000/10, or $25,000. An annual saving of $25,000 − $19,000, or $6000, can be realized by making the replacement. The cost of the new equipment is $40,000, but the sale of the existing property would provide $6000; therefore, it would be necessary to invest only $34,000 to bring about an annual saving of $6000. Since this represents a return greater than 10 percent, the existing equipment should be replaced.

Book Values and Unamortized Values

In the preceding example, the book value of the existing property was $15,000 at the time of the proposed replacement. However, this fact was given no consideration in determining whether the replacement should be made. The book value is based on past estimates, but the decision as to the desirability of making a replacement must be based on present conditions. The difference between the book value and the net realizable value at any time is commonly designated as the *unamortized value*.

Although unamortized values have no part in a replacement study, they must be accounted for in some manner. One method for handling these capital losses or gains is to charge them directly to the profit-and-loss account for the current operating period. When a considerable loss is involved, this method may have an unfavorable and unbalanced effect on the profits for the current period. Many companies protect themselves against such unfavorable effects by building up a surplus reserve fund to handle unamortized values. This fund is accumulated by setting aside a certain sum in each accounting period. When losses due to unamortized values occur, they are charged against the accumulated fund. In this manner, unamortized values can be handled with no serious effects on the periodic profits.

Investment on which the Replacement Comparison Is Based

The advisability of making a replacement is usually determined by the rate of return that can be realized from the necessary investment. It is, therefore, important to consider the amount of the investment. *The difference between the total cost of the replacement property and the net realizable value of the existing property equals the necessary investment.* Thus, the correct determination of the investment involves only consideration of the present capital outlay required if the replacement is made.

In replacement studies, the *net realizable value* of an existing property should be assumed to be the market value. Although this may be less than the actual value of the property as far as the owner is concerned, it still represents the amount of capital that can be obtained from the old equipment if the replacement is made. Any attempt to assign to an existing property a value greater than the net realizable value tends to favor replacements which are uneconomical.

An Extreme Situation to Illustrate Result of Replacement Economic Analysis — EXAMPLE 8-9

A new manufacturing unit has just been constructed and put into operation by a midwestern company. The basis of the manufacturing process is a special computer for control (designated as the OVT computer) as developed by the research department. The plant has now been in operation for less than one week and is performing according to expectations. However, a new computer (designated as the NTR computer) has just become available on the market. This new computer can easily be installed to replace the present computer and will provide the same process control at considerably less annual cash expense because of reduced maintenance and personnel costs. However, if the new computer is installed, the present computer is essentially worthless because there is no other use for it.

The pertinent economic information relative to the two computers is shown here:

	OVT computer	NTR computer
Capital investment, $	2,000,000	1,000,000
Estimated economic life, years	10	10
Salvage value at end of economic life, $	0	0
Annual cash expenses, $	250,000	50,000

What recommendation should be made relative to replacing the present $2,000,000 computer with the new computer?

■ Solution

The before-tax annual cost of continuing to operate the OVT computer is the cash expense plus depreciation, or $250{,}000 + 2{,}000{,}000/10 = \$450{,}000$.

The total capital investment for the OVT computer is completely lost if the NTR computer is installed; thus, the total investment in the NTR is $1,000,000. The after-tax savings based on a 35 percent tax rate will be $(450{,}000)(0.65) - [50{,}000 + (1{,}000{,}000/10)]0.65$, or $195,000. The rate of return on this investment is $(195{,}000/1{,}000{,}000)(100)$, or 19.5 percent. This after-tax return is most likely satisfactory to the company for a safe, though unnecessary, investment. If so, then the replacement should be recommended. If the return is not satisfactory, then the possibility of a corporate capital gains tax reduction on the capital loss for the OVT computer should be investigated and factored into the calculation.

Because of the reduced costs for the NTR computer, a profitability evaluation method that includes the time value of money will tend to favor the replacement even more than the rate of return method on investment used in the solution of this example.

■

PRACTICAL FACTORS IN ALTERNATIVE-INVESTMENT AND REPLACEMENT ANALYSIS

The discussions to this point have presented the theoretical viewpoint of alternative-investment and replacement analyses. However, certain practical considerations also influence the final decision. In many cases, the amount of available capital is limited. Accordingly, it may be desirable to accept the smallest investment that will give the necessary service and provide the required return. Although a greater investment might be better on a theoretical basis, the additional return would not be worth the extra risks involved when capital must be borrowed or obtained from some other outside source.

A second practical factor that should be considered is the accuracy of the estimates used in determining the rates of return. All risk factors should be given careful consideration before any investment is made, and the risk factors should receive particular attention before an investment greater than that absolutely necessary is accepted.

The economic conditions existing at the particular time have an important practical effect on the final decision. In depression periods or in times when economic conditions are very uncertain, it may be advisable to refrain from investing any more capital than is absolutely necessary. The tax situation for the corporation can also have an effect on the decision.

Many intangible factors enter into the final decision on a proposed investment. Sometimes it may be desirable to impress the general public by the external appearance of a property or by some unnecessary treatment of the final product. These advertising benefits would probably receive no consideration in a theoretical economic analysis, but they certainly could influence management's final decision in choosing one investment over another.

Personal whims or prejudices, desire to better a competitor's rate of production or standards, the availability of excess capital, and the urge to expand an existing plant are other practical factors that may be involved in determining whether a particular investment will be made. Theoretical analyses of alternative investments and replacements can be used to obtain a dollar-and-cents indication of what should be done about a proposed investment. The final decision depends on these theoretical results plus practical factors determined by existing circumstances.

NOMENCLATURE

A_j = cash flow in year j, dollars
A_x = nonmanufacturing fixed-capital investment, dollars
$-b$ = year in which investment is made, dimensionless
c_{oj} = all product cost except depreciation in year j, dollars
d_j = depreciation charge in year j, dollars
DCFR = discounted cash flow rate, percent/100
i = interest rate based on length of one interest period, percent/100
i' = annual rate of inflation, percent/100
j = year under consideration, dimensionless
m_{ar} = minimum acceptable annual rate of return, percent/100
N = evaluation period for economic analysis, years

N_p = annual net profit, dollars; subscript p,ave refers to average annual net profit; subscript p,j indicates net profit in year j
NPW = net present worth, dollars
p = price of product, dollars/kg
P_r = product rate, kg/yr
PBP = payback period, years
PWF = present worth factor, most forms are dimensionless
r = nominal interest rate for compounding other than annual, percent/100
r_{ma} = minimum acceptable rate of return for continuous compounding that corresponds to m_{ar}, percent/100
r' = nominal rate of interest for continuous compounding that corresponds to i', percent/100
rec_j = recovery of working capital and physical assets in year j, dollars
R_n = net return, dollars; subscript ave designates average net return
ROI = return on investment, percent or percent/100
s_j = value of sales in year j, dollars
$\mathcal{T}$ = total capital investment, dollars
V = manufacturing capital investment, dollars

Greek Symbol

Φ = income tax rate, percent/100

PROBLEMS

8-1 What total amount of funds before taxes will be available 10 years from now if $10,000 is placed in a savings account earning an interest rate of 6 percent compounded monthly? How many years will be required for this amount to double at the same interest rate compounded semiannually? What is the shortest time in years for the doubling to occur if continuous compounding is available?

8-2 A proposed chemical plant will require a fixed-capital investment of $10 million. It is estimated that the working capital will be 25 percent of the total investment. Annual depreciation costs are estimated to be 10 percent of the fixed-capital investment. If the annual profit will be $3 million, determine the percent return on the total investment and the payout period.

8-3 An investigation of a proposed investment has been made. The following result has been presented to management: The payback period is 5 years. Annual depreciation is 10 percent per year of the fixed-capital investment; and fixed-capital investment is 85 percent of total capital investment. Using this information, determine the rate of return on the investment.

8-4 Two pumps are being considered for pumping water from a reservoir. Installed cost and salvage value for the two pumps are given below:

	Pump A	Pump B
Installed cost	$20,000	$25,000
Salvage value	2,000	4,000

Pump A has a service life of 4 years. Determine the service life of pump B at which the two pumps are competitive if the annual effective interest rate is 15 percent. Competitiveness refers to the requirement that the installed cost of the pumps plus the amount that must be invested at the time of installation so that sufficient interest will be earned over the service life (when added to the salvage value) to replace the pumps at the original cost.

8-5 A heat exchanger has been designed, and insulation is being considered for the unit. The insulation can be obtained in thicknesses of 0.025, 0.051, 0.076, or 0.102 m. The following data have been determined for the differzent insulation thicknesses:

	0.025 m	0.051 m	0.076 m	0.102 m
kJ/s energy saved	88	102	108	111
Cost for installed insulation	$8,000	$10,100	$11,100	$11,500
Annual fixed charges, % of installed cost	10	10	10	10

What thickness of insulation should be used? The value of heat is $1.50/GJ. An annual after-tax return of 15 percent on the fixed-capital investment is required for any capital utilized in this type of investment. The income tax rate is 35 percent/yr. The exchanger operates for 300 days/yr.

8-6 A design engineer is evaluating two pumps for handling a corrosive solution. The information on the pumps is the following:

	Pump A	Pump B
Installed cost	$15,000	$22,000
Service life, years	2	5

Determine the annual interest rate at which the two pumps are competitive. Neglect salvage value. See Prob. 8-4 for the definition of competitiveness. Which pump would you recommend?

8-7 A company must purchase one reactor to be used in an overall operation. Four reactors have been designed, all of which are equally capable of giving the required service. The following data apply to the four designs:

	Design 1	Design 2	Design 3	Design 4
Fixed-capital investment	$10,000	$12,000	$14,000	$16,000
Sum of after-tax operating and fixed costs per year (all other costs are constant)	3,000	2,800	2,350	2,100

If the company demands a 15 percent return after taxes on any unnecessary investment, which of the four designs should be accepted?

8-8 In the design of a chemical plant, the following expenditures and revenues are estimated after the plant has achieved its desired production rate:

Total capital investment	$10,000,000
Working capital	$1,000,000
Annual sales	$8,000,000/yr
Annual expenditures	$2,000,000/yr

Assuming straight-line depreciation over a 10-year project analysis period, determine

a. The return on the investment after taxes

b. The payback period

8-9 An existing warehouse is worth \$500,000, and the average value of the goods stored in the warehouse is \$400,000. The annual insurance rate on the warehouse is 1.1 percent, and the insurance rate on the stored goods is 0.95 percent. If a proposed sprinkling system is installed in the warehouse, both insurance rates will be reduced to three-quarters of the original rate. The installed sprinkler system will cost \$20,000, while the additional annual cost of maintenance, inspection, and taxes will be \$300. The required depreciation period for the entire investment in the sprinkler system is 20 years. The capital necessary to make the investment is available. The operation of the warehouse is now providing an 8 percent return on the original investment. Give reasons why you would or would not recommend installing the sprinkler system.

8-10 A proposed chemical plant has the following projected revenues and operating expenses in millions of dollars

Year	Annual revenue	Annual operating expenses (excluding depreciation)
1	7.0	4.0
2	10.0	5.6
3	15.0	6.8
4	20.0	7.8
5	22.5	8.8
6	24.0	9.6
7	25.0	10.0

The fixed-capital investment for the plant is \$50 million with a working capital of \$7.5 million. Using a MACRS depreciation schedule with a class life of 5 years, determine

a. The annual cash flows

b. The net present worth, using a nominal discount rate of 15 percent

c. The DCFR

8-11 A power plant for generating electricity is part of a plant design proposal. Two alternative power plants with the necessary capacity have been suggested. One uses a boiler and steam turbine while the other uses a gas turbine. The following information applies to the two proposals:

	Boiler and steam turbine	Gas turbine
Initial investment, \$	600,000	400,000
Fuel costs per year, \$	160,000	230,000
Maintenance and repairs per year, \$	12,000	15,000
Insurance and taxes per year, \$	18,000	12,000
Depreciation recovery period, yr	20	10
Salvage value at end of service life, \$	0	0

All other costs are the same for either type of power plant. A 12 percent return is required on any investment. If one of these power plants must be accepted, which one should be recommended?

8-12 The facilities of an existing chemical company must be increased if the company is to continue in operation. There are two alternatives. One of the alternatives is to expand the present plant. The expansion would cost $130,000. Additional labor costs would be $150,000 per year, while additional costs for overhead, depreciation, property taxes, and insurance would be $60,000 per year.

The second alternative requires construction and operation of new facilities at a location about 50 mi from the present plant. This alternative is attractive because cheaper labor is available at this location. The new facilities would cost $200,000. Labor costs would be $120,000 per year. Overhead costs would be $70,000 per year. Annual insurance and property taxes would amount to 2 percent of the initial cost. All other costs except depreciation would be the same at each location. If the minimum acceptable return on any unnecessary investment is 9 percent per year after an income tax of 35 percent, determine the minimum recovery period for the facilities at the distant location for this alternative to meet the required incremental return. The salvage value should be assumed to be zero, and straight-line depreciation may be used.

8-13 A chemical company is considering replacing a batch reactor with a continuous reactor. The old unit cost $40,000 when new 5 years ago, and depreciation has been charged on a straight-line basis using an estimated service life of 10 years with a final salvage value of $1000.

The new unit would cost $70,000. It would save $15,000 per year in expenses not including depreciation. The straight-line depreciation period is taken to be 10 years with a zero salvage value. All costs other than those for labor, insurance, taxes, and depreciation may be assumed to be the same for both units. The old unit can now be sold for $5000. Income tax is 35 percent per year. If the after-tax minimum acceptable return on any investment is 15 percent, should the replacement be made?

8-14 A project is being considered that requires $1,000,000 for fixed-capital investment and $100,000 for working capital. The fixed capital is depreciated on a straight-line basis to a book value of zero at the end of the fifth year. The annual revenue in those 5 years is $500,000. The total product cost not including depreciation is $100,000 annually. The discount rate is 10 percent, and the income taxation rate is 35 percent. Develop a spreadsheet that shows the annual cash flow, the discounted cash flow, and the net present worth for each year. Treat the investments as occurring in a lump sum at zero time. The revenues and expenses occur continuously and utilize continuous compounding.

Now assume an inflation rate of 5 percent on both the revenues and the expenses. Again, develop another spreadsheet that shows the annual cash flow, the discounted cash flow, and the net present worth for each year.

8-15 The owner of a small antifreeze plant has a small canning unit which cost him $5000 when he purchased it 10 years ago. The unit has completely depreciated, but the owner estimates that it will still give him good service for 5 more years. At the end of 5 years the unit will have a salvage value of zero. The owner now has an opportunity to buy a more efficient canning unit for $6000 having an estimated service life of 10 years and zero salvage value. This new unit would reduce annual labor and maintenance costs by $1000 and increase annual expenses for taxes and insurance by $100. All other expenses except depreciation would be unchanged. If the old canning unit can be sold for $600, what replacement return on his capital investment will the owner receive if he decides to make the replacement?

8-16 An engineer in charge of the design of a plant must choose either a batch or a continuous system. The batch system offers a lower initial outlay but, because of higher labor requirements, exhibits a higher operating cost. The cash flows relevant to this decision have been estimated as follows:

	Year		Discounted cash flow rate of return	Net present worth at 10%
	0	1–10 (after taxes)		
Batch system	−$20,000	$5600/year	25%	$14,400
Continuous system	−$30,000	$7650/year	22%	$17,000

Check the values given for the discounted cash flow rate of return and net present worth. If the company requires a minimum rate of return of 10 percent, which system should be chosen?

8-17 An oil company is offered a lease of a group of oil wells on which the primary reserves are close to exhaustion. The major condition of the purchase is that the oil company agree to undertake a water flood project at the end of 5 years to undertake possible secondary recovery. No immediate payment by the oil company is required. The relevant cash flows have been estimated as follows:

Year				Discounted cash flow rate of return	Net present worth at 10%
0	1–4	5	6–20		
0	$50,000	−$650,000	$100,000	?	$242,000

Continuous, constant cash flows were used except for the expenditure that occurs in one sum at the end of year 5. Continuous discounting at 10 percent per year was used for all cash flows. Check the net present worth value. Should the lease and flood arrangement be accepted? How should this proposal be presented to the company board of directors who understand and make it a policy to evaluate proposals by using the discounted cash flow rate of return method?

8-18 A process with a depreciable capital investment of $100 million is to be constructed over a 3-year period. At start-up, $20 million of working capital is required. The plant is expected to operate for 10 years. At full capacity expected for the third and subsequent years of operation, the sales revenues are projected to be $150 million per year, and the total operating expenses, excluding depreciation, are projected to be $100 million per year. During the first and second years of operation, the sales revenues are anticipated to be 50 and 75 percent of the sales revenues projected in the third and subsequent years, respectively. The operating expenses during the first and second years will be the same as in the third and subsequent years. Assume that the income tax rate is 35 percent. Using the third year as a basis, determine

a. The return on the investment after taxes

b. The payback period

8-19 Assuming that the construction of the plant in Prob. 8-18 requires investments of $20 million during the first year, $30 million in the second year, and $50 million during the third year, evaluate the annual net present worth and the total net present worth of the project. Assume that the construction costs are continuous throughout the 3 years of construction. Use continuous, constant cash flows and continuous discounting.

CHAPTER

9

Optimum Design and Design Strategy

In engineering process design, the criteria for optimality can ultimately be reduced to a consideration of costs or profits.[†] The factors affecting the economic performance of the design include the types of processing technique used, processing equipment used, arrangement and sequencing of the processing equipment used in the design, and the actual physical parameters for the equipment. Of course, the operating conditions for the equipment are also of prime concern and import. Thus the optimum for a process design is the most cost-effective selection, arrangement, sequencing of processing equipment, and operating conditions for the design.

Optimization of process design follows the general aspects and processes for optimization in general. First the optimization criteria must be established; this is done by using an objective function that is an economic performance measure. Next, the optimization problem must be defined. This requires the establishment of the various mathematical relations and limitations that describe the aspects of the design. These relations are used to explore their effects on the objective function, and then to determine the optimal values of the parameters that affect the economic performance of the design.

Although cost considerations and economic balances are the basis of most optimum designs, at times factors other than cost can determine the most favorable conditions. For example, in the operation of a catalytic reactor, an optimum operation temperature may exist for each reactor size because of equilibrium or reaction rate limitations. This particular temperature could be based on the maximum percentage conversion or on the maximum amount of final product per unit of time. Ultimately, however, cost variables need to be considered, and the development of an optimum operation design is usually merely one step in the determination of an optimum economic design.

[†]T. F. Edgar, D. M. Himmelblau, and L. S. Lasdon, *Optimization of Chemical Processes,* 2d ed., McGraw-Hill, New York, 2001, p. 4.

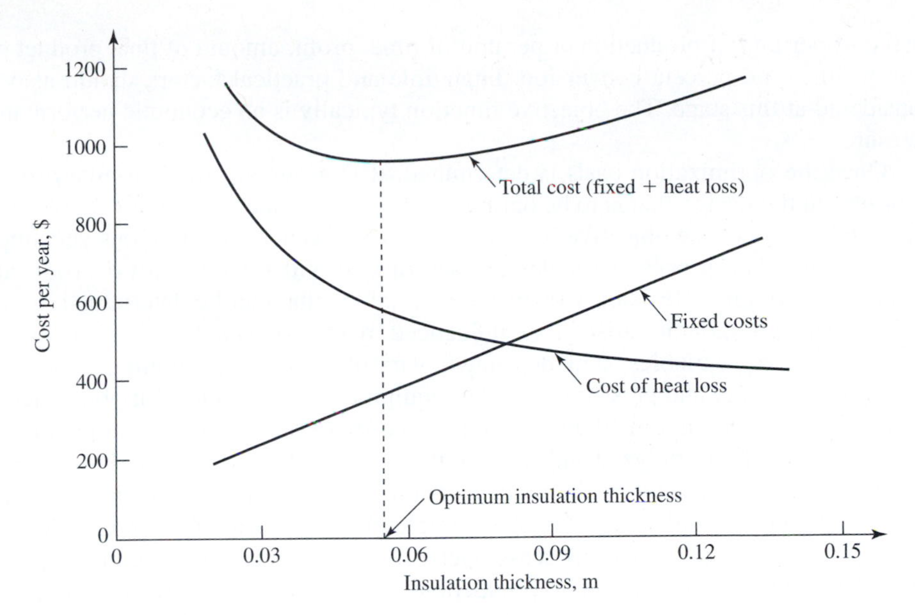

Figure 9-1
Illustration of the basic principle of an optimum design

When a design variable is changed, often some costs increase and others decrease. Under these conditions, the total cost may go through a minimum at one value of the particular design variable, and this value is the optimum value of that variable. An example illustrating the principles of an optimum economic design is presented in Fig. 9-1. In this simple case, the problem is to determine the optimum thickness of insulation for a given steam pipe installation. As the insulation thickness is increased, the annual fixed costs increase, the cost of heat loss decreases, and all other costs remain constant. Therefore, as shown in Fig. 9-1, the sum of the costs goes through a minimum, and this determines the optimum insulation thickness.

Although profitability or cost is generally the basis for optimization, practical and intangible factors usually need to be included as well in the final investment decision. Such factors are often difficult or impossible to quantify, and so decision maker judgment must weigh such factors in the final analysis.

Optimum design analysis begins with the establishment of the optimization task. This requires the selection of an economic criterion that is to be the objective function. Next, the process is examined to determine variables and constraints that affect process performance and the objective function. A process model with appropriate cost and economic data is then used, guided by an extremum-seeking strategy, to find the optimum conditions.

DEFINING THE OPTIMIZATION PROBLEM

The first phase in the development of an optimum design is to determine the objective function that is to be maximized or minimized as well as the structural and parametric variables and constraints that affect the objective function. Such factors would be total cost per unit of production or per unit of time, profit, amount of final product per unit of time, and percent conversion. Intangible and practical factors should also be considered at this stage. The objective function typically is an economic performance measure.

Once the optimization basis is determined, it is necessary to determine *process variables* in the design that is to be optimized. Process variables are the variables that affect the values of the objective functions. These are generally numerous yet simple and concrete. The process variables are scrutinized and divided into *decision* and *dependent* variables. Decision variables are variables that can be determined, or set, while dependent variables arise or are influenced by *process constraints*. Examples of process decision variables, and dependent variables include operating conditions such as temperature and pressure as well as equipment specification variables such as the number of trays in a distillation column. These are the topologic and parametric variables that will be analyzed and set for optimum design performance. Process constraints are the constraints on the process. These are sometimes so obvious they are overlooked, but having these defined completely is critical. Common examples of process constraints include process operability limits, reaction chemical species dependence, or product purity and production rate. In addition to domain considerations (linearity), variables, objective functions, and constraints are further qualified according to their range (continuous or discrete). Continuous variables may vary continuously over a certain range. These include such variables as temperature and flow rates that can be set to vary smoothly from zero and up. Conversely, discrete factors are limited to a discrete group of values. An example of discrete variables is set-position switches that can be set to a fraction of a maximum throughput value. An especially common subset of discrete variables is that of variables that can take on only integer values. An example of integer variables would be the number of process equipment pieces, where fractions of equipment are meaningless. It is important to note another commonly encountered subclassification of integer discrete variables—binary-type variables. These are variables that are limited to two possible values, most often off/on, closed/open, and yes/no. It is worthwhile to note that noninteger discrete variables can be described as the product of a value and a group of integers and are often represented thus for programming and analysis purposes. (A discrete distribution group of 1, $\frac{3}{4}$, $\frac{1}{2}$, and $\frac{1}{4}$ can be described as the group 4, 3, 2, and 1 multiplied by $\frac{1}{4}$.)

The large number of variables that are generally involved in any design can often be reduced through screening analysis of a variable's relative impact on design performance and costs.† Further simplification may also be possible through suboptimization of sections of the overall design, either as a basis for subsequent overall design optimization or for direct analysis.

†Variable effects of screening practices are described in any number of experimental analysis and statistics texts, e.g., R. L. Mason, R. F. Gunst, and J. L. Hess, *Statistical Design and Analysis of Experiments,* J. Wiley, New York, 1989, and G. E. Box, W. G. Hunter, and J. S. Hunter, *Statistics for Experiments,* J. Wiley, New York, 1978, and D. C. Montgomery, *Design and Analysis of Experiments,* 5th ed., J. Wiley, New York, 2001.

SELECTING AN OBJECTIVE FUNCTION

The quantity to be maximized or minimized in an optimization is designated as the *objective function*. The definition of this objective function is fundamental to the success of the optimization effort. The objective function is nearly always an economic function. In Chap. 8, five economic criteria for rating investments were discussed. Each of the five has advantages and disadvantages for judging the economic performance of an investment. What is the "best" investment strategy? On the basis of the profitability analysis discussed earlier, it is proposed that the best strategy is one that favors those investments that meet or exceed an established minimum acceptable rate of return for each investment, but that also ensures that no investment is made which does not meet that acceptable return standard. In particular, any incremental investment must meet the minimum acceptable return standard. When several discrete investment alternatives are being compared, an effort should be made first to identify the smallest investment that meets the minimum acceptable rate of return standard. This should be followed by comparing the return on the incremental investment for each alternative to the best previous alternative to determine whether the incremental investment meets the prescribed standard.

The methodologies of optimization are based on obtaining a maximum or minimum in the objective function. For optimization purposes, the methods of return on investment, payback period, and discounted cash flow rate of return are not ideal, since it may be possible to invest more than indicated at the maximum or minimum of these measures and still earn a rate that is equal to or above the return standard. The other two profitability methods—net return and net present worth—are recommended as the basis for the objective function in optimization problems. Of these two, the net present worth method is more strongly recommended because it includes the time value of money.

Structural Optimization

Optimization is often divided into two separate fields for convenience, parametric and structural optimization. Topologic optimization concerns itself with equipment selection alternatives, arrangement, and inter- and intraconnectivity. This translates to such issues as potential arrangements of processing steps, considerations of alternative processing techniques, and considerations of alternative types of both individual and groupings of processing equipment.

Structural optimization consists in finding one or more "best" flowsheets for a process. Considerable attention already has been given to the subject of developing flowsheets in Chap. 4. Candidate operation diagrams are identified by using heuristic or algorithmic procedures. Equipment is selected by primarily heuristic methods. The arrangement and sequencing of the equipment can be done heuristically, algorithmically, or by a combination of the two. These procedures can yield a relatively small set of flowsheets that satisfy the requirements of the process. Selection of one or several "best" flowsheets is done by economic comparisons of these candidates. This entire procedure constitutes topographic optimization, but since the first three steps were covered earlier, only the final step of economic comparisons is considered.

Ideally, topographic and parametric optimization would be conducted simultaneously. However, the methodologies and the computing power are not yet adequate to do this for any but simple processes. When the number of feasible flowsheets is on the order of 10 or fewer, and reasonable operating conditions can be specified for each flowsheet, then direct economic comparisons can be made. An appropriate objective function, such as net present worth, is to be maximized, or a cost function is to be minimized. If the number of flowsheets and their parameters are sufficiently small, the operating parameters of each flowsheet can be optimized by MINLP methodology. These optimized flowsheets can then be compared to determine which is best economically. When the number of feasible flowsheets is large, it is possible to eliminate infeasible and noneconomic flowsheets simultaneously by using the accelerated branch and bound method.† This variation of the branch and bound method discussed later in this chapter utilizes peculiar characteristics of chemical processes to significantly reduce the scope of the search. It does not, however, optimize the parameters for individual flowsheets, so the method might eliminate flowsheets that would prove acceptable if they were optimized.

Typically, a few near optimal flowsheets are obtained by establishing several feasible flowsheets and then comparing their net present worths or annualized costs. One or more of these is then subjected to parametric optimization, and the most economically attractive one is selected.

Parametric Optimization

Parametric optimization deals with process operating variables and equipment design variables other than those strictly related to structural concerns. Some of the more obvious examples of such decisions are operating conditions, recycle ratios, and stream properties such as flow rates and compositions. In addition to the decision variables common to most processing equipment, every type and variant of process equipment may have from one to several unique decision variables. Thus, parametric optimization problems often involve hundreds of decision variables, including many that are common to multiple pieces of equipment such as operating pressure and temperature. Optimization of all decision variables is therefore not realistic. Instead, it is more efficient to analyze the magnitude of the effects these decision variables have on process performance and economics and to select only the more influential variables. In this way a balance is struck between the increased difficulty of high-variable-number optimization and optimization accuracy.

Variable Screening and Selection

The selection of the more influential key decision variables for optimization can be made by analyzing the principles underlying the process and consulting optimization

†F. Friedler, J. B. Varga, E. Feher, and L. T. Fan, in *State of the Art in Global Optimization, Computational Methods and Applications,* C. A. Floudas and P. M. Pardalos, eds., Kluwer Academic Publishers, Norwell, MA, 1996, p. 609.

literature. However, this type of selection is qualitative and therefore can lead to overlooking influential variables or including unnecessary variables and thus needlessly complicating the optimization requirements. A more quantitative selection is possible if process models are available. Process models can then be used to analyze the relative effects of decision variables on process performance using statistical factor screening practices.† The model analysis variable selection route is more involved and time-consuming, but it does lead to quantifiable results. It also has the added benefit of revealing variable interaction effects, which are difficult to determine qualitatively. Further, this method does not require a complete model and can use partial process or single-equipment-piece models.

Much can be learned about the most significant economic aspect of a design by examining the equipment and operating costs associated with a base-case design. This will reveal those areas of the design that have the more significant equipment and operating costs. It is these design areas that then offer the greatest opportunity for affecting costs and improving profitability and in which optimization efforts have the greatest impact.

SUBOPTIMIZATION

Simultaneous optimization of the many parameters present in a chemical process design can be a daunting task, even with a computer-based system. This is so because of the large number of variables that can be present in both integer and continuous form, the nonlinearity of the property prediction relationships and performance models, and the frequent ubiquity of recycle. It is therefore common to seek out suboptimizations for some of the variables, so as to reduce the dimensionality of the problem. To suboptimize requires taking a subset of the entire design and optimizing it without regard to the rest of the process. It is possible to do this if the subproblem can be optimized without influencing the rest of the process. Although this may never be entirely true, there are instances for which it is valid in a practical, economic sense. It is common to optimize subsystems whose primary function is to receive feeds with fixed states (temperature, pressure, composition, and flow rate) from the rest of the process and deliver products back to the process with fixed states. Examples are pumps, compressors, heat exchange networks, and separation trains. As long as the fixed states of the inputs and outputs of these systems *do not change significantly* as a result of the optimization applied to the rest of the process, the suboptimization is effectively independent. Note that it is still possible that the suboptimization may have a significant impact on utility needs, for example, in which case the changes in the cost of supplying the utilities should be included in the suboptimization.

†Statistical factor screening practices are discussed in most higher-level statistics and experimental analysis texts, including these: R. L. Mason, R. F. Gunst, and J. L. Hess, *Statistical Design and Analysis of Experiments,* J. Wiley, New York, 1989, G. E. Box, W. G. Hunter, and J. S. Hunter, *Statistics for Experiments,* J. Wiley, New York, 1978, and D. C. Montgomery, *Design and Analysis of Experiments,* 5th ed., J. Wiley, New York, 2001.

Optimization of parameters in a chemical process design is usually attempted for a fixed plant capacity. When this is the case, it is common not to include the sales revenue in the analysis and to minimize the annual costs (costs taken as positive in such a case) rather than to maximize the net present worth. In fact, only those costs that actually change with respect to the optimization variable need to be considered. This is quite appropriate; but to do it in a manner consistent with the net present worth methodology, the variable costs must be included in the form in which they appear in the expression for net present worth, as given by Eq. (8-4).

Practical and Intangible Considerations

The quantitative methods of optimization presented in this chapter are highly useful. However, there are practical and intangible factors that may be difficult to quantify or to include in mathematical optimization methods yet which can significantly impact the decision of what is the best design. Such factors are weighed into investment decisions by methods other than those discussed in this chapter.

Practical considerations are realities and uncertainties that may be difficult to include in an optimization procedure or that require special treatment or judgment. Realities such as discrete sizes for equipment and auxiliaries and the lower cost of off-the-shelf equipment can be included in integer methods of mathematical programming. Other realities, such as the benefits of early entry of a product into a market, financing windows of opportunity and limitations, and fabrication and construction lead times may be difficult to program, but can be very important to the success of a project. Uncertainties include such factors as future economic and market conditions, dependability of raw material and intermediate supplies, and unproven technologies. These and other realities and uncertainties may be factored into investment decisions by sensitivity analysis, by adjustments to the minimum acceptable rate of return, and by decision maker judgment.

Intangible considerations are factors that are not quantifiable. Some examples are corporate image, investors' perceptions, community and labor relations, and compatibility with corporate product lines and policies. Such factors may be important in corporate planning and decision making, and the inclusion of such considerations in investment decisions is one of the main responsibilities of top corporate officers.

PROGRAMMING OPTIMIZATION PROBLEMS

The term *programming* as used in design optimization literally means optimization.† Application programming is the practice of assembling the mathematical relationships that describe a design process and its economic behavior and then using these relationships to explore and optimize the design.

†S. J. Wright, in *5th International Conference on Foundations of Computer-Aided Process Design,* M. F. Malone, J. T. Trainham, and B. Carnaham, eds., American Institute of Chemical Engineers (CACHE), New York, 2000.

There are no generalized systematic programming practices that can be applied to all designs. Instead, a number of programming methods have been developed for the various kinds of optimization situations most commonly encountered. The types of programming are categorized according to a number of classifications.[†] The first differentiation is by the range of the mathematical relations of the problem, separating programming into continuous and discrete. This categorization yields two optimization situations, *linear integer programming* (LIP) and *mixed-integer linear programming* (MILP). These are situations where all or some of the variables required for the optimization solution must be integers, or can be represented by integers as previously stated.

Next comes the question of the need for constraints in the programming formulation. Real engineering optimization situations are by necessity constrained, and therefore most programming approaches are constrained. However, for analysis and optimization purposes, many problems are treated as unconstrained, and constrained solution methods are often extensions of unconstrained approaches. Optimization problems combining discrete or continuous and constrained characteristics give rise to *stochastic programming*. Stochastic programming methods are particularly useful for dealing with uncertain processes that exhibit a certain degree of uncertainty or to evaluate process flexibility. They are also in general less applicable to process design optimization.

A further subclass of continuous constrained programming is *bound constrained programming*. These are situations where some variables or relations are artificially bound. A useful development of these methods is called *branch and bound* (BB). This is one of the most used programming approaches for both linear and nonlinear integer and mixed-integer situations and is used in some form or capacity in most commercial software for these applications. The essence of the BB method is to convert all evaluation variables to binary form, resulting in a set of sequential binary linear programming subproblems, or branches. If either of these subproblems has a noninteger solution, it is further converted to two subproblems. The latter often is termed *relaxation*. Integer solutions in turn are evaluated to see if they produce better objective function values than the best achieved to this point. This is the bounding part of BB.[‡] The optimal value of the subproblems is the lower bound of the original problem. If this value is equal to or greater than the best value so far, then the current branching is not the path to the optimum and the search retraces to previous variable value investigation paths. Nonfeasible subproblems are, of course, disregarded.

Another widely used continuous constrained optimization type is *linear programming*. Linear programming is widely used because of both its relative simplicity and wide availability of good general-purpose linear optimization algorithms and

---

[†]NEOS Guide, www-fp.mcs.anl.gov/otc/Guide/OptWeb/index.html.

[‡]Dedicated programming publications such as these can provide further guidelines to subproblem creation, relaxation, and bounding: G. L. Nemhauser and L. A. Wosley, *Integer and Combinatorial Optimization,* J. Wiley, New York, 1988, or I. E. Grossman and J. Hooker, *5th International Conference on Foundations of Computer-Aided Process Design,* M. F. Malone, J. T. Trainham, and B. Carnaham, eds., American Institute of Chemical Engineers (CACHE), New York, 2000.

practices, and because many real-world situations can be converted to linear behavior without too much difficulty or loss of accuracy. Further, the reduction to linear programming often does not result in acuity losses, especially in cases where model and data accuracy is inexact or not well known.†

The other widely used continuous constrained optimization approach is *nonlinear programming* (NLP). This category of programming is the most commonly encountered situation combining nonlinear objective functions or constraints. These are generally demanding and difficult to implement, but are the most useful. The relations of the various programming approaches are illustrated in Fig. 9-2. Together, linear and nonlinear programming, especially in the integer (LIP and NLIP) and mixed-integer (MILP and MINLP) forms, accounts for the large majority of all optimization approaches. These are therefore expanded upon later in the chapter.

Figure 9-2
Programming approach methodology, divisions, and relations

†T. F. Edgar, D. M. Himmelblau, and L. S. Lasdon, *Optimization of Chemical Processes,* 2d ed., McGraw-Hill, New York, 2001, p. 223.

Linear Programming

One strategy for simplifying the approach to a programming problem is based on expressing the constraints and the objective function in a linear mathematical form. The *straight-line,* or linear, expressions are stated mathematically as

$$av_1 + bv_2 + \cdots + jv_j + \cdots + nv_n = b \tag{9-1}$$

where the coefficients $a, \ldots, n$ and b are known values and $v_1, \ldots, v_n$ are unknown variables. With two variables, the result is a straight line on a two-dimensional plot, while a plane in a three-dimensional plot results for the case of three variables. Similarly, for more than three variables, the geometric result is a hyperplane. The general procedure mentioned in the preceding paragraph is designated as *linear programming.* It is a mathematical technique for determining optimum conditions for allocation of resources and operating capabilities to attain a definite objective. It is also useful for analysis of alternative uses of resources or alternative objectives.

As an example to illustrate the basic methods involved in linear programming for determining optimum conditions, consider the following simplified problem. A brewery has received an order for 400 liters of beer with the constraints that the beer must contain 4 percent alcohol by volume, and it must be supplied immediately. The brewery wishes to fill the order, but does not have 4 percent beer in stock. It is decided to mix two beers now in stock to give the desired final product. One of the beers in stock (beer A) contains 4.5 percent alcohol by volume and is valued at \$0.09 per liter. The other beer in stock (beer B) contains 3.7 percent alcohol by volume and is valued at \$0.07 per liter. Water (W) can be added to the blend at no cost. What volume combination of the two beers in stock with water, including at least 40 liters of beer A, will give the minimum ingredient cost for the 400 liters of 4 percent beer?

This example is greatly simplified because only a few constraints are involved and there are only three independent variables, that is, volume of beer A (V_A), volume of beer B (V_B), and volume of water (V_W). When a large number of possible choices are involved, the optimum set of choices may be far from obvious, and a solution by linear programming may be the best way to approach the problem. A step-by-step rational approach is needed for linear programming. This general rational approach is outlined in the following with application to the blending example cited.

Analytic Approach to Problems Involving Linear Programming A systematic rationalization of a problem being solved by linear programming can be broken down into the following steps:

1. *A systematic description of the limitations or constraints.* For the brewery example, the constraints are as follows:

   a. Total volume of product is 400 liters, or

$$V_W + V_A + V_B = 400 \tag{9-2}$$

   b. Product must contain 4 percent alcohol, or

$$0.0V_W + 4.5V_A + 3.7V_B = 4.0(400) \tag{9-3}$$

c. Volume of water and beer B must be zero or greater than zero, while volume of beer A must be 40 liters or greater; that is,

$$V_W \geq 0 \qquad V_B \geq 0 \qquad V_A \geq 40 \qquad \text{or} \qquad V_A - S = 40 \tag{9-4}$$

where S is the *slack variable*.

2. *A systematic description of the objective.* In the brewery example, the objective is to minimize the cost of the ingredients; that is, the objective function is

$$C = \text{cost} = \text{a minimum} = 0.0V_W + 0.09V_A + 0.07V_B \tag{9-5}$$

3. *Combination of the constraint conditions and the objective function to choose the best result out of many possibilities.* One way to do this would be to use an intuitive approach whereby every reasonable possibility would be considered to give ultimately, by trial and error, the best result. This approach would be effective for the brewery example because of its simplicity. However, the intuitive approach is not satisfactory for most practical situations, and linear programming can be used. The computations commonly become so involved that a computer is required for the final solution. If a solution is so simple that a computer is not needed, linear programming will probably not be needed. To illustrate the basic principles, the brewery example is solved in the following by linear programming including intuitive solution, graphical solution, and computer solution.

Previous relations yield the following linearized basic equations:

$$V_W + V_A + V_B = 400$$

$$4.5V_A + 3.7V_B = 1600$$

$$0.09V_A + 0.07V_B = C = \text{minimum value}$$

where C is the objective function.

Combining the first two relations

$$V_A = 150 + 4.625V_W \tag{9-6}$$

we see that the optimum must fall on the line represented by this relation. Combining the first relation with the objective function gives

$$V_A = 50C + 3.5V_W - 1400 \tag{9-7}$$

Intuitive Approach to Problems Involving Linear Programming It can be seen intuitively, from Eqs. (9-6) and (9-7), that the minimum value of the objective function C occurs when V_W is zero. Therefore, the optimum value of V_A from Eq. (9-6) is 150 liters, and the optimum value of V_B from Eq. (9-2) is 250 liters.

Graphical Approach to Problems Involving Linear Programming Figure 9-3 is the graphical representation of this problem. Line OE represents the overall constraint placed on the problem by Eqs. (9-2), (9-3), and (9-4). The parallel dashed lines

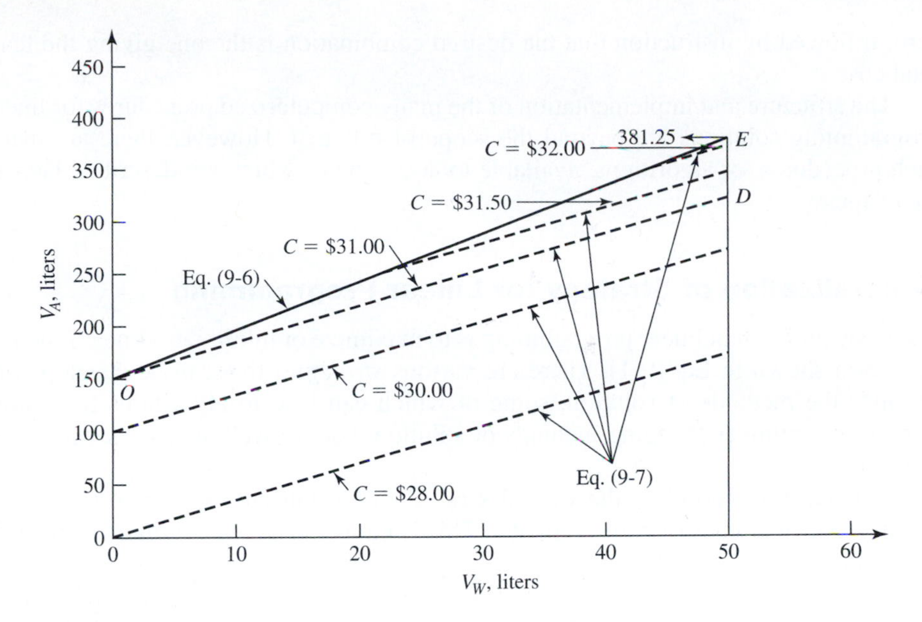

Figure 9-3
Graphical representation of linear programming solution based on a brewery example

represent possible conditions of cost. The goal of the program is to minimize cost (i.e., minimize C) while still remaining within the constraints of the problem. The minimum value of C that still meets the constraints occurs for the line OD, and the optimum must be at point O. Thus, the recommended blend is no water, 150 liters of A, 250 liters of B, for a total cost C of \$31.00 for 400 liters of blend.

Computerized Approach to Problems Involving Linear Programming Although the simplicity of this problem makes it trivial to use a computer for solution, the following is presented to illustrate the basic type of reasoning involved in developing a computer program for the linearized system.

An iterative procedure must be used for the computer solution to permit the computer to make calculations for repeated possibilities until the minimum objective function C is attained. In this case, there are four variables (V_A, V_B, V_W and S) and three nonzero constraints (total volume, final alcohol content, and $V_A = 40 + S$). Because the number of real variables cannot exceed the number of nonzero constraints, one of the four variables must be zero.† Thus, one approach for a computer solution merely involves solving a 4×3 matrix with each variable alternatively being set equal to

†For proof of this statement, see any book on linear programming, for example, S. I. Gass, *Linear Programming: Methods and Applications,* 3d ed., McGraw-Hill, New York, 1969, p. 54.

zero, followed by instruction that the desired combination is the one giving the least total cost.[†]

The structure and implementation of the many computerized procedures for linear programming solutions are beyond the scope of this text. However, there are many such procedures, or algorithms, available today, some of which are described later in the chapter.

Generalization of Strategy for Linear Programming

The basic problem in linear programming is to maximize or minimize a linear function of a form shown in Eq. (9-1). There are various *strategies* that can be developed to simplify the methods of solution, some of which can lead to algorithms that allow rote or pure number-plugging methods of solution that are well adapted for machine solution.

In linear programming, the variables $v_1, \ldots, v_n$ are usually restricted (or can be transformed) to values of zero or greater. This is known as a *nonnegativity restriction* on v_j; that is,

$$v_j \geq 0 \qquad j = 1, 2, \ldots, n$$

Consider a simple two-dimensional problem such as the following: The objective function is to maximize

$$3v_1 + 4v_2 \tag{9-8}$$

subject to the linear constraints of

$$2v_1 + 5v_2 \leq 10 \tag{9-9}$$

$$4v_1 + 3v_2 \leq 12 \tag{9-10}$$

$$v_1 \geq 0 \tag{9-11}$$

$$v_2 \geq 0 \tag{9-12}$$

This problem and its solution are pictured graphically in Fig. 9-4, which shows that the answer is $3v_1 + 4v_2 = 11$. From Fig. 9-4 it can be seen that the linear constraints, in the form of inequalities, restrict the solution region to the crosshatched area. This solution region is a polygon designated as *convex* because all points on the line between any two points in the crosshatched region are in the set of points that

---

[†]For more about computerized procedures to optimization problems, see algorithm design and numerical optimization texts such as C. A. Floudas, *Deterministic Global Optimization: Theory, Methods, and Applications,* Kluwer Academic Press, Norwell, MA, 2000; and J. Nocedal and S. J. Wright, *Numerical Optimization,* Springer, New York, 1999.

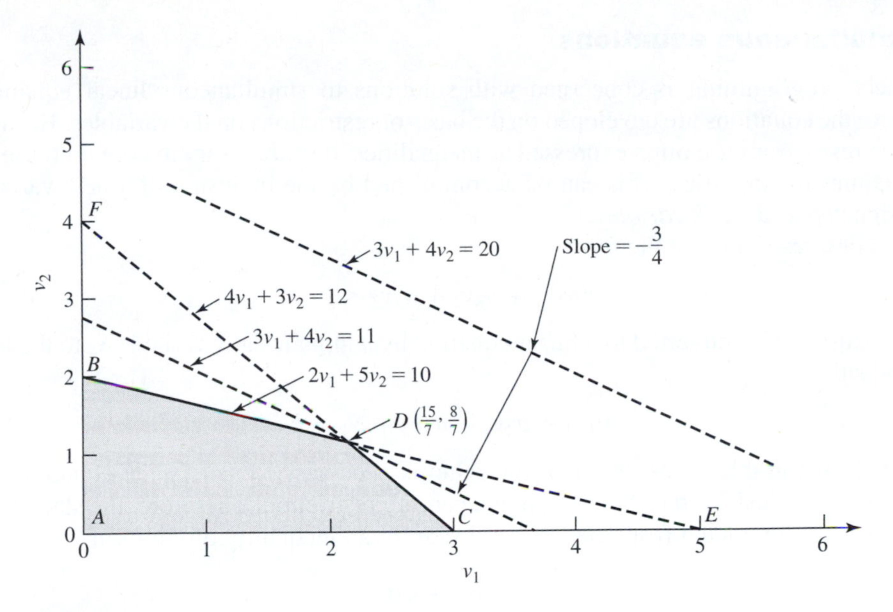

Figure 9-4
Graphical illustration of two-dimensional linear programming solution

satisfy the constraints. The set of the objective function is a family of lines with slope of $-\frac{3}{4}$. The maximum value of the objective function occurs for the line passing through the polygon vertex D. Thus, the maximum value of the objective function occurs for the case of $3v_1 + 4v_2 = 11$ at $v_1 = 15/7$ and $v_2 = 8/7$.

For the two-dimensional case considered in the preceding paragraph, one linear condition defines a line that divides the plane into two half-planes. For a three-dimensional case, one linear condition defines a set plane which divides the volume into two half-volumes. Similarly, for an n-dimensional case, one linear condition defines a hyperplane which divides the space into two half-spaces.

For the n-dimensional case, the region that is defined by the set of hyperplanes resulting from the linear constraints represents a *convex set* of all points that satisfy the constraints of the problem. If this is a bounded set, the enclosed space is a convex polyhedron; and for the case of monotonically increasing or decreasing values of the objective function, the maximum or minimum value of the objective function will always be associated with a vertex or *extreme point* of the convex polyhedron. This indicates that the linear programming solution for the model of inequality or equality constraints combined with the requested value for the objective function will involve determination of the value of the objective function at the extreme points of the set of all points that satisfy the constraints of the problem. The desired objective function can then be established by comparing the values found at the extreme points. If two extremes give the same result, then an infinite number of solutions exist as defined by all points on the line connecting the two extreme points.

Simultaneous equations†

Linear programming is concerned with solutions to simultaneous linear equations where the equations are developed on the basis of restrictions on the variables. Because these restrictions are often expressed as inequalities, it is necessary to convert these inequalities to equalities. This can be accomplished by the inclusion of a new variable designated as a *slack variable.*

For a restriction of the form

$$a_1v_1 + a_2v_2 + a_3v_3 \leq b \qquad (9\text{-}13)$$

the inequality is converted to a linear equation by adding a slack variable S_4 to the left-hand side:

$$a_1v_1 + a_2v_2 + a_3v_3 + S_4 = b \qquad (9\text{-}14)$$

The slack variable takes on whatever value is necessary to satisfy the equation and normally is considered as having a nonnegativity restriction. Therefore, the slack variable can be subtracted from the left-hand side for an inequality of the form

$$a_1v_1 + a_2v_2 + a_3v_3 \geq b \qquad (9\text{-}15)$$

to give

$$a_1v_1 + a_2v_2 + a_3v_3 - S_4 = b \qquad (9\text{-}16)$$

After the inequality constraints have been converted to equalities, the complete set of restrictions becomes a set of linear equations with n unknowns. The linear programming problem then will involve, in general, maximizing or minimizing a linear objective function for which the variables must satisfy the set of simultaneous restrictive equations with the variables constrained to be nonnegative. Because there will be more unknowns in the set of simultaneous equations than there are equations, there will be a large number of possible solutions, and the final solution must be chosen from the set of possible solutions.

If there are m independent equations and n unknowns with $m < n$, one approach is to choose arbitrarily $n - m$ variables and set them equal to zero. This gives m equations with m unknowns so that a solution for the m variables can be obtained. Various combinations of this type can be obtained so that the total possible number of solutions by this process becomes

$$\binom{n}{m} = \frac{n!}{m!(n-m)!} \qquad (9\text{-}17)$$

†Cases are often encountered in design calculations where a large number of design equations and variables are involved with long and complex simultaneous solution of the equations being called for. The amount of effort involved for the simultaneous solution can be reduced by using the *structural-array algorithm,* which is a purely mechanical operation involving crossing out rows for equations and columns for variables to give the most efficient order in which the equations should be solved. For details, see D. F. Rudd and C. C. Watson, *Strategy of Process Engineering,* J. Wiley, New York, 1968, pp. 45–49.

representing the total number of possible combinations obtainable by taking n variables m at a time. Another approach is to let $n - m$ combinations of variables assume any zero or nonzero value which results in an infinite number of possible solutions. Linear programming deals only with the situation in which the excess variables are set equal to zero.

Nonlinear programming

As stated previously, in principle, nonlinear programs are optimization situations where nonlinear behavior is exhibited by either or both the objective functions and the constraints. However, in general, nonlinear programming approaches concentrate on nonlinear objective functions. The formulation of the nonlinear problem therefore takes the following form:

$$\begin{array}{lll} \text{Minimize (or maximize):} & f'(\boldsymbol{v}) & \boldsymbol{v} = v_1, v_2, \ldots, v_n \\ \text{Subject to the constraints:} & h_i(\boldsymbol{v}) = b_i & i = 1, 2, \ldots, m \\ & g_j(\boldsymbol{v}) \leq c_j & j = 1, 2, \ldots, r' \end{array}$$

Again, f' is generally nonlinear, and h and g may also be nonlinear. All are usually assumed to be smooth, typically twice differentiable.

The programming approaches to solving these types of problems is more varied than those for linear problems, and a combination of methods is common. However, a number of major approaches account for most nonlinear programming and are representative of the methodologies employed in nonlinear programming. The major approaches are successive linear programming (SLP), successive quadratic programming (SQP), augmented Lagrangian methods or method of multipliers, and generalized reduced gradient (GRG) methods.

Nonlinear programming is beneficial, if not absolutely necessary, in situations where linear or polynomial models do not adequately describe a design. However, NLP methods are in general more limited in their applicability than the general LP methods. Also, unlike LP methods, most NLP methods cannot guarantee overall, or global, minimum solutions to design situations. Instead, most NLP methods only reveal optimum conditions for a limited set of variables and their ranges. This means that NLP methods are in general more effective in establishing optima for a subsection of a design than for the entire design. These design subsection, or local, optimum values can then be used as a basis for global design optimization.

Successive Linear Programming The essence of successive linear programming is programming linear approximations to nonlinear situations. Basically, this means that nonlinear functions and constraints are linearized by using Taylor series expansion:

$$g_j(\boldsymbol{v}) = g_j(\boldsymbol{v}_0 + \Delta\boldsymbol{v}) \cong g_j(\boldsymbol{v}_0) + \nabla g_j(\boldsymbol{v}_0)^T(\Delta\boldsymbol{v}) \tag{9-18}$$

where $\boldsymbol{v}_0$ is the set of initial values for $\boldsymbol{v}$ and $\Delta\boldsymbol{v}$ is the difference between the nonlinear relations and their linear approximations. The resultant linear program can then be solved, recalling that after linearization it is the $\Delta\boldsymbol{v}$'s that are the variables and that they are the changes from the initial variable values. There is also a need to limit the range of linearization, so as to maintain reasonable approximation of the original nonlinear problem. This is accomplished by using imposed upper and lower change variable bounds, sometimes called *step bounds*.

The linear problem resulting from the linearization is solved using LP methods, and the solution is compared to the best value achieved so far. For the first iteration this is the initial value set. If the new point is better than the best so far, the process is repeated until improved values are no longer obtained. New points that do not improve on the objective function may indicate either that they are within bounds of the optimum or that the bounds need to be altered. This means that the SLPs converge rapidly near the apex optima even though the value obtained does not need to satisfy equalities during each iteration. However, SLPs do not converge well for nonapex optima. Still the relative simplicity and usefulness of SLPs make this approach one of the most widely used methods for optimization.

Augmented Lagrangian Method This is a smooth type of penalty function method, utilizing Lagrange multipliers. The penalty function method relies on transforming constrained problems to a sequence of unconstrained problems. Basically this means that the objective function and its constraints are manipulated and combined to make a new unconstrained objective function, called a penalty function. For example,

$$\left.\begin{array}{ll}\text{Minimize:} & f(\boldsymbol{v})\\ \text{Subject to:} & \boldsymbol{g}(\boldsymbol{v}) \le 0\\ & \boldsymbol{h}(\boldsymbol{v}) = 0\end{array}\right\} \Rightarrow \text{minimize } P(f, \boldsymbol{g}, \boldsymbol{h}, r') \tag{9-19}$$

where r is a positive penalty parameter. Once the problem has been reformulated, the penalty function is minimized for increasing values of penalty parameters, approaching the original constrained problem optimum. It is possible to improve the performance of the penalty function method by formulating the penalty function to account for solutions that violate the original problem constraints. This is generally done by adding positive weighting terms that sum the extent of each constraint violation to the penalty function when constraints are violated.†

The augmented Lagrangian method introduces the Lagrange multiplier to the penalty function, as well as a quadratic penalty term, resulting in a function of the form

$$AL(\boldsymbol{v}, \lambda, r') = f'(\boldsymbol{v}) + \sum_{i=1}^{m} \lambda_i h_i(\boldsymbol{v}) + r \sum_{i=1}^{m} h_i^2(\boldsymbol{v}) \tag{9-20}$$

where λ_i are the Lagrange multipliers. Since optimum conditions require that the gradient of the Lagrangian function with respect to $\boldsymbol{v}$ at the optimum be zero, this method also weights the penalties toward the optimum.‡ This gives the augmented Lagrangian method a certain edge since it relies on exact penalty function use. However, the use of

---

†This method of weighting is in fact used in the common simplex algorithm.

‡T. Rapscák, in *Nonlinear Optimization and Related Topics,* G. Di Pillo and F. Gianessi, eds., Kluwer Academic Publishers, Norwell, MA, 1999, p. 351.

augmented Lagrangian methods does seem to be declining in favor of SLP, SQP, and GRG approaches.[†]

Successive Quadratic Programming This approach is similar to SLP approaches in that it approximates nonlinear programming problems with linear substitutes. However, in contrast to SLP methods, SQP approximations are made using successive solutions of quadratic objective functions. Further, the objective function used in SQP is a quadratic approximation of the Lagrange function, and the solution must meet a set of optimality conditions.

Here it is necessary to introduce a series of conditions that are necessary for optimality. These are a series of conditions necessary for NLP optimality, commonly called the *Kuhn-Tucker* or *Karuch-Kuhn-Tucker* (KTC) conditions. The basis for these is the concept that no allowable changes of problem variables can improve the objective function at a local optimum.[‡] The first condition is the linear dependence of gradients, or sometimes called the *balance of forces* condition. The algebraic form of this condition for a Lagrange function[§] is

$$\nabla L(v_{\min}, \mu_{\min}, \lambda_{\min}) = \nabla f'(v_{\min}) + \nabla g(v_{\min})\mu_{\min} + \nabla g(v_{\min})\lambda_{\min} = 0 \tag{9-21}$$

where μ is the weighting variable and the subscript min indicates values at the minimum of f'.

Conceptually, this means that all the movement or change tendencies of the objective function are balanced at the optimum. The second condition is that the optimum solution to a nonlinear programming problem must satisfy all the constraints of the problem. The third and fourth conditions deal with equality constraints and require that when equality constraints are ignored, inferring that they do not exist, the multiplier[¶] must be zero. Conversely, if an equality constraint does exist, then the multiplier can be positive. This means that in case of equality, the optimum must lie on the equality constraint; and in case of inequality, the influence of inequality constraints is one-directional. Last, additional constraint requirements are often necessary; a common use of this type is that the gradients of the constraints have to be linearly dependent.[*]

The quadratic linearization of the nonlinear problem begins much as in simple linearization where the Lagrangian function is also used. For a simple case the objective function and constraints would therefore be

$$\begin{aligned} &\text{Minimize:} && \nabla L^T + \tfrac{1}{2}\Delta \mathbf{v}^T \nabla_{\mathbf{v}}^2 L\, \Delta \mathbf{v} \\ &\text{Subject to:} && \mathbf{g} + \nabla \mathbf{g}\, \Delta \mathbf{v} = 0 \end{aligned} \tag{9-22}$$

where the Lagrangian is

$$L(\Delta \mathbf{v}, \Delta \lambda) = \nabla L^T \Delta \mathbf{v} + \tfrac{1}{2}\Delta \mathbf{v}^T\, \nabla_{\mathbf{v}}^2 L\, \Delta \mathbf{v} + \Delta \lambda^T(\mathbf{g} + \nabla \mathbf{g}\, \Delta \mathbf{v}) \tag{9-23}$$

[†]J. Nocedal and S. J. Wright, *Numerical Optimization,* Springer, New York, 1999.

[‡]J. Nocedal and S. J. Wright, *Numerical Optimization,* Springer, New York, 1999, pp. 328, 342.

[§]The Lagrange function is $L(\mathbf{v}, \mu, \lambda) = f(\mathbf{v}) + g(\mathbf{v})^T \mu + h(\mathbf{v})^T \lambda$, where the vectors μ and λ are set so that they "balance" Eq. (9-20).

[¶]In this case the multipliers for $g(v_{\min})$ and $h(v_{\min})$ are μ and λ, respectively.

[*]NEOS Guide, www-fp.mcs.anl.gov/otc/Guide/OptWeb/index.html.

and the optimality conditions become

$$\nabla_v L + \nabla_v^2 L \, \Delta \boldsymbol{v} + \nabla \boldsymbol{g}^T \Delta \boldsymbol{\lambda} = 0$$
$$\boldsymbol{g} + \nabla \boldsymbol{g} \, \Delta \boldsymbol{v} = 0 \tag{9-24}$$

Now these functions and equations which are based on the requirements for optimality, Eqs. (9-22) through (9-24), are successively solved linearly and iterated for $(\Delta \boldsymbol{v}, \Delta \boldsymbol{\lambda})$.

Inclusion of more and varying constraint situations entails added complexity to the equations presented, although the overall method is still similar.†

SQP offers the advantages of providing fairly quick convergence and not solving equalities at each iteration. It therefore forms the basis for many important practical optimization routines and software and is widely used. However, it does usually violate nonlinear constraints in intermediate iterations.

Generalized Reduced Gradient Methods GRG methods are based on the use of an objective function, bounds, and constraints only and avoid the use of penalty parameters by iteratively establishing and searching along curves that stay near proven feasible values. The variables in the problem are separated into three classes: fixed, basic, and superbasic. The fixed components are rigidly anchored to at least one of their bounds for the current iteration. The basic components of the problem are then expressed implicitly in terms of the fixed and superbasic components. The superbasic components are then allowed to vary in a direction that reduces the value of the objective function. The strategies for choosing the direction and magnitude of the allowed superbasic component variation are therefore integral to the performance of reduced gradient programming methods. They are derived from unconstrained optimization principles and include strategies such as steepest descent and quasi-Newtonian.‡

The nature of GRG methods makes them very robust and versatile; they are also useful in that they are able to follow feasible solutions, useful for global optimization. They do, however, need to satisfy equalities at each iteration and are often relatively slower than other NLP methods.

Mixed-Integer Programming A commonly occurring type of optimization problem involves both integer and continuous variables. These are very useful to many engineering applications, especially the various requirements of process design optimization, especially topology. Mixed-integer programming, both linear (MILP) and nonlinear (MINLP) methods, has therefore been of prime interest. The complexity involved in treating problems of mixed integers does, however, tend to make MIP methods involved and beyond the scope of this text.

---

†Descriptions of these more complex cases and their treatment are widely available in the optimization literature, including J. Nocedal and S. J. Wright, *Numerical Optimization,* Springer, New York, 1999; or J. S. Albuquerque, V. Gopal, G. H. Staus, L. T. Biegler, and B. E. Ydstie, *Comp. Chem. Eng.,* **21:** 853 (1997).

‡The strategies for choosing the direction and magnitude of the allowed superbasic component variation are critical to GRG performance. Further details on selecting and applying these are available in more advanced optimization literature, including P. E. Gill, W. Murray, and M. H. Wright, *Practical Optimization,* Academic Press, New York, 1981.

Table 9-1 Common methods for solving MIPs

MILP	MINLP
BB	BB Outer approximation Generalized benders decomposition Disjunctive programming

MIP methods follow very much the various programming methods described earlier, with the additional limitation of the constraint that some variables are limited to certain values. In general, MIPs follow one of the approaches for dealing with this limitation. The first is simply to optimize the continuous variables for the various integer variable combinations (i.e., optimize continuous variables for each set of possible integer variable values, then compare and select the best values). The other method is to relax the integer variables and solve the resulting continuous problem, as in BB methods. The solution is then examined at the various integer variable combinations. Some of the more popular methods for solving MIPs are listed in Table 9-1.[†]

Dynamic Programming

The concept of dynamic programming is based on converting an overall decision situation involving many variables into a series of simpler individual problems with each of these involving a small number of total variables. In its extreme, an optimization problem involving a large number of variables, all of which may be subject to constraints, is broken down into a sequence of problems with each of these involving only one variable. A characteristic of the process is that the determination of one variable leaves a problem remaining with one less variable. The computational approach is based on the principle of optimality, which states that an optimal policy has the property that, no matter what the initial state and initial decision, the remaining decisions must constitute an optimal policy with regard to the state resulting from the first decision.

The use of dynamic programming is pertinent for design in the chemical industry where the objective function for a complicated system can often be obtained by dividing the overall system into a series of stages. Optimizing the resulting simple stages can lead to the optimal solution for the original complex problem.

The general formulation for a dynamic programming problem, presented in a simplified form, is shown in Fig. 9-5. On the basis of the definitions of terms given in Fig. 9-5a, each of the variables v_{j+1}, v_j, and d_j may be replaced by vectors because there may be several components or streams involved in the input and output, and several decision variables may be involved. The profit or return P_j is a scalar which gives a measure of the contribution of stage j to the objective function.

[†]More details on the approach to MIP methods can be found in optimization literature, such as C. A. Floudas, *Deterministic Global Optimization: Theory, Methods, and Applications,* Kluwer Academic Publishers, Norwell, MA, 2000, p. 571.

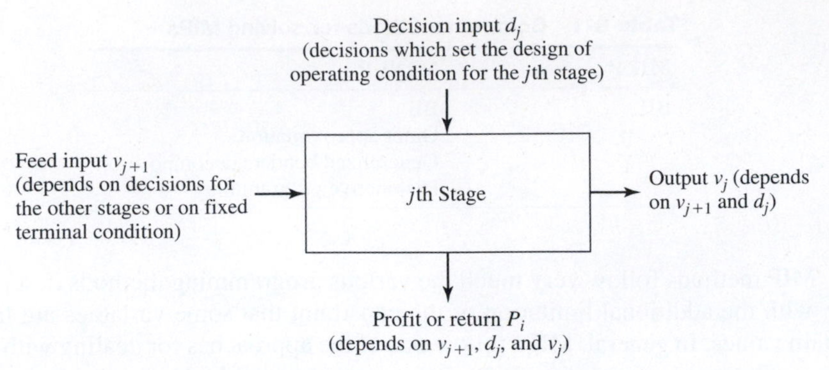

(a) General formulation for one stage in a dynamic programming model.

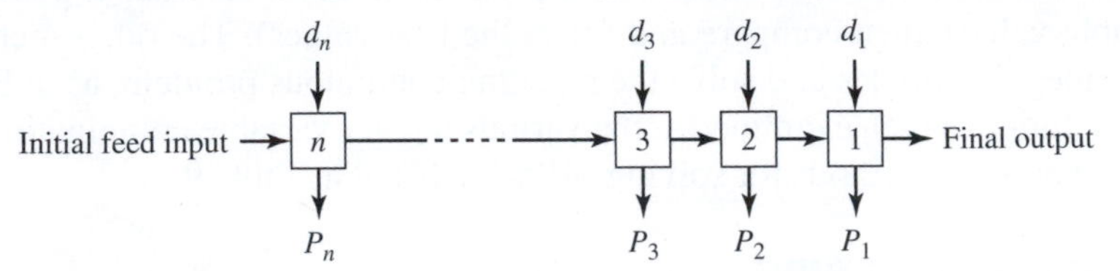

(b) A simple multistage process with n stages.

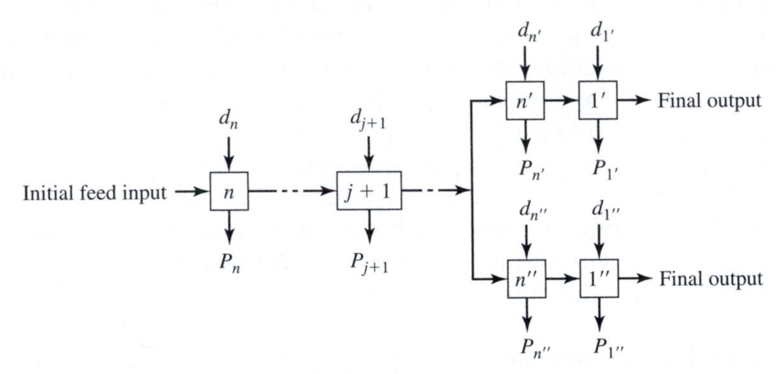

(c) A multistage process with separating branches.

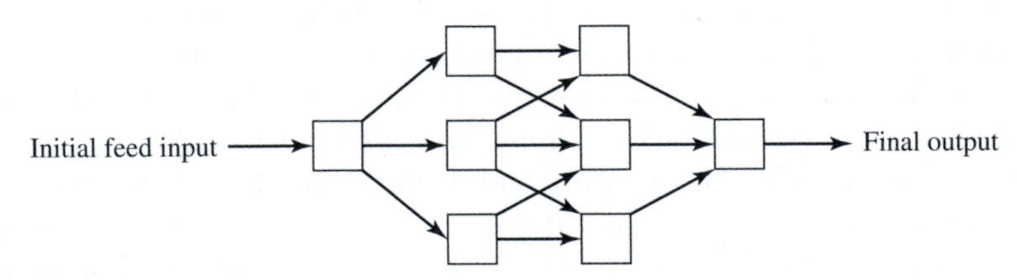

(d) A network of dynamic programming showing various paths.

Figure 9-5
Illustration of stages involved in dynamic programming

For the operation of a single stage, the output is a function of the input and the decisions, or

$$v_j = h_j(v_{j+1}, d_j) \tag{9-25}$$

Similarly, for the individual-stage objective function P_j

$$\text{P}_j = g_j(v_{j+1}, v_j, d_j) \tag{9-26}$$

or, on the basis of the relation shown as Eq. (9-25),

$$\text{P}_j = g_j(v_{j+1}, d_j) \tag{9-27}$$

For the simple multistage process shown in Fig. 9-5b, the process design strategy to optimize the overall operation can be expressed as

$$f'_j(v_{j+1}) = \max_{d_j}[g_j(v_{j+1}, d_j) + f'_{j-1}(v_j)] = \max_{d_j}[Q_j(v_{j+1}, d_j)] \tag{9-28}$$

for

$$v_j = h_j(v_{j+1}, d_j) \qquad j = 1, 2, \ldots, n\text{—subject to } f'_0 = 0$$

The symbolism $f'_j(v_{j+1})$ indicates that the maximum (or optimum) return or profit from a process depends on the input to that process, and the terms in the square brackets of Eq. (9-28) refer to the function that is being optimized. Thus, the expression $Q_j(v_{j+1}, d_j)$ represents the combined return from all stages through j and must equal the return from stage j, or $g_j(v_{j+1}, d_j)$, plus the maximum return from the preceding stages 1 through $j-1$, or $f'_{j-1}(v_j)$.

In carrying out the procedure for applying dynamic programming to the solution of appropriate plant design problems, each input v_{j+1} is considered as a parameter. Thus, at each stage, the problem is to find the optimum value of the decision variable d_j for all feasible values of the input variable. By using the dynamic programming approach involving n stages, a total of n optimizations must be carried out. This approach can be compared to the conventional approach in which optimum values of all the stages and decisions would be chosen by a basic probability combination analysis. Thus, the conventional method would have a computational effort that would increase approximately exponentially with the number of stages, while the dynamic programming approach can provide a great reduction in necessary computational effort because this effort would increase only about linearly with the number of stages. However, this advantage of dynamic programming is based on a low number of components in the input vector v_{j+1}, and dynamic programming rapidly loses its effectiveness for practical computational feasibility if the number of these components increases above 2 or 3.

A severe limitation of dynamic programming for chemical process design is that it is not practical for systems containing recycle.

OPTIMIZATION SOLUTION METHODOLOGIES

Solving and obtaining the optimum values for optimization problems is the last phase in design optimization. A number of general methods for solving the programmed optimization problems relate various relations and constraints that describe the process to their effect on the objective function. These examine the effect of variables on the objective function using analytic, graphical, and algorithmic techniques based on the principles of optimization and programming methods.

Procedure with One Variable

There are many cases in which the factor being minimized (or maximized) is an analytic function of a single variable. The procedure then becomes very simple. Consider the example presented in Fig. 9-1, where it is necessary to obtain the insulation thickness that gives the least total cost. The primary variable involved is the thickness of the insulation, and relationships can be developed showing how this variable affects all costs.

Cost data for the purchase and installation of the insulation are available, and the length of service life can be estimated. Therefore, a relationship giving the effect of insulation thickness on fixed charges can be developed. Similarly, a relationship showing the cost of heat lost as a function of insulation thickness can be obtained from data on the thermal properties of steam, properties of the insulation, and heat-transfer considerations. All other costs, such as maintenance and plant expenses, can be assumed to be independent of the insulation thickness.

The two cost relationships obtained might be expressed in a simplified form similar to the following:

$$\text{Fixed charges} = \phi(x) = ax + b \tag{9-29}$$

$$\text{Cost of heat loss} = \phi^i(x) = \frac{c}{x} + d \tag{9-30}$$

$$\text{Total variable cost} = C_T = \phi(x) + \phi^i(x) = ax + b + \frac{c}{x} + d \tag{9-31}$$

where a, b, c, and d are constants, C_T is the total variable cost or the objective function and x is the common variable (insulation thickness).

The *graphical* method for determining the optimum insulation thickness is shown in Fig. 9-1. The optimum thickness of insulation is found at the minimum point on the curve obtained by plotting total variable cost versus insulation thickness.

The slope of the total variable cost curve is zero at the point of optimum insulation thickness. Therefore, if Eq. (9-31) applies, the optimum value can be found *analytically* by merely setting the derivative of C_T with respect to x equal to zero and solving for x.

$$\frac{dC_T}{dx} = a - \frac{c}{x^2} = 0 \tag{9-32}$$

$$x = \left(\frac{c}{a}\right)^{1/2} \tag{9-33}$$

If the factor C_T does not attain a usable maximum or minimum value, the solution for the dependent variable will indicate this condition by giving an impossible result, such as infinity, zero, or the square root of a negative number.

The value of x shown in Eq. (9-33) occurs at a minimum, maximum, or a point of inflection. The second derivative of Eq. (9-31), evaluated at the given point, indicates if the value occurs at a minimum (second derivative greater than zero), maximum (second derivative less than zero), or point of inflection (second derivative equal to zero).

An alternative method for determining what type of point is involved is to calculate values of the objective function at points slightly greater and slightly smaller than the optimum value of the dependent variable.

The second derivative of Eq. (9-31) is

$$\frac{d^2C_T}{dx^2} = \frac{2c}{x^3} \tag{9-34}$$

If x represents a variable such as insulation thickness, its value must be positive. Since c must be positive, the second derivative at the optimum point must be greater than zero. The term $(c/a)^{1/2}$ represents the optimum value of x at the point where the total variable cost is a minimum.

Procedure with Two or More Variables

When two or more independent variables affect the objective function, the procedure for determining the optimum conditions may become rather tedious; however, the general approach is the same as when only one variable is involved.

Consider the case in which the total cost for a given operation is a function of the two independent variables x and y, or

$$C_T = \phi^{ii}(x, y) \tag{9-35}$$

By analyzing all the costs involved and reducing the resulting relationships to a simple form, the following function might be found for Eq. (9-35):

$$C_T = ax + \frac{b}{xy} + cy + d \tag{9-36}$$

where $a, b, c,$ and d are positive constants.

Graphical Procedure The relationship among $C_T, x,$ and y could be shown as a curved surface in a three-dimensional plot, with a minimum value of C_T occurring at the optimum values of x and y. However, the use of a three-dimensional plot is not practical for most engineering determinations.

The optimum values of x and y in Eq. (9-36) can be obtained graphically on a two-dimensional plot by using the method indicated in Fig. 9-6. In this figure, the objective function is plotted against one of the independent variables x, with the second variable y held at a constant value. A series of such plots is made with each dashed curve representing a different constant value of the second variable. As shown in Fig. 9-6, each of the curves ($A, B, C, D,$ and E) gives one value of the first variable x at the point where the total cost is a minimum. The solid curve NM represents the locus of all these minimum points, and the optimum value of x and y occurs at the minimum point on curve NM.

Similar graphical procedures can be used when there are more than two independent variables. For example, if a third variable z were included in Eq. (9-36), the first step would be to make a plot similar to Fig. 9-6 at one constant value of z. Similar plots

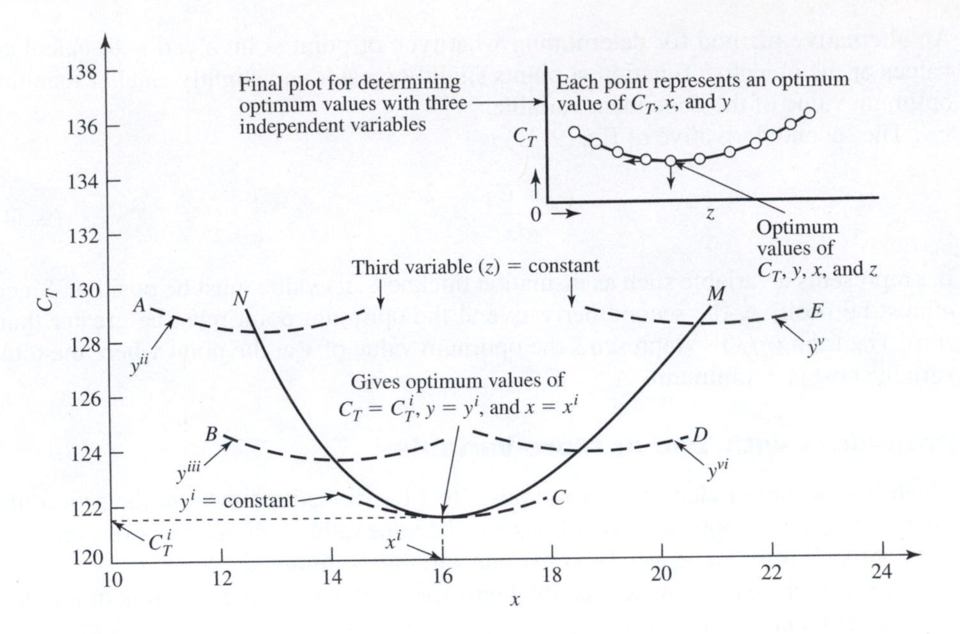

Figure 9-6
Graphical determination of optimum conditions with two or more independent variables

would then be made at other constant values of z. Each plot would give an optimum value of x, y, and C_T for a particular z. Finally, as shown in the insert in Fig. 9-6, the overall optimum value of x, y, z, and C_T could be obtained by plotting z versus the individual optimum values of C_T.

Analytical Procedure In Fig. 9-6, the optimum value of x is found at the point where $(\partial C_T/\partial x)_{y=y^i}$ is equal to zero. Similarly, the same results would be obtained if y were used as the abscissa instead of x. If this were done, the optimum value of y (that is, y^i) would be found at the point where $(\partial C_T/\partial y)_{x=x^i}$ is equal to zero. This immediately indicates an analytical procedure for determining optimum values.

Using Eq. (9-36) as a basis gives

$$\frac{\partial C_T}{\partial x} = a - \frac{b}{x^2 y} \tag{9-37}$$

$$\frac{\partial C_T}{\partial y} = c - \frac{b}{x y^2} \tag{9-38}$$

At the optimum conditions, both of these partial derivatives must be equal to zero; thus, Eqs. (9-37) and (9-38) can be set equal to zero, and the optimum values of $x = (cb/a^2)^{1/3}$ and $y = (ab/c^2)^{1/3}$ can be obtained by solving the two simultaneous equations. If more than two independent variables were involved, the same procedure would be followed, with the number of simultaneous equations being equal to the number of independent variables.

Determination of Optimum Values with Two Independent Variables — EXAMPLE 9-1

The following equation shows the effect of the variables x and y on the total cost for a particular operation:

$$C_T = 2.33x + \frac{11{,}900}{xy} + 1.86y + 10$$

Determine the values of x and y that will give the least total cost.

■ Solution

Analytical Method

The two partial derivatives required are

$$\frac{\partial C_T}{\partial x} = 2.33 - \frac{11{,}900}{x^2 y}$$

$$\frac{\partial C_T}{\partial y} = 1.86 - \frac{11{,}900}{x y^2}$$

At the optimum point,

$$0 = 2.33 - \frac{11{,}900}{x^2 y}$$

$$0 = 1.86 - \frac{11{,}900}{x y^2}$$

Solving simultaneously for the optimum values of x and y results in

$$x = 16 \qquad y = 20 \qquad C_T = 121.7$$

A check should be made to make certain the preceding values represent conditions of minimum cost.

$$\frac{\partial^2 C_T}{\partial x^2} = \frac{2(11{,}900)}{(16)^3(20)} = +0.290 \text{ at optimum point}$$

$$\frac{\partial^2 C_T}{\partial y^2} = \frac{2(11{,}900)}{16(20)^3} = +0.186 \text{ at optimum point}$$

Since the second derivatives are positive, the optimum conditions must occur at a point of minimum cost.

Graphical Method

The following constant values of y are chosen arbitrarily:

$$y^i = 20 \qquad y^{ii} = 32 \qquad y^{iii} = 26 \qquad y^{iv} = 15 \qquad y^v = 12$$

At each constant value of y, a plot is made of C_T versus x. These plots are presented in Fig. 9-6 as curves C, A, B, D, E, respectively. A summary of the results is presented in the following table:

y	Optimum x	Optimum C_T
$y^i = 20$	16.0	121.7
$y^{ii} = 32$	12.7	128.3
$y^{iii} = 26$	14.1	123.6
$y^{iv} = 15$	18.5	123.9
$y^v = 12$	20.7	128.5

One curve (*NM* in Fig. 9-6) through the various optimum points shows that the overall optimum occurs at

$$x = 16 \qquad y = 20 \qquad \text{where } C_T = 121.7$$

Note: In this case, a value of y was chosen that corresponded to the optimum value. Usually, it is necessary to interpolate or make further calculations to determine the final optimum conditions.

Comparison of Graphical and Analytical Methods In the determination of optimum conditions, the same final results are obtained with either graphical or analytical methods. Sometimes it is impossible to set up one analytical function for differentiation, and the graphical method must be used. If the development and simplification of the total analytical function require complicated mathematics, it may be simpler to resort to the direct graphical solution; however, each individual problem should be analyzed on the basis of the existing circumstances. For example, if numerous repeated trials are necessary, the extra time required to develop an analytical solution may be time well spent.

The graphical method has one distinct advantage over the analytical method. The shape of the curve indicates the importance of operating at or very close to the optimum conditions. If the maximum or minimum occurs at a point where the curve is flat with only a gradual change in slope, there will be considerable spread in the choice of the final conditions, and incremental cost analyses may be necessary. On the other hand, if the maximum or minimum is sharp, it may be essential to operate at the exact optimum conditions.

Breakeven Chart for Optimum Analysis of Production

In considering the overall costs or profits in a plant operation, one of the factors that has an important effect on the economic results is the fraction of total available time during which the plant is in operation. If the plant stands idle or operates at low capacity, certain costs, such as those for raw materials and labor, are reduced, but costs for depreciation and maintenance continue at essentially the same rate even though the plant is not in full use.

There is a close relationship between operating time, rate of production, and selling price. It is desirable to operate at a schedule that will permit maximum utilization of fixed costs while simultaneously meeting market sales demand and using the capacity of the plant production to give the best economic results. Figure 9-7 shows graphically how production rate affects costs and profits. The fixed costs remain constant while the total product cost, as well as the profit, increases with increased rate of production. The point where total product cost equals total income represents the breakeven point, and the optimum production schedule must be at a production rate higher than that corresponding to the breakeven point.

Experimental Design and Analysis of Process Simulations

The ability to create accurate process design simulations allows an additional approach to process optimization, referred to as *simulation experimentation optimization* (SEO).

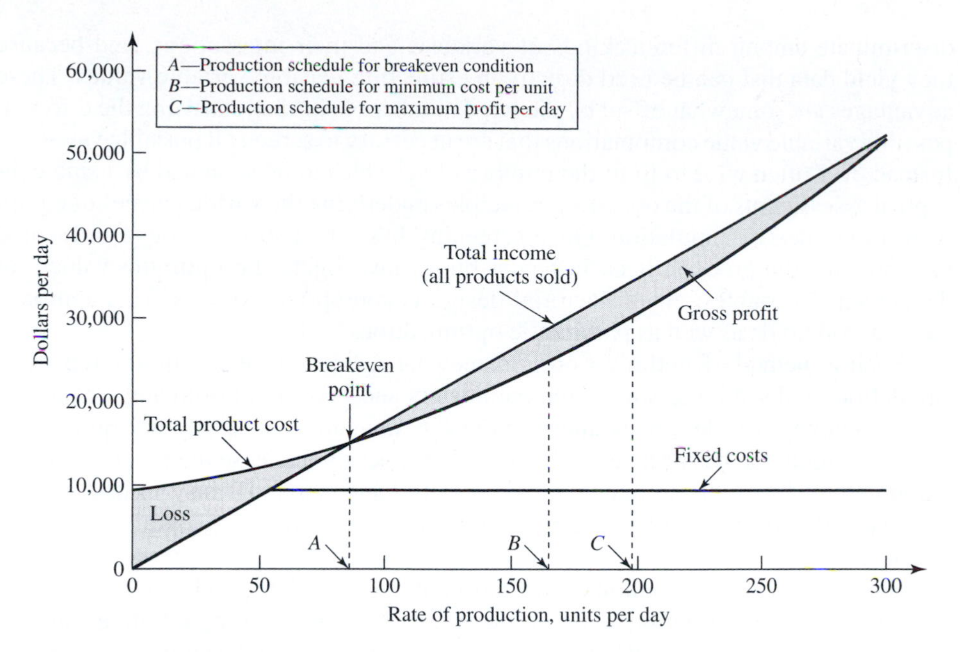

Figure 9-7
Breakeven chart for operating a production plant (based on the production schedule in Example 9-2)

This method essentially treats simulations as an experimental apparatus, for which experiments are designed for the various topologic and parametric aspects of the design. The simulation performance for those topologic and parametric aspects is then analyzed using established experimental analysis techniques. One of these methods, known as *factorial experimental design,*† is especially appropriate for SEO. Otherwise, SEO follows generally the same procedures as do other optimization solution methodologies.

Simulation experimentation optimization begins with a construction of the simulations; this is analogous to the process of establishing the mathematical performance of the design used in other optimization methods. As with other methods, suboptimization may be appropriate. Next, the simulation is used to screen out the more important variables for the design. Factorial design and analysis methods are particularly well suited for this step since they yield quantitative evaluation of variables and, are able to

†The principle underlying factorial design is that simulations are evaluated at two or more levels of the variables influencing the simulation. The simulation results are then compared to other simulation results for various combinations of variables at the different variable levels and analyzed. The analysis of these results and the variable levels that caused them provides quantitative evaluations of both the variable and variable combination influence on the results for the simulation.

Factorial experimental design and analysis and its use and implementation are described in greater detail by R. L. Mason, R. F. Gunst, and J. L. Hess, *Statistical Design and Analysis of Experiments,* J. Wiley, New York, 1989, and G. E. Box, W. G. Hunter, and J. S. Hunter, *Statistics for Experiments,* J. Wiley, New York, 1978, and D. C. Montgomery, *Design and Analysis of Experiments,* 5th ed., J. Wiley, New York, 2001.

discriminate among different kinds of variables and their interactions, and because they yield data that can be used directly in estimating feasible variable values. These advantages are somewhat offset by the large number of simulations using the different possible variable value combinations that are necessary to screen all possible variables. Instead, it is often wise to limit the number of variables to be screened by using conceptual assessments of the operating principles underlying the various pieces of equipment in the design simulation. Once screening has established the more influential variables for the process, it is then possible to investigate the optimum values for the screened variables; again, factorial design is appropriate for this. This approach includes topologic as well as parametric optimization.

Such a method of optimization is somewhat tedious since it requires repetitive simulations and setting of simulation parameters and structure. Fortunately, changing the simulation variable values and running the simulation generally are quite rapid once the simulation has been assembled. It is also possible to ease the repetitive simulation process by using external simulation control and manipulation routines that will automatically perform the necessary variable changes as well as record the results of the simulation.[†]

Simulation experimentation optimization offers a number of advantages. It is applicable to all types of optimization problems as well as being more intuitive and easy to understand and implement for those not experienced with other types of optimization. Further, the perceived added difficulty of constructing the simulation when compared to other optimization methods is partially offset since many other optimization methods also require simulations and it is required when testing the optimization results. Last, simulation experimentation optimization is capable of producing global optimization conditions.

Algorithm Solutions to Optimization Problems

An algorithm is simply an objective mathematical method for solving a problem and is purely mechanical so that it can be programmed for a computer. Solution of programming problems generally requires a series of actions that are iterated to a solution, based on a programming method and various numerical calculation methods. The input to the algorithm can be manual, where the relations governing the design behavior are added to the algorithm. They can also be integrated or set to interface with rigorous computer simulations that describe the design.

Use of algorithms thus requires the selection of an appropriate programming method, methods, or combination of methods as the basic principals for the algorithm function. It also requires the provision of the objective functions and constraints, either as directly provided relations or from computer simulation models. The algorithm then uses the basic programming approach to solve the optimization problem set by the objective function and constraints.

Algorithms can therefore be integrated parts of software, stand-alone optimization software with one or several built-in algorithms, or a general-purpose algorithm that

---

[†]Some optimization software programs use process simulations as the basis for optimization as well.

Table 9-2 Some commonly used algorithms[†]

System type	Algorithm	Notes
Linear		
	LINDO	
	CPLEX	
	OSL	
	LAPMS	
Nonlinear	Augmented Lagrangian	
	LANCELOT	Allows sparse matrix solution
	Successive quadratic	
	NPSOL	
	SNOPT	
	DONLP2	
	NLPQL	
	MINOS	Available at a nominal fee for educational purposes. Default for linear and nonlinear problems in GAMS software
	General reduced gradient	
	CONOPT	Allows sparse matrix solution
	LSGRG2	Allows sparse matrix solution
	GRG2	The most widely used GRG code, used in the Excel solver
MILP[‡]		
	LINDO	
	OSL/GAMS	
MINLP[‡]		
	DICOPT++/GAMS	
	MINOPT	An academic algorithm
	Extended cutting plane method	
	α-ECP	An academic algorithm

[†]The general algebraic modeling system (GAMS) is perhaps the most widely used optimization software and contains several algorithms including the following:

Linear programming problems—BDMLP, CONOPT, CPLEX, MINOS, OSL, SNOPT, XA, and XPRESS
Nonlinear programming problems—CONOPT, MINOS, SNOPT
Mixed-integer linear programming problems—BDMLP, CPLEX, OSL, XA, and XPRESS
Mixed-integer nonlinear programming problems—DICOPT and SBB

[‡]The branch and bound method is particularly useful to these applications. Several specialized algorithms are available that form the basis for many of the algorithms used for these types of problems.

can be customized. A description of the use and application of these algorithms for whatever purpose is well beyond the scope of this text;[†] instead, an example of the process of developing a linear programming algorithm in the form of the widely used simplex algorithm is provided to illustrate algorithm development. Also, Table 9-2 gives a brief summary of some of the more commonly used algorithms.

[†]A detailed guide to optimization methods and algorithms as well as available software is the NEOS Guide, which can be found at www-fp.mcs.anl.gov/otc/Guide/.

Linear Programming Algorithm Development To develop this form of approach for linear programming solutions, a set of linear inequalities which form the constraints are written in the form of "equal to or less than" equations as

$$\begin{aligned} a_{11}v_1 + a_{12}v_2 + \cdots + a_{1n}v_n &\le b_1 \\ a_{21}v_1 + a_{22}v_2 + \cdots + a_{2n}v_n &\le b_2 \\ \cdots\cdots\cdots\cdots\cdots\cdots\cdots\cdots & \\ a_{m1}v_1 + a_{m2}v_2 + \cdots + a_{mn}v_n &\le b_m \end{aligned} \tag{9-39}$$

or in general summation form as

$$\sum_{j=1}^{n} a_{ij}v_j \le b_i \qquad i = 1, 2, \ldots, m \tag{9-39a}$$

for

$$v_j \ge 0 \qquad j = 1, 2, \ldots, n$$

where i refers to rows (or equation number) in the set of inequalities and j refers to columns (or variable number).

As indicated earlier, these inequalities can be changed to equalities by adding a set of slack variables $v_{n+1}, \ldots, v_{n+m}$ (here v is used in place of S to simplify the generalized expressions), so that

$$\begin{aligned} a_{11}v_1 + a_{12}v_2 + \cdots + a_{1n}v_n + v_{n+1} &= b_1 \\ a_{21}v_1 + a_{22}v_2 + \cdots + a_{2n}v_n + v_{n+2} &= b_2 \\ \cdots\cdots\cdots\cdots\cdots\cdots\cdots\cdots\cdots & \\ a_{m1}v_1 + a_{m2}v_2 + \cdots + a_{mn}v_n + v_{n+m} &= b_m \end{aligned} \tag{9-40}$$

or, in general summation form,

$$\sum_{j=1}^{n} (a_{ij}v_j + v_{n+i}) = b_i \qquad i = 1, 2, \ldots, m \tag{9-40a}$$

for

$$v_j \ge 0 \qquad j = 1, 2, \ldots, n + m$$

In addition to the constraining equations, there is an objective function for the linear program which is expressed in the form of

$$z = \text{maximum (or minimum) of } c_1v_1 + c_2v_2 + \cdots + c_jv_j + \cdots + c_nv_n \qquad \text{(9-41)}^\dagger$$

where the variables v_j are subject to $v_j \ge 0\,(j = 1, 2, \ldots, n + m)$. Note that, in this case, all the variables above v_n are slack variables and provide no direct contribution to the value of the objective function.

---

†In a more compact form using matrix notation, the problem is to find the solution to $\mathbf{Av} = \mathbf{B}$ which maximizes or minimizes $\mathbf{z} = \mathbf{cv}$, where $\mathbf{v} \ge 0$.

Within the constraints indicated by Eqs. (9-39) or (9-40), a solution for values of the variables v_j must be found that meet the maximum or minimum requirement of the objective function given by Eq. (9-41). As demonstrated in previous examples, the solution to a problem of this sort must lie on an extreme point of the set of possible feasible solutions. For any given solution, the number of equations to be solved simultaneously must be set equal to the number of variables. This is accomplished by setting n (number of variables) minus m (number of equations) equal to zero and then proceeding to obtain a solution.

While the preceding generalization is sufficient for ultimately achieving a final solution, the procedure can be very inefficient unless some sort of special method is used to permit generation of extreme-point solutions in an efficient manner to allow a rapid and effective approach to the optimum condition. This is what the *simplex method* does.[†]

Simplex Algorithm The basis for the simplex method is the generation of extreme-point solutions by starting at any one extreme point for which a feasible solution is known and then proceeding to a neighboring extreme point. Special rules are followed that cause the generation of each new extreme point to be an improvement toward the desired objective function. When the extreme point is reached where no further improvement is possible, this will represent the desired optimum feasible solution. Thus, the simplex algorithm is an iterative process that starts at one extreme-point feasible solution, tests this point for optimality, and proceeds toward an improved solution. If an optimal solution exists, this algorithm can be shown to lead ultimately and efficiently to the optimal solution.

The stepwise procedure for the simplex algorithm is as follows (based on the optimum being a maximum):

1. State the linear programming problem in standard equality form.
2. Establish the initial feasible solution from which further iterations can proceed. A common method to establish this initial solution is to base it on the values of the slack variables, where all other variables are assumed to be zero. With this assumption, the initial matrix for the simplex algorithm can be set up with a column showing those variables that will be involved in the first solution. The coefficient for these variables appearing in the matrix table should be 1 with the rest of the column being 0.
3. Test the initial feasible solution for optimality. The optimality test is accomplished by the addition of rows to the matrix which give (*a*) a value of z_j for each column, where z_j is defined as the sum of the objective function coefficient for each solution variable (c_i corresponding to solution v_i in that row) times the coefficient of the constraining-equation variable for that column [a_{ij} in Eq. (9-40*a*)] [that is, $z_j = \sum_{i=1}^{n}(c_i a_{ij})$ when $(j = 1, 2, \ldots, n)$], (*b*) c_j [see Eq. (9-41)], and (*c*) $c_j - z_j$. If $c_j - z_j$ is positive for at least one column, then a better solution is possible.

[†]The simplex method and algorithm were first made generally available when published by G. B. Dantzig in *Activity Analysis of Production and Allocations,* T. C. Koopmans, ed., J. Wiley, New York, 1951, chap. 21. It is still widely used and forms the basis for many, if not most, current linear programming algorithms.

4. Iteration toward the optimal solution is accomplished as follows: Assuming that the optimality test indicates that the optimal program has not been found, the following iteration procedure can be used:
   a. Find the column in the matrix with the maximum value of $c_j - z_j$ and designate this column as k. The incoming variable for the new test will be the variable at the head of this column.
   b. For the matrix applying to the initial feasible solution, add a column showing the ratio b_i/a_{ik}. Find the minimum *positive* value of this ratio, and designate the variable in the corresponding row as the outgoing variable.
   c. Set up a new matrix with the incoming variable, as determined under (*a*), substituted for the outgoing variable, as determined under (*b*). The modification of the table is accomplished by matrix operations so that the entering variable will have a 1 in the row of the departing variable and 0s in the rest of that column. The matrix operations involve row manipulations of multiplying rows by constants and subtracting from or adding to other rows until the necessary 1 and 0 values are reached. This new matrix should have added to it the additional rows and column as explained under parts 3, 4*a*, and 4*b*.
   d. Apply the optimality test to the new matrix.
   e. Continue the iterations until the optimality test indicates that the optimum objective function has been attained.
5. Special cases:
   a. If the initial solution obtained by use of the method given in the preceding is not feasible, a feasible solution can be obtained by adding more artificial variables which must then be forced out of the final solution.
   b. Degeneracy may occur in the simplex method when the outgoing variable is selected. If there are two or more minimal values of the same size, the problem is degenerate, and a poor choice of the outgoing variable may result in cycling, although cycling almost never occurs in real problems. This can be eliminated by a method of multiplying each element in the rows in question by the positive coefficients of the kth column and choosing the row for the outgoing variable as the one first containing the smallest algebraic ratio.
6. The preceding method for obtaining a maximum as the objective function can be applied to the case when the objective function is a minimum by recognizing that maximizing the negative of a function is equivalent to minimizing the function.

OPTIMIZATION APPLICATIONS

Optimization Application: Optimum Production Rates in Plant Operation

The same principles used for developing an optimum design can be applied to determine the most favorable conditions in the operation of a manufacturing plant. One of the most important variables in any plant operation is the amount of product produced per unit of time. The production rate depends on many factors, such as the number of hours in operation per day, per week, or per month; the load placed on the equipment; and the sales market available. From an analysis of the costs involved under different situations and consideration of other factors affecting the particular plant, it is possible to determine an *optimum rate of production* or an *economic lot size*.

When a design engineer submits a complete plant design, the study ordinarily is based on a given production capacity for the plant. After the plant is put into operation, however, some of the original design factors will have changed, and the optimum rate of production may vary considerably from the original design capacity. For example, suppose a plant was designed originally for the batchwise production of an organic chemical, providing one batch every 8 h. After the plant has initiated production, cost data on the actual process are obtained, and tests with various operating procedures are conducted. It is determined that additional production per month can be achieved if the time per batch is reduced and the number of batches is increased. However, with the shorter batch time, not only is more labor required, but also the percent conversion of raw materials is reduced and steam and power costs increase. Here is an obvious case in which an economic balance can be used to establish the optimum production rate. Although the design engineer may have based the original recommendations on a similar type of economic balance, price and market conditions do not remain constant, and the operations engineer now has actual results on which to update the economic evaluation. The following analysis indicates the general method for determining economic production rates or lot sizes.

The total product cost per unit of time may be divided into the classifications of *operating costs* and *organization costs*. Operating costs depend on the rate of production and include expenses for direct labor, raw materials, power, heat, supplies, and similar items which are a function of the amount of material produced. Organization costs, on the other hand, are due to expenses for directive personnel, physical equipment, and other services or facilities which must be maintained irrespective of the amount of material produced. Organization costs are independent of the rate of production.

It is convenient to consider operating costs on the basis of one unit of production. When this is done, the operating costs can be divided into two types of expenses as follows: (1) minimum expenses for raw materials, labor, power, etc., that remain constant and must be paid for each unit of production as long as any material is produced; and (2) extra expenses due to increasing the rate of production. These extra expenses are known as *superproduction costs*. They become particularly important at high rates of production. Examples of superproduction costs are extra expenses caused by overload on power facilities, additional labor requirements, or decreased efficiency of conversion. Superproduction costs can often be represented in the following manner:

$$\text{Superproduction costs per unit of production} = m\mathcal{P}_r^n \tag{9-42}$$

where $\mathcal{P}_r$ is the rate of production in term of total units produced per unit of time, and m and n are constants.

Designating $\hbar$ as the operating costs which remain constant per unit of production and O_c as the organization costs per unit of time, the total product cost c_T per unit of production is

$$c_T = \hbar + m\mathcal{P}_r^n + \frac{O_c}{\mathcal{P}_r} \tag{9-43}$$

The following equations for various types of costs or profits are based on Eq. (9-43):

$$C_T = c_T Ᵽ_r = \left(\hbar + mᵽ^n_r + \frac{O_c}{Ᵽ_r} \right) Ᵽ_r \tag{9-44}$$

$$ᵽ_r = s - c_T = s - \hbar - mᵽ^n_r - \frac{O_c}{Ᵽ_r} \tag{9-45}$$

$$R' = ᵽ_r Ᵽ_r = \left(s - \hbar - mᵽ^n_r - \frac{O_c}{Ᵽ_r} \right) Ᵽ_r \tag{9-46}$$

where $ᵽ_r$ is the profit per unit of production, C_T the cost per unit time, R' the profit per unit of time, and s the selling price per unit of production.

Optimum Production Rate for Minimum Cost per Unit of Production Quite often there is a need to know the rate of production which will give the least cost on the basis of one unit of material produced. At this particular optimum rate, a plot of the total product cost per unit of production versus the production rate shows a minimum product cost; therefore, the optimum production rate must occur where $dc_T/dᵽ_o = 0$. An analytical solution for this case may be obtained from Eq. (9-44), and the optimum rate $Ᵽ_{r,o}$ giving the minimum cost per unit of production is found as follows:

$$\frac{dc_T}{dᵽ} = 0 = nmᵽ^{n-1}_{r,o} - \frac{O_c}{Ᵽ^2_{r,o}} \tag{9-47}$$

$$Ᵽ_{r,o} = \left(\frac{O_c}{nm} \right)^{1/(n+1)} \tag{9-48}$$

The optimum rate shown in Eq. (9-48) would, of course, give the maximum profit per unit of production if the selling price exceeds the total product cost.

Optimum Production Rate for Maximum Total Profit per Unit of Time In most business operations, the amount of money earned over a given time is much more important than the amount of money earned for each unit of product sold. Therefore, it is necessary to recognize that the production rate for maximum profit per unit of time may differ considerably from the production rate for minimum cost per unit of production.

Equation (9-46) presents the relationship between costs and profits. A plot of profit per unit of time versus production rate goes through a maximum. Equation (9-46), therefore, can be used to find an analytical value for the optimum production rate. When the selling price remains constant, the optimum rate providing the maximum profit per unit of time is

$$Ᵽ_{r,o} = \left[\frac{s - \hbar}{(n+1)m} \right]^{1/n} \tag{9-49}$$

Example 9-2 illustrates the preceding principles and shows the analytical solution for the situation presented in Fig. 9-7.

Determination of Profits at Optimum Production Rates — EXAMPLE 9-2

A plant produces small water pumps at the rate of P_r units per day. The variable costs per pump have been established to be $47.73 + 0.1P_r^{1.2}$. The total daily fixed charges are $1750, and all other expenses are constant at $7325 per day. If the selling price per pump is $173, determine

a. The daily profit at a production schedule giving the minimum cost per pump
b. The daily profit at a production schedule giving the maximum daily profit
c. The production schedule at the breakeven point

■ Solution

a. Total cost per pump is given by

$$C_T = 47.73 + 0.1P_r^{1.2} + \frac{1750 + 7325}{P_r}$$

At a production schedule for minimum cost per pump,

$$\frac{dC_T}{dP_r} = 0 = 0.12P_o^{0.2} - \frac{9075}{P_{r,o}^2}$$

$$P_o = 165 \text{ units/day for minimum cost per unit}$$

Daily profit at a production schedule for minimum cost per pump is

$$R' = \left[173 - 47.73 - 0.1(165)^{1.2} - \frac{9075}{165}\right]165 = \$4040$$

b. Daily profit is

$$R' = \left(173 - 47.73 - 0.1P_r^{1.2} - \frac{1750 + 7325}{P_r}\right)P_r$$

At a production schedule for maximum profit per day,

$$\frac{dR'}{dP_r} = 0 = 125.27 - 0.22P_{r,o}^{1.2}$$

$$P_{r,o} = 198 \text{ units/day for maximum daily profit}$$

Daily profit at a production schedule for maximum daily profit is

$$R'\left[173 - 47.73 - 0.1(198)^{1.2} - \frac{9075}{198}\right]198 = \$4440$$

c. Total profit per day at the breakeven point is given by

$$\left[173 - \left(47.73 + 0.1P_r^{1.2} + \frac{1750 + 7325}{P_r}\right)\right]P_r = 0$$

Solving the preceding equation for P_r gives

$$P_{\text{at breakeven point}} = 88 \text{ units/day}$$

■

Optimization Application: Cyclic Operations

Many processes are carried out by the use of cyclic operations that involve periodic shutdowns for discharging, cleanout, or reactivation. This type of operation occurs when the product is produced by a batch process or when the rate of production decreases with time, as in the operation of a plate-and-frame filtration unit. In a true batch operation, no product is obtained until the unit is shut down for discharging. In semicontinuous cyclic operations, product is delivered continuously while the unit is in operation, but the rate of delivery decreases with time. Thus, in batch or semicontinuous cyclic operations, the variable of total time required per cycle must be considered when determining optimum conditions.

Analyses of cyclic operations can be carried out conveniently by using the time for one cycle as a basis. When this is done, relationships similar to the following can be developed to express overall factors, such as total annual cost or annual rate of production:

$$\text{Total annual cost} = (\text{operating and shutdown costs per cycle})(\text{cycles per year}) + \text{annual fixed costs} \tag{9-50}$$

$$\text{Annual production} = (\text{production per cycle})(\text{cycles per year}) \tag{9-51}$$

$$\text{Cycles per year} = \frac{\text{operating and shutdown time used per year}}{\text{operating and shutdown time per cycle}} \tag{9-52}$$

Example 9-3 illustrates the general method for determining optimum conditions in a batch operation.

EXAMPLE 9-3 Determination of Conditions for Minimum Total Cost in a Batch Operation

An organic chemical is being produced by a batch operation in which no product is obtained until the batch is finished. Each cycle consists of the operating time necessary to complete the reaction plus an additional time of 1.4 h required for discharging and charging. The operating time per cycle is equal to $1.5P_b^{0.25}$ h, where P_b is the kilograms of product produced per batch. The operating costs during the operating period are \$20 per hour, while the costs during the discharge-charge period are \$15 per hour. The annual fixed costs C_F for the equipment vary with the size of the batch in the following manner:

$$C_F = 340P_b^{0.8} \text{ \$/yr}$$

Inventory and storage charges may be neglected. If necessary, the plant can be operated 24 h/day for 300 days/yr. The annual production is 10^6 kg of product. At this capacity, raw material and miscellaneous costs, other than those already mentioned, amount to \$260,000 per year. Determine the cycle time for conditions of minimum total cost per year.

■ **Solution**

$$\text{Cycles/year} = \frac{\text{annual production}}{\text{production/cycle}} = \frac{1{,}000{,}000}{P_b}$$

$$\text{Cycle time} = \text{operating time} + \text{shutdown time} = 1.5P_b^{0.25} + 1.4 \text{ h}$$

$$\frac{\text{Operating + shutdown costs}}{\text{cycle}} = 20\left(1.5P_b^{0.25}\right) + (15)(1.4)\ \$/\text{cycle}$$

$$\text{Annual fixed costs} = 340P_b^{0.8} + 260{,}000\ \$/\text{yr}$$

$$\text{Total annual costs} = \frac{\left[20\left(1.5P_b^{0.25}\right) + 15(1.4)\right]1{,}000{,}000}{P_b} + 340P_b^{0.8} + 260{,}000\ \$/\text{yr}$$

The total annual cost is a minimum where $d(\text{total annual cost})/dP_b = 0$.

Performing the differentiation, setting the result equal to zero, and solving for P_b give

$$P_{b,\text{optimum cost}} = 1630\ \text{kg/batch}$$

This same result could have been obtained by plotting total annual cost versus P_b and determining the value of P_b at the point of minimum annual cost.

For conditions of minimum annual cost and 10^6 kg/yr production,

$$\text{Cycle time} = 1.5(1630)^{0.25} + 1.4 = 11\ \text{h}$$

$$\text{Total time used per year} = \frac{11(1{,}000{,}000)}{1630} = 6750\ \text{h}$$

$$\text{Total time available per year} = 300(24) = 7200\ \text{h}$$

Thus, for conditions of minimum annual cost and a production of 10^6 kg/yr, not all the available operating and shutdown time would be needed.

■

Semicontinuous Cyclic Operations Such types of operation are often encountered in the chemical industry, and the design engineer should understand the methods for determining optimum cycle times in this type of operation. Although product is delivered continuously, the rate of delivery decreases with time due to scaling, collection of by-product, reduction in conversion efficiency, or similar causes. It becomes necessary, therefore, to shut down the operation periodically to restore the original conditions for high production rates. The optimum cycle time can be determined for conditions such as maximum amount of production per unit of time or minimum cost per unit of production.

Scale Formation in Evaporation A situation often encountered in cyclic operations involving evaporation is scale formation. During evaporation operations, solids often deposit on the heat-transfer surfaces, forming a scale. The continuous formation of the scale causes a gradual increase in the resistance to the flow of heat and, consequently, a reduction in the rate of heat transfer and rate of evaporation if the same temperature-difference driving forces are maintained. Under these conditions, the evaporation unit must be shut down and cleaned after an optimum operation time, and the cycle is repeated.

Scale formation occurs to some extent in all types of evaporators, but it is of particular importance when the feed mixture contains a dissolved material that has an inverted solubility. The expression *inverted solubility* indicates that the solubility decreases as the temperature of the solution is increased. For a material of this type, the solubility is lower near the heat-transfer surface where the temperature is the largest. Thus, any solid crystallizing out of the solution does so near the heat-transfer surface and is quite likely to form a scale on this surface. The most common scale-forming substances are calcium sulfate, calcium hydroxide, sodium carbonate, sodium sulfate, and calcium salts of certain organic acids.

When true scale formation occurs, the overall heat-transfer coefficient may be related to the time the evaporator has been in operation by the equation.[†]

$$\frac{1}{U^2} = a\theta_b + c \tag{9-53}$$

where a and c are constants for any given operation and U is the overall heat-transfer coefficient at any operating time θ_b since the beginning of the evaporation process.

If it is not convenient to determine the heat-transfer coefficients and the related constants as shown in Eq. (9-53), any quantity that is proportional to the heat-transfer coefficient may be used. Thus, if all conditions except scale formation are constant, the feed rate, production rate, and evaporation rate can each be represented in a form similar to Eq. (9-53). Any of these equations can be used as a basis for determining the optimum conditions. The general method is illustrated by the following treatment, which employs Eq. (9-53) as a basis.

If Q represents the total amount of heat transferred in the operating time θ_b, and A and ΔT represent, respectively, the heat-transfer area and the temperature-difference driving force, then the rate of heat transfer at any instant is

$$\frac{dQ}{d\theta_b} = UA\,\Delta T = \frac{A\,\Delta T}{(a\theta_b + c)^{1/2}} \tag{9-54}$$

The instantaneous rate of heat transfer varies during the time of operation, but the heat-transfer area and the temperature-difference driving force remain essentially constant. Therefore, the total amount of heat transferred during an operating time of θ_b can be determined by integrating Eq. (9-54) as follows:

$$\int_0^Q dQ = A\,\Delta T \int_0^{\theta_b} \left(\frac{1}{a\theta_b + c}\right)^{1/2} d\theta_b \tag{9-55}$$

$$Q = \frac{2A\,\Delta T}{a}[(a\theta_b + c)^{1/2} - c^{1/2}] \tag{9-56}$$

Cycle Time for Heat Transfer Another common concern in cyclic operations is the determination of the heat-transfer cycle time. There are two main approaches for

[†]W. L. McCabe and C. Robinson, *Ind. Eng. Chem.*, **16:** 478 (1924).

determining cycle times for heat transfer. The first is to find the cycle time that maximizes the heat transfer. Equation (9-56) can be used as a basis for determining the cycle time which will permit the maximum amount of heat transfer during a given period. Each cycle consists of an operating (or boiling) time of θ_b h. If the time per cycle for emptying, cleaning, and recharging is θ_c, the total time in hours per cycle is $\theta_t = \theta_b + \theta_c$. Therefore, if we designate the total time used for actual operation, emptying, cleaning, and refilling as H, the number of cycles during the total time used is given by $H/(\theta_b + \theta_c)$.

The total amount of heat transferred during time H is then given by

$$Q_H = \left(\frac{Q}{\text{cycle}}\right)\left(\frac{\text{cycles}}{H}\right)$$

Therefore,

$$Q_H = \frac{2A\,\Delta T}{a}[(a\theta_b + c)^{1/2} - c^{1/2}]\frac{H}{\theta_b + \theta_c} \tag{9-57}$$

Under ordinary conditions, the only variable in Eq. (9-57) is the operating time θ_b. A plot of the total amount of heat transferred versus θ_b shows a maximum at the optimum value of θ_b. Figure 9-8 presents a plot of this type. The optimum cycle time can also be obtained by setting the derivative of Eq. (9-57) with respect to θ_b equal to zero

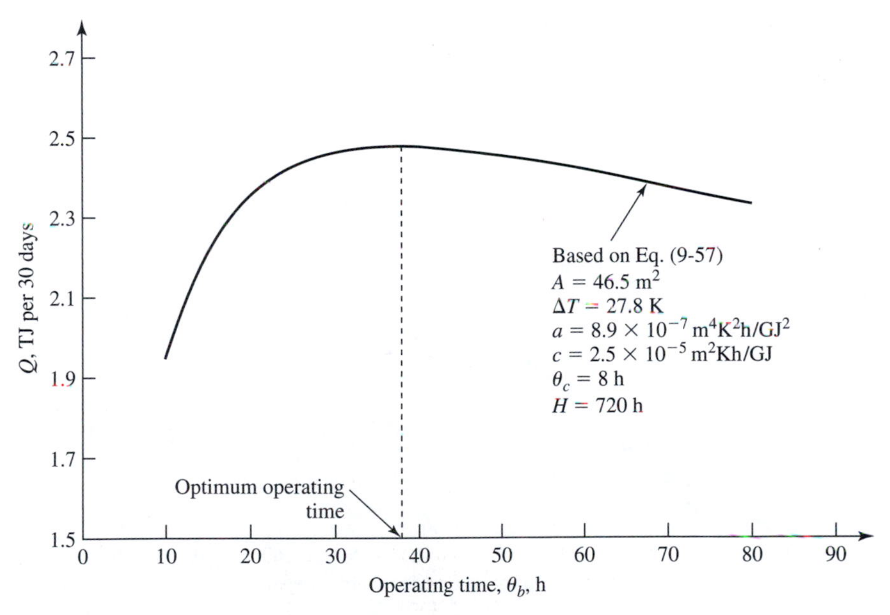

Figure 9-8
Determination of optimum operating time for maximum amount of heat transfer in evaporator with scale formation

and solving for θ_b. The result is

$$\theta_{b,\text{per cycle for maximum amount of heat transferred}} = \theta_c + \left(\frac{2}{a}\right)(ac\theta_c)^{1/2} \tag{9-58}$$

The optimum boiling time given by Eq. (9-58) shows the operating schedule necessary to permit the maximum amount of heat transfer. All the time available for operation, emptying, cleaning, and refilling should be used. For constant operating conditions, this same schedule would also give the maximum amount of feed consumed, product obtained, and liquid evaporated.

A third method for determining the optimum cycle time is known as the *tangential method for finding optimum conditions,* and this method is applicable to many types of cyclic operations. The method is illustrated for conditions of constant cleaning time θ_c in Fig. 9-9 where a plot of the amount of heat transferred as a function of boiling time is presented. Curve *OB* is based on Eq. (9-56). The *average* amount of heat transferred per unit of time during one complete cycle is $Q/(\theta_b + \theta_c)$. When the total amount of heat transferred during a number of repeated cycles is a maximum, the average amount of heat transferred per unit of time must also be a maximum. The optimum cycle time, therefore, occurs when $Q/(\theta_b + \theta_c)$ is a maximum.

The straight line CD' in Fig. 9-9 starts at a distance equivalent to θ_c on the left of the plot origin. The slope of this straight line is $Q/(\theta_b + \theta_c)$, with the values of Q and θ_b determined by the point of intersection between line CD' and curve *OB*. The maximum value of $Q/(\theta_b + \theta_c)$ occurs when line *CD* is tangent to curve *OB*, and the point

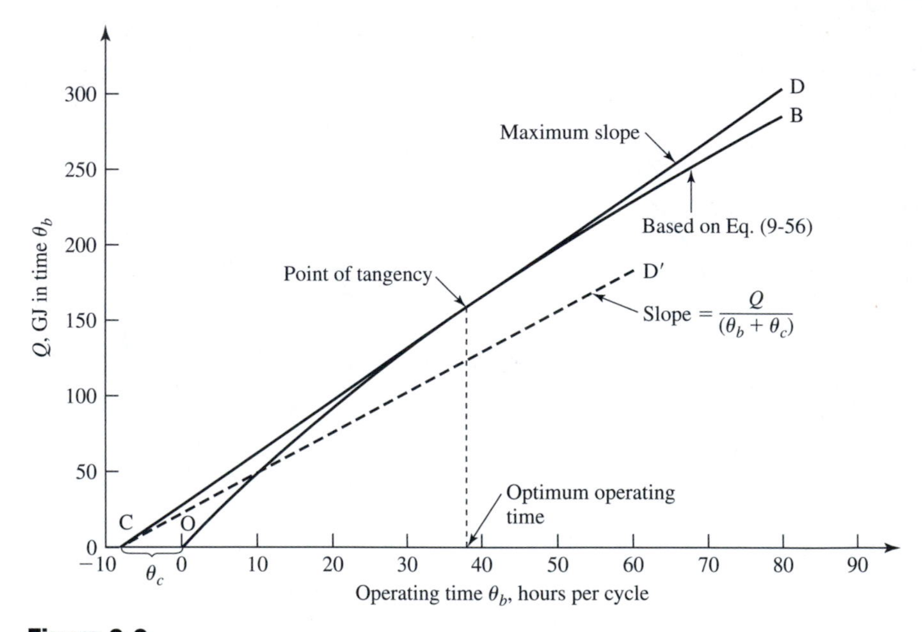

Figure 9-9
Tangential method for finding optimum operating time for maximum amount of heat transfer in evaporator with scale formation

of tangency indicates the optimum value of the boiling time per cycle for conditions of maximum amount of heat transfer.

Another approach commonly used for establishing heat-transfer cycle times involves the determination of cycle times to minimize the cost per unit of heat transfer. There are many different circumstances which may affect the minimum cost per unit of heat transferred in operations involving evaporation. One simple and commonly occurring case will be considered. It may be assumed that an evaporation unit of fixed capacity is available, and a definite amount of feed and evaporation must be processed each day. The total cost for one cleaning and inventory charge is assumed to be constant no matter how much boiling time is used. The problem is to determine the cycle time that will permit operation at the least total cost.

The total cost includes (1) fixed charges on the equipment and fixed overhead expenses; (2) steam, materials, and storage costs which are proportional to the amount of feed and evaporation; (3) expenses for direct labor during the actual evaporation operation; and (4) cost of cleaning. Since the size of the equipment and the amounts of feed and evaporation are fixed, the costs included in items 1 and 2 are independent of the cycle time. The optimum cycle time, therefore, can be found by minimizing the sum of the costs for cleaning and for direct labor during the evaporation.

If C_c represents the cost for one cleaning and S_b is the direct labor cost per hour during operation, the total variable costs during H h of operating and cleaning time must be

$$C_{T,\,\text{for}\,H} = (C_c + S_b\theta_b)\frac{H}{\theta_b + \theta_c} \tag{9-59}$$

Equations (9-57) and (9-59) may be combined to give

$$C_{T,\,\text{for}\,H} = \frac{aQ_H(C_c + S_b\theta_b)}{2A\,\Delta T[(a\theta_b + c)^{1/2} - c^{1/2}]} \tag{9-60}$$

The optimum value of θ_b for minimum total cost may be obtained by plotting C_T versus θ_b or by setting the derivative of Eq. (9-60) with respect to θ_b equal to zero and solving for θ_b. The result is

$$\theta_{b,\,\text{per cycle for minimum total cost}} = \frac{C_c}{S_b} + \frac{2}{aS_b}(acC_cS_b)^{1/2} \tag{9-61}$$

Equation (9-61) shows that the optimum cycle time is independent of the required amount of heat transfer Q_H. Therefore, a check must be made to make certain the optimum cycle time for minimum cost permits the required amount of heat transfer. This can be done easily by using the following equation, which is based on Eq. (9-57):

$$\theta_t = \theta_b + \theta_c = \frac{2AH'\Delta T}{aQ_H}[(a\theta_{b,\text{opt}} + c)^{1/2} - c^{1/2}] \tag{9-62}$$

where H' is the total time *available* for operation, emptying, cleaning, and recharging. If θ_t is equal to or greater than $\theta_{b,\text{opt}} + \theta_c$, the optimum boiling time indicated by

Eq. (9-61) can be used, and the required production can be obtained at minimum-cost conditions.

The optimum cycle time determined by the preceding methods may not fit into convenient operating schedules. Fortunately, as shown in Figs. 9-8 and 9-9, the optimum conditions usually occur where a considerable variation in cycle time has little effect on the objective function. It is possible, therefore, to adjust the cycle time to fit a convenient operating schedule without causing too much change in the final results.

The approach described in the preceding sections can be applied to many different types of semicontinuous cyclic operations. An illustration showing how the same approach is used for determining optimum cycle times in filter-press operations is presented in Example 9-4.

EXAMPLE 9-4 Cycle Time for Maximum Amount of Production from a Plate-and-Frame Filter Press

Tests with a plate-and-frame filter press, operated at constant pressure, have shown that the relation between the volume of filtrate delivered and the time in operation can be represented as

$$V_f^2 = 18(\theta_f + 0.11)$$

where V_f is the filtrate in cubic meters delivered during θ_f h of filtering time.

The cake formed in each cycle must be washed with an amount of water equal to one-sixteenth times the volume of filtrate delivered per cycle. The washing rate remains constant and is equal to one-fourth of the filtrate delivery rate at the end of the filtration. The time required per cycle for dismantling, dumping, and reassembling is 6 h. Under the conditions where the preceding information applies, determine the total cycle time necessary to permit the maximum output of filtrate during each 24 h.

■ Solution

Let θ_f represent the hours of filtering time per cycle.

$$\text{Filtrate delivered per cycle} = V_{f,\text{cycle}} = 4.24(\theta_f + 0.11)^{1/2}\ \text{m}^3$$

Rate of filtrate delivery at the end of the cycle is

$$(\text{Washing rate})(4) = \frac{dV_f}{d\theta_f} = \frac{4.24}{2}\frac{1}{(\theta_f + 0.11)^{1/2}}\ \text{m}^3/\text{h}$$

$$\text{Time for washing} = \frac{\text{volume of wash water}}{\text{washing rate}}$$

$$= \frac{\frac{1}{16}[4.24(\theta_f + 0.11)^{1/2}]}{\frac{1}{4}[(4.24/2)(\theta_f + 0.11)^{1/2}]}$$

$$= \frac{1}{2}(\theta_f + 0.11)\ \text{h}$$

$$\text{Total time per cycle} = \theta_f + \frac{1}{2}(\theta_f + 0.11) + 6 = 1.5\theta_f + 6.06$$

$$\text{Cycle per 24 h} = \frac{24}{1.5\theta_f + 6.06}$$

Filtrate delivered in 24 h is given by

$$\text{P}_{f,\text{cycle}}\,(\text{cycles per 24 h}) = 4.24(\theta_f + 0.11)^{1/2}\left(\frac{24}{1.5\theta_f + 6.06}\right)$$

At the optimum cycle time, the maximum output of filtrate per 24-h period is obtained from

$$\frac{d(\text{volume filtrate delivered per 24 h})}{d\theta_f} = 0$$

Performing the differentiation and solving for θ_f yield

$$\theta_{f,\text{opt}} = 3.8\text{ h}$$

Total cycle time necessary to permit the maximum output of filtrate is therefore

$$1.5\theta_f + 6.06 = 1.5(3.8) + 6.06 = 11.8\text{ h}$$

Optimization Application: Economic Pipe Diameter

A classic example showing how added refinements can be brought into an optimization analysis involves the development of methods used to determine the optimum economic pipe diameter in a flow system. The following analysis gives a detailed derivation to illustrate how simplified expressions for optimum conditions can be developed.

The investment for piping and pipe fittings can amount to an important part of the total investment for a chemical plant. It is necessary, therefore, to choose pipe sizes that provide a minimum total cost for pumping and fixed charges. For any given set of flow conditions, the use of an increased pipe diameter will cause an increase in the fixed charges for the piping system and a decrease in the pumping charges. Therefore, an optimum economic pipe diameter must exist. The value of this optimum diameter can be determined by combining the principles of fluid dynamics with cost considerations. The optimum economic pipe diameter is found at the point at which the sum of pumping costs and fixed charges based on the cost of the piping system is a minimum.

Pumping Costs For any given operating conditions involving the flow of an incompressible fluid through a pipe of constant diameter, the total mechanical energy balance can be reduced to the form

$$W_s = \frac{2\mathfrak{f}V^2L(1+J)}{D_i} + B \tag{9-63}$$

where W_s is the mechanical work added to the system from an external mechanical source, $\mathfrak{f}$ the Fanning friction factor, V the average linear velocity of the fluid, L the length of the pipe, J the frictional loss due to fittings and bends expressed as equivalent fractional loss in a straight pipe, D_i the inside diameter of the pipe, and B a constant that takes all other factors of the mechanical energy balance into consideration.

In the region of turbulent flow (Re > 2100), $\mathfrak{f}$ may be approximated for new steel pipes by the equation

$$\mathfrak{f} = \frac{0.04}{(\text{Re})^{0.16}} \tag{9-64}$$

where Re is the Reynolds number. If the flow is viscous (Re < 2100),

$$\mathfrak{f} = \frac{16}{\text{Re}} \tag{9-65}$$

By combining Eqs. (9-63) and (9-64) and applying the necessary conversion factors, the following equation can be obtained, representing the annual pumping cost when the flow is turbulent:

$$C_{\text{pumping}} = \frac{1.248 \times 10^{-4} q_f^{2.84} \rho^{0.84} \mu_c^{0.16} K(1+J)H_y}{D_i^{4.84} E} + B' \tag{9-66}$$

where C_{pumping} is the pumping costs in dollars per year per meter of pipe length when the flow is turbulent, q_f the fluid volumetric flow rate in m^3/s, ρ the fluid density in kg/m^3, μ_c the fluid viscosity in Pa·s, K the cost of electrical energy in \$/kWh, H_y the hours of operation per year, E the efficiency of the motor and pumps expressed as a fraction, and B' a constant independent of D_i.

Similarly, Eqs. (9-63) and (9-65) and the necessary conversion factors can be combined to give the annual pumping costs when the flow is viscous:

$$C'_{\text{pumping}} = \frac{0.0407 q_f^2 \mu_c K(1+J)H_y}{D_i^4 E} + B' \tag{9-67}$$

where C'_{pumping} is the pumping cost in dollars per year per meter of pipe length when the flow is viscous.

Equations (9-66) and (9-67) apply to incompressible fluids. In engineering calculations, these equations are also generally accepted for gases if the total pressure drop is less than 10 percent of the initial pressure.

Fixed Charges for Piping System For most types of pipe, a plot of the logarithm of the pipe diameter versus the logarithm of the purchase cost per meter of pipe is essentially a straight line. Therefore, the purchase cost for pipe may be represented by

$$c_{\text{pipe}} = X\left(\frac{D_i}{0.0254}\right)^n \tag{9-68}$$

where c_{pipe} is the purchase cost of new pipe per meter of pipe length, X the purchase cost of new 0.0254-m inside-diameter pipe per meter of pipe length, and n a constant whose value depends on the type of pipe selected.

The annual cost for the installed piping system may be expressed as

$$C_{\text{pipe}} = (1 + F)X\left(\frac{D_i}{0.0254}\right)^n K_F \tag{9-69}$$

where C_{pipe} is the cost for the installed piping system in dollars per year per meter of pipe length,† F the ratio of the total cost for fittings and installation to the purchase cost for the new pipe, and K_F the annual fixed charges including maintenance, expressed as a fraction of initial cost for the completely installed pipe.

Optimum Economic Pipe Diameter The total annual cost for the piping system and pumping can be obtained by adding Eqs. (9-66) and (9-69) or Eqs. (9-67) and (9-69). The only variable in the resulting total cost expressions is the pipe diameter. The optimum economic pipe diameter can be found by taking the derivative of the total annual cost with respect to pipe diameter, setting the result equal to zero, and solving for the inside diameter of the pipe. This procedure gives the following results:

For turbulent flow,

$$D_{i,\text{opt}} = \left[\frac{6.04 \times 10^{-4}(0.0254)^n q_f^{2.84} \rho^{0.84} \mu_c^{0.16} K(1 + J)H_y}{n(1 + F)XEK_F}\right]^{1/(4.84+n)} \tag{9-70}$$

For viscous flow,

$$D_{i,\text{opt}} = \left[\frac{0.1628\,(0.0254)^n q_f^2 \mu_c K(1 + J)H_y}{n(1 + F)XEK_F}\right]^{1/(4.0+n)} \tag{9-71}$$

The value of n for steel pipes is approximately 1.5 if the pipe diameter is 0.0254 m or larger and 1.0 if the diameter is less than 0.0254 m. Substituting these values in Eqs. (9-70) and (9-71) gives the following:

For turbulent flow in steel pipes,

$D_i \geq 0.0254$ m:

$$D_{i,\text{opt}} = q_f^{0.448} \rho^{0.132} \mu_c^{0.025} \left[\frac{1.63 \times 10^{-6} K(1 + J)H_y}{(1 + F)XEK_F}\right]^{0.158} \tag{9-72}$$

$D_i < 0.0254$ m:

$$D_{i,\text{opt}} = q_f^{0.487} \rho^{0.144} \mu_c^{0.027} \left[\frac{1.53 \times 10^{-5} K(1 + J)H_y}{(1 + F)XEK_F}\right]^{0.171} \tag{9-73}$$

For viscous flow in steel pipes,

$D_i \geq 0.0254$ m:

$$D_{i,\text{opt}} = q_f^{0.364} \mu_c^{0.182} \left[\frac{4.39 \times 10^{-4} K(1 + J)H_y}{(1 + F)XEK_F}\right]^{0.182} \tag{9-74}$$

---

†Pump cost could be included if desired; however, in this analysis, the cost of the pump is considered as essentially invariant with pipe diameter.

$D_i < 0.0254$ m:

$$D_{i,\text{opt}} = q_f^{0.40} \mu_c^{0.20} \left[\frac{4.14 \times 10^{-3} K(1+J)H_y}{(1+F)XEK_F} \right]^{0.20} \tag{9-75}$$

The exponents in Eqs. (9-72) through (9-75) indicate that the optimum diameter is relatively insensitive to most of the terms. It is possible to simplify the equations further by substituting average numerical values for some of the less critical terms. The following values are applicable under ordinary industrial conditions:

$K = \$0.05/\text{kWh}$
$J = 0.35$, or 35 percent
$H_y = 8760$ h/yr
$E = 0.50$, or 50 percent
$F = 1.4$
$K_F = 0.20$, or 20 percent
$X = \$2.43$ per meter for a 0.0254-m D_i steel pipe

Substituting these values into Eqs. (9-72) through (9-75) gives the following simplified results:

For turbulent flow in steel pipes,

$D_i \geq 0.0254$ m:

$$D_{i,\text{opt}} = 0.363 q_f^{0.45} \rho^{0.13} \mu_c^{0.025} = \frac{0.363 \dot{m}^{0.45} \mu_c^{0.025}}{\rho^{0.32}} \tag{9-76}$$

where $\dot{m}$ is the mass flow rate, kg/s.

$D_i < 0.0254$ m:

$$D_{i,\text{opt}} = 0.49 q_f^{0.49} \rho^{0.14} \mu_c^{0.027} \tag{9-77}$$

For viscous flow in steel pipes,

$D_i \geq 0.0254$ m:

$$D_{i,\text{opt}} = 0.863 q_f^{0.36} \mu_c^{0.18} \tag{9-78}$$

$D_i < 0.0254$ m:

$$D_{i,\text{opt}} = 1.33 q_f^{0.40} \mu_c^{0.20} \tag{9-79}$$

Depending on the accuracy desired and the type of flow, Eqs. (9-70) through (9-79) may be used to estimate optimum economic pipe diameters.† The simplified Eqs. (9-76) through (9-79) are sufficiently accurate for design estimates under ordinary

†This type of approach was first proposed by R. P. Genereaux, *Ind. Eng. Chem.*, **29:** 385 (1937) and *Chem. Met. Eng.*, **44**(5): 281 (1937), and B. R. Sarchet and A. P. Colburn, *Ind. Eng. Chem.*, **32:** 1249 (1940).

plant conditions, and as shown later in Table 9-4, the diameter estimates obtained are usually on the safe side in that added refinements in the calculation methods generally tend to result in smaller diameters. A nomograph based on these equations is presented in Chap. 12.

Analysis Including Tax Effects and Return on Investment The preceding analysis clearly neglects a number of factors that may have an influence on the optimum economic pipe diameter, such as return on investment, cost of pumping equipment, taxes, and the time value of money. If the preceding development of Eq. (9-70) for turbulent flow is refined to include the effects of taxes and the cost of capital (or return on investment) plus a more accurate expression for the frictional loss due to fittings and bends, the result is:[†]

For turbulent flow,

$$\frac{D_{\text{opt}}^{4.84+n}}{1+0.794L_e'D_{\text{opt}}} = \frac{7.61\times 10^{-5}YK\dot{m}^{2.84}\mu_c^{0.16}\{[1+(a'+b')M](1-\Phi)+ZM\}}{n(1+F)X'[Z+(a+b)(1-\Phi)]E\rho^2} \tag{9-80}$$

where D_{opt} is the optimum economic inside diameter in m, X' the purchase cost of new pipe per meter of pipe length when the inside pipe diameter is expressed in meters based on $c_{\text{pipe}} = X'(D/0.0254)^n$ in \$/m, L_e' the frictional loss due to fittings and bends, expressed as equivalent pipe length in pipe diameters per unit length of pipe divided by the optimum pipe diameter, $\dot{m}$ the mass flow rate in kg/s, M the ratio of total cost for pumping installation to yearly cost of pumping power required, Y the days of operation per year, a the fraction of initial cost of the installed piping system for annual depreciation, a' the fraction of initial cost of the pumping installation for annual depreciation, b the fraction of initial cost of the installed piping system for annual maintenance, b' the fraction of initial cost of pumping installation for annual maintenance, Φ the fractional rate of income tax, and Z the acceptable fractional rate of return (or cost of capital before taxes) on the incremental investment.

By using the values given in Table 9-3 for turbulent flow in steel pipes, Eq. (9-80) simplifies to

$$D_{\text{opt}} = \frac{0.38(1+1.865\,D_{\text{opt}})^{0.158}\dot{m}^{0.45}\mu_c^{0.025}}{(1+F)^{0.158}\rho^{0.32}} \tag{9-81}$$

Sensitivity of Results The simplifications made in obtaining Eqs. (9-76) to (9-79) and Eq. (9-81) illustrate an approach that can be used for approximate results when certain variables appear in a form where relatively large changes in them have little effect on the final results. The variables appearing in Table 9-3 and following Eq. (9-75) are relatively independent of pipe diameter, and each is raised to a small power for the final determination of diameter. Thus, the final results are not particularly sensitive to

[†]A similar equation was presented by J. H. Perry and C. H. Chilton, eds., *Perry's Chemical Engineers' Handbook*, 5th ed., McGraw-Hill, New York, 1973, p. 5–32. See Prob. 9-15 for derivation of Eq. (9-80) and comparison to form of equation given in above-mentioned reference.

Table 9-3 Values of variables used to obtain Eq. (9-81)†

Turbulent flow, steel pipe diameter 0.025 m or larger			
Variable	Value used	Variable	Value used
n	1.5	Φ	0.35
L'_e	2.35	Z	0.20
Y	328	X	2.43
K	0.05	$a+b$	0.18
μ_c	1×10^{-3}	$a'+b'$	0.40
M	0.8	E	0.50

The variable F is a function of diameter and can be approximated by $F = 3 + 0.23/D_{\text{opt}}$.
†For turbulent flow in a steel pipe with $D \geq 0.0254$ m.

Table 9-4 Comparison of optimum economic pipe diameter estimated from Eqs. (9-76) and (9-81)†

$D_{i,\text{opt}}$, m			
By Eq. (9-81)	By Eq. (9-76)	$\dot{m}$, kg/s	ρ, kg/m^3
0.21	0.25	204	3200
0.12	0.14	2.04	32
0.071	0.091	5.68	560
0.036	0.049	0.204	32

†Turbulent flow, schedule 40 steel pipe, approximate 4.5-m spacing of fittings.

the variables listed in Table 9-3, and the practical engineer may decide that the simplification obtained by using the approximate equations is worth the slight loss in absolute accuracy.

Table 9-4 shows the extent of change in optimum economic diameter obtained by using Eq. (9-81) versus Eq. (9-76) and illustrates the effect of bringing in added refinements as well as changes in values of some of the variables.

Optimization Application: Cooling Water Flow Rate

If a condenser, with water as the cooling medium, is designed to carry out a given duty, the cooling water may be circulated at a high rate with a small change in water temperature or at a low rate with a large change in water temperature. The temperature of the water affects the temperature-difference driving force for heat transfer. Use of an increased amount of water, therefore, will cause a reduction in the necessary amount of heat-transfer area and a resultant decrease in the original investment and fixed charges. On the other hand, the cost for the water will increase if more water is used. An economic balance between conditions of high water rate and low surface area, and low water rate and high surface area, indicates that the optimum flow rate of cooling water occurs at the point of minimum total cost for cooling water and equipment fixed charges.

Consider the general case in which heat must be removed from a condensing vapor at a given rate designated by $\dot{q}$ in kJ/s. The vapor condenses at a constant temperature of T_{cond}, and cooling water is supplied at a temperature T_1. The rate of heat transfer can be expressed as

$$\dot{q} = \dot{m} C_p (T_2 - T_1) = UA\,\Delta T_{\text{log mean}} = \frac{UA(T_2 - T_1)}{\ln[(T_{\text{cond}} - T_1)/(T_{\text{cond}} - T_2)]} \qquad (9\text{-}82)$$

where $\dot{m}$ is the flow rate of cooling water, C_p the heat capacity of the cooling water, T_2 the temperature of the cooling water leaving the condenser, U the constant overall heat-transfer coefficient determined at optimum conditions, and A the area of heat transfer. Solving Eq. (9-82) for the flow rate of cooling water gives

$$\dot{m} = \frac{\dot{q}}{C_p(T_2 - T_1)} \qquad (9\text{-}83)$$

The design conditions set the values of $\dot{q}$ and T_1, and the heat capacity of water generally can be approximated as 4.2 kJ/(kg·K). Therefore, Eq. (9-83) shows that the flow rate of the cooling water is fixed if the temperature of the water leaving the condenser is fixed. Under these conditions, the optimum flow rate of cooling water can be found directly from the optimum value of T_2.

The annual cost for cooling water is $\dot{m} H_y C_w$. From Eq. (9-83),

$$\dot{m} H_y C_w = \frac{\dot{q} H_y C_w}{C_p(T_2 - T_1)} \qquad (9\text{-}84)$$

where H_y designates the number of hours of condenser operation per year and C_w the cooling water cost, assumed to be directly proportional to the amount of water supplied.† The annual fixed charges for the condenser are $AK_F C_A$, where K_F is the annual fixed charge including maintenance expressed as a fraction of the initial cost for the installed equipment and C_A the installed cost of the condenser per square meter of heat-transfer area. All other costs are assumed constant, so the total annual variable cost for cooling water plus fixed charges is

$$\text{Total annual variable cost} = \frac{\dot{q} H_y C_w}{C_p(T_2 - T_1)} + AK_F C_A \qquad (9\text{-}85)$$

Substituting for A from Eq. (9-82) gives

$$\text{Total annual variable cost} = \frac{\dot{q} H_y C_w}{C_p(T_2 - T_1)} + \frac{\dot{q} K_F C_A \ln[(T_{\text{cond}} - T_1)/T_{\text{cond}} - T_2]}{U(T_2 - T_1)} \qquad (9\text{-}86)$$

The only variable in Eq. (9-86) is the temperature of the cooling water leaving the condenser. The optimum cooling water rate occurs when the total annual cost is a minimum.

†Cooling water is assumed to be available at a pressure sufficient to handle any pressure drop in the condenser; therefore, any cost due to pumping the water is included in C_w.

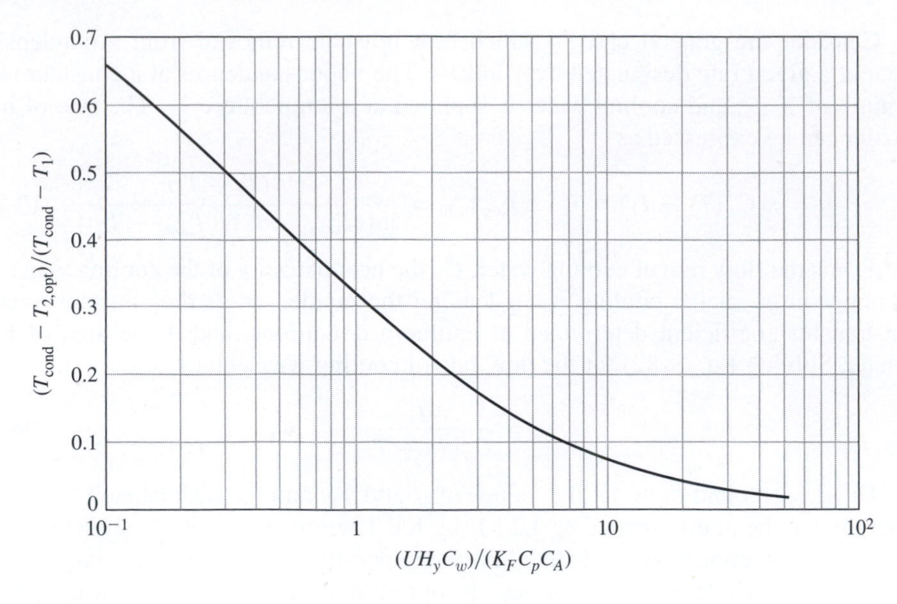

Figure 9-10
Solution of Eq. (9-87) for use in evaluating optimum flow rate of cooling medium in condenser

Thus, the corresponding optimum exit temperature can be found by differentiating Eq. (9-86) with respect to T_2 (or, more simply, with respect to $T_{cond} - T_2$) and setting the result equal to zero. When this is done, the following equation is obtained:

$$\frac{T_{cond} - T_1}{T_{cond} - T_{2,opt}} - 1 + \ln\frac{T_{cond} - T_{2,opt}}{T_{cond} - T_1} = \frac{UH_yC_w}{K_FC_pC_A} \tag{9-87}$$

The optimum value of T_2 can be obtained by either using a simple iterative computer routine or utilizing the graphical solution of Fig. 9-10 to solve Eq. (9-87). The optimum flow rate of the cooling water is then obtained with Eq. (9-83).

EXAMPLE 9-5 Optimum Cooling Water Flow Rate in Condenser

A condenser for a distillation unit must be designed to condense 2300 kg of vapor per hour. The effective condensation temperature for the vapor is 77°C. The heat of condensation for the vapor is 465 kJ/kg. The cost of the cooling water at 21°C is \$2.56 per 100 m^3. The overall heat-transfer coefficient at the optimum conditions may be taken as 0.284 kJ/(m^2·s·K). The cost for the installed heat exchanger is \$380 per square meter of heat-transfer area, and annual fixed charges including maintenance are 20 percent of the initial investment. The heat capacity of the water may be assumed to be constant at 4.2 kJ/(kg·K). If the condenser is to operate 6000 h/yr, determine the cooling water flow rate in kilograms per hour for optimum economic conditions.

■ Solution

$$U = 0.284 \text{ kJ/(m}^2\cdot\text{s}\cdot\text{K)}$$
$$H_y = 6000 \text{ h/yr}$$
$$K_F = 0.20$$
$$C_p = 4.2 \text{ kJ/(kg}\cdot\text{K)}$$
$$C_A = \$380/\text{m}^2$$

$$C_w = \frac{2.56}{100(1000)} = \$0.0000256/\text{kg}$$

$$\frac{UH_y C_w}{K_F C_p C_A} = \frac{(0.284)(6000)(3600)(0.0000256)}{(0.2)(4.2)(380)} = 0.492$$

The optimum exit temperature may be obtained by solving Eq. (9-87) with a computer routine as 53.1°C. Use of Fig. 9-10 gives similar results.

By Eq. (9-83), at the optimum economic conditions,

$$\dot{m} = \frac{\dot{q}}{C_p(T_{2,\text{opt}} - T_1)} = \frac{2300(465)}{4.2(53.1 - 21)} = 7930 \text{ kg/h water}$$

■

Optimization Application: Distillation Reflux Ratio

The design of a distillation unit is ordinarily based on specifications giving the degree of separation required for a feed supplied to the unit at a known composition, temperature, and flow rate. The design engineer must determine the size of the column and the reflux ratio necessary to meet the specifications. As the reflux ratio is increased, the number of theoretical stages required for the given separation decreases. An increase in reflux ratio, therefore, may result in lower fixed charges for the distillation column and greater costs for the reboiler heat duty and condenser coolant.

The optimum reflux ratio occurs at the point where the sum of fixed charges and operating costs is a minimum. As a rough approximation, the optimum reflux ratio generally is between 1.1 and 1.3 times the minimum reflux ratio. Example 9-6 illustrates the general method for determining the optimum reflux ratio in distillation operations.

Determination of Optimum Reflux Ratio — EXAMPLE 9-6

A sieve-plate distillation column is being designed to handle 318 kg mol of feed per hour. The unit is to operate continuously at a total pressure of 1 atm. The feed contains 45 mol percent benzene and 55 mol percent toluene, and the feed enters at its boiling temperature. The overhead product from the distillation tower must contain 92 mol percent benzene, and the bottoms must contain 95 mol percent toluene. Determine the following:

a. The optimum reflux ratio as mols of liquid returned to tower per mol of distillate product withdrawn.

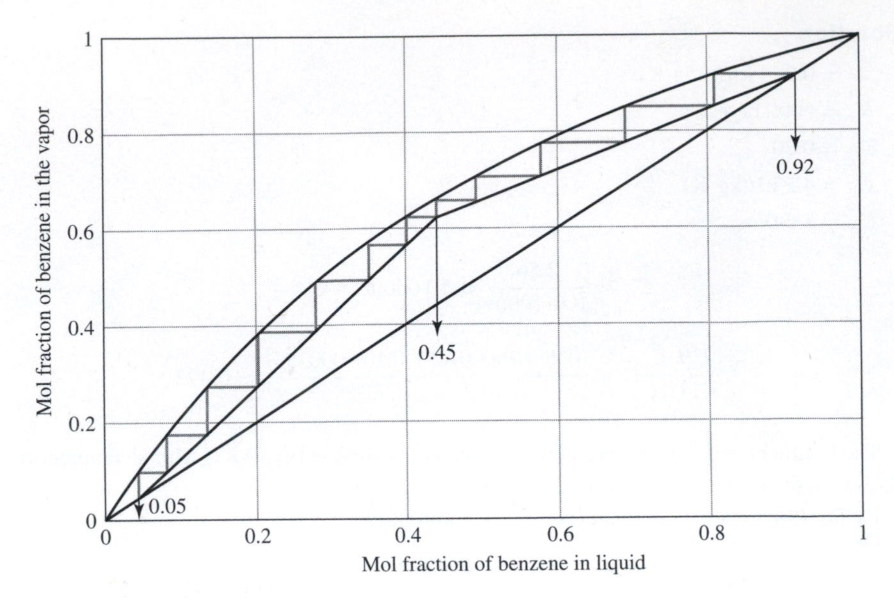

Figure 9-11
Equilibrium diagram for benzene-toluene mixture at total pressure of 1 atm (McCabe-Thiele method for determining number of theoretical plates)

b. The ratio of the optimum reflux ratio to the minimum reflux ratio.

c. The percent of the total variable cost due to steam consumption at the optimum conditions.

The following data apply:

Vapor-liquid equilibrium data for benzene-toluene mixtures at atmospheric pressure are presented in Fig. 9-11.

The molal heat capacity for liquid mixtures of benzene and toluene in all proportions may be assumed to be 1.67×10^5 J/(kg mol·K).

The molal heat of vaporization of benzene and toluene may be taken as 3.2×10^4 kJ/kg mol.

Effects of change in temperature on heat capacity and heats of vaporization are assumed negligible. Heat losses from the column are negligible. Effects of pressure drop over the column may be neglected.

The overall coefficient of heat transfer is 0.454 kJ/(m^2·s·K) in the reboiler and 0.568 kJ/(m^2·s·K) in the condenser.

The boiling temperature is 94°C for the feed, 81.7°C for the distillate, and 108°C for the bottoms. The temperature-difference driving force in the reflux condenser may be based on an average cooling water temperature of 32.2°C, and the change in cooling water temperature is 27.8°C for all cases. Saturated steam at 415 kPa is used in the reboiler. At this pressure, the temperature of the condensing steam is 144.8°C, and the heat of condensation is 2.13×10^3 kJ/kg. No heat-saving devices are used.

The column diameter is to be based on a maximum allowable vapor velocity of 0.76 m/s at the top of the column. The overall plate efficiency may be assumed to be 70 percent. The unit is to operate 8500 h/yr.

Cost Data

$$\text{Steam} = \$3.31 \text{ per } 10^3 \text{ kg}$$

$$\text{Cooling water} = \$0.0238 \text{ per } 10^3 \text{ kg}$$

The sum of costs for piping, insulation, and instrumentation can be estimated to be 60 percent of the cost for the installed equipment. Annual fixed charges amount to 15 percent of the total cost for installed equipment, piping, instrumentation, and insulation.

The following costs for the sieve-plate distillation column, condenser, and reboiler are for installed equipment and include delivery and erection costs. To simplify the calculations, all equipment costs may be interpolated.

Sieve-plate distillation column		Condenser, tube-and-shell heat exchanger		Reboiler, tube-and-shell heat exchanger	
Diameter, m	$/plate	Heat-transfer area, m^2	Cost, $	Heat-transfer area, m^2	Cost, $
1.50	2640	75	21,100	90	37,200
1.75	3170	95	24,600	130	45,700
2.0	3910	110	27,000	165	52,700
2.25	4730	130	29,800	200	59,100
2.5	5680	150	32,250	240	65,200

■ Solution

The variable costs involved are the cost of column, cost of reboiler, cost of condenser, cost of steam, and cost of cooling water. Each of these costs is a function of the reflux ratio, and the optimum reflux ratio occurs at the point where the sum of the annual variable costs is a minimum. The total variable cost will be determined at various reflux ratios, and the optimum reflux ratio will be found by the graphical method. A sample calculation for a reflux ratio of 1.5 is outlined below.

Annual Cost for Distillation Column

The McCabe-Thiele simplifying assumptions apply for this case, and the number of theoretical plates can be determined by the standard graphical method shown in Fig. 9-11. The slope of the enriching line is $1.5/(1.5 + 1) = 0.6$. From Fig. 9-11, the total number of theoretical stages required for the separation is 12.1.

The actual number of plates is $(12.1 - 1)/0.7 = 16$.

The mols of distillate D per hour and the mols of bottoms B per hour are obtained by a benzene material balance from

$$0.45(318) = 0.92D + 0.05(318 - D)$$

$$D = 146 \text{ kg mol/h of distillate}$$

$$B = 318 - 146 = 172 \text{ kg mol/h of bottoms}$$

The number of mols of vapor per hour at the top of the column is

$$146(1 + 1.5) = 365 \text{ kg mol/h}$$

Apply the ideal gas law, using the allowable vapor velocity of 0.76 m/s at the top of the column.

$$0.76 = \frac{365(22.4)(355)(4)}{3600(273.1)(\pi D^2)}$$

$$D = 2.22 \text{ m}$$

Cost per plate for this diameter is $4630.

$$\text{Annual cost for distillation column} = (4630)(16)(1 + 0.6)(0.15) = \$17,780$$

Annual Cost for Condenser

The rate of heat transfer per hour in the condenser is given by

$$\dot{q}_c = (\text{kg mol/h condensed})(\text{molal heat of condensation}) = (365)(3.2 \times 10^4) = 11.68 \times 10^6 \text{ kJ/h}$$

The heat-transfer area required for condensation is

$$A = \frac{\dot{q}_c}{U\Delta T} = \frac{11.68 \times 10^6}{0.568(81.7 - 32.2)(3600)} = 115.4 \text{ m}^2$$

$$\text{Annual cost for condenser} = (27,750)(1 + 0.6)(0.15) = \$6660$$

Annual Cost for Reboiler

The rate of heat transfer in the reboiler can be determined with a total energy balance around the distillation column. Use the temperature of the liquid feed as the energy reference level. The energy balance is then

$$\dot{q}_r + (318)(94 - 81.7)(1.67 \times 10^2) = 11.68 \times 10^6 + (172)(108 - 81.7)(1.67 \times 10^2)$$

$$\dot{q}_r = 11.782 \times 10^6 \text{ kJ/h}$$

The heat-transfer area required for reboiling is

$$A = \frac{\dot{q}_r}{U\Delta T} = \frac{11.782 \times 10^6}{(0.454)(144.8 - 108)(3600)} = 195.9 \text{ m}^2$$

$$\text{Annual cost for reboiler} = (58,350)(1 + 0.6)(0.15) = \$14,000$$

Annual Cost for Cooling Water

The rate of heat transfer in the condenser is 11.68×10^6 kJ/h.

$$\text{Annual cost for cooling water} = \frac{(11.68 \times 10^6)(0.0238)(8500)}{(4.2)(27.8)(1000)} = \$20,230$$

Annual Cost for Steam

$$\text{Annual cost for steam} = \frac{(11.782 \times 10^6)(3.31)(8500)}{(2.13 \times 10^3)(1000)}$$
$$= \$155{,}630$$

Total Annual Variable Cost at a Reflux Ratio of 1.5

$$\text{Cost} = 17{,}780 + 6660 + 14{,}000 + 20{,}230 + 155{,}630$$
$$= \$214{,}300$$

With the use of a simple computer routine, the previous calculations can be repeated for different reflux ratios. The results are presented in the following table.

Reflux ratio	No. of actual plates required	Column diameter, m	Annual cost, $					Total annual cost, $
			Column	Condenser	Reboiler	Cooling water	Steam	
1.14	∞	2.04	∞	6060	12,830	17,350	133,000	∞
1.2	29	2.07	28,930	6190	13,090	17,830	136,700	202,740
1.3	21	2.13	21,450	6320	13,380	18,600	142,900	202,650
1.4	18	2.16	19,180	6480	13,740	19,420	149,200	208,020
1.5	16	2.22	17,780	6660	14,000	20,230	155,630	214,300
1.7	14	2.35	17,140	6960	14,700	21,880	167,600	228,280
2.0	13	2.44	16,880	7390	15,550	24,300	186,000	250,120

a. The data presented in the preceding table are plotted in Fig. 9-12. The minimum total cost per year occurs at a reflux ratio of 1.25. Thus, the optimum reflux ratio is 1.25.

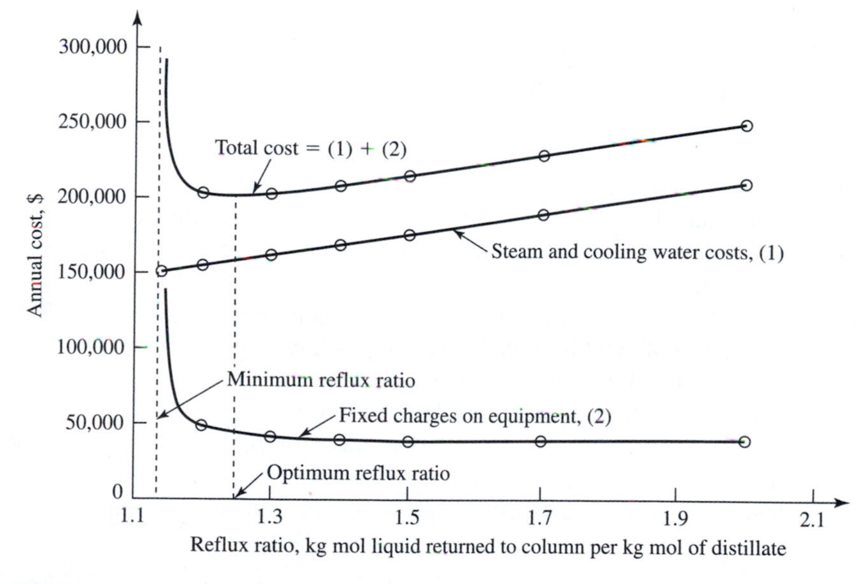

Figure 9-12
Optimum reflux ratio in distillation operation

b. For conditions of minimum reflux ratio, the slope of the enriching line in Fig. 9-11 is 0.532.

$$\frac{\text{Minimum reflux ratio}}{\text{Minimum reflux ratio} + 1} = 0.532$$

$$\text{Minimum reflux ratio} = 1.14$$

$$\frac{\text{Optimum reflux ratio}}{\text{Minimum reflux ratio}} = \frac{1.25}{1.14} = 1.1$$

c. At the optimum reflux ratio

$$\text{Annual steam cost} = \$139{,}600$$

$$\text{Total annual variable cost} = \$201{,}200$$

$$\text{Percent of variable cost due to steam consumption} = \frac{139{,}600}{201{,}200}(100) = 69.4\%$$

■

Optimization Application: Pinch Technology Analysis

Pinch Technology Concept Based on thermodynamic principles, pinch technology offers a systematic approach to optimum energy integration in a process. The improvements in the process associated with this technique are not due to the use of advanced unit operations, but to the generation of a heat integration scheme. One of the key advantages of pinch technology over conventional design methods is the ability to set an *energy target* for the design. The energy target is the minimum theoretical energy demand for the overall process.

The principal objective of this technology is to match cold and hot process streams with a network of exchangers so that demands for externally supplied utilities are minimized. Pinch technology establishes a temperature difference, designated as the pinch point, that separates the overall operating temperature region observed in the process into two temperature regions. Once a pinch point has been established, heat from external sources must be supplied to the process only at temperatures above the pinch and removed from the process by cooling media only at temperatures below the pinch. Such a methodology will maximize the heat recovery in the process with the establishment of a heat exchanger network based on pinch analysis principles. The best design for an energy-efficient heat exchanger network will result in a tradeoff between the energy recovered and the capital costs involved in this energy recovery.

The success of pinch technology has led to more inclusive ideas of *process integration* in which chemical processes are examined for both mass and energy efficiency. Even though process integration is a relatively new technology, its importance in process design is continuing to grow as processes become more complex.[†]

[†]N. Hallale, *Chem. Eng. Prog.*, **97**(7): 30 (2001).

As noted above, one concept of pinch analysis is to set energy targets prior to the design of the heat exchanger network. Targets can be set for the heat exchanger network without actually having to complete the design.[†] Energy targets can also be set for the utility heat duties at different temperature levels such as refrigeration and steam heat supply levels. Pinch analysis provides the thermodynamic rules to ensure that the energy targets are achieved during the heat exchanger network design.

Pinch Technology Analysis The starting point for a pinch technology analysis is to identify in the process of interest all the process streams that need to be heated and all those that need to be cooled. This means identifying the streams, their flow rates and thermal properties, phase changes, and the temperature ranges through which they must be heated or cooled. This can be accomplished after mass balances have been performed and temperatures and pressures have been established for the process streams. Energy quantities can be calculated conveniently by using a simulation program or by traditional thermodynamic calculations. Some heat duties may not be included in the network analysis because they are handled independently of the integration. For example, distillation column reboiler heating and condenser cooling may be treated independently of the rest of the heat duties. However, such independent duties should always be considered for inclusion in the network.

All the process streams that are to be heated, their temperatures, and enthalpy change rates corresponding to their respective temperature changes or phase changes are then tabulated. The enthalpy change rate for *each* stream is obtained from

$$\Delta \dot{H} = \dot{m} C_p \Delta T = CP \Delta T \tag{9-88}$$

where $\Delta \dot{H}$ is the enthalpy change rate, $\dot{m}$ the mass flow rate, C_p the heat capacity, ΔT the temperature change in the stream, and CP the heat capacity rate defined as the $\dot{m} C_p$ product. The enthalpy change rates are then added over each temperature interval that includes one or more of the streams to be heated. The resulting values allow plotting of the temperature versus enthalpy rate to provide a composite curve of all the streams that require a heat source. The same information and procedures are followed to develop a composite curve of the streams to be cooled. The resulting diagram, shown in Fig. 9-13, is designated as a *composite diagram* for the heat integration problem. The actual steps involved in preparing such a diagram are presented in Example 9-7.

It must be recognized that while each temperature is a fixed value on the vertical axis, enthalpy change rates are relative quantities. Enthalpy changes rather than absolute enthalpies are calculated via thermodynamic methods. Thus, the horizontal location of a composite line on the diagram is arbitrarily fixed. For the purposes of pinch technology analysis, the composite curve for streams to be cooled is located so as to be to the left, at every temperature, of the composite curve for those streams to be heated. Fixing the location of the composite curves with respect to one another with the use of a preselected value of $\Delta T_{\min}$ completes the composite diagram. The location

---

[†]R. Smith, *Chemical Process Design*, McGraw-Hill, New York, 1995.

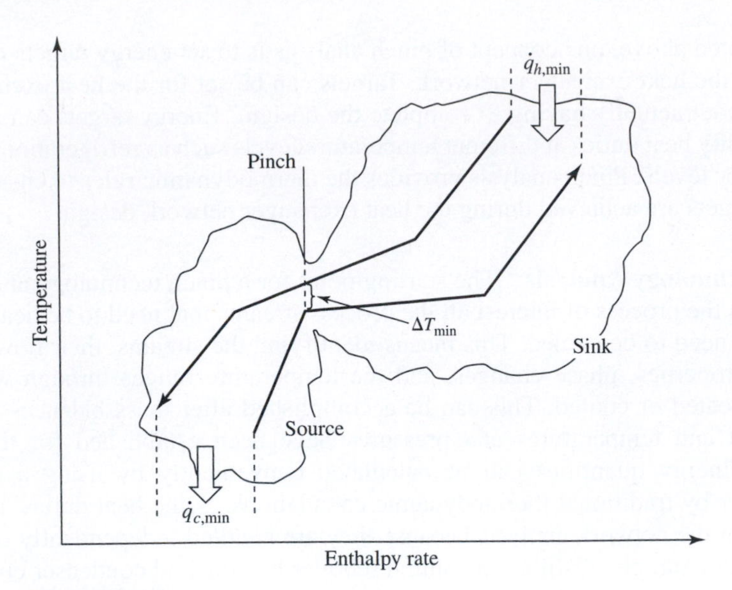

Figure 9-13
Composite diagram prepared for pinch technology analysis

of $\Delta T_{\min}$ on the composite diagram is where the two curves most closely approach each other in temperature, when measured in a vertical direction. On the first plotting of these curves, the vertical distance will rarely equal the preselected $\Delta T_{\min}$. This deficiency is remedied by moving one of the two curves *horizontally* until the distance of closest vertical approach matches the preselected $\Delta T_{\min}$. This can be done graphically or by calculation. All these steps can be accomplished readily with a spreadsheet, provided adequate thermodynamic property values are available.

The optimum value for $\Delta T_{\min}$ is generally in the range of 3 to 40°C for heat exchange networks, but is unique for each network and needs to be established before the pinch technology analysis is completed. If no cooling media are required below about 10°C, the optimum $\Delta T_{\min}$ is often in the range of 10 to 40°C. For a given $\Delta T_{\min}$, the composite curves define the utility heating and cooling duties.

The composite curves show the overall profiles of heat availability and heat demand in the process over the entire temperature range. These curves represent the cumulative heat sources and heat sinks in the process. The overlap between the two composite curves indicates the maximum quantity of heat recovery that is possible within the process.[†] The overshoot of the hot composite curve represents the minimum quantity of external cooling $\dot{q}_{c,\min}$ required, and the overshoot of the cold composite curve represents the minimum quantity of external heating $\dot{q}_{h,\min}$ required for the process.

Note that the composite curves can be used to evaluate the overall tradeoff between energy and capital costs. An increase in $\Delta T_{\min}$ causes the energy costs to

---

[†]B. Linnhof and D. R. Vredevelt, *Chem. Eng. Prog.*, **80**(7): 33 (1984).

increase, but also provides larger driving forces for heat transfer and accompanying reduced capital costs.

Construction of a Composite Diagram — EXAMPLE 9-7

Figure 9-14 shows a process flowsheet in which two reactant streams, each at 20°C, are to be heated to 160°C and fed to a reactor. It has been decided to mix the two streams before heating them, since both reactants need to be heated to the same temperature and will be mixed in the reactor anyway. Mixing these feed streams before they enter the reactor reduces the number of heat exchangers required from two to one. Since the reaction is slightly endothermic, the product stream leaves the reactor at 120°C. After further heating the reactant stream to 260°C, it is sent to a distillation column to recover the product. The liquid distillate product from the column is at 180°C and must be cooled to 20°C for storage. The bottom product from the column is cooled from 280 to 60°C. Although hot and cold utilities could be used for all the heating and cooling requirements, there clearly is an opportunity for savings in heat exchange since it is apparent from Fig. 9-14 that the process streams are, for the most part, within overlapping temperature ranges.

Since the reboiler temperature is too high for heat exchange with any of the process streams, it becomes an independent heat exchange problem. A hot oil utility stream available at 320°C and cooled to 310°C is to be used for heating the reboiler, as well as any process heating loads not met by process-process exchange. The cost of the hot oil is $2.25/GJ. The condenser temperature is in a range that could be used to heat some of the process streams; however, for brevity purposes it will not be included in the present analysis. The cooling utility is cooling water available at 10°C with an allowable temperature rise of 10°C and a cost of $0.25/GJ.

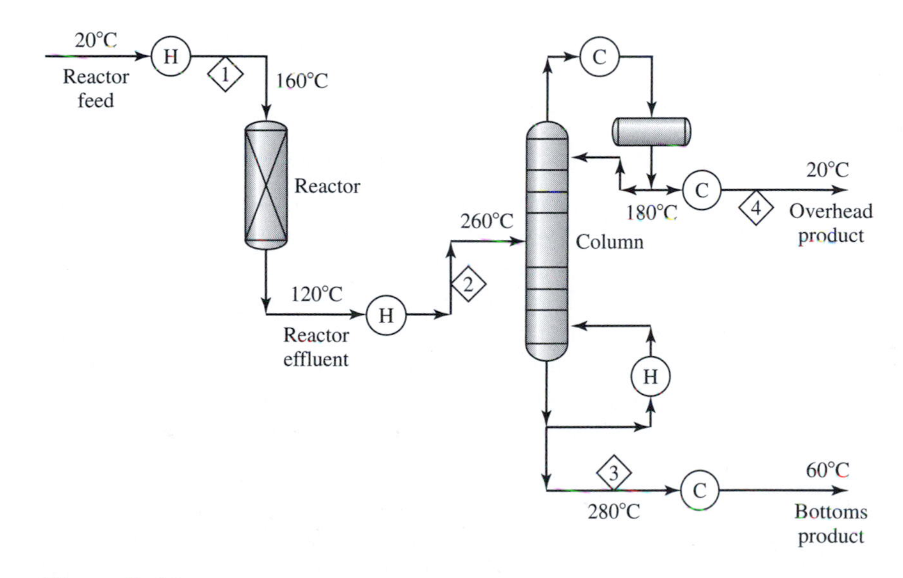

Figure 9-14

Process flowsheet diagram applicable for Example 9-7

(Reprinted with permission from *Handbook of Energy Efficiency,* F. Krieth and R. E. West, eds., Fig. 15.6, p. 597. Copyright © 1997 CRC Press, Boca Raton, Florida.)

Data for the process streams are provided in the following table.[†‡]

Stream Number	Stream description	Temperature interval, °C		Heat capacity rate, kJ/(s·°C) $CP = \dot{m}C_p$	Enthalpy change rate, kJ/s $CP(T_{out} - T_{in})$
Streams to be heated		In	Out		
1	Reactor feed	20	160	50	7,000
2	Reactor effluent	120	260	55	7,700
				Total	14,700
Streams to be cooled					
3	Bottom product	280	60	30	−6,600
4	Overhead product	180	20	40	−6,400
				Total	−13,000

a. Find the annual cost if only hot and cold utilities are used to supply all the heating and cooling for the process streams. Construct a composite diagram for this process.

b. Reconstruct the composite diagram to achieve a ΔT_{min} of 20°C.

c. Construct the balanced composite diagram for the process with a ΔT_{min} of 20°C, and find the minimum utility duties and the annual cost for the utilities after the heat integration.

■ Solution

a. The total enthalpy change rate provided in the table can be used directly to obtain the cost of providing all the heating and cooling requirements with only the hot and cold utilities.

$$\text{Hot utility cost} = \left(\frac{2.25}{10^6}\right)(14{,}700) = \$0.0331/\text{s}$$

$$\text{Cold utility cost} = \left(\frac{0.25}{10^6}\right)(13{,}000) = \$0.00325/\text{s}$$

$$\text{Total utility cost} = \$0.0363/\text{s, or } \$1.03 \times 10^6/\text{yr (at 90\% operating factor)}$$

A review of the stream information in the table shows that only stream 1 is to be heated over the temperature interval from 20 to 120°C; between 120 and 160°C, streams 1 and 2 are to be heated; and from 160 to 260°C, only stream 2 is to be heated. These four temperatures and the corresponding stream numbers and values for the heat capacity rates are entered into the temperature interval table in the next page. Where there is more than one stream in an interval, the sum of the heat capacity rate values for all the streams is entered. The same procedure is followed for the streams to be cooled. The enthalpy change rate is obtained for each interval by multiplying the total heat capacity value by the temperature interval, and this product is entered into the sixth column of the table.

[†]See Fig. 9-14 to identify these streams.

[‡]In these tabulations, it is assumed that the heat capacity is independent of temperature. A consequence of this assumption is that the temperature versus enthalpy change rate curves are linear. In reality, the heat capacity generally tends to increase with temperature, resulting in some curvature in the curves. Determination of these curves using temperature-dependent heat capacity relationships would provide better results. However, the use of temperature-dependent properties does not change the heat integration method; only the details of the computations are changed.

Enthalpy change rates are fixed by selecting a baseline value for the enthalpy change rate at one stream temperature. A starting enthalpy change rate of 5000 kJ/s at 20°C is selected for the streams to be heated, while a value of 15,000 kJ/s at 280°C is chosen for the streams to be cooled. These values are arbitrary, selected only for graphical convenience since there is no unique composite diagram until a ΔT_{min} has been implemented. The enthalpy change rates in the table are added to the initial enthalpy change rate values to yield the enthalpy rate values tabulated with the corresponding temperatures.

Initial temperature interval table

Stream number†	Required temperature interval, °C		Heat capacity rate CP, kJ/(s·°C)	Enthalpy change rate, kJ/s	Initial enthalpy selection	
					X	Y
Streams to be heated	**In**	**Out**			5,000	20
1	20	120	50	5,000	10,000	120
1 & 2	120	160	105	4,200	14,200	160
2	160	260	55	5,500	19,700	260
			Total	14,700		
Streams to be cooled					15,000	280
3	280	180	30	−3,000	12,000	180
3 & 4	180	60	70	−8,400	3,600	60
4	60	20	40	−1,600	2,000	20
			Total	−13,000		
Utilities						
Hot oil	320	310		−3,900		
Cooling water	10	20		2,200		

†See Fig. 9-14 to identify these streams.

The sets of temperature versus enthalpy rate values that have been established for the streams that are to be cooled and those that are to be heated are plotted in Fig. 9-15. This is a composite diagram for the heat integration problem. It is apparent from the figure that the closest vertical approach of the two curves occurs at an enthalpy change rate of 10,000 kJ/s. This is the pinch point for the two composite curves and occurs where the temperature of the streams that are to be heated is 120°C and the temperature of the streams that are to be cooled is about 153°C. This ΔT_{min} of 33°C is simply a consequence of the starting enthalpy rates that were initially chosen.

b. To achieve a ΔT_{min} of 20°C, one of the curves must be moved horizontally to bring the two curves closer together. One way to do this is to move the curve representing the streams that are to be cooled to the right, so that a temperature of 140°C is intercepted at an enthalpy rate of 10,000 kJ/s. In Fig. 9-15, the slope of that portion of the curve at the pinch point is obtained from

$$\frac{180 - 60}{12{,}000 - 3600} = 0.01428$$

and the intercept is

$$180 - (0.01428)(12{,}000) = 8.57$$

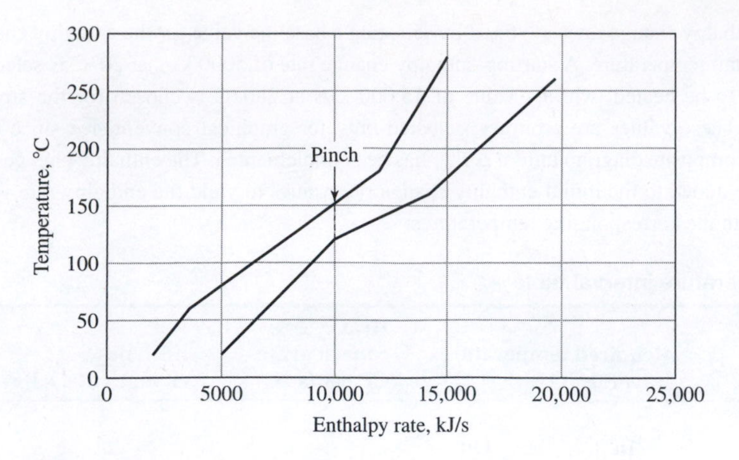

Figure 9-15
Composite diagram with 33°C approach temperature

At 140°C, the enthalpy rate now is

$$\frac{140 - 8.57}{0.01428} = 9204 \text{ kJ/s}, \quad \text{or} \quad \sim 9200 \text{ kJ/s}$$

This value of 9200 must be increased to 10,000 kJ/s to make the ΔT at the pinch point equal to 20°C. Therefore, 800 kJ/s must be added to every enthalpy change rate value associated with the streams to be heated. This action changes the enthalpy rate to 15,800 kJ/s. The revised temperature interval table is shown below.

Revised temperature interval table

Stream number†	Required temperature interval, °C		Heat capacity rate, kJ/(s·°C)	Enthalpy change rate, kJ/s		Revised enthalpy selection	
						X	*Y*
Streams to be heated	In	Out				5,000	20
1	20	120	50		5,000	10,000	120
1 & 2	120	160	105		4,200	14,200	160
2	160	260	55		5,500	19,700	260
				Total	14,700		
Streams to be cooled						15,800	280
3	280	180	30		−3,000	12,800	180
3 & 4	180	60	70		−8,400	4,400	60
4	60	20	40		−1,600	2,800	20
				Total	−13,000		
Utilities							
Hot oil	320	310			−3,900	15,800	310
						19,700	320
Cooling water	10	20			2,200	2,800	10
						5,000	20

†See Fig. 9-14 to identify these streams.

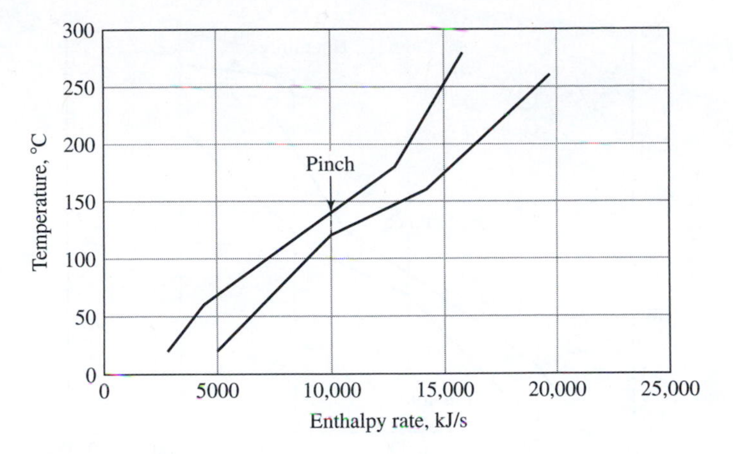

Figure 9-16
Composite diagram with 20°C approach temperature

These values are plotted in Fig. 9-16, the composite diagram for this problem for a $\Delta T_{\min}$ of 20°C.

c. It is clear from the composite diagram of Fig. 9-16 that above the cold stream temperature of about 190°C there is no hot stream curve above the cold stream curve. Since all heat transfer is vertical on a composite diagram, there is no process stream available to heat the cold stream from 190 to 260°C with an enthalpy change rate of about 3900 kJ/s. Therefore, a hot utility must be used to provide this heat. In fact, this quantity of heat needed is the *minimum hot utility requirement* for the problem as defined in Fig. 9-16, with its temperatures, heat duties, and specified $\Delta T_{\min}$ of 20°C. Similarly, below a hot stream temperature of about 70°C there is no cold process stream available to cool the hot process streams. Thus, a cold utility must be used to remove this heat. The corresponding $\Delta \dot{H}$ of about 2200 kJ/s is the *minimum cold utility requirement* for the problem as defined. For this process with a $\Delta T_{\min}$ of 20°C, various heat exchanger networks can be devised which require more hot and cold utilities, but no network that will require less utilities. Only by decreasing $\Delta T_{\min}$ can these heat duties be reduced, and then not reduced beyond the values corresponding to a $\Delta T_{\min}$ of zero. If the curves for the required heating and cooling utilities are included in the composite diagram and all heating and cooling loads are satisfied, the diagram is called a *balanced composite diagram,* as shown in Fig. 9-17.

Examination of Fig. 9-17 shows that at the cold end of the network, the process stream temperature is 20°C and the cooling water enters at 10°C, for an approach temperature difference of only 10°C. This is a consequence of the available water inlet temperature and the required outlet temperature for the warm stream. While this temperature difference appears to violate the $\Delta T_{\min}$ specification of 20°C, it actually does not because $\Delta T_{\min}$ applies only to process stream heat exchanges and not to utility process heat exchanges. Nonetheless, it might be useful to determine whether the 20°C outlet temperature for the warm stream is necessary. If that temperature can be increased, the area of the cooler can be reduced.

The minimum hot utility requirement is 3900 kJ/s—the difference between the highest enthalpy rate of 19,700 kJ/s for the streams that are to be heated and the highest enthalpy rate of

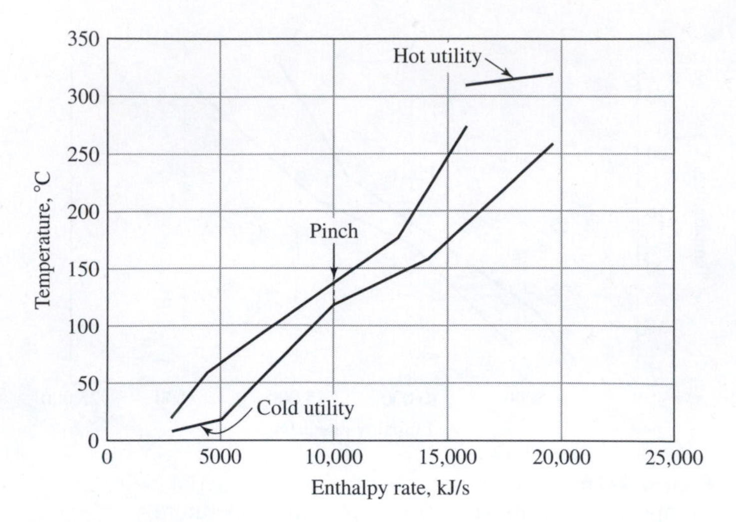

Figure 9-17
Balanced composite diagram

15,800 kJ/s for the streams that are to be cooled. The minimum cold utility requirement is 2200 kJ/s—the difference between the lowest enthalpy rate of 5000 kJ/s for the streams that are to be heated and the lowest enthalpy rate of 2800 kJ/s for the streams that are to be cooled. The total utility cost for these duties is

$$\left(\frac{2.25}{10^6}\right)(3900) + \left(\frac{0.25}{10^6}\right)(2200) = \$0.00933/\text{s}$$

providing a utility savings of \$0.0270/s or $\$7.66 \times 10^5$/yr (at 90 percent operating factor) compared to using utilities for all the heating and cooling duties. This savings does not come without a cost, however, since it requires purchasing, installing, operating, and maintaining the heat exchangers needed for the process-process heat exchange. Whether this is worthwhile depends upon an economic analysis of the savings and costs.

Pinch Technology Guidelines Pinch technology includes several principles that offer guidance in constructing a feasible and near optimal heat exchanger network:

1. Do not transfer heat across the pinch point; the pinch point divides the heat exchanger network into two distinct regions.
2. Do not use a hot utility below the pinch point.
3. Do not use a cold utility above the pinch point.
4. Heat transfer always takes place from a higher to a lower temperature.
5. No process-process heat exchanger should have an approach temperature less than the specified ΔT_{min}.
6. Minimize the number of heat exchangers that are needed.
7. Avoid loops in the heat integration system.

A few comments on these guidelines may be helpful. For example, any process-stream heat that is transferred from one side of the pinch point to the other side of the pinch point only increases the requirements for both utilities. This generally results in a nonoptimal network design and should be recommended only if it can be economically justified. Also, using a hot utility below the pinch point or a cold utility above the pinch point only increases the requirement for each utility and thereby nearly always leads to a nonoptimal design. Violation of the $\Delta T_{\min}$ guideline will change the structure of the heat integration system and will probably move the solution away from optimal conditions. Note also that the optimal network is usually the one that uses the least number of heat exchangers to meet the problem needs. Using more than the minimum number is usually not optimal. The *loop* referred to in the guidelines infers that there is an energy route that could be followed that leads back to the starting point. This can happen, for example, when the same two streams exchange heat in more than one heat exchanger or when a utility is used where it is not needed. Each loop adds an unnecessary heat exchanger to the network.

The minimum number of heat exchangers required for a given composite diagram can be obtained from the diagram. The number of exchangers required is given by[†]

$$N_E = N_s - 1 \tag{9-89}$$

where N_E is the number of heat exchangers and N_s the total number of streams exchanging heat. This rule must be applied to each separate section of heat transfer, generally four in most network problems. These are identified from a balanced composite diagram by drawing a vertical line at the pinch point, another vertical line where the hot utility is initially required, and a third vertical line where the cold utility initially is required. This divides the diagram into four distinct sections which are, moving from left to right on the diagram, the cold utility section, the process exchange section below the pinch (also designated as the *source section*), the process exchange section above the pinch (also designated as the *sink section*), and the hot utility section. In each section, the sum of the enthalpy change rate values for all the streams in the section that are to be heated will match the sum obtained for all the streams in that same section that are to be cooled, except the latter will have a minus sign. Moreover, because the two curves are developed to have only one pinch point, there will be process stream matches that are consistent with the specified $\Delta T_{\min}$.

In each of the four sections, the number of streams participating in the heat exchange, including any utility streams, is counted. Labeling line segments on the diagram with stream names or numbers expedites this counting. Within any one section, each stream is counted only once; but each stream is counted in every section in which it appears. The minimum number of heat exchangers needed in a section is then obtained directly with Eq. (9-89). The resulting four values, when added, give the minimum total number of heat exchangers needed for the overall network. By emphasizing the goal of minimizing the number of heat exchangers used, the design engineer should be able to develop a heat integration network by using the minimum number of heat

[†]E. C. Hohmann, Ph.D. thesis, University of Southern California, Los Angeles, 1971.

exchangers. Such a network should be at least near optimal for the problem posed. It is virtually certain that a network using more than the minimum number of exchangers will not be optimal.

Identifying an Optimal Heat Exchange Network There is not a unique network for any but a two-stream heat exchange problem. So the design engineer needs both insight and creativity, in addition to described procedures that identify an appropriate network among the many possibilities. A network is developed one section at a time. Since the minimum number of heat exchangers already has been established, the task now becomes one of identifying which streams go to which exchangers. For each heat exchanger, a heat balance must be satisfied. If it is assumed that there are negligible heat gains or losses from the exchanger, the heat balance equation is

$$\Delta \dot{H} = 0 = [CP(T_{\text{out}} - T_{\text{in}})]_{\text{hot stream}} + [CP(T_{\text{out}} - T_{\text{in}})]_{\text{cold stream}} \qquad (9\text{-}90)$$

Some specific guidelines useful in finding good heat exchange matches are given below:

1. At the pinch point, each stream that is to be heated must enter or leave an exchanger at the pinch point, cold composite temperature; and each stream that is to be cooled must enter or leave an exchanger at the pinch point, hot composite temperature.
2. Start the analysis of exchangers in the sink and source sections at the pinch point where all temperatures are fixed.
3. A point of discontinuity in a composite curve indicates the addition or removal of a stream, or the onset of a phase change. The stream that is being added or removed must enter or leave an exchanger at the temperature where the discontinuity occurs.
4. If there are only two streams in a section, they both go to the one exchanger that is reserved for the section.
5. If there are three streams in a section, the stream with the largest change in enthalpy should be split across two exchangers to satisfy the heat duties for each of the other two streams.
6. If there are four streams in a section, three heat exchangers will be required. If three streams are either heated or cooled, then the fourth stream is split into three flows to satisfy the heat duties from the other three streams. If there are two streams that are to be heated and two streams that are to be cooled, a convenient way to allocate the streams to exchangers is to prepare a new composite diagram and use this to make the allocations. (This is demonstrated in Example 9-8.)
7. If there are more than four streams in a section, attempt to follow guideline 6. The use of a computer-based algorithm is recommended for these more complicated cases.[†]

[†]Two examples of this type of software are ASPEN PLUS-PINCH and PRO/II-LNGHY.

8. If the matches between heat exchange duties result in more than the minimum number of exchangers being required, try other matches. Look for loops and eliminate them.
9. If a discontinuity occurs in a process stream curve within a utility section, it may be possible by means of the adjacent process section to meet the duty of the stream by leaving the curve at the discontinuity and still not violate the ΔT_{min}. Doing so reduces the required number of exchangers by 1 without changing the utility requirements and many times is an economical choice.

Establishing Minimum Number of Heat Exchangers and Recommended Heat Exchange Network — EXAMPLE 9-8

Refer to the process heating and cooling problem presented in Example 9-7. In this problem,

a. Determine the minimum number of heat exchangers required for a ΔT_{min} of 20°C.
b. Establish a heat exchanger network meeting the process requirements with the minimum required number of exchangers, as evaluated in part (a).
c. Reevaluate the number of exchangers required in the cold utility section.
d. Recommend a heat exchanger network for this process.

■ Solution

a. The minimum number of heat exchangers required for the problem is determined with the aid of Fig. 9-18 and Eq. (9-89). On the graph, vertical lines are drawn to divide the curves into four independent exchange sections: the cold utility section, the process exchange section below the pinch point, the process exchange section above the pinch point, and the hot utility section.

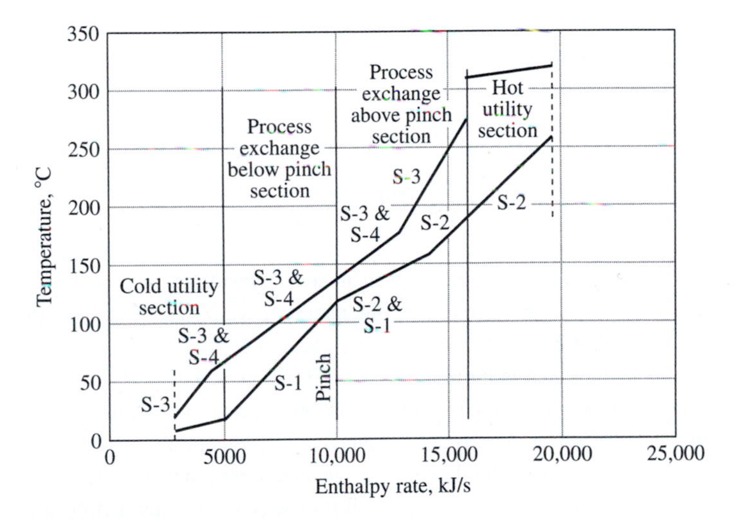

Figure 9-18
Balanced composite diagram with 20°C approach temperature

The lines on the diagram are labeled with the streams that are represented. The total number of hot and cold streams in each section is then counted and decreased by 1 to provide the number of heat exchangers required in those sections. Doing this in the cold utility section shows that three streams and two exchangers are required. In the process section below the pinch, there are three streams; thus, two exchangers are required. In the process section above the pinch, there are four streams and three exchangers. Since there are two streams in the hot utility section, one exchanger is needed. The minimum total number of exchangers required therefore is eight.

b. Starting with the cold utility section, and referring to Fig. 9-18, streams are matched to obtain the desired heat exchange. In this section, two exchangers are needed; one is for stream 4 and the cooling water, and the other is for stream 3 and the cooling water. These exchangers are designated as C-1 and C-2.

The inlet temperatures of the process streams to the cold utility section in C-1 and C-2 are determined by energy balances made in the process section below the pinch. Consideration should be given to cooling stream 3 to the required 60°C by using a process stream, and this will be done in part (c).

Now proceed to the process section below the pinch point shown in Fig. 9-18. In this section two heat exchangers are required. Stream 1 with an inlet temperature of 20°C and an exit temperature of 120°C, will be split between two exchangers, E-1 and E-2. Both stream 3 and stream 4 will be cooled from 140°C to an exit temperature T_{out}, given by an energy balance over the two exchangers

$$50(120 - 20) + (30 + 40)(T_{out} - 140) = 0$$

$$T_{out} = \frac{-5000 + 9800}{70} = 68.6^\circ\text{C}$$

This is the temperature of streams 3 and 4 leaving E-1 and E-2 and the inlet temperature to coolers C-1 and C-2.

An energy balance for exchanger E-1, between streams 1 and 3, gives the fraction of stream 1 that must be used in this exchanger:

$$50x(120 - 20) + 30(68.6 - 140) = 0$$

$$x = \frac{2142}{5000} = 0.428$$

where x is the fraction of stream 1 sent to exchanger E-1 resulting in a heat duty of 2142 kJ/s for E-1. The fraction of stream 1 sent to exchanger E-2 is 0.572. The heat duty for E-2 is 0.572(50)(120 − 20), or 2860 kJ/s.

The section above the pinch point requires three exchangers. The problem here is to match up the four streams by using only three exchangers. At the pinch point, both of the streams that are to be cooled must exit from the heat exchanger at their pinch point temperature of 140°C while both of the streams that are to be heated must enter the heat exchanger at their pinch point temperature of 120°C. A simple way to illustrate the problem is to plot all four streams on a temperature versus enthalpy graph. Begin by setting up an energy balance table for all the individual streams in the section above the pinch point, as shown here.

Streams in the section above the pinch	Temperature interval, °C		Heat capacity rate, kJ/(s·°C)	Enthalpy change rate, kJ/s	Enthalpy rate, kJ/s	
Streams to be heated	In	Out			In	Out
S-1	120	160	50	2000	10,000	12,000
S-2	120	189.1	55	3800	10,000	13,800
				Total 5800		
Streams to be cooled						
S-3	280	140	30	−4200	14,200	10,000
S-4	180	140	40	−1600	11,600	10,000
				Total −5800		

All the values in the first four columns of the table are directly available from the given data, except the exit temperature for stream 2. This value may be estimated from Fig. 9-18 or calculated from an energy balance. The total enthalpy change per second of 5800 kJ/s for streams 1 and 2 must equal that of streams 3 and 4. Since the enthalpy change rate of stream 2 is known to be 2000 kJ/s, that for stream 1 must be the difference between 5800 kJ/s and 2000 kJ/s, or 3800 kJ/s. The exit temperature of stream 2 can be calculated from

$$55(T_{\text{out}} - 120) = 3800$$

$$T_{\text{out}} = 189.1°\text{C}$$

The values in the preceding table are then used to plot the individual stream values, as shown in Fig. 9-19. The latter can now be used to match streams to form a heat exchange network, for this section. Three different matches will be developed.

Match 1

In Fig. 9-20, a vertical line is drawn from the upper end of line S-4 to line S-1. All the 1600 kJ/s heat duty of stream S-4 will be transferred to stream S-1 in heat exchanger E-3. The exit temperature of stream S-1 can be obtained from the graph or calculated by an energy balance as

$$1600 = 50(T_{\text{out}} - 120)$$

$$T_{\text{out}} = \frac{1600}{50} + 120 = 152°\text{C}$$

Next, a vertical line is drawn from the upper end of line S-2 to line S-3. All the 3800 kg/s heat duty of stream 2 will be supplied by stream 3 in heat exchanger E-4. The exit temperature of stream 3 leaving exchanger E-4 can be obtained from the graph or calculated by another energy balance as

$$3800 = 30(T_{\text{out}} - 140)$$

$$T_{\text{out}} = \frac{3800}{30} + 140 = 266.7°\text{C}$$

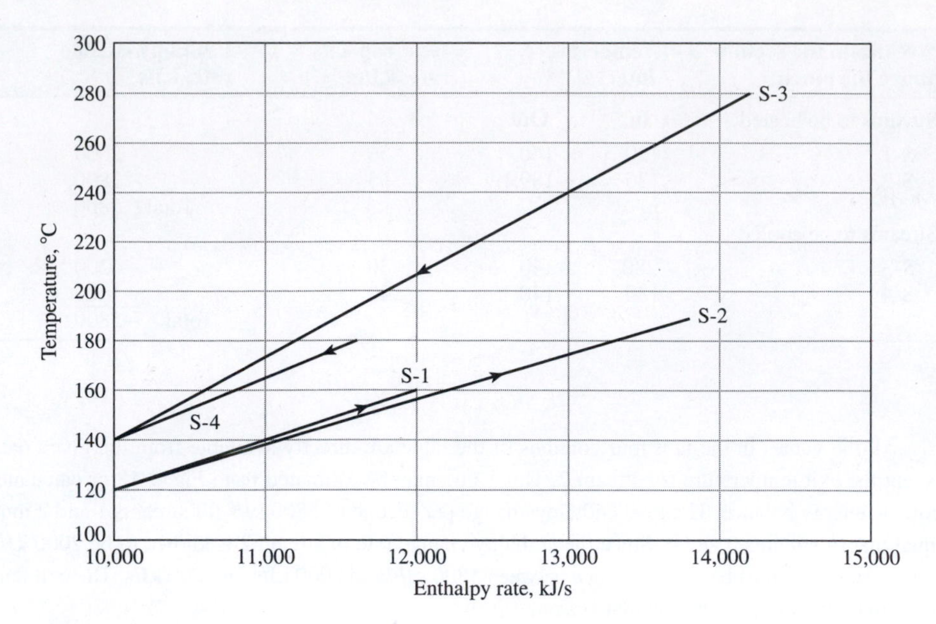

Figure 9-19
Enthalpy rates for the four streams in section above pinch

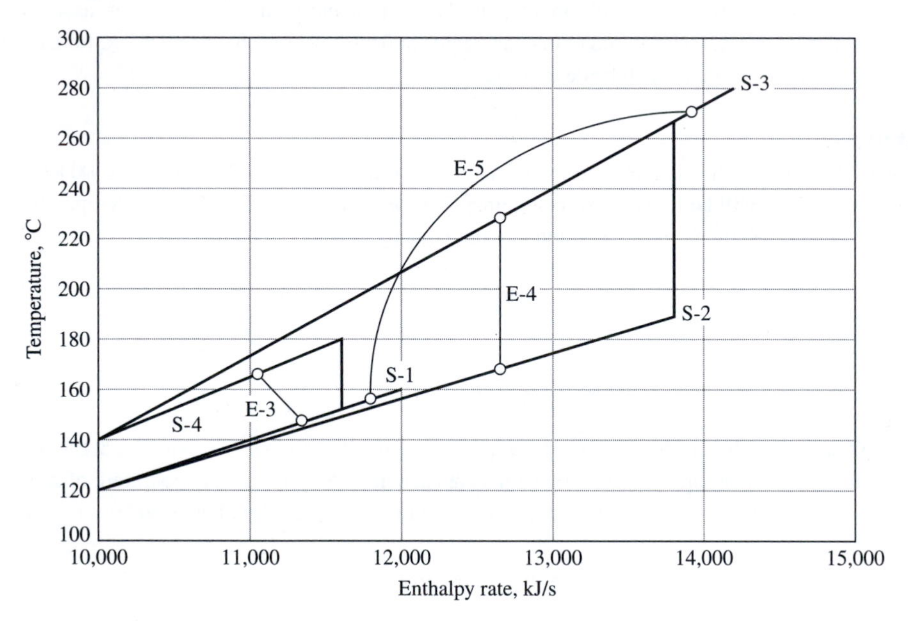

Figure 9-20
Possible heat exchange for match 1

The remaining 400 kJ/s heat duty of stream 3 not removed in exchanger 4 is removed in exchanger E-5 by the 400 kJ/s heat duty that is available from stream 1. This arrangement of the three heat exchangers is shown in Fig. 9-23c.

The minimum number of heat exchangers for this section has been met. Since all temperature differences are 20°C or greater, an acceptable network has been developed. Note, however, that E-5 has a much smaller heat duty than in the other two heat exchangers. As a consequence, consideration should be given to the elimination of E-5 in this match. This can be accomplished with a small increase in heating and cooling utilities, provided the savings in heat exchanger and associated costs are sufficient to justify such a change. This choice can only be made by an economic analysis.

Match 2

Note that in match 1 the higher-temperature end of stream S-3 is used to heat a much cooler portion of stream S-1. To alleviate this heat exchange inefficiency, some modification to match 1 is suggested. Begin by leaving exchanger E-3 as provided in match 1, but providing the remaining duty for the upper end of stream S-1 by heat duty from the section of stream S-3 immediately above that of stream S-1 in heat exchanger E-4, as shown in Fig. 9-21. This provides a much closer temperature match between the two streams.

Now use the two remaining sections of stream S-3 to meet the heat duties of stream S-2; but this requires two exchangers, E-5 and E-6. Even though the heat loads all balance and the temperature differences are above the $\Delta T_{\min}$, this match is not a good choice because an extra heat exchanger is required. In fact, this match has created a loop that can be verified by starting at E-6 in Fig. 9-21, moving down line S-2 to E-5, and following the E-5 curve to line S-3 and then back to E-6. This path

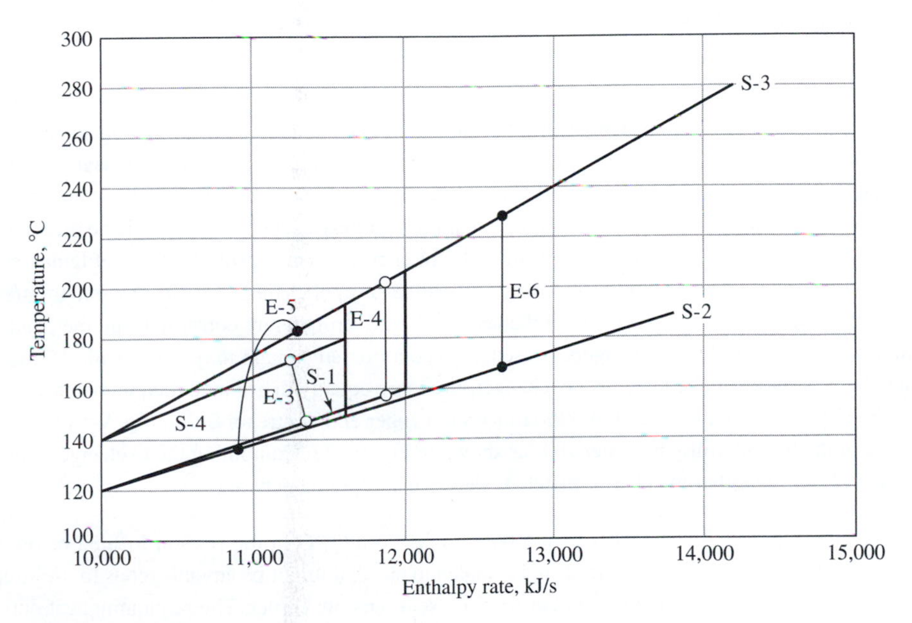

Figure 9-21
Possible heat exchange match 2

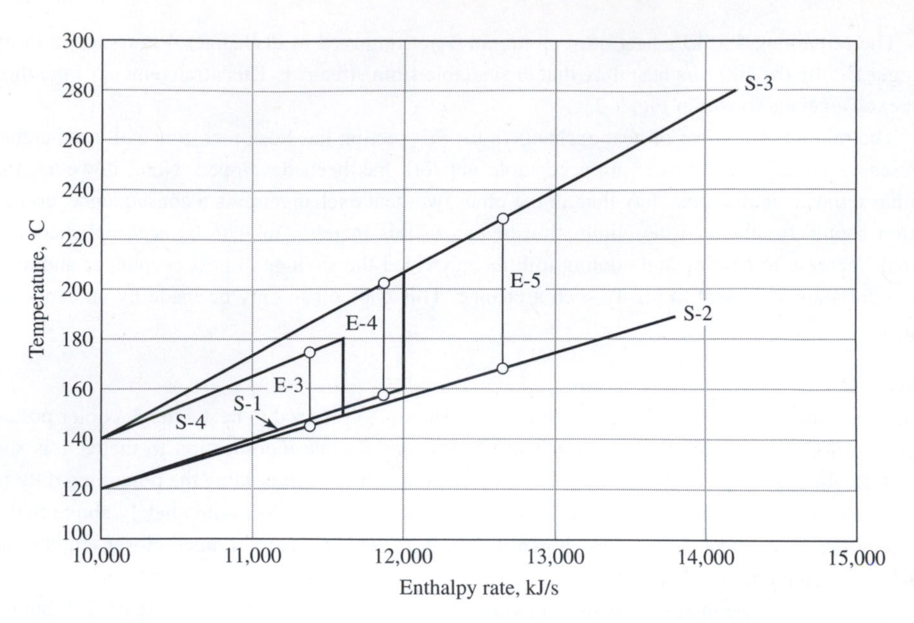

Figure 9-22
Possible heat exchange match 3

is marked with the darkened circles in Fig. 9-21. Since this violates one of the heat integration guidelines, this match should be rejected.

Match 3

With reference to Fig. 9-22, match streams S-4 and S-2 in heat exchanger E-3. This match removes the total heat duty of 1600 kJ/s from stream S-4. The exit temperature of stream S-2 in heat exchanger E-3 can be obtained from the graph or calculated by an energy balance as 149.1°C. Now match all of stream S-1 with the low-temperature end of stream S-3 in heat exchanger E-4. This heat exchange utilizes all the 2000 kJ/s heat duty of stream S-1. An energy balance provides an inlet temperature into E-4 of 206.7°C. The remaining heat duty of stream S-3 supplies the 2200 kJ/s required by stream S-2 in exchanger E-5. As in match 1, all the heat loads and temperature constraints are met, and the minimum number of heat exchangers is used. An economic advantage may occur with this match since the heat duties in the three heat exchangers are more equally distributed than in match 1.

Finally, the unmet heat load of 3900 kJ/s on the upper end of stream S-2 in Fig. 9-18 is met by the use of the hot oil utility in heater H-1, as shown in Fig. 9-23e. Combining the exchangers shown in Fig. 9-23, using either match 1 or match 3, yields an acceptable network.

c. Streams S-3 and S-4 in part (b) were both cooled to 68.6°C in the section located below the pinch. Since stream S-3 only needs to be cooled to 60°C, it might be advantageous to use stream S-1 to cool stream S-3 to 60°C rather than to use a separate cooler. The remaining heat duty in stream S-1 could be used to partially cool stream S-4. Further cooling of stream S-4 to 20°C would be accomplished with cooling water. No additional cooling water would be required, but one less cooler would be needed. For this option, exchanger E-3 needs to be reevaluated with

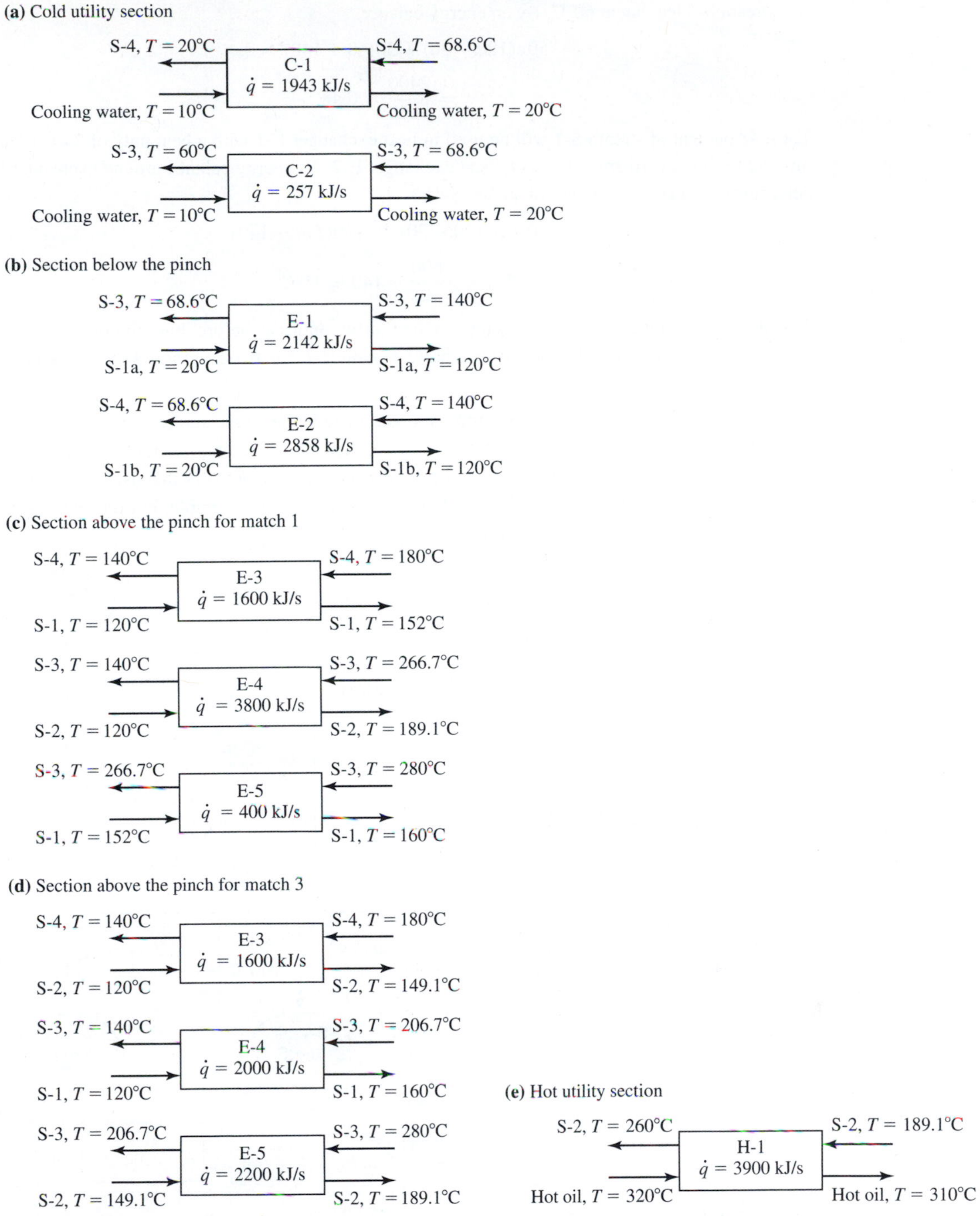

(a) Cold utility section
S-4, T = 20°C
S-4, T = 68.6°C
C-1
q̇ = 1943 kJ/s
Cooling water, T = 10°C
Cooling water, T = 20°C
S-3, T = 60°C
S-3, T = 68.6°C
C-2
q̇ = 257 kJ/s
Cooling water, T = 10°C
Cooling water, T = 20°C
(b) Section below the pinch
S-3, T = 68.6°C
S-3, T = 140°C
E-1
q̇ = 2142 kJ/s
S-1a, T = 20°C
S-1a, T = 120°C
S-4, T = 68.6°C
S-4, T = 140°C
E-2
q̇ = 2858 kJ/s
S-1b, T = 20°C
S-1b, T = 120°C
(c) Section above the pinch for match 1
S-4, T = 140°C
S-4, T = 180°C
E-3
q̇ = 1600 kJ/s
S-1, T = 120°C
S-1, T = 152°C
S-3, T = 140°C
S-3, T = 266.7°C
E-4
q̇ = 3800 kJ/s
S-2, T = 120°C
S-2, T = 189.1°C
S-3, T = 266.7°C
S-3, T = 280°C
E-5
q̇ = 400 kJ/s
S-1, T = 152°C
S-1, T = 160°C
(d) Section above the pinch for match 3
S-4, T = 140°C
S-4, T = 180°C
E-3
q̇ = 1600 kJ/s
S-2, T = 120°C
S-2, T = 149.1°C
S-3, T = 140°C
S-3, T = 206.7°C
E-4
q̇ = 2000 kJ/s
S-1, T = 120°C
S-1, T = 160°C
S-3, T = 206.7°C
S-3, T = 280°C
E-5
q̇ = 2200 kJ/s
S-2, T = 149.1°C
S-2, T = 189.1°C
(e) Hot utility section
S-2, T = 260°C
S-2, T = 189.1°C
H-1
q̇ = 3900 kJ/s
Hot oil, T = 320°C
Hot oil, T = 310°C

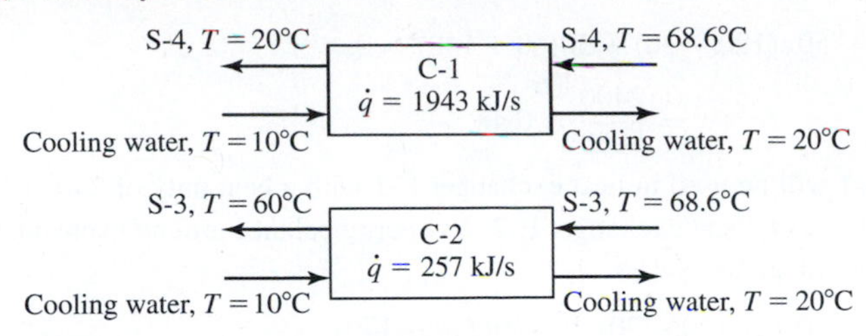

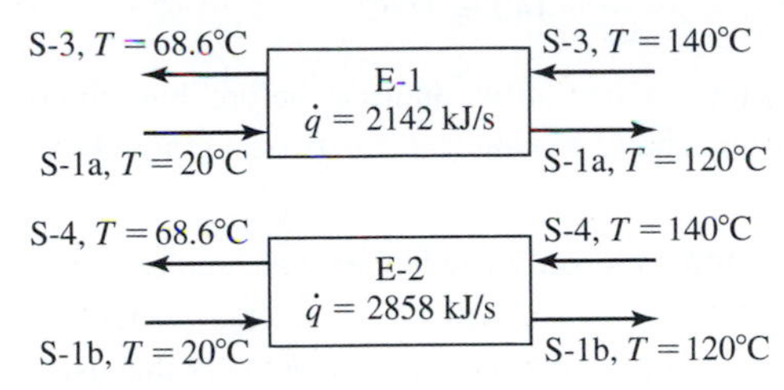

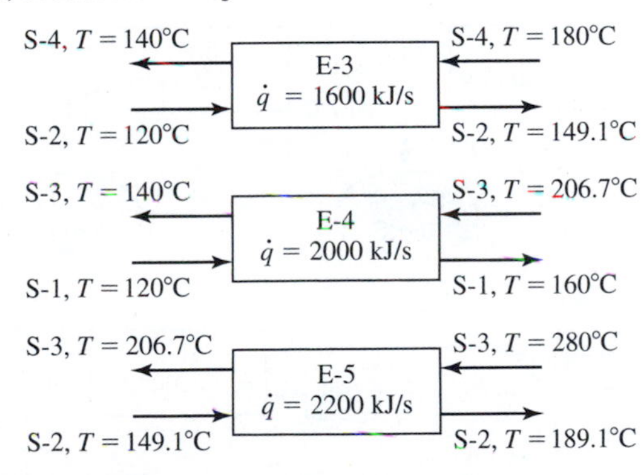

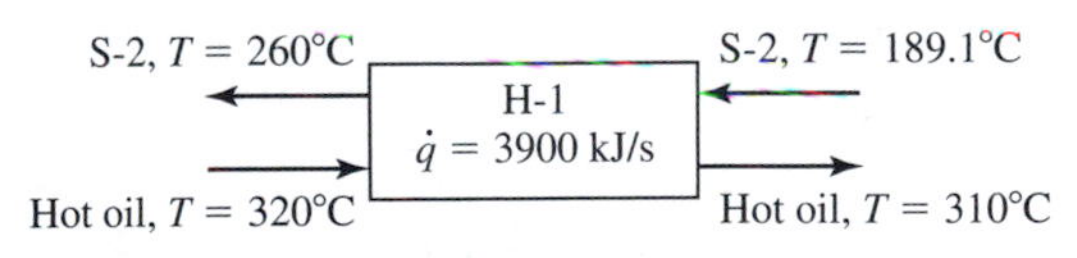

Figure 9-23
Development of heat exchanger network

stream S-3 leaving at 60°C. By an energy balance

$$50x(120 - 20) + 30(60 - 140) = 0$$

$$x = \frac{2400}{5000} = 0.48$$

Thus, 48 percent of stream S-1 will be used in heat exchanger E-1 with a heat duty of 2400 kJ/s. In turn, 52 percent of stream S-1 goes to heat exchanger E-2. An energy balance around exchanger E-2 determines the exit temperature of stream S-1 as

$$50(0.52)(120 - 20) = -40(T_{\text{out}} - 140)$$

$$T_{\text{out}} = \frac{2600}{-40} + 140 = 75°\text{C}$$

This flow arrangement results in the same cooling utility duties as before, but with only one cooler rather than two as in match 1. One cooler is more economical than two and should be selected for the proposed flowsheet.

d. A recommended network for this problem could include the single cooler C-1, exchangers E-1 and E-2 corresponding to the single-cooler case, match 1 or match 3 for the section above the pinch point, and heater H-1 for the hot utility section. The final network that uses one cooler and the results from match 3, as shown in Fig. 9-24, is probably preferable, but the final network selection depends upon an economic analysis.

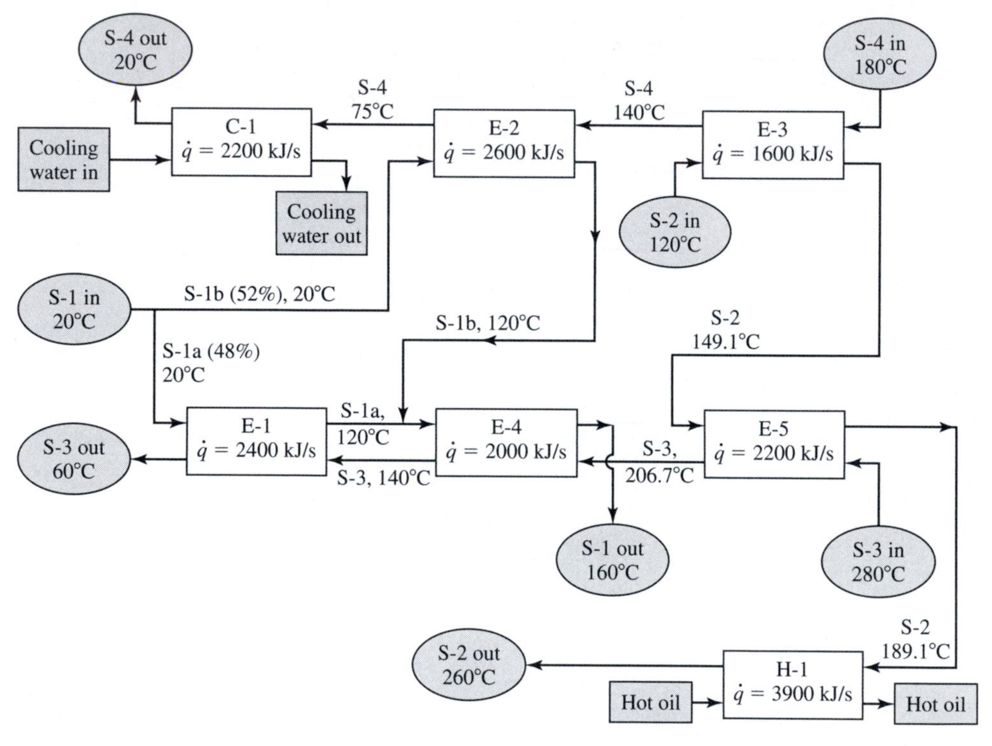

Figure 9-24
Recommended heat exchanger network

Optimization of Heat Exchange Networks In Example 9-8, a heat exchange network was recommended that appeared to be close to optimal based on qualitative observations. However, finding the feasible networks and then finding that network which maximizes the net present worth, or minimizes the present worth of all variable costs, for the application would make a preferable selection. This could be done with a computer-based, heat exchanger network synthesis tool, or with a good optimization program. The optimal network in Example 9-8 has not yet been determined, however, because the example was prepared with a predetermined value of $\Delta T_{\min}$. The particular value selected may not yield a global optimum. Therefore, it is necessary to repeat the process for other values of $\Delta T_{\min}$ until a global optimum is obtained. Once again, this can be done with a computer-based, heat exchanger-network synthesis tool; or it can be done with a single-variable directed search such as a five-point or Golden section search. It should be clear from the foregoing simple example why a good computer-based method is desirable for such an optimization.

NOMENCLATURE

a = constraint equation coefficient; subscript i designates coefficient corresponding to variable i; subscript j designates coefficient corresponding to variable j; cost equation coefficient

A = heat-transfer area, m^2

b = constant

B = constant used for calculation of mechanical work energy balance, J/kg

B' = constant in pump cost equation

c = cost coefficient; subscript j designates constant for variable j

c_{pipe} = purchase cost of new pipe per meter of pipe length, \$/m

c_T = total cost per unit of production, \$/unit

C = cost, \$; subscript F designates annual fixed costs for equipment, \$/yr; subscript c the cost for one cleaning; subscript "pumping" designates pumping costs as dollars per year per meter of pipe length; subscript w designates cooling water costs; subscript A designates installed cost of the condenser per square meter of heat-transfer area; subscript "pipe" designates installed cost of piping system per year per meter of pipe length; superscript prime designates viscous flow

C_p = heat capacity, kJ/(kg·°C), kJ/(kg·K), kJ/(kg mol·°C), or kJ/(kg mol·K)

C_T = total cost, \$; cost per unit of time, \$/s

CP = heat capacity rate, which equals $\dot{m}C_p$, kJ/(s·°C) or kJ/(s · K)

d = constant decision variable; subscript denotes jth stage

D_i = inside pipe diameter, m

E = efficiency of motors and pumps, dimensionless

f' = objective function in cost per time or per unit, \$/s or \$/unit; subscript j designates objective function for stage j

$\mathfrak{f}$ = Fanning friction factor, dimensionless

F = ratio of total cost for fittings and installation to purchase cost for new pipe

g = inequality constraint function in terms of variables v_j and constants c_j; subscript j designates constraint j; bold type indicates vector of functions

h = equality constraint function in terms of variables v_j and constants b_i; subscript i designates constraint i; bold type indicates vector of functions

$\hbar$ = operating costs which remain constant per unit of production, \$/unit

H = total time used for actual operation, emptying, cleaning, and refilling equipment, s; superscript prime designates total time *available* for operation, emptying, cleaning, and recharging; subscript y designates yearly operation time, h

J = frictional loss due to fittings and bends expressed as equivalent fractional loss in a straight pipe, dimensionless

K = electric energy cost, \$/kWh

K_F = annual fixed charges including maintenance, expressed as a fraction of the initial cost for installed equipment, dimensionless

L = length of pipe, m; reflux flow rate, kg mol/s

L'_e = frictional loss due to fittings and bends, expressed as equivalent pipe length in pipe diameters per unit length of pipe divided by pipe diameter, m^2

$\dot{m}$ = mass flow rate, kg/s

m = constant for various equations; number of equations

n = constant for various equations; number of variables

N_E = number of heat exchangers

N_s = total number of streams exchanging heat

O_c = organization costs per unit of time, \$/time

ᵽ$_r$ = profit per unit of production, \$/unit

P = objective function; objective function in cost per time or per unit, \$/s or \$/unit

Ᵽ = profit or return, \$; subscript j designates jth stage

Ᵽ$_b$ = mass of product produced per batch, kg per batch

Ᵽ$_f$ = filtrate delivered during θ_f filtering time, m^3/h

Ᵽ$_r$ = rate of production, total units produced per unit of time; subscript o designates optimum value; subscript cycle denotes per one cycle

$\dot{q}$ = rate of heat transfer or heat duty, kJ/s or kW; subscript r designates reboiler duty; subscript c designates condenser or cooling duty; subscript min refers to minimum duty

q_f = fluid volumetric flow rate, m^3/s

Q = total amount of heat transferred in a given time, kJ; subscript H designates heat transferred in time H h

Q_j = function for dynamic programming indicating combined return from all stages through j

r = positive penalty parameter used in objective function reformulation, dimensionless; prime denotes number of inequalities

R' = profit per unit of time, \$/time

s = selling price per unit of production, \$/unit

S = slack variable

S_b = direct labor cost per hour during operation, \$

T = temperature, K; subscript i designates temperature of stream i; subscript cond designates condensation temperature

U = overall heat-transfer coefficient, $kJ/(s \cdot m^2 \cdot K)$

v = variable; subscript j designates variable j; subscript 0 designates initial variable value; subscript min designates value at minimum objective function value; slack variable; bold type indicates vector of values

V = volume, m^3; subscript A designates species A volume; subscript B designates species B volume; subscript W designates water volume; linear velocity, m/s

W_s = mechanical work added to system from an external mechanical source, kJ/kg

x = independent variable; insulation thickness, m; fraction of stream sent to exchanger

y = independent variable

X = purchase cost for new pipe per meter of length if pipe diameter is 0.0254 m, \$/m

$\Delta \dot{H}$ = enthalpy change rate, kJ/s

ΔT = temperature difference, K or °C; subscript 1 designates the difference at one point; subscript 2 designates the difference at another point; subscript min designates the smaller of ΔT_2 and ΔT_1; subscript log mean designates the logarithmic mean of two temperature differences

z = objective function; subscript j corresponds to variable j

Z = fractional rate of return

Greek Symbols

λ = Lagrange multiplier, dimensionless; subscript i designates multiplier for variable i; subscript min designates value at minimum objective function value

μ = weighting variable, dimensionless; subscript i designates weighting variable i; subscript min designates value at minimum objective function

μ_c = fluid viscosity, Pa·s

ϕ = cost function, \$ per time; superscript i designates heat loss

Φ = income tax rate, fraction/yr

ρ = fluid density, kg/m^3

θ = operating time, s; subscript b designates boiling time; subscript c designates recharging time; subscript i designates total time for boiling and recharging time; subscript f designates filtering time; subscript t designates total time per complete cycle; subscript opt denotes optimum value

∇ = gradient operator

PROBLEMS

9-1 A multiple-effect evaporator is to be used for evaporating 200,000 kg of water per day from a salt solution. The total initial cost for the first effect is \$18,000, and each additional effect costs \$15,000. The service life of the evaporator is estimated to be 10 years, and the salvage or scrap value at the end of the life period may be assumed to be zero. The straight-line depreciation method is used. Fixed charges minus depreciation are 15 percent yearly, based on the first cost of the equipment. Steam costs \$0.0033/kg. Annual maintenance charges are 5 percent of the initial equipment cost. All other costs are independent of the number of effects. The unit will operate 300 days/yr. If the kilograms of water evaporated per kilogram of steam equals 0.85(number of effects), determine the optimum number of effects for minimum annual cost.

9-2 Determine the optimum economic thickness of insulation that should be used under the following conditions: Saturated steam is being passed continuously through a steel pipe with an outside diameter of 0.273 m. The temperature of the steam is 480 K, and the steam is valued at \$0.004/kg. The pipe is to be insulated with a material that has a thermal conductivity of 5.2×10^{-2} W/m·k. The cost of the installed insulation per meter of pipe length is \$$180I_t$, where I_t is the thickness of the insulation in meters. Annual fixed charges including maintenance amount to 20 percent of the initial installed cost. The total length of the pipe is 300 m, and the average temperature of the surroundings may be taken as 295 K. Heat-transfer resistances due to the steam film, scale, and pipe wall are negligible. The air-film coefficient at the outside of the insulation may be assumed constant at 11.4 W/(m^2·K), for all insulation thicknesses.

9-3 An absorption tower containing wooden grids is to be used for absorbing SO_2 in a sodium sulfite solution. A mixture of air and SO_2 will enter the tower at a rate of 33 m^3/s, temperature of 400 K, and pressure of 111 kPa. The concentration of SO_2 in the entering gas is specified, and a given fraction of the entering SO_2 must be removed in the absorption tower. The molecular weight of the entering gas mixture may be assumed to be 29.1. Under the specified design conditions, the number of transfer units necessary varies with the superficial gas velocity as follows:

$$\text{Number of transfer units} = 1.05G_s^{0.18}$$

where G_s is the entering gas velocity in kg/m^2·s based on the cross-sectional area of the empty tower. The height of a transfer unit is constant at 4.5 m. The cost for the installed tower is \$40/m^3 of inside volume, and annual fixed charges amount to 20 percent of the initial cost. Variable operating charges for the absorbent, blower power, and pumping power are represented by the following equation:

$$\text{Total variable operating costs, \$/s} = 9.78 \times 10^5 G_s^2 + \frac{0.1098}{G_s} + \frac{0.0244}{G_s^{0.8}}$$

The unit is to operate 8000 h/yr. Determine the height and diameter of the absorption tower at conditions of minimum annual cost.

9-4 Derive an expression for the optimum economic thickness of insulation to put on a flat surface if the annual fixed charges per square meter of insulation are directly proportional to the thickness, (a) neglecting the air film and (b) including the air film. The air-film coefficient of heat transfer may be assumed constant for all insulation thicknesses.

9-5 A continuous evaporator is operated with a given feed material under conditions in which the concentration of the product remains constant. The feed rate at the start of a cycle after the tubes have been cleaned has been found to be 5000 kg/h. After 48 h of continuous operation, tests have shown that the feed rate decreases to 2500 kg/h. The reduction in capacity is due to true scale formation. If the downtime per cycle for emptying, cleaning, and recharging is 6 h, how long should the evaporator be operated between cleanings in order to obtain the maximum amount of product per 30 days?

9-6 A solvent extraction operation is carried out continuously in a plate column with gravity flow. The unit is operated 24 h/day. A feed rate of 40 m^3/day must be handled 300 days/yr. The allowable velocity per square meter of cross-sectional tower area is 12.2 m^3 of combined solvent and charge per hour. The annual fixed costs for the installed equipment can be predicted from the following equation:

$$C_F = 8800F_{sf}^2 - 51{,}000F_{sf} + 110{,}000 \text{ \$/yr}$$

where F_{sf} is the cubic meters of solvent per cubic meter of feed. Operating and other variable costs depend on the amount of solvent that must be recovered, and these costs are \$1.41 for each cubic meter of solvent passing through the tower. What tower diameter should be used for optimum conditions of minimum total cost per year?

9-7 Prepare a plot of optimum economic pipe diameter versus the flow rate of fluid in a steel pipe under the following conditions:

Costs and operating conditions ordinarily applicable in industry may be used.

The flow of the fluid may be considered to be in the turbulent range.

The viscosity of the fluid may range from 0.1 to 20 cP.

Express the diameters in meters and use inside diameters.

The plot should cover a diameter range of 2.5×10^{-3} to 2.5 m.

Express the flow rate in kg/s.

The plot should cover a flow rate range of 1×10^{-3} to 10 kg/s.

The plot should be presented on log-log coordinates.

One curve on the plot should be presented for each of the following fluid densities: 1600, 800, 160, 16, 1.6, 0.16, and 0.016 kg/m^3.

9-8 For the conditions indicated in Prob. 9-7, prepare a log-log plot of fluid velocity in meters per second versus optimum economic pipe diameter in meters. The plot should cover a fluid velocity range of 0.1 to 30 m/s and a pipe diameter range of 0.01 to 0.25 m.

9-9 A continuous evaporator is being used to concentrate a scale-forming solution of sodium sulfate in water. The overall coefficient of heat transfer decreases according to

$$U^{-2} = 6.88 \times 10^{-5}\theta_b + 0.186$$

where U = overall coefficient of heat transfer, (kJ/s)/(m^2·K) and θ_b = time in operation, s. The only factor which affects the overall coefficient is the scale formation. The liquid enters the evaporator at the boiling point, and the temperature and heat of vaporization are constant. At the operating conditions, 2300 kJ is required to vaporize 1 kg of water, the heat-transfer area is 37 m^2, and the temperature-difference driving force is 40°C. The time required to shut down, clean, and get back on stream is 4 h for each shutdown, and the total cost for this cleaning operation is $100 per cycle. The labor costs during operation of the evaporator are $20 per hour. Determine the total time per cycle for minimum total cost under the following conditions:

a. An overall average of 30,000 kg of water per 24-h day must be evaporated during each 30-day period.

b. An overall average of 37,000 kg of water per 24-h day must be evaporated during each 30-day period.

9-10 An organic chemical is produced by a batch process. In this process, chemicals X and Y react to form chemical Z. Since the reaction rate is very high, the total time required per batch has been found to be independent of the amounts of the materials, and each batch requires 2 h, including time for charging, heating, and dumping. The following equation shows the relation between the kilograms of Z produced (kg_Z) and the kilograms of X (kg_X) and Y (kg_Y) supplied:

$$kg_Z = 1.5(1.1kg_X kg_Z + 1.3kg_Y kg_Z - kg_X kg_Y)^{0.5}$$

Chemical X costs $0.2/kg. Chemical Y costs $0.1/kg. Chemical Z sells for $1.75 per kg. If one-half of the selling price for chemical Z is due to costs other than for raw materials, what is the maximum profit obtainable per kilogram of chemical Z?

9-11 Derive an expression similar to Eq. (9-87) for finding the optimum exit temperature of cooling water from a heat exchanger when the temperature of the material being cooled is not constant. Designate the true temperature-difference driving force by $F_G \, \Delta T_{lm}$, where F_G is a correction factor with a value dependent on the geometric arrangement of the passes in the exchanger. Use primes to designate the temperature of the material that is being cooled.

9-12 Under the following conditions, determine the optimum economic thickness of insulation for a 0.038-m (1½-in. standard) pipe carrying saturated steam at 800 kPa. The line is in use continuously. The covering specified is light carbonate magnesia, which is marketed only in 0.025-m increments (0.025 m, 0.05 m, 0.075 m, etc.). The cost of the installed insulation may be approximated as $700 per cubic meter of insulation. Annual fixed charges are 20 percent of the initial investment, and the heat of the steam is valued at $1.42/GJ. The temperature of the surroundings may be assumed to be 300 K.

L. B. McMillan [*Trans. ASME,* **48:** 1269 (1926)] has presented approximate values of optimum economic insulation thickness versus the group $(kb_c H_y \Delta T/a_c)^{0.5}$, with pipe size as a parameter.

k = thermal conductivity of insulation, (kJ/s)/(m·K)

b_c = cost of heat, $/kJ

H_y = operation per year, s/yr

ΔT = overall temperature-difference driving force, K

a_c = cost of insulation, ($/m^3)/yr

The following data are based on the results of McMillan, and these data are applicable to the conditions of this problem:

$\left(\frac{kb_c H_y \Delta T}{a_c}\right)^{0.5}$	**Optimum economic thickness of insulation, for a nominal pipe diameter of**			
	0.013 m ($\frac{1}{2}$ in.)	**0.025 m (1.0 in.)**	**0.051 m (2.0 in.)**	**0.102 m (4.0 in.)**
0.03	—	0.010	0.013	0.015
0.06	0.020	0.024	0.028	0.033
0.09	0.030	0.036	0.041	0.048
0.15	0.047	0.053	0.062	0.074
0.24	0.070	0.079	0.091	0.109
0.37	0.097	0.109	0.124	—

9-13 A catalytic process uses a catalyst which must be regenerated periodically because of reduction in conversion efficiency. The cost for one regeneration is constant at $800. This figure includes all shutdown and start-up costs as well as the cost for the actual regeneration. The feed rate to the reactor is maintained constant at 70 kg/day, and the cost for the feed material is $5.50 per kilogram. The daily costs for operation are $300, and fixed charges plus general overhead costs are $100,000 per year. Tests on the catalyst show that the yield of product as kilograms of product per kilogram of feed during the first day of operation with the regenerated catalyst is 0.87, and the yield decreases as $0.87/\theta_D^{0.25}$, where θ_D is the time in operation expressed in days. The time necessary to shut down the unit, replace the catalyst, and start-up the unit is negligible. The value of the product is $31.00 per kilogram, and the plant operates 300 days/yr. Assuming no costs are involved other than those mentioned, what is the maximum annual profit that can be obtained under these conditions?

9-14 Derive the following equation for the optimum outside diameter of insulation on a wire for maximum heat loss:

$$D_{\text{opt}} = \frac{2k_m}{(h_c + h_r)_c}$$

where k_m is the mean thermal conductivity of the insulation and $(h_c + h_r)_c$ is the combined convection and radiation heat-transfer coefficient. The values of k_m and $(h_c + h_r)_c$ can be considered as constants independent of temperature level and insulation thickness.

9-15 Derive Eq. (9-80) for the optimum economic pipe diameter, and compare this to the equivalent expression presented as Eq. (5-90) in *Perry's Chemical Engineers' Handbook,* 5th ed. (J. H. Perry and C. H. Chilton, eds., McGraw-Hill, New York, 1973, p. 5–32).

9-16 Using a direct partial derivative approach for the objective function, instead of the Lagrangian multiplier $L(x, y) = xy + \lambda(x^2 + y^2 - 10)$, determine the optimum values of x and y in this equation.

9-17 Find the values of $x, y,$ and z that minimize the function $x + 2y^2 + z^2$ subject to the constraint that $x + y + z = 1$, making use of the Lagrangian multiplier.

9-18 To continue the operation of a small chemical plant at the same capacity, it will be necessary to make some changes on one of the reactors in the system. The decision has been made by management that the unit must continue in service for the next 12 years, and the company policy is that no unnecessary investments are made unless at least an 8 percent rate of return (end-of-year compounding) can be obtained after income taxes. Two possible ways for making a satisfactory change in the reactor are as follows:

1. Make all the critical changes now at a cost of \$5800 so the reactor will be satisfactory to use for 12 years.
2. Make some of the changes now at a cost of \$5000 which will permit operation for 8 years, and then make changes costing \$2500 to permit operation for the last 4 years.

   a. Which alternative should be selected if no inflation is anticipated over the next 12 years?

   b. Which alternative should be selected if inflation at a rate of 7 percent (end-of-year compounding) is assumed for all future costs?

9-19 As noted in Example 9-8 on heat integration, the heat exchange network that was developed was based on a predetermined value of $\Delta T_{\min}$ of 20°C. However, this particular value may not yield a global optimum. Therefore, it is necessary to investigate other values of $\Delta T_{\min}$ until this global optimum is obtained.

In this evaluation, use the network that was recommended with a single cooler C-1, exchangers E-1 and E-2 corresponding to the single cooler, match 3 network for the section above the pinch point, and heater H-1 for the hot utility section. Use the cost data for the hot oil and cooling water given in Example 9-8. Use an overall heat transfer coefficient of 1 kW/m^2·K for each exchanger, including utilities. Heat exchanger purchase price = \$1,000$A^{0.6}$, where A is the heat exchange area in m^2. No single exchanger should exceed 500 m^2 in area. Use straight-line depreciation for 7 years, with no salvage value. Use an income tax rate of 35 percent per year. As the objective function, use the following relation that is based on the net return, Eq. (8-3*c*)

$$\text{Minimize:}\quad \text{annual cost, } C_T = (\text{annual cost of heating oil and cooling water})\,(1 - \Phi) + (\text{annual depreciation})\,(1 - \Phi) + m_{ar}(T_{hx})$$

where T_{hx} is the total investment in the heat exchanger system assumed to be five times the exchanger purchase price. The minimum acceptable rate of return m_{ar} is 15 percent per year after income tax.

Obtain the optimal $\Delta T_{\min}$ for this network. Plot the results of the optimization process to show what effect a small change in the $\Delta T_{\min}$ value has on the economic analysis.

CHAPTER

10

Materials and Fabrication Selection

As chemical process plants turn to higher temperatures and flow rates to boost yields and throughputs, selection of construction materials takes on added importance because these severe conditions intensify corrosive action. Fortunately, a broad range of materials is now available for corrosive service. However, this apparent abundance of materials also complicates the task of selecting the "best" material because, in many cases, a number of alloys and plastics will have sufficient corrosion resistance for a particular application. Final choice cannot be based simply on selecting a suitable material from a corrosion table, but must be based on a sound economic analysis of competing materials.

The purpose of this chapter is to provide the design engineer with a working knowledge of some of the major forms and types of materials available, what they offer, and how they are specified. With this background, the engineer can consult a materials specialist during the early stages of the design to establish the materials that best meet the process conditions.

FACTORS CONTRIBUTING TO CORROSION

Corrosion is a complex phenomenon that may involve one or more forms. Corrosion of metals specifically applies to chemical or electrochemical attack. The corrosion is often confined to the metal surface, but sometimes it may occur along grain boundaries or other metal weaknesses because of a difference in resistance to attack or local electrolytic action.

The deterioration of plastics and other nonmetallic materials is essentially physiochemical rather than electrochemical and results in swelling, crazing, cracking, softening, and/or decomposition. Nonmetallic materials can either deteriorate rapidly when exposed to a particular environment or be practically unaffected.

There are both localized and structural aspects of corrosion associated with metals. Corrosive action on a localized level occurs at areas such as crevices, intergranular boundaries, oxygen-deficient cells, or stress-activated surfaces. Corrosive action can also be initiated by galvanic action, impingement, cavitation, or hydrogen attack. In galvanic corrosion, for example, an electric potential can be set up between two different metals in contact with a conducting fluid. Table 10-1 shows the galvanic series of various metals. The table should be used with caution, since exceptions to this series in actual use are possible. However, as a general rule, when dissimilar metals are used in contact with each other and are exposed to an electrically conducting solution,

Table 10-1 Galvanic series for metals and alloys

Corroded end (anodic, or least noble)
Magnesium
Magnesium alloys
Zinc
Aluminum alloys
Aluminum
Al-clad
Cadmium
Mild steel
Cast iron
Ni-resist
13% Chromium stainless (active)
50-50 Lead-tin solder
18-8 Stainless type 304 (active)
18-8-3 Stainless type 316 (active)
Lead
Tin
Muntz metal
Naval brass
Nickel (active)
Inconel 600 (active)
Yellow brass
Admiralty brass
Aluminum bronze
Red brass
Copper
Silicon bronze
70-30 Cupronickel
Nickel (passive)
Inconel 600 (passive)
Monel 400
18-8 Stainless type 304 (passive)
18-8-3 Stainless type 316 (passive)
Silver
Graphite
Gold
Platinum
Protected end (cathodic, or most noble)

combinations of metals that are as close as possible in the galvanic series should be selected. Galvanic corrosion can be prevented by insulating the metals from each other.

The corrosion rate of most metals is affected by pH. Metals such as aluminum and zinc dissolve rapidly in either acidic or basic solutions while noble metals are not appreciably affected by pH. Oxidizing agents often are powerful accelerators of corrosion. In many cases the oxidizing power of a solution may be the most important single property associated with the corrosion activity. For example, over a pH range from 4 to 10 the corrosion rate of acid-soluble metals such as iron is controlled by the rate of transport of oxidizer (usually dissolved oxygen) to the metal surface. Note, however, that oxidizing agents may accelerate the corrosion of one class of metals and retard the corrosion of another class of metals by the formation of a protective oxide on the surface. This property of chromium is responsible for the principal corrosion-resisting characteristics of the stainless steels.

The rate of corrosion tends to increase with rising temperature. In addition, temperature has a secondary effect through its influence on the solubility of air (oxygen), the principal oxidizer influencing corrosion. The rate of corrosion quite often will also increase with an increase in the velocity of the corrosive solution relative to the metallic surface.

Structural corrosion, on the other hand, includes at least three different types of material damage. Graphitic corrosion usually involves gray cast iron in which the metal is converted to graphite mixed with iron corrosion products. Dezincification is corrosion of a brass alloy containing zinc in which the principal product of corrosion is metallic copper. In biological corrosion the metabolic activity of microorganisms can either directly or indirectly cause deterioration of a metal by corrosion processes.

Biological activity in a fluid can produce a corrosive environment, create electrolytic concentration cells on the metal surface, alter the resistance of surface films, have an influence on the rate of anodic or cathodic reaction, or alter the environmental composition. Microorganisms associated with corrosion are either aerobic or anaerobic. The former species readily thrive in an oxygen-containing environment while the latter species thrive in an environment essentially devoid of atmospheric oxygen.

The manner in which many of these microorganisms proceed with their chemical processes is quite complicated. As a group, these destructive microorganisms can grow over a pH range of about 0 to 11. The bacteria involved may use almost any available organic carbon molecule as well as a number of inorganic elements or ions as sources of energy. The nutritional requirements apparently range from the very simple to the very complex. Most, however, fall between these extremes and require a limited number of organic molecules, moderate temperatures, moist environment, and a pH close to 7.

Most chemical engineers are not, and do not need to become, experts in the details of biological corrosion. What is required, however, is to recognize that this type of corrosion process is active, that process equipment and structures are at risk, and that there are numerous tools available to detect and monitor microbiologically influenced corrosion.

COMBATING CORROSION

In view of the problems that are associated with corrosion, it is necessary for the design engineer to make material selection decisions that most economically fulfill the process requirements. The best source of guidance for such a decision is well-documented experience from an identical process unit. In the absence of such data, other data sources must be used. The data from such alternative sources must be properly evaluated, taking into account the differences that exist between the conditions under which such data were obtained and the actual conditions in the proposed unit. Such a review may suggest a minimization of the corrosion process with a change in the operating environment, use of an inhibitor, cathodic or anodic protection, substitution of less corrosive materials, or the use of a permissible corrosion rate in the design of certain equipment if the corrosion is anticipated and properly allowed for in the thickness calculation.

Simple changes in an environment may make an appreciable difference in the corrosion of metals and need to be considered as a means of combating corrosion. Oxygen is an important factor, and its removal or addition can result in marked changes in corrosion. The treatment of boiler feedwater to remove oxygen greatly reduces the corrosiveness of the water on the chamber and piping. Corrosiveness of acid fluids to stainless alloys, on the other hand, may be reduced by aeration because of the formation of passive oxide films. Reduction in temperature is often beneficial with a reduction in corrosion. Modification of pH values to less acidic conditions also can result in less corrosion. Elimination of moisture in a system can and frequently does minimize corrosion of metals, and this environmental alteration should always be considered.

The use of inhibitors as additives to corrosive environments to decrease corrosion of metals is another means of combating corrosion. Inhibitors are effective when they can influence the cathode- or anode-area reactions. Typical examples of inhibitors for minimizing corrosion of iron and steel in aqueous solutions are the chromates, phosphates, and silicates.

Cathodic protection is widely used in the protection of underground pipes and tanks from external soil corrosion and in water systems to protect water storage tanks and offshore structures. Two methods of providing cathodic protection are presently available. Both methods depend on making the metal that is to be protected the cathode in the electrolyte involved.

The use of nonmetallic castings and lining materials in combination with steel or other materials has often proved to be an economical solution for combating corrosion. Organic coatings of many kinds are used. The most dependable barrier linings include flake-glass-reinforced resin systems, elastomers, and plasticized plastic systems. Special glasses can be bonded to steel, providing an impervious liner with a thickness of 1.5 to 2.5 mm.

The cladding of steel with an alloy is another approach to the corrosion problem. For mild environments, steel can also be coated with zinc, cadmium, aluminum, or lead. Tests may be required when it becomes necessary to determine the most economical selection for a particular environment.

PROPERTIES OF MATERIALS

Materials used in the process industry may be divided into two general classifications, namely, *metals* and *nonmetals.* Pure metals and metallic alloys are included under the first classification. Because of the many types of metallic products used in industry, basic specifications have been developed to define the chemistry and properties of such materials. The American Iron and Steel Institute (AISI) has set up a series of standards for steel products.† However, even the relatively simple product descriptions provided by AISI and shown in Table 10-2 must be used carefully. For instance, the AISI 1020 carbon steel does not refer to all 0.20 percent carbon steels. AISI 1020 is part of the numerical designation system defining the chemical composition of certain "standard steels" used primarily in bar, wire, and some tubular steel products. The system generally does not apply to sheet, strip, plate, or structural material. One reason is that the chemical composition ranges of standard steels are unnecessarily restrictive for many applications.

Carbon-steel plates for reactor vessels are a good example. This application generally requires a minimum level of mechanical properties, weldability, formability, and toughness as well as some assurance that these properties will be uniform throughout. A knowledge of the detailed composition of the steel alone will not ensure that these requirements are met. Even welding requirements for plate can be met with far less restrictive chemical compositions than would be needed for the same type of steel used in bar stock suitable for heat treating to a minimum hardness or tensile strength.

Ferrous Metals and Alloys

Steel Carbon steel is the most commonly used material in the chemical process industry despite its somewhat limited corrosion resistance. This type of steel is routinely used with most organic chemicals and neutral or basic aqueous solutions at moderate temperatures. Because of its availability, low cost, and ease of fabrication, carbon steel is used even in situations with corrosion rates of 0.13 to 0.5 mm/yr by specifying an added thickness to ensure achievement of the desired service life.

Low-alloy steels contain one or more alloying components to improve the mechanical and corrosion-resistant properties over those of carbon steel. Nickel increases toughness and improves low-temperature properties and corrosion resistance. Chromium and silicon improve hardness, abrasion resistance, corrosion resistance, and resistance to oxidation. Molybdenum provides added strength for higher-temperature operation.

Stainless Steel There are more than 70 standard types of stainless steel and many special alloys. These steels are produced in the wrought form (AISI types) and as cast

†Specifications and codes on materials have also been established by the Society of Automotive Engineers (SAE), the American Society of Mechanical Engineers (ASME), and the American Society for Testing and Materials (ASTM).

Table 10-2 AISI standard steels[†]

Carbon steel AISI series designations	Nominal composition or range[‡]
10XX	Nonresulfurized carbon steels with 44 compositions ranging from 1008 to 1095. Manganese ranges from 0.30 to 1.65%; if specified, silicon is 0.10 max. to 0.30 max., each depending on grade. Phosphorus is 0.040 max., sulfur is 0.050 max.
11XX	Resulfurized carbon steels with 15 standard compositions. Sulfur may range up to 0.33%, depending on grade.
B11XX	Acid Bessemer resulfurized carbon steels with three compositions. Phosphorus generally is higher than 11XX series.
12XX	Rephosphorized and resulfurized carbon steels with five standard compositions. Phosphorus may range up to 0.12% and sulfur up to 0.35%, depending on grade.
13XX	Manganese, 1.75%. Four compositions from 1330 to 1345.
40XX	Molybdenum, 0.20 or 0.25%. Seven compositions from 4012 to 4047.
41XX	Chromium, to 0.95%; molybdenum, to 0.30%. Nine compositions from 4118 to 4161.
43XX	Nickel, 1.83%; chromium to 0.80%; molybdenum, 0.25%. Three compositions from 4320 to E4340.
44XX	Molybdenum, 0.53%. One composition, 4419.
46XX	Nickel, to 1.83%; molybdenum to 0.25%. Four compositions from 4615 to 4626.
47XX	Nickel, 1.05%; chromium, 0.45%; molybdenum to 0.35%. Two compositions, 4718 and 4720.
48XX	Nickel, 3.50%; molybdenum, 0.25%. Three compositions from 4815 to 4820.
50XX	Chromium, 0.40%. One composition, 5015.
51XX	Chromium to 1.00%. Ten compositions from 5120 to 5160.
5XXXX	Carbon, 1.04%; chromium to 1.45%. Two compositions, 51100 and 52100.
61XX	Chromium to 0.95%, vanadium to 0.15% min. Two compositions, 6118 and 6150.
86XX	Nickel, 0.55%; chromium, 0.50%; molybdenum 0.20%. Twelve compositions from 8615 to 8655.
87XX	Nickel, 0.55%; chromium, 0.50%; molybdenum, 0.25%. Two compositions, 8720 and 8740.
88XX	Nickel, 0.55%; chromium, 0.50%; molybdenum, 0.35%. One composition, 8822.
92XX	Silicon, 2.00%. Two compositions, 9255 and 9260.
50BXX	Chromium to 0.50%, also containing boron. Four compositions from 50B44 to 50B60.
51BXX	Chromium to 0.80%, also containing boron. One composition, 51B60.
81BXX	Nickel, 0.30%; chromium, 0.45%; molybdenum, 0.12%, also containing boron. One composition, 81B45.
94BXX	Nickel, 0.45%; chromium, 0.40%; molybdenum, 0.12%, also containing boron. Two compositions, 94B17 and 94B30.

[†]When a carbon or alloy steel also contains the letter L in the code, it contains from 0.15 to 0.35 percent lead as a free-machining additive, that is, 12L14 or 41L40. The prefix E before an alloy steel, such as E4340, indicates the steel is made only by electric furnace. The suffix H indicates an alloy steel made to more restrictive chemical composition than that of standard steels and produced to a measured and known hardenability requirement; for example, 8630H or 94B30H.XX indicates nominal carbon content within range.

[‡]For a detailed listing of nominal composition or range, see *Perry's Chemical Engineers' Handbook,* 7th ed., R. H. Perry and D. W. Green, eds., McGraw-Hill, New York, 1997.

alloys (ACI[†] types). In the former, there are three groups of stainless alloys, namely, *martensitic, ferritic,* and *austenitic.*

The martensitic alloys contain 12 to 20 percent chromium with controlled amounts of carbon and other alloys. These alloys can be hardened by heat treatment, which can increase the tensile strength from 550 to 1380 MPa. Corrosion resistance is poor relative to that for austenitic stainless steels. This relegates martensitic steels to environments that are only mildly corrosive.

Ferritic stainless steels contain 15 to 30 percent chromium with 0.1 percent or less carbon content. The strength of ferritic stainless can be increased by cold working but not by heat treatment. Corrosion resistance is good, although ferritic alloys experience considerable corrosion with reducing acids such as HCl. However, mildly corrosive solutions and oxidizing media are handled with minimum damage.

Austenitic stainless steels are the most corrosion-resistant of the three groups of stainless steels. These steels contain 16 to 26 percent chromium and 6 to 22 percent nickel. Carbon content is maintained below 0.08 percent to minimize carbide precipitation. These alloys can be work-hardened, but heat treatment will not cause hardening. Tensile strength in the annealed condition is about 585 MPa, but work-hardening can increase this to 2000 MPa. The addition of molybdenum to the austenitic alloy, as in type 316, increases the corrosion resistance and improves high-temperature strength.

Although fabrication operations on stainless steels are more difficult than on standard carbon steels, all types of stainless steel can be fabricated successfully.[‡] The properties of types 430F, 416, 410, 310, 309, and 303 make these materials particularly well suited for machining or other fabricating operations. In general, machinability is improved if small quantities of phosphorus, selenium, or sulfur are present in the alloy.

A preliminary approach to the selection of the stainless steel required for a specific application is to classify the various types according to alloy content, microstructure, and major characteristics. Table 10-3 outlines the information according to the classes of stainless steels—austenitic, martensitic, and ferritic. Table 10-4 presents characteristics and typical applications of various types of stainless steel while Table 10-5 indicates resistance of stainless steels to oxidation in air.

In addition to the three groups of stainless steels noted above, many cast stainless alloys are available and widely used in pumps, valves, and fittings. These casting alloys are defined under the ACI system where the letter L refers to corrosion-resistant alloys, the letter H refers to a series of heat-resistant grades of ACI cast alloys, and the letters PH refer to precipitation-hardening stainless steels. Essentially, these cast stainless alloys contain chromium and nickel with added alloying agents such as copper, aluminum, beryllium, molybdenum, nitrogen, and phosphorus.

A group of mostly proprietary alloys with somewhat better corrosion resistance than the stainless steels are known as medium alloys. One member of this group is the

---

[†]Alloy Casting Institute.

[‡]For a detailed discussion of machining and fabrication of stainless steels, see "Selection of Stainless Steels," *Bulletin* OLE 11366, Armco Steel Corporation, Middletown, OH 45042; and "Fabrication of Stainless Steel," *Bulletin* 031478, United Steel Corporation, Pittsburgh, PA 15230.

Table 10-3 Classification of stainless steels by alloy content and microstructure

Stainless steels	Chromium types	Martensitic	Hardenable (types 403, 410, 414, 416, 416Se, 420, 431, 440A, 440B, 440C)
		Ferritic	Nonhardenable (types 405, 430, 430F, 430Se, 442, 446)
	Chromium-nickel types	Austenitic	Nonhardenable, except by cold working (types 201, 202, 301, 302, 302B, 303, 303Se, 304, 304L, 305, 308, 309, 309S, 310, 310S, 314, 316, 316L, 317, 321, 347, and 348)
			Strengthened by aging or precipitation-hardening (types 17-14 CuMo, 17-10P, HNM)
		Semiaustenitic	Precipitation-hardening (PH 15-7 Mo, 17-7 PH, AM 355)
		Martensitic	Precipitation-hardening (17-4 PH, 15-5 PH, stainless W)

20 alloy which was originally developed to meet the need for a material with greater resistance to hot sulfuric acid than the available stainless steels. Other members of the medium-alloy group are Incoloy 825 and Hastelloy G-3 and G-30.

Materials in another group designated as high alloys all contain large percentages of nickel. For example, Hastelloy B contains 61 percent nickel and 28 percent molybdenum. Oxidizing acids and salts rapidly corrode this alloy, but it provides unusually high resistance to all concentrations of hydrochloric acid at all temperatures in the absence of oxidizing agents. Hastelloy C, on the other hand, with a reduction in the nickel/molybdenum content to permit additions of chromium, iron, tungsten, and trace amounts of several key elements, provides an alloy with structural strength and good corrosion resistance under conditions of high temperatures. These characteristics permit this alloy to be used in the form of valves, piping, heat exchangers, and various types of vessels. Other types of Hastelloys and Chlorimets are also available for use under special corrosive conditions.

Table 10-4 Stainless steels most commonly used in the chemical process industries†

Type§	Composition, %			Other significant elements‡	Major characteristics	Properties	Applications
	Cr	Ni	C max.				
301	16.00–18.00	6.00–8.00	0.15		High work-hardening rate combines cold-worked high strength with good ductility	Good structural qualities	Structural applications, bins and containers
302	17.00–19.00	8.00–10.00	0.15		Basic, general-purpose austenitic type with good corrosion resistance and mechanical properties	General-purpose	Heat exchangers, towers, tanks, pipes, heaters, general chemical equipment
303	17.00–19.00	8.00–10.00	0.15	S 0.15 min	Free-machining modification of type 302; contains extra sulfur	Type 303Se is also available for parts involving extensive machining	Pumps, valves, instruments, fittings
304	18.00–20.00	8.00–12.00	0.08		Low-carbon variation of type 302, minimizes carbide precipitation during welding	General-purpose. Also available as 304L with 0.03% carbon to minimize carbide precipitation during welding	Perforated blow-pit screens, heat exchanger tubing, preheater tubes
305	17.00–19.00	10.00–13.00	0.12		Higher heat and corrosion resistance than type 304	Good corrosion resistance	Funnels, utensils, hoods
308	19.00–21.00	10.00–12.00	0.08		High Cr and Ni produce good heat and corrosion resistance. Used widely for welding rod	In order of their numbers, these alloys show increased resistance to high-temperature corrosion. Types 308S, 309S, and 310S are also available for welded construction (applies to 308, 309, 310)	Welding rod, more ductile welds for type 430
309	22.00–24.00	12.00–15.00	0.20		High strength and resistance to scaling at high temperatures		Welding rod for type 304, heat exchangers, pump parts
310	24.00–26.00	19.00–22.00	0.25		Higher alloy content improves basic characteristics of type 309		Jacketed high-temperature, high-pressure reactors, oil-refining still tubes
314	23.00–26.00	19.00–22.00	0.25	Si 1.5–3.0	High silicon content	Resistant to oxidation in air to 1100°C	Radiant tubes, carburizing boxes, annealing boxes
316	16.00–18.00	10.00–14.00	0.08	Mo 2.00–3.00	Mo improves general corrosion and pitting resistance and high-temperature strength over that of type 302	Resistant to high pitting corrosion. Also available as 316L for welded construction	Distillation equipment for producing fatty acids, sulfite paper processing equipment
317	18.00–20.00	11.00–15.00	0.08	Mo 3.00–4.00	Higher alloy content improves basic advantages of type 316	Type 317 has the highest aqueous corrosion resistance of all AISI stainless steels	Process equipment involving strong acids or chlorinated solvents
321	17.00–19.00	9.00–12.00	0.08	Ti 5 × C, min.	Stabilized to permit use in 420–870°C range without harmful carbide precipitation	Stabilized with titanium and columbium-tantalum, respectively, to permit their use for large welded structures which cannot be annealed after welding (applies to 321, 347)	Furnace parts in presence of corrosive fumes
347	17.00–19.00	9.00–13.00	0.08	Cb-Ta 10 × C, min.	Characteristics similar to type 321. Stabilized by Cb and Ta		Like 302 but used where carbide precipitation during fabrication or service may be harmful, welding rod for type 321

403	11.50–13.50		0.15	Si 0.50 max.	Version of type 410 with limited hardenability but improved fabricability	Not highly resistant to high-temperature oxidation in air	Steam turbine blades
405	11.50–14.50		0.08	Al 0.10–0.30	Version of type 410 with limited hardenability but improved weldability	Good weldability and cladding properties	Tower linings, baffles, separator towers, heat exchanger tubing
410	11.50–13.50		0.15		Lowest-cost general-purpose stainless steel	Wide use where corrosion is not severe	Bubble-tower parts for petroleum refining, pump rods and valves, machine parts, turbine blades
416	12.00–14.00		0.15	S 0.15 min.	Sulfur added for free-machining version of type 410. Type 416Se also available	The most free-machining type of martensitic stainless	Valve stems, plugs, gates, useful for screws, bolts, nuts, and other parts requiring considerable machining during fabrication
420	12.00–14.00		0.15 min.		Similar to type 410 but higher carbon produces higher strength and hardness	High-spring temper	Utensils, bushings, valve stems, and wear-resisting parts
430	14.00–18.00		0.12		Most popular of nonhardening chromium types. Combines good corrosion resistance (to nitric acid and other oxidizing media)	Good heat resistance and good mechanical properties. Also available in type 430F	Chemical and processing towers, condensers. Furnace parts such as retorts and low stressed parts subject to temperatures up to 800°C. Type 430 nitric acid storage tanks, furnace parts, fan scrolls. Type 430F pump shafts, instrument parts, valve parts
431	15.00–17.00	1.25–2.50	0.20		High yield point	Very resistant to shock	Products requiring high yield point and resistance to shock
442	18.00–23.00		0.25		High-chromium nonhardenable type	High-temperature uses where high-sulfur atmospheres make presence of nickel undesirable	Fume furnaces, flare stacks, materials in contact with high-sulfur atmospheres
446	23.00–27.00		0.20		Similar to type 442 but Cr increased to provide maximum resistance to scaling. Especially suited to intermittent high temperatures	Excellent corrosion resistance to many liquid solutions; fabrication difficulties limit its use primarily to high-temperature applications. Useful in high-sulfur atmospheres	Burner nozzles, stack dampers, boiler baffles, furnace linings, glass molds

[†]Adapted from Biennial Material of Construction reports published regularly in *Chemical Engineering* and from tabulations in *Perry's Chemical Engineers' Handbook,* 7th ed., McGraw-Hill, New York, 1997. See the latter reference for mechanical property data such as yield strength, tensile strength, percent elongation, and hardness.
[‡]For a detailed listing of nominal composition or range, see the latest issue of *Data on Physical and Mechanical Properties of Stainless and Heat-Resisting Steels,* Carpenter Steel Company, Reading, PA 19603.
[§]In general, stainless steels in the 300 series contain large amounts of chromium and nickel; those in the 400 series contain large amounts of chromium and little or no nickel; those in the 500 series contain low amounts of chromium and little or no nickel; in the 300 series, except for type 309, the nickel content can be 10 percent or less if the second number is 0 and greater than 10 percent if the second number is 1; in the 400 series, an increase in the number represented by the last two digits indicates an increase in the chromium content.

Table 10-5 Resistance of stainless steels to oxidation in air

Maximum temperature, °C	Stainless steel type
650	416
700	403, 405, 410, 414
800	430F
850	430, 431
900	302, 303, 304, 316, 317, 321, 347, 348, 17-14 CuMo
1000	302B, 308, 442
1100	309, 310, 314, 329, 446

Nonferrous Metals and Alloys

Nickel and Its Alloys Nickel exhibits high corrosion resistance to most alkalies. Nickel-clad steel is used extensively for equipment in the production of caustic soda and alkalies. The strength and hardness of nickel are almost as great as those of carbon steel, and the metal can be fabricated easily. In general, oxidizing conditions promote the corrosion of nickel, and reducing conditions retard it.

Monel, an alloy of nickel containing 67 percent nickel and 30 percent copper, is often used in the food industries. This alloy is stronger than nickel and has better corrosion resistance properties than either copper or nickel. Another important nickel alloy is Inconel (77 percent nickel and 15 percent chromium). The presence of chromium in this alloy increases its resistance to oxidizing conditions.

Aluminum and Its Alloys The lightness and relative ease of fabrication of aluminum and its alloys are factors favoring the use of these materials. Aluminum resists attack by acids because a surface film of inert hydrated aluminum oxide is formed. This film adheres to the surface and offers good protection unless materials which can remove the oxide, such as halogen acids or alkalies, are present. Aluminum alloys generally have a lower corrosion resistance than the pure metal. The Al-clad alloys have been developed to overcome this shortcoming by metallurgically bonding a pure aluminum layer on the core alloy.

The corrosion resistance of aluminum and its alloys tends to be very sensitive to trace contamination. Very small amounts of metallic mercury, heavy-metal ions, or chloride ions can frequently cause rapid failure under conditions which otherwise would be fully acceptable.

Copper and Its Alloys Copper is relatively inexpensive, possesses fair mechanical strength, and can be fabricated easily into a wide variety of shapes. Although it shows little tendency to dissolve in nonoxidizing acids, it is readily susceptible to oxidation. Copper is resistant to atmospheric moisture or oxygen because a protective coating composed primarily of copper oxide is formed on the surface. The oxide, however, is soluble in most acids, and thus copper is not a suitable material for construction when it must contact any acid in the presence of either oxygen or oxidizing agents. Copper exhibits good corrosion resistance to strong alkalies, with the exception of ammonium hydroxide. At room temperature it can handle sodium and potassium hydroxide of all concentrations. It resists most organic solvents as well as aqueous solutions of organic acids.

Copper alloys, such as brass, bronze, admiralty, and Muntz metals, can exhibit better corrosion resistance and better mechanical properties than pure copper. In general, high-zinc alloys should not be used with acids or alkalies owing to the possibility of dezincification. Most of the low-zinc alloys are resistant to hot dilute alkalies.

Lead and Alloys Pure lead has low creep and fatigue resistance, but its physical properties can be improved by the addition of small amounts of silver, copper, antimony, or tellurium. Lead-clad equipment is in common use in many chemical plants. The excellent corrosion resistance properties of lead are caused by the formation of protective surface coatings. If the coating is one of the highly insoluble lead salts, such as sulfate, carbonate, or phosphate, good corrosion resistance is obtained. Little protection is offered, however, if the coating is a soluble salt, such as nitrate, acetate, or chloride. As a result, lead shows good resistance to sulfuric acid and phosphoric acid, but is susceptible to attack by either acetic or nitric acid.

Tantalum The physical properties of tantalum are similar to those of mild steel, with the exception that its melting point (2996°C) is much higher. It is ordinarily used in the pure form and readily fabricated into many different shapes. The corrosion resistance properties of tantalum resemble those of glass. The metal is attacked by hydrofluoric acid, by hot concentrated alkalies, and by materials containing free sulfur trioxide. It is resistant to all other acids and is often used for equipment involving contact with hydrochloric acid.

Titanium Because of its strength and medium weight, titanium has become increasingly important as a construction material. Corrosion resistance is superior in oxidizing and mild reducing media. Its general resistance to seawater is excellent. Titanium is resistant to nitric acid at all concentrations except with fuming nitric. The metal also resists ferric chloride, cupric chloride, and other hot chloride solutions. Disadvantages of forming and welding are limits for its general use.

Zirconium Commercial-grade zirconium resembles titanium from a fabrication standpoint. It also has excellent resistance to reducing environments. Oxidizing agents frequently cause accelerated attack. Zirconium resists all chlorides except ferric and cupric. Zirconium alloys are severely damaged when exposed to sulfuric acid at concentrations above about 70 percent.

Inorganic Nonmetals

Glass, stoneware, brick, and cement materials are common examples of inorganic nonmetals used as materials of construction. Many of the nonmetals have low structural strength. Consequently, they are often used in the form of linings or coatings bonded to metal supports. For example, glass-lined equipment has many applications in the chemical industries.

Glass and Glassed Steel Glass has excellent resistance and is subject to attack only by hydrofluoric acid and hot alkaline solutions. It is particularly suitable for processes which have critical contamination levels. A disadvantage is its brittleness and damage by thermal shock. On the other hand, glassed steel combines the corrosion resistance of glass with the working strength of steel. Nucerite is a ceramic-metal composite made in a manner similar to glassed steel and resists corrosive hydrogen chloride gas, chlorine, or sulfur dioxide at 650°C. Its impact strength is 18 times that of safety glass, and the abrasion resistance is superior to that of porcelain enamel.

Stoneware and Porcelain Materials of stoneware and porcelain are about as resistant to acids and chemicals as glass, but with the advantage of greater strength. This is offset somewhat by poor thermal conductivity and susceptibility to damage by thermal shock. Porcelain enamels are used to coat steel, but the enamel has slightly poorer chemical resistance, because of the presence of surface imperfections.

Brick and Cement Materials Brick-lined construction can be used for many severely corrosive conditions, where high alloys would fail. Acid-proof refractories can be used up to 900°C.

A number of cement materials are used with brick. Standard are phenolic and furane resins, polyesters, sulfur, silicate, and epoxy-based materials. Carbon-filled polyesters and furans are good against nonoxidizing acids, salts, and solvents. Silica-filled resins should not be used against hydrofluoric or fluorosilicic acids. Sulfur-based cements are limited to 95°C, while resins can be used to about 175°C. The sodium silicate-based cements provide good protection against acids to 400°C.

Organic Nonmetals

In comparison with metallic materials, the use of organic nonmetallics is limited to relatively moderate temperatures and pressures. Plastics, for example, are less resistant to mechanical abuse and have high expansion rates, low strengths (thermoplastics), and only fair resistance to solvents. However, they are lightweight, are good thermal and electrical insulators, are easy to fabricate and install, and have low friction factors.

Plastics Generally, plastics have excellent resistance to weak mineral acids and are unaffected by inorganic salt solutions—areas where metals are not entirely suitable. Since plastics do not corrode in the electrochemical sense, they offer another advantage over metals: Most metals are affected by slight changes in pH, minor impurities, or oxygen content, while plastics will remain resistant to these same changes.

One of the most chemical-resistant plastic commercially available today is tetrafluoroethylene, or TFE (Teflon). This thermoplastic is practically unaffected by all alkalies and acids except fluorine and chlorine gas at elevated temperatures and molten metals. It retains its properties up to 260°C. Chlorotrifluoroethylene or CTFE (Kel-F) also possesses excellent corrosion resistance to almost all acids and alkalies up to 175°C. FEP, a copolymer of tetrafluorethylene and hexafluoropropylene, has similar properties to TFE except that it is not recommended for continuous exposures at temperatures above 200°C. Also, FEP can be extruded on conventional extrusion equipment, while TFE parts must be fabricated by complicated *powdered-metallurgy* techniques. Polyvinylidene fluoride, or PVF_2 (Kynar), has excellent resistance to alkalies and acids to 150°C. Perfluoroalkoxy, or PFA, on the other hand, can tolerate temperatures up to 300°C while exhibiting the general properties and chemical resistance of FEP.

Polyethylene is the lowest-cost plastic commercially available. Mechanical properties are generally poor, particularly above 50°C, and pipe must be fully supported. Carbon-filled grades are resistant to sunlight and weathering.

Unplasticized polyvinyl chlorides (type I) have excellent resistance to oxidizing acids except when concentrated, and to most nonoxidizing acids. Resistance is also good when exposed to weak and strong alkaline solutions. Resistance to chlorinated hydrocarbons is not good, but can be greatly improved with the substitution of a polyvinylidene known as Saran.

Acrylonitrile butadiene styrene (ABS) polymers have good resistance to nonoxidizing and weak acids but are not satisfactory with oxidizing acids. Upper temperature limit is about 65°C. Resistance to weak alkaline solutions is excellent. They are not satisfactory with aromatic or chlorinated hydrocarbons but have good resistance to aliphatic hydrocarbons.

Chlorinated polyether can be used continuously up to 125°C and intermittently up to 150°C. Chemical resistance is between that of polyvinyl chloride and that of the fluorocarbons. Dilute acids, alkalies, and salts have no effect on the chemical resistance. Hydrochloric, hydrofluoric, and phosphoric acids can be handled at all concentrations up to 105°C. Sulfuric acid over 60 percent and nitric over 25 percent cause degradation, as do aromatics and ketones.

Acetals have excellent resistance to most organic solvents but are not satisfactory for use with strong acids and alkalies. Cellulose acetate butyrate is not affected by dilute acids and alkalies or gasoline, but chlorinated solvents cause some swelling. Nylons resist many organic solvents but are attacked by phenols, strong oxidizing agents, and mineral acids.

The chemical resistance of polypropylene is about the same as that of polyethylene, but it can be used at 120°C. Polycarbonate is a relatively high-temperature plastic. It can be used up to 150°C. Resistance to mineral acids is good. Strong alkalies slowly decompose it, but mild alkalies do not. It is partially soluble in aromatic solvents and soluble in chlorinated hydrocarbons.

Polyphenylene sulfide is resistant to aqueous inorganic salts and bases. In fact, it has no known solvents below 190 to 205°C, and its mechanical properties are unaffected by exposure to air at 235°C. Polysulfone is highly resistant to mineral acid, alkali and salt solutions, detergents, oils, and alcohols up to a temperature of 175°C. Damage is sustained with exposure to such solvents as ketones, chlorinated hydrocarbons, and aromatic hydrocarbons. Polyphenylene oxide displays good resistance to aliphatic solvents, acids, and bases, but poor resistance to esters, ketones, and aromatic or chlorinated solvents.

Polyamide or polyimide polymers are resistant to aliphatic, aromatic, and chlorinated or fluorinated hydrocarbons as well as to many acidic and basic systems, but are degraded by high-temperature caustic exposures.

Among the thermosetting materials are phenolic plastics filled with carbon, graphite, and silica. Relatively low cost, good mechanical properties, and chemical resistance (except against strong alkalies) make phenolics popular for chemical equipment. Furan plastics, filled with asbestos, have much better alkali resistance than phenolic asbestos. They are more expensive than the phenolics but also offer somewhat higher strengths.

General-purpose polyester resins, reinforced with fiberglass, have good strength and good chemical resistance, except to alkalies. Some special materials in this class, based on bisphenol, are more alkali-resistant. Temperature limit for polyesters is 95°C. The general area of fiberglass-reinforced plastic (FRP) represents a rapidly expanding application of plastics for processing equipment, but fabrication standards still need to be formalized.

Epoxies reinforced with fiberglass have very high strengths and resistance to heat. Chemical resistance of the epoxy resin is excellent in nonoxidizing and weak acids but poor with strong acids. Alkaline resistance is excellent in weak solutions. Chemical resistance of epoxy-glass laminates may be affected by any exposed glass in the laminate.

Phenolics, general-purpose polyester glass, Saran, and CAB (cellulose acetate butyrate) are adversely affected by alkalies. Thermoplastics generally show poor resistance to organics.

Rubber and Elastomers† Natural and synthetic rubbers are used as linings or as structural components for equipment in the chemical industries. By adding the proper ingredients, natural rubbers with varying degrees of hardness and chemical resistance can be produced. Hard rubbers are chemically saturated with sulfur. The vulcanized products are rigid and exhibit excellent resistance to chemical attack by dilute sulfuric acid and dilute hydrochloric acid.

Natural rubber is resistant to dilute mineral acids, alkalies, and salts; but oxidizing media, oils, benzene, and ketones will attack it. Chloroprene or neoprene rubber is resistant to attack by ozone, sunlight, oils, gasoline, and aromatic or halogenated solvents. Styrene rubber has chemical resistance similar to that of natural rubber. Nitrile rubber is known for resistance to oils and solvents. Butyl rubber's resistance to dilute mineral acids and alkalies is exceptional; resistance to concentrated acids, except nitric and sulfuric, is good. Silicone rubbers, also known as polysiloxanes, have outstanding resistance to high and low temperatures as well as against aliphatic solvents, oils, and greases. Chlorosulfonated polyethylene, known as Hypalon, has outstanding resistance to ozone and oxidizing agents except fuming nitric and sulfuric acids. Oil resistance is good. Fluoroelastomers (Viton A, Kel-F) combine excellent chemical and high-temperature resistance. The polyvinyl chloride elastomer (Koroseal) was developed to overcome some of the limitations of natural and synthetic rubbers. This elastomer exhibits excellent resistance to mineral acids and petroleum oils.

Carbon and Graphite‡ Generally, impervious graphite is completely inert to all but the most severe oxidizing conditions. This property, combined with excellent heat transfer, has made impervious carbon and graphite very popular in heat exchangers, as brick lining, and in pipe and pump systems. One limitation of these materials is low

†Further information on chemical resistance of rubbers and elastomers is presented in *Perry's Chemical Engineers' Handbook,* 7th ed., R. H. Perry and D. W. Green, eds., McGraw-Hill, New York, 1997, Tables 28-26 and 28-27, respectively.

‡Properties available in Table 28-29, loc. cit.

tensile strength. Threshold oxidation temperatures are 350°C for carbon and 400°C for graphite.

Wood[†] This material of construction, while fairly inert chemically, is readily dehydrated by concentrated solutions and consequently shrinks badly when subjected to the action of such solutions. It also has a tendency to slowly hydrolyze when in contact with hot acids and alkalies.

Low- and High-Temperature Materials

The extremes of low and high temperatures used in many recent chemical processes have created some unusual problems in fabrication of equipment.[‡] For example, some metals lose their ductility and impact strength at low temperatures, although in many cases, yield and tensile strengths increase as the temperature is decreased. It is important in low-temperature applications to select materials resistant to shock. Usually a minimum Charpy value is specified at the operating temperature. Ductility tests are performed on notched specimens since smooth specimens usually show amazing ductility. Table 10-6 provides a brief summary of metals and alloys recommended for low-temperature use.

Among the most important properties of materials at the other end of the temperature spectrum are creep, rupture, and short-time strengths. Stress rupture is another important consideration at high temperatures since it relates stress and time to produce rupture. Ferritic alloys are weaker than austenitic compositions, and in both groups molybdenum increases strength. Higher strengths are available in Inconel, cobalt-base Stellite 25, and iron-base A286. Other properties that become important at high temperatures include thermal conductivity, thermal expansion, ductility, alloy composition, and stability.

Table 10-6 Metals and alloys for low-temperature process use

ASTM specification and grade	Recommended minimum service temperature, °C
Carbon and alloy steels:	
T-1	−45
A 201, A 212, flange or firebox quality	−45
A 203, grades A and B (2.25% Ni)	−60
A 203, grades D and E (3.50% Ni)	−100
A 353 (9% Ni)	−195
Copper alloys, silicon bronze, 70-30 brass, copper	−195
Stainless steel types 302, 304L, 304, 310, 347	−255
Aluminum alloys 5052, 5083, 5086, 5154, 5356, 5454, 5456	−255

[†]Chemical resistance given in Table 28-31, loc. cit.

[‡]See R. M. McClintock and H. P. Gibbons, *Mechanical Properties of Structural Materials at Low Temperatures*, National Bureau of Standards, June 1960, K. D. Timmerhaus and T. M. Flynn, *Cryogenic Process Engineering*, Plenum Press, New York, 1989.

Table 10-7 Alloys for high-temperature process use†

Alloys	Nominal composition, %				Max. temp., °C
	Cr	Ni	Fe	Other	
Ferritic steels:					
Carbon steel			bal.		480
$2\frac{1}{4}$ chrome	$2\frac{1}{4}$		bal.	Mo	
Type 502	5		bal.	Mo	620
Type 410	12		bal.		700
Type 430	16		bal.		850
Type 446	27		bal.		1100
Austenitic steels:					
Type 304	18	8	bal.		900
Type 321	18	10	bal.	Ti	
Type 347	18	11	bal.	Cb	
Type 316	18	12	bal.	Mo	
Type 309	24	12	bal.		1100
Type 310	25	20	bal.		1100
Type 330	15	35	bal.		
Nickel-base alloys:					
Nickel		bal.			
Incoloy	21	32	bal.		1100
Hastelloy B		bal.	6	Mo	
Hastelloy C	16	bal.	6	W, Mo	
60/15	15	bal.	25		
Inconel	15	bal.	7		1100
80/20	20	bal.			
Hastelloy X	22	bal.	19	Co, Mo	
Multimet	21	20	bal.	Co	
Rene 41	19	bal.	5	Co, Mo, Ti	
Cast irons:					
Ductile iron			bal.	C, Si, Mg	
Ni-resist, D-2	2	20	bal.	Si, C	
Ni-resist, D-4	5	30	bal.	Si, C	
Cast stainless (ACI types):					
HC	28	4	bal.		
HF	21	11	bal.		
HH	26	12	bal.		
HK	26	20	bal.		
HT	15	35	bal.		
HW	12	bal.	28		
Superalloys:					
Inconel X	15	bal.	7	Ti, Al, Cb	
A 286	15	25	bal.	Mo, Ti	
Stellite 25	20	10	Co-base	W	1100
Stellite 21 (cast)	27.3	2.8	Co-base	Mo	
Stellite 31 (cast)	25.2	10.5	Co-base	W	

†*Perry's Chemical Engineers' Handbook,* 7th ed., R. H. Perry and D. W. Green, eds., McGraw-Hill, New York, 1997.

Actually, in many cases strength and mechanical properties become of secondary importance in process applications, compared with resistance to corrosive surroundings. All common heat-resistant alloys form oxides when exposed to hot oxidizing environments. Whether the alloy is resistant depends upon whether the oxide is stable and forms a protective film. Thus, mild steel is seldom used above 500°C because of excessive scaling rates. Higher temperatures require chromium. This is evident not only from Table 10-5, but also from Table 10-7, which lists the important commercial alloys for high-temperature use.

Gasket Materials

Metallic and nonmetallic gaskets of many different forms and compositions are used in industrial equipment. The choice, as shown in Table 10-8, of a gasket material depends on the corrosive action of the chemicals that may contact the gasket, the location of the gasket, and the type of gasket construction.[†] Other factors of importance are the cost of the materials, pressure and temperature involved, and frequency of opening the joint.

TABULATED DATA FOR SELECTING MATERIALS OF CONSTRUCTION

Table 10-8 presents information on the corrosion resistance of some common metals, nonmetals, and gasket materials. Table 10-9 presents similar information for various types of plastics. These tables can be used as an aid in selecting materials to be used in the various components of the design. However, no single table can take into account all the factors that can affect corrosion. Temperature level, concentration of the corrosive agent, presence of impurities, physical methods of operation, and slight alterations in the composition of the construction material can affect the degree of corrosion resistance. The final selection of a material of construction, therefore, may require reference to manufacturers' bulletins and consultation with individuals who are experts in the particular field of application.[‡]

SELECTION OF MATERIALS

The chemical engineer responsible for the selection of materials of construction must have a thorough understanding of all the basic process information available. This knowledge of the process can then be used to select materials of construction in a logical

[†]Further information on chemical resistance of gaskets is given in *Perry's Chemical Engineers' Handbook,* 7th ed., R. H. Perry and D. W. Green, eds., McGraw-Hill, New York, 1997, Table 28-28.

[‡]Up-to-date information on various aspects of materials of construction is presented in the Biennial Materials of Construction Reports published by *Chemical Engineering.* See also "Current Literature on Materials of Construction," *Chem. Eng.,* **95**(15): 69 (1988).

Table 10-8 Corrosion resistance of construction materials†

Code designation for corrosion resistance

A = Acceptable, can be used successfully
C = Caution, resistance varies widely depending on conditions; used when some corrosion is permissible
X = Unsuitable
Blank = Information lacking

Code designation for gasket materials‡

a = Asbestos, white (compressed or woven)
b = Asbestos, blue (compressed or woven)
c = Asbestos (compressed and rubber-bonded)
d = Asbestos (woven and rubber-frictioned)
e = GR-S or natural rubber
f = Teflon

	Metals								Nonmetals					
			Stainless steel											
Chemical	Iron and steel	Cast iron (Ni-resist)	18-8	18-8 Mo	Nickel	Monel	Red brass	Aluminum	Industrial glass	Carbon (Karbate)	Phenolic resins (Haveg)	Acrylic resins (Lucite)	Vinylidene chloride (Saran)	Acceptable nonmetallic gasket materials
Acetic acid, crude	C	C	C	C	C	C	C	A	A	A	A	A	C	b, c, d, f
Acetic acid, pure	X	X	C	A	C	A	X	A	A	A	A	X	X	b, c, d, f
Acetic anhydride	C	C	A	A	A	A	X	A	A	A	A	X	C	b, c, d, f
Acetone	A	A	A	A	A	A	A	A	A	A	C	X	C	a, e, f
Aluminum chloride	X	C	X	X	C	C	A	A	A	A	A	...	A	a, c, e, f
Aluminum sulfate	X	C	C	A	C	C	X	A	A	A	A	A	A	a, c, d, e, f
Alums	X	C	C	A	C	A	X	A	A	A	A	A	A	a, c, d, e, f
Ammonia (gas)	A	A	C	A	A	A	X	C	A	...	A	...	C	a, f
Ammonium chloride	C	A	C	C	A	A	C	C	A	A	A	A	A	b, c, d, e, f
Ammonium hydroxide	A	A	A	A	C	C	X	C	A	...	A	A	C	a, c, d, f
Ammonium phosphate, monobasic	X	C	A	A	...	C	X	X	A	A	A	...	...	b, c, d, e, f
Ammonium phosphate, dibasic	C	A	A	A	...	A	C	C	A	A	A	...	...	a, c, d, e, f
Ammonium phosphate, tribasic	A	A	A	A	A	A	X	C	A	A	A	...	...	a, c, d, e, f
Ammonium sulfate	C	A	C	C	A	A	C	A	A	A	A	A	A	b, c, d, e, f
Aniline	A	A	A	A	...	A	X	...	A	A	C	...	C	a, f
Benzene, benzol	A	A	A	A	A	A	A	A	A	A	A	...	C	a, f
Boric acid	X	C	A	A	A	A	C	A	A	A	A	...	A	a, c, d, e, f
Bromine	X	C	C	C	C	C	C	...	A	C	X	...	X	b, f

Calcium chloride	C	A	C	C	A	A	C	C	A	A	A	A	A	b, c, d, e, f
Calcium hydroxide	A	A	A	A	...	A	C	...	...	...	...	...	...	a, c, d, e, f
Calcium hypochlorite	X	C	C	A	C	C	C	C	A	A	C	...	C	b, c, d, f
Carbon tetrachloride	C	C	C	A	A	A	C	C	A	A	A	A	C	a, f
Carbonic acid	C	C	A	A	A	A	C	A	A	A	A	A	A	a, e, f
Chloracetic acid	X	...	X	X	C	C	X	C	A	...	A	...	...	b, f
Chlorine, dry	A	A	C	A	A	A	A	A	A	A	A	...	X	b, e, f
Chlorine, wet	X	X	X	X	X	X	X	X	A	C	A	...	X	b, e, f
Chromic acid	C	C	C	C	C	C	X	C	A	X	X	X	A	b, f
Citric acid	X	C	C	A	C	A	C	A	A	A	A	A	A	b, c, d, e, f
Copper sulfate	X	C	A	A	C	C	X	X	A	A	A	X	...	b, c, d, e, f
Ethanol	A	A	A	A	A	A	A	A	A	A	A	...	A	a, c, e, f
Ethylene glycol	A	A	A	A	A	A	A	A	A	A	A	A	C	a, c, e, f,
Fatty acids	C	C	A	A	A	A	C	A	A	A	A	...	A	a, e, f
Ferric chloride	X	X	X	C	X	X	X	X	A	C	A	...	A	b, e, f
Ferric sulfate	X	X	C	A	C	C	X	C	A	C	A	...	A	b, c, e, f
Ferrous sulfate	C	A	A	A	A	A	C	C	A	A	A	...	C	
Formaldehyde	C	C	A	A	A	A	C	A	A	A	A	...	A	a, c, e, f
Formic acid	X	...	C	C	C	C	X	X	A	A	A	...	A	b, c, e, f
Glycerol	A	A	...	A	A	A	A	A	A	A	C	A	C	a, c, e, f
Hydrocarbons (aliphatic)	A	A	A	A	A	A	A	A	A	A	A	A	C	a, c, d, f
Hydrochloric acid	X	X	X	X	C	C	X	X	A	A	A	A	C	b, c, d, f
Hydrofluoric acid	C	X	X	X	C	C	X	X	X	A	C	...	C	b, f
Hydrogen peroxide	C	...	C	C	C	C	C	A	A	A	A	A	C	a, e, f
Lactic acid	X	C	C	A	C	C	A	C	A	A	A	...	...	a, b, c, d, e, f
Magnesium chloride	C	C	C	C	A	A	C	C	A	A	A	...	A	b, c, e, f
Magnesium sulfate	A	A	A	A	A	A	A	A	A	...	A	...	A	b, c, e, f
Methanol	A	A	A	A	A	A	A	A	A	A	A	...	A	a, c, e, f
Nitric acid	X	C	C	C	X	X	X	C	A	C	C	...	C	b, f
Oleic acid	C	C	A	A	A	A	C	A	A	A	A	...	A	a, e, f
Oxalic acid	C	C	C	C	C	A	C	C	A	A	A	...	...	b, c, d, e, f

(Continued)

Table 10-8 *Continued*

Chemical	Metals								Nonmetals					Acceptable nonmetallic gasket materials
	Iron and steel	Cast iron (Ni-resist)	Stainless steel 18-8	Stainless steel 18-8 Mo	Nickel	Monel	Red brass	Aluminum	Industrial glass	Carbon (Karbate)	Phenolic resins (Haveg)	Acrylic resins (Lucite)	Vinylidene chloride (Saran)	
Phenol (carbolic acid)	C	A	C	A	A	A	C	A	A	A	C	A	C	a, f
Phosphoric acid	C	C	C	A	C	C	X	X	C	A	A	...	A	b, c, f
Potassium hydroxide	C	C	A	A	A	A	X	X	...	...	...	...	C	a, e, f
Sodium bisulfate	X	C	A	A	A	A	C	C	A	A	A	...	A	b, c, d, e, f
Sodium carbonate	A	A	A	A	A	A	C	C	C	A	A	X	...	a, c, d, e, f
Sodium chloride	A	A	C	C	A	A	C	C	A	A	A	...	...	a, c, d, e, f
Sodium hydroxide	A	A	A	A	A	A	C	X	C	A	A	A	C	a, c, d, f
Sodium hypochlorite	X	C	C	A	C	C	C	X	A	C	X	...	A	b, c, d, f
Sodium nitrate	A	A	A	A	A	A	C	A	A	A	A	...	...	b, c, d, e, f
Sodium sulfate	A	A	A	C	A	A	A	A	A	A	A	...	A	a, c, d, e, f
Sodium sulfide	A	A	C	A	A	A	X	X	C	A	A	...	...	a, e, f
Sodium sulfite	A	A	A	A	A	A	C	C	A	A	A			
Sodium thiosulfate	C	...	A	A	A	A	C	C	A	A	A	...	A	a, c, d, e, f
Stearic acid	C	A	A	A	A	A	C	A	A	A	A	...	...	a, e, f
Sulfur	A	C	C	C	C	C	C	C	A	A	A	...	...	a, e, f
Sulfur dioxide	C	C	C	C	C	C	C	A	A	A	A	...	A	a, f
Sulfuric acid, 98% to fuming	A	C	X	C	X	X	X	C	A	X	X	X	C	b, f
Sulfuric acid, 75–95%	A	C	X	X	X	C	X	X	A	C	X	X	C	b, f
Sulfuric acid, 10–75%	X	C	X	X	C	C	X	X	A	A	C	C	A	b, f
Sulfuric acid, <10%	X	C	X	C	C	C	C	C	A	A	C	A	A	a, b, c, e, f
Sulfurous acid	X	...	C	A	X	X	C	C	A	A	A	...	C	b, c, d, e, f
Trichloroethylene	C	A	C	A	A	A	C	C	A	A	A	...	C	a, f
Zinc chloride	C	C	C	X	A	A	X	C	...	...	A	A	...	b, c, d, e, f
Zinc sulfate	C	A	A	A	A	A	C	C	...	...	...	...	...	b, c, d, e, f

[†]From miscellaneous sources. For additional details, see *Perry's Chemical Engineers' Handbook,* 7th ed., R. H. Perry and D. W. Green, eds., McGraw-Hill, New York, 1997, D. A. Hansen and R. B. Puyear, *Materials Selection for Hydrocarbon and Chemical Plants,* Marcel Dekker, New York, 2000, and P. A. Schweitzer, *Corrosion Resistance Tables,* Marcel Dekker, New York, 1990.

[‡]Asbestos is listed because many seals still in operation contain this unacceptable material.

Table 10-9 Typical property and chemical resistance data for selected plastics†

Plastics[a]	Specific gravity	Tensile strength MPa	Modulus of elasticity, tension MPa × 10^2	Impact strength, Izod[b] J	Maximum temp. (no load) °C	HDT[c] at 1.75 MPa °C	Chemical resistance[d]: Weather resistance	Weak acid	Strong acid	Weak alkali	Strong alkali	Solvents
Alkyds												
Glass-filled	2.12–2.15	28–66	138–193	0.8–14	230	200–260	R	A	A	A	A	A
Mineral-filled	1.60–2.30	21–62	34–207	0.4–0.7	150–230	180–260	R	R	A	A	D	A
Asbestos-filled	1.65	31–48		0.6–0.7	230	160	R	R	S	R	S	R
Synthetic fiber-filled	1.24–2.10	31–48	138	0.7–6.1	150–220	120–220	R	R	S	R	S	A
Alkyl diglycol carbonate	1.30–1.40	34–41	21	0.3–0.5	100	60–90	R	R	A[e]	R	R–S	R
Diallyl phthalates												
Glass-filled	1.61–1.78	41–76	97–152	0.5–20	150–200	165–280	R	R	S	R–S	S	R
Mineral-filled	1.65–1.68	34–62	83–152	0.4–1	150–200	160–280	R	R	S	R–S	S	R
Asbestos-filled	1.55–1.65	48–55	83–152	0.5–0.7	150–200	160–280	R	R	S	R–S	S	R
Epoxies (*bis*-A)												
No filler	1.06–1.40	28–90	15–36	0.3–1.4	120–260	45–260	R	R	A	R	S	R–S
Graphite-fiber	1.37–1.38	1280–1380	814–827				S	R	R	R	R	R–S
Mineral-filled	1.6–2.0	34–103		0.4–0.5	150–260	120–260	S	R	R	R	R	R–S
Glass-filled	1.7–2.0	69–207	207	14–41	150–260	120–260	S	R	R–S	R	R	R–S
Epoxies (novolac): no filler	1.12–1.24	34–76	15–36	0.4–0.9	200–260	230–260	R	R	R	R	R	R
Epoxies (cycloaliphatic), no filler	1.12–1.18	69–121	34–48		250–290	260–290	R	R	R–A	R	R–A	R
Melamines												
Cellulose-filled	1.45–1.52	34–62	76	0.3–0.5	120	130	S	R–S	D	R	D	R
Flock-filled	1.50–1.55	48–62		0.5–0.7	120	130	S	R–S	D	R	D	R–S
Asbestos-filled	1.70–2.0	34–48	138	0.4–0.5	120–200	130	S	R–S	D	S	S	R
Fabric-filled	1.5	55–76	97–110	0.8–1.4	120	150	S	R	D	R	A	R–S
Glass-filled	1.8–2.0	34–69	165	0.8–24	150–200	200	S	R	D	R	R–S	R
Phenolics												
Wood-flour-filled	1.34–1.45	34–62	55–117	0.3–0.8	150–180	150–190	S	R–S	S–D	S–D	A	R–S
Asbestos-filled	1.45–2.00	31–52	69–207	0.3–0.5	180–260	150–260	S	R–S	S–D	S–D	A	R–S
Mica-filled	1.65–1.92	38–48	172–345	0.4–0.5	120–150	150–180	S	R–S	S–D	S–D	A	R–S
Glass-filled	1.69–1.95	34–124	131–228	0.4–24	180–290	150–320	S	R–S	S–D	S–D	A	R–S
Fabric-filled	1.36–1.43	21–62	62–97	1.1–11	100–120	120–170	S	R–S	S–D	S–D	A	R–S
Polybutadienes, very high vinyl (no filler)	1.00	55	14	1.5	260	—	S	R	R	R	R	R
Polyesters												
Glass-filled BMC	1.7–2.3	28–69	110–172	2.0–22	150–180	200–230	R–E	R–A	S–A	S–A	S–D	A–D
Glass-filled SMC	1.7–2.1	55–138	110–172	11–30	150–180	200–230	R–E	R–A	S–A	S–A	S–D	A–D
Glass-cloth reinforced	1.3–2.1	172–345	131–310	7–41	150–180	200–230	R–E	R–A	S–A	S–A	S–D	A–D
Silicones												
Glass-filled	1.7–2.0	28–45	69–103	4–20	320	320	R–S	R–S	R–S	S	S–A	R–A
Mineral-filled	1.8–2.8	28–41	90–124	0.4–0.5	320	320	R–S	R–S	R–S	S	S–A	R–A
Ureas												
Cellulose-filled	1.47–1.52	38–90	69–103	0.3–0.5	80	130–140	S	R–S	A–D	S–A	D	R–S
Urethanes, no filler	1.1–1.5	1–69	7–69	7	90–120	—	R–S	S	A	S	S–A	R–S

(Continued)

Table 10-9 *Continued*

Thermosets[a]	Specific gravity	Tensile strength MPa	Modulus of elasticity, tension MPa × 10^2	Impact strength, Izod[b] J	Maximum temp. (no load) °C	HDT[c] at 1.75 MPa °C	Weather resistance	Chemical resistance[d] Weak acid	Strong acid	Weak alkali	Strong alkali	Solvents
ABS												
GP	1.05–1.07	41	21	8	70–90	90–95	R–E	R	A[e]	R	R	A/R
High-impact	1.01–1.06	33	17	10	60–100	85–100	R–E	R	A[e]	R	R	A/R
Heat-resistant	1.06–1.08	51	27	3.0	90–110	110–115	R–E	R	A[e]	R	R	A/B
Trans.	1.07	39	20	7.1	55	75	R–E	R	A[e]	R	R	A/R
	1.20	41	22	3.4	55–80	90	R–E	R	A[e]	R	R	A/R
Acetals												
Homopolymers	1.42	69	36	1.9	90	125	R	R	A	R	A–D	R
Copolymers	1.41	61	28	1.6–2.2	100	110	R	R	A	R	R	R
Acrylics												
GP	1.11–1.19	39–76	16–32	0.4–3.1	55–110	75–100	R	R	A[e]	R	A	A/R
High-impact	1.12–1.16	40–55	16–23	1.1–3.1	60–90	75–90	R	R	A[e]	R	R	A/R
	1.21–1.28	55–86	24–33	0.4–0.5	50–90	70–95	R	R	A[e]	R	A	A/R
Cast	1.18–1.28	62–86	26–34	0.5–2.0	60–90	70–100	R	R	A[e]	R	A	A/R
Multipolymer	1.09–1.14	41–55	21–30	1–4	75–80	85–90	E	R	A[e]	R	S	A[f]
Cellulosics												
Acetate	1.23–1.34	21–55	7–18	1.5–9	60–105	45–90	S	S	D	S	D	D–S
Butyrate	1.15–1.22	21–48	5–12	4–14	60–105	45–95	S	S	D	S	D	D–S
E cellulose	1.10–1.17	21–55	3–24	2.3–9.5	45–85	45–90	S	S	D	R	S	D
Nitrate	1.35–1.40	48–55	13–15	7–9	60	60–70	E	S	D	S	D	D
Propionate	1.19–1.22	28–45	8–12	2.3–13	70–105	45–110	S	S	D	S	D	D–S
Chloropolyether	1.4	37	10	0.5	140		R–S	R	A[e]	R	R	R
Fluoropolymers												
FEP	2.14–2.17	17–27	3–5		208		R	R	R	R	R	R
PTFE	2.1–2.3	7–28	2.6–4.5	3.4–5.4	290		R	R	R	R	R	R
CTFE	2.10–2.15	32–39	12–14	4.7–4.9	180–200		R	R	R	R	R	S[g]
PVF_2	1.77	50	12	5.2	150	90	S	R	A[h]	R	R	R
ETFE	1.68	45–48	14–17		150	70	R	R	R	R	R	
Methylpentene	0.83	23–25	10–13	1.3–5.2	135		E	R	A[h]	R	R	A
Nylons												
6/6	1.13–1.15	62–83	27	27	80–150	65–105	R	R	A	R	R	R–D
6	1.14	86		1.6	80–170	60–70	R	R	A	R	R	R–A[i]
6/10	1.07	49	19	2.2	80		R	R	A	R	R	R–A[i]
8	1.09	27		>22			R	R	A	R	R	R–A[i]
12	1.01	45–59	12–14	1.6–5.7	80–125	50–55	R	R	A	R	R	R–A[i]
Copolymers	1.08–1.14	52–76		2–26	80–120	55–180	R	R	A	R	R	R–A[i]

Polyesters												
PET	1.37	72		1.1	80	85	R	R	A[e]	R	A	R–A[i]
PBT	1.31	55–57	25	1.6–1.8	140	55	R	R	R	R	A	R
PTMT	1.31	57		1.4	130	50	R	R	R	R	A	R
Copolymers	1.2	50		1.4		70						
Polyaryl ether	1.14	52	22	14	120	150	E	R	R	R	R	A
Polyaryl sulfone	1.36	90	26	2.7	260	275	Darkens	R	R	R	R	R
Polybutylene	0.910	26	1.8		105	55	E	R	A[e]	R	R	
Polycarbonate	1.2	62	24	16–22	120	130–140	R	R	A[e]	A	A	A
PC-ABS	1.14	57	26	14	105	105	R–E	R	A[e]	R	S	A
Polyethylenes												
LD	0.91–0.93	6–17	1.4–1.9		80–100	30–40	E	R	A[e]	R	R	R
HD	0.95–0.96	20–37		0.5–19	80–120	45–55	E	R	R–A[e]	R	R	R
HMW	0.95	17	7			40–80	E	R	A[e]	R	R	R
Polypropylenes												
GP	0.90–0.91	33–38	11–15	0.5–3.0	105–150	50–60	E	R	A[e]	R	R	R
High-impact	0.90–0.91	21–34	9	2–16	95–120	50–60	E	R	A[e]	R	R	A
Polystyrenes												
GP	1.04–1.07	41–50	31	0.4	65–80	80–105	S	R	A[e]	R	R	D
High-impact	1.04–1.07	20–32	20–28	0.9–1.4	60–80	80–100	S	R	A[e]	R	R	D
Polysulfone	1.24	70	25	1.6	150	175	S	R	R	R	R	R–A
Polyurethanes	1.11–1.25	31–58	0.7–24		90		R–S	S–D	S–D	S–D	S–D	R
Vinyl, rigid	1.3–1.5	34–55	21–34	0.7–27	65–80	55–80	R	R	R–S	R	R	R–A
Vinyl, flexible	1.2–1.7	7–28		0.7–27	60–80		S	R	R–S	R	R	R–A
Rigid CPVC	1.49–1.58	52–62	25–32	1.4–7.6	110	95–115	R	R	R	R	R	R
PVC-acrylic	1.30–1.35	38–45	19–23	20		80	R	R	S	R	R	A
PVC-ABS	1.10–1.21	18–41	6–23	14–20			S	R	R–S	R	R	R–D
SAN	1.08	69–83	34–39	0.5–0.7	60–95	90–105	S–E	R	A	R	R	A

[a]All values at room temperature unless otherwise listed
[b]Notched samples
[c]Heat deflection temperature
[d]R = resistant; A = attacked; S = slight effects; E = embrittles; D = decomposes
[e]By oxidizing acids
[f]By ketones, esters, and chlorinated and aromatic hydrocarbons
[g]Halogenated solvents cause swelling
[h]By fuming sulfuric
[i]Dissolved by phenols and formic acid
†Modified from *Plastics Engineering Handbook of the Society of the Plastics Industry,* 5th ed., Kluwer Academic Publishers, Dordrecht, The Netherlands, 1991, and P. A. Schweitzer, *Mechanical and Corrosion Resistant Properties of Plastics and Elastomers,* Marcel Dekker, New York, 2001.

manner. A brief plan for studying materials of construction is as follows:

1. Preliminary selection—experience, manufacturer's data, special literature, general literature, availability, safety aspects, preliminary laboratory tests
2. Laboratory testing—reevaluation of apparently suitable materials under process conditions
3. Interpretation of laboratory results and other data—effect of possible impurities, excess temperature, excess pressure, agitation, and presence of air in equipment; avoidance of electrolysis; fabrication method
4. Economic comparison of apparently suitable materials—material and maintenance cost, probable life, cost of product degradation, and liability to special hazards
5. Final selection

In making an economic comparison, the engineer is often faced with the question of where to use high-cost claddings or coatings over relatively cheap base materials such as steel or wood. For example, a column requiring an expensive alloy-steel surface in contact with the process fluid may be constructed of the alloy itself or with a cladding of the alloy on the inside of a carbon-steel structural material. Other examples of commercial coatings for chemical process equipment include baked ceramic or glass coatings, flame-sprayed metal, hard rubber, and many organic plastics. The durability of coatings is sometimes questionable, particularly where abrasion and mechanical wear conditions exist. As a general rule, if there is little economic incentive between a coated type versus a completely homogeneous material, a selection should favor the latter material, mainly on the basis of better mechanical stability.

Economics Involved in Selection†

First cost of equipment or material often is not a good economic criterion for comparing alternate materials of construction for chemical process equipment. Any cost estimation should include the following items:

1. Total equipment or materials costs
2. Installation costs
3. Maintenance costs, amount and timing
4. Service life
5. Replacement costs
6. Downtime costs
7. Cost of inhibitors, control facilities required to achieve estimated service life
8. Depreciation and taxes
9. Time value of money
10. Inflation

†For further information, see the material in NACE Publication 3C14, Item No. 24182, "Economics of Corrosion," September, 1994.

When these factors are considered, cost comparisons bear little resemblance to first costs. Table 10-10 presents a typical analysis of comparative costs for alternative materials when based on return on investment. One difficulty with such a comparison is the uncertainty associated with *estimated life.* Well-designed laboratory and plant tests can at least give order-of-magnitude estimates. Another difficulty arises in estimating the annual maintenance cost. This can only be predicted from previous experience with the specific materials.

Table 10-10 Alternative investment comparison

	Material *A*	Material *B*	Material *C*
Purchased cost	$25,000	$30,000	$35,000
Installation cost	15,000	20,000	25,000
Total installed cost	40,000	50,000	60,000
Additional cost over material *A*		10,000	20,000
Estimated life, years	4	8	10
Estimated maintenance cost/year	5,000	4,500	3,000
Annual replacement cost (installed cost/estimated life)	10,000	6,250	6,000
Total annual cost	15,000	10,750	9,000
Annual savings vs. cost for material *A*		4,250	6,000
Tax on savings, 34%		1,445	2,040
Net annual savings		2,805	3,960
Return on investment over material *A*, % (net savings/additional cost over material *A*) 100		28.05	19.8
Return on investment of material *C* versus *B*, %			11.5

The comparison could be extended by the use of compounding interest methods as outlined in Chap. 7 to show the value of money to a company. Without compounding, Table 10-10 indicates that both material *B* and *C* are better than material *A* and material *C* is marginally better than material *B* as determined by the annual return on investment method. However, depending on the time value of money to a company, this may not always be true.

FABRICATION OF EQUIPMENT

Fabrication expenses account for a large fraction of the purchased cost for equipment. A chemical engineer, therefore, should be acquainted with the methods for fabricating equipment, and the problems involved in the fabrication should be considered when equipment specifications are prepared.

Many of the design and fabrication details for equipment are governed by various codes, such as the ASME codes. These codes can be used to indicate definite specifications or tolerance limits without including a large amount of descriptive restrictions. For example, fastening requirements can often be indicated satisfactorily by merely stating that all welding should be in accordance with the ASME code.

Methods of Fabrication

The exact methods used for fabrication depend on the complexity and type of equipment being prepared. In general, however, the following steps are involved in the complete fabrication of major pieces of chemical equipment such as tanks, autoclaves, reactors, towers, and heat exchangers:

1. Layout of materials
2. Cutting to correct dimensions
3. Forming into desired shape
4. Fastening
5. Testing
6. Heat-treating
7. Finishing

Layout The first step in the fabrication is to establish the layout of the various components on the basis of detailed instructions prepared by the fabricator. Flat pieces of the metal or other constructional material involved are marked to indicate where cutting and forming are required. Allowances must be made for losses caused by cutting, shrinkage due to welding, or deformation caused by the various forming operations. Once the equipment takes shape, location of various outlets and attachments can be made.

Cutting Several methods can be used for cutting the laid-out materials to the correct size. *Shearing* is the cheapest method and is satisfactory for relatively thin sheets. The edge resulting from a shearing operation may not be usable for welding, and the sheared edges may require an additional grinding or machining treatment. *Burning* is often used for cutting metals. This method can be employed to cut and, simultaneously, prepare a beveled edge suitable for welding. *Sawing* can be used to cut metals that are in the form of flat sheets. However, sawing is expensive and is used only when the heat effects from burning would be detrimental.

Forming After the construction materials have been cut, the next step is to form them into the desired shape. This can be accomplished by various methods, such as by rolling, bending, pressing, bumping (i.e., pounding), or spinning on a die. In some cases, heating may be necessary to carry out the forming operation. Because of work-hardening of the material, annealing may be required before forming and between stages during the forming.

When the shaping operations are finished, the different parts are assembled and fitted for fastening. When the fitting is complete and all edges are correctly aligned, the main seams can be tack-welded in preparation for the final fastening.

Fastening Riveting can be used for fastening operations, but electric welding is far more common and gives superior results. The quality of a weld is very important, because the ability of equipment to withstand pressure or corrosive conditions is often limited by the conditions along the welds. Although good welds may be stronger than the metal that is fastened together, design engineers usually assume a weld is not perfect and employ weld efficiencies of 80 to 95 percent in the design of pressure vessels.

In some cases, fastening can be accomplished by the use of various solders. Screw threads, packings, gaskets, and other mechanical methods are also used for fastening various parts of equipment.

Testing All welded joints can be tested for concealed imperfections by X-rays, and code specifications usually require X-ray examination of main seams. Hydrostatic tests can be conducted to locate leaks. Sometimes, delicate tests, such as a helium probe test, are used to check for very small leaks.

Heat-Treating After the preliminary testing and necessary repairs are completed, it may be necessary to heat-treat the equipment to remove forming and welding stresses, restore corrosion resistance properties to heat-affected materials, and prevent stress corrosion conditions. A low temperature may be adequate, or the particular conditions may require a full anneal followed by a rapid quench.

Finishing The finishing operation involves preparing the equipment for final shipment. Sandblasting, polishing, and painting may be necessary. Final pressure tests at $1\frac{1}{2}$ to 2 or more times the design pressure are conducted together with other tests as demanded by the specified code or requested by the inspector.

PROBLEMS

10-1 For each of the following materials of construction, prepare an approximate plot of temperature versus concentration in water for sulfuric acid and for nitric acid, showing conditions of generally acceptable corrosion resistance.

a. Stainless steel type 302
b. Stainless steel type 316
c. Karbate
d. Haveg

10-2 A process for sulfonation of phenol requires the use of a 11.4-m^3 storage vessel. It is desired to determine the most suitable material of construction for this vessel. Interest will be charged on the installed cost at a rate of 10 percent/yr.

The life of the storage vessel is calculated by dividing the corrosion allowance of 3.175 mm by the estimated corrosion rate. The equipment is assumed to have a salvage value of 10 percent of its original cost at the end of its useful life.

For the case in question, corrosion data indicate that only a few corrosion resistance alloys will be suitable:

Vessel type	Installed cost	Average corrosion rate, mm/yr
Nickel-clad	$ 88,000	0.5
Monel-clad	105,000	0.25
Hastelloy B	198,000	0.114

Determine which material of construction should be used with appropriate justification for the selection.

10-3 A new plant requires a large rotary vacuum filter for the filtration of zinc sulfite from a slurry containing 1 kg of zinc sulfite solid per 20 kg of liquid. The liquid contains water, sodium sulfite, and sodium bisulfite. The filter must handle 8000 kg of slurry per hour. What additional information is necessary to design the rotary vacuum filter? How much of this information could be obtained from laboratory or pilot-plan tests? Outline the method for converting the test results to the conditions applicable in the final design.

10-4 A manhole plate for a reactor is to be 0.05 m thick and 0.5 m in diameter. It has been proposed that the entire plate be made of stainless steel type 316. The plate will have 18 bolt-holes, and part of the face will need to be machined for close gasket contact. If the base price for stainless steel type 316 in the form of industrial plates is $10.00 per kilogram, estimate the purchased cost for the manhole plate.

10-5 Six tanks of different construction materials and six different materials to be stored in these tanks are listed in the following columns:

Tanks	*Materials*
Brass-lined	20% hydrochloric acid
Carbon steel	10% caustic soda
Concrete	75% phosphoric acid for food products
Nickel-lined	98% sulfuric acid
Stainless steel type 316	Vinegar
Wood	Water

All tanks must be used, and all materials must be stored without using more than one tank for any one material. Indicate the material that should be stored in each tank.

10-6 A cylindrical storage tank is to have an inner diameter of 3.6 m and a length of 11 m. The seams will be welded ($E = 0.90$), and the material of construction will be plain carbon steel (0.15 percent C). The maximum working pressure in the tank will be 790 kPa, and the maximum temperature will be 25°C. On the basis of the equations recommended by the API-ASME Code for Unified Pressure Vessels presented in Table 12-10, estimate the required wall thickness.

10-7 A proposal has been made to use stainless-steel tubing as part of the heat-transfer system in a nuclear reactor. High temperatures and very high rates of heat transfer will be involved. Under these conditions, temperature stresses across the tube walls will be high, and the design engineer must choose a safe wall thickness and tube diameter for the heat-transfer system. List in detail all information and data necessary to determine if a proposed tube diameter and gauge number would be satisfactory.

10-8 What materials of construction should be specified for the thiophane process-described in Prob. 3-16. Note the extremes of temperatures and corrosion which are encountered in this process because of the regeneration step and the presence of H_2S and caustic.

10-9 Liquid chlorine is to be transferred from a chlorine storage container by pressurizing with dry chlorine gas. What materials of construction should be selected for this transfer process? What corrosion effects may be anticipated? How might they be minimized?

10-10 Review the flow diagram for the manufacture of nitric acid by the ammonia oxidation process as presented in *Plant Design and Economics for Chemical Engineers.*[†] For each piece of equipment, specify the materials of construction that would most economically provide a service life of 10 years.

[†]M. S. Peters and K. D. Timmerhaus, *Plant Design and Economics for Chemical Engineers,* 4th ed., McGraw-Hill, New York, 1991, p. 21.

CHAPTER

11

Written and Oral Design Reports

A successful engineer not only must be able to apply theoretical and practical principles in the development of ideas and methods, but also must have the ability to express the results clearly and convincingly in written and oral presentations. During the course of a design project, the engineer may be requested to prepare numerous written and oral reports. These may involve feasibility investigations, environmental impact studies, progress reports, and project conclusions and recommendations. The decision on the advisability of continuing the design project will often be made on the basis of the material presented in the written and oral communications. The value of the engineer's work is measured to a large extent by the results given in these various reports covering the design study and the manner in which these results are presented.

The essential purpose of any written or oral communication is to succinctly pass on information to others who may have a need for the results and recommendations. The design engineer preparing either a written or an oral report should never forget the words *to others*. The abilities, functions, and needs of the intended reader or listener should be kept in mind constantly during the preparation of such communications. The success of this communication transfer depends to a large extent on the type, detail, and scope of information that the recipients expect, whether they be fellow chemical engineers, other engineers, managers, executives, clients, or the general public. Both the focus and the appropriate level of detail will vary with the background of the recipient. Some questions that each design engineer should ask before starting, preparing, or finalizing such a communication are listed here:

What is the purpose of this communication?

What audience will read or listen to this communication?

What is the function of this audience?

What background information does the audience have?

What is the technical level that the audience will understand?

What response is desired from this audience?

The answers to these questions indicate the type of information that should be presented, the quantity of detail required, and the most satisfactory method of presentation.

WRITTEN REPORTS

A design report, generally, presents the results of a design effort to decision makers in an organization and aids them in their assigned responsibilities. Similar to the design process, design reports are of varying scope, detail, and complexity. Order-of-magnitude studies generally entail relatively short reports whose purpose is to convince the decision maker whether to continue with the process evaluation or to abandon the project in favor of other, more profitable opportunities. Design reports based on preliminary design, definition, and detailed estimates generally are addressed to more technical audiences, such as project engineers and engineering managers. This group will have a greater interest in the details of the design to commit the company to develop a more detailed design, to modify a process to increase its yield, or to reduce the environmental impact.

Reports can be designated as *formal* and *informal*. Formal reports are often prepared as research, development, or design reports. These reports present results in considerable detail, and the design engineer is generally allowed considerable leeway in selecting the type of presentation to be made. Informal reports include memoranda, letters, progress notes, survey-type results, and similar items in which the major purpose is to present a result without including detailed information. Stereotyped forms are often used for informal reports, such as those for sales, production, calculations, progress, analyses, or summary of economic relations.

Although many general rules can be applied to the preparation of written reports, it should be realized that each individual company has its own guidelines and specifications. A stereotyped form shows exactly what information is desired, and detailed instructions are often provided for preparing other types of informal reports. Many companies also have standard outlines that must be followed for formal reports. In general, these outlines are similar to the one shown in the next section.

Organization of a Written Report

The organization of a formal report requires a careful sectioning and the use of subheadings in order to maintain a clear and effective presentation.[†] To a lesser extent, the

[†]Many books and articles have been written on effective technical writing. For example, see M. Gunter, *Guide to Managerial Communications,* 3d ed., Prentice-Hall, Englewood Cliffs, NJ, 1991; H. F. Ebel, C. Bliefert, and W. E. Russey, *The Art of Scientific Writing,* VCH Publishers, New York, 1987; and the review in *Chem. & Eng. News,* **66**(48): 34 (1988).

same type of sectioning is valuable for informal reports. The following discussion applies to formal reports, but by deleting or combining appropriate sections, the same principles can be applied to the organization of any type of report.

A complete design report consist of several independent parts, with each succeeding part giving greater detail on the design and its development. A covering *letter of transmittal* is usually the first item in any report. After this come the *title page,* the *table of contents,* and an *abstract* or *summary* of the report. The *body* of the report is next and includes essential information, presented in the form of discussion, graphs, tables, and figures. The *appendix* at the end of the report gives detailed information which permits complete verification of the results shown in the body of the report. Tables of data, sample calculations, and other supplementary material are included in the appendix. A typical outline for a design report is given in Table 11-1.

Letter of Transmittal The purpose of a letter of transmittal is to refer to the original instructions or developments that have made the report necessary. The letter should be brief, but it can call the reader's attention to certain pertinent sections of the report or give definite results which are particularly important, and it can affect the bottom line of a company. The writer should express any personal opinions in the letter of transmittal rather than in the report itself. Personal pronouns and an informal business style of writing may be used.

Title Page and Table of Contents In addition to the title of the report, a title page usually indicates other basic information, such as the name and organization of the person (or persons) submitting the report and the date of submittal. A table of contents may not be necessary for a short report of only six or eight pages, but for longer reports it is a convenient guide for the reader and indicates the scope of the report. The titles and subheadings in the written text should be shown as well as the appropriate page numbers. Indentations can be used to indicate the relationships of the various subheadings. A list of tables, figures, and graphs should be presented separately at the end of the table of contents.

Summary The summary is probably the most important part of a report, since it is referred to most frequently and is often the only part of the report that is read. Its purpose is to give the reader the entire contents of the report in one or two pages. It covers all phases of the design project, but it does not go into detail on any particular phase. All statements must be concise and give a minimum of general qualitative information. The aim of the summary is to present precise quantitative information and final conclusions with no unnecessary details.

The following outline shows what should be included in a summary:

1. A statement introducing the reader to the subject matter
2. What was done and what the report covers
3. How the final results were obtained
4. The important results including quantitative information, major conclusions, and recommendations

Table 11-1 Typical outline for a written design report

I. Letter of transmittal
   A. Indicates why the report has been prepared
   B. Gives essential results that have been *specifically requested*

II. Title page
   A. Includes title of report, name of individual to whom report is submitted, name and organization of the writer, and date

III. Table of contents
   A. Indicates location and title of figures, tables, and all major sections

IV. Summary
   A. Briefly presents essential results and conclusions in a clear and precise manner

V. Body of report
   A. Introduction
      1. Presents a brief discussion to explain what the report is about and the reason for the report; no results are included
   B. Previous work
      1. Discusses important results obtained from literature surveys and other previous work
   C. Discussion
      1. Outlines method of attack on project and gives design basis
      2. Includes graphs, tables, and figures that are essential for understanding the discussion
      3. Discusses technical matters of importance
      4. Indicates assumptions made and their justification
      5. Indicates possible sources of error
      6. Gives a general discussion of results and proposed design
   D. Final recommended design with appropriate data
      1. Drawings of proposed design
         a. Qualitative flowsheets
         b. Quantitative flowsheets
         c. Combined-detail flowsheets
      2. Tables listing equipment and specifications
      3. Tables giving material and energy balances
      4. Process economics including costs, profits, and return on investment
   E. Conclusions and recommendations
      1. Presented in greater detail than that provided in the summary
   F. Acknowledgment
      1. Acknowledges important assistance of others who are not listed as preparing the report
   G. Table of nomenclature
      1. Sample units should be shown
   H. References to literature (bibliography)
      1. Gives complete identification of literature sources referred to in the report

VI. Appendix
   A. Sample calculations
      1. One example should be presented and explained clearly for each type of calculation
      2. Computer programs developed or used in the calculations
   B. Derivation of equations essential to understanding the report but not presented in detail in the main body of the report
   C. Tables of data employed with reference to sources
   D. Results of laboratory tests
      1. If laboratory tests were used to obtain design data, the experimental data, apparatus and procedure description, and interpretation of the results may be included as a special appendix to the design report.

An ideal summary can be completed on one printed page. If the summary must be longer than two pages, it may be advisable to precede the summary by an *abstract,* which merely indicates the subject matter, what was done, and a brief statement of the major results.

Body of the Report The first section in the body of the report is the *introduction.* It states the purpose and scope of the report and indicates why the design project originally appeared to be feasible or necessary. The relationship of the information presented in the report to other phases of the company's operations can be covered, and the effects of future developments may be worthy of mention. References to *previous work* can be discussed in the introduction, or a separate section can be presented dealing with literature survey results and other previous work.

Modern bibliographic software can be used to advantage to obtain information from current reference materials and allow the writer to file such data, sorted in a coherent manner, and provide easy accessibility. Whether the information is in print, a PDF file, or an electronic citation, bibliographic software simplifies the organization of reference material and increases the efficiency of report preparation. Besides this benefit, one of the most attractive features of using bibliographic software is the ability to create in-text citations, bibliographies, footnotes, and endnotes for the report.

A description of the methods used for developing the proposed design is presented in the next section under the heading of *discussion.* Here the writer shows to the reader the methods used in reaching the final conclusions. The validity of the methods must be made apparent, but the writer should not present an annoying or distracting amount of detail. Any assumptions or limitations on the results should be discussed in this section.

The next section presents the *recommended design,* complete with figures and tables giving all the necessary qualitative and quantitative data. An analysis of the cost and profit potential of the proposed process should accompany the description of the recommended design.

The body of a design report often includes a section giving a detailed discussion of all *conclusions* and *recommendations.* When applicable, sections covering *acknowledgment, table of nomenclature,* and *literature references* may be added.

Appendix To make the written part of a report more readable, the details of calculation methods, experimental data, reference data, certain types of derivations, and similar items are often included as separate appendixes to the report and identified by a table of contents. This information is thus available to anyone who wishes to make a complete check on the work; yet the descriptive part of the report is not made ineffective because of excess information.

Preparing the Report

The physical process of preparing a report can be divided into the following steps:

1. Define the subject matter, scope, and intended audience.
2. Prepare a skeleton outline and then a detailed outline.

3. Write the first draft.
4. Edit and improve the first and any successive drafts before preparing the final draft.
5. Check the final draft carefully, print the report, and proofread the final report.

To accomplish each of these steps successfully, the writer must make certain the initial work on the report is started soon enough to allow a thorough job and still meet any predetermined deadline date. Many of the figures, graphs, and tables, as well as some sections of the report, can be prepared while the design work is in progress.

Word processing, spreadsheet, and graphics software are modern tools that can assist the writer to produce a better final report. However, many of these software programs have limitations that the writer must recognize when they are used. Such software problems arise because of programming errors, misunderstandings by software developers of the rules of writing, or differences between technical writing and other forms of written communication.

Presenting the Results

Accuracy and logic must be maintained throughout any report. The writer has a moral responsibility to present the facts accurately and not mislead the reader with incorrect or dubious statements. If approximations or assumptions are made, their effect on the accuracy of the results should be indicated. For example, a preliminary plant design might show that the total investment for a proposed plant is $5,500,000. This is not necessarily misleading as to the accuracy of the result, since only two significant figures are indicated. On the other hand, a proposed investment of $5,554,328 does not make sense, and the reader knows at once that the writer did not use any type of logical reasoning in determining the accuracy of the results.

The style of writing in technical reports should be simple and straightforward. Although short sentences are preferred, variation in the sentence length is necessary in order to avoid a disjointed staccato effect. The presentation must be convincing, but it must also be devoid of distracting and unnecessary details. Flowery expressions and technical jargon are often misused by technical writers in an attempt to make their writing more interesting. Certainly, an elegant or forceful style is sometimes desirable, but the technical writer must never forget that the major purpose of the report is to present information clearly and understandably.

Subheadings and Paragraphs The use of effective and well-placed subheadings can improve the readability of a report. The sections and subheadings follow the logical sequence of the report outline and permit the reader to become oriented and prepared for a new subject.

Paragraphs are used to cover one general thought. A paragraph break, however, is not nearly as definite as a subheading. The length of paragraphs can vary over a wide range, but any thought worthy of a separate paragraph should require at least two sentences. Long paragraphs are a strain on the reader, and the writer who consistently uses paragraphs longer than 10 to 12 printed lines will have difficulty holding the reader's attention.

Tables The effective use of tables can save many words, especially if quantitative results are involved. Tables are included in the body of the report only if they are essential to the understanding of the written text. Any type of tabulated data that is not directly related to the discussion should be located in the appendix.

Every table requires a title, and the headings for each column should be self-explanatory. If numbers are used, the correct units must be shown in the column heading or with the first number in the column. A table should never be presented on two pages unless the amount of data makes a break absolutely necessary.

Graphs In comparison with tables, which present definite numerical values, graphs serve to show trends or comparisons. The interpretation of results is often simplified for the reader if the tabulated information is presented in graphical form. This can be done quite easily with one of many computer software programs that are presently available to portray information in a graphical form.

If possible, the experimental or calculated points on which a curve is based should be shown on the plot. These points can be represented by large dots, small circles, squares, triangles, or some other identifying symbol. The most probable smooth curve can be drawn on the basis of the plotted points, or a broken line connecting each point may be more appropriate. In any case, the curve should not extend through the open symbols representing the data points. If extrapolation or interpolation of the curve is doubtful, the uncertain region can be designated by a dotted or dashed line.

The ordinate and the abscissa must be labeled clearly, and any nomenclature used should be defined on the graph or in the body of the report. If numerical values are presented, the appropriate units are shown immediately after the labels on the ordinate and abscissa. Restrictions on the plotted information should be indicated on the graph itself or with the title.

The title of the graph must be explicit but not obvious. For example, a log-log plot of temperature versus the vapor pressure of pure glycerol should not be entitled "Log-Log Plot of Temperature versus Vapor Pressure for Pure Glycerol." A much better title, although still somewhat obvious, would be "Effect of Temperature on Vapor Pressure of Pure Glycerol."

Some additional suggestions for the preparation of graphs follow:

1. The independent or controlled variable should be plotted as the abscissa, and the variable that is being determined should be plotted as the ordinate.
2. Permit sufficient space between grid elements to prevent a cluttered appearance (ordinarily, one to two grid lines per centimeter are adequate).
3. Use coordinate scales that give good proportionment of the curve over the entire plot, but do not distort the apparent accuracy of the results.
4. The values assigned to the grids should permit easy and convenient interpolation.
5. If possible, the label on the vertical axis should be placed in a horizontal position to permit easier reading.
6. Unless families of curves are involved, it is advisable to limit the number of curves on any one plot to three or less.

7. The curve should be drawn as the heaviest line on the plot, and the coordinate axes should be heavier than the grid lines.

Illustrations Flow diagrams, photographs, line drawings of equipment, and other types of illustrations may be a necessary part of a report. They can be inserted in the body of the text or included in the appendix. Complete flow diagrams, prepared on oversize paper, and other large drawings are often folded and inserted in an envelope at the end of the report.

Process simulation software is generally used to produce a flowsheet. The resulting flowsheet, however, may not be a true process flow diagram if the software did not follow the accepted conventions regarding equipment symbols, line crosses, labels, etc. Additionally, the process simulated occasionally is not the actual process, particularly when no calculations were required by the process simulator for certain process units such as storage tanks. Under such conditions the process unit may not be included as part of the process flow diagram. Nevertheless, the flowsheet developed by a process simulator should be included in the appendix since it provides a visual portrayal of the simulation process.

Nomenclature If many different symbols are used repeatedly throughout a report, a table of nomenclature, showing the symbols, meanings, and sample units, should be included in the report. Each symbol can be defined when it first appears in the written text. If this is not done, a reference to the table of nomenclature should be given with the first equation.

Ordinarily, the same symbol is used for a given physical quantity regardless of its units. Subscripts, superscripts, and lower- and uppercase letters can be employed to give special meanings. The nomenclature should be consistent with common usage.

References to the Literature Bibliographic software enables the writer to collect cutting-edge information from database services on the web or a CD-ROM and to search various libraries across the world via the Internet and automatically download the reference information in a very short time. To cite this information properly in the report, it will be necessary to include the Internet locator (URL) so that the writer can go back to the file to obtain the reference author, publication source, publication date, and other desired information.

The original sources of any literature referred to in the report should be listed at the end of the body of the report. References are usually tabulated and numbered in alphabetical order on the basis of the first author's surname, although the listing can be based on the order of appearance in the report.

When a literature reference is cited in the written text, the last name of the author is mentioned and the bibliographical identification is shown by a superscript number after the author's name or at the end of the sentence. A number enclosed in square brackets may be used in place of the raised number, if desired.

The bibliography should give the following information:

1. For journal articles: (*a*) authors' names, followed by initials; (*b*) journal, abbreviated to conform to the List of Periodicals as established by *Chemical Abstracts;* (*c*) volume number, sometimes shown in bold; (*d*) issue number, if necessary;

(*e*) page number, which is preceded by a colon; and (*f*) date (in parentheses). The title of the article is usually omitted. Issue number is omitted if paging is on a yearly basis. The date is sometimes included with the year in place of the issue number. Here are some examples:

McCormick, J. E., *Chem. Eng.,* **95**(13): 75 (1988).

McCormick, J. E., *Chem. Eng.,* **95:** 75 (Sept. 26, 1988).

Seibert, A. F., and Fair, J. R., *Ind. Eng. Chem.,* **32:** 2213 (1993).

2. For single publications such as books or pamphlets: (*a*) authors' names, followed by initials; (*b*) title, in italics for books and pamphlets, in roman for theses; (*c*) edition (if more than one has appeared); (*d*) volume (if there is more than one); (*e*) publisher; (*f*) place of publication; and (*g*) year of publication; (*h*) the chapter or page number. Titles of theses are often omitted.

   Humphrey, J. L., and Keller, II, G. E., *Separation Process Technology,* McGraw-Hill, New York, 1997.

   Al-Rabiah, A. A., Ph.D. thesis in chemical engineering, University of Colorado, Boulder, 2001.

3. For unknown or unnamed authors, alphabetize by the journal or organization publishing the information.

   Chem. Eng., **95**(13): 26 (1988).

4. For patents: (*a*) patentees' names, followed by initials, and assignee (if any) in parentheses; (*b*) country granting patent and numbers; and (*c*) date issued.

   Fenske, E. R. (to Universal Oil Products Co.), U.S. Patent 3,249,650, May 3, 1986.

5. For unpublished information: (*a*) *in press* means formally accepted for publication by the indicated journal or publisher; (*b*) the use of *private communication* and *unpublished data* is not recommended unless absolutely necessary, because the reader may find it impossible to locate the original material.

   Morari, M., *Chem. Eng. Prog.,* in press, 1988.

Sample Calculations The general method used in developing the proposed design is discussed in the body of the report, but detailed calculation methods are not presented in this section. Instead, sample calculations are given in the appendix. One example should be shown for each type of calculation, and sufficient detail must be included to permit the reader to follow each step. The particular conditions chosen or the sample calculations must be designated. The data on which the calculations are based should be listed in detail at the beginning of the section, even though these same data may be available through reference to one of the tables presented with the report.

Word Processing The arrival of the word processor has had a major impact on the manner in which written design reports are prepared. Since various sections of the text can be cut and pasted with ease, it has become much easier to add new sections as the process design is modified and then add the previously prepared sections usually with minor alternations. For technical writing, Word, WordPerfect, and LaTex are three of the most commonly used word processors.

Since there are generally many cross-references between sections of a lengthy design report, it becomes helpful to add headers to the pages that identify the section numbers and titles. Additionally, an index can be of considerable assistance when the reader is interested in a specific topic of the report. To add headers and create an index, desktop publishing packages such as PageMaker are quite useful.

Mechanical Details The final report should be submitted in a neat and businesslike form. Formal reports are usually bound with a heavy cover, and the information shown in the title page is repeated on the cover. If paper fasteners are used for binding in a folder, the pages should be attached only to the back cover.

The report should be printed on a good grade of paper with a margin of at least 2.5 cm on all sides. Normally, only one side of the page is used; and all material is double-spaced except the letter of transmittal, footnotes, and long quotations. Starting with the summary, all pages including graphs, illustrations, and tables should be numbered in sequence.

Short equations can sometimes be included directly in the written text if the equation is not numbered. In general, however, equations are centered on the page and given a separate line, with the equation number appearing at the right-hand margin of the page. Explanation of the symbols used can be presented immediately following the equation.

Proofreading and Checking Before final submittal, the completed report should be read carefully and checked for printout errors, consistency of data quoted in the text with those presented in the table and graphs, grammatical errors, spelling errors, and similar obvious mistakes. If numerous corrections or changes are necessary, the appearance of the report must be considered, and some sections will need to be redone with the aid of the word processor.

With the present availability of computers, no report should be submitted that has not been subjected to a spell checker. However, no spell checker is perfect since such computer software can only check the spelling of a word but cannot determine whether it is the word that the writer intended. For example, if the writer uses the term *principal* rather than the term *principle,* the spell checker will not be able to detect the error.

Grammar checkers have improved over the past decade, but those that are based on simple rules still cause problems in the reediting of technical communications. This is particularly true in long sentences involving written possessive constructions and subject/verb agreement.

Abbreviations Time and space can be saved by the use of abbreviations, but the writer must be certain the reader knows what is meant. Unless the abbreviation is standard, the meaning should be explained the first time it is used. The following rules are generally applicable for the use of abbreviations.

1. Abbreviations are acceptable in tables, graphs, and illustrations when space limitations make them desirable.
2. Abbreviations are normally acceptable in the text only when preceded by a number, such as 3 cm/s (3 centimeters per second).

3. Periods may be omitted after abbreviations for common scientific and engineering terms, except when the abbreviation forms another word (e.g., in. for inch).
4. The plural of an abbreviation is the same as the singular (kilograms are abbreviated as kg).
5. The abbreviation for a noun derived from a verb is formed by adding *n* (concentration becomes concn).
6. The abbreviation for the past tense is formed by adding d (concentrated becomes concd).
7. The abbreviation for the participle is formed by adding g (concentrating becomes concg).

Examples of accepted abbreviations are shown in Appendix A.

Rhetoric Correct grammar, punctuation, and style of writing are obvious requirements for any report. Many engineers, however, submit unimpressive reports because they do not concern themselves with the formal style of writing required in technical reports. This section deals with some of the restrictions placed on formal writing and presents a discussion of common errors.

Personal Pronouns The use of personal pronouns should be avoided in technical writing. Many writers eliminate the use of personal pronouns by resorting to the passive voice. This is certainly acceptable, but, when applicable, the active voice gives the writing a less stilted style. For example, instead of saying "We designed the absorption tower on the basis of . . . ," a more acceptable form would be "The absorption tower was designed on the basis of . . ." or "The basis for the absorption tower design was . . . ."

The pronoun *one* is sometimes used in technical writing. In formal writing, however, it should be avoided or, at most, employed only occasionally.

Tenses Both past and present tenses are commonly used in report writing; however, tenses should not be switched in one paragraph or in one section unless the meaning of the written material requires the change. General truths that are not limited by time are stated in the present tense, while references to a particular event in the past are reported in the past tense (e.g., "The specific gravity of mercury is 13.6." "The experiment was performed . . .").

Diction Contractions such as *don't* and *can't* are seldom used in technical writing, and informal or colloquial words should be avoided. Humorous or witty statements are out of place in a technical report, even though the writer may feel they are justified because they can stimulate interest. Too often, the reader will be devoid of a sense of humor, particularly when engrossed in the serious business of digesting the contents of a technical report. A good report is made interesting by clarity of expression, skillful organization, and the significance of its contents.

Singular and Plural Many writers have difficulty determining whether a verb should be singular or plural. This is especially true when a qualifying phrase separates the subject and its verb. For example, "A complete list of the results *is* (not *are*) given in the appendix."

Certain nouns, such as *number* and *series,* can be either singular or plural. As a general rule, the verb should be singular if the subject is viewed as a unit ("The number of engineers in the United States *is* increasing") and plural if the terms involved are considered separately ("A number of the workers *are* dissatisfied"). Similarly, the following sentences are correct: "*Thirty thousand gallons was* produced in *two* hours." When the subject is a number with a unit of measure, it always takes a singular verb. "The tests show that *18* batches *were* run at the wrong temperature."

Dangling Modifiers The technical writer should avoid dangling modifiers that cannot be associated directly with the words they modify. For example, the sentence "Finding the results were inconclusive, the project was abandoned" is incorrect because *finding* modifies *the project.* It could be rewritten correctly as "Finding the results were inconclusive, the investigators abandoned the project."

Poor construction caused by dangling modifiers often arises from retention of the personal viewpoint, even though personal pronouns are eliminated. The writer should analyze the work carefully and make certain the association between a modifying phrase and the words referred to is clear.

Compound Adjectives Nouns are often used as adjectives in scientific writing. This practice is acceptable; however, the writer must use it in moderation. A sentence including "a centrally located natural gas production plant site is . . ." should certainly be revised. Prepositional clauses are often used to eliminate a series of compound adjectives.

Hyphens are employed to connect words that are compounded into adjectives, for example, *hot-wire anemometer* and a *high-pressure line,* but no hyphen appears in *highly sensitive element.*

Split Infinitives The use of split infinitives is acceptable in some types of writing, but they should be avoided in technical reports. A split infinitive bothers many readers and frequently results in misplaced emphasis. Instead of "The supervisor intended to carefully check the data," the sentence should read, "The supervisor intended to check the data carefully."

That versus Which Many technical writers tend to overwork the word *that.* Substitution of the word *which* for *that* is often acceptable even though a strict grammatical interpretation would require repetition of *that.* The general distinction between the pronouns *that* and *which* can be stated as follows: *That* is used when the clause it introduces is necessary to define the meaning of its antecedent; *which* introduces some additional or incidental information.

Comments on Common Errors

1. The word *data* is usually plural. Say "data are," not "data is."
2. *Balance* should not be used when *remainder* is meant.
3. Use *different from* instead of *different than.*

4. The word *farther* refers to distance, and *further* indicates "in addition to."
5. *Affect,* as a verb, means "to influence." It should never be confused with the noun *effect,* which means "result."
6. *Due to* should be avoided when *because of* can be used.
7. Use *fewer* when referring to a countable amount and *less* when referring to a non-countable quantity or degree.

Checklist for the Final Report

Before submitting the final draft, the writer should make a critical analysis of the report. Following is a list of questions the writer should ask when evaluating the report. These questions cover the important considerations in report writing and can serve as a guide for both experienced and inexperienced writers.

1. Does the report fulfill the purpose?
2. Will it be understandable to the principal readers?
3. Does the report attempt to cover too broad a subject?
4. Is sufficient information presented?
5. Are too many details included in the body of the report?
6. Are the objectives stated clearly?
7. Is the reason for the report indicated?
8. Is the summary concise? Is it clear? Does it give the important results, conclusions, and recommendations? Is it a true summary of the entire report?
9. Is there an adequate description of the work done?
10. Are the important assumptions and the degree of accuracy indicated?
11. Are the conclusions and recommendations valid?
12. Are sufficient data included to support the conclusions and recommendations?
13. Have previous data and earlier studies in the field been considered?
14. Is the report well organized?
15. Is the style of writing readable and interesting?
16. Has the manuscript been rewritten and edited ruthlessly?
17. Is the appendix complete?
18. Are tables, graphs, and illustrations presented in a neat, readable, and organized form? Is all necessary information shown?
19. Has the report been proofread? Are pages, tables, and figures numbered correctly?
20. Is the report ready for submittal on time?

ORAL REPORTS

Oral presentations are often used in addition to written reports to communicate the results of a design investigation. When done well, such presentations can communicate information more rapidly than the written word and can provide immediate response from the audience to the speaker as well as between members of the audience. However, maintaining audience interest in an oral presentation is more difficult than maintaining the interest of a reader of a written report since the latter can quickly skip over the written material that is of lesser interest and continue to those items of greater interest.

Oral presentation of a design report is somewhat different from preparation of a written report. In the short time available for the presentation, the speaker or speakers will need to cover certain independent parts of the written version in a cohesive and informative sequence that can capture the attention of the audience. To achieve this goal, the presentation and the accompanying visuals need to focus the audience's attention on the areas of the design study that the speaker and the audience feel are key items.

Organization and Presentation of an Oral Design Report

The presentation is initiated with a visual that identifies the study and the individuals involved in that study. In this introduction, the objectives of the study are presented. This is generally followed by a brief overview of the proposed process and the alternative designs that were considered. With that background, the presenter can then review the key areas of the proposed process. Any innovative designs that enhance the profitability of the process can be highlighted in the economic analysis that follows. Any environmental and safety issues that are associated with the process and their effect on the plant employees and the public should be noted. The presentation normally closes with a summary of the process design and recommended actions. A typical outline for an oral presentation of a design report is given in Table 11-2.

Table 11-2 Typical outline for an oral design report

I. Title of presentation
   A. Identify report and the presenter(s)

II. Outline of presentation (optional)

III. Introduction of the design problem
   A. Orients audience to the design problem
   B. Outlines objective of study

IV. Overview of proposed process
   A. If applicable, brief overview of alternative processes considered
   B. Brief review of proposed process and its justification
   C. Selective review of key areas in proposed process
   D. Highlighting of innovative areas

V. Economic analysis of the process design
   A. Present results of the analysis
   B. Economic effect of assumptions and uncertainties

VI. Recognition of other factors related to the design
   A. Environmental effects
   B. Safety and health concerns
   C. Site location

VII. Summary and recommendations

VIII. Acknowledgments

Oral Presentation One of the most difficult tasks that the presenter of an oral report encounters is that of summarizing the most important features of an extensive written report. For this reason, it is incumbent on the speaker to identify these features and place them in an order that makes sense to audience members who have just been introduced to the details of this design study. It is important to realize that an audience cannot absorb technical material in the same manner that they can when reading a report. For example, a detailed material and energy balance will not be readily understood by a general audience. Also, what is effectively communicated in a table in the report will create little interest on the part of the audience given the time available to digest the material. Such information could better be communicated orally by using a graph or even a pie chart. For all these reasons, it is imperative that the speaker rehearse the presentation, possibly in the presence of colleagues, to obtain constructive comments that can be used to improve the oral communication effect of the formal presentation. If no audience is available, modern technology has made it possible to use a videotape system to provide a record of such an oral communication that permits the speaker to make a self-evaluation of both the presentation and the supporting visuals.

Visual Aids Most oral presentations of a technical nature are developed around the use of visual aids. Transparencies, slides, computer projections, and videos can be used as visual aids. Each has its advantages, and the presenter should consider all options before deciding how best to present the material.

The overhead projector with transparencies is the visual aid most widely used by presenters for displaying key concepts, graphs, and figures that accompany a design project. The use of transparencies provides considerable latitude for the presenter to rearrange the order of the visuals or to change or mark up a particular transparency before or during a presentation. The use of slides, on the other hand, is limited to rearrangement before the presentation and does not provide the opportunity for changes during the presentation. However, an oral presentation accompanied by this type of visual generally gives an impression of a more professional presentation.

With the availability of LCD pads and computer projection devices, more and more presenters are opting to use computerized projection facilities. To prepare and display the images, several software packages have been developed, including Powerpoint. This software is capable of displaying animated sequences, halftones, and videos. In most cases the quality of the display is measurably improved relative to the use of transparencies and slides.

Regardless of the visual aid selected for the presentation, great care must be exercised in preparing the visual. There are numerous guidelines concerning the font size of the lettering and the density of information to use on visuals. If the lettering is too small or the density of information too large, audience interest will diminish, making the presentation considerably less effective. In preparing any type of visual, the best approach is to imagine what the audience response would be to the visual. As noted by Turton et al.,[†] "No visual should ever be used in a presentation unless it has been

---

[†]R. Turton, R. C. Bailie, W. B. Whiting, and J. A. Shaewitz, *Analysis Synthesis, and Design of Chemical Processes,* Prentice-Hall, Upper Saddle River, NJ, 1998.

previewed with similar equipment, in a similar room, from the most remote position that an audience member can be located."

PROBLEMS

11-1 Obtain the guidelines for preparing written reports from a company that employs engineers, and compare the company's guidelines with those presented in this chapter. Provide explanations for any differences between the two.

11-2 If your department has guidelines for preparing laboratory reports, compare the department guidelines with those presented in this chapter. Why would the two guidelines show some key differences?

11-3 Review a design report prepared by a colleague. After quickly reviewing the report, answer the following questions:

What was the objective of the study?

Is the design study clearly and succinctly presented?

What are the key conclusions and recommendations?

Is the report sufficiently persuasive to accept the recommendations?

11-4 Run a spell checker and grammar checker on a completed design report prepared by the writer to determine if improvements can be made.

11-5 Investigate the features of a software package used to prepare graphs, and list those features that could lead to poor visual aids.

11-6 Prepare a skeleton outline and a detailed outline for a final report on the detailed-estimate design of a distillation unit. The unit is to be used for recovering methanol from a by-product containing water and methanol. In the past, this by-product has been sold to another chemical plant, but the head of the engineering development group feels that the recovery should be accomplished within the company. The report will be reviewed by the head of the design group and then will be submitted to plant management for final approval.

CHAPTER

12

Materials-Handling Equipment—Design and Costs

The design of equipment and systems used for the safe handling of materials is encountered in nearly every type of plant design. The most common process for transferring liquid materials uses pumps and piping systems. Compressors, fans, blowers, conveyors, chutes, and hoists are examples of other types of equipment used extensively to handle and transfer gaseous and solid materials. Other types of special equipment may further change the characteristics of the materials with the use of blenders, mixers, kneaders, crushers, and grinders. Collection and storage of these materials are still another phase of this materials-handling process.

In all these cases the design engineer must decide which type of equipment is best suited for achieving the objectives of the process design to prepare equipment specifications that will satisfy the operational demands of the process under reasonable cost conditions. Therefore, design principles, practical problems of operation, and cost effects are all involved in the final choice of materials transfer, handling, and treatment equipment.

BASIC CONCEPTS OF FLUID TRANSPORT

A key factor involved in the design of systems providing the transport of fluids from one location to another is the quantity of power that is required for this particular operation. For example, mechanical energy may be necessary to overcome frictional resistance, changes in elevation, changes in internal energy, and other resistances encountered in the flow system.

The various forms of energy can be related to the total energy balance or the total mechanical energy balance. On the basis of 1 kg of fluid flowing under steady-state conditions between the inlet and exit locations in the system, the total energy balance can be written in differentiated form as

$$Z_1 g + p_1 v_1 + \frac{V_1^2}{2\alpha} + u_1 + Q + W = Z_2 g + p_2 v_2 + \frac{V_2^2}{2\alpha} + u_2 \qquad (12\text{-}1)$$

and the mechanical energy balance as

$$Z_1 g - \int_1^2 v\,dp + \frac{V_1^2}{2\alpha} + W_o = Z_2 g + \frac{V_2^2}{2\alpha} + \sum F \qquad (12\text{-}2)$$

where Z is the vertical distance above an arbitrarily chosen datum plane, g the local gravitational acceleration, p the absolute pressure, v the specific volume of the fluid, V the average fluid velocity, α the correction factor to account for the use of the average velocity, u the internal energy of the fluid, Q the heat energy transmitted across the fluid boundary from an outside source, W the total shaft work provided from an outside source, W_o the mechanical work provided from an outside source,† and F the mechanical energy loss due to friction. The value for α is usually assumed to be 1.0 if the flow is turbulent and 0.5 if the flow is viscous. Equations (12-1) and (12-2) are sufficiently general for treatment of almost any flow problem and are the basis for many design equations that apply for a number of simplified conditions.

Evaluation of the $\int_1^2 v\,dp$ term in Eq. (12-2) may be difficult if a compressible fluid is flowing through the system, because the exact path of the compression or expansion is often unknown. For noncompressible fluids, however, the specific volume v remains essentially constant, and the integral term reduces simply to $v(p_2 - p_1)$. Consequently, the total mechanical energy balance is especially useful and easy to apply when the flowing fluid can be considered as noncompressible.

Newtonian Fluids

For a Newtonian fluid in a smooth pipe, dimensional analysis relates the friction drop per unit length of pipe to the pipe diameter, density of the fluid, and average fluid velocity through two dimensionless groups, namely, the Fanning friction factor,‡ given by

$$\mathfrak{f} = \frac{D\,\Delta p}{2\rho^2 V L} \qquad (12\text{-}3a)$$

and the Reynolds number, given by

$$\mathrm{Re} = \frac{DV\rho}{\mu} \qquad (12\text{-}3b)$$

The friction factor in smooth pipes is a function of the Reynolds number. However, in rough pipes, the relative roughness ε/D also affects the friction factor. Figure 12-1

†The mechanical work W_o is equal to the total shaft work W minus the amount of energy transmitted to the fluid as a result of pump friction or pump inefficiency. When W_o is used in the mechanical energy balance, pump friction is not included in the term for the mechanical energy loss due to friction F.

‡The Fanning friction factor should not be confused with the Darcy friction factor that is 4 times the Fanning friction factor value.

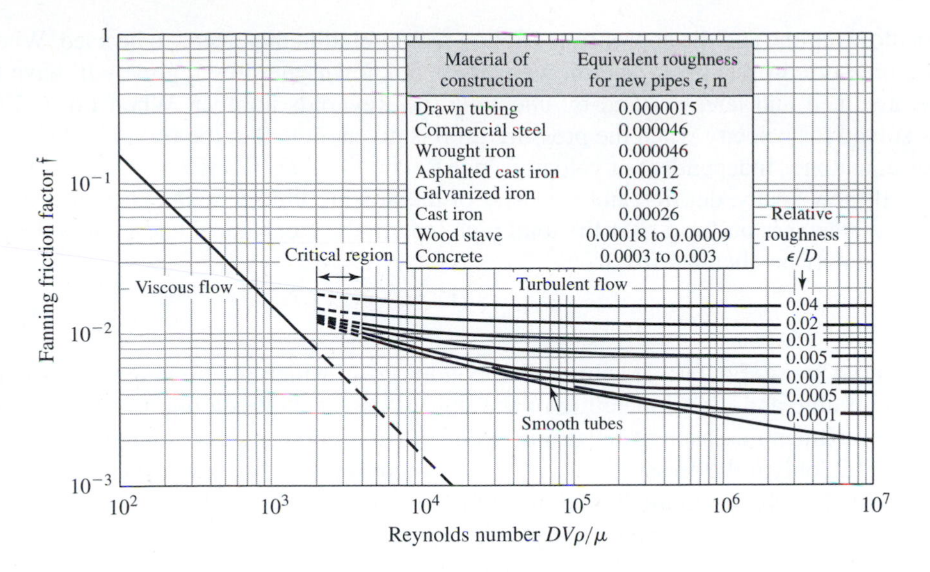

Figure 12-1
Fanning friction factors for long, straight pipes. [Based on L. F. Moody, *Trans. ASME.* **66:** 671–684 (1944).]

presents a plot of the friction factor as a function of the Reynolds number and ε/D. Representative values for the surface roughness factor ε associated with various materials are indicated in the figure.

Below a critical Reynolds number of 2100, the flow of a fluid in a pipe is laminar, and the Fanning friction factor is given by the Hagen-Poiseuille equation

$$\mathfrak{f} = \frac{16}{\text{Re}} = \frac{16\mu}{DV\rho} \qquad \text{Re} \leq 2100 \tag{12-4}$$

A transition from laminar flow to turbulent flow occurs over the range $2100 < \text{Re} < 4000$. Since there is doubt as to which type of flow is predominant in this range, safe design practice favors the assumption of turbulent flow in this transitional region.

For turbulent flow in smooth pipes, the Blasius equation provides a reasonably accurate friction factor over a wide range of Reynolds number as given by

$$\mathfrak{f} = \frac{0.079}{(\text{Re})^{0.25}} \qquad 4000 < \text{Re} < 10^5 \tag{12-5}$$

The Colebrook relation gives a good approximation of the friction factor for rough pipes over the entire turbulent flow range:

$$\frac{1}{\mathfrak{f}^{1/2}} = -4 \log\left[\frac{\varepsilon}{3.7D} + \frac{1.256}{\text{Re}(\mathfrak{f})^{1/2}}\right] \qquad \text{Re} > 4000 \tag{12-6}$$

Many pipe flow problems either include determination of the pressure drop given the flow rate (or velocity) or the calculation of the flow rate (or velocity) for a given pressure drop. When the flow rate is given, the Reynolds number is determined to identify the flow regime and the appropriate friction factor relation that needs to be used. When the pressure drop is given and the velocity is unknown, the flow regime will have to be assumed and later verified by checking the Reynolds number. When Eq. (12-6) is solved for velocity when the pressure drop is given, note that the right-hand side of the equation is independent of velocity since $\text{Re}(\mathfrak{f})^{1/2} = (D^{3/2}/\mu)(\rho\,\Delta p/2L)$.

If the velocity, density, and viscosity of the flowing fluid remain constant and the pipe diameter is uniform over the total pipe length, the mechanical energy loss due to friction may be obtained from

$$F = \frac{2\mathfrak{f}V^2L}{D} \tag{12-7}$$

In a strict sense, Eq. (12-7) is limited to conditions in which the flowing fluid is noncompressible and the temperature of the fluid is constant. When one is dealing with compressible fluids, such as air, steam, or any gas, it is good engineering practice to use Eq. (12-7) only if the pressure drop over the system is less than 10 percent of the initial pressure. If a change in the fluid temperature occurs, Eq. (12-7) should not be used in the form indicated unless the total change in the fluid viscosity is less than approximately 50 percent based on the maximum viscosity.[†‡] If Eq. (12-7) is used when pressure changes or temperature changes are involved, the best accuracy is obtained by using the linear velocity, density, and viscosity of the fluid as determined at the average temperature and pressure. Exact results for compressible fluids or nonisothermal flow can be obtained from the Fanning equation by integrating the differential expression, taking all changes into consideration.

For turbulent flow in a conduit of noncircular cross section, an equivalent diameter can be substituted for the circular-section diameter, and the equations for circular pipes can be applied without introducing too large an error. This equivalent diameter is defined as 4 times the hydraulic radius R_H, where the hydraulic radius is the ratio of the cross-sectional flow area to the wetted perimeter. When the flow is viscous, substitution of $4R_H$ for D does not give accurate results, and the exact expressions relating frictional pressure drop and velocity can be obtained only for certain conduit shapes.

Non-Newtonian and Bingham Fluids

For flow of non-Newtonian fluids, the Fanning friction factor design chart (Fig. 12-2) developed by Dodge and Metzner is often used.[§] In this chart, n' is the slope of a plot of $D\,\Delta p/(4L)$ versus $8V/D$ on logarithmic coordinates, and γ is a general relationship

[†]Overall effects of temperature on the friction factor are more important in the laminar-flow range where $\mathfrak{f}$ is directly proportional to the viscosity than in the turbulent-flow range where $\mathfrak{f}$ is approximately proportional to $\mu^{0.16}$.

[‡]For heating or cooling of fluids, a temperature gradient must exist from the pipe wall across the flowing fluid. A simplified design procedure for this case is as follows: When the temperature and viscosity changes must be taken into consideration, the friction factor for use in Eq. (12-7) should be taken as the isothermal friction factor (Fig. 12-1) based on the arithmetic-average temperature of the fluid divided by a correction factor ϕ, where $\phi = 1.1(\mu_a/\mu_w)^{0.25}$ when $\text{Re} < 2100$ or $\phi = 1.02(\mu_a/\mu_w)^{0.14}$ when $\text{Re} > 2100$, where μ_a is the viscosity of the fluid at the average bulk temperature and μ_w the viscosity of the fluid at the temperature of the pipe wall.

[§]B. F. Dodge and A. B. Metzner, *AIChE J.*, **5:** 189 (1959).

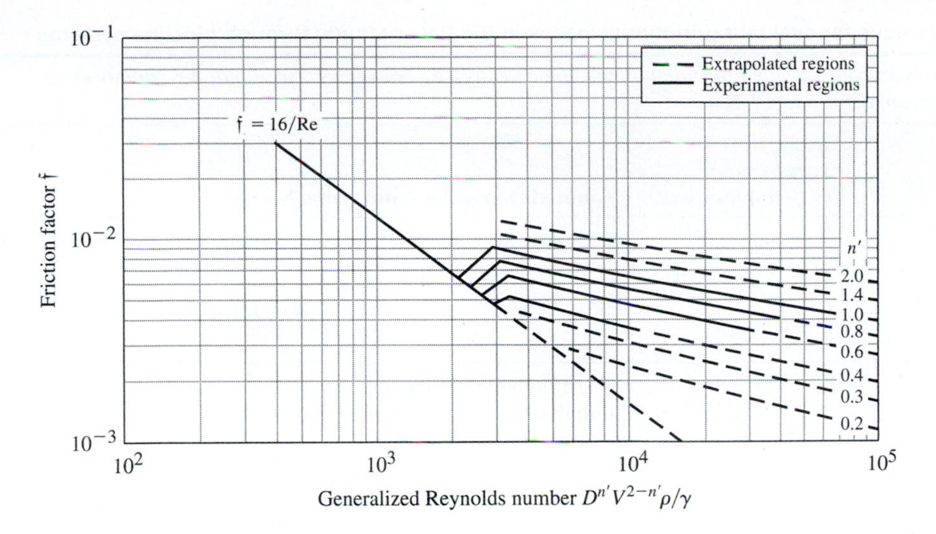

Figure 12-2
Fanning friction factor for non-Newtonian flow where $n' = d\ \ln(D\,\Delta p/4L)/d\ \ln(8V/D)$. [Based on B. F. Dodge and A. B. Metzner, *AIChE J.*, **5:** 189 (1959).]

for the shear rate of the fluid given by

$$\gamma = \frac{8V}{D}\frac{1+3n'}{4n'} \tag{12-8}$$

By plotting capillary viscometry information in this manner, the data can be used directly for pressure drop calculations. For such calculations, the flow rate and diameter determine the fluid velocity from which $8V/D$ is determined and $D\Delta p/(4L)$ is read from the plot.

For Bingham plastic materials in turbulent flow, it is generally assumed that the wall stresses greatly exceed the yield stress, so that the friction factor–Reynolds number relationship for Newtonian fluids applies. This is equivalent to setting $n' = 1$ in the Dodge-Metzner method.

Vacuum Flow

For gas flows under high vacuum conditions, the continuum hypothesis used in previous flow calculations is no longer appropriate. Accordingly, vacuum flow is generally described with such flow variables as S and Q', the pumping speed and throughput, respectively. The pumping speed is the actual volumetric flow rate of the gas through a flow cross section, and throughput is the product of the pumping speed and the absolute pressure. The mass flow rate $\dot{m}$ is directly related to the throughput, using the ideal gas law, as

$$\dot{m} = \frac{M}{RT}Q' \tag{12-9}$$

where M is the molecular weight of the gas. For a given mass flow rate, throughput is independent of pressure. The relation between throughput and pressure drop across a

Table 12-1 Expressions for evaluating frictional losses in the flow of fluids through pipeline systems

For noncircular, cross-sectional area and turbulent flow, replace D by $4R_H = 4$ (cross-sectional flow area/wetted perimeter) for an approximate frictional loss.

Friction caused by	General expression for frictional loss	Limited expressions and remarks
Flow through long, straight pipe of constant cross-sectional area	$dF = \frac{2\mathfrak{f}V^2\,dL}{D}$	For cases in which fluid is essentially noncompressible and temperature is constant $F = \frac{2\mathfrak{f}V^2 L}{D}$
Sudden enlargement	$F_e = \frac{(V_1 - V_2)^2}{2\alpha}$	The following values for α may be used in design calculations: turbulent flow, $\alpha = 1$; streamline flow, $\alpha = 0.5$.
Sudden constriction	$F_c = \frac{K_c V_2^2}{2\alpha}$	The following values for α may be used in design calculations: turbulent flow, $\alpha = 1$; streamline flow, $\alpha = 0.5$. For $A_2/A_1 < 0.715$, $K_c = 0.4(1.25 - A_2/A_1)$. For $A_2/A_1 > 0.715$, $K_c = 0.75(1 - A_2/A_1)$. For conical or rounded shape, $K_c = 0.05$.
Fittings, valves, etc.	$F = \frac{2\mathfrak{f}V^2 L_e}{D}$	See L_e/D values below

Fitting	L_e/D per fitting (dimensionless)
45° elbows	15
90° elbows, std. radius	32
90° elbows, medium radius	26
90° elbows, long sweep	20
90° square elbows	60
180° close-return bends	75
180° medium-radius return bends	50
Tee (used as elbow, entering run)	60
Tee (used as elbow, entering branch)	90
Couplings	Negligible
Unions	Negligible
Gate valves, open	7
Globe valves, open	300
Angle valves, open	170
Water meters, disk	400
Water meters, piston	600
Water meters, impulse wheel	300

Table 12-1 *Continued*

Sharp-edged orifice	$F_p = -\Delta p_f$	$\dfrac{D_o}{D}$	$\dfrac{\Delta p_f(100)}{\Delta p \text{ across orifice}}$, %
		0.8	40
		0.7	52
		0.6	63
		0.5	73
		0.4	81
		0.3	89
		0.2	95
		D; D_o; Measured Δp across orifice	
Rounded orifice	$F_p = \dfrac{(V_o - V_2)^2}{2\alpha}$	The following values for α may be used in design calculations: turbulent flow, $\alpha = 1$; streamline flow, $\alpha = 0.5$ (V_o, V_2)	
Venturi	$F_p = -\Delta p_f$	$-\Delta p_f = \frac{1}{8}$ to $\frac{1}{10}$ of total pressure drop from upstream section to venturi throat	

flow element is designated in terms of the conductance C_c. Conductance has dimensions of volume per time and is defined as

$$C_c = \frac{Q'}{\Delta p} \tag{12-10}$$

For a vacuum pump with a pumping speed S_p and evacuating a gas from a vacuum vessel through a connecting pipe with a conductance C_{cp}, the pumping speed at the vessel is

$$S = \frac{S_p C_{cp}}{S_p + C_{cp}} \tag{12-11}$$

Frictional Losses Encountered in Pipelines

If the cross-sectional area of a pipe changes gradually to a new cross-sectional area, the disturbances to the flow pattern can be so small that the amount of mechanical energy lost as friction due to the change in cross section is negligible. However, if the change is sudden, an appreciable amount of mechanical energy can be lost as friction. Similarly, the presence of bends, fittings, valves, orifices, or other installations that disturb the flow pattern can cause frictional losses. All these effects must be included in the friction term appearing in the total mechanical energy balance. Recommended expressions for evaluating the important types of frictional losses are presented in Table 12-1.

Power Requirements for Transport of Liquids and Gases

For noncompressible fluids, Eq. (12-2) can be reduced to

$$W_o = g\,\Delta Z + \Delta\left(\frac{V^2}{2\alpha}\right) + \Delta(pv) + \sum F \qquad (12\text{-}12)$$

Since the individual terms in Eq. (12-12) can be evaluated directly from the physical properties of the system and the associated flow conditions, the design engineer can apply this equation to many liquid-flow systems without making any major assumptions. Example 12-1 illustrates the application of Eq. (12-12) to determine the power requirements for the given pumping operation.

EXAMPLE 12-1 Application of Mechanical Energy Balance to Noncompressible Flow Systems

Water is pumped from a large reservoir into an open tank using a standard steel pipe with an inside diameter of 0.0525 m. The reservoir and the tank are open to the atmosphere. The difference in vertical elevation between the water surface in the reservoir and the discharge point at the top of the tank is 21.3 m. The length of the pipeline is 305 m. Two gate valves and three standard 90° elbows are part of the piping system. The efficiency of the pump is 40 percent and includes the losses at the entrance and exit of the pump. If the flow rate of the water is to be maintained at 3.15×10^{-3} m^3/s and the water temperature remains constant at 16°C, estimate the power rating of the motor required to drive the pump.

■ Solution

Use the mechanical energy balance between locations 1 (surface of the water in the reservoir) and 2 (outside at the discharge point). By rearranging Eq. (12-2), the mechanical work required is given by

$$W_o = g(Z_2 - Z_1) + \left(\frac{V_2^2}{2\alpha} - \frac{V_1^2}{2\alpha}\right) + (p_2 v_2 - p_1 v_1) + \sum F$$

The selection of points 1 and 2 results in $V_2^2/(2\alpha) = 0$, $V_1^2/(2\alpha) = 0$, and $p_2 = p_1$. Since water can be assumed to be noncompressible, $v_2 = v_1$, $p_2 v_2 = p_1 v_1$, and $Z_2 - Z_1 = 21.3$ m.

$$g(Z_2 - Z_1) = (21.3)(9.806) = 208.7 \text{ N·m/kg}$$

Now determine the friction factor.

$$\text{Average velocity in the pipe} = \frac{3.15 \times 10^{-3}}{(\pi/4)(0.0525)^2} = 1.455 \text{ m/s}$$

$$\mu_{H_2O}(16°\text{C}) = 1.12 \times 10^{-3} \text{ Pa·s} \qquad \rho_{H_2O}(16°\text{C}) = 997 \text{ kg/m}^3$$

$$\text{Re} = \frac{DV\rho}{\mu} = \frac{0.0525(1.455)(997)}{1.12 \times 10^{-3}} = 68{,}000$$

$$\frac{\varepsilon}{D} = \frac{4.57 \times 10^{-5}}{0.0525} = 0.00087$$

The friction factor from Fig. 12-1 is estimated as 0.0057. The total L_e for fittings and valves is

$$L_e = 2(7)(0.0525) + 3(32)(0.0525) = 5.8 \text{ m}$$

Friction due to flow through pipe and all fittings is

$$F = \frac{2\mathfrak{f}V^2(L + L_e)}{D}$$

$$= \frac{2(0.0057)(1.455)^2(305 + 5.8)}{0.0525} = 143.4 \text{ N}\cdot\text{m/kg}$$

Friction due to contraction and enlargement (from Table 12-1) is

$$F = \frac{0.5(1.455)^2}{2(1)} + \frac{(1.455 - 0)^2}{2(1)} = 1.6 \text{ N}\cdot\text{m/kg}$$

$$\sum F = 143.4 + 1.6 = 145.0 \text{ N}\cdot\text{m/kg}$$

The theoretical mechanical energy required from the pump is $208.7 + 145.0 = 353.7$ N·m/kg.

$$P = \frac{353.7(3.15 \times 10^{-3})(997)}{0.4} = 2780 \text{ W} \simeq 3 \text{ kW}$$

A software package designed to establish pumping requirements indicates that a 3.5-kW motor would be needed. Assuming 85 percent efficiency for the motor, the software value for W_o is also $\simeq$3 kW. ■

The use of the mechanical energy balance is not recommended for compressible fluids when large pressure drops in the flow are involved. In such cases, the total energy balance should be used if the required data are available. A rearrangement of the total energy balance given in Eq. (12-1) for the work involved in the fluid system is given by

$$W = g\,\Delta Z + \Delta h + \Delta\left(\frac{V^2}{2\alpha}\right) - Q \tag{12-13}$$

where h is the enthalpy and is equal to $u + pv$. When Eq. (12-13) is applied in design calculations, information must be available for determining the change in enthalpy over the range of temperature and pressure involved. Example 12-2 shows how Eq. (12-13) can be used to calculate the pumping requirement when a compressible fluid is involved.

EXAMPLE 12-2 Application of the Total Energy Balance to a Compressible Flow System

Nitrogen gas is flowing under turbulent conditions at a constant rate through a long, straight, and horizontal pipe with an inside diameter of 0.0525 m. At the upstream location (point 1), the temperature of the nitrogen gas is 21°C, the pressure is 103 kPa, and the linear velocity of the gas is 15 m/s. At the downstream location (point 2), the temperature of the nitrogen gas is 60°C and the pressure is 345 kPa. An external heater is located between points 1 and 2, and 23.2 kJ is transferred from the heater to each kilogram of the flowing gas. No other heat sources or losses occur between the gas and the surroundings. Estimate the work input to the gas in kJ/kg supplied by the compressor located between points 1 and 2.

■ Solution

At the operating conditions specified, nitrogen can be considered an ideal gas with a mean heat capacity of 1.044 kJ/(kg·°C). The total energy balance between points 1 and 2 for the horizontal system is

$$W = (h_2 - h_1) + \left(\frac{V_2^2}{2} - \frac{V_1^2}{2}\right) - Q$$

where $V_1 = 15$ m/s and $V_2 = 15(333)(103)/(294)(345) = 5.07$ m/s. For an ideal gas, the enthalpy change of the nitrogen is given by

$$h_2 - h_1 = C_p(T_2 - T_1) = (1.044)(60 - 21) = 40.7 \text{ kJ/kg}$$

$$Q = 23.2 \text{ kJ/kg}$$

$$W = 40.7 + \left[\frac{(5.07)^2}{2} - \frac{(15)^2}{2}\right]10^{-3} - 23.2 = 17.4 \text{ kJ/kg}$$

When the data necessary for application of the total energy balance are not available, the engineer may be forced to use the mechanical energy balance for design calculations, even though compressible fluids and large pressure drops are involved. Example 12-3 illustrates the general method for applying the total mechanical energy balance under these conditions.[†]

EXAMPLE 12-3 Application of the Mechanical Energy Balance to a Compressible Flow System with a Large Pressure Drop

Air at a flow rate of 0.115 kg/s is transferred through a straight, horizontal steel pipe with an inside diameter of 0.0525 m and a length of 1000 m. A compressor is located at the upstream end of the pipe with the air entering the compressor through a 0.0525-m pipe. The pressure at the downstream end of the pipe is 135 kPa while the temperature is 21°C. If the air pressure in the pipe at the entrance to the compressor is 170 kPa and the temperature is 27°C, determine (1) the pressure in the pipe as the air

[†]For additional discussion and methods for integrating the total mechanical energy balance when the flow of a compressible fluid and high pressure drop are involved, see *Perry's Chemical Engineers' Handbook,* 7th ed., R. H. Perry and D. W. Green, eds., McGraw-Hill, New York, 1997.

leaves the compressor and (2) the mechanical energy added to the air by the compressor in N·m/kg, assuming that the compression operation is isothermal.

■ Solution

Since the heat exchanged between the surroundings and the system is unavailable, the total energy balance cannot be used to solve this problem. However, an approximate solution can be obtained by using the total mechanical energy balance.

Designate point 1 as the entrance to the compressor, point 2 as the exit from the compressor, and point 3 as the downstream end of the pipe. Under these conditions, the total mechanical energy balance for the system between points 2 and 3, written in differential form, is given by

$$\int_2^3 v\,dp + \int_2^3 \frac{V\,dV}{\alpha} = -\int_2^3 \frac{2\mathfrak{f}V^2\,dL}{D} \tag{1}$$

Since the mass velocity G in kg/(m^2·s) is constant,

$$V = Gv$$

$$dV = G\,dv$$

Eliminating V and dV from Eq. (1) and dividing by v^2 give

$$-\int_2^3 \frac{dp}{v} = \int_2^3 \frac{G^2\,dv}{\alpha v} + \int_2^3 \frac{2\mathfrak{f}G^2\,dL}{D} \tag{2}$$

Assume that air will behave as an ideal gas at the pressures involved. Then

$$v = \frac{RT}{Mp} \tag{3}$$

where R is the ideal gas constant, M the molecular weight, and T the absolute temperature. Substituting Eq. (3) into Eq. (2) and integrating give

$$\frac{M\left(p_2^2 - p_1^2\right)}{2RT_{avg}} = \frac{G^2}{\alpha}\ln\frac{v_3}{v_2} + \frac{2\mathfrak{f}_{avg}G^2L}{D} \tag{4}$$

The T_{avg} term represents the average absolute temperature between points 2 and 3, and temperature variations up to 20 percent from the average absolute value will introduce only a small error in the final result. The error introduced by using a constant $\mathfrak{f}_{avg}$ (based on average temperature and pressure) instead of the exact integrated value is not important unless pressure variations are considerably greater than those involved in this problem.

If the pump operation is isothermal, $T_2 = 27°\text{C}$ and $T_{avg} = (21 + 27)/2 + 273$, or 297 K. (If the pump operation were adiabatic, a different value for T_{avg} would have been necessary.)

$$\mu(297\text{ K}) = 0.018\text{ cP} = 1.8 \times 10^{-5}\text{ Pa·s}$$

$$G = \frac{0.115/0.0525^2}{\pi/4} = 53.15\text{ kg/(m}^2\text{·s)}$$

$$\text{Re} = \frac{DG}{\mu} = \frac{0.0525(53.15)}{1.8 \times 10^{-5}} = 155{,}000$$

$$\alpha = 1 \qquad \frac{\varepsilon}{D} = \frac{4.57 \times 10^{-5}}{0.0525} = 0.00087$$

From Fig. 12-1, $\mathfrak{f}_{avg} = 0.0052$.

Since air is assumed to behave as an ideal gas,

$$\frac{v_3}{v_2} = \frac{T_3 p_2}{T_2 p_3} = \frac{294 p_2}{300(135)} = \frac{p_2}{137.7}$$

Substituting values in Eq. (4) gives

$$\frac{29 \times 10^6}{2(8.31 \times 10^3)(297)}\left[p_2^2 - (135)^2\right] = (53.15)^2 \ln\frac{p_2}{137.7} + \frac{2(0.0052)(53.15)^2(1000)}{0.0525}$$

With the aid of Mathcad, the pressure of the air as it leaves the pump is

$$p_2 = 337 \text{ kPa}$$

The mechanical energy added by the pump can be determined by making a total mechanical energy balance between points 1 and 3.

$$\int_1^3 v\,dp + \int_1^3 \frac{V\,dV}{\alpha} = \int_1^3 dW_o - \int_1^3 \delta F \tag{5}$$

The friction term, by definition, includes all the friction except that occurring in the pump. Thus,

$$\int_1^3 \delta F = \int_2^3 \delta F$$

and

$$\int_1^2 v\,dp + \int_2^3 v\,dp + \int_1^2 \frac{V\,dV}{\alpha} + \int_2^3 \frac{V\,dV}{\alpha} = W_o - \int_2^3 \delta F \tag{6}$$

Subtracting Eq. (1) from Eq. (6) gives

$$\int_1^2 v\,dp + \int_1^2 \frac{V\,dV}{\alpha} = W_o$$

The value of $\int_1^2 v\,dp$ depends on the conditions or path followed in the pump, and the integral can be evaluated if the necessary p–v relationships are known. Although many pumps and compressors operate near adiabatic conditions, the pump operation will be assumed to be isothermal in this example.

For an ideal gas and isothermal compression,

$$\int_1^2 v\,dp = \frac{RT}{M}\ln\frac{p_2}{p_1}$$

$$V_1 = Gv_1 = \frac{53.15(22.4)(300)(101.3)}{29(273)(170)} = 26.88 \text{ m/s}$$

$$V_2 = \frac{V_1 p_1}{p_2} = \frac{26.88(170)}{337} = 13.56 \text{ m/s}$$

$$W_o = \left(\frac{RT}{M}\right)\ln\frac{p_2}{p_1} + \frac{V_2^2}{2} - \frac{V_1^2}{2}$$

$$= \left[\frac{(8.31 \times 10^3)(300)}{29}\right]\ln\frac{337}{170} + \frac{(13.56)^2}{2} - \frac{(26.88)^2}{2}$$

$$= 58{,}600 \text{ N}\cdot\text{m/kg}$$

Mechanical energy added to the air by the compressor, assuming isothermal operation, is

$$P = (58{,}600)(0.115) = 6740 \text{ N}\cdot\text{m/s} = 6740 \text{ W}$$

The total power supplied to the compressor can be determined if the isothermal efficiency of the compressor is available.

PIPING IN FLUID TRANSPORT PROCESSES

The American National Standards Institute (ANSI) and the American Petroleum Institute (API) have established detailed standards for the most widely used components of piping systems. Lists of these standards as well as specifications for pipe and fitting materials can be found in the ANSI B31 code sections. Many of these standards contain pressure-temperature ratings that will be of assistance to engineers in their design function. Even though safety is the basic consideration of the code, design engineers are cautioned that the code is not a design handbook and does not eliminate the need for engineering judgment.

Some of the specific requirements for pumping systems have been included in the Occupational Safety and Health Administration (OSHA) requirements. A few of the more significant requirements of ANSI B31.3 have been summarized and are included in this section since they relate directly to the minimum requirements for the selection and design of piping systems.

Selection of Piping Materials

General aspects that need to be evaluated when selecting piping materials are (1) possible exposure to fire with respect to the loss in strength or combustibility of the pipe and supports; (2) susceptibility of the pipe to brittle failure or thermal shock failure when exposed to fire; (3) ability of thermal insulation to protect the pipe from fire; (4) susceptibility of the pipe and joints to corrosion or adverse electrolytic effect; (5) suitability of packing, seals, gaskets, and lubricants used on joints and connections; (6) refrigeration effect during sudden loss of pressure with volatile fluids; and (7) compatibility with the fluid handled.

Specific material precautions and/or alternatives that need to be considered when selecting piping materials are presented in Table 12-2. Iron and steel pipes are specified according to wall thickness by a standard formula for schedule number as designated by the American Standards Association (ASA). The schedule number is defined

Table 12-2 Specific material precautions for piping systems[†]

Metal piping materials	Possible material precautions and alternatives
Metal	
Iron (cast, malleable, and high-silicon)	Lack of ductility and sensitivity to thermal and mechanical shock.
Carbon steel and several low-alloy steels	Embrittlement when handling alkaline or strong caustic fluids; conversion of carbides to graphite during long-time exposure at temperature above 427°C; hydrogen damage when exposed to hydrogen or aqueous acid solutions; deterioration when exposed to hydrogen sulfide. (Consider silicon-killed carbon steel for temperatures above 480°C.)
High-alloy stainless steels	Stress corrosion cracking of austenitic stainless steels exposed to chlorides and other halides; intergranular corrosion of austenitic stainless steel after long-term exposure to temperatures between 427 and 871°C unless stabilized or low-carbon grades are used; intercrystalline attack of austenitic stainless steels on contact with molten lead and zinc; brittleness of ferritic stainless steels at room temperature after use at temperatures above 315°C.
Nickel and nickel-base alloys	Grain boundary attack of both materials not containing chromium when exposed to sulfur at temperatures above 315°C; grain boundary attack of other materials containing chromium at temperatures above 595°C under reducing conditions and above 760°C under oxidizing conditions; stress-corrosion cracking of 70Ni-30Cu alloy in HF vapor.
Aluminum and aluminum alloys	Corrosion from concrete, mortar, lime, plaster, and other alkaline materials; intergranular attack of alloys 5154, 5087, 5083, and 5456 when exposed to temperatures above 65°C.
Copper and copper alloys	Loss of zinc from brass alloys; stress-corrosion cracking of copper-based alloys; unstable acetylide formation when exposed to acetylene.
Titanium and titanium alloys	Deterioration of titanium and its alloys above 315°C.
Zirconium and zirconium alloys	Deterioration of zirconium and its alloys above 315°C.
Nonmetal piping materials	
Thermoplastics	Minimum and maximum operating temperatures limits between −34 and 210°C for thermoplastic pipe materials and −198 and 260°C for thermoplastics used in linings;[‡] special precautions when used for transporting compressed air or other gases.
Reinforced thermosetting resins	Minimum and maximum operating temperature limits between 29 and 93°C.[‡]
Borosilicate glass and impregnated graphite	Lack of ductility and sensitivity to thermal and mechanical shock.

[†]Modified from data presented by M. P. Boyce, in *Perry's Chemical Engineers' Handbook,* 7th ed., R. H. Perry and D. W. Green, eds., McGraw-Hill, New York, 1997, p. 10-68, as extracted from the *Chemical Plant and Petroleum Refinery Piping Code,* ANSI B31.3.

[‡]Recommended temperature limits for specific thermoplastics and reinforced thermosetting resins are given in Tables 10-15, 10-16, and 10-17 of *Perry's Chemical Engineers' Handbook,* 7th ed., R. H. Perry and D. W. Green, eds., McGraw-Hill, New York, 1997.

by the ASA as the approximate value of

$$\frac{1000\,p_s}{S_s} = \text{schedule number} \qquad (12\text{-}14)$$

where S_s is the safe working stress and p_s the safe working pressure, defined by

$$p_s = \frac{2S_s t_m}{D_m} \qquad (12\text{-}14a)$$

Here t is the minimum wall thickness in m, D_m the mean diameter in m, and p_s and S_s in kPa. For temperatures up to 120°C, the recommended safe working stress is 62,000 kPa for lap-welded steel pipe and 49,000 kPa for butt-welded steel pipe.† Thus, if the schedule number is known, the safe working pressure can be estimated directly from Eq. (12-14). Ten schedule numbers are in use at present.

Pipe sizes are based on the approximate diameter and are reported as nominal pipe sizes. Although the wall thickness varies depending on the schedule number, the outside diameter of any pipe having a given nominal size is constant and independent of the schedule number. This permits the use of standard fittings and threading tools on pipes of different schedule numbers. A table showing outside diameters, inside diameters, and other dimensions for pipes of different diameters and schedule numbers is presented in Appendix D.

A number of the piping materials are available in the form of tubing. Although pipe specifications are based on standard nominal sizes, tubing specifications are based on the actual outside diameter with a designated wall thickness. Conventional systems, such as the Birmingham wire gauge (BWG), are used to indicate the wall thickness. Common designations of tubing dimensions are given in Appendix D.

Threaded fittings, flanges, valves, flowmeters, steam traps, and many other auxiliaries are used in piping systems to connect sections of pipe, change the directions of flow, or obtain desired conditions in a flow system. Flanges are usually employed for piping connections when the pipe diameter is 0.075 m or larger, while screwed fittings are commonly used for smaller sizes. In the case of cast-iron pipe used as underground water lines, bell-and-spigot joints are ordinarily employed rather than flanges.

The auxiliaries in piping systems must have sufficient structural strength to resist the pressure or other strains encountered in the operation, and the design engineer should provide a wide safety margin when specifying the ratings of these auxiliaries. Fittings, valves, steam traps, and similar items are often rated on the basis of safe operating pressure.

Design of Piping Systems

Various items need to be considered by the design engineer when developing the design for a piping system. The overriding factor in all these design considerations is the provision of protective measures as required to ensure the safe operation of the

†For allowable stresses at other temperatures and for other piping materials, see M. P. Boyce, in *Perry's Chemical Engineers' Handbook,* 7th ed., R. H. Perry and D. W. Green, eds., McGraw-Hill, New York, 1997, Table 10-49.

proposed piping system. General aspects to be evaluated should include (1) the hazardous properties of the fluid, (2) the quantity of fluid that could be released by a piping failure, (3) the effect of a failure on overall plant safety, (4) evaluation of the effects of a reaction of the fluid with the environment, (5) the probable extent of human exposure to all aspects of the piping failure, and (6) the inherent safety of the piping system by virtue of materials of construction, methods of fabrication, and history of service reliability.

These safety considerations must also be exercised in such design items as (1) selecting piping materials and pipe sizes, (2) checking effects of temperature level and temperature changes on thermal expansion, freezing, and insulation requirements, (3) ensuring flexibility in the piping system to withstand physical and thermal shocks, (4) establishing adequate support structures for the system, and (5) providing a system configuration that is easy to install, inspect, and maintain. This section relates to several of these design items.

Dynamic Effects The pipe stresses resulting from thermal expansion or contraction must be considered in any piping system design. For example, if the temperature of the pipe changes from 10 to 315°C, the increase in length for each 100 m of pipe will be 0.4 m for steel pipe and 0.6 m for brass pipe. This amount of thermal expansion could easily cause a pipe or wall to buckle if the pipe were fastened firmly at each end with no allowances for expansion. The necessary flexibility for the piping system can be provided by the use of expansion loops, bellows, slip joints, and other flexible expansion devices.

The piping design must provide for possible impact from the effect of high winds, earthquakes, discharge reactions, and vibrations from piping arrangement and support. For example, water hammer in the piping system may cause extreme stresses at bends in pipelines. In steam lines, this effect can be minimized by eliminating liquid pockets in the line through the use of steam traps and sloping of the line in the direction of flow. Quick-opening or quick-closing valves may cause water hammer, and valves of this type may require protection by the use of expansion or surge chambers.

Ambient Effects If cooling of the fluid in the system results in a vacuum, the design must provide for the additional pressure difference experienced by the system, or a vacuum breaker may need to be installed. Provision must be made for thermal expansion of fluid trapped between or in closed valves. Nonmetallic or nonmetallic-lined pipe may require protection when the ambient temperature exceeds the design temperature.

The possibility of solidification of the fluid should not be overlooked in the design of a piping system. Insulation, steam tracing, and sloping the line to drain the valves are methods for handling this type of problem.

Pipe Sizing Specification of the pipe diameter to be used in a given piping system will depend on the economic factors involved. Theoretically, the optimum pipe diameter recommended is the one that provides the minimum total cost of both the annual pumping power and the fixed charges for a particular piping system. Some general

Table 12-3 Recommended economic velocities for sizing steel pipes

Turbulent flow	
Type of fluid	**Recommended velocity range, m/s†**
Water or fluid similar to water	1–3
Low-pressure steam (250 kPa)	15–30
High-pressure steam (750 kPa)	30–60
Air (250–500 kPa)	15–30

Viscous flow			
	Recommended velocity range, m/s†		
Nominal pipe diameter, m (in.)	$\mu = 50$ **cP**	$\mu = 100$ **cP**	$\mu = 1000$ **cP**
0.0254 (1)	0.5–1	0.3–0.6	0.1–0.2
0.0508 (2)	0.75–1.07	0.5–0.75	0.15–0.24
0.102 (4)	1.07–1.5	0.75–1.07	0.24–0.36
0.203 (8)	—	1.2–1.5	0.4–0.55

†These values apply for motor drives. Multiply indicated velocities by 0.6 to give recommended velocities when steam turbine drives are used.

recommendations for use in design estimates of pipe diameters are presented in Table 12-3.

The derivation of equations for determining optimum economic pipe diameters is presented in Chap. 9. The following simplified equations can be used for making these design estimates:

For turbulent flow (Re $>$ 2100) and $D_i \geq 0.0254$ m,

$$D_{i,\text{opt}} = 0.363 \dot{m}_v^{0.45} \rho^{0.13} \tag{12-15}$$

For viscous flow (Re $<$ 2100) and $D_i \leq 0.0254$ m,

$$D_{i,\text{opt}} = 0.133 \dot{m}_v^{0.40} \mu_f^{0.20} \tag{12-16}$$

where $D_{i,\text{opt}}$ is the optimum pipe diameter in m, $\dot{m}_v$ the volumetric flow rate in m^3/s, ρ the fluid density in kg/m^3 and μ_f the fluid viscosity in Pa·s.

The preceding equations are the basis for the nomograph presented in Fig. 12-3 that can be used for estimating the optimum diameter of steel pipe under normal process conditions. Equations (12-15) and (12-16) should not be used when the flowing fluid is steam because the derivation of these equations makes no allowance for the effects of pressure on the condition of the fluid. Equation (12-15) is limited to conditions in which the viscosity of the fluid is between 0.2 and 20 cP.

The constants in Eqs. (12-15) and (12-16) are based on average cost and operating conditions. When unusual conditions exist or when a more accurate determination of the optimum diameter is desired, other relationships given in the prior reference should be used.

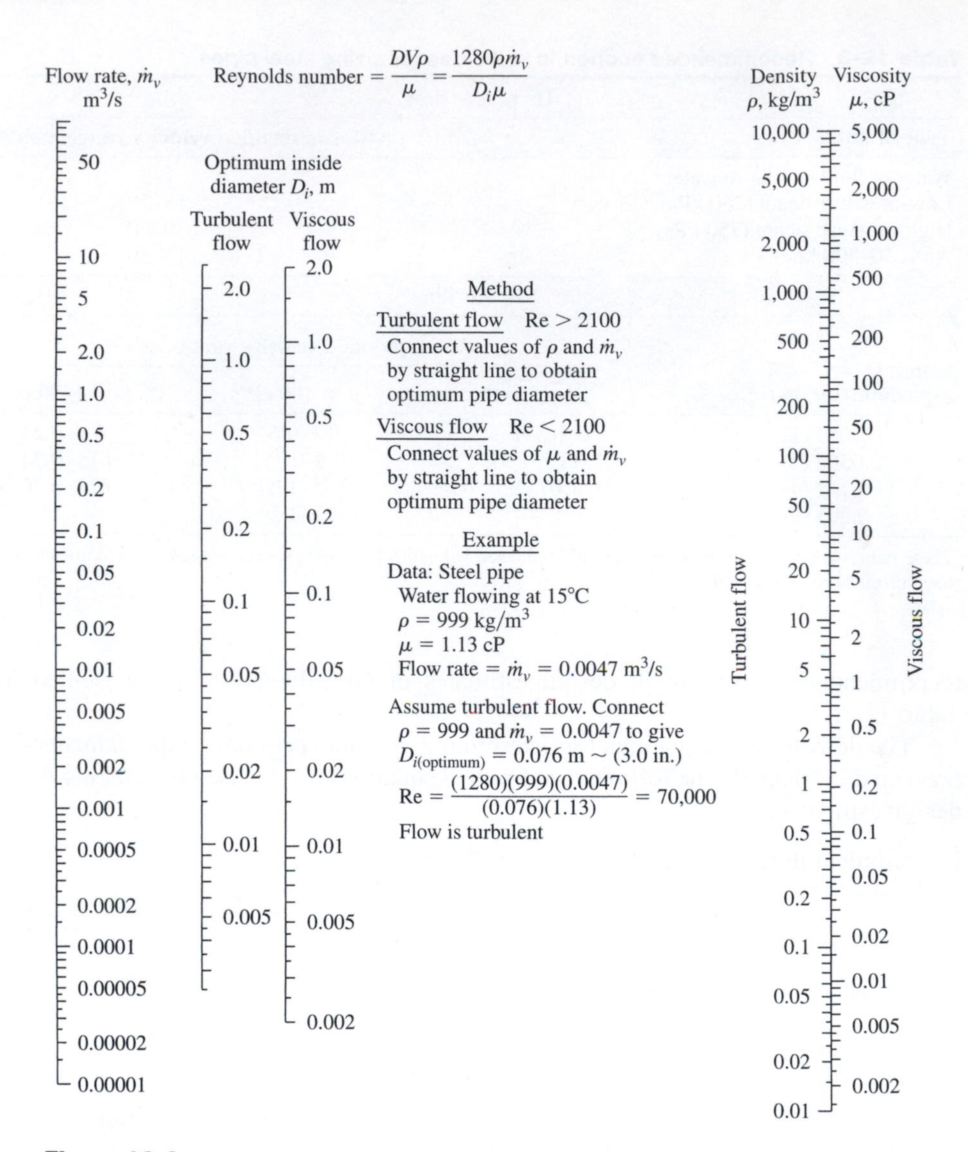

Figure 12-3
Nomograph for estimation of optimum economic pipe diameters with turbulent or viscous flow based on Eqs. (12-15) and (12-16)

Costs for Piping and Piping System Auxiliaries

Piping is a major item in the cost of all types of process plants. These costs in a fluid-process plant can run as high as 80 percent of the purchased equipment cost or 25 percent of the fixed-capital investment. There are essentially two basic methods for preparing piping cost estimates: the percentage of installed equipment method and the material and labor takeoff method. Several variations of each method have appeared in the literature.

The percentage of installed equipment method, as described in Chap. 6, is a quick procedure for preliminary or order-of-magnitude type of cost estimates. In the hands of experienced estimators it can be a reasonably accurate method, particularly on repetitive-type units. It is not recommended on alteration jobs or on projects where the total installed equipment is less than $150,000.

The material and labor takeoff method is the recommended method for definitive estimates where accuracy within 10 percent is required. To prepare a cost estimate by this method usually requires piping drawings and specifications, material costs, fabrication and erection labor costs, testing costs, auxiliaries, supports, and painting requirements. The takeoff from the drawings must be made with the greatest possible accuracy because it is the basis for determining material and labor costs. In the case of revisions to existing facilities, thorough field study is necessary to determine job conditions and their possible effects.

Although accurate costs for pipes, valves, and piping system auxiliaries can be obtained only by direct quotations from manufacturers, the design engineer can often make satisfactory estimates from data such as those presented in Figs. 12-4 through 12-6 for various types of pipes, Figs. 12-7 through 12-11 for different types of valves, and Fig. 12-12 for insulation and painting costs. The cost of materials presented in these figures covers the types of equipment most commonly encountered in industrial operations.

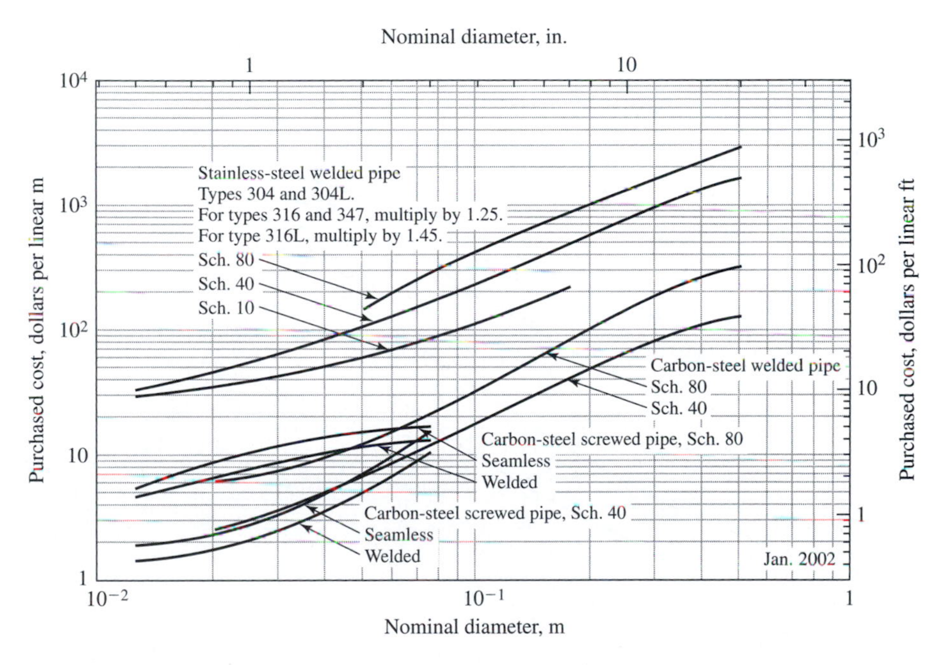

Figure 12-4
Purchased cost of welded and screwed pipe per unit length

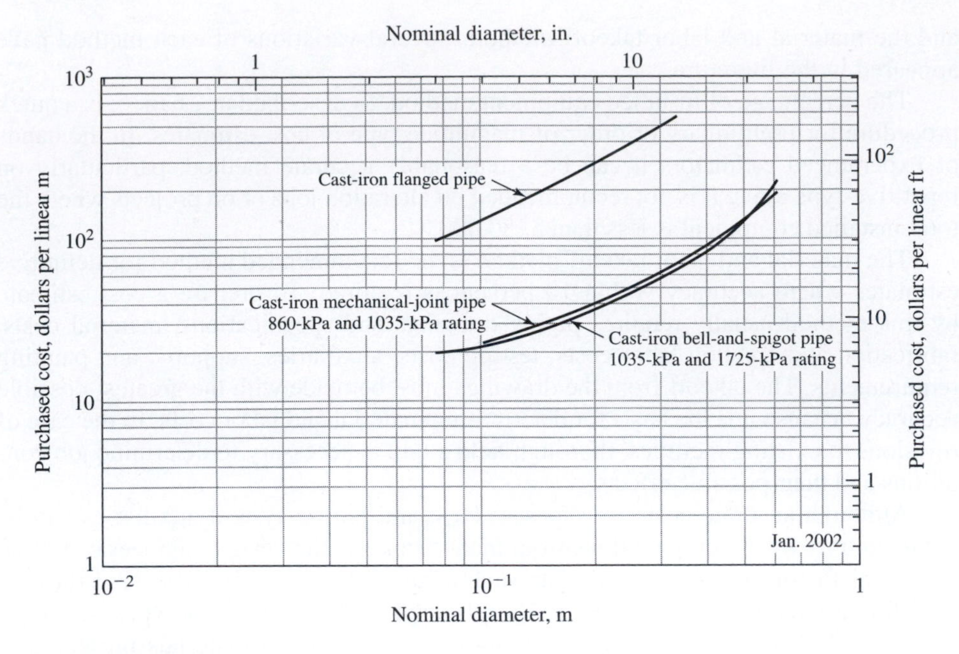

Figure 12-5
Purchased cost of cast-iron pipe per unit length

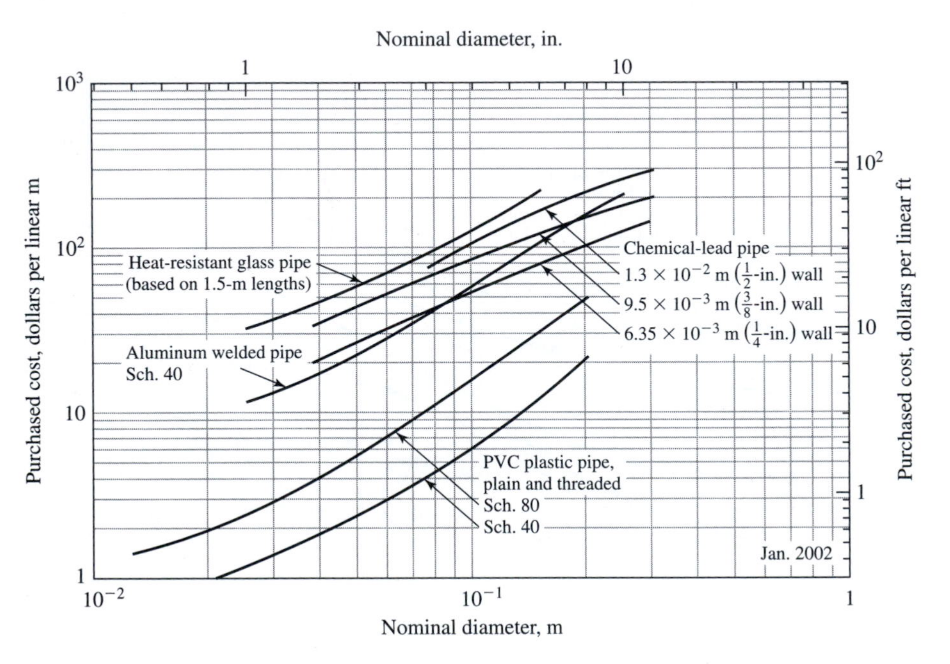

Figure 12-6
Purchased cost of specialty pipe per unit length

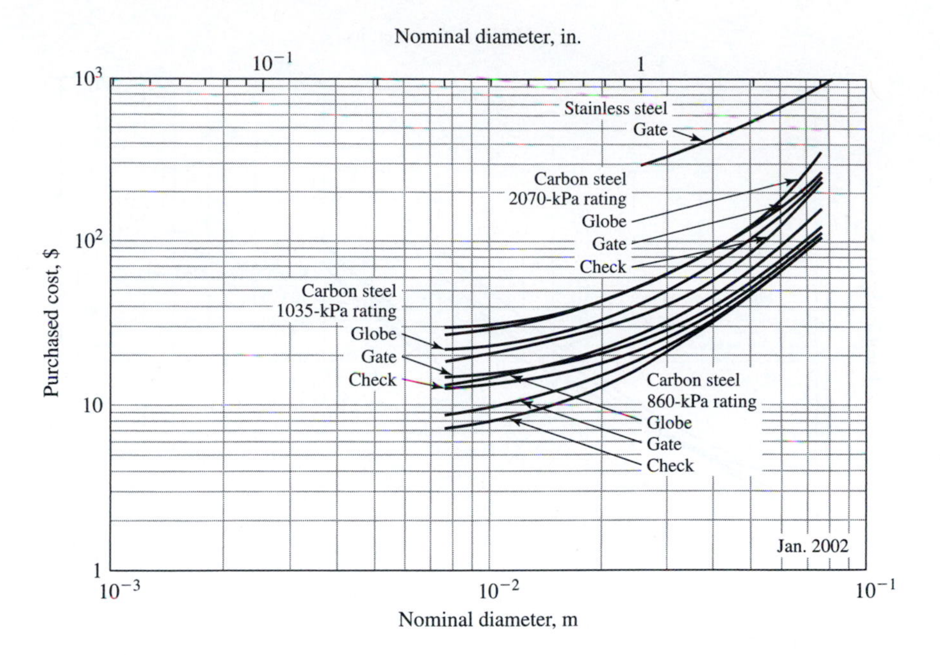

Figure 12-7
Purchased cost of screwed valves for water, oil, and gas

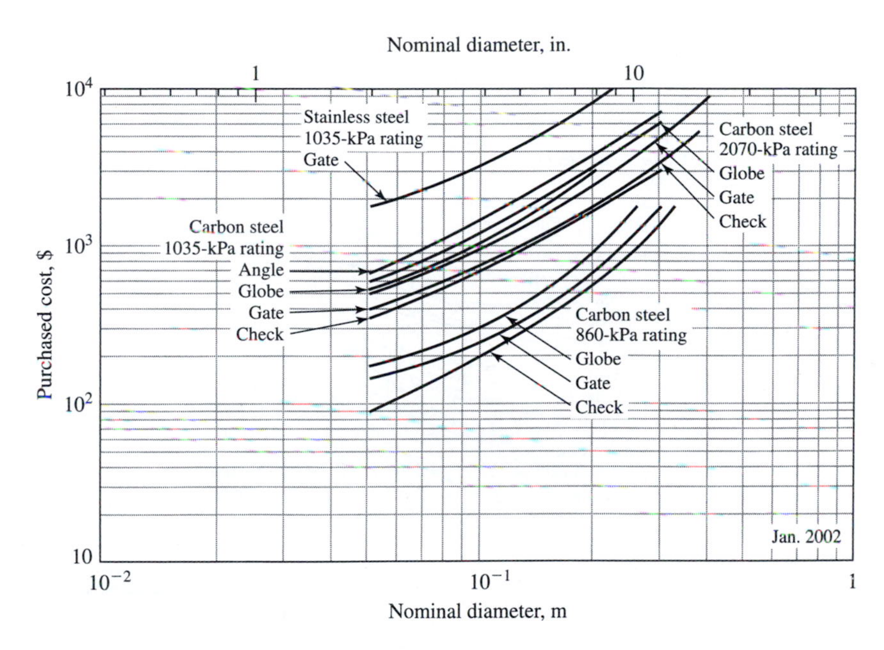

Figure 12-8
Purchased cost of flanged valves for water, oil, and gas

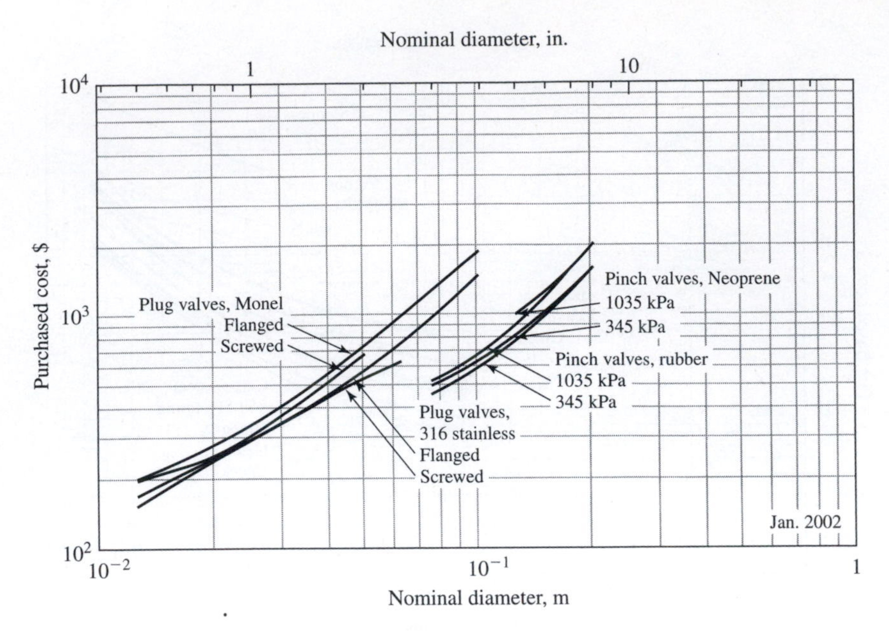

Figure 12-9
Purchased cost of plug and pinch valves

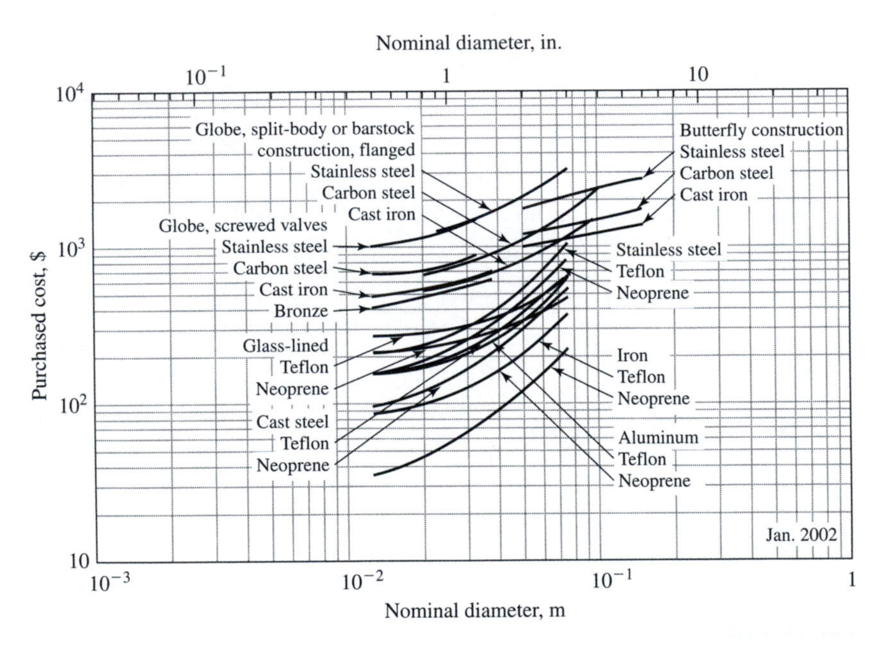

Figure 12-10
Purchased cost of diaphragm valves

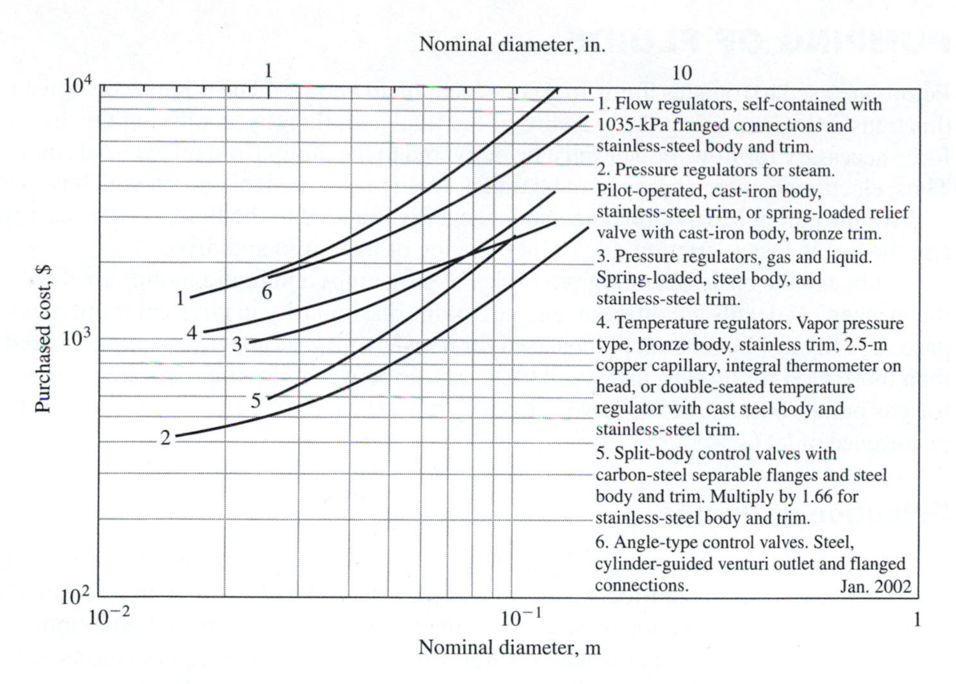

Figure 12-11
Purchased cost for control and relief valves

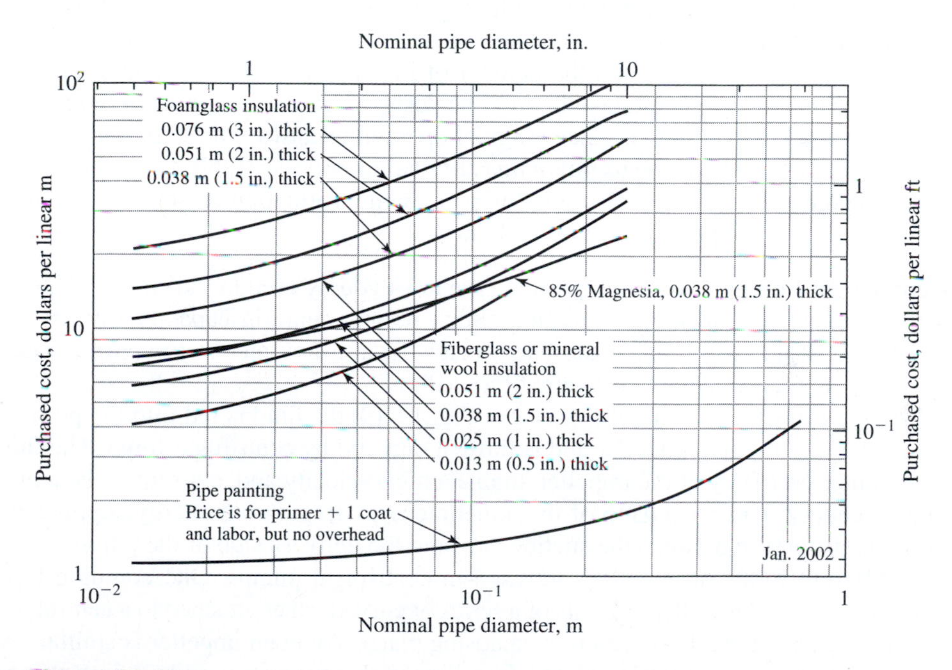

Figure 12-12
Purchased cost of pipe insulation and pipe painting per unit length. Insulation price includes cost of standard covering

PUMPING OF FLUIDS

Pumps are used to transfer fluids from one location to another. The pump accomplishes this transfer by increasing the pressure of the fluid and, thereby, supplying the driving force necessary for flow. Power must be delivered to the pump from an external source. Thus, electric or steam energy may be transformed to mechanical energy which is used to drive the pump. Most of this mechanical energy is added to the fluid as work energy, and the rest is lost as friction due to inefficiency of the pump and drive.

Although the basic operating principles of gas pumps and liquid pumps are similar, the mechanical details are different because of the dissimilarity in physical properties of gases and liquids. In general, pumps used for transferring gases operate at higher speeds than those used for liquids, and smaller clearances between moving parts are necessary for gas pumps because of the lower viscosity of gases and the greater tendency for the occurrence of leaks.

Selection of Pumps

Selection of a pump for a specific service requires knowledge of the liquid to be handled, the total dynamic head required, the suction and discharge heads, and, in most cases, the temperature, viscosity, vapor pressure, and density of the fluid. Special attention will need to be given to those cases where solids are contained in the liquid. Additionally, liquid corrosion characteristics will require the use of special materials.

The different types of pumps commonly employed in industrial operations can be classified as (1) centrifugal pumps (including turbine and axial pumps), (2) positive displacement pumps, (3) jet pumps, and (4) electromagnetic pumps. Because of the wide variety of pump types and the number of factors which determine the selection of any one type for a specific installation, the design engineer will want to reduce this choice as quickly as possible. Since range of operation is always an important consideration, Fig. 12-13 should be useful in this elimination process. Table 12-4 can provide additional guidelines in this selection. Supplemental information on some of the more widely used pump types is summarized below.

Centrifugal Pumps This pump is the type most widely used in the chemical industry for transferring all kinds of liquids. Such pumps range in capacity from 0.5 to 2×10^4 m^3/h and can provide discharge heads from a few meters to approximately 4.9×10^3 m (equivalent to a pressure of 48 MPa).

In the centrifugal pump, illustrated in Fig. 12-14, the fluid is fed into the pump at the center of a rotating impeller and is thrown outward by centrifugal force. The fluid at the outer periphery of the impeller attains a high velocity and, consequently, a high kinetic energy. The conversion of this kinetic energy to pressure energy supplies the pressure difference between the suction side and the delivery side of the pump.

Different forms of impellers are used in centrifugal pumps. One common type, known as a closed impeller, consists of a series of curved vanes attached to a central hub which extends outward between two enclosing plates. An open impeller is similar, except that there are no enclosing plates. Impellers of this type are used in volute pumps, which are the simplest form of centrifugal pumps.

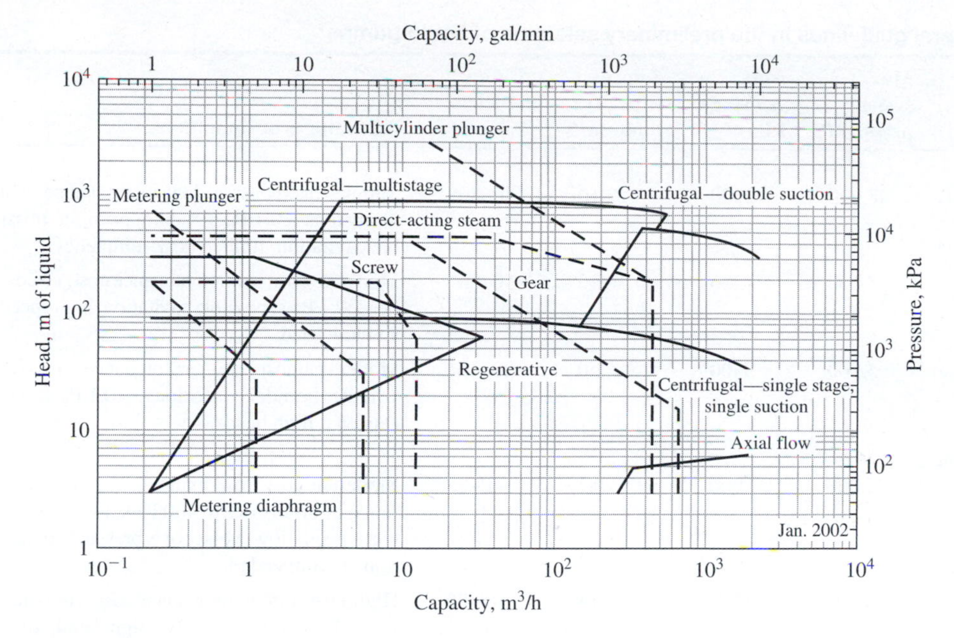

Figure 12-13
Normal operating ranges of commercially available pumps.
Solid lines: use left ordinate, head. Broken lines: use right ordinate, pressure.

For an ideal centrifugal pump, the speed of the impeller N_r in revolutions per minute should be directly proportional to the fluid discharge rate $\dot{m}_v$, or

$$\frac{N_{r,1}}{N_{r,2}} = \frac{\dot{m}_{v,1}}{\dot{m}_{v,2}} \tag{12-17}$$

The head, or pressure difference, produced by the pump is a function of the kinetic energy developed at the point of release from the impeller. The head developed by the pump is directly proportional to the square of the impeller speed. Since the power required for such a pump is directly proportional to the product of the head and the flow rate, the following relationship results:

$$\frac{\text{Power}_1}{\text{Power}_2} = \frac{\dot{m}_{v,1}^3}{\dot{m}_{v,2}^3} = \frac{N_{r_1}^3}{N_{r_2}^3} \tag{12-18}$$

The preceding equations apply for the ideal case in which there are no friction, leakage, or recirculation losses. In any real pump, however, these losses do occur, and their magnitude can be determined only by actual tests. As a result, characteristic curves are usually supplied by pump manufacturers to indicate the performance of any particular centrifugal pump, as shown in Fig. 12-15. The design engineer must recognize that for any fixed rotational speed the pump will operate along this curve and very inefficiently at any other operational point not on the curve. For example, at a flow of 54.5 m^3/h (240 gal/min) the pump can generate a 23.2-m (76-ft) head. If the head is increased to 26.5 m (87 ft), 45.4 m^3/h (200 gal/min) will be delivered. It is not possible to reduce the

Table 12-4 General guidelines in the preliminary selection of liquid pumps[†]

Type of pump	Max. system press., kPa	Max. Δp/stage, kPa	Approx. capacity limit, m^3/s	Pump efficiency range, %	Advantages or limitations
Centrifugal					
Centrifugal (radial)	48,000	2000	10	40–80*	Simple, inexpensive, low maintenance cost, viscosities <0.1 Pa·s, require priming, possible cavitation, limited peak efficiency.
Axial	35,000	200	10	50–85*	Moderate cost, low maintenance cost, viscosities <0.1 Pa·s, possible cavitation, high speed, and low head.
Regenerative (turbine)	5,000	3500	<1.0	20–40*	Moderate cost and maintenance, handles volatile liquids, viscosities <0.1 Pa·s, low capacity, high head.
Positive displacement					
Gear	35,000	20,000	0.1	40–85	Moderate cost, low maintenance cost, wide range of viscosities to 400 Pa·s, low capacity, high head, low noise, overpressure protection recommended.
Lobe	35,000	1700	0.1	40–85	High cost, low maintenance cost, viscosities <1.0 Pa·s, low capacity, high head, overpressure protection recommended.
Piston (circumferential)	35,000	1700	0.04	40–85	Moderate cost, low maintenance cost, wide range of viscosities to 400 Pa·s, low capacity, low noise, overpressure protection recommended.
Screw	20,000	2000	0.1	40–70	High cost, moderate maintenance cost, wide range of viscosities to 1000 Pa·s, low capacity, high head, and moderate noise.
Sliding vane	35,000	15,000	0.1	40–85	Moderate cost, low maintenance cost, wide range of viscosities to 100 Pa·s, low capacity, high head, no abrasive liquids.
Reciprocating					
Piston	100,000	15,000	0.03	50–90	High cost, high maintenance cost, very high head, low capacity, viscosities to 400 Pa·s, pulsating flow, fluid leakage, noisy.
Diaphragm	35,000	7,000	0.006	20–50	High cost, moderate maintenance cost, very low capacity, low viscosities, pulsating flow, no fluid leakage, failure potential.
Momentum transfer					
Jet	35,000	100	1.0	5–20	Low cost, low maintenance cost, low capacity, low head, low viscosities, contamination of fluid, can handle corrosive fluids.

[†]Modified from data presented by G. D. Ulrich, *A Guide to Chemical Engineering Process Design and Economics,* J. Wiley, New York, 1984, D. R. Woods, *Process Design and Engineering Practice,* Prentice-Hall, Englewood Cliffs, NJ, 1995, R. H. Perry and D. W. Green, *Perry's Chemical Engineers' Handbook,* 7th ed., McGraw-Hill, New York, 1997, S. M. Walas, *Chemical Process Equipment—Selection and Design,* Butterworth-Heinemann, Newton, MA, 1990, and M. S. Peters and K. D. Timmerhaus, *Plant Design and Economics for Chemical Engineers,* 4th ed., McGraw-Hill, New York, 1991.

*Independent of viscosity up to 0.05 Pa·s.

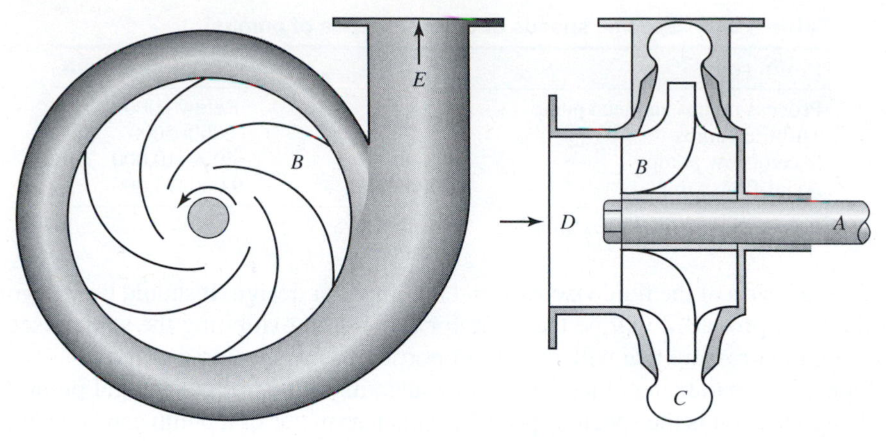

Figure 12-14
Simple centrifugal pump. A. shaft, B. impeller, C. stationary casing, D. suction pipe, and E. discharge pipe.

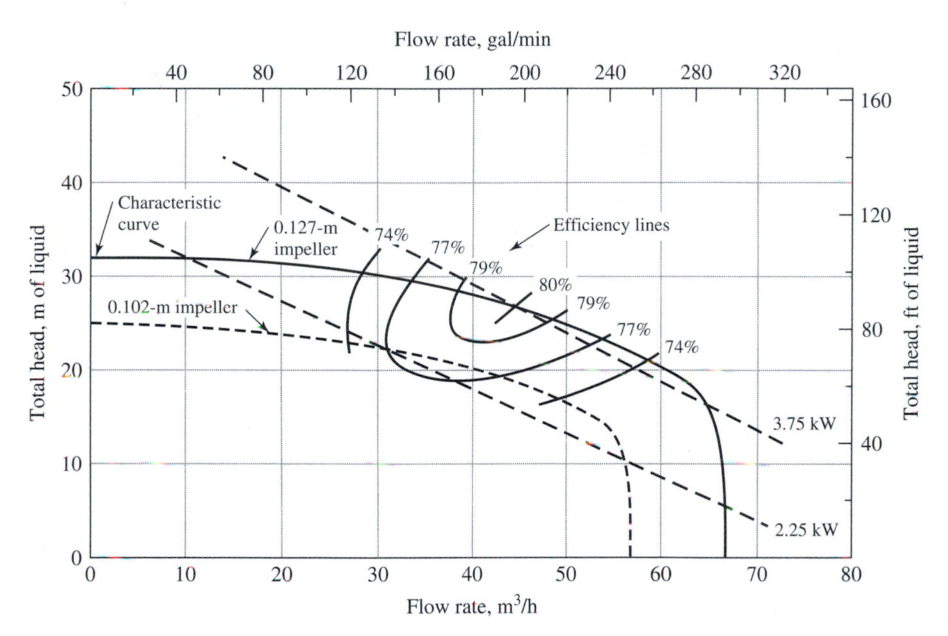

Figure 12-15
Characteristic curve of a centrifugal pump operating at a constant speed of 3450 r/min

capacity to 45.4 m^3/h (200 gal/min) for a head of 23.2 m (76 ft) unless the discharge is throttled so that a head of 26.5 m (87 ft) is actually generated in the pump. This action would result in a reduction in the efficiency of the pump. Note that the characteristic curve of a pump can be changed with the use of variable-speed drives such as steam turbines.

Note that the head developed in a pump depends on the velocity of the fluid, and this, in turn, depends upon the capability of the impeller to transfer energy to the fluid.

Table 12-5 Specific speeds of different types of pumps[†]

Pump type	Range of specific speeds
Process pumps and feed pumps	Below 2000
Turbine pumps	2000–5000
Mixed-flow pumps	4000–10,000
Axial-flow pumps	9,000–15,000

[†]*Perry's Chemical Engineers' Handbook,* 7th ed., p. 10-75.

This is a function of the fluid viscosity and the impeller design. It should be recognized that the head produced will be the same for any liquid exhibiting the same viscosity. However, the pressure rise will vary in proportion to the specific gravity.

One of the parameters that is useful in selecting a type of centrifugal pump for a certain application is the specific speed N_s. This parameter of a pump can be evaluated based on its design speed, rate of flow, and head generated from

$$N_s = \frac{286.4 N_r \dot{m}_v^{0.5}}{H^{0.75}} \tag{12-19}$$

where N_r is the revolutions per minute of the impeller, $\dot{m}_v$ the flow rate in m^3/s, and H the head in N·m/kg. Specific speed is a parameter that defines the speed at which impellers of geometrically similar design have to be run to discharge 6.31×10^{-5} m^3/s against a 0.3048-m head. Specific speeds of different centrifugal pumps are given in Table 12-5. In general, pumps with a low specific speed have a low flow capacity while pumps with a high specific speed have a high capacity.

Process pumps are typically single-stage, pedestal-mounted pumps with single-suction overhung impellers. These pumps are designed for ease in dismantling and accessibility. With appropriate material selection, these pumps can transfer corrosive or otherwise difficult to handle liquids. Most pump manufacturers provide both horizontal and vertical process pumps built to an ANSI standard. The horizontal pumps are available for capacities up to 900 m^3/h while the vertical in-line pumps have capacities up to 320 m^3/h. Both types of pumps can operate with heads up to 120 m.

Double-suction, single-stage pumps are used for general water supply, circulation service, and chemical service with noncorrosive liquids. These units are available for capacities from about 6 to more than 10,000 m^3/h and fluid heads up to 300 m.

Higher heads than can be generated by a single impeller require the use of multistage centrifugal pumps. In these units the impellers are in series, and the total head generated is the summation of the heads of the individual impellers. Pressures as high as 20 to 40 MPa can be achieved. Such units are used as deep-well pumps, high-pressure water supply pumps, boiler feed pumps, and fire pumps.

Axial-flow pumps, on the other hand, are essentially very high-capacity, low-head units. Often these pumps are designed for flows greater than 450 m^3/h with liquid heads of 10 m or less. Such pumps are widely used in closed-loop circulation systems in which the pump casing essentially becomes an elbow in the line.

Energy losses caused by turbulence at the point where the liquid path changes from radial flow to tangential flow in the pump casing can be decreased by using so-called turbine pumps. With this type of centrifugal pump, the liquid flows from the impeller through a series of fixed vanes forming a diffusion ring. The change in direction of the

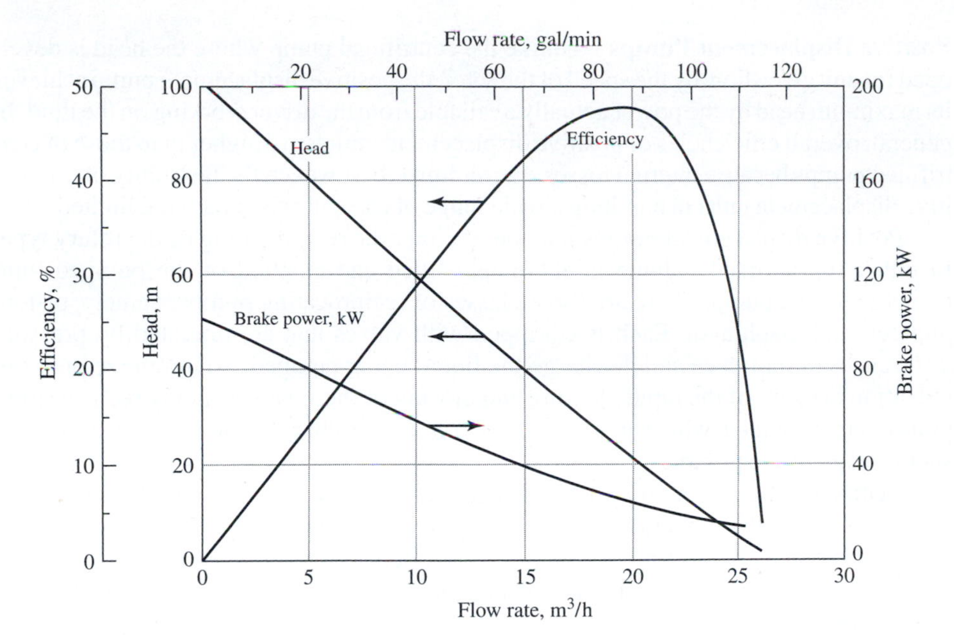

Figure 12-16
Characteristic curves of a regenerative pump

fluid is more gradual than in a volute pump, and a more efficient conversion of kinetic energy to pressure energy is obtained. Such units are available in capacities from 20 m^3/h and above with liquid heads up to 30 m per stage.

Regenerative pumps are also referred to as turbine pumps because of the shape of the impeller. Such units employ a combination of mechanical impulse and centrifugal force to produce heads greater than 100 m at capacities below 20 m^3/h. These units are very useful when small flow quantities of low-viscosity liquids must be transferred at higher pressures than are normally available with other types of centrifugal pumps. Figure 12-16 shows typical characteristic curves for a regenerative pump.

Still another centrifugal pump that needs to be mentioned is the canned-motor pump. Since the cavity containing the motor rotor and the pump casing are interconnected, the motor bearings operate in the process liquid and all seals are eliminated. Standard single-stage canned-motor pumps are available for flows up to 160 m^3/h and heads up to 75 m. These units are being used in installations handling toxic or hazardous liquids where leakage is an environmental and economic problem.

The major advantages of a centrifugal pump are simplicity, low initial cost, low maintenance expense, uniform (nonpulsating) flow, quiet operation, and adaptability to use either motor or turbine drivers. This type of pump also can handle liquids with large amounts of solids since there are no close metal-to-metal fits within the pump and there are no valves involved in the pump operation. On the other hand, centrifugal pumps generally cannot be operated at high heads and have efficiency problems with handling highly viscous fluids. The maximum efficiency for a given pump is only available over a fairly narrow range of operating conditions. Finally, this type of pump is subject to air binding and often must be primed.

Positive Displacement Pumps Unlike the centrifugal pump where the head is developed for any given flow by the speed of the rotor, the positive displacement pump achieves its maximum head by the power actually available from the driver working on the fluid. In general, overall efficiencies of positive displacement pumps are higher than those of centrifugal pumps because internal losses are minimized. However, the flexibility of the positive displacement units in handling a wide range of capacities is somewhat limited.

Positive displacement pumps may be of either the reciprocating or the rotary type. In either type a fixed volume is alternately filled and emptied of the pumped fluid by action of the pump. There are three classes of reciprocating pumps, namely, piston, plunger, and diaphragm. Each is equipped with valves that are operated by pressure differences to introduce and discharge the liquid being pumped. To minimize pressure pulsation because of the rapid opening and closing of these valves, many reciprocating pumps are provided with surge chambers on the discharge side as well as on the suction side of the pump.

Reciprocating pumps may be either single-cylinder or multicylinder in design. The latter have all cylinders in parallel for increased capacity. Piston-type pumps can be classified as single-acting or double-acting, depending on whether energy is supplied to the fluid on both the forward and backward strokes of the piston. Plunger pumps are always single-acting.

The simplex double-acting pumps and the duplex double-acting pumps are the most widely used piston pumps. Both pumps may be direct-acting or power-driven. The duplex type differs from the simplex type in the use of two cylinders whose operation is coordinated.

Plunger pumps differ from piston pumps because they have one or more constant-diameter plungers moving back and forth through packing glands while displacing liquid from cylinders in which there is considerable radial clearance. Simplex plunger pumps are quite often used as metering or proportioning pumps. When used in this capacity, these pumps can be designed to measure or control the flow of a liquid to within a deviation of ± 2 percent with capacities up to 11 m^3/h and pressures as high as 70 MPa.

Diaphram pumps perform similarly to piston and plunger pumps, but the reciprocating driving member is a flexible diaphram. The principal advantage of such a pump is the elimination of all packing and seals exposed to the liquid being pumped. Units are available to withstand pressures up to 7 MPa and capacities greater than 25 m^3/h. In view of possible diaphram failure, the design engineer should realistically appraise the consequences that could result from such a failure.

Reciprocating pumps, in general, have the advantage of being able to deliver fluids against high pressures and operate with good efficiency over a wide range of operating conditions. For example, the volumetric efficiency, defined as the ratio of the actual fluid displacement of a piston pump to the theoretical fluid displacement, is usually in the range of 70 to 95 percent. Likewise, the pressure efficiency, defined as the ratio of the theoretical to the actual pressure required by the driver to pump the fluid, varies from about 50 percent for small pumps up to 90 percent for large pumps.

Rotary positive displacement pumps combine a rotary motion with a positive displacement of the fluid. Two intermeshing gears are fitted into a casing with a sufficiently close spacing to provide an effective seal between each tooth space. As the gears rotate in opposite directions, fluid is trapped in each tooth space and transferred to the discharge side of the pump. Similar results can be obtained by using a rotating cam or separately driven impellers containing several lobes.

A modification of the gear pump is the screw pump. In the two-rotor version, the liquid being pumped progresses axially in the cavities formed by the meshing screw threads. Screw pumps can achieve pressures as high as 7 MPa.

No priming is required for rotary positive displacement pumps. These units are well adapted for pumping highly viscous fluids. A constant rate of delivery is obtained, and the fluid may be delivered at high pressures. Because of the small clearance that must be maintained between interacting surfaces, this type of pump should not be used with nonlubricating fluids or with fluids containing solid particles.

Jet Pumps This type of pump utilizes the momentum of one fluid to transport the desired fluid. The ejector is designed for those operations in which the head to be attained is low and less than the head of the fluid used for pumping. The steam injector, on the other hand, is a special type of jet pump utilizing steam in which the fluid being pumped is discharged at the same pressure as that of the steam being used. The efficiency of an ejector is generally only a few percent, and the head developed is also low except in special situations. However, in steam injectors used in boiler feed operations in which the energy of the steam is recovered, efficiency can be close to 100 percent. Dilution of the fluid being pumped needs to be considered when the use of an ejector is considered.

Electromagnetic Pumps All electromagnetic pumps utilize the principle that a conductor in a magnetic field, carrying a current that flows at right angles to the direction of the field, has a force exerted on it. Since these types of pumps are used to move fluids that exhibit electrical conductivity properties, the force suitably directed in the fluid manifests itself as a pressure if the fluid is suitably contained.

Design Procedures for Pumps

The amount of useful work that a pump performs is the product of the flow rate handled by the pump and the total pressure differential measured across the device. The latter quantity is usually expressed in terms of an equivalent height of the fluid being pumped under adiabatic conditions and is defined simply as the *head.* Thus, the shaft work of the pump in kilowatts is given by

$$W_o = \frac{H\dot{m}_v\rho}{10^3} \tag{12-20a}$$

where H is the total dynamic head (column of liquid) in N·m/kg, $\dot{m}_v$ the volumetric flow rate in m^3/s, and ρ the liquid density in kg/m^3. When the total dynamic head is expressed in pascals, the shaft work of the pump in kilowatts is given by

$$W_o = \frac{H\dot{m}_v}{10^3} \tag{12-20b}$$

Power input to any type of pump is obtained by dividing the power output by the efficiency of the pump. Figure 12-17 gives values that are suitable for design estimates of centrifugal pump efficiencies. Since pump and driver efficiencies must both be considered when total power costs are determined, necessary design data on the efficiency of electric motors are presented in Fig. 12-18.

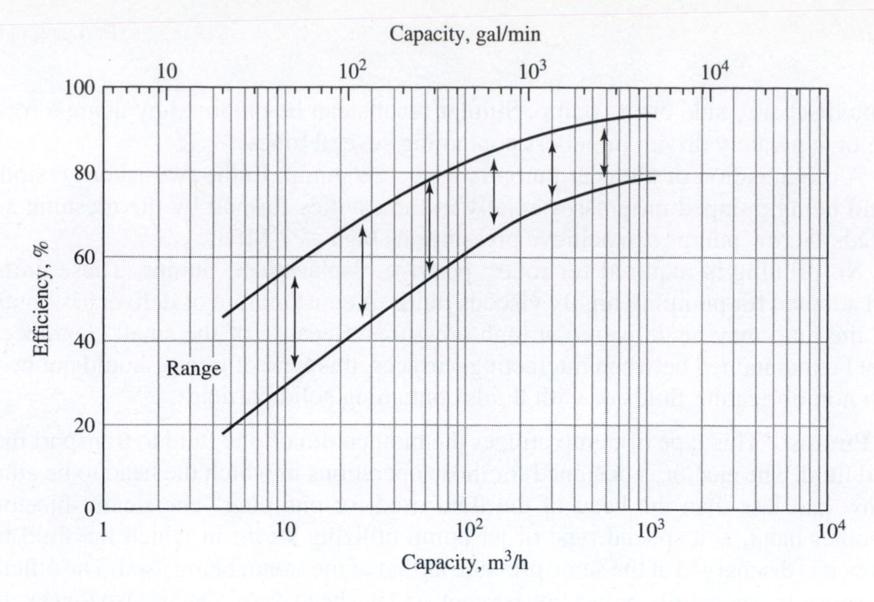

Figure 12-17
Efficiencies of centrifugal pumps

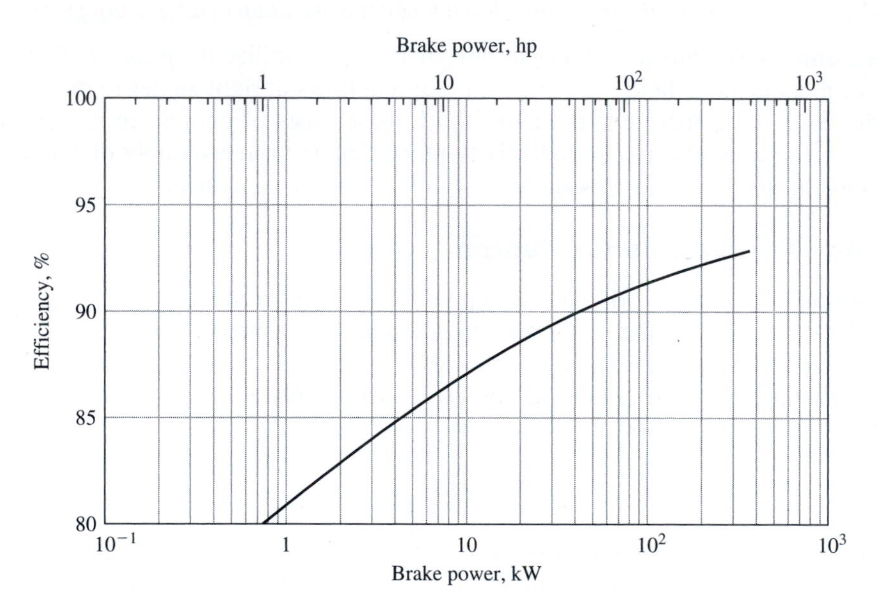

Figure 12-18
Efficiencies of three-phase motors

Costs for Pumps and Motors

Figures 12-19 through 12-24 provide approximate costs for different types of pumps and motors that can be used for preliminary design estimates; firm estimates should be based on vendors' quotations.

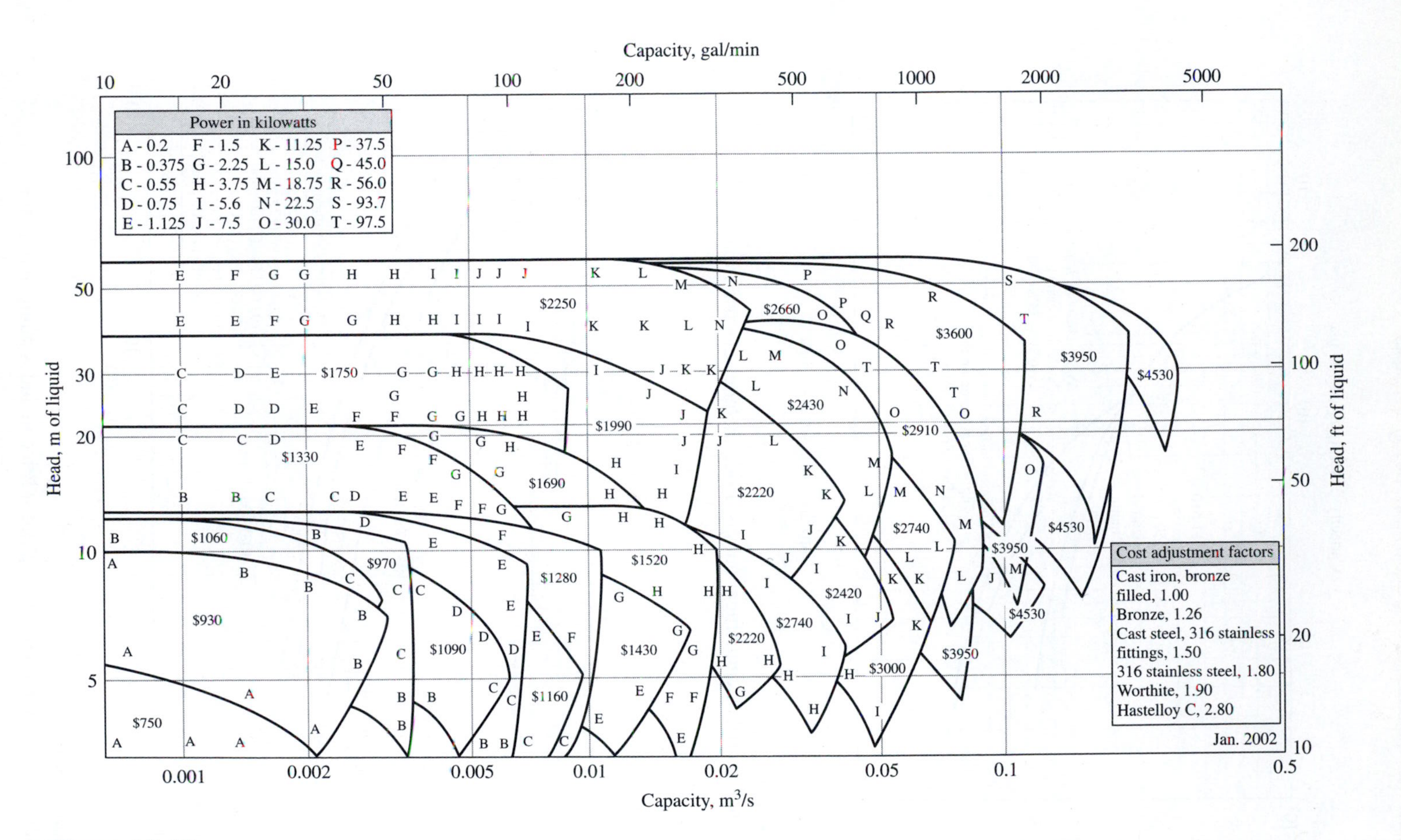

Figure 12-19
Pumps: general-purpose centrifugal (single- and two-stage, single-suction). Price includes pump, steel base, and coupling, but no motor. Letters located within selection blocks approximate the pump power requirement in kilowatts.

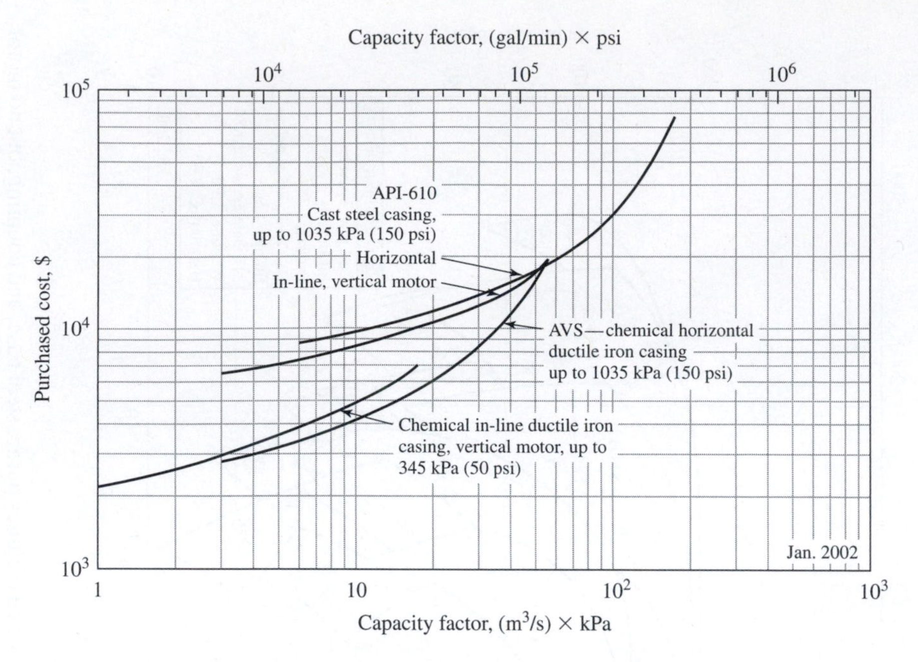

Figure 12-20
Purchased cost of centrifugal pumps. Price includes electric motor.

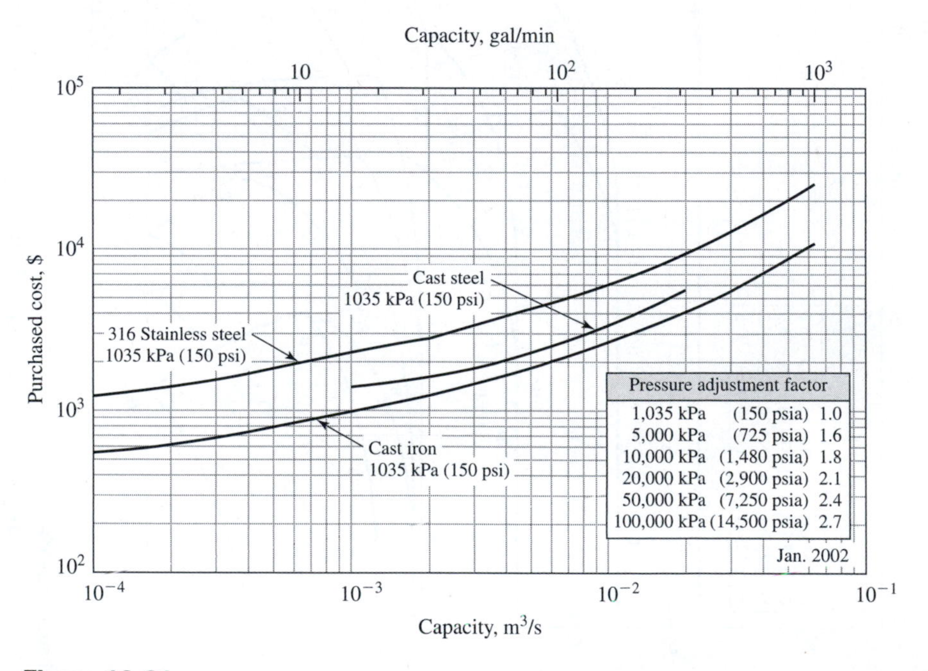

Figure 12-21
Purchased cost of reciprocating pumps. Price includes pump and motor.

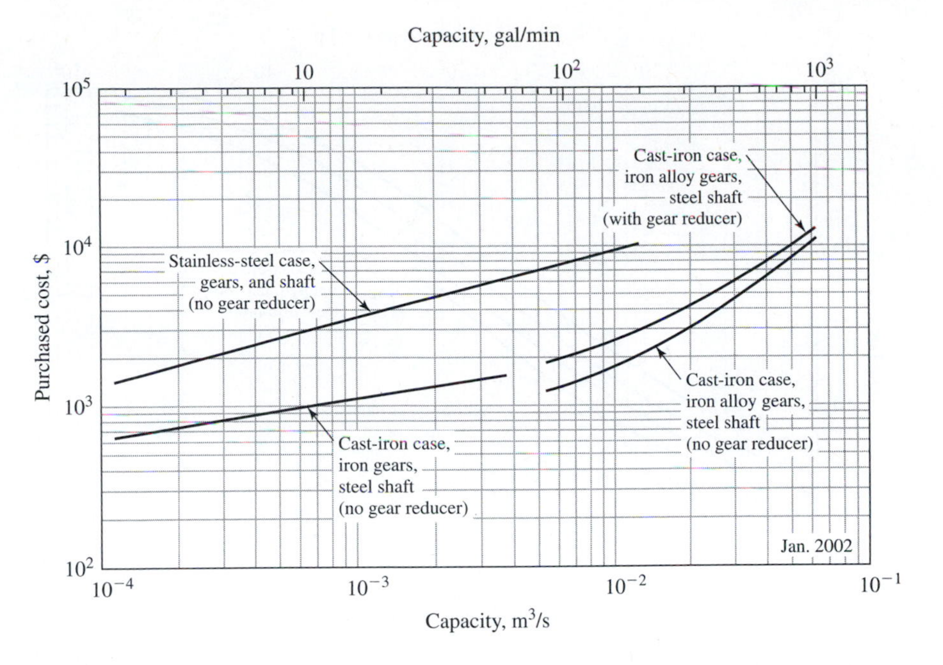

Figure 12-22
Purchased cost of gear pumps, 690-kPa (100 psig) discharge pressure. Cost includes pump, baseplate, and V-belt drive, but no motor.

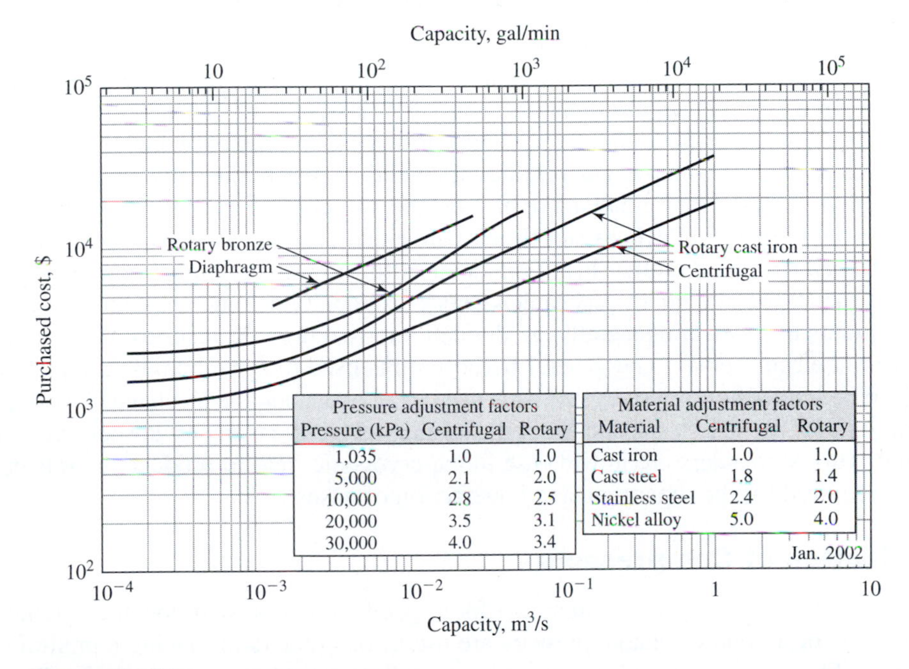

Pressure adjustment factors

Pressure (kPa)	Centrifugal	Rotary
1,035	1.0	1.0
5,000	2.1	2.0
10,000	2.8	2.5
20,000	3.5	3.1
30,000	4.0	3.4

Material adjustment factors

Material	Centrifugal	Rotary
Cast iron	1.0	1.0
Cast steel	1.8	1.4
Stainless steel	2.4	2.0
Nickel alloy	5.0	4.0

Figure 12-23
Purchased cost of diaphragm, centrifugal, and rotary pumps

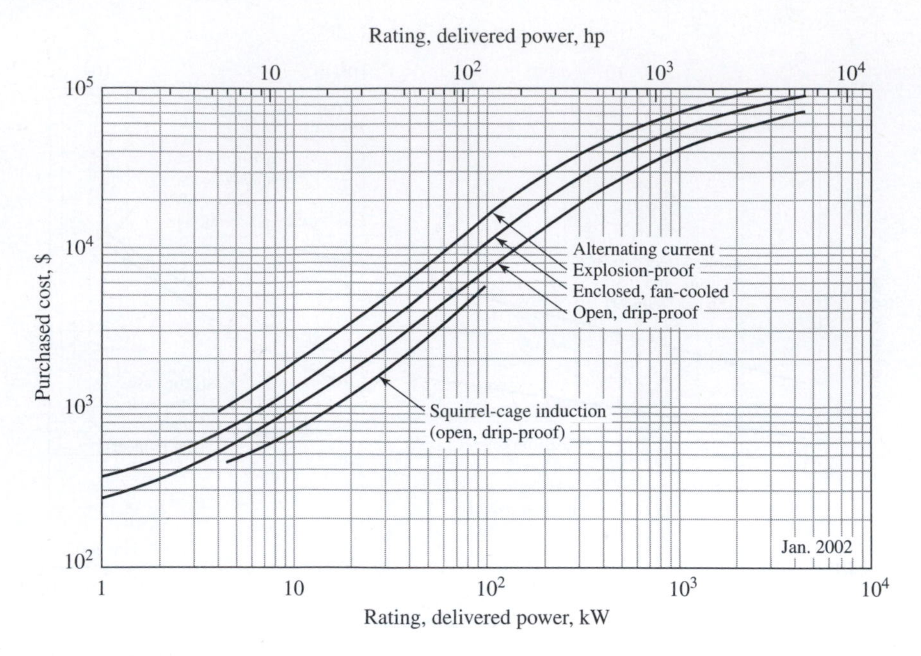

Figure 12-24
Purchased cost of electric motors

COMPRESSION AND EXPANSION OF FLUIDS

Compressors are used to transfer large volumes of gas while increasing the pressure of the gas from an inlet condition to pressures as high as 300 MPa. Even though there are many types of compressors, they are generally classified into two major categories, namely, continuous-flow compressors and positive displacement compressors. Similarly, fans are used to move gas volumes at conditions where the delivery pressure differential is no more than 3.5 kPa. For blowers the pressure differential is slightly greater but no more than 10 kPa. Fans and blowers are either centrifugal or axial-flow units.

Expanders, on the other hand, are devices for converting the pressure energy of a gas or vapor stream to mechanical work as the fluid undergoes an expansion. Expanders using steam generally are divided into two broad categories: those used to generate electric power and general-purpose units used to drive pumps, compressors, etc. When other fluids besides steam are used in an expander, the mechanical work produced generally is a by-product since the primary objective is to provide cooling of the fluid. Turboexpanders are in wide use in the cryogenic field to produce the refrigeration required for the separation and liquefaction of gases.

Selection of Compressors

In selecting the appropriate compressor for a specific process condition, the volumetric capacity, head, and discharge pressure are useful parameters in making a preliminary choice. Figure 12-25 provides operating boundaries using these parameters for the four types of compressors discussed in this section, namely, centrifugal, axial, reciprocating,

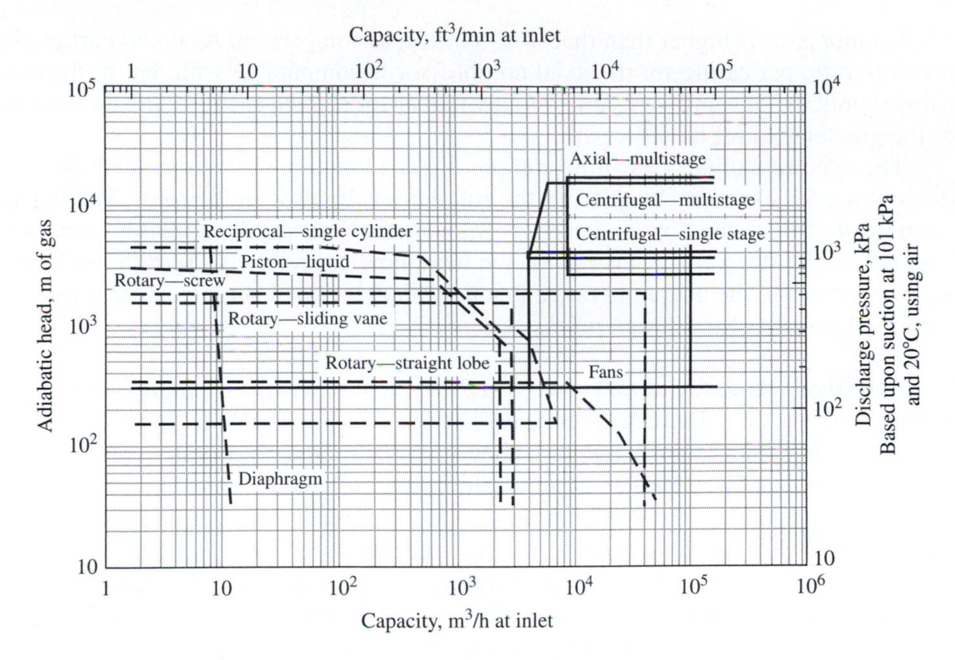

Figure 12-25
Compressor coverage chart based on the normal range of operation of commercially available types shown. Solid lines: use left ordinate, head. Broken lines: use right ordinate, pressure.

and rotary. The selection process must also recognize process problems with certain gases at elevated temperatures that could create a potential explosion hazard or with the admission of small amounts of lubricating oil or water that would contaminate the process gas stream. Finally, for continuous process operation, a high degree of equipment reliability will be required since frequent shutdowns cannot be tolerated.

Centrifugal and Axial Compressors Both centrifugal and axial-flow units are continuous-flow compressors. Centrifugal compressors, in general, are used for higher pressure ratios and lower flow rates compared to lower-stage pressure ratios and higher flow rates in axial compressors. The pressure ratio in a single-stage centrifugal compressor used in the petrochemical industry is about 1.2 : 1. The pressure ratio of axial-flow compressors is between 1.05 : 1 and 1.15 : 1. Because of the low pressure ratios for each stage, a single compressor may include a number of stages in one casing to achieve the desired overall pressure ratio.

The operating range of a centrifugal compressor is between that of *surge,* which is the lower flow limit of stable operation, and *choke,* which is the maximum flow through the compressor at a given operating speed. This operating range is reduced as the pressure ratio per stage is increased or the number of stages is increased. To increase the surge-to-choke operating range, most centrifugal compressors use backward-curved impellers that provide very low pressure ratios as well as higher efficiencies.

Axial-flow compressors are mainly used as compressors for turbines. The use of multiple stages permits overall pressure increases up to 30 : 1. The efficiency of an axial compressor is higher than that of a centrifugal compressor. As noted earlier, the pressure ratio per casing for the axial compressor is comparable with that of the centrifugal unit; the flow rates are considerably higher for a given casing diameter because of the greater area for the flow path.

The operation of the axial-flow compressor is a function of the rotational speed of the low-aspect stationary blades and the rotation of the flow in the rotor. The blades are used to diffuse the flow and convert the velocity increase to a pressure increase. Because of the steep characteristics of the head/flow capacity curve exhibited by the axial compressor, the operating range is much smaller with a surge point that often is within 10 percent of the design point.

Positive Displacement Compressors These type of units are essentially volume gas movers with variable discharge pressures. This classification includes both reciprocating and rotary compressors. Reciprocating compressors are generally used when a high-pressure head is requested at a low flow rate. Such units are available in either single-stage or multistage types. Single-acting, air-cooled and water-cooled air compressors are available in sizes up to 75 kW with pressure levels as high as 24 MPa. However, because of the difficulty in preventing gas leakage and lubricating oil contamination, these units are seldom used for compression of other gases than air. The compressors that are used for compressing gases use a crosshead that connects the piston rod and the connecting rod to provide a straight-line motion for the piston rod and the use of simple packing.

Intercoolers are inserted between stages in multistage units. These heat exchangers remove the heat of compression, thereby reducing the volume of gas going to the next stage and reducing the power required for the compression. More importantly, the cooling maintains the temperature within safe operating limits.

Suction and discharge valve losses begin to exert significant effects on the actual internal compression ratio of many reciprocating compressors when piston speeds greater than 25 m/s are permitted. The obvious effects noted are higher operating temperatures and correspondingly higher power requirements. The less obvious effects are increased valve problems that can contribute to larger maintenance costs experienced by such compressors.

The four types of rotary positive displacement compressors include the straight-lobe, screw, sliding-vane, and liquid-piston machines. The volume displaced in any of these units can only be varied by changing the rotor speed or through bypassing and wasting some of the capacity of the units. The straight-lobe units are available for pressure differentials up to about 80 kPa and capacities up to 2.5×10^4 m^3/h. Higher pressures can be obtained by using multiple units operated in series. The rotary screw compressor can handle capacities up to about 4.2×10^4 m^3/h at pressure ratios of 4:1 or higher. Unlike the straight-lobe rotary compressor, the screw compressor uses the rotation of two rotors to cause axial progression of successive sealed cavities of gas through the unit. The sliding-vane compressors are available for operating pressures up to 850 kPa with capacities up to 3.4×10^3 m^3/h. Pressure ratios per stage are generally limited to 4:1. Since lubrication of the vanes is required, the discharge gas stream will contain lubricating oil.

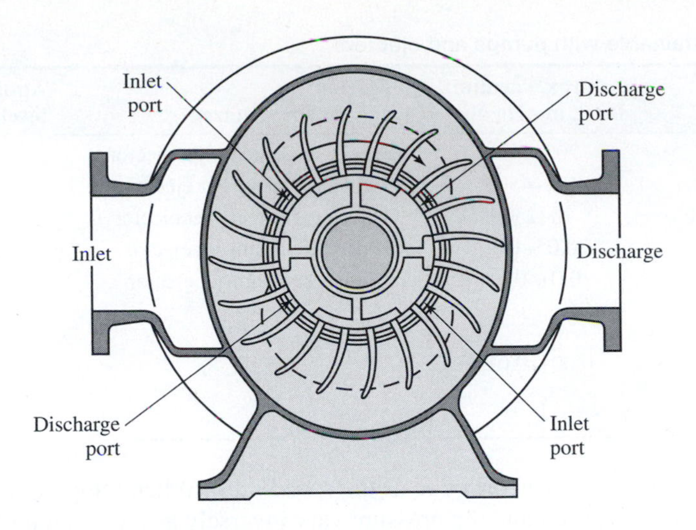

Figure 12-26
Liquid-piston type of rotary compressor.

The liquid-piston type of compressor is very useful when hazardous gases are being compressed. They have also found wide application as vacuum pumps for wet-vacuum service. In this type of unit, illustrated in Fig. 12-26, the liquid serves as the sealing fluid. As the vaned impeller rotates, centrifugal force drives the sealing liquid against the walls of the pump housing, causing the gas to be successively drawn into the vane cavities and expelled against the discharge pressure. A separator is generally required in the discharge line to minimize carryover of the entrained liquid. This type of compressor is available as single-stage units for pressure differentials up to about 520 kPa in the smaller sizes and capacities up to 6.8×10^3 m^3/h when used with a lower pressure differential.

Selection of Fans and Blowers

Fans and blowers are used to move large volumes of air or gas through ducts, conveying fine material in a gas stream, removing fumes, providing air for drying and condensing towers, and other high-flow, low-pressure applications. The disk-type fan, similar to a household fan, and the propeller-type fan are both axial-flow fans. In the centrifugal fan, the gas enters in an axial direction but is discharged in a radial direction. These blowers use either a radial, forward-curved blade or a backward-curved blade. The forward-curved blade blowers provide the largest exit gas velocity, but with a low discharge pressure. To obtain a higher discharge pressure, the backward-curved blade blower is used. The radial blade blower is a compromise between these two blowers.

The performance of centrifugal blowers varies with changes in conditions such as temperature, density, and speed of the gas being handled. If the blower is not operating at stated conditions specified by a vendor, corrections must be made to account for these variations. For example, when the speed of the blower varies, (1) capacity varies directly as the speed ratio, (2) pressure varies as the square of the speed ratio, and

Table 12-6 Vacuum levels attainable with pumps and ejectors

Pump equipment required	Approx. vacuum level, mmHg abs	Ejector equipment required	Approx. vacuum level, mmHg abs
Multistage centrifugal pump	200–760	Single-stage steam-jet ejector	70–760
Rotary (water-sealed) pump	100–450	Two-stage steam-jet ejector	10–100
Reciprocating (dry vacuum) pump	10–250	Three-stage steam-jet ejector	2–25
Rotary (oil-sealed) pump	0.05–15	Four-stage steam-jet ejector	0.2–2.5
Single-stage diffusion pump with discharge to rotary pump or steam-jet ejector	0.01–0.1	Five-stage steam-jet ejector	0.05–0.5
Multistage diffusion pump with discharge to rotary pump or steam-jet ejector	0.001–0.01		

(3) power required varies as the cube of the speed ratio. When temperature of the gas varies, the power required and the pressure vary inversely as the absolute temperature (assuming that the speed and capacity are constant). When the density of the gas varies, the power required and the pressure vary directly as the density (again, assuming speed and capacity are constant).

Selection of Vacuum System Equipment

Development of vacuum systems generally requires the use of either mechanical vacuum pumps or steam-jet ejectors. Table 12-6 indicates vacuum levels attainable with such equipment.

The equipment listed in Table 12-6 has been discussed earlier with the exception of the diffusion pump and the ejector. In the diffusion pump, a liquid with a very low vapor pressure is vaporized in the pump and ejected at high velocity in a downward direction, thereby trapping molecules that have come from the evacuated space. The trapped molecules are removed through the discharge line by a backup pump such as an oil-sealed rotary pump. The diffusion pump vapor is condensed and returned to the pump. The ultimate vacuum that can be attained by diffusion pumps depends to a degree on the condensation temperature of the pump liquid.

The ejector, on the other hand, is a simplified type of compressor or vacuum pump that has no moving parts. In vacuum systems, the unit consists essentially of a nozzle which discharges a high-velocity jet across a suction chamber that is connected to the space to be evacuated. The gas from the evacuated space is entrained by the high-velocity gas or steam and carried into a venturi-shaped diffuser which converts the velocity energy to pressure energy.

Steam-jet ejectors can be connected in parallel to handle quantities of gas or in series to develop lower levels of vacuum, as shown in Table 12-6. The capacity of steam-jet ejectors is usually reported as kilograms per hour rather than on a volume basis. For design purposes it is often necessary to make a rough estimate of the steam requirements for various ejector capacities and conditions. The data provided in Table 12-7 can be used for this purpose.

The choice of the most suitable type of ejector depends on the steam pressure available, the maximum water temperature allowed, the flow capacity required, and the level

Table 12-7 **Average consumption in kg/h of 775-kPa steam in a steam-jet ejector to provide various levels of vacuum†**

Capacity, kg/h gas-vapor mixture handled‡	Wt% net dry air (noncondensibles) in gas-vapor	Suction pressure, mmHg absolute									
		12.5		25		37.5	50	75	100		150
		3-stage	2-stage	3-stage	2-stage	2-stage	2-stage	2-stage	2-stage	1-stage	1-stage
4.5	100	33	45	27	32	26	23	19	17	26	16
4.5	70	27	38	21	27	22	19	16	14	29	17
4.5	40	20	31	15	21	17	15	12	10	31	18
4.5	10	11	20	7	13	10	8	6	5	34	19

†From R. H. Perry and D.W. Green, *Perry's Chemical Engineer's Handbook,* 4th ed., McGraw-Hill, New York, 1963. Reproduced with permission of The McGraw-Hill Companies.
‡For larger and smaller capacities, steam consumption is approximately proportional to capacity.

of vacuum desired. Ejectors are relatively inexpensive initially, are easy to operate, but require large amounts of steam. Since they have no moving parts, they provide long life, sustained efficiency, and low maintenance cost. Practically any type of gas or vapor can be handled by ejectors. They can even handle wet or dry gas mixtures as well as gases containing sticky or solid material such as dust or chaff.

Selection of Turbines, Expanders, and Other Drivers

There are several types of steam turbines available. The single-stage, single-valve turbine is the simplest turbine in which the steam, after passing through the governor valve, expands through nozzles, where it gains velocity and momentum to drive the turbine wheels by impulse action against the blades. The majority of the applications are below 1100 kW, operating with steam pressures up to 4100 kPa. To obtain power outputs to 4500 kW, the multistage, single-valve turbine is widely used. Exhaust pressures of the expanded steam are at low condensing pressures of 7 to 14 kPa absolute.

The major variables that affect steam turbine selection include (1) the power and speed of the device that is to be driven, (2) the pressure and temperature of the steam available and its cost, (3) the quantity of steam required based on turbine efficiency (as determined by stage and valve options), and (4) the control and safety features required. Logically, the first step in the selection process is to make an estimate of the steam flow at various steam pressures by using Fig. 12-27 for a rough estimate of the efficiency. (Vendors of steam turbines prefer to work with the pressure levels proposed by an ASME-IEEE standards committee, as shown in this figure.) Note that 538°C is a good upper limit for steam turbines without experiencing a sharp increase in cost because of the requirement for more expensive materials.

The gas turbine in the simple cycle mode consists of a compressor (axial or centrifugal) that compresses the air, a combustor that heats the air, and a turbine that expands the combustion gases with a resultant power production. The power required to compress the air varies from 40 to 60 percent of the total power produced by the turbine. The axial turbine is the most widely used gas turbine and is more efficient than a radial-flow turbine in most operational regions. Axial-flow turbine efficiencies range between 88 and 92 percent. However, radial-flow turbines are used because of lower production costs, partly because the nozzle blading does not require as much airfoil design. Radial-flow turbines are more robust, but due to cooling restrictions are used for lower turbine inlet temperatures.

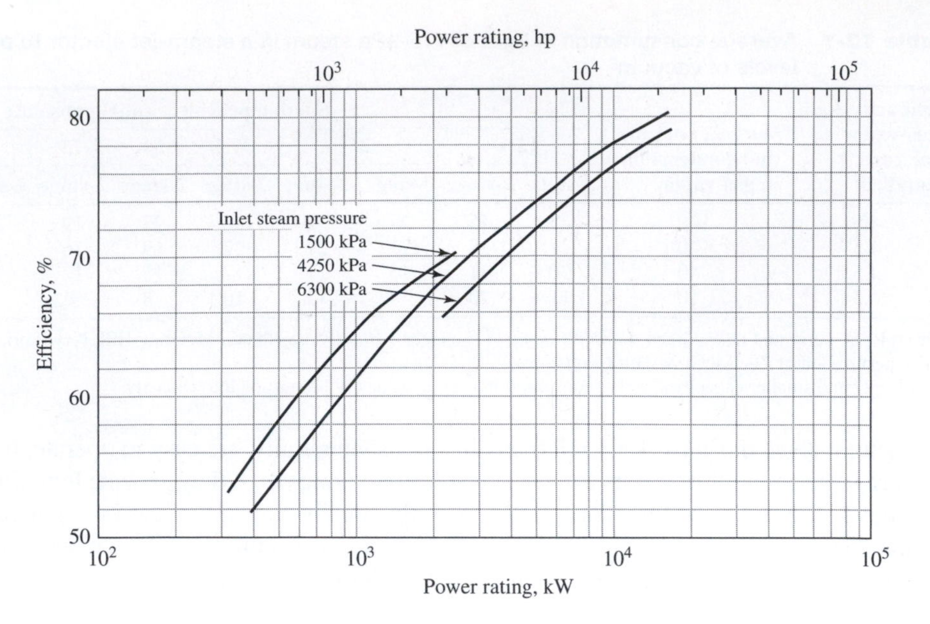

Figure 12-27
Approximate efficiency for multistage turbines

The effect on performance due to inlet and exit conditions of the gas turbine is significant. An increase in the inlet temperature of about 3°C will reduce the design power produced by 2 percent, and a reduction in the inlet pressure by 7 kPa reduces the power output by approximately 3 percent. Additionally, a 1 percent decrease in compressor efficiency reduces the overall thermal efficiency by 0.5 percent and the power output by 2 percent. A 1 percent change in the turbine efficiency will produce a change of about 3 percent in turbine power output and 0.75 percent change in the overall thermal efficiency. Thus, careful control needs to be exercised if the design predictions are to be achieved.

Radial-flow turboexpanders were developed primarily for the production of low temperatures, but they can also be used as power recovery devices. These expanders normally are single-stage with combination impulse-reaction blades. These units often operate with small or moderate streams which dictate a rather high rotational speed up to 500,000 r/min. Because of their heavy-duty construction, they are resistant to abuse and provide high reliability. Such units can provide efficiencies between 75 to 88 percent, calculated on the basis of isentropic rather than polytropic expansion.

Several established operating limitations for turboexpanders without special design features are an enthalpy drop in the expanding gas of 93 to 116 kJ/kg per stage of expansion and a maximum rotor tip speed of 305 m/s. Commercial turboexpanders are available for inlet pressures up to 20.7 MPa and inlet temperatures from near absolute zero to 538°C. The permissible liquid condensation in the expanding stream may be 50 weight percent or more in the discharge, provided the turboexpander has been designed to handle condensation.

A potential application for power recovery exists whenever a large flow of gas is throttled to a lower pressure or when high-temperature process streams are available to vaporize a secondary liquid. When such conditions exist, they should be reviewed to determine whether the use of a turboexpander is justified. In such cases a turboexpander can be used to drive a pump, compressor, or electric generator and recover a large fraction of the energy that is lost. In applications of this type, consideration should be given to the temperature drop in the expander that could lead to possible condensation or formation of solids.

In most process applications, particularly those involving less than 100 kW of power, the electric motor generally is the driver selected. Low capital cost, very low maintenance cost, and high reliability are strong arguments favoring such a choice.

Design Procedures for Compressors

To properly specify any type of compressor requires certain information about the operating conditions—the type of gas, its pressure, temperature, and molecular weight. It also requires knowledge about the corrosive properties of the gas so that proper material selection can be made. Gas fluctuation due to process instabilities must also be pinpointed so that the compressor can operate without surging.

Expressions for the theoretical power requirements of gas compressors can be obtained from basic equations of thermodynamics. Since most compressors operate along a polytropic path approaching the adiabatic, compressor calculations are generally based on the adiabatic curve given by the equation of state $pv^k = \text{constant}$, where k is the ratio of specific heat at constant pressure to that at constant volume. The adiabatic head for a compressor is expressed as

$$H_{ad} = R'T_1\left(\frac{k}{k-1}\right)\left[\left(\frac{p_2}{p_1}\right)^{(k-1)/k} - 1\right] \tag{12-21}$$

where H_{ad} is the adiabatic head in N·m/kg, R' the gas constant in kJ/(kg·K), T_1 the inlet gas temperature in K, k the heat capacity ratio, p_1 the inlet pressure in kPa, and p_2 the discharge pressure in kPa. The power required for the compression is equal to the product of the adiabatic head and the mass flow rate of gas handled. Thus, the adiabatic power for single-stage compression is given by

$$P_{ad} = \dot{m}H_{ad} = \dot{m}R'T_1\left(\frac{k}{k-1}\right)\left[\left(\frac{p_2}{p_1}\right)^{(k-1)/k} - 1\right] \tag{12-22a}$$

or

$$P_{ad} = 2.78 \times 10^{-4}\dot{m}_{v,1}p_1\left(\frac{k}{k-1}\right)\left[\left(\frac{p_2}{p_1}\right)^{(k-1)/k} - 1\right] \tag{12-22b}$$

where P_{ad} is the adiabatic power in kW, $\dot{m}$ the mass flow rate in kg/s, and $\dot{m}_{v,1}$ the volumetric gas flow rate at the compressor inlet in m^3/h. The adiabatic discharge temperature is obtained from

$$T_2 = T_1\left(\frac{p_2}{p_1}\right)^{(k-1)/k} \tag{12-23}$$

As noted earlier, the work of compression under ideal conditions occurs at constant entropy. The actual process of compression is polytropic, as given by the equation of state $pv^n = \text{constant}$. The expressions defining single-stage polytropic head and polytropic power of compression are similar to those developed for adiabatic head and compression power. Thus,

$$H_{\text{poly}} = Z_c R T_1 \left(\frac{n}{n-1}\right)\left[\left(\frac{p_2}{p_1}\right)^{(n-1)/n} - 1\right] \tag{12-24}$$

and

$$P_{\text{poly}} = \dot{m} Z_c R T_1 \left(\frac{n}{n-1}\right)\left[\left(\frac{p_2}{p_1}\right)^{(n-1)/n} - 1\right] \tag{12-25}$$

The addition of an average compressibility factor Z_c in Eqs. (12-24) and (12-25) attempts to account for any deviations from the ideal gas law that are present during the compression process. For adiabatic multistage compression, assuming equal division of compression work between the stages and intercooling of the gas after each stage back to the original inlet temperature, Eq. (12-22*b*) is modified to

$$P_{\text{ad}} = 2.78 \times 10^{-4} N_{\text{st}} \dot{m}_{v,1} p_1 \left(\frac{k}{k-1}\right)\left[\left(\frac{p_2}{p_1}\right)^{(k-1)/kN_{\text{st}}} - 1\right] \tag{12-26}$$

$$T_2 = T_1 (p_2/p_1)^{(k-1)/(kN_{\text{st}})} \tag{12-27}$$

where N_{st} is the number of stages involved in the compression, T_1 the temperature of the gas at the compressor inlet, and T_2 the temperature of the gas at the compressor discharge, both in K. If the compression cycle approaches isothermal conditions where the equation of state $pv =$ constant prevails, a simple expression for the compression of an ideal gas is given by

$$P_{\text{iso}} = 2.78 \times 10^{-4} \dot{m}_{v,1} p_1 \ln \frac{p_2}{p_1} \tag{12-28}$$

where P_{iso} is the single-stage isothermal compression in kW.

Compression efficiencies are generally expressed as isentropic efficiencies, that is, on the basis of an adiabatic and reversible process. Isothermal efficiencies are sometimes quoted, and design calculations are simplified when isothermal efficiencies are used. In either case, the efficiency is defined as the ratio of the power required for the ideal process to the power actually consumed. The adiabatic efficiency can be represented in terms of the total pressure change as

$$\eta_{\text{ad}} = \frac{(p_2/p_1)^{(k-1)/k} - 1}{(p_2/p_1)^{(n-1)/n} - 1} \tag{12-29}$$

Similarly, the polytropic efficiency, often labeled the hydraulic efficiency, is obtained from

$$\eta_{\text{poly}} = \frac{(k-1)/k}{(n-1)/n} \tag{12-30}$$

and is the limiting value of the isentropic efficiency as the pressure ratio approaches unity. Table 12-8 provides typical adiabatic efficiencies observed for some of the compressors discussed in this section.

Table 12-8 Typical efficiencies for various types of compressors

Type of compressor	Typical efficiency range, %
Centrifugal	70–80
Axial	80–85
Reciprocating	60–80
Rotary	60–80
Rotary (vacuum)	40–60
Ejector	25–30

Design Procedures for Turbines and Expanders

The design of turbines and expanders uses the same theoretical principles as were used in the design of compressors. Either a simplified mechanical energy equation or an overall energy balance can be utilized. Disregarding potential and kinetic energy effects, the energy relation developed in Example 12-2 can be written as

$$(h_2 - h_1) - Q - W = 0 \tag{12-31}$$

where h_1 and h_2 are the specific enthalpies of the entering and exiting streams, respectively, through the expansion device. Since adiabatic operation is generally the case, the equation reduces to the familiar form of

$$W = -(h_1 - h_2) \tag{12-32}$$

where the energy recovered is positive as defined in Eq. (12-31). The principal difficulty in applying Eq. (12-32) is the undetermined value for h_2, the exit enthalpy. This difficulty is normally overcome by evaluating the ideal power recovery when the expansion process is assumed to follow an isentropic path. Thus,

$$W_{\text{ad}} = -(h_1 - h_{2,s}) \tag{12-33}$$

where $h_{2,s}$ is the exit enthalpy at the exit pressure but at the inlet entropy. Since $W = \eta W_i$, the actual power recovery is given by

$$P = -\eta \dot{m}(h_1 - h_{2,s}) \tag{12-34}$$

and the actual exit enthalpy can be obtained from

$$h_2 = h_1 - \eta(h_1 - h_{2,s}) \tag{12-35}$$

Energy recovery follows directly from Eq. (12-32). This is very convenient if an enthalpy-entropy table is available.

Determination of Energy Recovery Using a Steam Turbine — EXAMPLE 12-4

A paper mill has 2 kg/s of steam available at 4240 kPa and 400°C which is currently throttled to 690 kPa for use in drying paper. If this steam were sent through a turbine prior to the drying process, how much energy could be recovered?

■ Solution

The steam tables show that inlet steam has the following conditions:

$$h_1 = 3204 \text{ kJ/kg} \qquad s_1 = 6.73 \text{ kJ/(kg·K)}$$

At the outlet conditions under isentropic expansion to 690 kPa, the ideal exit enthalpy would be 2770 kJ/kg. The ideal isentropic energy recovery is

$$P_{ad} = -2.0(3204 - 2770) = -868 \text{ kW}$$

Figure 12-27 indicates that an efficiency of approximately 60 percent might be expected for this power rating. Thus, the actual energy recovered is 521 kW. The actual enthalpy is obtained with the use of Eq. (12-35):

$$h_2 = 3204 - 0.6(3204 - 2770) = 2944 \text{ kJ/kg}$$

A return to the steam tables indicates that under these actual enthalpy and pressure conditions, the exit temperature will be 246°C with an entropy value of 7.09 kJ/(kg·K).

An alternate analysis can be made by using the mechanical energy balance presented in Eq. (12-2). Assuming polytropic expansion of a real gas, the energy recovery can be estimated from a relationship that is similar to Eq. (12-25), namely,

$$P_{poly} = \dot{m} Z_c R T_1 \left(\frac{n}{n-1}\right)\left[\left(\frac{p_2}{p_1}\right)^{(n-1)/n} - 1\right] \tag{12-36}$$

where Z_c is the compressibility factor for the gas. The present availability of computer software that can continuously monitor the pressure-volume-temperature relationship of a gas during the expansion process not only has greatly improved the ease with which estimates of energy recovery can be made but also has increased the accuracy of the result.

Note that if condensation occurs during the expansion of the gas, there is a reduction in the expander efficiency. In such instances, the general rule is to set the efficiency equal to the product of the noncondensate value and the vapor fraction. Thus if 5 weight percent of the steam had condensed in Example 12-4, the efficiency assumed would have been (0.95)(0.60), or 0.57.

Devices driven by high-pressure liquids in actuality are not expanders since the specific volume of the liquid does not change appreciably. Energy recovery in a liquid turbine can be estimated from a simplified form of the mechanical energy balance, assuming constant density. Thus

$$P = \frac{\eta \dot{m}(p_2 - p_1)}{\rho} \tag{12-37}$$

EXAMPLE 12-5 Evaluation of Energy Recovery Using a Liquid Turbine

Determine the energy recovered from expanding 2 kg/s of water initially at an inlet pressure of 4240 kPa and a temperature of 25°C to an outlet pressure of 690 kPa.

■ Solution

Assuming a turbine efficiency of 0.6 and a liquid density of 1000 kg/m^3, the energy recovery from this expansion is given by Eq. (12-37), where $p_2 - p_1 = -3550$ kPa:

$$P = -0.6(2)(3550)/10^3 = -4.26 \text{ kW}$$

The exit temperature can be estimated for this adiabatic system from an energy balance.

$$P = \dot{m}(h_2 - h_1) = \dot{m}C_p(T_2 - T_1)$$

$$T_2 - T_1 = \frac{-4.26}{2(4.19)} = -0.5^\circ\text{C}$$

$$T_2 = 24.5^\circ\text{C}$$

This example shows that the energy available from a compressed liquid is significantly smaller than that available from the same mass of gas and subjected to the same pressure reduction. As a consequence, new high-pressure vessels are tested hydrostatically for safety reasons with compressed water instead of compressed air.

Costs for Compressors, Fans, Blowers, and Expanders

Figures 12-28 through 12-30 provide cost data for various types of compressors. Figures 12-31 and 12-32 give cost data for fans and blowers, while Fig. 12-33 provides such data for steam-jet ejectors. Cost data for expanders and internal combustion engines are presented in Figs. 12-34 and 12-35 and for variable-speed drives in Figs. 12-36 and 12-37. Cost data from these figures are adequate for most preliminary design estimates.

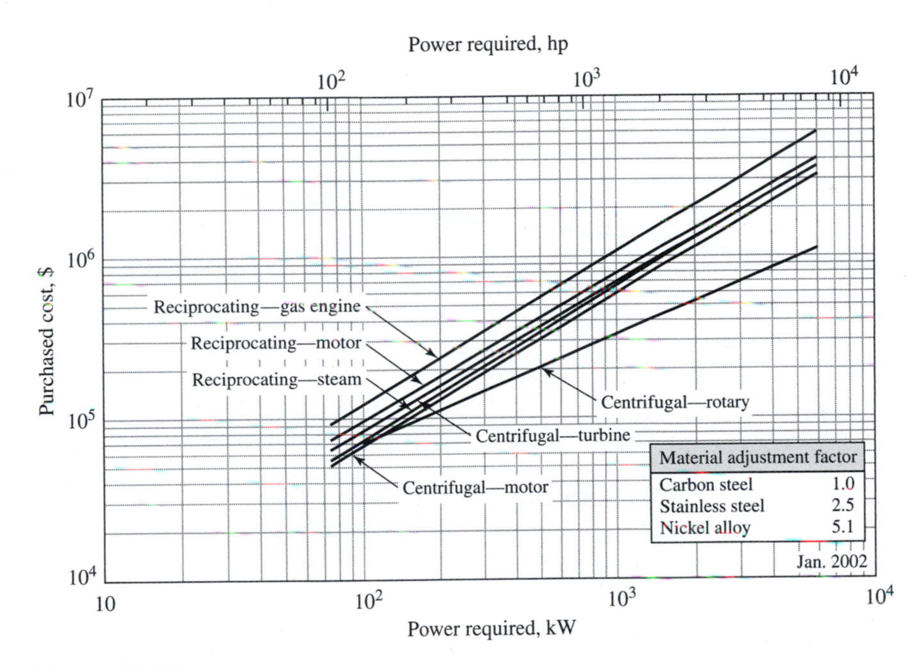

Figure 12-28
Purchased cost of compressors. Price includes drive, gear mounting, baseplate, and normal auxiliary equipment; operating pressure to 7000 kPa (1000 psig).

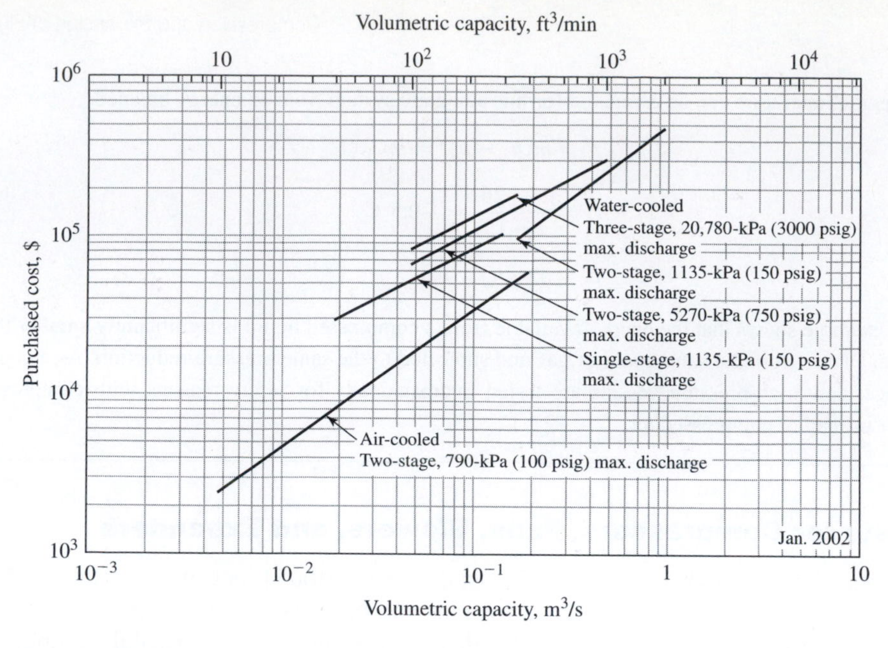

Figure 12-29
Purchased cost of reciprocating compressors

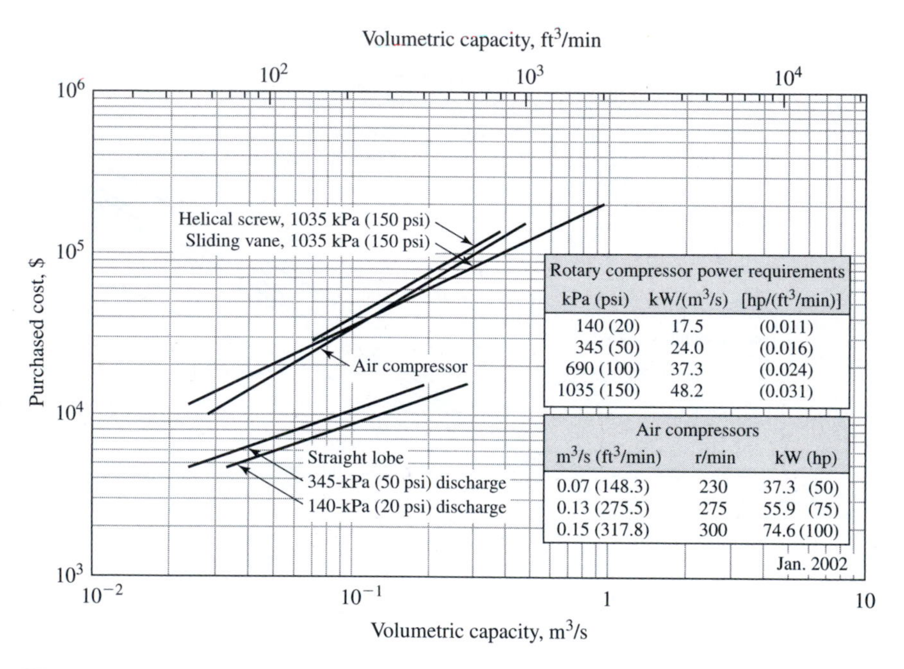

Rotary compressor power requirements		
kPa (psi)	kW/(m³/s)	[hp/(ft³/min)]
140 (20)	17.5	(0.011)
345 (50)	24.0	(0.016)
690 (100)	37.3	(0.024)
1035 (150)	48.2	(0.031)

Air compressors		
m³/s (ft³/min)	r/min	kW (hp)
0.07 (148.3)	230	37.3 (50)
0.13 (275.5)	275	55.9 (75)
0.15 (317.8)	300	74.6 (100)

Figure 12-30
Purchased cost of single-stage rotary compressors and air compressors. Prices are for completely packaged compressor units (freight and installation costs not included). The straight lobe prices also exclude aftercooler, trap, and controls.

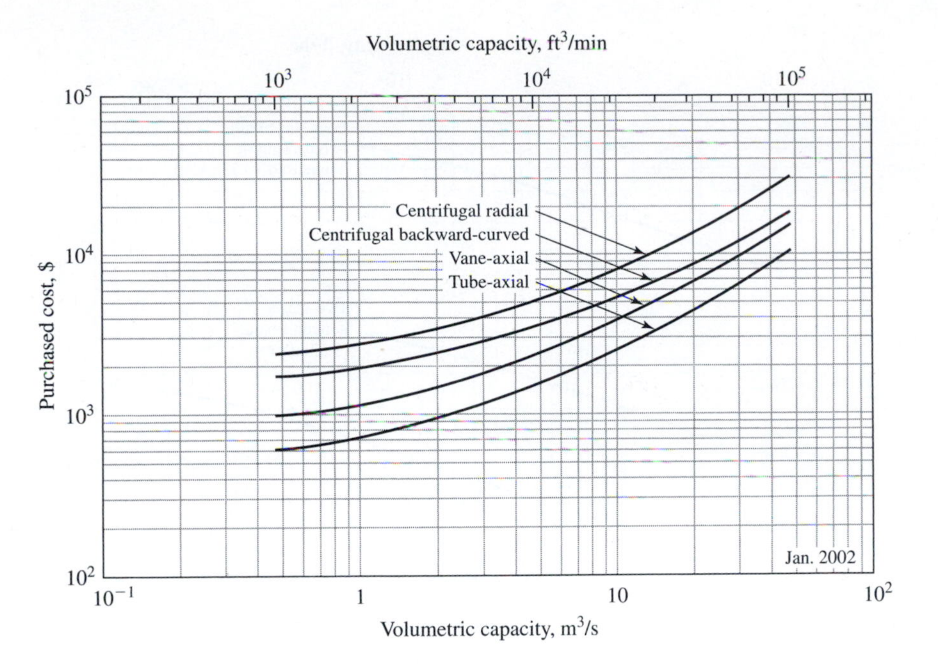

Figure 12-31
Purchased cost of centrifugal fans with electric drives

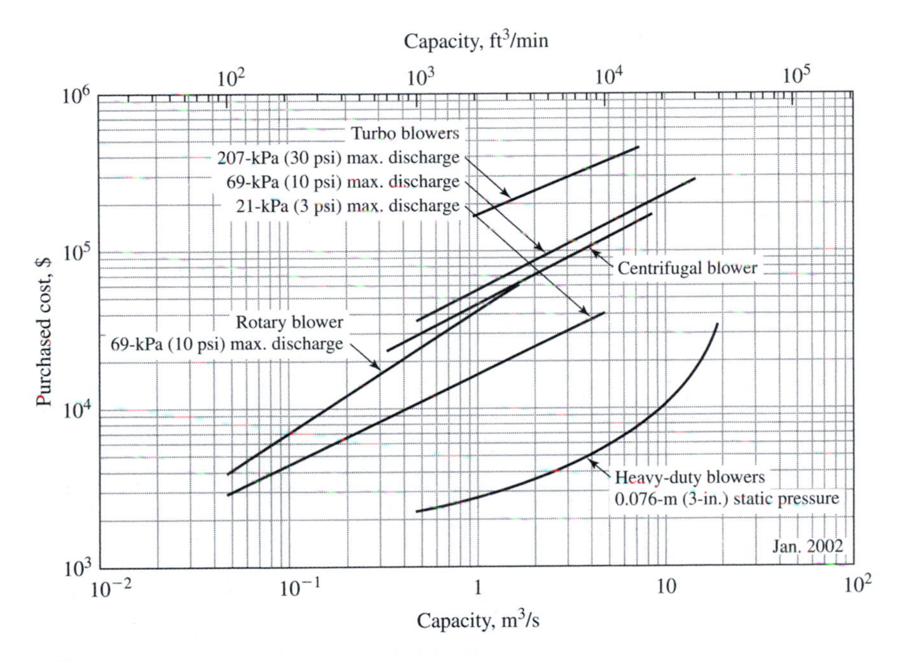

Figure 12-32
Purchased cost of blowers (heavy-duty, industrial type)

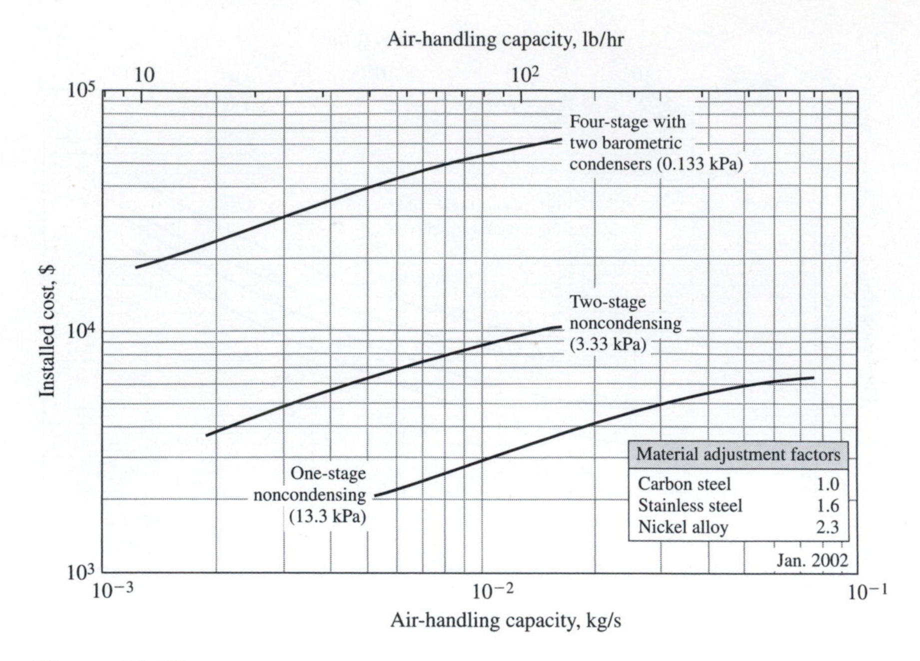

Figure 12-33
Purchased cost of steam-jet ejectors for a specified steam requirement of 1.26×10^{-1} kg/s (1000 lb/h)

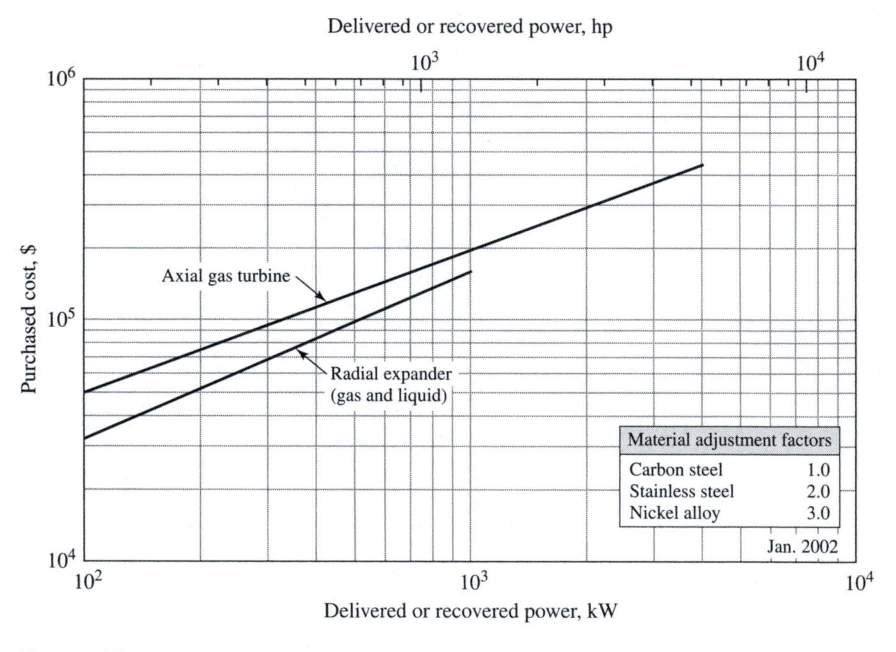

Figure 12-34
Purchased cost of turbines and expanders

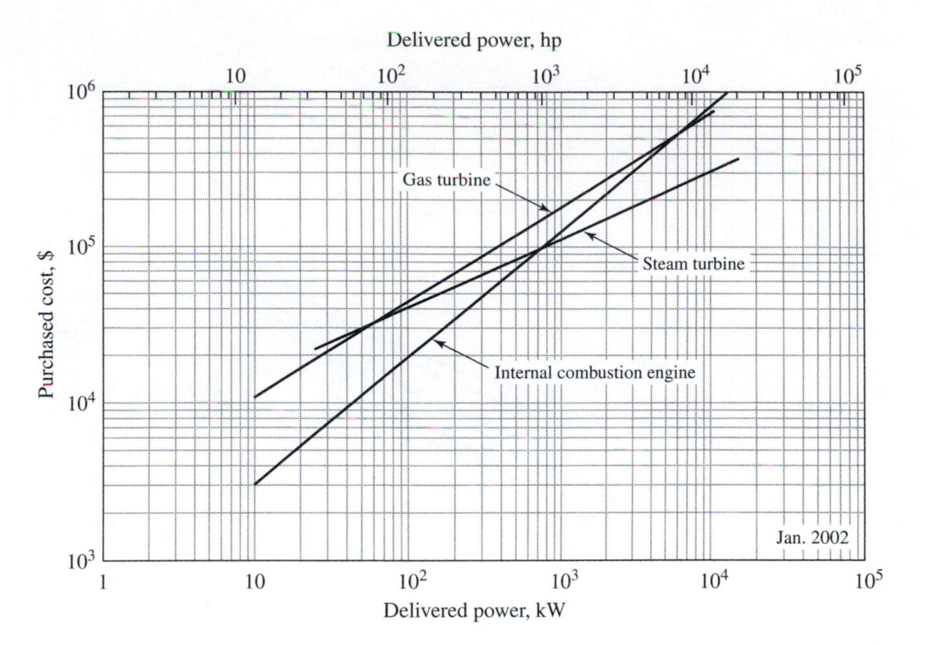

Figure 12-35
Purchased cost of turbine and internal combustion engine drivers

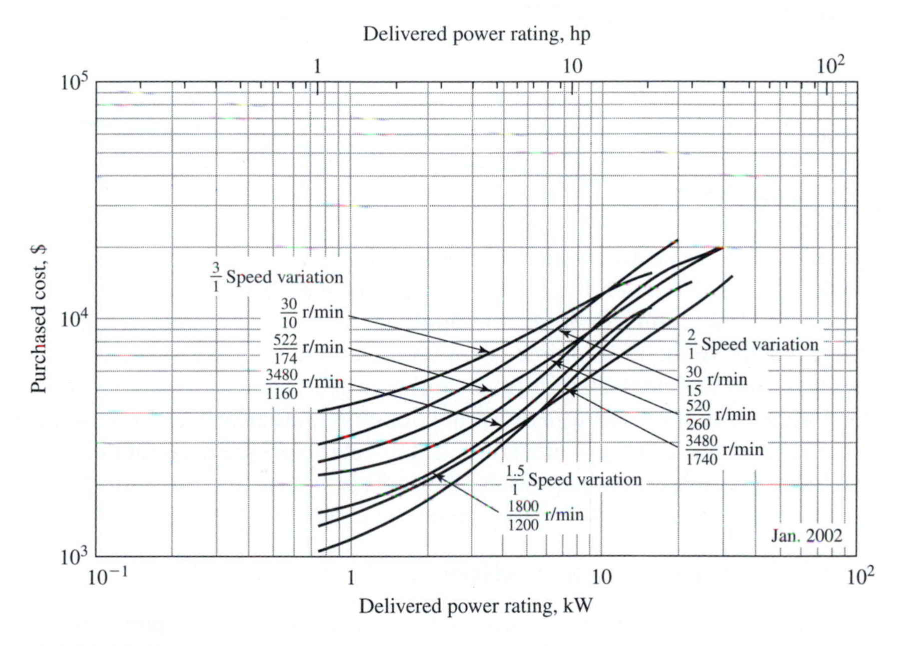

Figure 12-36
Purchased cost of variable-speed drives. Price includes handwheel control with a built-in indicator and TEFC motors.

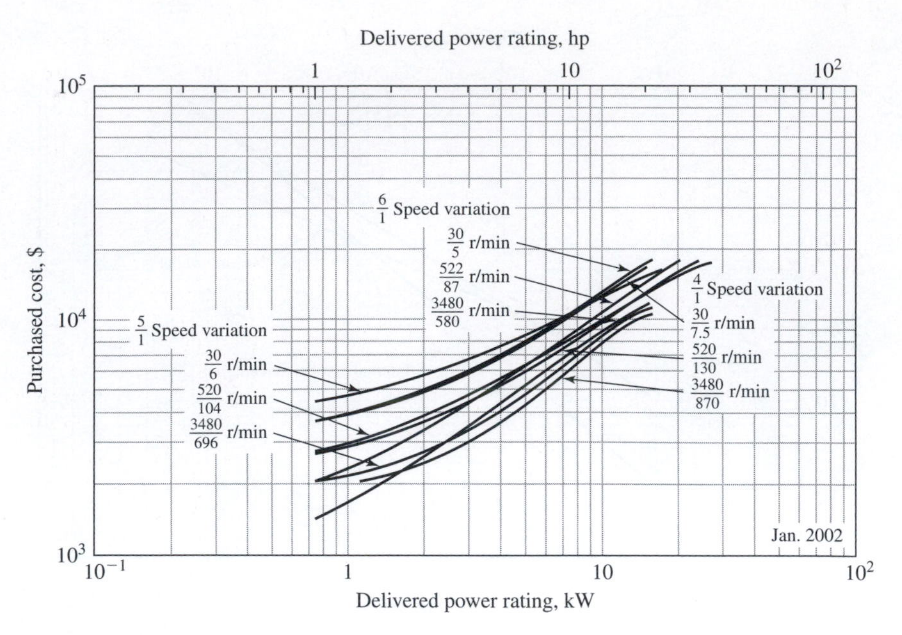

Figure 12-37
Purchased cost of variable-speed drives. Price includes handwheel control with a built-in indicator and TEFC motors.

AGITATION AND MIXING OF FLUIDS

A comparison of agitators and mixers indicates some unique differences between these two types of equipment. Besides mixing of fluids, agitation devices can suspend solid particles in fluids, disperse gases, emulsify liquids, and even enhance heat transfer between a fluid and a solid surface or increase mass transfer between phases. Mixers, on the other hand, involve the subdivision and blending of separate fluids so that microscopic diffusion or shear will lead to more complete homogeneity.

Mixing of low-viscosity immiscible fluids is done efficiently with devices that create turbulence by relative fluid motion or by transfer through various flow turbulators. As viscosity and/or immiscibility of the fluids increases, more mechanical energy is necessary to achieve the level of intensity required. Generally, the suspension of solid particles in a fluid will necessitate an even higher degree of agitation. Propeller turbine agitators, because of their controllability and flexibility, are nearly always used for this mixing task. Even more energy-intensive mechanical mixers are used for mixing highly viscous liquids, pastes, and solid powders. It is not surprising, therefore, that the type of mixing device selected depends very much on the characteristics of the given feed.

Selection of Agitators and Mixers

As noted above, mixing of low-viscosity fluids can often be accomplished within the pipe transporting the fluids to be mixed. For example, gases can be mixed through differential fluid motion or injection of one stream into another. For high-intensity mixing of gases, injection of one stream at sonic velocity into a low-velocity stream is common practice and requires an absolute pressure in the motive gas about twice that of the other stream. Pressure differentials for mixing of liquids to achieve comparable turbulence are considerably higher. With the mixing of more-viscous liquids, the pressure differential for a given intensity of mixing increases linearly with the square of the viscosity. Thus, fluid-jet mixing becomes impractical for mixing liquids with viscosities greater than 0.01 Pa·s. The cost and maintenance of a jet mixer usually are negligible compared with those of the supportive equipment.

An orifice plate in a pipe can be used to promote mixing of gases and nonviscous liquids. This restriction induces turbulence and recirculation similar to that of the fluid jet. The ratio of orifice to pipe diameter ranges from 0.5 to 0.2 and creates a pressure drop of 5 to 30 kPa, respectively, in typical pipelines. The power consumption of this restriction is essentially the mechanical energy required to overcome the added pressure drop in the line.

Motionless mixers are devices that subdivide, rotate, and recombine various elements of viscous liquids, slurries and pastes. The mixer is often a series of twisted metal ribbons inserted inside a section of pipe. The low capital, maintenance, and operating costs as well as the mixing characteristics and low pressure drops of these devices make them preferable to orifice plates in many applications.

Gas spargers are often used in the mixing of corrosive liquids in a tank or where mild agitation is required for gas-liquid contact. These units utilize perforated tubes or porous elements immersed near the bottom of the tank through which the gases bubble and rise to the surface. The recommended gas rate is 0.004 m^3/s per square meter of tank cross section for mild agitation, 0.008 m^3/s for moderate agitation, and 0.02 m^3/s for intense agitation. Gas spargers can be used with liquids as viscous as 1 Pa·s.

The use of a centrifugal pump either installed directly in the pipeline or inserted in an external pipe loop that provides fluid recirculation to a tank can provide mixing, dispersion, or emulsification through fluid interaction. Pressure drops, mixing intensities, and power consumption for various types of pump agitation are similar to those for the nonmechanical modes that they simulate. As a rule of thumb, a fluid volume equal to that maintained within the tank should be circulated through the external loop to provide adequate mixing. This rule can be employed to determine mixing times in batch operations or residence times in continuous operations. The power consumption and capital costs are similar to those for centrifugal pumps.

Propeller or turbine mixers, however, still are the most widely used mechanical agitators for low- to medium-viscosity fluids. These units consist basically of a motorized rotating impeller immersed in the liquid contained within a tank. Propeller agitators often use three-bladed marine-type propellers and are employed extensively in small-scale, flexible operations. These units are characterized by rotational speeds that seldom permit the propeller diameter to exceed 1.5 m. Agitation of large tanks or vessels can be achieved by inserting one or more propeller units through the sidewalls. The power requirements of propeller agitators range from laboratory size to more than 50 kW depending on the vessel size and the viscosity of the fluid.

Turbine impellers generally are much larger than propeller agitators with impellers operating at lower rotational speeds. Turbines employ a variety of impeller designs and are more flexible and efficient than propeller-driven agitators for a number of special applications. Generally, turbine impellers are classified as either radial flow or axial flow. The type of flow obtained is directly related to the configuration of the impeller blades. In the radial impeller the vertical flat-bladed impeller discharges the liquid at high velocity in the radial direction. The axial impeller uses pitched blades that direct the liquid to flow downward parallel to the shaft and then upward along the vessel wall. Axial impellers are primarily used for the suspension of solids, dispersion of miscible liquids, heat-transfer enhancement, and promotion of chemical reactions. Radial impellers are excellent for gas dispersion.

Most turbine-agitated vessels are designed with a ratio of tank diameter to impeller diameter ranging from 2 to 4. The impeller diameter normally does not exceed 5 m, restricting the tank diameter to around 20 m. Larger capacities will require the use of multiple vessels or multiple agitators. The impeller is normally located about one-fourth to one-third of the tank diameter above the bottom of the tank. The liquid height in the tank ranges from 0.75 to 1.5 times the tank diameter. For higher liquid levels, two or more impellers are mounted on the same shaft. Baffles are often installed to increase fluid agitation in the tank.

Power consumption for propeller and turbine agitators generally falls within a range of 0.03 to 0.2 kW/m^3 for mild agitation, 0.2 to 0.5 kW/m^3 for vigorous agitation, and 0.5 to 2 kW/m^3 for intense agitation. These ranges are for overall power consumption and are valid for viscosities up to about 25 Pa·s. At higher viscosities, special impellers are generally used.

For mixing materials that have viscosities greater than 200 Pa·s, such as pastes and polymers, simple agitation normally is insufficient. Such materials will require simultaneous squeezing, dividing, and folding to achieve homogeneous mixing and necessitate increasing power requirements as the material becomes more viscous. Figure 12-38 shows the type of mixing equipment used and the power consumption of these devices with increasing viscosity.

The kneader with its separation, folding, and compression action similar to a kneading operation can process up to 1.0 kg/s of heavy, stiff, and gummy material. The viscosity range covered by this device is between 200 and 2000 Pa·s. Power consumption varies from 10 kW/(kg/s) for low-viscosity pastes to 150 kW/(kg/s) for polymers.

Extruders, used for plastic fabrication, can also be used as mixers. In such devices, a screw of variable pitch and diameter rotates inside a tapered cylinder and extrudes the mixed product from the tapered end. Extrusion action is used for materials that require high shear under pressure to provide improved homogeneity. Extruders can handle materials over the same viscosity range processed by kneaders as well as higher-viscosity polymers and elastomers. Power consumption may approach 1000 kW/(kg/s) for those materials exhibiting a viscosity of 10,000 Pa·s.

Other types of mixers used with high-viscosity semisolids include rotor mixers, muller mixers, and mixing rolls. The rotor mixer is a variation of the auger conveyor using rotors with propellerlike blades that cut and mix as the material is conveyed. Rotor mixers are most effective with modestly viscous nonsticky pastes, soft lumpy solids, and dry powders. Power consumption is below that required for kneaders, but the mixing intensity is also less.

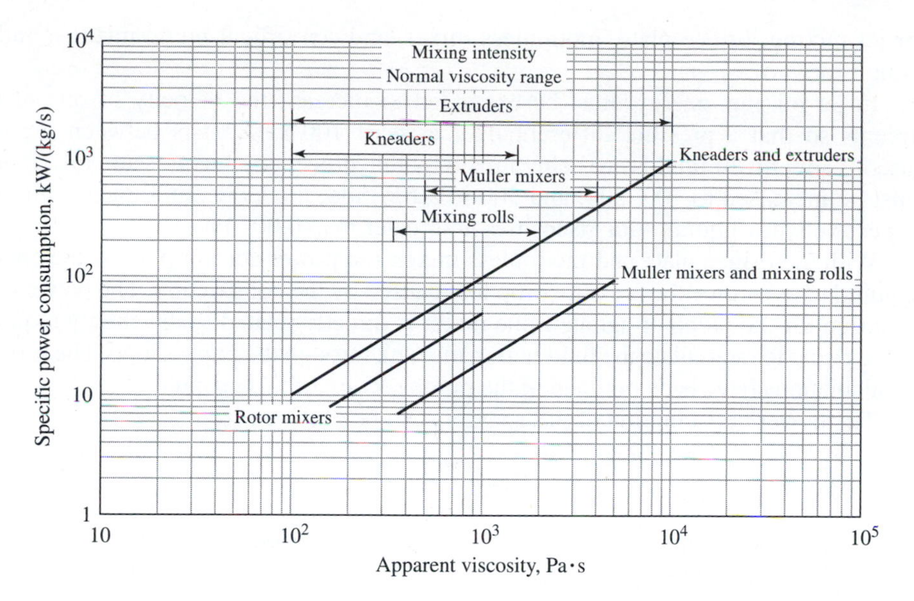

Figure 12-38
Normal power consumption ranges for mixers processing high-viscosity pastes and polymers

Muller mixers operate with wheels rolling over the material, crushing and rubbing the material in an action similar to that of a mortar and pestle. The mixers can be used in both batch and continuous operation where the latter operation can provide reasonable residence times and steady flow. Muller mixers are useful for uniformly coating granular solids with liquid and the mixing of some dry powders.

In mixing rolls, material is forced through a narrow space between two or more rotating rollers that may rotate at different speeds to create shear as well as compression. Because of their simple mechanical design, mixing rolls are relatively low in capital cost and are selected for those applications for which their mixing characteristics are adequate. However, mixing rolls are not as versatile for compounding operations as those associated with kneaders and extruders.

The ribbon mixer is another variation of auger conveying where concentric, double-helical counterrotating ribbons disperse and blend free-flowing powders. Such action on a more intense scale can come from using certain mills, normally used for size reduction. For example, drum or vibratory pebble and jet mills can be used for continuous mixing and blending of friable and free-flowing solids, while hammer, cage, and attrition mills are often used to blend mixtures of sticky and gummy materials. For mixing purposes, power consumption of these devices is lower and capacity higher than when they are used strictly as grinding equipment.

Design Procedures for Agitators and Mixers

In agitators and mixers, the most important design variables are power consumption and residence time required to accomplish the mixing process. The power consumption for jet mixing, orifice plate, motionless mixer, and centrifugal pump mixer is rather straightforward.

In jet mixing, energy must be expended to increase the velocity of one of the streams so that a pressure differential of at least 100 kPa exists between the two gaseous streams. If liquids are being mixed by this procedure, the pressure differential must be increased to 100, 200, and 300 kPa for mild, moderate, and intense mixing, respectively, for liquids with viscosities no greater than 0.001 Pa·s.

With the orifice plate and motionless mixer, the power consumption is the energy required to overcome the pressure drop loss caused by the constriction. The power consumption for the pump or agitated-line mixer is directly related to the energy required by the pump to recirculate the fluid within the tank through the bypass loop. These types of calculations have been outlined in the pump section of this chapter.

The power consumption of gas spargers can be obtained from

$$P = \frac{\dot{m}_G \eta_c \, \Delta p}{\rho_V} \tag{12-38}$$

where $\dot{m}_G$ is the gas mass flow rate, η_c the overall compressor efficiency, Δp the compressor differential pressure, and ρ_V the gas density. To determine Δp, assume 10 to 30 kPa as the pressure drop through the sparger, and then add the static pressure exerted by the fluid in the tank.

Mixing with propeller or turbine agitators is considerably more complex. The power consumption depends on the flow pattern in the mixer and on the geometric dimensions of the equipment. These variables, through dimensional analysis, provide the relation

$$N_{\text{Po}} = \mathfrak{f}'(\text{Re, Fr}, S_1, \ldots, S_n) \tag{12-39}$$

where N_{Po} is the power number defined by $P/(N_r^3 D_a^5 \rho)$ and is a function of the Froude number expressed as $D_a N_r^2/g$, and the shape factors $S_1, \ldots, S_n$ are based on the dimensions of the mixer, as shown in Fig. 12-39. In these three expressions, P is the power consumption, N_r the impeller rotation per unit time, D_a the diameter of the impeller, ρ the density of the fluid, μ the viscosity of the fluid, and g the acceleration of gravity. The Froude number in Eq. (12-39) becomes a factor only when the Reynolds number is greater than 300. When it is a factor, the Froude number is related to the power number by the exponential expression

$$N_{\text{Po}}/\text{Fr}^m = \mathfrak{f}'(\text{Re}, S_1, \ldots, S_n) = \phi \tag{12-40}$$

where ϕ is referred to as the power function and the exponent m is empirically related to the Reynolds number by the relation

$$m = \frac{a - \log \text{Re}}{b} \tag{12-41}$$

Constants a and b must be evaluated experimentally for each new set of shape factors.

Figure 12-40 presents a typical plot of the power function as a function of the Reynolds number for a flat-bladed turbine mixer with six blades.† When baffled with

---

†J. H. Rushton, E. W. Costich, and H. J. Everett, *Chem. Eng. Prog.*, **46:** 395, 467 (1950).

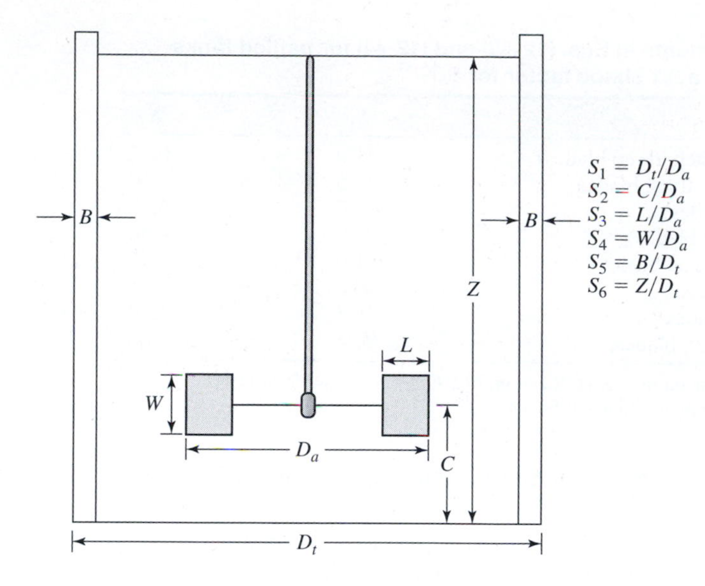

Figure 12-39
Turbine mixer dimensions and associated shape factors used for evaluating power requirements for the mixer

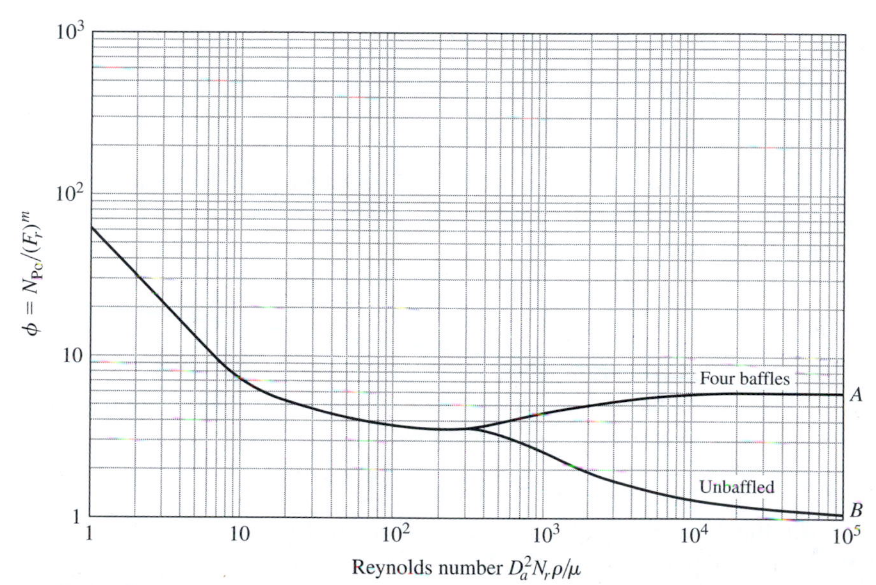

Figure 12-40
Relation between the power function ϕ and the Reynolds number for a six-blade turbine mixer. Constants a and b in Eq. (12-41) for this mixer have been evaluated as 1.0 and 40.0, respectively.

four baffles, curve A applies and $N_{Po} = \phi$; without baffles, curve B applies. The shape factors that apply to both curves are $S_1 = 3.0$, $S_2 = 1.0$, $S_3 = 0.25$, and $S_6 = 1.0$. In addition, for curve A, $S_5 = 0.1$. The constants a and b in Eq. (12-41) for this specific set of shape factors are 1.0 and 40.0, respectively. Power function curves for other

Table 12-9 Constants in Eqs. (12-43) and (12-44) for baffled tanks with a 0.1 shape factor for S_5†

Type of impeller	K_L	K_T
Propeller (square pitch, three blades)	41.0	0.32
Propeller (pitch 2:1, three blades)	43.5	1.00
Turbine (six flat blades)	71.0	6.30
Turbine (six curved blades)	70.0	4.80
Turbine (six arrowhead blades)	71.0	4.00
Fan turbine (six blades)	70.0	1.65
Flat paddle (two blades)	36.5	1.70
Shrouded turbine (six blades)	97.5	1.08

†Reprinted with permission from J. H. Rushton, *Ind. Eng. Chem.*, 44, 2931 (1952). Copyright © 1952 American Chemical Society.

shape factors are available in the literature for both propeller and turbine agitators and mixers.

The power consumption that is delivered to the fluid is obtained by combining Eq. (12-40) and the definition of the power number, resulting in

$$P = \phi \mathrm{Fr}^m N_r^3 D_a^5 \rho \tag{12-42}$$

When the Froude number is not a factor, the power relation reduces to

$$P = \phi N_r^3 D_a^5 \rho \tag{12-42a}$$

Further simplifications can be made at both low and high Reynolds numbers. For Reynolds numbers less than 10, the flow is laminar and density no longer is a factor. Equation (12-40) becomes

$$N_{\mathrm{Po}}\mathrm{Re} = \mathfrak{f}'_L(S_1, \ldots, S_m) = \frac{P}{N_r^2 D_a^3 \mu} = K_L \tag{12-43}$$

In baffled tanks at Reynolds numbers greater than 10,000, the power function is independent of the Reynolds number, and viscosity is no longer a factor. In this flow range, the flow is turbulent and Eq. (12-39) becomes

$$N_{\mathrm{Po}} = \mathfrak{f}'_T(S_1, \ldots, S_n) = K_T \tag{12-44}$$

Values for the constants K_L and K_T are given in Table 12-9 for various types of impellers operating in tanks provided with four baffles.

Residence time for mixing is a function of the viscosity of the fluid, the volume to be mixed, and indirectly the power consumed. For either a propeller or a turbine impeller, mixing miscible liquids and solutions with moderate agitation, the mixing time for a batch vessel can be estimated from†

$$\theta_r = 12{,}000 \left(\frac{\mu V'}{P}\right)^{0.5} \left(\frac{V'}{1.0\ \mathrm{m}^3}\right)^{0.2} \tag{12-45}$$

†W. L. McCabe and J. C. Smith, *Unit Operations of Chemical Engineering*, 3d ed., McGraw-Hill, New York, 1976.

where μ is the viscosity in Pa·s, P the power in W, and V' the volume being mixed in the tank in m^3. Equation (12-45) estimates that batch mixing of 0.01 m^3 of water-based solutions at room temperature with a 0.5-kW mixing device requires a mixing time of approximately 21 s. For continuous-flow tanks, a residence time equal to this batch mixing time is adequate.

The task of developing mixer specifications for propeller and turbine mixers operating at viscosity levels less than 0.1 Pa·s can be greatly simplified by the use of a spreadsheet. The procedures for developing such a spreadsheet have been outlined in considerable detail by Drury and Gates.† This spreadsheet illustrates the interrelationship of mixer performance criteria exhibited in Eq. (12-39) and how these criteria vary with scale-up. For tank capacities from 1 to 40 m^3 the spreadsheet provides the required tank and impeller diameters, liquid level in the tank, impeller speed, power consumption, torque, Reynolds number, turnover ratio, and residence time for mild, medium, and intense mixing. The latter are defined as the equivalent of average bulk-fluid velocities of 0.09, 0.18, and 0.27 m/s, respectively. The spreadsheet developed by these authors assumes a fixed set of shape factors that are most often used to obtain optimum mixing. For example, the ratio of the height of the fluid to the tank diameter has been set at 1.0, and the ratio of the impeller diameter to the tank diameter at either 0.25 or 0.35.

The design of mixers for high-viscosity applications or for non-Newtonian fluids requires considerable expertise. Assistance with such designs is provided by Dickey.‡

Determination of Power Requirements for a Mixer — EXAMPLE 12-6

A turbine mixer has been selected to mix concentrated sodium hydroxide with water to produce a 50% caustic soda solution at a temperature of 65°C and a pressure of 101.3 kPa. The vertical tank is 1.8 m in diameter and filled to a depth of 1.8 m. A 0.6-m-diameter flat-blade turbine with six blades, located centrally in the tank and 0.6 m above the bottom of the tank, rotates at 90 r/min. The tank is unbaffled. What is the power requirement for this mixer?

■ Solution

Estimation of the power function requires calculation of the Reynolds and Froude numbers. Data from *Perry's Chemical Engineers' Handbook* indicate that the viscosity and density of a 50 weight percent caustic solution at 65°C are 12 cP and 1498 kg/m^3, respectively. The shape factors for the turbine mixer are $S_1 = 3.0$, $S_2 = 1.0$, and $S_6 = 1.0$. Since the shape factors match those for curve B of Fig. 12-40, this figure can be used to simplify the calculation. The Reynolds and Froude numbers for the caustic solution are

$$\text{Re} = \frac{D_t^2 N_r \rho}{\mu} = \frac{(0.6)^2(90/60)(1498)}{12(10^{-3})} = 67{,}400$$

$$\text{Fr} = \frac{D_t N_r^2}{g} = \frac{(0.6)(90/60)^2}{9.8} = 0.137$$

†S. F. Drury and L. E. Gates, *Chem. Eng.*, **108**(2): 62 (2001). The spreadsheet can be downloaded directly from www.sharpemixers.com.

‡D. S. Dickey, *Chem. Eng.*, **107**(5): 68 (2000).

The power consumption can be obtained by using Eq. (12-42). This requires evaluation of the exponent *m* from Eq. (12-41). The constants *a* and *b* for the unbaffled six-blade turbine mixer in Fig. 12-40 are 1.0 and 40.0, respectively. Thus,

$$m = \frac{a - \log \text{Re}}{b} = \frac{1.0 - \log 67{,}400}{40} = -0.096$$

From Fig. 12-40 using curve *B* for an unbaffled tank with a Reynolds number of 67,400, the power function ϕ is 1.06. The power required can now be determined from Eq. (12-42) as

$$P = (1.06)(0.137)^{-0.096}\left(\frac{90}{60}\right)^3 (0.6)^5(1498)(10^{-3}) = 0.504 \text{ kW}$$

Conservatively, a 0.75-kW motor should be recommended.

What is the power requirement if four baffles, each with a width of 0.02 m ($S_5 = 0.1$), are added to the tank to mix the caustic soda? Since all the shape factors are the same as those associated with curve *A* of Fig. 12-40, the same procedure as followed above can be used to determine the power consumption. At a Reynolds number of 67,400, the power function ϕ is 6.0. At these conditions, the Froude number no longer is a factor, and Eq. (12-42*a*) can be used.

$$P = (6.0)\left(\frac{90}{60}\right)^3 (0.6)^5(1498)(10^{-3}) = 2.35 \text{ kW}$$

The power consumption for the baffled tank could also have been estimated by using Eq. (12-44) and Table 12-9. Rearrangement of Eq. (12-44) results in

$$P = K_T N_r^3 D_a^5 \rho = (6.3)\left(\frac{90}{60}\right)^3 (0.6)^5(1498)(10^{-3}) = 2.47 \text{ kW}$$

Similar results are obtained by using a software package designed for mixing various kinds of miscible fluids.

■

Costs for Agitators and Mixers

The costs for jet mixers, orifice plates, and gas spargers are quite small compared with the costs of other process equipment. Generally, their cost is the cost of the pumps and compressors that are needed to pressurize the fluid or overcome the pressure drop generated with the insertion of the mixer. When a vessel or tank is involved as with a gas sparger or pump-mixing loop, the cost of the vessel or tank is included in the cost tabulation. For an agitated-line mixer, assume a cost of 1.5 times that of the centrifugal pump of the same capacity.

Estimated costs for motionless mixers, propeller and turbine agitators, rotary and ribbon blenders, kneaders, mullers, and extruders are presented in Figs. 12-41 through 12-49. Costs for milling equipment, used as mixers, can be found later in the chapter by using Figs. 12-68 through 12-74.

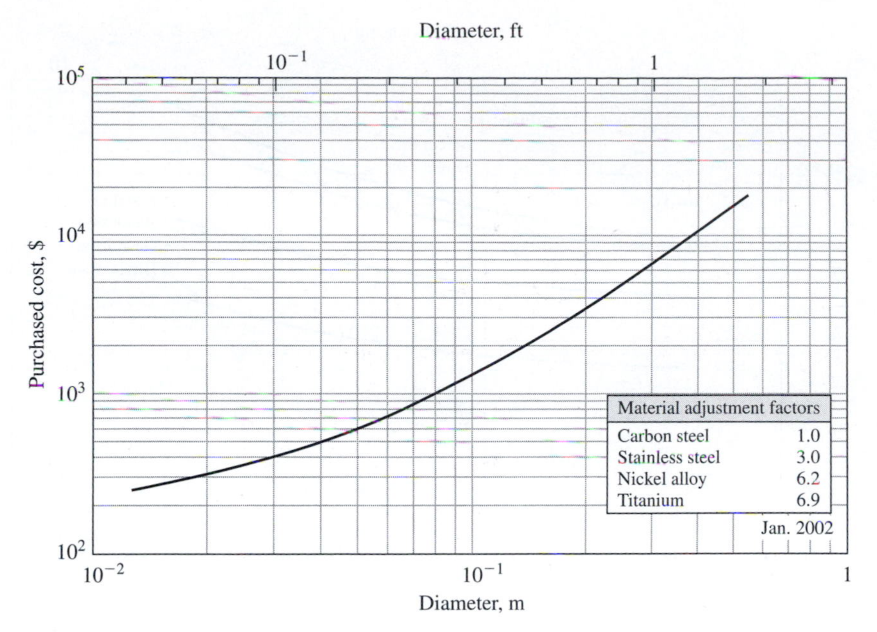

Figure 12-41
Purchased cost of motionless mixers

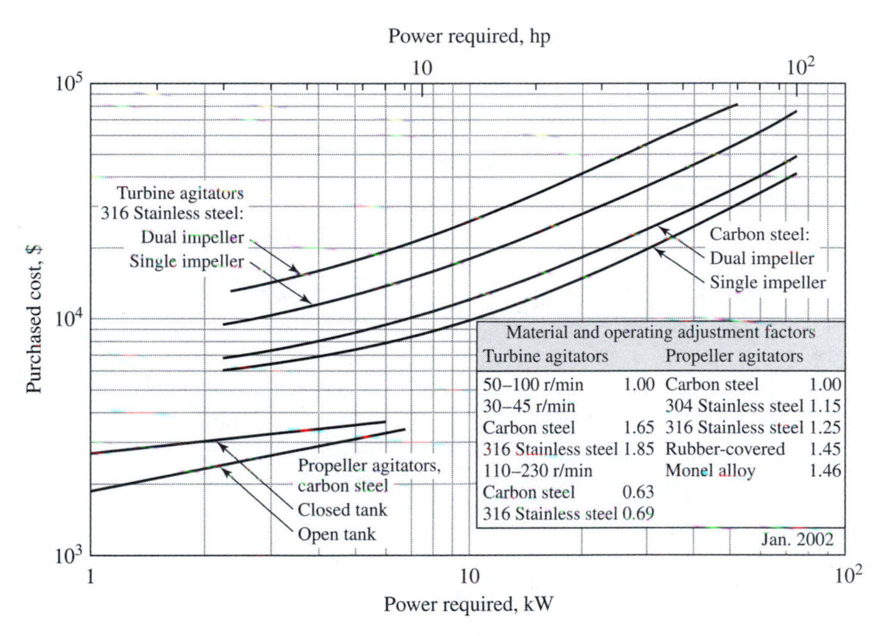

Figure 12-42
Purchased cost of turbine and propeller agitators

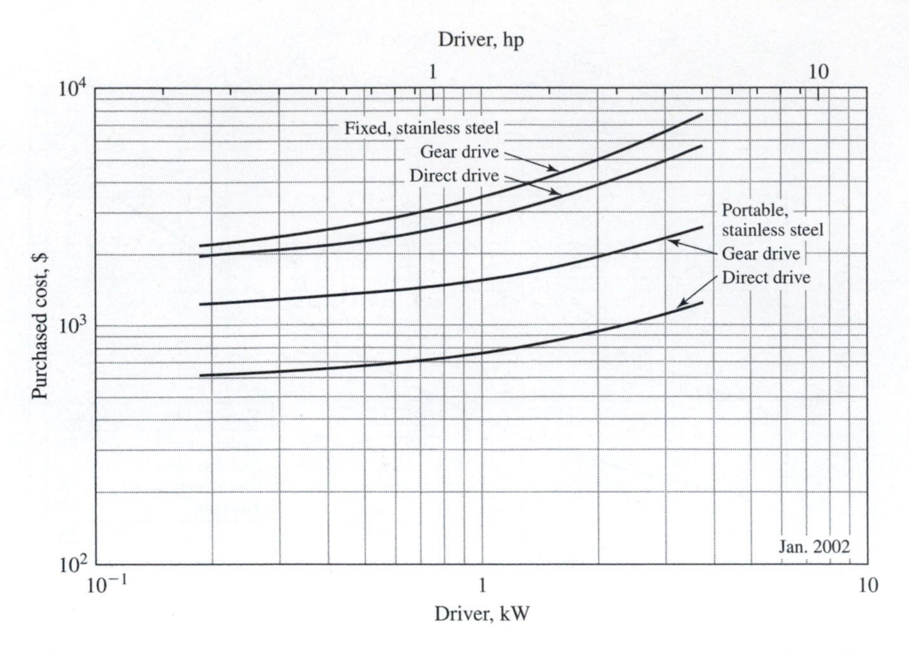

Figure 12-43
Purchased cost of propeller mixers

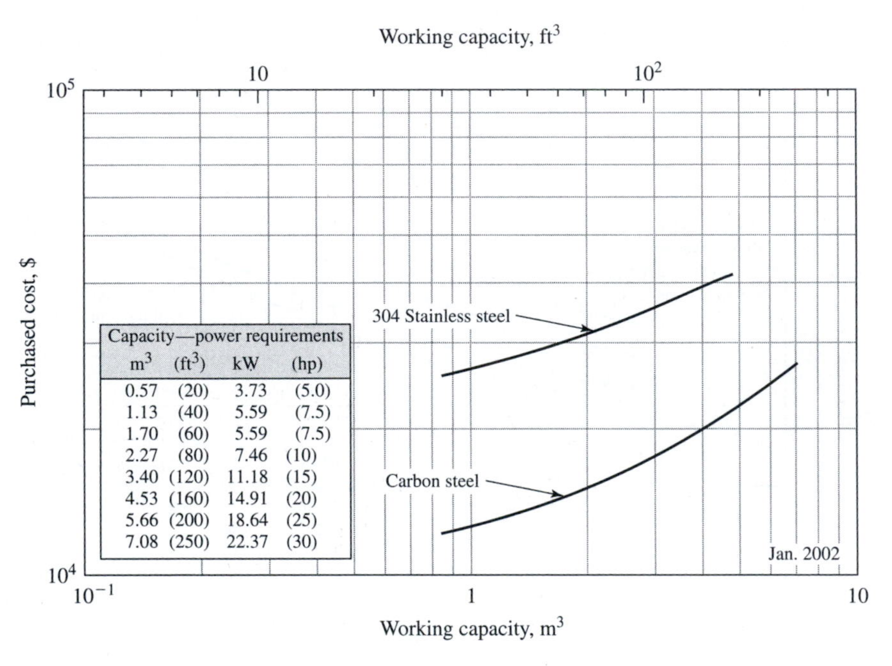

Capacity—power requirements			
m^3	(ft^3)	kW	(hp)
0.57	(20)	3.73	(5.0)
1.13	(40)	5.59	(7.5)
1.70	(60)	5.59	(7.5)
2.27	(80)	7.46	(10)
3.40	(120)	11.18	(15)
4.53	(160)	14.91	(20)
5.66	(200)	18.64	(25)
7.08	(250)	22.37	(30)

Figure 12-44
Purchased cost of double-cone rotary blenders. Price does not include motor.

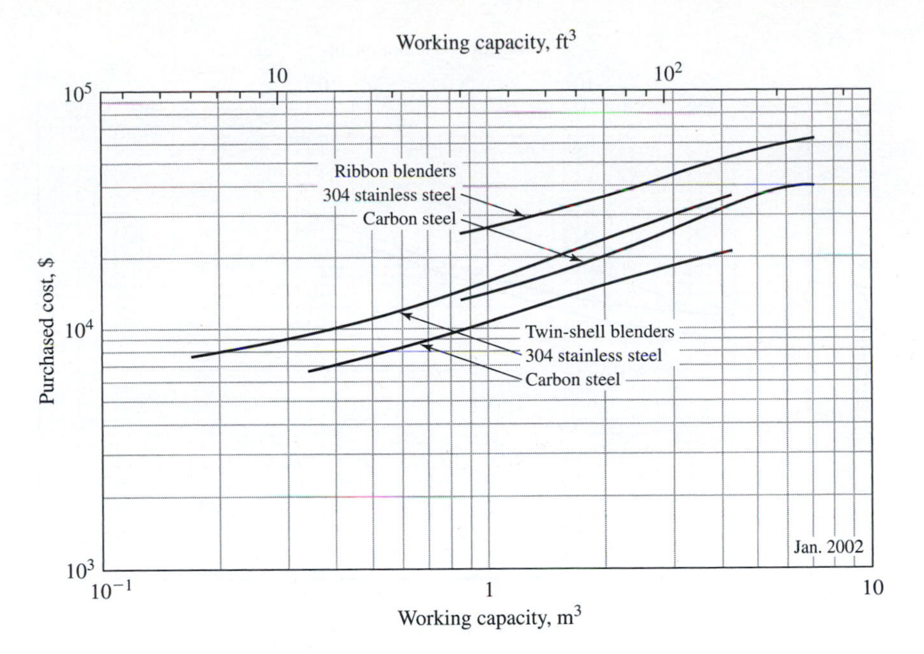

Figure 12-45
Purchased cost for ribbon blenders and twin-shell blenders

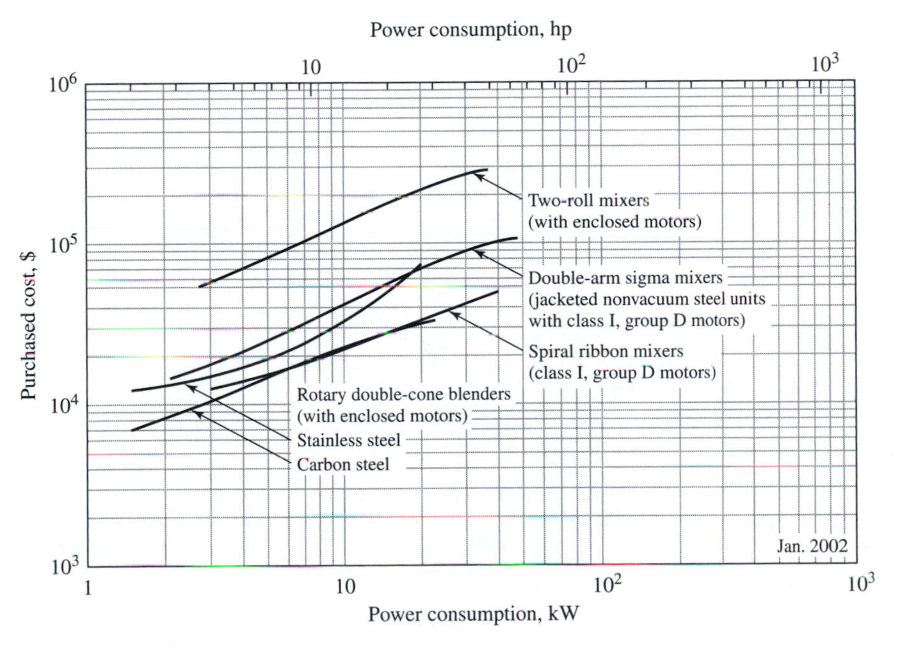

Figure 12-46
Purchased cost for low-power consumption mixers and blenders

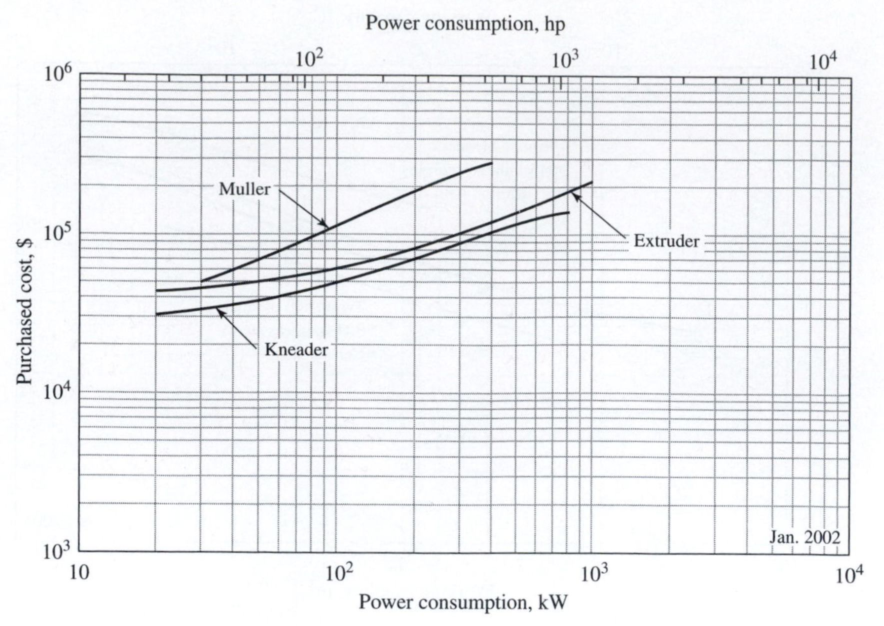

Figure 12-47
Purchased cost of mixers, including motors, for heavy-power consumption processes

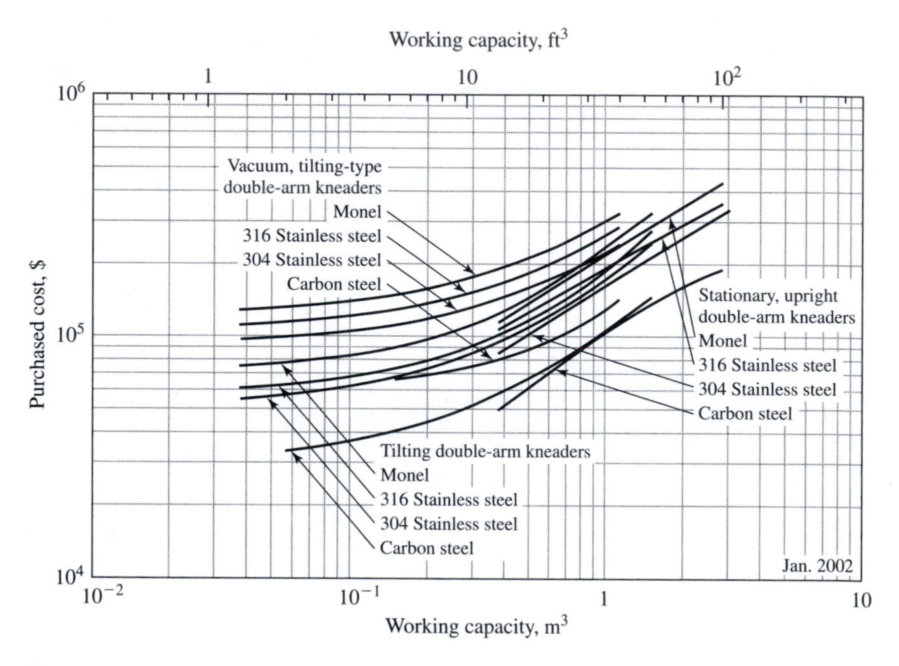

Figure 12-48
Purchased cost of kneaders. Price includes machine, jacket, gear reducer and drive, cover, nozzle, and agitator. Motor and starter are not included in purchase price.

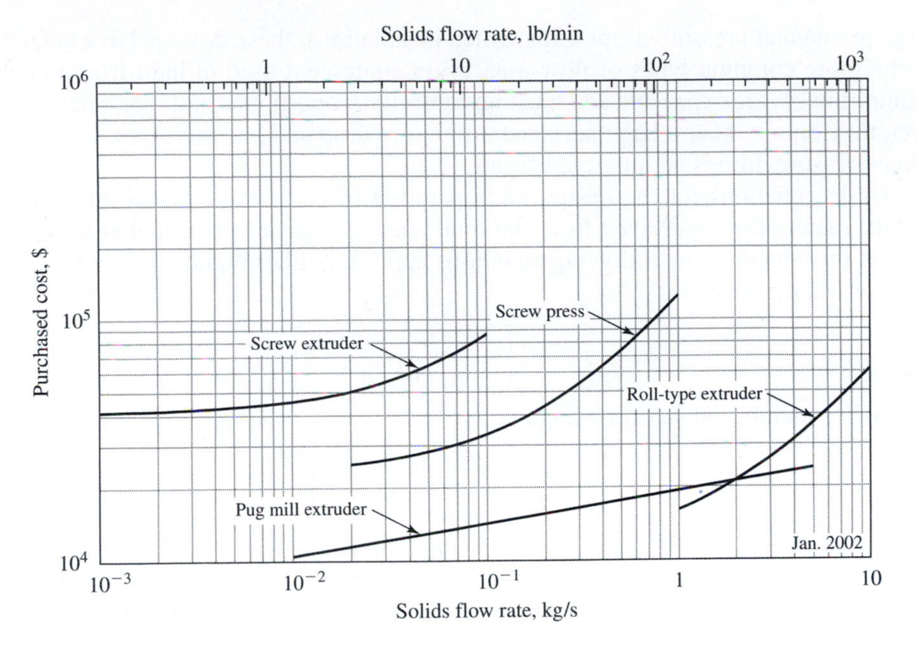

Figure 12-49
Purchased cost of solids extrusion and consolidation equipment

FLOW MEASUREMENT OF FLUIDS

The transfer and storage of liquids and gases involve a comprehension of the properties and behavior of these fluids. Such fluids are acted upon by many forces that result in changes in pressure, temperature, stress, and strain. A fluid is said to be isotropic when the relationship between the components of stress and those of the strain rate is the same in all directions. To completely describe the flow regime requires a knowledge of the pressures and temperatures existing throughout the transfer system. However, for preliminary design estimates, average pressures and temperatures are normally used to obtain the flow rate of the liquid, the key parameter required for specifying pipe, pump, and compressor sizes or capacities.

Rotameters, orifice meters, venturi meters, and displacement meters are used extensively in industrial operations for measuring the rate of fluid flow. Other flow-measuring devices such as weirs, anemometers, pitot tubes, and wet-test meters are also useful.† The rotameter and the orifice meter are the most widely used of these measuring instruments. The rotameter indicates the flow rate with a plummet supported in a slightly tapered tube by a fluid stream that has been throttled by a constriction. Precision-bore tubing for rotameters contributes to the higher cost of these units. Orifice meters, on the other hand, are the least expensive and most flexible of the various types of devices for measuring flow rates. Despite the inherent disadvantage of

†Detailed descriptions of various types of flow-measuring equipment and derivations of related equations are presented in detail in essentially all texts dealing with chemical engineering principles. See R. H. Perry and D. W. Green, eds., *Perry's Chemical Engineers' Handbook,* 7th ed., McGraw-Hill, New York, 1997.

large permanent pressure drops with orifice installations, these devices have been one of the more common types of flow-measuring equipment used in industrial practice. Venturi meters are expensive and must be carefully proportioned and fabricated. However, they do not incur a large permanent pressure drop and are, therefore, very useful when pressure drop is an important factor.

Basic equations for the design and operation of rotameters, orifice meters, and venturi meters can be derived from the total energy balances presented at the beginning of this chapter. The following equations apply for either liquids or gases.

For rotameters

$$\dot{m}_v = YC_d A_c \left[\frac{2V_p' g(\rho_p - \rho)/\rho}{A_p(1-\beta^4)}\right]^{1/2} \tag{12-46}$$

For orifice meters and venturi meters

$$\dot{m}_v = YC_d A_c \left[\frac{2(p_1 - p_2)/\rho}{1-\beta^4}\right]^{1/2} \tag{12-47}$$

where $\dot{m}_v$ is the volumetric flow rate in m^3/s, Y the expansion factor (dimensionless), C_d the coefficient of discharge (dimensionless), A_c the cross-sectional flow area at the minimum cross-sectional flow area in m^2, A_p the cross-sectional area in the upstream pipe before the constriction in m^2, V_p' the volume of the plummet in m^3, g the local gravitational acceleration in m/s^2, ρ_p the density of the plummet in kg/m^3, ρ the average density of the fluid in kg/m^3, β the ratio of the throat diameter to the pipe diameter (dimensionless), p_1 the static pressure in the upstream pipe before the constriction in kPa, and p_2 the static pressure at the minimum cross-sectional flow area in kPa. For the flow of liquids, the expansion factor Y is unity. For the flow of gases, the expansion factor that allows for the change in gas density as it expands adiabatically from p_1 to p_2 is given by

$$Y = \left[r^{2/k}\left(\frac{k}{k-1}\right)\left(\frac{1-r^{(k-1)/k}}{1-r}\right)\left(\frac{1-\beta^4}{1-\beta^4 r^{2/k}}\right)\right]^{1/2} \tag{12-48}$$

for venturi meters and flow nozzles, where $r = p_2/p_1$ and k is the ratio of the specific heat of the gas at constant pressure to the specific heat at constant volume. Values of Y obtained from Eq. (12-48) are shown in Fig. 12-50 as a function of r, k, and β.

The value of the coefficient of discharge C_d for orifice meters depends on the properties of the flow system, the ratio of the orifice diameter to the upstream diameter, and the location of the pressure-measuring taps. Values of C_d for sharp-edged orifice meters are presented in Fig. 12-51. These values apply strictly for pipe orifices with throat taps, in which the downstream pressure tap is located one-third of one pipe diameter from the downstream side of the orifice plate and the upstream tap is located one pipe diameter from the upstream side. However, within an error of about 5 percent, the values of C_d indicated in Fig. 12-51 may be used for manometer taps located anywhere between the orifice plate and the hypothetical throat taps.

Venturi meters usually have a tapered entrance with an interior total angle of 25 to 30° and a tapered exit with an interior angle of 7°. Under these conditions, the value of the coefficient of discharge may be assumed to be 0.98 if the Reynolds number based on conditions in the upstream section is greater than 5000.

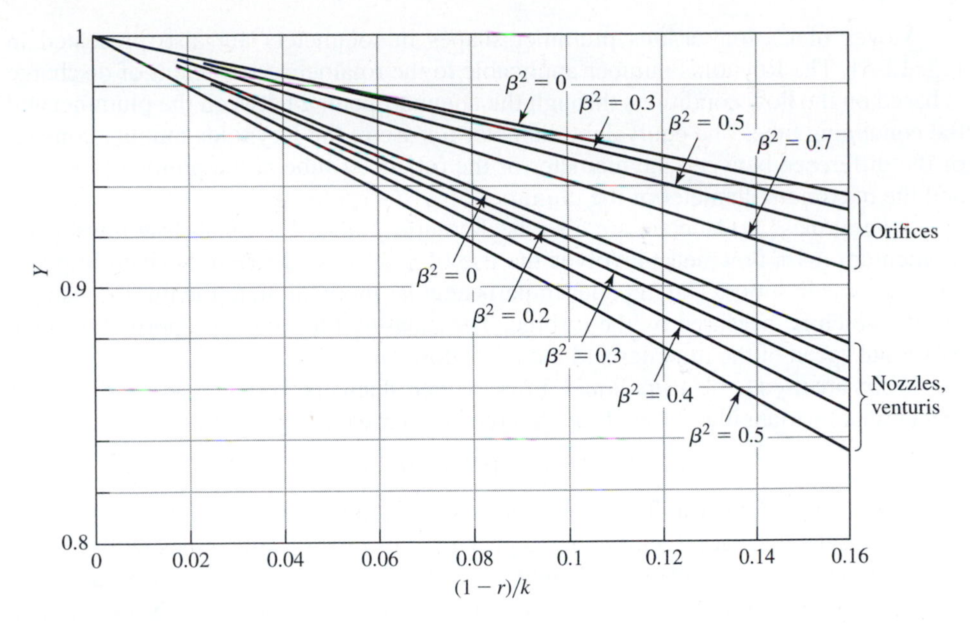

Figure 12-50
Values of expansion factor Y for orifices, nozzles, and venturis

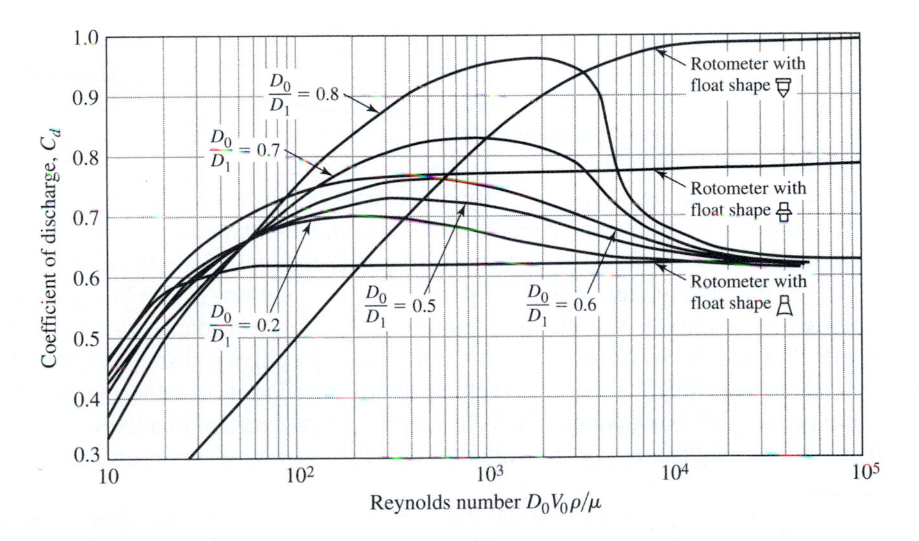

Figure 12-51
Coefficients of discharge for square-edged orifices with centered circular openings and for rotameters. (Subscript 0 indicates "at orifice or at constriction," and subscript 1 indicates "at upstream section.")

Values of C_d for various plummet shapes in rotameters are also presented in Fig. 12-51. The Reynolds number applicable to the rotameter coefficient of discharge is based on the flow conditions through the annular opening between the plummet and the containing tube. The equivalent diameter for use in the Reynolds number consists of the difference between the diameter of the rotameter tube at the plummet location and the maximum diameter of the plummet.

Several mass flowmeters are also used commercially. The axial-flow transverse-momentum mass flowmeter involves the use of axial flow through a driven impeller and a turbine in series. The impeller imparts angular momentum to the fluid, causing a torque which is measured by the turbine. The measured torque is proportional to the rotational speed of the impeller and the mass flow rate.

Weirs are used to measure liquid flows in open channels. Weirs are generally either rectangular or triangular. Flow through a rectangular weir is given by

$$\dot{m}_v = 0.415(L - 0.2H_o)H_o^{1.5}(2g)^{1/2} \tag{12-49}$$

where $\dot{m}_v$ is the volumetric flow rate, L the length of the rectangular weir, H_o the weir head, and g the local acceleration due to gravity. This relationship agrees within 3 percent of experimental results if the length is greater than $2H_o$, the velocity of approach is 0.6 m/s or less, the weir head is not less than 0.09 m, and the height of the crest above the bottom of the channel is at least $3H_o$.

STORAGE AND CONTAINMENT OF FLUIDS

Storage of liquid materials is commonly accomplished in industrial plants by the use of cylindrical, spherical, or rectangular tanks. These tanks may be constructed of wood, concrete, fiber-reinforced plastic, or metal. The last is the most common material of construction, although use of fiber-reinforced plastics has increased during the past decade. The design of storage vessels involves consideration of details such as wall thickness, size and number of openings, shape of heads, necessary temperature and pressure controls, and corrosive action of the contents.

The same principles of design apply for other types of tanks, including pressure vessels such as those used for chemical reactors, mixers, and distillation columns. For these cases, the shell is often designed and its cost estimated separately with the other components, such as tray assemblies, agitators, linings, and packing units, also being handled separately. Process pressure vessels are normally designed in accordance with the *ASME Boiler and Pressure Vessel Code.*[†] These units are usually cylindrical shells capped with an elliptical or hemispherical head at each end with installation in either a vertical or horizontal position. A major concern in the design is to make certain that the walls of the vessel are sufficiently thick to permit safe use under all operating conditions.

[†]The *ASME Boiler and Pressure Vessel Code* is published by the ASME Boiler and Pressure Vessel Committee, American Society of Mechanical Engineers, New York, with a new edition coming out every 3 years. Section VIII of the code deals specifically with pressure vessels with the basic rules being given in Division 1 and alternative rules being presented in Division 2. For a summary of Section VIII, Division 1 of this ASME code, see R. H. Perry and D. W. Green, eds., *Perry's Chemical Engineers' Handbook,* 7th ed., McGraw-Hill, New York, 1997, pp. 10-144 to 10-148.

Design Procedures for Pressure Vessels

The necessary wall thickness for metal vessels is a function of the ultimate tensile strength or yield point of the metal at the operating temperature, the operating pressure, the diameter of the tank, and the joint or welding efficiencies during fabrication.† Table 12-10 presents a summary of design relationships based on the *ASME Boiler and Pressure Vessel Code* as specified in Section VIII, Division 1.

Costs for Tanks, Pressure Vessels, and Storage Equipment

Numerous approaches have been developed over the years which provide methods for obtaining tank and pressure vessel costs based on detailed estimates of the costs for individual components in the fabrication of such equipment, such as materials, nozzles, manholes, and support skirts as well as labor and overhead. Final installed cost can be obtained by applying factors to account for freight, labor, materials, engineering, and overhead related to shipping the unit to the plant and installing it ready for use. These methods take into account the materials of construction to be used as well as operating temperature and pressure. However, the time needed to make such cost estimates cannot be justified when only a preliminary design is required.

For making a preliminary cost estimate for tanks and pressure vessels, it is preferable to estimate their weight and apply a unit cost in dollars per kilogram of material required. The unit f.o.b. cost of carbon steel and 304 stainless steel was determined to vary as the -0.34 power of the weight. Since many stainless-steel vessels frequently use considerable carbon steel in the form of support skirts, brackets, flanges, bolts, etc., it is necessary to assume that each kilogram of carbon steel is equivalent to approximately 0.4 kg of stainless steel.

Weights of tanks and pressure vessels are obtained by calculating the cylindrical shell and heads separately and then adding the weights of the nozzles, manholes, and skirts or saddles. If information on the weight of this tank hardware is not available, increase the weight calculated for that required by the tank shell and the two heads by 15 percent for vessels to be installed in a horizontal position and by 20 percent for vessels to be installed in a vertical position. Steel density can be assumed to be 7833 kg/m^3.

The weight of the formed head of such tanks can be approximated by calculating the area of the blank disk of metal used for forming the heads. The required diameter of the blank disk can be obtained by multiplying the actual outside diameter of the shell by the appropriate factor given in Table 12-11. These factors include an allowance for the flange that is necessary for welding purposes.

Cost for pressure vessels in January 2002 (including nozzles, manholes, saddles or skirts, but no internals such as trays or agitators) as dollars per kilogram weight of the fabricated unit f.o.b. with carbon steel as the cost basis is given by

$$\text{Cost} = 73(W_v)^{-0.34} \tag{12-50}$$

where W_v is the total calculated weight of the vessel in kg. This relation is applicable over a weight range from 400 to 50,000 kg. The cost factor to convert from carbon

†In the design of vacuum vessels, the ratio of length to diameter must also be considered.

Table 12-10 Design equations and data for pressure vessels based on the *ASME Boiler and Pressure Vessel Code*†

Recommended design equations for vessels under internal pressure	Limiting conditions
For cylindrical shells	
$t = \dfrac{Pr_i}{SE_J - 0.6P} + C_c$	$t \leq \dfrac{r_i}{2}$ or $P \leq 0.385SE_J$
$t = r_i \left(\dfrac{SE_J + P}{SE_J - P} \right)^{1/2} - r_i + C_c$	$t > \dfrac{r_i}{2}$ or $P > 0.385SE_J$
For spherical shells	
$t = \dfrac{Pr_i}{SE_J - 0.2P} + C_c$	$t \leq 0.356r_i$ or $P \leq 0.665SE_J$
$t = r_i \left(\dfrac{2SE_J + 2P}{2SE_J - P} \right)^{1/3} - r_i + C_c$	$t > 0.356r_i$ or $P > 0.665SE_J$
For ellipsoidal head	
$t = \dfrac{PD_a}{2SE_J - 0.2P} + C_c$	0.5 (minor axis) $0 = 0.25D_a$
For torispherical (spherically dished) head	
$t = \dfrac{0.885PL_a}{SE_J - 0.1P} + C_c$	r = knuckle radius = 6% of inside crown radius and is not less than $3t$
For hemispherical head Same as for spherical shells with $r_i = L_a$	

Properties of vessel heads (include corrosion allowance in variables)	2:1 Ellipsoidal	Hemi-spherical	Standard ASME torispherical
Capacity as volume in head, m^3	$\dfrac{\pi D_a^3}{24}$	$\dfrac{2}{3}\pi L_a^3$	$0.9\left(\dfrac{2\pi L_a^2}{3}\text{IDD}\right)$
IDD = inside depth of dish, m	$\dfrac{D_a}{4}$	L_a	$L_a - [(L_a - r)^2 - (L_a - t - r)^2]^{1/2}$
Approximate weight of dished portion of head, kg	$\rho_m \dfrac{\pi(nD_a + t)^2 t}{4}$	$\rho_m \left(2\pi L_a^2 t\right)$	$\rho_m \dfrac{\pi(\text{OD} + \text{OD}/24 + at)^2 t}{4}$

Table 12-10 *Continued*

Joint efficiencies	Metal	Temp., °C	S, kPa
	Recommended stress values		
For double-welded butt joints	Carbon steel	−29 to 343	94,500
If fully radiographed = 1.0	(SA-285, Gr. C)	399	82,700
If spot-examined = 0.85		454	57,200
If not radiographed = 0.70			
In general, for spot examined	Low-alloy steel	−29 to 427	94,500
If electric resistance weld = 0.85	for resistance to	510	75,800
If lap-welded = 0.80	H_2 and H_2S	565	34,500
If single-butt-welded = 0.60	(SA-387, Gr.12C1.1)	649	6,900
	High-tensile steel	−29 to 399	137,900
	for heavy-wall	454	115,800
	vessels	510	69,000
	(SA-302, Gr.B)	538	42,750
	High-alloy steel		
	for cladding and	−29	128,900
	corrosion resistance	343	77,200
	Stainless 304	427	72,400
	(SA-240)	538	66,900
	Stainless 316	−29	128,900
	(SA-240)	345	79,300
		427	75,800
		538	73,100
	Nonferrous metals		
	Copper	38	46,200
	(SB-11)	204	20,700
	Aluminum	38	15,900
	(SB-209, 1100-0)	204	6,900

Nomenclature for Table 12-10

a = 2 for thicknesses <0.0254 m and 3 for thicknesses ≥ 0.0254 m
C_c = allowance for corrosion, m
D_a = major axis of an ellipsoidal head, before corrosion allowance is added, m
E_J = efficiency of joints expressed as a fraction
IDD = inside depth of dish, m
L_a = inside radius of hemispherical head or inside crown radius of torispherical head, before corrosion allowance is added, m
n = 1.2 for $D \leq 1.55$ m, 1.21 for $D = 1.55$–2.0 m, 1.22 for $D = 2.0$–2.7 m, and 1.23 for $D > 2.7$ m
OD = outside diameter, m
P = maximum allowable internal pressure, kPa (gauge)
r = knuckle radius, m
r_i = inside radius of shell, before corrosion allowance is added, m
S = maximum allowable working stress, kPa
t = minimum wall thickness, m
ρ = density of metal, kg/m^3

†See the latest *ASME Boiler and Pressure Vessel Code* for further details.

Table 12-11 Factors to obtain diameters of blank disks required for three types of formed heads

Type of head	Ratio D/t†	Blank diameter factor
ASME head	>50	1.09
	30–50	1.11
	20–30	1.15
Ellipsoidal head	>20	1.24
	10–20	1.30
Hemispherical head	>30	1.60
	18–30	1.65
	10–18	1.70

†D is the head diameter and t the nominal minimum head thickness.

Table 12-12 Cost factors to account for internal pressure levels of vessels†

Pressure level, kPa	Cost factor	Pressure level, kPa	Cost factor
Up to 425	1.0 (basis)	5,500	3.8
775	1.3	6,150	4.0
1450	1.6	6,850	4.2
2100	2.0	10,200	5.4
2800	2.4	13,600	6.5
3450	2.8	20,300	8.8
4150	3.0	27,000	11.3
4800	3.3	33,800	13.8

†If the data are available, it is much better to use the design equations presented in Table 12-10 to obtain the necessary wall thickness based on the stress value at the operating temperature, in place of using the given pressure factors since there is a critical interrelationship among material of construction, operating pressure, and operating temperature in establishing the design and cost of pressure vessels.

steel as the material of construction for the fabricated unit to one of 304 stainless steel ranges between 2.0 and 3.5, for 316 stainless steel between 2.3 and 4.3, for Monel between 4.5 and 9.8, and for titanium between 4.9 and 10.6.

To estimate the purchased cost of pressure vessels that operate with internal pressures greater than the upper limit of 425 kPa allowed by Eq. (12-50), Table 12-12 provides still another set of cost factors to account for this internal pressure change. For example, the cost of a pressure vessel subjected to an pressure of 2100 kPa will be twice the cost of a vessel of the same capacity operating below the 425-kPa limit.

In general, the minimum wall thickness, not including allowances for corrosion, for any metal plate subject to pressure should not be less than 2.4 mm for welded or brazed construction and not less than 4.8 mm for riveted construction except that the thickness of walls for unfired steam boilers should not be less than 6.35 mm. A corrosion allowance of 0.25 to 0.38 mm annually or about 3 mm for a 10-year life is a reasonable value.

maximum angle of inclination is considerably less than the angle of repose of the solid, normally ranging from 15 to 20° with a maximum of 30°. However, inclination of the belt reduces its transport capacity by 5 to 10 percent.

Temperature, chemical activity, and hours of continuous service play an important role in belt selection. Depending upon the operating conditions, belts are fabricated from a variety of elastomers, special rubbers, cotton, and asbestos fibers. Belt width and belt speed are functions of bulk density and lump size of the material being transported. Even though a lower initial conveyor cost can generally be obtained by using the narrowest belt for a given lump size and operating it at the maximum speed, it may be more economical on a long-term basis to use a wider belt with fewer plies to achieve the necessary tensile strength with good belt troughing characteristics. The transport of abrasive or very lumpy material also requires a belt width larger than that recommended for transporting fine or granular material.

Continuous-Flow Conveyors These conveyors employ a chain-supported conveying element that is transversely pulled through the solid being transported. The conveying action of the various designs of continuous-flow conveyors varies with the type of conveying flight used. Because of the conveying principle involved, these units are totally enclosed and provide a relatively high capacity per unit of cross-sectional area.

The fabrication costs associated with the heavy-duty casing and chain-supported conveying element make these units relatively expensive. However, these units require little operating space, have low maintenance, may travel in several directions with only a single drive, are self-feeding, and can feed and discharge at several points during the solids transport. These factors may often compensate for what appears to be a high cost per length of travel. In fact, this type of conveyor is widely used in the chemical industry whenever there is considerable rehandling of the solid or there are requirements for many feed and discharge points of the solid.

The most common chain conveyor is the apron conveyor. These units are available in a wide variety of sizes for both horizontal and inclined transport. The typical design is a series of pans mounted between two strands of roller chains, with pans overlapping to minimize solid losses. Pan design may vary according to material requirements. The main application for these conveyors is the feeding of solids at controlled rates.

Pneumatic Conveyors Solids ranging from fine powders to 6.35-mm pellets and bulk densities of 15 to 3200 kg/m^3 can be suspended in a high-velocity airstream and moved in a pneumatic conveyor both vertically and horizontally over distances of 100 m. The capacity of the pneumatic conveying system depends on the solid bulk density, the diameter and length of the transport system, and the energy content of the air throughout the conveying system. Minimum capacity for pneumatic conveyors is attained when the energy of the conveying air is just sufficient to transport the solid material through the system without any loss due to solids settling. Pneumatic conveyors are generally classified into the five types noted below.

In the pressure-controlled pneumatic conveyor, the solid material is dropped into an airstream where it is suspended until the airstream reaches the receiving vessel where the solid once again is separated from the air. Pressure systems are used for free-flowing solids of any particle size up to 6.35 mm, and solid flow rates exceed

2.5 kg/s. These systems are preferred when one source of solids must supply several receivers.

The advantage of using a vacuum system is that all the pumping energy is used to move the solid. Otherwise, operation of a vacuum system is similar to that for the pressure system. Vacuum systems are generally used when solid flows do not exceed 2 kg/s and the conveying length is no more than 300 m. These units are widely used for moving finely divided solid materials.

The pressure vacuum system combines the advantages of the previous two systems to induce solids in storage onto the conveyor and then transports the solids in suspension to the receiver. Vacuum is used in the first step followed by the use of pressure in the second step. The most typical application is combined bulk vehicle unloading and solids transport to product storage.

Another type of pneumatic conveyor is the fluidized system. These conveyors generally convey prefluidized, finely divided, non-free-flowing solids over short distances. Fluidizing is accomplished in a chamber in which air is passed through a porous membrane located at the bottom of the conveyor. At the point of incipient fluidization, the solid particles take on the characteristics of free flow. Prefluidizing reduces the volume of air required per unit mass of solids transported, and this, in turn, reduces the power requirement. The characteristics of the rest of this system are similar to those of regular pressure- or vacuum-type conveyors.

Still another type of pneumatic conveying is the blow tank in which pressurized air is introduced into a pressure vessel that is used for storage of powders and granules. If the solid material is free-flowing, it will flow through a valve at the bottom of the vessel and move through a short conveying line depending on the solid characteristics. The blow tank principle can be used to feed regular pneumatic conveyors.

Vibrating Conveyors Most vibrating conveyors are essentially directional-throw units which consist of a spring-supported horizontal pan that is vibrated by a direct-connected eccentric arm and rotating eccentric weights. The motion imparted to the solid material particles may vary, but its purpose is to throw the material upward and forward so that it will travel along the conveyor path in a series of short hops.

The solids capacity of vibrating conveyors is determined by the magnitude of trough displacement, frequency of this displacement, slope of trough, angle of throw, and the ability of the solid to receive and transmit through its mass the directional throw of the trough. To be properly conveyed, the solid should have a high friction factor with the pan and with itself. Since there are so many variables that affect the performance of vibrating conveyors, there are no simple relationships for determining the capacity and the power requirements for these conveying units. Available data are generally the result of actual experiments that are quantified with empirical equations.

Bucket Elevators The use of bucket elevators is the simplest and most dependable approach for transporting nonsticky solids in a vertical direction. The conveyor can be visualized as a belt conveyor with the belt replaced by a series of buckets. The latter are linked together to form a continuous chain that moves up and down between rotating top and bottom sprocket wheels. Bucket elevators can be enclosed for dust control but cannot operate leak-free for operation with controlled environments or under reduced pressures. Bucket elevators are common in mineral processing where vertical lifting of abrasive and lumpy material is encountered.

General Design Procedures for Solids Transport Equipment

Even though most of the more meaningful design procedures for estimating the capacity and power requirements for solids transport equipment are proprietary and only available from the vendors of such equipment, it is still possible to arrive at a reasonable estimate of these design quantities with a few simplified relationships that are supported with some generalized solids transport data. This approach will be used to provide some guidance in the design of some of the solids transport equipment discussed in the previous section.

The power required to operate a screw conveyor combines the power necessary to drive the unit empty and the power necessary to move the solid material. The first component is a function of conveyor length, speed of rotation, and friction in the conveyor bearings. The second is a function of the total weight of material conveyed per unit time, conveyed length, and depth to which the trough is loaded. The latter power component is, in turn, a function of the friction between the solid material and the rotating conveyor surface. Since these factors are highly dependent on the mechanical design of the conveyor, no specific relationship would be applicable. However, a preliminary estimate of the power required for horizontal movement of solids in a screw conveyor is given by

$$P = \frac{C_{sc}\dot{m}_s L}{114} \tag{12-51}$$

where P is the power in kW, $\dot{m}_s$ the mass flow rate of the solid in kg/s, and L the length of the conveyor in m. The constant C_{sc} accounts for both power components noted above and the type of solid being conveyed. For free-flowing nonabrasive solids, free-flowing slightly abrasive solids, and free-flowing highly abrasive solids, the values of C_{sc} are 1.3, 2.5, and 4.0, respectively. These three types of solids are identified in one classification scheme as class I, class IIX, and class III, respectively. This classification scheme is also used in Table 12-15, which provides typical capacities and operating conditions for several different diameter horizontal screw conveyors. For inclined conveyors, the additional power to attain this lift must be added to the total power requirement.

Belt conveyor design first should accurately determine the weight per cubic meter and the lump size of the solid that is to be conveyed. For example, for a 0.6-m belt, uniform lump size can range up to about 0.1 m. For each 0.15-m increase in belt width, the lump size can increase by about 0.05 m. The larger the lump size, the greater the possibility of it falling off the belt or rolling back on inclines.

Power to drive a belt conveyor is the summation of the power to move the empty belt and that to move the solid material horizontally and vertically. As with most other conveying systems, it is advisable to obtain relationships from a vendor in making these calculations. A typical equation obtained from one of the manufacturers of belt

Table 12-15 **Capacity and operating conditions for horizontal screw conveyors**[†]

	Maximum capacity for economical service[‡]			
Screw diameter, m	Class I,[§] m^3/h (r/min)	Class IIX,[§] m^3/h (r/min)	Class III,[§] m^3/h (r/min)	Maximum power at 100 r/min, kW
0.15	10.6 (165)	2.1 (60)	0.7 (60)	3.75
0.25	76.5 (140)	18.8 (50)	5.7 (50)	7.5
0.40	158.6 (120)	44.5 (45)	12.0 (45)	18.6
0.51	283.2 (105)	87.9 (40)	22.1 (40)	29.8

[†]Data provided by FMC Corporation, Material Handling Systems Division.
[‡]These classifications cover a broad list of materials that can be conveyed in screw compressors. Care must be given to highly corrosive or highly aerated materials.
[§]Class I material: free-flowing nonabrasive solid, occupying 45% of the conveyor trough area; class IIX: free-flowing, slightly abrasive solid, occupying 30% of the conveyor trough area; class III: free-flowing, highly abrasive solid, occupying 15% of the conveyor trough area.

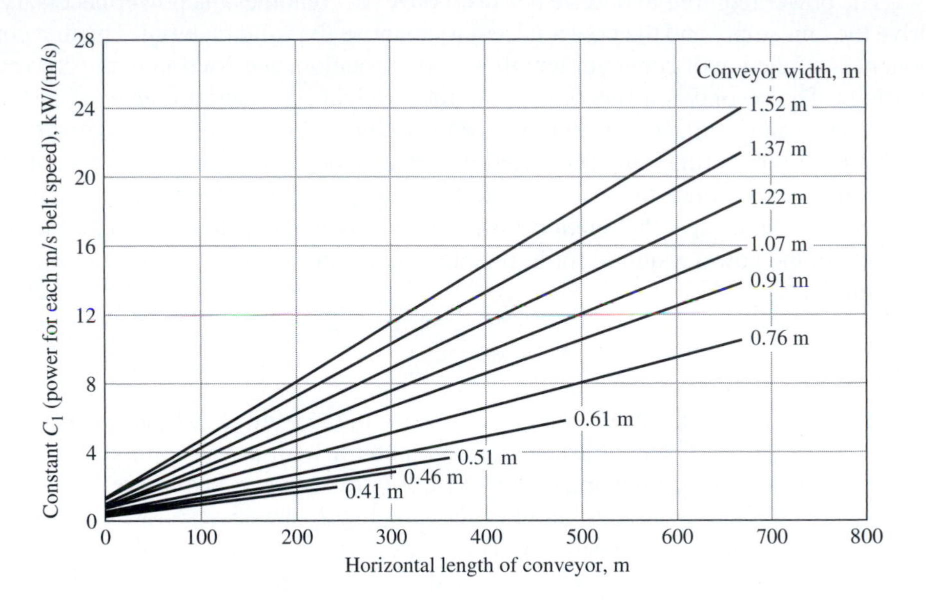

Figure 12-58
Power required to move an empty belt conveyor

conveyors is

$$P = P_{\text{empty}} + P_{\text{horizontal}} + P_{\text{vertical}} \tag{12-52a}$$

or

$$P = C_1 V_{\text{BS}} + 0.0295\left(0.4 + \frac{L}{91.42}\right)\dot{m} + 9.69 \times 10^{-3}\,\Delta Z\dot{m} \tag{12-52b}$$

where P is the total power requirement in kW, C_1 is a constant obtained from Fig. 12-58 for specific belt widths and lengths, V_{BS} the belt speed in m/s, L the horizontal length of the conveyor in m, and $\dot{m}$ the mass flow rate of the solid material in kg/s. Table 12-16

Table 12-16 Design data for 45° troughed belt conveyors†

Belt width, m	Belt speed, m/s‡	Capacity, kg/s, at indicated slope angle§				Power req'mt., kW/L (horizontal movement)	Power req'mt., kW/ΔZ (vertical movement)
		0°	10°	20°	30°		
0.46	0.51	13.6	15.1	17.4	19.8	0.0171	0.142
	(1.27)	33.8	37.8	43.5	49.5	0.0431	0.348
0.61	0.51	24.7	29.0	33.3	37.8	0.0250	0.250
	(1.52)	74.2	87.0	99.9	113.4	0.0744	0.749
0.91	0.51	58.1	70.2	80.3	91.0	0.0389	0.597
	(2.03)	232	281	321	364	0.1560	2.384
1.22	0.51	111	129	148	167	0.0744	1.141
	(2.03)	444	516	592	668	0.297	4.577
1.52	0.51	182	206	235	266	0.122	1.870
	(2.29)	818	927	1058	1197	0.548	8.420
1.83	0.51	259	301	343	388	0.232	3.573
	(2.29)	1166	1355	1543	1745	1.042	16.165

†Data attributed to Conveyor Equipment Manufacturers Association and Fairfield Engineering Company.
‡Normal belt speed for each belt width shown in parentheses.
§Density of the material conveyed, 1600 kg/m^3. Slope angle is angle of repose of material.
For inclined conveyors, add vertical lift power to horizontal movement power. For terminals, multiply the total power by the following factors: 0 to 15.2 m, 1.20; 15.2 to 30.5 m, 1.10; 30.5 to 45.7 m, 1.05. Add tripping power if a tripper device is used.

provides another method for estimating the power requirements of a belt conveyor. These two approaches are presented in Example 12-7.

Belt selection depends on the driveshaft power developed and the required tensile strength of the belt material. Since various combinations of width and ply thickness will provide the required strength, final selection is dictated by lump size, troughability of the belt, and ability of the belt to support the load between idlers. Capacities of flat-belt conveyors can be increased with the use of sidewalls. Further increases in capacity can also be achieved by the use of troughed belts. The 20° troughing belt is the most common and can increase the capacity by about 33 percent over the belt with 0° troughing. Consideration of these many aspects makes it necessary to use an empirical approach to arrive at a belt selection that meets all requirements.

EXAMPLE 12-7 Determination of the Power Requirement for a Belt Conveyor

A cement clinker solid with a bulk density of 1600 kg/m^3 is to be transported a horizontal distance of 365 m up an incline of 5°. A troughed belt conveyor has been selected to handle 101 kg/s of this material with a running angle of repose of 19°. Assuming a normal belt speed, what is the minimum belt width that should be selected? What is the power requirement for this solids transport?

■ Solution

Examination of Table 12-16 indicates that a 0.61-m troughed belt conveyor with an indicated slope angle of 20° and operating at a normal belt speed of 1.52 m/s has a capacity of 99.9 kg/s. Thus, an increase of less than 1 percent in belt speed would meet the transport requirements. A 0.91-m belt conveyor would be too large.

Actual length L_a of the belt conveyor is given by

$$L_a = \frac{365}{\cos 5^\circ} = 366.4 \text{ m}$$

Lift of the conveyor is

$$\Delta Z = 365 \tan 5^\circ = 31.9 \text{ m}$$

With Eq. (12-52*b*) and Fig. 12-58, where C_1 = 4.6 kW/(m/s), the power requirement is evaluated (neglecting the 1 percent increase in belt speed) as

$$P = 4.6(1.52) + 0.0295\left(0.4 + \frac{365}{91.42}\right)(101) + (9.69 \times 10^{-3})(31.9)(101)$$
$$= 51.3 \text{ kW}$$

The actual power requirement will be at least 10 to 20 percent more to account for the losses in the drive gear and motor. Now estimate the power requirements for the identical belt conveyor, using the data and notations from Table 12-16.

$$P = [0.0744(365) + 0.749(31.9)](1.05) = 53.5 \text{ kW}$$

In view of the estimated power values in Table 12-16 and the interpolation performed in Fig. 12-58, this close agreement between the two methods would not have been anticipated. ■

The power requirement associated with pneumatic conveyors is a summation of the work of compression of the air and the frictional losses that develop because of the air and solids flow through the transport line. The work of compression of the air is calculated by using either Eq. (12-22*a*) or Eq. (12-22*b*). Friction losses are evaluated separately for the air and the solid being transported. For each of these, frictional losses come from the transport line itself and from elbows, line restrictions, and receiving equipment. Generally, it is assumed that the linear velocities of the air and the solid suspension are the same. Expressions for evaluating the friction losses of the air in the transport line are given in Table 12-1. Since the airflow normally is at a high Reynolds number, the friction factor for the air may be assumed constant at a value of 0.015.

The work of moving the solid at a specified rate is due to the kinetic energy gain of the solid at the conveyor entrance, the lift provided to the solid through a change in elevation, friction in the transport line, and friction in any elbows or line restrictions. The gain in kinetic energy is N·m/s given by

$$W_{\mathrm{KE}} = \left(\frac{V^2}{2}\right) \tag{12-53}$$

The work involved in elevating the solid is evaluated from

$$W_L = \Delta Z\, g \tag{12-54}$$

where ΔZ is the change in elevation and g is the local acceleration due to gravity. The coefficient of sliding friction $\mathfrak{f}_s$ of the solid equals the tangent of the angle of repose, which is between 30 and 45° for most solids. The work involved with the sliding friction in the transport line in N·m/s is

$$W_{\mathrm{sf}} = \mathfrak{f}_s L g \tag{12-55}$$

where L is the length of the pneumatic conveyor. Friction in the elbows is increased because of centrifugal force so that the additional work in N·m/s is represented by

$$W_e = 1.155\mathfrak{f}_s V^2 \tag{12-56}$$

The total friction power required in kW therefore is

$$P_f = (W_{\mathrm{air}})\dot{m}_a + (W_{\mathrm{KE}} + W_L + W_{\mathrm{sf}} + W_e)\dot{m}_s \tag{12-57}$$

and the total power requirement is

$$P = 0.001\left(\frac{W_c \dot{m}_a}{\eta} + P_f\right) \tag{12-58}$$

where η is the efficiency of the blower and the units for P are $\mathrm{kW/(kg/s)_{solid}}$. Pressure drop in the transport line in kPa is obtained from the friction power, the total flow rate, and the density of the mixture

$$\Delta p = \frac{0.001\, P_f \rho_m}{\dot{m}_a + \dot{m}_s} \tag{12-59}$$

The specific air rate, often referred to as *saturation,* is defined as the ratio of the air required to the mass of solids transported in $(m^3/s)/(kg/s)$. In this determination of the specific air rate, the velocity of the air is normally evaluated at atmospheric pressure for comparative purposes.

EXAMPLE 12-8 Determination of Power Requirements for a Pneumatic Conveyor

A 0.154-m-diameter tube is used by a pneumatic conveyor to transport a finely crushed material with a bulk density of 1000 kg/m^3 at a solids transport rate of 2.5 kg/s. The 100-m conveyor tube contains two 90° long sweep elbows and provides a lift of 15 m. Air for the transport is available from another source at a velocity of 73.4 m/s, at a temperature of 38°C, a pressure of 101.3 kPa, and a density of 1.132 kg/m^3. After compression and cooling of the gas, the inlet conditions to the pneumatic conveyor system are 38°C and 186 kPa. What is the theoretical power requirement for this solids transport system?

■ Solution

For a quick estimate, assume that the pressure drop through the conveyor is about 40 kPa and that the temperature remains constant during the conveying process. This pressure loss assumption will need to be verified.

Determine the flow rate of the air from

$$\dot{m}_a = VA\rho = (73.4)\left(\frac{\pi}{4}\right)(0.154)^2(1.132) = 1.55 \text{ kg/s}$$

Average density of the air at 38°C and an average pressure of 186 – 40/2 or 166 kPa is

$$\rho_a = (1.132)\left(\frac{166}{101.3}\right) = 1.855 \text{ kg/m}^3 \qquad \text{(assuming an ideal gas)}$$

Average density of the mixture is given by

$$\rho_m = \frac{\dot{m}_a + \dot{m}_s}{\dot{m}_a/\rho_a + \dot{m}_s/\rho_s}$$

$$= \frac{1.55 + 2.5}{1.55/1.855 + 2.5/1000} = 4.83 \text{ kg/m}^3$$

Average velocity of the air in the conveyor is

$$V = (73.4)\left(\frac{101.3}{166}\right) = 44.8 \text{ m/s}$$

This is greater than the minimum velocity of 37 m/s (even at the inlet conditions) required to suspend the solids.[†] The power of compression of the air from 101.3 kPa and 38°C to 186 kPa, using

[†]See *Perry's Chemical Engineers' Handbook,* 7th ed., R. H. Perry and D. W. Green, eds., McGraw-Hill, New York, 1997, Table 21-13.

Eq. (12-22*a*), is

$$P_c = RT\left(\frac{k}{k-1}\right)\left[\left(\frac{p_2}{p_1}\right)^{(k-1)/k} - 1\right]\dot{m}_a$$

$$= \left(\frac{8314}{28.9}\right)(311)\left(\frac{1.4}{0.4}\right)\left[\left(\frac{186}{101.3}\right)^{0.4/1.4} - 1\right](1.55)(10^{-3}) = 92.0\text{ kW}$$

Now determine the frictional contributions of the air, using the expressions listed in Table 12-1.

$$P_a = \left(F_c + F_e + \sum F\right)\dot{m}_a$$

$$F_c = \frac{K_c V^2}{2\alpha} \qquad \text{assume} \qquad \frac{A_2}{A_1} = 0.75 \qquad K_c = 0.2 \qquad \alpha = 0.5$$

$$F_e = \frac{V^2}{2\alpha} \qquad \alpha = 0.5$$

$$\sum F = \frac{\mathfrak{f}(V^2/2)\left(L + \sum L_e\right)}{D} \qquad L = 100\text{ m} \qquad \mathfrak{f} = 0.015$$

$$\sum L_e = 2(20D) = 2(20)(0.154) = 6.16\text{ m}$$

$$P_a = \left[\frac{(44.8)^2}{2}\right]\left[\frac{0.2}{0.5} + \frac{1}{0.5} + \frac{0.015(100 + 6.16)}{0.154}\right](1.55)(10^{-3})$$

$$= 19.8\text{ kW}$$

For the frictional contributions of the solid, assume that the coefficient of sliding friction is 1.0. The frictional contributions are

$$P_s = (W_{KE} + W_L + W_{sf} + W_e)\dot{m}_s$$

$$= \left[\frac{V^2}{2} + \Delta Z g + \mathfrak{f}_s L g + 2(1.155)\mathfrak{f}_s V^2\right]\dot{m}_s$$

$$= \left[\frac{(44.8)^2}{2} + 15(9.8) + 1(100)(9.8) + 2(1.155)(1)(44.8)^2\right](2.5)(10^{-3})$$

$$= 16.9\text{ kW}$$

The total frictional contribution is

$$P_f = P_a + P_s = 19.8 + 16.9 = 36.7\text{ kW}$$

The total power requirement is

$$P = P_c + P_f = 92.0 + 36.7 = 128.7\text{ kW}$$

The calculated pressure drop is given by

$$\Delta p = \left(\frac{P_f}{\dot{m}_a + \dot{m}_s}\right)\rho_m$$

$$= \left(\frac{36.7}{1.55 + 2.5}\right)(4.83) = 43.8 \text{ kPa}$$

Thus, the assumed pressure drop is relatively close to the calculated value, and a recalculation would have little effect on the theoretical power requirements in this example. The frictional contribution can also be estimated by using the nomographs presented in *Perry's Chemical Engineers' Handbook.*[†] For this example, the frictional contribution is estimated as 54 kW and the pressure drop across the pneumatic conveyor as 35 kPa. This compares with calculated values of 36.7 kW and 43.8 kPa, respectively. The variation in results between the two methods is on the order of 20 to 30 percent.

Software that simulates the operation of pneumatic conveyors shows nearly a 10 percent higher power requirement than was obtained in the above calculations. This reduction is attributed to higher inefficiencies experienced within the conveying system.

■

For estimating purposes, specifications and power requirements for centrifugal discharge and continuous bucket elevators are also available in this reference.[‡] Determination of the capacity and power requirements for vibrating or oscillating conveyors is best accomplished with the use of Fig. 12-59, since there are no simple expressions to cover all the variables that affect their ability to transport solids.

Costs for Solids Transport Equipment

The purchased costs for screw, belt, apron, bucket, roller, pneumatic, and vibratory conveyors are presented in Figs. 12-60 through 12-64. Equipment costs for chutes, gates, and hoists are available from Figs. 12-65 and 12-66.

HANDLING OF SOLIDS

Solids, in general, are more difficult to handle in processing operations than liquids or gases. Solids can exist in many forms. They may be abrasive, fragile, dusty, explosive, plastic, or sticky. The use of solids in chemical processing generally requires some form of size reduction. Some of the purposes of size reduction are to free a desired component for subsequent separation, prepare the material for subsequent separation, subdivide the material so that it can be blended more easily with other components, meet size requirements of an end product, or even prepare wastes for storage and recycling.

[†]G. J. Raymus, in *Perry's Chemical Engineers' Handbook,* 7th ed., R. H. Perry and D. W. Green, eds., McGraw-Hill, New York, 1997, Fig. 21-13, pp. 21–22.

[‡]Loc. cit., Tables 21-8 and 21-9.

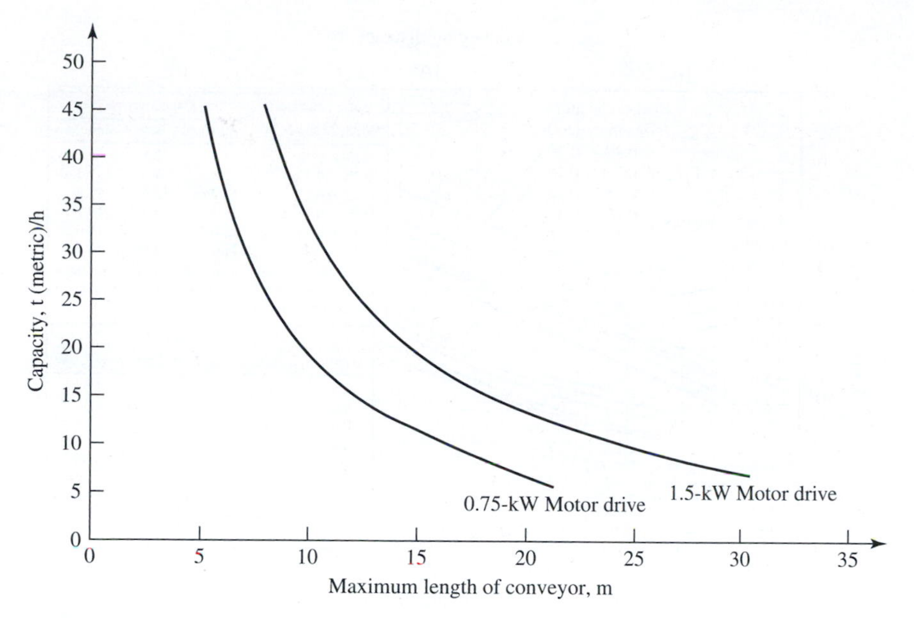

Figure 12-59
Capacity of a leaf-spring mechanical oscillating conveyor with a fixed motor drive. *(Courtesy FMC Corporation Materials Handling Systems Division.)*

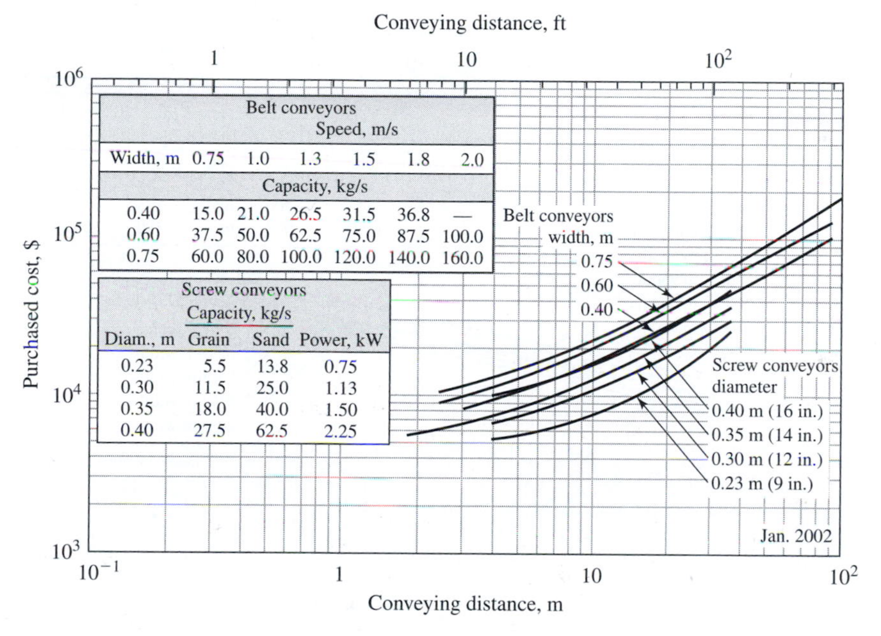

Figure 12-60
Purchased cost of screw and belt conveyors

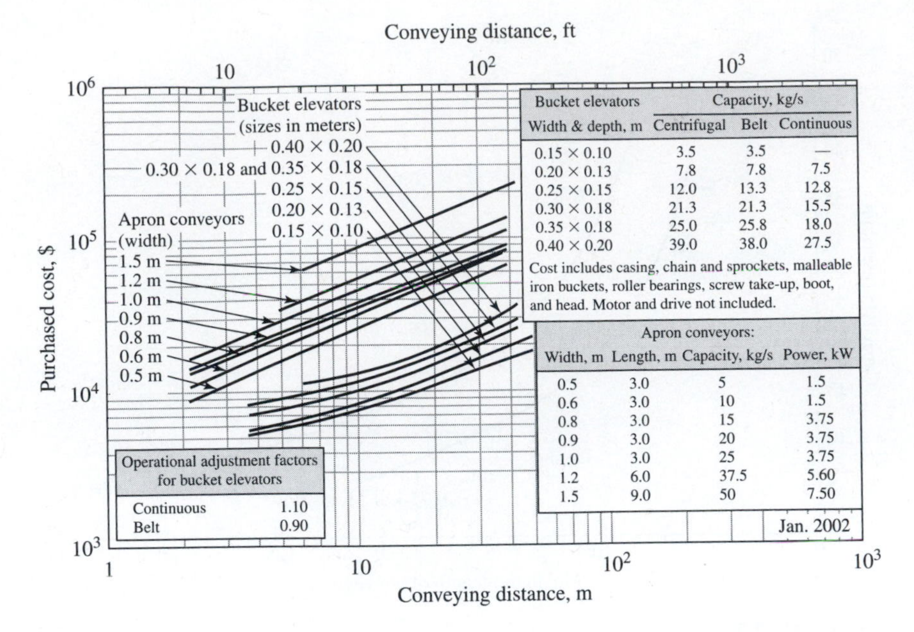

Figure 12-61
Purchased cost of apron conveyors and bucket elevators

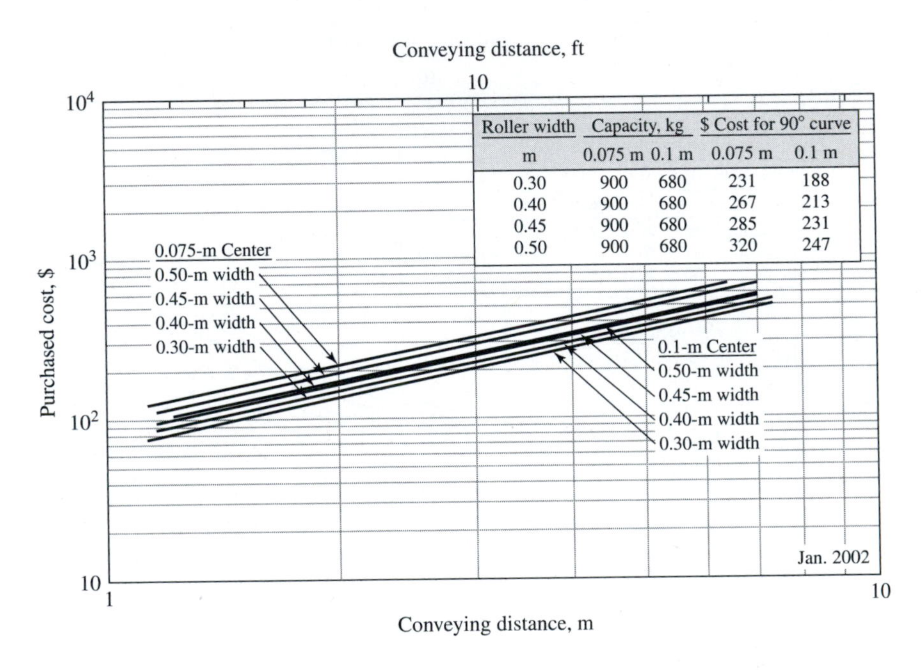

Figure 12-62
Purchased cost of roller conveyors

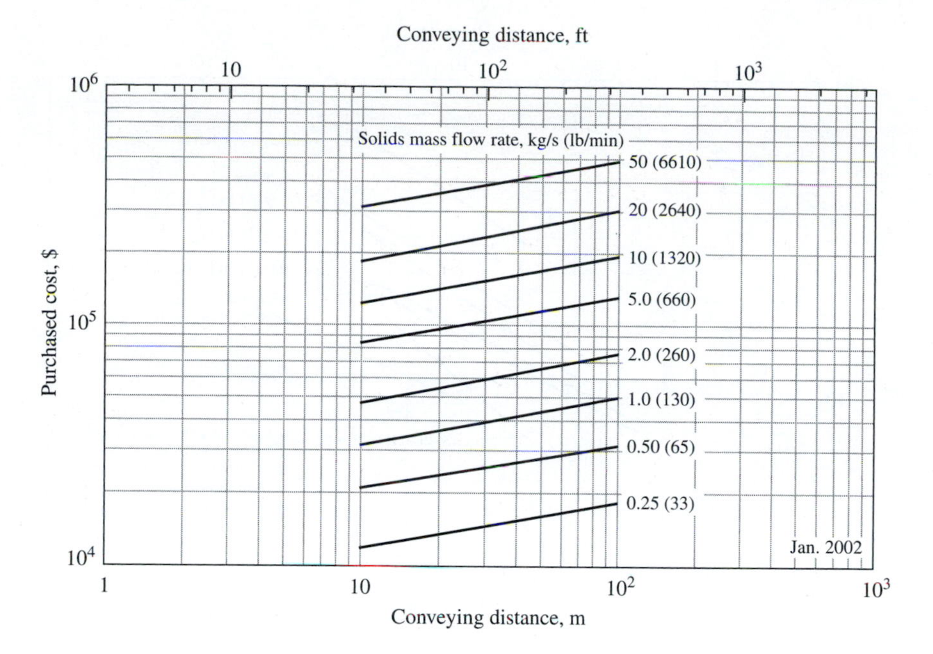

Figure 12-63
Purchased cost of pneumatic solids-conveying equipment. Drives are included.

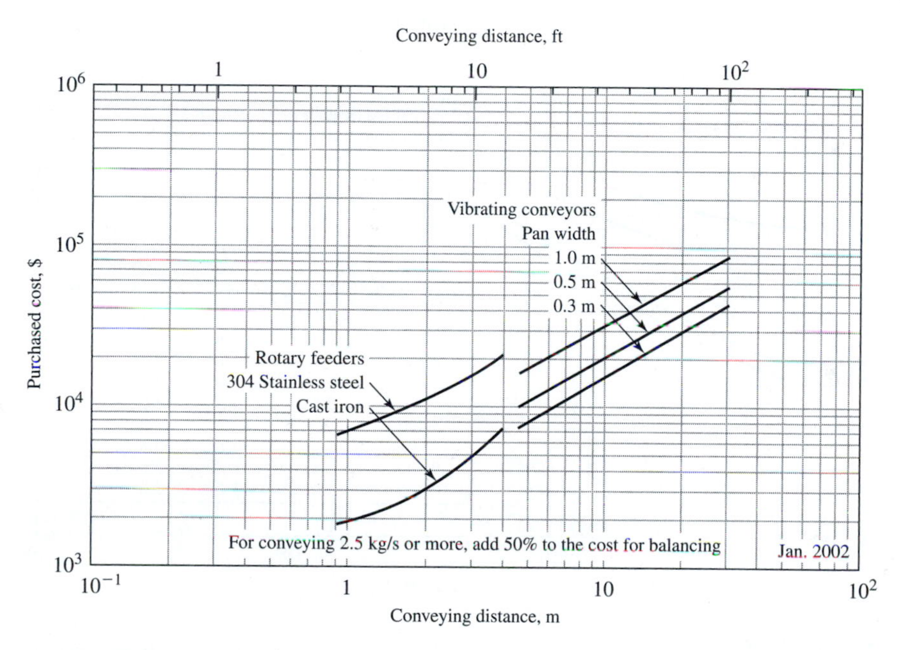

Figure 12-64
Purchased cost of rotary feeders and vibrating conveyors

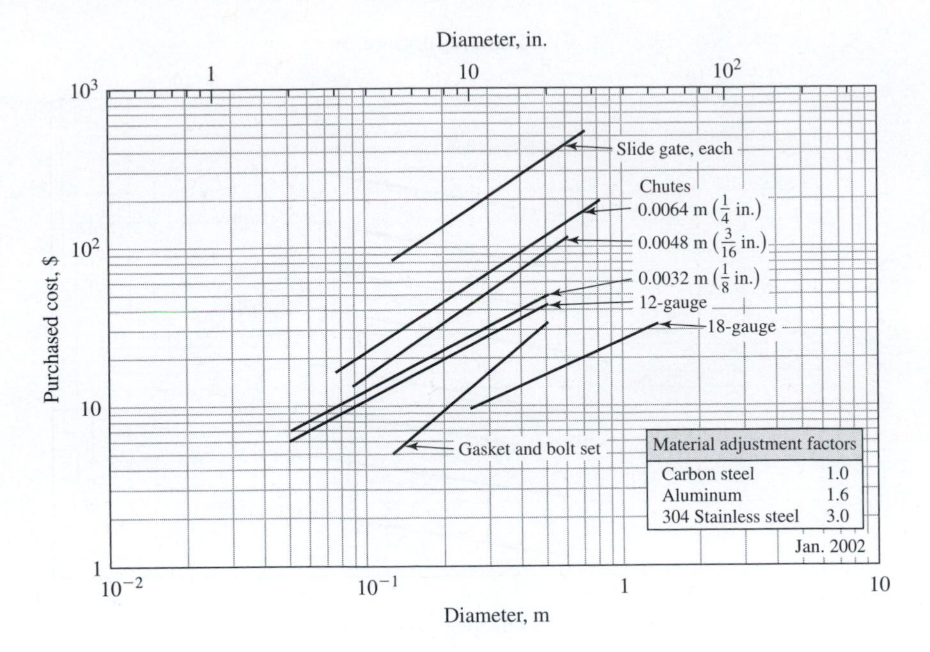

Material adjustment factors	
Carbon steel	1.0
Aluminum	1.6
304 Stainless steel	3.0

Figure 12-65
Purchased cost of chutes and gates. Price for flexible connections not included.

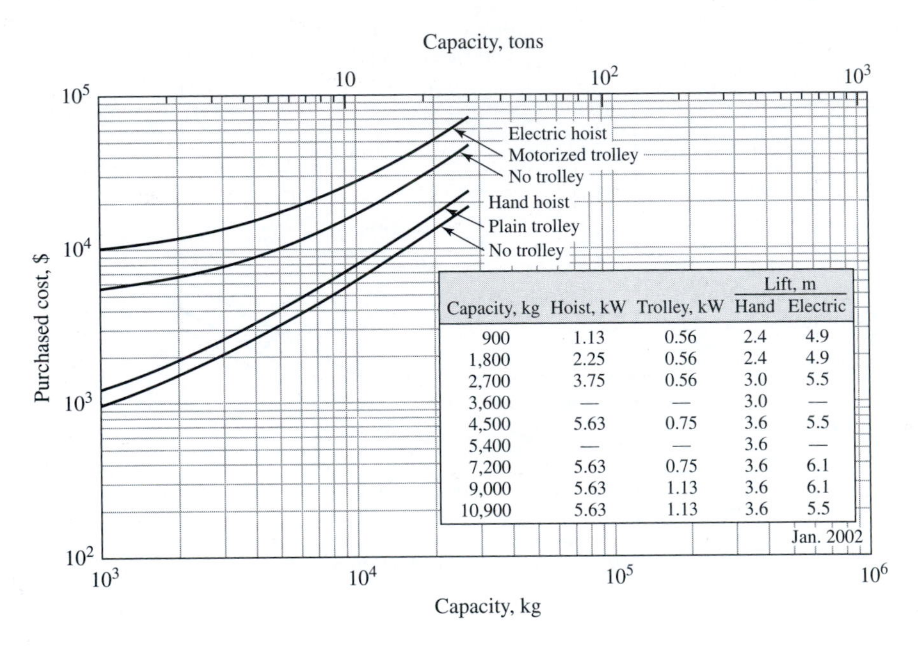

Capacity, kg	Hoist, kW	Trolley, kW	Lift, m Hand	Lift, m Electric
900	1.13	0.56	2.4	4.9
1,800	2.25	0.56	2.4	4.9
2,700	3.75	0.56	3.0	5.5
3,600	—	—	3.0	—
4,500	5.63	0.75	3.6	5.5
5,400	—	—	3.6	—
7,200	5.63	0.75	3.6	6.1
9,000	5.63	1.13	3.6	6.1
10,900	5.63	1.13	3.6	5.5

Figure 12-66
Hoisting equipment. Extra costs include acid-resistant construction, dust-tight construction, power reels, and chain container.

Selection of Solids-Handling Equipment

A wide variety of size reduction equipment is available. The equipment involved is generally classified according to the manner in which forces are applied to achieve the size reduction. These forces are applied between two solid surfaces, impact at one surface or between particles, shear action of the surrounding medium, or through nonmechanical introduction of energy.

A number of general principles govern the selection of crushers. When the solid contains a predominant amount of material that tends to be cohesive, any type of repeated pressure crusher will result in the fines to pack and clog the outlet of the crushing zone. Under these situations, impact breakers are more suitable, provided the solid is not too hard and abrasive.

When the solid is not hard but cohesive, toothed roll mills give satisfactory performance. With harder solids (8 to 10 moh), jaw and gyratory crushers are required. Jaw crushers are less prone to clogging than gyratory crushers. For secondary crushing, the high-speed conical-head gyratory crusher is preferred except when sticky material precludes its use. When a wide size distribution is to be avoided, a compression-type crusher is normally recommended.

A useful guide for the selection of size reduction equipment based on feed size and hardness of the solid being handled is presented in Table 12-17. It should be emphasized that the selection criteria listed are a guide and that exceptions can be found in practice. Further support of these guidelines is provided in the following summary of the specific types of equipment noted in this table.

Size Reduction Crushers The jaw and gyratory crushers are used in the crushing of hard materials and usually are followed by other crushers for further size reductions. Jaw crushers are usually rated by the dimensions of their feed area and can accommodate the same size of solid material as that accepted for a gyratory crusher. A comparison between these two crushers indicates that the jaw crushers have lower capital and maintenance costs, similar installation costs, but lower throughput than the gyratory crushers. Power consumption for gyratory crushers is lower than that of jaw crushers and makes the former a preferred choice when capacities of 225 kg/s or higher are required.

The use of impact crushers, such as heavy-duty hammer crushers and rotor impact breakers, is best suited for the breaking and crushing of tough, fibrous, or sticky materials. In hammer mills, solids are broken by impact of the high-speed hammers combined with shearing and attrition between hammers and anvils. Rotor impactors, on the other hand, depend solely on impact for size reduction. This type of impactor requires less power than a hammer crusher, but is limited to nonabrasive solids that fracture easily by impaction. Cage impactors are employed by medium-scale operations where relatively large reduction ratios are required. In applications where either compression or impaction crushers can handle the size reduction requirement, impact crushers are generally selected because capital costs are one-half to one-third of those for jaw and gyratory crushers.

Even though jaw and gyratory crushers have replaced many roll crushers because of the poor wear characteristics of these devices, roll crushers are still commonly used

Table 12-17 Guidelines for selection of size reduction equipment[†]

Size reduction operation	Hardness of solid[‡]	Range of particle size,[§] mm				Open circuit reduction ratio[¶]	Equipment type generally applicable for this specified size reduction
		Feed		Product			
		Max.	Min.	Max.	Min.		
Crushing							
Primary	Hard	1500	300	500	100	3 to 1	Jaw and gyratory crushers
		500	100	125	25	4 to 1	
Secondary	Hard	125	25	25	5	5 to 1	Jaw, gyratory, impact, roll and pan crushers
		35	6	5	0.8	7 to 1	
	Soft	1500	100	150	10	10 to 1	Impact, roll and pan crushers, disk mills, rotary cutters
Grinding							
Pulverizing							
Coarse	Hard	5.0	0.8	0.5	0.08	10 to 1	Roll and pan crushers; rotary cutters; tumbling, vibratory, and ring-roller mills
Fine	Hard	1.2	0.15	0.08	0.01	15 to 1	Tumbling, vibratory, ring-roller, high-speed, colloid, and jet mills
Disintegration							
Coarse	Soft	12	1.6	0.6	0.08	20 to 1	Disk and ring-roller mills
Fine	Soft	4	0.5	0.08	0.01	50 to 1	Ring-roller, high-speed, colloid, and jet mills

[†]Modified from data presented by R. H. Snow, in *Perry's Chemical Engineers' Handbook,* 7th ed., R. H. Perry and D. W. Green, eds., McGraw-Hill, New York, 1997, Table 20-7, p. 20-23, D. R. Woods, *Process Design and Engineering Practice,* Prentice-Hall, Englewood Cliffs, NJ, 1995, and G. D. Ulrich, *A Guide to Chemical Engineering Process Design and Economics,* J. Wiley, New York, 1984.
[‡]Hard (8–10 moh); soft (1–3 moh).
[§]85% by weight smaller than the size given.
[¶]Higher reduction ratios achievable for closed-circuit operations.

for both primary and secondary crushing of friable solids. The roll press has achieved commercial success in the cement industry. It is used for fine crushing, replacing the function of a coarse ball mill or a tertiary crusher.

The dry pan crusher, on the other hand, is appropriate for crushing medium-hard and soft materials and minerals. High reduction ratios with low power and maintenance are features of this type of crusher.

Size Reduction Mills The disk or attrition mill is a modern counterpart of the early buhrstone mill. Although its efficiency is relatively low, this type of mill is excellent for size reduction of tough or resilient particles. An advantage of these devices is that they can be heated or cooled and can be used for blending as well.

Ball, pebble, rod, tube, or bead mills are all classified as tumbling or media mills. In each of these devices a grinding medium such as metal balls, rods, or pebbles is used to produce a fine or extra-fine grinding of hard and abrasive powders. These mills can be operated either wet or dry. Wet grinding, when permissible, yields a finer powder with a power saving of about 30 percent.

Capacities for pebble mills are 30 to 50 percent of the capacity of the same size of ball mill with steel grinding media and liners. The tube mill is essentially a longer ball mill that uses smaller steel or ceramic balls and produces a finer powder. Rod mills are not efficient for fine grinding but are widely used for coarse or intermediate grinding. Often they are employed prior to the ball mill. In general, tumbling mills are relatively inexpensive with low operating costs. Their efficiency is good, but because of the fine particles handled, specific power consumption is high.

The selection of a ball or rod mill grinding device is based on experimental laboratory data and scaled up on the basis that production is proportional to energy input. Without experimental data, performance is based on published data for similar types of material, expressed in terms of either grindability or energy requirement. Newer methods of sizing these mills and determining operating conditions for optimum performance are based on computer solutions of the grinding equations with values of rate and breakage functions determined from laboratory or full-scale tests.

Fine Grinding Mills Primary applications of vibratory mills are in fine milling of medium to hard materials, generally in dry form, producing particle sizes of 1 μm or finer. Grinding increases with residence time, active mill volume, energy density, and vibration frequency. Larger vibratory amplitudes are more favorable for communition than higher frequency. Advantages of vibratory mills include simple construction and low capital cost, attainment of very fine product size, large reduction ratio in a single pass, small space and weight requirements, and low cost of maintenance. Disadvantages are limited mill size and thus limited capacity, vibration of supporting structures, and often high-noise output, particularly when run dry.

The ring-roller mill with internal air classification is used for the large-capacity fine grinding of most of the softer nonmetallic minerals, since this type of mill is more energy-efficient than a ball or hammer mill. Materials with a moh hardness up to and including 5 are handled economically with these units.

Hammer mills operated at high speeds are used for fine pulverizing and disintegration. The grinding action results from impact and attrition between the particles being ground, the mill housing, and the grinding elements. The fineness of the product can be regulated by a change in rotor speed, feed rate, or the clearance between the rotating hammers and grinding plates. In general, the feed must be nonabrasive with a moh hardness of 1.5 or less.

In contrast to peripheral hammers, pin mills have pin breakers in the grinding circuit. These may be on a rotor with stator pins between circular rows of pins on the rotor disk, or they may be on rotors operating in opposite directions and thus obtain an increased differential in speed.

To disrupt lightly bonded clusters or agglomerates, a new aspect of fine grinding is required. Colloid and dispersion mills employed for this purpose operate on the principle of high-speed fluid shear. This action produces dispersed droplets with a size range of about 3 to 5 μm. The concentration of energy in mills of this type is high, resulting in a considerable amount of heating. This can be reduced with the use of a cooling-water jacket. Unfortunately, energy requirements for this physical disruption differ significantly with the type of materials involved so that other devices are used to accomplish this task. These devices include high-speed stirrers, turbine mixers, bead mills, and vibratory mills.

In jet mills, particles are either conveyed or intercepted by a fine high-velocity stream, resulting in a high energy release and a high order of turbulence. The latter causes the particles to grind upon themselves with eventual rupture. Since not all particles are fully ground, it is necessary to carry out a classifying operation with the return of the oversize particles for further grinding. Most of these mills use the energy of the flowing fluid stream to effect a centrifugal classification.

Rotary Cutters The device most commonly employed to cut soft and fibrous materials into smaller dimensions is a rotary knife cutter. With appropriate mechanical design, rotary cutters are the most economical means of cutting rods into pellets or dicing sheets into squares or cubes.

General Design Procedures for Solids-Handling Equipment

The science associated with size reduction has been developed to such a point that the theoretical energy required to fracture a solid can be evaluated quite accurately with present computer software. However, this theoretical power often is less than 2 percent of that required by the crusher or grinder to accomplish the communition purpose. The balance of the power is utilized to overcome friction or is dissipated as heat.

The specific power required for communition increases as the solid particles become smaller. Thus, fine grinders require more power per unit of product than primary crushers. Since power consumption also depends on the hardness of the solid being handled, none of the size reduction equipment referenced in Table 12-17 is suitable for all types of solids. Note that the power relationship presented in Table 12-18 for each device applies only over the hardness range specified. The designer is cautioned that these relationships are approximate and should be used for preliminary design work only when actual operating data are not available.

Table 12-18 Design criteria for size reduction equipment under closed-circulation operation[†]

Equipment type	Solid hardness, moh	Max. capacity, kg/s	Max. reduction ratio R^*	Relationship for power requirement, kW[‡]
Jaw crusher	8–10	400 (coarse) 200 (intermediate)	8	$P = 3\dot{m}_s^{0.88} R^*$
Gyratory crusher	8–10	4000 (coarse) 400 (intermediate)	8	$P = 2.5\dot{m}_s^{0.88} R^*$
Impact crusher	1–3	400	35	$P = 1.0\dot{m}_s^{0.88} R^*$
Roll crusher	8–10 4–7	125	16	$P = 0.6\dot{m}_s R^*$ $P = 0.3\dot{m}_s R^*$
Disk or attrition mill	1–3	15	15	$P = 10\dot{m}_s$ to $50\,\dot{m}_s$
Tumbling mills				
Rod mill	8–10	50	15	$P = 0.007\dot{m}_s/D_{pp}$
Ball mill	8–10	15	20	$P = 0.008\dot{m}_s/D_{pp}$
Vibratory mill	8–10	0.1	30	$P = 40\dot{m}_s/D_{pp}^{0.3}$
Ring-roll grinder	4–7	15	15	$P = 0.3\dot{m}_s R^*$
Hammer mill	1–3	2	50	$P = 40\dot{m}_s \ln R^*$
Jet mill	8–10	1	50	1–10 kg air (800 kPa/kg solid)
Rotary cutter	1–3	50	50	$P = 100\dot{m}_s$ to $500\dot{m}_s$

[†]Modified from data presented by G. D. Ulrich, *A Guide to Chemical Engineering Process Design and Economics,* J. Wiley, New York, 1984, R. H. Snow, in *Perry's Chemical Engineers' Handbook,* 7th ed., R. H. Perry and D. W. Green, eds., McGraw-Hill, New York, 1997, and vendor literature.
[‡]Power P in kW, solids mass flow rate $\dot{m}_s$ in kg/s, reduction ratio R^* dimensionless, and product diameter of particle D_{pp} in m.

A more semitheoretical relationship is also available which assumes that the power required to form particles of size D_{pp} from large feed particles is proportional to the square root of the surface-to-volume ratio of the product. This relationship requires the use of W_i, a work index term defined as the total energy per kg needed to reduce the particle size of the feed such that 80 percent of the product passes through a 100-μm screen. Thus, if 80 percent of the feed has a mesh size of D_{pf} and 80 percent of the product has a mesh size of D_{pp}, the relationship can be expressed as

$$P = 0.01\dot{m}_s W_i \left(\frac{1}{D_{pp}^{1/2}} - \frac{1}{D_{pf}^{1/2}} \right) \tag{12-60}$$

where P is the required power in kW, $\dot{m}_s$ the mass flow rate of the solids in kg/s, W_i the work index of the feed, and D_{pp} and D_{pf} the particle sizes of the product and the feed in m, respectively.

Table 12-19 provides typical work indices for several materials that generally require crushing and grinding. These indices do not vary greatly for different size reduction devices of the same general type and apply to wet grinding. For dry grinding, the power calculated from Eq. (12-60) should be increased by 33 percent.

Table 12-19 Work indices for grinding specific materials

Material	Density, kg/m^3	Work index, kJ/kg
Cement clinker	3140	53.26
Cement (raw material)	2665	41.62
Coal	1395	51.48
Coke	1305	59.91
Granite	2655	59.91
Gravel	2655	63.60
Iron ore	3520	50.85
Limestone	2655	50.45
Quartz	2645	53.74
Shale	2625	62.85
Slate	2565	56.63

EXAMPLE 12-9 Power Requirement for Secondary Crushing of Limestone

What is the power required to crush 25.3 kg/s of limestone if 80 percent of the feed passes through a 0.051-m screen and 80 percent of the product passes through a 0.0032-m screen?

■ Solution

The work index for limestone from Table 12-19 is 50.45. From Eq. (12-60) the power requirement for dry grinding is

$$P = (0.01)(25.3)(50.45)\left[\frac{1}{(0.0032)^{1/2}} - \frac{1}{(0.051)^{1/2}}\right](1.33) = 225 \text{ kW}$$

Compare this result with that obtained from using the relationship provided in Table 12-18. Since secondary crushing of a hard material is involved, the roll crusher is adequate for the task with a reduction ratio of 15.9.

$$P = 0.6\dot{m}_s R^*$$
$$= 0.6(25.3)(15.9) = 241 \text{ kW}$$

The comparison is reasonable considering all the assumptions made in the development of the two relationships.

The power requirement for crushing the limestone has also been evaluated with software specifically developed for this purpose. In this case, the power requirements are about 10 percent lower than those obtained above.

Costs for Solids-Handling Equipment

The purchased cost of various types of primary and secondary crushers is provided in Fig. 12-67. Costs for purchased or installed grinders, cutters, and pulverizers are presented in Figs. 12-68 through 12-73. Costs for cutters and disintegrators are given in Fig. 12-74.

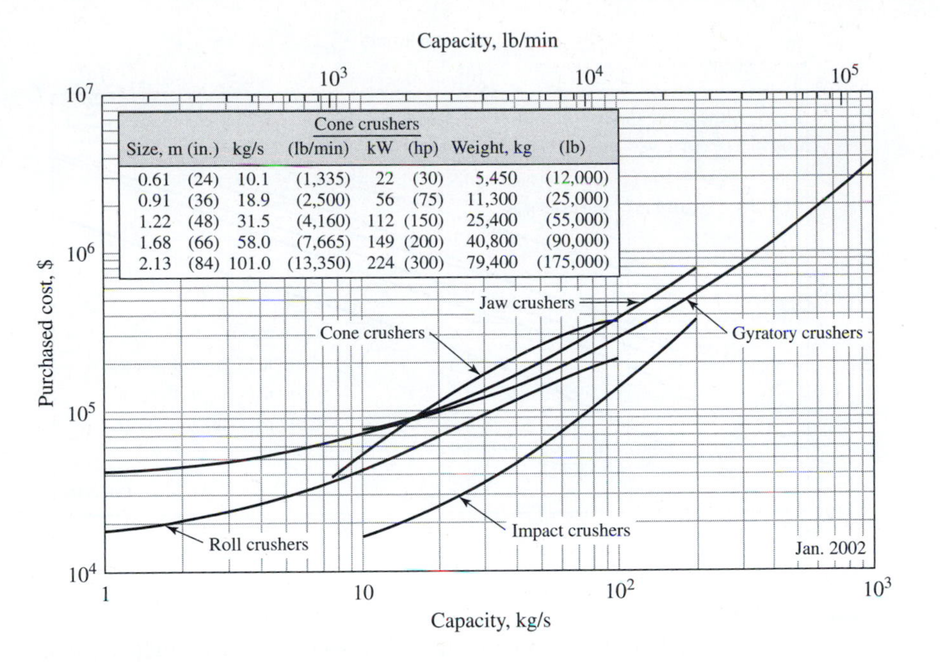

Size, m	(in.)	kg/s	(lb/min)	kW	(hp)	Weight, kg	(lb)
Cone crushers							
0.61	(24)	10.1	(1,335)	22	(30)	5,450	(12,000)
0.91	(36)	18.9	(2,500)	56	(75)	11,300	(25,000)
1.22	(48)	31.5	(4,160)	112	(150)	25,400	(55,000)
1.68	(66)	58.0	(7,665)	149	(200)	40,800	(90,000)
2.13	(84)	101.0	(13,350)	224	(300)	79,400	(175,000)

Figure 12-67
Purchased costs of crushers. Price includes motor and drive.

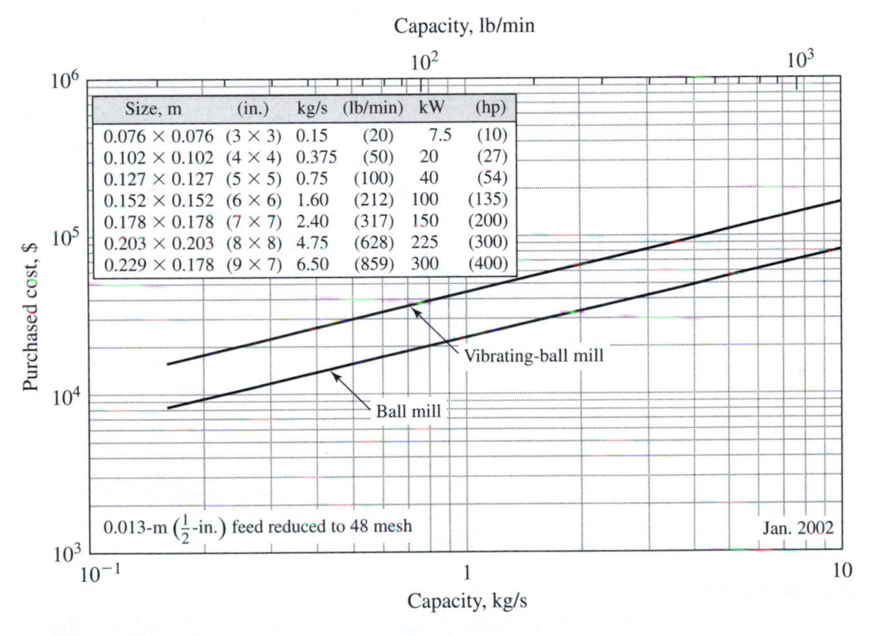

Size, m	(in.)	kg/s	(lb/min)	kW	(hp)
0.076 × 0.076	(3 × 3)	0.15	(20)	7.5	(10)
0.102 × 0.102	(4 × 4)	0.375	(50)	20	(27)
0.127 × 0.127	(5 × 5)	0.75	(100)	40	(54)
0.152 × 0.152	(6 × 6)	1.60	(212)	100	(135)
0.178 × 0.178	(7 × 7)	2.40	(317)	150	(200)
0.203 × 0.203	(8 × 8)	4.75	(628)	225	(300)
0.229 × 0.178	(9 × 7)	6.50	(859)	300	(400)

Figure 12-68
Purchased cost of ball grinders. Price includes liner, motor, drive, and guard.

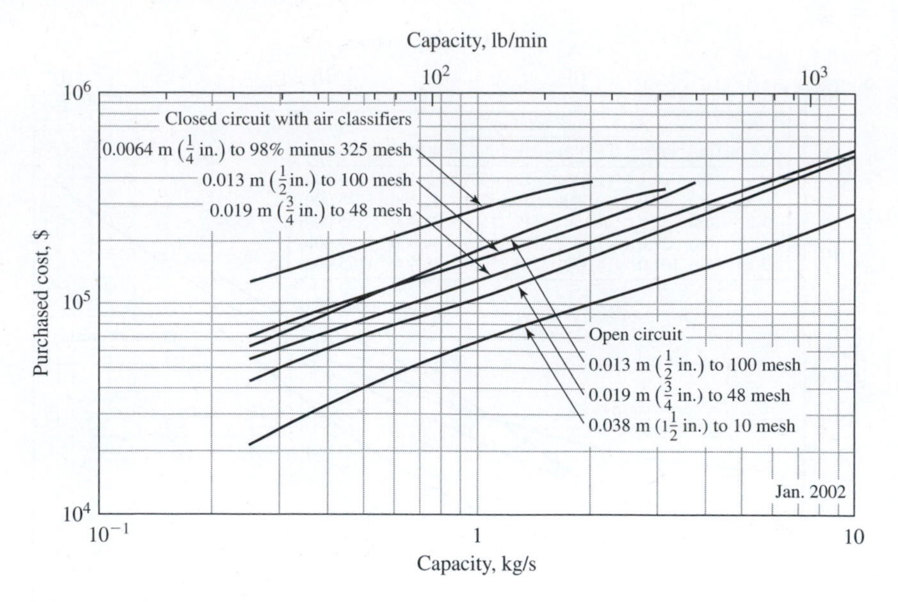

Figure 12-69
Purchased cost of ball mill dry grinding. Includes installation, classifier, motors, drives, and an allowance for foundations and erection. Does not include freight, auxiliary equipment, or handling equipment.

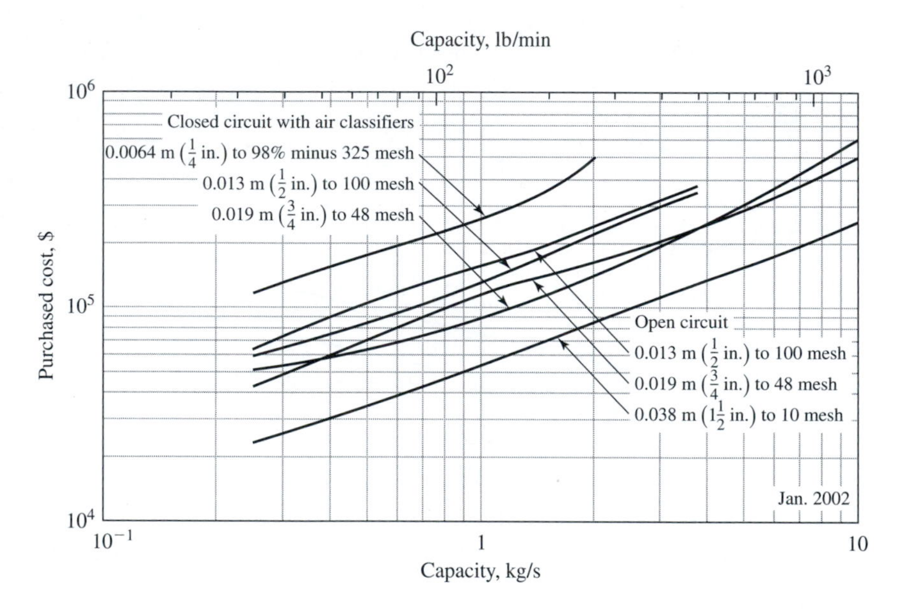

Figure 12-70
Purchased cost of ball mill wet grinding. Includes installation, classifier, motors, drives, and an allowance for installation. Does not include freight, auxiliary equipment, or handling equipment.

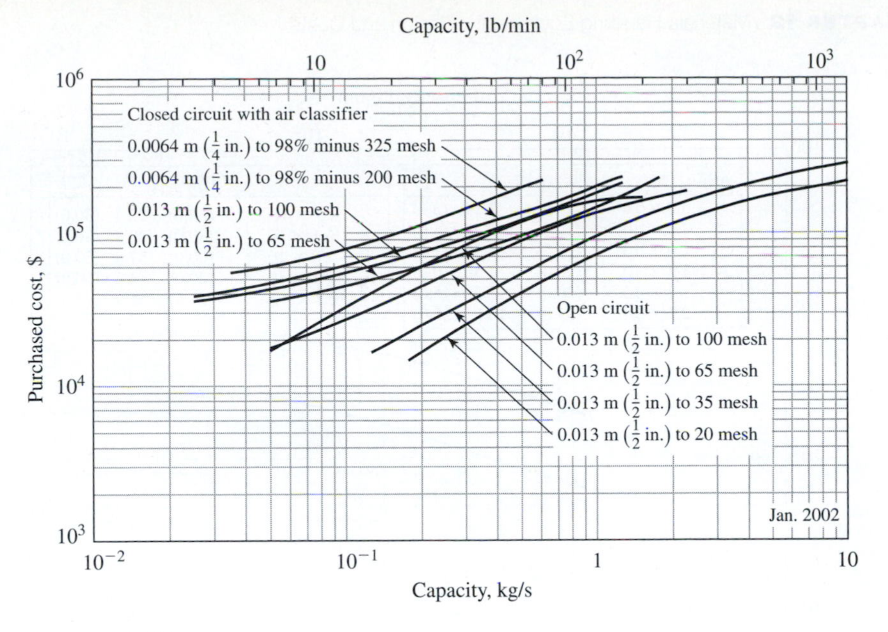

Figure 12-71
Purchased cost of AG/SAG mill dry grinders. Price includes installation, classifier, motors, and drives. Does not include freight, auxiliary equipment, or handling equipment.

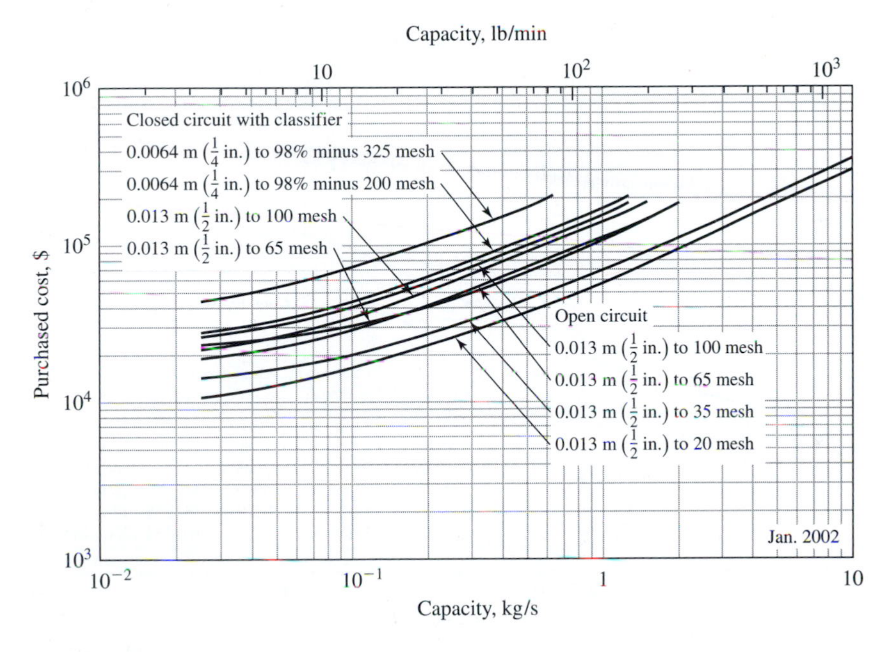

Figure 12-72
Purchased cost of AG/SAG mill wet grinders. Price includes installation, classifier, motors, and drives. Does not include freight, auxiliary equipment, or handling equipment.

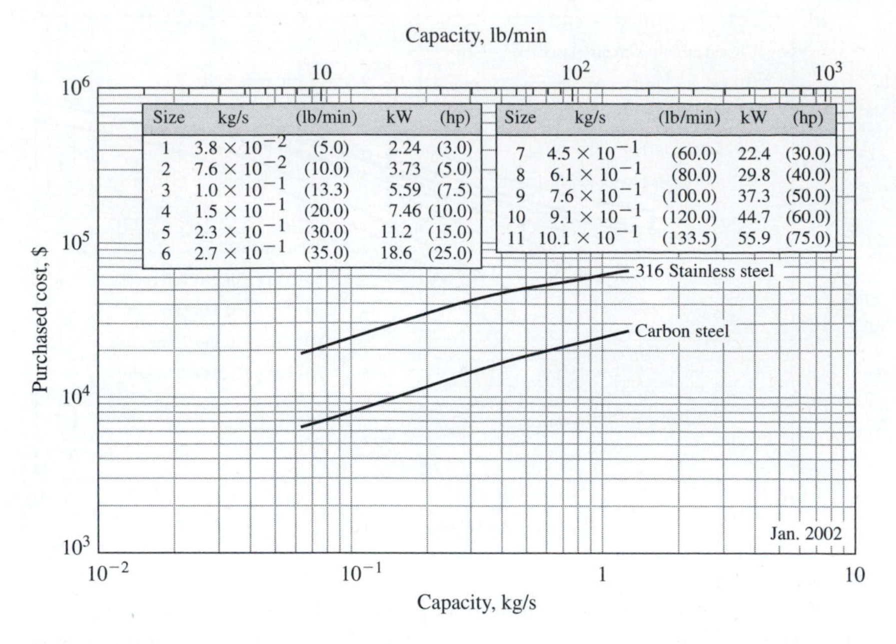

Size	kg/s	(lb/min)	kW	(hp)
1	3.8×10^{-2}	(5.0)	2.24	(3.0)
2	7.6×10^{-2}	(10.0)	3.73	(5.0)
3	1.0×10^{-1}	(13.3)	5.59	(7.5)
4	1.5×10^{-1}	(20.0)	7.46	(10.0)
5	2.3×10^{-1}	(30.0)	11.2	(15.0)
6	2.7×10^{-1}	(35.0)	18.6	(25.0)

Size	kg/s	(lb/min)	kW	(hp)
7	4.5×10^{-1}	(60.0)	22.4	(30.0)
8	6.1×10^{-1}	(80.0)	29.8	(40.0)
9	7.6×10^{-1}	(100.0)	37.3	(50.0)
10	9.1×10^{-1}	(120.0)	44.7	(60.0)
11	10.1×10^{-1}	(133.5)	55.9	(75.0)

Figure 12-73
Purchased cost of pulverizers. Price does not include motor and drive.

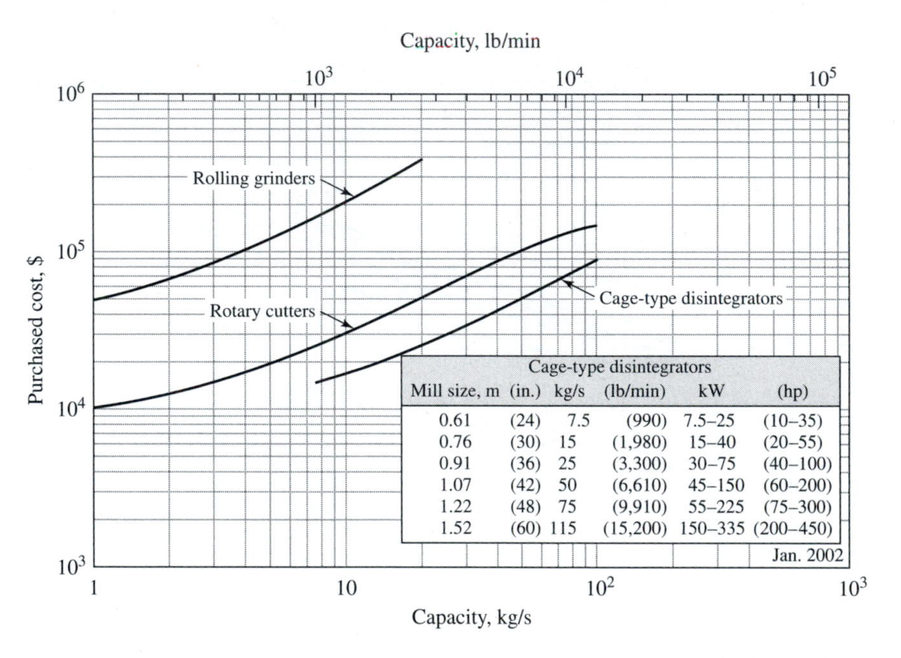

Cage-type disintegrators					
Mill size, m	(in.)	kg/s	(lb/min)	kW	(hp)
0.61	(24)	7.5	(990)	7.5–25	(10–35)
0.76	(30)	15	(1,980)	15–40	(20–55)
0.91	(36)	25	(3,300)	30–75	(40–100)
1.07	(42)	50	(6,610)	45–150	(60–200)
1.22	(48)	75	(9,910)	55–225	(75–300)
1.52	(60)	115	(15,200)	150–335	(200–450)

Figure 12-74
Purchased cost of grinders, cutters and disintegrators. Price includes motor, drive, and guard.

NOMENCLATURE

a = constant in Eq. (12-41), dimensionless

A_c = cross-sectional flow area at minimum cross-sectional area in three flowmeters, m^2

A_p = cross-sectional area in upstream pipe before flow constriction in three flowmeters, m^2

b = constant in Eq. (12-41), dimensionless

C_c = conductance in a vacuum system defined by Eq. (12-10), m^3/s

C_{cp} = conductance in a pipe connecting to a vacuum system, m^3/s

C_d = coefficient of discharge, dimensionless

C_{sc} = constant relating power components in a screw conveyor, dimensionless

C_1 = constant in Eq. (12-52*b*) obtained from Fig. 12-58, dimensionless

D = diameter, m; subscript a refers to diameter of impeller; subscript i refers to the inside diameter; subscript i,opt refers to the optimum inside diameter; subscript m refers to the mean diameter; subscript o designates the orifice diameter; subscript t designates the inside vessel diameter of a mixer

D_{pf} = solid particle size in the feed, m

D_{pp} = solid particle size in the product, m

E_J = efficiency of welded joints, expressed as a fraction

$\mathfrak{f}$ = Fanning friction factor, dimensionless; subscripts L, T, and s refer to laminar, turbulent, and sliding friction factors, respectively

$\mathfrak{f}'$ = function term, dimensionless

F = mechanical energy loss due to friction, N·m/kg

F_c = friction loss due to sudden contraction, N·m/kg

F_e = friction loss due to sudden expansion, N·m/kg

Fr = Froude number defined as V^2/gD, dimensionless

g = local gravitational constant, m/s^2

G = mass velocity, $kg/s \cdot m^2$

h = enthalpy, kN·m/kg or kJ/kg; subscripts 1 and 2 refer to the inlet and outlet conditions, respectively; subscript 2,s designates exit conditions while maintaining constant entropy from the inlet to the exit of the expander

H = head of fluid, N·m/kg; subscript ad refers to adiabatic head

H_o = weir head in an open channel, m

k = ratio of specific heat at constant pressure to specific heat at constant volume, dimensionless

K_c = constant in expression for evaluating friction due to sudden constriction of fluid (see Table 12-1), dimensionless

K_L = constant in Eq. (12-43) for laminar mixing, dimensionless

K_T = constant in Eq. (12-44) for turbulent mixing, dimensionless

L = length of pipe or unit, m; subscript e designates actual length; subscript e refers to equivalent length

m = constant in Eq. (12-40) and defined in Eq. (12-41), dimensionless

$\dot{m}$ = mass flow rate, kg/s; subscript a designates mass flow rate of air; subscript G designates mass flow rate of gas; subscript s designates mass flow rate of solid

$\dot{m}_v$ = volumetric flow rate, m^3/s

M = molecular weight, kg/kg mol

n = constant used in polytropic compression, dimensionless

n' = slope of $D\,\Delta p/(4L)$ versus $8V/D$ for non-Newtonian fluids on logarithmic coordinates, dimensionless

N_r = speed of impeller, r/min

N_{Po} = power number in mixing, dimensionless

N_{st} = number of stages involved in a compression, dimensionless

p = absolute pressure, kPa; subscripts s refers to the safe working pressure

Δp_f = pressure drop due to friction loss, kPa

P = power requirement, kW; subscripts *ad*, *iso*, and *poly* refer to the power required for adiabatic, isothermal, and polytropic compression; subscript *empty* designates power required when belt conveyor is not loaded; subscript *horizontal* designates additional power required when belt conveyor is loaded for horizontal operation; subscript *vertical* designates additional power required when belt conveyor is loaded for vertical operation

Q = heat transmitted to a system from an external source, kW

Q' = product of pumping speed and absolute pressure, (m^3/s)(kPa)

r = ratio of p_2/p_1, dimensionless

R = universal gas constant, kJ/(kg mol·K)

R^* = reduction ratio, dimensionless

R_H = hydraulic radius, m

Re = Reynolds number, defined by Eq. (12-3*b*), dimensionless

S = pumping speed, m^3/s

S_p = pumping speed of vacuum pump, m^3/s

S_s = safe working stress for tubes, kPa

t = minimum wall thickness of tube, m

T = temperature, °C or K

u = internal energy, kN·m/kg, or kJ/kg

v = specific volume, m^3/kg

V = velocity, m/s; subscript BS designates belt speed

V' = volume, m^3; subscript p refers to volume of plummet

W = shaft work, N·m/kg; subscript *ad* refers to work under isentropic compression; subscript e refers to centrifugal work required in flow through elbows; subscript KE refers to kinetic energy gain at conveyor entrance; subscript L refers to work in elevating solids; subscript *sf* refers to work due to sliding friction between solid particles; subscript s refers to the total work associated with solid transport; subscript a refers to work with air

W_i = work index, kJ/kg

W_o = mechanical work provided to a system from an external source, kJ/kg

W_v = weight of a storage or pressure vessel, kg

Y = expansion factor for certain flowmeters, dimensionless

Z = vertical distance above an arbitrarily chosen datum plane, m

Z_c = compressibility factor, dimensionless

Greek Symbols

α = correction coefficient to account for use of average velocity, dimensionless

β = ratio of throat diameter to pipe diameter for certain flowmeters, dimensionless

γ = general relationship for shear rate of a fluid, s^{-1}

ε = equivalent roughness, m

η_{ad} = adiabatic efficiency, dimensionless; subscript *c* refers to compressor efficiency; subscript poly refers to polytropic efficiency

θ_r = residence time, s

μ = viscosity, Pa·s

ρ = density, kg/m^3; subscript *m* designates mean density; subscript V refers to vapor; subscript *p* refers to density of the plummet

ϕ = power function in mixing, dimensionless

PROBLEMS

12-1 Estimate the size of the motor necessary to pump a lean oil to the top of an absorption tower operating at a pressure of 445 kPa. The oil is to be pumped from an open tank with a liquid level 3 m above the floor through 46 m of pipe with an inside diameter of 0.078 m. There are five 90° elbows in the line, and the top of the tower is 9.1 m above the floor level. A flow of 2.7 kg/s of lean oil is required. The viscosity of the oil is 15 cP, and its density is 857 kg/m^3. Assume that the efficiency of the pumping system including the motor is 40 percent.

12-2 What is the pressure loss when 2.14 kg/s of pure benzene at 40°C flows through a 21-m length of straight pipe with an inside diameter of 0.0409 m? The pipeline contains six 90° elbows, one tee used as an elbow (equivalent resistance equal to 60 pipe diameters), one globe valve, and one gate valve. The density of the benzene is 849 kg/m^3, and its viscosity at 40°C is 5×10^{-4} Pa·s.

12-3 A condenser is to be supplied with 0.0095 m^3/s of cooling water at 25°C. What diameter of Schedule 40 pipe should be recommended for this application?

12-4 Liquid benzene at 38°C with a vapor pressure of 26.4 kN/m^2 and a density of 860 kg/m^3 is to be pumped at a rate of 0.0025 m^3/s from a storage tank to a discharge location 3 m above the liquid level in the tank. The pump with a mechanical efficiency of 60 percent is 1 m above the liquid level. The storage tank is at atmospheric pressure. The pressure at the end of the discharge line is 445 kPa absolute. The inside diameter of the pipe used to transfer the benzene is 0.0409 m. The frictional pressure drops in the suction line and the discharge line has been evaluated as 3.45 and 37.9 kN/m^2, respectively. Determine the head developed by the pump and the total power requirement. If the pump manufacturer specifies a required net positive head of 3.0 m, will such a pump be applicable for this service?

12-5 A centrifugal pump delivers 0.0063 m^3/s of water at a temperature of 15°C when the impeller speed is 1800 r/min and the pressure drop across the pump is 138 kPa. If the speed of the impeller is reduced to 1200 r/min, estimate the water delivery rate and the head that is developed by the pump if pump operation is ideal.

12-6 A preliminary estimate of the total cost for a completely installed pumping system is required for a certain design project. In this system 15.75 kg/s of cooling water at 15.5°C is to be provided using a 305-m pipeline. It has been estimated that the theoretical power requirement (100 percent efficiency) for the pump will be 7.5 kW. Using the following data, estimate the total installed cost of the pumping system:

Materials of construction—carbon steel

Number of fittings (equivalent to tees)—40

Number of valves (gate)—4

Insulation (85 percent magnesia)—0.038 m thick

Pump—centrifugal with no standby

Motor—ac, enclosed, three-phase, 1800 r/min

12-7 A two-stage steam jet is to be used on a large vacuum system. It is estimated that 10 kg of air must be removed from the system each hour. The vapors being removed will contain water vapor at a pressure equivalent to the equilibrium vapor pressure of water at 15°C. If a suction pressure of 50 mmHg absolute is to be maintained by the steam jet, estimate the kilograms of steam per hour that will be required to operate the jet.

12-8 A reciprocating compressor has been selected to compress 0.085 m^3/s of 25°C methane gas from 100 to 6200 kPa. Assume that the heat capacity remains constant at 2433 J/(kg·K) over the temperature range of compression and that the ratio of C_p/C_v remains constant at 1.31. How many stages of compression should be used? What is the power requirement if the compressor exhibits an efficiency of 80 percent? What is the exit temperature of the gas at the exit from the first stage? If the temperature of the cooling water can only increase by 15°C, how much water is required for the intercoolers and aftercooler to ensure that the compressed gas from each stage after cooling is again returned to 25°C?

12-9 With a software package that uses a suitable equation of state and a procedure for estimating thermodynamic properties, repeat Prob. 12-8 and compare the results. Where should the largest differences occur between these two calculations?

12-10 Hydrogen gas at a temperature of 20°C and an absolute pressure of 1380 kPa undergoes compression to 4140 kPa. If the mechanical efficiency of the compressor is 55 percent on the basis of an isothermal and reversible operation, determine the kilograms of hydrogen that can be compressed per second when the power supplied to the compressor is 224 kW. Assume that the kinetic energy effects can be neglected. Also evaluate the mechanical efficiency of the compressor on the basis of an adiabatic and reversible operation. Assume that a single-stage compressor is used, and the ratio of C_p/C_v remains constant at 1.4.

12-11 A centrifugal fan is used to remove stagnant air in a large hangar at a pressure of 100 kPa and a temperature of 32°C and discharge it externally at a pressure of 103 kPa and a velocity of 40 m/s. Determine the power requirement needed for the removal of 4.7 std m^3/s of air. The efficiency of the fan can be assumed to be 65 percent.

12-12 What power input in kW will be required to mix a rubber latex compound with a viscosity of 1.2×10^5 cP and a density of 1121 kg/m^3 in the same mixer that was used in Example 12-6? Would there be any difference in the power requirement if baffles were added to the mixing tank?

12-13 What power will be required to mix an aqueous solution of 50% NaOH in a baffled tank, 2 m in diameter? The mixing will be performed in the vertical tank filled to a height of 2 m by a disk turbine with six flat blades. The turbine is 0.67 m in diameter and positioned 0.67 m above the bottom of the tank. The turbine blades are 0.134 m wide and turn at 90 r/min. The solution has a viscosity of 0.012 Pa·s and a density of 1500 kg/m^3.

12-14 A spherical carbon-steel tank with an inside diameter of 9 m will be subjected to a working absolute pressure of 310 kPa and a temperature of 27°C. All the welds are butt-welded with a backing strip. Assuming no corrosion allowance is required, what is the required wall thickness of the tank? Estimate the cost of the steel for this tank if the cost of steel sheet is $1.10 per kilogram. On the basis of the data presented in Fig. 12-52, determine the fraction of the purchased cost of the tank that is due to the cost for the steel.

12-15 Air at 15°C and 275 kPa is admitted to the entrance of a horizontal steel pipe with an inside diameter of 0.0779 m. The entering velocity is 15 m/s. Spherical particles with a 60-mesh average particle size are picked up by the airstream immediately downstream from the entrance to the pipe; the weight ratio of the solid particles to the air is 4 : 1, and the density of the particles is 2690 kg/m^3. If the pipe is 50 m in length, what is the pressure loss in the pipe?

12-16 An available crusher has been accepting hard rock with a volume-surface mean diameter of 0.069 m and providing a product with a volume-surface mean diameter of 5×10^{-3} m. The power required for crushing 10,000 kg/h of this specific rock is 6.35 kW. What would be the power consumption if the capacity were reduced to 9000 kg/h with the same feed characteristics but with a reduction in the volume-surface mean diameter of the product to 4×10^{-3} m? Assume that the mechanical efficiency of the unit will remain unchanged.

CHAPTER

13

Reactor Equipment—Design and Costs†

Chemical reaction engineering is one of two fields unique to chemical engineering, the other being separation processes. The large variety of chemical reactions and reactors and the general methods of analysis available in reactor design provide many opportunities for creativity. The complex nature of chemical reaction engineering often requires extensive and repetitive calculations that can only be carried out by sophisticated computer models now available to the design engineer for the analysis of most reactions and the selection and arrangement of reactor types.

To produce a desired product, a reaction path is required that indicates the reaction or reactions necessary to convert the starting materials to the desired product. The reaction path selected as part of the design process may need to be established by a research and development program, or be prespecified, or be limited by the availability of specific starting materials. The reaction path combined with the reactor performance, in turn, determines many other process conditions, such as the nature, quantity, and purity of feed streams, operating temperatures and pressures, catalysts, safety issues, and product separation and purification requirements. Once a reaction path has been defined, a complete process flowsheet is developed, as discussed in Chap. 4.

The process involving the design and selection of chemical reactors and reactor systems commences in the following manner. First, the reactor type or types to be evaluated must be established. Then the reaction mixture or the catalyst volume is determined by solving the appropriate design equation(s), utilizing kinetic and thermodynamic data.

The volume establishment step of the design process can present problems because the required kinetic data may not be available and may require experimental determination. From the reaction mixture or catalyst volume, plus ullage or void volume, the dimensions of the reactor are established. This is followed by a determination of the heating and cooling requirements established from reactor/reaction mass and energy balances. The materials of construction for the required equipment are then established,

†This chapter was prepared in part by B. Halevi, Computer Consultant, H. Barr Inc., Boulder, CO.

as discussed in Chap. 10, from the character of the species in the reaction mixture and the temperature and pressure. All this information is then used to estimate the equipment and utility costs for the reactor system. Once the various possible designs and their associated costs are identified, a determination of the optimal design can be made. A more graphical description of chemical reaction design is illustrated in flowchart form in Fig. 13-1.

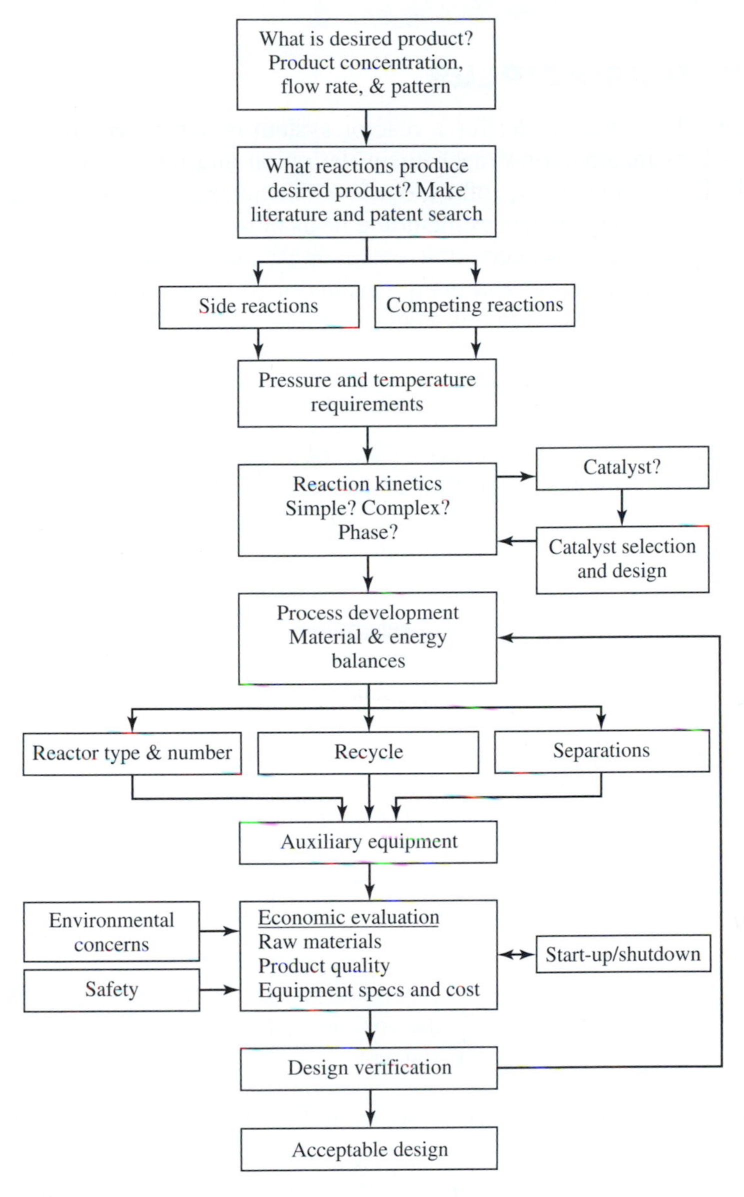

Figure 13-1
Reaction design and evaluation scheme flowchart

The central role of the reactor in so many chemical processes indicates that the reaction path, reactor selection, and design may play a key role in the complete process design, as well as in the overall economics of the process. It is therefore possible that the reaction path and reactor type selection may be the key variables in the structural optimization of a process.

REACTOR PRINCIPLES

The basic mathematical model for a reactor system is developed from (1) reaction rate expressions incorporating mechanism definition and temperature functionality; (2) material balances including inflow, outflow, reaction rates, mixing effects, and diffusion effects; (3) energy balances including heats of reaction, heat transfer, and latent and sensible heat effects; (4) economic evaluations; and (5) special constraints on the design system. The material and energy balances, shown graphically in Fig. 13-2, are developed from the fundamental conservation equations based on a differential volume of the reacting system.

The material balance is given by

$$\begin{pmatrix}\text{Rate of}\\ \text{accumulation of}\\ \text{species } i \text{ in the}\\ \text{volume element}\end{pmatrix} = \begin{pmatrix}\text{rate of inward}\\ \text{flow of species } i\\ \text{into the volume}\\ \text{element}\end{pmatrix} - \begin{pmatrix}\text{rate of outward}\\ \text{flow of species } i\\ \text{from the volume}\\ \text{element}\end{pmatrix} + \begin{pmatrix}\text{rate of species}\\ i \text{ generation}\\ \text{in the volume}\\ \text{element}\end{pmatrix} \tag{13-1}$$

The energy balance is expressed as

$$\begin{pmatrix}\text{Rate of energy}\\ \text{accumulation}\\ \text{in the volume}\\ \text{element}\end{pmatrix} = \begin{pmatrix}\text{rate of inward}\\ \text{energy flow into}\\ \text{the volume}\\ \text{element}\end{pmatrix} - \begin{pmatrix}\text{rate of outward}\\ \text{energy flow from}\\ \text{the volume}\\ \text{element}\end{pmatrix} + \begin{pmatrix}\text{rate of energy}\\ \text{generation in}\\ \text{the volume}\\ \text{element}\end{pmatrix} \tag{13-2}$$

An overall form for the energy balance given in Eq. (13-2) is, for a completely mixed reactor,

$$\sum_{\text{species}} m_i C_{p_i} \frac{dT}{d\theta} = \sum_{\text{species}} \dot{m}_i \int_{T_{i,\text{in}}}^{T_{i,\text{out}}} C_{p_i}\, dT + V_R \sum (-\Delta H_{\text{rxn}}) r_i + P \tag{13-3}$$

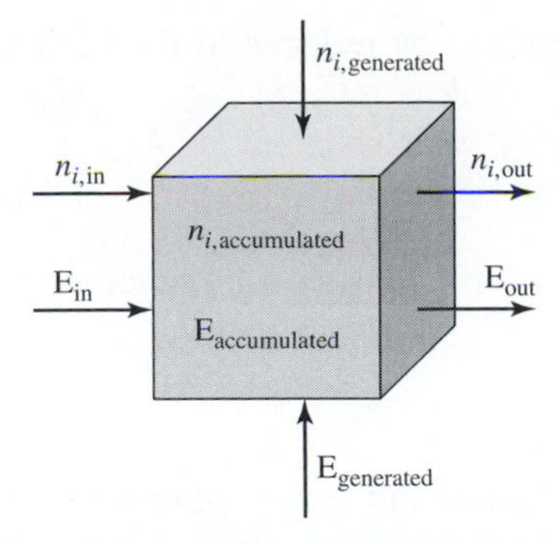

Figure 13-2
Material and energy balances applied to a differential volume of the reacting system.

where m_i is the density of species i, C_{p_i} the heat capacity of species i, $\dot{m}_i$ the mass flow rate of species i, V_R the reactor volume, ΔH_{rxn} the heat generated by the reaction per unit of reaction, r_i the rate of reaction per unit volume, and P the power supplied to the reactor. More explicit and useful forms for the general material balance expression are developed later in the chapter for the major reactor types.

There are many types of chemical reactors that operate under various conditions such as batch, flow, homogeneous, heterogeneous, and steady-state. Thus, one general mathematical description that would apply to all types of reactors would be extremely complex. The general approach for reactor design, therefore, is to develop the appropriate mathematical model that will describe the specific reaction system for the type of reactor under consideration. For example, if the reaction system is to be evaluated for steady-state operating conditions, the time-differential terms on the left-hand sides of Eqs. (13-1) and (13-2) are zero, resulting in a simplified mathematical model.

It is convenient to use different forms for expressing the rate of generation of species i and the accompanying rate of energy generation due to the heat of reaction that is associated with the selected reaction system. For example, if a batch reactor of volume V_R involving a homogeneous reaction is under consideration, the convenient form for expressing the rate of reaction, based on component i as a reactant, is

$$r_i = \frac{1}{V_R}\frac{dN_i}{d\theta} = k_i(c_a, c_b, \ldots, c_i, \ldots, c_{n-1}, c_n) \tag{13-4}$$

where N_i is the amount of reactant i present in the reactor at time θ, k_i the reaction rate constant for component i, and $c_a, c_b, \ldots, c_n$ the concentrations of species $a, b, \ldots, n$ in the reactor at time θ, respectively. On the other hand, for a flow reactor where concentration varies throughout the reactor, it is usually more convenient to base the rate on a differential volume of the reactor, and a commonly used rate form is given by

$$r_i = \text{div}(\mathbf{U}c_i) \tag{13-5}$$

where $\mathbf{U}$ is the velocity vector with normal x, y, and z directed velocity elements using Cartesian coordinates. For the case of a plug-flow reactor with z representing the length of the reactor in the direction of flow, Eq. (13-5) reduces to the familiar form of

$$r_i = \frac{d(U_z c_i)}{dz} \tag{13-6}$$

Similar simplified expressions can be developed for back-mix tank reactors and other types of reactor systems.

Reactor Types

The common, idealized designations for types of reactors are *batch, plug-flow,* and *back-mix* or *continuous stirred tank*. In an idealized batch reactor, the reactants initially are fully mixed and no reaction mixture is removed during the reaction period. Complete mixing is assumed during the reaction so that all the reactor contents are at the same temperature and concentration during the reaction process. The composition (and often the temperature) changes with time. The idealized plug reactor is a tubular reactor in which the reacting fluid moves through the tube with no back mixing or radial concentration gradients. Conditions are at steady state so that the concentration as well as the temperature profile along the length of the reactor does not change with time. An idealized back-mix flow reactor is equivalent to a continuous stirred-tank reactor (CSTR) where the contents of the reactor are completely mixed so that the complete contents of the reactor are at the same concentration and temperature as the product stream. Since the reactor is designed for steady state, the flow rates of the inlet and outlet streams, as well as the reactor conditions, remain unchanged with time. These three basic types of reactors, represented schematically in Fig. 13-3, form the basis for all reactor designs, with modifications to meet specific needs. In reality, very few reactors can fulfill the requirements for ideality, and the design engineer therefore must generally design for nonideal reactors. While the reasons for nonideal reactor behavior are

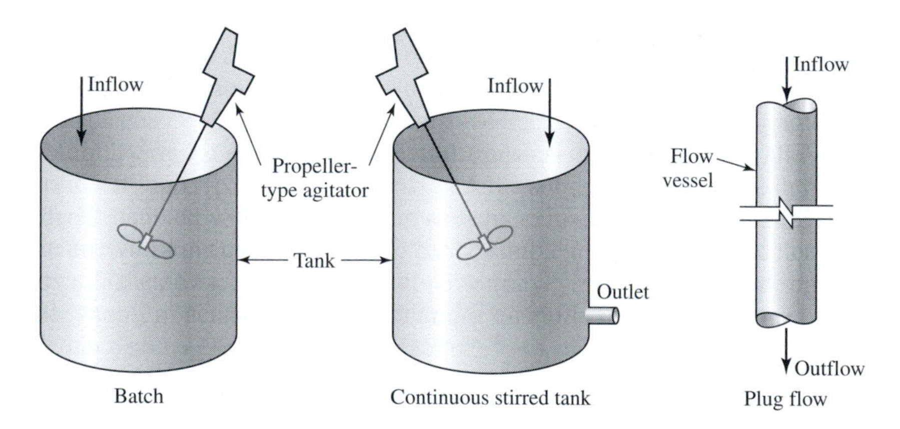

Figure 13-3
Schematics for the three basic reactor types: batch, continuous stirred tank, and plug flow

many, they are generally well recognized and the design engineer can account for the nonidealities. Another useful concept in the design of chemical reactor systems is that of recycling some of the reactor product and mixing it with the feed stream entering the reactor. Recycling often can improve the overall performance of the system, particularly if the reacted and nonreacted species in the reactor product stream are separated, and unreacted reactants are recycled to the reactor inlet.

Space Velocity and Space Time

Flow reactor analysis often utilizes two concepts, *space velocity* and *space time*. Space velocity is defined as the ratio of the volumetric feed rate to the volume of the reactor, which permits determination of the number of reactor volumes of feed that can be treated during a specified time period. The expression for this concept is given by

$$\text{Space velocity} = \frac{F_i v_{iF}}{V_R} = \frac{F_i}{V_R c_{iF}} \tag{13-7}$$

where F_i is the feed rate of component i, v_{iF} the volume of the feed per mol of component i in the feed, and c_{iF} the concentration of component i in the feed as mols of i per unit volume and therefore is the inverse of v_F. Space time is simply the inverse of space velocity, or the time required to process one reactor volume of feed.

Batch Reactors

For batch reactors, Eq. (13-4) can be rearranged to provide the basic design equation as†

$$\int_0^{\theta} d\theta = \theta = \int_{N_{io}}^{N_{ie}} \frac{dN_i}{V_R r_i} \tag{13-8}$$

where N_{ie} is the mols of component i remaining at the end of reaction time θ and N_{io} the mols of component i at the start of the reaction. In terms of conversions, X_{ie} represents the total fractional conversion of component i during the reaction time θ with N_{io} being a constant,

$$N_i = N_{io} - N_{io}X_i = N_{io}(1 - X_i) \qquad \text{and} \qquad dN_i = -N_{io}\,dX_i \tag{13-9}$$

Substituting Eq. (13-9) into Eq. (13-8) gives

$$\theta = N_{io} \int_0^{X_{ie}} \frac{dX_i}{-V_R r_i} \tag{13-10}$$

Equation (13-10) represents the basic design equation to be used with batch reactors. If the reactor volume V_R remains constant during the entire reaction, integration of the equation can be simplified by removing the V_R term from under the integral sign and recognizing that N_{io}/V_R is the concentration of the reactant at the start of the reaction.

†Note that $-r_i$ as used in this chapter represents the reaction rate expressed as the rate of reaction of reactant i, while r_i represents the rate of generation of species i.

Another useful version of the batch reactor performance expression gives the volume required to produce a given conversion in a given time

$$V_R = \frac{N_{io}}{\theta} \int_0^{X_{ie}} \frac{dX_i}{-r_i} \tag{13-11}$$

Tubular Plug-Flow Reactors

Tubular reactors are often designed for steady-state operation on the basis of idealized plug flow so that Eq. (13-6) applies, or

$$z = \int_{(U_z c_i)_o}^{(U_z c_i)_e} \frac{d(U_z c_i)}{r_i} \tag{13-12}$$

To simplify this expression, designate the constant cross-sectional area of the flow reactor as A_R, the total volume of the reactor to point z along the length of the reactor as V_R so that $V_R = zA_R$, and the constant feed rate of component i as F_{io}. When both sides of Eq. (13-12) are multiplied by A_R, it is apparent that the term $A_R U_z c_i$ is actually the mols of component i flowing over a particular point z in the reactor per unit time. This term is equivalent to $(F_{io} - F_{io}X_{iz})$, where X_{iz} is the fractional conversion of component i in the entering feed at any distance z along the reactor length. Since $d(F_{io} - F_{io}X_{iz})$ is $-F_{io}dX_{iz}$, Eq. (13-12) can be rewritten as

$$\int_0^{zA_R} d(zA_R) = \int_0^{V_R} dV_R = V_R = F_{io} \int_o^{X_{iz}} \frac{dX_{iz}}{-r_i} \tag{13-13a}$$

or

$$\frac{V_R}{F_{io}} = \int_0^{X_{iz}} \frac{dX_{iz}}{-r_i} \tag{13-13b}$$

As an alternate expression for plug-flow reactors, the more general case can be considered in which the reactant i enters the reactor already partly converted with an initial conversion, designated as X_{io}, replacing the zero conversion assumed initially in Eq. (13-11). The resulting design equation equivalent to Eq. (13-13*b*) is

$$\frac{V_R}{F_{io}} = \int_{X_{io}}^{X_{iz}} \frac{dX_{iz}}{-r_i} \tag{13-14}$$

Equations (13-13*b*) and (13-14) are in a form often used for tubular, idealized, plug-flow reactors when the reaction rate is based on the volume of the reactor. When the reaction is heterogeneous, such as a reaction occurring on the surface of a catalyst, it is common practice to base the reaction rate on the mass of the catalyst rather than on the volume of the reactor by substituting r_{ic} for r_i. The resulting equivalent design equation is

$$\frac{W_c}{F_{io}} = \int_{X_{io}}^{X_{iz}} \frac{dX_{iz}}{-r_{ic}} \tag{13-15}$$

where W_c is the mass of the catalyst in the reactor and r_{ic} the rate of the reaction as mols of component i converted per unit time per unit mass of catalyst.

Back-Mix Reactors

For the steady-state operation of idealized back-mix reactors, Eq. (13-1) becomes

$$\begin{pmatrix}\text{Rate of accumulation}\\ \text{of species } i \text{ in the}\\ \text{volume element}\end{pmatrix} = 0 = \begin{pmatrix}\text{inflow rate of}\\ \text{species } i-\text{outflow}\\ \text{rate of species } i\end{pmatrix} + \begin{pmatrix}\text{rate of}\\ \text{generation of}\\ \text{species } i\end{pmatrix} \tag{13-16a}$$

or

$$0 = F_{io} - F_{io}(1 - X_{ie}) + (r_i V_R) \tag{13-16b}$$

Therefore,

$$-F_{io}X_{ie} = r_i V_R \tag{13-17a}$$

which upon rearrangement becomes

$$\frac{V_R}{F_{io}} = \frac{X_{ie}}{-r_i} \tag{13-17b}$$

where X_{ie} is the final conversion based on the fraction of entering material i converted and V_R the total volume of the back-mix reactor. Because X_{ie} is the same everywhere in the reactor, the reaction rate must be evaluated at the exit conditions.

For the case in which the feed stream component, on which the conversion is based, enters the reactor with part of the material already converted, the initial or entering conversion becomes X_{io} instead of zero, as was the case for Eqs. (13-17*a*) and (13-17*b*). Under such conditions, Eq. (13-17*b*) assumes the more general form of

$$\frac{V_R}{F_{io}} = \frac{X_{ie} - X_{io}}{-r_i} \tag{13-18}$$

Nonideality of Reactors

The concept of ideal reactors, while useful, is misleading since few, if any, reactors and reactor systems perform ideally. All reactor nonideality can be traced to imperfect distribution of material and energy within the reactor due to mass and energy transport limitations of the reacting species, reacted species, nonreactive species, and catalytic species. Imperfect mass transport can result in altered performance, when the contact between reactive species is not idea, leading to reduced reaction rates, an increase in undesired reactions, and fouling of the reactor system. Deficient energy transport leads to a thermal distribution within the reactor system and subsequently to a range of kinetic performances due to the thermal variations. These thermally caused kinetic variations lead to changes in reaction rates, both increased and decreased, that are based on the specific kinetics of the system and can be minor, major, or even catastrophic with extremely exothermic and autocatalytic reactions. Since the reactions are limited to a certain temperature range, temperature variations in a reactor caused by inadequate energy transport can lead to unforeseen reactions that can be harmful or dangerous.

The mass and energy transport limitations can be divided into two molecular regimes—those that involve molecular-level transport and those that involve greater than molecular size transport. The larger-scale transport (or macrotransport) imperfections are due to mechanical design issues of reactors and include incomplete mixing and inadequate cooling and heating. These sources of reactor nonideality can be controlled to a certain extent, as they are all artificially influenced and therefore can be adjusted to reduce nonideality, subject to economic considerations. This can be accomplished through adjustment of the heat transfer in the reactor system and changes in reactor agitation. Molecular-level transport issues are due to inherent transport issues that cannot be influenced economically by mechanical design because of their scale. These types of transport issues are mainly encountered in fluid-solid systems, such as systems involving fluid reactive species and solid catalysts, although they can also occur in immiscible fluid systems. Some control over microscale transport is possible through catalyst morphologic design, though to a much lesser extent than is possible for larger-scale sources of nonideality.[†]

Accounting for Reactor Nonideality

Models representing real chemical reactors become prohibitively complex as the number and accuracy of the design factors are increased. It is, however, possible to reduce the complexity of nonideal chemical reactor design by combining the estimation of reactor nonideality and flow patterns and then using combinations of ideal reactor models to emulate the nonideality of the actual reactor. Presently, a number of software packages permit the analysis and design of several standard types of real reactors. The quality of these packages is increasing, and with increased performance of computers it may soon be possible to design real reactor systems with an ease similar to that experienced in the design of ideal reactors.

Residence Time Distribution

The residence time distribution (RTD) of reactants in a flow reactor is paramount to the performance of reactors, particularly as it relates to reactor flow and its effect on reactor nonideality. Flow patterns in reactors produce a distribution of particle residence times that results in a distribution of reactant conversions. An understanding of reactor residence time distribution is therefore necessary when designing and selecting reactors for a chemical reaction system.

Residence time distribution investigations are usually performed using easily detected inert tracer materials, referred to as tracers. The latter enter the reactor system in the same manner as reactants enter the system and use the same flow rate and carrier fluid. The reactor inlet and exit are monitored, and the collected tracer response data are used to calculate a residence time distribution pattern for the reactor system. This

[†]The design of catalysts is beyond the scope of this text. For a more complete study of catalyst design, consult catalysis texts such as H. F. Rase, *Handbook of Commercial Catalysts: Heterogeneous Catalysts,* CRC Press, Boca Raton, FL, 2000.

residence time distribution pattern can then be applied to the potential reactants and design of the system can proceed.

It is, however, important to note that tracer use does not give a unique model of flow patterns within a reactor system since several reactor arrangements may yield similar tracer responses. This duplication of tracer response is demonstrated in the case of a series of continuous stirred tank reactors producing a residence time distribution pattern much like that of a single plug-flow reactor. The inability of tracers to predict a unique flow pattern limits the use of tracer-predicted residence time distribution to simple first-order reactions with relatively constant kinetic constants as well as a guideline for cases with more complex reaction kinetics.

Nonideal Reactor Emulation Using Ideal Reactor Combinations

The use of ideal reactor models to emulate nonideal reactors is both simple and useful. The methodology of fitting an ideal reactor system to an actual reactor begins with an examination of the results of flow tests carried out in the design reactor. Ideal reactor models are then arranged in combinations that produce the same flow test results. Further tests can be run to confirm that the ideal reactor combination chosen produces the required flow patterns. It is, however, important to remember that the solution to a flow pattern is not unique and that it is therefore important to evaluate multiple prospective arrangement schemes. The disadvantage of using combinations of ideal reactor models to create real reactor mock-ups is that the true flow patterns are not known, and the number of reactor arrangements and selection possibilities makes the analysis complex. The advantage of using a nonideal mock-up procedure is that it allows for nonideality consideration while avoiding the extreme complexity of directly modeling nonideal reactors.

Emulation of a Real Plug-Flow Reactor Using Ideal Reactor Combinations — EXAMPLE 13-1

A reaction is underway in a plug-flow reactor 200 m in length and 0.02 m in diameter with a dilute aqueous feed stream at a temperature of 50°C and a volumetric flow rate of 0.00052 m^3/s. The experimental conversion for this reaction is lower than estimated. Use a combination of ideal reactors to create a mock-up model of the real reactor.

■ Solution

The lower conversion is probably associated with imperfect plug flow in the reactor because of non-plug flow at the reactor inlet due to entrance effects. This experimental situation can be simulated by using a combination of an ideal continuous stirred tank reactor representing the mixed-flow region of the inlet region and an ideal plug-flow reactor representing the fully developed flow region in the reactor itself.

The entrance region volume can be determined by using an entrance length relation given by Dombrowski et al.[†]

$$\frac{L_{\text{entrance}}}{D} = 0.370e^{-0.148\text{Re}} + 0.0550\,\text{Re} + 0.260$$

[†]N. Dombrowski, E. A. Foumeny, S. Ookawara, and A. Riza, *Can. J. Chem. Engr.,* **71:** 472 (1993).

Since the feed is dilute, assume that it has waterlike properties: $\rho = 1000\ \text{kg/m}^3$ and $\mu = 0.00055\ \text{kg/(m·s)}$ in determining the Reynolds number for the reactor

$$\text{Re} = \frac{\rho V D}{\mu} = \frac{4\rho \dot{m}_v}{\pi D \mu} = \frac{4(1000)(0.00052)}{\pi(0.02)(0.00055)} = 60{,}190$$

Now calculate the entrance length

$$L_{\text{entrance}} = 0.02[0.370e^{(-0.148)(60{,}190)} + 0.0550(60{,}190) + 0.260] = 66.2\ \text{m}$$

This calculated entrance length is a significant fraction of the overall reactor length, indicating that the assumption of imperfect plug flow can be a major contributor to the lower than expected conversion in the system. The entrance length is then used to calculate the volume of the continuous stirred tank reactor V_{CSTR} that is directly related to the nonplug flow character of the inlet portion of the reactor

$$V_{\text{CSTR}} = A_{\text{PFR}} L_{\text{entrance}} = \frac{\pi}{4} D^2 L_{\text{entrance}} = \frac{\pi}{4}(0.02)^2(66.2\ \text{m}) = 0.021\ \text{m}^3$$

Next calculate the plug flow reactor volume V_{PFR} that will account for the plug-flow portion of the simulated reactor

$$V_{\text{PFR}} = A_{\text{PFR}}(L_{\text{reactor}} - L_{\text{entrance}}) = \frac{\pi}{4} D^2 (L_{\text{reactor}} - L_{\text{entrance}})$$

$$= \frac{\pi}{4}(0.02)^2(200 - 66.2) = 0.042\ \text{m}^3$$

The reactor performance can now be simulated with a continuous stirred tank reactor representing the reactor inlet region and a plug-flow reactor representing the plug-flow region of the reactor. The performance of the entrance region, represented by a CSTR, can be determined with the use of Eq. (13-18)

$$\frac{V_{\text{CSTR}}}{F_{io}} = \frac{X_{\text{CSTR,out}} - X_{\text{initial}}}{-r_i} = \frac{0.021}{0.00052} = 40\ \text{s}$$

so that $X_{\text{CSTR,out}} = 40(-r_i) + X_{\text{initial}}$. Now use Eq. (13-14) to account for the plug-flow region in the reactor, substituting the final conversion expression for the inlet region of the reactor for the initial conversion in the general performance expression for plug flow, to find the overall ideal reactor mock-up performance expression for this system:

$$\frac{V_{\text{PFR}}}{F_{io}} = \frac{0.042}{0.00052} = 81\ \text{s} \qquad \text{thus} \qquad \int_{X_{\text{CSTR,out}}}^{X_{\text{final}}} \frac{dX_{iz}}{-r_i} = 81\ \text{s}$$

In combination with the reaction rate, these two expressions determine the actual performance of the reactor in the convenient form of a general ideal reactor performance expression.

■

Recycle Reactors

Under some circumstances it is possible to improve reactor performance through a partial recycling of the reactor exit stream back to the reactor inlet, as illustrated in

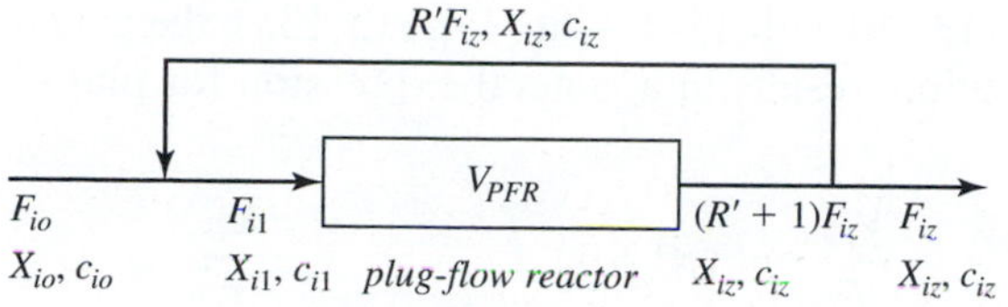

$R'F_{iz}, X_{iz}, c_{iz}$

F_{io} X_{io}, c_{io} — F_{i1} X_{i1}, c_{i1} — V_{PFR} plug-flow reactor — $(R'+1)F_{iz}$ X_{iz}, c_{iz} — F_{iz} X_{iz}, c_{iz}

$$\textit{Recycle ratio} = R' = \frac{\text{volume of fluid returned to the reactor entrance}}{\text{volume of fluid leaving the system}}$$

Figure 13-4
Recycle operation ratios and relation schematic

Fig. 13-4. Such recycle is useful in autocatalytic reactions, where products increase the reaction rate, and in auto-thermal reactions, where heat of reaction increases the rate. The main examples of the former are fermentation reactions, and of the latter are highly exothermal reactions.†

Beginning with Eq. (13-13*a*), the generalized equation for plug-flow reactors, the design engineer must account for the fact that the feed stream to the reactor depends on the recycled product stream as well as the fresh feed stream to the system. A simple material balance yields

$$F_{i1} = \begin{pmatrix}\text{reactant entering} \\ \text{in recycle stream}\end{pmatrix} + \begin{pmatrix}\text{reactant entering} \\ \text{in fresh feed}\end{pmatrix}$$

or analytically as

$$F_{i1} = R'F_{iz} + F_{io} = R'F_{io} + F_{io} = (R'+1)F_{io} \tag{13-19}$$

with $X_{io} = 0$ the conversion X_{i1} of reactant i of the mixed feed stream is given by

$$X_{i1} = \frac{1 - c_{i1}/c_{io}}{1 + \varepsilon_1 c_{i1}/c_{io}} \tag{13-20}$$

where ε_1 is the fractional change in volume of the system between no conversion and complete conversion of component i. By adding the recycle and fresh feed streams, the reactor feed concentration becomes

$$c_{i1} = c_{io}\frac{1 + R' - R'X_{iz}}{1 + R' + \varepsilon_1 R'X_{iz}} \tag{13-21}$$

Combining Eqs. (13-20) and (13-21) gives

$$X_{i1} = \left(\frac{R'}{1+R'}\right)X_{iz} \tag{13-22}$$

†O. Levenspiel, *Chemical Reaction Engineering,* 3d ed., J. Wiley, New York, 1999.

Substitution of Eqs. (13-19) and (13-22) into Eq. (13-13*a*), the general plug flow reactor performance equation, results in a general expression for plug-flow reactors with recycle as

$$\frac{V_{\mathrm{PFR}}}{F_{io}} = (R' + 1) \int_{R' X_{iz}/(1+R')}^{X_{iz}} \frac{dX_{iz}}{-r_i} \tag{13-23}$$

It is interesting to note that for a zero recycle ratio this result reduces this system to a plug-flow reactor, and for infinite recycle ratio the result becomes that for a continuous stirred tank reactor.

EXAMPLE 13-2 Recycle Operation of a Plug-Flow Tubular Reactor Saponification Reaction

Ethyl acetate is a harmful chemical used in process operations. Disposal of ethyl acetate (EtAc) can be simplified by reacting it with sodium hydroxide (NaOH), also a harmful substance, to produce ethyl alcohol (EtOH) and sodium acetate (NaAc), which are environmentally much more acceptable. Experimental results show that the reaction and reaction rate at 40°C are[†]

$$\mathrm{EtAc}(aq) + \mathrm{NaOH}(aq) \longrightarrow \mathrm{NaAc}(aq) + \mathrm{EtOH}(aq)$$

$$-r_{\mathrm{EtAc}} = k c_{\mathrm{EtAc}} c_{\mathrm{NaOH}} \qquad k = 0.5\ \mathrm{m^3/kg\ mol \cdot s}$$

Both ethyl acetate and sodium hydroxide aqueous solutions are fed to a reactor at a volumetric flow rate of 0.0005 m^3/s and a concentration of 0.1 kg mol/m^3. The reactor is 100 m in length with a diameter of 0.05 m. Compare the conversion of this system at recycle ratios of 0, 1, and 5. Assume negligible volume changes, fully developed flow, and isothermal operations.

■ Solution

The first step is to convert the concentration-based reaction expression given above to a conversion-based reaction rate expression. This results in

$$-r_{\mathrm{EtAc}} = k c_{\mathrm{EtAc}} c_{\mathrm{NaOH}} = k[c_{\mathrm{EtAc},1}(1 - X_{\mathrm{EtAc},z})][c_{\mathrm{NaOH},1}(1 - X_{\mathrm{NaOH},z})]$$

Since both reactants are fed at the same concentration and have identical first orders for the reaction expression, this expression can be converted to the following rate expression:

$$-r_{\mathrm{EtAc}} = k c_{\mathrm{EtAc},1}^2 (1 - X_{\mathrm{EtAc},z})^2 = k c_{\mathrm{EtAc},o}^2 \frac{(1 - X_{\mathrm{EtAc},z})^2}{1 + R'(1 - X_{\mathrm{EtAc},z})^2}$$

$c_{\mathrm{EtAc},o} = 0.05$ kg mol/m^3 after the two feeds mix.

Combining the conversion-based rate expression with the reactor performance equation, Eq. (13-23), and substituting the correct flow rate and volume relations provide the expression

$$\frac{k(c_{\mathrm{EtAc},o})^2 (\pi/4) D_{\mathrm{PFR}}^2 L_{\mathrm{PFR}}}{c_{\mathrm{EtAc},o}(\dot{m}_{v,\mathrm{EtAc}} + \dot{m}_{v,\mathrm{NaOH}})} = (R' + 1) \int_{R'(X_{\mathrm{EtAc},z})/(1+R')}^{X_{\mathrm{EtAc},z}} \frac{[1 + R'(1 - X_{\mathrm{EtAc},z})]^2 dX_{\mathrm{EtAc},z}}{(1 - X_{\mathrm{EtAc},z})^2}$$

where $\dot{m}_{v,\mathrm{EtAc}}$ and $\dot{m}_{v,\mathrm{NaOH}}$ are the ethyl acetate and sodium hydroxide feed volumetric flow rates, respectively.

[†]B. Halevi, University of Colorado, *Chemical Engineering Department,* private communication, 1999.

Replacement of known values and integration results in

$$\frac{(\pi/4)(0.05)^2(100)(0.5)(0.05)^2}{(0.0005+0.0005)(0.05)} = 4.90$$

$$= \frac{1}{1-X_{\mathrm{EtAc},z}} - \frac{1}{1-(R'/1+R')(X_{\mathrm{EtAc},z})}$$

$$-2R'\ln\left[\frac{(1-X_{\mathrm{EtAc},z})}{1-(R'/1+R')X_{\mathrm{EtAc},z}}\right] + (R')^2 X_{\mathrm{EtAc},z}(1-R'/1+R')$$

This equation can be solved directly for $R' = 0$ to give $X_{\mathrm{EtAc},z} = 0.831$. Solving it iteratively gives $X_{\mathrm{EtAc},z} = 0.596$ and 0.131 for $R' = 1$ and 5, respectively. These results demonstrate that recycle without separation is undesirable for this non-autocatalytic reaction.

DEVELOPMENT OF CHEMICAL REACTION RATE EXPRESSIONS

It is normally necessary to use a simplified or empirical expression for the reaction rate r_i in terms of constants and concentrations of reactant and product that can be assumed from the stoichiometry of a proposed reaction mechanism or developed purely empirically on the basis of experimental data. One of the key components of the rate expression is the specific rate constant k which must almost always be determined directly from laboratory data, although some theoretical expressions do exist.

The most common form of presenting a rate constant is in the form of the Arrhenius equation as

$$k = Ae^{-E_a/RT} \tag{13-24}$$

where k is the specific rate constant with appropriate units to fit the rate equation, A the frequency factor with units identical to those of k, E_a the activation energy with units that make E_a/RT dimensionless, R the ideal gas law constant, and T the absolute temperature.

It is worthwhile to note that since the reaction constant is dependent on the temperature, the reaction rate is also dependent on the temperature. The effect of temperature on the reaction coefficient and reaction rate can be substantial for even small temperature variations. The sensitivity of reaction rates to temperature variation is due to the dependence of the Arrhenius rate coefficient on the exponent of the negative inverse of the reaction temperature. This dependence is illustrated in Example 13-3 with the gas-phase degradation of dinitrogen pentoxide at temperatures of 293 and 303 K.

Effect of Temperature on the Reaction Coefficient and Rate — EXAMPLE 13-3

The degradation of dinitrogen pentoxide to nitrogen oxide and oxygen occurs according to the following reaction, with an energy of activation of 1×10^5 kJ/kg mol.[†]

[†]S. S. Zumdahl, *Chemistry,* D.C. Heath, Lexington, MA, 1986, pp. 501–503.

$$2N_2O_5(g) \longrightarrow 4NO_2(g) + O_2(g)$$

The reaction is assumed to be taking place at a temperature of 303 K. What would happen to the reaction rate if the reaction actually took place at 293 K?

■ **Solution**

To compare the reaction rates, compare the reaction coefficients by assuming Arrhenius behavior for the reaction coefficients and setting up the ratio given by

$$\frac{k_{293\text{ K}}}{k_{303\text{ K}}} = \frac{A\ e^{-E_a/293R}}{A\ e^{-E_a/303R}} = e^{\frac{-Ea}{R}\left(\frac{1}{293}-\frac{1}{303}\right)} = e^{\frac{-10^5(1.126\times10^{-4})}{8.314}} = 0.26$$

Thus, a 10°C decrease in temperature has reduced the reaction rate to 26 percent of what it would have been at 303 K (nearly a 75 percent reduction). It is important to be certain that the rate equation and the rate constant used in the design equations are valid for the reactor temperature, pressure, and concentration conditions used in the reaction process.

Types of Reactions

Elementary Reactions The simplest type of reactions are those reactions in which the observed reaction kinetics match the overall reaction stoichiometry. This type of reaction of one or more molecules of one or more species to form one or more molecules of another species is often represented by either of the following general expressions:

$$-r_A = kc_A^a c_B^b \cdots c_N^n \tag{13-25a}$$

or

$$-r_A = kp_A^a p_B^b \cdots p_N^n \tag{13-25b}$$

where the c_i and p_i terms are the concentrations and partial pressures of species involved in the reaction, respectively. Equations (13-25*a*) and (13-25*b*) typify the elementary reaction rate for the general reaction

$$aA + bB + \cdots + nN \xrightarrow{k} oO + pP + \cdots + zZ \tag{13-26}$$

where the uppercase letters represent chemical species and the lowercase letters represent their coefficients. Equations (13-25*a*) and (13-25*b*) are related to Eq. (13-26) in that the uppercase and lowercase letters in these three expressions are the same, designating stoichiometric coefficients and species from Eq. (13-26) and relating their effect on the reaction rate as demonstrated in Eqs. (13-25*a*) and (13-25*b*). An example of an elementary reaction is the collision of two gas-phase chlorine oxide molecules to produce chlorine and oxygen as follows:†

$$2ClO(g) \longrightarrow Cl_2(g) + O_2(g)$$

Following the general formula for the rate expression, the reaction rate for this reaction

†J. B. Russell, *General Chemistry,* 2d ed., McGraw-Hill, New York, 1991.

would be

$$-r_{ClO} = kc_{ClO}^2 = 2r_{Cl_2}$$

Reaction Order The concept commonly used in rate expressions is the one that refers to the order dependence of a reaction on one or more of the reactants. There are two uses for this concept. The first applies to the overall order of the reaction by referring to the sum of the orders of all the reactants with respect to a particular product. The overall order of a reaction is simply the summation of the stoichiometric coefficients of the species, represented by the lowercase exponents in Eqs. (13-25*a*) and (13-25*b*), and the prefixes in Eq. (13-26). The value of the overall reaction order is that it gives a general trend for reaction dependence on reactant availability.

The second application of reaction order is to desegregate the order of the reaction with respect to a particular species. Again, this is established from specie stoichiometric coefficients, represented by lowercase prefixes in Eq. (13-26). This segregated order of a reaction is of value as it permits direct prediction of reaction performance in response to a change in reactant supply. While not complex, understanding of reaction order is important since it is widely used. Example 13-4 illustrates the various types of reaction order, their establishment, and use.

Establishment of Reaction Order — EXAMPLE 13-4

For the following reactions, establish the overall reaction order, the order with respect to specie, and qualitatively show how the reaction rate would respond to an increase or decrease in the availability of one of the reactants.

Reaction rate	Overall order	Species order	Behavior
$r_{C_2H_6} = kc_{C_2H_4}$	1	First order with respect to C_2H_4	Reaction rate is proportional to methane concentration
$r_{Cl_2} = kc_{ClO}^2$	2	Second order with respect to ClO	Reaction rate is proportional to square of ClO concentration
$r_{HI} = kc_{H_2}c_{I_2}$	2	First order with respect to H_2 and I_2	Reaction rate is proportional to hydrogen and iodine concentration
$r_{NOBr} = kc_{NO}^2 c_{Br_2}$	3	Second order with respect to NO, first order with respect to Br_2	Reaction rate is proportional to bromine concentration and proportional to square of nitrogen oxide concentration

Nonelementary Reactions Such reactions do not exhibit reaction rates that match their stoichiometry. However, it is often possible to represent nonelementary reaction rates as a group of elementary reactions to simplify the analysis and provide a qualitative design for nonelementary reactions. An example of a nonelementary reaction is the gas-phase formation of hydrogen bromide as established by Bodenstein's classic research.[†] The reaction

$$H_2 + Br_2 \longrightarrow 2HBr$$

was found to proceed via the following set of elementary reactions:

$$Br_2 \longrightarrow 2Br^*$$
$$Br^* + H_2 \xrightarrow{k_1} HBr + H^*$$
$$H^* + Br_2 \xrightarrow{k_2} HBr + Br^*$$
$$H^* + HBr \xrightarrow{k_3} H_2 + Br^*$$
$$Br^* + Br^* \longrightarrow Br_2$$
$$H^* + H^* \longrightarrow H_2$$

Utilizing the assumption that the concentration of the atomic forms Br^* and H^* are constant, these reactions result in the following nonelementary rate expression:

$$r_{HBr} = k_1 c_{Br^*} c_{H_2} + k_2 c_{H^*} c_{Br_2} - k_3 c_{H^*} c_{HBr} = \frac{k_4 c_{H_2} c_{Br_2}^{1.5}}{k_2 c_{Br_2} + k_3 c_{HBr}}$$

where k_4 is a combination of k_1 and other coefficients. In addition to providing an explanation for experimentally determined reaction rates, representing a nonelementary reaction with a system of elementary reaction steps can provide a better understanding of the requirements of the reaction, resulting in increased performance through appropriate reactor design. It is, however, important to note that the elementary reaction scheme used to represent a nonelementary reaction is not necessarily unique.

Contrasting elementary and nonelementary reactions leads to an interesting observation. While in elementary reactions there is only conversion of reactants, a nonelementary reaction often displays both the reaction of reactants to form products and the reaction of products to form reactants. This is an example of what designers designate as irreversible and reversible reactions, respectively.

Irreversible Reactions Those reactions in which the reaction of reactants to form products is the only reaction observed are designated as *irreversible reactions.* Design for irreversible reactions is simple since such reactions only create specific products. This type of reaction is demonstrated in the homogeneous formation of nitrogen oxide, where the only reaction is the conversion of nitrous oxide and oxygen to form nitrogen oxide.[‡]

$$2NO(g) + O_2(g) \longrightarrow 2NO_2(g)$$

Reversible Reactions Reactions that involve both the formation of products from the reactants and a simultaneous formation of reactants from the products are known as *reversible reactions.* An example of such a reversible reaction is the formation of

[†]M. Bodenstein, *Zeitschrift für Physikalische Chemie* (Leipzig), **22:** 1 (1897).

[‡]L. D. Schmidt, *The Engineering of Chemical Reactions,* Oxford University Press, New York, 1998, p. 26.

nitrous oxide from oil and gas furnaces†

$$O_2(g) + N_2(g) \rightleftharpoons 2NO(g)$$

where both the formation of nitrous oxide from nitrogen and oxygen and the dissociation of nitrous oxide back to oxygen and nitrogen occur. The complexity of reversible reactions requires much more careful design to maximize the production of the required product while minimizing the reversion of the product back to the reactants. The techniques available for maximizing reversible reactions are based on the principle of maximizing the ratio of the forward, reactant-producing product, to the reverse, product-producing reactant. The factors that affect this ratio are the activation energies of the forward and reverse reactions and the entropy of the products versus those of the reactants. Running the reversible reaction at an appropriate temperature, removing the product from the reaction vessel shortly after formation, and operating at certain pressures all favor the desired forward, product-producing reaction, while reducing the undesired reverse reaction. Example 13-5 demonstrates the evaluation of a reversible reaction.

Effects of Temperature and Composition on Reversible Reactions — EXAMPLE 13-5

The proposed mechanism for NO emissions from oil and gas-fired furnaces is based on three reversible reactions with the following rate coefficients:‡

1. $O_2 + M \underset{k_{1,R}}{\overset{k_{1,F}}{\rightleftharpoons}} 2O^* + M \qquad k_{1,F} = 5.5 \times 10^{14} T^{1/2} e^{-59,400/T} \qquad k_{1,R} = 2.2 \times 10^{10} T^{1/2}$

2. $N_2 + O^* \underset{k_{2,R}}{\overset{k_{2,F}}{\rightleftharpoons}} NO + N^* \qquad k_{2,F} = 1.3 \times 10^{11} e^{-37,943/T} \qquad k_{2,R} = 3.1 \times 10^{10} e^{-166/T}$

3. $O_2 + N^* \underset{k_{3,R}}{\overset{k_{3,F}}{\rightleftharpoons}} NO + O^* \qquad k_{3,F} = 6.43 \times 10^{6} T e^{-3145/T} \qquad k_{3,R} = 1.55 \times 10^{6} T e^{-19,445/T}$

where M is any molecule in the system, $k_{1,R}$ is in (m^3/kg mol)^2/s, and all other k units are in (m^3/kg mol)/s. Analyze each reaction for temperature and concentration dependence, assuming rudimentary reaction orders.

■ Solution

To compare temperature dependence of the reactions, simply compare the rate coefficients for the forward and reverse reactions.

1. $$\frac{k_{1,F}}{k_{1,R}} = \frac{5.5 \times 10^{14} T^{1/2} e^{-59,400/T}}{2.2 \times 10^{10} T^{1/2}} = 2.5 \times 10^{4} e^{-59,400/T} \text{ m}^3/\text{ kg mol}$$

2. $$\frac{k_{2,F}}{k_{2,R}} = \frac{1.3 \times 10^{11} e^{-37,943/T}}{3.1 \times 10^{10} e^{-166/T}} = 4.2 e^{-37,777/T}$$

3. $$\frac{k_{3,F}}{k_{3,R}} = \frac{6.43 \times 10^{6} T e^{-3145/T}}{1.55 \times 10^{6} T e^{-19,445/T}} = 4.15 e^{16,300/T}$$

---

†C. D. Holland and R. G. Anthony, *Fundamentals of Chemical Reaction Engineering,* Prentice-Hall, Englewood Cliffs, NJ, 1989, pp. 184–5.

‡Adapted from C. D. Holland and R. G. Anthony, loc. cit.

The rate coefficient relations indicate that higher temperatures increasingly favor the forward reactions for the first two reversible reactions. On the other hand, lower temperatures increasingly favor the forward reaction for the third reversible reaction.

To compare the concentration or pressure dependence of the reactions, examine the order of the reactions.

1. Forward reaction: first in O_2, first in M; reverse reaction: second in O^*, first in M
2. Forward reaction: first in N_2, first in O^*; reverse reaction: first in NO, first in N^*
3. Forward reaction: first in O_2, first in N^*; reverse reaction: first in NO, first in O^*

For reaction 1, increased O_2 will increase the forward rate while decreased O^* will reduce the reverse reaction. Similarly, in reaction 2, increased N_2 and O^* will increase the forward rate while decreased NO and N^* will reduce the reverse reaction; and in reaction 3, increased O_2 and N^* will increase the forward rate, and decreased NO and O^* will reduce the reverse reaction.

■

Catalytic Reactions For fluid-phase reactions, solid catalysts are most frequently used, because solids are convenient for contacting gases and liquids, easy to fabricate, place in a reactor, and separate from the reaction mixture. The Langmuir-Hinshelwood (L-H) concept of such reactions is that molecules of at least one reactant must be adsorbed on the catalyst surface, there to react with another reactant either gaseous or adsorbed, and then the reaction product(s) must desorb from the catalyst surface back to the gas phase.[†] Any one of these three steps can be much slower than the others and is therefore called the *rate-limiting* or *rate-controlling* step. A general form of rate expression for such reactions is

$$r_i = \frac{(\text{kinetic term})(\text{driving force})}{(\text{adsorption term})} \tag{13-27}$$

Specific versions of this equation, for various rate-limiting steps and reaction orders, are available.[‡]

An example of this type of reaction is the synthesis of methanol from syngas (a gas containing carbon monoxide plus hydrogen and usually including some carbon dioxide and water vapor)

$$CO(g) + 2H_2(g) \rightleftharpoons CH_3OH(g)$$

This reaction is usually carried out in a reactor containing a packed bed of solid catalyst particles. This reversible reaction has been found to fit a surface-reaction-rate-controlled form where one adsorbed carbon monoxide reacts with two adsorbed hydrogen molecules.[§] Adsorption of carbon monoxide, hydrogen, and carbon dioxide are all significant.

[†]J. Langmuir, *J. Am. Chem. Soc.,* **40:** 1361 (1918), and J. Hinshelwood, *Kinetics of Chemical Change,* Oxford Press, Cambridge, UK, 1940.

[‡]For a table of the various forms of fluid-solid catalysis rate expressions, see *Perry's Chemical Engineers' Handbook,* 7th ed., R. H. Perry and D. W. Green, eds., McGraw-Hill, New York, 1997, Tables 7-2 and 7-3, pp. 7-12 and 7-13.

[§]P. Villa, P. Forazatti, G. Buzzi-Ferraris, G. Garone, and J. Pasquon. *Ind. Eng. Chem. Process Des. & Dev.,* **24:** 12 (1985), and K. J. Smith, *Ind. Eng. Chem. Res.,* **26:** 400 (1987).

$$r_{CH_3OH} = \frac{p_{CO}p_{H_2}^2 - p_{CH_3OH}/K}{(A_a + B_a p_{CO} + C_a p_{H_2} + D_a p_{CO_2})^2}$$

where p_i is partial pressure of species i, K the reaction equilibrium constant, A_a a constant, and B_a, C_a, and D_a adsorption coefficients.

The rate expressions for homogeneous catalytic reactions follow the same general format as for noncatalyzed reactions.

Enzymatic Reactions Another common type of catalysis is found in biochemical reactions that use biochemical catalysts, called *enzymes.* Such reactions generally involve a reactant and an enzyme that combine reversibly to form a complex. The reactant portion of the complex is then quickly transformed to the product species, at which point the complex splits and the enzyme is released and freed to complex again. The apparent simplicity of enzymatic reactions carries over to the rate expression for enzyme-catalyzed reactions[†]:

$$\text{Reactant} + \text{enzyme} \longrightarrow \text{complex} \longrightarrow \text{product} + \text{enzyme}$$

$$r_{\text{product}} = \frac{k_E c_{\text{enzyme,total}} c_{\text{reactant}}}{K_{\text{MM}} + c_{\text{reactant}}} \tag{13-28}$$

where k_E is the enzymatic reaction rate coefficient, K_{MM} the enzyme-product dissociation constant (commonly referred to as the Michaelis-Menten constant), $c_{\text{enzyme,total}}$ the total concentration of the complexed and uncomplexed enzyme, and c_{reactant} the concentration of the reactant.

Biological Reactions Some of the most widely utilized chemical processes involve biological reactions. Biological reactions are found in a variety of applications, ranging from long-traditional industries such as the fermentation of sugars into alcoholic beverages to newer uses such as the production of penicillin and the recent advances in gene therapy and bioengineered therapeutics. Biological reactions can be compared to a catalyzed, reversible reaction, much like enzymatic reactions in that they depend on reactant, product, and catalytic species concentration. In fact, many biological reactions are combinations of enzymatic reactions carried out in biological species. Such reactions do, however, require an additional level of design to account for the inherent sensitivity of biological organisms. Therefore, the general format of biological reaction rate expressions is very similar to the general enzyme reaction rate expression, but must be adjusted to account for the limitations of the biological components. The general process involved in biological reactions is that a feed, often composed of several substances, is consumed by biological organisms that in turn multiply and secrete a number of other substances. It is important to note that the secretions produced in the reaction may adversely affect organism growth, either of the secreting species or of another species present. It is further important to note that biological organisms usually

[†]L. Michaelis and M. L. Menten, *Biochemische Zeitschrift,* **49:** 333 (1913).

require a period of adjustment to new environments, after which they begin or resume their regular living cycle. Once the biological organisms have readjusted to a new environment, the biological reaction process proceeds via a general biological reaction rate expression with the form:[†]

$$\text{Feed} \xrightarrow{\text{biological organism}} \text{biological organism} + \text{secretion}$$

$$r_{\text{BO},i} = k_{\text{BO}}\left(1 - \frac{c_{\text{sec},i}}{c^*_{\text{sec},i}}\right)^n \frac{c_{\text{feed},i}\,c_{\text{BO},i}}{c_{\text{feed},i} + c_M} \tag{13-29}$$

where $r_{\text{BO},i}$ is the rate of production of biological species i, k_{BO} the reaction rate constant in secretion-free surroundings, $c_{\text{sec},i}$ the concentration of secretion species i, $c^*_{\text{sec},i}$ the concentration of secretion species i at which microbe growth ceases, n the order of secretion poisoning, $c_{\text{feed},i}$ the concentration of feed species i, $c_{\text{BO},i}$ the concentration of biological organism species i, and c_M the concentration of feed species i when the growth rate of the biological organism i is 50 percent of the maximum growth rate (often called the *Monod coefficient*[‡]). In biological reactions the desired reaction product may be the actual organism, such as yeast, secretions of the organisms as in the case of penicillin, or a component within the organisms as in the case of DNA produced for gene therapy. The nature of the desired product, secretion or microbe, determines which component of the reaction must be maximized and thereby controls the operational requirements. An example of a biological reaction is given in Example 13-6.

EXAMPLE 13-6 Growth of Bacterial Biomass

Nielsen and Valladsen[§] provide experimental data for the aerobic growth of *Aerobacter aerogenes* with glycerol as the carbon source. Schematically, the reaction is

$$\text{Glycerol} + O_2 \xrightarrow{\textit{Aerobacter aerogenes}} \textit{Aerobacter aerogenes} + CO_2 + H_2O$$

There is no secretion poisoning in this reaction, so the rate expression is

$$-r_g = k\frac{c_g c_A}{c_g + c_M}$$

where subscript g refers to glycerol, and subscript A to *Aerobacter aerogenes*. The reported data are: $k = 5.05 \times 10^{-4}\ \text{s}^{-1}$, $c_M = 0.01$ kg glycerol/m^3, and the yield coefficient $Y = 0.55$ kg *A. aerogenes* (dry weight)/kg glycerol.

For an aerated CSTR with a volume of 20 m^3, and a feed rate of 0.003 m^3/s of aqueous glycerol at a concentration of 10 kg/m^3, find the steady state concentration of glycerol and of *A. aerogenes* in the reactor for these conditions utilizing no recycle.

[†] O. Levenspiel, *Chemical Reaction Engineering,* 3d ed., J. Wiley, New York, 1999, p. 645.

[‡] J. Monod, *Ann. Rev. of Microbiology,* **3:** 271 (1949).

[§] J. Nielsen and J. Villadsen, *Bioreaction Engineering Principles,* Plenum Press, New York, 1994.

■ Solution

The CSTR-design equation, Eq. (13-17*b*), will be used. First, notice that the *A. aerogenes* concentration is given by $c_A = Y(c_{g,in} - c_g)$, then substitute this into Eq. (13-17*b*)

$$\frac{V_R}{m_v} = \frac{c_{g,in} - c_g}{(-r_g)} = \frac{c_{g,in} - c_g}{k\frac{c_g c_A}{c_g + c_M}} = \frac{20}{0.003} = \frac{10 - c_g}{5.05 \times 10^{-4}\frac{c_g[0.55(10 - c_g)]}{c_g + 0.01}}$$

and solve the expression for c_g to give a value of 0.0117 kg/m^3 for c_g.

The bacterial concentration is given by

$$c_A = Y(c_{g,in} - c_g) = 0.55(10 - 0.0117) = 5.5 \text{ kg/m}^3.$$

■

REACTION AND REACTOR PERFORMANCE

The mechanisms involved in reactions depend on the appropriate spatial position of the reacting species, the correct attitude of the species, and the energy level of the reacting species so that a chemical or physical change between the reactants is possible. Unique products are therefore not common, and the products of a reaction will often react themselves under the conditions that led to their production in the first place. It is therefore advantageous to develop a procedure to enable the design engineer to selectively improve the desired product production over competing, selective, or product-consuming reactions. The great variety in reaction types as well as the performance requirements make a simple process optimization procedure highly unlikely. However, several courses of action can be pursued for improving and determining an optimal arrangement for a desired chemical synthesis operation. The optimization guidelines are basically a general set of practices that help the design engineer select a design that best fills the project requirements. These guidelines include an approach for choosing reactor types and arrangements of multiple reactors to facilitate the required product production and a broad strategy for determining the best operating conditions for those situations where other reactions compete or hinder the desired reaction. The goal of these methods is always to maximize the desired reaction relative to competing reactions through control of the temperature, pressure, concentration, and reactor design.

Parallel Reactions

Parallel reactions are reactions that utilize one or more of the reactants but form products other than the desired ones. These parallel, or side, reactions consume valuable reactant, require separation from the desired product, can produce substances that react with the desired product, may physically impede the production of the desired product by changing concentrations, or interfere with process equipment. While complete extinction of parallel reactions is generally not possible, careful design can minimize them to the point of near nonexistence, or at least reduce them to such a level that their presence does not affect the overall reaction process. Controlling reaction mechanism-influencing factors such as temperature, concentration, pressure, and available species makes this possible. Examining the possible reaction rate coefficients will reveal the reaction temperature dependence, allowing for comparison of competing reactions and the possibility of selective enhancement through temperature control. This analysis will not always lead to elimination of parallel reactions, but often does allow for selectivity improvement.

Examination of the reaction rate expression reveals the temperature or pressure dependence of the reaction rate. Comparison of the concentration or pressure dependence of the possible parallel reactions permits increased selectivity by operation of the reactor at a concentration or pressure that promotes the formation of the desired product over that of parallel reaction by-products. Finally, parallel reactions can occur because of the presence of substances, reactant type or catalytic, that do not affect the desired reaction but lead to parallel reactions. Any such interfering substances should be evaluated and eliminated, if possible.

Series Reactions

Series reactions are reactions where the desired product of a reaction will itself react under the conditions that led to its creation. This is very similar to the process observed in reversible reactions, except here the product-consuming reaction does not lead to more reactant, but rather consumes some of the reactant to produce undesired species. Analysis of series reactions follows much the same methodology as for parallel reactions, that is, by examination and control of operational temperature, concentration, and pressure.

Systems of Identical Multiple Reactors

A number of basic principles and relations govern the performance of connected multiples of the three basic reactor designs. These relations are useful since multiple smaller reactors are often more advantageous than single reactors because of reduced capital cost, reduced operating cost, redundancy, system reliability, and utilization of used reactor equipment.

Plug-Flow Reactors for Any Reaction Kinetics Order Equation (13-14) can be applied to systems of multiple, equal-volume reactors to produce a generalized expression for N connected plug-flow reactors in series. This adaptation results in the following expression:

$$\frac{V_R}{F_{i0}} = \sum_{n=1}^{N} \frac{V_n}{F_{i0}} = \frac{V_1 + V_2 + \cdots + V_N}{F_{i0}} = \int_{X_0}^{X_1} \frac{dX_{iz}}{-r_i} + \int_{X_1}^{X_2} \frac{dX_{iz}}{-r_i} + \cdots + \int_{X_{N-1}}^{X_N} \frac{dX_{iz}}{-r_i}$$

$$= \int_{X_0}^{X_N} \frac{dX_{iz}}{-r_i} \qquad (13\text{-}30)$$

This expression demonstrates that the performance of a single plug-flow reactor is equivalent to that of multiple reactors in series of equal total volume, regardless of the reaction order.

Similarly, for N reactors in parallel, each with volume V_R/N and each receiving F_{i0}/N of the feed

$$\frac{V_R/N}{F_{i0}/N} = \frac{V_R}{F_{i0}} = \int_{X_0}^{X_N} \frac{dX_{i,z}}{-r_i} \tag{13-30a}$$

That is, these N plug flow reactors in parallel give the same result as does one plug flow reactor of the same total volume and total feed. This is true for any series and parallel combination of plug flow reactors, providing that all streams that mix have the same composition.†

Continuous Stirred Tank Reactors with First-Order Kinetics Applying Eq. (13-18) to a series of N completely mixed equal-volume reactors in series gives

$$\left(\frac{V_R}{F_{i0}}\right)_n = \frac{X_n - X_{n-1}}{-r_i} \tag{13-31}$$

and

$$\left(\frac{c_0 V_R}{F_{i0}}\right)_n = \theta_{r,n} = \frac{c_0(X_n - X_{n-1})}{-r_i} = \frac{c_0[(1 - c_n/c_0) - (1 - c_{n-1}/c_0)]}{kc_n} = \frac{c_{n-1} - c_n}{kc_n} \tag{13-32}$$

where $\theta_{r,n}$ is the residence time or the time a fluid element remains in reactor n. Further,

$$\frac{c_{n-1}}{c_n} = 1 + k\theta_{r,n} = \frac{1 - X_{n-1}}{1 - X_n} = 1 + k\left(\frac{c_0 V_R}{F_{i0}}\right)_n \tag{13-33}$$

For equal-size reactors with a volume of V_R, θ_r is the same for all reactors.

$$\frac{c_0}{c_N} = \frac{1}{1 - X_N} = \frac{c_0}{c_1}\frac{c_1}{c_2}\cdots\frac{c_{N-1}}{c_N} = (1 + k\theta_r)^N \tag{13-34}$$

Applying this relation to the entire system results in the relation

$$\theta_{N\text{ reactors}} = N\theta_r = \frac{N}{k}\left[\left(\frac{c_0}{c_N}\right)^{1/N} - 1\right] \tag{13-35}$$

When $N \to \infty$, Eq. (13-35) reduces to the first-order plug-flow reactor equation

$$\theta_{r,p} = \frac{1}{k}\ln\left(\frac{c_0}{c_N}\right) \tag{13-36}$$

and demonstrates that plug-flow reactor performance can be approximated by a large number of mixed flow reactors in series.

Results for a 2nd order reaction are available also.‡

†O. Levenspiel, *Chemical Reaction Engineering,* 3d ed., J. Wiley, New York, 1999.

‡O. Levenspiel, *Chemical Reaction Engineering,* 3d ed., J. Wiley, New York, 1999, pp 124–125.

Batch Reactors for Any Reaction Order Kinetics Inspection of the batch reactor performance provided in Eq. (13-10) and the assumption of reactor volume constancy result in a relation that demonstrates that the performance of multiple batch reactors equals the performance of a single reactor of the same total volume.

$$\theta_r V_R = \theta_r \sum_{n=1}^{N} V_n = \theta_r (V_1 + V_2 + \cdots + V_N)$$
$$= \frac{N_{io}}{N}\left(\int_0^{X_{ie}} \frac{dX_i}{-r_i} + \int_0^{X_{ie}} \frac{dX_i}{-r_i} + \cdots + \int_0^{X_{ie}} \frac{dX_{iz}}{-r_i}\right) = N_{io}\int_0^{X_{ie}} \frac{dX_i}{-r_i} \qquad (13\text{-}39)$$

REACTOR AND CATALYST EQUIPMENT

Selection of Catalyst

A key factor in many chemical processes is the use of a catalyst, since an appropriate catalyst can reduce residence time requirements, provide greater selectivity for the desired product, and thereby reduce investment and operating costs. There is no single path of finding a catalyst for a particular chemical reaction; experience and imagination are perhaps the main ingredients in the search. Although catalyst selection is both an art and a science, trial and error experimentation and scientific analysis provide guidance for the selection.

The single most important characteristic of a catalyst is its chemical constituency. Usually one element or compound is the key species in catalyst performance. Platinum and related metallic elements such as palladium and ruthenium are known to catalyze hydrocarbon oxidation and hydrogenation reactions. Zinc and zinc chloride catalyze the chlorination of organic compounds. Strong acids, such as sulfuric in aqueous solution catalyze a number of organic synthesis reactions, such as formation of bisphenol A from phenol and acetone. Zeolites, crystalline aluminosilicates, catalyze the cracking of petroleum fractions. The chemical and patent literature can serve as a valuable guide in the selection of a species to catalyze a particular reaction.

Given the chemical nature of a catalyst, the next choice is the physical form in which the catalyst is to be used. Catalysts are virtually always either liquids, solids in liquid solution or suspension, or solids. Catalysts in liquid solution are most common for liquid-phase reactions, but solid catalysts have some applications for such reactions also. For gaseous reactions, solid-phase catalysts are by far the most frequent choice, but liquid-phase catalysts find some uses. Once the chemical composition and phase have been selected, then factors such as catalyst concentration and support must be established.

The physical structure of solid catalysts for gas-phase reactions is very important. The catalytic substance is usually located on the surface of the solid. Therefore, large surface areas of the solid are usually required to allow sufficient exposure of the catalyst to the reactants. Areas on the order of 100 m^2 per g of catalyst are not uncommon. Such large areas are obtained by using solids containing micropores and mesopores of the order of nanometers in size. Typically, these catalyst particles are made from more or less inert solid supports, such as alumina, silica and aluminosilicates. The active catalyst species are then deposited on the surface of the support.

The transport of reactants to and product from reaction sites within a catalyst particle can limit the effective utilization of the catalyst. Such effects can be predicted based on the catalyst particle size and the relative values of the reaction rate constant and the diffusion coefficient. These effects can be at least partially compensated for by changing the catalyst particle size, the catalyst loading on the support, and the pore structure of the particles, which in turn affects the diffusion coefficient.

Another important factor in solid catalyst behavior is the stability of the catalyst to impurities and to degradation over time in the reaction atmosphere. There are a number of substances, sulfur for example, that decrease catalyst activity by being strongly adsorbed on or reacting with reaction sites. Under many organic reaction conditions, catalyst particles become coated with elemental carbon deposits thereby suffering reduced activity. The ability either to regenerate the activity of the catalyst or to protect the catalyst from exposure to potential poisons is essential to many catalytic processes. One of the main reasons for the removal of lead anti-knock compounds from gasoline was because lead is a poison for the catalysts used in automotive catalytic converters.

Sufficient physical strength to withstand the handling and abrasion to which catalysts are usually subjected is another important characteristic. Abnormal operating conditions, especially temperatures that are too high, can cause loss of mechanical strength, or decrease catalytic activity.

Catalysis remains an important area of research in chemistry and chemical engineering. Because chemical reactions involve changes at the electron level, electronic materials such as semiconductors present an intriguing opportunity for further developments.

Types of Reactors

Chemical reactors come in the form of vessels or tanks for batch reactors or back-mix flow reactors, as cylinders for fluidized-bed reactors, or as single or multiple tubes inside a cylindrical container for plug-flow reactors, particularly when special needs exist for temperature control. High pressure and extreme temperatures as well as corrosive action of the materials involved can introduce complications in the design of such systems.

Tank reactors must be designed with closed ends that can withstand the operating conditions. Cooling or heating can be provided either by internal or external coils or pipes, or by jacketing, as shown in Fig. 13-5.

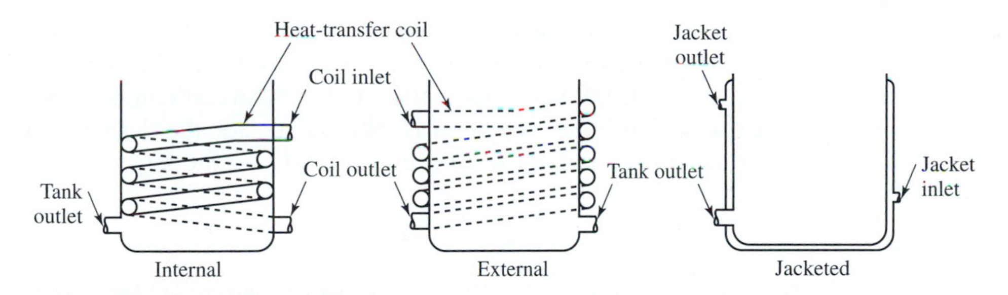

Figure 13-5
Typical systems used for heating or cooling reactors

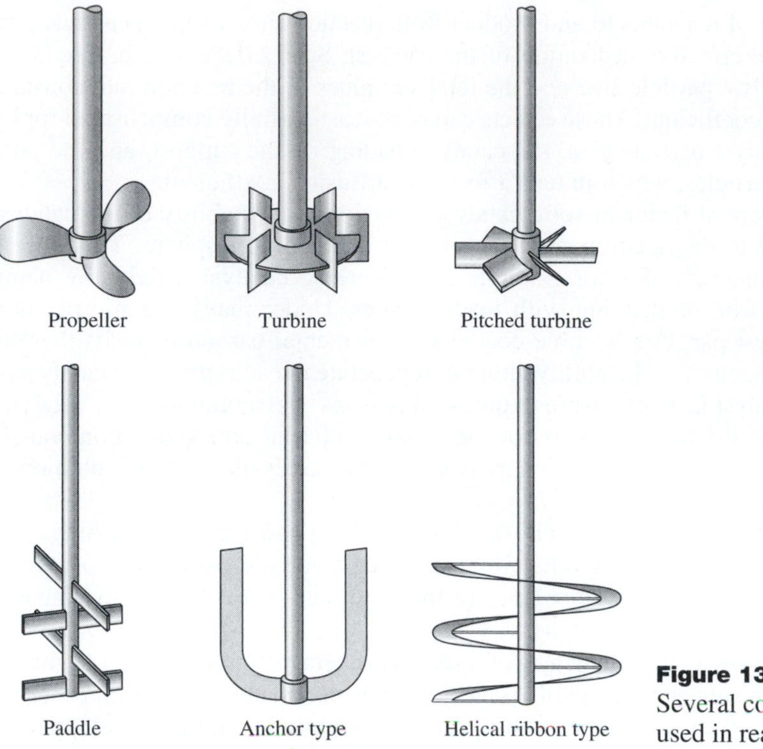

Figure 13-6
Several common agitators used in reactors

Vents, pressure releases, sampling inlets, seals, and auxiliaries for agitation, if needed, must be provided, as well as adequate foundations, supports, and facilities for personnel access to handle operations, inspection, and maintenance. Special attention to agitation equipment and operation is required in stirred tank reactor design, as correct agitation is critical to reactor performance. The key criteria in establishing correct agitation are the type of agitator and baffling used, and the agitation speed. The agitator type and speed plus baffling determine the flow pattern within the reactor. There is an extensive choice of agitation equipment that can meet a wide range of operational and mixing requirements. Figure 13-6 illustrates some of the most widely used types of agitation equipment.†

Some flow reactors are designed as large cylinders closed at both ends through which the reactant flows, while others are designed in the form of multiple tubes located in parallel with the reactant flowing inside the tubes. In both configurations, the simple assumption of ideal plug-flow operation is often made for the design even though it is clear that some back mixing will occur, especially in the large-cylinder type of flow units.

---

†Other types of agitation equipment and relations describing their operation are provided in Chap. 12 or are available in equipment handbooks.

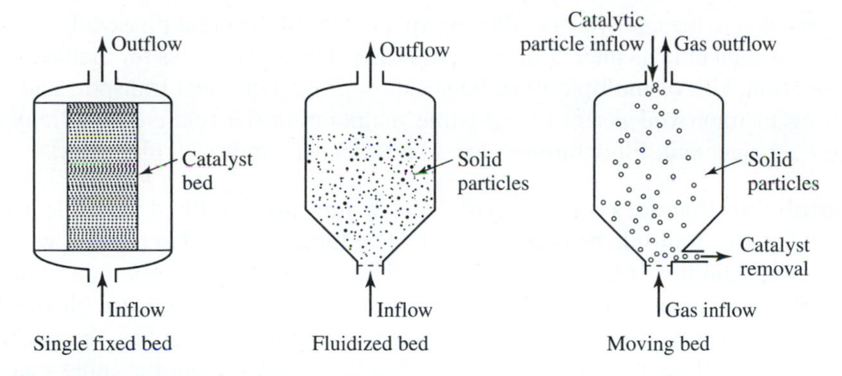

Figure 13-7
General solid-fluid reactor types

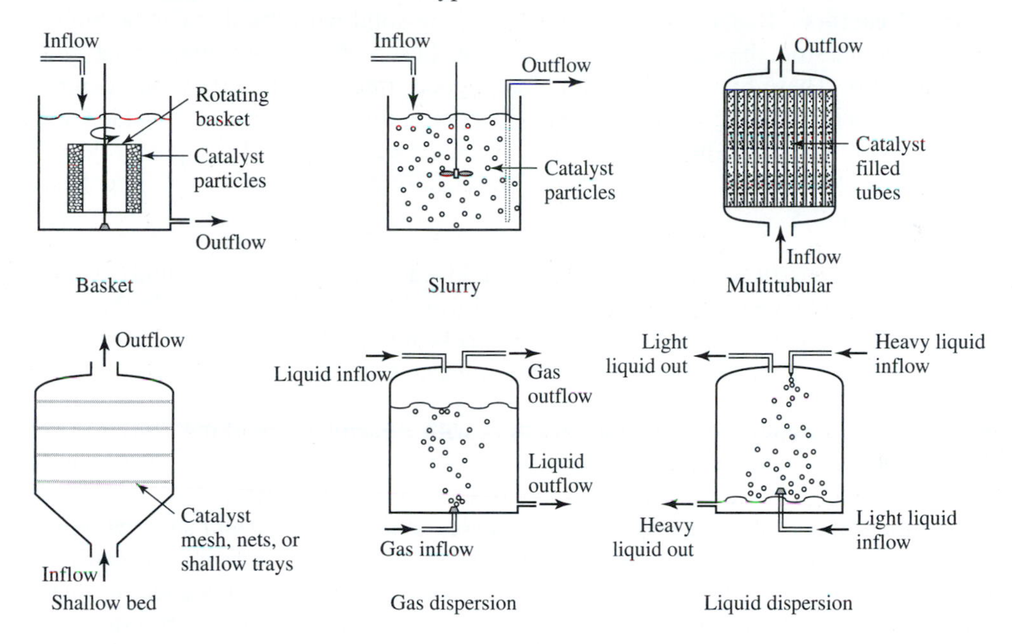

Figure 13-8
Specialized catalytic reactor types

Use of catalysts requires modifications to basic reactor design in order to account for mass and energy transport issues arising from catalysis. These modifications result in a number of basic reactor design variations that include *fixed-bed reactors, multitubular reactors, slurry reactors, moving bed reactors, fluidized-bed reactors, thin-bed reactors, dispersion reactors,* and *film reactors*. Several of these reactor configurations are shown in Figs. 13-7 and 13-8.

Fixed-Bed Reactors These reactors are solid-catalyst-containing vessels. Their design can lead to high pressure drops. These units are generally used in heterogeneous catalysis, where the catalyst and reacting species are of different phases. The major advantage of such units is their simplicity and ease of catalyst access for maintenance and regeneration. Use of multiple fixed beds can improve both heat transport and control, resulting in improved performance while maintaining the relative simplicity of this reactor arrangement. Flow through fixed beds tends to approach plug flow.

Multitubular Reactors These types of reactors are modified multiple fixed-bed units, where the multiple beds are catalyst-filled tubes arranged in parallel with a heat-conducting fluid flowing outside the tubes. These reactors offer good thermal control and uniform residence time distribution, but experience increased complexity as well as catalyst inaccessibility. Catalyst access is somewhat simplified by the packed tube arrangement, although packing and removing the catalyst from the tubes can still be difficult.

Slurry Reactors Reactions of slurries containing solid particles that can be physically separated from the suspension fluid are often best performed in agitated tank-type fluid reactors. These reactors offer simplicity, good transport properties, and control while sacrificing nothing in catalyst access since catalyst particles can be added and removed continuously. There is, however, an increased element of equipment degradation due to particle impingement on the fluid-handling equipment, such as impellers, nozzles, and pipes.

Moving Bed Reactors These units are also fluid reactors used where the fluid contains solid particles that can be physically separated from the suspension fluid. In this case however, the slurry travels through the reactor in essentially plug flow. Again, simplicity, access, and control are good with a uniform residence time distribution.

Fluidized-Bed Reactors These are reactors with a gas-phase working fluid that require gas flow around and past fine particles at a rate sufficient to fluidize the particles suspended within the reactor. There are considerable operating difficulties associated with initiating and running fluidized-bed reactors due to the flow and suspension issues. Further, these types of reactors have large residence time distributions because of the ease of backflow in the gas and approach CSTR behavior. The advantages of these reactors are their ability to process fine particles and suitability to high reaction rate processes.

Moving-Bed Reactors A subset of the fluidized-bed reactor, moving-bed reactors operate by creating a suspension of particles through fluid impingement. The usefulness of this type of reactor is limited, but it does allow for processing of particles that are too large to be handled in fluidized-bed reactors.

Thin- or Shallow-Bed Reactors These designations are reserved for reactors where the reactant fluid flows through catalyst meshes or thin beds. These are simple reactors particularly suitable for fast reactions that require good control, where catalyst access is important for purposes of catalyst reactivation or maintenance, or where large heats of reaction are involved.

Dispersion Reactors These types of reactors are fluid-containing vessels that allow dispersion of liquid and gas-phase reactants by bubbling the latter through the liquid or dripping the liquid into the gas stream or into a less dense liquid, to achieve increased contact area and reaction performance. Even though these reactors are simple and inexpensive reactors, they require careful planning due to their sensitivity to flow behavior.

Film Reactors A reactor design that maximizes contact area for gas/liquid reactions is the film reactor that brings together a gas and liquid as a thin film over a solid support. This type of reactor offers an added benefit of increased thermal control via the solid support. Such an arrangement also allows for complex phase-dependent reactions in which solid, liquid, and gas phases are all involved.

Selection of Reactors

The selection of the best reactor type for a given process is subject to a number of major considerations. Such design aspects, for example, include (1) temperature and pressure of the reaction; (2) need for removal or addition of reactants and products; (3) required pattern of product delivery (continuous or batchwise); (4) catalyst use considerations, such as the requirement for solid catalyst particle replacement and contact with fluid reactants and products; (5) relative cost of the reactor; and (6) limitations of reactor types as discussed in the previous section. Other considerations such as available space, safety, and related factors can be important and should not be overlooked. The resulting complex set of reactor physical requirements is often possible to achieve by using multiple reactor types, in which such considerations of cost, safety, and related concerns become the determining considerations in selecting a reactor.†

It is important to note that while explicit guidelines for reactor selection are not available, there are some general rules of thumb that can be followed in the selection process of an appropriate reactor for a given reaction.‡ These are briefly summarized here:

1. For conversions up to 95 percent of equilibrium the performance of five or more CSTRs connected in series approaches that of a PFR.
2. CSTRs are usually used for slow liquid-phase or slurry reactions.
3. Batch reactors are best suited for small-scale production, very slow reactions, those which foul, or those requiring intensive monitoring or control.
4. The typical size of catalytic particles is approximately 0.003 m for fixed-bed reactors, 0.001 m for slurry reactors, and 0.0001 m for fluidized-bed reactors.
5. Larger pores in catalytic particles favor faster, lower-order reactions; conversely, smaller pores favor slower, higher-order reactions.

†For more about reactor selection, see any advanced reactor text or handbook as well as industrial catalyst texts.

‡Summarized from selection criteria provided by J. M. Douglas, *Conceptual Design of Chemical Processes,* McGraw-Hill, New York, 1988, L. T. Biegler, I. E. Grossmann, and A. W. Westerberg, *Systematic Methods of Chemical Process Design,* Prentice-Hall, Upper Saddle River, NJ, 1997, and W. D. Seider, J. D. Seader, and D. R. Lewin, *Process Design Principles,* J. Wiley, New York, 1999.

Design of Reactor Systems

Generally speaking, a number of connected reactors are often more economical than any single reactor for achieving the same conversion. This is especially true when designing a new process using existing equipment. It is therefore worthwhile to establish which reactor combinations and/or arrangements are optimal for a prescribed product specification. The choice of reactors and their arrangement is based on the same criteria used for single reactor selection, with the added complexity of determining the optimal reactor arrangement.

Generally, for any nth-order elementary reaction, where n is positive, the reactors should be connected in series. Further, reactant concentrations should be kept as low as possible for reaction orders less than 1 and as high as possible for reaction orders greater than 1. This simple rule is unfortunately not applicable to more complex, nonelementary reactions where the reaction rate has at least one minimum or maximum value. These more kinetically complex reaction systems must be analyzed on a case-by-case basis that can often result in several reactor combinations and arrangements of similar performance. Additionally, different conversions of the same reaction system may call for different reactor combinations and arrangements. Nonelementary reactions therefore require a combinatorial approach to find the optimal solution that is tedious but straightforward when using modern computing techniques. In the case of ideal reactors it is possible to do this analysis manually by observing and examining the various combinations either rigorously or graphically. The graphical analysis approach is often useful as it is easy, quick, and reasonably accurate, if done correctly. In this method, reactor combinations are analyzed by graphing the inverse rate versus the conversion for the reaction, so that a reaction can be analyzed even if the rate expression is unknown. The required reactor volumes that are necessary to achieve a specified conversion can then be found since these reactor volumes are directly proportional to the area under the reaction rate–conversion curve plot for the reaction in plug-flow reactors, and to the rectangular area formed by the product of the conversion range and inverse-rate corresponding to the exit conversion for continuous stirred tank reactors. This procedure is demonstrated here.

EXAMPLE 13-7 Selection of Reactor and Reactor Combinations by Graphical Method

The autocatalytic reaction A $\rightharpoonup$ B has the isothermal rate versus conversion behavior shown in the table; the rate values are in kg mol/(m^3·s).

$-r_A$	X_A	$-r_A$	X_A	$-r_A$	X_A
5.43×10^{-3}	0	8.58×10^{-3}	0.35	1.40×10^{-2}	0.70
5.77×10^{-3}	0.05	9.21×10^{-3}	0.40	1.48×10^{-2}	0.75
6.14×10^{-3}	0.10	9.90×10^{-3}	0.45	1.52×10^{-2}	0.80
6.55×10^{-3}	0.15	1.06×10^{-2}	0.50	1.51×10^{-2}	0.85
6.99×10^{-3}	0.20	1.15×10^{-2}	0.55	1.36×10^{-2}	0.90
7.47×10^{-3}	0.25	1.23×10^{-2}	0.60	9.56×10^{-3}	0.95
8.00×10^{-3}	0.30	1.32×10^{-2}	0.65	2.59×10^{-3}	0.99

For single and combinations of two equal-volume reactors in series, use the graphical method to find the reactor volumes required to produce a 96 percent conversion of A.

■ Solution

Plot the inverse rate versus conversion from the table, as shown in Fig. 13-9. From this plot determine the graph area for reactors that result in the required conversion. The required volume is the graph area multiplied by the feed rate of A in kg mol/m^3; the graph areas will be used to represent the required reactor volumes.

Based on Eq. (13-14), the volume required for a PFR is represented in Fig. 13-9 by the shaded area under the inverse reaction-rate curve up to a conversion of 0.96. The required volume is proportional to this area of approximately 105 m^3·s/kg mol on the graph.

The graph area for a CSTR, from Eq. (13-17*b*), is given by the rectangular area with a height equal to the ordinate of the curve at the final conversion of 0.96, multiplied by the final conversion. The result, as shown in Fig. 13-10, is a volume proportional to the area of about 120 m^3·s/kg mol for one CSTR. So, for this reaction, the single CSTR requires about 14 percent more volume than does the single PFR for the same conversion.

Now, consider two equal-volume reactor combinations of PFRs and CSTRs in series, namely, PFR + PFR, CSTR + CSTR, CSTR + PFR, and PFR + CSTR. First, consider two PFRs in series. As shown by Eq. (13-30), PFRs in series give the same result as a single PFR of the same total volume. A total PFR volume is proportional to 105 m^3·s/kg mol as determined earlier, so two equal volume PFRs in series each would have half that volume. In fact, for any combination of 2 (or more) PFRs in series, the sum of the individual volumes must be the same total volume as found above to give the same conversion.

For other combinations of two equal volume reactors, a conversion exiting the first reactor is assumed (this then is the conversion entering the second reactor) and the volumes of the two reactors are found from the corresponding areas on the graph. This procedure is repeated until the two areas

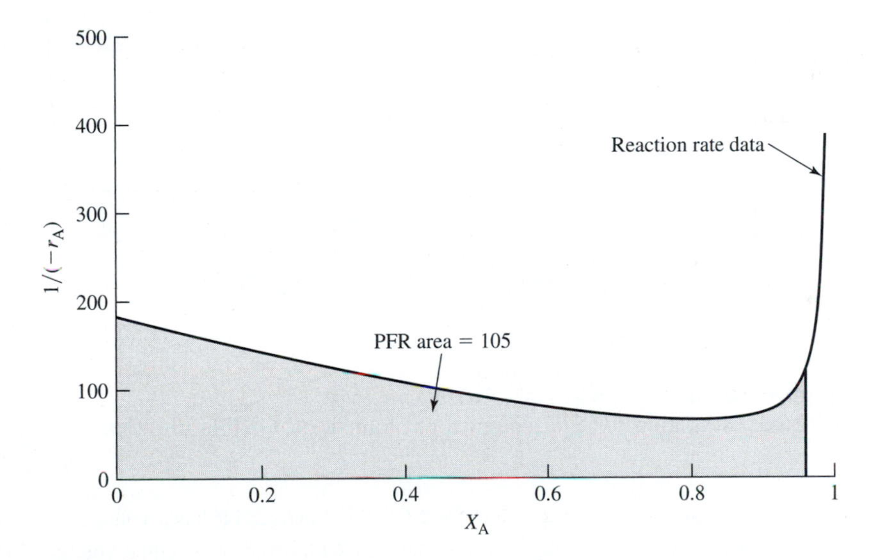

Figure 13-9
Graphic estimation of required reactor volume for a single PFR

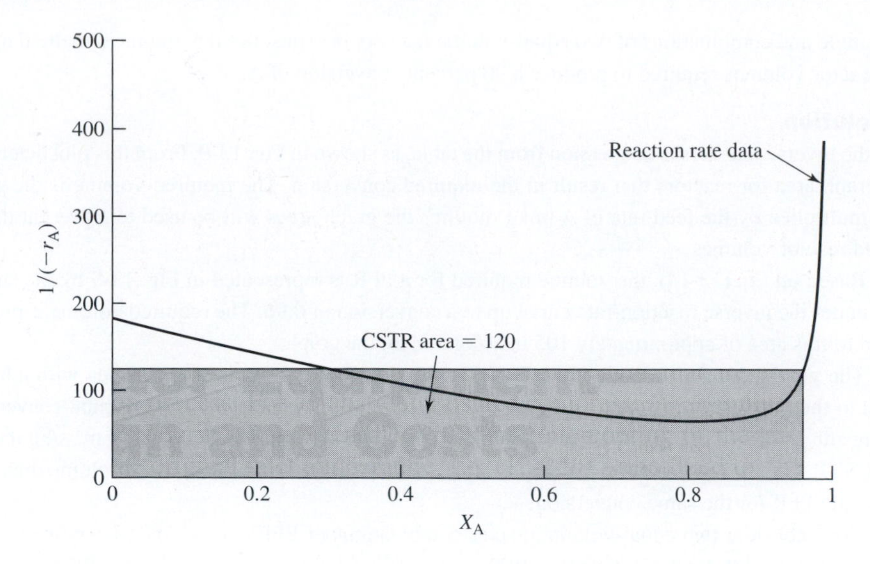

Figure 13-10
Graphic estimation of required reactor volume for a single CSTR

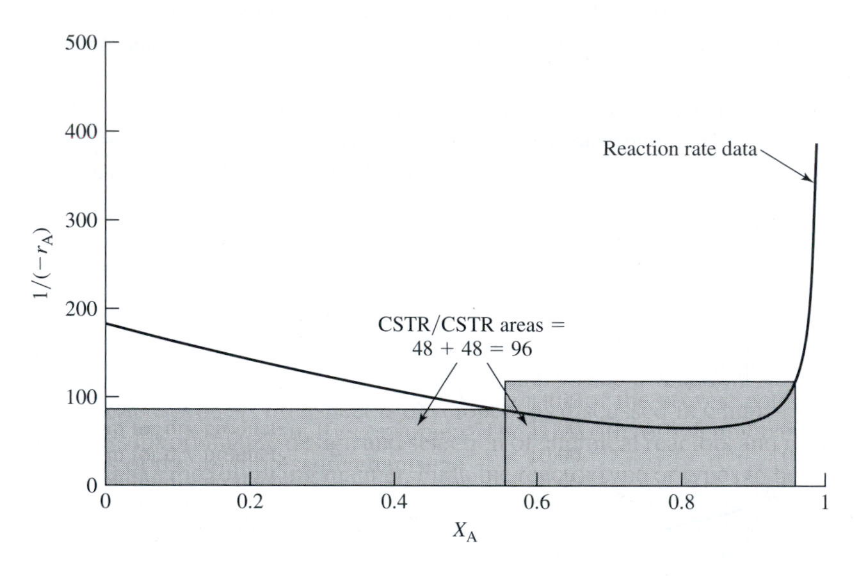

Figure 13-11
Graphic estimation of required reactor volume for 2 CSTRs in series

are equal. The results are shown in Fig. 13-11 for 2 CSTRs. Each CSTR has a volume proportional to the area of approximately 48 $m^3 \cdot s/kg$ mol. For one CSTR followed by an equal volume PFR, the result of the iteration is shown in Fig. 13-12. The required volume is proportional to the area of 45 $m^3 \cdot s/kg$ mol for each reactor. Finally, for a PFR followed by an equal volume CSTR, the required volume of each is proportional to the area of approximately 60 $m^3 \cdot s/kg$ mol, as shown in Fig. 13-13.

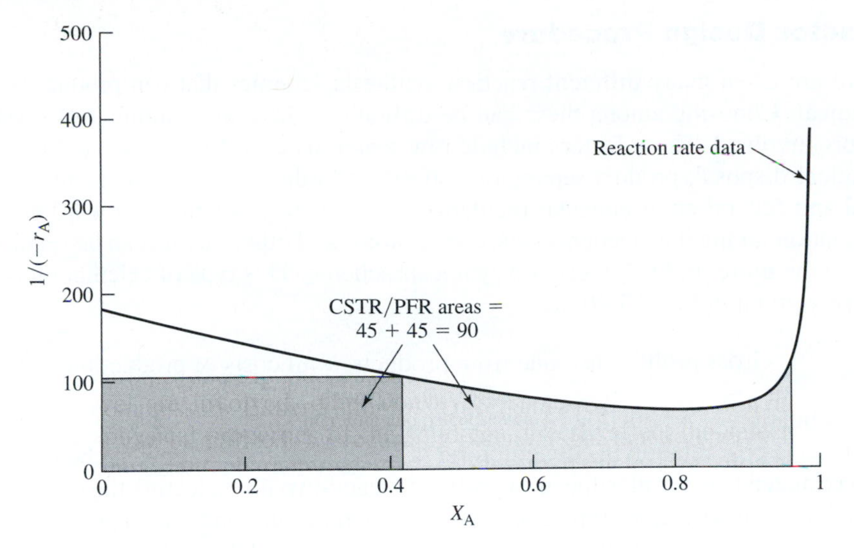

Figure 13-12
Graphic estimation of required reactor volume for a CSTR followed in series by a PFR

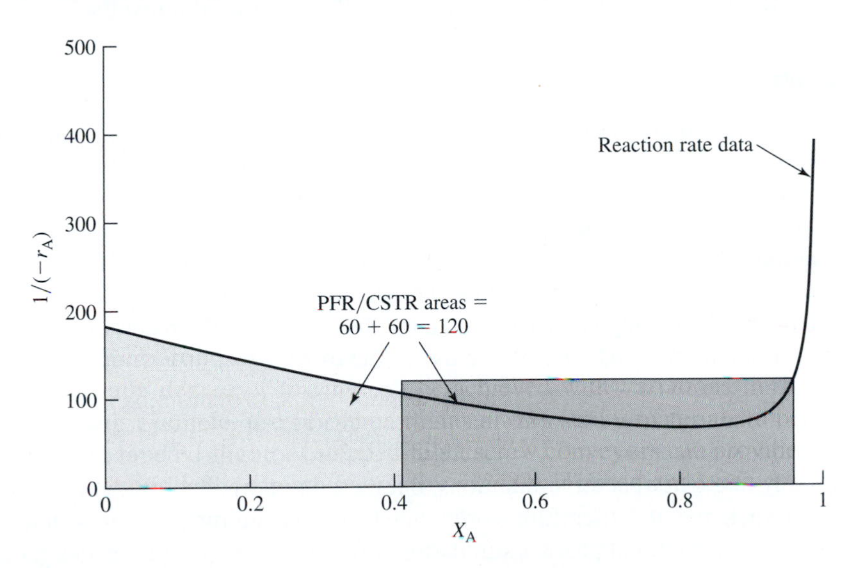

Figure 13-13
Graphic estimation of required reactor volume for a PFR followed in series by a CSTR

Thus, among these combinations, one CSTR followed by one PFR gives the smallest reactor volumes, although the combination of 2 CSTRs is not much different. It can be seen by examining Fig. 13-9 that the smallest total volume requirement for this reaction and conversion would be obtained by using 2 reactors of unequal volume, a CSTR with a final conversion equal to that which gives a minimum in the inverse rate curve (at between 0.80 and 0.85 conversion), followed by a PFR. The optimal combination would be the one that gives the minimum total annual cost of operation.

Reactor Design Procedure

There are often many different reaction synthesis schemes that can produce a given chemical. Choosing among these can be difficult as there are a number of economic factors involved. These factors include raw materials cost, by-product and unreacted chemical disposal, product separation, safety, and other less obvious factors such as local and federal environmental regulations. Fortunately, a straightforward algebraic calculation taking into account costs, conversions, and other factors can be set up to establish the more profitable reaction synthesis scheme. This type of calculation is qualitatively given in Eq. (13-40) as

$$\text{Gross profit} = \text{revenue from products} - \text{all costs of production} \qquad (13\text{-}40)$$

The estimation of production costs is discussed in Chap. 6.

Once a cost relation in equation form is developed, the relation can be minimized or maximized to determine the most profitable reaction path. Selecting the most suitable reactor can, however, be further complicated if multiple reaction steps are required. Again, the most cost-effective scheme must be determined through optimization of the process conditions. Use of software at this level of complication is highly recommended, although tedious manual calculations will lead to similar results.

Software

Software is particularly helpful in reaction process design because of the complex calculations required to simulate and optimize chemical reactors and the many possible equipment arrangements in determining the most economic reaction system and conditions.[†] However, the large number of reactor variations permits use of only a small number of explicit models, limiting the accuracy of off-the-shelf software packages.[‡] There are methods of increasing the accuracy and complexity of mainstream software packages, through experimental data obtained from the reactor manufacturer, from in-house experiments, or with the use of the mock-up approach discussed earlier in the chapter.

There are many more elaborate and accurate reactor simulation software packages available today as well as add-on modules and custom software. These can often result in extremely accurate reactor simulations, but these are complex, often unwieldy, and relatively expensive. It is therefore worthwhile to at least attempt reactor system analysis using general chemical process simulation software. If mainstream packages do not fulfill the requirements, it is often advisable to consult reaction specialists, who are more familiar with both the availability and the use of more specialized reactor software. Contacting specialists is advisable when it is determined that the system is sufficiently complex that it cannot be simulated by general software.

[†]See Chap. 5 for more details about availability, use, and limitations of chemical process design simulation software.

[‡]The most widely used general software packages currently available are from Aspen Technologies, Hyprotech, and ChemCAD, respectively.

Costs for Reactor Equipment

The design and costing of reactor vessels are handled in a manner similar to that for regular mixing and pressure vessels, as described in Chap. 12. The most reliable results are obtained by direct fabricator quotations or by the assistance of a fabricator representative who is an expert in the design of reactor vessels.

One critical item in the design of reactor vessels, in addition to establishing the type and size of the reactor by the procedures outlined earlier, is to select the correct materials of construction and wall thicknesses to handle the given operating conditions. Table 12-10 gives design equations and stress values for various materials of construction that can be applied to the design of reactor vessels, and Tables 12-11 and 12-12 present additional information that can be helpful in the design and costing of reactor vessels.† Costs for tubular flow reactors with internal tubes can often be estimated by considering the reactor unit as equivalent to a heat exchanger, so that the data presented in Chap. 14 can be used for preliminary cost estimates. Costs for kettle reactors are given in Fig. 13-14, while Fig. 13-15 gives cost data for jacketed and agitated reactors. Figure 13-16 gives the cost information for autoclaves. Costing of reactors is illustrated in the following two examples.

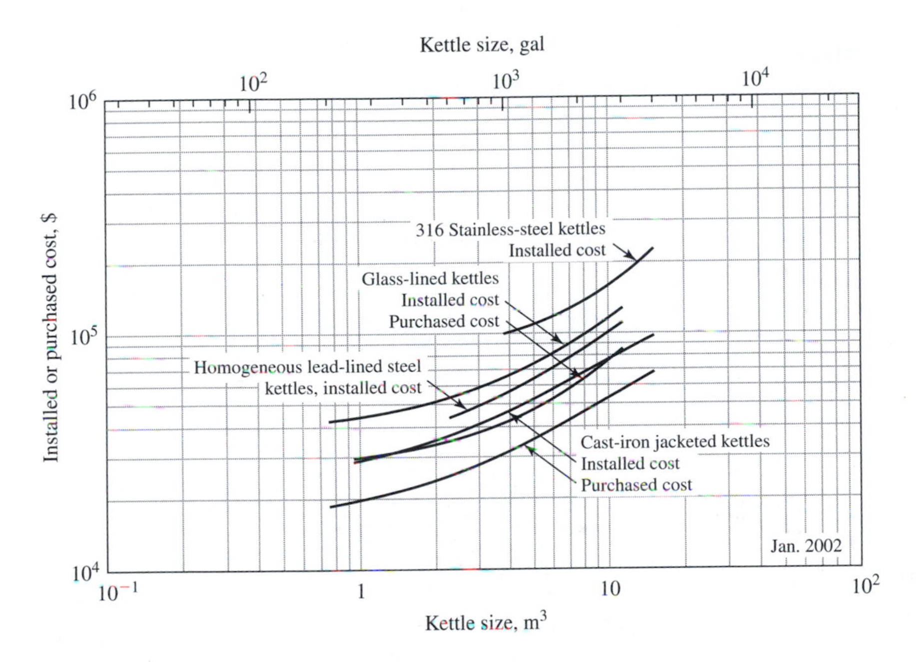

Figure 13-14
Installed or purchased cost of kettles. Cost includes kettle, jacket, agitator, thermometer well, drive and support, manhole cover, and stuffing box.

†See also the information presented in Chap. 15 on costs for trayed and packed vessels and the cost information given in Chap. 12 for tanks, pressure vessels, and storage equipment.

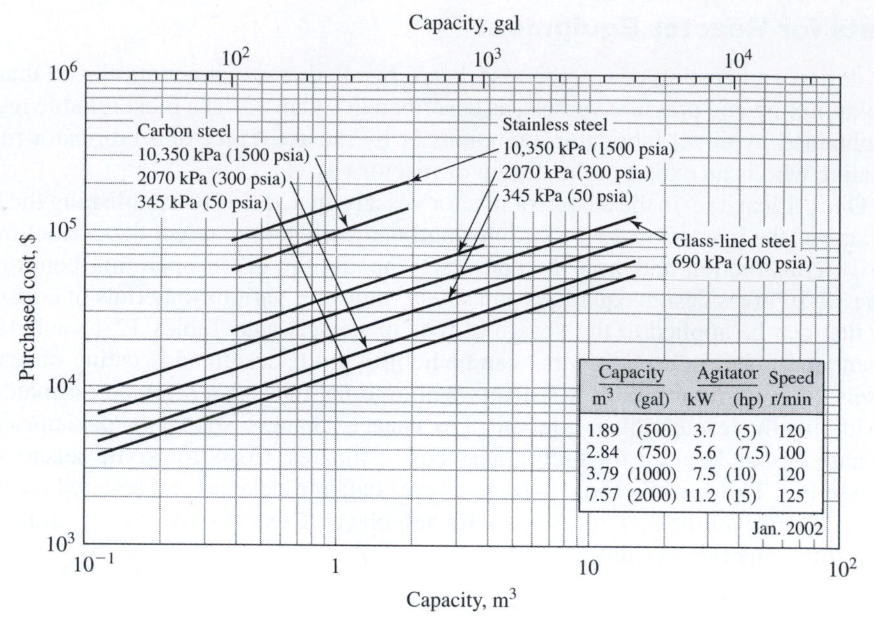

Figure 13-15
Purchased cost of jacketed and stirred reactors

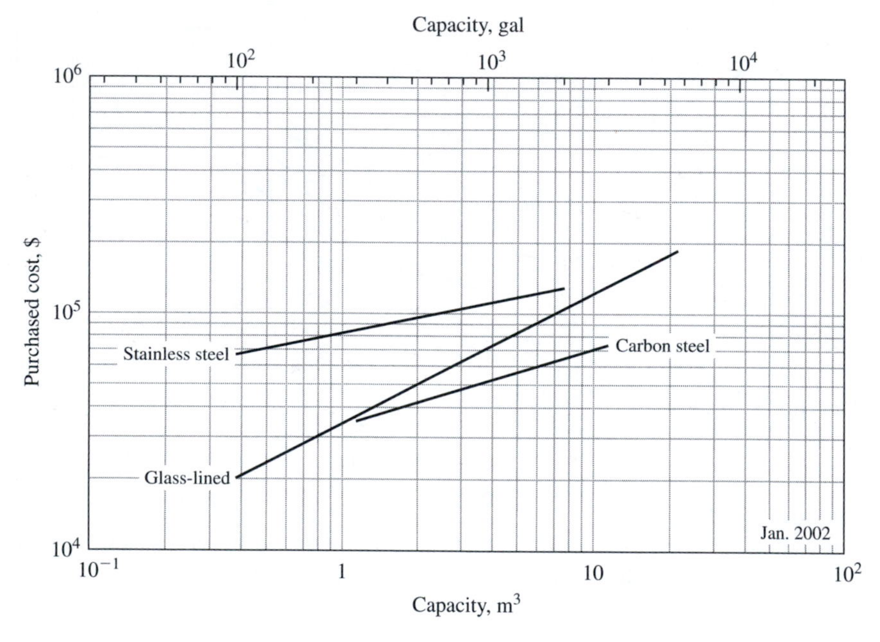

Figure 13-16
Purchased cost of autoclaves

EXAMPLE 13-8 Selection, Design, and Costing of a Reactor for the Production of Styrene

Determine the cost of a reactor required to annually produce 100,000 t (metric) of styrene, assuming 8000 operating h/yr. A literature search reveals that current production of styrene generally entails dehydrogenation of ethylbenzene using an iron oxide/potassium oxide catalyst. Besides the dehydrogenation step, there are two competing reactions that must be considered in this production process. The three reactions are

$$C_6H_5C_2H_5 \rightleftharpoons C_6H_5CHCH_2 + H_2 \tag{1}$$

$$C_6H_5C_2H_5 \longrightarrow C_6H_6 + C_2H_4 \tag{2}$$

$$C_6H_5C_2H_5 + H_2 \longrightarrow C_6H_5CH_3 + CH_4 \tag{3}$$

Further, the rates of these reactions are as follows:†

$$-r_1 = \rho_{\text{bulk}} A_1 \exp\left(\frac{-E_{A_1}}{RT}\right)\left(p_{\text{EB}} - \frac{p_{\text{S}} p_{\text{H}}}{K_1}\right)$$

$$-r_2 = \rho_{\text{bulk}} A_2 \exp\left(\frac{-E_{A_2}}{RT}\right) p_{\text{EB}}$$

$$-r_3 = \rho_{\text{bulk}} A_3 \exp\left(\frac{-E_{A_3}}{RT}\right) p_{\text{EB}}\, p_{\text{H}}$$

where

Reaction	**E_{A_i}, kJ/kg mol**	**A_i, $\frac{\text{kg mol/m}^3\cdot\text{s}}{(\text{kg catalyst/m}^3)(\text{kPa})^n}$**	***n***
(1)	9.086×10^4	0.03263	1
(2)	2.079×10^5	5.548	1
(3)	9.148×10^4	1.33	2

and $\rho_{\text{bulk}} = 1260$ and K_1 in kilopascals is given by the relation

$$K_1 = 101.3e^{\left(-17.34 - \frac{1.302 \times 10^4}{T} + 5.051 \ln T - 4.931 \times 10^{-3} T + 1.302 \times 10^{-6} T^2 - 2.314 \times 10^{-10} T^3\right)}$$

Subscripts EB, S, and H refer to ethylbenzene, styrene, and hydrogen, respectively. All rates are in (kg mol/s)/m^3 (of reactor volume).

■ Solution

The operating conditions for the reaction must be established first. This can be done by examining the temperature dependence of the three reactions by plotting their reaction rate constant-temperature dependence. Such a plot shows that the rate constants for the styrene- and benzene-producing reactions are significantly greater than the rate constant for the toluene at higher temperatures. Note that the styrene-producing reaction rate constant is considerably greater than the benzene-producing reaction rate

†Adapted from J. D. Snyder and B. Subramaniam, *Chem. Eng. Sci.*, **49:** 5585 (1994), J. G. P. Sheel and C. M. Crowe, *Can. J. Chem. Eng.*, **47:** 183 (April 1969).

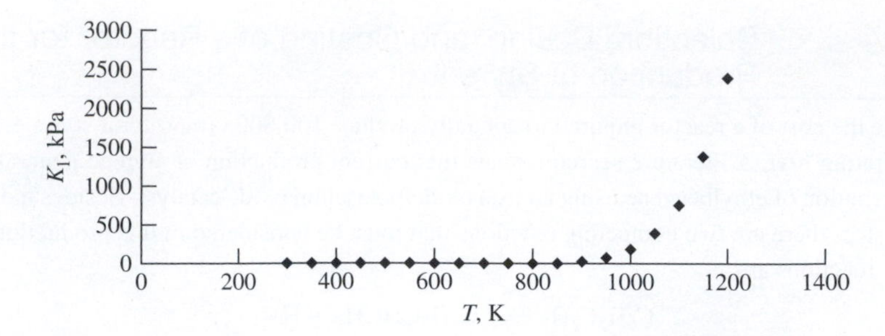

Figure 13-17
Temperature dependence of K_1

constant at lower temperatures. However, the benzene-producing reaction rate constant shows a more rapid increase with increase in temperature. This leads to the general observation that while higher temperatures increase the rate of formation of styrene, they also lead to an increase for the side reaction formation of benzene, which is undesirable.

Next examine the other temperature-dependent factor in the reaction kinetics, the ethylbenzene/styrene/hydrogen equilibrium constant K_1. This plot (Fig. 13-17) demonstrates that the equilibrium constant increases exponentially with increasing temperature. Thus temperatures above 900 K favor the production of styrene.

Next, analyze the pressure dependence of the reactions by examining the pressure dependence of the three reaction rates as depicted in the rate expressions. The desired styrene-producing reaction is both proportional to the ethylbenzene pressure and negatively proportional to the product of the hydrogen and styrene pressures. Therefore, maintaining a high pressure for the ethylbenzene reactant while keeping the product pressure low will help promote the desired reaction. Maintaining low product pressures may be difficult, but fortunately the reaction rate is negatively proportional to the product of the hydrogen and styrene pressures divided by the equilibrium constant. This means that the reaction can operate at high pressures, as long as the temperature is high enough to increase the equilibrium constant to such a level that it overrides the effect of high product pressures. The benzene-producing reaction displays first-order kinetics, where the reaction rate is proportional to the pressure of the ethylbenzene reactant. This indicates that lower pressures of ethylbenzene will reduce the side formation of benzene. The toluene-producing reaction displays second-order kinetics, where the reaction rate is proportional to the pressures of the two reactants, ethylbenzene and hydrogen. This indicates that low pressures of ethyl benzene and hydrogen are required for reducing the production of toluene. Combining the pressure dependence observations of the three reactions encountered in this process leads to a preference for low pressures to reduce the formation of the undesired by-products of benzene, and toluene.

Combining all the temperature and pressure dependency observations points to a number of operating condition recommendations. First, the desired reaction favors higher temperatures, but this also favors the undesired side reactions. A balance of these factor leads to an operating temperature greater than 900 K, but not above 1000 K. The analysis also indicates that maintaining low product concentrations and pressures is desirable since this increases the styrene reaction rate while not adversely affecting the side reactions. An operating pressure is selected to avoid external leakage into the reactor system while providing sufficient pressure potential to ensure adequate material flow.

The final consideration is the need to supply heat to the endothermic reactions taking place in the reactor. The analysis suggests that a steam feed to the reactor could not only supply heat, but also lower the relative pressures of species, thereby promoting styrene production.

Based upon this analysis and the conditions for an isothermal, packed bed process, the following conditions are selected:

Pressure: 150 kPa
Temperature: 930 K
Steam to organic feed molar ratio = 8

Equipment Selection

The reaction is hydrogenation of a gas-phase reactant on a solid catalyst. The isothermal European process† utilizes a fixed catalyst bed in the tubes of a reactor resembling a shell-and-tube heat exchanger. A molten mixture of the salts sodium, lithium and potassium carbonate supplies heat.

Reactor Material Selection

Containment of the molten salt on the shell side of the reactor is the main concern. It is recommended that type 304 stainless steel be used for the tubes and the shell of the reactor. Above about 1000 K more expensive alloys must be used, and this is one of the key considerations in the selection of 930 K as the reactor temperature.

Reactor Sizing

First, it is decided that for purposes of illustrating a preliminary sizing, reactions 2 and 3 need not be included in the calculations. As can be seen from the rate equations, they are slow relative to reaction 1 so long as reaction 1 does not approach equilibrium too closely. It is known that typically about 97 percent of the ethylbenzene is converted to styrene, so this value will be included in the material balances for styrene. A design target of 70 percent conversion of ethylbenzene-per pass through the reactor is selected.

The design production rate of 100,000 t (metric)/8000 hours is

$$\frac{10^8}{8000(3600)} = 3.472 \text{ kg styrene/s}$$

The fresh feed rate of ethylbenzene with 97 percent selectivity for styrene, and an assumed 98 percent recovery and recycle of ethylbenzene is

$$\frac{3.472(106)}{(0.97)(0.98)(104)} = 3.723 \text{ kg ethylbenzene/s}$$

And, with a 70 percent conversion per pass through the reactor, the required feed rate of ethylbenzene to the reactor is

$$\frac{3.723}{0.70} = 5.318 \text{ kg/s} = 0.05017 \text{ kg mol/s}$$

†This process, offered by Lurgi, Montedison and Denggendarfer, is described in M. Howe-Grant, ed., *Kirk-Othmer Encyclopedia of Chemical Technology,* J. Wiley, New York, 1994.

The equilibrium constant for reaction 1 at 930 K is given by

$$K_1 = 101.3\exp[-17.34 - 1.302 \times 10^4(930)^{-1} + 5.051 \ln 930$$
$$-4.931 \times 10^{-3}(930) + 1.302 \times 10^{-6}(930)^2 - 2.314 \times 10^{-10}(930)^3]$$
$$= 63.863 \text{ kPa}$$

The rate constant for reaction 1 at 930 K is given by

$$k = \rho_{\text{bulk}} A_1 e^{-E_1/RT} = 1260(0.03263)\exp\left(\frac{-90{,}860}{8.314(930)}\right)$$
$$= 3.23 \times 10^{-4} \text{ kg mol/(m}^3 \text{ reactor) (kPa)(s)}$$

Eq. (13-14), the design equation for ideal plug flow, is

$$\frac{V_R}{F_{i0}} = \int_0^{X_e} \frac{dX_i}{(-r_i)}$$

In order to integrate this equation and solve for the reactor volume needed, the reaction rate must be expressed in terms of the conversion. The partial pressure of each component equals the total pressure multiplied by the component mol fraction. Because the number of mols changes in reaction 1, the number of mols and the mol fractions must be expressed in terms of conversion. On the basis of one mol of ethylbenzene fed, and only ethylbenzene and steam in the feed, the total number of mols of ethylbenzene, styrene, and hydrogen is $(1 + X)$, where X is the conversion at any point. The total mols of steam is constant at 8, and the total mols of gas is $(9 + X)$. The mol fractions of the reacting species are thus given by

$$y_{\text{EB}} = (1 - X)/(9 + X)$$
$$y_{\text{S}} = X/(9 + X)$$
$$y_{\text{H}} = X/(9 + X)$$

The reaction rate can now be written as

$$-r_{\text{EB}} = 3.23 \times 10^{-4}\left[p_T \frac{(1 - X)}{9 + X} - p_T^2 \frac{X^2}{(9 + X)^2 K_1}\right]$$

where p_T the total pressure equals 150 kPa. Substituting into the reactor design equation gives

$$\frac{V_R}{0.05017} = \frac{1}{3.23 \times 10^{-4}(150)} \int_0^{0.70} \frac{(9 + X)^2 dX}{(1 - X)(9 + X) - 150X^2/63.863}$$

This equation can be integrated graphically or with a package integration routine. It has been integrated here using a spreadsheet. Integration has been done for various increments of X until the value of the integral changes by less than 0.01 percent for successive increments. The result is that the value of the integral is 16.65 and the reactor volume is given by

$$V_R = \frac{(0.05017)(16.65)}{3.23 \times 10^{-4}(150)} = 17.24 \text{ m}^3$$

Estimation of Reactor and Catalyst Costs

A tubular reactor with the catalyst inside the tubes is to be used, so the reactor volume must equal the inside volume of the tubes. By selecting a tube diameter and length, the volume per tube is calculated as follows:

Pick a tube of outside diameter 0.0508 m (2 in.), which has an inner cross sectional area of 0.00157 m^2. This is a large size for heat exchanger tubing, but a large size is desirable for good catalyst distribution and minimal wall effects. Pick a tube length of 6.09 m (20 ft). The inside volume of one tube is thus

$$0.00157(6.09) = 0.00956 \text{ m}^3$$

The number of tubes needed to hold the catalyst is

$$17.24/(0.00956) = 1803 \text{ tubes}$$

The outside surface area per tube length is 0.159 m^2/m, or 0.973 m^2 per tube. Therefore, the heat transfer area available is (0.973)(1803) = 1754 m^2. Referring to Fig. 14-17, the cost for a shell-and-tube, floating head exchanger for this service can be estimated. Notice that the maximum exchanger size on the figure is 1000 m^2; therefore, two exchangers of 880-m^2 areas each should be used. The graph gives about $61,000 as the cost for a base size, lowest pressure, carbon steel exchanger. A multiplier of 3 is suggested to account for the use of 304 for both the tubes and shell. Adjustments for the tube diameter (0.0508 rather than 0.019 m), tube length (6.08 rather than 4.88 m), and operating pressure (150 versus 690 kPa), are respectively obtained from Figs. 14-21 (extrapolate to estimate a value of 1.75/0.91), 14-22 (0.96), and 14-23 (0.93/0.96). Thus the purchased cost of the 3 exchangers is estimated to be

$$\$61{,}000(2)(3)(1.75/0.91)(0.96)(0.93/0.96) = \$655{,}000$$

This result will be compared with the estimate for a U-tube heat exchanger, based on Fig. 14-17. The maximum area for a stainless steel exchanger is 350 m^2, thus 5 exchangers of 350-m^2 area each will be used. The cost for one stainless steel exchanger from the graph is $75,000. So the total cost for 5 exchangers, including diameter, length and pressure corrections is

$$\$75{,}000(5)(1.75)(0.93) = 610{,}000$$

This shows that the estimate for the floating head exchanger is not unreasonable. The rest of the analysis is done for the floating head case, because the higher estimate is probably more realistic.

Next, the capability of the heat exchanger to add the heat of reaction, 280,000 kJ per kg mol, is checked. The rate of heat transfer needed is

$$280{,}000(0.035) = 9.8 \times 10^3 \text{ kJ/s}$$

Assuming a $\Delta T_{\text{log mean}}$ of 40 K, and an overall heat transfer coefficient of 0.1 kW/$m^2 \cdot$K, it is estimated that the heat exchanger can handle

$$\dot{q} = UA_s \Delta T_{\text{log mean}} = 0.1(1760)(40) = 7.0 \times 10^3 \text{ kJ/s}$$

This suggests that the heat exchanger is undersized for the heat transfer needed.

As an alternative, the floating head design is repeated using 0.038 m ($1\frac{1}{2}$ inch) tubes, which have more heat-transfer area per unit volume. The result is to use 3 exchangers of the required volume with 825-m^2 areas each. The rate of heat transfer predicted is 9.9×10^3 kJ/s, so this design appears to be satisfactory. The total purchased cost of the 3 floating head units is estimated to be $735,000, about 12 percent more than for the 0.0508-m tubes, but the exchanger is more satisfactory, so this design is recommended.

Finally, the cost of the catalyst is estimated. The catalyst price of $6.60 per kg (January 2002) is based on quotations from a number of catalyst manufacturers.

$$\text{Cost of catalyst} = 17.24(1260)(6.60) = \$143{,}000.$$

■

EXAMPLE 13-9 Cost of a Pressure Vessel Reactor

Consider the case where a reactor is to be designed and a preliminary cost estimate (January, 2002) is to be made for the installed reactor for the following conditions:

The reactor will be cylindrical with a 2.74 m inside diameter and a length of 9.15 m. It will be constructed of carbon steel and will operate at a temperature of 670°K and an internal pressure of 690 kPa guage. It will require double-welded butt joints and will be spot-examined by the radiograph technique. It will be operated as a flow reactor in a horizontal position with the reacting materials flowing through it. There are no major heating or cooling needs, and, therefore, no special internal accessories are needed. Heads for the reactor will be hemispherical and the same thickness as the shell. The corrosion allowance for the wall thickness is 0.0032 m.

■ Solution

Use Table 12-10. For carbon steel at 670 K, the stress value is 82,000 kPa and the joint efficiency is 0.85 for the given fabrication conditions as shown in Table 12-10.

The minimum wall thickness is (see Table 12-10 for equations and nomenclature)

$$\text{Thickness} = \frac{P(\text{inside radius})}{SE_J - 0.6P} + C_c$$

$$= \frac{\overbrace{(690)}^{P}\left[\overbrace{(2.74)(0.5)}^{\text{inside radius}}\right]}{\underbrace{(82{,}000)}_{S}\underbrace{(0.85)}_{E_J} - (0.6)\underbrace{(690)}_{P}} + \overbrace{0.0032}^{C_c}$$

$$= 0.017 \text{ m}$$

The vessel weight, with density of steel = 7840 kg/m^3, is (see Table 12-10)

$$\text{Weight of shell} = \pi(2.74)(9.15)(0.017)(7840) = 10{,}500 \text{ kg}$$

$$\text{Weight of two heads} = (2\pi)\left[\left(\frac{2.74}{2}\right)^2 (0.017)(7840)(1.6)\right](2) = 5{,}000 \text{ kg}$$

Total weight including 15 percent increase for nozzles, manholes, and saddles = (15,500) (1.15) = 17,800 kg.

F.o.b. cost for reactor based on Eq. 12-50 = (cost per kg) (17,800) = 147(17,800)$^{-0.34}$ × (17,800) = \$94,000.

■

SUMMARY

Chemical reaction process design is generally encountered in most chemical processes. The complexity of some chemical reactions and their dependence on mass and energy transport can lead to considerable difficulties in the design of reactive chemical processes. These complexities make a simple optimization routine impossible, although a general approach to designing and optimizing chemical reaction processes greatly helps in establishing the possible designs that fulfill the process requirements, the optimal design, and the best operating conditions.

Idealized reactor models can be adapted for use in most situations. While these basic designs are idealized and often are not accurate enough for proper design, there are ways of increasing their usefulness through careful analysis and the use of software. At the same time, the kinetics of the reaction drive the operational requirements of the design to maximize the required reaction and discourage any unwanted reactions. This requires careful study of the reaction kinetics and mechanism, followed by the selection of temperature, pressure, concentration, and material contact operating conditions. Further improvement of reactions is often possible through the utilization of catalytic substances. These often improve the economics of a reaction, but require additional equipment design and investigation of operating conditions to fulfill catalytic requirements and thus achieve an optimal design.

NOMENCLATURE

A_a = constant in catalytic reaction rate expressions, dimensionless

A_R = constant cross-sectional area of a flow reactor, m^2

A_s = heat exchanger surface area, m^2

B_a = adsorption coefficient in catalytic reaction rate expressions, dimensionless

c = concentration, kg mol/m^3 or kg/m^3; subscript i designates concentration of species i; subscript BO,i designates concentration of biological organism species i; subscript enzyme,total designates total concentration of the complexed and uncomplexed enzyme; subscript feed,i designates feed concentration of species i; subscript iF designates concentration of component i in the feed as mols of i per unit volume; subscript io designates concentration of component i in the stream feeding to a recycle system; subscript iz designates the concentration of component i in the stream leaving the reactor in a recycle system; subscript $i1$ designates concentration of component i in the stream feeding to the reactor in a recycle system; subscript M designates concentration of feed species i where the biological organism i reproduces at one-half the maximum growth rate (commonly referred to as the Monod constant); subscript N designates concentration in reactor N; subscript reactant designates concentration of reactant; subscript sec,i designates concentration of secretion species i

$c^*_{sec,i}$ = concentration of secretion species i at which microbe growth ceases, kg mol secretion/m^3

C_a = adsorption coefficient in catalytic reaction rate expressions, dimensionless

C_{pi} = heat capacity of species i, kJ/kg·K or kJ/kg mol·K

D_a = adsorption coefficient in catalytic reaction rate expressions, dimensionless

E_i = energy, kJ

F_i = flow rate of feed component i, kg mol/s; subscript io designates feed rate of component i; subscript $i1$ designates rate of component i in the feed plus recycle stream of a recycle system

ΔH_{rxn} = heat generated by the reaction per mol reacted, kJ/kg mol

k = reaction rate constant, units vary with order of reaction; subscript i designates reaction rate constant for species i; subscript BO designates reaction rate constant in secretion-free surroundings; subscript E designates enzymatic reaction rate coefficient

K = reaction equilibrium constant, specified units or dimensionless; subscript e denotes reaction number

K_{MM} = enzyme-product dissociation constant (commonly referred to as the Michaelis-Menten constant), kg/m^3

m_i = mass of i in reactor, kg

$\dot{m}_{v,i}$ = volumetric flow rate of species i, m^3/s

n = reactor number exponent on secretion poisoning term, dimensionless

n_i = number of mols of species i, kg mol

N = number of reactors, dimensionless

N_i = amount of reactant i present at time θ in a volume V_R, kg mol

N_{ie} = amount of component i remaining at the end of reaction time θ, kg mol

N_{io} = amount of component i at the start of the reaction, kg mol

p_i = partial pressure of the species i, kPa

P = mechanical power to reactor, kJ/s

p_T = total pressure, kPa

$\dot{q}$ = rate of heat transfer, kJ/s

r_i = rate of reaction of species i, units vary with order of reaction, of the form kg mol/s; subscript BO,i designates rate of production of biological species i; subscript ic designates rate of the reaction as kilogram mols of component i converted per unit of time per kilogram mass of catalyst

R' = recycle ratio, dimensionless

R = ideal gas constant, kJ/kg mol·K

T = temperature, K; subscript i designates species i; subscript i,in designates species i entering the reactor; subscript i,out designates temperature of species i leaving the reactor

$\Delta T_{\text{log mean}}$ = log mean temperature difference, K

U = velocity vector with normal x, y, and z directed velocity elements using Cartesian coordinates

U = overall heat transfer coefficient, $kJ/s{\cdot}m^2{\cdot}K$

v_F = volume of the feed per mol of component i in the feed, m^3

V_R = volume of reactor, m^3; subscript PFR designates the volume of a plug-flow reactor; subscript CSTR designates the volume of a continuous stirred tank reactor

W_c = mass of catalyst in the reactor, kg

X = fractional conversion of species, dimensionless; subscript i designates conversion of species i; subscript ie represents the overall conversion based on the fraction of entering material i converted; subscript io designates the initial conversion of species i; subscript iz designates conversion of species i at point z along the length of the reactor; subscript N designates the conversion in reactor N

y_i = mol fraction of component i, dimensionless

z = length of the reactor in the direction of flow, m

Greek Symbols

ε = catalyst void fraction, dimensionless

ε_i = fractional change in volume of the system between no conversion and complete conversion of component i, dimensionless

ρ_{bulk} = bulk density of the catalyst, kg catalyst/m^3 reactor volume

θ = time, s; subscript r refers to residence time; subscript N reactor designates residence time of N reactors; subscript r,n designates residence time in nth reactor; subscript r,p designates residence time of plug-flow reactor

PROBLEMS

13-1 A reaction is carried out utilizing four equal-volume CSTRs connected in series. The flow through the system is 0.01 m^3/s, and the CSTRs each have a volume of 3 m^3. This reactor system has been in operation for more than 5 years, and the engineer supervising the system has noticed an 8 percent drop in the conversion for the system since it was first fabricated. Investigate the possibility that the decrease in performance is due to equipment malfunction, using an average fluid density of 870 kg/m^3 and a mean viscosity of 0.00062 Pa·s.

a. There is a possibility that the performance decrease is due to the nonmixing of the reactants in the bottom of the reactors. For the case where all the reactors provide equal performance, what would be the actual volume of the reactors that is being actually utilized in the reaction process?

b. Is it possible that the decrease in performance is due to one of the mixers completely malfunctioning, and if so, which one?

13-2 Triphenyl methyl ether is produced from the reaction of triphenyl methyl chloride with methanol, both dissolved in benzene, according to the following reaction:

$$CH_3OH + (C_6H_5)_3CCl \longrightarrow (C_6H_5)_3COCH_3 + HCl$$

The reaction is second-order for methanol and first-order for triphenyl methyl chloride with a rate constant of 4.48×10^{-3} $(m^3/kg\ mol)^2/s$ at 298 K, 101 kPa.[†] Feed concentrations are 0.11 kg mol/m^3 for the triphenyl methyl chloride and one-half that concentration for the methanol.

a. For 50 percent conversion of methanol utilizing a single CSTR, calculate the required reactor space time.

b. What reactor type, CSTR, PFR, or batch, would you recommend for this reaction?

13-3 The irreversible reaction of bromine cyanide with methylamine produces cyanamide ($CNNH_2$) and methylbromide. The reaction is first-order with respect to both reactants and has a reaction constant of 2.22 $(m^3/kg\ mol)/s$ at 283 K.[‡] The reactants have feed aqueous concentrations of 0.1 and 0.2 kg mol/m^3 for bromine cyanide and methylamine, respectively. Each reactant is fed to the reactor at a rate of 0.1 m^3/s. The volume of the CSTR used is 2 m^3.

a. What is the conversion of each reactant for this system?

b. Assuming a recycle stream composed of all but 10 mol percent of each of the reactants leaving the reactor, 20 percent of the water, and 20 percent of the cyanamide product, what is the conversion for each reactant?

13-4 Acrylic acid is commercially produced by catalytically oxidizing propylene. The oxidation of propylene does, however, lead to two side reactions forming acetic acid and carbon dioxide. The reactions arising from the oxidation of propylene are

$$C_3H_6 + \frac{3}{2}O_2 \longrightarrow C_3H_4O_2 + H_2O \qquad (1)$$

[†]C. G. Swain, *J. Am. Chem. Soc.*, **70:** 1119 (1948).

[‡]R. O. Griffith, R. S. Jobin, and A. McKeown, *Trans. Far. Soc.*, **34:** 316 (1938).

$$C_3H_6 + \frac{5}{2}O_2 \longrightarrow C_2H_4O_2 + CO_2 + H_2O \qquad (2)$$

$$C_3H_6 + \frac{9}{2}O_2 \longrightarrow 3CO_2 + 3H_2O \qquad (3)$$

where $-r_i = A_i \exp(-E_{A_i}/RT) p_{\text{propylene}}\, p_{\text{oxygen}}$ and

Reaction	E_{A_i}, kJ/kg mol	A_i, (kg mol/m^3)(kPa)$^{-2}$(s)$^{-1}$
(1)	3590	4.42×10^1
(2)	5980	5.03×10^4
(3)	4780	2.45×10^2

Establish general guidelines for operating temperature, pressure, and concentrations to maximize the desired reaction and minimize the side reactions.

13-5 The thermal, gas-phase dimerization of butadiene follows the following kinetics:

$$-r_{d-\text{butadiene}} = kc^2_{\text{butadiene}} \qquad \text{where } k = 5.4 \times 10^7 \exp\left(\frac{-100{,}250}{RT}\right) \text{ m}^3/(\text{kg mol}\cdot\text{s})$$

where R is 8.314 kJ/kg mol·K.

A current setup employs three CSTRs arranged in series, with volumes of 8, 6, and 4 m^3, respectively. The reaction may be assumed to be isothermal at 500 K, and 101 kPa. Butadiene is fed at a rate of 7 m^3/day.

a. Estimate the total conversion of butadiene through the three reactors.

b. Size a batch reactor to achieve the same conversion as the system described above. Assume the batch reactor operates 50 percent of the time.

13-6 An aqueous glucose stream is to be catalytically hydrogenated to sorbitol in a slurry tank reactor. Feeds to the reactor are an aqueous glucose stream with a concentration of 2.6 kg mol/m^3 at a flow rate of 10^{-3} m^3/s and a 120 percent of stoichiometric hydrogen gas stream. The reactor is to be operated isothermally at 423 K, where the reaction is observed to follow the kinetics shown here:

$$-r_{\text{glucose}} = kc^{0.6}_{\text{hydrogen}}\, c_{\text{glucose}} \qquad k_{423\text{ K}} = 3.76 \times 10^{-4}\ (\text{kg mol/m}^3)^{-0.6}\text{s}^{-1}$$

The reaction is conducted at 10,000 kPa, the solubility of hydrogen may be assumed to be 5 kg mol/m^3 at this pressure.

The actual kinetics of the reaction require adjustment to account for catalyst mass-transfer effects. Neglecting these effects, calculate the conversion for a 2-m^3 CSTR.

13-7 Dilute propylene oxide is to be catalytically hydrolyzed to propylene glycol in an adiabatic PFR according to the kinetics

$$-r_{\text{propylene oxide}} = kc_{\text{propylene oxide}} \qquad k = 4.71 \times 10^9 \exp\left(\frac{-63{,}010}{RT}\right) \text{ s}^{-1}$$

where R is in kJ/kg mol·K. The reaction is conducted isothermally at 300 K. The feed consists of a 10 weight percent aqueous stream of propylene oxide at 300 K with a flow rate of 0.01 m^3/s. Water stream containing 0.1 weight percent aqueous sulfuric acid (the catalyst) at 300 K is added at a flow rate of 0.01 m^3/s.

Size the reactor to achieve 90 percent conversion.

13-8 The rate of fermentation of glucose to ethanol by *Saccharomyces cerevisiae* at 303 K is given by:[†]

$$-r_{\text{glucose}} = k\left(1 - \frac{c_{\text{ethanol}}}{c^*_{\text{ethanol}}}\right)^{0.52} \frac{c_{\text{glucose}}\, c_{\textit{Saccharomyces cerevisiae}}}{c_{\text{glucose}} + c_M}$$

where $k = 1.53 \times 10^{-3}$ kg/m^3·s, $c^*_{\text{ethanol}} = 93$ kg/m^3, and $c_M = 1.7$ kg/m^3.
The cell yield is 0.06 kg *S. cerevisiae* produced per kg of glucose consumed.

a. Evaluate the time required for 95 percent conversion of glucose in a batch reactor with an initial charge of 10 kg/m^3 glucose and initial cell concentration of 0.01 kg/m^3 in a 10-m^3 batch reactor filled to 70 percent.

b. Determine the required flow rate to produce the same glucose conversion for a PFR of equal volume with the same glucose-feed concentration, but with 0.9 kg/m^3 of *S. cerevisiae* in the feed.

c. Size and cost a single CSTR as well as three CSTRs of equal volume connected in series to produce the required conversion for the flow rate calculated from part (b).

13-9 Based on the information provided in Prob. 13-8 and the availability of two PFRs with volumes of 1 and 2 m^3 and 1 : 100 diameter-to-length ratios and two CSTRs with volumes of 1 and 2 m^3 as well:

a. What is the best arrangement for the two PFRs in series?

b. What is the best arrangement for the two CSTRs in series?

c. What is the best arrangement using any two PFRs and CSTRs of equal volume?

d. What is the best arrangement for all four reactors together?

e. What is the conversion for the arrangements listed in parts (a), (b), (c), and (d)?

13-10 The packed-bed catalytic syngas production of methanol is a reversible reaction as follows:[‡]

$$CO + 2H_2 \rightleftharpoons CH_3OH \qquad \Delta H_{298} = -89.98 \text{ kJ/kg mol}$$

The rate expression is given by

$$-r_{CO} = \frac{\bar{p}_{CO}\,\bar{p}^2_{H_2} - \bar{p}_{CH_3OH}/K}{(A + B\bar{p}_{CO} + C\bar{p}_{H_2})^2}, \text{ (kg·mol·CO/kg catalyst·min)}$$

The coefficients at 506 K are given by

$$K = 3.18 \times 10^{-7}\,(\text{kPa})^{-2} \text{ (reaction equilibrium constant)}$$
$$A = 23{,}400\,(\text{kPa})^{3/2}(\text{kg mol/kg·min})^{-1/2}$$
$$B = 126\,(\text{kPa})^{1/2}(\text{kg mol/kg·min})^{-1/2}$$
$$C = 47\,(\text{kPa})^{1/2}(\text{kg mol/kg·min})^{-1/2}$$

The reaction is to be conducted in a packed bed reactor at 506 K and 5065 kPa. The catalyst has a bulk density of 653 kg/(m^3 of reactor volume). The following assumptions may be made: the reaction takes place isothermally (it is exothermic, but externally cooled),

---

[†]R. Miller and M. Melick, *Chem. Eng.*, **94**(2): 112 (1987).

[‡]K. J. Smith, *Ind. Eng. Chem. Res.*, **26:** 401 (1987), and P. Villa, P. Forazatti, G. Buzzi-Ferraris, G. Garone, and I. Pasquon, *Ind. Eng. Chem. Process Des. & Dev.*, **24:** 12 (1985).

carbon monoxide and hydrogen are fed in a stoichiometric ratio, fugacity coefficients are 1.0, and the water-gas-shift reaction may be neglected.

a. Calculate the carbon monoxide conversion at equilibrium.

b. For a carbon monoxide conversion equal to 90 percent of the equilibrium conversion, calculate the catalyst to feed-rate ratio in (kg catalyst)(min)/(kg mol CO).

c. For the conversion of part (b) and a reactor production rate of 50 t (metric)/h of methanol, select, and estimate the size and cost of a suitable reactor.

13-11 Propionic acid is used in the production of a number of valuable chemical products, including herbicides, food flavorings, preservatives, and animal feed. A proposed process uses *Propionibacterium acidipropionici* bacteria to convert the aqueous lactose in sweet whey to propionic acid. (Sweet whey is a by-product of cheese making and does not have many commercial uses; thus, it is generally sent to waste with the associated waste disposal costs.[†])

$$-r_{\text{lactose}} = k_l\left(1 + \frac{c_{\text{propionic acid}}}{c^*_{\text{propionic acid}}}\right)^n \left(\frac{c_{\text{lactose}}\, c_{\textit{Propionibacterium acidipropionici}}}{c_{\text{lactose}} + c_M}\right)$$

$$k_l = 0.901\ \text{h}^{-1} \qquad c_M = 32.5\ \text{kg/m}^3 \qquad c^*_{\text{propionic acid}} = 4.4218\ \text{kg/m}^3 \quad n = -1$$

$$c_{\text{lactose(initial)}} = 50\ \text{kg/m}^3 \qquad c_{\textit{Propionibacterium acidipropionici}\ \text{(initial)}} = 0.1\ \text{kg/m}^3$$

The yield of *P. acidipropionici* is 0.145 kg dry cells/kg of lactose, and that of propionic acid is 0.307 kg/kg of lactose. Since the bacteria are very sensitive to temperature, a constant temperature of 308 K must be maintained. The reaction constants provided are those for 308 K.

Calculate the time required in a batch reactor to reduce the lactose concentration to 20 kg/m^3.

Plot the lactose, *P. acidipropionici,* and propionic acid concentrations with time.

13-12 The average catalytic converter contains approximately 0.05 kg of platinum-group metals contained within a 5 kg ceramic honeycomb block. Recovery of the platinum-group metals in spent catalytic converters is desirable for purposes of reducing waste disposal costs, reducing the amount of toxic heavy metals sent to landfills, and selling the valuable metals that can be recovered from the catalytic converters. A proposed recovery process complexes a 1 part catalyst to two parts water by weight aqueous slurry of the catalyst with excess cyanide that is obtained from sodium cyanide ionization. The actual recovery of the platinum metal occurs via heating to decompose the complexed metal cyanides and remaining sodium cyanide which precipitates the platinum. The destruction of the species occurs as follows:[‡]

$$-r_{\text{NaCN destruction}} = k_1 c_{\text{NaCN}} \qquad k_1 = 3.78 \times 10^8 \exp\left(\frac{-11{,}320}{T}\right)\ \text{s}^{-1}$$

$$-r_{\text{metal-cyanide}} = k_2 c_{\text{metal-cyanide}} \qquad k_2 = 5.0 \times 10^7 \exp\left(\frac{-13{,}720}{T}\right)\ \text{s}^{-1}$$

[†]C. Coronado, J. E. Botello, and F. Herrera, *Biotechnol. Prog.,* **17:** 669 (2001).

[‡]Adapted from P. M. Brown, *Chem. Eng. Ed.,* **28:** 266 (1994).

Assume a 99.99 percent conversion of the metal complex, while processing 12 catalytic converters per hour. A reactor space time of 4 h has been selected.

a. Select a type of ideal reactor that is best suited to perform the precipitation step.
b. Determine the required operating temperature and pressure for the reactor.
c. Size the reactor and make construction material recommendations.
d. Estimate the reactor cost.

13-13 Consider the case of a reactor design dealing with a unit that is to be used to chlorinate benzene with ferric chloride as the catalyst.† A semibatch reactor is to be used containing an initial charge of pure benzene into which dry chlorine gas is fed at a steady state. The reaction vessel is operated at a constant pressure of 202 kPa and is equipped with coils to maintain a constant temperature of 55°C. The liquid is agitated to give a homogeneous mixture. The unit contains a reflux condenser which returns all vaporized benzene and chlorobenzenes to the reactor while allowing the generated hydrogen chloride gas and excess chlorine gas to leave the system. Monochlorobenzene, dichlorobenzene, and trichlorobenzene are produced through successive irreversible reactions as follows, where the k's represent the forward rate constants for the equations as written:

$$C_6H_6 + Cl_2 \xrightarrow{k_1} C_6H_5Cl + HCl$$
$$C_6H_5Cl + Cl_2 \xrightarrow{k_2} C_6H_4Cl_2 + HCl$$
$$C_6H_4Cl_2 + Cl_2 \xrightarrow{k_3} C_6H_3Cl_3 + HCl$$

The rate of each of these reactions is first order with respect to each reactant.

The feed rate of the dry chlorine is 1.4 mol of chlorine/(h) kg/kg mol of initial benzene charge). The following rate constants are estimated values for the catalyst used at 320 K:

$$k_1 = 31.8(\text{kg mol/m}^3)^{-1}\,\text{h}^{-1}$$
$$k_2 = 3.99(\text{kg mol/m}^3)^{-1}\,\text{h}^{-1}$$
$$k_3 = 0.131(\text{kg mol/m}^3)^{-1}\,\text{h}^{-1}$$

There is negligible liquid or vapor holdup in the reflux condenser.

Volume changes in the reacting mixture are negligible, and the volume of liquid in the reactor remains constant at 0.091 m^3/kg mol of initial benzene charge.

Hydrogen chloride has a negligible solubility in the liquid mixture.

The chlorine gas dissolves completely in the liquid phase until such time that the liquid-phase concentration of chlorine reaches the solubility limit of 0.12 kg mol of chlorine per kg mol of original benzene. After that point sufficient chlorine dissolves to keep the liquid-phase concentration of chlorine constant and the remainder goes to recovery. Operationally the chlorine addition rate may be reduced at this time, so long as the liquid-phase concentration remains constant, but this does not affect the solution to this problem.

Find the reaction times at which the concentrations of the mono-, di-, and trichlorobenzene reach their maximum values.

†Adapted from information and methods presented by R. B. MacMullin, *Chem. Eng. Progr.,* **44:** 183 (1948) and A. Carlson, *Instr. and Control Systems,* **38(4):** 147 (1965).

CHAPTER

14

Heat-Transfer Equipment—Design and Costs

Modern heat-transfer equipment includes not only simple concentric-pipe exchangers but encompasses complex surface exchangers with thousands of square meters of heating area. Between these two extremes are the conventional shell-and-tube exchangers, coil heaters, bayonet heaters, extended-surface finned exchangers, plate exchangers, condensers, evaporators, furnaces, and many other, more specialized heat-transfer units. Intelligent selection of heat-transfer equipment requires an understanding of the basic theories of heat transfer that are incorporated in the sophisticated computer programs presently being used to perform the required design calculations. This chapter presents an outline of heat-transfer theory and the associated design calculation methods, together with an analysis of the general factors that must be considered in the selection of heat-transfer equipment.

The shell-and-tube exchanger is the most commonly used type of heat exchanger in the process industry, accounting for at least 60 percent of all heat exchangers in use today. Figure 14-1 shows design details of a conventional two-pass exchanger of the shell-and-tube type. Such exchangers are designed according to TEMA (Tubular Exchanger Manufacturers Association) specifications and are available for virtually any capacity and operating condition, from high vacuums to ultrahigh pressures, from very low to very high temperatures, and for many temperature and pressure differences between the fluids being heated and cooled. Materials of construction generally provide the only limitations on the operating conditions since they can be designed to meet other conditions of vibration, heavy fouling, highly viscous fluids, erosion, corrosion, toxicity, multicomponent mixtures, etc. Shell-and-tube exchangers may be fabricated from a variety of metal and nonmetal materials with surface areas available from 0.1 to 100,000 m^2. However these heat exchangers generally exhibit an order of magnitude less surface area per unit volume than compact heat exchangers and require considerable space, weight, and support structure.

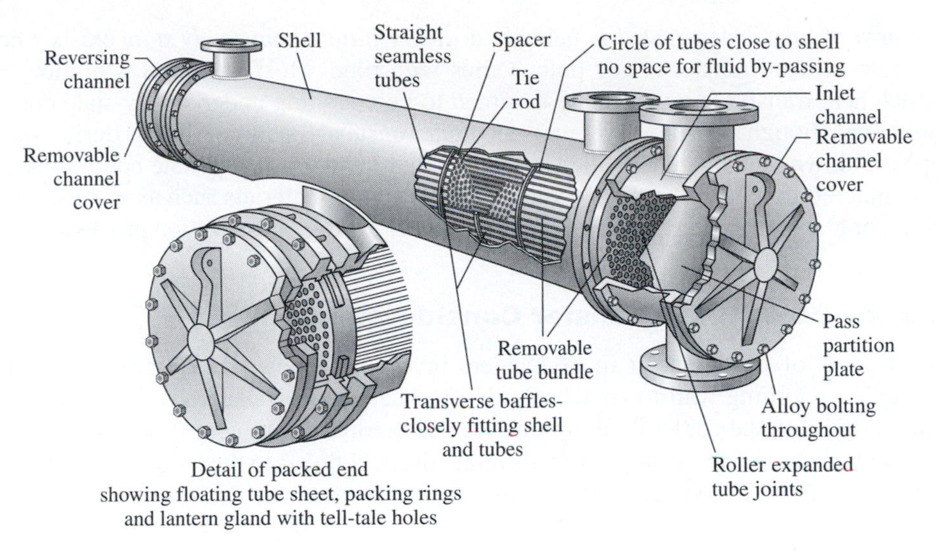

Figure 14-1
Two-pass shell-and-tube heat exchanger showing construction details. (*Ross Heat Exchanger Division of American-Standard.*)

Compact heat exchangers possess advantages since such units can operate with lower energy requirements and permit lower fluid inventory as well as improved process design, plant layout, and processing conditions when compared with conventional shell-and-tube heat exchangers. Compact heat exchangers may be designed to operate at temperatures as high as 800°C and pressures as high as 20 MPa, although present applications do not involve the combined use of high temperature and high pressure. Basic types of gas-to-gas compact heat exchangers are plate-fin, tube-fin, and all prime surface recuperators and regenerators. Liquid-to-liquid and liquid-to-phase-change compact heat exchangers include gasketed and welded plate-to-frame, welded plate (without frames), spiral plate, and dimple plate heat exchangers. Fouling is one of the major problems encountered in many compact heat exchangers partly due to hindered access for cleaning purposes. An exception to the cleaning difficulty is the gasketed plate heat exchangers which permit easy cleaning. With a large frontal area exchanger, flow maldistribution can also lead to a loss in exchanger efficiency. Unlike the shell-and-tube exchangers, there is no recognized standard or accepted practice available to designers of compact heat exchangers.

BASIC THEORY OF HEAT TRANSFER IN EXCHANGERS

Heat can be transferred from a source to a receiver by *conduction, convection,* or *radiation*. In many cases, the exchange occurs by a combination of two or three of these mechanisms. When the rate of heat transfer remains constant and is unaffected by time, the flow of heat is designated as being in a *steady state;* an *unsteady state* exists when the rate of heat transfer at any point varies with time. Most industrial operations in which heat transfer is involved are assumed to be operating under steady-state conditions even though such processes may encounter unsteady-state conditions during start-up, cooldown, and surge conditions. On the other hand, unsteady-state conditions are encountered in batch processes, cooling and heating of materials such as metals, polymers, or glasses, and certain types of regeneration, curing, or activation processes.

Steady-State Heat-Transfer Considerations

Most cases of heat transfer in exchangers involve the flow of heat from one fluid through a retaining wall to another fluid. The heat that is transferred flows from the warmer fluid to the colder fluid through several thermal resistances in series. By analogy with electrical conduction, we may relate the heat flux $\dot{q}$ to the temperature driving force $T_h - T_c$ by introducing a total resistance to heat transfer R_{tot}, defined by

$$R_{\text{tot}} = \frac{T_h - T_c}{\dot{q}} \tag{14-1}$$

The total resistance is comprised of the resistances due to convective heat transfer in the two fluids, the resistance due to the fouling that occurs with time on either wall surface, and the resistance to the transfer of heat through the retaining wall. The resistances associated with convective heat transfer in the hot and cold fluids are the reciprocals of the convective heat transfer coefficients h_h and h_c, respectively. The wall resistance is given by x/k_w (assuming a planar wall), where x is the thickness of the wall and k_w is the thermal conductivity of the wall material. Convention calls for carrying out heat-transfer calculations in terms of the reciprocal of R_{tot} and defining that term as U, the overall heat-transfer coefficient. With the appropriate substitutions, the expression for R_{tot} in terms of $1/U$ becomes

$$R_{\text{tot}} = \frac{1}{U} = \frac{1}{h_h} + R_{h,f} + \frac{x}{k_w} + R_{c,f} + \frac{1}{h_c} \tag{14-2}$$

where $R_{h,f}$ and $R_{c,f}$ are the thermal fouling resistances for the hot and cold streams, respectively. Since the more normal configuration for heat transfer involves one fluid flowing inside of a tube and the other fluid flowing on the outside of the tube, an adjustment must be made to Eq. (14-2) to account for the difference in heat-transfer area per unit length of tube for the inside or outside areas of the tube. Thus, the overall heat-transfer coefficient is conventionally based on either the inside wall area A_i, the outside wall area A_o, or infrequently on the mean area A_m. The three overall heat-transfer coefficients are related as

$$\frac{1}{U_i A_i} = \frac{1}{U_o A_o} = \frac{1}{U_m A_m} \tag{14-3}$$

and make no distinction where the hot and cold streams are located.

When the outside wall area A_o is used as the reference area for the heat transfer, Eq. (14-2) becomes

$$\frac{1}{U_o} = \frac{A_o}{h_i A_i} + \frac{A_o}{h_{i,d} A_i} + \frac{A_o x_w}{k_w A_{m,w}} + \frac{1}{h_o} + \frac{1}{h_{o,d}} \tag{14-4}$$

where h_i and h_o are the individual heat-transfer coefficients evaluated for the clean inside and outside surfaces of the tube, respectively, while $h_{i,d}$ and $h_{o,d}$ are the individual heat-transfer coefficients that account for the dirt and scale accumulation that results in the fouling of the two surfaces. Note that the latter two coefficients are the reciprocals of the fouling resistances associated with these surfaces. When the inside surface of the tube is used as the reference area, the relation becomes

$$\frac{1}{U_i} = \frac{1}{h_i} + \frac{1}{h_{i,d}} + \frac{A_i x_w}{k_w A_{m,w}} + \frac{A_i}{h_o A_o} + \frac{A_i}{h_{o,d} A_o} \tag{14-4a}$$

where A_i is the internal surface area, A_o the external surface area, and $A_{m,w}$ the mean surface area of the wall given by $A_i = \pi D_i L$, $A_o = \pi D_o L$, and $A_{m,w} = [\pi L(D_o - D_i)]/\ln(D_o/D_i)$, respectively. The units for U are normally expressed in $W/m^2{\cdot}K$.[†]

Before we discuss the details of calculating values for the individual heat-transfer coefficients, let us return to Eq. (14-1) which indicates that the net rate of heat transfer per unit area is a function of the reciprocal of the thermal resistance and the temperature driving force between the two fluids. Even though the temperature driving force generally varies over the length of the exchanger, as shown in Fig. 14-2 for a number of single-pass exchangers, Eq. (14-1) may be simplified to obtain the rate of heat transfer, assuming that the overall heat-transfer coefficient remains constant, using the expression

$$\dot{q} = AU\,\Delta T_{o,m} \tag{14-5}$$

where $\Delta T_{o,m}$ is the mean overall temperature difference between the two fluids. As noted above, the overall heat-transfer coefficient must be related to either the inside surface area, the outside surface area, or the mean surface area of the tube or pipe. Assuming no change in phase for either fluid, constant heat capacity for either fluid, and negligible external heat loss from the exchanger, the rate of heat transfer is also given by

$$\dot{q} = -(\dot{m} C_p \,\Delta T)_h = (\dot{m} C_p \,\Delta T)_c \tag{14-6}$$

where $\dot{m}$ is the mass rate of flow of either the hot or cold streams and C_p is the heat capacity of these two streams, respectively.

When the temperature profiles displayed in Fig. 14-2 are essentially linear, the mean temperature $\Delta T_{o,m}$ can be expressed in terms of the temperature differences at both ends of the exchanger as

$$\Delta T_{o,m} = \Delta T_{o,\text{log mean}} = \frac{\Delta T_1 - \Delta T_2}{\ln(\Delta T_1/\Delta T_2)} \tag{14-7}$$

where ΔT_1 is the larger temperature difference occurring at either end of the exchanger

[†]To convert units for the heat-transfer coefficient from $W/m^2{\cdot}K$ to English units of $Btu/h{\cdot}ft^2{\cdot}{}^\circ F$, multiply by 0.176.

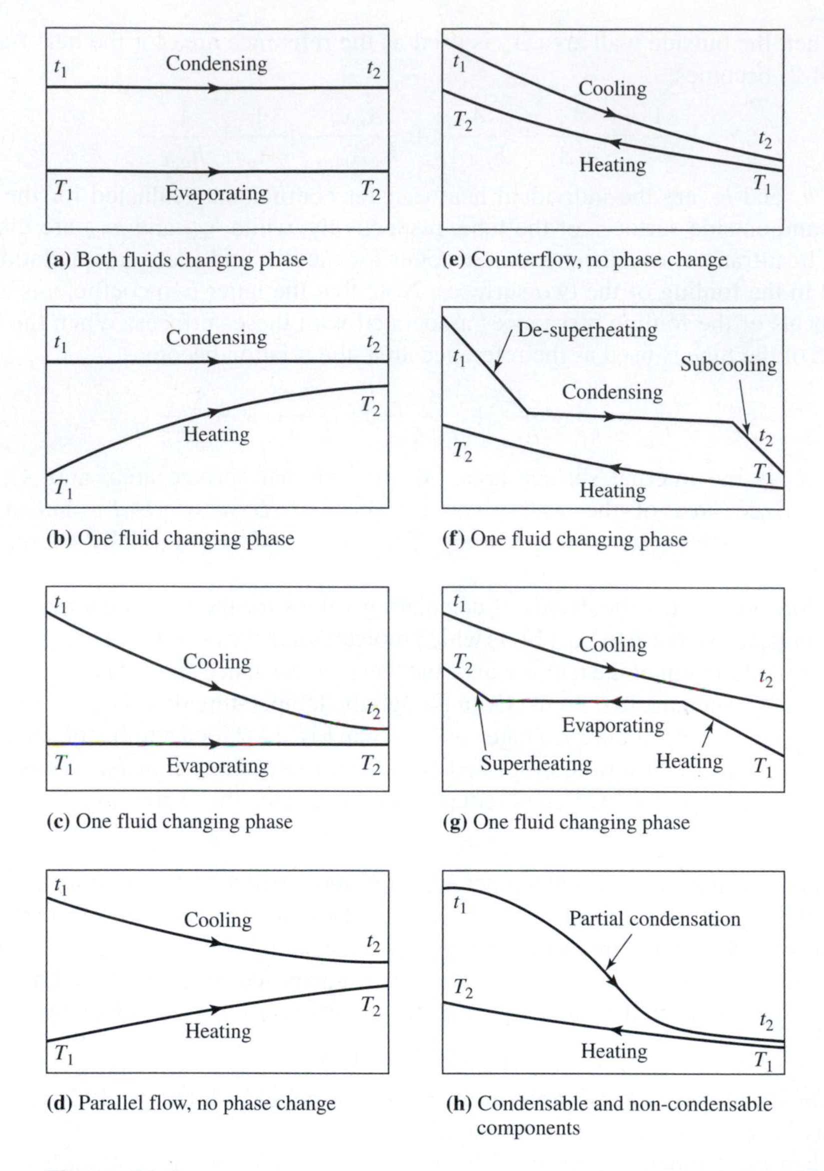

Figure 14-2
Various temperature profiles developed in heat exchangers for parallel or countercurrent flow, with one or two phases present (temperature t_1 and T_1 are the inlet temperatures for the hot and cold streams, respectively, and t_2 and T_2 are the exit temperatures for the hot and cold streams, respectively).

and ΔT_2 is the smaller temperature difference present at either end of the exchanger. The temperature profiles in an exchanger are linear when the heat capacities of the fluids are constant over the temperature range observed in the exchanger or when a phase change at constant temperature occurs within the exchanger. For design calculations, an

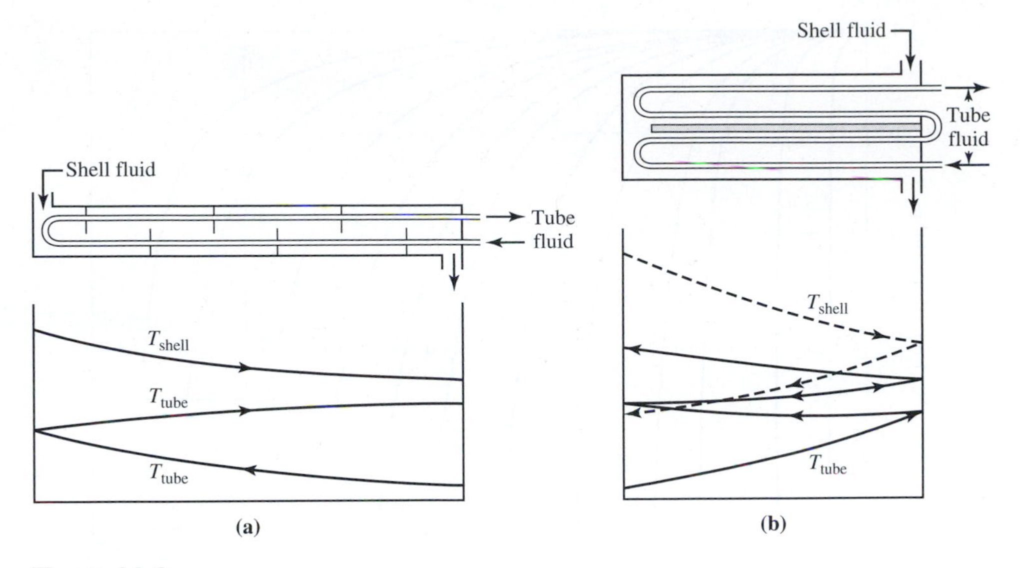

Figure 14-3
Typical temperature profiles developed in multipass heat exchangers with (a) one shell pass, two tube passes, and (b) two shell passes, four tube passes.

arithmetic average temperature difference can be used in place of the log mean value if the ratio of $\Delta T_1/\Delta T_2$ does not exceed 2.0.

When the temperature profiles for the single-pass heat exchangers exhibit several different linear sections throughout the unit (as shown in Fig. 14-2f and g), the exchanger can be treated as an exchanger with several linear sections, each characterized with its own log-mean temperature difference. Intermediate temperatures within the exchanger may be established with an appropriate stepwise energy balance either with a lengthy hand calculation or with the aid of various computer programs. The latter are particularly useful for cases that involve one or two continuously curved temperature profiles, as shown in Fig. 14-2h.

Temperature profiles within the exchanger become even more complex when multipass heat exchangers are used to achieve greater compactness of the equipment, as shown in Fig. 14-3 for two simple shell-and-tube exchangers. Assuming a constant overall heat-transfer coefficient, constant heat capacity of both fluids, and negligible heat loss from the exchanger, modified values for $\Delta T_{o,m}$ can be obtained by integrating the basic relations for convective heat transfer or, more simply, by utilizing plots to provide appropriate correction factors that are used in the relation

$$\Delta T_{o,m} = F \Delta T_{o,\text{log mean}} \tag{14-8}$$

where F is the correction factor to account for the type of multipass arrangement that has been selected. Figure 14-4 is a plot used to obtain the correction factor for a multipass exchanger with one shell and two or more even-numbered tube passes.[†]

[†]Plots for the other multipass heat exchanger configurations are provided by J. Taborek in *Heat Exchanger Design Handbook,* G. F. Hewitt, ed., Begell House, New York, 1998, Secs. 1.5.2 and 1.5.3.

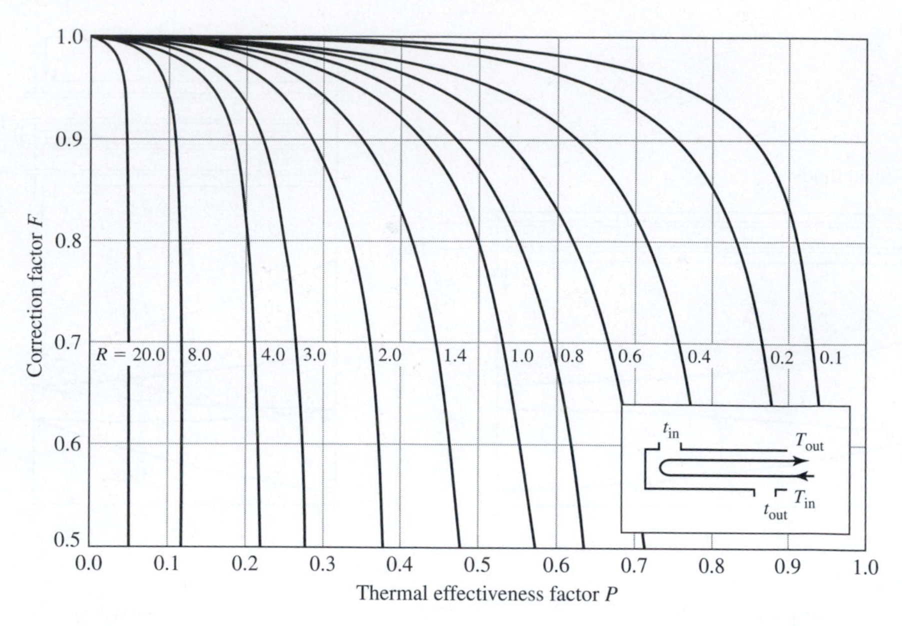

Figure 14-4
Chart for determining correct mean temperature-difference driving force for an exchanger with one shell pass and two or more even-numbered tube passes (T and t are interchangeable). Correction factor F is based on the $\Delta T_{\text{log mean}}$ for counterflow. If F is less than 0.7, operation of the exchanger may not be practical.

The correction factor for this specific multipass arrangement may also be calculated from

$$F = (R^2 + 1)^{1/2} \ln\left(\frac{1 - P}{1 - RP}\right) \Big/ (R - 1) \ln \frac{2 - P[(R + 1) - (R^2 + 1)^{1/2}]}{2 - P[(R + 1) + (R^2 + 1)^{1/2}]} \tag{14-9}$$

where R and P are defined as

$$R = \frac{T_{h,\text{in}} - T_{h,\text{out}}}{T_{c,\text{out}} - T_{c,\text{in}}} = \frac{(\dot{m}C_p)_c}{(\dot{m}C_p)_h} \tag{14-9a}$$

$$P = \frac{T_{c,\text{out}} - T_{c,\text{in}}}{T_{h,\text{in}} - T_{c,\text{in}}} \tag{14-9b}$$

In heat exchanger applications, it is desirable to have a value of 0.85 or higher for the correction factor. Values for the correction factor are only minimally decreased when using multipass heat exchangers with additional tube passes. For example, the value of F for a one-shell–eight-tube pass exchanger is only about 2 percent less than the value of the correction factor for a one-shell–two-tube pass exchanger for the same operating conditions.

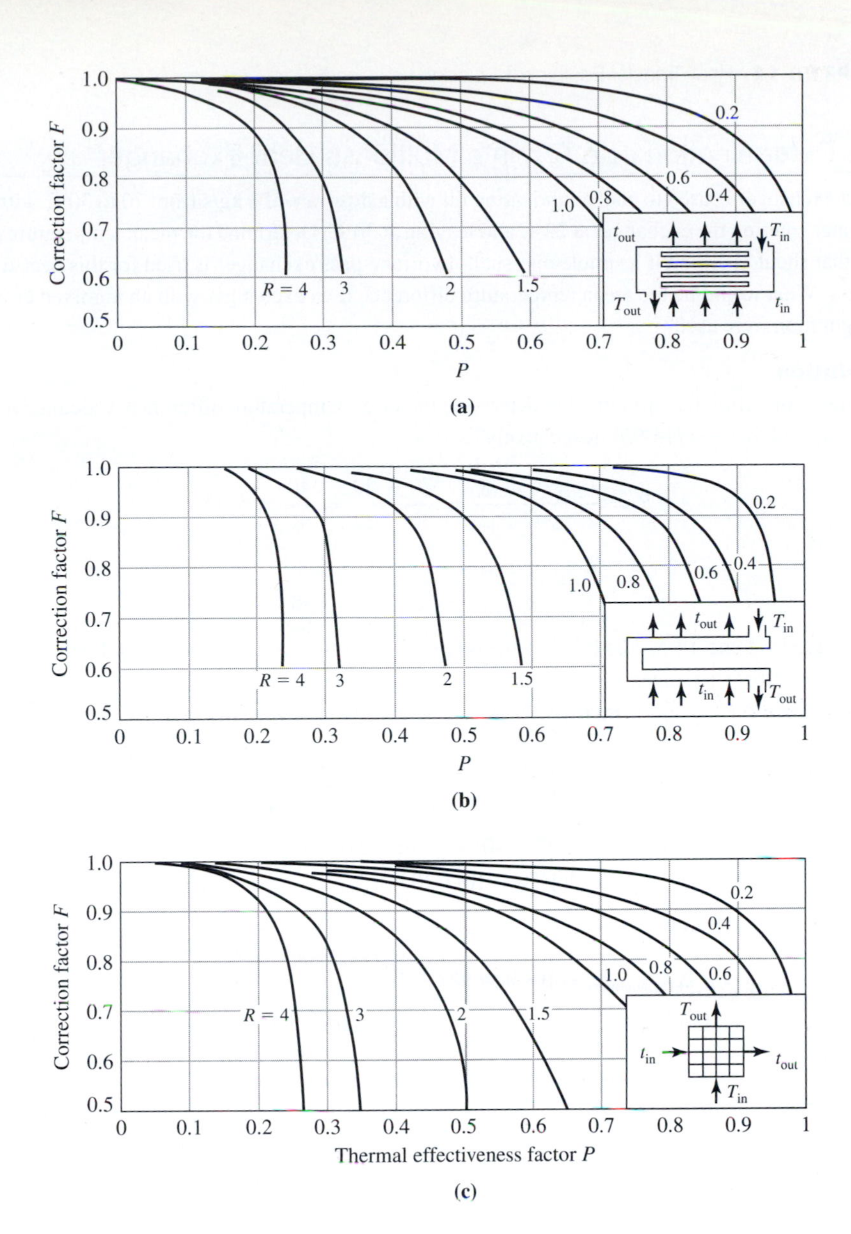

Figure 14-5
Temperature driving force correction factor for crossflow heat exchangers: (a) one shell pass, one or more parallel rows of tubes; (b) two shell passes, two rows of tubes (for more than two passes, use $F = 1$); (c) one shell pass, both fluids unmixed. (T and t are interchangeable.)

Multipass exchangers with cross-flow across the tubes exhibit somewhat higher correction factors than multipass countercurrent exchangers. A plot for obtaining the correction factor for an unmixed cross-flow exchanger is presented in Fig. 14-5. Note that it is immaterial whether the hot fluid flows on the shell or the tube side. The use of the correction factor is illustrated in Example 14-1.

EXAMPLE 14-1 Determination of $\Delta T_{o,m}$ in a Multipass Heat Exchanger

A heat exchanger is used to cool a lubricating oil with a flow rate of 6 kg/s from 70 to 30°C with cooling water entering the exchanger at 22°C and leaving at 30°C. Determine the mean temperature difference that should be used if a simple one-shell–two-tube pass exchanger is used for this heat-transfer process. What would be the mean temperature difference if an exchanger with an unmixed crossflow configuration were used?

■ Solution

Use the F correction factor method to determine the mean temperature difference. Calculate R and P from Eqs. (14-9*a*) and (14-9*b*), respectively.

$$R = \frac{T_{h,\text{in}} - T_{h,\text{out}}}{T_{c,\text{out}} - T_{c,\text{in}}} = \frac{70 - 30}{30 - 22} = \frac{40}{8} = 5.0$$

$$P = \frac{T_{c,\text{out}} - T_{c,\text{in}}}{T_{h,\text{in}} - T_{c,\text{in}}} = \frac{30 - 22}{70 - 22} = \frac{8}{48} = 0.167$$

From Figs. 14-4 and 14-5, F is 0.83 for the one-shell–two-tube pass exchanger and 0.91 for the unmixed cross-flow exchanger, respectively.

Calculate $\Delta T_{o,\text{log mean}}$ from Eq. (14-7).

$$\Delta T_{o,\text{log mean}} = \frac{(T_{h,\text{in}} - T_{c,\text{out}}) - (T_{h,\text{out}} - T_{c,\text{in}})}{\ln[(T_{h,\text{in}} - T_{c,\text{out}})/(T_{h,\text{out}} - T_{c,\text{in}})]}$$

$$= \frac{(70 - 30) - (30 - 22)}{\ln[(70 - 30)/(30 - 22)]} = 19.88°\text{C}$$

Determine $\Delta T_{o,m}$ from Eq. (14-8).

$$\Delta T_{o,m} = F\,\Delta T_{o,\text{log mean}} = 0.83(19.88) = 16.5°\text{C} \qquad \text{(one-shell–two-tube pass)}$$

$$\Delta T_{o,m} = 0.91(19.88) = 18.1°\text{C} \qquad \text{(unmixed cross-flow)}$$

The calculations show that using the one-shell–two-tube pass exchanger is somewhat marginal since the F factor is slightly below the 0.85 value recommended above. To increase the F factor value above 0.85, either use the unmixed cross-flow configuration with an F factor of 0.91, or go to a two-shell exchanger and any even number of tube passes with an increase in the F factor to 0.97.

■

If the overall heat-transfer coefficient and the local temperature difference vary throughout the exchanger, one could envision a heat exchanger with a number of smaller exchangers in series and assume that the coefficient of heat transfer varies linearly with the temperature in each of these small exchangers. In one common situation where the mass flow rate and heat capacities are constant, heat losses from the exchanger are negligible, and there is no partial phase change in either stream, Eq. (14-5) can be modified to include the above conditions and written as

$$\dot{q} = A\frac{U_1\,\Delta T_{o,1} - U_2\,\Delta T_{o,2}}{\ln(U_1\,\Delta T_{o,1}/U_2\,\Delta T_{o,2})} \qquad (14\text{-}10)$$

Note that values of $\dot{q}$ and A in Eq. (14-10) only apply to each individual section of the exchanger between the limits indicated by subscripts 1 and 2. Consequently, the total value of $\dot{q}$ and A for the entire heat exchanger must be obtained by summing the areas for each of the individual sections. The procedure is shown in Example 14-2.

Evaluation of the Heat-Transfer Area When U and ΔT Vary Within the Exchanger — EXAMPLE 14-2

A countercurrent heat exchanger is used to remove 1000 kW in the condensation of an organic mixture from 206 to 146°C. Cooling is provided by vaporizing another organic mixture that enters the exchanger at 96°C and leaves at 145°C. Temperatures of each stream within the exchanger have been calculated as a function of the heat removed from the hot stream and transferred to the cold stream. To further simplify the problem, overall heat-transfer coefficients have been evaluated at these same temperature locations within the exchanger and are available for the solution. Determine the total surface area that will be required in this exchanger, using Eq. (14-10). The available data for Example 14-2 and the incremental areas for each of the 10 sections in the exchanger are given below.

Available data					Calculated values		
$\dot{q}$, kW	T_h, °C	T_c, °C	U, W/m²·K		$\Delta T_{o,1}$, °C	$\Delta T_{o,2}$, °C	A_i, m²
0	206	145	705				
				Sec. 1	61	66	2.024
100	204	138	855				
				Sec. 2	66	70	1.653
200	200	130	925				
				Sec. 3	70	70	1.508
300	195	125	970				
				Sec. 4	70	72	1.426
400	190	118	1005				
				Sec. 5	72	71	1.378
500	184	113	1025				
				Sec. 6	71	70	1.491
600	178	108	1030				
				Sec. 7	70	66	1.435
700	171	105	1020				
				Sec. 8	66	61	1.560
800	163	102	1000				
				Sec. 9	61	55	1.760
900	155	100	960				
				Sec. 10	55	50	2.022
1000	146	96	925				

■ Solution

Solve Eq. (14-10) for the area. Substitute the values for T_h, T_c, and U in this equation, and solve for the required area in section 1 of the exchanger. Repeat the calculations for all 10 sections, and sum the areas.

$$A_{1-2} = \dot{q}_{1-2} \Big/ \frac{U_1\,\Delta T_{o,1} - U_2\,\Delta T_{o,2}}{\ln(U_1\,\Delta T_{o,1}/U_2\,\Delta T_{o,2})}$$

Assume section 1 involves 100 kW of energy transferred from the hot stream to the cold stream. Thus,

$$A_{1-2} = 100{,}000 \Big/ \frac{855(66) - 705(61)}{\ln[(855)(66)/(705)(61)]} = 2.024 \text{ m}^2$$

The calculation is repeated for all 10 sections in a spreadsheet, and the results are recorded in the table. The required area is the sum of the last column

$$A_T = \sum A_i = 16.26 \text{ m}^2$$

If, in this example, only the inlet and exit temperatures of the two streams had been used with an average value for the overall coefficient of heat transfer, the required area would have been evaluated from

$$A_T = \frac{\dot{q}_T}{U_m \, \Delta T_{o,\text{log mean}}}$$

where

$$\Delta T_{o,\text{log mean}} = \frac{(206 - 145) - (146 - 96)}{\ln[(206 - 145)/(146 - 96)]} = 55.3°\text{C}$$

$$A_T = \frac{1 \times 10^6}{947.3(55.3)} = 19.1 \text{ m}^2$$

These differing results show that temperature and overall heat-transfer coefficient variations within a heat exchanger need to be considered carefully before conventional techniques are blindly applied.

■

The solution of Example 14-2 could also have been performed graphically. Figure 14-6 shows the variation of $1/(U \Delta T_o)$ as a function of the heat transferred per unit time as the fluid moves through the exchanger. The required exchanger area is given by the crosshatched area displayed on the figure. From this plot one may derive a rather general method for calculating the required area for a simple heat exchanger based on the knowledge of the local temperature differences and overall heat-transfer coefficients. Many modern computer programs for heat exchanger design carry out this numerical integration to determine the area required for a specified heat transfer.

Alternative Approaches to Heat Exchanger Performance

Earlier we described the use of the F correction factor to obtain the overall mean temperature by using Eq. (14-8). Evaluation of F requires knowledge of the inlet and outlet temperatures of both streams and the exchanger configuration. However, often there is a need to determine the exit temperatures of the two fluids in a given heat exchanger configuration when only the inlet temperatures are known. The solution of such a problem using the F correction method requires making successive estimates of the outlet temperatures and proceeding with an iterative process. The latter can often be avoided by using alternative approaches involving the concepts of heat exchanger effectiveness E and number of transfer units NTU.

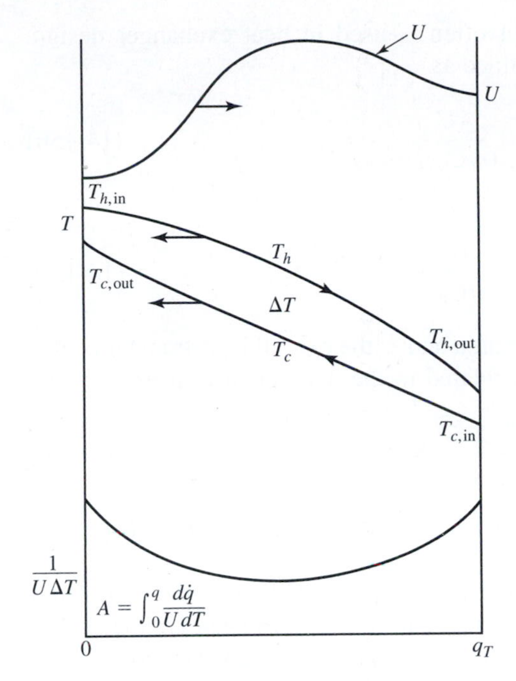

Figure 14-6
Graphical method for single heat exchanger design

The effectiveness of a heat exchanger is defined as

$$E = \frac{\dot{q}}{\dot{q}_{\text{max}}} \tag{14-11}$$

where $\dot{q}$ and $\dot{q}_{\text{max}}$ are the actual and the maximum amounts of heat that can be transferred per unit time in an exchanger. The maximum amount of heat transfer between two streams in a countercurrent flow heat exchanger occurs when the outlet temperature of the stream with the lowest $\dot{m}C_p$ product approaches the inlet temperature of the other stream. Thus

$$\dot{q}_{\text{max}} = (\dot{m}C_p)_{\text{min}}\,\Delta T_{o,\text{max}} \tag{14-12}$$

where $\Delta T_{o,\text{max}} = T_{h,\text{in}} - T_{c,\text{in}}$. Making the substitution into Eq. (14-11) yields

$$E = \frac{T_{c,\text{out}} - T_{c,\text{in}}}{T_{h,\text{in}} - T_{c,\text{in}}} \tag{14-13}$$

when the cold fluid has the minimum value of $\dot{m}C_p$. When the hot fluid has the minimum value of $\dot{m}C_p$, then E is defined as

$$E = \frac{T_{h,\text{in}} - T_{h,\text{out}}}{T_{h,\text{in}} - T_{c,\text{in}}} \tag{14-14}$$

The effectiveness of conventional heat exchangers generally is between 0.4 and 0.8 depending on the configuration.

The number of transfer units concept often is used in heat exchanger design. Values for the hot and cold streams are defined as

$$\mathrm{NTU}_h = \frac{AU}{(\dot{m}C_p)_h} \tag{14-15a}$$

and

$$\mathrm{NTU}_c = \frac{AU}{(\dot{m}C_p)_c} \tag{14-15b}$$

respectively, where A is the heat exchanger area and U the overall heat-transfer coefficient. The heat exchanger effectiveness is related to the number of transfer units by

$$E = \mathrm{NTU}_{\min} \frac{\Delta T_{o,m}}{\Delta T_{\max}} = \mathrm{NTU}_{\min}\theta \tag{14-16}$$

where $\mathrm{NTU}_{\min}$ is the NTU for the stream with the minimum value of $\dot{m}C_p$ and θ is the ratio of $\Delta T_{o,m}/\Delta T_{\max}$.

With these definitions note that E and θ are functions of the number of transfer units and the exchanger configuration. The Engineering Sciences Data Unit (ESDU)† has developed plots which relate these different variables. A typical plot for a shell-and-tube heat exchanger with one shell and even-numbered tube passes is shown in Fig. 14-7 while a plot for a cross-flow heat exchanger with both fluids unmixed is shown in Fig. 14-8. The values for R and P in both figures were defined earlier in Eqs. (14-9*a*) and (14-9*b*), respectively. Thus, for given values of R and NTU_c, it is possible to obtain θ and P from these plots. The lower part of the combined plot permits evaluation of F as a function of P with R as the parameter. The combination of these plots permits direct extrapolation of F to determine whether the design meets the criteria of $F \geq 0.85$. Additional plots for other exchanger configurations are available in the literature.‡

The steps involved in obtaining the two outlet temperatures for a specified multipass exchanger when the total surface area, overall heat-transfer coefficient, stream flow rates with corresponding heat capacities, and both inlet temperatures are known, are as follows:

1. Determine R from the ratio $(\dot{m}C_p)_c/(\dot{m}C_p)_h$.
2. Evaluate NTU_c from $AU/(\dot{m}C_p)_c$.
3. Use the upper part of the charts similar to Figs. 14-7 and 14-8 to obtain a value for P for the calculated values of R and NTU_c.

---

†Engineering Sciences Data Unit, ESDU data item 86018, ESDU International Plc, London, 1986.

‡J. Taborek, in *Heat Exchanger Design Handbook,* G. F. Hewitt, ed., Hemisphere Publishing, New York, 1983, Sec. 1.5.

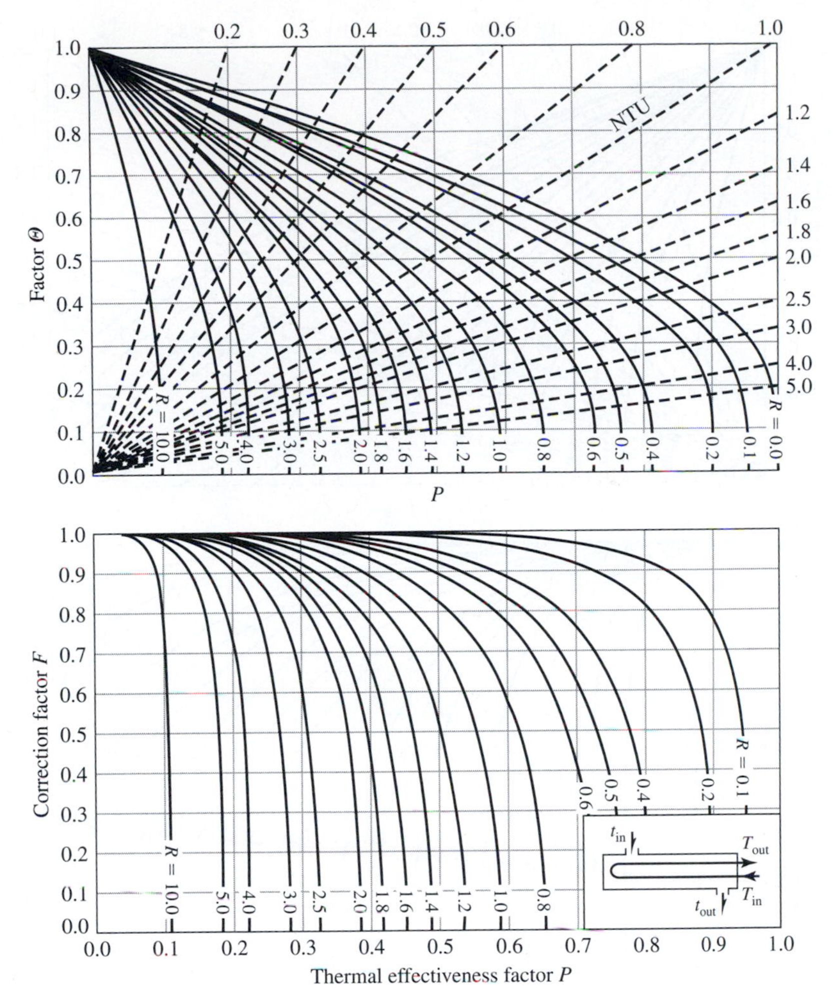

Figure 14-7
F-*θ*-NTU-*P* chart for a two-pass shell-and-tube heat exchanger. The same chart can be used for four, six, eight, . . . passes. (*T* and *t* are interchangeable.) (From *Heat Exchanger Design Handbook,* 1983. Reprinted by permission of Begell House Inc., Publishers.)

4. Rearrange Eq. (14-9*b*) to calculate $T_{c,\text{out}}$ from

$$T_{c,\text{out}} = T_{c,\text{in}} + P(T_{h,\text{in}} - T_{c,\text{in}})$$

5. Determine $T_{h,\text{out}}$ from a rearrangement of Eq. (14-9*a*)

$$T_{h,\text{out}} = T_{h,\text{in}} - R(T_{c,\text{out}} - T_{c,\text{in}})$$

where $T_{c,\text{out}}$ was calculated in step 4.

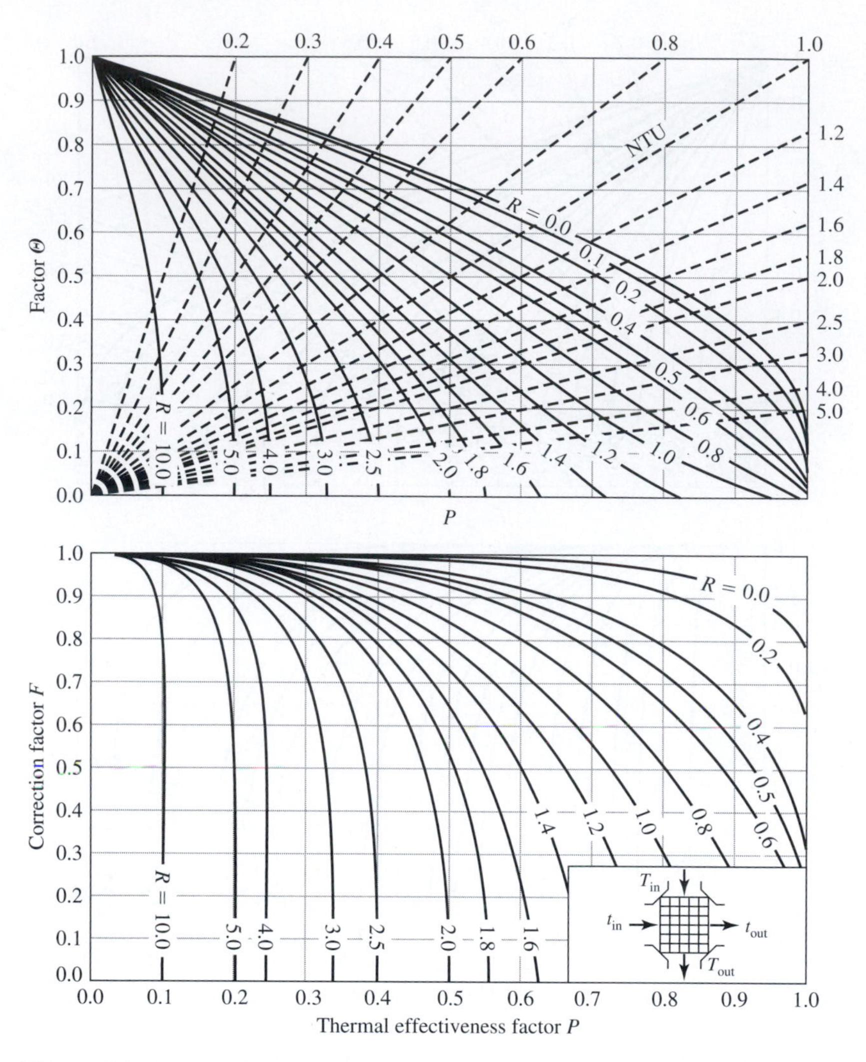

Figure 14-8
F-θ-NTU-P chart for a cross-flow heat exchanger with no lateral mixing of either stream. (T and t are interchangeable.) (From *Heat Exchanger Design Handbook,* 1983. Reprinted by permission of Begell House Inc., Publishers.)

DETERMINATION OF HEAT-TRANSFER COEFFICIENTS

Exact values of convection heat-transfer coefficients for a given situation can only be obtained by experimental measurements under those same operating conditions. Approximate values, however, can be obtained for use in design by employing correlations based on general experimental data. A number of correlations that are particularly useful in design work are presented in the following sections. In general, the relationships applicable to turbulent conditions are more accurate than those for laminar conditions. Film coefficients obtained from the correct use of equations in the turbulent flow range will ordinarily be within ±20 percent of the true experimental value; but values determined for viscous flow conditions or for condensation, boiling, natural convection, and those for shell sides of heat exchangers may be in error by more than 100 percent. Because of the inherent inaccuracies in the methods for estimating film coefficients, some design engineers prefer to use overall coefficients based on past experience, while others include a large safety factor in the form of fouling factors or fouling coefficients.

Film Coefficients for Fluids Flowing Inside of Pipes and Tubes (No Phase Change)

The following equations are based on the correlations presented by Sieder and Tate:†

For viscous flow ($D_iG_i/\mu < 2100$),

$$h_i = 1.86\frac{k}{D_i}\left(\frac{D_iG_i}{\mu}\right)^{1/3}\left(\frac{C_p\mu}{k}\right)^{1/3}\left(\frac{D_i}{L}\right)^{1/3}\left(\frac{\mu}{\mu_w}\right)^{0.14}$$

$$= 1.86\left(\frac{4\dot{m}_iC_p}{\pi kL}\right)^{1/3}\left(\frac{\mu}{\mu_w}\right)^{0.14} \tag{14-17}$$

For turbulent flow above the transition region ($D_iG_i/\mu > 10{,}000, 0.7 < \Pr < 160$ and $L/D_i > 10$)

$$h_i = 0.23\frac{k}{D_i}\left(\frac{D_iG_i}{\mu}\right)^{0.8}\left(\frac{C_p\mu}{k}\right)^{1/3}\left(\frac{\mu}{\mu_w}\right)^{0.14} \tag{14-18}$$

For the transition region ($2100 < D_iG_i/\mu < 10{,}000$), see Fig. 14-9.

In Eqs. (14-17) and (14-18), D_i is the inside diameter of the tube, G_i the mass velocity inside the tube, C_p the heat capacity of the fluid at constant pressure, k the thermal conductivity of the fluid, μ the viscosity of the fluid (subscript w indicates evaluation at the wall temperature), L the heated length of the straight tube, and $\dot{m}$ the mass rate of flow per tube. Physical properties k, C_p, and μ are evaluated at the average bulk temperature of the fluid.

Equation (14-18) is recommended for general use, bearing in mind that the standard deviation of error, according to Hewitt et al., is approximately 13 percent.‡ Equations (14-17) and (14-18) are plotted in Fig. 14-9 to facilitate their solution and to indicate values to use in the transition zone.

In some references a constant of 0.027 is used in place of the 0.023 value used in Eq. (14-18). The 0.023 constant is more appropriate for water, organic fluids, and gases at moderate ΔT values, while the 0.027 constant is more applicable to heat transfer of viscous liquids, particularly when the viscosities at the wall and in the bulk fluid show a significant difference in value.

---

†E. N. Sieder and G. E. Tate, *Ind. Eng. Chem.*, **28:** 1429 (1936).

‡G. F. Hewitt, G. L. Shires, and T. R. Bott, *Process Heat Transfer,* CRC Press, Boca Raton, FL, 1994, Sec. 2.4.7.3.

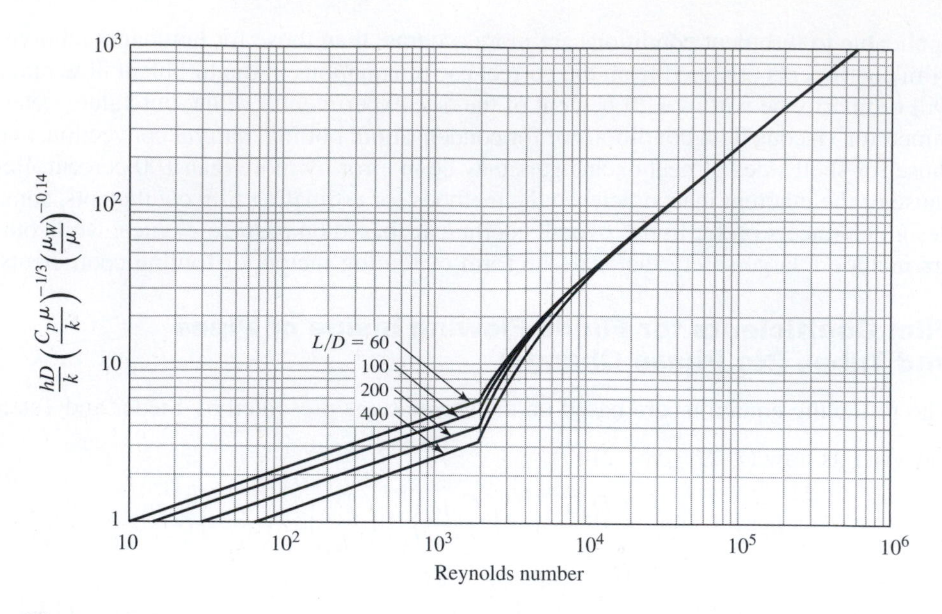

Figure 14-9
Simple procedure for estimating film coefficients for fluids flowing in pipes and tubes [based on Eqs. (14-17) and (14-18)].

The use of computers has made it possible to develop a more accurate and more widely applicable correlation. Such a correlation was developed by Guielinski and summarized by Hewitt.† This correlation incorporates a semitheoretical basis with the Prandtl analogy to skin friction in terms of the Darcy friction factor f_D. (The latter is 4 times the Fanning friction factor.) The correlation has been shown to cover a range in Reynolds number from 2300 to 1,000,000 and a range in Prandtl number from 0.6 to 2000.

$$h_i = \frac{k(f_D/8)(\text{Re} - 1000)\text{Pr}}{D_i[1 + 12.7(f_D/8)^{1/2}](\text{Pr}^{2/3} - 1)}\left[1 - \left(\frac{D_i}{L}\right)^{2/3}\right] \tag{14-19}$$

where

$$f_D = \frac{1}{(1.82 \log \text{Re} - 1.64)^2} \tag{14-19a}$$

All the properties in Eq. (14-19) are evaluated at the bulk fluid conditions. For viscous liquids, the right-hand side of Eq. (14-19) is multiplied by the correction factor $(\text{Pr}_b/\text{Pr}_w)^{0.11}$, where the subscripts b and w refer to the bulk and wall temperature conditions, respectively. Similarly, for gases, the right-hand side of the equation is multiplied by the correction factor $(T_b/T_w)^{0.45}$, using the absolute temperatures of the bulk fluid and the wall.

†G. F. Hewitt, ed., *Handbook of Heat Exchanger Design,* Begell House, New York, 1992, Sec. 2.5.1.

Occasionally, a situation is encountered in which a fluid is transmitted through a conduit having a noncircular cross section. The heat-transfer coefficients for turbulent flow in this case can be determined by using the same equations that apply to pipes and tubes if the pipe diameter D appearing in these equations is replaced by an equivalent diameter D_e. Best results are obtained if this equivalent diameter is taken as 4 times the hydraulic radius, where the hydraulic radius is defined as the cross-sectional flow area divided by the *heated* perimeter. For example, if heat is being transferred from a fluid in a center pipe to a fluid flowing through a circular annulus, the film coefficient for the outside surface of the inner pipe would be based on the following equivalent diameter D_e:

$$D_e = 4 \times \text{hydraulic radius} = 4\left[\frac{\pi D_2^2/4 - \pi D_1^2/4}{\pi D_1}\right] = \frac{D_2^2 - D_1^2}{D_1} \qquad (14\text{-}20)$$

where D_1 and D_2 are the inner and outer diameters of the annulus, respectively.

The difference between the hydraulic radii for heat transfer and for fluid flow should be noted. In Chap. 12, the correct equivalent diameter for evaluating friction due to the fluid flow in the annulus was shown to be 4 times the cross-sectional flow area divided by the *wetted* perimeter, or $4(\pi D_2^2/4 - \pi D_1^2/4)/(\pi D_2 + \pi D_1) = D_2 - D_1$.

Film Coefficients for Fluids Flowing Outside of Pipes and Tubes (No Phase Change)

Heat transfer flow often involves flow over bluff bodies of finite dimensions such as cylinders, spheres, or airfoil shapes. Heat transfer to and from cylinders is particularly important because of the application to cross-flow in heat exchangers. Since the heat transfer in cross-flow is linked to the dynamic behavior of the fluid, it has been necessary to develop suitable correlations based essentially on experimental data. One such correlation developed for cross-flow across a single cylinder or sphere is that developed by Churchill and Bernstein.†

$$\frac{h_o D_o}{k} = 0.3 + \frac{0.62\,\text{Re}^{1/2}\text{Pr}^{1/3}}{[1 + (0.4/\text{Pr})^{2/3}]^{1/4}}\left[1 + \left(\frac{\text{Re}}{28{,}200}\right)^{5/8}\right]^{4/5} \qquad (14\text{-}21)$$

where the Reynolds number is defined as $D_o V \rho/\mu$, based on the approach velocity V and the outside diameter D_o of the cylinder or sphere.

In process plants, heat exchangers are designed so that one of the fluids flows through a bank of tubes and is heated or cooled by the cross-flow of another fluid. The outside heat-transfer coefficient for the first row of tubes is similar to that for a single tube in cross-flow, as noted in Eq. (14-21). However, in subsequent rows of tubes, the heat transfer is enhanced by the increased local velocity and additional turbulence created by the preceding tubes. The average heat-transfer coefficient for the fluid flowing between the tubes has been found to correlate with the maximum velocity between the tubes V_{max} (rather than the approach velocity), the fluid properties, and the geometry of the tube bank. Various empirical correlations based on experimental data have been published. Equation (14-22) is recommended because of its relative simplicity and the

†S. W. Churchill and J. Bernstein, *J. Heat Transfer,* **99:** 300 (1977).

ease with which it can be incorporated into a computer code. However, it should be recognized that correlations of this type are not as accurate as those developed for internal flow in tubes and have a standard deviation of at least 25 percent for laminar flow and 15 percent for turbulent flow.

The majority of cross-flow tube banks have geometries that are designated either as in-line square pitch arrays, in which tubes in successive rows are in line with the direction of fluid flow, or as staggered triangular pitch arrays, in which tubes in successive rows are staggered so that the tubes in any given row occupy a position in line with the midpoint of the open space between the tubes in the previous row. (See Fig. 14-40 for details.) The empirical correlation that applies to both geometries is of the form

$$\frac{h_o D_o}{k} = a\mathrm{Re}^m \mathrm{Pr}^{0.34} F_1 F_2 \tag{14-22}$$

where a and m are correlation constants tabulated in Table 14-1, the cross-flow Reynolds number is given by

$$\mathrm{Re} = \frac{D_o V_{\max} \rho}{\mu} \tag{14-22a}$$

the correction factor F_1, for wall-to-bulk physical property variations, is evaluated from

$$F_1 = \left(\frac{\mathrm{Pr}_b}{\mathrm{Pr}_w}\right)^{0.26} \qquad \text{for Pr} < 600 \tag{14-22b}$$

and the correction factor F_2, which accounts for the effect of the number of tube rows in either the in-line or the staggered configuration, is presented in Table 14-2. For

Table 14-1 Values of a and m for use with Eq. (14-22)[†]

Range of Re number	In-line banks $1.2 < P_T/D_o < 4$		Staggered banks $1 < P_T/D_o < 4$	
	a	m	a	m
$10 - 3\times10^2$	0.742	0.431	1.309	0.360
$3\times10^2 - 2\times10^5$	0.211	0.651	0.273	0.635
$2\times10^5 - 2\times10^6$	0.116	0.700	0.124	0.700

[†]ESDU, Convective heat transfer during crossflow of fluids over plain tube banks, ESDU International plc, London, Data Item No. 73031, Nov. 1973 unamended. Reprinted by permission.

Table 14-2 Correction factor F_2 as a function of tube rows in Eq. (14-22)[†]

Number of tube rows	F_2, in-line banks (Re > 2×10^3)	F_2, staggered banks (Re > 10^2)
3	0.86	0.85
4	0.90	0.90
6	0.94	0.95
8	0.98	0.985
10^+	0.99	0.99

[†]ESDU, Convective heat transfer during crossflow of fluids over plain tube banks, ESDU International plc, London, Data Item No. 73031, Nov. 1973 unamended. Reprinted by permission.

either tube configuration V_{max} is given by

$$V_{max} = \frac{\dot{m}_T}{\rho N_T (P_T - D_o) L} \tag{14-22c}$$

where $\dot{m}_T$ is the total mass flow rate of the fluid, ρ the fluid density, N_T the number of tubes in the tube bank, P_T the tube pitch (as defined in Fig. 14-40), D_o the outside tube diameter, and L the tube length. Equation (14-22*c*) assumes that the sum of the spaces between tubes in a given transverse row is equal to the sum of the tube diameters in a row. For example, in an in-line square array, the gap at each edge of the tube bundle is equal to one-half the transverse tube spacing. Excessive gaps at the edge of tube bundles should be avoided to minimize bypass flow.

Film Coefficients and Overall Coefficients for Various Heat-Transfer Situations

Table 14-3 briefly summarizes heat transfer coefficients and fouling resistances for sensible heat transfer in tubular exchangers, while Table 14-4 provides information on these same two quantities for condensation and vaporization. Since there is considerable latitude between the lower and upper values of many of these coefficients, considerable care must be exercised in their use. Accordingly, design engineers often elect to use overall heat-transfer coefficients directly without attempting to evaluate the individual film coefficients. When this is the case, the engineer must predict an overall coefficient on the basis of past experience with equipment and materials similar to those involved in the new study. Design values of overall heat-transfer coefficients for many of the situations commonly encountered by design engineers are listed in Table 14-5.

Table 14-3 General range of heat-transfer coefficients and fouling resistances for sensible heat transfer in tubular exchangers[a]

Fluid conditions	h, W/m²·K[b]	h_d^{-1}, m²·K/W
Water, liquid	$5\times10^3 - 1\times10^4$	$1\times10^{-4} - 2.5\times10^{-4}$
Light organics,[c] liquid	$1.5\times10^3 - 2\times10^3$	$1\times10^{-4} - 2\times10^{-4}$
Medium organics,[d] liquid	$7.5\times10^2 - 1.5\times10^3$	$1.5\times10^{-4} - 4\times10^{-4}$
Heavy organics,[e] liquid, heating	$2.5\times10^2 - 7.5\times10^2$	$2\times10^{-4} - 1\times10^{-3}$
Heavy organics,[e] liquid, cooling	$1.5\times10^2 - 4\times10^2$	$2\times10^{-4} - 1\times10^{-3}$
Very heavy organics,[f] liquid, heating	$1\times10^2 - 3\times10^2$[g]	$4\times10^{-4} - 3\times10^{-3}$
Very heavy organics,[f] liquid, cooling	$6\times10 - 1.5\times10^2$[g]	$4\times10^{-4} - 3\times10^{-3}$
Gas,[h] $p = 100$–200 kPa	$8\times10 - 1.2\times10^2$	$0 - 1\times10^{-4}$
Gas,[h] $p = 1$ MPa	$2.5\times10^2 - 4\times10^2$	$0 - 1\times10^{-4}$
Gas,[h] $p = 10$ MPa	$5\times10^2 - 8\times10^2$	$0 - 1\times10^{-4}$

[a]Data adapted from *Heat Exchanger Design Handbook*, G. F. Hewitt, ed., Begell House, New York, 1998, Sec. 3.1.4.
[b]Coefficients based on clean surface area in contact with the fluid and an allowable pressure drop of about 50–100 kPa for the fluid.
[c]Hydrocarbons through C_8, gasoline, light alcohols, ketones, etc.; $\mu < 0.5 \times 10^{-3}$ Pa·s
[d]Absorber oil, hot gas oil, kerosene, and light crudes; $0.5 \times 10^{-3} < \mu < 2.5 \times 10^{-3}$ Pa·s
[e]Lube oils, fuel oils, cold gas oil, heavy and reduced crudes; $2.5 \times 10^{-3} < \mu < 5.0 \times 10^{-2}$ Pa·s
[f]Tars, asphalts, greases, polymer melts, etc.; $\mu > 5.0 \times 10^{-2}$ Pa·s
[g]Estimation of coefficients is approximate and depends strongly on the temperature difference.
[h]Air, nitrogen, oxygen, carbon dioxide, light hydrocarbon mixtures, etc.

Table 14-4 General range of heat-transfer coefficients and fouling resistances for condensation and vaporization processes[a]

Fluid conditions	h, W/m^2·K[b]	h_d^{-1}, m^2·K/W
Steam, dropwise condensation	$5\times10^4 - 1\times10^5$	$0 - 1\times10^{-4}$
Steam, condensation, $p = 10$ kPa	$8\times10^3 - 1.2\times10^4$	$0 - 1\times10^{-4}$
Steam, condensation, $p = 100$ kPa	$1\times10^4 - 1.5\times10^4$	$0 - 1\times10^{-4}$
Steam, condensation, $p = 1$ MPa	$1.5\times10^4 - 2.5\times10^4$	$0 - 1\times10^{-4}$
Light organics,[c] condensation, $p = 10$ kPa	$1.5\times10^3 - 2\times10^3$	$0 - 1\times10^{-4}$
Light organics,[c] condensation, $p = 100$ kP	$2\times10^3 - 4\times10^3$	$0 - 1\times10^{-4}$
Light organics,[c] condensation, $p = 1$ Mpa	$3\times10^3 - 7\times10^3$	$0 - 1\times10^{-4}$
Medium organics,[d] condensation, $p = 100$ kPa, pure or narrow condensing range	$1.5\times10^3 - 4\times10^3$	$1\times10^{-4} - 3\times10^{-4}$
Heavy organics,[e] narrow condensing range, $p = 100$ kPa	$6\times10^2 - 2\times10^3$	$2\times10^{-4} - 5\times10^{-4}$
Light organic mixtures,[c] medium condensing range, $p = 100$ kPa	$1\times10^3 - 2.5\times10^3$	$0 - 2\times10^{-4}$
Medium organic mixtures,[d] medium condensing range, $p = 100$ kPa	$6\times10^2 - 1.5\times10^3$	$1\times10^{-4} - 4\times10^{-4}$
Heavy organic mixtures,[e] medium condensing range, $p = 100$ kPa	$3\times10^2 - 6\times10^2$	$2\times10^{-4} - 8\times10^{-4}$
Water, vaporization, $p < 500$ kPa ΔT superheat max. = 25°C	$3\times10^3 - 1\times10^4$	$1\times10^{-4} - 2\times10^{-4}$
Water, vaporization, 0.5 MPa $< p <$ 10 MPa ΔT superheat max. = 20°C	$4\times10^3 - 1.5\times10^4$	$1\times10^{-4} - 2\times10^{-4}$
Light organics,[c] vaporization, $p < 2$ MPa, ΔT superheat max. = 20°C	$1\times10^3 - 4\times10^3$	$1\times10^{-4} - 2\times10^{-4}$
Light organic mixtures,[c] vaporization, $p < 2$ MPa, ΔT superheat max. = 15°C, narrow boiling range	$7.5\times10^2 - 3\times10^3$	$1\times10^{-4} - 3\times10^{-4}$
Medium organics,[d] vaporization, $p < 2$ MPa, ΔT superheat max. = 20°C	$1\times10^3 - 3.5\times10^3$	$1\times10^{-4} - 3\times10^{-4}$
Medium organic mixtures,[d] vaporization, $p < 2$ MPa, ΔT superheat max. = 15°C, narrow boiling range	$6\times10^2 - 2.5\times10^3$	$1\times10^{-4} - 3\times10^{-4}$
Heavy organics,[e] vaporization, $p < 2$ MPa, ΔT superheat max. = 20°C	$7.5\times10^2 - 2.5\times10^3$	$2\times10^{-4} - 5\times10^{-4}$
Heavy organic mixtures,[e] vaporization, $p < 2$ MPa, ΔT superheat max. = 15°C, narrow boiling range	$4\times10^2 - 1.5\times10^3$	$2\times10^{-4} - 8\times10^{-4}$
Very heavy organic mixtures,[g] vaporization, $p < 2$ MPa, ΔT superheat max. = 15°C, narrow boiling range	$3\times10^2 - 1\times10^3$[f]	$2\times10^{-4} - 1\times10^{-3}$

[a]Data adapted from *Heat Exchanger Design Handbook,* G. F. Hewitt, ed., Begell House, New York, 1998, Sec. 3.1.4.
[b]Coefficients based on clean surface area in contact with the fluid and an allowable pressure drop of about 50–100 kPa for the fluid.
[c]Hydrocarbons through C_8, gasoline light alcohols, ketones, etc.; $\mu < 0.5 \times 10^{-3}$ Pa·s
[d]Absorber oil, hot gas oil, kerosene, and light crudes; $0.5 \times 10^{-3} < \mu < 2.5 \times 10^{-3}$ Pa·s
[e]Lube oils, fuel oils, cold gas oil, heavy and reduced crudes; $2.5 \times 10^{-3} < \mu < 5.0 \times 10^{-2}$ Pa·s
[f]Estimation of coefficients is approximate.
[g]Tars, asphalts, greases, polymer melts, etc.; $\mu > 5.0 \times 10^{-2}$ Pa·s

Table 14-5 Approximate design values of overall heat-transfer coefficients

The following values of overall heat-transfer coefficients are based primarily on results obtained in ordinary engineering practice. The values are approximate because variation in fluid velocities, amount of noncondensable gases, viscosities, cleanliness of heat-transfer surfaces, type of baffles, operating pressure, and similar factors can have a significant effect on the overall heat-transfer coefficients. The values are useful for preliminary design estimates or for rough checks on heat-transfer calculations.

The upper range of overall coefficients given for coolers may also be used for condensers, while the upper range of overall coefficients given for heaters may also be used for evaporators.

Hot fluid	Cold fluid	U_d, W/m²·K	h_d^{-1}, m²·K/W
Coolers			
Water	Water	1250 – 2500	2×10^{-4}
Methanol	Water	1250 – 2500	2×10^{-4}
Ammonia	Water	1250 – 2500	2×10^{-4}
Aqueous solutions	Water	1250 – 2500	2×10^{-4}
Light organics†	Water	375 – 750	6×10^{-4}
Medium organics‡	Water	250 – 600	6×10^{-4}
Heavy organics§	Water	25 – 375	6×10^{-4}
Gases	Water	10 – 250	6×10^{-4}
Water	Brine	500 – 1000	6×10^{-4}
Light organics†	Brine	200 – 500	6×10^{-4}
Heaters			
Steam	Water	1000 – 3500	2×10^{-4}
Steam	Methanol	1000 – 3500	2×10^{-4}
Steam	Ammonia	1000 – 3500	2×10^{-4}
Steam	Aqueous solutions:		
	$\mu < 2 \times 10^{-3}$ Pa·s	1000 – 3500	2×10^{-4}
	$\mu > 2 \times 10^{-3}$ Pa·s	500 – 2500	2×10^{-4}
Steam	Light organics†	500 – 1000	6×10^{-4}
Steam	Medium organics‡	250 – 500	6×10^{-4}
Steam	Heavy organics§	30 – 300	6×10^{-4}
Steam	Gases	25 – 250	6×10^{-4}
Utility fluid (e.g., Dowtherm)	Gases	20 – 200	6×10^{-4}
Utility fluid (e.g., Dowtherm)	Heavy organics§	30 – 300	6×10^{-4}
Exchangers (no phase change)			
Water	Water	1400 – 2850	2×10^{-4}
Aqueous solutions	Aqueous solutions	1400 – 2850	2×10^{-4}
Light organics†	Light organics†	300 – 425	6×10^{-4}
Medium organics‡	Medium organics‡	100 – 300	6×10^{-4}
Heavy organics§	Heavy organics§	50 – 200	6×10^{-4}
Heavy organics§	Heavy organics§	150 – 300	6×10^{-4}
Light organics†	Heavy organics§	50 – 200	6×10^{-4}

†Hydrocarbons through C_8, gasoline, light alcohols, ketones, etc.; $\mu < 0.5 \times 10^{-3}$ Pa·s
‡Absorber oil, hot gas oil, kerosene, and light crudes; $0.5 \times 10^{-3} < \mu < 2.5 \times 10^{-3}$ Pa·s
§Lube oils, fuel oils, cold gas oil, heavy and reduced crudes; $2.5 \times 10^{-3} < \mu < 5.0 \times 10^{-2}$ Pa·s

DETERMINATION OF PRESSURE DROP IN HEAT EXCHANGERS

It may seem strange to include pressure drop considerations in the design of heat exchangers. However, it has become increasingly apparent that increased fluid velocities generally result in larger heat-transfer coefficients and thus less required heat-transfer area and consequently lower heat exchanger costs for a given rate of heat transfer. On the other hand, increased fluid velocities lead to increased pressure drop through the exchanger, which can result in higher operating costs. In fact, often there is a specified maximum pressure drop that can be tolerated in a heat exchanger to keep the annual operating cost at minimum levels. Thus, most heat exchanger design calculations also require an evaluation to determine whether the pressure drop limitation has been exceeded.

The major source of pressure drop in a heat exchanger is friction encountered by the fluid as it flows through either the shell or the tubes of the exchanger. Other pressure drops occur because of friction due to sudden expansion, sudden contraction, or reversal in the direction of flow of the fluids. Changes in vertical head and kinetic energy can influence the pressure drop, but these two effects are ordinarily relatively small and can be neglected in most design calculations.

Tube-Side Pressure Drop

It is convenient to express the pressure drop for heat exchangers in a form similar to the Fanning equation as presented in Chap. 12. Because the transfer of heat is involved, a factor must be included for the effect of temperature change on the friction factor. Under these conditions, the pressure drop on the tube side of a heat exchanger may be expressed as

$$\Delta p_i = \frac{2\beta_i \mathfrak{f}_i G_i^2 L n_p}{\rho_i D_i \Phi_i} \qquad (14\text{-}23)$$

where the subscript i refers to the inside of the tube at the bulk temperature. The Fanning friction factor $\mathfrak{f}_i$ for isothermal flow is based on the arithmetic-average bulk temperature of the fluid. For fully developed turbulent flow in a smooth pipe, the following correlations for $\mathfrak{f}_i$ are recommended:

$$\mathfrak{f}_i = 0.079\ \text{Re}^{-0.25} \qquad \text{for Re} \leq 2100 \qquad (14\text{-}23a)$$

$$\mathfrak{f}_i = 0.046\ \text{Re}^{-0.2} \qquad \text{for Re} > 2100 \qquad (14\text{-}23b)$$

The correction factor Φ_i for nonisothermal flow is equal to $1.1(\mu_i/\mu_w)^{0.25}$ when the Reynolds number is less than 2100 and $1.02(\mu_i/\mu_w)^{0.14}$ when the Reynolds number is greater than 2100; μ_i is the viscosity at the arithmetic-average bulk temperature of the fluid, and μ_w the viscosity of the fluid at the average temperature of the inside-tube wall surface. The n_p term recognizes if more than one tube pass is involved with a length of L.

The correction factor β_i to account for the added friction due to sudden contraction, sudden expansion, and reversal of flow direction for the fluid is given by

$$\beta_i = 1 + \frac{F_e + F_c + F_r}{2\mathfrak{f}_i G_i^2 L/(\rho_i^2 D_i \Phi_i)} \tag{14-23c}$$ †

or

$$\beta_i = 1 + \frac{0.51 K_1 n_p \, \Delta T_{f,i} (\mu_i/\mu_w)^{0.28}}{(T_{i,\text{in}} - T_{i,\text{out}})\text{Pr}_i^{2/3}} \tag{14-23d}$$ ‡

where $\Delta T_{f,i}$ is the temperature difference across the film located inside the tube.

Shell-Side Pressure Drop

The pressure drop due to friction when a fluid is flowing parallel to and outside of tubes can be calculated in the normal manner described in Chap. 12 by using a mean diameter equal to 4 times the hydraulic radius of the system and by including all friction effects due to contraction and expansion. In heat exchangers, however, the fluid flow on the shell side is usually across the tubes, and many types and arrangements of baffles may be used. As a result, no single explicit equation can be given for evaluating pressure drop on the shell side of these types of heat exchangers. A detailed procedure for making such an evaluation is presented in the shell-and-tube heat exchanger analysis.

For the case of flow directly across tubes, either in-line or staggered, the following equation can be used to approximate the pressure drop due to friction based on the outside diameter of the tubes at the bulk temperature of the fluid:

$$\Delta p_o = \frac{2B_o \mathfrak{f}' N_{\text{tr}} G_s^2}{\rho_o} \tag{14-24}$$

where B_o is a correction factor to account for friction due to reversal in the direction of flow, recrossing of tubes, and variation in the cross section. When the flow is across unbaffled tubes, B_o is assumed to be equal to unity, while across baffled tubes B_o is estimated as equal to the number of tube crossings. In Eq. (14-24), $\mathfrak{f}'$ is a modified friction factor and N_{tr} the number of tube rows over which the shell fluid flows. The friction factor is a function of the shell-side Reynolds number and is influenced by the configuration of the tube array. For the normal case where the external Reynolds number $D_o G_s/\mu_o$ is between 2000 and 40,000, the friction factor may be represented as§

$$\mathfrak{f}' = b_o \left(\frac{D_o G_s}{\mu_o} \right)^{-0.15} \tag{14-24a}$$

---

†The friction due to sudden expansion F_e may be obtained from $F_e = (V_1 - V_2)^2/2$. The friction due to sudden contraction F_c is given by $F_c = K_c V_2^2/2$, where V_1, V_2, and K_c are defined in Table 12-1. The friction due to reversal of flow direction F_r depends on the details of the exchanger construction, but a good estimate of the friction term for design work is $F_r = 0.5V_2^2(n_p - 1)/(2n_p)$.

‡K_1 is obtained from $(1 - S_i/S_H)^2 + K_c + 0.5(n_p - 1)/n_p$, where S_i/S_H is the ratio of the total inside-tube cross-sectional area per pass to the header cross-sectional area per pass.

§For flows with higher Reynolds numbers, see R. H. Perry and D. W. Green, *Perry's Chemical Engineers' Handbook,* 6th ed., McGraw-Hill, New York, 1984, pp. 5-50 to 5-53.

where b_o has been approximated by Grimson for in-line tubes as†

$$b_o = 0.044 + \frac{0.08x_L}{(x_T - 1)^{(0.43-1.13/x_L)}} \tag{14-24b}$$

and for staggered tubes as

$$b_o = 0.23 + \frac{0.11}{(x_T - 1)^{1.08}} \tag{14-24c}$$

In Eqs. (14-24*b*) and (14-24*c*), x_T is the dimensionless ratio of pitch (i.e., the tube center-to-center distance) transverse to flow to the outside tube diameter while x_L is the dimensionless ratio of pitch parallel to flow to the outside tube diameter. Best results are obtained if x_T is between 1.5 and 4.0. For design purposes the range can be extended down to 1.25.

EXAMPLE 14-3 Estimation of Film Coefficient and Pressure Drop Inside Tubes in a Shell-and-Tube Exchanger

A horizontal shell-and-tube heat exchanger with two tube passes and one shell pass is being used to heat 9 kg/s of 100% ethanol from 15 to 65°C at atmospheric pressure. The ethanol passes through the inside of the tubes, and saturated steam at 110°C condenses on the shell side of the tubes. The tubes are steel with an OD of 0.019 m and a Birmingham wire gauge (BWG) of 14. The exchanger contains a total of 100 tubes (50 tubes per pass). The ratio of total inside-tube cross-sectional area per pass to header cross-sectional area per pass S_i/S_H can be assumed to be 0.5. Estimate the film coefficient for the ethanol and the pressure drop due to friction through the tube side of the exchanger.

■ Solution

Equations (14-18) and (14-23) can be used to determine the heat-transfer coefficient and pressure drop inside the tubes, respectively. Note that the average wall temperature will need to be evaluated to determine the effect of viscosity on the heat-transfer coefficient. This can be accomplished with a simple estimate based on the thermal resistances in the films associated with the ethanol and condensing steam.

Physical property data

At an average temperature of $(15 + 65)/2 = 40°C$

μ for ethanol at 40°C = 0.0009 Pa·s

k for ethanol at 40°C = 1.63×10^{-4} kJ/s·m·K

C_p for ethanol at 40°C = 2.594 kJ/kg·K

ρ for ethanol at 40°C = 785 kg/m^3

k_w for steel = 0.045 kJ/s·m·K

†E. D. Grimson, *Trans. ASME,* **59:** 583 (1937); **60:** 381 (1938).

Exchanger configuration

Tube outside diameter	$D_o = 0.019$ m
Tube inside diameter	$D_i = 0.015$ m
Tube wall thickness	$t_w = 0.002$ m
Flow area per tube	$A_c = 0.000177\ \text{m}^2$
Inside surface area/m	$A_i = 0.047\ \text{m}^2\text{/m}$ tube length

First, determine the mass velocity in each tube.

$$G_i = \frac{\dot{m}_T}{A_c} = \frac{9}{0.000177(50)} = 1017\ \text{kg/s·m}^2$$

The Reynolds and Prandtl numbers are

$$\text{Re} = \frac{D_i G_i}{\mu} = \frac{0.015(1017)}{0.0009} = 16{,}950$$

$$\text{Pr} = \frac{C_p \mu}{k} = \frac{2.594(0.0009)}{0.000163} = 14.32$$

Equation (14-18), for evaluation of the average heat-transfer coefficient, applies under these conditions. The only remaining unknown in the relation is μ_w, and this requires the temperature of the tube wall. As a first approximation, assume the following:

ΔT across ethanol film = 70% of average total ΔT, or $0.7(110 - 40) = 49°\text{C}$

ΔT across steam film = 10% of average total ΔT, or $0.10(110 - 40) = 7°\text{C}$

Wall temperature = $40 + 49 = 89°\text{C}$

μ_w of ethanol at 89°C = 0.0004 Pa·s

$$\frac{\mu}{\mu_w} = \frac{0.0009}{0.0004} = 2.25$$

From Eq. (14-18)

$$h_i = 0.023\left(\frac{k}{D_i}\right)\text{Re}^{0.8}\text{Pr}^{1/3}\left(\frac{\mu}{\mu_w}\right)^{0.14}$$

$$= (0.023)\left(\frac{0.163}{0.015}\right)(16{,}950)^{0.8}(14.32)^{1/3}(2.25)^{0.14} = 1650\ \text{W/m}^2\text{·K}$$

Now check on the ΔT assumptions. From Table 14-3 a fouling coefficient of 5000 W/m²·K is adequate for the ethanol. No fouling coefficient will be used for the steam.

From Table 14-4, the steam film coefficient can be estimated conservatively as 1.0×10^4 W/m²·K. The overall heat-transfer coefficient can now be estimated from Eq. (14-4*a*):

$$\frac{1}{U_i} = \frac{1}{h_i} + \frac{1}{h_{i,d}} + \frac{D_i \ln(D_o/D_i)}{2k_w} + \frac{D_i}{h_o D_o}$$

$$= \frac{1}{1650} + \frac{1}{5000} + \frac{0.015 \ln(0.019/0.015)}{2(45)} + \frac{0.015}{10{,}000(0.019)}$$

$$= 0.00061 + 0.0002 + 0.00004 + 0.00008 = 0.00093\ \text{m}^2\text{·K/W}$$

$$\text{Percent } \Delta T \text{ across ethanol film} = \frac{0.00061}{0.00093}(100) = 65.6\% \text{ of total } \Delta T$$

$$\text{Percent } \Delta T \text{ across steam film} = \frac{0.00008}{0.00093}(100) = 8.6\% \text{ of total } \Delta T$$

The ΔT assumptions are sufficiently adequate to proceed.

In the determination of the pressure drop due to friction in a new tube or pipe, assume that U is constant over the length of the exchanger.

$$U_i = \frac{1}{0.00093} = 1075 \text{ W/m}^2\cdot\text{K}$$

$$\Delta T_{o,\text{log mean}} = \frac{(110 - 15) - (110 - 65)}{\ln[(110 - 15)/(110 - 65)]} = 66.9°\text{C}$$

From Eq. (14-6)

$$\dot{q} = (\dot{m}C_p\,\Delta T)_c = 9(2594)(65 - 15) = 1.17 \times 10^6 \text{ W}$$

By Eq. (14-5)

$$A_i = \dot{q}/U_i\,\Delta T_{o,\text{log mean}} = \frac{1.17 \times 10^6}{1075(66.9)} = 16.3 \text{ m}^2$$

Length per tube is then

$$L = \frac{16.3}{0.047(100)} = 3.47 \text{ m}$$

To obtain the pressure drop inside the tube, use Eq. (14-23) and Eqs. (14-23*d*) and (14-23*b*) to evaluate β_i and $\mathfrak{f}_i$, respectively.

$$\beta_i = 1 + \left[0.51 K_1 n_p\,\Delta T_{f,i}(\mu_i/\mu_w)^{0.28}/(T_{i,\text{out}} - T_{i,\text{in}})\text{Pr}_i^{2/3}\right]$$

$$K_l = \left(1 - \frac{S_i}{S_H}\right)^2 + K_c + \frac{0.5(n_p - 1)}{n_p}$$

where (from Table 12-1) $K_c = 0.4(1.25 - S_i/S_H) = 0.4(1.25 - 0.5) = 0.3$ and

$$K_1 = (1 - 0.5)^2 + 0.3 + \frac{0.5(2 - 1)}{2} = 0.8$$

$$\mathfrak{f}_i = 0.046\,\text{Re}^{-0.2} = 0.046(16{,}950)^{-0.2} = 0.0066$$

$$\beta_i = 1 + \frac{0.51(0.8)(2)(0.656)(110 - 40)(2.25)^{0.28}}{(65 - 15)(14.32)^{2/3}} = 1.16$$

$$\Delta P_i = \frac{2\beta_i \mathfrak{f}_i G_i^2 L n_p}{\rho_i D_i \Phi_i} \qquad \Phi_i = 1.02\left(\frac{\mu}{\mu_w}\right)^{0.14}$$

$$= \frac{2(1.16)(0.0066)(1017)^2(3.47)(2)}{785(0.015)(1.02)(2.25)^{0.14}}$$

$$= 8169 \text{ Pa}$$

The pressure drop due to friction through the tube side of the exchanger is slightly greater than 8 kPa. A software program utilizing the given data determines that the pressure drop is 6 kPa. ■

SELECTION OF HEAT EXCHANGER TYPE

One of the more important actions taken by the design engineer in arriving at a satisfactory solution for a specific heat exchange is the careful selection of the heat exchanger type that should be used. In the chemical industry the preferred choice in the past has been the shell-and-tube heat exchanger for which there are well-established codes and standards prepared by TEMA (Tubular Exchanger Manufacturers Association) and ASME (American Society for Mechanical Engineers). However, to attain higher thermal efficiencies while minimizing capital costs, the use of other types of heat exchangers has received considerable attention in both the chemical industry and other manufacturing industries.

The selection process normally includes a number of factors, all of which are related to the heat-transfer application. These factors include, but are not limited to, the following items:

1. Thermal and hydraulic requirements
2. Material compatibility
3. Operational maintenance
4. Environmental, health, and safety considerations and regulations
5. Availability
6. Cost

Any heat exchanger selected must be able to provide a specified heat transfer, often between a fixed inlet and outlet temperature, while maintaining a pressure drop across the exchanger that is within the allowable limits dictated by process requirements or economics. The exchanger should be able to withstand the stresses due to fluid pressure and temperature differences. The material or materials selected for the exchanger must be able to provide protection against excessive corrosion. The propensity for fouling in the exchanger must be evaluated to assess the requirements for periodic cleaning. The exchanger must meet all the safety codes. Potential toxicity levels from all fluids must be assessed, and appropriate types of heat exchangers selected to eliminate or minimize the human and environmental effects in the event of an accidental leak or failure of the exchanger. Finally, to meet construction time constraints and project cost controls, the design engineer may have to select a heat exchanger based on a standard design used by the fabricator to maintain timeliness and reduce costs.

Note that heat exchangers often do not operate at the conditions for which they were designed. For example, if fouling allowances are included in the design, the heat exchanger will be overdesigned initially and underdesigned prior to shutdown for periodic cleaning. As a consequence, the outlet conditions from the exchanger will vary, with possible downstream effects. Modern quality assurance procedures make it mandatory to inspect the delivered heat exchanger to determine whether the original design specifications have been followed.

Key Heat Exchanger Types Available

As noted earlier, heat exchangers are normally classified as either a shell-and-tube type or a compact type of exchanger. A short description of several key types of heat exchangers follows.

Double-Pipe and Multiple Double-Pipe Exchangers The double-pipe heat exchanger in its earliest form is the simplest device for exchanging heat between two fluids, consisting of a "tube inside of a tube" with suitable inlet and outlet connections for both fluids. Modern industrial designs of the double-pipe exchanger use modular sections with a U-tube in a "hairpin" arrangement. The flow in each section is essentially countercurrent. However, double-pipe heat exchangers using normal piping have only limited industrial use, except when the required heat transfer is small or high pressures exist for both fluid streams. To make these units useful, fins are added to the outside surface of the inner tube. With such an addition, the fluid with a low heat-transfer coefficient, such as a high-viscosity fluid or a gas, can be located on the fin side, resulting in a relative area enlargement of 5 to 20. Since the fin surface is relatively inexpensive compared to the primary surface of the tube, very cost-effective designs can be established.

A modification of the simple double-pipe arrangement involves the use of multiple tubes in a U-bend, as shown in Fig. 14-10. This design is essentially a mixture of the double-pipe and shell-and-tube heat exchangers. The requirement of a tube sheet partially reduces the advantage of double-pipe construction, but much larger surface areas can be achieved with such an arrangement.

Shell-and-Tube Heat Exchangers The types of shell-and-tube exchangers shown in Fig. 14-11 are the three most commonly used types, but many other configurations are possible. Such configurations are generally described in terms of the TEMA standard nomenclature, in which the first letter describes the head type at the front end; the second letter, the shell type; and the third letter, the head type at the rear end of the exchanger.

The *fixed-tube type* of exchanger is probably used more often than any other type. The construction is simple and economical. The inside of the tubes can be cleaned mechanically or chemically, while the outside of the tubes requires chemical cleaning.

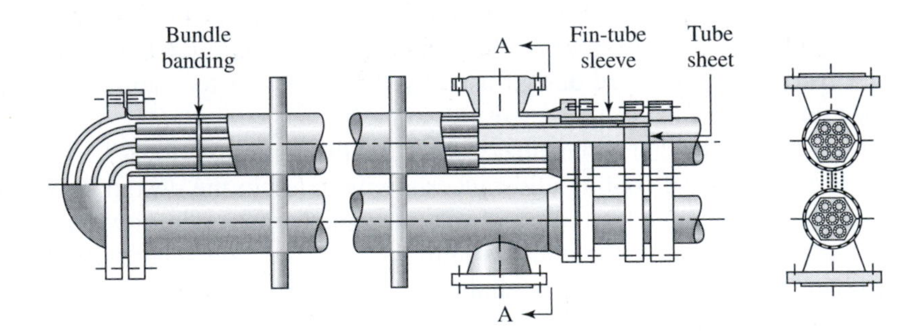

Figure 14-10
Multitube unit with six longitudinally finned tubes

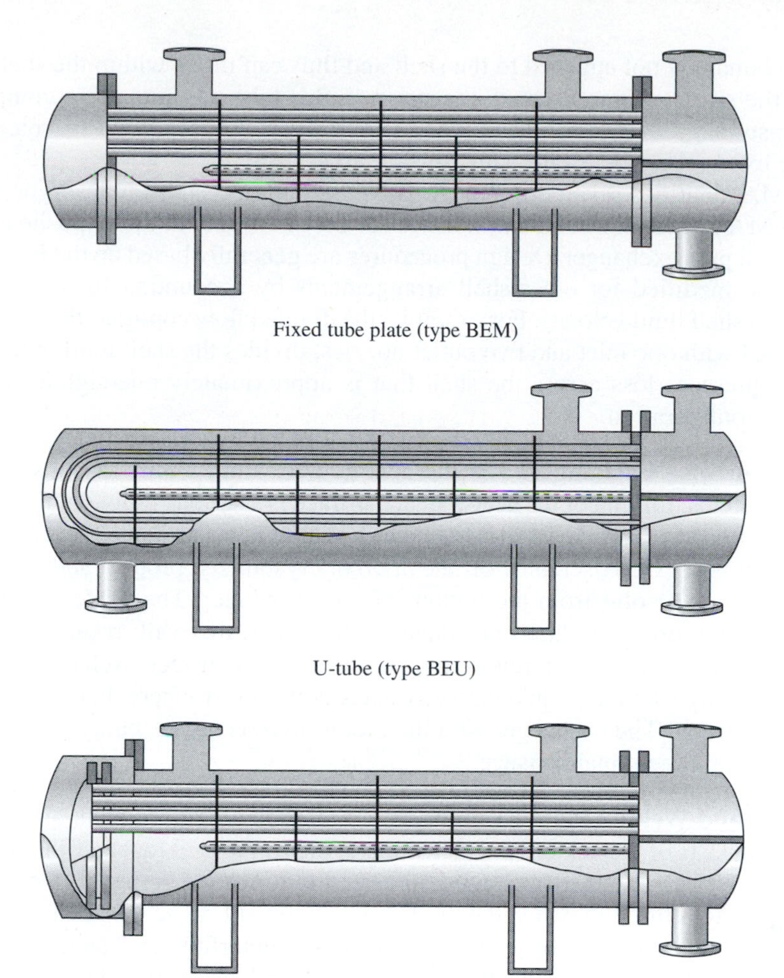

Figure 14-11
Shell-and-tube heat exchanger design types. (*Courtesy of TEMA.*)

An expansion bellows is often used to accommodate excessive stresses caused by thermal expansion.

The *U-tube type* of exchanger is able to absorb the stresses due to thermal expansion. However, the use of the U-tube makes cleaning of the inside surface of the tubes much more difficult. This type of exchanger is similarly priced to the fixed-tube type when operating at low pressure, but shows a significant saving at high operating pressures. A major disadvantage of the U-tube design is that complete counterflow across the tubes does not occur except when a type-F shell is designated since the latter is a two-pass shell with a longitudinal baffle.

The *floating-head type* of exchanger is a more robust exchanger capable of handling both high temperatures and high pressures. It gets its name because one end of the tube bundle is not attached to the shell and thus can move within the shell to take care of the stresses from thermal expansion. Since fabrication is more complex and time-consuming, the floating-head type is approximately 25 percent more expensive than the fixed-tube type for the same surface area.

TEMA identifies seven different shell configurations. The simplest type, referred to as TEMA E, has entry and exit nozzles for the shell fluid at the opposite ends of a single-shell pass exchanger. Design procedures are generally based on the E-type shell, but can be modified for other shell arrangements by accounting for the respective changes in shell fluid velocity. For example, the divided flow configuration, designated as TEMA J with one inlet and two outlet nozzles, divides the shell fluid in half with a resultant pressure loss across the shell that is approximately one-eighth that of the TEMA E configuration.

Scraped-Surface Exchangers In the heat transfer from highly viscous or crystallization systems, fouling of the heat-transfer surface is a serious problem. For such systems it is often necessary to use a heat exchanger in which a rotating blade periodically moves over the surface, scraping off the dried or crystallized product and allowing the latter to exit as a solid from the bottom of the exchanger. The blade action creates shearing of the dried product immediately adjacent to the wall, resulting in locally high-heat transfer rates. As a rule of thumb, the scraped-surface exchanger should be considered only when the liquid viscosity exceeds 1 Pa·s or where there is heavy fouling or deposition. The exchanger with the motor drive is heavy, bulky, and expensive and requires frequent maintenance.

Gasketed and Welded Plate Exchangers A gasketed plate or plate-and-frame exchanger consists of a stack of corrugated metal plates pressed together in a frame and sealed at the plate edges by a compressible gasket to form a series of interconnected narrow passages through which the fluids flow. Figure 14-12 schematically shows the plate arrangement for four plates in a single-pass counterflow exchanger of this type. The hot and cold fluids flow in alternate passages, and heat is transferred through the

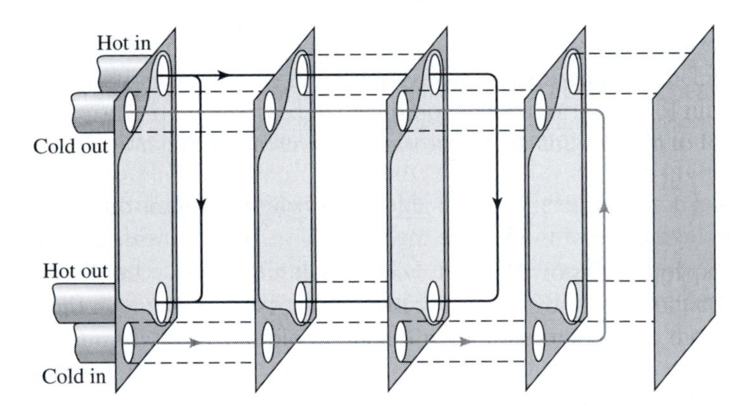

Figure 14-12
Plate arrangement in a single-pass, counterflow heat exchanger

thin corrugated plate separating the two fluids with relatively low thermal resistance. The plates are corrugated to improve mechanical rigidity, control plate spacing, and increase heat-transfer area.

The nature of the design permits rapid optimization to provide a highly efficient heat-transfer performance, although often at a relatively high pressure drop. A major advantage of the gasketed plate exchanger is that its overall surface area can be increased simply by adding more plates. The gaskets used as a seal between each plate are generally butyl and/or nitrile rubber. Other compressible materials include neoprene, Hypalon, and Viton. The relative integrity of the sealant material to high temperatures and pressures is the major limitation of the gasketed plate exchanger.

The sealing problems noted with the gasketed plate exchanger can be overcome by welding the edges of the plates. The operating pressures of the welded plate exchanger are normally limited to around 3 MPa. The plate areas of the welded plate exchanger are often much larger than those for the gasketed plate exchanger, to minimize the periphery-to-area ratio and thereby reduce the welding cost associated with each plate. Typical plates in such units can have a length of 10 m and a width of 1.5 m. Nevertheless, the purchased cost per unit area for the welded plate exchanger can be as much as 20 to 35 percent higher than that for the gasketed plate exchanger.

A maximum differential pressure of 3 MPa between the hot and cold fluids can be a limitation with this type of exchanger. Also, since the plates are welded around the periphery, only chemical cleaning can be used to remove fouling deposits. This limitation may rule out many applications.

Spiral Plate and Tube Exchangers To further reduce the cost of plate exchangers, the plates can be coiled to form countercurrent flow passages in a configuration shown in Fig. 14-13. The spiral surfaces are separated by raised protrusions and sealed with

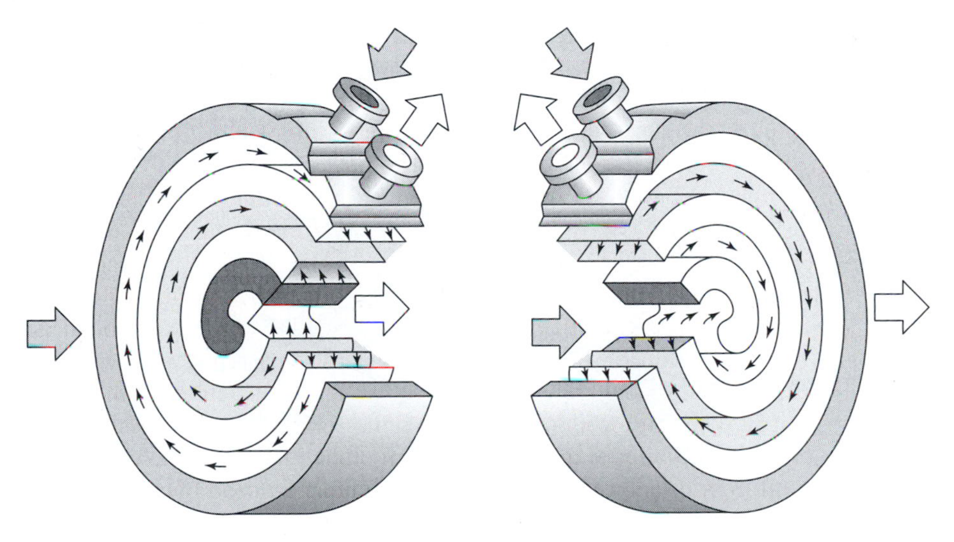

Figure 14-13
Flow paths in a spiral heat exchanger. (*Courtesy of Alfa-Laval Inc.*)

two end plates. The cold fluid enters at the periphery and flows toward the center, while the hot fluid enters at the center of the exchanger and flows countercurrent to the cold stream. The end plates can be removed easily for cleaning purposes. Spiral plate exchangers are suitable for small-capacity service with viscous, fouling, and corrosive fluids, but have limitations similar to those for flat plate exchangers. Since the colder fluid enters at the periphery and is located in the center passage, there is an effective blanket of cooler liquid surrounding the spiral assembly. This configuration often minimizes or eliminates the need for external insulation.

A modification that overcomes the pressure limitations of plate exchangers is to use a spiral of adjacent tubes. Since adjacent tubes are in direct contact with each other, the shell-side fluid must flow in the spiral gap between turns of the tube coil. The overall heat-transfer coefficients for spiral tube exchangers generally are somewhat higher than conventional exchangers because of the increased turbulence created by the centrifugal flow of the fluid. The latter also produces a scrubbing effect that reduces fouling.

The spiral tube exchanger is extremely compact and can handle a variety of fluids, but capacities are limited. The shell side is accessible for cleaning purposes. The spiral configuration of the tubes, however, does not permit mechanical cleaning of the tubes. Where such limitations are acceptable, spiral tube exchangers are good choices for small-scale applications.

Compact Exchangers For many applications there is a premium in using a heat exchanger that is small in both size and weight. Often this need is met with plate-fin heat exchangers that are fabricated by stacking alternate layers of corrugated, high-uniformity die-formed metallic sheets (fins) between flat separator plates to form individual flow passages. The edges of each layer are sealed with solid metal bars. An expanded view of the elements of one layer of a plate-fin heat exchanger before brazing is shown in Fig. 14-14. The flow passages usually have integral welded headers

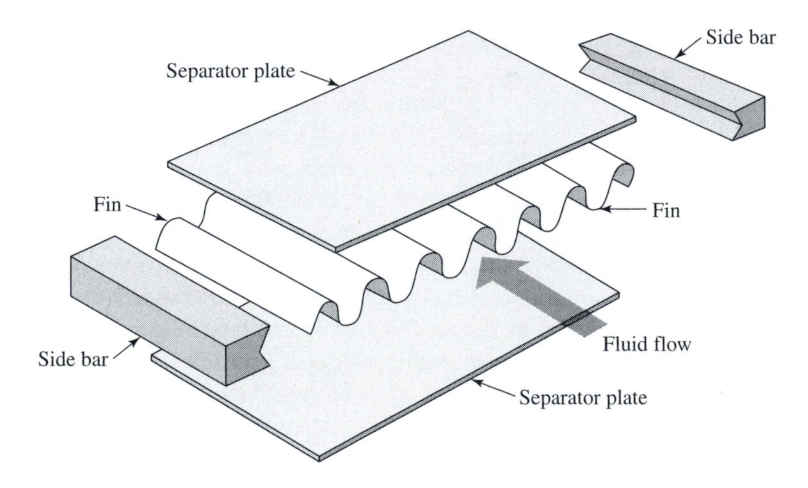

Figure 14-14
Exploded view of one layer of a typical plate-fin heat exchanger before brazing

which are connected to form the overall heat exchanger. Plate-fin exchangers are about 9 times as compact as conventional shell-and-tube heat exchangers while providing the same surface area, weigh less than conventional heat exchangers, and withstand design pressure up to 6 MPa for temperatures between −270 and 800°C.

The corrugated sheets, representing the finned portion of the exchanger, are fabricated in plain, wavy, perforated, and multientry or laced configurations. The plain (straight) fins offer both the lowest heat-transfer coefficients and the lowest pressure drop; conversely, the more complex fins provide the higher heat-transfer coefficients but also the higher pressure drops.

Many different types of flow patterns are available in plate-fin heat exchangers. To provide either multipass or multistream arrangements, suitable internal seals, distributors, and external headers are used. The multipass cross-flow arrangement is comprised of several cross-flow passages, assembled in counterformation to provide effective mean temperature differences in the approaches, and thereby more closely approximates the conditions obtained with countercurrent operations. This type of construction is often used in gas-gas and gas-liquid applications. Countercurrent flow arrangements should be used when very high thermal effectivenesses, above 95 percent, are required in a heat exchanger.

Gas-to-Gas Exchangers A major application of this type of exchanger is seen in the recovery of energy from combustion gases to preheat furnace air. There are many different configurations for this type of exchanger; however, only two will be discussed briefly.

In the plain tube gas-to-gas heat exchanger, the hot gas flows past the tubes in the tube bank in either a single-pass (cross-flow) or a multipass configuration. The normal maximum operating temperature for such units is 250°C.

Another type of gas-to-gas exchanger establishes a convection bank with polished, dimple-ribbed stainless-steel tubes. In a typical unit, the flat-sided tubes are mounted horizontally and sealed with an elastomer to perforated end plates that are formed into a flange plate and an enclosure. The units are compact and lightweight and can be used to recover heat from exhaust gases up to 250°C.

Air-Cooled Exchangers This type of exchanger is sometimes referred to as a *fin-fan* unit because the tubes have external fins and fans are employed to force or draw air through the tube banks in a cross-flow arrangement. The latter arrangement makes these units less efficient than a countercurrent exchanger. The penalty for cross-flow often can be decreased through the use of multiple tube-side passes.

The use of fins is required to increase the overall heat-transfer coefficient. The extra investment in finned tubing is partially compensated for by the fact that no shell is required and no cooling water (including pumps and associated piping) is necessary. In addition, shell-side fouling generally is not a problem, and tube cleaning is relatively easy.

Condensers A wide range of equipment is used to meet the condensation requirements of industrial processes. Condensing equipment can be classified into four generic types, namely, *tubular, plate-type, air-cooled,* and *direct contact*. In the process industry, tubular condensers are normally conventional shell-and-tube types with condensation either inside the tube or in the shell depending on the process requirements, particularly with respect to fouling tendencies. The plate type includes both plate-and-frame exchangers and plate-fin exchangers. Condensation in the air-cooled condensers occurs within the tubes while in the direct-contact condensers the coolant is brought in direct contact with the condensing vapor, eliminating the need for a heat-transfer surface as in the other generic types.

Evaporators Heat exchangers used for concentration and crystallization are labeled as evaporators. Those designed for vapor production are designated as vapor generators. There are many types of evaporators including *film, in-tube boiling, shell-side boiling, flash,* and *direct-contact* types. Selection of the most suitable type is very dependent on the properties of the evaporating fluid. For example, falling and climbing film evaporators are more appropriate for concentrating heat-sensitive liquids since overheating is minimized. On the other hand, some of the in-tube evaporators and direct-contact evaporators are more suitable for concentrating corrosive or scaling fluids. Many of the evaporators mentioned can also be used as vapor generators.

Preliminary Selection of Heat Exchanger Types

Tables 14-6 and 14-7 provide brief summaries of key criteria that the design engineer needs to consider when making preliminary selection of heat exchanger, condenser, and evaporator types for a particular application. The tables provide information on pressure and temperature limits as well as ranges of surface area normally available from vendors of such equipment. (It should be recognized that area limitations can generally be alleviated by using two or more heat exchangers in parallel involving only additional piping and installation costs.) The table also lists potential compatibility problems between the fluids and various components of the exchanger relative to fluid leakage and possible equipment failure. Additional information on the various types of equipment listed is provided in this section and in the literature.[†]

Note that in many of the exchanger types, the material of construction used in the exchanger is the only limitation in the heating or cooling of many fluids. Generally, where seals are required to maintain separation between the two fluids, the nature of the seals or gaskets will be the major fluid limitation. Heat exchangers that have limited accessibility, such as welded plate exchangers and compact heat exchangers, should not be used with fluids that either are corrosive or have a known fouling characteristic. Understandably, if several possible types of heat-transfer equipment meet the thermal and hydraulic requirements, show compatibility with the fluids involved, and indicate no serious problems with maintenance, safety, health, and protection of the environment, then preliminary selection of the type of heat-transfer equipment recommended will generally be based on economics.

Costs of Heat Exchangers

Some of the major factors that influence the cost of heat-transfer equipment include the following: heat-transfer area, tube diameter and thickness, tube length, pressure of fluids, materials of construction, baffling requirements, special design features (finned surface,

---

[†]G. F. Hewitt, ed., *Heat Exchanger Design Handbook,* Begell House, New York, 1998, Part 3.

Table 14-6 Criteria for the preliminary selection of the appropriate heat exchanger type[†]

Exchanger type	Max. pressure, approx. range, MPa	Temperature, approx. range, °C	Normal area, approx. range, m^2	Fluid velocities, (shell/tube), m/s	Fluid limitations	Key features
Double-pipe Multiple-pipe	30 (shell) 140 (tube)	−100 – ~600	0.25 – 20 10 – 200	Liq. (2–3)/(2–3) Gas (10–20)/(10–20)	Materials of construction	Modular construction, small scale
Shell-and-tube	30	−200 – 600+	3 – 1000	Liq. (1–2)/(2–3) Gas (5–10)/(10–20)	Materials of construction	Very adaptable, many types
Scraped-wall	~0.11	Up to 200	2 – 20	Liq. (1–2)/(1–2)	Liquids solidifying on hot surface	For viscous, crystallization systems
Gasketed plate	0.1–2.5	−25 – 175	1 – 2500	Liq. (1–2)/(1–2) Gas (5–10)/(5–10)	Limited to gasket material, avoid gas flow	Modular construction, minimal $/area cost
Welded plate	3	>400	1 – 2500	Liq. (1–2)/(1–2) Gas (5–10)/(5–10)	Materials of construction, avoid fouling fluids	Δp between fluids <3 MPa
Spiral plate	2	Up to 300	10 – 200	Liq. (1–2)/(1–2) Gas (5–10)/(5–10)	Materials of construction	For viscous, corrosive fluids
Spiral tube	50	350	1 – 50	Liq. (2–3)/(2–3) Gas (5–10)/(5–10)	Materials of construction	Adaptable, low maintenance
Compact	3–10	−270 – 80 with Al −270 – 800 with ss	10 – 30,000	Gas (2–5)/(2–5)	Materials of construction, no corrosive fluids	Large area/volume, can operate with small ΔT
Gas-to-gas	~0.11	Up to 250	6–100 for low temperatures 1200–3000 for regenerators	Gas (5–10)/(5–10)	Materials of construction	Many types, some for corrosive gases
Air-cooled	Variable on tube side	Variable on tube side	6 – 20,000	Liq. (NA)/(2–3) Gas (3–6)/10–20)	Materials of construction	Use for heat rejection, standardized design

[†]Modified from data presented by G. F. Hewitt, G. L. Shires, and T. R. Bott, *Process Heat Transfer,* CRC Press, Boca Raton, FL, 1994, Sec. 4.3, and G. D. Ulrich, *A Guide to Chemical Engineering Design and Economics,* J. Wiley, New York, 1984.

Table 14-7 Criteria for the preliminary selection of appropriate condenser or evaporator types†

Equipment type	Normal area, approx. range, m^2	Approx. maximum fluid viscosity, Pa·s	Fluid compatibility	Equipment suitability
Condensers				
Vertical shell side	3–200	1.0	Condense distillation fluids as thermosiphon reboiler	Shell-side venting and condensate removal problems
Vertical tube side	3–200	1.0	Condense organic vapors	Design is very flexible and efficient, cleaning problems
Horizontal shell side	3–1000	2.0	Condense organic and refrigerant vapors	Widely used, pressure drop controlled with shell selection
Horizontal tube side	3–1000	2.0	Condense high-pressure vapors	Avoid multipass tube arrangements
Air cooled	6–2000	2.0	Condense organic and refrigerant vapors	High capacity, eliminates coolant, nonuniformity in air distribution
Gasketed plate	1–1200	1.0	Temperature limitation based on gasket choice	Units have lower $/area cost, appropriate gaskets must be selected
Plate-fin	10–150	0.01	Fluids must be very clean with no fouling, can operate with small ΔT's between fluids	Typically used in cryogenic processes, large area/volume ratio
Direct contact	NA	2.0	Can handle corrosive and fouling vapors	Solids content of fluid dictates choice of direct-contact spray or submerged contact
Vaporizers				
Jacketed vessel	7–30	0.01	Can vaporize heat-sensitive fluids, materials of construction permit corrosive fluids	Easy cleaning, medium $/area cost, requires little vertical space

Direct contact	7–160	0.01	Can vaporize heat-sensitive fluids, materials of construction permit corrosive fluids	Low \$/area cost, reasonable capacity, requires little vertical space
In-tube evaporators				
Falling film	30–300	1.0	Can vaporize heat-sensitive fluids, but not scale-producing liquids and slurries	Operates with small temperature difference, limited vaporization capacity
Long tube, vertical	100–10,000	1.0	Avoid severe scaling fluids	Requires large vertical space, high-velocity circulation, high capacity, and high efficiency
Short tube, vertical (Callandria, basket)	20–300	0.01	Works well with conventional, low-viscosity fluids	Unit is compact, relatively inexpensive, and efficient; has limited capacity
Forced circulation, vertical	20–2000	2.0	Can operate with low- and high-viscosity fluids and slurries	Large capacity, multiple-effect use, requires vertical space, medium \$/area cost
Forced circulation, horizontal	20–2000	2.0	Can operate with low- and high-viscosity fluids and slurries	Large capacity, multiple-effect use, medium \$/area cost
Agitated film (scraped-wall)	2–20	100	Low-viscosity fluids unacceptable, materials of construction permit use of corrosive fluids	Small capacity, variable power requirement, high \$/area cost
Shell-side evaporators				
Shell-and-tube, horizontal	3–1000	1.0	Avoid severe scaling or fouling fluids	Large capacity, medium \$/area cost

†Modified from data presented by G. F. Hewitt, G. L. Shires, and T. R. Bott, *Process Heat Transfer,* CRC Press, Boca Raton, FL, 1994, Secs. 15.7 and 17.3, G.D. Ulrich, *A Guide to Chemical Engineering Design and Economics,* J. Wiley, New York, 1984, and S. M. Walas, *Chemical Process Equipment,* Butterworth-Heinemann, Newton, MA, 1988, pp. 204–211.

U bends, removable bundles, etc.), and supports and auxiliaries. Certain equipment manufacturers specialize in specific types of heat exchangers and often are able to provide lower-cost quotes on their units than other equipment fabricators. Accordingly, price quotes should be obtained from several manufacturers to obtain a firm purchase price.

Figures 14-15 and 14-16 provide average cost data for single and multiple double-pipe heat exchangers, respectively. Average cost data for U-tube, fixed-tube, and floating-head shell-and-tube heat exchangers are presented in Figs. 14-17 through 14-20. The relative effect of tube diameter, tube length, and operating pressure on the purchased cost of shell-and-tube heat exchangers is shown in Figs. 14-21 through 14-23. If two floating heads are used in a single shell in place of one floating head, the exchanger cost will be increased by approximately 30 percent while a decrease in cost of about 10 to 15 percent occurs if fixed tube sheets or U tubes are used instead of a floating head. Table 14-8 lists relative costs for tube-and-shell heat exchangers on the basis of the construction material used for the tubes and the exchanger. Figure 14-24 shows the costs of heat exchangers with two types of stainless-steel tubes relative to all-carbon-steel construction.

The average cost data for welded and gasketed plate exchangers, scraped-wall and spiral heat exchangers, flat and spiral plate exchangers, and air-cooled heat exchangers are provided in Figs. 14-25 through 14-28. The costs for vent, spray, and barometric condensers are given in Figs. 14-29 through 14-31. Costs for vaporizers, forced-circulation evaporators, long-tube evaporators, and agitated falling-film evaporators are presented in Figs. 14-32 through 14-34. Information on purchased cost of small immersion heaters is given in Figs. 14-35 and 14-36 while Figs. 14-37 and 14-38 present costs for process furnaces and direct-fired furnaces.

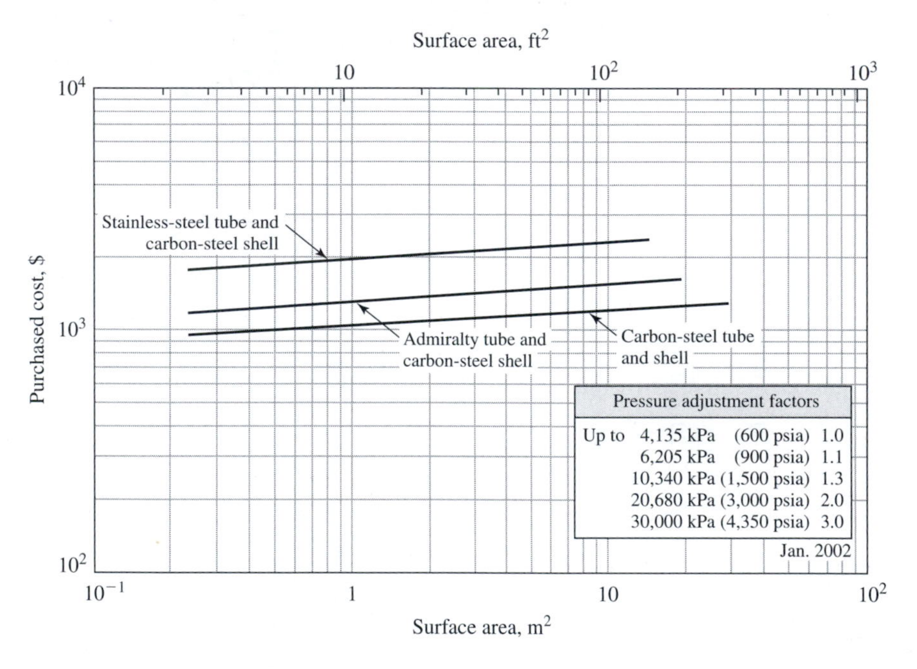

Figure 14-15
Purchased cost of double-pipe heat exchangers

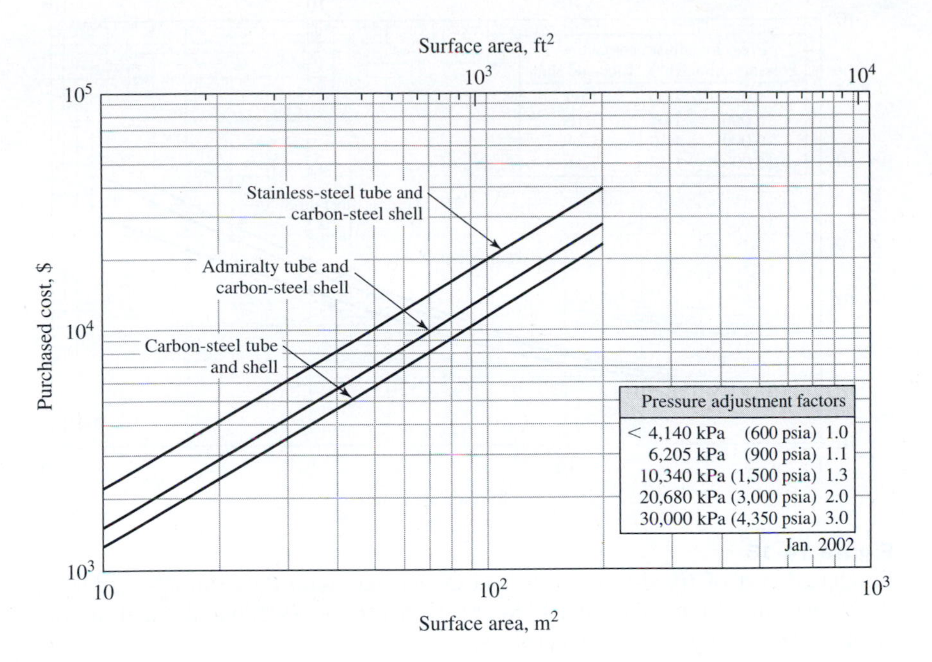

Figure 14-16
Purchased cost of multiple-pipe heat exchangers

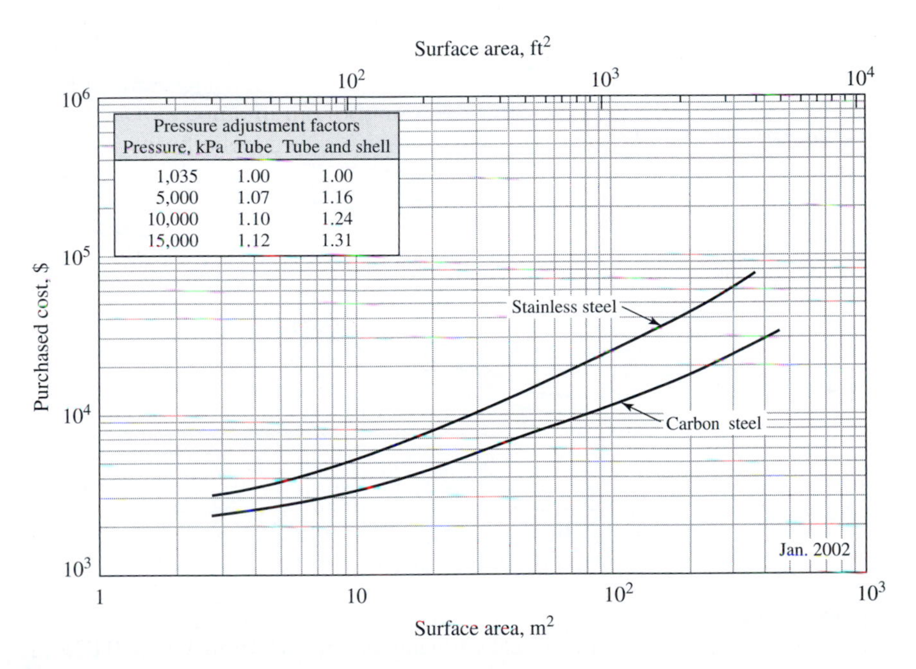

Figure 14-17
Purchased cost of U-tube heat exchangers with 0.0254-m (1-in.) OD tubes × 0.0254-m (1-in.) square pitch and 4.88-m (16-ft) bundles operating at 1035 kPa (150 psia)

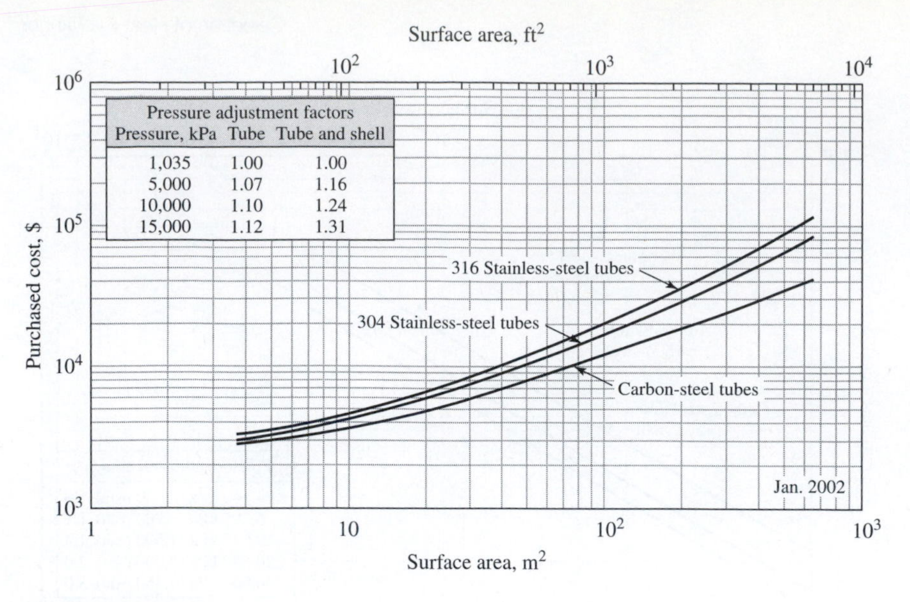

Figure 14-18

Purchased cost of fixed-tube-sheet heat exchangers with 0.019-m ($\frac{3}{4}$-in.) OD × 0.025-m (1-in.) square pitch and 4.88- or 6.10-m (16- or 20-ft) bundles and carbon-steel shell operating at 1035 kPa (150 psia)

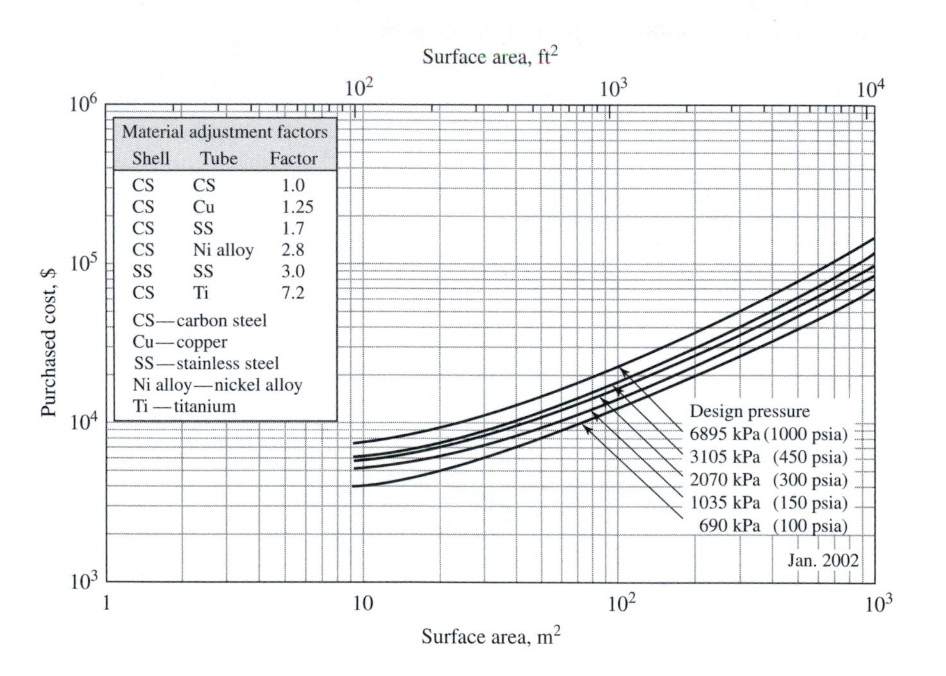

Figure 14-19

Purchased cost of floating-head heat exchangers with 0.019-m OD × 0.025-m ($\frac{3}{4}$-in. OD × 1-in.) square pitch and 4.88-m (16-ft) bundles of carbon-steel construction

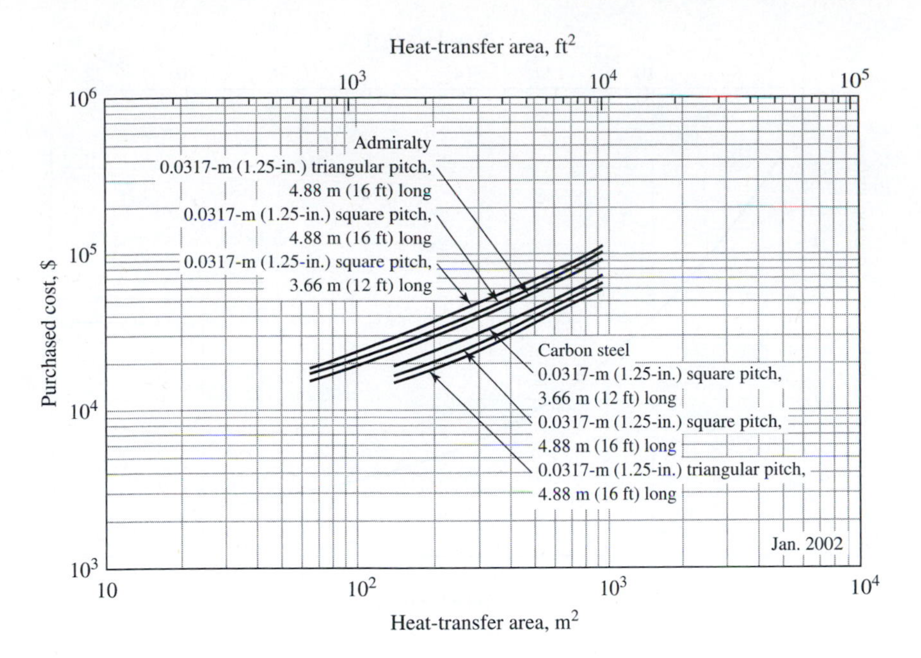

Figure 14-20
Purchased cost of finned-tube floating-head exchangers at 1035 kPa (150 psi). Cost is for 0.0254-m (1-in.) OD Turfin tubes.

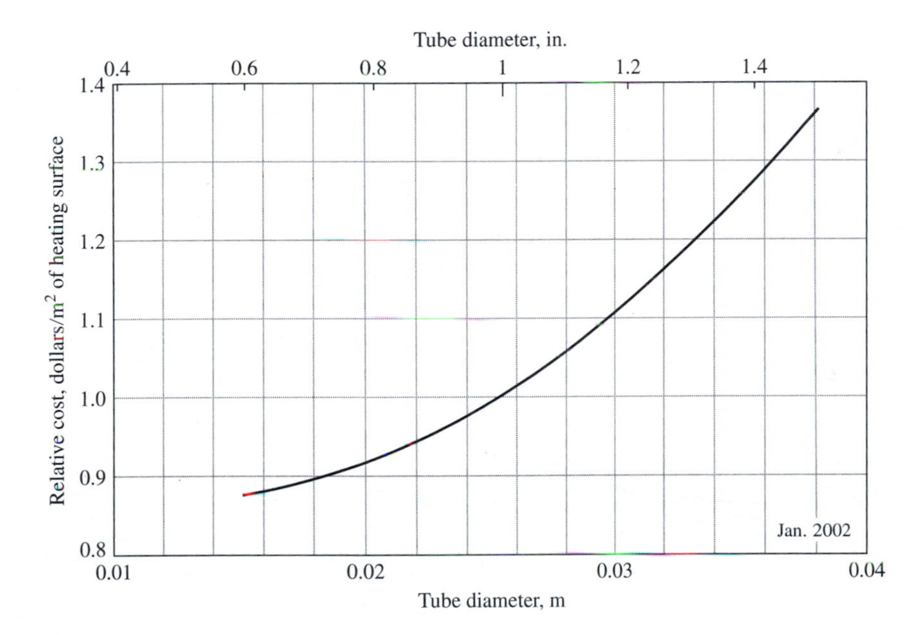

Figure 14-21
Effect of tube diameter on cost of conventional shell-and-tube heat exchangers

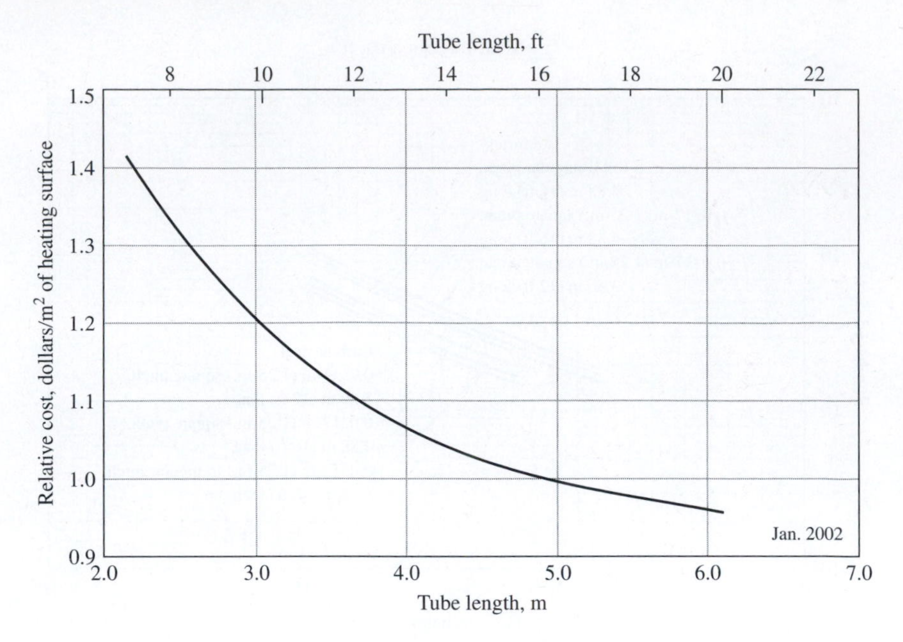

Figure 14-22
Effect of tube length on cost of conventional shell-and-tube heat exchangers

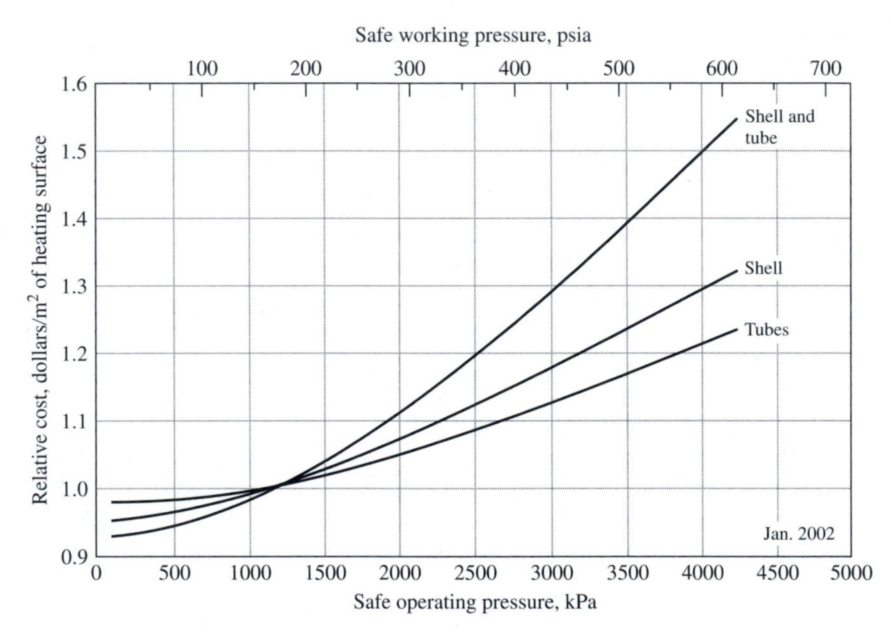

Figure 14-23
Effect of operating pressure on cost of conventional shell-and-tube heat exchangers

Table 14-8 Relative costs of heat exchanger tubing and heat exchangers with 150 m^2 of surface†

Materials	Tubing	Heat exchanger
Zirconium, seamless	25.1	7.7
Hastelloy, C-276, welded	18.2	5.9
Zirconium, 20 BWG, seamless	15.8	5.2
Inconel, 625, welded	15.1	5.0
Carpenter, 20 CB3, welded	8.6	3.3
Incoloy, 825, welded	7.6	3.0
Monel, 400, seamless	7.5	3.0
Titanium, welded	6.8	2.8
E-Brite, 26-1, welded	5.2	2.4
Titanium, 20 BWG, welded	3.6	2.3
316L Stainless steel, welded	3.2	2.2
Cu/Ni of 70/30, seamless	2.9	2.0
Cu/Ni of 70/30, welded	2.4	1.8
304L Stainless steel, welded	2.2	1.8
Carbon steel	1.0	1.0

Basis: Tubing is 0.0254-m OD × 16 BWG, except as noted. Heat exchangers are TEMA-type BEM with 0.019-m OD × 16 BWG × 6.1-m welded tubing and mild carbon-steel shells.
†Estimates will be conservative for smaller exchangers (< 100 m^2) and too low for larger exchangers (> 1000 m^2).

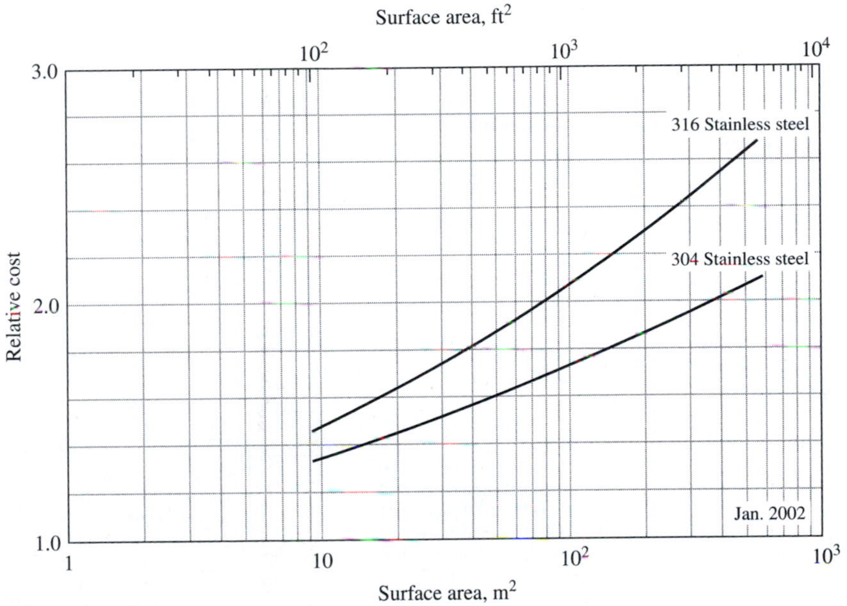

Figure 14-24
Cost of heat exchangers with stainless-steel tubes relative to all-carbon-steel construction

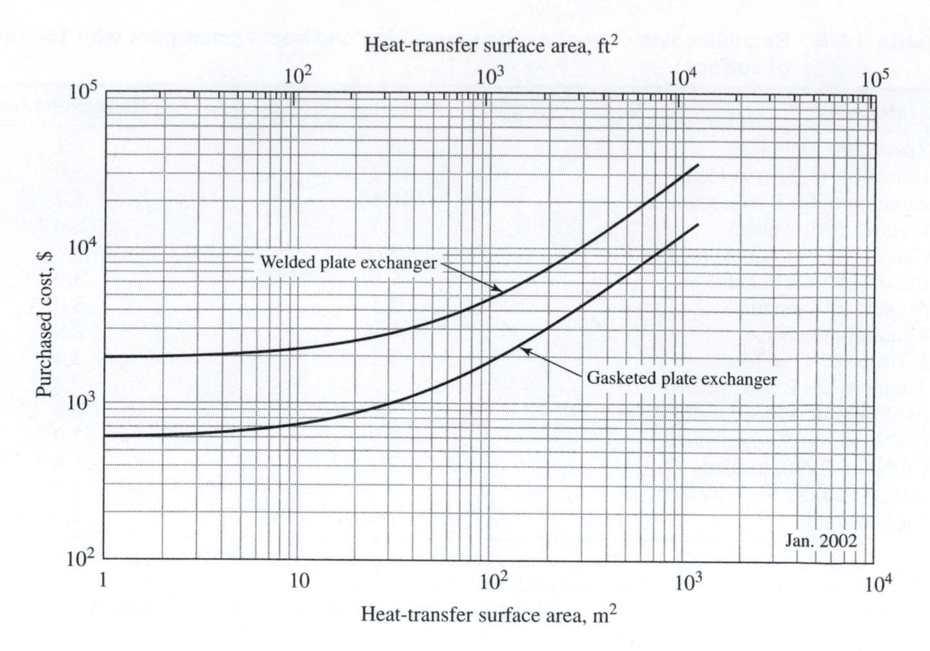

Figure 14-25
Purchased cost of gasketed and welded plate exchangers

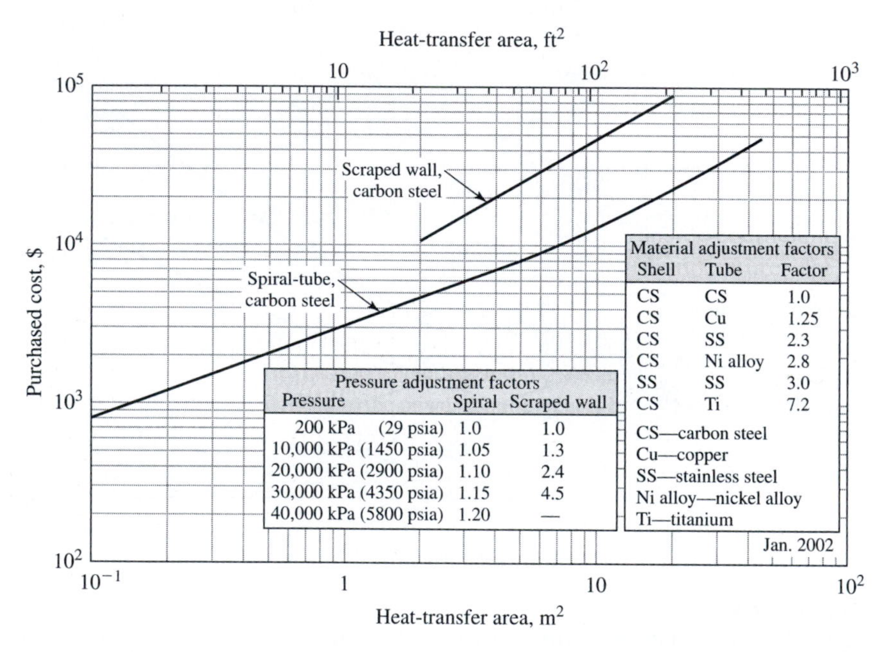

Pressure adjustment factors		
Pressure	Spiral	Scraped wall
200 kPa (29 psia)	1.0	1.0
10,000 kPa (1450 psia)	1.05	1.3
20,000 kPa (2900 psia)	1.10	2.4
30,000 kPa (4350 psia)	1.15	4.5
40,000 kPa (5800 psia)	1.20	—

Material adjustment factors		
Shell	Tube	Factor
CS	CS	1.0
CS	Cu	1.25
CS	SS	2.3
CS	Ni alloy	2.8
SS	SS	3.0
CS	Ti	7.2

CS—carbon steel
Cu—copper
SS—stainless steel
Ni alloy—nickel alloy
Ti—titanium

Figure 14-26
Purchased cost of scraped wall and spiral tube heat exchangers

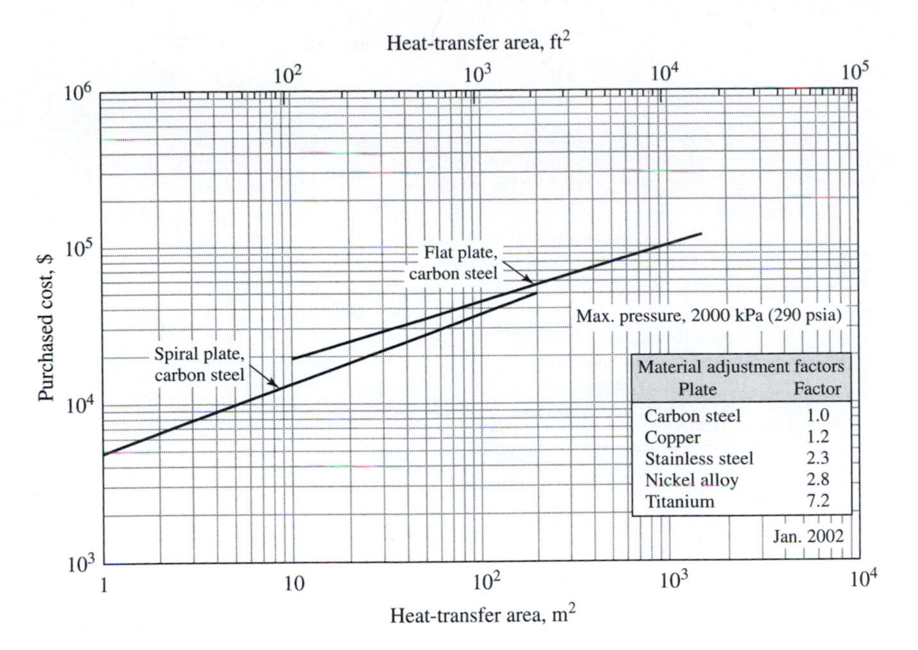

Figure 14-27
Purchased cost of spiral and flat-plate heat exchangers

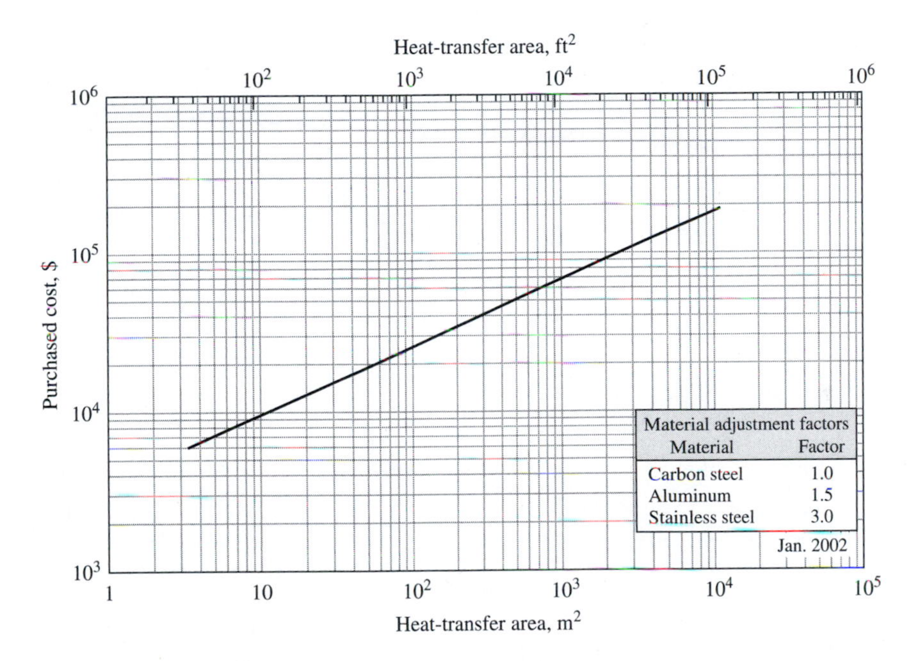

Figure 14-28
Purchased cost of air-cooled heat exchangers. Area is outside area of equivalent bare tubes, excluding fins.

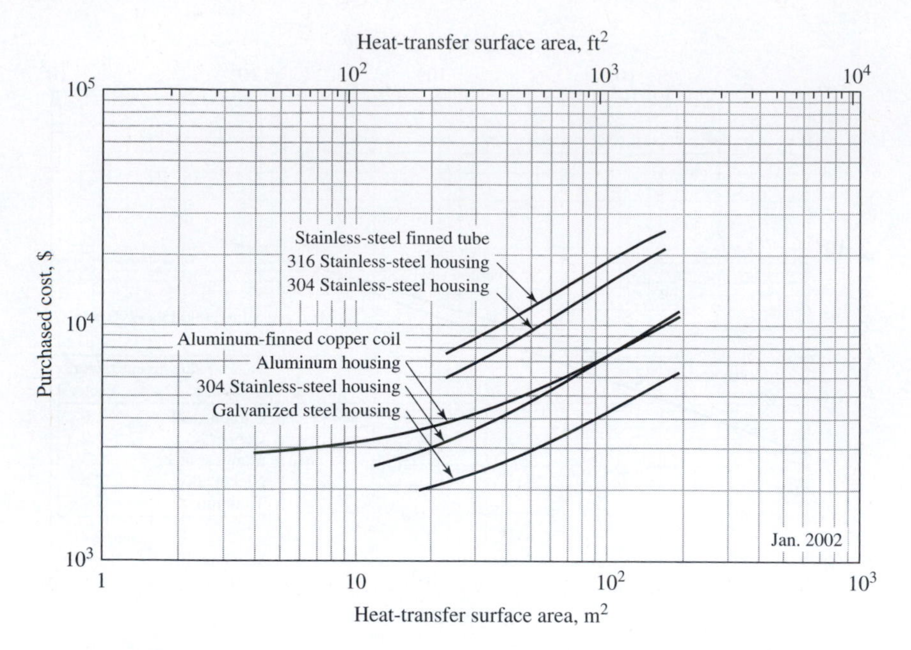

Figure 14-29
Purchased cost of tank vent condensers

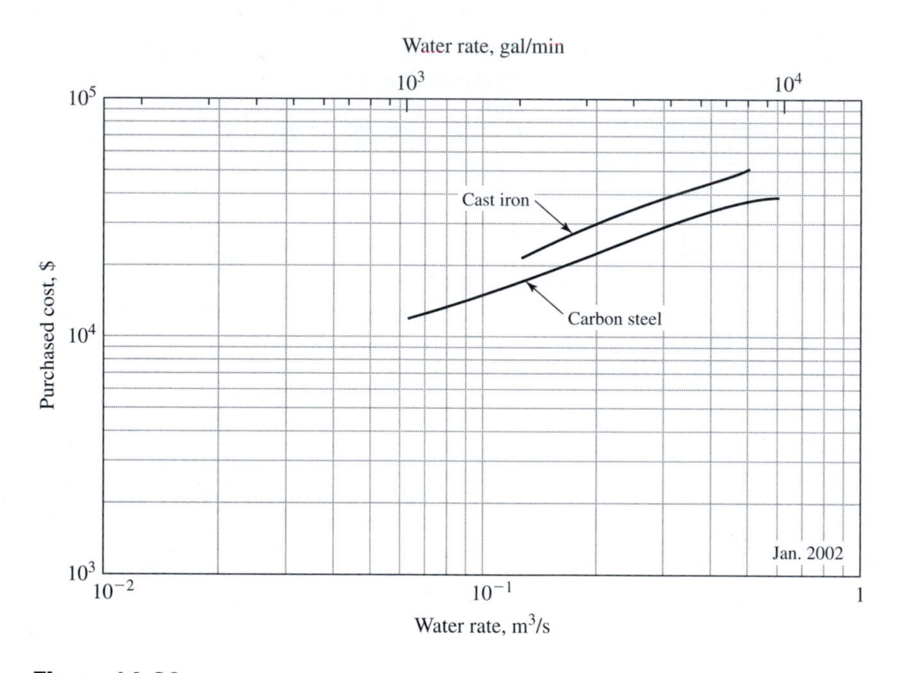

Figure 14-30
Purchased cost of multijet spray-type condensers. Basic construction consists of a shell, water nozzle, case and plate, and spray-type nozzles.

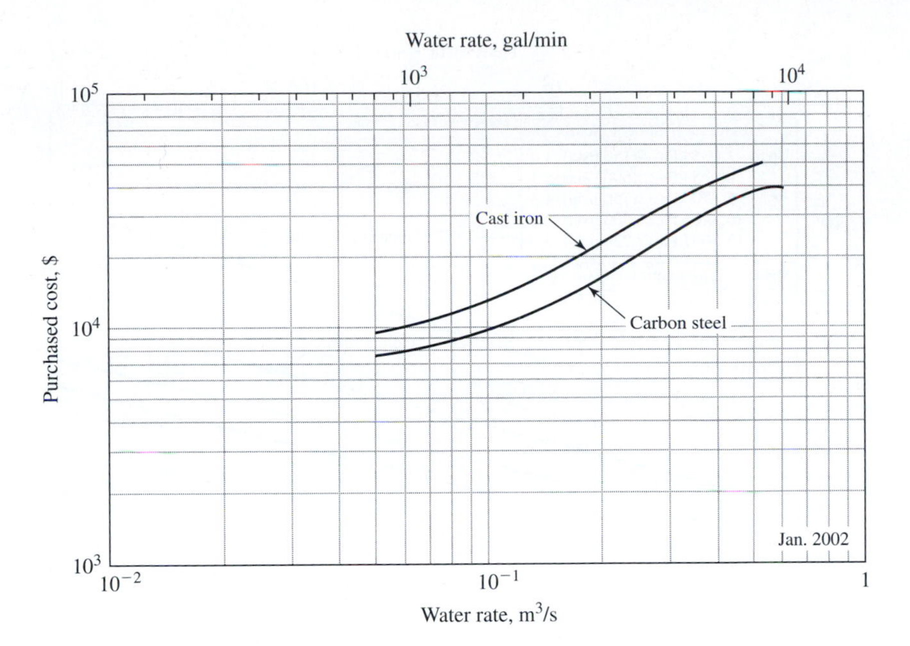

Figure 14-31
Purchased cost of multijet barometric condensers

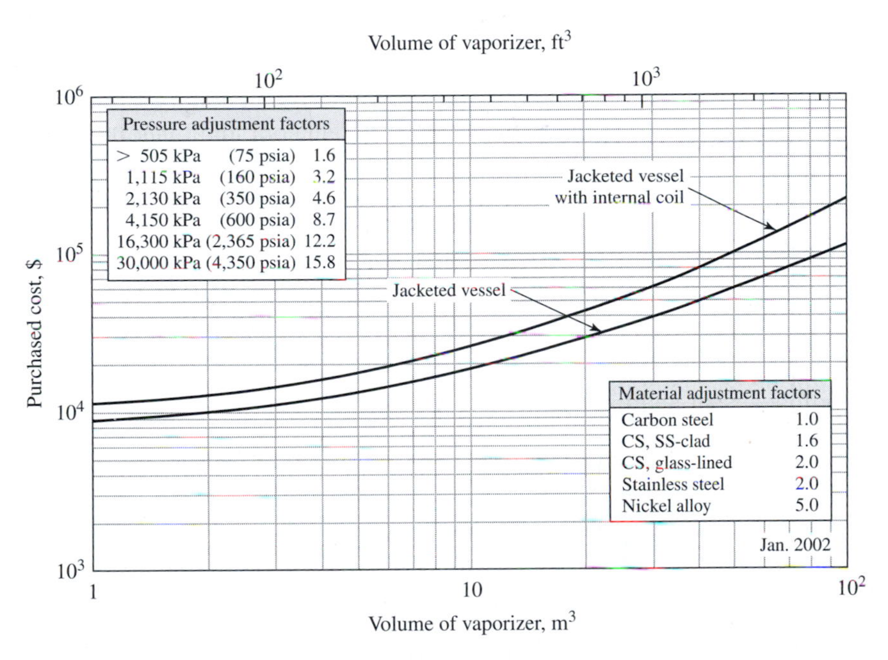

Figure 14-32
Purchased cost of vaporizers

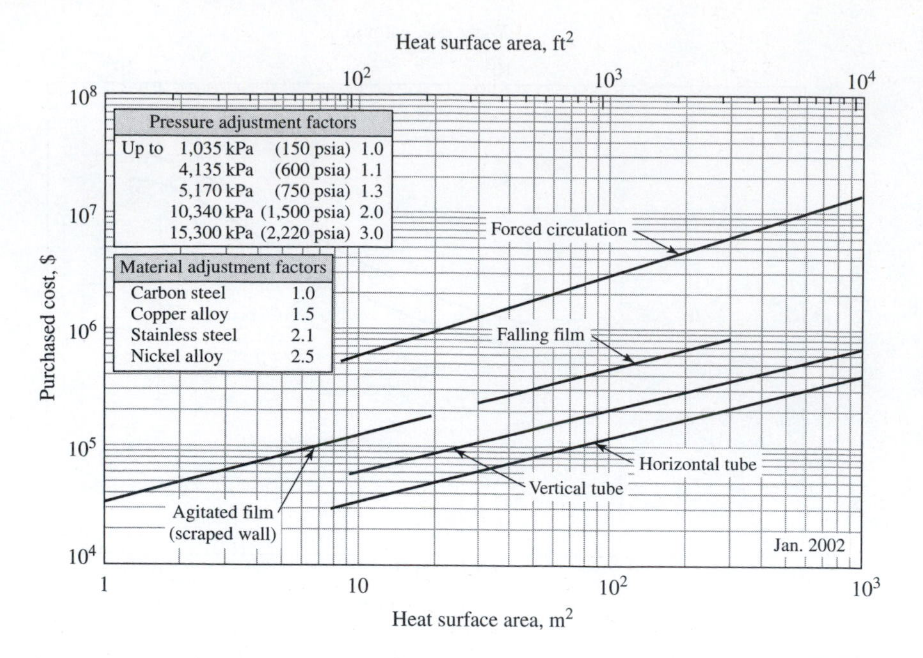

Figure 14-33
Purchased cost of single-effect evaporators. Prices for multiple effects are proportional to the number of effects required.

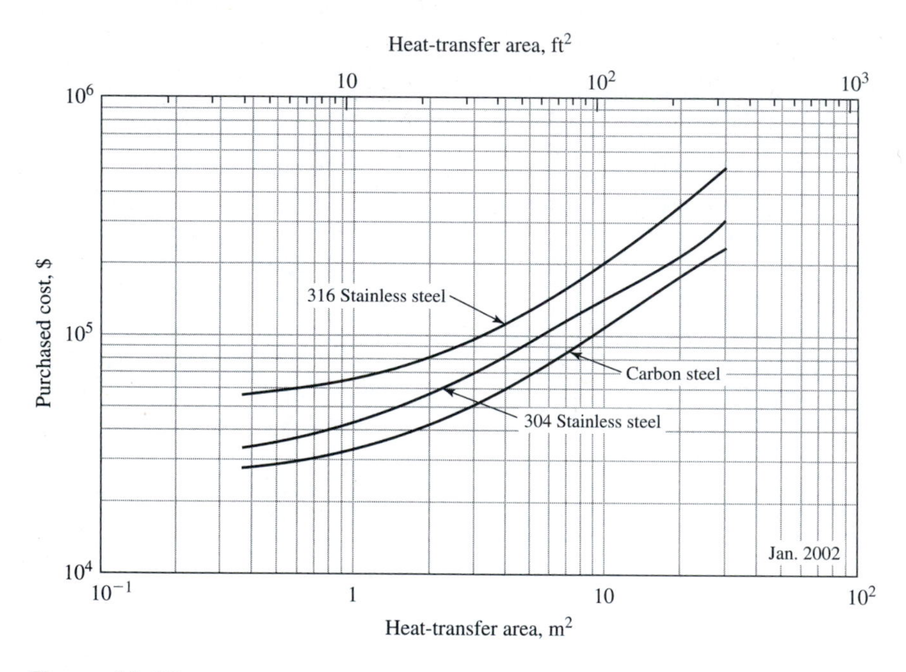

Figure 14-34
Purchased cost of agitated falling-film evaporators, complete with motor and drive

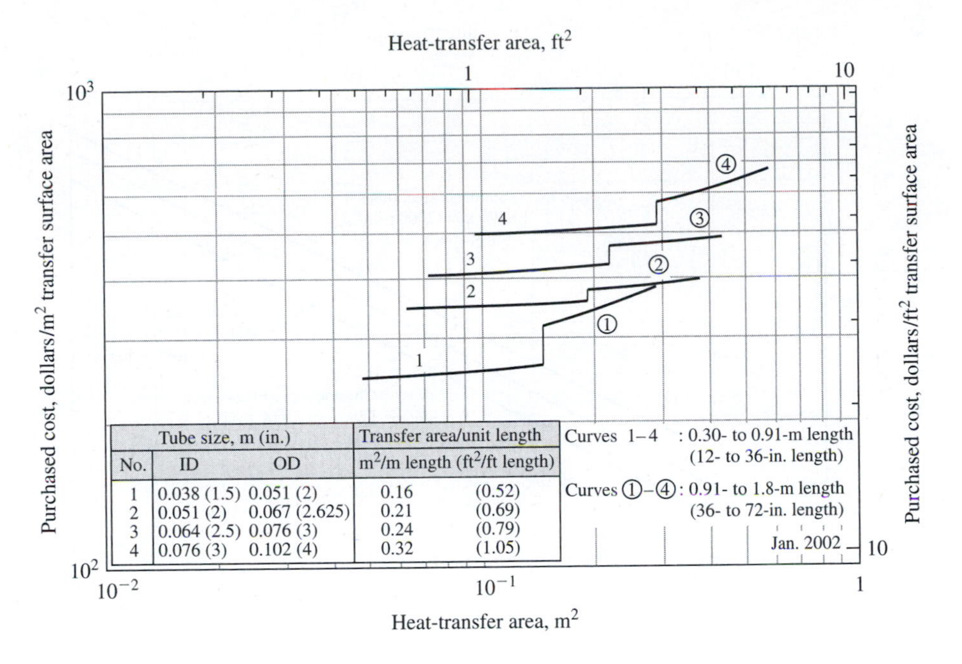

Tube size, m (in.)			Transfer area/unit length	
No.	ID	OD	m²/m length	(ft²/ft length)
1	0.038 (1.5)	0.051 (2)	0.16	(0.52)
2	0.051 (2)	0.067 (2.625)	0.21	(0.69)
3	0.064 (2.5)	0.076 (3)	0.24	(0.79)
4	0.076 (3)	0.102 (4)	0.32	(1.05)

Figure 14-35
Purchased cost of bayonet heaters

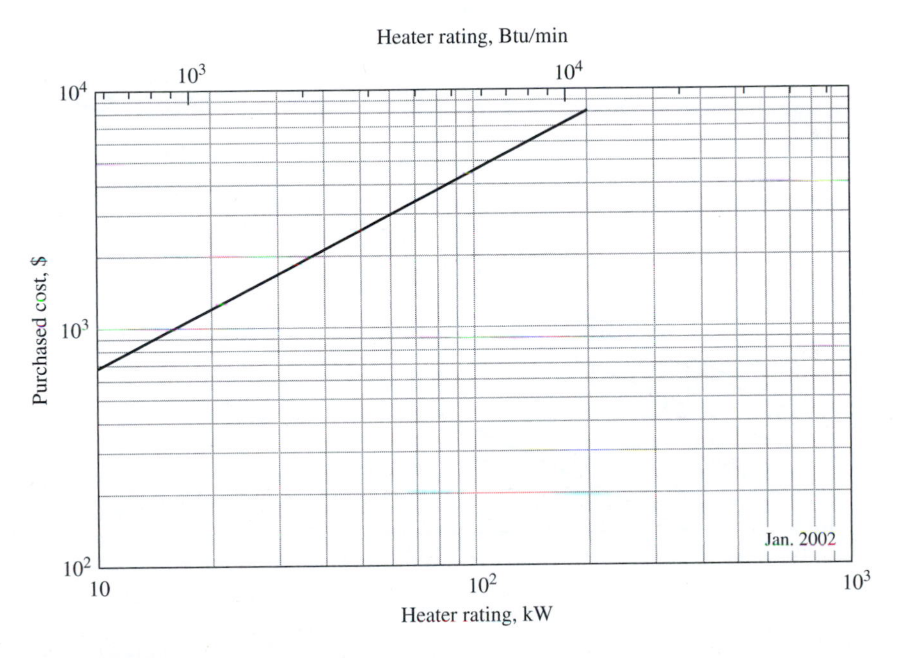

Figure 14-36
Purchased cost of electric immersion heaters

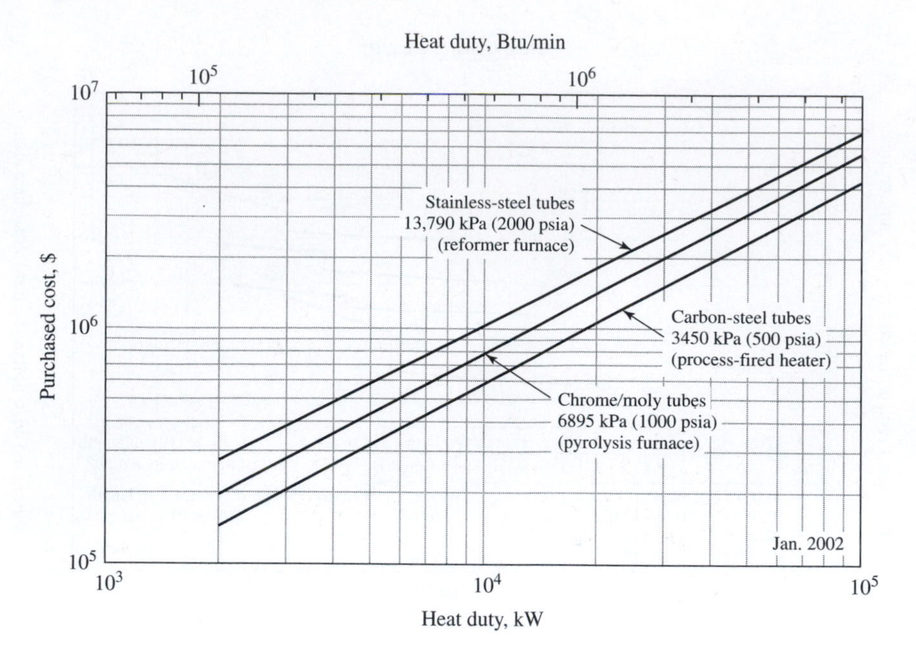

Figure 14-37
Purchased cost of process furnaces, box type with horizontal radiant tubes

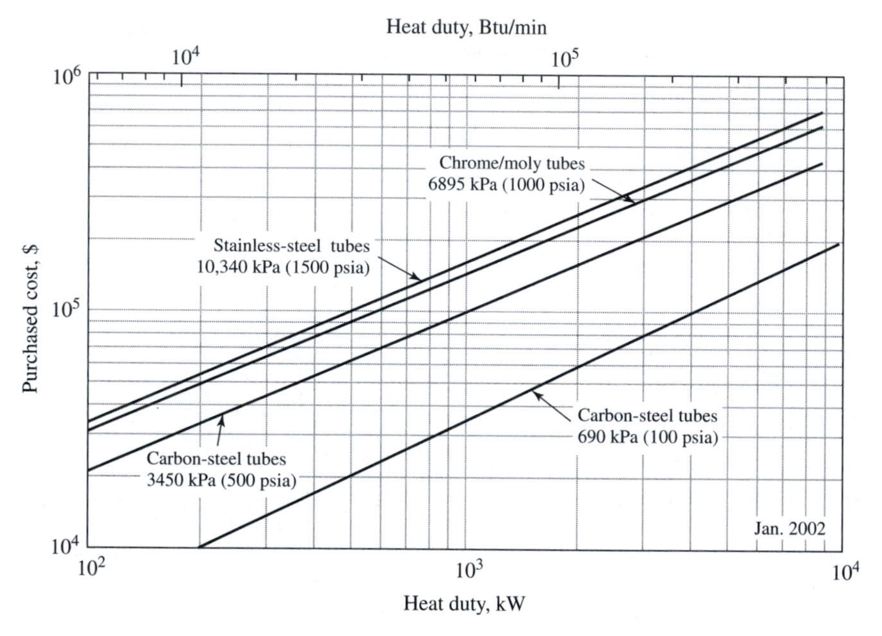

Figure 14-38
Purchased cost of direct-fired heaters, cylindrical type with vertical tubes

Even though many of the cost data presented have been obtained from a large number of vendors, these data must be regarded as approximate because of the many possible variations in heat-transfer equipment. Accordingly, the cost data should only be used for preliminary estimates when making economic comparisons.

Preliminary Selection of an Appropriate Type of Heat Exchanger — EXAMPLE 14-4

A heat exchanger is required for cooling a medium organic, as defined in Table 14-5, with a flow rate of 1.0 kg/s and a pressure of 4 MPa. The organic, with a heat capacity of 2 kJ/kg·K, enters the exchanger at 60°C and leaves at 30°C. Water, with a heat capacity of 4.2 kJ/kg·K, serves as the coolant and enters the exchanger at 20°C with a flow rate of 2.86 kg/s. Evaluate the alternative heat exchangers listed in Table 14-6, and make a preliminary selection based on exchanger suitability and economics.

■ Solution

Review the heat exchanger types and select those that could meet the stated requirements. Determine the heat requirements, exit water temperature, and the log mean temperature difference. With an average overall heat-transfer coefficient from Table 14-5, evaluate the required area for each type of heat exchanger and compare the purchased costs, using the cost charts.

Table 14-6 indicates that only the double-pipe and shell-and-tube exchangers are suitable candidates for further evaluation. The pressure is too high for the gasketed plate, welded plate, and spiral plate exchangers. Compact heat exchangers can be eliminated because of pressure limitations.

The heat load is obtained from

$$\dot{q} = (\dot{m}C_p)_h(T_{h,\text{in}} - T_{h,\text{out}})$$
$$= 1(2)(60 - 30) = 60 \text{ kW}$$

while the water outlet temperature is given by

$$T_{c,\text{out}} = T_{c,\text{in}} + \frac{\dot{q}}{(\dot{m}C_p)_c}$$
$$= 20 + \frac{60}{2.86(4.2)} = 25^\circ\text{C}$$

Use Eq. (14-7) to determine $\Delta T_{o,\text{log mean}}$:

$$\Delta T_{o,\text{log mean}} = \frac{\Delta T_1 - \Delta T_2}{\ln(\Delta T_1/\Delta T_2)} = \frac{(T_{h,\text{in}} - T_{c,\text{out}}) - (T_{h,\text{out}} - T_{c,\text{in}})}{\ln[(T_{h,\text{in}} - T_{c,\text{out}})/(T_{h,\text{out}} - T_{c,\text{in}})]}$$
$$= \frac{(60 - 25) - (30 - 20)}{\ln[(60 - 25)/(30 - 20)]} = 19.95^\circ\text{C}$$

From Table 14-5 the overall heat-transfer coefficient involving a medium organic and water is between 375 and 750 W/m^2·K. Assume an average value of 562 W/m^2·K.

The surface required by the double-pipe exchanger is then

$$A = \frac{\dot{q}}{U_{\text{avg}}\,\Delta T_{o,\text{log mean}}} = \frac{60 \times 10^3}{562(19.95)} = 5.35 \text{ m}^2$$

For the shell-and-tube exchanger we will select a one-shell and two-tube pass design. This will require a correction to the log mean temperature difference as presented in Eq. (14-8). This requires calculation of parameters R and P given in Eqs. (14-9a) and (14-9b).

$$R = \frac{T_{h,\text{in}} - T_{h,\text{out}}}{T_{c,\text{out}} - T_{c,\text{in}}} = \frac{60 - 30}{25 - 20} = 6$$

$$P = \frac{T_{c,\text{out}} - T_{c,\text{in}}}{T_{h,\text{in}} - T_{c,\text{in}}} = \frac{25 - 20}{60 - 20} = 0.125$$

Figure 14-4 provides a correction factor of 0.93. Thus, $\Delta T_{o,m}$ is

$$\Delta T_{o,\text{mean}} = F\,\Delta T_{o,\text{log mean}} = 0.93(19.95) = 18.55^\circ\text{C}$$

The required area for the shell-and-tube exchanger, assuming the same average overall heat-transfer coefficient, is then

$$A = \frac{\dot{q}}{U_{\text{avg}}\,\Delta T_{o,m}} = \frac{60 \times 10^3}{562(18.55)} = 5.75 \text{ m}^2$$

The purchased costs of these two heat exchangers, assuming carbon-steel shell and tubes, are obtained from Figs. 14-15 and 14-17. (Since the high-pressure fluid will flow inside the tube, a pressure correction must be made to the cost of the shell-and-tube exchanger.)

Cost of double-pipe exchanger = \$1150
Cost of shell-and-tube exchanger = 1.07(2750) = \$2940

The double-pipe heat exchanger is probably the best choice for this small high-pressure heat-transfer application since the modular construction of this type of exchanger provides a significant cost advantage.

DESIGN OF KEY HEAT EXCHANGER TYPES

In the previous section different types of heat exchangers were described, and their individual advantages were highlighted. This section seeks to outline techniques from which satisfactory designs can be made for these different types. The basis of such designs is the capability of establishing the thermohydraulic performance or *rating* of the heat-transfer equipment. By successive iterations, the specification for the type of heat exchanger selected is determined to meet the given process requirements. In the past this iterative procedure has been carried out by hand, but nowadays such a procedure is predominantly performed with commercial software or custom programs and algorithms. These are often proprietary codes developed by the design industry, large processing companies, and international research organizations such as HTRI (Heat Transfer Research, Inc.) or HTFS (Heat Transfer and Fluid Flow Service).

Since there are many alternative designs that would meet a specific heat duty, it is necessary to optimize the design. The process by which the design engineer moves from one design to a more satisfactory one and its relationship to the total design process are shown in Fig. 14-39. The critical feature of the design process is the design modification step. If it is done by hand, the design modification is largely intuitive and depends on the design engineer's experience and insight. However, if the design process is carried out with a fully integrated computer program, the success of the process is heavily dependent on the skill and care with which the design logic has been constructed. Rapid

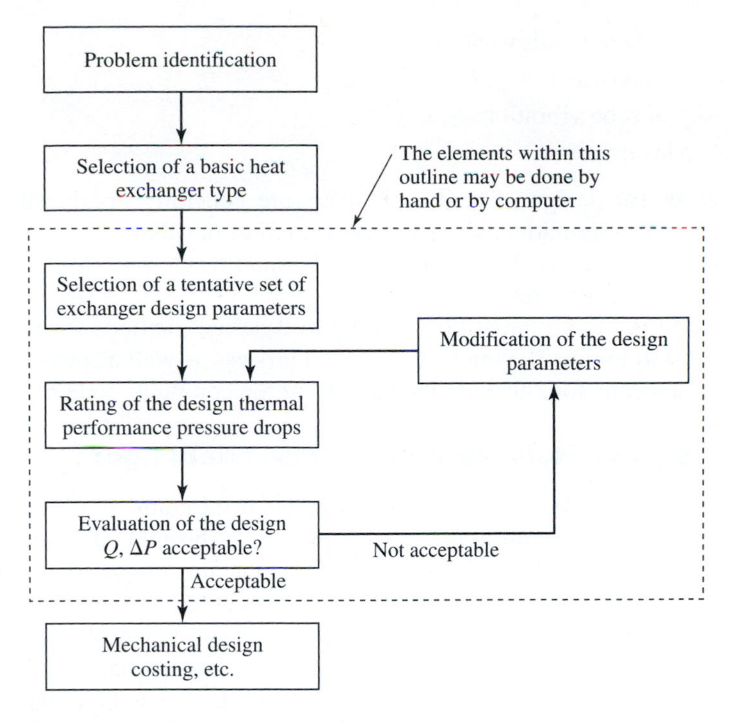

Figure 14-39
Structure for the process of heat exchanger design logic

convergence to an acceptable design is less important than being certain that the logical structure does not eliminate possible design options by imposing a poorly framed decision point. Even a modest-size design program may have 40 separate logical decisions that need to be made, leading to 2^{40}, or 1.1×10^{12}, different paths through the logic. Since it is impossible to check all these possibilities, great care and conservatism must be exercised in spelling out these decision points.

For a typical design, the following parameters and constraints are usually given:

Fluids used and their properties
Inlet and exit fluid temperatures
Fluid flow rates
Operating pressure
Allowable pressure drop
Fouling resistances

For these given specifications, a design procedure will select designs that satisfy Eq. (14-5). In this design process, the exchanger configuration will also need to be evaluated against other specified constraints, such as

Maximum and minimum fluid velocities
Maximum and minimum temperatures

Necessary corrosion allowances
Materials of construction
Propensity of tube vibrations
Special codes involved

Computer codes for design of heat exchangers are generally organized to vary the design parameters systematically to identify configurations that satisfy the constraints listed above. As with many design situations, the final result is usually a compromise between all the competing requirements.

In this section, the principal goal will be to describe methods for thermal rating that can be used to initiate the interactive design process as well as provide the design engineer with a means for checking the validity of the computer code being used.

Double-Pipe and Multiple Double-Pipe Exchangers

The double-pipe heat exchanger is generally operated in the countercurrent mode. When both fluids are single-phase throughout the exchanger, the heat transfer can be evaluated by using Eqs. (14-4) through (14-7) and recognizing that these equations assume a constant heat capacity and constant overall heat-transfer coefficient. Equations (14-17) and (14-18) permit evaluation of the individual heat-transfer coefficients for laminar and turbulent flow, respectively. Note that for the annulus an equivalent diameter D_e is required, as given by Eq. (14-20). Tube and shell-side pressure drops can be obtained by using modified forms of Eqs. (14-23) and (14-24), respectively.

When the inlet and outlet temperatures and fluid flow rates are known, estimation of the required surface area of the exchanger is quite straightforward since the heat-transfer rate can be obtained directly from Eq. (14-6) and the overall heat-transfer coefficient and log mean temperature difference calculated from Eqs. (14-4) and (14-7), respectively. However, if the area is known while the fluid outlet temperatures are unknown, the basic equations will need to be rearranged to provide a noniterative solution using the temperature ratio parameters R and P as defined in Eqs. (14-9*a*) and (14-9*b*), respectively. Such rearrangement leads to

$$\frac{\dot{q}}{UA} = \Delta T_{o,\text{log mean}} = \frac{(T_{h,\text{in}} - T_{c,\text{in}})P(R-1)}{\ln[(1-P)/(1-PR)]} \tag{14-25}$$

$$\frac{(\dot{m}C_P)_c}{UA} = \frac{R-1}{\ln[(1-P)/(1-PR)]} \tag{14-25a}$$

and

$$\frac{(\dot{m}C_p)_h}{UA} = \frac{(R-1)/R}{\ln[(1-P)/(1-PR)]} \tag{14-25b}$$

When U, A, R, and either $\dot{m}_c$ or $\dot{m}_h$ are known, Eq. (14-25*a*) or (14-25*b*) can be used to determine P, which in turn permits evaluation of $T_{c,\text{out}}$ and finally $T_{h,\text{out}}$.

When the pressure drop available for moving the fluids is limited or the length of the double-tube exchanger is excessive, a variety of different multiple-tube arrangements are feasible. For example, when the pressure drop available for a cold fluid is limited, a parallel/series arrangement can be selected. Here the hot fluid flows in series through the shell side of the two double-pipe heat exchangers while the cold fluid is divided equally into two streams for parallel flow in the two tube sections of the exchanger. This reduces the pressure drop on the tube side to approximately one-fourth of the original value at the expense of lowering the outlet temperature of the cold fluid. A new effective temperature difference term[†] for the parallel/series arrangement needs to be used that considers the asymmetry of heat transfer in the two units.

Enhanced shell-side heat transfer is often achieved by attaching longitudinal fins to the outer surface of the inner tube. The fins increase the surface area for heat transfer while reducing the cross-sectional flow area on the shell side, resulting in an increase in the fluid velocity. The net effect is an increase in both heat transfer and pressure drop.

Since a temperature gradient normally is present along the extended height of the fin, the heat-transfer effectiveness of the fin surface is less than that observed for the bare surface of the tube. Jakob[‡] has shown that the effectiveness η_f of the fin area for straight longitudinal and pin fins is given by

$$\eta_f = \frac{\tanh ML_f}{ML_f} \tag{14-26}$$

where $M = (h/k_f\delta)^{1/2}$, and h is the heat-transfer coefficient, k_f the thermal conductivity of the fin, δ the ratio of the volume of the fin to the surface area of the fin, that is, $\delta = V_f/A_f$, and L_f the height of the fin.

The effective heat transfer area $\eta_o A_o$ may then be defined as

$$\eta_o A_o = A_p + \eta_f A_f = A_o - A_f(1 - \eta_f) \tag{14-27}$$

where η_o is the overall surface effectiveness, A_o the total heat-transfer area, A_p the outside area of the inner pipe not covered with fins, and A_f the fin area. Solving for the overall surface effectiveness results in

$$\eta_o = 1 - \frac{A_f}{A_o}(1 - \eta_f) \tag{14-28}$$

The overall heat-transfer coefficient for the finned double-tube heat exchanger may be obtained by modifying Eq. (14-4) to include the correction for the effective area of the fins, as given by

$$\frac{1}{U_1} = \frac{A_1}{h_1(\eta_f A_f + A_p)} + \frac{D_1 \ln(D_1/D_2)}{2k_w} + \frac{D_1}{h_2 D_2} \tag{14-29}$$

where the subscripts 1 and 2 refer to the outside and inside areas and diameters of the tube, respectively, rather than the total area of the tube and fins. Example 14-5

[†]D. Q. Kern, *Process Heat Transfer,* McGraw-Hill, New York, 1986.

[‡]M. Jakob, *Heat Transfer,* J. Wiley, New York, 1949.

illustrates the procedure for estimating the length of a double-pipe exchanger required for a specified heat duty.

EXAMPLE 14-5 Determination of Surface Area and Length for a Double-Pipe Heat Exchanger

It is desired to preheat 4 kg/s of Dowtherm from 10 to 70°C with a hot water condensate that is to be cooled from 95 to 60°C. Two double-pipe heat exchangers, one without fins and one with fins on the outer surface of the inner tube, are being considered for this heat-transfer process. Determine the area and exchanger length anticipated for these two exchangers fabricated with copper tubes and fins. Dimensions are as follows:

Tube dimension	$D_1 = 0.0483$ m	$D_2 = 0.0408$ m
Shell dimension	$D_{os} = 0.0889$ m	$D_s = 0.0779$ m
Fin height	$L_f = 0.0127$ m	
Fin thickness	$t_f = 0.0027$ m	
Number of fins	$N_f = 24$ longitudinal fins	

To simplify the calculations, assume that each copper fin is attached individually to the outside surface of the inner copper tube, and thus each fin occupies 0.0027 m^2 of the bare surface tube area for each 1-m length of tube.

■ Solution

Determine the heat duty and the mass flow rate of the water. Obtain the individual heat-transfer coefficients for the two fluids and the overall heat-transfer coefficient. With this information calculate the area and length of the exchanger required. For the exchanger with fins, the latter calculation can be made only after the effectiveness of the fin area has been established.

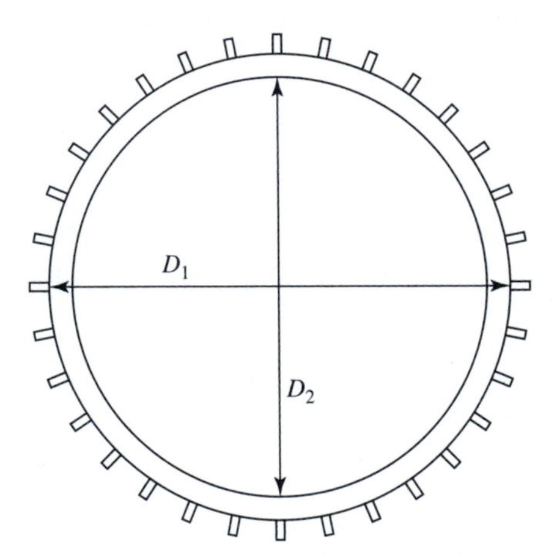

In this calculation, assume that the fluid properties at the average bulk temperature for each fluid are adequate. A more precise calculation would require using the mean temperature between the average fluid temperature and the surface temperature of the wall. Also, to further simplify the calculation, assume negligible resistance on the hot and cold surfaces due to fouling. Consider first the double-pipe exchanger without the finned surface. The fluid properties of each stream at the average temperature are obtained from *Perry's Chemical Engineers' Handbook.*

Property	Dowtherm at 40°C	Water at 77°C
Heat capacity C_p	1.622×10^3 J/kg·K	4.198×10^3 J/kg·K
Thermal conductivity k	0.138 J/s·m·K	0.668 J/s·m·K
Viscosity μ	2.70×10^{-3} Pa·s	3.72×10^{-4} Pa·s
Density ρ	1.044×10^3 kg/m³	9.74×10^2 kg/m³

The total rate of heat transfer is obtained with Eq. (14-6) using the flow rate of Dowtherm through the annulus.

$$\dot{q} = (\dot{m}C_p\,\Delta T)_c = 4(1.622 \times 10^3)(70 - 10) = 3.89 \times 10^5\ \text{W}$$

The mass flow rate of water, after rearranging Eq. (14-6), is

$$\dot{m}_h = \frac{\dot{q}}{(C_p\,\Delta T)_h} = \frac{3.89 \times 10^5}{(4.198 \times 10^3)(95 - 60)} = 2.65\ \text{kg/s}$$

Evaluate h_c and h_h, using Eq. (14-18), assuming the flow is turbulent.

$$h = 0.023\left(\frac{k}{D}\right)\text{Re}^{0.8}\text{Pr}^{1/3}$$

The Reynolds and Prandtl numbers for the Dowtherm in the annulus are obtained from

$$\text{Re}_c = \frac{D_e V_c \rho_c}{\mu_c} \qquad \text{Pr}_c = \left(\frac{C_p\mu}{k}\right)_c$$

where D_e, the mean hydraulic diameter for the annulus, is given by

$$D_e = D_s - D_1 = 0.0779 - 0.0483 = 0.0296\ \text{m}$$

Calculation of the average fluid velocity requires the cross-sectional area of the annulus, given by

$$S_c = \frac{\pi}{4}\left(D_s^2 - D_1^2\right) = \frac{\pi}{4}[(0.0779)^2 - (0.0483)^2] = 2.934 \times 10^{-3}\ \text{m}^2$$

Thus

$$V_c = \frac{\dot{m}_c}{S_c\rho_c} = \frac{4}{(2.934 \times 10^{-3})(1.044 \times 10^3)} = 1.306\ \text{m/s}$$

and

$$\text{Re}_c = \frac{0.0296(1.306)(1.044 \times 10^3)}{2.70 \times 10^{-3}} = 1.495 \times 10^4$$

$$\text{Pr}_c = \frac{(1.622 \times 10^3)(2.70 \times 10^{-3})}{0.138} = 31.73$$

Repeating the calculations for the Reynolds and Prandtl numbers for the water inside the tube gives

$$D_e = D_2 = 0.0408 \text{ m}$$

$$S_i = \frac{\pi}{4} D_2^2 = \frac{\pi}{4}(0.0408)^2 = 1.307 \times 10^{-3} \text{ m}^2$$

$$V_h = \frac{\dot{m}_h}{S_i \rho_h} = \frac{2.65}{(1.307 \times 10^{-3})(9.74 \times 10^2)} = 2.08 \text{ m/s}$$

$$\text{Re}_h = \frac{0.0408(2.08)(9.74 \times 10^2)}{3.72 \times 10^{-4}} = 2.22 \times 10^5$$

$$\text{Pr}_h = \frac{(4.198 \times 10^3)(3.72 \times 10^{-4})}{0.668} = 2.34$$

The individual heat-transfer coefficients can now be calculated as

$$h_c = (0.023)\left(\frac{0.138}{0.0296}\right)(1.495 \times 10^4)^{0.8}(31.73)^{1/3} = 7.42 \times 10^2 \text{ W/m}^2\cdot\text{K}$$

$$h_h = (0.023)\left(\frac{0.668}{0.0408}\right)(2.22 \times 10^5)^{0.8}(2.34)^{1/3} = 9.46 \times 10^3 \text{ W/m}^2\cdot\text{K}$$

Since most of the thermal resistance occurs in the cold fluid, the tube wall temperatures will be close to those of the hot fluid, namely, the water. The overall heat-transfer coefficient is calculated by using Eq. (14-4) in terms of diameters D_1 and D_2, except that the fouling resistances are neglected.

$$\frac{1}{U_c} = \frac{1}{h_c} + \frac{D_1}{2k_w} \ln \frac{D_1}{D_2} + \frac{D_1}{h_h D_2}$$

where $D_1/D_2 = 0.0483/0.0408 = 1.184$.

The thermal conductivity of copper at 77°C is 379 J/s·m·K. Substitution gives

$$\frac{1}{U_c} = \frac{1}{7.42 \times 10^2} + \frac{0.0483 \ln(1.184)}{2(379)} + \frac{1.184}{(9.46 \times 10^3)}$$

$$= 1.484 \times 10^{-3} \text{ m}^2\cdot\text{K/W}$$

$$U_c = 674 \text{ W/m}^2\cdot\text{K}$$

The log mean temperature difference is obtained by using Eq. (14-7):

$$\Delta T_{o,\text{log mean}} = \frac{(T_{h,\text{in}} - T_{c,\text{out}}) - (T_{h,\text{out}} - T_{c,\text{in}})}{\ln[(T_{h,\text{in}} - T_{c,\text{out}})/(T_{h,\text{out}} - T_{c,\text{in}})]}$$

$$= \frac{(95 - 70) - (60 - 10)}{\ln[(95 - 70)/(60 - 10)]} = 36.067 \text{ K}$$

The area of the exchanger based on the outside tube area can now be calculated from

$$A_c = \frac{\dot{q}}{U_c \, \Delta T_{o,\text{log mean}}} = \frac{3.89 \times 10^5}{674(36.067)} = 16.0 \text{ m}^2$$

The length of the double-pipe heat exchanger with no fins is then

$$L = \frac{A_c}{\pi D_1} = \frac{16.0}{\pi(0.0483)} = 105.4 \text{ m}$$

The only way to accommodate this length would be to use several smaller exchangers in series. It would be better either to adjust the tube parameters or to add fins to the outside surface of the inner tube. The latter choice is the one investigated below.

The addition of fins changes the cross-sectional area and fluid velocity in the annulus. The new cross-sectional area is given by

$$S_c = \frac{\pi}{4}\left(D_s^2 - D_1^2\right) - N_f(L_f t_f)$$

where N_f is the number of fins attached to the outside surface of the inner tube, L_f the length of the fin, and t_f the thickness of the fin.

$$S_c = \frac{\pi}{4}[(0.0779)^2 - (0.0483)^2] - 24(0.0127)(2.7 \times 10^{-3}) = 2.107 \times 10^{-3}\ \text{m}^2$$

The equivalent diameter D_e is defined as 4(cross-sectional area)/wetted perimeter P':

$$D_e = \frac{4S_c}{P'} = \frac{4(2.107 \times 10^{-3})}{\pi(0.0779) + \pi(0.0483) + 2(24)(0.0127)}$$
$$= 8.377 \times 10^{-3}\ \text{m}$$

The fluid velocity of the cold fluid is

$$V_c = \frac{\dot{m}_c}{S_c \rho_c} = \frac{4}{(2.107 \times 10^{-3})(1.044 \times 10^3)} = 1.82\ \text{m/s}$$

The new Reynolds number in the annulus is

$$\text{Re}_c = \frac{D_e V_c \rho_c}{\mu_c} = \frac{(8.377 \times 10^{-3})(1.82)(1.044 \times 10^3)}{2.70 \times 10^{-3}}$$
$$= 5895$$

The new heat-transfer coefficient in the annulus is then

$$h_c = (0.023)\left(\frac{0.138}{8.377 \times 10^{-3}}\right)(5895)^{0.8}(31.73)^{1/3} = 1245\ \text{W/m}^2\cdot\text{K}$$

Now determine the fin efficiency η_f from Eq. (14-26).

$$h_f = \frac{\tanh ML_f}{ML_f}$$

where $M = (h_c/k_f\delta)^{1/2}$

$$\delta = \frac{V_f}{A_f} \qquad \delta = \frac{L_f t_f L}{2L_f L + t_f L} \qquad \text{set } L = 1\ \text{m}$$

$$= \frac{0.0127(0.0027)}{2(0.0127) + 0.0027} = 1.22 \times 10^{-3}\ \text{m}$$

$$M = \left[\frac{1245}{379(1.22 \times 10^{-3})}\right]^{1/2} = 51.88$$

$$ML_f = 51.88(0.0127) = 0.6589$$

$$\eta_f = \frac{\tanh 0.6589}{0.6589} = 0.877$$

Since the heat-transfer coefficient for the warm fluid inside the tube has not changed, calculate U_c from Eq. (14-29)

$$\frac{1}{U_c} = \frac{A_1}{h_c(\eta_f A_f + A_p)} + \frac{D_1 \ln(D_1/D_2)}{2k_w} + \frac{D_1}{h_h D_2}$$

where

$$A_f = 24(2L_f + t_f)L = 24[2(0.0127) + 0.0027] = 0.6744 \text{ m}^2/\text{m}$$

$$A_p = (\pi D_1 - 24t_f)L = \pi(0.0483) - 24(0.0027) = 0.0869 \text{ m}^2/\text{m}$$

Now calculate the new overall heat-transfer coefficient.

$$\frac{1}{U_c} = \frac{\pi(0.0483)}{1245[0.877(0.6744) + 0.0869]} + \frac{0.0483 \ln(1.184)}{2(379)} + \frac{1.184}{9.46 \times 10^3}$$

$$= 3.156 \times 10^{-4} \text{ m}^2\cdot\text{K/W}$$

$$U_c = 3169 \text{ W/m}^2\cdot\text{K}$$

The area of the exchanger based on the outside tube area is then

$$A_c = \frac{\dot{q}}{U_c \, \Delta T_{o,\text{log mean}}} = \frac{3.89 \times 10^5}{3169(36.067)} = 3.40 \text{ m}^2$$

and the tube length is

$$L = \frac{A_c}{\pi D_1} = \frac{3.40}{\pi(0.0483)} = 22.4 \text{ m}$$

Note that the addition of the 24 copper fins reduced the required length of the exchanger by a factor of 4.7. However, it is still too long to be accommodated in one exchanger and will require several exchangers in series.

The calculations have been simplified by neglecting possible fouling effects. Inclusion of these effects will increase the lengths of both exchangers since the overall heat-transfer coefficient will decrease with time due to surface fouling. Note that the fluid velocity of the water in both exchangers is high, and this will contribute to considerable pressure loss in the tubes. This deficiency can be corrected by selecting a slightly larger tube diameter.

■

Shell-and-Tube Exchangers

Extensive studies over several decades have been made to simulate the heat transfer and pressure drop that occur in a shell-and-tube heat exchanger. Evaluation of these parameters for the fluid flowing inside the tube is relatively simple, and several approaches have been described earlier in this chapter. However, because of the complex flow conditions that are encountered on the shell side of this widely used heat exchanger, careful analysis must be made with respect to tube layout and spacing of the tube bundle,

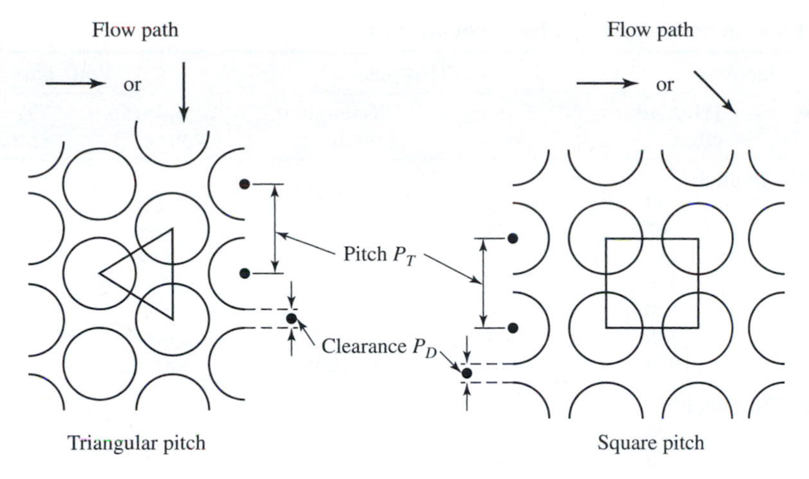

Figure 14-40
Conventional tube-plate layouts in shell-and-tube heat exchangers

baffle selection, fluid flow across the tube bundle, fluid leakage across various shell components, ineffective heat-transfer areas, and effects of various fluid conditions.

With respect to tube layout in the tube bundle, the tubes are commonly laid out on a square pattern or on a triangular pattern, as shown in Fig. 14-40. (Either pattern may also be rotated to meet specific exchanger requirements.) The shortest center-to-center distance between adjacent tubes in either arrangement is defined as the *tube pitch* P_T, while the shortest distance between any two tubes is designated as the *tube clearance* P_D. In most shell-and-tube exchangers, the tube pitch for economic reasons is maintained between 1.25 and 1.50 times the tube diameter. SAE recommends that the tube clearance not be less than one-fourth of the tube diameter with a minimum clearance of no less than 0.0048 m.

Although a square pitch pattern has the advantage of easier external cleaning, the triangular pitch pattern is often preferred because it permits the use of more tubes and hence results in more surface area in a given shell diameter. Table 14-9 provides the number of tubes that can be placed within an exchanger with conventional tube sizes and tube pitches.

The standard length of tubes in shell-and-tube heat exchangers is normally 2.44, 3.66, or 4.88 m. These tubes are available in a variety of different diameters and wall thicknesses. Exchangers with small-diameter tubes are less expensive per square meter of heat-transfer surface than those with large-diameter tubes because a given surface area may be arranged in a tube bundle requiring a smaller shell diameter; however, the small-diameter tubes are more difficult to clean. A tube diameter of 0.019- or 0.0254-m OD is the most common, but outside diameters up to 0.038 m are used in many industrial installations. Tube wall thickness is usually specified by the Birmingham wire gauge, and variations from the normal thickness may be ± 10 percent for “average wall” tubes and $+22$ percent for “minimum wall” tubes. Normal pipe sizes apply to the shell for diameters up to 0.710 m. In general, a shell thickness of 0.0095 m is used for

Table 14-9 Number of tubes in conventional tube sheet layouts

	One-pass		Two-pass		Four-pass	
Shell ID, m	Square pitch	Triangular pitch	Square pitch	Triangular pitch	Square pitch	Triangular pitch
0.019-m-OD tubes on 0.0254-m pitch						
0.203	32	37	26	30	20	24
0.305	81	92	76	82	68	76
0.387	137	151	124	138	116	122
0.540	277	316	270	302	246	278
0.635	413	470	394	452	370	422
0.787	657	745	640	728	600	678
0.940	934	1074	914	1044	886	1012
0.0254-m-OD tubes on 0.0317-m pitch						
0.203	21	21	16	16	14	16
0.305	48	55	45	52	40	48
0.387	81	91	76	86	68	80
0.540	177	199	166	188	158	170
0.635	260	294	252	282	238	256
0.787	406	472	398	454	380	430
0.940	596	674	574	664	562	632

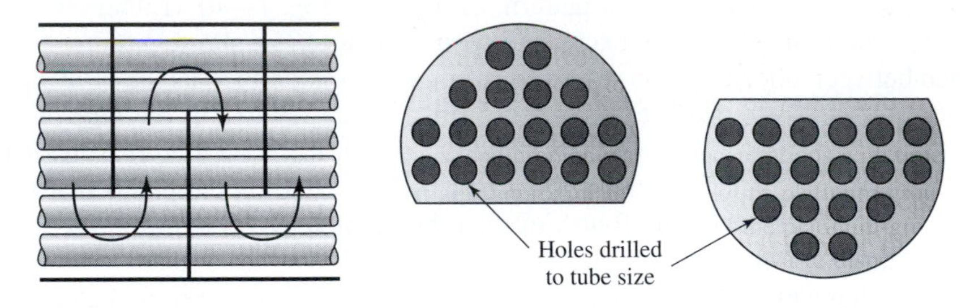

Figure 14-41
Segmented baffles

shell diameters between 0.305 and 0.710 m unless the shell fluids are extremely corrosive or the operating pressure on the shell side exceeds 2200 kPa.

The function of the baffle in a shell-and-tube heat exchanger is to direct the flow across the tube bundle as well as to support the tubes from sagging and possible vibration. Note that the presence of baffles in the shell side of such exchangers increases the pressure drop on the shell side; however, the advantage of better mixing of the fluid with increased turbulence more than offsets the pressure drop disadvantage. In general, baffles should not be spaced closer than one-fifth of the inner shell diameter or farther apart than that specified by the TEMA code for the tube material used.

The most common type of baffle used in shell-and-tube heat exchangers is the segmental baffle, illustrated in Fig. 14-41. Many segmental baffles have a baffle height that is 75 percent of the inside diameter of the shell. This arrangement is designated as

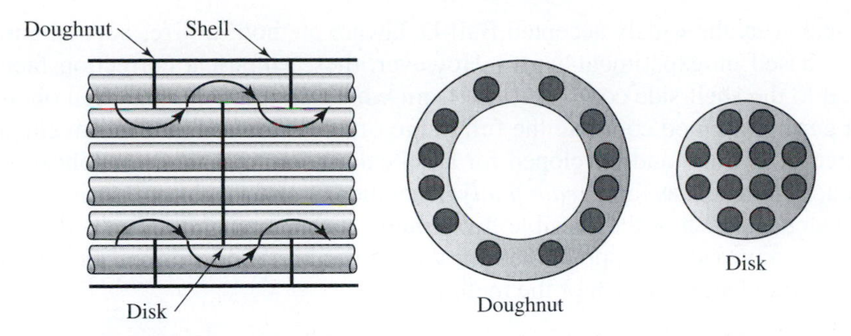

Figure 14-42
Disk and doughnut baffles

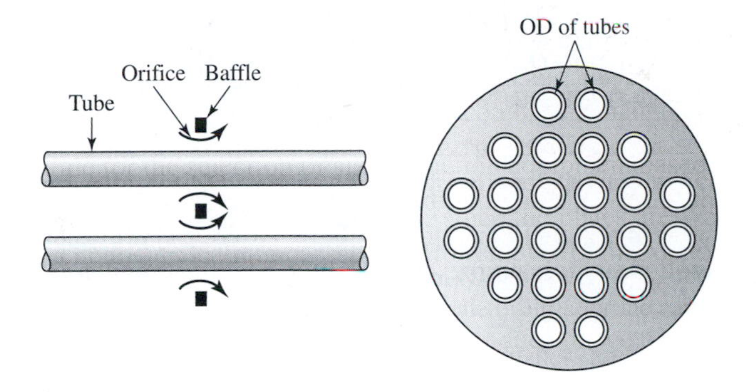

Figure 14-43
Orifice baffles

a segmental baffle with a 25 percent cut. The ratio of baffle spacing to baffle cut is a critical design parameter for determining a realistic pressure drop on the shell side of the exchanger.

Other types of baffles include the disk-and-doughnut baffle and the orifice baffle, shown in Figs. 14-42 and 14-43. Use of the disk-and-doughnut baffle can often result in a 50 to 60 percent reduction in the pressure drop through the shell. Other baffle types providing reduced pressure drops include the triple segmental and the "no tubes in the window" arrangement.

The thickness of the baffles should be at least twice the thickness of the tube walls and is generally in the range of 0.003 to 0.006 m. With respect to tube sheets, 0.022 m is ordinarily considered as a minimum thickness in industrial exchangers.

There are three recognized methods for determining heat-transfer and pressure drop information on the shell side of these exchangers. Design engineers will note that the Kern method[†] is the simplest approach and uses equations analogous to the equations used for flow in tubes. This method, however, does not adequately account for fluid leakages occurring within the shell side of the exchanger. To account for such

[†]D. Q. Kern, *Process Heat Transfer,* McGraw-Hill, New York, 1986.

fluid leakages, the widely accepted Bell-Delaware method[†] utilizes several correction factors based on experimental data. However, these empirical correction factors are limited to the shell-side configurations from which the database has been obtained. A more generic method covering the full range of possible shell-side arrangements was initiated by Tinker[‡] and developed for hand calculation by Wills and Johnston.[§] This approach is known as the *stream analysis* method in which separate fluid streams are designated for each of the possible flow routes of the fluid in the shell side of the exchanger. To provide an appreciation for these three methods, this section summarizes the procedural steps in each of the methods.

Kern Method The correlation for the shell-side heat-transfer coefficient developed by Kern, and based solely on the operation of an industrial exchanger with a baffle cut of 25 percent, is given by

$$h_s = 0.36\frac{k}{D_e}\left(\frac{D_e G_s}{\mu}\right)^{0.55}\left(\frac{C_p\mu}{k}\right)^{0.33}\left(\frac{\mu}{\mu_w}\right)^{0.14} \tag{14-30}$$

where G_s is the shell-side mass velocity obtained from

$$G_s = \frac{\dot{m}_T}{S_m} \tag{14-31}$$

Here $\dot{m}_T$ is the total mass flow on the shell side, and S_m the cross-flow area measured along the centerline of flow in the shell. The value for S_m is obtained from

$$S_m = \frac{D_s P_D L_B}{P_T} \tag{14-32}$$

where D_s is the inside diameter of the shell, P_T the tube pitch, P_D the tube clearance or distance between adjacent tubes, and L_B the distance between adjacent baffles. The equivalent diameter D_e for the square pitch arrangement is defined as

$$D_e = \frac{4\left(P_T^2 - 0.25\pi D_o^2\right)}{\pi D_o} \tag{14-33}$$

where D_o is the outside diameter of the tubes. For the triangular pitch arrangement D_e is defined as

$$D_e = \frac{4\left(0.86 P_T^2 - 0.25\pi D_o^2\right)}{\pi D_o} \tag{14-34}$$

[†]K. J. Bell, "Cooperative Research Program on Shell and Tube Heat Exchangers," *University of Delaware Eng. Exp. Station Bulletin 5* (June 1963).

[‡]T. Tinker, *Proceeding of General Discussion on Heat Transfer,* Institute of Mechanical Engineers and American Society for Mechanical Engineers, New York, 1951.

[§]M. J. N. Wills and D. Johnston, *22nd National Heat Transfer Conference, HTD,* vol. 36, American Society for Mechanical Engineers, New York, 1984.

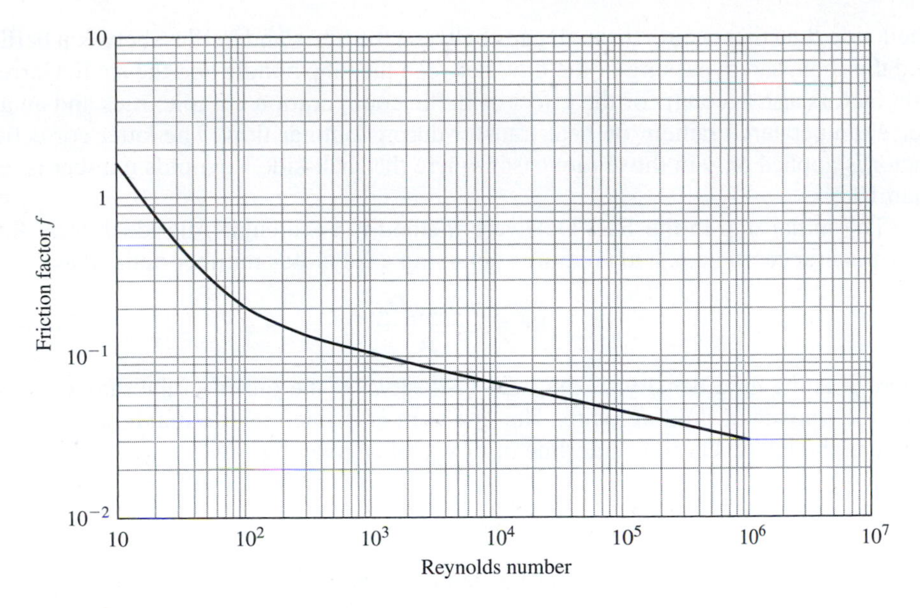

Figure 14-44
Plot of friction factor f as a function of shell-side Reynolds number

The pressure loss Δp_s observed on the shell side of the heat exchanger, assuming no fluid leakage, is given by the simple relation

$$\Delta p_s = \frac{4 f G_s^2 D_s (N_B + 1)}{2 \rho D_e (\mu / \mu_w)_s^{0.14}} \tag{14-35}$$

where the friction factor f is obtained from Fig. 14-44 as a function of the shell-side Reynolds number and N_B is the number of baffles specified for the heat exchanger. The $N_B + 1$ value accounts for the fluid passes across the tube bundle. The number of baffles in the shell is given by

$$N_B = \frac{L_s}{L_B + t_b} - 1 \tag{14-36}$$

where L_s is the length of the exchanger, L_B the spacing between baffles, and t_b the thickness of the baffle.

The maximum number of tubes located across the centerline of the tube bundle is determined from

$$N_T = \frac{D_s - \Delta_b}{P_T} = \frac{D_{OTL}}{P_T} \tag{14-37}$$

where Δ_b is the bundle-to-shell diametral clearance recommended by TEMA and $D_s - \Delta_b$ is the tube bundle diameter, designated as D_{OTL}.

Bell-Delaware Method This method applies various correction factors to the "ideal" cross-flow heat-transfer and pressure drop evaluations to account for the leakage of the shell-side fluid that occurs through gaps between the tubes and baffles, between baffles and the shell, and bypassing of the flow between the tube bundle and the shell. Correction factors also account for the effect of baffle configuration complexities and an adverse temperature gradient on heat-transfer during laminar flow. The latter correction factor is applied only in those rare cases where the shell-side Reynolds number is less than 100.

The initial step in the Bell-Delaware method is to calculate the ideal cross-flow heat-transfer coefficient. This requires obtaining a Reynolds number defined as

$$\mathrm{Re} = \frac{\rho V_{\max} D_o}{\mu} \tag{14-38}$$

where ρ is the fluid density, μ the fluid viscosity, D_o the outside tube diameter, and $V_{\max}$ the maximum fluid velocity. The last term is defined as the maximum velocity between the tubes near the centerline of flow and is given by

$$V_{\max} = \frac{\dot{m}_T}{\rho S_m} \tag{14-39}$$

where S_m is the cross-flow area near the centerline and is defined for square and triangular pitch arrangements as

$$S_m = L_B\left[D_s - D_{OTL} + \frac{(D_{OTL} - D_o)(P_T - D_o)}{P_T}\right] \tag{14-40}$$

Definitions of S_m for rotated square pitch and rotated triangular pitch arrangements are given in the Bell reference mentioned earlier.

The actual shell-side heat-transfer coefficient is obtained from

$$h_s = h_i J_C J_L J_B \tag{14-41}$$

where h_i is the ideal cross-flow heat-transfer coefficient calculated from Eq. (14-22) by using the appropriate values for the constant and exponents provided in Tables 14-1 and 14-2. The correction factors J_C, J_L, and J_B are evaluated with relationships developed by Bell as described below.

The correction factor J_C that accounts for baffle configuration is a function of the fraction of tubes assumed to be in cross-flow. This fraction is calculated from the relation

$$F_c = \frac{1}{\pi}\left[\pi + \frac{2(D_s - 2L_c)}{D_{OTL}}\sin\left(\cos^{-1}\frac{D_s - 2L_c}{D_{OTL}}\right) - 2\cos^{-1}\frac{D_s - 2L_c}{D_{OTL}}\right] \tag{14-42}$$

where L_c is the baffle cut distance obtained from the product $B_c D_s/100$. Here B_c is the percentage baffle cut, normally between 15 and 45 percent of the inside shell diameter. Figure 14-45 is used to obtain J_C.

The correction factor J_L is related to the shell-to-baffle and tube-to-baffle leakage areas S_{sb} and S_{tb}, respectively. These two leakage areas are evaluated from

$$S_{sb} = D_s\left(\frac{\Delta_{sb}}{2}\right)\left[\pi - \cos^{-1}\left(1 - \frac{2L_c}{D_s}\right)\right] \tag{14-43a}$$

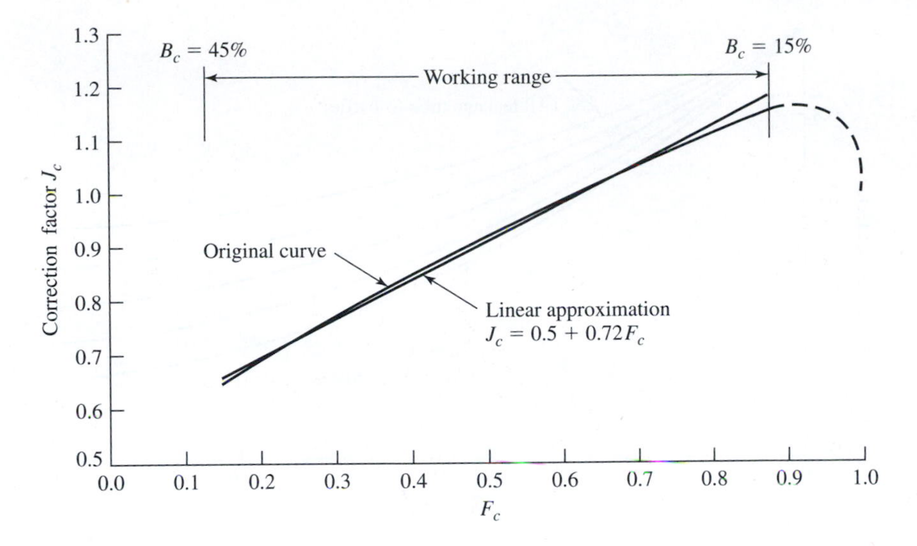

Figure 14-45
Correction factor for the influence of baffle configuration. (From R. H. Perry and D. W. Green, *Perry's Chemical Engineers' Handbook, 4th ed.,* McGraw-Hill, New York, 1963. Reproduced with permission of The McGraw-Hill Companies.)

and

$$S_{tb} = \pi D_o \left(\frac{\Delta_{tb}}{2} \right) N_T \frac{1 + F_c}{2} \tag{14-43b}$$

where Δ_{sb} and Δ_{tb} are the shell-to-baffle and tube-to baffle diametral clearances, respectively, and J_L is related to the ratio $(S_{sb} + S_{tb})/S_m$ in Fig. 14-46 with $S_{tb}/(S_{sb} + S_{tb})$ as the parameter. The cross-flow area S_m is obtained directly from Eq. (14-40).

The correction factor J_B, which accounts for the bypass in the bundle-to-shell gap, is expressed as a function of F_{bp}, the fraction of the cross-flow area available for bypass flow. Here F_{bp} is defined as

$$F_{bp} = \frac{L_B}{S_m}(D_s - D_{OTL}) \tag{14-44}$$

To minimize the bypass flow in the bundle-to-shell gap, pairs of sealing strips are added to the shell side. These affect the heat-transfer rate and need to be included when obtaining the correction factor J_B. Figure 14-47 shows J_B as a function of the ratio S_b/S_m with N_{ss}/N_c as the parameter. Here N_{ss} is the number of pairs of sealing strips, and N_c is the number of cross rows as estimated from

$$N_c = \frac{D_s}{P_{TP}}\left(1 - \frac{2L_c}{D_s}\right) \tag{14-45}$$

where P_{TP} is the distance between tube rows in the direction of flow and is equal to P_T for square tube arrangements and $0.866P_T$ for triangular tube arrangements.

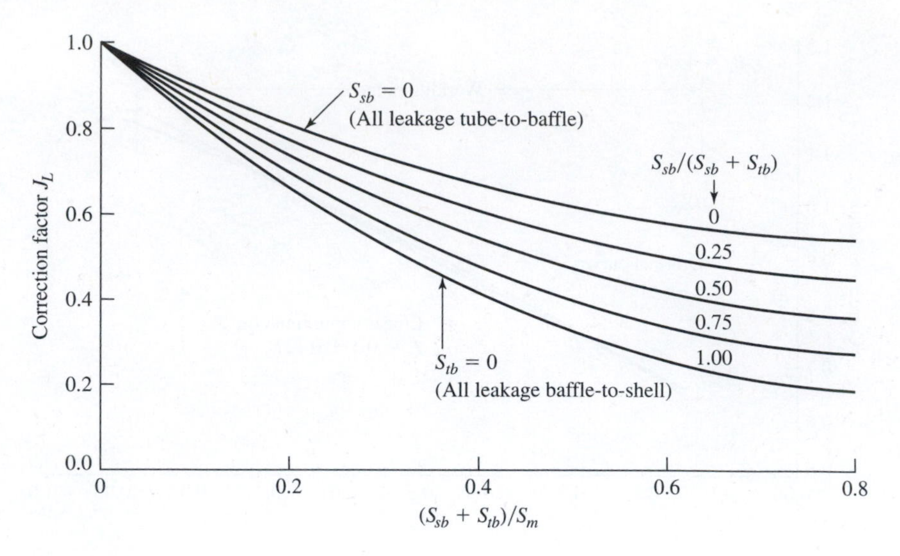

Figure 14-46
Correction factor for the effect of tube-to-baffle and baffle-to-shell leakage for calculating heat-transfer coefficients in shell-and-tube heat exchangers. (From R. H. Perry and D. W. Green, *Perry's Chemical Engineers' Handbook, 7th ed.,* McGraw-Hill, 1999. Reproduced with permission of The McGraw-Hill Companies.)

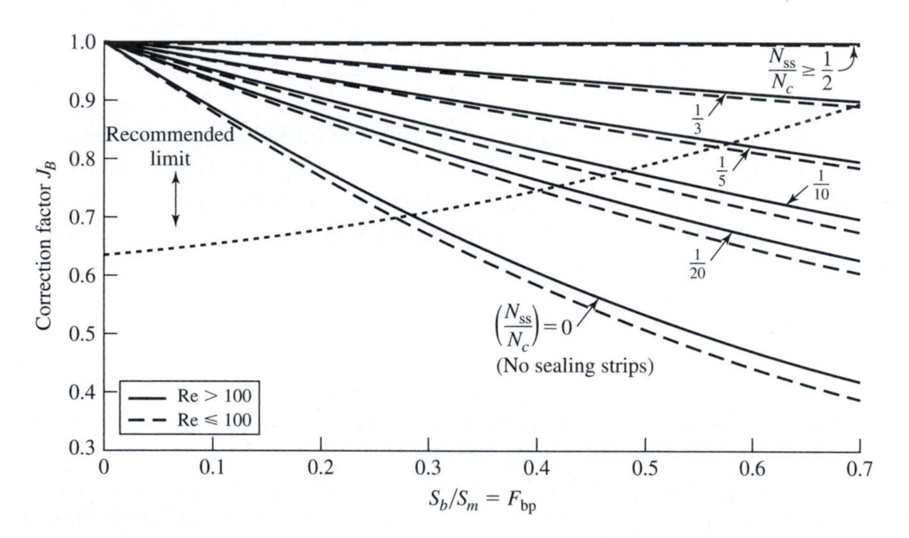

Figure 14-47
Correction factor for the effect of bypass on the heat-transfer coefficient in shell-and-tube heat exchangers. N_{ss} is the number of pairs of sealing strips and N_c is the number of cross-flow rows. (From R. H. Perry and D. W. Green, *Perry's Chemical Engineers' Handbook, 7th ed.,* McGraw-Hill, 1999. Reproduced with permission of The McGraw-Hill Companies.)

Calculations for the shell-side pressure drop utilizing the Bell-Delaware method are similar to those made for the heat-transfer calculations. First the pressure drop for an ideal cross-flow is determined, and then correction factors are applied to account for baffle leakage and bypass. Separate calculations are made for the cross-flow and window zones, thereby eliminating the configuration correction.

The ideal cross-flow pressure drop is calculated from

$$\Delta p_c = (K_a + N_c K_f)\left(\frac{\rho V_{max}^2}{2}\right) \tag{14-46}$$

where the constant K_a is related to the pressure losses of the fluid at the inlet and exit of the tube bundle and is generally approximated with a value of 1.5. (See Table 12-1 for additional details.) The constant K_f accounts for the friction and momentum losses as the fluid passes between each successive row of tubes in the bundle. Values for this constant are given in Table 14-10 as a function of the Reynolds number for the shell-side fluid. The values for V_{max} and N_c are obtained from Eqs. (14-39) and (14-45), respectively.

The pressure drop for the "ideal" window zone between baffles is given by Bell as

$$\Delta p_w = \frac{26\dot{m}_T \mu}{\rho(S_m S_w)^{1/2}}\left(\frac{N_{cw}}{P_T - D_o} + \frac{L_c}{D_w^2}\right) + \frac{\dot{m}_T^2}{S_m S_w \rho} \qquad \text{for Re} \le 100 \tag{14-47a}$$

$$\Delta p_w = \frac{(2 + 0.6N_{cw})\dot{m}_T^2}{2S_m S_w \rho} \qquad \text{for Re} > 100 \tag{14-47b}$$

Table 14-10 Values of K_f for square and triangular tube configurations

	K_f (Square pitch)			K_f (Triangular pitch)		
Re	**1.25**	**1.50**	**2.0**	**1.25**	**1.50**	**2.0**
10^1	21.7	9.4	4.0	29.0	12.0	6.0
10^2	2.35	1.2	0.65	3.20	1.80	1.25
10^3	0.48	0.38	0.24	0.99	0.70	0.60
10^4	0.42	0.34	0.24	0.51	0.43	0.38
10^5	0.29	0.25	0.18	0.29	0.22	0.20
10^6	0.27	0.25	0.18	0.23	0.18	0.17

The 1.25, 1.50, and 2.0 values in Table 14-10 are the pitch-to-diameter ratios for the specific tube arrangement. Interpolation relationships for a tube pitch of 1.25 have been developed by Zukauskas and Ulinskas as follows:

For in-line square arrays

$$K_f = 0.267 + \frac{0.249 \times 10^4}{\text{Re}} - \frac{0.927 \times 10^7}{\text{Re}^2} + \frac{0.100 \times 10^{11}}{\text{Re}^3} \qquad \text{for } 2 \times 10^3 < \text{Re} < 2 \times 10^6$$

For staggered arrays

$$K_f = 0.245 + \frac{0.339 \times 10^4}{\text{Re}} - \frac{0.984 \times 10^7}{\text{Re}^2} + \frac{0.133 \times 10^{11}}{\text{Re}^3} - \frac{0.599 \times 10^{13}}{\text{Re}^4} \qquad \text{for } 10^3 < \text{Re} < 10^6$$

A. Zukauskas and R. Ulinskas, in *Heat Exchanger Design Handbook,* Hemisphere Publishing, Washington, 1983, Sec. 2.2.4.

Here N_{cw} is the number of effective cross-flow rows in the window zone, defined as

$$N_{cw} = \frac{0.8L_c}{P_{TP}} \tag{14-48}$$

The window flow area S_w is defined as

$$S_w = \frac{D_s^2}{4}\left[\cos^{-1} D_B - D_B\left(1 - D_B^2\right)^{1/2}\right] - \frac{N_T}{8}(1 - F_c)\pi D_o^2 \tag{14-49}$$

where D_B is defined as $(D_s - 2L_c)/D_s$ in radians and N_T is the total number of tubes in the tube bundle. The equivalent diameter in the window zone D_w is given by

$$D_w = \frac{4S_w}{4[(\pi/2)N_T(1 - F_c)D_o + 2\cos^{-1} D_B]} \tag{14-50}$$

The pressure drop across the shell side is then given as

$$\Delta p_s = [(N_B - 1)\,\Delta p_c\, R_B + N_B\,\Delta p_w]R_L + 2\Delta p_c R_B\left(1 + \frac{N_{cw}}{N_c}\right) \tag{14-51}$$

where the correction factors R_B and R_L are obtained from Figs. 14-48 and 14-49, respectively, and N_B is the number of baffles required as evaluated from Eq. (14-36).

Wills and Johnston Method The stream analysis method has formed the basis for many modern computer programs. A simplified version of this method can serve as a means of verifying the results obtained with various commercial computer programs.

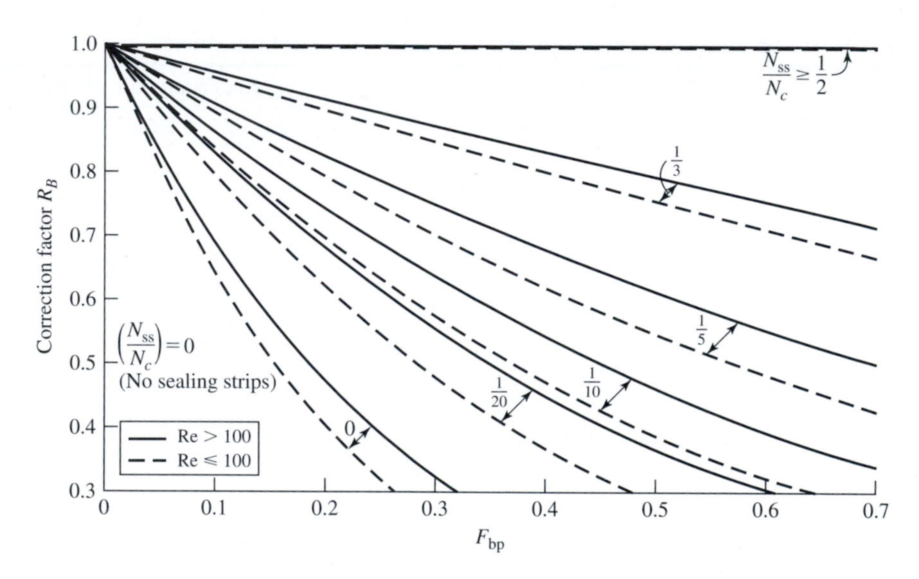

Figure 14-48

Correction factor R_B for the influence of bypass on pressure drop. (From R. H. Perry and D. W. Green, *Perry's Chemical Engineers' Handbook, 7th ed.*, McGraw-Hill, 1999. Reproduced with permission of The McGraw-Hill Companies.)

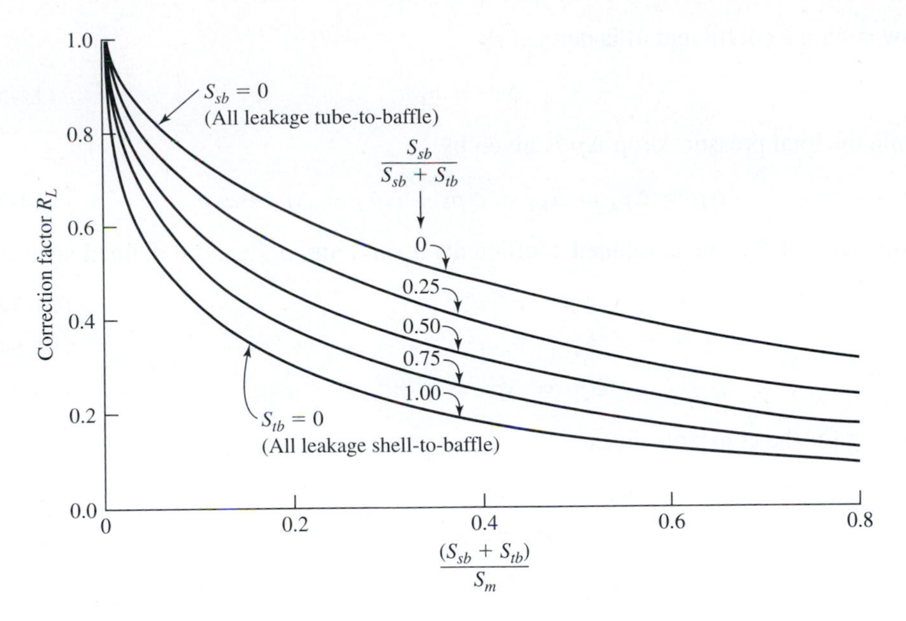

Figure 14-49
Correction factor R_L for the influence of tube-to-baffle and shell-to-baffle leakage on the pressure drop. (From R. H. Perry and D. W. Green, *Perry's Chemical Engineers' Handbook, 7th ed.,* McGraw-Hill, 1999. Reproduced with permission of The McGraw-Hill Companies.)

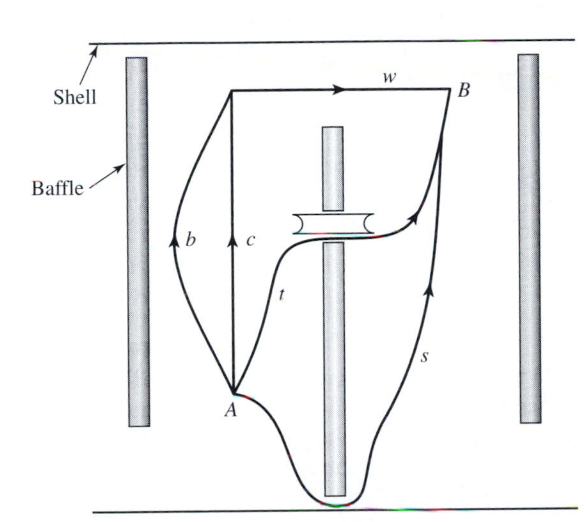

Figure 14-50
Flow streams in the Wills and Johnston method. (From "Flow Streams in the Wills and Johnston Method," M. J. Wills and D. Johnston in 22nd National Heat Transfer Conference, *Heat Transfer Division,* vol. 36, ASME, New York, 1984. Reprinted by permission of the American Society of Mechanical Engineers.)

Figure 14-50 illustrates the flow paths considered when the fluid in the shell flows from A to B. These include leakages between the tubes and the baffle in path t and between the baffle and shell in path s. Some of the flow passes over the tubes in cross-flow in path c, and part of it bypasses the bundle in path b. The cross-flow and bypass streams combine and form stream w that flows through the window zone. For each flow stream a coefficient n_i is defined as

$$\Delta p_i = n_i \dot{m}_i^2 \tag{14-52}$$

while the total pressure drop Δp is given by

$$\Delta p = \Delta p_s = \Delta p_t = \Delta p_b + \Delta p_w = \Delta p_c + \Delta p_w \tag{14-53}$$

From Eq. (14-52) the combined coefficients n_a, n_p, and n_{cb} can be defined such that

$$\Delta p = n_a \dot{m}_w^2 \tag{14-54a}$$

$$\Delta p = n_p \dot{m}_T^2 \tag{14-54b}$$

$$\Delta p_c = \Delta p_b = n_{cb} \dot{m}_w^2 \tag{14-54c}$$

Solving for the constants gives

$$n_a = n_w + n_{cb} \tag{14-55a}$$

$$n_p = \left(n_a^{-1/2} + n_s^{-1/2} + n_t^{-1/2}\right)^{-2} \tag{14-55b}$$

$$n_{cb} = \left(n_c^{-1/2} + n_b^{-1/2}\right)^{-2} \tag{14-55c}$$

For simplification, Wills and Johnston assumed that n_s, n_t, n_w, and n_b were constant, independent of flow rate, and depended only on the internal geometry of the exchanger. The cross-flow stream coefficient n_c is the only coefficient allowed to vary as a function of the cross-flow Reynolds number and is determined from the nominal cross-flow pressure loss correlations. Because n_c is not constant, an iterative solution is required.

To initiate the shell-side calculations for a specified exchanger, the fixed values of n_s, n_t, n_w, and n_b are calculated from the relationships given below. The shell-to-baffle leakage resistance coefficient is given by

$$n_s = \frac{0.036(2t_b/\Delta_{sb}) + 2.3(2t_b/\Delta_{sb})^{-0.177}}{2\rho S_s^2} \tag{14-56}$$

Where S_s is the shell-to-baffle leakage area obtained from

$$S_s = \pi \left(D_s - \frac{\Delta_{sb}}{2}\right)\left(\frac{\Delta_{sb}}{2}\right) \tag{14-57}$$

Here Δ_{sb} is the shell-to-baffle diametral clearance while t_b is the baffle thickness. The tube-to-baffle clearance resistance coefficient n_t is given by

$$n_t = \frac{0.036(2t_b/\Delta_{tb}) + 2.3(2t_b/\Delta_{tb})^{-0.177}}{2\rho S_t^2} \tag{14-58}$$

where Δ_{tb} is the tube-to-baffle diametral clearance and S_t is the tube-to-baffle leakage area, given by

$$S_t = N_T \pi \left(D_o + \frac{\Delta_{tb}}{2}\right)\left(\frac{\Delta_{tb}}{2}\right) \tag{14-59}$$

The window flow resistance coefficient n_w is obtained from

$$n_w = \frac{1.9e^{0.6856 S_w/S_m}}{2\rho S_w^2} \tag{14-60}$$

where S_m is the cross-flow area as defined in Eq. (14-40) for square and triangular pitch arrays and S_w is the window flow area as given by Eq. (14-49). The bypass flow resistance coefficient n_b is evaluated from

$$n_b = \frac{a(D_s - 2L_c)/P_{TP} + N_{ss}}{2\rho S_b^2} \tag{14-61}$$

where D_s is the internal diameter of the shell, L_c the baffle cut length, P_{TP} the spacing between tube rows in the direction of flow, N_{ss} the number of sealing strips, and S_b the bypass flow area, defined as

$$S_b = (\Delta_b + \Delta_{pp})L_B \tag{14-62}$$

where Δ_b is the tube bundle-to-shell diametral clearance and Δ_{pp} the clearance encountered with an in-line pass partition. Any pass partition within the shell can give rise to an additional bypass area when these are parallel with the flow. The constant a in Eq. (14-61) is 0.266 for the square tube configuration and 0.133 for the triangular, rotated triangular and rotated square tube configurations.

As noted above, the cross-flow resistance n_c varies with the flow rate. Rather than use the method proposed by Wills and Johnston, a simpler method has been proposed by Hewitt et al.[†] The cross-flow pressure drop based on Eq. (14-52) is related to n_c by

$$\Delta p_c = n_c \dot{m}_c^2 \tag{14-63}$$

It is also related to the maximum velocity between the tubes near the centerline of flow by Eq. (14-46). When one replaces V_{max} with $\dot{m}_c/\rho S_m$, where S_m is the flow area near the centerline, solution of the previous two equations leads to the following relation for n_c:

$$n_c = \frac{K_a + N_c K_f}{2\rho S_m^2} \tag{14-64}$$

The number of cross rows N_c is estimated from Eq. (14-45), and the parameter K_f is obtained from Table 14-10 as a function of the Reynolds number for square or triangular tube configurations. Since the Reynolds number, given by $D_o \dot{m}_c / S_m \mu$, is also related to $\dot{m}_c$, it is initially necessary to estimate its value from $\dot{m}_c = F_{cr} \dot{m}_T$, where F_{cr} is the fraction of flow that is in cross-flow over the tube bundle. (An initial estimate of 0.5 for F_{cr} is reasonable.) With this information, n_c, n_a, and n_p can be determined. To verify the estimate of F_{cr}, a new value of F_{cr} is calculated from the relationship

$$F_{cr} = \frac{(n_p/n_a)^{1/2}}{1 + (n_c/n_b)^{1/2}} \tag{14-65}$$

[†]G. F. Hewitt, G. L. Shires, and T. R. Bott, *Process Heat Transfer,* CRC Press, Boca Raton, FL, 1994.

With this value for F_{cr}, the calculations for n_c, n_{cb}, n_a, and n_p are repeated to provide a new value for F_{cr}. The procedure is repeated until a convergence value for F_{cr} is obtained. The pressure loss per baffle space can now be calculated using the converged value of n_p given in Eq. (14-55*b*).

The flow fractions for each flow stream considered in this analysis are calculated from the following relationships:

$$F_s = \left(\frac{n_p}{n_s}\right)^{1/2} \tag{14-66}$$

$$F_t = \left(\frac{n_p}{n_t}\right)^{1/2} \tag{14-67}$$

$$F_b = \frac{(n_p/n_a)^{1/2}}{1 + (n_b/n_c)^{1/2}} \tag{14-68}$$

$$F_w = F_b + F_{cr} \tag{14-69}$$

Note that the sum of the flow fractions should equal unity, namely,

$$F_{cr} + F_s + F_t + F_b = 1 \tag{14-70}$$

The pressure drop per baffle space is given by Eq. (14-54*b*), and the total shell-side pressure drop is then

$$\Delta p_s = (N_B + 1)\Delta p \tag{14-71}$$

where Δp is the pressure drop per baffle. For exchangers with less than 10 baffles, Wills and Johnston developed a method to include the additional pressure drop effect for the end baffles. Since the effect is small for most large shell-and-tube heat exchangers, this correction procedure is not presented.

Estimation of the shell-side heat-transfer coefficient generally is performed by assuming that the coefficient corresponds to the value obtained at the centerline using $\dot{m}_c$ rather than $\dot{m}_T$ in Eq. (14-22).

EXAMPLE 14-6 Estimation of Heat-Transfer Coefficient and Pressure Drop on the Shell Side of a Shell-and-Tube Exchanger Using the Kern, Bell-Delaware, and Wills-Johnston Methods

A shell-and-tube exchanger with one shell and one tube pass is being used as a cooler. The cooling medium is water with a flow rate of 11 kg/s on the shell side of the exchanger. With an inside diameter of 0.584 m, the shell is packed with a total of 384 tubes in a staggered (triangular) array. The outside diameter of the tubes is 0.019 m with a clearance between tubes of 0.00635 m. Segmental baffles with a 25 percent baffle cut are used on the shell side, and the baffle spacing is set at 0.1524 m. The length of the exchanger is 3.66 m. (Assume a split backing ring, floating heat exchanger.)

The average temperature of the water is 30°C, and the average temperature of the tube walls on the water side is 40°C. Under these conditions, estimate the heat-transfer coefficient for the water and the pressure drop on the shell side, using the Kern, Bell-Delaware, and Wills and Johnston methods.

■ Solution

The procedures for all three methods have been outlined briefly in the shell-and-tube section. Appendix D provides the following data for water:

	30° C	35° C	40° C
Physical property data			
Thermal conductivity k, kJ/s·m·K	0.000616	0.000623	0.000632
Heat capacity C_p, kJ/kg·K	4.179	4.179	4.179
Viscosity μ, Pa·s	0.000803	0.000724	0.000657
Density ρ, kg/m^3	995	995	995
Exchanger configuration			
Shell internal diameter		$D_s = 0.584$ m	
Tube outside diameter		$D_o = 0.019$ m	
Tube pitch (triangular)		$P_T = 0.0254$ m	
Number of tubes		$N_T = 384$	
Baffle spacing		$L_B = 0.1524$ m	
Shell length		$L_s = 3.66$ m	
Bundle-to-shell diametral clearance†		$\Delta_b = 0.035$ m	
Shell-to-baffle diametral clearance†		$\Delta_{sb} = 0.005$ m	
Tube-to-baffle diametral clearance†		$\Delta_{tb} = 0.0008$ m	
Thickness of baffle†		$t_b = 0.005$ m	
Sealing strips per cross-flow row†		$N_{ss}/N_c = 0.2$	

†Items consistent with recommendations by J. Taborek, in *Heat Exchanger Design Handbook,* Hemisphere Publishing, Washington, 1983, Sec. 3.3.5.

Kern Method

Determine the flow area at the shell centerline. The gap between tubes P_D is given as 0.00635 m. The cross-flow area along the centerline of flow in the shell is given by Eq. (14-32).

$$S_s = \frac{D_s P_D L_B}{P_T} = \frac{0.584(0.00635)(0.1524)}{0.0254} = 0.02225 \text{ m}^2$$

Determine D_e from Eq. (14-33).

$$D_e = \frac{4\left(P_T^2 - \pi D_o^2/4\right)}{\pi D_o} = \frac{4[(0.0254)^2 - (\pi/4)(0.019)^2]}{\pi(0.019)} = 0.02423 \text{ m}$$

The mass flow rate G_s is

$$G_s = \frac{\dot{m}_T}{S_s} = \frac{11}{0.02225} = 494.4 \text{ kg/m}^2\text{·s}$$

To obtain the heat-transfer coefficient at an average water-film temperature requires evaluation of the Reynolds and Prandtl numbers.

$$\text{Re} = \frac{D_e G_s}{\mu_f} = \frac{0.02423(494.4)}{0.000724} = 16{,}550$$

$$\text{Pr} = \left(\frac{C_p \mu}{k}\right)_f = \frac{4.179(0.000724)}{0.000623} = 4.86$$

From Eq. (14-30)

$$h_s = 0.36\left(\frac{k}{D_e}\right)\text{Re}^{0.55}\text{Pr}^{0.33}\left(\frac{\mu}{\mu_w}\right)^{0.14}$$

$$= 0.36\left(\frac{0.623}{0.02423}\right)(16{,}550)^{0.55}(4.86)^{0.33}\left(\frac{0.000803}{0.000657}\right)^{0.14}$$

$$= 3369\ \text{W/m}^2\cdot\text{K}$$

Calculate the pressure drop on the shell side, assuming no effect for any type of fluid leakage. The number of baffles on the shell side is obtained from Eq. (14-36).

$$N_B = \frac{L_s}{L_B + t_b} - 1 = \frac{3.66}{0.1524 + 0.005} - 1 = 22.2 \text{ or } 22$$

For a shell-side Reynolds number of 16,550, Fig. 14-44 provides a value of 0.062 for the friction factor. The pressure drop is obtained from Eq. (14-35) as

$$\Delta p_s = \frac{4\mathfrak{f} G_s^2 D_s(N_B + 1)}{2\rho D_e(\mu/\mu_w)_s^{0.14}}$$

$$= \frac{4(0.062)(494.4)^2(0.584)(22+1)}{2(995)(0.02423)(0.000803/0.000657)^{0.14}} = 16{,}420\ \text{Pa}$$

Bell-Delaware Method

The first step in this method is to calculate the ideal cross-flow heat-transfer coefficient. Calculate $V_{\max}$ from Eq. (14-39) and obtain S_m from Eq. (14-40) to substitute into Eq. (14-22).

$$S_m = L_B\left[D_s - D_{OTL} + \frac{(D_{OTL} - D_o)(P_T - D_o)}{P_T}\right] \qquad \text{where } D_{OTL} = D_s - \Delta_b = 0.549$$

$$= 0.1524\left[0.035 + \frac{(0.549 - 0.019)(0.0254 - 0.019)}{0.0254}\right] = 0.0255\ \text{m}^2$$

$$V_{\max} = \frac{\dot{m}_T}{\rho S_m} = \frac{11}{995(0.0255)} = 0.4335\ \text{m/s}$$

$$\text{Re} = \frac{\rho V_{\max} D_o}{\mu} = \frac{995(0.4335)(0.019)}{0.000803} = 10{,}205$$

$$\text{Pr} = \frac{C_p \mu}{k} = \frac{4.179(0.000803)}{0.000616} = 5.449$$

The ideal heat-transfer coefficient is given by

$$h_i = \frac{k}{D_o} a\,\text{Re}^m\,\text{Pr}^{0.34} F_1 F_2$$

where constants a and m are obtained from Table 14-1 for a staggered tube array, F_1 from Eq. (14-22b), and F_2 from Table 14-2.

$$h_i = \left(\frac{0.616}{0.019}\right)(0.273)(10{,}205)^{0.635}(5.449)^{0.34}\left(\frac{5.449}{4.345}\right)^{0.26}(0.99)$$

$$= 5807\ \text{W/m}^2\cdot\text{K}$$

The actual shell-side heat-transfer coefficient is obtained from Eq. (14-41). This requires obtaining values for J_C, J_L, and J_B using the appropriate correction factors to account for baffle configuration, leakage, and bypass. Equation (14-42) permits calculation of F_c

$$F_c = \frac{1}{\pi}\left[\pi + \frac{2(D_s - 2L_c)}{D_{OTL}} \sin\left(\cos^{-1}\frac{D_s - 2L_c}{D_{OTL}}\right) - 2\cos^{-1}\frac{D_s - 2L_c}{D_{OTL}}\right]$$

For a baffle cut of 25 percent

$$L_c = 0.25D_s = 0.25(0.584) = 0.146 \text{ m}$$

$$\frac{D_s - 2L_c}{D_{OTL}} = \frac{0.584 - 2(0.146)}{0.549} = 0.5318$$

$$F_c = \frac{1}{\pi}[\pi + 2(0.5318)\sin(\cos^{-1} 0.5318) - 2\cos^{-1} 0.5318] = 0.6437$$

From Fig. 14-45

$$J_C = 0.55 + 0.72F_c = 0.55 + 0.72(0.6437) = 1.013$$

To obtain J_L, calculate the leakage areas S_{sb} and S_{tb} from Eqs. (14-43*a*) and (14-43*b*), respectively.

$$S_{sb} = D_s\left(\frac{\Delta_{sb}}{2}\right)\left[\pi - \cos^{-1}\left(1 - \frac{2L_c}{D_s}\right)\right]$$

$$= (0.584)\left(\frac{0.005}{2}\right)\left\{\pi - \cos^{-1}\left[1 - \frac{2(0.146)}{0.584}\right]\right\} = 0.003058 \text{ m}^2$$

$$S_{tb} = \pi D_o\left(\frac{\Delta_{tb}}{2}\right) N_T \frac{1 + F_c}{2}$$

$$= \pi(0.019)\left(\frac{0.0008}{2}\right)(384)\left(\frac{1 + 0.6437}{2}\right) = 0.007535 \text{ m}^2$$

The correction factor J_L is obtained from Fig. 14-46, utilizing S_{sb} and S_{tb}.

$$\frac{S_{sb} + S_{tb}}{S_m} = \frac{0.003058 + 0.007535}{0.0255} = 0.4154$$

$$\frac{S_{sb}}{S_{sb} + S_{tb}} = \frac{0.003058}{0.003058 + 0.007535} = 0.2887$$

Figure 14-46 provides a value of 0.56 for J_L.

To obtain the correction factor J_B for bypass in the bundle-shell gap, obtain F_{bp}, the fraction of the cross-flow area available for bypass flow, with Eq. (14-44).

$$F_{\text{bp}} = \frac{L_B}{S_m}(D_s - D_{OTL}) = \frac{0.1524}{0.0255}(0.035) = 0.2092$$

Note that $F_{\text{bp}} = S_b/S_m$, and Fig. 14-47 can be used to obtain a J_B value of 0.935 when $N_{\text{ss}}/N_c = 0.2$. The corrected heat-transfer coefficient from Eq. (14-41) is then

$$h = h_i J_c J_L J_B$$

$$= 5807(1.013)(0.56)(0.935) = 3080 \text{ W/m}^2\cdot\text{K}$$

Evaluation of the pressure drop using the Bell-Delaware method is similar to the process for obtaining the heat-transfer coefficient. The ideal cross-flow pressure drop through one baffle space is obtained with the use of Eq. (14-46).

$$\Delta p_c = (K_a + N_c K_f)\left(\frac{\rho V_{\max}^2}{2}\right) \qquad \text{assume } K_a = 1.5$$

$$N_c = \frac{D_s}{P_{TP}}\left(1 - \frac{2L_c}{D_s}\right) \qquad P_{TP} = 0.866 P_T, \text{ for triangular array}$$

$$= \frac{0.584}{0.866(0.0254)}\left[1 - \frac{2(0.146)}{0.584}\right] = 13.27$$

A value of 0.495 for K_f is obtained by using the following relation given in the footnote of Table 14-10:

$$K_f = 0.245 + \frac{0.339 \times 10^4}{\text{Re}} - \frac{0.984 \times 10^7}{\text{Re}^2} + \frac{0.133 \times 10^{11}}{\text{Re}^3} - \frac{0.599 \times 10^{13}}{\text{Re}^4}$$

$$\Delta p_c = [1.5 + 13.27(0.495)](995)\frac{(0.4335)^2}{2} = 754 \text{ Pa}$$

Calculate the window zone pressure loss from Eq. (14-47*b*). First, determine the window flow area S_w from Eq. (14-49).

$$S_w = \frac{D_s^2}{4}\left[\cos^{-1} D_B - D_B\left(1 - D_B^2\right)^{1/2}\right] - \frac{N_T}{8}(1 - F_c)\pi D_o^2$$

where

$$D_B = \frac{D_s - 2L_c}{D_s} = 1 - \frac{2L_c}{D_s} = 1 - \frac{2(0.146)}{0.584} = 0.5$$

$$S_w = \frac{(0.584)^2}{4}\{\cos^{-1} 0.5 - 0.5[1 - (0.5)^2]^{1/2}\} - \left(\frac{384}{8}\right)(1 - 0.6437)\pi(0.019)^2$$

$$= 0.03298 \text{ m}^2$$

Next, calculate the number of effective cross-flow rows in the window zone from Eq. (14-48).

$$N_{cw} = \frac{0.8 L_c}{P_{TP}} = \frac{0.8(0.146)}{0.866(0.0254)} = 5.31$$

Now calculate the window zone pressure drop for Re > 100.

$$\Delta p_w = \frac{(2 + 0.6 N_{cw})\dot{m}_T^2}{2 S_m S_w \rho}$$

$$= [2 + 0.6(5.31)]\frac{(11)^2}{2(0.0255)(0.03298)(995)} = 375 \text{ Pa}$$

Finally, estimate the leakage and bypass correction factors R_B and R_L. To obtain R_B, use the calculated values of F_{bp} and $N_{\text{ss}}/N_c = 0.2$ with Fig. 14-48. This gives a value of 0.82 for R_B. For R_L use the area ratio values of $(S_{sb} + S_{tb})/S_m$ and $S_{sb}/(S_{sb} + S_{tb})$ with Fig. 14-49 to obtain a value of 0.365 for R_L.

The pressure drop across the shell is given by Eq. (14-51).

$$\Delta p_s = [(N_B - 1)\,\Delta p_c R_B + N_B \Delta p_w]R_L + 2\,\Delta p_c R_B\left(1 + \frac{N_{cw}}{N_c}\right)$$

$$= [(22 - 1)(754)(0.82) + 22(375)](0.365) + 2(754)(0.82)\left(1 + \frac{5.31}{13.27}\right)$$

$$= 7750 + 1731 = 9481 \text{ Pa}$$

Wills and Johnston Method

The heat-transfer coefficient calculated in this method is similar to that used in the Bell-Delaware method except that the value of the Reynolds number is estimated from $\dot{m}_c = F_{cr}\dot{m}_T$. To determine F_{cr} requires evaluating the flow stream resistance coefficients in Fig. 14-50 as defined in Eqs. (14-55a) through (14-55c), (14-56), (14-58), (14-60), and (14-61).

Calculate the shell-to-baffle resistance coefficient n_s, using Eqs. (14-56) and (14-57).

$$S_s = \pi\left(D_s - \frac{\Delta_{sb}}{2}\right)\left(\frac{\Delta_{sb}}{2}\right) = \pi\left(0.584 - \frac{0.005}{2}\right)\left(\frac{0.005}{2}\right) = 0.004567 \text{ m}^2$$

$$n_s = \frac{0.036(2t_b/\Delta_{sb}) + 2.3(2t_b/\Delta_{sb})^{-0.177}}{2\rho S_s^2}$$

$$= \frac{0.036(2)(0.005)/0.005 + 2.3[2(0.005)/0.005]^{-0.177}}{2(995)(0.004567)^2}$$

$$= 50.75$$

Calculate the tube-to-baffle clearance resistance coefficient n_t from Eqs. (14-58) and (14-59).

$$S_t = N_T\pi\left(D_o + \frac{\Delta_{tb}}{2}\right)\left(\frac{\Delta_{tb}}{2}\right)$$

$$= 384\pi(0.019 + 0.0004)(0.0004) = 0.00936 \text{ m}^2$$

$$n_t = \frac{0.036(2t_b/\Delta_{tb}) + 2.3(2t_b/\Delta_{tb})^{-0.177}}{2\rho S_t^2}$$

$$= \frac{0.036(2)(0.005/0.0008) + 2.3[2(0.005/0.0008)]^{-0.177}}{2(995)(0.00936)^2}$$

$$= 11.02$$

Calculate the window flow resistance coefficient n_w from Eq. (14-60).

$$n_w = \frac{1.9e^{0.6856\,S_w/S_m}}{2\rho S_w^2}$$

where $S_m = 0.0255$ m^2 and $S_w = 0.03298$ m^2 from the Bell-Delaware calculations.

$$n_w = \frac{1.9\exp[0.6856(0.03298/0.0255)]}{2(995)(0.03298)^2} = 2.13$$

The bypass flow resistance coefficient n_b is calculated from Eqs. (14-61) and (14-62).

$$S_b = (\Delta_b + \Delta_{pp})L_B \qquad \text{assume } \Delta_{pp} \cong 0$$

$$= (0.035 + 0)(0.1524) = 0.00533 \text{ m}^2$$

$$N_{ss} = N_c \frac{N_{ss}}{N_c} = 13.27(0.2) = 2.65 \cong 3$$

$$n_b = \frac{a(D_s - 2L_c)/P_{TP} + N_{ss}}{2\rho S_b^2}$$

Since $N_c = (D_s/P_{TP})(1 - 2L_c/D_s)$, this can be rearranged and simplified to

$$n_b = \frac{aN_c + N_{ss}}{2\rho S_b^2} \qquad \text{where } a = 0.133 \text{ for triangular arrays}$$

$$= \frac{0.133(13.27) + 3}{2(995)(0.00533)^2} = 84.2$$

For a first approximation assume that the fraction F_{cr} of the flow that is in cross-flow over the bundle is 0.5 to initiate a calculation for the flow resistance coefficient n_c. For an F_{cr} of 0.5

$$\text{Re} = \frac{D_o \dot{m}_T F_{cr}}{S_m \mu}$$

$$= \frac{0.019(11)(0.5)}{0.0255(0.000803)} = 5103$$

The flow resistance coefficient n_c is evaluated by using Eq. (14-64) where K_f is obtained from the relation given in Table 14-10 for a triangular tube array with $10^3 < \text{Re} < 10^6$.

$$K_f = 0.245 + \frac{0.339 \times 10^4}{\text{Re}} - \frac{0.984 \times 10^7}{\text{Re}^2} + \frac{0.133 \times 10^{11}}{\text{Re}^3} - \frac{0.599 \times 10^{13}}{\text{Re}^4}$$

$$K_f(\text{Re} = 5103) = 0.6227$$

Calculate n_c, n_{cb}, n_a, and n_p to determine a new value for F_{cr}.

$$n_c = \frac{K_a + N_c K_f}{2\rho S_m^2} \qquad \text{assume } K_a = 1.5$$

$$= \frac{1.5 + 13.27(0.6227)}{2(995)(0.0255)^2} = 7.55$$

$$n_{cb} = \left(n_c^{-1/2} + n_b^{-1/2}\right)^{-2}$$

$$= (7.55^{-1/2} + 84.2^{-1/2})^{-2} = 4.47$$

$$n_a = n_w + n_{cb} = 2.13 + 4.47 = 6.60$$

$$n_p = \left(n_a^{-1/2} + n_s^{-1/2} + n_t^{-1/2}\right)^{-2}$$

$$= [(6.60)^{-1/2} + (50.75)^{-1/2} + (11.02)^{-1/2}]^{-2} = 1.47$$

Now calculate a new F_{cr} with Eq. (14-65).

$$F_{cr} = \frac{(n_p/n_a)^{1/2}}{1 + (n_c/n_b)^{1/2}}$$

$$= \frac{(1.47/6.60)^{1/2}}{1 + (7.55/84.2)^{1/2}} = 0.363$$

Repeat the above calculations beginning with the Reynolds number evaluation to determine a new value for F_{cr} until a convergence value for F_{cr} is obtained. The iteration results are shown below.

	Iteration attempts			
	1	**2**	**3**	**4**
F_{cr} (initial)	0.50	0.363	0.355	0.354
Re	5103	3705	3618	3614
K_f	0.6227	0.6729	0.676	0.676
n_c	7.55	8.06	8.09	8.09
n_{cb}	4.47	4.70	4.72	4.72
n_a	6.60	6.83	6.85	6.85
n_p	1.47	1.47	1.474	1.474
F_{cr} (calc.)	0.363	0.355	0.354	0.354

The iteration establishes F_{cr} at a value of 0.354 and fixes the Reynolds number for this calculation of the heat-transfer coefficient from Eq. (14-22) with constants a and m listed in Table 14-1, and F_1 and F_2 obtained from Eq. (14-22a) and Table 14-2, respectively.

$$h = \frac{k}{D_o} a \text{Re}^m \text{Pr}^{0.34} F_1 F_2$$

$$= \frac{0.616}{0.019}(0.273)(3614)^{0.635}(5.449)^{0.34}\left(\frac{5.449}{4.345}\right)^{0.26}(0.99)$$

$$= 3004 \text{ W/m}^2\cdot\text{K}$$

For the pressure drop calculation determine the various flow fractions.

Equation (14-66) for shell-to-baffle leakage flow:

$$F_s = \left(\frac{n_p}{n_t}\right)^{1/2} = \left(\frac{1.474}{50.75}\right)^{1/2} = 0.1704$$

Equation (14-67) for tube-to-baffle leakage:

$$F_t = \left(\frac{n_p}{n_t}\right)^{1/2} = \left(\frac{1.474}{11.02}\right)^{1/2} = 0.3657$$

Equation (14-68) for bypass flow:

$$F_b = \frac{(n_p/n_a)^{1/2}}{1 + (n_b/n_c)^{1/2}}$$

$$= \frac{(1.474/6.85)^{1/2}}{1 + (84.2/8.09)^{1/2}} = 0.1098$$

Check on the flow fractions that should equal unity.

$$F_s + F_t + F_b + F_{cr} \equiv 1.000$$

$$0.1704 + 0.3657 + 0.1098 + 0.3540 = 0.9999 \qquad \text{good check}$$

Calculate the total pressure drop per baffle on the shell side, using Eq. (14-54*b*).

$$\Delta p = n_p \dot{m}_T^2 = (1.474)(11)^2 = 178.4 \text{ Pa}$$

The total shell-side pressure drop is given by

$$\Delta p_s = (N+1)\Delta p = (22+1)(178.4) = 4103 \text{ Pa}$$

A comparison of the results for the shell-side heat-transfer coefficient and shell-side pressure drop from the three methods as well as from a widely used computer program is shown below:

Method	h, W/m²·K	Δp, Pa
Kern	3,369	16,420
Bell-Delaware	3,080	9,481
Wills-Johnston	3,004	4,103
Computer (CC-Therm)	3,035	4,155

Note that the Kern method provides higher values for the heat-transfer coefficient and pressure drop on the shell side. The Bell-Delaware and Wills-Johnston methods provide similar results for the heat-transfer coefficient.

Plate Exchangers

Gasketed and welded plate heat exchangers are used when a compact, flexible, liquid heat exchange system is required that encounters temperatures up to 250°C and pressures up to 2.5 MPa. Most plate heat exchangers are fitted with gasket seals that have to be compatible with the fluid being processed. Plate heat exchangers with welded joints may be used with toxic or highly inflammable liquids, but such units can experience difficult maintenance problems.

The fabricators of plate heat exchangers offer a wide range of plate corrugations designed to enhance heat transfer. The most popular types, described in detail by Cooper and Usher,[†] are the intermating and chevron configurations. The latter type generally provides the greater heat-transfer enhancement.

The total heat transferred between the two fluids passing through a plate heat exchanger is given by Eq. (14-5). Care must be taken in defining the projected area through which heat is being transferred as

$$A = N_p L W \tag{14-72}$$

where N_p is the number of plates and $N_p + 1$ the number of flow passages, L the plate height, and W the plate width. The overall heat-transfer coefficient obtained from

[†]A. Cooper and J. D. Usher, in *Heat Exchanger Design Handbook,* Hemisphere Publishing, New York, 1983, Sec. 3.7.

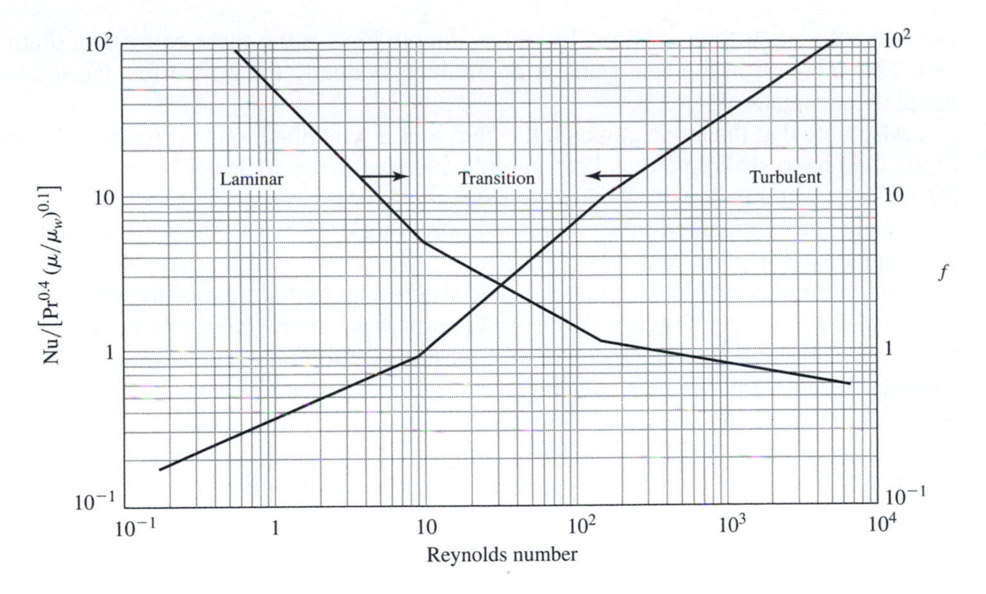

Figure 14-51
Performance characteristics of a small plate with chevron corrugations. (From A. Cooper and J. D. Usher, in *Heat Exchanger Design Handbook,* 1983. Reprinted by permission of Begell House Inc., Publishers.)

Eq. (14-4) must be based on this projected area of the exchanger. Figure 14-51 developed by Cooper and Usher provides curves for the heat-transfer coefficient and friction factor as a function of the Reynolds number for a small plate exchanger using the chevron configuration with an angle of 30° to the direction of flow in the exchanger. The dimensionless parameters are defined as

$$\text{Re} = \frac{D_e V \rho}{\mu} \tag{14-72a}$$

$$\text{Nu} = \frac{h D_e}{k} \tag{14-72b}$$

$$f = \frac{\Delta p}{4(L/D_e)(\rho V^2/2)} \tag{14-72c}$$

where V is the mean velocity of the fluid in the space between the plates, h the individual heat-transfer coefficient between the plate surface and the fluid, Δp the frictional pressure drop between the ends of the plate with a length L and an equivalent diameter D_e, defined as twice the axial spacing b observed between adjacent plate surfaces. This corresponds to 4 times the volume of the free space divided by the sum of the projected areas of the two adjacent plates. The fluid properties ρ, μ, and k are evaluated at the mean fluid temperature.

Note in Fig. 14-51 that the Reynolds number at which transition from laminar flow occurs is less than that for smooth pipes, and the transition region is considerably larger. Also, the heat-transfer coefficients in the transition and turbulent regions are considerably larger than for flow through a smooth pipe of the same equivalent diameter. This increase in the heat-transfer coefficients is due to the flow effect caused by the plate corrugations.

Also note that the angle at which the chevrons are inclined to the direction of flow of the fluid has a striking effect. For example, increasing the angle of the chevron from 30° to 75° essentially doubles the heat-transfer coefficient. This increase in heat-transfer coefficient is accompanied by an increase in pressure drop across the plate associated with an increase in pumping costs.

The thermal performance methods described in the first section of the chapter are also applicable to plate heat exchangers. One of these approaches was the effective mean temperature difference method with its F correction factor that is a function of parameters R, P, and the heat exchanger configuration. The other approach analyzed the heat exchanger in terms of two parameters, namely, the heat-transfer effectiveness E and the number of transfer units NTU. Graphs and tables for F and E as functions of R and NTU for various plate exchangers are available in the literature.[†]

One of the problems associated with plate and frame exchanger design is that of matching the heat-transfer requirement while making the best use of the available pressure drop. Note that the heat-transfer area A and the total cross-sectional flow area S for each fluid are not independent, as shown in Eq. (14-72*d*).

$$A = N_p L W = N_p b W \frac{L}{b} = 2S\frac{L}{b} \tag{14-72d}$$

where S is defined as $N_p bW/2$. Thus, the ratio of surface area to cross-sectional flow area can only be increased by an increase in the L/b ratio. Since spatial and construction constraints limit the variation in this ratio, other configurations must be used to obtain significant increases in both surface area and fluid velocity. To make the best use of the available pressure drop and to increase the heat-transfer coefficients, a multipass arrangement is often selected in which both streams are redirected several times through the heat exchanger in countercurrent fashion. A simple example is the two-pass/two-pass arrangement where the total surface area is the same as that of a single-pass plate and frame exchanger with the same number of plates, but the fluid velocities and fluid path lengths are doubled. The overall effect of such a change results in a large increase in the heat-transfer coefficient but at the expense of a significant increase in the pressure drop.

EXAMPLE 14-7 Estimation of the Number of Plates Required in a Plate and Frame Exchanger for a Specific Duty

In Example 14-5 a double-pipe exchanger was used to preheat 4 kg/s of Dowtherm from 10 to 70°C with a hot water condensate that was cooled from 95 to 60°C. For the same specific heat duty, determine the number of plates required for a single-pass counterflow plate and frame exchanger, as

[†]R. K. Shah and W. W. Foote, in *Heat Transfer Equipment Design,* R. K. Shah, E. C. Subarao, and R. A. Mashelkar, eds., Hemisphere Publishing, New York, 1988.

described previously in Fig. 14-12. Assume that each mild stainless-steel plate [$k_w = 45$ J/s·m·K] has a length of 1.0 m and a width of 0.25 m with a spacing between the plates of 0.005 m. Also, estimate the pressure drop of the hot water stream as it flows through the exchanger.

The performance characteristics for the chevron configuration selected for the plates are shown in Fig. 14-51. For Re > 100, Nu and $\mathfrak{f}$ can be represented by the following relationships:

$$\text{Nu} = 0.4\ \text{Re}^{0.64}\ \text{Pr}^{0.4} \qquad \mathfrak{f} = 2.78\ \text{Re}^{-0.18}$$

Properties of each fluid at the mean temperature in the exchanger are provided in Example 14-5.

■ Solution

To avoid an iterative calculation because of the interdependency between the heat-transfer area and the total flow area, use the NTU approach to determine the NTU_{min} required, noting that $\text{NTU}_{\text{min}} = UA/(\dot{m}C_p)_{\text{min}}$. The area of the plate and frame exchanger can be calculated once the overall heat-transfer coefficient has been evaluated.

For a single-pass configuration with N_p plates and $N_p + 1$ flow passages, solution of the problem can be simplified mathematically by assuming n flow passages and $n - 1$ plates, since flow velocities involve flow passages and not plates. With this modification, the heat-transfer surface area of the exchanger in terms of n is

$$A = (n-1)LW = (n-1)(1)(0.25) = 0.25(n-1)\ \text{m}^2$$

The flow area for each stream with $n/2$ flow passages is given by

$$S = \frac{n}{2}Wb = \frac{n}{2}(0.25)(0.005) = (6.25 \times 10^{-4})n$$

From Example 14-5, the mass flow rate of the water is 2.65 kg/s. The velocity of the water is then

$$V_h = \frac{\dot{m}_h}{\rho_h S} = \frac{2.65}{(9.74 \times 10^2)(6.25 \times 10^{-4})n} = \frac{4.35}{n}\ \text{m/s}$$

This permits evaluation of the Reynolds number in terms of n where D_e is defined as equal to $2b$, or $2(0.005) = 0.01$ m.

$$\text{Re}_h = \frac{D_e V_h \rho_h}{\mu_h} = \frac{0.01(4.35/n)(9.74 \times 10^2)}{3.72 \times 10^{-4}} = \frac{1.139 \times 10^5}{n}$$

This indicates that Re > 100 and the correlation for Nu can be used.

$$h_h = (0.4)\left(\frac{k_h}{D_e}\right)\text{Re}^{0.64}\text{Pr}^{0.4} \qquad \text{Pr} = 2.34 \text{ from Example 14-5}$$

$$= (0.4)\left(\frac{0.668}{0.01}\right)\left(\frac{1.139 \times 10^5}{n}\right)^{0.64}(2.34)^{0.4}$$

$$= \frac{6.467 \times 10^4}{n^{0.64}}\ \text{W/m}^2\cdot\text{K}$$

The same calculations are repeated for the cold stream.

$$V_c = \frac{\dot{m}_c}{\rho_c} = \frac{4.0}{(1.044 \times 10^3)(6.25 \times 10^{-4})n} = \frac{6.13}{n} \text{ m/s}$$

$$\text{Re}_c = \frac{D_e V_c \rho_c}{\mu_c} = \frac{0.01(6.13/n)(1.044 \times 10^3)}{2.70 \times 10^{-3}} = \frac{2.37 \times 10^4}{n}$$

This also indicates that Re > 100

$$h_c = (0.4)\left(\frac{k_c}{D_e}\right)\text{Re}^{0.64}\text{Pr}^{0.4} \qquad \text{Pr} = 31.73 \text{ from Example 14-5}$$

$$= (0.4)\left(\frac{0.138}{0.01}\right)\left(\frac{2.37 \times 10^4}{n}\right)^{0.64}(31.73)^{0.4} = \frac{1.388 \times 10^4}{n^{0.64}} \text{ W/m}^2\cdot\text{K}$$

The overall heat-transfer coefficient can now be determined in terms of n. Since the surface areas on either side of the plate are the same, no correction for area is required in Eq. (14-4) or (14-4*a*). Assume a thickness of the plate x_w of 0.0032 m.

$$\frac{1}{U} = \frac{1}{h_h} + \frac{x_w}{k_w} + \frac{1}{h_c}$$

$$= \frac{n^{0.64}}{6.467 \times 10^4} + \frac{0.0032}{45} + \frac{n^{0.64}}{1.388 \times 10^4}$$

$$= 8.751 \times 10^{-5} n^{0.64} + 7.11 \times 10^{-5} \text{ m}^2\cdot\text{K/W}$$

A NTU_{min} for the cold stream with a minimum $\dot{m}C_p$ is defined as

$$\text{NTU}_{\text{min}} = \frac{UA}{(\dot{m}C_p)_{\text{min}}} = \frac{T_{c,\text{out}} - T_{c,\text{in}}}{F\,\Delta T_{o,\text{log mean}}}$$

For a single-pass counterflow plate and frame exchanger, $F = 1$. From Example 14-5, $\Delta T_{o,\text{log mean}} = 36.067$ K. Thus,

$$\text{NTU}_{\text{min}} = \frac{70 - 10}{36.067} = 1.664$$

To satisfy the other NTU definition of $UA/(\dot{m}C_p)_{\text{min}}$ in terms of n results in the relation

$$1.664 = \left(\frac{1}{8.751 \times 10^5 n^{0.64} + 7.11 \times 10^{-5}}\right)\left(\frac{0.25(n-1)}{4.0(1.622 \times 10^3)}\right)$$

This equation can be solved quickly with a computer to indicate that $n = 51$. Thus 50 plates are required to meet the heat-transfer needs to preheat 4 kg/s of Dowtherm from 10 to 70°C. A software program that includes possible fouling effects indicates that 59 plates would be required for the heat exchange.

Now estimate the pressure drop in the water stream. The water velocity and Reynolds number are

$$V_h = \frac{4.35}{51} = 0.0853 \text{ m/s}$$

$$\text{Re}_h = \frac{1.139 \times 10^5}{51} = 2233$$

Since Re > 100,

$$\mathfrak{f} = 2.78\,\text{Re}^{-0.18} = 2.78(2233)^{-0.18} = 0.694$$

Neglecting friction due to entrance and exit losses as well as temperature effects on the viscosity between the wall and the bulk fluids, a rearrangement of Eq. (14-72*c*) provides the relation for determining the pressure drop as

$$\Delta p = 4\mathfrak{f}\left(\frac{L}{D_e}\right)\frac{\rho_h V_h^2}{2}$$

$$= 4(0.694)\left(\frac{1}{0.01}\right)\frac{(9.74 \times 10^2)(0.0853)^2}{2} = 984\ \text{N/m}^2 = 984\ \text{Pa}$$

Since the entrance and exit friction losses will be small, the pressure loss per plate is small, and a new configuration with modified dimensions should be considered.

Compact Exchangers

One of the most widely used compact heat exchangers is the plate-fin heat exchanger. The construction principle of such an exchanger was illustrated earlier in Fig. 14-14. By vertically stacking many of these finned layers on top of each other and allowing the two fluids to flow through alternating layers with appropriate headers, a high rate of heat transfer per unit volume is achieved. The method of assembly results in robust construction even when relatively thin sheet metal is used for both the separator plates and the corrugated fins. An added feature for this type of exchanger is that surface geometry and flow area for both streams can be specified to obtain the optimum combination.

When designing plate-fin heat exchangers, a geometry and surface arrangement should be selected that results in a UA product that has the right magnitude to satisfy Eq. (14-5). The overall UA product for a plate-fin exchanger can be related to the individual hA product values for the hot and cold streams by

$$UA = \frac{\sum(hA)_h \sum(hA)_c}{\sum(hA)_h + \sum(hA)_c} \tag{14-73}$$

where h is the individual heat-transfer coefficient and A_h and A_c are the effective heat-transfer areas for the hot and cold streams, respectively, as defined by Eq. (14-27).

The Colburn correlation is generally used to express the heat-transfer coefficients for single-phase flow in the form

$$h = j_t C_p G \text{Pr}^{-2/3} \tag{14-74}$$

where values of j_t, a modified Colburn factor that includes the effect of the fluid properties, are generally obtained experimentally as a function of the Reynolds number. These curves are available from fabricators of such heat exchangers.

Pressure drop relations for single-phase flow in plate-fin heat exchangers are generally expressed in terms of a modified friction factor f_t as

$$\Delta p = f_t\left(\frac{L}{D_e}\right)\left(\frac{G^2}{\rho}\right) \tag{14-75}$$

Table 14-11 **Empirical heat-transfer and associated pressure-drop correlations for brazed, aluminum plate-fin heat exchangers†**

Type and flow conditions	Heat transfer correlations	Pressure-drop/length correlations
Straight fins $H = 7.87$ mm $W = 0.15$ mm 492 fins/m $500 < \text{Re} < 10^4$ Gas phase only	$h = 0.0291\text{Re}^{-0.24}\text{Pr}^{-2/3}C_pG$	$\Delta p/\Delta L = (0.0099 + 40.8\text{Re}^{-1.033})G^2/\rho D_e$
Wavy fins $H = 9.53$ mm $W = 0.20$ mm 591 fins/m $300 < \text{Re} < 10^4$ Gas phase only	$h = 0.085\text{Re}^{-0.265}\text{Pr}^{-2/3}C_pG$	$\Delta p/\Delta L = (0.0834 + 23.6\text{Re}^{-0.862})G^2/\rho D_e$
Herringbone fins $H = 10.82$ mm $W = 0.15$ mm 472 fins/m $400 < \text{Re} < 10^4$ Gas phase only	$h = 0.555(\text{Re} + 500)^{-0.482}\text{Pr}^{-2/3}C_pG$	$\Delta p/\Delta L = (7.04\text{Re}^{-0.547})G^2/\rho D_e$

†Fluid properties for all equations are at the mean film temperature $T_m = (T_b + T_w)/2$, where T_b is the bulk temperature of the fluid and T_w the wall or separator plate temperature. Equation (14-75*a*) is to be used to obtain D_e. The $\Delta p/\Delta L$ pressure drop relation accounts for the friction pressure drop and does not include any pressure drops that are associated with entrance and exit losses as well as pressure drops caused by flow acceleration or deceleration due to fluid heating or cooling.

where D_e is given by

$$D_e = 4\left(\frac{A_c L}{A_w}\right) \tag{14-75a}$$

and where A_c is the free-flow cross-sectional area, L the length of a fluid passage, and A_w the wetted surface area. Friction factor values as a function of the Reynolds number must be obtained from fabricators for their various fin geometries. Proprietary methods for determining the pressure drop in inlet and exit nozzles, distributors, and headers are also available from such vendors. Table 14-11 provides heat-transfer and associated pressure-drop correlations for three typical fin configurations used in plate-fin heat exchangers.

EXAMPLE 14-8 Evaluation of Design Parameters for a Plate-Fin Exchanger Required for Gas/Gas Heat Exchange at Low Temperatures

A warm helium gas with a mass flow rate of 0.3 kg/s is to be cooled from 300 K and 0.202 MPa to 120 K in a brazed aluminum plate-fin heat exchanger with a countercurrent flow of cold helium gas initially at 110 K and 0.101 MPa. The mass flow rate of the cold stream is also 0.3 kg/s. The exchanger utilizes herringbone fins, 10.82 mm in height, 0.15 mm in thickness, with 472 fins per meter. The total free-flow area available at the entrance of the fin channels for each gas stream is 0.03 m^2. Assume an average thermal conductivity of 150 W/m·K for the aluminum fins and an average wall (separator plate) temperature of 200 K between the hot and cold helium streams. Evaluate the overall heat-transfer coefficient, the number of heat-transfer units for the heat exchanger, the exit temperature of the cold gas, and the length of heat exchanger required if the overall effectiveness of the exchanger cannot be less than 0.85.

■ Solution

The exit temperature of the cold gas stream can be obtained by using a relationship that is derived from the definition for heat exchanger effectiveness. With the exit temperature, the heat-transfer coefficient, NTU, surface area, and length of the heat exchanger required can be evaluated. A schematic of the entrance to the fin channels for each gas stream is shown below.

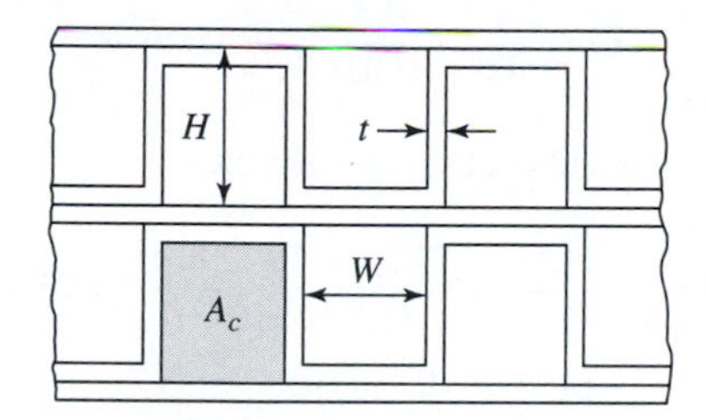

$$H = 1.082 \times 10^{-2}\ \text{m}$$

$$W = \frac{1}{472} - 0.15 \times 10^{-3} = 1.97 \times 10^{-3}\ \text{m}$$

$$D_e = \frac{4A_c L}{A_w} \qquad \text{(assume } L = 1\text{)}$$

$$= \frac{4(1.082 \times 10^{-2} - 1.5 \times 10^{-4})(1.97 \times 10^{-3})}{2(1.082 \times 10^{-2} - 1.5 \times 10^{-4}) + 2(1.97 \times 10^{-3})}$$

$$= 3.33 \times 10^{-3}\ \text{m}$$

To achieve an overall exchanger effectiveness of 0.85, the exit temperature of the cold stream can be obtained from Eq. (14-13) by assuming that the cold fluid has the minimum value of $\dot{m}C_p$.

$$E = 0.85 = \frac{T_{c,\text{out}} - T_{c,\text{in}}}{T_{h,\text{in}} - T_{c,\text{in}}}$$

$$T_{c,\text{out}} = 110 + 0.85(300 - 110) = 271.5\ \text{K}$$

$$T_{c,\text{avg}} = \frac{271.5 + 110}{2} = 190.8\ \text{K}$$

Since the average wall temperature was 200 K, the average film temperature of the cold helium gas is essentially 195 K. The average bulk temperature of the warm helium gas is $(300 + 120)/2$, or 210 K, giving an average film temperature of 205 K. Properties of helium gas at these temperatures and pressures are as follows:

	195 K	205 K
C_p, J/kg·K	5.2×10^{-3}	5.2×10^{-3}
k, W/m·K	0.1134	0.1173
μ, Pa·s	1.471×10^{-5}	1.52×10^{-5}
Pr	0.674	0.674

Because the heat capacity values are the same for the cold and hot streams, the approximation of assuming that $(\dot{m}C_p)_c = (\dot{m}C_p)_{\min}$ is still reasonable so that the exit temperature of the cold helium gas is 271.5 K.

The mass flow rate per unit area for both helium gas streams is

$$G = \frac{\dot{m}}{A_c} = \frac{0.3}{0.03} = 10 \text{ kg/m}^2\cdot\text{s}$$

The Reynolds numbers for the hot and cold streams are then

$$\text{Re}_h = \frac{D_e G}{\mu} = \frac{(3.33 \times 10^{-3})(10)}{(1.52 \times 10^{-5})} = 2190$$

$$\text{Re}_c = \frac{(3.33 \times 10^{-3})(10)}{1.471 \times 10^{-5}} = 2264$$

Since $400 < \text{Re} < 10^4$ applies to both streams, the correlation for herringbone fins given in Table 14-11 is adequate

$$h = 0.555(\text{Re} + 500)^{-0.482}\text{Pr}^{-2/3}C_p G$$

$$h_h = 0.555(2190 + 500)^{-0.482}(0.674)^{-2/3}(5.2 \times 10^3)(10) = 834 \text{ W/m}^2\cdot\text{K}$$

$$h_c = 0.555(2264 + 500)^{-0.482}(0.674)^{-2/3}(5.2 \times 10^3)(10) = 824 \text{ W/m}^2\cdot\text{K}$$

The heat-transfer coefficients for the hot and cold sides of the heat exchanger are nearly identical, indicating that the wall temperature should be equidistant between the hot and cold temperatures of 210 and 190.8 K. This is approximately 200 K, matching the assumed wall temperature.

Evaluate the overall heat-transfer coefficient, assuming that $A_h = A_c$ and the separator plate has a thickness that is 3 times that of the fin.

$$\frac{1}{U} = \frac{1}{h_h} + \frac{x_w}{k_w} + \frac{1}{h_c} = \frac{1}{834} + \frac{3(1.5 \times 10^{-4})}{150} + \frac{1}{824} = 2.42 \times 10^{-3} \text{ m}^2\cdot\text{K/W}$$

$$U = 414 \text{ W/m}^2\cdot\text{K}$$

Equation (14-16) relates the heat exchanger effectiveness with the minimum number of transfer units. By appropriate substitutions for $\Delta T_{o,m}$ and $\Delta T_{\max}$ in terms of $(\dot{m}C_p)_{\min}$ and $(\dot{m}C_p)_{\max}$, an effectiveness-NTU expression for countercurrent heat exchange can be developed as

$$E = \frac{1 - \exp[-\text{NTU}(1 - C_{\min}/C_{\max})]}{1 - (C_{\min}/C_{\max})\exp[-\text{NTU}(1 - C_{\min}/C_{\max})]}$$

where $C_{\min} = (\dot{m}C_p)_{\min}$ and $C_{\max} = (\dot{m}C_p)_{\max}$. When $C_{\min} = C_{\max}$, as in this example, the equation reduces to the relation

$$E = \frac{\text{NTU}}{1 + \text{NTU}}$$

Thus, with $E = 0.85$

$$\text{NTU} = \frac{E}{1 - E} = \frac{0.85}{1 - 0.85} = 5.667$$

Since $\text{NTU} = UA/(\dot{m}C_p)_{\min}$,

$$A = (\dot{m}C_p)_{\min}\frac{\text{NTU}}{U} = (0.3)(5.2 \times 10^3)\frac{5.667}{414} = 21.35 \text{ m}^2$$

where the area per unit length of exchanger is obtained from a rearrangement of Eq. (14-75*a*).

$$\frac{A}{L} = \frac{4A_c}{D_e} = \frac{4(0.03)}{3.33 \times 10^{-3}} = 36.04 \text{ m}^2/\text{m}$$

The required length of the exchanger therefore is

$$L = \frac{A}{A/L} = \frac{21.35}{36.04} = 0.59 \text{ m}$$

(Comparison of results using appropriate software was not attempted.)

Air-Cooled Exchangers

There has been a steady increase in the use of air to replace more expensive fluids as the cooling medium with the use of air-cooled heat exchangers. In these heat-transfer units, the fluid to be cooled flows inside a bundle of externally finned tubes while forced air passes over the outside of the tubes in a cross-flow arrangement. Various flow configurations and fin designs are used in these units.

The basic expression for the total rate of heat transfer in a cross-flow air-cooled heat exchanger is given by Eq. (14-5) except that the area is the total external surface area of the tubes without the fins and $U_{e,o}$, the enhanced overall heat-transfer coefficient based on this same area, is obtained from

$$\frac{1}{U_{e,o}} = \frac{1}{h_{e,o}} + \frac{D_o}{2k_w}\ln\frac{D_o}{D_i} + \frac{1}{h_i}\frac{D_o}{D_i} + R_e \tag{14-76}$$

where $h_{e,o}$ is the enhanced heat-transfer coefficient for the air film based on the outside diameter D_o of the tube, k_w the thermal conductivity of the tube, h_i the coefficient of heat transfer for the fluid being cooled based on the inside diameter of the tube, and R_e the thermal contact resistance between the attached fins and tube. The enhanced air-side heat-transfer coefficient is evaluated from the relation

$$h_{e,o} = \frac{h_o(\eta_f A_f + A_p)}{A_T} \tag{14-77}$$

where A_f is the external area of the fins, A_p the external surface tube area not covered by the fins, and A_T the total area of the fins plus the tube area not covered by the fins. Equation (14-26) can be used to determine η_f.

Calculation of $\Delta T_{o,m}$ for cross-flow exchangers is best performed by using the appropriate F-NTU-P charts and following the accompanying procedures to obtain the correction factor F used in Eq. (14-8). Derivation of these charts for air-cooled cross-flow heat exchangers assumes complete mixing of the air. Because of this assumption, the only effect on the thermal performance of air-cooled heat exchangers occurs when more than one row of tubes is used in the exchanger. Thus, separate charts have been developed for exchangers that require more than one row of tubes. Separate charts have also been developed for countercurrent flow of the cooled fluid for two-, three-, and four-pass arrangements.[†]

If outlet temperatures from the air-cooled heat exchanger are unknown, the procedure is to use the NTU values for the air along with the parameter R to identify the operating point to obtain a value for P from which the outlet temperature can be evaluated. If all fluid temperatures are known, the operating point can be identified with the R and P values. The required area can be calculated by using the corresponding NTU value associated with the airflow across the tubes.

Condensers

Condensation may occur in a variety of modes generally designated as *filmwise, homogeneous, dropwise,* and *immiscible liquid*. Many condensers operate with filmwise condensation, but the design engineer must be aware of the other modes. For instance, dropwise condensation could be encountered in the early stages of operation of a condenser, resulting in higher heat-transfer rates than those originally predicted.

Both horizontal and vertical tube heat-transfer surfaces are used for condensation. In the horizontal arrangement, the condensation can occur either outside or inside the tubes. In the former case, the tubes are normally placed in tube bundles with the coolant flowing inside the tubes. The average heat-transfer coefficient $\bar{h}_o$ for laminar filmwise condensation on the outside surface of a horizontal tube is given by

$$\bar{h}_o = 0.725\left[\frac{k_L^3 \rho_L(\rho_L - \rho_V) g \lambda_C}{\mu_L D_o (T_{\text{sat}} - T_w)}\right]^{1/4} \qquad (14\text{-}78)$$

where k_L, ρ_L, and μ_L are the thermal conductivity, density, and viscosity of the condensate, respectively, ρ_V the density of the vapor, g the local gravitational acceleration, λ_C the heat of condensation, and T_{sat} the temperature of the saturated condensate. Note that solution of this equation generally will require an iteration to determine T_w, the temperature of the wall. If the condensate is saturated steam, Eq. (14-78) can be simplified to the relation

$$\bar{h}_o = \frac{3100}{D_o^{1/4}(T_{\text{sat}} - T_w)^{1/3}} \qquad (14\text{-}78a)$$

[†]J. Taborek, in *Heat Exchanger Design Handbook,* G. F. Hewitt, ed., Begell House, New York, 1998, Sec. 1.5.3.

For a vertical row of N_V tubes in which the condensate flows from the top tube to the tubes directly below due to gravity effects, the average heat-transfer coefficient should be modified by the expression

$$\bar{h}_N = \bar{h}_o N_V^{-1/4} \tag{14-78b}$$

where $\bar{h}_o$ is the average heat-transfer coefficient for the first horizontal tube as determined from either Eq. (14-78) or Eq. (14-78*a*).

For condensation on the inside of horizontal tubes under stratifying conditions, a relation similar to Eq. (14-78) can be used by replacing the coefficient with a variable coefficient that changes with the angle subtended by the condensate drainage region that exists at the bottom of the tube. This coefficient varies from 0.725 for low condensate drainage to 0.90 for high condensate drainage, with 0.80 being a reasonable initial estimate.

If the condensation on the inside of the horizontal tube is shear-controlled, an annular film condensate is formed on the inside wall of the tube. The local heat-transfer coefficient for turbulent annular flow is given by

$$h_{a,i} = \frac{C_{pL}(\rho_L \tau_o)^{1/2}}{T_{\delta+}} \tag{14-79}$$

where τ_o is the shear stress defined as the product of $D_i/4$ and the frictional pressure gradient dp_F/dz. The latter term is obtained from

$$\frac{dp_F}{dz} = \frac{2\dagger G^2}{D_i \rho_L} \tag{14-79a}$$

while $T_{\delta+}$, the external film temperature for $\mathrm{Re}_f > 1483$, is estimated from

$$T_{\delta+} = 5\left[\mathrm{Pr}_L + \ln(1 + 5\mathrm{Pr}_L) + 0.5\ln\left(\frac{0.0504\mathrm{Re}_f^{0.875}}{30}\right)\right] \tag{14-79b}$$

where the Reynolds number for the film with a liquid flow quality of x is obtained from

$$\mathrm{Re}_f = \frac{G(1-x)D_i}{\mu_L} \tag{14-79c}$$

Between the pure stratifying and annular flow conditions there is an intermediate region where both mechanisms may apply. Since the modified flux velocity of the vapor is the determining factor, Palen et al.[†] have suggested that stratifying conditions exist when $V_V^* \leq 0.5$ and annular conditions prevail when $V_V^* \geq 1.5$. The modified flux velocity for the vapor inside the tube is defined as

$$V_V^* = \frac{V_V \rho_V^{1/2}}{[gD_i(\rho_L - \rho_V)]^{1/2}} \tag{14-80}$$

[†] J. W. Palen, G. Trebes, and J. Taborek, *Heat Transfer Eng.*, **1**(2): 47 (1979).

For the intermediate region ($0.5 < V_V^* < 1.5$), they recommend the interpolation formula

$$\bar{h} = h_{a,i} + (V_V^* - 1.5)(h_{a,i} - h_{\mathrm{st},i}) \tag{14-80a}$$

where $h_{a,i}$ and $h_{\mathrm{st},i}$ are the annular and stratifying heat-transfer coefficients, respectively.

The average heat-transfer coefficient $\bar{h}_o$ for laminar filmwise condensation on the outside of a vertical tube for pure saturated vapors operating under conditions of $4\dot{m}/(\pi D_o \mu_L) < 2000$ is given by

$$\bar{h}_o = 1.47 \left(\frac{\pi D_o k_L^3 \rho_L^2 g}{4 \dot{m} \mu_L} \right)^{1/3} \tag{14-81}$$

where the thermal conductivity, density, and viscosity are those of the condensate at the average condensate temperature.

The discussion to this point has only considered condensation of a pure component. However, in many practical applications, there can be two or more components in the inlet stream to the condenser. The complexity of the multicomponent diffusion equations that need to be solved simultaneously has led to the development of approximate methods for condenser design. One such development is the Silver-Bell-Ghaly (SBG) method[†] that assumes the vapor phase temperature is similar to the temperature of the equilibrium condensation curve throughout the condenser and that all heat removed from the condensing/cooling process will be absorbed by the coolant.

For a countercurrent flow, the heat removed from an increment of condenser area includes the heat extracted due to the latent heat of condensation and the cooling of both the liquid and vapor phases. The removal of heat in this element of area is expressed in differential form as

$$d\dot{q}_c = U_c(T_i - T_c)\, dA \tag{14-82}$$

where U_c is the overall heat-transfer coefficient from the gas/liquid interface at temperature T_i to the coolant at temperature T_c. The sensible heat flux from the vapor is given by

$$d\dot{q}_V = h_V(T_E - T_i)\, dA \tag{14-83}$$

where T_E is the equilibrium temperature and h_V the gas-to-interface heat-transfer coefficient that may require a correction at high rates of condensation to account for the influence of mass transfer on the heat transfer as originally derived by Ackermann.[‡] Elimination of T_i between Eqs. (14-82) and (14-83) and setting $z = d\dot{q}_V/d\dot{q}_c$ result in a differential equation that can be integrated to obtain the required condenser area from

$$A = \int_o^{\dot{q}} \frac{(1 + U_c z/h_V)\, d\dot{q}}{U_c(T_E - T_c)} \tag{14-84}$$

[†]K. J. Bell and M. A. Ghaly, *AIChE Symp. Series,* **69**(131): 72 (1972).

[‡]G. Ackermann, *Forsch. Ing. Wis., VDI Forschungsheft,* **8**(382): 1 (1937).

Since z is a function of $\dot{q}$ and therefore also with T_E, Eq. (14-84) can be integrated once a condensation curve has been developed and the variations of h_V and U_c throughout the condenser have been established. This method is the basis for many current computer design programs that involve multicomponent vapor condensation.

Evaporators

As noted earlier, there are many types of evaporators and vapor generators. For example, a typical falling-film evaporator used for concentrating a solution consists of a bank of vertical tubes mounted within a cylindrical shell with the liquid entering at the top of the unit and flowing down the inside of the tubes, with a hot fluid flowing on the shell side serving as the heat source. The generation of vapor takes place at the surface of the downward-flowing liquid. The latent heat of vaporization is supplied by conduction across the film from the tube wall to the liquid surface. If no noncondensible gases are present in the liquid, the surface temperature of the liquid is assumed to approach the saturation temperature of the liquid. Determination of the heat-transfer coefficient can then be made after calculating the film thickness δ.

At low liquid flow rates, the flow is laminar, and the film thickness can be calculated by using hydrodynamic relationships. At higher liquid velocities, waves appear on the liquid surface, and the heat-transfer coefficient can no longer be calculated theoretically but must be evaluated with empirical correlations. At still higher liquid velocities, the flow is turbulent, and an empirical relationship involving both the Reynolds number and the Prandtl number is more appropriate.†

The film thickness for a falling-film evaporator operating with conditions of smooth laminar flow and low vapor velocity is given by

$$\delta' = \left[\frac{3\mu_L \Gamma}{g\rho_L(\rho_L - \rho_V)}\right]^{1/3} \tag{14-85}$$

where Γ is the fluid feed rate per unit length of tube periphery. The heat transfer from the wall to the liquid surface is controlled by conduction through the liquid film. Thus, the local heat-transfer coefficient that relates the heat transfer to the temperature driving force can be developed from Eq. (14-85), resulting in

$$h = \left[\frac{k_L^3 g(\rho_L - \rho_V)}{3\mu_L \Gamma}\right]^{1/3} \tag{14-86}$$

Equation (14-86) can also be expressed in terms of the Reynolds number as

$$h = 1.10\left[\frac{k_L^3 g\rho_L(\rho_L - \rho_V)}{\mu_L^2 \mathrm{Re}}\right]^{1/3} \tag{14-86a}$$

where the Reynolds number is equal to $4\Gamma/\mu_L$.

---

†D. A. Labuntsov, *Teploenergetika,* **4**(7): 72 (1957).

For wavy laminar flow (30 < Re < 1800), the local heat-transfer coefficient can be evaluated from the empirical relation

$$h = 0.756\text{Re}^{-0.22}\left[k_L^3 g\rho_L(\rho_L - \rho_V)\mu_L^2\right]^{1/3} \tag{14-87}$$

Above a Reynolds number of approximately 1800, the falling film becomes turbulent and the local heat-transfer coefficient may be obtained from

$$h = 0.023\text{Re}^{0.25}\text{Pr}^{0.5}\left[\frac{k_L^3 g\rho_L(\rho_L - \rho_V)}{\mu_L^2}\right]^{1/3} \tag{14-88}$$

Obtaining an average heat-transfer coefficient for the three operating conditions listed above will require determination of the local heat-transfer coefficient at several locations along the vertical tube followed by an averaging of these values. Note that correlations of local heat-transfer coefficients derived from condensing film experiments are also applicable to evaporating films.

A climbing film evaporator operates in the convective boiling heat-transfer mode with the liquid entering the bottom of a vertical bank of tubes that are heated from the shell side to generate a vapor product. Liquid flow rates are kept low to ensure boiling within the tubes while vapor velocities are high enough to maintain a climbing film of liquid over most of the tube length. Since liquid and vapor are present in the tubular evaporator, a more complex correlation for the heat-transfer coefficient is required, namely,

$$h = \frac{k_L}{D_i}(1.3 + 128\,D_i)\text{Pr}_L^{0.9}\text{Re}_L^{0.23}\text{Re}_V^{0.34}\left(\frac{\rho_L}{\rho_V}\right)^{0.25}\left(\frac{\mu_V}{\mu_L}\right) \tag{14-89}$$

where D_i is the inside tube diameter and the liquid and vapor Reynolds numbers are calculated on the basis of each fluid flowing by itself in the tube. The climbing film mode of operation can be found in both long-tube and short-tube vertical evaporators. Estimation of the heat-transfer coefficient in either type of evaporator is similar since performance primarily is a function of temperature level, temperature difference, and viscosity.

Horizontal-tube evaporators operating with partially or fully submerged heating surfaces provide behavior that is similar to that of the short-tube vertical evaporators. Heat-transfer coefficients for liquids vaporizing inside horizontal tubes under constant-heat-flux conditions can be obtained from the empirical relationship

$$h = 1.59\left(\frac{k_L^3\rho_L^2 g}{\mu_L^2\text{Re}_f}\right)^{1/3} \tag{14-90}$$

where the Reynolds number for the film flow was defined earlier as equal to $4\Gamma/\mu_L$.

OPTIMUM DESIGN OF HEAT EXCHANGERS

Two types of quantitative problems are commonly encountered by the design engineer when dealing with heat-transfer calculations. The first type is illustrated in this chapter by Examples 14-3 through 14-8 in which all the design variables are set, and the calculations involve only a determination of the indicated nonvariant quantities. The conditions specified in these examples normally give low pressure drops through the exchangers. Alternative conditions could be specified which would give higher pressure drops and less heat-transfer area. By selecting various conditions, the engineer could ultimately arrive at a final design that would minimize the sum of the annual variable costs and the annual operating costs of the exchanger.

General Case

The design of most heat exchangers involves initial conditions in which the following variables are known:

Process-fluid rate of flow

Allowed temperature change of process fluid

Inlet temperature of utility fluid (for cooling or heating)

With this information, the engineer must prepare a design for the optimum exchanger that will meet the required process conditions. Ordinarily, the following results must be determined:

Heat-transfer area

Exit temperature and flow rate of utility fluid

Number, length, diameter, and arrangement of tubes

Pressure drops for the hot and cold fluids

The variable annual costs of importance are the fixed charges on the equipment, the cost for the utility fluid, and the power cost for pumping both fluids through the exchanger. The total annual cost for optimization, therefore, can be represented by:

$$C_T = A_o K_F C_{A_o} + \dot{m}_u H_y C_u + A_o E_i H_y C_i + A_o E_o H_y C_o \tag{14-91}$$

where C_T is the total annual variable cost for the heat exchanger and its operation, C_{A_o} the installed cost of the heat exchanger per unit of outside-tube heat-transfer area, C_u the cost of the utility fluid, C_i the cost of supplying 1 N·m to pump the fluid flowing through the inside of the tubes, C_o the cost of supplying 1 N·m to pump the fluid flowing through the shell side of the unit, A_o the outside tube area, K_F the annual fixed charges including maintenance, expressed as a fraction of the installed cost of the exchanger, $\dot{m}_u$ the hourly flow rate of the utility fluid, H_y the hours of exchanger operation per year, E_i the power loss experienced inside the tubes in Nm/s per unit of outside tube area, and E_o the power loss experienced outside of the tubes in Nm/s per unit of outside tube area.

An optimum design could be developed by using Eq. (14-91) with a laborious procedure of trial and error, taking all possible variations into consideration. This time-consuming procedure can be reduced somewhat by making some reasonable design assumptions listed below and using the method of partial derivatives as described by Peters and Timmerhaus.[†]

[†]M. S. Peters and K. D. Timmerhaus, *Plant Design and Economics for Chemical Engineers,* 4th ed., McGraw-Hill, New York, 1991, pp. 627–635.

Under ordinary circumstances, the effect of tube diameter on total cost at the optimum operating conditions is not great, and a reasonable choice of tube diameter, wall thickness, and tube spacing can be specified at the start of the design. Similarly, the number of tube passes can usually be specified. If a change in phase of one of the fluids occurs (e.g., the utility fluid is condensing steam), solution of Eq. (14-91) for optimum conditions can often be simplified. Where there is no change in phase, the solution can become complex, because the velocities and resulting power costs and heat-transfer coefficients can vary quite independently over a wide range of values. The only satisfactory means of arriving at an optimization of a shell-and-tube heat exchanger under such conditions requires the iterative use of available computer optimization programs that follow the TEMA codes. Example 14-9 demonstrates such an optimization procedure using a commercial computer program.

EXAMPLE 14-9 Development of the Optimum Design for a Shell-and-Tube Heat Exchanger

A gas under an average pressure of 1010 kPa (10 atm) with properties equivalent to those of air must be cooled from 65.5 to 37.8°C (150 to 100°F). Cooling water is available at a temperature of 21.1°C (70°F). Use of a shell-and-tube floating-head heat exchanger with cooling water as the utility fluid has been proposed. On the basis of the following data and specifications, determine the optimum tube length, number of tubes, and installed cost for the optimum exchanger which will handle 2.52 kg/s (20,000 lb/h) of high-pressure gas.

Exchanger specifications

Steel shell-and-tube exchanger with cross-flow baffling
Cooling water on the shell side of the exchanger
One tube pass and countercurrent flow
Tubes staggered with a tube pitch of 1.25
Tube dimensions: OD = 0.0254 m, ID = 0.0198 m

Costs

Use the January 1990 purchased cost data for heat exchangers presented in Fig. 15-13 (page 616 of *Plant Design and Economics for Chemical Engineers,* 4th ed.) to permit a direct comparison between the hand calculation and computer solutions of this same design problem.
Installation is 15 percent of the purchased equipment cost.
Annual fixed charges, including maintenance, are 20 percent of the installed cost.
Cost for the cooling water or utility fluid is \$0.02 per 10^3 kg (\$0.009 per 10^3 lb) excluding pumping costs.
Cost of energy required to force the cooling water and the pressurized gas through the exchanger (including the effect of pump and motor efficiency) is \$0.04/kWh.

General information

Fouling coefficient for the cooling water is 8500 W/m^2·K [1500 Btu/h·ft^2·°F].
Fouling coefficient for the high-pressure gas is 11,350 W/m^2·K [2000 Btu/h·ft^2·°F].

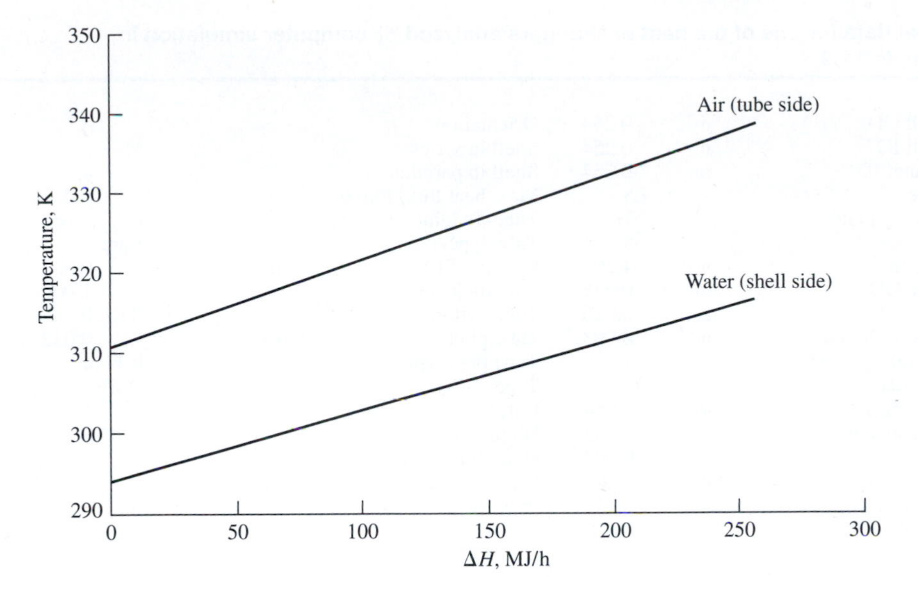

Figure 14-52
Heating and cooling curves for Example 14-9

At the optimum conditions, flow in the tube and shell side is turbulent.

The exchanger operates for 7000 h/yr.

■ Solution

The optimum design of this shell-and-tube heat exchanger, using Eq. (14-91), was presented with a trial-and-error hand calculation as Example 5 in Chap. 15 of the 4th edition of this text. This solution has been greatly simplified by using available computer software to provide a series of heat exchanger designs for different tube lengths and plotting C_T as a function of these tube lengths to determine the optimum exchanger design. This optimum obtained from the computer solution using the same 1990 cost data is compared to the optimum solution obtained from the hand calculations.

Based on the available data, a widely used computer simulation program provided the following results. Figure 14-52 provides the temperature profile in the exchanger as a function of the heat transferred. Table 14-12 shows the tabulated analysis typically provided by the computer program for one of the many heat exchangers examined in this optimization process. The information provided is for a shell-and-tube heat exchanger with 36 tubes, an outside tube diameter of 0.0254 m (1.0 in.), and a length of 4.267 m (14 ft). The pressure drops on the shell and tube sides are 3151 Pa (0.46 psia) and 11,521 Pa (1.671 psia), respectively.

The computer-aided design results for the shell-and-tube heat exchangers optimized at four tube lengths, using a staggered tube array with a tube pitch of 1.25 and an outside tube diameter of 0.0254 m, are summarized in Table 14-13.

To obtain the total annual cost for the heat exchangers tabulated in Table 14-13, a simple computer program was developed to solve Eq. (14-91) incorporating the design data summarized on page 743.

Table 14-12 **Actual data for one of the heat exchangers analyzed by computer simulation in Example 14-9**

General data

Shell:	Shell OD	m	0.254	Orientation		H
	Shell ID	m	0.254	Shell in series		1
	Bonnet ID	m	0.254	Shell in parallel		1
	Type		AES	Max. heat flux, J/kmol		0.0
	Imping. plate		No	Tube-Ts joint		Expanded
Tubes:	Number		36	Tube type		Bare
	Length	m	4.267	Free Int. Fl area	m^2	0.00
	Tube OD	m	0.025	Fin efficiency		1.000
	Tube ID	m	0.020	Tube pattern		TRI30
	Tube wall thk.	m	0.003	Tube pitch	m	0.032
	No. tube passes		1	Tube pass type		Ribbon
Baffles:	Number		77	Type		DSEG
	Spacing	m	0.054	Cut, %		31
	Inlet spacing	m	0.063	Direction		Hor.
Clearances:	Baffle	m	0.003	Outer tube limit	m	0.222
	Tube hole	m	0.001	Outer tube clearance	m	0.032
	Bundle top space	m	0.016	Pass part clearance	m	0.000
	Bundle bottom space	m	0.000			

Overall data

Area:	Area total	m^2	12.26	% Excess		0.94
	Area required	m^2	12.03	*U* calc.	$W/m^2 \cdot K$	310.49
	Area, effective	m^2	12.14	*U* service	$W/m^2 \cdot K$	307.59
	Area clean	m^2	11.17	*U* clean	$W/m^2 \cdot K$	334.34
	Area per shell	m^2	12.14	Heat duty	MJ/h	2.59E+002
Temperatures:	Weight LMTD	K	19.31	Corr. factor 1.00	Corr LMTD K	19.31
Resistances:	Shell-side film coeff.				$W/m^2 \cdot K$	2010.62
	Shell-side fouling factor				$m^2 \cdot K/W$	0.00012
	Tube wall resistance				$m^2 \cdot K/W$	0.00006
	Tube-side fouling factor				$m^2 \cdot K/W$	0.00009
	Tube-side film coeff./outside of wall				$W/m^2 \cdot K$	414.12
	Reference factor					1.279

Shell-side data

Cross-flow vel.	m/s	1.1E-001	End zone vel. 7.0E-002	Window vel.	1.2E-001
Film coeff.	$W/m^2 \cdot K$	2010.62	Reynolds no.	7547	
Allow press. drop.	Pa	34473.80	Calc. press. drop	Pa	3150.91
Inlet nozzle size	m	0.03	Press. drop/in nozzle	Pa	89.63
Outlet nozzle size	m	0.03	Press. drop/out nozzle	Pa	834.27
Interm. nozzle size	m	0.00	Mean temperature	K	305.37
ρV^2	$kg/m \cdot s^2$	880.1	Press. drop (dirty)	Pa	3406.01

Stream analysis

SA factors:	*A* 9.28	*B* 46.84	*C* 31.08	*E* 12.80	*F* 0.00
Ideal cross-vel.	m/s	0.10	Ideal window vel	m/s	0.09
H Xflow	$W/m^2 \cdot K$	3017.55	H window	$W/m^2 \cdot K$	1491.74

Tube-side data

Film coeff.	$W/m^2 \cdot K$	529.56	Reynolds no.		229,132
Allow press. drop	Pa	34,473.80	Calc. press. drop	Pa	11,521.144
Inlet nozzle size	m	0.152	Press. drop/in nozzle	Pa	944.582
Outlet nozzle size	m	0.152	Press. drop/out nozzle	Pa	599.844
Interm. nozzle size	m	0.000	Mean temperature	K	308.694
Velocity	m/s	21.30			

Table 14-13 Summary of exchanger designs optimized at four tube lengths

Tube length, m	2.438	3.048	3.658	4.267
Tubes required, N_T	300	130	58	36
h (shell-side), W/m^2·K	589	875	1536	2010
h (tube-side), W/m^2·K	97	190	367	530
U (overall), W/m^2·K	66	122	278	311
Area, m^2	58.4	31.6	17	12.3
Δp (shell-side), Pa	965	1123	1910	3151
Δp (tube-side), Pa	1662	2253	5120	11521

Design Data

$$A_o = \pi D_o N_T L \text{ m}^2 \qquad K_f = 0.2$$

$$C_{A_o} = (1 + 0.15)(\text{purchased cost/m}^2)$$

$$\dot{q} = (\dot{m} C_p \, \Delta T)_o \qquad \dot{m}_u = \frac{\dot{q}}{(C_p \, \Delta T)_u} \qquad \dot{m} = 2.52 \text{ kg/s}$$

$$H_y = 7000 \text{ h}$$

$$C_u = \frac{\$0.02}{10^3 \text{ kg}} = \$2.0 \times 10^{-5}/\text{kg}$$

$$C_i = C_o = \$0.04/\text{kWh} = \$1.11 \times 10^{-8}/\text{N·m}$$

$$E_i = \frac{-\Delta p_i \, \dot{m}_i}{\rho_o A_o} \text{ N·m/s·m}^2$$

$$E_o = \frac{-\Delta p_o \, \dot{m}_o}{\rho_o A_o} \text{ N·m/s·m}^2$$

The total annual cost for the shell-and-tube heat exchangers optimized at the four tube lengths is as follows:

Tube length, m (ft)	Total annual cost, $
2.438 (8)	2792
3.048 (10)	2139
3.658 (12)	1928
4.267 (14)	2183

A plot of these values as a function of tube length indicates that the most appropriate length is close to 3.65 m (12 ft). A comparison of the optimum heat exchanger obtained from the hand calculation presented in the 4th edition and the computer-aided calculation is given here:

Exchanger details	Hand calculation	Computer-aided calculation
Tube length, m (ft)	3.26 (10.7)	3.65 (12)
Number of tubes	45	58
Area required, m^2 (ft^2)	11.7 (126)	17 (183)
U_o, W/m^2·K (Btu/h·ft^2·°F)	312 (55)	233 (41)
Installed cost, $	4930	6050

The differences in the results for the two solutions require some further analysis. Note that the difference in the number of tubes required is related to the overall heat-transfer coefficient, and this is related to the tube and shell stream velocities. The controlling individual heat-transfer coefficient is that on the tube side. The velocities of the tube-side fluid were 16.6 m/s for the hand calculation and 13.1 m/s for the computer-aided solution. By adjusting the tube-side velocity in the computer solution to 16.6 m/s, the overall heat-transfer coefficient increases from 233 to 282 $W/m^2 \cdot K$, and the number of tubes required decreases to 48 tubes in the 3.65-m shell-and-tube exchanger. This reduction in tubes will decrease the installed cost of the exchanger and increase the annual operating costs. The installed cost of this 48-tube exchanger is $5750.

The principal reason why the computer solution do not come closer to the hand calculation is that the software closely approximates the Bell-Delaware method, which accounts for various shell, tube, and baffle leakages that are not considered in the Kern method. Additionally, the software follows TEMA guidelines and uses available industrial tube shell diameters and tube installations. Finally, note that the installed cost of the heat exchanger in the hand calculation was approximately 10 percent too low, and a recalculation was suggested. Thus, even without the above analysis, the computer solution quite rapidly provided an acceptable solution that could be used in a preliminary design effort. ■

GENERAL METHODS FOR DESIGN OF HEAT EXCHANGERS

The procedures used for developing the design of heat exchangers vary with the type of problem and the preference of the designer. For example, some engineers prefer to develop the design of a heat exchanger based on past experience in which they assume the configuration of an exchanger and use a computer software program to determine whether the exchanger obtained from this computer printout will be able to handle the process requirements under reasonable conditions. If the design does not meet the requirements, various modifications are made to the design and analyzed with the computer program until a suitable design is achieved.

An alternative approach is to base the design on optimum economic considerations, using methods described in Chap. 9. No matter which approach is used, the general method of attack for a given set of conditions consists of the following steps:

1. Determine the rates of flow and rate of heat transfer necessary to meet the given conditions.
2. Select the type of heat exchanger to be used, and establish the basic equipment specifications.
3. Evaluate the overall heat-transfer coefficient and the film coefficients, if necessary. In many cases, fluid velocities and leakage rates must be determined in order to obtain accurate heat-transfer coefficients.
4. Evaluate the temperature driving force throughout the heat exchanger.
5. Determine the required area of heat transfer and the exchanger dimensions.
6. Analyze the results to determine if all dimensions, pressure drops, capital and operating costs, and other design details are satisfactory.

7. If the analysis in step 6 indicates that the heat exchanger does not meet the desired requirements, the specifications given in step 2 are inadequate. New specifications must be established, and steps 3 through 6 need to be repeated until a satisfactory design is achieved.

NOMENCLATURE

a = constant in Eq. (14-22), tabulated in Table 14-1, dimensionless

A = area of heat transfer, m^2; subscript c designates surface area for cold fluid or cross-sectional flow area of tube; subscript f designates area of tube fins; subscript h designates surface area for hot fluid; subscript i designates inside area; subscript m,w designates mean area of tube wall; subscript o designates outside area or overall area of tube and fins; subscript p designates tube area not covered by fins; subscript T designates total area

b_o = constant in Eq. (14-24*a*) for evaluating shell-side friction factor, dimensionless

B_c = percentage baffle cut in a shell-and-tube heat exchanger used to obtain baffle cut distance, dimensionless

B_o = correction factor in Eq. (14-24) to account for friction due to reversal of flow direction, recrossing of tubes, and variation in cross section, dimensionless

C_{A_o} = installed cost of heat exchanger per unit of outside-tube heat-transfer area, dollars/m^2

C_i = cost for supplying 1 N·m to pump fluid through inside of tubes, dollar/N·m

C_p = heat capacity, J/kg·K or kJ/kg·K

C_T = total annual variable cost for heat exchanger and its operation, dollar/yr

C_u = cost of utility fluid, dollar/kg

D = diameter, m; subscripts i and o designate inside and outside diameters, respectively

D_e = equivalent diameter equal to 4 × hydraulic radius, m

D_{OTL} = outside diameter of tube bundle, m

D_s = inside diameter of heat exchanger shell, m

E = effectiveness of a heat exchanger as defined in Eq. (14-11), dimensionless

E_i = power loss due to fluid flow inside heat exchanger tubes per unit of outside tube area, N·m/s

E_o = power loss due to fluid flow outside heat exchanger tubes per unit of outside tube area, N·m/s

f = Fanning friction factor, dimensionless; subscript i designates friction factor for isothermal flow; prime designates a modified friction factor for shell-side flow given by Eq. (14-24*a*); subscript t refers to a modified friction factor

f_D = Darcy friction factor, dimensionless

F = correction factor to account for type of multipass heat exchanger arrangement selected and obtained from Fig. 14-4

F_1 = correction factor for wall-to-bulk physical property variations evaluated from Eq. (14-22*b*), dimensionless

F_2 = correction factor to account for number of tube rows for different tube configurations presented in Table 14-2, dimensionless

F_b = fraction of flow stream b given by Eq. (14-68), dimensionless

F_{bp} = fraction of cross-flow area available for bypass flow given by Eq. (14-44), dimensionless

F_c = friction factor accounting for sudden contraction of a fluid in Eq. (14-23*c*), dimensionless; fraction of tubes assumed to be in cross-flow in the Bell-Delaware method as determined from Eq. (14-42), dimensionless

F_{cr} = fraction of flow that is in cross-flow over tube bundle in Wills and Johnston method given by Eq. (14-65), dimensionless

F_e = friction factor accounting for sudden expansion of a fluid in Eq. (14-23*c*), dimensionless

F_r = friction factor accounting for reversal of flow direction in Eq. (14-23*c*), dimensionless

F_s = fraction of flow in stream *s* given by Eq. (14-66), dimensionless

F_t = fraction of flow in stream *t* given by Eq. (14-67), dimensionless

F_w = fraction of flow in stream *w* given by Eq. (14-69), dimensionless

g = local gravitational acceleration, m/s^2

G = mass velocity, kg/m^2·s; subscript *i* designates mass velocity inside tube; subscript *s* designates mass velocity on shell side

h = film coefficient of heat transfer, W/m^2·K; subscript *ai* designates annular flow coefficient in film condensation; subscripts *c* and *h* refer to cold and hot streams, respectively; subscript *d* is the coefficient accounting for fouling and subscripts *i* and *o* refer to inside and outside surfaces of the tube, respectively; subscript *e, o* designates enhanced air-side coefficient evaluated from Eq. (14-77); subscript *i* refers to ideal cross-flow coefficient used in Eq. (14-41); subscript *s* designates actual shell-side coefficient obtained from Eq. (14-41); subscript st designates stratified flow coefficient in film condensation

$\bar{h}_o$ = average coefficient of heat transfer for laminar filmwise condensation on outside surface of a horizontal tube, W/m^2·K; subscript *N* designates average coefficient when a vertical row of N_V tubes is involved in condensation

H_y = hours of operation per year, h/yr

j_t = modified Colburn factor used to determine heat-transfer coefficients, dimensionless

J = correction factors applied to ideal heat-transfer coefficient, dimensionless; subscript *C* accounts for baffle configuration; subscript *L* refers to shell-to-baffle and tube-to-baffle leakage correction; subscript *B* accounts for bypass in bundle-to-shell gap

k = thermal conductivity, W/m·K; subscript *w* designates values at temperature of tube wall; subscript *f* designates value for fin material; subscripts *L* and *V* refer to liquid and vapor, respectively

K_a = constant related to pressure losses of fluid at inlet and exit of tube bundle, dimensionless

K_c = constant in expression for evaluating friction due to sudden contraction of fluid (see Table 12-1), dimensionless

K_f = constant that accounts for friction and momentum losses when fluid passes between successive rows of tubes in a bundle, given in Table 14-10, dimensionless

K_F = annual fixed charges, including maintenance, expressed as a fraction of initial cost for a completely installed heat exchanger unit, dimensionless

K_1 = constant in Eq. (14-23*d*), dimensionless

L = length of heat exchanger or plate height in a plate heat exchanger, m; subscript *B* designates the spacing between baffles; subscript *C* designates baffle cut distance obtained from $B_c D_s/100$; subscript *f* designates height of a fin used to enhance heat-transfer area; subscript *s* refers to shell length

m = constant in Eq. (14-22) and tabulated in Table 14-1, dimensionless

$\dot{m}$ = mass flow rate, kg/s; subscripts c and h refer to the flow rates on cold and hot streams, respectively; subscript T designates the total mass flow rate; subscript w refers to flow rate in stream w of Wills and Johnston method; subscript u designates the hourly flow rate of utility fluid

M = identity used in evaluating heat-transfer effectiveness of fin area and defined as $[h/k_f\delta]^{1/2}$, m^{-1}

n = number of flow passages in a plate heat exchanger, dimensionless

n_i = flow stream constant used in Wills and Johnston method, dimensionless; subscripts a, b, c, cb, p, s, t, and w are flow stream constants involved

n_p = number of tube passes, dimensionless

N_B = number of baffles specified for heat exchanger, dimensionless

N_c = number of cross rows estimated from Eq. (14-45), dimensionless

N_{cw} = number of effective cross-flow rows in window zone from Eq. (14-48), dimensionless

N_f = number of fins attached to outside surface of inner tube, dimensionless

N_p = number of plates in a plate heat exchanger, dimensionless

N_{tr} = number of tube rows over which shell fluid flows, dimensionless

N_T = total number of tubes in exchanger defined as product of number of tubes per pass and number of tube passes, dimensionless

N_V = number of rows of tubes in a vertical tier, dimensionless

NTU = number of transfer units, dimensionless; subscript c and h refer to cold and hot streams, respectively

NTU_{min} = subscript min refers to minimum number of transfer units required

p = absolute pressure, kPa

P = parameter defined by Eq. (14-9b), dimensionless

P' = wetted perimeter in a flow passage, m

P_D = tube clearance, m

P_T = tube pitch, m

P_{TP} = distance between tube rows in direction of flow, m

Pr = Prandtl number defined as $C_p\mu/k$, dimensionless; subscripts c and h refer to cold and hot streams, respectively; subscripts b and w refer to physical properties at temperature of bulk fluid and wall, respectively; subscript L refers to liquid physical properties

Δp = pressure drop exhibited by a fluid stream, kPa; subscript c designates ideal cross-flow pressure drop evaluated from Eq. (14-46); subscript s designates pressure loss on shell side of heat exchanger; subscript w designates pressure drop for ideal window zone between baffles as evaluated from Eq. (14-47a)

$\dot{q}$ = rate of heat transfer, kJ/s or kW; subscript c refers to rate of heat removal during condensation; subscript max refers to maximum rate of heat transfer that can be achieved

R = parameter defined by Eq. (14-9a), dimensionless

R_B, R_L = correction factors used in Eq. (14-51) to obtain pressure drop across shell side, dimensionless

R_{tot} = total resistance to heat transfer, $m^2 \cdot K/kW$ or $m^2 \cdot K/(kJ/s)$; subscripts c, f and h, f refer to thermal fouling resistances of cold and hot streams, respectively

Re = Reynolds number, defined as DG/μ, dimensionless; subscripts c and h refer to cold and hot streams, respectively; subscripts L and V refer to liquid and vapor portions of a two-phase fluid, respectively; subscript f designates film conditions

R_e = thermal resistance between an attached fin and tube, $m^2 \cdot K/W$

S = cross-sectional flow area, m^2; subscript b designates bypass flow area given by Eq. (14-62); subscript c designates cross-sectional flow area when fins are used or refers to cold stream; subscript H designates header cross-sectional area per pass; subscript i designates cross-sectional area inside tubes per pass; subscript m refers to cross-flow area near centerline of tube bundle; subscripts s and t designate shell-to-baffle and tube-to-baffle leakage areas in Wills and Johnston method, respectively; subscripts sb and tb designate shell-to-baffle and tube-to-baffle leakage areas in Bell-Delaware method, respectively; subscript w designates window flow area for both methods

t_b = thickness of a baffle, m

t_f = thickness of a fin, m

T = absolute temperature, K; subscript b refers to average bulk temperature; subscripts c and h refer to cold and hot streams, respectively; subscripts c,in and h,in refer to inlet cold and hot streams, respectively; subscripts c,out and h,out refer to exit cold and hot streams, respectively; subscript w refers to wall temperature; in general, subscript 1 refers to entering temperature and subscript 2 refers to leaving temperature; subscripts E and sat refer to equilibrium and saturated temperatures, respectively; subscript δ^+ designates external film temperature

ΔT = temperature-difference driving force, K; subscripts o,m and o,log mean refer to arithmetic mean and logarithmic mean temperature differences, respectively

U = overall coefficient of heat transfer, $W/m^2 \cdot K$; subscript d indicates that a dirt or fouling factor is included; subscripts i and o designate that coefficient is based on inside or outside area of tube, respectively; subscript m indicates that coefficient is based on mean area; subscript e,o refers to an enhanced overall coefficient

V = velocity of fluid, m/s; subscripts c and h refer to cold and hot streams, respectively; subscript max designates maximum fluid velocity allowed between tubes near centerline of flow; subscript V designates vapor velocity

V_V^* = modified flux velocity for vapor inside a tube during condensation process, dimensionless

w = plate width in a plate heat exchanger, m

x = length of conduction path, m; subscript w refers to thickness of pipe wall

x_L = ratio of pitch parallel to flow to tube diameter, dimensionless

x_T = ratio of pitch transverse to flow to tube diameter, dimensionless

z = defined as $d\dot{q}_v/d\dot{q}_c$ and used in Eq. (14-84), dimensionless

Greek Symbols

β = coefficient of volumetric expansion, K^{-1}

β_i = correction factor to account for added friction due to sudden contraction, sudden expansion, and flow reversal, dimensionless

Γ = fluid feed rate per unit length of tube periphery, kg/s·m

δ = ratio of volume of fin to surface area of fin, m

δ' = film thickness for a falling-film evaporator, m

η	= effectiveness of heat- or mass-transfer operation, dimensionless; subscript f designates effectiveness of fin area; subscript o designates overall surface effectiveness of a finned tube
θ	= ratio of $\Delta T_{o,m} / \Delta T_{\max}$, dimensionless
λ_c	= heat of condensation, J/kg
μ	= fluid viscosity, Pa·s; subscripts L and V refer to liquid and vapor stream, respectively; subscript w refers to conditions at temperature of tube wall
ρ	= fluid density, kg/m^3; subscripts L and V refer to liquid and vapor stream, respectively
τ_o	= shear stress for fluid, kPa
Φ_i	= correction factor for nonisothermal flow, dimensionless
Δ_b	= bundle-to-shell diametral clearance, m
Δ_{sb}	= shell-to-baffle diametral clearance, m
Δ_{tb}	= tube-to-baffle diametral clearance, m

PROBLEMS

14-1 A thin-walled copper tube with an external diameter of 0.01 m used for transporting a liquid at 100°C needs to be insulated to minimize the heat loss to the surrounding air at 15°C. What is the heat loss per meter of uninsulated tubing, assuming that the inside heat-transfer coefficient is quite high and the heat-transfer coefficient from the tube to the surrounding air is 10 W/m^2·K? What is the effect on the heat loss with each addition of a 0.005-m layer of insulation with a thermal conductivity of 0.1 W/m·K? What would be the optimum thickness of insulation that should be used? Make appropriate assumptions to simplify the calculations.

14-2 At an average film temperature of 350 K, what are the individual heat-transfer coefficients when the fluid flowing in a 0.0254-m inside diameter tube is air, water, or oil? Each fluid in this comparison exhibits a Reynolds number of 5×10^4. How would the pressure drop vary for each fluid? The properties of the three fluids at 350 K are as follows:

	Air	Water	Oil
Density, kg/m^3	0.955	973	854
Viscosity, Pa·s	2×10^{-5}	3.72×10^{-4}	3.56×10^{-2}
Thermal conductivity, W/m·K	0.030	0.668	0.138
Heat capacity, J/kg·K	1050	4190	2116

14-3 Heat is transferred between a hot gas and a liquid with the gas flowing on the outside of a finned tube bundle and the liquid flowing inside the tube. The tube has an inside diameter of 0.022 m and an outside diameter of 0.028 m which serves as the base for the external fins. The circular fins have an outside diameter of 0.040 m, a thickness of 0.002 m, and a spacing between fins of 0.008 m. The heat-transfer coefficient for the gas side is 200 W/m^2·K, and that for the liquid side is 1500 W/m^2·K. Both the tube and the fin material have a thermal conductivity of 25 W/m·K. Determine the overall heat-transfer coefficient based on the outside surface area.

14-4 Estimate the temperature of saturated steam that is required in the shell of a new (i.e., no scale deposit) single-pass shell-and-tube heat exchanger when water on the tube side, for either the new design or the original design that included scale formation, is heated from 21 to 65.5°C at a flow rate of 15.75 kg/s. The exchanger contains 60 steel tubes, each with an inside diameter of 0.0186 m and an outside diameter of 0.0254 m. The unit is designed with sufficient tube area to permit the heating of the water as specified. In the original design, an h_d of 8250 W/m^2·K was used to allow for scaling on the water side of the tube. The film coefficient for the steam was taken as 11,400 W/m^2·K, and the temperature of the saturated steam required to account for the heating of the water was found to be 143.3°C. No scaling factor for the shell side was used in the original design, and it can be neglected in the new design.

14-5 A saturated organic fluid with a latent heat of vaporization of 200 kJ/kg and a flow rate of 2 kg/s is to be vaporized at a constant saturation temperature of 90°C. The hot fluid used to vaporize the organic fluid enters the evaporator at a temperature of 200°C and leaves at a temperature of 120°C. The heat capacity of the hot fluid may be assumed to remain constant at 2.2 kJ/kg·K over the specified temperature range. If the average overall heat-transfer coefficient is 400 W/m^2·K, determine the required flow rate of the hot fluid, the value of ΔT_m, and the heat-transfer surface area required.

14-6 A heat exchanger with tubes inserted in a rectangular duct is used to heat 0.914 kg/s of air initially at 15°C. A high flow rate of hot water in the tubes maintains the temperature of the tube wall at 70°C. The rectangular duct has a height and width of 0.25 and 0.5 m, respectively. The tubes with an outside diameter of 0.0104 m and a length of 1.0 m are inserted in the duct opening in a staggered array with a 60° triangular pitch. This configuration permits the placement of 7 successive rows of tubes with each row consisting of 8 tubes with a pitch of 0.0313 m. With this type of heat exchanger, what is the exit temperature of the air?

14-7 Condensing steam is to be used to heat water in a single-pass tubular heat exchanger. The water flow through the horizontal tubes is turbulent, and the steam condenses dropwise on the outside of the tubes. The water flow rate, the inlet and exit temperatures of the water, the condensing temperature of the steam, and the available tube-side pressure drop (neglecting entrance and exit losses) are all specified. To determine the optimum exchanger design, it is desired to establish how the total required area of the exchanger varies with the tube diameter selected. Assuming that the water flow remains turbulent and the thermal resistances of the tube wall and the steam condensate film are negligible, determine the effect of tube diameter on the total area required for the exchanger.

14-8 A tubular gas heater is to be designed to heat 0.075 kg/s of a gas from 21 to 88°C with steam condensing on the outside surface of the carbon-steel tubes at 104.4°C. The inside diameter of the tubes is 0.023 m. The velocity of the gas in the tubes is to be maintained near the optimal value. Determine the number of tubes in parallel that will be required and the length of each tube.

Before initializing fabrication of this heat exchanger, some concerns were raised by the design engineer relating to the accuracy of the relationship used in the software package for the above calculation. Accordingly, it has been decided to fabricate the exchanger with a 20 percent safety factor. The design engineer proposes to obtain this factor of safety by adding 20 percent more tubes with the same diameter and length as used in the original design. A colleague, however, indicates that the 20 percent safety factor should be obtained by increasing the tube diameters by 20 percent while keeping the same number of tubes and the same tube length. Discuss the advantages and disadvantages of these two recommendations. Based on this examination, how should the exchanger be fabricated to provide a 20 percent factor of safety?

14-9 A heat exchanger is to be constructed by forming copper tubing into a coil and placing the latter inside an insulated steel shell. In this exchanger, water will flow inside the tubing, and a hydrocarbon vapor at a rate of 0.126 kg/s will be condensing on the outside surface of the tubing. The inside and outside diameters of the tube are 0.0127 and 0.0152 m, respectively. Inlet and exit temperatures for the water are 10 and 32°C, respectively. The heat of condensation of the hydrocarbon at a condensing temperature of 88°C is 335 kJ/kg, and the heat-transfer coefficient for the condensing vapor is 1420 $W/m^2 \cdot K$. Heat losses from the shell may be neglected. What length of copper tubing will be required to accomplish the desired heat transfer?

14-10 A heat exchanger with two tube passes has been recommended for cooling distilled water from 33.9 to 29.4°C. The proposed unit contains 160 tubes, each with a 0.0191-m outside diameter, 18 BWG, and 4.876-m in length. The tubes are laid out on a 0.0238-m triangular pitch within a 0.387-m inside diameter shell. Twenty-five percent cut segmental baffles, spaced 0.3 m apart, are located within the shell. Cooling water at a temperature of 24°C and a flow rate of 2.0 m/s will be used in the tubes to provide the cooling. Under these conditions, the fouling coefficients for the distilled water and the cooling water can be assumed to be 11,360 and 5680 $W/m^2 \cdot K$, respectively. The pressure drop on either the tube side or the shell side may not exceed 69 kPa. Would the recommended design be adequate for cooling 22 kg/s of distilled water?

14.11 Use the Bell-Delaware method to calculate the shell-side heat-transfer coefficient and pressure drop for the cooling of a light hydrocarbon with a mass flow rate of 5.5 kg/s. Physical properties for the light hydrocarbon are as follows: $\rho_L = 730$ kg/m^3; $k_L = 0.13$ W/m·K; $C_{p,L} = 2470$ J/kg·K; $\mu_L = \mu_w = 4 \times 10^{-6}$ Pa·s. The dimensions and internal configurations for the shell-and-tube heat exchanger are listed below.

Inside tube diameter	$D_i = 0.0206$ m
Outside tube diameter	$D_o = 0.0254$ m
Inside shell diameter	$D_s = 0.54$ m
Tube pitch (square)	$P_T = 0.03175$ m
Shell length	$L_s = 4.267$ m
Number of tubes	$N_T = 158$
Baffle spacing	$L_B = 0.127$ m
Baffle thickness	$t_b = 0.005$ m
Diametral clearances	
Tube-to-baffle	$\Delta_{tb} = 0.0008$ m
Shell-to-baffle	$\Delta_{sb} = 0.005$ m
Bundle-to-shell	$\Delta_b = 0.035$ m
Sealing strips/cross-flow row	$N_{ss}/N_c = 0.2$
Number of tube passes	$n = 4$

14.12 A heat exchanger with seven steel tubes placed within a shell is used to cool pure ethyl alcohol flowing inside the tubes from 65.5 to 37.8°C. The inside and outside diameters of the tubes are 0.0203 and 0.0254 m, respectively. The inside diameter of the shell is 0.127 m. Water at 21°C enters the shell side of the exchanger and flows countercurrent to the ethyl alcohol. It is desired to cool 6.3 kg/s of ethyl alcohol with 12.6 kg/s of cooling water. Under the following conditions, determine the total annual pumping costs for the fluids in the tube and shell side of the exchanger.

Conditions and assumptions:

The exchanger operates for 7200 h/yr.

The specific heat of the ethyl alcohol may be assumed to be constant at 0.6. For water, the value may be assumed to be 1.0.

The specific gravity of ethyl alcohol may be assumed to be constant at 0.77. For water, the value may be assumed to be 1.0.

Contraction, expansion, and friction losses can be included by increasing the straight-section frictional pressure drop by 20 percent.

Scale formation is assumed to be negligible.

No baffles are present in the shell side of the exchanger, and the flow on that side can be considered to be parallel to the flow in the tubes.

The outside of the shell is insulated, with negligible heat loss from the shell.

The efficiency of both pumps is 60 percent.

The cost of power is fixed at $0.08/kWh.

14-13 Steam is condensed in a one-shell-pass and two-tube-pass exchanger with 129 tubes per pass. The diameter of the brass tubes is 0.0159 m, 18 gauge, and the length of the tubes is 1.83 m. Estimate the overall heat-transfer coefficient and the kilograms per second of steam that will be condensed at a saturation temperature of the steam of 51.7°C if cooling water operating between 21 and 35°C is used as the coolant. Assume an average velocity for the water inside the tubes of 1.22 m/s.

14.14 Saturated isopentane condenses at 127°C on the outside surface of a copper tube through which water flows as the coolant. The tube has an inside and outside diameter of 0.025 and 0.030 m, respectively. At a position along the tube at which the water is at a temperature of 21°C, determine the average condensation heat-transfer coefficient and the rate of condensation. In the calculation, assume that the thermal resistance of the copper tube is negligible, that the outside wall temperature of the tube is uniform, and that the physical properties of the bulk fluid can be used in the calculation of the tube-side heat-transfer coefficient. Appropriate physical properties for both fluids are given here:

	Isopentane (127°C)	Water (30°C)
Density of liquid, kg/m^3	551	998
Density of gas, kg/m^3	14.1	—
Thermal conductivity, W/m·K	0.0082	0.599
Viscosity, Pa·s	1.285×10^{-4}	9.82×10^{-4}
Heat of condensation, kJ/kg	283.8	—
Heat capacity, kJ/kg·K	—	4.184

14-15 The overhead vapor from the C_2 splitter in Fig. 3-13 is partially condensed in E-601. The process conditions for the vapor entering the condenser are:

Temperature	−30.1°C
Pressure	2944 kPa
Flow rate into condenser	
CH_4	3×10^{-3} kg/s
C_2H_4	64.53 kg/s
C_2H_6	6.26×10^{-2} kg/s

A shell-and-tube exchanger has been selected for this heat-transfer process to condense 73.5 percent of the overhead vapor. Use an appropriate software package (based on TEMA guidelines) to obtain the overall heat-transfer coefficient and the area required for the condensation if the tubes used have an outside diameter of 0.0127 m and an inside diameter of 0.0094 m. Assuming that the maximum length of the tubes is 3.05 m, how many tubes will be required and what shell diameter is recommended? Propylene at −46°C serves as the coolant for the condensation process.

14-16 Air used in a catalytic oxidation process is to be heated from 15 to 270°C before entering the oxidation chamber. The heating is accomplished with the use of product gases, which cool from 380 to 200°C. A steel one-pass shell-and-tube exchanger with cross-flow on the shell side has been proposed. The average absolute pressure on both the tube side and the shell side is 1010 kPa, with the hot gases being sent through the tubes. The flow rate for the air has been set at 1.9 kg/s. The inside and outside diameters for the tubes are 0.0191 and 0.0254 m, respectively. The tubes will be arranged in line with a square pitch of 0.0381 m. The exchanger operates for 8000 h/yr. The properties of the hot product gases can be considered identical to those of air. The cost data for the exchanger are given in Fig. 14-19.

Installation costs are 15 percent of the purchased cost, and annual fixed charges including maintenance are 20 percent of the installed cost. The cost for power delivered is $0.12/kWh. Under these conditions, determine the most appropriate tube length and the purchased cost for the optimum heat exchanger.

CHAPTER

15

Separation Equipment—Design and Costs

Since the separation of most mixtures into their constituent chemical species is not a spontaneous process, separation will require the expenditure of energy or the use of external forces. Many techniques are available for separating either homogeneous or heterogeneous mixtures. This chapter examines several of the more widely used separation techniques.

When the feed mixture is a homogeneous, single-phase solution, a second immiscible phase must often be developed or added before separation of the chemical species can be achieved. The second phase is created by the use of an energy-separating agent and/or the addition of a mass-separating agent. Some of the more important separation processes besides *distillation* and *absorption* include *azeotropic* and *extractive distillation, stripping, extraction,* and *crystallization.* Design procedures for most of these separation operations have been incorporated as mathematical models into widely used commercial computer-aided chemical process simulation and design programs for steady-state operations.

The use of microporous and nonporous membranes as semipermeable barriers in the separation process has been receiving considerable attention with numerous applications. Separation techniques that involve some form of barrier in the separation process include *microfiltration, ultrafiltration, nanofiltration, osmosis, gas separation,* and *pervaporation.* On the other hand, removal of certain components may be accomplished more successfully with the use of solid mass-separating agents. Separation processes that operate in this fashion are *adsorption, chromatography,* and *ion exchange.* Finally, the use of external force fields can sometimes be used to provide the separation driving force between dissimilar molecules and ions. The use of this concept leads to another list of separation processes including *centrifugation, thermal diffusion, electrolysis, electrophoresis,* and *electrodialysis.*

When the mixture is heterogeneous, it is often more practical to use some mechanical process based on gravity, centrifugal force, pressure reduction, or electric and/or magnetic field to separate the phases. Further separation techniques, if required, can then be applied to each phase. Some of the methods used for the separation of heterogeneous mixtures include *settling and sedimentation, flotation, centrifugation, drying, evaporation,* and *filtration.*

There are two measures, besides economics, that provide a good insight with respect to the overall effect of a separation operation. One measure is the recovery percentage of key products obtained from the feed mixture. The other measure of separation is product purity. Both measures have served as guides for design engineers in the selection of suitable separation processes required in industrial processes. These measures have become even more critical as more recent separations involve the recovery of very high-cost products requiring particularly high purities from temperature-sensitive dilute feed streams as, for example, in many biotech and pharmaceutical separations.

SELECTION OF SUITABLE SEPARATION PROCESSES

The selection of the best separation process often must be made from a number of possible separation processes. The important factors to be considered in making the selection include exploitable property differences, feed and product conditions, and characteristics of the separation process. The most important feed conditions are flow rate and composition, particularly of the key components to be recovered or separated. Generally, the more dilute the key components are in the feed mixture, the greater is the cost of separation.† However, some separation operations, such as those based on the use of barriers or solid agents, perform very well even when feeds are dilute in the component that is to be recovered. The most important product condition is the product purity of the component or components that are recovered.

A guide to exploitable property differences and the separation processes utilizing these property differences is provided in Table 15-1. In reviewing the suggested separation processes that are based on these property differences, the designer must have some understanding of the general regions of applicability for these processes. For example, selecting chromatography to separate components from an industrial multicomponent mixture would be unrealistic even though a large exchange equilibrium might exist between the components. However, it would make some sense to use the latter separation process to establish the presence of undesirable impurities in the feed or product stream.

The achievement of very pure products either requires large differences in certain properties between the components that are to be separated or requires a large number of stages for the separation. It is important to consider not only molecular properties but also the bulk thermodynamic and transport properties. A good source of such data with estimation methods for many bulk properties is provided by Reid,

†G. E. Keller, *AIChE Monogr. Ser.,* **83:** 17 (1987).

Table 15-1 Property differences associated with various separation key processes[†]

Exploitable property differences	Related separation processes
Vapor pressure	Distillation Absorption and stripping Drying Evaporation
Solubility	Crystallization Leaching
Distribution coefficient	Solvent extraction Adsorption
Exchange equilibrium	Ion exchange Chromatography
Surface activity	Froth flotation
Molecular geometry	Membrane Dialysis
Molecular kinetic energy	Mass diffusion Thermal diffusion
Electrical field	Precipitation
Particle size	Filtration Screening Settling Sedimentation
Particle size and density	Centrifugation Classification Thickening Decantation Scrubbing

[†]Modified from data presented by D. R. Woods, *Process Design and Engineering Practice,* Prentice-Hall, Englewood Cliffs, NJ, 1995.

Prausnitz, and Poling.[†] Both molecular and bulk properties are given by Daubert and Danner.[‡]

Generally, when a separation process is well understood, the process can readily be designed using a mathematical model and/or scaled up to handle industrial flow rates directly from laboratory data. Other separation processes may require pilot plant data before a final design can be made with some assurance of operational success. Operations based on a barrier separation are more expensive to stage than those based on the creation or addition of a second phase. Since some separation processes are limited to a maximum size, the use of parallel units will become necessary. The choice of single or parallel units can have an effect on the economics of the separation process.

[†]R. C. Reid, J. M. Prausnitz, and B. E. Poling, *The Properties of Gases and Liquids,* 4th ed., McGraw-Hill, New York, 1987.

[‡]T. E. Daubert and R. P. Danner, *Physical and Thermodynamic Properties of Pure Chemicals—Data Compilation,* DIPPR-AIChE, Hemisphere, New York, 1989.

For example, doubling the capacity of a single unit will normally increase the capital investment by about 50 percent while the use of parallel units to handle the doubling in capacity will require an additional investment of 100 percent.

Tables 15-2 and 15-3 present some guidelines to assist the design engineer in identifying one or more feasible processes that would be appropriate for the separation of homogeneous and heterogeneous mixtures, respectively. The tables outline the normal feed conditions and the separating agent, force field, or gradient that will be involved in the separation process under consideration. Additional information in the table indicates the approximate concentration range of key components that may be encountered in the feed for each of the listed separation processes as well as an estimate of the range in flow capacity of the equipment available for laboratory to commercial use associated with these processes. Finally, advantages and disadvantages listed for each of the separation processes can provide insights for recognizing whether such alternative processes could be utilized to perform the required separation. The next section provides additional information on several of the more widely used separation processes identified in these tables.

Guidelines for the Separation Process Selection

Distillation The creation or addition of another phase in distillation is obtained by the repeated vaporization and condensation of the fluid. The separation process exploits the differences in vapor pressure of the key components in the mixture to initiate the separation. The advantages of distillation are its simple flowsheet, low capital investment, and low risk. The separation process has the ability to handle wide ranges of feed concentrations and throughputs while producing a high-purity product.

This separation process can easily be staged. It is possible to have more than 100 theoretical stages in one distillation column. Accordingly, relative volatilities between key components to be separated can be as low as 1.2, and the separation can still be achieved. In addition, mass-transfer rates between phases are generally quite high, and phase contacting and disengaging are easily accomplished so that equipment dimensions needed to obtain high degrees of separation are usually not excessive.

The equipment requirements for distillation are small compared to those for most other separation processes. All that is generally required is a column, a reboiler, a condenser, and some relatively small ancillary equipment. By comparison, separation processes based on the use of mass-separating agents typically involve two or more separate mass-transfer zones, or require semibatch operation, or do both to perform the same separation.

The economies of scale, as noted earlier, favor distillation for large-scale separations. Process information on many of the other separation processes indicates that these have higher scale-up factors. Thus, the economic improvement associated with higher throughputs is not normally as pronounced with alternative separation processes as it is with distillation. This lowered potential for cost reduction is due to the added flowsheet complexity and/or the necessity of using equipment in parallel to meet the increased throughput. Membrane separations often encounter this design limitation.

Table 15-2 General guidelines in the process selection for separating homogenous mixtures†

Separation processes for homogeneous mixtures	Feed condition and approx. mass fraction of key component	Separating agent, force field, or gradient	Normal feed capacity limits set by equip., kg/s	Key process advantages	Key process limitations
Distillation	Liquid and/or vapor $10^{-1} - 0.95$	Heat transfer	$10^{-2} - 100$	Simple flowsheet, low capital investment, easily scalable, most widely used	Needs adequate relative volatility and thermal stability of components, low energy efficiency
Azeotropic distillation	Liquid and/or vapor (dictated by azeotropic conc.)	Liquid entrainer plus heat transfer	$10^{-4} - 60$	Breaks azeotrope, operates at normal temperature and pressure, easily scalable	System more complex, requires recovery of entrainer, increased capital investment
Extractive distillation	Liquid and/or vapor $10^{-3} - 0.95$	Liquid solvent plus heat transfer	$10^{-4} - 60$	Handles low-relative-volatility mixtures, operates at normal temperature and pressure, easily scalable	Requires inexpensive solvent with high recovery, system more complex, increased capital investment
Absorption	Vapor $10^{-3} - 0.95$	Liquid absorbent	$10^{-4} - 50$	Good for recovery of soluble gases, medium capital investment	Requires low-cost solvent and recovery system, cross-contamination
Stripping	Liquid $10^{-3} - 0.75$	Stripping vapor	$10^{-4} - 50$	Strips volatile component from liquid, simple operating conditions	Requires recovery of volatile component, involves cross-contamination
Extraction	Liquid	Liquid solvent	$10^{-4} - 50$	Great flexibility in selective solvents, easily scalable	Mutual solubilities of other components, involves cross-contamination, solvent loss
Crystallization	Liquid	Heat transfer	$10^{-4} - 10$	Single processing step at low temperatures, low energy requirement, large capacity variation	Staging not easy, may require parallel units, limited to crystal-forming components

Membrane separation		Pressure gradient plus:		Good for bulk separation, clean air/water purification, some trace minerals, wide variety of membranes with range of selectivities	Low to moderate feed rates, requires chemically stable membrane, not easily staged, needs repressurization between stages, limited to non fouling fluids
Microfiltration	Liquid	Microporous membrane	$10^{-5} - 1$		
Ultrafiltration	Liquid $3 \times 10^{-3} - 4 \times 10^{-3}$	Macroporous membrane			
Reverse osmosis	Liquid $7.5 \times 10^{-4} - 4 \times 10^{-2}$	Nonporous membrane	$10^{-5} - 5$		
Pervaporation	Liquid $5 \times 10^{-4} - 0.75$	Nonporous membrane	$10^{-5} - 1$		
Adsorption	Vapor or liquid $2 \times 10^{-3} - 2 \times 10^{-1}$	Solid adsorbent	$10^{-4} - 30$	High selectivity for low-concentration stream, good for gas purification	Requires recovery sequencing, needs high selectivity and easy regeneration
Electrical field and gradient separation	Vapor or liquid	Either centrifugal, electric force field or thermal gradient	Generally very low	Effective when electrical or gradient forces enhance the separation	Generally restricted to low flow rates, high equipment costs/unit capacity

[†] Modified from data presented by J. D. Seader and E. J. Henley, *Separation Process Principles,* J. Wiley, New York, 1988, G. D. Ulrich, *A Guide to Chemical Engineering Design and Economics,* J. Wiley, New York, 1984, J. C. Humphrey and G. E. Keller, II, *Separation Process Technology,* McGraw-Hill, New York, 1997, and D. R. Woods, *Process Design and Engineering Practice,* Prentice-Hall, Englewood Cliffs, NJ, 1995.

Table 15-3 General guidelines in the process selection for separating heterogeneous mixtures†

Separation processes for heterogeneous mixtures	Feed condition and approx. mass fraction of key component	Separating agent, force field, or gradient	Normal feed capacity limits set by equip., kg/s	Key process advantages	Key process limitations
Settling and sedimentation	Liquid or gas with solid 10^{-2}– 0.4	Gravitational field	$10^{-5} - 2\times10^{1}$	Separation based on gravitational field, large capacities, low capital and maintenance costs	Poor filtrate clarity, high filtrate loss, requires large differences in density, poor with slime mixtures
Flotation	Liquid and solid $5 \times 10^{-4} - 4 \times 10^{-3}$	Adsorbent gas and surface activity	$10^{-5} - 4\times10^{-5}$	Useful in solid-solid separation, large capacity range	Separation based on selective gas bubble attachment, high operating cost to attain suitable particle size
Centrifugation	Solid-solid or liquid-solid $4 \times 10^{-1} - 0.3$	Centrifugal force field	$10^{-6} - 10^{-2}$	High separation efficiency	High capital and energy costs, limited capacity, discharge difficult
Drying	Wet solid 0.5 – 0.99	Gas and/or heat transfer	$5 \times 10^{-4} - 2$	Final drying of solids once most, liquid has been removed	Staging inconvenient, may require parallel units, high energy requirement
Evaporation	Liquid	Heat transfer	$10^{-1} - 30$	Use when relative volatility >20	Low thermal efficiency
Condensation	Vapor	Heat transfer	$10^{-1} - 30$	Recovery of less volatile components in vapor or condensation of vapor	Appropriate coolant required
Filtration	Liquid-solid or gas-solid $10^{-2} - 0.75$	Pressure gradient plus porous filter	$10^{-6} - 10^{-4}$	Good solid-liquid separation of fibers, pulps, granules and slimes; low capital cost	High operating cost, often batch operation, poor with toxic or hazardous materials

†Modified from data presented by J. D. Seader and E. J. Henley, *Separation Process Principles,* J. Wiley, New York, 1988, G. D. Ulrich, *A Guide to Chemical Engineering Design and Economics,* J. Wiley, New York, 1984, J. C. Humphrey and G. E. Keller, II, *Separation Process Technology,* McGraw-Hill, New York, 1997, and D. R. Woods, *Process Design and Engineering Practice,* Prentice-Hall, Englewood Cliffs, NJ, 1995.

On the other hand, distillation can become very expensive when the relative volatility between key components is decidedly below 1.2 over part of or all the distillation operating curve in the column. In fact, some separations can involve azeotrope-forming systems in which the relative volatility for the key components actually becomes less than unity. In cases of low relative volatility, other separation processes should be examined to attain less expensive options.

Fixed composition and product requirements for purity can sometimes create additional problems. For example, separation of a high-boiling component from a 10 to 20 percent or less concentration in the feed will increase the energy costs per unit of desired product and make it excessive. Also, when the boiling range of one set of components overlaps the boiling range of another set of components from which it must be separated, distillation will require a number of columns to complete the separation. In such a case, another separation process might exploit other property differences and achieve the separation in fewer steps.

Distillation can also be at a disadvantage when extremes of temperature and pressure are encountered. For temperatures less than $-40°C$ or greater than $250°C$, the cost for the added refrigeration or heating involved escalates rapidly from typical cooling water and steam costs. If required operating pressures are less than about 2 kPa, column size and vacuum costs also escalate rapidly. Likewise, pressures greater than 5 MPa will result in an escalation of column costs.

Earlier we noted that distillation provides a relatively low scale-up factor. Unfortunately, it does not scale down very well. Thus, alternative processes that may not compete economically in the separation of large production rates may compete quite well when production rates are quite low.

Finally, distillation may be unsuitable when undesirable reactions occur at column temperatures. This can result in significant product loss or in the formation of by-products that can further complicate the separation process. For example, if only a small fraction of the feed reacts to form a by-product that precipitates on the tray openings, fouling may require intermittent cleaning with considerable loss in production or the requirement of a duplicate separation system to maintain continuous operation during the cleanup period.

Absorption A widely used alternative to distillation for the separation of a solute from a gas stream is absorption. In this separation process, the gas mixture is contacted with a liquid solvent which preferentially absorbs one or more components from the gas stream. Liquid flow rate, temperature, and pressure are the variables that must be set in this separation process.

The *absorption factor* L/KV for a relatively dilute component in the gas phase determines how readily that component will absorb in the liquid phase. Here L and V are the liquid and vapor mass flow rates, respectively, and K is the vapor/liquid equilibrium value for that component. When the absorption factor for a specified component is large, its absorptivity in the liquid is increased. When the absorption factor is increased by increasing the liquid flow rate, the number of trays in the column required to obtain a specified separation decreases. However, at high values of the absorption factor, an increase in the liquid flow rate achieves diminishing returns.

Accordingly, an economic optimum for process design has been established, namely, $1.2 < L/KV < 2.0$. A value of 1.4 is often used.†

In an absorber, the transfer of solute from the gas to the liquid phase can bring about a heating effect that will generally lead to a temperature rise down the column. If the component being absorbed by the liquid is dilute, the heating effect will be small, resulting in a small temperature rise. However, with a large absorption the temperature rise down the column can be significant. Unfortunately, an increase in temperature decreases the solubility of the absorbent and reduces the separating efficiency of the absorption process. To minimize the effect of the temperature rise in absorbers, the liquid flowing down the column is sometimes cooled through heat exchange with cooling water or refrigeration at intermediate locations in the column.

Greater solubility of the desired component or components can be obtained by operating the absorber at higher pressures. However, operating at higher pressure tends to be more expensive, particularly if this requires the use of a gas compressor. Thus, economically there is an optimal pressure for the absorption process.

Stripping Once a solute is dissolved in a solvent, it is often necessary to separate the solute from the absorbent in a stripping operation to recycle the solvent back to the absorber. The *stripping factor* KV/L should be large, to permit stripping the solute and concentrating the latter in the vapor phase. For a stripping column, the stripping factor should be in the range of $1.2 < KV/L < 2.0$ and economically is often close to 1.4. With strippers a temperature decrease is noted as the liquid flows down the column. The reasoning is analogous to that developed earlier for absorbers. Thus, to enhance the stripping process requires an increase in operating temperature and/or a decrease in column pressure.

Extraction Liquid-liquid extraction is a process often selected to separate azeotropes and components with overlapping boiling points. The extraction process requires the addition of a solvent in which one or more of the components in the mixture are soluble. Components in the liquid mixture are separated based on their different solubilities in the solvent.

The extraction process can offer energy savings, particularly when solvent recovery is relatively easy. The process can be operated at low to moderate temperatures for recovery of thermally sensitive products in the food and pharmaceutical industries. However, the requirement of a solvent increases the complexity of the process. The latter consists of an extractor plus at least another distillation column to recover the solvent. If the solvent has some miscibility in both components that are being separated, a second distillation column will be required to complete the recovery of the solvent.

One of the more recent advances in liquid-liquid extraction has been the use of a supercritical fluid such as carbon dioxide as a solvent. Supercritical solvents offer enhanced transport properties because solutes diffuse more rapidly through a supercritical solvent than through a liquid solvent. Supercritical fluids also provide other desirable solvent properties, with equilibrium ratios and separation factors that are

†J. M. Douglas, *Conceptual Design of Chemical Processes,* McGraw-Hill, New York, 1988.

appreciably larger than with normal solvents. Since the solvent is in the supercritical state, it can be recovered as a gas by simply reducing the pressure and changing the temperature.

The primary disadvantage of supercritical fluid extraction is that the extractor must be maintained at a pressure sufficient to keep the solvent in the supercritical state. Since supercritical fluid extraction generally requires operating pressures between 7 and 35 MPa, the capital and operating costs are higher. Nevertheless, supercritical fluid extraction has found favor in the purification of high-boiling specialty products where molecular distillation previously had been used. In such applications this separation process has the advantage because multiple stages can be used, whereas in molecular distillation staging is difficult and expensive.

Crystallization Under certain operating conditions, dissolved materials in solution are recoverable in solid form by precipitation upon cooling, removal of the solvent, or the addition of precipitating agents. The design of crystallizers is based on the knowledge of phase equilibria, solubilities, rates and amounts of nuclei generated, and rates of crystal growth. Each crystallization system is unique in most of these design attributes, and these are not predictable. Since theoretical advances have been slow in this separation process, pilot-plant evaluation is normally required to establish critical design parameters.

Various crystallizers presently employed in a number of industrial processes have been identified by Walas.[†] For example, batch crystallizers are primarily used in the production of fine chemicals and pharmaceuticals at a rate of 0.0015 to 0.15 kg/s. On the other hand, crystallizers are also of the continuous type utilizing circulating evaporators and/or circulating cooling crystallizers. Many of these crystallizers provide some control of crystal size with the aid of special design features.

Membrane Separation This type of separation is accomplished with the use of specially prepared membranes that selectively permit one or more components of a feed stream to pass through the barrier while retarding the passage of other components in the feed stream. The stream passing through the membrane is referred to as the *permeate* while the stream retained by the membrane is the *retentate*. In most membrane processes, a pressure difference between the feed and permeate streams provides the driving force for the separation.

Membrane processes are classified by the size of particles or molecules being separated as defined in Table 15-4. Filtration, discussed under heterogeneous separation, was not included in this classification since the barrier in filtration separates particles with dimensions greater than 5×10^{-6} m.

Since membrane processes can separate at the molecular and very small particle levels, a large number of difficult separations are now possible. Membrane processes, except for pervaporation, generally do not require a phase change to make a separation. Energy requirements, as a rule, will be low unless energy is required to increase the pressure of the feed stream to obtain a suitable pressure driving force. Suitable

[†]S. M. Walas, *Chemical Process Equipment,* Butterworth-Heinemann, Newton, MA, 1988.

Table 15-4 Size classification for membranes

Membrane process	Separation mechanism	Pore or inter-molecular size, m	Transport regime
Microfiltration	Size exclusion	$5\times10^{-8}-5\times10^{-6}$	Macropores
Ultrafiltration	Size exclusion	$2\times10^{-9}-5\times10^{-9}$	Mesopores
Nanofiltration	Size exclusion	$<2\times10^{-9}$	Micropores
Reverse osmosis	Solution/diffusion	$<1\times10^{-9}$	Molecular
Gas separation	Solution/diffusion	$<1\times10^{-9}$	Molecular
Pervaporation	Solution/diffusion	$<1\times10^{-9}$	Molecular

membranes with high selectivity for various components are available or can be fabricated from a large number of polymers and inorganic media. Membrane processes generally are fairly simple with no complex control schemes and require little ancillary equipment except for pumps and compressors.

Membrane processes, however, also have their limitations. For example, product purity can be a problem since the process seldom can produce two pure products. Improvement in product purity would require staging such as used in distillation. Generally, it is difficult to stage membrane processes. This means that membranes being used for a given separation must have much higher selectivities than would be necessary for distillation with respect to relative volatilities.

Many membranes selected for their high selectivity can exhibit incompatibilities with one or more components in the feed stream. For example, many polymers expand and cause damage to the structural integrity of the membrane when exposed to aromatics and various oxygenated solvents. Likewise, membranes generally cannot operate much above room temperature, to avoid damage to the membrane.

Finally, membrane processes generally become uneconomical for large flow rates when compared to those of distillation and absorption. This conclusion is reached because larger flow rates require the replication of identical membrane units which do not enjoy the economies of scale-up. There are, of course, some notable exceptions to this rule, including seawater desalination and large-scale hydrogen recovery.

Adsorption The selective adsorption of components on an adsorbent followed by removal of these components through regeneration forms the basis for adsorption. In most cases, the forces that bind adsorbates to adsorbents are weaker than the forces that bind atoms together. This fact allows the adsorption process to be reversed by either raising the temperature of the absorbent or reducing the concentration or partial pressure of the adsorbate. Sometimes, both effects are utilized.

An adsorbent is another example of using a mass-separating agent to facilitate a separation. However, because the mass-separating agent is a solid, the ability to move it from one location to another is rather remote. Therefore the functions of adsorbing and desorbing are generally limited to changing the thermodynamic conditions which a fixed adsorbent bed experiences as a function of time. However, by creating a desorption or regeneration step in the overall process, the adsorbent generally can be reused many times, thereby creating a highly economic design feature. The downside of the separate regeneration step is that the overall process is cyclic in time. Such periodic operation can introduce a great deal of process complexity, which can in some cases result in unacceptably high capital costs for the separation.

Adsorption processes find applicability throughout a broad band of industries in both bulk separations and gas purification. When adsorption processes are used in bulk separation applications, the design engineers must recognize that large quantities of heat are liberated during the adsorption process and that conversely large quantities of heat are consumed during the desorption process. For gas adsorptions, this release of energy is often 2 to 3 times the heat of vaporization. For liquid absorptions, this energy release is somewhat smaller. For gas purification, which involves adsorption of small amounts of gas, this energy release can usually be neglected. However, for bulk separations, the energy release becomes an important issue in the process design of such equipment.

Chromatographic processes have a technological similarity to adsorption processes since both use selective concentration on a solid surface to effect a separation. The chromatographic separations generally involve either laboratory work or the separation of very small, high-value chemicals for use by the pharmaceutical and biotechnology industries.

Ion exchange is also similar to adsorption in that solid particles are used in the separation process and regeneration is necessary. However, a chemical reaction is involved. For example, in water softening, the solid organic polymer in its sodium form removes calcium ions from the water and replaces them chemically with sodium ions. After prolonged use, the polymer is regenerated to remove the calcium ions.

External Field or Gradient Separation As noted earlier, separations utilizing external force fields or thermal gradients can be used to take advantage of the response of molecules and ions to such forces. Since the use of these separation processes is limited to a few specialized industrial applications, the emphasis in this chapter will not be directed to these specialized separation processes.

Settling and Sedimentation In settling processes, solid particles or liquid drops are separated from a fluid by gravitational forces acting on the particles or drops. The fluid can be a liquid or a gas. Flash drums are often used to separate a vapor-liquid mixture by gravity. In such a separation the velocity of the vapor within the drum must be less than the settling velocity of the liquid drops. In a simple gravity settler for removing a dispersed liquid phase from another liquid phase, the horizontal velocity of the fluid within the settler must be low enough to allow the low-density droplets to rise to the interface and the high-density droplets to move away from the interface and coalesce.

The prime function of a sedimentation device is to produce a more concentrated slurry. In one type of thickener the slurry is fed to a cone-shaped tank with a slowly revolving rake to scrape and move the thickened slurry toward the center of the cone for removal. Often in such operations a flocculating agent is added to the slurry to aid in the settling process.

Flotation In this process of separating specific solid-solid mixtures, gas bubbles are generated in a liquid and become attached to selected solid particles, allowing them to rise to the liquid surface. The solid particles are removed from the surface by an overflow weir or a mechanical scraper. The separation of the solid particles is dependent on differing surface properties which permit preferential attachment to the gas bubbles. A number of additives can be added to meet various requirements of the flotation process. For example, modifiers control the pH of the liquid-solid mixture, activators increase the activity of the solid surface, and frothers create a stable froth and assist in the separation. The bubbles of gas can be generated by gaseous dispersion, dissolution, or electrolysis of the liquid.

Centrifugation Separation by applying a centrifugal force field was noted earlier under external field separation. It is mentioned again since it is a useful separation tool when gravity separations of heterogeneous mixtures may be too slow because of the close densities of the particles and the fluid, the small particle size of the particle resulting in low settling velocities, or the formation of a stable emulsion. Use of centrifugal force increases the force acting on the particle and permits more rapid separation times. Besides solids separation from liquids, it is also sometimes used to separate two liquids with different densities by centrifugation. In this case, the denser fluid occupies the outer periphery of the centrifuge, since the centrifugal force is greater on the denser liquid.

Drying The main purpose of this separation process is to remove liquid from a liquid-solid system to produce a dry solid. Generally, the liquid removed is water, but drying can also be used to separate organic liquids from solids. Most dryers obtain the heat required to vaporize the liquid from the solid by employing a series of gas-solid contacting devices. The specific type of contacting device selected is based on the condition of the feed, the desired form of the product, and the temperature sensitivity of the solid. For example, if the feed is a dilute mixture of solids in a liquid, the dryer selected to remove the liquid could be either a single-drum or a spray dryer. On the other hand, for a moist solid the drying choice could be a tunnel, screw, or rotary dryer. The variety of equipment available for drying service is quite large, particularly when additional requirements of product form and heat sensitivity need to be considered.

Drying devices are generally subdivided into two groups: those that depend on mechanical means for gas-solid contact and those that depend on fluid motion. Within each group there is a second subdivision, namely, those that utilize either direct or indirect heating. Direct heating occurs when hot gases or combustion products come in direct contact with the solid being dried. In indirect heating, the energy transfer is through a conductive surface that separates the process stream from the heating media.

In the processing of solutions or slurries, successful heating devices provide a transition zone at the entrance, atomize the fluid, or premix it with recycled solids to enhance the flow. From an energy conservation viewpoint, total evaporization of the liquid from the solution or slurry should be avoided unless other separating or concentrating techniques are impractical or inappropriate. For gummy pastes, a pretreatment step is often necessary to convert the paste to a form that it can be successfully processed. With dusty materials, special equipment may be required to maintain a dust-free environment. Heat-sensitive solids will require dryers with precise temperature control and may even require drying under vacuum conditions.

Evaporation In separation technology, evaporators may be considered as liquid dryers or concentrators that separate solvents from a solution by evaporation. In most cases the solvent is water. Evaporation, as a separation process, is distinct from distillation in that the solute is nonvolatile and a high degree of separation can be achieved in one stage.

Evaporators are essentially reboilers with adequate provisions for separating liquid and vapor phases and for removing solids when they are precipitated or crystallized from the solution. However, because of the varied fluid property characteristics of the solutions that need to be concentrated by evaporation of the solvent, there are presently many specialized types of evaporators in use. For those evaporators employing tubes to circulate the heating fluid, the tubes may be horizontal or vertical, long or short. The liquid may be inside or outside of the tubes, and circulation may be natural or forced.

The forced circulation type of heater is the most versatile, particularly for viscous and fouling operation, but such units are more expensive than the natural circulation evaporators. Long-tube vertical evaporators, with either natural or forced convection, are the most widely used units. The tubes in these units range from 0.019 to 0.063 m in diameter with a length of 4 to 12 m. Thus, there is a wide range in capacity for this type of evaporator.

Since vaporization is energy-intensive, thermal economy is a major consideration. If a stream must be concentrated, separation processes such as precipitation and filtration or reverse osmosis that do not require a heat input should be considered. If there is no practical alternative to evaporation, then it may be possible to improve on the thermal economy by recovering and reusing the latent heat of the vaporized solvent. In practice, individual evaporators are staged successively so that each unit operates at a lower pressure than the preceding one. In this way, the vapor from one stage can be used as a source of heat to evaporate the liquid in the next stage. There are a number of different arrangements available when multistage systems are being considered for performing the separation process. Such arrangements make it possible to recover energy even from low-level process steam that is used as the heat source.

Filtration In this mode of separation, solid particles in a liquid or gas are removed by passing the mixture through a porous medium that retains the particles and passes the fluid. Filtration occurs by either cake filtration in which the particles are retained on the surface of the filter medium or depth filtration in which the particles are retained within the filter medium. The filter medium in cake filtration is generally a cloth of natural or artificial fibers or even metal. The filter cloth most often is supported between plates in an enclosure. To separate solid particles from a gas, the filter cloth is arranged in the shape of a large thimble and is designated as a bag filter. Filtration of a slurry is best done with either a porous belt or drum that rotates through the slurry to remove particles by direct contact. If purity of the filter cake is of no concern, filter aids can be added prior to the filtration step to assist with the filtration process.

Rather than using a cloth, the deep-bed filters use a granular medium in which the solid particles are trapped within the medium. The trapped particles can be released by periodically reversing the flow of the filtrate. Approximately 3 percent of the throughput is required for the periodic backwashing.

EXAMPLE 15-1 Identification of Feasible Options to Obtain a High-Purity Propylene Product from a Propane/Propylene Mixture

This separation process is presently carried out using a 150-180 stage distillation tower with a large reflux ratio. Operating conditions of such a separation require the use of two columns in series, as detailed in Example 15-2. Because of the large capital investment and operating costs for achieving this separation, what are some other feasible separation options that could be used to obtain a high-purity propylene product similar to that achieved with the conventional separation process?

■ Solution

An evaluation of the properties associated with the two components is necessary to determine what properties might be exploited to effect a separation. Key properties are listed below.

Property	Propane	Propylene
Molecular weight, kg/kg mol	44.096	42.081
Normal melting point, K	85.5	87.9
Normal boiling point, K	231.1	225.4
Critical temperature, K	369.8	364.8
Critical pressure, kPa	4250	4610
Dipole moment, debyes	0.0	0.4
Van der Waal's volume, m^3/kg mol	0.03757	0.03408
Van der Waal's area, m^2/kg mol $\times 10^{-8}$	5.59	5.060
Acentric factor	0.152	0.142

The only property listed above that might be exploited is the dipole moment. Because of the asymmetric location of the double bond in propylene, its dipole moment is considerably greater than that of propane. A review of the literature reveals that the use of facilitated transport membranes with silver ions selectively transporting the propylene across the membrane has provided a promising separation process.[†] The economic benefits of using a hybrid system that combines the separation aspects of the membrane systems coupled with a considerably smaller distillation column have been examined by other investigators.[‡] One configuration investigated by these investigators was shown in the design problem presented in Chap. 3.

Another separation process that exploits the polarity difference between propylene and propane uses extractive distillation with a polar solvent, such as furfural or an aliphatic nitrile, to reduce the volatility of the propylene.[§] The literature also reports that either silica gel or a zeolite can be used to selectively recover the propylene by adsorption.[¶]

■

[†]R. D. Hughes, J. A. Mahoney, and E. F. Steigelmann, in *Recent Developments in Separation Science,* vol. 9, N. Li and J. M. Calo, eds., CRC Press, Cleveland, Ohio, 1982, p. 173.

[‡]A. A. Al-Rabiah, K. D. Timmerhaus, and R. D. Noble, in *Proceedings 5th World Congress of Chemical Engineering,* American Institute of Chemical Engineers, New York, 1996, vol. 4, pp. 335–340.

[§]U.S. Patent 2,588,056, March 1952.

[¶]C. M. Shu, S. Kulvaronan, M. E. Findley, and A. T. Liapis, *Separation Technology,* **1:** 18 (1990).

EQUIPMENT DESIGN AND COSTS FOR SEPARATING HOMOGENEOUS MIXTURES

As noted earlier, the separation of homogeneous mixtures requires the formation or addition of another phase. The transfer of mass from one phase to another is involved in the operations of distillation, azeotropic distillation, extractive distillation, absorption, stripping, extraction, and crystallization. The principal function of the associated equipment used for these operations is to permit efficient contact between the phases. Even though many different types of equipment have been developed for these separation processes, only a few of the more widely used separation devices can be included in this equipment design and cost summary.

SEPARATION BY DISTILLATION

There are two basic approaches used in distillation column design, namely, *design* and *rating*. The former approach involves the design of a new column to determine the column diameter and height required to achieve a specified separation. This approach utilizes stage-to-stage calculations for determining the number of equilibrium stages. The rating approach involves the retrofit of an existing column in which the column diameter and height are fixed and the flow capacity and separation are to be determined. Since the rating approach has convergence advantages, it is most often used in computer algorithms which do stage-to-stage calculations.

In the design of a new column or rating of an existing one, many design engineers obtain final designs and price quotes from equipment vendors. However, to evaluate such vendor responses, a preliminary design should be made. Accordingly, the procedures involved in the design of a distillation column are summarized in Fig. 15-1 and briefly outlined with supporting equations in this section of the chapter.

Distillation Design Procedures for Columns with Sieve Trays

Designation of Design Bases The first step in the design procedure (see Fig. 15-1) is to establish the composition and physical properties of the feed and products, the feed rate, and any special limitations placed on the separation. The latter could include maximum temperature and pressure drop restrictions, presence of toxic materials or reactive components, etc.

Selection of Design Variables The second step involves selecting adequate design variables. The operating pressure is generally the first variable to be addressed. An increase in operating pressure is often reflected by an increase in separation difficulty, an increase in reboiler and condenser temperatures, an increase in vapor density, and a decrease in the latent heat of vaporization. As the pressure is lowered, these effects are reversed. The lower limit is often set by the desire to avoid vacuum operation and use of external refrigeration for the condenser because of capital and operating cost penalties. It is usually adequate, if process constraints permit, to fix the operating pressure at as low a pressure above ambient that permits cooling water or air cooling to be used in the condenser. In other words, the operating pressure should be selected so that the bubble point of the overhead product is at least 5 to 10°C above the summer cooling water temperature, or to atmospheric pressure if the latter would introduce vacuum operation.

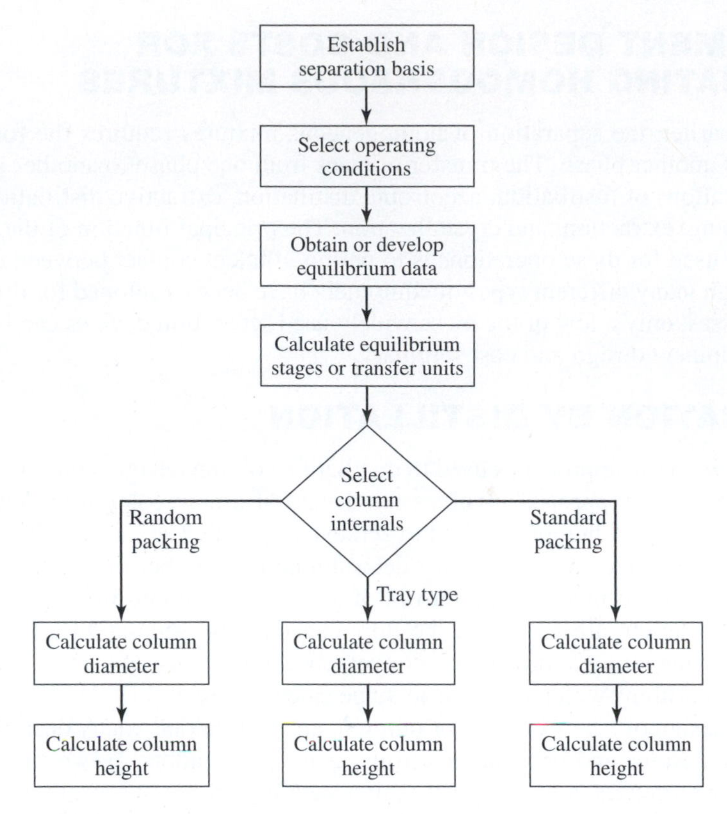

Figure 15-1
Procedural steps involved in the design of a distillation column

The second variable that needs to be set for the distillation process is the reflux ratio. With an increase in reflux ratio from its minimum, the capital cost decreases initially because of a reduction in the number of plates required for the separation. However, the utility costs increase as more reboiling and condensation duties are required. If the capital costs involved in the separation process are annualized and combined with the annual cost of the utilities, an optimal reflux ratio can be obtained. Most designers select a ratio of actual reflux to minimum reflux of at least 1.2 or above, except in special cases such as ultralow-temperature separation, since a small error in the design data or a small change in the operating conditions might lead to a process that does not meet the original design specifications.

A design variable of lesser importance is the vapor/liquid condition of the feed. If the feed is subcooled, the number of trays in the rectifying section is decreased and the number of trays in the stripping section is increased. This feed condition requires more heat in the reboiler and less cooling in the condenser. Partially vaporized feed reverses these effects. For a given separation, the feed conditions can be optimized.

Establishment of Physical Equilibria Data Once the bases and operating conditions have been established, the appropriate phase equilibria must be determined. Either experimental or predicted phase equilibria can be used. Since most experimental vapor/liquid equilibria are for binary mixtures, and not for multicomponent mixtures, it is necessary to use the data for binary pairs and combine these data with a model to predict multicomponent behavior. The Wilson, NRTL, and UNIQUAC models[†] use this approach to predict vapor/liquid equilibria. When no binary data are available, the UNIFAC[‡] model can be used for the prediction based only on functional groups. In either case, most commercial simulators are capable of providing reasonably accurate thermodynamic bases to predict phase equilibria.

Determination of Number of Equilibrium Stages Even though shortcut and stage-by-stage methods for calculating the number of equilibrium stages are available in many commercial computer software programs, it is necessary to understand the fundamentals involved to determine whether the computer results are realistic. Of the shortcut methods, the Fenske-Underwood-Gilliland method is the most widely used. To determine the number of equilibrium stages, first the minimum number of stages and the minimum reflux must be evaluated. The minimum number of stages N_{min} is obtained from the Fenske relation[§]

$$N_{\text{min}} = \frac{\ln[(x_{LK}/x_{HK})_D(x_{HK}/x_{LK})_B]}{\ln(\alpha_{LK/HK})_{\text{av}}} \tag{15-1}$$

where x_{LK} is the mol fraction of the light key, x_{HK} the mol fraction of the heavy key, $(\alpha_{LK/HK})_{\text{av}}$ the average geometric relative volatility of the light key to the heavy key, while the subscripts D and B refer to the distillate and bottom products, respectively. The geometric average value for the relative volatility is calculated by using the dew point temperature of the assumed overhead product and the bubble point temperature of the assumed bottoms product. Thus,

$$(\alpha_{LK/HK})_{\text{av}} = [(\alpha_{LK/HK})_D(\alpha_{LK/HK})_B]^{1/2} \tag{15-1a}$$

To determine the minimum reflux ratio requires Eqs. (15-2) and (15-3) developed by Underwood,[¶] namely,

$$\sum_{i=1}^{n} \frac{\alpha_i x_{F,i}}{\alpha_i - \Theta} = 1 - \bar{q} \tag{15-2}$$

where α_i is the average geometric relative volatility of component i in the mixture relative to the heavy key, $x_{F,i}$ the mol fraction of component i in the feed, and $\bar{q}$ the mols of saturated liquid on the feed tray per mol of feed. The value of Θ is determined by trial and error and lies between the relative volatilities of the two key components. The minimum reflux R_{min} is obtained from

$$R_{\text{min}} + 1 = \sum_{i=1}^{n} \frac{\alpha_i x_{D,i}}{\alpha_i - \Theta} \tag{15-3}$$

[†]G. M. Wilson, *J. Am. Chem. Soc.,* **86:** 127 (1964), H. Renon and J. M. Prausnitz, *AIChE Journal,* **14:** 135 (1968), J. M. Prausnitz et al., *Computer Calculations for Multicomponent Vapor-Liquid and Liquid-Liquid Equilibria,* Prentice-Hall, Englewood Cliffs, NJ, 1980.

[‡]A. Fredenslund et al., *Vapor-Liquid Equilibria Using UNIFAC,* Elsevier, Amsterdam, Netherlands, 1977.

[§]M. R. Fenske, *Ind. Eng. Chem.,* **24:** 482 (1932).

[¶]A. J. V. Underwood, *Chem. Eng. Prog.,* **44:** 603 (1948).

where n is the number of individual components in the feed and $\alpha_{D,i}$ the mol fraction of component i in the distillate. Gilliland[†] related the number of equilibrium stages N as a function of the number of equilibrium stages and the minimum reflux ratio with a plot that was transformed by Eduljee[‡] into the relation

$$\frac{N - N_{\min}}{N+1} = 0.75\left[1 - \left(\frac{R - R_{\min}}{R+1}\right)^{0.566}\right] \tag{15-4}$$

where R is the operating reflux selected by the designer. Distillate and bottoms distributions of any nonkey components in the feed may be evaluated after the minimum number of stages has been calculated from

$$\frac{x_{D,i}}{x_{B,i}} = (\alpha_i)_{\text{av}}^{N_{\min}} \frac{(x_{HK})_D}{(x_{HK})_B} \tag{15-5}$$

where $x_{B,i}$ is the mol fraction of component i in the bottoms and $(\alpha_i)_{\text{av}}$ is the average geometric relative volatility of component i relative to the heavy key as given by Eq. (15-1a). The Kirkbride[§] method is used to determine the ratio of trays above and below the feed point.

$$\log \frac{N_D}{N_B} = 0.206 \log\left\{\left(\frac{B}{D}\right)\left(\frac{x_{HK}}{x_{LK}}\right)_F \left[\frac{(x_{LK})_B}{(x_{HK})_D}\right]^2\right\} \tag{15-6}$$

where B and D are the mol flow rates of the bottoms and distillate, respectively, and N_D and N_B are the number of equilibrium stages above and below the feed tray, respectively.

Selection of Column Internals At this point the designer must make a selection based on performance and cost whether a tray, random packing, or structural packing is best for the separation process being considered. Typically, trays are favored when the operating pressure and liquid flow rate are high and when the column diameter is large. Random packings are more often recommended when the column diameter is small, corrosion and foaming are present, or batch columns are to be used. Structural packings, on the other hand, are considered for low-pressure and vacuum operation. Additionally, they are often selected when low pressure drop across the column is required or low liquid holdup is desired.

Sieve, valve, and bubble-cap trays are examples of traditional crossflow trays. The development of newer trays over the past decade to provide improved performance has been significant. Descriptions of some of these newer trays, including the Nye,™ Max-Frac,™ some of the improved multiple-downcomer trays, such as ECMB and EEMD, as well as the Ultra-Frac,™ P-K,™ and Trutna trays, are presented by Humphrey and Keller.[¶]

Figure 15-2 provides a cross-sectional view of a traditional distillation column in operation showing an example of a sieve, valve, and bubble-cap tray. Of these tray types, the sieve tray is the choice in many distillation separations since its tray fundamentals

[†]E. R. Gilliland, *Ind. Eng. Chem.,* **32:** 1220 (1940).

[‡]H. E. Eduljee, *Hydrocarbon Proc.,* **54:** 120 (1975).

[§]C. G. Kirkbride, *Petrol. Refiner.,* **23:** 32 (1944).

[¶]J. L. Humphrey and G. E. Keller, II, *Separation Process Technology,* McGraw-Hill, New York, 1997.

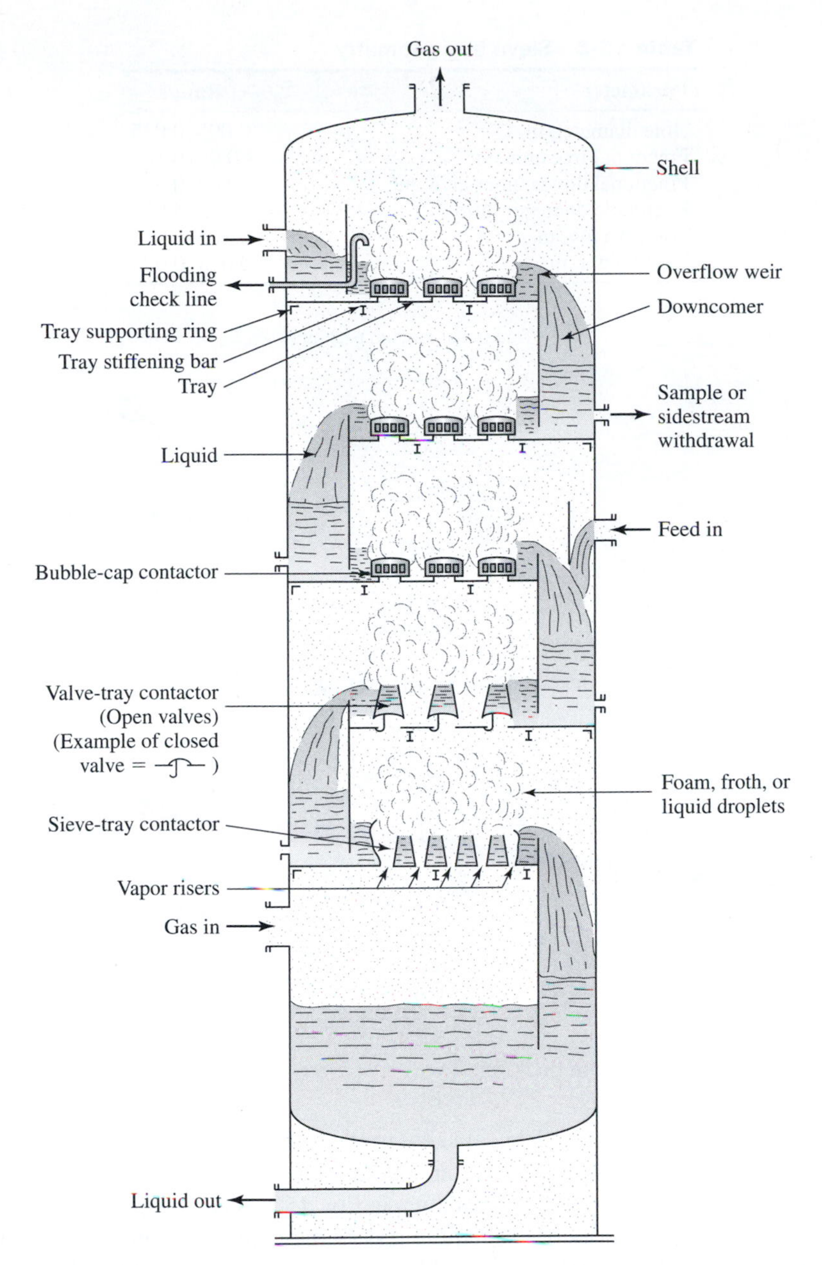

Figure 15-2
Cross-sectional view of a finite-stage contactor column in operation showing an example of a sieve tray, a valve tray, and a bubble-cap tray

are well established, entailing low risk. In addition, the trays are low in cost relative to many other tray types while handling wide variations in flow rates. Overall tray efficiencies of sieve trays generally are between 60 and 85 percent. Disadvantages of these trays are their higher pressure drops and lower capacities relative to some of the newer

Table 15-5 Sieve tray geometry

Parameter	Range
Hole diameter, m	0.005–0.025
Fractional free area, m^2	0.06–0.16
Fractional downcomer area, m^2	0.05–0.30
Pitch/hole diameter ratio	2.5–4.0
Tray spacing, m	0.305–0.915
Weir height, m	0.025–0.075

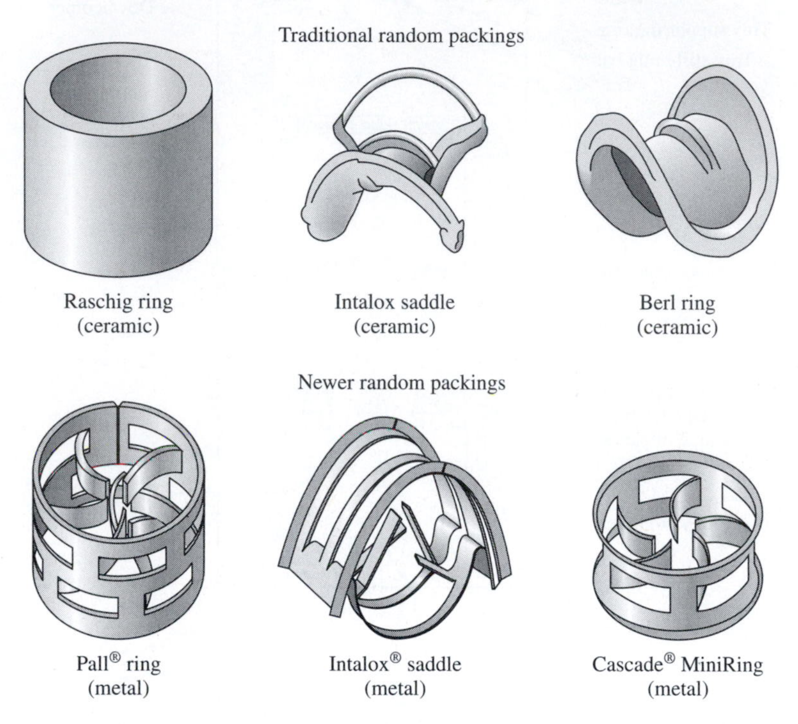

Figure 15-3
Examples of random packings

trays and structured packings. Typical sieve tray geometries, including guidelines for hole dimensions, free areas, and downcomer areas, are provided in Table 15-5.

Examples of random packings are shown in Fig. 15-3. The newer random packings have thinner elements and therefore require less of the column volume than traditional packings. A variety of materials are available for random packings with both advantages and disadvantages. For example, stoneware is susceptible to attack by alkali and hydrofluoric acid. When metals are used, there is a concern for wettability of the surface and the possibility of high corrosion rates. Plastic packings are lightweight, are easy to install, provide a low pressure drop per theoretical stage, are low-cost, and corrosion is not a problem. However, such packings can exhibit wettability problems and normally experience an upper temperature range. The geometric characteristics and the height

Table 15-6 Geometry and efficiency of several random packings†

Type of packing	Void fraction	Surface area per volume, m^2/m^3	Approx. HETP, m
25-mm Ceramic Raschig rings	0.73	190	0.6–0.12
25-mm Ceramic Intalox saddles	0.78	256	0.5–0.9
25-mm Ceramic Berl saddles	0.69	259	0.6–0.9
25-mm Plastic Pall rings	0.90	267	0.4–0.5
25-mm Metal Pall rings	0.94	207	0.25–0.3
50-mm Norpac®	0.94	102	0.45–0.6
50-mm Highflow® rings	0.93	108	0.4–0.6

†Additional values are available in M. S. Peters and K. D. Timmerhaus, *Plant Design and Economics for Chemical Engineers,* 4th ed., McGraw-Hill, New York, 1991, p. 690.

Table 15-7 Geometry and HETP of several structured packings†

Type of packing	Void fraction	Surface area per volume, m^2/m^3	Approx. HETP, m
Intalox 2T (Norton)	0.96	213	0.2–0.3
Flexipac® 1 (Koch)	0.91	558	0.2–0.3
Flexipac® 2 (Koch)	0.93	249	0.3–0.4
Gempak® 4A (Glitsch)	0.91	525	0.2–0.3
Gempak® 2A (Glitsch)	0.93	262	0.3–0.4
Sulzer BX (Sulzer)	0.90	499	0.2–0.3

†Additional values are available in H. Z. Kister, *Distillation Design,* McGraw-Hill, New York, 1992, p. 446.

equivalent to a theoretical plate (HETP) of several random packings are presented in Table 15-6.

Structured packings with a geometric structure of corrugated sheets positioned in a parallel arrangement are fabricated to fit the dimensions of the column. When placed in the column, successive elements are oriented at 90° from each other, as shown in Fig. 15-4. The structured packing occupies about 60 to 70 percent of the column volume while the remaining volume is used for flow distribution and phase disengagement. Structured packings are available in sheet or gauze form and in a variety of materials. The geometric characteristics and HETP of structured packings are highlighted in Table 15-7.

Structured packings can provide higher mass-transfer efficiencies and capacities than traditional tray contactors for vacuum and low-pressure applications. The effect of this increased efficiency and capacity is a smaller required column size. The performance of structured packings remains unaffected even at gas flow rates as low as 10 percent of the design load. In the case of absorption, another possible benefit of this packing is a reduction in the solvent/gas feed ratio that again permits the use of smaller equipment. However, disadvantages of structured packings are higher costs relative to traditional trays and the difficulty of maintaining good liquid and vapor distribution throughout the column.

The choice between use of a tray column or a packed column for a given mass-transfer operation should, theoretically, be based on a detailed cost analysis for the two

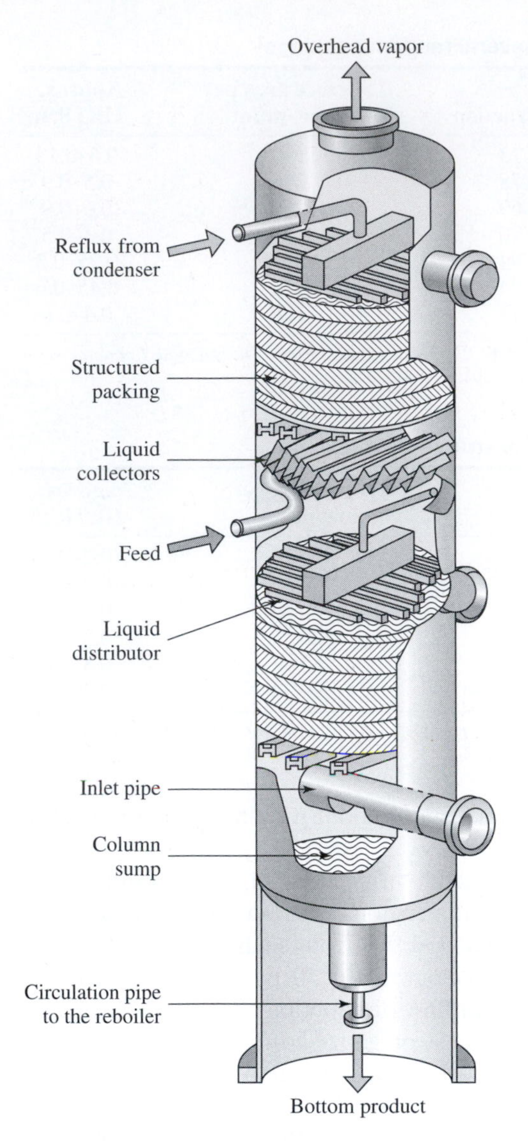

Figure 15-4
Distillation column equipped with structured packing

types of contactors. Thus, the optimum economic design for each type would be developed in detail, and the final choice would be based on a consideration of costs and profits at the optimum conditions. In many cases, however, the decision can be made on the basis of a qualitative analysis of the relative advantages and disadvantages, eliminating the need for a detailed cost comparison. The following general advantages and disadvantages of tray and packed columns should be considered when a choice must be made between the two types of contactors:

1. Stage efficiencies for packed towers must be based on experimental tests with each type of packing. The efficiency varies not only with the type and size of packing,

but also with the fluid rates, fluid properties, column diameters, operating pressure, and, in general, extent of liquid dispersion over the available packing surface.

2. Because of liquid dispersion difficulties in packed columns, the design of tray columns is considerably more reliable and requires a lower safety factor when the ratio of liquid mass velocity to gas mass velocity is low.
3. Tray columns can be designed to handle wide ranges of liquid rates without flooding.
4. If the operation involves liquids that contain dispersed solids, use of a tray column is preferred because the plates are more accessible for cleaning.
5. Tray columns are preferred if interstage cooling is required to remove heats of reaction or solution because cooling coils can be installed on the plates, or the liquid-delivery line from plate to plate can be passed through an external cooler.
6. The total weight of a dry tray column is usually less than that of a packed column designed for the same duty. However, if liquid holdup during operation is taken into account, both types of columns have about the same weight.
7. When large temperature changes are involved, as in distillation operations, tray columns are often preferred because thermal expansion or contraction of the equipment components may crush the packing.
8. Design information for tray columns is generally more readily available and more reliable than that for packed columns.
9. Random-packed columns generally are not designed with diameters larger than 1.5 m, and diameters of commercial tray columns are seldom less than 0.67 m.
10. Packed columns prove to be less expensive and easier to construct than tray columns if highly corrosive fluids must be handled.
11. Packed columns are usually preferred if the liquids have a large tendency to foam.
12. The amount of liquid holdup is considerably less in packed columns.
13. The pressure drop through packed columns may be less than the pressure drop through tray columns designed for the same duty. This advantage, plus the fact that the packing serves to lessen the possibility of column wall collapse, makes packed columns particularly desirable for vacuum operation.

Diameter Evaluation for Columns with Sieve Trays Determination of the column diameter first requires calculation of the net vapor (gas) velocity at flood conditions V_{nf} in the column from

$$V_{nf} = C_{sb}\left(\frac{\sigma}{20}\right)^{0.2}\left(\frac{\rho_L - \rho_V}{\rho_V}\right)^{0.5} \tag{15-7}$$

where C_{sb} is the Souders and Brown factor at flood conditions in m/s, σ is the surface tension in dyne/cm, and ρ_L and ρ_V are the densities of the liquid and vapor streams in the column, respectively. The Souders and Brown factor is obtained from Fig. 15-5 after specifying a reasonable tray spacing. Standard tray spacings for large-diameter columns are generally either 0.46 or 0.61 m, but 0.3- and 0.91-m spacings are also used.

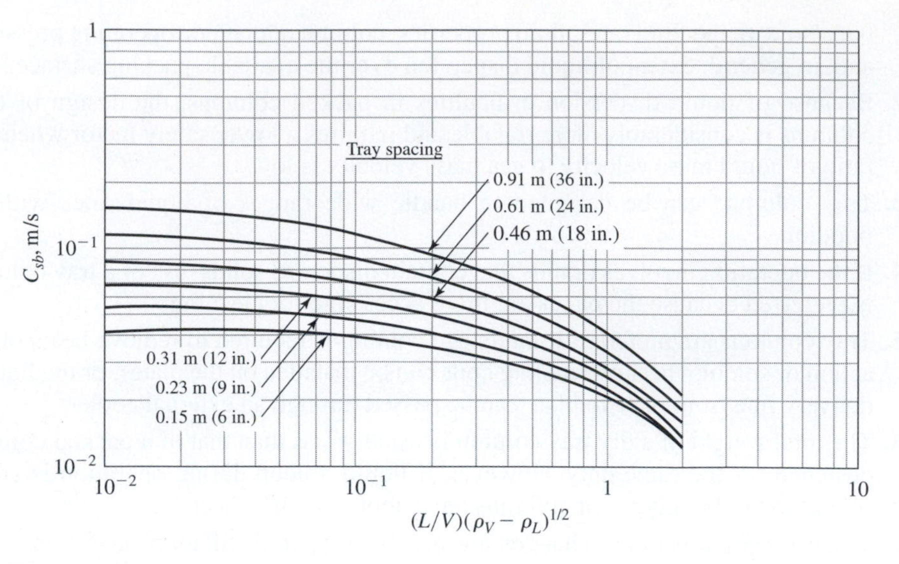

Figure 15-5
Chart for estimating values of C_{sb} (±10 percent) in Eq. (15-7). [*Adapted from J. R. Fair, Petro/Chem. Eng.,* ***33****(10): 45 (1961) with permission.*]

The actual vapor velocity V_n is selected by assuming that it is 50 to 90 percent of the net vapor velocity at flood conditions. (The lower and upper values are only selected when some unusual flow conditions are expected in the separation process.) The net column area is then obtained from

$$A_n = \frac{\dot{m}'_V}{V_n} \tag{15-8}$$

and

$$A_c = A_n + A_d \tag{15-8a}$$

where A_n is the net column area used in the separation process, A_d the downcomer area, A_c the cross-sectional area of the column, $\dot{m}'_V$ the volumetric flow rate of the vapor, and V_n the actual vapor velocity. The column diameter is then calculated from $D = (4A_c/\pi)^{1/2}$.

Height Evaluation of Columns with Sieve Trays Conversion of the equilibrium stages to actual stages requires the use of an overall tray efficiency. The most rigorous method begins with point efficiencies and then converts these efficiencies to an overall tray efficiency. Since this approach is not practical for multicomponent mixtures, a simple analytical expression, developed by Lockett,† is recommended when no supporting experimental data are available, namely,

$$E_o = 0.492[\mu_L(\alpha_{LK/HK})_{av}]^{-0.245} \tag{15-9}$$

†M. J. Lockett, *Distillation Design Fundamentals,* Cambridge University Press, Cambridge, United Kingdom, 1986.

where E_o is the overall tray efficiency, $(\alpha_{LK/HK})_{av}$ the average relative volatility between the light and heavy keys, and μ_L the liquid viscosity of the feed mixture. Since the plate spacing was selected in a prior design step, all the information is now available to determine the height of the column exclusive of the height required for phase disengagement and required internal hardware. Thus,

$$H_c = (N_{act} - 1)H_s + \Delta H \tag{15-10}$$

where H_c is the actual column height, N_{act} the actual number of trays, H_s the plate spacing, and ΔH the additional height required for column operation as noted above.

Determination of Diameter and Height of a Distillation Column — EXAMPLE 15-2

A sieve tray distillation column is to be used to separate a propane/propylene mixture. The multicomponent feed to the column is completely vapor at a temperature of 43.3°C, a pressure of 1668 kPa, and a composition as shown below.

Component	$\dot{m}$, kg/s	Component	$\dot{m}$, kg/s
Acetylene	1.3×10^{-6}	MAPD†	5.4×10^{-4}
Ethylene	1.3×10^{-6}	1,3-Butadiene	3.7×10^{-4}
Ethane	3.3×10^{-4}	*i*-Butene	1.9×10^{-4}
Propylene	6.41	1-Butene	1.2×10^{-4}
Propane	4.62		

†MAPD refers to the C_3 acetylenes (C_3H_4), methylacetylene and propadiene.

It is desired to obtain a product with a propylene concentration of 99.6 mol percent. Determine the diameter and height of the column if a tray spacing of 0.762 m is specified.

■ Solution

A review of the feed composition indicates that the separation is essentially a separation between propane and propylene. However, use of the McCabe-Thiele procedure could introduce considerable inaccuracies because of the difficulty in the graphical evaluation. Therefore, the relations for separating multicomponent mixtures will be used with a number of assumptions to minimize the calculations.

The first step is a material balance around the column. Since the other components in the feed are negligible relative to the two key components, assume only these two components in the material balance.

$$F = B + D$$

$$Fx_{(C_3H_6)_F} = Bx_{(C_3H_6)_B} + Dy_{(C_3H_6)_D}$$

$$Fx_{(C_3H_8)_F} = Bx_{(C_3H_8)_B} + Dy_{(C_3H_8)_D}$$

where $F = 0.257$ kg mol/s, $x_{(C_3H_6)_F} = 0.5924$, and $x_{(C_3H_6)_B} = 1 - x_{(C_3H_8)_B}$, also

$$y_{(C_3H_6)_D} = 0.996 \qquad y_{(C_3H_8)_D} = 0.004 \qquad B = F - D$$

Solving these equations simultaneously provides the following results:

$$D = 0.1527 \text{ kg mol/s} \qquad B = 0.1043 \text{ kg mol/s}$$

$$x_{(C_3H_6)_B} = 0.0022 \qquad x_{(C_3H_8)_B} = 0.9978$$

To obtain the relative volatility between the light key (propylene) and the heavy key (propane), assume an average temperature of 43.3°C or 316.4 K. From vapor pressure data provided by the *International Critical Tables,* the relative volatility of propylene to propane at this temperature is 1.136. A slightly more accurate value could have been obtained with Eq. (15-1*a*) by using the bubble point temperature of the bottoms and the dew point temperature of the distillate to calculate the geometric mean of the relative volatility.

Determine the minimum number of stages with Eq. (15-1).

$$N_{\min} = \frac{\ln[(x_{LK}/x_{HK})_D(x_{HK}/x_{LK})_B]}{\ln \alpha_{LK/HK}}$$

$$= \frac{\ln[(0.996/0.004)(0.9978/0.0022)]}{\ln 1.136} = 91.2 \text{ stages}$$

To obtain $R_{\min}$, use Eqs. (15-2) and (15-3).

$$\sum_{i=1}^{n} \frac{\alpha_i x_{F,i}}{\alpha_i - \Theta} = 1 - \bar{q} \qquad \text{let } q = 0 \text{ since feed is all vapor}$$

$$\frac{1.136(0.5924)}{1.136 - \Theta} + \frac{1(0.4076)}{1 - \Theta} = 1 \qquad \Theta = 1.0555$$

$$R_{\min} + 1 = \sum_{i=1}^{n} \frac{\alpha_i x_{D,i}}{\alpha_i - \Theta}$$

$$= \frac{1.136(0.996)}{1.1336 - 1.0555} + \frac{1(0.004)}{1 - 1.0555} = 13.98$$

$$R_{\min} = 12.98 \qquad \text{assume } R = 1.2R_{\min}$$

$$R = 1.2(12.98) = 15.58$$

The actual number of theoretical stages is determined with Eq. (15-4).

$$\frac{N - N_{\min}}{N + 1} = 0.75\left(1 - \frac{R - R_{\min}}{R + 1}\right)^{0.566}$$

$$\frac{N - 91.2}{N + 1} = 0.75\left(1 - \frac{15.58 - 12.98}{15.58 + 1}\right)^{0.566}$$

$$N - 91.2 = (N + 1)(0.487)$$

$$N = 179 \text{ theoretical stages}$$

The location of the feed is obtained from Eq. (15-6).

$$\log \frac{N_D}{N_B} = 0.206 \log\left\{\left(\frac{B}{D}\right)\left(\frac{x_{HK}}{x_{LK}}\right)_F \left[\frac{(x_{LK})_B}{(x_{HK})_D}\right]^2\right\}$$

$$= 0.206 \log\left[\left(\frac{0.1043}{0.1527}\right)\left(\frac{0.4076}{0.5924}\right)\left(\frac{0.0022}{0.004}\right)^2\right]$$

$$N = N_D + N_B$$

Solving these two equations results in

$$N_D = 107 \text{ theoretical stages} \qquad N_B = 72 \text{ theoretical stages}$$

Thus, the feed enters the column 72 theoretical stages above the bottom stage.

Equation (15-7) is used to obtain the net vapor velocity at flood conditions. Normally, the net vapor velocity should be obtained at both ends of the column to determine the maximum column diameter that will be required. In this design, the maximum diameter will occur at the top of the column, and only this calculation is pursued. Equation (15-7) requires information on the liquid and vapor densities as well as the surface tension. Since the distillate is essentially pure propylene, the liquid and vapor densities of this component are required. Liquid density is obtained from Table 2-30 of *Perry's Chemical Engineers' Handbook,* 7th edition, as 11.309 kg mol/m^3 or 476 kg/m^3. Vapor density is obtained with the aid of the ideal gas relationship as 27 kg/m^3. The surface tension for propylene has been estimated as 6 dyne/cm. Once the function $(L/V)(\rho_V/\rho_L)^{0.5}$ has been established, Fig. 15-5 can be used to obtain C_{sb}.

$$\left(\frac{L}{V}\right)\left(\frac{\rho_V}{\rho_L}\right)^{0.5} = \left(\frac{15.58}{16.58}\right)\left(\frac{27}{476}\right)^{0.5} = 0.224$$

From Fig. 15-5, $C_{sb} \cong 0.075$ m/s.

With Eq. (15-7)

$$\begin{aligned} V_{nf} &= C_{sb}\left(\frac{\alpha}{20}\right)^{0.2}\left(\frac{\rho_L - \rho_V}{\rho_V}\right)^{0.5} \\ &= (0.075)\left(\frac{6}{20}\right)^{0.2}\left(\frac{476 - 27}{27}\right)^{0.5} = 0.240 \text{ m/s} \end{aligned}$$

Assume 80 percent of flooding.

$$V_n = 0.8V_{nf} = 0.8(0.240) = 0.192 \text{ m/s}$$

From Eq. (15-8),

$$\begin{aligned} A_n &= \frac{\dot{m}_V}{V_n} \\ &= \frac{(16.58)(0.1527)(42.08)/27}{0.192} = 20.55 \text{ m}^2 \end{aligned}$$

Assume the downcomer occupies 15 percent of the cross-sectional area of the column. Thus,

$$A_c = \frac{A_n}{0.85} = \frac{20.55}{0.85} = 24.2 \text{ m}^2$$

and

$$D = \left(\frac{4A_c}{\pi}\right)^{1/2} = \left[\frac{4(24.2)}{\pi}\right]^{1/2} = 5.55 \text{ m}$$

Since the number of theoretical stages is too large for one column, two columns will be required, each with 90 theoretical stages and a column diameter of 5.6 m.

■

EXAMPLE 15-3 Computer Determination of Diameter and Height of a Distillation Column

With the data provided in Example 15-2 for separating a propane/propylene mixture, use one of the commercially available simulators to determine the required diameter and height of the distillation column that will provide the same high-purity product as specified.

■ Solution

The operation of the propane/polymer distillation column was simulated with the *SCDS* (Chemstations) model using the Soave-Redlich-Kwong equation of state. The reflux ratio used will also be $1.2R_{min}$. The feed location has been optimized to provide the lowest number of trays required. A tray spacing of 0.762 m was specified to maintain the flood percent below 80 percent (similar to that selected in Example 15-2). For this separation, the relative volatility varies between 1.132 and 1.14. It was assumed that the separation required with this low a relative volatility and high product purity would necessitate a two-column arrangement. This was verified in the simulation results.

The temperature profile of the propane/propylene two-column system as a function of the theoretical stage number prepared by the simulation program is shown in Fig. 15-6 to emphasize the small temperature changes that occur from one stage to the next. The design data provided by the program are shown here.†

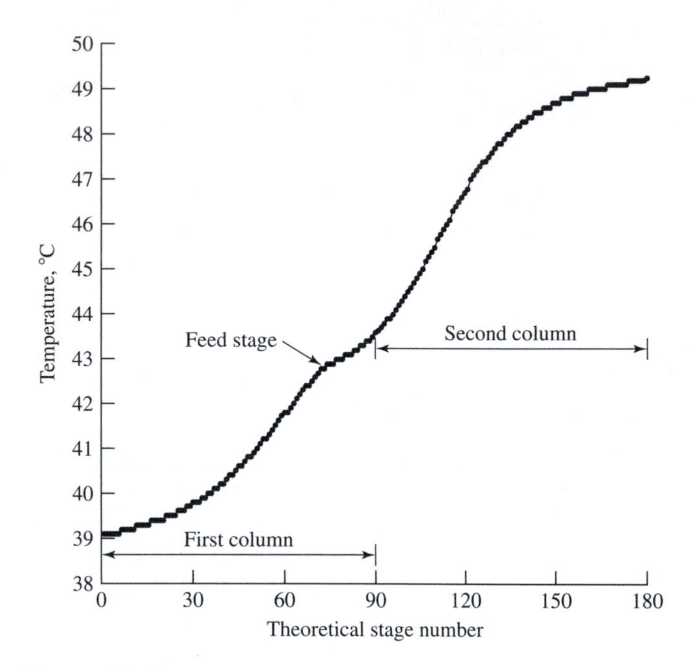

Figure 15-6
Temperature profile for the propylene/propane two-column system as a function of theoretical stage number. (From R. E. Treybal, *Mass Transfer Operation,* McGraw-Hill, 1980. Reproduced with permission of The McGraw-Hill Companies.)

Selected design parameters	Design data from simulation
Number of columns	2
Number of theoretical trays	180 (valve trays)
Average tray efficiency	0.85
Tray spacing	0.762 m
Column internal diameter	5.33 m
Column reflux ratio	15.563
Column cross-sectional area	21.86 m^2
Column active area	17.85 m^2
Percentage of flood at bottom stage	74.7%
Percentage of flood at top stage	79.4%
Pressure drop per tray	0.77 kPa per tray

It is apparent that the results from the computer simulation provide excellent agreement with the hand calculations. Generally such good agreement is not obtained; rather, there can be as much as a ±25 percent difference in the height and diameter calculations, particularly with the separation of multicomponent mixtures exhibiting a large degree of nonideality. In such cases, a good simulation program will provide a much better approach to actual operating conditions.

Distillation Design Procedures for Columns with Random Packing

As noted in Fig. 15-1, the initial procedures for designing a new distillation column are similar when either a trayed column or a packed column is selected for the separation process. Differences in the design calculations become apparent when the column diameter and height need to be established. Accordingly, the following will describe the design procedures associated with the diameter and height aspects of those columns utilizing random packing.

Diameter Evaluation for Columns with Random Packing In contrast to distillation columns with trays, performance in packed columns is strongly affected by both liquid and vapor rates in the column. Not only is flow limited, but at high throughputs the gas flow impedes the liquid flow, which can eventually lead to flooding of the column. Thus, packed columns also operate at a vapor velocity that is 70 to 90 percent of the flooding velocity. A simplified shortcut method that has been the standard of the industry for several decades utilizes the generalized pressure drop correlation chart originally developed by Sherwood[‡] and improved by several other investigators.[§] One

[†]Considerable additional design parameters are provided by this simulation program, including extensive design data on liquid and vapor velocities, weir details, downcorner specifications, etc.

[‡]R. E. Treybal, *Mass Transfer Operations,* 3d ed., McGraw-Hill, New York, 1980, p. 195.

[§]For example, see M. Leva, *Chem. Eng. Prog.,* **88**(1): 65 (1992).

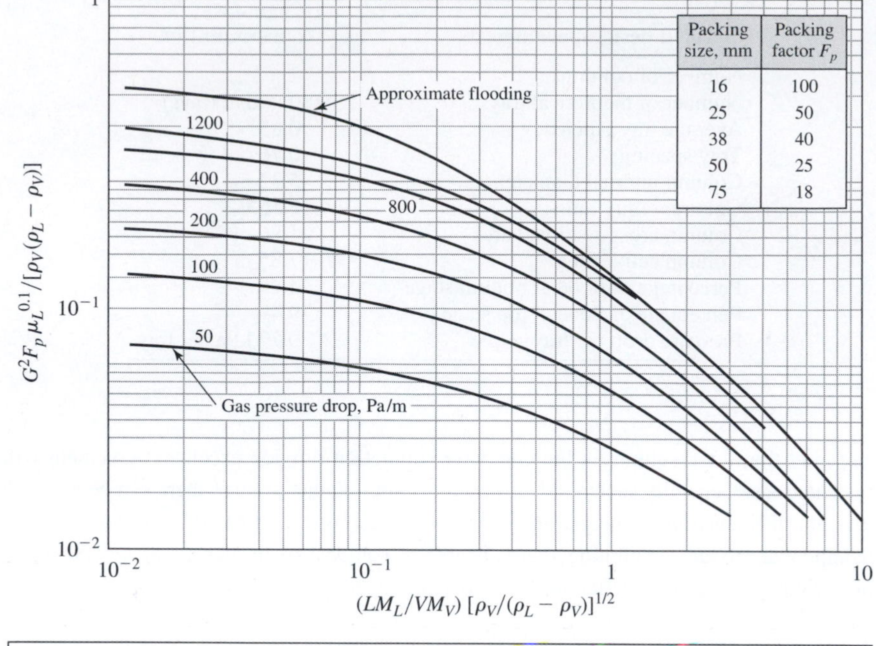

where

$G = VM_V/A_c$, superficial gas mass flux, kg/(s·m^2)
F_p = packing factor, dimensionless; see insert
μ_L = liquid viscosity, Pa·s
ρ_V = gas density, kg/m^3
ρ_L = liquid density, kg/m^3
L = liquid flow rate, kg mol/s
A_c = column cross-sectional area, m^2
M_V = molecular weight of vapor, kg/kg mol
M_L = molecular weight of liquid, kg/kg mol
V = gas flow rate, kg mol/s

Figure 15-7
Flooding and pressure drop correlations for packed columns. (From R. E. Treybal, *Mass Transport Operations,* 3d ed., McGraw-Hill, New York, 1980, p. 195 with permission of McGraw-Hill.)

version of this correlation chart, shown in Fig. 15-7, permits the designer to estimate the cross-sectional area of the column after selecting a recommended pressure drop per unit height of packing. Recommended pressure drops in packed columns for atmospheric and high-pressure separations range from 400 to 600 Pa/m, for vacuum operation between 4 and 50 Pa/m, and for absorption and stripping columns between 200 and 400 Pa/m.

The parameters involved in this estimation of the cross-sectional area A_c are L, V, ρ_L, ρ_V, μ_L, and F_p. In Fig. 15-7, L and V are the liquid and vapor mass flow rates, respectively, ρ_L and ρ_V are the liquid and vapor densities, respectively, μ_L is the liquid viscosity in Pa·s, and F_p is the packing factor associated with the random packing selected. The latter parameter is normally available from the supplier of the packing. If such information is not available, an approximation relating packing size (diameter in mm) to F_p for some of the more widely used packings is displayed in the upper right-hand corner of Fig. 15-7. Raschig rings and Berl saddles are excluded from this approximation.

The cross-sectional area for the column can be evaluated directly from the value obtained for the ordinate. The column diameter again is given by $D = (4A_c/\pi)^{1/2}$.

Height Evaluation for Columns with Random Packing Determination of the height of packing required in a column to achieve a specific separation involves use of either the height of a transfer unit (HTU) or the height equivalent to a theoretical plate (HETP) approach. To obtain the total height using the first approach also requires one to evaluate the number of transfer units (NTU) to satisfy the relation

$$Z = (\text{HTU})(\text{NTU}) \tag{15-11}$$

where Z is the total height of the mass-transfer zone. HTU and NTU are defined as

$$\text{HTU} = \frac{\dot{m}_v''}{(K_G a_e) A_c} \tag{15-12}$$

and

$$\text{NTU} = \int_{y_2}^{y_1} \frac{dy}{y - y^*} \tag{15-13}$$

where $\dot{m}_v''$ is the molar flow rate of vapor, K_G the overall mass-transfer coefficient, a_e the area of interfacial contact between the liquid and vapor phase per unit volume of contactor, A_c the cross-sectional area of the column, y the mol fraction of the component in the vapor phase, and y^* the mol fraction of the component in the vapor phase that would be in equilibrium with that component in the liquid phase.

To include the contribution of both the gas- and liquid-phase resistances, the HTU is defined as

$$\text{HTU} = \text{HTU}_V + \lambda \text{HTU}_L = \frac{V_V}{k_V a_e} + \lambda \frac{V_L}{k_L a_e} \tag{15-14}$$

where V_V and V_L are the superficial velocities of the vapor and liquid, respectively, k_V and k_L the individual mass-transfer coefficients for the vapor and liquid expressed in units of velocity, respectively, and λ the ratio of the slope of the equilibrium line to the slope of the operating line L/V.

A number of design procedures for evaluating the HTU for packed columns are available in the literature. For example, the approach by Bravo and Fair[†] involves estimating the individual mass-transfer coefficients and an *effective* interfacial area for mass transfer. The mass-transfer coefficients established by these investigators, in units of m/s, are obtained from

$$k_V = \frac{0.119 a_p D_V \text{Re}_V^{0.7} \text{Sc}_V^{0.333}}{(a_p D_p)^2} \tag{15-15}$$

[†]J. L. Bravo and J. R. Fair, *Ind. Eng. Chem. Proc. Des. & Dev.*, **21:** 162 (1982).

and

$$k_L = 3.72 \times 10^{-5}\left(\frac{\mu_L g}{\rho_L}\right)^{0.333}\left(\frac{\mathrm{Re}_L a_p}{a_w}\right)^{0.667}\mathrm{Sc}_L^{-0.5}(a_p D_p)^{0.4} \tag{15-16}$$

where

$$\frac{a_p}{a_w} = \left\{1 - \exp\left[-1.45\,\mathrm{Re}_L^{0.1}\mathrm{Fr}_L^{-0.05}\mathrm{We}_L^{0.2}\left(\frac{\sigma_c}{\sigma}\right)^{0.75}\right]\right\}^{-1} \tag{15-17}$$

$$\mathrm{Re}_V = \frac{1488\,\rho V_V}{a_p \mu_V} \tag{15-18}$$

$$\mathrm{Re}_L = \frac{1488\,\rho_L V_L}{a_p \mu_L} \tag{15-19}$$

$$\mathrm{Fr}_L = \frac{a_p V_L^2}{g} \tag{15-20}$$

$$\mathrm{We}_L = \frac{453.2\rho_L V_L^2}{a_p \sigma} \tag{15-21}$$

$$\mathrm{Sc}_L = \frac{0.000672\,\mu_L}{\rho_L D_L} \tag{15-22}$$

Note that subscripts L and V refer to the liquid and vapor phases. Other terms with their units include the packing surface area a_p m^2/m, the respective diffusion coefficients D_V and D_L m^2/s, the packing diameter D_p m, the area of wetted packing a_w m^2, the respective densities ρ_L and ρ_V kg/m^3, the respective viscosities μ_L and μ_V cP, the surface tension σ dyne/cm, and the critical surface tension σ_c, which is assumed to be 61 dyne/cm for ceramic packing, 75 dyne/cm for structured packings, and 33 dyne/cm for PEB packing. The terms Fr, We, and Sc refer to the Froude, Weber, and Schmidt dimensionless numbers, respectively.

Determination of the individual mass-transfer coefficients now permits evaluation of the effective area a_e for mass transfer from

$$a_e = a_p \sigma^{0.5}(\mathrm{Ca}_L \mathrm{Re}_V)^{0.392}\mathrm{HTU}^{0.4} \tag{15-23}$$

where the dimensionless capillary number Ca_L is obtained from

$$\mathrm{Ca}_L = \frac{0.304\,V_L \mu_L}{\sigma} \tag{15-23a}$$

The HTU_V, HTU_L, and HTU can now be determined from Eq. (15-14).

Even though the HTU approach is more rigorous, many designers elect to use the HETP approach because it provides a comparison with the number of theoretical stages determined with trayed columns. The relationship between HETP and HTU is given by

$$\mathrm{HETP} = \mathrm{HTU}\left(\frac{\ln \lambda}{\lambda - 1}\right) \tag{15-24}$$

and the total height of the packing is obtained from

$$Z = (\text{HETP})N \tag{15-25}$$

where N is the number of equilibrium stages required. Since evaluation of HETP generally involves determining the HTU, the preferred method is to obtain HETP values from experimental or plant data. For example, Kister[†] presents HETP values for a wide range of packing used in industrial separations. If such data are not available, the same author has also provided some helpful rules of thumb for predicting the HETP values for columns with random packings in terms of the column diameter obtained earlier, namely,

$$\text{HETP} = \begin{cases} D & \text{for } D \leq 0.5\,\text{m} & (15\text{-}26a) \\ 0.5D^{0.3} & \text{for } D > 0.5\,\text{m} & (15\text{-}26b) \\ D^{0.3} & \text{for absorption columns with } D > 0.5\,\text{m} & (15\text{-}26c) \end{cases}$$

For vacuum distillation, it is recommended that an extra 0.15 m be added to these predicted HETP values.

Distillation Design Procedures for Columns with Structured Packing

Diameter Evaluation for Columns with Structured Packing A procedure similar to that used for determining the diameter of a column with random packing may be used to obtain the column diameter with structured packing. If Fig. 15-7 is used to obtain the cross-sectional area of the column, the designer will need to obtain a suitable packing factor from the vendor for the structured packing selected.

Another procedure, developed by Kister and Gill,[‡] involves calculating the pressure drop over a unit length of the column under flood conditions. A correlation, similar in format to Fig. 15-7, is used to determine the flood velocity in columns with structured packing. The operating velocity in the column is designated as a fraction of the flood velocity. The cross-sectional area of the column is then obtained by dividing the volumetric flow rate of the vapor by the operating velocity. Column diameter is obtained from $D = (4A_c/\pi)^{1/2}$.

Height Evaluation for Columns with Structured Packing The total height of packing required to achieve a specific separation again necessitates determination of HTU and HETP. Several options are available. A rule of thumb for quick estimation of HETP for structured packing has been presented by Harrison and France[§]

$$\text{HETP} = \frac{9.29}{a_p} + 0.10 \tag{15-26d}$$

[†]H. Z. Kister, *Distillation Design,* McGraw-Hill, New York, 1992.

[‡]H. Z. Kister and D. R. Gill, *Chem. Eng. Prog.,* **87**(2): 32 (1991).

[§]M. E. Harrison and J. J. France, *Chem. Eng.*, **96**(4): 121 (1989).

where a_p is the packing surface area per unit volume in m^2/m^3 and the HETP obtained is in meters. A more accurate approach would be to use the interpolation or extrapolation of packing efficiency data of Kister noted in the previous section.

Rating of Distillation Equipment

As noted earlier, rating a distillation column can entail either examining performance under new operating conditions or determining whether a retrofit of an existing column with perhaps improved column internals could prove to be economically attractive. For example, Humphrey and Seibert[†] have shown that in vacuum distillation, the replacement of sieve trays with structured packing with its increased capacity is a good return on the investment.

The steps involved in the rating of a distillation column are not unlike those used in the design of a new distillation column. The first step must carefully detail the objective of the rating study and provide precise information about the column including existing internal hardware. Accurate information must also be available about feed composition, rates, and physical properties. Equilibrium data must be obtained from experimental or literature sources. If none are available, such data will have to be developed by using commercially available software programs.

If the internal configuration of the column is not to be modified, the next step is to calculate the number of equilibrium stages required for the new design specifications. However, if the objective is to increase the capacity or the purity of the product, the internal configuration will probably have to be changed. For example, improved purity could be achieved by removing sieve trays and replacing them with commercial structured packing with smaller spacings between packing elements. Calculation of the equilibrium stages can be initiated once the column modifications have been finalized.

Next, the designer must establish if the diameter of the existing column is adequate to handle the new vapor and liquid rates. The procedures for evaluating the required diameter of the column are similar to those described for the design of a new column. Likewise, the procedures for evaluating the required height of the column are similar to those discussed earlier. If the resulting values of column diameter and height match those for the existing column, the proposed separation can be achieved. Obviously, if the required diameter and height are larger than those available for the existing column, modifications will have to be made either in the separation objectives or in the column configuration and operation.

Rigorous Design Methods for Multicomponent Distillation

The preceding section has outlined shortcut methods for the design of distillation columns using either trays or packing. Since these methods included many simplifying assumptions to reduce the calculation efforts, the results often can only be considered as reasonable estimates. To obtain a more accurate picture of the separation process requires a rigorous, stage-by-stage evaluation. The advent of high-speed digital computers

---

[†]J. L. Humphrey and A. F. Seibert, "Separation Technologies—Advances and Priorities," Final Rept., Gas Research Institute, ID No. 89/0005, February 1989.

has made possible solutions of equilibrium-stage models for multicomponent, multistage distillation-type columns to an exactness limited only by the accuracy of the phase equilibria and enthalpy data used. Time and cost requirements for obtaining such solutions are low compared with the cost of hand calculations.

Over the past several decades, methods have also become available that, with appropriate computer programs, can accurately solve most distillation problems quickly and efficiently. Some of the more widely used models, namely, the BP (boiling-point) method for distillation, the SR (sum rates) method for absorption and stripping, the ISR (isothermal sum rates) method for liquid-liquid extraction, the SC (simultaneous correction) procedures for distillation of wide-boiling mixtures, and the IO (inside-out) method that reduces the computational time of earlier methods are detailed in Seader and Henley.[†]

Many of the simulators presently available offer various options which allow the user to select not only a particular algorithm for stage calculations, but also the model or approach to determine the vapor/liquid equilibria. Some simulators also allow the user to incorporate tray efficiencies as a function of composition, so that the actual number of trays can be established as well. Again, users of such simulators must have a good understanding of distillation fundamentals to appreciate the limitations of the methods and models used, and to recognize errors should they occur.

The deficiencies associated with the equilibrium stage models enumerated above have been recognized for some time, and a new model has been developed for distillation and other staged processes by Krishnamurty and Taylor.[‡] The feature of the RBD (rate-based design) method is that component material and energy balances for both the liquid and vapor phases, together with mass- and energy-transfer rate equations, and equilibrium equations for the phase interface are solved directly to determine the actual separation.

In this design method, calculations are performed on an incremental basis as one proceeds from stage to stage through the column. The uncertainties in the computations using average tray efficiencies in prior models based on individual components are avoided. A major advantage of this approach is that it replaces the two-step method of first determining the number of equilibrium stages and then using an average tray or packing efficiency to determine the required column height. A review of this method, as well as a comparison with the traditional equilibrium approach, has been presented by Seader.[§]

Design Procedures for Other Distillation Processes

Batch Distillation Most distillation processes operate in a continuous fashion, but there is a growing interest in batch distillation, particularly in the food, pharmaceutical, and biotechnology industries. The advantage of this separation process is that the distillation unit can be used repeatedly, after cleaning, to separate a variety of products.

[†]J. D. Seader and E. J. Henley, *Separation Process Principles,* J. Wiley, New York, 1998.

[‡]R. Krishnamurty and R. Taylor, *AIChE J.,* **31**(3): 449 (1985).

[§]J. D. Seader, in *Perry's Chemical Engineers' Handbook,* 7th ed., R. H. Perry and D. W. Green, eds., McGraw-Hill, New York, 1997, pp. 13-52 to 13-54.

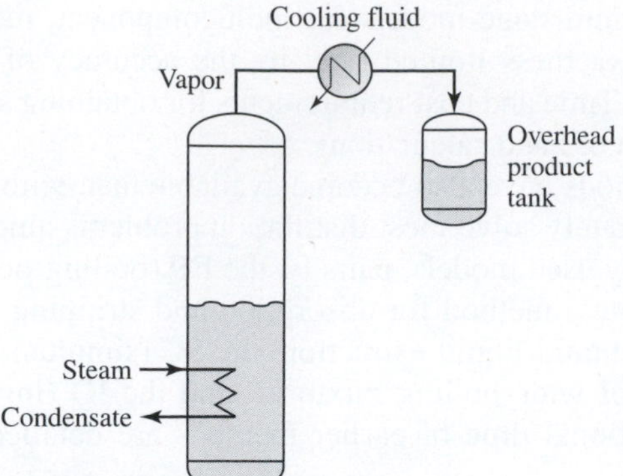

Figure 15-8
Simple batch distillation

The unit generally is quite simple, but because concentrations are continually changing, the process is more difficult to control.

In single-stage distillation, like that shown in Fig. 15-8, it is assumed that the vapor and liquid are in equilibrium. If L' is the number of mols of liquid in the column at any given time and dL' is the differential amount vaporized, a material balance on component i yields

$$L'x_i = (L' - dL')(x_i - dx_i) + (y_i + dy_i)dL' \tag{15-27}$$

Neglecting products of differentials and integrating over the change in L' from the initial to the final condition and from the initial to the final concentration of liquid establishes the Rayleigh equation for batch distillation, given as

$$\ln \frac{L'_2}{L'_1} = \int_{x_{i1}}^{x_{i2}} \frac{dx_i}{y_i - x_i} \tag{15-28}$$

where L'_1 and L'_2 are the initial and final mols of liquid in the column, respectively; x_{i1} and x_{i2} are the initial and final mol fractions of component i in the column, respectively; and x_i and y_i are the mol fractions of component i in the liquid and vapor, respectively, during the batch distillation process. If the mixture to be separated is a binary one and equilibrium data are available, the vapor mol fraction can be expressed in terms of x_i and α_{ij}, the relative volatility between components i and j. For a constant relative volatility between these two components, Eq. (15-28) can be integrated to give

$$\ln \frac{L'_2}{L'_1} = \frac{1}{\alpha_{ij} - 1} \ln \frac{x_{i2}(1 - x_{i1})}{x_{i1}(1 - x_{i2})} + \ln \frac{1 - x_{i1}}{1 - x_{i2}} \tag{15-29}$$

From this relation the designer can estimate the quantity of liquid that must be distilled to change the composition of the remaining liquid from x_{i1} to x_{i2}.

Azeotropic Distillation A separation more sophisticated than simple distillation is required when the components to be separated form an azeotrope. If the composition of the azeotrope is sensitive to pressure, this property can be used to carry out the

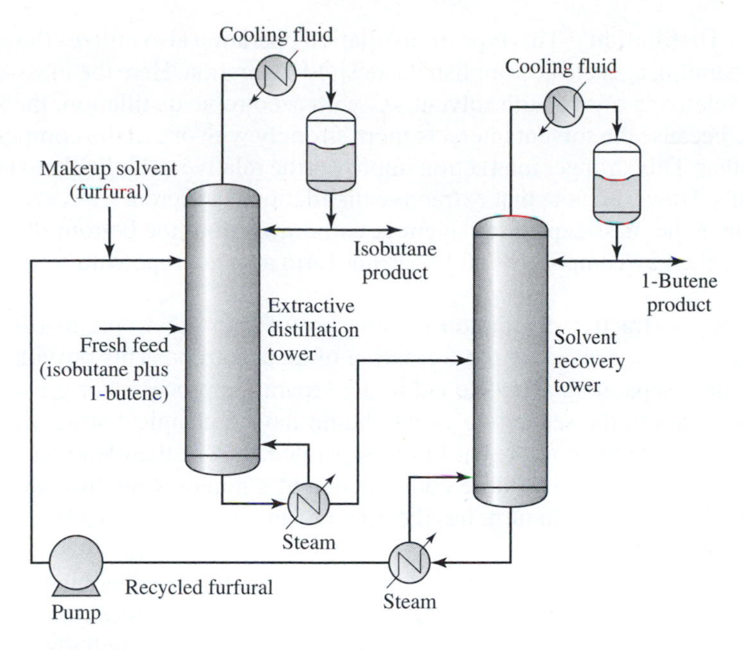

Figure 15-9
Example of extractive distillation separating 1-butene from isobutane

separation provided no material decomposition occurs. A change in azeotropic composition of at least 5 percent with a change in pressure is generally required.†

The design process for separating an azeotrope sensitive to pressure changes requires the use of two distillation columns in which the azeotrope product from the high-pressure column is fed to the low-pressure column, as shown in Fig. 15-9. The azeotrope product from this column is recompressed and added to the fresh feed of the high-pressure column. A major problem in using the pressure change scheme is that the smaller the change in azeotropic composition with change in pressure, the larger is the recycle back to the high-pressure column.

If the azeotrope is not sensitive to changes in pressure, it may be possible to add another component to the mixture to favorably change the relative volatility of the key components. This technique is also useful when a mixture is difficult to separate because of a fairly low relative volatility between the key components. The addition of a relatively volatile mass-separating agent to the mixture that is to be separated forms either a low-boiling binary azeotrope with one of the key components or, more often, a tenary azeotrope containing both key components. Separation by the latter process is only feasible if condensation of the overhead vapor results in two liquid phases, one of which contains the majority of one of the key components and the other contains most of the added component. This type of a separation process will require the use of a decanter located after the overhead condenser to separate the two liquid phases.

†C. D. Holland, S. E. Gallum, and M. J. Lockett, *Chem. Eng.*, **88**(3): 185 (1981).

Extractive Distillation This type of distillation operation also utilizes the addition of a mass-separating agent to accomplish the desired separation. Here the mass-separating agent is a relatively nonvolatile solvent. As with azeotropic distillation, the separation is possible because the solvent interacts more strongly with one of the components than with the other. This stronger interaction improves the relative volatility between the key components. However, note that extractive distillation is different from azeotropic distillation since the mass-separating agent is withdrawn from the bottom of the column with one of the key components and does not form an azeotrope with any of the other components.

Generally, extractive distillation is more useful than azeotropic distillation since the former does not depend on the formation of an azeotrope. This provides a greater choice of mass-separating agents to aid in the separation process. For greatest success with the separation, the separating agent should have a chemical structure similar to that of the less volatile component being separated. It will then tend to form a near ideal mixture with the less volatile component and a nonideal mixture with the more volatile component. This, in turn, has the effect of increasing the volatility of the more volatile component.

In those cases where the flow rate of the solvent can be varied within limits, a larger solvent flow rate often results in a better separation. However, since the solvent is part of the reboil in both the extraction and stripping columns, there will also be a greater energy demand with a larger solvent flow rate. Additionally, higher solvent flow rates increase the distillate temperature.

During the separation process there will inevitably be losses if a mass-separating agent is used. Even if these losses in the agent are insignificant in terms of the cost of the agent, they could create environmental problems later in the process. The best way to solve the effluent problems caused by the loss of mass-separating agents is to eliminate them from the design unless practical difficulties and excessive cost dictate their use.

Equipment Costs for Tray and Packed Columns

The purchased cost for tray and packed columns can be divided into the following components: (1) cost for column shell, including heads, skirts, manholes, and nozzles; (2) cost for internals, including trays and accessories, packing, supports, and distributor plates; and (3) cost for auxiliaries, such as platforms, ladders, handrails, and insulation.

The cost for fabricated column shells is estimated on the basis of either weight or the diameter and height of the column. The former estimate is often preferred to more accurately account for the effect of pressure on the wall thickness of the column shell. Figure 15-10 presents the cost based on column weight and Fig. 15-11 on the basis of diameter and height. This cost is for the column shell only, without trays, packing, or connections.[†] The cost for installed connections, such as manholes and nozzles, may be approximated from the data presented in Fig. 15-12.

[†]See also the information presented in Chap. 12 on costs for tanks, pressure vessels, and storage equipment along with information on design methods for tanks and pressure vessels.

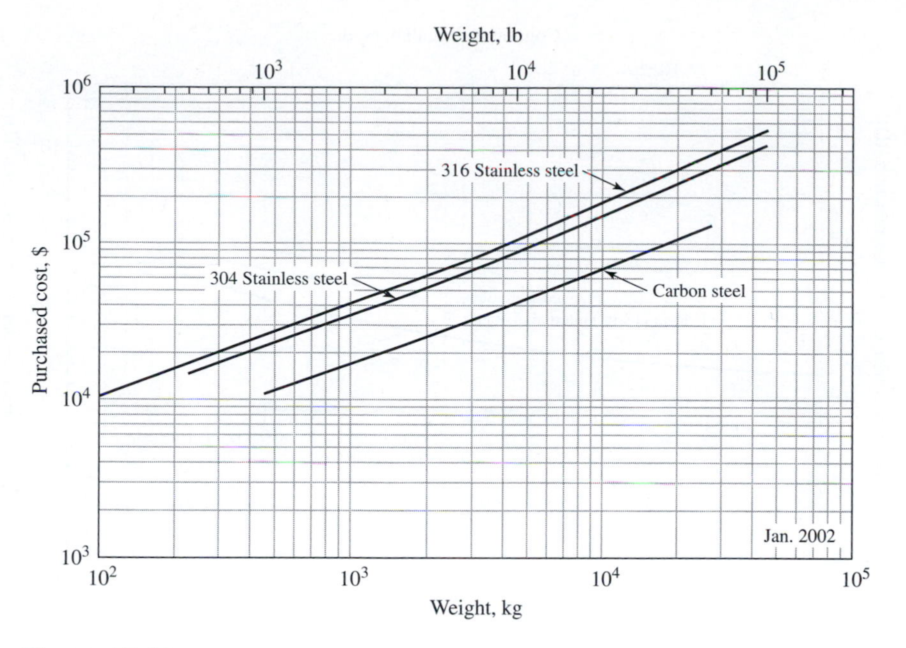

Figure 15-10
Purchased cost of columns and towers. Costs are for shell with two heads and skirt, but without trays, packing, or connections.

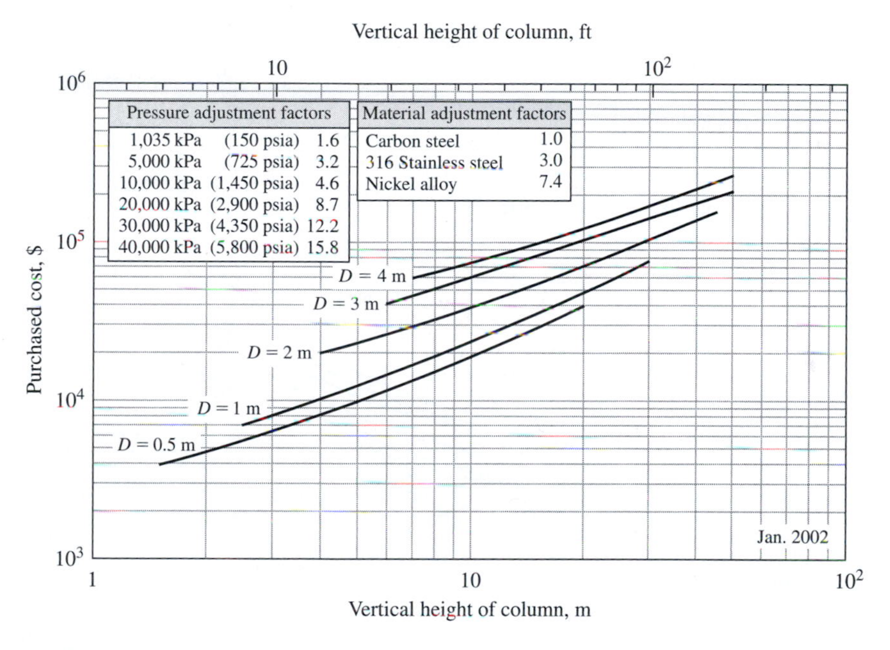

Figure 15-11
Purchased cost of vertical columns. Price does not include trays, packing, or connections.

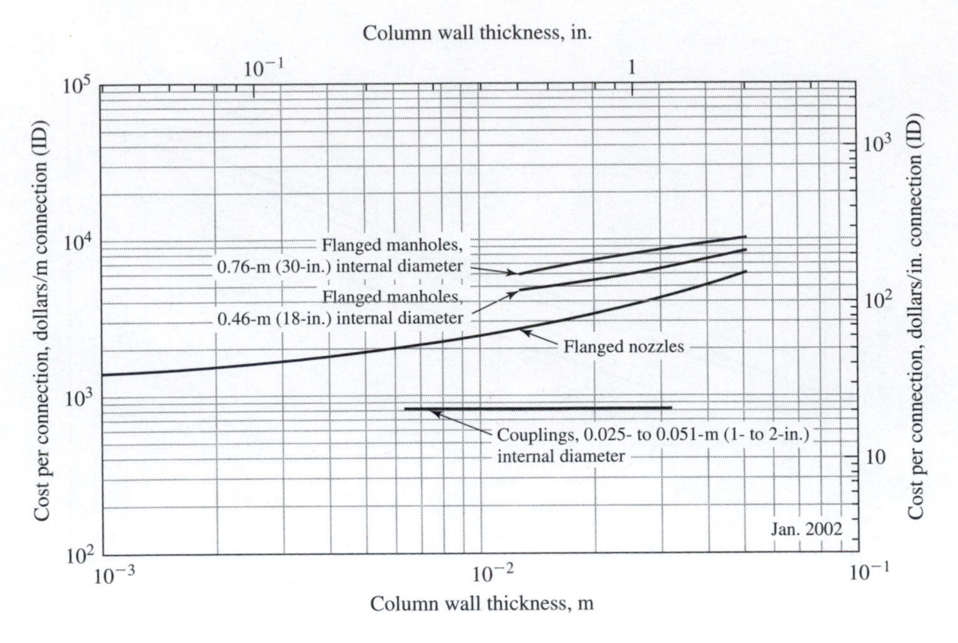

Figure 15-12

Installed cost of steel column connections. Values apply to 136-kg (300-lb) connections. Multiply costs by 0.9 for 68-kg (150-lb) connections and by 1.2 for 272-kg (600-lb) connections.

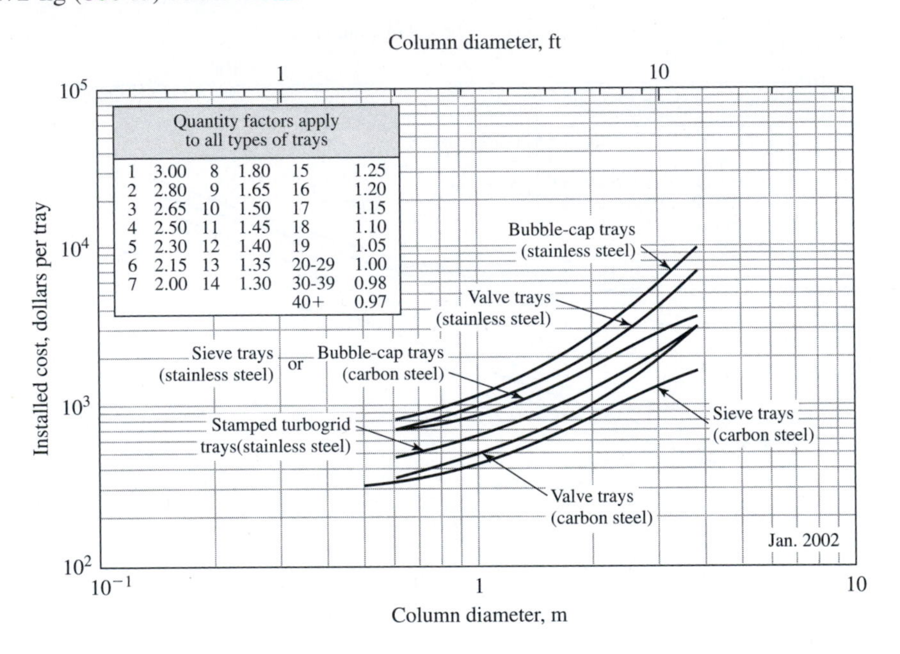

1	3.00	8	1.80	15	1.25
2	2.80	9	1.65	16	1.20
3	2.65	10	1.50	17	1.15
4	2.50	11	1.45	18	1.10
5	2.30	12	1.40	19	1.05
6	2.15	13	1.35	20-29	1.00
7	2.00	14	1.30	30-39	0.98
				40+	0.97

Figure 15-13

Purchased cost of trays in tray columns. Price includes tray deck, bubble caps, risers, downcomers, and structural-steel parts.

Tray costs are shown in Fig. 15-13 for conventional installations. The purchased costs for slotted-ring and high-efficiency saddle packings including column internal supports and distributors are given in Fig. 15-14. For rough estimates, the cost for a distributor plate in a packed column may be assumed to be the same as that for one bubble-cap tray. The complete installed costs of various types of distillation columns are presented in Fig. 15-15. Figure 15-16 provides complete installed costs for packed columns. Both cost estimates cover the costs of all components normally associated with both types of columns, as outlined above. The installed cost of several industrial-type insulations, all with aluminum jackets, for columns and tanks may be approximated from Fig. 15-17.

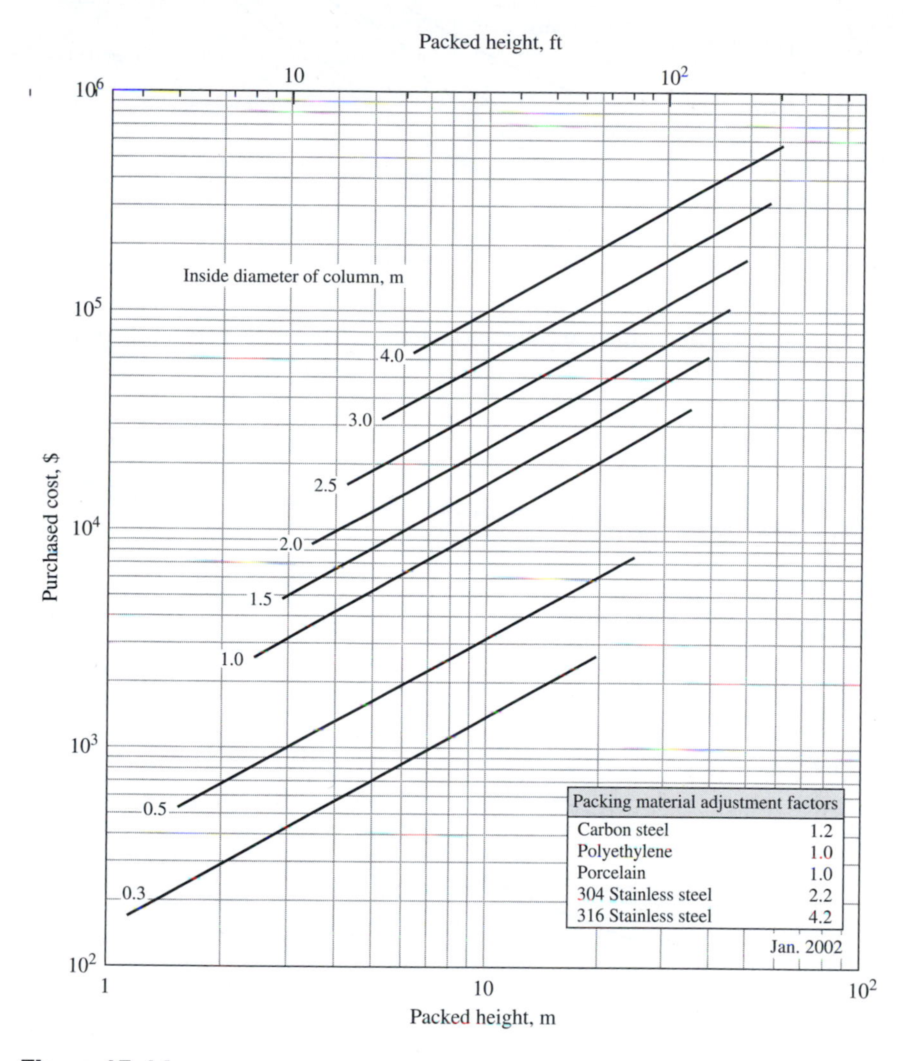

Figure 15-14
Purchased cost of stacked-ring and high-efficiency saddle packing (price includes column internal support and distribution)

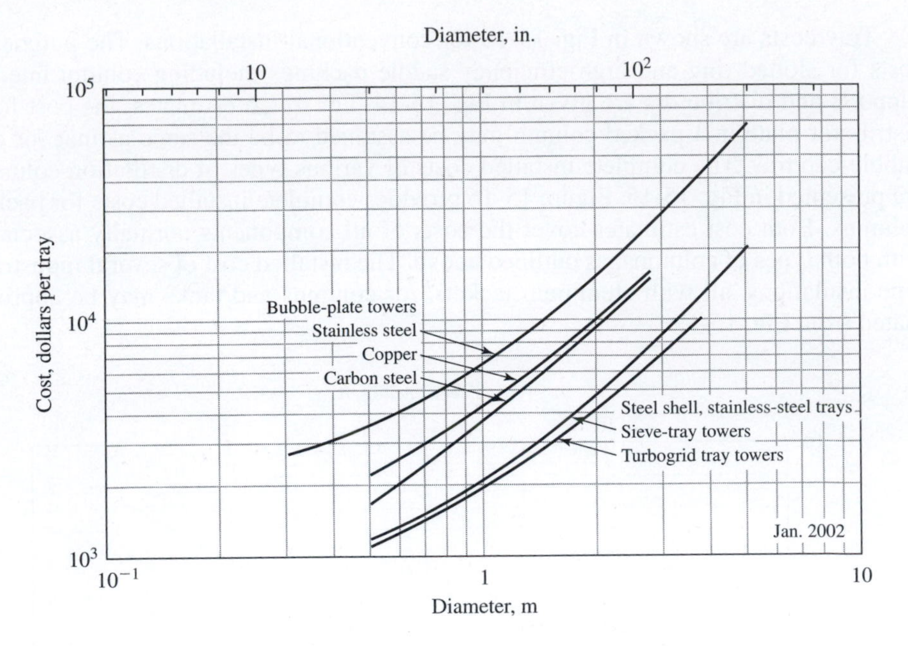

Figure 15-15
Purchased cost of distillation columns including installation and auxiliaries

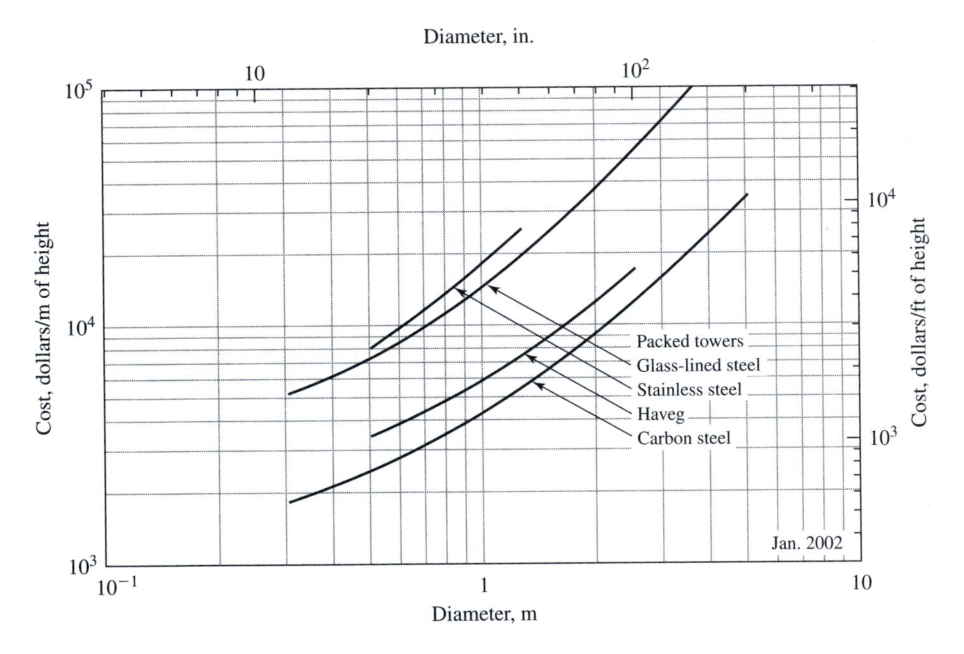

Figure 15-16
Purchased cost of packed columns including installation and auxiliaries

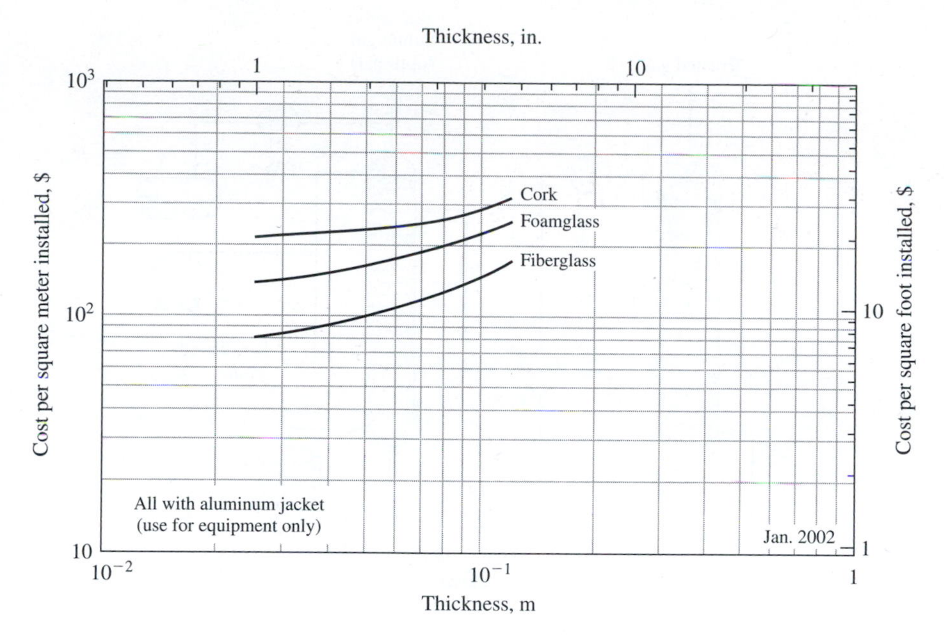

Figure 15-17
Installed cost of various industrial insulations for columns, towers, and tanks

SEPARATION BY ABSORPTION AND STRIPPING

Absorption and stripping processes are similar to the distillation process. In the latter, the mass transfer is between the liquid and the vapor in each stage throughout the column. In absorption and stripping, the mass transfer occurs between a gas phase and the liquid phase on each stage throughout the column. In absorption the solute, or component to be absorbed, is transferred from the gas phase to the liquid phase. In stripping, the reverse is true; that is, the transfer is from the liquid phase to the gas phase.

Absorbers are often coupled with strippers to provide recovery of the adsorbent, as shown in Fig. 15-18. Since the stripping process is not perfect, the absorbent recovered contains components that were present in the fresh feed to the absorber. When water is used as the absorbent, it is many times more common to recover the absorbent by distillation than by stripping.

The equipment used for absorption and stripping is very similar to that used for distillation except that no reboiler or condenser is required in the process. The separation process is conducted with trayed columns, packed columns, as well as spray towers and centrifugal contactors. The column internals for trayed columns preferably include sieve and valve trays. The packed towers use random and structured packing similar to the geometric configurations noted earlier in Figs. 15-3 and 15-4.

In most applications the choice of the absorber is between a trayed column and a packed column. The latter is generally favored when the requirements indicate a column diameter less than 0.65 m and a packed height of less than 6 m. Packed towers should also be considered if the separation process involves corrosive fluids or if there

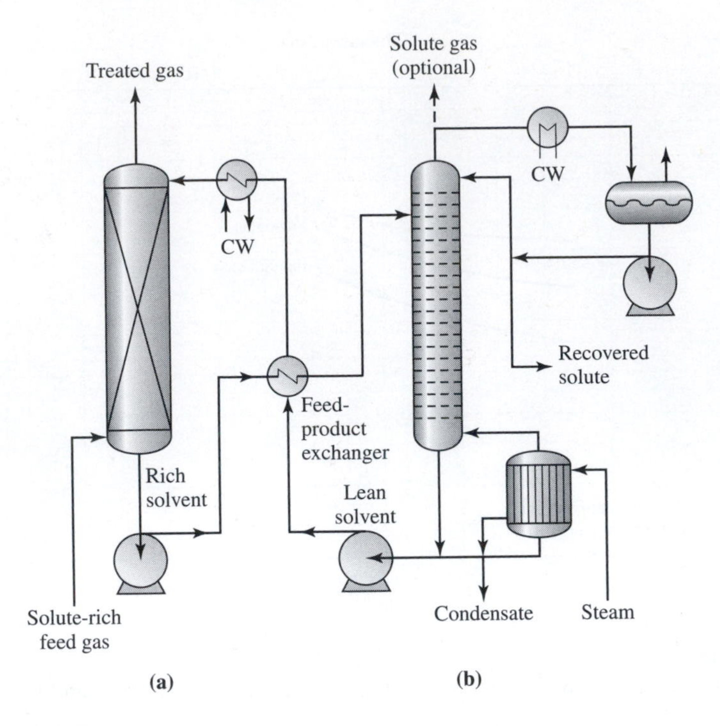

Figure 15-18
Gas absorber using a solvent regenerated by stripping: (a) absorber; (b) stripper

is a need for a minimum pressure drop across the column, as in vacuum or near ambient pressure operation. The use of a spray tower may be advantageous if the solute is very soluble in the liquid phase and only one or two stages are required for the separation process. Otherwise, trayed columns, which can be designed and scaled up more easily, are normally preferred.

The absorbent selected is one that should have a high solubility for the solute and have a low volatility to reduce absorbent losses and to facilitate its recovery in the stripper. If there is a choice, preference should be given to an absorbent with a low viscosity to achieve lower pressure drops and higher mass- and heat-transfer rates in the column. Additionally, the absorbent should be nontoxic, nonflammable, and noncorrosive to facilitate its safe use. The most widely used absorbents are water, hydrocarbon oils, and aqueous solutions of acids and bases. The most common stripping agents are air, steam, inert gases, and hydrocarbon gases.

Most absorbers generally are operated at pressures greater than ambient pressure and temperatures near ambient. This operating decision minimizes the stage requirements and/or the absorbent flow rate, which in turn reduces the equipment volume required to accommodate the gas flow. On the other hand, the operating pressure should be low and the temperature high for a stripper to reduce the stage requirements and/or the flow rate of the stripping agent. However, care must be exercised in the use of these guidelines. For example, an absorber should not be operated at a pressure and/or a temperature that would condense the feed gas. Similarly, a stripper should not operate with pressure and temperature conditions that would vaporize the liquid feed.

In design problems involving flow rates for the absorbent or the stripping agent and the number of stages required, there is always a tradeoff between the two parameters at flow rates greater than the minimum value. The next section briefly outlines an analytical method that can be used to determine the minimum flow rate and the tradeoff involved for a mixture that is dilute in the solute. For this situation, the absorption or stripping process is essentially isothermal, and an energy balance can be neglected. The separation process becomes complicated with concentrated mixtures and necessitates the use of computer-aided methods. Numerous literature sources detailing the design procedures for absorption and stripping are available.[†]

Design Procedures for Trayed Columns Separating Dilute Solutes

Initiation of the design procedures for separations by either absorption or stripping is very similar to the procedures outlined for distillation design. This includes establishing the objectives of the separation; identifying feed components, concentrations, and flow rates; providing operating conditions whenever these are known; selecting a suitable solvent; and obtaining appropriate equilibrium data. The objectives of the separation generally will specify the percentage of the solute in the gas stream that is to be removed in the separation process. Guidelines for selecting reasonable process conditions and a suitable solvent have been discussed previously. Appropriate equilibrium data can be obtained or generated with the use of the literature references cited in the distillation design procedures section.

Determination of Number of Equilibrium Stages Graphical methods for determining equilibrium stages become very time-consuming when more than a few stages are required or the best operating conditions are to be determined. Under such conditions, an algebraic method is preferred for manual calculations if an appropriate simulator is not available.

The Kremser method for single-section cascades, as modified by Edmister,[‡] is applicable for absorption and stripping of dilute mixtures. The fraction of a solute i absorbed is given by

$$\frac{A_i^{N+1} - A_i}{A_i^{N+1} - 1} = \text{solute fraction absorbed} \tag{15-30}$$

while the fraction of solute i stripped is

$$\frac{S_i^{N+1} - S_i}{S_i^{N+1} - 1} = \text{solute fraction stripped} \tag{15-31}$$

where A_i and S_i are the solute absorption and stripping factors for component i,

[†]For example see, S. Diab and R. N. Maddox, *Chem. Eng.*, **89**(26): 38 (1982), and J. D. Seader and E. J. Henley, *Separation Process Principles*, J. Wiley, New York, 1998.

[‡]W. C. Edmister, *AIChE J.*, **3:** 165 (1957).

respectively, and N is the number of equilibrium stages required for the specified solute fraction either absorbed or stripped. The absorption and stripping factors at low solute concentrations are conveniently defined in terms of L and V, the entering liquid and vapor flow rates into the column, respectively, and K_i, the vapor/liquid equilibrium ratio for the solute i, as

$$A_i = \frac{L}{K_i V} \tag{15-32}$$

and

$$S_i = \frac{K_i V}{L} \tag{15-33}$$

Estimating K_i values in either of these two equations is a function of temperature, pressure, and liquid-phase composition. For dilute solute mixtures and essentially ambient pressure, a number of common expressions for K_i can be estimated from

$$K_i = \frac{p_i^S}{p} \quad \text{ideal solution and subcritical temperature} \tag{15-34a}$$

$$K_i = \frac{\gamma_{i,L}^{\infty} p_i^S}{p} \quad \text{nonideal solution and subcritical temperature} \tag{15-34b}$$

$$K_i = \frac{H_i}{p} \quad \text{solute at supercritical temperature} \tag{15-34c}$$

$$K_i = \frac{p_i^S}{x_i^* p} \quad \text{slightly soluble solute at subcritical temperature} \tag{15-34d}$$

where p_i^S is the vapor pressure of the solute, p the pressure of the gas, $\gamma_{i,L}^{\infty}$ the activity coefficient of the liquid at infinite dilution, H the Henry's law constant, and x_i^* the mol fraction of the solute under equilibrium conditions. Once the absorption and stripping factors are estimated, Eqs. (15-30) and (15-31) can be used to determine the fraction of the solute absorbed or stripped when the number of equilibrium stages is known.

EXAMPLE 15-4 Determination of Number of Theoretical Stages Required to Strip Wastewater Containing Volatile Organic Compounds to an Acceptable Level

A water recovery stripper using dry air at 15°C and 103 kPa as the stripping gas is to remove 99.9+ percent of the benzene and ethylbenzene in a wastewater stream. The flow rates of the air and wastewater streams are 2.41 and 0.0475 m^3/s, respectively. The stripping is to be done at 20°C and 103 kPa. Necessary thermodynamic data from API tabulations are listed here.

Volatile organic component	Concentration in wastewater, mg/L	Vapor pressure at 20°C, kPa	Solubility in water at 20°C, mol fraction
Benzene	150	10.55	0.00041
Ethylbenzene	20	1.03	0.000035

■ Solution

Since the solute concentration is dilute, the Kremser relation may be used and applied independently to both organic impurities. Absorption of air by the water and stripping of water by the air will be considered negligible. The K value can be computed with either Eq. (15-34b) or Eq. (15-34d). Since solubility data are available, use the latter equation. For benzene

$$K_i = \frac{p_i^S}{x_i^* p} = \frac{10.55}{0.00041(103)} = 250$$

for ethylbenzene

$$K_i = \frac{1.03}{0.000035(103)} = 285$$

The vapor and liquid rates are obtained from

$$V = \frac{2.41}{22.4(273/288)} = 0.102 \text{ kg mol/s} \qquad \text{assuming ideal gas}$$

$$L = \frac{0.0475(10^3)}{18.02} = 2.636 \text{ kg mol/s} \qquad \rho_L = 1000 \text{ kg/m}^3$$

The stripping factors for benzene and ethylbenzene can now be calculated individually with Eq. (15-33):

$$S = \frac{K_i V}{L} = \frac{250(0.102)}{2.636} = 9.67 \qquad \text{for benzene}$$

$$S = \frac{K_i V}{L} = \frac{285(0.102)}{2.636} = 11.03 \qquad \text{for ethylbenzene}$$

With the aid of Eq. (15-31) the fraction stripped in each theoretical stage can be evaluated with a spreadsheet computer program, to provide the following results:

Organic stripped	Percent stripped as a function of stages required			
	Stage 1	Stage 2	Stage 3	Stage 4
Benzene	90.62	99.04	99.90	99.96
Ethylbenzene	91.68	99.25	99.93	99.99

Thus, three theoretical stages should be sufficient to achieve a 99.9+ percent recovery for the two volatile organic compounds. The actual height of the tower will depend on the stage efficiency provided by the stripping column under the specified operating conditions. In addition, various other airflow rates should be investigated to make certain that flooding or weeping will not occur in the column. ■

Determination of Stage Efficiency and Column Height To determine the actual number of plates required for an absorption or stripping process, the number of theoretical stages must be adjusted by a stage efficiency E_o. For well-designed trays and flow rates near the capacity limit, E_o depends mainly on the physical properties of the liquid and vapor streams. Values of stage efficiency are normally predicted either by comparison with performance data for industrial columns or by using empirical correlations or semitheoretical relationships.

A widely used empirical equation, based on the observation that overall stage efficiencies depend primarily on the average molar viscosity, is given by

$$E_o = 19.2 - 57.8 \log \mu_L \tag{15-35}$$

where μ_L is the average viscosity in cP. This relation unfortunately provides inadequate results for absorbers and strippers when applied to components exhibiting wide ranges in volatility or equilibrium ratios. To minimize some of these deficiencies, a more general empirical correlation has been suggested

$$\log E_o = 1.597 - 0.199 \log \frac{KM_L\mu_L}{\rho_L/16.02} - 0.0896\left(\log \frac{KM_L\mu_L}{\rho_L/16.02}\right)^2 \tag{15-36}$$

where K is the equilibrium ratio, M_L the molecular weight of the liquid, μ_L the viscosity of the liquid in cP, and ρ_L the density of the liquid in kg/m^3. This correlation must also be used with caution.

The use of semitheoretical models to arrive at tray efficiencies for each component on each tray and then to average these to obtain an overall stage efficiency is much more complex and requires adequate software. A number of commercial simulation programs are available for such calculations.

Note that the above correlations only apply to separation devices in which discrete stages are present in the column. Thus, they should not be applied to packed towers or continuous-contact devices. In these units, the efficiency becomes part of an equipment and system-dependent parameter such as the HETP.

The actual number of trays is given by

$$N_{\text{act}} = \frac{N}{E_o} \tag{15-37}$$

where N_{act} is the actual number of trays and N the theoretical stages or trays. The actual column height is given by Eq. (15-10).

Evaluation of Tray Diameter Determination of the diameter of an absorber or stripper column with trays involves the same procedural steps that were outlined earlier for trayed distillation columns. This entails calculating the vapor velocity at flooding by using Fig. 15-5 for either sieve or bubble-cap trays and a specified tray spacing. (Figure 15-5 appears to be conservative for valve trays.) The actual velocity at the top and bottom of the column is assumed to be between 70 and 90 percent of the net vapor velocity at flood conditions. The net and total column areas at both locations are obtained with Eqs. (15-8) and (15-8*a*), respectively. The larger of the top or bottom diameter is used for the entire column.

Design Procedures for Packed Columns Separating Dilute Solutes

Since packed columns are continuous differential contacting devices, they are best analyzed by mass-transfer considerations rather than by the equilibrium stage concept described for trayed columns. The rate of mass transfer for absorption and stripping in a packed column is normally expressed in terms of a volumetric mass-transfer coefficient ka, where a represents the area of mass transfer per unit volume of packed bed. At steady state and no chemical reaction within the column, the rate of solute mass transfer across the gas-phase film must equal the rate across the liquid-phase film. If the system is dilute with respect to the solute, the rate of mass transfer per unit volume of packed bed may be written in terms of the mol fraction driving forces in each of the two phases or in terms of a partial-pressure driving force in the gas phase and a concentration driving force in the liquid phase.

Determination of the packed height of a column most commonly involves the overall gas-phase coefficient since the liquid usually has a strong affinity for the solute so that resistance to mass transfer is mostly in the gas phase. This requires obtaining the height of a transfer unit HTU and the number of transfer units NTU as defined by Eqs. (15-12) and (15-13). The approach outlined to determine the HTU for a packed column used in distillation service is also adequate for use in a packed column for absorption and stripping service.

To obtain the height of packing required, the number of transfer units needs to be established, as noted in Eq. (15-11). For absorption and stripping with dilute solute mixtures, the integral in Eq. (15-13) can be eliminated. This is done by using the linear equilibrium condition $y^* = Kx$ to eliminate y^* and using the linear solute material balance operating line to eliminate x. The resulting relationship after integration and the substitution of $L/KV = A$ becomes

$$\text{NTU} = \frac{\ln\{[(A-1)/A][(y_{\text{in}} - Kx_{\text{in}})/(y_{\text{out}} - Kx_{\text{in}})] + 1/A\}}{(A-1)/A} \tag{15-38}$$

The total height of the column is the sum of the height required for the packing and the additional height required for phase disengagement and flow distribution hardware.

The generalized pressure drop correlation shown in Fig. 15-7 can also be used once a suitable pressure drop per unit height of packing between 200 and 400 Pa/m has been selected to determine the cross-sectional area of the absorber or stripper. The column diameter is obtained from the cross-sectional area.

Design Procedures for Packed Columns Separating Concentrated Solutes

The design procedures associated with absorbing or stripping dilute solute mixtures cannot be used as solute concentrations are increased since Eq. (15-13) cannot be integrated analytically. There are several factors that make the separation process more complex. In absorption, for example, if there is appreciable absorption, the mass flux rate decreases from the bottom to the top of the absorber. Also, as solute is transferred from the gas to the liquid phase, there is a general heating effect with an increase in temperature down the column. Thus, isothermal operation does not exist, resulting in a continual change in vapor/liquid equilibrium and the requirement for material and energy balances over the length of the column.

The most common approach to determine the height of the packed column under these circumstances is to use a computer program that will stepwise move through the column to establish the equilibrium and operating conditions. The equilibrium conditions will be established with appropriate thermodynamic data or correlations. The operating conditions will be determined after convergence is attained with heat and material balances. Again, a number of commercial computer programs are available to simulate the absorption or stripping processes for such operating conditions.

SEPARATION BY EXTRACTION

In liquid/liquid extraction, components of a liquid are separated according to their preferential distribution in a suitable solvent. Intimate contact between the two liquid phases is made to achieve the maximum approach to equilibrium. After contacting the phases, provision is made to separate the two liquid phases. The extraction is accomplished either with simple extractors in which one phase is dispersed in the other under countercurrent flow conditions or in mechanical agitators where dispersion is enhanced by mechanical agitation and shear.

The simplest extraction system involves the *solute* (the material to be extracted), the *solvent,* and the *carrier* (the nonsolute portion of the feed mixture). In extraction the designer must also differentiate between the light phase, the heavy phase, the dispersed phase, the continuous phase, and the feed, raffinate, and extract phases.

An extraction system always includes at least another separation device to recover the solvent from the extract phase. For example, Fig. 15-19 shows an industrial extraction system where the trayed extraction column is provided with reflux in the form of extract product from the distillation column to reduce the equilibrium limitations placed on the purity of the extracted material. Note that this system utilizes a low-boiling solvent that is recovered as the distillate bottoms from the distillation column.

Equipment Selection for Liquid/Liquid Extraction

Because of the wide diversity of applications utilizing liquid/liquid extraction, a wide variety of separation devices have been developed. Some equipment is similar to that

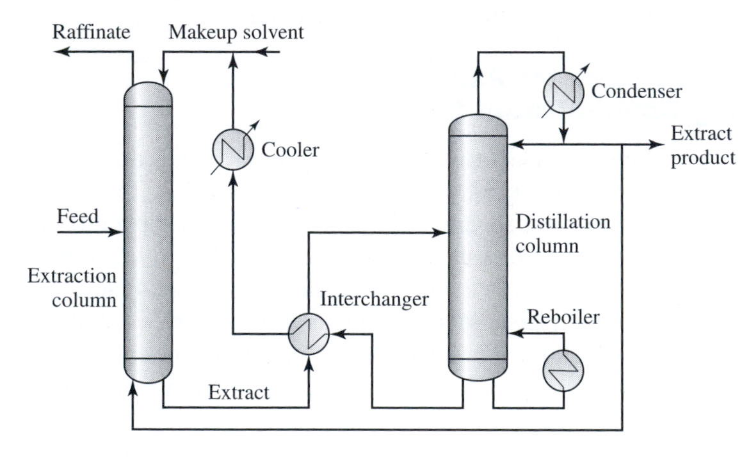

Figure 15-19
Industrial extraction system

Table 15-8 General guidelines affecting extraction equipment selection

Equipment class	Advantages	Limitations
Continuous counterflow contactors, no mechanical agitation	Low capital cost, low operating costs, simple construction	Can exhibit low efficiency, limited capacity with small density differences, flow rate restrictions, poor scale-up
Continuous counterflow contactors, with mechanical agitation	Medium capital cost, good dispersion, capability of many stages, easy scale-up	Limited capacity with small density differences, flow rate restrictions, inadequate for emulsifying systems
Mixed-settlers	Good contacting, high efficiency, capability of many stages, wide flow rates, good scale-up	High capital cost, high operating costs, large liquid holdup, may require interstage pumping and storage
Centrifugal extractors	Low holdup volume, short holdup time, small solvent requirement, handle low-density differences between phases	High capital cost, high operating costs, high maintenance costs, limited number of stages per unit

used for distillation, absorption, and stripping. However, such devices can be inefficient unless the liquid viscosities are low and the differences in phase density are high. For this reason, mechanically agitated devices are often better choices. In either case, the design engineer needs to evaluate the number of theoretical stages that will be required for the separation and to determine the dimensions of the equipment selected for the extraction.

Extractors normally are grouped into four different classes: (1) continuous counterflow contactors, (2) continuous counterflow contactors with mechanical agitation, (3) mixer-settlers, and (4) centrifugal extractors. Spray, packed, and trayed columns are utilized as continuous counterflow contactors. Examples of extraction equipment that provide external mechanical agitation to increase the dispersion between liquids include such devices as the Scheibel and Oldshue-Rushton columns, the rotating-disk contactor (RDC), the asymmetric rotating-disk (ARD) contactor, the Kuhni column, the Karr reciprocating-plate column, and the Graesner extractor. The Lurgi column is an example of the mixer-settler extractors, and the Podbielniak extractor is one of several industrial centrifugal extractors. With such an array of extraction equipment available, the design engineer will need to consider the advantages and limitations associated with each class of equipment. Table 15-8 presents some general guidelines prepared by Seader and Henley[†] that can provide assistance in this selection process. Reissinger and Schroeder[‡] have compared a number of industrial column-type extractors for maximum liquid capacity and maximum column design. Their results are tabulated in Table 15-9. Liquid capacities per unit of cross-sectional area are highest with the Karr extractor and lowest with the Graesser extractor.

[†]J. D. Seader and E. J. Henley, *Separation Process Principles,* J. Wiley, New York, 1998.

[‡]K. H. Reissinger and J. Schroeder, in *Encyclopedia of Chemical Processing and Design,* vol. 21, J. J. McKetta et al., eds., Marcel Dekker, New York, 1984.

Table 15-9 Maximum loading and diameter of commercial column-type extractors

Commercial column type	Approx. maximum liquid loading, $m^3/m^2 \cdot h$	Maximum column diameter, m
Lurgi	30	8.0
Pulsed packed	40	3.0
Pulsed sieve tray	60	3.0
Scheibel	40	3.0
RDC	40	8.0
ARD	25	5.0
Khuni	50	3.0
Karr	100	1.5
Graesser	<10	7.0

Above data apply under conditions of (1) high interfacial surface tension (30–40 dyne/cm), (2) viscosity of approximately 1 cP, (3) volumetric phase ratio of 1 : 1, and (4) phase-density difference of approximately 0.6 g/10^{-6} m^3.

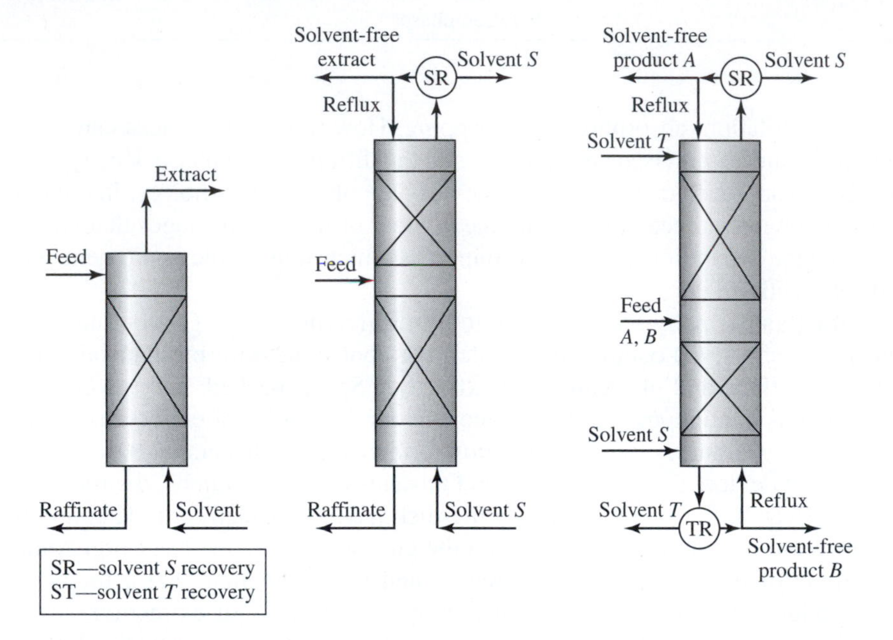

Figure 15-20
Common liquid-liquid extraction cascade configurations: (a) single-section cascade; (b) two-section cascade; (c) dual solvent with two-section cascade

Design Procedures for Liquid/Liquid Extraction

The design for a liquid/liquid extractor is more involved than the design for a vapor/liquid operation because of complications introduced by the two liquid phases. Several different column-type arrangements, as shown in Fig. 15-20, are used depending upon the difficulty of the separation. The single-section cascade is similar to that used in absorption and stripping and is used in transferring to the solvent a certain fraction of the solute in the feed. The two-section cascade is similar to distillation with a feed between the two sections. Depending on solubility considerations, it is often possible to attain a good separation between components in the feed. If this is not possible, then it may be necessary to utilize a dual solvent arrangement with two sections. For this arrangement involving a minimum of four components (two in the feed and two solvents), the calculations become very tedious and will require the use of rather sophisticated computer simulation programs that are presently available. (Seader and Henley adequately discuss the solution methods used by such programs with applications related to liquid/liquid extraction.†) Note that the columns in Fig. 15-20 are shown with packed sections; any of the other extraction devices discussed earlier can be substituted.

Many factors may influence the separation process involved in extraction. The more important factors include the following:

Feed flow rate, composition, temperature, and pressure

Type of extraction configuration required (single-section, two-section, or dual solvent and two-section configurations)

Desired degree of recovery of solute in a single section or degree of separation of feed in a two-section configuration

Choice of liquid solvent and its properties

Operating temperature, pressure, and associated phase equilibria

Interfacial tension and phase-density difference between the two liquids

Emulsification and scum formation tendency

Extractor type selected

Selection of the solvent is a major decision in extraction. Again, a number of criteria need to be considered in this selection. First, a high value for the partition ratio is desired to permit a lower solvent-to-feed ratio separation process. The *partition ratio* is defined as the ratio of the weight fraction of the solute in the extract phase to the weight fraction of the solute in the raffinate phase. To expedite shortcut calculations, the partition ratio K' is often defined in Bancroft coordinates as

$$K' = \frac{Y}{X} \tag{15-39}$$

where Y is the weight ratio of solute to extraction solvent in the extract phase and X is the weight ratio of solute to feed solvent in the raffinate phase. The slope of the equilibrium line m in Bancroft coordinates is then

$$m = \frac{dY}{dX} \tag{15-40}$$

For low concentrations in which the equilibrium line is linear, $K' = m$.

Another desirable criterion in solvent selection is the need for a high value of the selectivity or separation factor β^o, to lower the number of equilibrium stages that will be required for the separation. The selectivity between two components in the feed is

†J. D. Seader and E. J. Henley, *Separation Process Principles,* J. Wiley, New York, 1998.

given by the ratio of the two partition ratios, namely,

$$\beta^{o}_{\text{solute/carrier}} = \frac{K'_{\text{solute}}}{K'_{\text{carrier}}} \tag{15-41}$$

Selection of a solvent also dictates the need for maintaining a large difference in density between the extract and raffinate phases to permit higher capacities in extraction devices using gravity in the phase separation, and providing a low viscosity to improve the rate of mass transfer and a low interfacial tension to aid in the dispersion of the phases. Similar to the selection of solvents for absorption, the more normal criteria include, for example, low cost and availability of the solvent, easy recovery of the solvent, nontoxicity, nonflammability, and noncorrosiveness of the solvent, as well as stability and compatibility of the solvent over the operating conditions of the extraction process. Obviously, solvent selection is generally a compromise among all the properties listed above. However, initial consideration is generally given first to selectivity and environmental concerns, followed by capacity requirements.

Liquid/liquid equilibria are either obtained from literature sources or generated by computer simulation using binary phase equilibria. Several large tabulations of experimental data are available for liquid/liquid systems. Among these are those by Sorenson and Arlt[†] and by Wisniak and Tamir.[‡] *Perry's Chemical Engineers' Handbook* provides a selected list of partition ratios for ternary systems.[§]

Once ternary phase equilibria data are available in weight percent, a spreadsheet can rapidly organize the data in the form required for developing either a triangular or a right triangular ternary diagram. Figure 15-21 illustrates these two types of ternary systems. In a Type I system, the carrier and the solvent are essentially immiscible, while the carrier-solute and the solvent-solute pairs are miscible. In Type II, there are immiscibilities between two pairs, namely, solvent-solute and solvent-carrier. In both liquid/liquid equilibria systems, the tie lines shown in the two-phase region connect the extract and raffinate equilibrium compositions and thus can be used to determine K' values and separation factors. There is also a Type III, which is characterized by immiscibilities between all three pairs.

Countercurrent Equilibrium Stage Calculations The countercurrent configuration is generally employed in multiple-stage extractions. In such configurations, the feed flows from one stage to the next while interacting with the solvent flowing in the opposite direction. Figure 15-22 shows the graphical stepwise procedure for determining equilibrium stages using the ternary diagram. For a given feed F and solvent B, values of the extract E and raffinate R can be obtained graphically as shown. The minimum solvent rate occurs when the extract and solvent are in equilibrium. If the solvent-to-feed ratio is increased, stages can be stepped off graphically, using the difference point

---

[†]J. M. Sorenson and W. Arlt, *Liquid-Liquid Equilibrium,* Chem. Data Series, vol. 5 (Part 1, Binary Systems), 1979, Part 2, Ternary Systems, 1980; Part 3, Ternary and Quaternary Systems, 1980; Supplement 1, 1987, by E. A. Macedo and P. Rasmussen, DECHEMA, Frankfurt, Germany.

[‡]J. Wisniak and A. Tamir, *Liquid-Liquid Equilibrium and Extraction: A Literature Source Book,* vol. 1, 1980; vol. 2, 1980; Supplement 1, 1987; Elsevier, New York.

[§]*Perry's Chemical Engineers' Handbook,* 7th ed., R. H. Perry and D. W. Green, eds., McGraw-Hill, New York, 1997, Table 15-5.

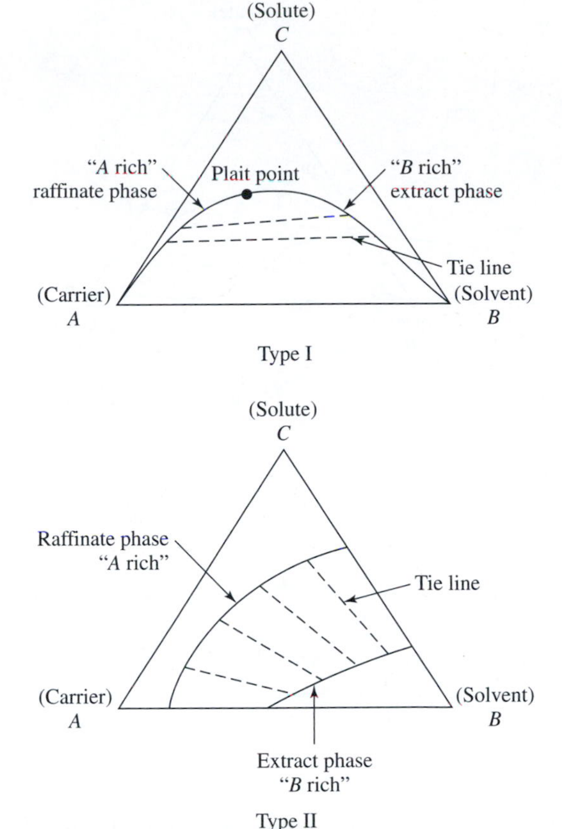

Figure 15-21
Liquid-liquid equilibria for Type I and Type II systems

Δ as the pivot. Note, however, that a tradeoff exists between the number of equilibrium stages and the solvent/feed flow rate ratio.

Multicomponent systems containing four or more components are difficult to display graphically. Nevertheless, process design calculations can often be made for the extraction of the component with the lowest partition ratio and can use this component to establish a ternary system. The components with higher partition ratio values will be extracted even more thoroughly from the raffinate than the solute selected for the ternary system. Computer calculations should be used to reduce the tedious hand calculations.

Algebraic methods for Type I ternaries, utilizing a modified Kremser equation similar to that used in absorption and stripping, are useful for quickly determining the number of equilibrium stages. Such relationships, again, are valid only when the concentration of the solute in the feed stream is dilute. One form of the Kremser relation in Bancroft coordinates, when the extraction factor $\varepsilon \neq 1.0$, is given by

$$N = \ln\left[\frac{X_F - Y_S/K'_S}{X_R - Y_S/K'_S}\left(1 - \frac{1}{\varepsilon}\right) + \frac{1}{\varepsilon}\right]\left(\frac{1}{\ln \varepsilon}\right) \tag{15-42}$$

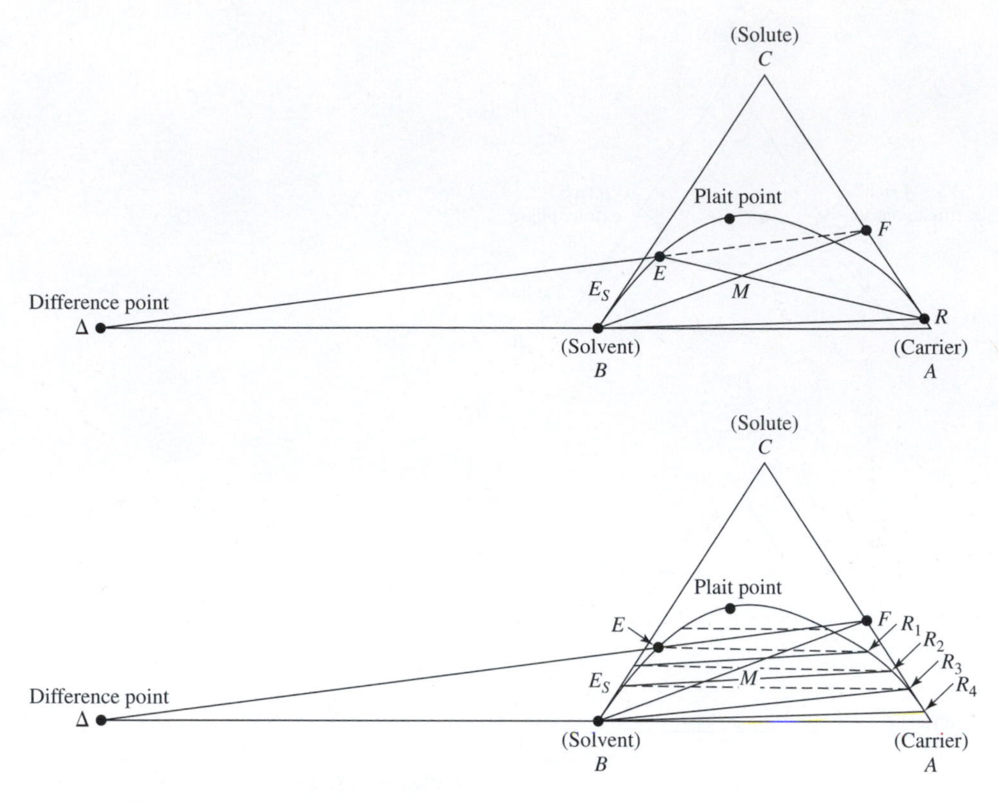

Figure 15-22
Triangular diagram for determining equilibrium stages in liquid-liquid equilibria extraction

and when $\varepsilon = 1.0$, this reduces to

$$N = \frac{X_F - Y_S/K'_S}{X_R - Y_S/K'_S} - 1 \tag{15-42a}$$

where the subscripts F, R, and S refer to the feed, raffinate, and solvent, respectively, while the extraction factor is defined as

$$\varepsilon = \frac{m}{F'/S'} \tag{15-42b}$$

where m is the slope of the equilibrium line and F'/S' the slope of the operating line. When the equilibrium line is not straight, Treybal recommends that a geometric mean value of m be used.† The geometric mean of the slope of the equilibrium line at the concentration leaving the feed stage m_1 and at the raffinate concentration leaving the raffinate stage m_R is given by $(m_1 m_R)^{1/2}$.

†R. E. Treybal, *Liquid Extraction,* 2d ed., McGraw-Hill, New York, 1963.

Equilibrium Stage Calculation for Extraction of a Low-Concentration Solute — EXAMPLE 15-5

A 10 kg/min feed contains 10 weight percent acetic acid (Ac) in water and is to be extracted with a 20 kg/min recycle flow of methyl isobutyl ketone (MIBK) that contains 0.1 weight percent acetic acid and 0.01 weight percent water. It is desired to reduce the acetic acid concentration in the raffinate down to 1 percent on a weight basis. How many equilibrium stages will be required for the separation? Liquid/liquid equilibrium data for the acetic acid–water–MIBK ternary system have been obtained by Sherwood et al.† and can be represented by two phase equilibria relations, namely,

$$Y = 0.93X^{1.10} \qquad \text{for } 0.0299 < X < 0.2708$$

$$Y = 1.27X^{1.29} \qquad \text{for } 0.2708 < X < 0.8065$$

■ Solution

From a material balance establish values for F' and S'. Use the equilibrium data to calculate X and m values. From the latter determine ε. Then use Eq. (15-42) to establish the equilibrium stages required.

$$F' = 10(1 - 0.1) = 9 \text{ kg water/min}$$

$$X_F = \frac{0.1}{0.9} = 0.111 \text{ kg Ac/kg water}$$

$$X_R = \frac{0.01}{0.99} = 0.01 \text{ kg Ac/kg water}$$

$$S' = 20(1 - 0.001 - 0.0001) = 19.98 \text{ kg MIBK/min}$$

$$Y_S = \frac{0.01}{19.98} = 0.0005 \text{ kg Ac/kg MIBK}$$

If we make a material balance around the multistage extractor and assume that $S' = E'$ and $F' = R'$, then

$$Y_e = \frac{FX_F + S'Y_S - R'X_R}{E'}$$

$$= \frac{9(0.111) + 19.98(0.0005) - 9(0.01)}{19.98}$$

$$= 0.046 \text{ kg Ac/kg MIBK}$$

The correlation of the equilibrium data indicates that $Y = 0.93X^{1.10}$ at low values of X. Check the value of X_1.

$$X_1 = \left(\frac{0.046}{0.93}\right)^{1/1.10} = 0.065 \qquad \text{and correlation can be used}$$

$$m_1 = \frac{dY}{dX} = (0.93)(1.10)(0.065)^{0.1} = 0.778 \qquad \text{at } X = 0.065$$

At $X = 0.0299$,

$$Y = (0.93)(0.0299)^{1.10} = 0.196$$

†T. K. Sherwood et al., *Ind. Eng. Chem.*, **31:** 599 (1939).

This will permit obtaining the slope of the equilibrium line at the raffinate end of the extractor

$$m_R = \frac{dY}{dX} = K'_S = \frac{0.0196}{0.0299} = 0.655$$

The geometric mean value of m is obtained from

$$m = (m_1 m_R)^{1/2} = [0.778(0.655)]^{1/2} = 0.714$$

Equation (15-42b) establishes the extraction factor

$$\varepsilon = \frac{mS'}{F'} = (0.714)\left(\frac{19.98}{9}\right) = 1.585$$

The required equilibrium stages are obtained by using Eq. (15-42)

$$N = \ln\left[\frac{0.111 - 0.0005/0.655}{0.01 - 0.0005/0.655}\left(1 - \frac{1}{1.585}\right) + \frac{1}{1.585}\right]\left(\frac{1}{\ln 1.585}\right) = 3.5 \text{ stages}$$

Use of a computer program indicates a requirement of 3.2 stages.

To determine the height of the extraction unit will require evaluation of the height equivalent to a theoretical stage.

Countercurrent Mass-Transfer Unit Calculations The mass-transfer unit concept is used to more rigorously represent the interaction of fluids in a differential contactor. For the recovery of low concentrations of solute from an extraction feed, the assumption of a straight operating line with an intercept of zero is often valid. Under these conditions, the number of mass-transfer units based on the overall raffinate phase composition N_{oR} can be determined by using a modified form of Eq. (15-42). When $\varepsilon \neq 1.0$,

$$N_{oR} = \ln\left[\frac{X_F - Y_S/m}{X_R - Y_S/m}\left(1 - \frac{1}{\varepsilon}\right) + \frac{1}{\varepsilon}\right] \Big/ \left(1 - \frac{1}{\varepsilon}\right) \tag{15-43}$$

When $\varepsilon = 1.0$, the number of mass-transfer units is identical to the number of equilibrium stages obtained with Eq. (15-42a).

Equations (15-44) and (15-45) show the effect on the solute concentration in the raffinate when the solvent/feed ratio is changed while the number of mass-transfer units is kept constant. When $\varepsilon \neq 1.0$,

$$\frac{X_R - Y_S/m}{X_F - Y_S/m} = \frac{1 - 1/\varepsilon}{e^{N_{oR}(1-1/\varepsilon)} - 1/\varepsilon} \tag{15-44}$$

and when $\varepsilon = 1.0$,

$$\frac{X_R - Y_S/m}{X_F - Y_S/m} = \frac{1}{N_{oR} + 1} \tag{15-45}$$

Solution of these last two equations indicates that the solute concentration in the raffinate changes very little when $\varepsilon > 5.0$. This occurs because mass-transfer from the

Table 15-10 Approximate HETS values for several types of extractors

Extractor type	HETS, m	Superficial velocity of dispersed and continuous phases, m/h
Sieve column	0.5–3.6	27–75
Random packing		
Pall rings	0.6–1.5	12–40
Raschig rings	0.6–1.5	12–40
Intalox saddles	0.7–2.1	12–40
Structured packing		
Sulzer SMV	0.6–1.3	65–90
Sulzer SX	0.5–1.2	65–90
Intalox 2T	1.0–1.6	65–90
Mechanically aided		
Pulsed sieve column	0.8–1.3	25–35
Pulsed packed column	0.2–1.3	17–23
Karr (recip. – plate)	0.2–0.3	20–50
RDC	0.3–0.5	7–30
Scheibel	0.2–0.4	6–14
Graessner	0.1–0.2	1–2
Centrifugal	0.1–0.2	75–110

Modified from data presented by J. Stichlmair, *Chemi-Ingenieur-Technik,* **52:** 253 (1980), G. D. Ulrich, *A Guide to Chemical Engineering Process Design and Economics,* J. Wiley, New York, 1984, J. L. Humphrey and G. E. Keller, II, *Separation Process Technology,* McGraw-Hill, New York, 1997, and D. R. Woods, *Process Design and Engineering Practice,* Prentice-Hall, Englewood Cliffs, NJ, 1995.

raffinate phase limits the performance, which is true even in staged equipment with insufficient residence time.

Evaluation of HETS and HTU The height equivalent to a theoretical stage is defined as

$$\text{HETS} = \frac{Z}{N} \tag{15-46}$$

where Z is the height of the column and N the number of theoretical stages required. HETS values are generally obtained from actual laboratory data. Table 15-10 provides performance data for a number of commercial extraction devices noted earlier.

Similarly, the height of a transfer unit based on the raffinate composition is expressed as

$$\text{HTU}_{oR} = \frac{Z}{N_{oR}} \tag{15-47}$$

where N_{oR} is the number of mass-transfer units required to achieve the desired separation based on the raffinate phase composition. Here again, contributions to the HTU based on the raffinate composition arise from the resistance to mass transfer in the raffinate phase and the resistance to mass transfer in the extract phase, as shown by

$$\text{HTU}_{oR} = \text{HTU}_R + \text{HTU}_E/\varepsilon \tag{15-48}$$

Here HTU_R and HTU_E are defined in terms of the raffinate rate R and the extract rate

E, respectively, as

$$\text{HTU}_R = \frac{R}{A_c k_r a} \tag{15-49a}$$

$$\text{HTU}_E = \frac{E}{A_c k_e a} \tag{15-49b}$$

where A_c is the cross-sectional area of the extraction column, k_r and k_e the mass-transfer coefficients in the raffinate and extract phases, respectively, and a the interfacial (droplet) mass-transfer area per column volume. Predictions of both mass-transfer coefficients and mass-transfer areas of droplets are quite complex. Only limited success has been achieved in correlating extractor performance with basic principles. As a consequence, rather than attempt tedious hand calculations, approximate results can be obtained through the use of a number of commercially available computer programs.

Equipment Costs for Liquid/Liquid Extraction Equipment

Since the equipment utilized in this separation process uses trayed and packed columns similar to those used in distillation, the costs for such columns can be obtained by using the cost data provided in the distillation section.

SEPARATION USING MEMBRANES

In membrane separations the two products, namely, the permeate and retentate, are usually miscible, the separating agent is a semipermeable barrier, and a sharp separation is generally not achievable. Thus membrane separations differ from the more common separation processes discussed earlier. However, the replacement of such separation processes with membrane separations many times has the potential to reduce the energy requirements for the separation. This replacement, though, requires a high mass-transfer flux, defect-free, long-life membranes, and fabrication of membranes into compact, economical modules of high surface area per unit volume.

Selection of Membrane Types

Nearly all industrial membranes are fabricated from natural or synthetic polymers. Natural polymers include wool, rubber, and cellulose. Some of the more widely used synthetic polymers include polyimide, polysulfone, polystyrene, polyisoprene, polycarbonate, polytetrafluorethylene, cellulose triacetate, and aromatic polyamide. These are glassy, rubbery, or crystalline polymers that are normally fabricated into membrane shapes identified as flat asymmetric or thin-composite sheet, tubular, or hollow fiber. The more common membrane modules are the plate and frame, spiral-wound, four-leaf spiral-wound, hollow fiber, tubular, and monolith. Table 15-11 provides a comparison of several of these membrane modules.

The application of polymer membranes is generally limited to temperatures below about 150°C and to the separation of mixtures that are chemically inert. Operation at higher temperatures or with chemically active mixtures requires the use of inorganic membranes such as microporous ceramics, carbon, or dense metals.

Table 15-11 Typical characteristics of four membrane modules

Membrane module	Packing density, m^2/m^3	Fouling resistance	Cleaning ease	Relative cost	Principal application†
Plate and frame	30–500	Good	Good	High	RO, PV, UF, MF
Spiral-wound	200–800	Moderate	Fair	Low	RO, GP, UF, MF
Hollow fiber	500–9000	Poor	Poor	Low	RO, GP, UF
Tubular	30–200	Very good	Very good	High	RO, UF

†RO = reverse osmosis, PV = pervaporation, GP = gas permeation, UF = ultrafiltration, MF = microfiltration

General Design Concepts for Membrane Separation

To be effective in the separation of a mixture of chemical components, the membrane must provide a high permeance and a high permeance ratio for the two components being separated. The *permeance* for a given component diffusing through a membrane is defined as the flow rate of that component per unit cross-sectional area of the membrane per unit driving force per unit thickness. This is analogous to the mass-transfer coefficient used in several separation processes discussed earlier.

The molar transmembrane flux N_i for component i is given by

$$N_i = \frac{P_{M_i}}{t_M} \Delta\Phi_i \tag{15-50}$$

where P_{M_i} is the permeance of component i, t_M the thickness of the membrane, and $\Delta\Phi_i$ the driving force of component i across the membrane. This relation can also be written as

$$N_i = \bar{P}_{M_i} \Delta\Phi_i \tag{15-50a}$$

where $\bar{P}_{M_i}$ is the permeability of component i.

For a given application, the calculation of the required membrane surface area is generally based on laboratory data for the selected membrane. Although permeation can occur by one or more of the transport mechanisms discussed in this section, all are consistent with either Eq. (15-50) or Eq. (15-50*a*). For example, the mechanisms for the transport of liquid or gas molecules through a porous membrane can involve bulk flow through the pores, diffusion through the pores, restricted diffusion through the pores, and solution diffusion through dense membranes.

In terms of the bulk flow of a gas through a porous membrane, subjected to a pressure difference, Eq. (15-50) takes the form of

$$N_i = \frac{P_{M_i}}{t_M}(p_F - p_P) \tag{15-51}$$

where p_F and p_P are the pressures on the feed and permeate surfaces of the membrane, respectively. Here P_{M_i} is given by

$$P_{M_i} = \frac{\rho\varepsilon^3}{4.16(1-\varepsilon)^2 a_v^2 \mu} \tag{15-51a}$$

Here ε is the porosity (void fraction) of the membrane, a_v the pore surface area per unit volume of membrane, ρ the density of the gas, and μ the viscosity of the gas. The porosity of the membrane is obtained from

$$\varepsilon = n\pi \frac{d_p^2}{4} \tag{15-51b}$$

where n is the number of pores per unit cross-section of the membrane and d_p the pore diameter. When the pores are assumed to be cylindrical, a_v is simply equal to $6/d_p$.

If the bulk flow is in the laminar region (Re < 2100), which is almost always the case for flow in small-diameter pores, the bulk-flow flux with the above assumption reduces to

$$N_i = \frac{\rho\varepsilon^3(p_F - p_P)}{4.16(1-\varepsilon)^2 a_v^2 t_M \mu} \tag{15-52}$$

Equation (15-52) can be used as a first approximation to obtain the pressure drop for flow through a porous membrane with a tortuosity factor greater than 2. For gas flow, the density may be taken as the arithmetic average of the densities at the feed and permeate surfaces of the membrane.

In liquid diffusion, the solute transmembrane flux is obtained from a modified form of Fick's law as

$$N_i = D_i\left(\frac{\varepsilon}{\tau}\right)(c_F - c_P)K_{r,i} \tag{15-53}$$

where D_i is the diffusivity of solute i in the feed mixture, ε the porosity of the membrane defined in Eq. (15-51b), τ the tortuosity factor, c_F and c_P the concentrations of solute i in the membrane pores at the feed and permeate surfaces of the membrane, respectively, and $K_{r,i}$ a restrictive factor that accounts for the effect of pore diameter when the ratio of molecular diameter d_m to pore diameter d_p exceeds about 0.01. The restrictive factor may be evaluated from

$$K_{r,i} = \left(1 - \frac{d_m}{d_p}\right)^4 \qquad \text{for } \frac{d_m}{d_p} \le 1 \tag{15-53a}$$

In general, transmembrane fluxes for liquids through microporous membranes are very small because diffusivities themselves are very low.

In gas diffusion, the rate of component diffusion can also be expressed in terms of Fick's law. If pressure and temperature on either side of the membrane are equal and the pressure is low to validate use of the ideal gas law, the transmembrane flux relation can be written as

$$N_i = \left(\frac{D_{e,i}}{RTt_M}\right)(\bar{p}_F - \bar{p}_P) \tag{15-54}$$

where $D_{e,i}$ is the effective diffusivity, including both ordinary diffusion and Knudsen diffusion, and $\bar{p}_F$ and $\bar{p}_P$ are the partial pressures of component i in the membrane pores at the feed and permeate surfaces of the membrane, respectively. When Knudsen flow predominates, as it usually does in the micropores of many membranes, the permeability ratio of the two components being separated is inversely proportional to the ratio of the molecular weights of the two components. Except for gaseous components with widely differing molecular weights, the permeability ratio is not very large. This indicates that separation by gaseous diffusion without an accompanying driving force in most cases is not very promising.

The transport of components through nonporous membranes, however, is the mechanism of membrane separators for reverse osmosis, gas permeation, and pervaporation. Liquid diffusivities are several orders of magnitude less than gas diffusivities, and diffusivities of solutes in solids are a few orders of magnitude less than diffusivities in liquids. Thus, differences between diffusivities in gases and solids are extremely large.

Because of the low separation efficiencies obtained with simple liquid or gas diffusion, most membrane systems utilize solution diffusion separations. The solution diffusion model developed by Lonsdale et al.† is most often used to design membrane separators. This model is based on Fick's law for diffusion through solid nonporous membranes subjected to a driving force in which the concentrations are those of the solute dissolved in the membrane. The concentrations in the membrane are related to the concentrations or partial pressures in the fluid adjacent to the membrane faces by assuming thermodynamic equilibrium at the fluid/membrane interfaces.

Comparisons of typical solute profiles for liquid mixtures and for gas mixtures with porous and nonporous membranes are shown in Fig. 15-23. For porous membranes the concentration and partial pressure profiles are continuous from the bulk feed side to the bulk permeate side. In nonporous membranes there are film resistances to mass transport at both surfaces of the membrane. Assuming that thermodynamic equilibrium exists at these two fluid/membrane interfaces, the concentrations in Fick's law can be related through the use of thermodynamic equilibrium partition coefficients K_i in the case of liquid mixtures and through the use of Henry's law H_i for gas mixtures.

When the mass-transfer resistances in the two boundary layer films are negligible, the solute transmembrane flux for liquid mixtures is given by

$$N_i = \frac{K_i D_i}{t_M}(c_{F,i} - c_{P,i}) \tag{15-55}$$

where D_i is the diffusivity of the solute in the membrane and the product $K_i D_i$ is the permeance $P_{M,i}$. In the solution diffusion model, K_i accounts for the solubility of the solute in the membrane, and D_i accounts for the diffusion in the membrane. Since D_i is generally very small, it is critical that the membrane material provide a large value for K_i and/or a small membrane thickness.

Using the same approach with similar assumptions, the solution diffusion model for gas mixtures provides a comparable relation for solute transmembrane flux of

$$N_i = \frac{H_i D_i}{t_M}(\bar{p}_{F,i} - \bar{p}_{P,i}) \tag{15-56}$$

†H. K. Lonsdale, U. Merten, and R. L. Riley, *J. Appl. Polym. Sci.*, **9:** 1341 (1965).

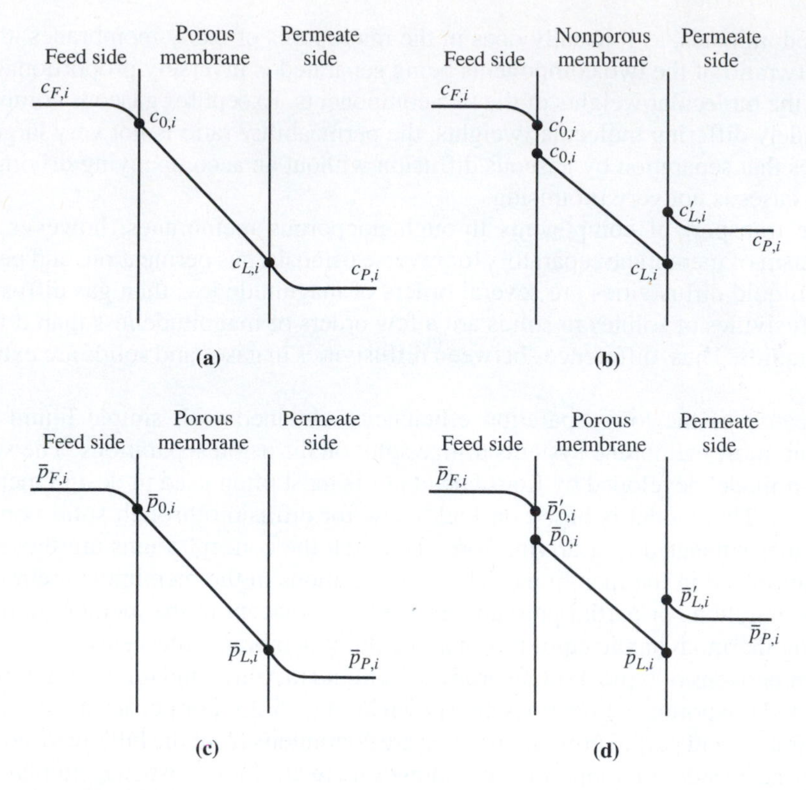

Figure 15-23
Concentration and partial pressure profiles for solute transport through membranes. Liquid mixture with (a) a porous and (b) a nonporous membrane; gas mixture with (c) a porous and (d) a nonporous membrane.

where H_i is Henry's law constant and $H_i D_i$ the permeance. The latter depends on both the solubility of the solute in the membrane and the diffusivity of that component in the feed.† An acceptable rate of transport through the membrane can only be achieved by using a very thin membrane and a high pressure difference across the membrane.

The ideal nonporous membrane exhibits a high permeance for the penetrant molecule and a high separation factor or selectivity between the components to be separated. The separation factor is defined as

$$a^*_{i,j} = \frac{y_{P,i}\; y_{P,j}}{x_{R,i}\; x_{R,j}} \tag{15-57}$$

where $y_{P,i}$ is the mol fraction of component i in the permeate leaving the membrane and $x_{R,i}$ the mol fraction of component i in the retentate on the feed side of the membrane.

†An extensive tabulation of solubility and diffusivity values is given by J. Bandrup and E. H. Immergut, eds., *Polymer Handbook,* 3d ed., J. Wiley, New York, 1989.

The subscripts i and j refer to the components being separated. (Note that y and x have no equilibrium implications.) When the film mass-transfer resistances are negligible on both sides of the membrane and the permeate pressure is very small compared to the feed pressure, the ideal separation factor can be simplified to

$$\alpha^*_{i,j} = \frac{H_i D_i}{H_j D_j} = \frac{P_{M_i}}{P_{M_j}} \tag{15-58}$$

The transport of components i and j through the membrane, because of perfect mixing, can be written as

$$y_{P,i} n_P = A_M \bar{P}_{M_i}(x_{R,i} p_R - y_{P,i} p_P) \tag{15-59a}$$

and

$$y_{P,j} n_P = A_M \bar{P}_{M_j}(x_{R,j} p_R - y_{P,j} p_P) \tag{15-59b}$$

where A_M is the membrane area normal to the flow, n_p the mols in the permeate, $\bar{P}_M$ the membrane permeability, and p_R and p_P the pressures at the retentate and permeate sides of the membrane, respectively.

When the permeate pressure is not negligible, an expression for $\alpha_{i,j}$ can be expressed in terms of the pressure ratio p_P/p_F as

$$\alpha_{i,j} = \alpha^*_{i,j} \frac{x_{R,i}(\alpha_{i,j} - 1) + 1 - \alpha_{i,j}(p_P/p_F)}{x_{R,i}(\alpha_{i,j} - 1) + 1 - p_P/p_F} \tag{15-60}$$

where $x_{R,i}$ is the mol fraction in the retentate on the feed side of the membrane corresponding to the partial pressure $\bar{p}_{F,i}$ in Fig. 15-23. Equation (15-60) is an implicit equation for $\alpha_{i,j}$ in terms of the p_P/p_F ratio and x_i that can be solved for $\alpha_{i,j}$ by using the formula for a quadratic equation. Generally, the permeabilities are high and the separation factors are low in rubbery polymers. The opposite is true for glassy polymers. For a given feed composition, however, the separation factor limits the degree of separation that can be attained. A listing of ideal separation factors is available in the *Membrane Handbook*.[†]

Discussions to this point have assumed a flow pattern with perfect mixing of both the retentate and permeate streams. Three other flow patterns, as shown in Fig. 15-24, are also used in membrane separation processes. These are the flow patterns of countercurrent, cocurrent, and crossflow; all assume no fluid mixing and are comparable to the idealized flow patterns used in heat exchanger design. For a given type of equipment discussed earlier, it is not always obvious which idealized flow pattern to assume. For example, hollow-fiber membrane separation systems may be designed to approximate countercurrent, cocurrent, or crossflow patterns.

Solution methods for all four flow patterns are presented by Walawender and Stern[‡] under the assumptions of a binary feed with constant p_P/p_F ratio and constant ideal separation factor. Exact analytical solutions are available for the perfect mixing

[†]W. S. W. Ho and K. K. Sirkar, *Membrane Handbook,* Van Nostrand Reinhold, New York, 1992.

[‡]W. P. Walawender and A. Stern, *Separation Sci.,* **7:** 553 (1972).

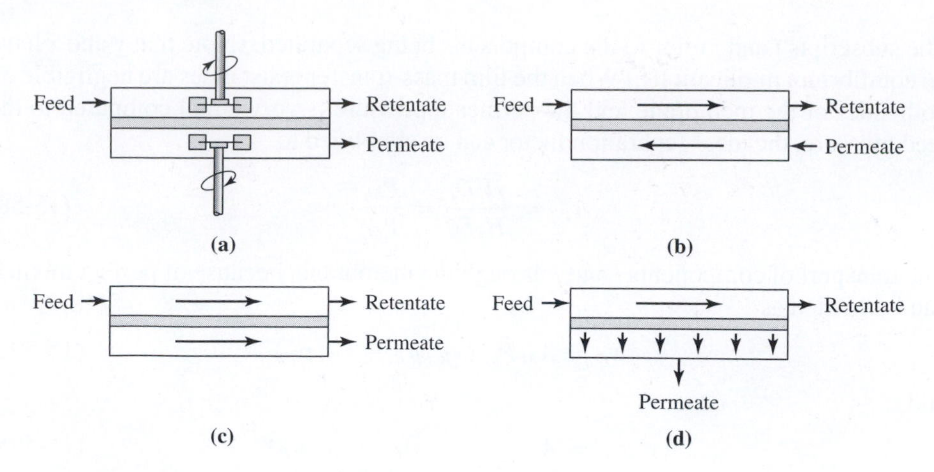

Figure 15-24
Idealized flow patterns in membrane modules: (a) perfect mixing; (b) countercurrent flow; (c) cocurrent flow; (d) cross-flow

and the crossflow case. Numerical solutions of ordinary differential equations with appropriate computer codes are required for the countercurrent and crossflow cases.

The extent to which a feed mixture can be separated in a single stage is determined by the separation factor α. The latter depends on the existing flow pattern, the permeability ratio, the driving force for mass transfer, and the *stage cut* Θ, defined as the ratio of the mols in the permeate to the mols in the feed n_P/n_F. To achieve a higher degree of separation will require going to a countercurrent cascade of stages. For a cascade, additional factors that affect the degree of separation of the feed are the number of stages and the *recycle ratio,* defined as the ratio of permeate recycle rate to permeate product rate. Studies have shown that it is best to manipulate the stage cut and reflux rate at each membrane stage so that the compositions of the two streams entering each stage will be the same.

Generally, it is assumed that the pressure drop on the retentate side of the membrane in a cascade arrangement is negligible. Thus, only the permeate must be pumped when it is a liquid, or compressed when it is a gas, in order to be transferred to the next stage. In the case of gas permeation, compression costs can be quite high. Therefore, membrane cascades for gas permeation are often limited to just two or three stages. Common recycle cascades normally involve a two-stage stripping cascade, a two-stage enriching cascade, or a two-step enriching cascade with an additional premembrane stage.

General Design Procedures for Membrane Separation

Since design procedures for countercurrent and cocurrent flow patterns necessitate developing a series of differential equations requiring computer solution, the design procedures discussed will be limited to the perfect mixing and crossflow patterns in membrane modules. For the perfect mixing flow pattern, the initial step is to establish the operating conditions of temperature, pressure, and flow rate of the feed. Using the guidelines presented earlier, select the most appropriate membrane for the separation, noting the need for a high separation factor; that is, the membrane must possess a high permeance and a high permeance ratio for the two components being separated by the membrane.

Next, a material balance must be made equating the molar flow rate of each component in the feed to the sum of the molar flow rates in the permeate and retentate streams. Upon substitution of the stage cut definition into the material balance equations, the resulting equations can be expressed in terms of the mol fraction in the retentate streams.

Since both fluid sides are well mixed, use Eq. (15-58) to obtain the ideal separation factor $\alpha^*_{i,j}$. The latter can be used in Eq. (15-60) to obtain the actual separation factor if the permeate pressure is not negligible. Solution of Eq. (15-59) should provide the required membrane area for the desired stage cut Θ.

In the crossflow pattern, the feed flows across the upstream membrane surface in plug flow with no longitudinal mixing. In the analytical solution developed by Naylor and Backer,[†] the pressure ratio and the ideal separation factor are assumed constant. The film mass-transfer coefficients on both sides of the membrane are assumed to be negligible. If a differential element within the membrane is considered, the local mol fractions in the retentate and the permeate are x_i and y_i, respectively, and the penetrant molar flux is dn/dA_M.

The local separation factor can be obtained with Eq. (15-60) in terms of the local x_i, p_P/p_F, and $\alpha^*_{i,j}$. An alternative expression for the local permeate composition is given by

$$\frac{y_i}{1 - y_i} = \frac{\alpha^*_{i,j}(x_i - y_i p_P/p_F)}{1 - x_i - (1 - y_i)(p_P/p_F)} \tag{15-61}$$

A material balance around the differential element within the membrane for component i results in

$$y_i\, dn = d(nx_i) = n\, dx_i + x_i\, dn \tag{15-62}$$

Equation (15-57) can be used to eliminate the y_i term in Eq. (15-62) to give

$$\frac{dn}{n} = \frac{[1 + (\alpha_{i,j} - 1)x_i]\, dx_i}{x_i(\alpha_{i,j} - 1)(1 - x_i)} \tag{15-63}$$

When the pressure ratio p_P/p_F is small, the separation factor remains relatively constant over much of stage cut range. With this assumption, Eq. (15-63) can be integrated from the local point within the differential element to the final retentate flow rate and composition. The integration results in

$$n = n_R(1 - \Theta)\left(\frac{x_i}{x_{R,i}}\right)^{1/(x_{i,j}-1)}\left(\frac{1 - x_{R,i}}{1 - x_i}\right)^{\alpha_{i,j}/(\alpha_{i,j}-1)} \tag{15-64}$$

[†]R. W. Naylor and P. O. Backer, *AIChE J.*, **1:** 95 (1955).

The mol fraction of component i in the final permeate is obtained by integrating the following relation:

$$y_{P_i} = \int_{x_{F_i}}^{x_{R_i}} \frac{y_i}{\Theta n_F} \, dn \tag{15-65}$$

This integral in terms of n can be transformed into an integral in x_i by utilizing Eqs. (15-57), (15-61), and (15-64). Integration of the resulting integral for component i gives

$$y_{P_i} = x_{R,i}^{1/(1-\alpha)} \left(\frac{1-\Theta}{\Theta}\right) \left[(1 - x_{R,i})^{\alpha/(\alpha-1)} \left(\frac{x_{F,i}}{1 - x_{F,i}}\right)^{\alpha/(\alpha-1)} - x_{R,i}^{1/(1-\alpha)} \right] \tag{15-66}$$

where the $\alpha_{i,j}$ represented as α can be estimated from Eq. (15-60) by assuming $x_i = x_{F_i}$. The differential rate of mass transfer of component i across the membrane is then given by

$$y_i \, dn = (P_{M_i}/t_M)(x_i p_F - y_i p_P) \, dA_M \tag{15-67}$$

from which the total membrane area can be obtained by integration of the relation

$$A_M = \int_{x_{R,i}}^{x_{F,i}} \frac{y_i t_M}{P_{M_i}(x_i p_F - y_i p_P)} \, dn \tag{15-68}$$

EXAMPLE 15-6 Separation of Air by Gas Permeation

A low-density thin-film polyethylene membrane with a thickness of 0.1μm is proposed to separate 37,000 m^3/h of air into an enriched oxygen stream and an enriched nitrogen stream. The air is compressed, cooled, and purified before entering the membrane at 25°C and 10 atm. The pressure on the permeate side of the membrane is 1 atm. Assume perfect mixing on both sides of the membrane so that compositions are uniform and equal to the exit compositions. Neglect pressure drop and mass-transfer resistances on both sides of the membrane. Composition of the feed air stream may be assumed to be 21 mol percent oxygen and 79 mol percent nitrogen. What membrane surface area will be required for a stage cut (fraction of feed permeated) of 0.4?

■ Solution

The *Polymer Handbook* provides the following data for oxygen and nitrogen at 25°C when low-density polyethylene is used as a membrane material:

	O_2	N_2
Diffusivity D_i, cm^2/s	4.62×10^{-7}	3.2×10^{-7}
Henry's law constant H_i, cm^3 (STP)/(cm^3·Pa)	4.72×10^{-7}	2.28×10^{-7}

As outlined earlier in this section, initiate the solution by obtaining the molar flow rate of the feed

$$n_p = \frac{37{,}000(273/298)}{22.4} = 1513 \text{ kg mol/h}$$

Calculate the permeance of oxygen and nitrogen, noting that $P_{M,i} = D_i H_i$.

$$P_{M,O_2} = (4.62 \times 10^{-7})(4.72 \times 10^{-7}) = 2.18 \times 10^{-13}\ \text{cm}^3\,(\text{STP})\cdot\text{cm/cm}^2\cdot\text{s}\cdot\text{Pa}$$

$$P_{M,N_2} = (3.2 \times 10^{-7})(2.28 \times 10^{-7}) = 7.296 \times 10^{-14}\ \text{cm}^3\,(\text{STP})\cdot\text{cm/cm}^2\cdot\text{s}\cdot\text{Pa}$$

The permeability of O_2 in appropriate units is

$$\bar{P}_{M,O_2} = P_{M,O_2}/t_M = (2.18 \times 10^{-13})(3600)(10^2)/(22{,}400)(10^3)(1 \times 10^{-7})$$
$$= 3.50 \times 10^{-8}\ \text{kg mol}\cdot\text{m/m}^2\cdot\text{h}\cdot\text{Pa}$$

Similarly,

$$\bar{P}_{M,N_2} = 1.17 \times 10^{-8}\ \text{kg mol}\cdot\text{m/m}^2\cdot\text{h}\cdot\text{Pa}$$

A material balance for the oxygen can be made in terms of the compositions of the feed, retentate, and permeate as

$$x_{F,O_2} n_F = x_{R,O_2} n_R + y_{P,O_2} n_p$$

Since the stage cut Θ is defined as n_P/n_F, $(1 - \Theta) = n_R/n_F$. This can be substituted into the oxygen material balance to obtain the mol fraction in the retentate

$$x_{R,O_2} = \frac{x_{F,O_2} - y_{P,O_2}\Theta}{1 - \Theta} \tag{1}$$

Upon appropriate substitution for y_{P,N_2} and x_{R,N_2} the separation factor, as defined in Eq. (15-57), becomes

$$\alpha^*_{O_2,N_2} = \left(\frac{y_{P,O_2}}{x_{R,O_2}}\right)\left(\frac{1 - y_{P,O_2}}{1 - x_{R,O_2}}\right) \tag{2}$$

Since the permeate pressure is not negligible relative to the pressure, Eq. (15-60) is used to obtain α_{O_2,N_2}. Note that the ratio of p_P/p_F is 0.1, and $\alpha^*_{O_2,N_2}$ is obtained from Eq. (15-58) as

$$\alpha^*_{O_2,N_2} = \frac{2.18 \times 10^{-13}}{7.296 \times 10^{-14}} = 2.99$$

Equation (15-60) can be simplified to

$$\alpha_{O_2,N_2} = 2.99\frac{x_{R,O_2}(\alpha_{O_2,N_2} - 1) + 1 - 0.1\alpha_{O_2,N_2}}{x_{R,O_2}(\alpha_{O_2,N_2} - 1) + 1 - 0.1} \tag{3}$$

For a stage cut of 0.4, we now have Eqs. (1), (2), and (3) with three unknowns, namely, α_{O_2,N_2}, y_{P,O_2}, and x_{R,O_2}. The simplest solution is to solve the three equations simultaneously with a computer program such as Mathcad. Such a solution results in the following values:

$$x_{R,O_2} = 0.146 \qquad y_{P,O_2} = 0.306 \qquad \alpha_{O_2,N_2} = 2.573$$

The membrane area can now be solved by using Eq. (15-59*a*)

$$A_M = \frac{y_{P,O_2} n_p}{\bar{P}_{M,O_2}(x_{R,O_2} p_R - y_{P,O_2} p_P)}$$

$$= \frac{0.306(1513)(0.4)}{3.50 \times 10^{-8}[0.146(1.013 \times 10^6) - 0.306(1.013 \times 10^5)]}$$

$$= 45{,}600\ \text{m}^2$$

Even though the oxygen concentration in the permeate has increased from 21 to 30.6 mol percent and the nitrogen concentration in the retentate from 79 to 85.4 mol percent, it is apparent that this membrane separation is uneconomical based on the required membrane area. The low-density polyethylene membrane is unsuitable for this separation. To achieve an economical enrichment will require a membrane with a much higher separation factor and a crossflow or countercurrent flow pattern. ■

Range of Costs for Several Membrane Systems

Since membranes can be fabricated in different sizes and thicknesses, there is quite a range in the cost of the membrane and the module. Costs for four widely used membrane modules are listed in Table 15-12.

SEPARATION BY ADSORPTION

In an adsorption process, molecules, atoms, or ions in a gas or liquid diffuse to the surface of a solid where they attach to that surface and are held there by weak intermolecular forces. The adsorbed solutes are referred to as the *adsorbate* while the solid material is the *adsorbent*. To achieve a very large surface area for adsorption per unit volume, highly porous solid particles with small-diameter, interconnected pores are used.

During adsorption and ion exchange, the adsorbent becomes essentially saturated with the adsorbate and requires regeneration by desorbing the adsorbate. The adsorbing and regeneration are generally carried out in a cyclical manner.

Adsorption processes are classified as either bulk separation or purification depending upon the concentration of the adsorbate in the liquid or gas stream being treated. A distinction is made since considerable quantities of energy can be liberated when large amounts of a component are adsorbed on a surface and likewise can be required when that component desorbs. For gas adsorptions, this release of energy may be as much as or more than the heat of vaporization. In liquid adsorptions this energy liberation is somewhat less. For purification, because of adsorbate concentrations of less than 1 weight percent, this release is not too important. However, for bulk separations, the heat release becomes an important parameter in the design process.

Table 15-12 Costs for membrane modules†

Membrane module	Cost, $\$/m^2$
Plate and frame	250–400
Spiral-wound	25–100
Hollow fiber	10–20
Tubular	250–400
Capillary	25–100
Ceramic	1000–1600

†R. W. Baker, *Membrane Technology and Applications,* McGraw-Hill, New York, 2000.

Selection of Sorbent for Separation by Adsorption

For commercial applications, a sorbent should exhibit a number of suitable characteristics. Among these are high selectivity to enable sharp separations, high adsorption capacity, chemical and thermal stability, free-flowing tendency, and relatively low cost. Most solids are able to adsorb components from liquids and gases. However, only a few have sufficient selectivity and capacity to serve as candidates for commercial adsorbents. Actually, only four types are predominantly used, and these are activated carbon, molecular sieve zeolites, silica gel, and activated alumina. Typical physical properties for these four adsorbent types are given in Table 15-13.

Activated carbon is the most widely used sorbent, and it may be in the form of a fine powder or as hard pellets, granules, cylinders, spheres, or fibers. A carbon molecular sieve has also been commercialized which exhibits a very narrow pore size distribution compared to the normally activated carbons.

Molecular sieve zeolites (MSZ) are highly crystalline aluminosilicate structures. Varying structures can be produced by changing the silicon/aluminum ratio in the feed solution and in the crystallization and drying conditions. Since zeolites consist of highly regular channels and cages, they are selective for polar and hydrophilic components. Very strong bonds are formed with water, carbon dioxide, hydrogen sulfide, and other similar components, while weaker bonds are created with organic components. Note should be made of a molecular sieve which consists almost entirely of silica. In contrast to the regular molecular sieve zeolites, these zeolites are hydrophobic, and their separation behavior more closely resembles that of the activated carbons. However, because of the higher thermal stability of these silica molecular sieves, such sorbents can be regenerated under much more severe conditions than those for regenerating activated carbons.

Silica gels provide an intermediate choice between highly hydrophilic and highly hydrophobic surfaces. These adsorbents are most widely used to remove water from various gases, but hydrocarbon removal is also possible.

Since activated alumina also has a high affinity for water, these adsorbents are often used in drying applications of gases. As with silica gels, the water bond with the alumina surface is not as strong as those bonds associated with molecular sieves and thus requires milder temperatures in the regeneration process. Activated alumina can provide a higher ultimate capacity for water than the typical molecular sieve, but the latter will have a higher capacity at low partial pressures of water. If both adsorbent capacity and high sorbent removal are important, it would be possible to use the adsorbents in series by using the alumina to remove the majority of the water followed by the molecular sieve to achieve the desired moisture content in the gas.

The necessity for highly reactive adsorbents to selectively remove undesirable organics has led to the development of biosorbents and affinity adsorbents. The former sorbs the organic molecule and during regeneration oxidizes the molecule to carbon dioxide, water, and other products, if such atoms exist in the original molecules. Affinity adsorbents are extremely selective materials used to recover specific biomaterials or organic molecules from a complex mixture of these molecules. The use of these adsorbents is not economically practical except in the recovery of high-priced biomaterials and pharmaceuticals.

Table 15-13 Typical properties for commercial adsorbents†

Type of adsorbent	Pore diameter, nm	Particle density, kg/m^3	Particle porosity,‡	Surface area $10^5\ m^2/kg$	H_2O cap. wt. %, 25°C, 4.6 mmHg	CO_2 cap. wt. %, 25°C, 250 mmHg	Reactivation temp., °C
Carbon, activated							
Small pore	1–2.5	500–900	0.4–0.6	4–12	1	5	—
Large pore	>3	600–800	—	2–6	—	7	—
Molecular sieve	0.2–1.0	980	—	4	—	NA	—
Molecular sieve							
3A	0.3	670–740	0.2	7	20	NA	>350
4A	0.4	660–720	0.3	7	23	13	
5A	0.5	670–720	0.4	6.5	21	15	
13X	0.8	610–710	0.5	6	25	16	
Mordenite	0.3–0.4	720–800	0.25	7	9	6	
Chabazite	0.4–0.5	640–720	0.35	6.5	16	12	
Silica gel							
Small pore	2.2–2.6	1000	0.47	8	11	3	130–280
Large pore	10–15	620	0.71	3.2	—	NA	130–280
Activated							
Alumina	1–7.5	800	0.50	3.2	7	2	150–315

†Modified from data presented by J. L. Humphrey and G. E. Keller, II, *Separation Process Technology*, McGraw-Hill, New York, 1997, J. D. Seader and E. J. Henley, *Separation Process Principles,* J. Wiley, New York, 1988, and G. E. Keller, R. A. Anderson, and C. M. Yon, in *Handbook of Separation Process Technology,* R. W. Rousseau, ed., J. Wiley, New York, 1987, chap. 12.

‡Particle porosity defined as 1 − (bulk density/particle density).

Basic Adsorption Cycles

As distinct from distillation, adsorption processes can operate with several different physical arrangements and cycles. Four basic cycles with combinations of these are considered in this section.

In the temperature-swing cycle, shown in Fig. 15-25, a feed stream containing a small quantity of an adsorbate at a partial pressure $\bar{p}_1$ is sent through the adsorption bed at a temperature T_1. The initial equilibrium loading on the adsorbent is X_1, generally expressed in units of weight of adsorbate per weight of adsorbent. After equilibrium between the adsorbate and the adsorbent is reached, regeneration requires raising the temperature of the bed to T_2. Desorption occurs as more feed is sent through the bed and a new equilibrium loading of X_2 is established. The net theoretical removal capacity of the adsorption bed is given by the difference between X_1 and X_2. When the

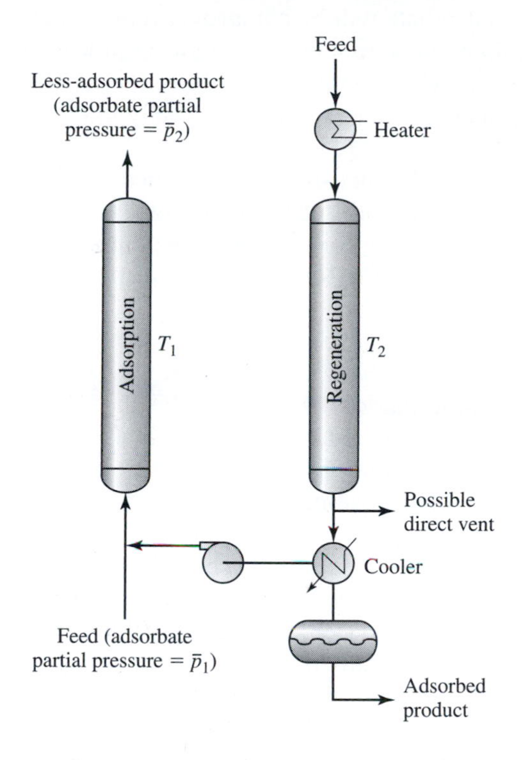

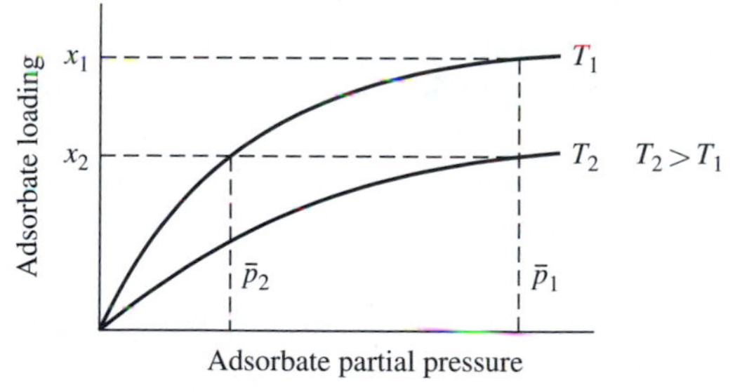

Figure 15-25
Temperature-swing cycle

bed is again cooled to T_1, purification of the gas stream can be reestablished. Typical adsorbate removal in a temperature-swing cycle is generally in excess of 1 kg per 100 kg of adsorbent.

Since the regeneration time associated with the temperature-swing cycle is relatively long, the cycle is used almost exclusively to remove small concentrations of adsorbates from feed streams. This allows the on-stream time to be a significant fraction of the total cycle time of the separation process. Additionally, temperature-swing cycles can require a substantial quantity of energy per unit of adsorbate. However, if the adsorbate concentration in the feed is small, the energy cost per unit of feed processed can be reasonable.

In the inert-purge cycle, the adsorbate during regeneration is removed by sending a nonadsorbing gas containing no adsorbate through the bed. This procedure lowers the partial pressure of the adsorbate and initiates the desorption process. If sufficient pure purge gas is directed through the bed, the adsorbate will be completely removed and adsorption can be resumed. Cycle times, in contrast to temperature-swing processes, are often only a few minutes rather than hours. However, because adsorption capacity is reduced as adsorbent temperatures rise, inert-purge processes are usually limited to 1 to 2 kg of adsorbate per 100 kg of adsorbent.

The displacement purge cycle differs from the inert-purge cycle since the cycle uses a gas or liquid in the regeneration step that adsorbs about as strongly as the adsorbate. In this way, desorption is facilitated both by adsorbate partial pressure or concentration reduction in the fluid and by competitive adsorption of the displacement medium. Typical cycle times are usually a few minutes.

The use of two different types of purge fluid negatively affects the contamination level of the less-adsorbed product. However, since the heat of adsorption of the displacement purge fluid is approximately equal to that of the adsorbate, as the two exchange places on the adsorbent, the net heat generated or consumed is essentially negligible and thus maintains nearly isothermal conditions throughout the cycle. The latter fact makes it possible to increase the adsorbate loading of the displacement purge cycle over that for the inert-purge cycle.

As noted in Fig. 15-26, the partial pressure of an adsorbate can be reduced by lowering the total pressure of the gas. The lower the total pressure of the regeneration step, the greater the adsorbate loading during the adsorption step. The time required to load, depressurize, regenerate, and repressurize a bed generally is only a few minutes. Thus, even though practical adsorbate loadings are almost always less than 1 kg of adsorbate per 100 kg of adsorbent, to minimize thermal gradient problems, the very short times make a pressure-swing cycle a viable option for bulk-gas separations.

It is quite common in a pressure-swing cycle to use a fraction of the less-adsorbed gas product as a low-pressure purge gas. Often the purge flow is in the opposite direction from that for the feed flow. The simplest version of this process is designated as *pressure-swing adsorption* (PSA). Rules for the minimum fraction of less-adsorbed gas product required for displacing the adsorbate have been developed by Skarstrom.†

†C. W. Skarstrom, in *Recent Developments in Separation Science,* vol. 2, N. N. Li, ed., CRC Press, Cleveland, Ohio, 1972, p. 95.

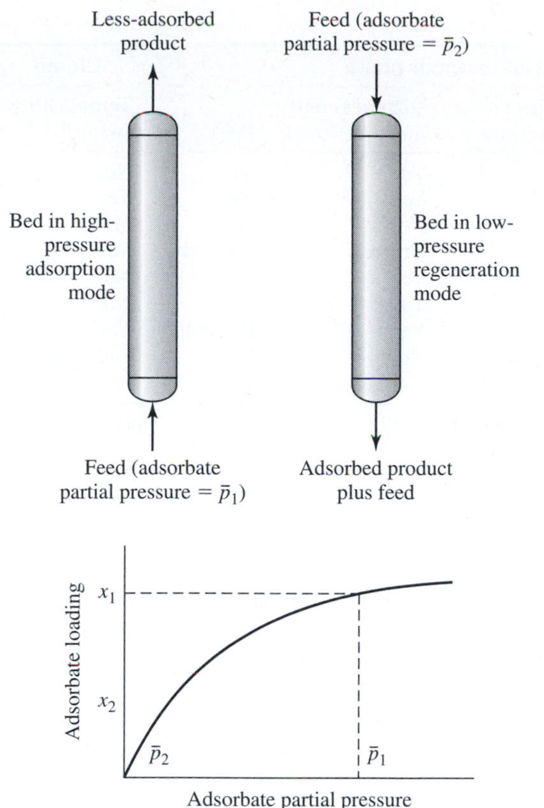

Figure 15-26
Pressure-swing cycle

Selection of Appropriate Adsorption Cycle

To assist in the selection of the appropriate adsorption cycle, Keller et al.[†] have developed a simple matrix that can rapidly serve as a guide for making this choice with essentially no calculations. In Table 15-14, nine process conditions are given that can be used to characterize the desired adsorption cycle. All that is necessary is to determine every process condition that applies to the separation being considered. By moving across the matrix, the applicability of a cycle for each specific process condition will indicate one of four designations. Obviously, a "no" designation eliminates that adsorption cycle from further consideration. A separation indicating a "yes" for each process condition would deserve strong consideration. The other two designations can be used to rank the adsorption cycles if more than one cycle is still a possibility. This selection procedure should be used with some caution since capital and energy costs can vary from one location to the next and can affect the final choice.

[†]G. E. Keller, II, R. A. Anderson, and C. M. Yon, in *Handbook of Separation Process Technology,* R. W. Rousseau, ed., J. Wiley, New York, 1987, chap. 12.

Table 15-14 Matrix for adsorption cycle selection†

Process condition	Gas or vapor phase				Liquid
	Temperature swing	Inert purge	Displacement purge	PSA	Temperature swing‡
Feed: gas or vapor	Yes	Yes	Yes	Yes	No
Feed: liquid, vaporize <200°C	Unlikely	Yes	Yes	Yes	Yes
Feed: liquid, vaporize >200°C	No	No	No	No	Yes
Adsorbate concentration in feed, <3%	Yes	Yes	Unlikely	Unlikely	Yes
Adsorbate concentration in feed, 3–10%	Yes	Yes	Yes	Yes	No
Adsorbate concentration in feed >10%	No	Yes	Yes	Yes	No
High product purity required	Yes	Yes	Yes	Possible§	Yes
Thermal regeneration required	Yes	Yes	No	No	Unlikely
Difficult purge/adsorbate separation	Possible¶	Unlikely	Unlikely	NA	Possible¶

†Based on the information provided by G. E. Keller, II, R. A. Anderson, and C. M. Yon, in *Handbook of Separation Process Technology,* R. W. Rousseau, ed., J. Wiley, New York, 1987, chap. 12, p. 670, with permission.
‡Includes powdered, fixed-bed and moving-bed processes.
§Requires very high adsorption pressure and low desorption pressure.
¶Only if adsorbate does not need to be recovered.

General Design Concepts for Separation by Adsorption

A dynamic phase equilibrium is established in adsorption for the distribution of the adsorbate between the fluid and the adsorbent surface. This equilibrium is generally expressed in terms of the partial pressure or the concentration of the adsorbate in the fluid and the adsorbate loading on the adsorbent. Unfortunately, no acceptable theory has been developed to estimate adsorption equilibria. Thus, if such data are not available, it will be necessary to obtain experimental equilibrium data for the specific adsorbate or mixture of adsorbates and the absorbent material of interest. The adsorption isotherms obtained from these experimental studies will provide an upper limit on the adsorption of the adsorbate from a given fluid mixture on a specific adsorbent for a designated set of conditions.

Experimental adsorption isotherms are classified into five types, as shown in Fig. 15-27. The type I isotherm corresponds to unimolecular adsorption, generally applicable to gases at temperatures above their critical temperature. The type II isotherm is associated with multimolecular adsorption and is observed for gases at temperatures below their critical temperature and pressures below but approaching the saturation temperature. Both types I and II are desirable isotherms that exhibit strong adsorption characteristics. The type III isotherm is undesirable since its adsorption effect is low except at high pressures. The adsorption isotherms for types IV and V indicate that the multimolecular adsorption observed in types II and III is augmented by capillary condensation. Note that a hystersis phenomenon can occur in the multimolecular adsorption regions of types IV and V.

Experimental adsorption data for 18 different pure gases adsorbing on various adsorbents are summarized by Valenzuela and Meyers.† Their analysis showed that

†D. P. Valenzuela and A. L. Meyers, *Adsorption Equilibrium Data Handbook,* Prentice-Hall, Englewood Cliffs, NJ, 1989.

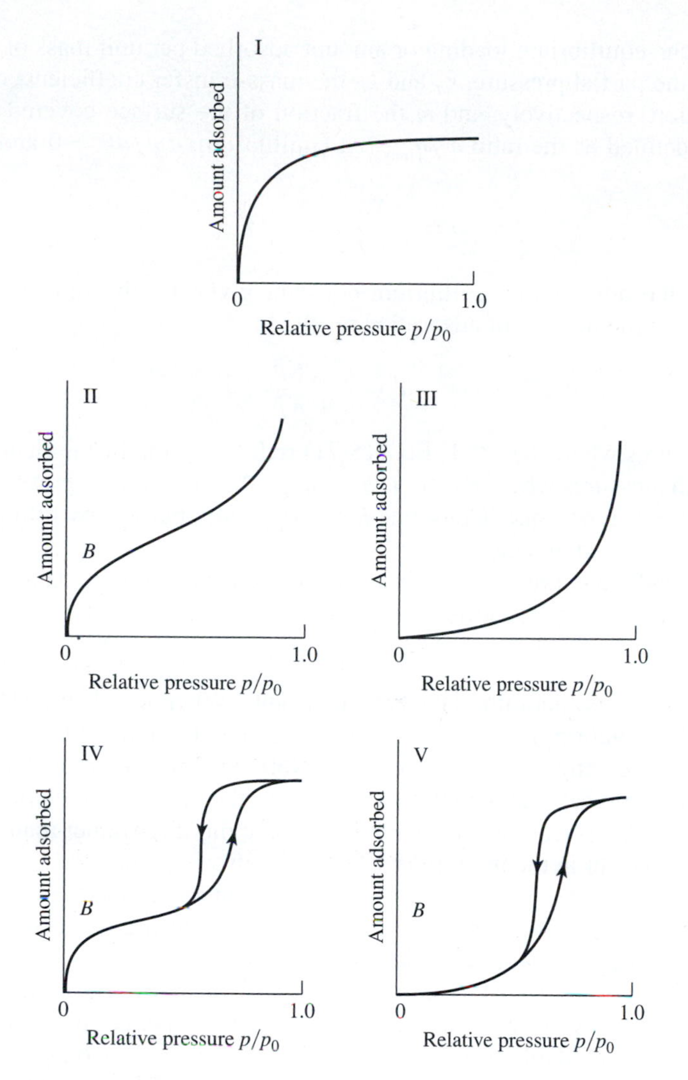

Figure 15-27
Brunauer's five types of adsorption isotherms

adsorption isotherms for these gases varied considerably with different adsorbents and that adsorbate loading varied essentially inversely with their molecular weight. For practical applications, however, the classical equations of Langmuir and Freundlich are still the best because of their simplicity and ability to correlate type I isotherms.

The Langmuir relation is derived from simple mass action kinetics, assuming chemisorptions. The net rate of adsorption is the difference between the rates of adsorption and desorption, given by

$$\frac{dq'}{d\theta} = k_a \bar{p}(1 - \omega) - k_d \omega \tag{15-69}$$

where q' is the equilibrium loading or amount adsorbed per unit mass of adsorbent, θ the time, $\bar{p}$ the partial pressure, k_a and k_d the mass-transfer coefficients of adsorption and desorption, respectively, and ω the fraction of the surface covered by adsorbed molecules, defined as the ratio $q'/q'_{\max}$. At equilibrium, $dq'/d\theta = 0$ and Eq. (15-69) reduces to

$$\omega = \frac{K\bar{p}}{1 + K\bar{p}} = q'/q'_{\max} \tag{15-70}$$

where K is the adsorption equilibrium constant given by the ratio k_a/k_d. Solving Eq. (15-70) for the net rate of adsorption results in

$$q' = \frac{q'_{\max} K\bar{p}}{1 + K\bar{p}} \tag{15-71}$$

At low pressures where $K\bar{p} \ll 1$, Eq. (15-71) reduces to the linear Henry's law form while at high pressures where $K\bar{p} \gg 1$, $q' = q'_{\max}$. At intermediate pressures, the equation is nonlinear in pressure. Constants K and $q'_{\max}$ are obtained by fitting Eq. (15-71) directly to experimental data.

The Freundlich expression that is used to correlate adsorption data is given by the empirical and nonlinear equation

$$q' = k(\bar{p})^{1/n} \tag{15-72}$$

where k and n are temperature-dependent constants. When $n = 1$, Eq. (15-72) reduces to Henry's law relation. Experimental data can generally be fitted to Eq. (15-72) with a nonlinear curve-fitting computer program. Note that the Langmuir relation predicts an asymptotic limit for q' at high pressure, whereas the Freundlich relation does not.

Commercial applications of adsorption involve mixtures rather than pure gases. If only one component in the mixture is selectively adsorbed, then the adsorption of that component is estimated from its pure gas adsorption as outlined above. However, if two or more components in the mixture are adsorbed during the adsorption process, the estimation becomes more complex and will require an extension of the Langmuir equation as developed by Markham and Benton.† In this development, the authors assumed that the only effect of the adsorption of one component from the mixture was a reduction in the available surface area for the adsorption of the other components.

For a binary gas mixture in which components A and B are being adsorbed on the surface of an adsorbent, the equilibrium loading of each component is given by

$$q'_A = \frac{(q'_A)_{\max} K_A \bar{p}_A}{1 + K_A\bar{p}_A + K_B\bar{p}_B} \tag{15-73a}$$

$$q'_B = \frac{(q'_B)_{\max} K_B \bar{p}_B}{1 + K_A\bar{p}_A + K_B\bar{p}_B} \tag{15-73b}$$

where $(q_i)_{\max}$ is the maximum loading of component i when it covers the entire adsorbent surface. Equations (15-73*a*) and (15-73*b*) can be extended to a multicomponent

†E. C. Markham and A. F. Benton, *J. Am. Chem. Soc.*, **53:** 497 (1931).

mixture with

$$q_i' = \frac{(q_i')_{\max} K_i \bar{p}_i}{1 + \sum_j K_j \bar{p}_j} \tag{15-74}$$

Similarly, the Freundlich equation can be combined with the Langmuir equation to provide another relation for adsorption of components from gas mixtures, namely,

$$q_i' = \frac{(q_i'')_{\max} K_i (\bar{p}_i)^{1/n_i}}{1 + \sum_j K_j (\bar{p}_j)^{1/n_i}} \tag{15-75}$$

where $(q_i'')_{\max}$ is the maximum loading that may differ from the $(q_i')_{\max}$ value obtained for monolayer adsorption. Even though Eqs. (15-74) and (15-75) lack thermodynamic consistency, their simplicity along with relatively reasonable results (±20 percent) encourages their continued use. Improved results may be obtained by the procedures outlined by Yang.[†]

Determination of Specific Adsorption from a Binary Gas Mixture — EXAMPLE 15-7

With the use of the extended Langmuir equation, predict the specific adsorption volumes (STP) from a vapor mixture of 69.6 mol percent CH_4 and 30.4 mol percent CO at 20°C and a total pressure of 2510 kPa. Experimental adsorption isotherms for the pure components and for the mixture have been presented by Ritter and Yang.[‡] At 20°C, the constants in the Langmuir relation, Eq. (15-70), are

	CH_4	CO
$(q_i')_{\max}$, cm^3 (STP)/g	133.4	126.1
K, kPa^{-1}	1.99×10^{-3}	9.05×10^{-4}

■ Solution

Use Eqs. (15-73*a*) and (15-73*b*) to predict the specific adsorption volumes (STP).

$$\bar{p}_{CH_4} = y_{CH_4} P = 0.696(2510) = 1747 \text{ kPa}$$

$$\bar{p}_{CO} = 0.304(2510) = 763 \text{ kPa}$$

From Eq. (15-73*a*)

$$q_{CH_4}' = \frac{133.4(1.99 \times 10^{-3})(1747)}{1 + (1.99 \times 10^{-3})(1747) + (9.05 \times 10^{-4})(763)} = 89.8 \text{ cm}^3 \text{ (STP)/g}$$

Likewise,

$$q_{CO}' = \frac{126.1(9.05 \times 10^{-4})(763)}{1 + (1.99 \times 10^{-4})(1747) + (9.05 \times 10^{-4})(763)} = 16.8 \text{ cm}^3 \text{ (STP)/g}$$

[†] R. T. Yang, *Gas Separation by Adsorption Processes,* Butterworths, Boston, 1987.

[‡] J. A. Ritter and R. T. Yang, *Ind. Eng. Chem. Res.,* **26:** 1679 (1987).

The total volume adsorbed per gram of adsorbent is 89.8 + 16.8, or 106.6 cm^3(STP)/g. The CH_4 concentration in the adsorbate has been increased to 84.2 mol percent while the CO concentration has been reduced to 15.8 mol percent. This compares reasonably well with the experimental concentrations of 86.7 and 13.3 mol percent for the CH_4 and CO concentrations in the adsorbate, respectively, obtained by Ritter and Yang.

■

Among the design considerations for adsorption processes besides adsorbent characteristics are the manner in which compositions change as a function of distance through an adsorbent bed and the position and movement of the temperature front in the bed. First, consider the adsorbate composition profile as a function of time inside a bed. Figure 15-28a illustrates the movement of the adsorbate concentration wave

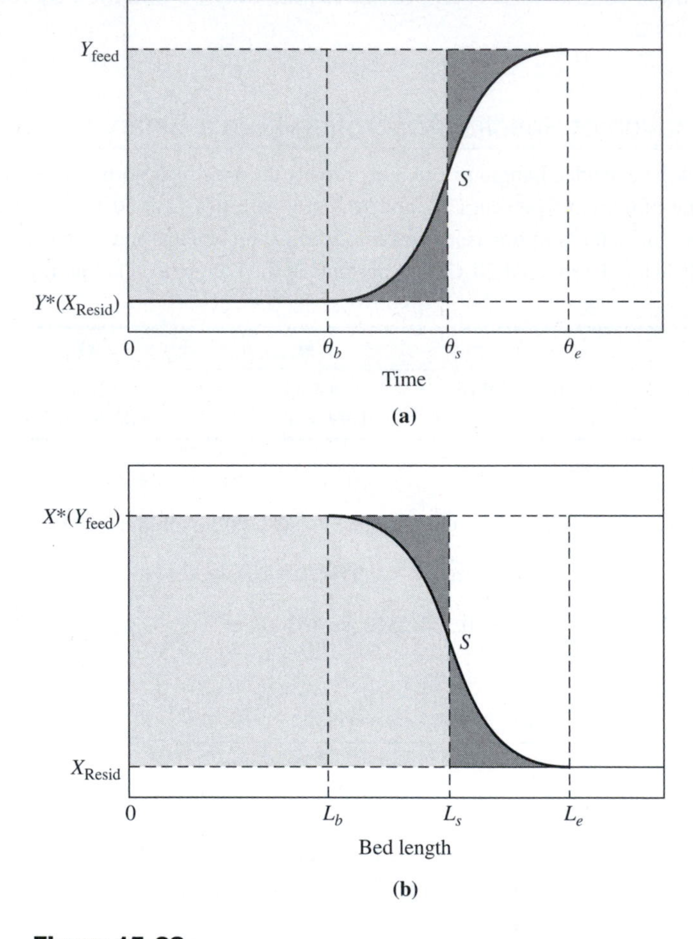

Figure 15-28
Time trace of adsorbate concentration in an adsorber effluent (a) and adsorbate loading along the axis of an adsorber during adsorption (b)

through the bed as well as the outlet concentration of adsorbate as a function of time for a bed operating essentially in an isothermal manner. There are three distinct zones within the bed. The first zone, nearest to the inlet, is where the adsorbent is fully saturated and in equilibrium with the adsorbate concentration in the feed. Farther down the bed is a zone in which the adsorbate concentration on the adsorbent approximates an S shape. In this zone, labeled as the mass-transfer zone (MTZ), most of the adsorption occurs. Beyond the MTZ is a zone in which no adsorption presently is underway since most of the adsorbate in the feed has been removed in the first two zones.

If the adsorbate is concave downward (see Fig. 15-28b) and the MTZ is reasonably isothermal, the S-shaped curve is generally established soon after adsorption is initiated, and it maintains this shape throughout the bed. Breakthrough is indicated when the leading point of the S curve reaches the end of the bed. At this point, the adsorption process normally is terminated, even though a certain fraction of the adsorbent is less than fully loaded, and the desorption process is initiated.

The width of the S curve or, more appropriately, the mass-transfer zone directly affects the operation of an adsorbent bed. As the ratio of the width of the S curve to the bed length increases, the bed use per cycle decreases, and more adsorbent will be required to process a given feed rate. The continual widening of the MTZ, unfortunately, is typical of that obtained with a linear adsorption isotherm. However, when the isotherm is either a Langmuir or a Freundlich type, the widening of the MTZ rapidly diminishes within the bed and an asymptotic or constant pattern front is developed in which the MTZ is also constant. For gas feeds, the MTZ for a well-designed adsorption system will often be between 0.3 and 1 m in length, but can be as small as 25 mm. The MTZ for liquid feeds will generally be larger because of the much higher viscosities of liquids. For a constant pattern front, Cooney[†] presents an approximate but rapid method using the Langmuir and Freundlich isotherms to estimate concentration profiles and breakthrough curves when mass-transfer and equilibrium parameters are available.

On the other hand, Collins[‡] has developed a technique for determining the length of a full-scale adsorbent bed, when the constant-front assumption is valid, directly from breakthrough curves obtained in small-scale laboratory experiments. In this approach the adsorbent bed is considered to be the sum of two sections. In the first section, the adsorbent is in equilibrium with the adsorbate concentration in the feed and is designated as the *length of the equilibrium section* (LES). In the second section, the adsorbate loading is zero and is specified as the *length of the unused bed* (LUB). The length of this section depends on the width of the MTZ and the shape of the concentration profile within the MTZ. The total required bed length L_B is

$$L_B = \text{LES} + \text{LUB} \tag{15-76}$$

Since the LUB is one-half the width of the MTZ for a constant pattern front, it may be determined from the experimental breakthrough curve from the relation

$$\text{LUB} = \frac{\theta_s - \theta_b}{\theta_s} L_e \tag{15-77}$$

[†]D. O. Cooney, *Chem. Eng. Comm.*, **91:** 1 (1990).

[‡]J. J. Collins, *Chem. Eng. Prog. Symp. Series,* **63**(74): 31 (1967).

where θ_s is the breakthrough time required to attain the midpoint concentration between the inlet and exit concentrations in the MTZ, θ_b the breakthrough time required to attain the desired exit concentration, and L_e the length of the experimental bed. For a cylindrical adsorbent bed, the LES value can be determined by assuming an ideal adsorber where LUB $= 0$ and making a solute mass balance. The resulting relation is

$$\text{LES} = \frac{c'_F G\theta}{q_F \rho_b} \tag{15-78}$$

where c'_F is the concentration of the adsorbate in the feed, G the mass flow rate per cross-sectional area of the bed, θ the time for breakthrough for which the LES is evaluated, q_F the loading of the adsorbent during each cycle, and ρ_b the density of the bed.

The length-to-diameter ratio of vertical cylindrical adsorption beds is usually greater than 1.5. Higher L/D values may be necessary for liquid feeds to minimize the problem with a large MTZ. To minimize pressure drop through adsorbers, it may be necessary to use either a slanted or a horizontal adsorption unit.

EXAMPLE 15-8 Use of Laboratory Data to Determine Actual Adsorber Height

Estimate the adsorber height required for a commercial adsorber to remove the water vapor from a nitrogen gas stream from 1440 to 15 ppm (by volume). Assume that the adsorber will be operated at the same conditions of temperature, pressure, inlet mass flow rate, and inlet and exit adsorbate loading of the adsorbent as experienced by the laboratory unit. Consider a time of 30 h for breakthrough to occur. The data provided by Collins† for the laboratory unit will be used.

Laboratory data are as follows: Temperature = 20°C, pressure = 593 kPa, laboratory adsorber height = 0.268 m, mass flow rate = 144.5 kg mol/h·m^2, inlet adsorbate loading = 1 kg water/100 kg adsorbent, exit adsorbate loading = 21.5 kg water/100 kg adsorbate, inlet adsorbate concentration = 1440 ppm (by volume), density of adsorbent bed = 713 kg adsorbent/m^3. The breakthrough data for removal of water from nitrogen with a fixed bed of 4-Å molecular sieves is presented below.

c_{exit}, ppm (by volume)	Time, h	c_{exit}, ppm (by volume)	Time, h
<1	0–9	650	10.8
1	9.0	808	11.0
4	9.2	980	11.25
9	9.4	1115	11.5
33	9.6	1235	11.75
80	9.8	1330	12.0
142	10.0	1410	12.5
238	10.2	1440	12.8
365	10.4	1440	13.0
498	10.6		

†J. J. Collins, loc. cit.

■ **Solution**

Obtain the adsorber height by using Eq. (15-78) to determine LES and Eq. (15-77) to determine LUB after graphically or numerically integrating the breakthrough curve to establish θ_s. Thus,

$$c'_F = \frac{1440(18)}{10^6} = 0.02592 \text{ kg water/kg mol N}_2$$

$$\text{LES} = \frac{0.02592(144.5)(30)}{(0.215 - 0.01)(713)} = 0.769 \text{ m}$$

From the breakthrough curve data

$$\theta_e(1440 \text{ ppm}) = 12.8 \text{ h}$$

$$\theta_b(15 \text{ ppm}) = 9.45 \text{ h}$$

By graphical integration of the breakthrough curve data $\theta_s = 10.9$ h. Numerical integration of the data could also have been performed by using the relation

$$\theta_s = \int_o^{\theta_e} \left(1 - \frac{c}{c_F}\right) d\theta$$

From Eq. (15-77)

$$\text{LUB} = \frac{10.9 - 9.45}{10.9}(0.268) = 0.036 \text{ m}$$

The height of the commercial adsorber is then

$$L_B = 0.769 + 0.036 = 0.805 \text{ m}$$

This provides a satisfactory L/D ratio for the adsorber. ■

Costs for Adsorption Equipment

Most adsorption beds containing adsorbent particles are vertically oriented, cylindrical vessels. The cost for such vessels is similar to those given earlier for fabricating the shell of a distillation column.

EQUIPMENT DESIGN AND COSTS FOR SEPARATING HETEROGENEOUS MIXTURES

When the feed mixture is a heterogeneous one, separation is generally best effected by taking advantage of the specific differences that exist between physical properties of the phases involved. The type of system requiring separation, namely, a gas-liquid or a solid-liquid system, has a major impact on what options are available to the design engineer. A brief enumeration of a few of these options is presented in the following section.

The equipment usually available to separate gas-liquid mixtures includes cyclones, knockout drums, venturis, and spray towers. The most common is the knockout drum that provides a sufficiently low gas velocity to allow the liquid droplets to settle out by gravity. For low liquid concentrations, the separation can be enhanced with the insertion of baffles.

Separation of gas-solid systems is best accomplished with the use of cyclones, electrostatic precipitators, scrubbers, and bag filters. Particle size is a major variable since for both wet scrubbers and electrostatic precipitators almost any solid particle loading can be handled. Most of these solids removal devices are about 90 to 99 percent efficient in mass removal. However, as the particle size to be removed decreases, the energy requirement for the solids removal increases.

Liquid-liquid systems can be separated by decanters, hydrocyclones, and deep-bed filtration. These devices utilize either decantation or gravity settling, often augmented with the addition of coalescence promoters as the particle size becomes smaller. Combustion and high-gradient magnetic separation are options for very dilute and very small particle systems.

For separating solid-solid mixtures, a variety of options are available. One of the options is leaching that dissolves one of the components and leaves the other insoluble components to be removed by a solid-liquid separation technique. Other options require that the desired component be reduced to a particle that can be separated physically from the undesired particles. Generally, this requires crushing the solid material to a particle size that permits formation of separate particles for the components involved. The resulting solid mixture can then be separated physically by exploiting various differences in solid properties. For example, froth flotation exploits the differences in surface wettability by removing one of the solid components by attachment to gas bubbles that are forced through the liquid containing the solid mixture. Other options exploit differences in density, magnetic, or electrostatic properties and are identified as classifiers, concentrators, or separators, respectively. Classifiers include hydrocyclones, wet and dry screens, as well as rake and spiral separation devices. Jigs, tables, and sluices serve as concentrators and exploit differences in the settling velocities of the particles to effect the separation.

For solid-liquid separations, a wide range of options exist, namely, settlers, filters, screens, dewatering presses, dryers, evaporators, and leachers. Some of the options are based on differences in settling characteristics, others exploit the differences in particle size relative to the size in the separation barrier, while still others exploit differences in vapor pressure. For filters and screens the overall class of filter is based on the size of the particle; within a class, the concentration of solids in solution aids in the selection of the type of filter recommended.

Guidelines were presented earlier in Table 15-3 for characterizing various types of equipment used in the separation of heterogeneous mixtures with reference to normal feed conditions and the separating agent or force field involved in the separation. Any advantages or limitations of such equipment were noted. This section provides equipment selection and design procedures for two of the more important separation processes identified in the table. Further details on many of the other separation devices identified in the table are given by Walas[†] and Woods.[‡]

[†]S. M. Walas, *Chemical Process Equipment—Selection and Design,* Butterworth Heineman, Newton, MA, 1988.

[‡]D. R. Woods, *Process Design and Engineering Practice,* Prentice-Hall, Englewood Cliffs, NJ, 1995.

SEPARATION BY DRYING

Drying of solids encompasses two fundamental and simultaneous processes involving the transfer of heat to evaporate the liquid and the transfer of liquid and vapor within the solid and vapor from the surface. The factors governing the rates of these two processes determine the drying rate. The methods of heat transfer in commercial dryers may utilize convection, conduction, radiation, or a combination of these three mechanisms. Mass transfer within the solid, however, occurs because of a concentration gradient that is established in the drying process which is dependent on the characteristics of the solid. Removal of the vapor from the surface will be dictated by the existing flow conditions provided by the dryers.

Evaluation of the drying process may be based either on the internal mechanisms of liquid and vapor flow or on the effect of external conditions of temperature, humidity, gas flow rate, state of solid subdivision, etc., on the drying rate of the solid. The latter procedure, even though it is less fundamental, is generally followed since the results have greater application in equipment selection and design.

Equipment Selection for Drying Solids

The key factors to consider in the preliminary selection of dryers include the properties of the material being dried, drying characteristics of the material, flow quantities and flow conditions, product quality, and possible dust and solvent recovery. The design engineer should select those dryers that appear best suited for handling the wet material and the dry product to provide a product with the desired physical properties. This preliminary selection can be made with the aid of Table 15-15, which classifies a number of commercial dryers on the basis of the materials handled and lists some advantages and limitations of these dryers. Following preliminary selection of suitable types of dryers, a quick evaluation of the size and cost should be made for each type to eliminate those that are uneconomical. Procedures for this evaluation are given following a brief discussion of the types of equipment involved.

In the tunnel dryer, wet material on trays or on a conveyor belt passes through a tunnel in which hot gas is being circulated. The flow of hot gas can be either countercurrent or cocurrent. This type of dryer is generally used when the wet material is not free-flowing.

The rotary dryer is a cylindrical shell, mounted slightly downward, that is slowly rotated. Wet material, fed at the higher end of the rotating shell, slowly flows down the shell due to gravitational forces. A hot gas flowing either countercurrent or cocurrent provides the heat source to vaporize the liquid. Rotary dryers are usually used when the wet material is free-flowing. With appropriate modifications, rotary dryers can be employed to process most conventional solids. Large capacities, flexibility, and good efficiencies make these devices useful for drying many solids. Rotary vacuum dryers are suitable only for nonsticking materials.

Vertical tower contactors are similar to conventional liquid/vapor distillation columns, but are more complex mechanically. Stirrers and wipers are required to mix,

Table 15-15 General guidelines to assist drying equipment selection[†]

Equipment types considered	Solids drying compatibility	Equipment advantages	Equipment limitations
Mechanically aided			
Tunnel dryer (continuous)	Sludges, pastes, large solids, heat-sensitive materials	Can process many materials, conveyors simplify handling, good temperature control	High labor requirement, vacuum impractical, unsuitable for dusty materials
Rotary dryer	Fine, granular or fibrous materials	High capacity, flexibility, good efficiency	Unsuitable for slurries, pastes, sludges, or heat-sensitive materials
Vacuum dryer	Fine flowing powders, granular or fibrous solids	Handles heat sensitive and nonsticking materials	Unsuitable for slurries, gummy pastes, sludges, batch operation, high cost
Tower contactor	Fine, free-flowing solids	Operation with controlled process and temperature feasible	Limited capacity, structured constraints, high investment
Vibrating conveyor contactor	Fine, free-flowing powders, small granular solids	Provides fluid bed action for particles up to 0.15 mm, high heat transfer	Limited to particles that can be fluidized
Drum dryer (indirect heating)	Slurries or solutions, flowing pastes and sludges	Provides drying of liquid solutions impractical by other means, high maintenance cost	Drying limited to solutions adhering to a drum, low capacity, high energy costs
Screw conveyor (indirect heating)	Gummy and sticky materials, sludges	Dries difficult-to-handle materials, operates under pressure and vacuum	Low capacity
Fluid-activated			
Fluid bed	Fine, free-flowing powders, very small granules or fibrous solids	High capacity, low cost, high heat transfer, uniform internal temperatures	Limited to solids that can be fluidized and not fractured in high-velocity gas streams
Spouted bed	Uniform granular solids	Handles granular solids up to 1 mm in diameter, relatively high capacity, good heat transfer	Limited to solids that can be agitated with a jet
Pneumatic conveyor	Very fine, free-flowing solids	High capacity, low processing time, good for heat-sensitive materials	Drying dependent on gas moisture and enthalpy content, generally requires recycling
Spray dryer	Slurries or solutions	Excellent for heat-sensitive materials, light and porous product	Handles only liquid mixtures that can be sprayed, verify pressure effects on liquid and solid

[†]Summarized from data presented by C. G. Moyers and G. W. Baldwin, in *Perry's Chemical Engineers' Handbook,* 7th ed., R. H. Perry and D. W. Green, eds., McGraw-Hill, New York, 1997, Sec. 12, S. W. Walas, *Chemical Process Equipment,* Butterworth-Heineman, Newton, MA, 1990, G. D. Ulrich, *A Guide to Chemical Engineering Design and Economics,* J. Wiley, New York, 1984, and D. R. Woods, *Process Design and Engineering Practice,* Prentice-Hall, Englewood Cliffs, NJ, 1995.

expose, and transport the solid as it slowly moves, tray by tray, down the column. The capacity of the units is limited because of mechanical and structural constraints. The additional complexity is reflected in the high cost of the equipment.

The vibrating conveyor provides gas/solid contact through mechanical and fluid-agitated contactors. In this device the hot gas stream, with the aid of mechanical vibration, agitates the solid particles and places them in fluidized suspension. This fluidization provides the contacting efficiencies of a fluid bed for solids that normally could not have been handled in a conventional fluid bed.

A drum dryer utilizes an internally heated rotating drum as the drying source. The partially submerged drum rotates through a bath of liquid to form a film that solidifies while the drum is rotating. The resulting solid is removed from the drum by a scraper. Drum dryers are suitable for drying slurries or free-flowing pastes, but are limited to rather low throughputs.

The screw conveyor, operating with indirect heating, moves the solid forward in an auger-type fashion. Although limited in capacity, the dryer efficiently handles solids that are gummy, sticky, or otherwise difficult to dry. The device is adaptive to either a controlled atmosphere or a moderate vacuum.

Fluid beds used for drying of solids are characterized by intimate gas-solid contact, high capacity, mechanical simplicity, low cost, and uniform internal temperatures. In this fluid-activated dryer, efficient fluidization is accomplished with countercurrent flow of the gas and the finely divided solids. Gummy materials are rarely handled and only after preforming. Because of the high fluid agitation, friable or fragile materials generally generate dust that must be removed with cyclones, filters, or scrubbers and therefore should be avoided.

The spouted bed dryer is essentially a combination of the fluid bed and gravity contactor. In this device, the solid is lifted by a high-velocity gas jet in a vertical column and then recirculated by gravity in the surrounding annular space. The spouted bed dryer is a complement to the fluidized bed dryer and is used when solids are too large for efficient fluidization. Spouted bed dryers are frequently used for drying uniform granular solids.

A flash contactor, also designated as a pneumatic conveyor contactor, entrains a very fine solid, similar to dust, in a high-velocity gas stream. As with the fluid bed, gas-solid contact is efficient. However, unlike in the fluid bed, the solid and gas remain together over most of the processing time. Because of limited equipment size, contact times are on the order of a second or less. Thus, materials that can dry quickly are required. If more processing time is required, part of the dried product may be recycled to improve the drying rate.

In the spray drier, the liquid or slurry is sprayed as fine droplets into a hot gas stream. Care must be exercised that the liquid or slurry must be able to withstand the pressures required for the droplet formation. The product from this dryer generally takes the form of light and porous particles. Since vaporization of the liquid is rapid in this dryer, the temperature of the product remains low even in the presence of the hot gas. This makes this dryer a good choice for dried products that are sensitive to thermal decomposition.

General Design Procedures for Drying

Because of the large number of equipment types in gas-solid contacting devices used for the drying of solids, it is not surprising that a specific, reliable, and consistent design procedure is available. Not only each type of equipment but also every application for each type must be evaluated separately. Nevertheless, despite the variety and complexity of gas-solid contacting equipment, some general guidelines are available to establish limits on equipment size and energy consumption required for a predesign and cost estimate.

The design of gas-solid contactors involves the interaction of fluid mechanics, heat and mass transfer, and possibly chemical kinetics. As with most chemical process designs, analysis can be pursued directly by utilizing conventional material balances, energy balances, and rate equations. As a general rule, the following procedure can be employed to provide approximate equipment dimensions for the gas-solid contactor selected to achieve the specified drying requirements.[†]

In most mechanically aided drying equipment, average velocities of solids are rather limited, as shown in Table 15-16. Gas flows, temperature, and composition, on the other hand, are easily varied and can be adjusted to meet specific drying requirements. The continuity equation can be used to determine the average flow rate of the solid for a longitudinal flow contactor

$$\dot{m}_S = \rho_S V_S A_S \tag{15-79}$$

where ρ_S is the density of the solid, V_S the average flow velocity of the solid, and A_S the solids cross-sectional area in the dryer. The mass of liquid removed from the solid can be established with a material balance based on the inlet and exit conditions of the solid as detailed by process requirements.

The quantity of energy crossing the gas-solid interface is also dictated by the inlet and exit conditions of the solid. Since the drying process generally involves removal of liquids that are physically bound, the heats of solution or adsorption are negligible and only the latent heats need to be included in the energy balance. Heat losses from the gas-solid contacting equipment range from 5 to 15 percent of that supplied to the drying process. In most applications, the mechanical work in the energy balance is negligible compared with the heat and enthalpy fluxes. However, it must be evaluated to determine the utility costs associated with the process. At this time the gas flow rate and inlet temperature must be fixed to define the inlet and outlet enthalpies. For a first trial, a representative value may be selected from Table 15-16. These operating conditions can be revised later to arrive at equipment dimensions that match those available from vendors.

Besides providing an adequate flow capacity, the gas-solid contactor must maintain an adequate solid-gas residence time to accomplish the degree of drying required.

[†]Based on recommendations by G. D. Ulrich, *A Guide to Chemical Engineering Process Design and Economics*, J. Wiley, New York, 1984.

Table 15-16 Normal operating conditions associated with drying equipment[†]

Type of drying equipment	Approx. dimensions[‡]			Solids velocity, m/s	% A_c used by solids[‡]	Gas velocity, m/s	Avg. heat-transfer coeff., J/s·m²·K	Avg. G,[‡] kg/s·m²	Thermal eff., %	Approx. power reqm't., kW
	D, m	L, m	L/D							
Mechanically aided										
Tunnel dryer (continuous)	0.5–4	5–20	5–10	0.006–0.2	10–20	1–5	30–50	—	20–50	0.035–12.5
Rotary dryer (direct)	1–3	4–20	4–6	0.02–0.1	10–15	0.3–1.0	$60\ G^{0.67}$	0.5–5.0	55–75	8–27
Vacuum dryer	0.5–3	1.5–12	3–4	—	50–65	—	40–60	Batch	60–80	2–27
Tower contactor	2–10	2–20	1–2	0.006–0.8 (solid flow, kg/s)	NA	0.6–3	40–70	—	55–75	1–300
Vibrating conveyor contactor	0.3–2	3–20	3–10	2–10	15–20	—	$10^3\ (G^{0.67})$ (volume)	—	—	0.1–1.3
Drum dryer	0.6–3	0.6–5	1–8	0.0003–0.07 (solid flow, kg/s)	NA	—	10^2–10^3	—	—	—
Screw conveyor	0.1–0.7	1–6	3–10	0.01–0.1	40–50	—	15–60	0.05–0.5	30–60	0.01–2.3
Fluid-activated										
Fluid bed	1–10	0.3–15	0.3–15	—	20–30	0.1–2	$10^3\ (G^{0.67})$ (volume)	—	50–80	0.3–1200
Spouted bed	0.2–1	0.5–4	1–10	—	20–30	—	$10^2\ (G^{0.67})$ (volume)	—	50–80	0.08–16
Pneumatic conveyor	0.2–1	10–30	>50	—	5–10	—	10^3–2×10^3 (volume)	—	55–75	3–230
Spray dryer	2–10	8–25	4–5	—	NA	0.2–3	—	—	40–60	—

[†]Summarized from data presented by G. D. Ulrich, *A Guide to Chemical Engineering Design and Economics,* J. Wiley, New York, 1984, D. R. Woods, *Process Design and Engineering Practice,* Prentice-Hall, Englewood Cliffs, NJ, 1995, and C. G. Moyers and G. W. Baldwin, in *Perry's Chemical Engineers' Handbook,* 7th ed., R. H. Perry and D. W. Green, eds., McGraw-Hill, New York, 1997, Sec. 12.

[‡]D = diameter, m; L = length, m; G = mass flux, kg/s·m², A_c = cross-sectional area of dryer, m².

In dryers the solids residence time θ_r is the critical parameter and is controlled by both diffusion and heat-transfer rates. In tunnel, tower contactors, and drum dryers, the rate of drying depends on the solids porosity, temperature difference, moisture content, and bed thickness. This is generally represented by the relation

$$\frac{dX}{d\theta_r} = \frac{k_{d'} X (T_G - T_{wb})}{t_s} \tag{15-80}$$

where X is the weight ratio of liquid to dry solid, T_G the dry-bulb temperature of the gas, T_{wb} the wet-bulb temperature of the gas, t_s the thickness of the solid bed, and $k_{d'}$ the drying rate constant. Integration of Eq. (15-80) results in an expression for the residence time of the solid given by

$$\theta_r = \frac{t_s \ln(X_o / X_i)}{k_d (T_G - T_{wb})} \tag{15-81}$$

when an arithmetic average of the terminal temperature differences is used. Values of k_d, for materials generally dried in static bed units, range from about 10^{-6} m/K·s for dense nonporous solids to 5×10^{-6} m/K·s for more granular or fibrous materials. Since the process of drying is complex, Eq. (15-81) should be used with considerable caution. Whenever possible, laboratory drying data should be used for improved reliability in the design of such dryers.

Heat transfer is generally assumed to limit vapor transport in those dryers that provide agitated beds or intimate gas-solid contact. In those types of equipment, an approximate residence time can be evaluated from an energy balance on the solids stream plus the normal convective heat-transfer relation given in Chap. 14, namely,

$$q = UA\,\Delta T_{\text{log mean}} \tag{15-82}$$

or

$$q = U' v_d\,\Delta T_{\text{log mean}} \tag{15-82a}$$

where U is the overall heat-transfer coefficient based on the area and U' the overall heat-transfer coefficient based on the volume of the dryer. The $\Delta T_{\text{log mean}}$ is based on the differences between the dry-bulb gas temperatures and the wet-bulb temperatures at the inlet and exit of the dryer. (For the drying of hygroscopic materials, the wet-bulb temperature should be replaced by the saturation temperature.) Once the area or volume of the dryer has been established from Eq. (15-82) or (15-82*a*), respectively, the residence time can be evaluated from

$$\theta_r = \frac{v_d f_S \rho_S}{\dot{m}_S} \tag{15-83}$$

where v_d is the volume of the dryer and f_S the fraction of the dryer occupied by the solid. For dryers with continuous solid-gas interactions, such as in fluid beds, Eq. (15-83) can be used directly. For example, dryer volume can be determined once the residence time, average velocity of the solid, and density of the solid are known.

For a particular type of dryer configuration and a required residence time, there are a variety of terminal temperatures and flow rates that will satisfy the mass and energy balances. However, the heat-transfer characteristics of the dryer uniquely establish these parameters. As noted above, the logarithmic-mean temperature driving force applies.

In this simplified design procedure for dryers, the analysis is normally overspecified because of some of the arbitrary assumptions that were made. One assumption is that the gas stream did not become saturated during its passage through the drier. It is important to check that saturation or near saturation has not occurred. Other restrictive assumptions may result in a design that either does not meet configurations normally available from vendors or is expensive to operate. At this point, the design should be reevaluated to determine whether other assumptions are justified or other equipment alternatives should be considered.

Costs for Typical Dryers

The purchased and/or installed costs of dryers are presented in Figs. 15-29 through 15-35. Costs are available for tray, pan, rotary, drum, vibratory conveyor, fluid-bed, and spray dryers. The cost data cover both vacuum and atmospheric systems.

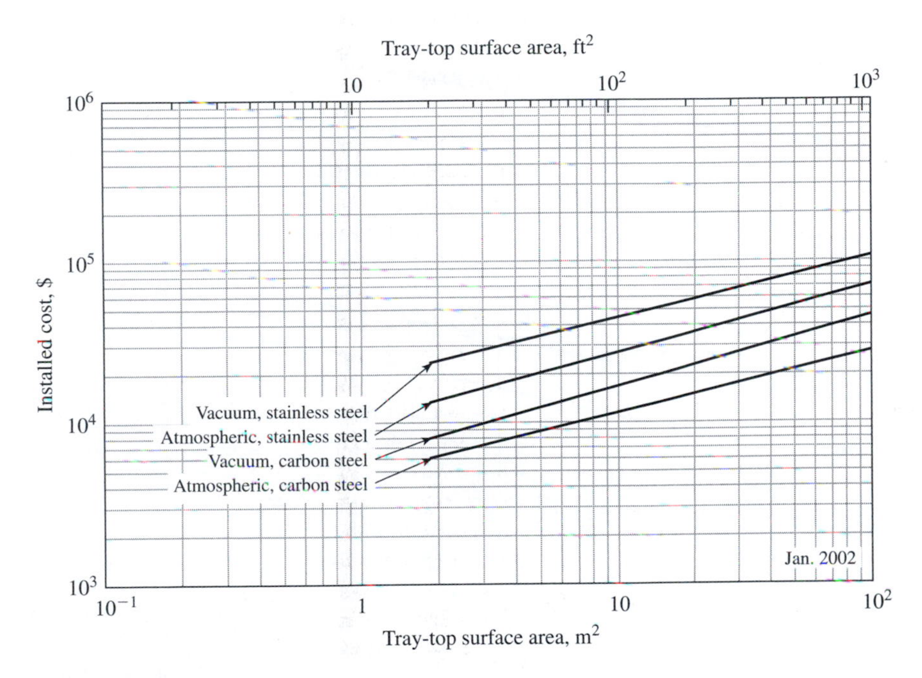

Figure 15-29
Installed cost of tray dryers

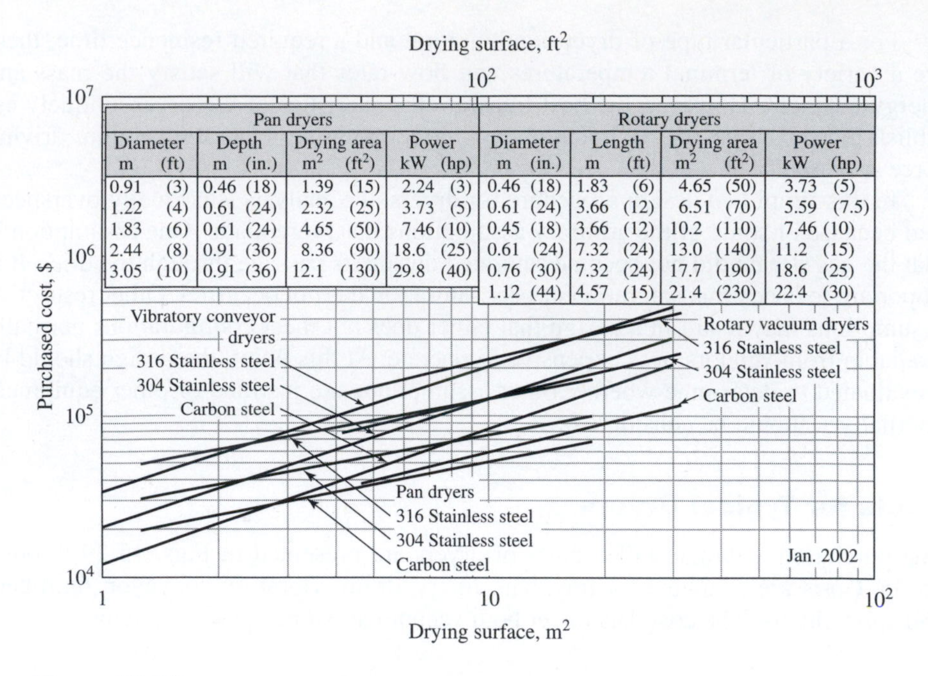

Pan dryers: Diameter m (ft)	Depth m (in.)	Drying area m^2 (ft^2)	Power kW (hp)
0.91 (3)	0.46 (18)	1.39 (15)	2.24 (3)
1.22 (4)	0.61 (24)	2.32 (25)	3.73 (5)
1.83 (6)	0.61 (24)	4.65 (50)	7.46 (10)
2.44 (8)	0.91 (36)	8.36 (90)	18.6 (25)
3.05 (10)	0.91 (36)	12.1 (130)	29.8 (40)

Rotary dryers: Diameter m (in.)	Length m (ft)	Drying area m^2 (ft^2)	Power kW (hp)
0.46 (18)	1.83 (6)	4.65 (50)	3.73 (5)
0.61 (24)	3.66 (12)	6.51 (70)	5.59 (7.5)
0.45 (18)	3.66 (12)	10.2 (110)	7.46 (10)
0.61 (24)	7.32 (24)	13.0 (140)	11.2 (15)
0.76 (30)	7.32 (24)	17.7 (190)	18.6 (25)
1.12 (44)	4.57 (15)	21.4 (230)	22.4 (30)

Figure 15-30
Purchased cost of drying equipment. Price includes motor and drive.

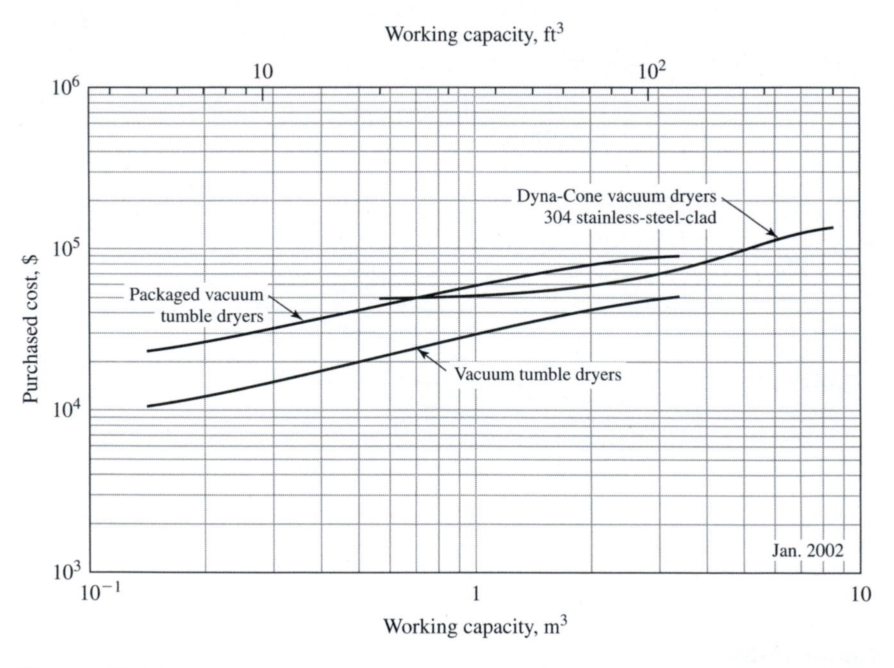

Figure 15-31
Purchased cost of tumble and Dyna-Cone vacuum dryers

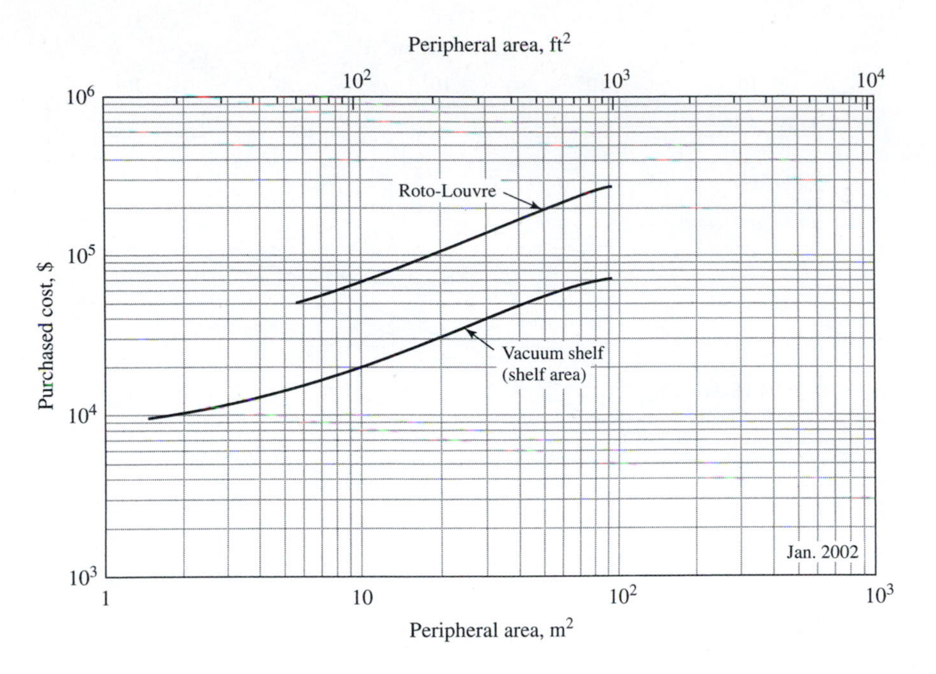

Figure 15-32
Purchased cost for Roto-Louvre and vacuum shelf dryers, carbon-steel construction. Price includes auxiliaries (motor, drive, fan, vacuum pump, condenser, and receiver).

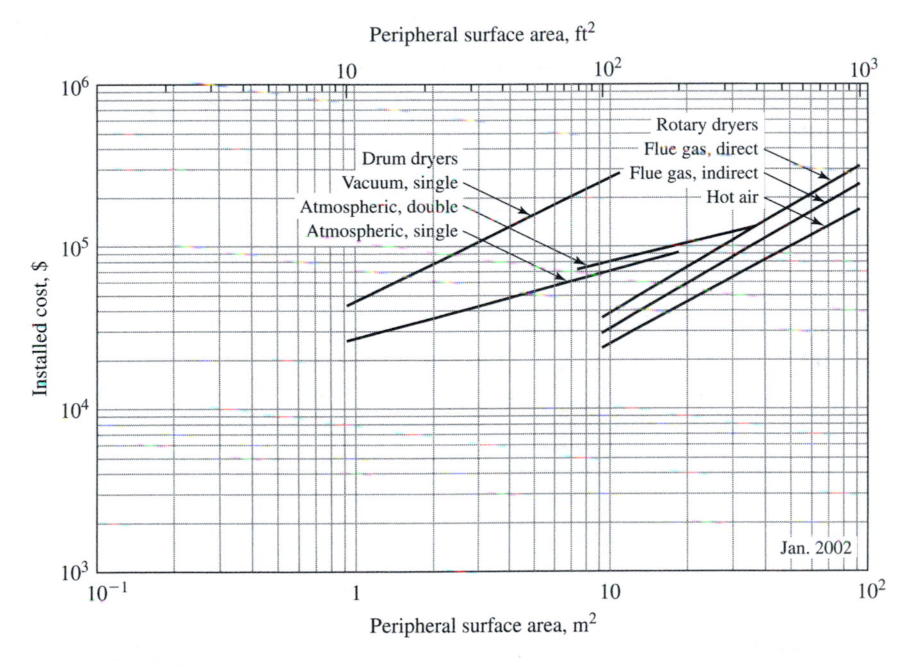

Figure 15-33
Installed cost of drum and rotary dryers

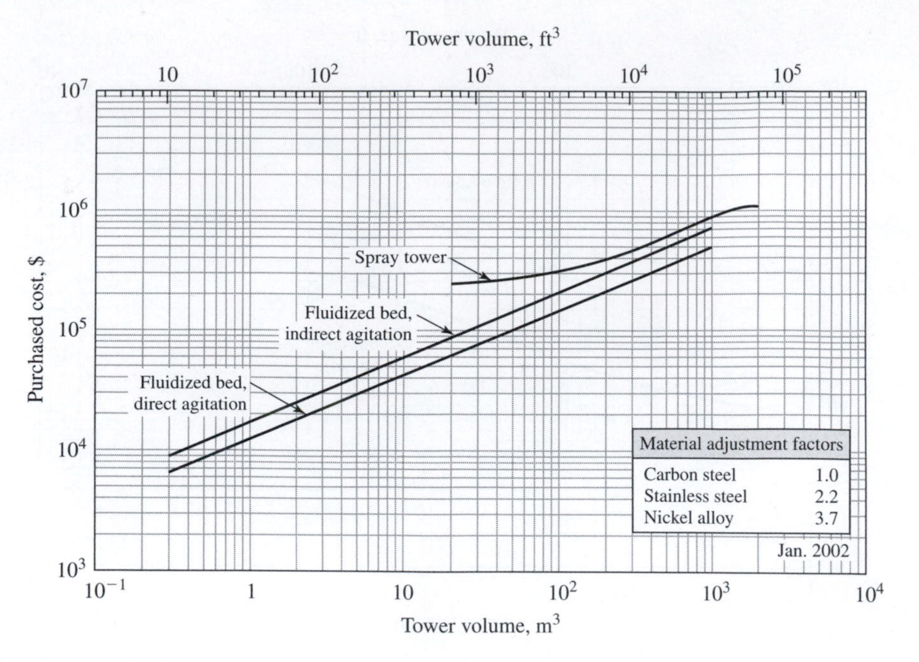

Figure 15-34
Purchased cost of fluidized-bed and spray tower contactors

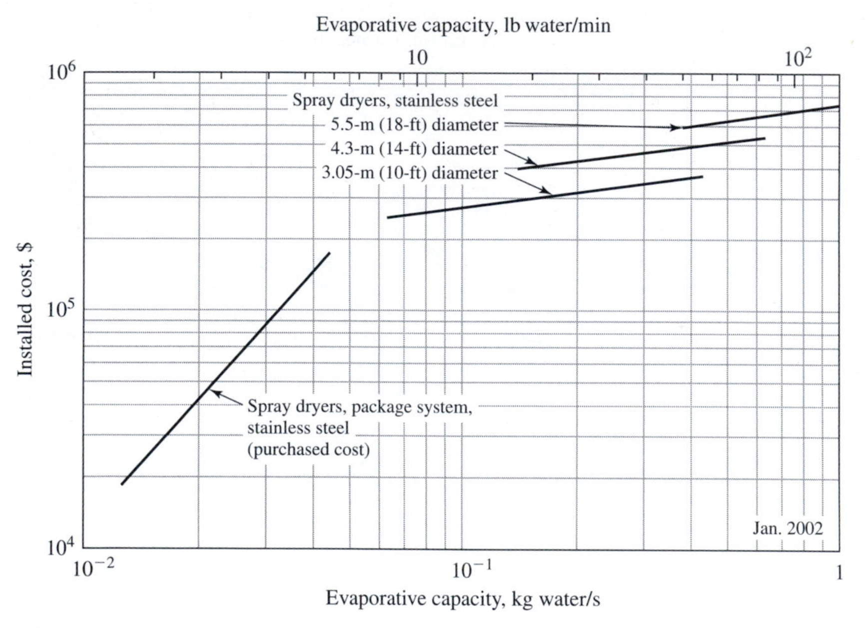

Figure 15-35
Purchased and installed cost of spray dryers

Preliminary Design of a Dryer Using Simplified Procedures — EXAMPLE 15-9

A granular solid contains 4 kg of water per kilogram of dry solids and enters the dryer at 20°C and leaves at 30°C . A final moisture content of 0.05 kg of water per kilogram of dry solids is to be achieved in this drying process. Air available at 20°C and 60 percent relative humidity is available and can be preheated to as high as 100°C before entry into the dryer. The wet solids enter the dryer at a rate of 0.01 kg/s with a solid density of 8.0 kg/m^3. The heat capacity for air is 1 kJ/kg·°C while that for the solid is 2.0 kJ/kg·°C. Determine whether a rotary dryer would be a good selection for the drying of this granular solid.

■ Solution

Use the information provided in Table 15-16 to develop a preliminary design. The equipment diameter can be obtained with the aid of Eq. (15-79); assume a solids velocity of 0.01 m/s and a solids occupancy within the dryer of 12.5 percent.

$$A_c = \frac{\pi D^2}{4} = \frac{A_S}{0.125} = \frac{\dot{m}_S/\rho_S V_S}{0.125}$$

$$= \frac{0.01/(8)(0.01)}{0.125} = 1\ \text{m}^2$$

$$D = \left(\frac{4A_c}{\pi}\right)^{1/2} = \left[\frac{(4)(1)}{\pi}\right]^{1/2} = 1.13\ \text{m}$$

From a mass balance, the water removed by evaporation is

$$\dot{m}_{H_2O} = \frac{0.01\ \text{kg wet solid}}{\text{s}}\left(\frac{1.0\ \text{kg dry solid}}{5.0\ \text{kg wet solid}}\right)\left[\frac{(4.0-0.05)\ \text{kg water}}{\text{kg dry solid}}\right]$$

$$= 0.0079\ \text{kg water/s}$$

From an energy balance, the total heat transferred from the gas to the solid, based on the solid stream, is given by

$$q_S = \dot{m}_{S,i}\left(\frac{1}{1+X_i}\right)C_{p_s}(T_{s,o}-T_{s,i}) + \dot{m}_{S,i}\left(\frac{X_i - X_o}{1+X_i}\right)(h_{V,T_{G,o}} - h_{L,20°C})$$

$$+\dot{m}_S\left(\frac{X_o}{1+X_i}\right)C_{p,H_2O}(T_{s,o}-T_{s,i})$$

where h is the enthalpy, subscripts i and o refer to the inlet and outlet conditions, respectively, and the subscript $V, T_{G,o}$ refers to the water vapor leaving at the outlet gas temperature. Substituting appropriate values gives

$$q_S = 0.01\left[\frac{1}{5}(2.0)(30-20) + \frac{4.95}{5.0}(h_{V,T_{G,o}} - 84) + \frac{0.05}{5.0}(4.19)(30-20)\right]$$

$$= 0.0442 + 0.0099(h_{V,T_{G,o}} - 84)$$

This energy requirement must be provided from the sensible heat in the air. Thus, a balance on the gas with a 10 percent loss through the dryer walls can be expressed as

$$q_S = 0.9\dot{m}_G C_{p,G}(T_{G,i} - T_{G,o})$$

Since a rotary dryer provides agitation and exposure of the solid to the countercurrent air flow, it can be assumed that the residence time is controlled by heat transfer rather than by mass transfer. The overall heat-transfer coefficient from the gas to the solid can be approximated from Table 15-16 to be $U = 60G^{0.67}$ J/s·m^2·K, where G is the mass flux based on the internal surface area of the dryer. Thus,

$$q_S = UA\Delta T_{\text{log mean}} = (60 \times 10^{-3})G^{0.67}\pi DL\Delta T_{\text{log mean}} \qquad G = \frac{4\dot{m}_G}{\pi D^2}\frac{1}{1-0.125}$$

$$= 60 \times 10^{-3}\left(\frac{4\dot{m}_G}{0.875\pi D^2}\right)^{0.67}\pi DL\,\Delta T_{\text{log mean}} = 0.2326\,\dot{m}_G^{0.67}L\,\Delta T_{\text{log mean}}$$

Since the amount of water evaporated is rather large, a mass flux rate of 3.0 kg/s·m^2 (midway between the recommended range for rotary dryers) will be selected. Thus,

$$\dot{m}_G = GA_c = 3.0\frac{\pi}{4}(1.13)^2(0.875) = 2.623\text{ kg/s}$$

Assume a gas outlet temperature of 40°C; the $h_{V,T_{G,o}}$ from the steam tables is 2573 kJ/kg. Then, from above

$$q_S = 0.0442 + 0.0099(2573 - 84) = 24.69\text{ kJ/s}$$

From this, $T_{G,i}$ can now be evaluated from

$$T_{G,i} = \frac{q_S}{0.9\,\dot{m}_G C_{p,G}} + T_{G,o}$$

$$= \frac{24.69}{0.9(2.623)(1.0)} + 40 = 10.5 + 40 = 50.5\text{°C}$$

With the aid of the psychrometric chart, the incoming air contains 0.009 kg of water/kg of dry air. From a material balance, the outlet air contains 0.0128 kg of water/kg of dry air. The inlet and outlet wet-bulb temperatures can also be obtained from the psychrometric chart as 22 and 28°C, respectively. The $\Delta T_{\text{log mean}}$ for these inlet and outlet conditions is given by

$$\Delta T_{\text{log mean}} = \frac{(50.5-28)-(40-22)}{\ln[(50.5-28)/(40-22)]} = 20.2\text{°C}$$

The dryer length can now be calculated from

$$L = \frac{q_S}{(0.2326)\dot{m}_G^{0.67}\,\Delta T_{\text{log mean}}}$$

$$= \frac{24.69}{(0.2326)(2.623)^{0.67}(20.2)} = 2.75\text{ m}$$

This provides an L/D ratio of approximately 2.5. The minimum L/D ratio for a rotary dryer is 4. Air temperatures and gas flow rates should be revised. Calculations should be repeated, setting the L/D ratio to a minimum value of 4. If the results are on the borderline, it may be necessary to select another type of dryer and repeat the general design procedure outlined above. In most cases, the dependability of the design will be controlled by the accuracy of the residence time calculations.

SEPARATION BY FILTRATION

Filtration uses a porous barrier to separate solid particles from a liquid. The barrier can be a cloth, screen, cartridge, or granular material. Straining, cake filtration, deep-bed filtration, and membrane filtration are included in this equipment category. Since the number of equipment options is large, it is necessary to examine each individual application to obtain the optimum equipment selection and design.

Selection of Filtration Equipment

Solid/liquid separation by filtration generally is divided into the four steps of pretreatment, thickening, separation, and posttreatment. Any separation operation can utilize one, two, three, or all four of these steps. Pretreatment involves either coagulation or flocculation to improve the separation operation. For dilute suspensions, thickening often is beneficial in reducing the cost of the separation. The separation step is normally accomplished with equipment such as cake filters, filtering centrifuges, or strainers. Posttreatment often is necessary whenever the clarity of the effluent stream, the dryness of the solid material, or the presence of solute in the mother liquor does not meet product specifications. Various types of settlers, centrifuges, or filters can be used to improve the clarity of the effluent while washing can be used to reduce the solute in the mother liquor.

Prior to selection of the appropriate device for this solid/liquid separation, it is essential to analyze the characteristics of the solid particles and the liquid involved in the system. Particle size is considered to be the most important parameter in the separation since it directly affects both the filterability and the settling rate of the particle. As a rule, the smaller the particle, the more difficult it is to separate. (A particle is considered large when its size exceeds 50 μm and small when it is less than 10 μm.) Solids concentration affects the type of separators that can be considered. For example, cake filters can handle a high solids content but will not work for dilute suspensions that do not form cakes.

With respect to the liquid phase, viscosity is the most important property since the filtration rate and sedimentation velocity are inversely proportional to the viscosity of the suspending fluid. Generally, direct measurement of the viscosity is necessary since it is affected by changes in temperature, liquid purity, and dissolved solids. Factors such as pH, toxicity, corrosivity, particle shape, and particle strength can also affect the separation process and must be considered in the design solution.

Filtration includes all separations that involve the use of a porous barrier to remove solid particles from a liquid. Based on the manner in which the particles are removed, filtration is divided into four basic types, designated as training, cake filtration, deep-bed filtration, and crossflow filtration. A schematic of the basic filtration mechanisms is shown in Fig. 15-36.

Straining refers to the use of a filter medium that has openings smaller than the solid particles. These devices are used inline often to remove solids larger than 150 μm from the suspension. Removal of the particles collected is either performed manually, or scrapers, brushes, and/or liquid backwash is used. When screens are used as the filter medium, vibration methods can be used for continuous discharge of the solids.

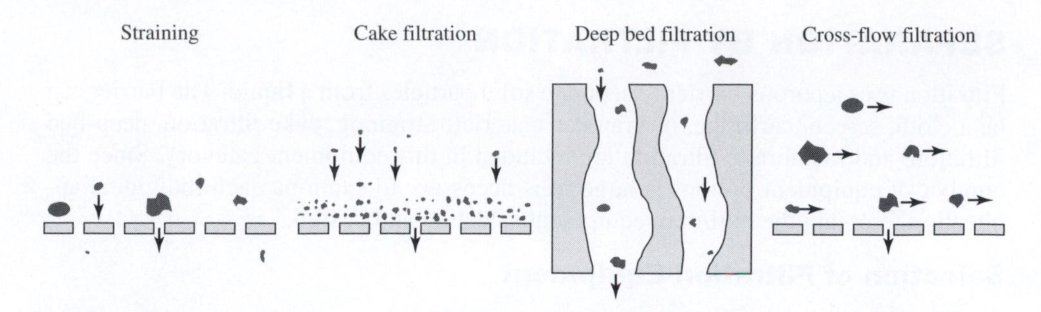

Figure 15-36
Simplified visualization of the semifiltration mechanism

Bag filters are similar to strainers, but the filter media are disposable synthetic bags. Depending on the medium used, bag filters can remove particles over a size range from 1 to 1000 μm. A number of cartridge filters, on the other hand, have higher particle removal efficiencies than bag filters, but are more expensive. However, neither bag nor cartridge filters are designed for high solids loadings. If the total suspended solids in the feed stream exceed 50 mg/l, either type of filter will require frequent solids removal, and other types of filtration equipment should be considered.

In cake filtration, the solids concentration is sufficiently large to form a cake on the separation barrier. Once the filter cake has been established, it becomes the primary filtration medium and can actually remove particles that are smaller than the openings of the original filter medium. Thus, the resistance to flow of the solid-liquid mixture is due to both the filter cake and the filter medium. As the filter cake grows thicker, the resistance to flow increases.

The driving forces used in cake filtration are gravity, vacuum, pressure, mechanical pressure, and centrifugal force. Gravity filters are only used for fast filtration of solid-liquid mixtures with average cake permeabilities greater than 10^{-9} m^2 and particle sizes greater than 1 mm. Gravity belt or gravity nutsche filters are commonly used.

Vacuum filters operate best handling slurries with particles greater than 30 μm and rapid drainage characteristics. Such filters generally require an average cake permeability between 10^{-9} and 10^{-11} m^2. Most vacuum filters are operated continuously and can therefore handle large flows. The most popular continuous types are the rotary drum, disk, and vacuum-belt filters. Other types of continuous vacuum filters include the table and pan filters. The nutsche and leaf filters are examples of batch-operated vacuum filters.

The driving force for pressure filtration is obtained from pump pressure or compressed gas with pressure up to 750 kPa. This type of filter generally is used for fine (1- to 70-μm) particles and slow-draining solids (average cake permeability of 10^{-11} to 10^{-15} m^2). Most pressure filters are batch-operated, but a few continuous types are also available. Pressure tank filters include vertical leaf, horizontal leaf, and candle filters. The filter elements of these units are located within the pressure vessel. The filter press, on the other hand, contains a series of plates that are pushed together with an external pressure to form chambers with the filtration area required for the application.

Several types of mechanical compression filters are available. They include high-pressure expression devices such as the diaphragm press and the tubular press. There are also low-pressure mechanical compression filters that operate continuously as belt and screw press units. Belt presses are often used for sludge dewatering by compression of the slurry between two belts moving around rollers of progressively decreasing size.

The capacities of most pressure filters are limited, due to their batch or semi-continuous operation. One misconception in pressure filtration is that a higher pressure will result in a higher filtration rate and a drier cake. This is not the case for highly compressible solids, where increased pressure actually has a negative effect on the filtration process. In such cases, filtration should be carried out by mechanical compression, rather than by pump pressure.

Another cake filtration device operates under centrifugal force. All filtering centrifuges can be enclosed for vapor- or gas-tight operation. Various types of batch and continuous operating filtering centrifuges are available. The vertical basket and the peeler centrifuge are examples of batch-operated units. Depending on the filter cloth used, these two types of centrifuges can filter fine particles down to about 3 μm. Continuous filtering centrifuges, on the other hand, can process large volumes of slurry but are limited to a lower particle size level of about 50 μm with the use of wedge wire as the filter medium. Pusher centrifuges, worm-screen centrifuges, and screen-bowl centrifuges are examples of continuous filtering centrifuges.

In deep-bed filtration the filter medium provides a specified filter depth, and particles are retained within the filter medium, rather than on the surface. This retention of particles relies more on adsorption than on filtering and can therefore remove particles smaller than the openings in the filter medium. One major type of deep-bed filter is the granular or sand filter which is widely used for water and wastewater treatment. In gravity-driven, sand-bed filters, the filtrate is periodically removed by backwash of the filter medium. These types of sand-bed filters are limited to low feed solids loadings of less than 70 mg/l. A high solids content is normally detrimental to the filter. However, with the use of continuous backwashing with an airlift tube, sand-bed filters are now available that can tolerate higher suspended solids up to 500 mg/l in the feed.

Traditional filtration media, such as filter cloth, wire mesh, and sintered metal, are used to remove solid particles down to 1 to 10 μm. Finer filtration can be achieved with membranes where pore sizes in certain polymers can be reduced to a few angstroms (10^{-10} m). Membrane filtration can be divided into microfiltration, ultrafiltration, and reverse osmosis. Ultrafiltration membranes often are rated by the molecular weight of the particles (2×10^3 to 10^6 molecular weight)[†] that can be separated. For even smaller molecules, such as ions, reverse osmosis is used.

Since membrane filters are designed to filter very fine particles, the traditional deadend filtration is inefficient and a high crossflow rate across the membrane surface is required to prevent cake formation. Filtration is continued in this manner until the circulating suspension becomes too thick to maintain the required crossflow rate. The final products are a clear filtrate and a thick suspension.

[†]K. Scott, *Handbook of Industrial Membranes,* Elsevier, Oxford, United Kingdom, 1995.

Membrane filters are designed to maximize the filtration area while maintaining a reasonable equipment configuration. The three most common configurations are the plate-and-frame, shell and hollow-fiber, and spiral-wound membrane systems. In each system, the suspension flows across the membrane, and the clean filtrate is collected on the permeate side of the membrane.

With the many types of filtration equipment identified in this section, there are no absolute selection techniques available to arrive at the best choice since there are so many factors involved. Nevertheless, Table 15-17 provides a few qualitative guidelines that can be used to narrow down some of the available choices. Additionally, an excellent tool in this selection process uses the commercial computer software p^c – SELECT developed by Wakeman and Tarleton.[†]

General Design Procedures for Separation by Filtration

The major factor in the design of filters is the cake resistance or cake permeability. Since the value of the cake resistance can only be determined on the basis of experimental data, laboratory or pilot-plant tests are almost always necessary to supply the information needed for a filter design. After the basic characteristics of the filter cake have been determined experimentally, the theoretical concepts of filtration can be used to establish the effects of changes in operating variables such as filtering area, slurry concentration, or pressure-difference driving force.

The rate at which filtrate is obtained in a filtering operation is governed by the materials making up the slurry and the physical conditions of the operation. The rate of filtrate delivery is inversely proportional to the combined resistance of the filter cake and the filtering medium, inversely proportional to the viscosity of the filtrate, and directly proportional to the available filtering area and the pressure-driving force. This rate of filtrate delivery can be expressed in equation form as

$$\frac{dv}{d\theta} = \frac{A\,\Delta p}{(R_K + R_F)\mu} \tag{15-84}$$

where v is the volume of filtrate delivered in time θ, A the area of the filtering surface, Δp the pressure drop across the filter, R_K the resistance of the filter cake, R_F the resistance of the filter medium, and μ the viscosity of the filtrate.

Cake resistance R_K varies directly with the thickness of the cake, and the proportionality can be expressed as

$$R_K = Cl \tag{15-85}$$

where C is a constant and l is the cake thickness at time θ.

It is convenient to express R_F in terms of a fictitious cake thickness l_F with resistance equal to that of the filter medium. Thus,

$$R_F = Cl_F \tag{15-86}$$

Designating w as the mass of dry cake solids per unit volume of filtrate, ρ_c as the cake density expressed as the mass of dry cake solids per unit volume of wet filter cake,

[†]R. J. Wakeman and E. S. Tarleton, *Filtration: Equipment Selection, Modelling and Process Simulation,* Elsevier Advanced Technology, Oxford, United Kingdom, 1999.

Table 15-17 Criteria for the selection and preliminary design of selected filtration equipment[†]

Filtration equipment, class and type	Typical equipment dimensions			Press. drop, kPa	Flow rate, m^3/s[‡]	Solids, wt% in cake	Equipment advantages	Equipment limitations
	L, m	D, or W, m	Area, m^2					
Strainer								
Screens (vibratory)	2–5	0.5–1.5	1–7.5		L_1	—	Low capital cost, low power reqm't., easy discharge	High maintenance, particle size limitations
Bag filters	<20	<50	$<3 \times 10^5$	0.5–2	G_1	—	High capacities, good cake dryness, easy discharge	Large space reqm't., nonwashability, high capital cost
Cartridge	1	1	<50	10–100	L_2, G_1	—	Good filtrate clarity and cake dryness, low capital cost	Poor washability, high labor costs, low capacity
Cake filtration								
Rotary drum (vac.) Single compart.	1.5–4	0.6–2	0.5–15	10–100	L_1	60–80	High capacities, good washability, low labor cost	High capital and labor costs, high power reqm't., poor filtrate clarity
Rotary drum (press.) Single compart.	1.5–4	0.6–2	0.5–15	50–300	L_1	50–70	High capacity, low labor cost, handles toxic materials	High capital and labor costs, high power reqm't., poor filtrate clarity
Rotary disk	1–4	2–6	10–600	10–100	L_1	40–60	High capacity, low labor cost, good washability, thick cakes	Poor filtrate clarity, poor cake dryness, limited washability
Table		2–8	1–60	<100	L_1	50–70	High capacity, low labor cost, good washability, thick cakes	Low cake comp., poor discharge, poor filtrate clarity, large space requirement
Tilting pan		6–25	6–200	<100	L_1	50–70	High capacity, low labor cost, good washability, thick cakes	High capital cost, poor filtrate clarity, large space reqm't.
Shell and leaf	2–5	0.5–2	0.1–300	10–600	L_1	50–70	Good filtrate clarity, low capital cost, high cake comp.	High labor costs, high filtrate loss, poor cake depth formation
Plate and frame	0.5–20	0.1–2	1–1000	10–1000	L_1	60–80	Low capital cost, high cake dryness, high cake comp.	Low capacity, high labor cost
Screw					L_1	80–90	High cake dryness, high cake comp., low labor cost	High capital cost, poor filtrate clarity
Roll					L_1	80–95	High cake dryness, high cake comp., low labor cost	High capital cost, poor filtrate clarity
Belt	0.5–15	0.2–3	0.2–100	<100	L_1	50–70	Med. capacity, low labor cost, thick cake formation	High capital cost, low cake comp., poor filtrate clarity, large space reqm't.
Deep-bed filtration								
Sand filter	<4	<12	$\pi D^2/4$	<5	L_2, G_2	10–20	High capacities, low capital cost, high filtrate clarity	Poor cake dryness, poor cake discharge, thin cake formation
Membrane								
Ultrafiltration	0.1–1		<80	<1000	L_1	—	Excellent filtrate clarity, filter very fine particles	High capital, labor and maintenance cost, high space reqm't.

[†]Modified from data presented by G. D. Ulrich, *A Guide to Chemical Engineering Design and Economics,* J. Wiley, New York, 1984, S. M. Walas, *Chemical Process Equipment—Selection and Design,* Butterworth-Heinemann, Newton, MA, 1988, R. J. Wakeman and E. S. Tarleton, *Filtration,* Elsevier Advanced Technology, Oxford, United Kingdom, 1999, D. R. Woods, *Process Design and Engineering Practice,* Prentice-Hall, Englewood Cliffs, NJ, 1995, and N. P. Cheremisinoff, *Liquid Filtration,* 2d ed., Butterworth-Heinemann, Newton, MA, 1998.

[‡]L_1 = liquid flow rate – for <1 wt % solids: $7 \times 10^{-6}A$ to $10^{-3}A$; for >5 wt % solids: $10^{-4}A$ to $3 \times 10^{-3}A$; for fibers or pulps: $0.001A$ to $0.015A$. L_2 = liquid flow rate – $0.001A$. G_1 = gas flow rate – $0.005A$ to $0.5A$; G_2 = gas flow rate – $0.1A$. (A is area in m^2.)

and v_F as the fictitious volume of filtrate per unit of filtering area necessary to lay down a cake of thickness l_F, the actual cake thickness plus the fictitious cake thickness is

$$l + l_F = \frac{w(v + Av_F)}{\rho_c A} \tag{15-87}$$

Equations (15-84) through (15-87) can be combined to give

$$\frac{dv}{d\theta} = \frac{A^2 \Delta p}{\alpha w(v + Av_F)\mu} \tag{15-88}$$

where α equals C/ρ_c and is known as the *specific cake resistance*. In the usual range of operating conditions, the value of the specific cake resistance can be related to the pressure difference by the empirical equation

$$\alpha = \alpha'(\Delta p)^s \tag{15-89}$$

where α' is a constant with its value dependent on the properties of the solid material and exponent s is a constant known as the *compressibility exponent* of the cake. The value of s would be zero for a perfectly noncompressible cake and unity for a completely compressible cake. For commercial slurries, the value of s is usually between 0.1 and 0.8.

The following general equation for the rate of filtrate delivery is obtained by combining Eqs. (15-88) and (15-89):

$$\frac{dv}{d\theta} = \frac{A^2(\Delta p)^{1-s}}{\alpha' w(v + Av_F)\mu} \tag{15-90}$$

This equation applies to the case of constant-rate filtration. For the more common case of constant-pressure-drop filtration, A, Δp, s, α', w, v_F, and μ can all be assumed to be constant with changes in v, and Eq. (15-90) can be integrated between the limits of 0 and v to give

$$v^2 + 2Av_F v = \frac{2A^2(\Delta p)^{1-s}\theta}{\alpha' w \mu} \tag{15-91}$$

Equations (15-90) and (15-91) are directly applicable for use in the design of batch filters. The constants α', s, and v_F must be evaluated experimentally, and then the general equations can be applied to conditions of varying A, Δp, v, θ, w, and μ. One point of caution is necessary, however. In the usual situations, the constants are evaluated experimentally in a laboratory or pilot-plant filter. These constants may be used to scale up to a similar filter with perhaps 100 times the area of the experimental unit. To reduce scale-up errors, the constants should be obtained experimentally with the same slurry mixture, same filter aid, and approximately the same pressure drop as are to be used in the final designed filter. Under these conditions, the values of α' and s will apply adequately to the larger unit. Fortunately, v_F is usually small enough for changes in its value due to scale-up to have little effect on the final results.

Example 15-10 illustrates the methods of determining the constants and applying them in the design of a large plate-and-frame filter.

Estimation of Filtering Area Required for a Specific Filtration System — EXAMPLE 15-10

A plate-and-frame filter press is to be used for removing the solid material from a slurry containing 80 kg/m^3 of solid-free liquid. The viscosity of the liquid is 1 cP, and the filter must deliver a minimum of 11.3 m^3 of solid-free filtrate over a continuous operating time of 2 h when the pressure-difference driving force over the filter unit is constant at 175 kPa. On the basis of the data obtained below with a small laboratory plate-and-frame filter press, estimate the total filter area required.

The following data were obtained with a total filtering area of 0.74 m^2.

Total volume of filtrate v processed in time θ, m^3	Filtration time θ required to process a volume of filtrate for several pressure-difference driving forces		
	$\Delta p_1 = 148$ kPa	$\Delta p_2 = 207$ kPa	$\Delta p_3 = 276$ kPa
0.144	0.34	0.25	0.21
0.226	0.85	0.64	0.52
0.283	1.32	1.00	0.81
0.340	1.90	1.43	1.17

The slurry and the filter medium were identical to those that were to be used in the large filter. The filtrate obtained was free of solid, and a negligible amount of liquid was retained in the filter cake.

■ Solution

An approximate solution can be obtained by interpolating for values of v at $\Delta p = 175$ kPa and then using two of these values to set up Eq. (15-91) in the form of two equations involving only the two unknowns v_F and $(\Delta p)^{1-s}/\alpha' w\mu$. By simultaneous solution, the values of v_F and $(\Delta p)^{1-s}/\alpha' w\mu$ can be obtained. The final required area can then be determined directly from Eq. (15-91). Because this method places considerable reliance on the precision of individual experimental measurements, a more involved procedure using all the experimental data is recommended.

The following procedures can be used to evaluate the constants v_F, s, and α'. Begin by rearranging Eq. (15-91) in the form

$$\frac{\theta \Delta p}{v/A} = 0.5\alpha' w(\Delta p)^s \left(\frac{v}{A}\right) + \alpha' w\mu v_F (\Delta p)^s$$

At constant Δp, a plot of $\theta \Delta p/(v/A)$ versus v/A should give a straight line with a slope equal to $0.5\alpha' w\mu(\Delta p)^s$ and an intercept at $v/A = 0$ of $\alpha' w\mu v_F (\Delta p)^s$. Figure 15-37 presents a plot of this type based on the experimental data for this problem. Anytime the same variable appears in both the ordinate and the abscissa of a straight-line plot, an analysis for possible misinterpretation should be made. In this case, the values of θ and Δp change sufficiently to make a plot of this type acceptable.

The following slopes and intercepts are obtained from Fig. 15-37:

Δp, kPa	Slope, h·kPa/m^2	Intercept, h·kPa/m
138	1227	11.1
207	1380	12.5
276	1505	13.6

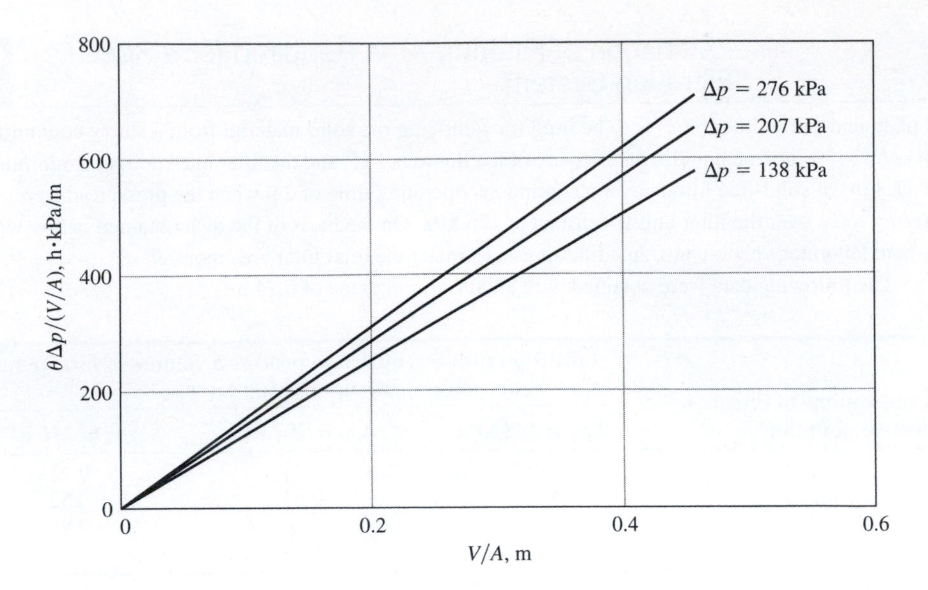

Figure 15-37
Plot used to evaluate constants for the filtrate rate equation in Example 15-10

Values of α' and s can now be obtained by simultaneous solution with any two of the three slope values presented in the preceding table. For example,

$$\alpha' = \frac{\text{slope}}{w\mu(\Delta p)^s} \qquad \text{where } \mu = 0.0353 \text{ kPa·m/h}$$

Select the filter runs at 138 and 207 kPa.

$$\frac{1227(2)}{80(0.0353)(138)^s} = \frac{1380(2)}{80(0.0353)(207)^s}$$
$$s = 0.3$$

Solving for α' gives $\alpha' = 198$.

On the basis of the Fig. 15-37 intercept for the 207-kPa data,

$$v_F = \frac{\text{intercept}}{\alpha' w\mu(\Delta p)^s}$$
$$= \frac{12.5}{198(80)(0.0353)(207)^{0.3}} = 4.51 \times 10^{-3} \text{ m}^3/\text{m}^2$$

Substitution of the constants in Eq. (15-91) provides the final equation to use in evaluating the total area needed for the large filter:

$$(11.3)^2 + 2A(4.51 \times 10^{-3})(11.3) = \frac{2A^2(175)^{1-0.3}(2)}{198(80)(0.0353)}$$
$$127.7 + 0.102A = 0.266A^2$$

Solving the relation, we see that the total area required is approximately 22.1 m^2. ■

A procedure to verify the reliability of α' and s involves taking the logarithm of the expressions for the slope and the intercept, as obtained in Example 15-10. This gives

$$\log(\text{slope}) = s \log \Delta p + \log \frac{\alpha' w \mu}{2} \tag{15-92a}$$

$$\log(\text{intercept}) = s \log \Delta p + \log(\alpha' w \mu v_F) \tag{15-92b}$$

A log-log plot of the slopes versus Δp should give a straight line with a slope of s and an intercept at $\log(\Delta p) = 0$ of $\log(\alpha' w \mu/2)$. This not only evaluates s and α', but also permits checking of the consistency of the experimental data.

Development of general design equations for continuous filters, such as rotary-drum or rotary-disk filters, follows the same line of reasoning as that presented in the development of Eq. (15-91). The following analysis is based on the design variables for a typical rotary vacuum filter of the type shown in Fig. 15-38.

It is convenient to develop the design equations in terms of the total area available for filtering service, even though only a fraction of this area is in direct use at any instant. Designate the total available area as A_D and the fraction of this area immersed in the slurry as, Ψ_f. The effective area of the filtering surface then becomes $A_D \Psi_f$, and Eq. (15-84) can be expressed in the following form:

$$\frac{dv}{d\theta} = \frac{A_D \Psi_f \Delta p}{(R_K + R_F)\mu} \tag{15-93}$$

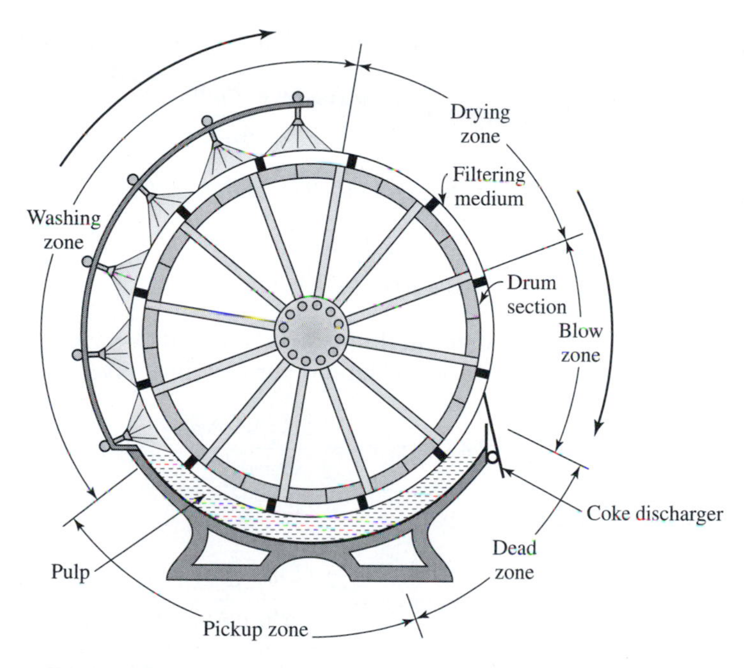

Figure 15-38
Cross-sectional end view of rotary vacuum-drum filter. (*Eimco Corporation.*)

Referring to Eqs. (15-84) and (15-85),

$$R_K + R_F = C(l + l_F) \tag{15-94}$$

With a continuous filter, the cake thickness at any given location on the submerged filtering surface does not vary with time. The thickness, however, does vary with location as the cake builds up on the filtering surface during passage through the slurry. The thickness of the cake leaving the filtering zone is a function of the slurry concentration, cake density, and volume of filtrate delivered per revolution. This thickness can be expressed by

$$l_{\text{leaving filter zone}} = \frac{wv_R}{\rho_c A_D} \tag{15-95}$$

where v_R is the volume of filtrate delivered per revolution and ρ_c the cake density, defined as the mass of dry cake solids per unit volume of wet filter cake leaving the filter zone.

An average cake thickness during the cake deposition period can be assumed to be one-half the sum of the thicknesses at the entrance and exit of the filtering zone. Since no appreciable amount of cake should be present on the filter when it enters the filtering zone,

$$l_{\text{avg}} = \frac{wv_R}{2\rho_c A_D} \tag{15-96}$$

By using the same procedure as was followed in the development of Eq. (15-87),

$$l + l_F = l_{\text{avg}} + l_F = \frac{w(v_R/2 + A_D \Psi_f v_F)}{\rho_c A_D} \tag{15-97}$$

Combination of Eqs. (15-93), (15-94), and (15-97), with $\alpha = C/\rho_c$, gives

$$\frac{dv}{d\theta} = \frac{2A_D^2 \Psi_f \Delta p}{\alpha w \mu (v_R + 2A_D \Psi_f v_F)} \tag{15-98}$$

Integration of Eq. (15-98) between the limits of $v = 0$ and $v = v_R$, and $\theta = 0$ and $\theta = 1/N_r$ where N_r is the number of revolutions per unit time, gives

$$v_R^2 + 2A_D \Psi_f v_F v_R = \frac{2A_D^2 \Psi_f \Delta p}{\alpha w \mu N_r} \tag{15-99}$$

or, by including Eq. (15-89),

$$v_R^2 + 2A_D \Psi_f v_F v_R = \frac{2A_D^2 \Psi_f (\Delta p)^{1-s}}{\alpha' w \mu N_r} \tag{15-100}$$

The constants in the preceding equations can be evaluated by a procedure similar to that described in Example 15-10. Equation (15-100) is often used in the following simplified forms, which are based on the assumptions that the resistance of the filter medium is negligible and the filter cake is noncompressible:

$$\text{Volume of filtrate per revolution } v_R = A_D\left(\frac{2\Psi_f \Delta p}{\alpha w \mu N_r}\right)^{1/2} \quad (15\text{-}101a)$$

$$\text{Volume of filtrate per unit time } v_R N_r = A_D\left(\frac{2\Psi_f N_r \Delta p}{\alpha w \mu}\right)^{1/2} \quad (15\text{-}101b)$$

$$\text{Weight of dry cake per unit time } v_R N_r w = A_D\left(\frac{2\Psi_f N_r w \Delta p}{\alpha \mu}\right)^{1/2} \quad (15\text{-}101c)$$

A vacuum pump must be supplied for the operation of a rotary vacuum filter, and the design engineer may need to estimate the size of the pump with its power requirement for a given filtration unit. Because air leakage into the vacuum system may supply a major amount of the air that passes through the pump, design methods for predicting air suction rates must be considered as approximate since they do not account for air leakage.

The rate at which air is drawn through the dewatering section of a rotary vacuum filter can be expressed in a form similar to Eq. (15-93) as

$$\frac{dv_a}{d\theta} = \frac{A_D \Psi_a \Delta p}{(R'_F + R'_K)\mu_a} \quad (15\text{-}102)$$

where v_a is the total volume of air at ambient conditions drawn through the filter cake in time θ, μ_a the viscosity of this air flow, and Ψ_a the fraction of total surface available for the air suction. The cake resistance R'_K is directly proportional to the cake thickness l, and the filter medium resistance R'_F can be assumed to be directly proportional to a fictitious cake thickness l'_f. Designating C' as the proportionality constant, we have

$$R'_K + R'_F = C'(l + l'_F) \quad (15\text{-}103)$$

If the cake is noncompressible, l must be equal to the thickness of the cake leaving the filtering zone. Therefore, by Eq. (15-95) and using the same procedure as was followed in the development of Eq. (15-87),

$$l + l'_F = \frac{w(v_R + A_D \Psi_a v'_F)}{\rho_c A_D} \quad (15\text{-}104)$$

where v'_F is the fictitious volume of filtrate per unit of air suction area necessary to provide a cake of thickness l'_F.

Combination of Eqs. (15-102) through (15-104) gives

$$\frac{dv_a}{d\theta} = \frac{A_D^2 \Psi_F \Delta p}{\beta' w \mu_a (v_R + A_D \Psi_a v'_F)} \quad (15\text{-}105)$$

where β' is C'/ρ_c and is designated as the specific air suction cake resistance.

Integration of Eq. (15-105) between the limits corresponding to $v_a = 0$ and $v_a = v_{ar}$, where v_{ar} designates the volume of air per revolution, gives

$$v_{ar} = \frac{A_D^2 \Psi_a \Delta p}{\beta' w (v_R + A_D \Psi_a v_F') \mu_a N_r} \tag{15-106}$$

If the cake is compressible, a rough correction for the variation in β' with a change in Δp can be made by use of the empirical equation

$$\beta' = \beta''(\Delta p)^{s'} \tag{15-106a}$$

where β'' and s' are constants. By neglecting the resistance of the filter medium, Eq. (15-106) can be simplified to

$$\text{Volume of air per revolution } v_{ar} = \frac{A_D^2 \Psi_a \Delta p}{\beta' w v_R \mu_a N_r} \tag{15-107}$$

$$\text{Volume of air per unit time } v_{ar} N_r = \frac{A_D^2 \Psi_a \Delta p}{\beta' w v_R \mu_a} \tag{15-108}$$

Equations (15-101*a*), (15-101*c*), and (15-108) can be combined to give

$$\text{Volume of air per unit time} = \frac{A_D \psi_a}{\beta' \mu_a} \left(\frac{\alpha \mu N_r \Delta p}{2 w \Psi_f} \right)^{1/2} \tag{15-109}$$

$$\frac{\text{Volume of air per unit time}}{\text{Weight of dry cake per unit time}} = \left(\frac{\Psi_a}{\Psi_f} \right) \left(\frac{\mu}{\mu_a} \right) \left(\frac{\alpha}{2\beta' w} \right) \tag{15-110}$$

If the constants in the preceding equations are known for a given filter system and the assumption of no air leakage is adequate, the total amount of suction air can be estimated. This value, combined with a knowledge of the air temperature and the pressures at the intake and discharge sides of the vacuum pump, can be used to estimate the power requirements of the vacuum pump.

EXAMPLE 15-11 Estimation of the Power Requirements for a Rotary Vacuum Filter

It is desired to use a rotary vacuum filter to separate a slurry containing 20 kg of water per kilogram of solid material. Tests on the rotary filter at the conditions to be used for the filtration have indicated that the dimensionless ratio of α/β' is 0.6 and 19 kg of filtrate (not including wash water) is obtained for each 21 kg of slurry. The temperature and pressure of the surroundings are 20°C and 1 atm, respectively. The pressure drop to be maintained by the vacuum pump is 35 kPa. The fraction of the drum area submerged in the slurry is 0.3, and the fraction of the drum area available for suction is 0.1. On the basis of the following assumptions, estimate the kilowatt rating of the motor necessary to operate the vacuum pump if the filter handles 23,000 kg/h of slurry.

Assumptions are as follows:

Resistance of the filter medium is negligible.

Effects of air leakage are included in the value given for α/β'. The value of β' is based on the temperature and pressure of the ambient air.

The filter removes all the solids from the slurry.

The vacuum pump operates isentropically with an overall efficiency of 50 percent for the pump and motor.

The ratio of C_p/C_v remains constant at a value of 1.4.

■ Solution

Since the value given for α/β' is applicable for the operating conditions for the filtration process and the resistance of the filter medium is negligible, Eq. (15-110) can be used to determine the volume of ambient air that should be removed during a unit time period.

The variables in Eq. (15-110) are as follows:

$$\psi_a = 0.1 \qquad \psi_f = 0.3 \qquad \alpha/\beta' = 0.6$$

$$\mu_{H_2O}(20°C) = 0.0347 \text{ kPa·m/h}$$

$$\mu_a(20°C) = 0.000635 \text{ kPa·m/h}$$

$$\rho_{H_2O}(20°C) = 998 \text{ kg/m}^3$$

$$w = \frac{1}{19/998} = 52.53 \text{ kg dry cake solids/m}^3 \text{ filtrate}$$

$$\text{Weight of dry cake/h} = (23{,}000)(1/21) = 1095 \text{ kg/h}$$

From Eq. (15-110),

$$\text{Volume of dry air/s} = \frac{1095(0.1)(0.0347)(0.6)}{0.3(0.000635)(2)(52.53)(3600)} = 0.0316 \text{ m}^3\text{/s}$$

The theoretical power required for an isentropic compression is given by

$$\text{Power} = \frac{k}{k-1} p_1 \dot{m}_{V,1} \left[\left(\frac{p_2}{p_1} \right)^{(k-1)/k} - 1 \right]$$

$$p_1 = 101.3 - 35 = 66.3 \text{ kPa}$$

$$p_2 = 101.3 \text{ kPa}$$

$$\dot{m}_{V,1} = \text{gas flow at vacuum pump inlet, m}^3\text{/s}$$

$$= \frac{0.0316(101.3)}{66.3} = 0.0483 \text{ m}^3\text{/s at } 20°\text{C and } 66.3 \text{ kPa}$$

$$\text{Power} = \left(\frac{1.4}{0.4} \right)(66.3)(0.0483) \frac{(101.3/66.3)^{0.4/1.4} - 1}{0.5} = 2.88 \text{ kW}$$

A 3-kW motor would be sufficient.

Costs for Filtration and Other Solids Separation Equipment

Information to permit estimation of the cost for various types of filtration equipment is presented in Figs. 15-39 through 15-42. Costs for solids separation by centrifugal force are given in Figs. 15-43 through 15-47 and by various types of dust collectors in Figs. 15-48 and 15-49.

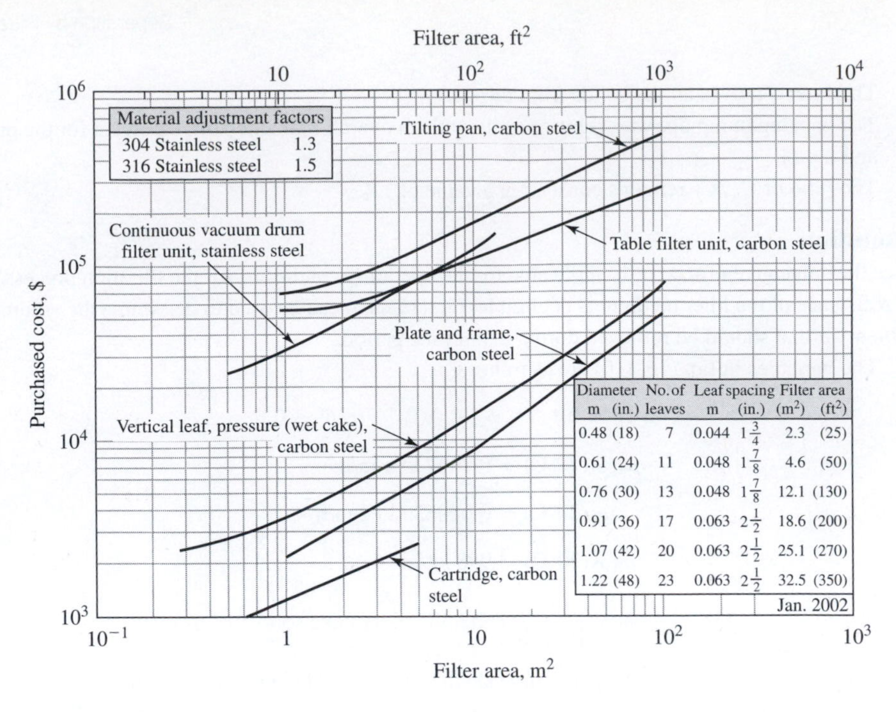

Diameter m (in.)	No. of leaves	Leaf spacing m (in.)	Filter area (m^2) (ft^2)
0.48 (18)	7	0.044 $1\frac{3}{4}$	2.3 (25)
0.61 (24)	11	0.048 $1\frac{7}{8}$	4.6 (50)
0.76 (30)	13	0.048 $1\frac{7}{8}$	12.1 (130)
0.91 (36)	17	0.063 $2\frac{1}{2}$	18.6 (200)
1.07 (42)	20	0.063 $2\frac{1}{2}$	25.1 (270)
1.22 (48)	23	0.063 $2\frac{1}{2}$	32.5 (350)

Figure 15-39
Purchased cost of fixed and continuous filters

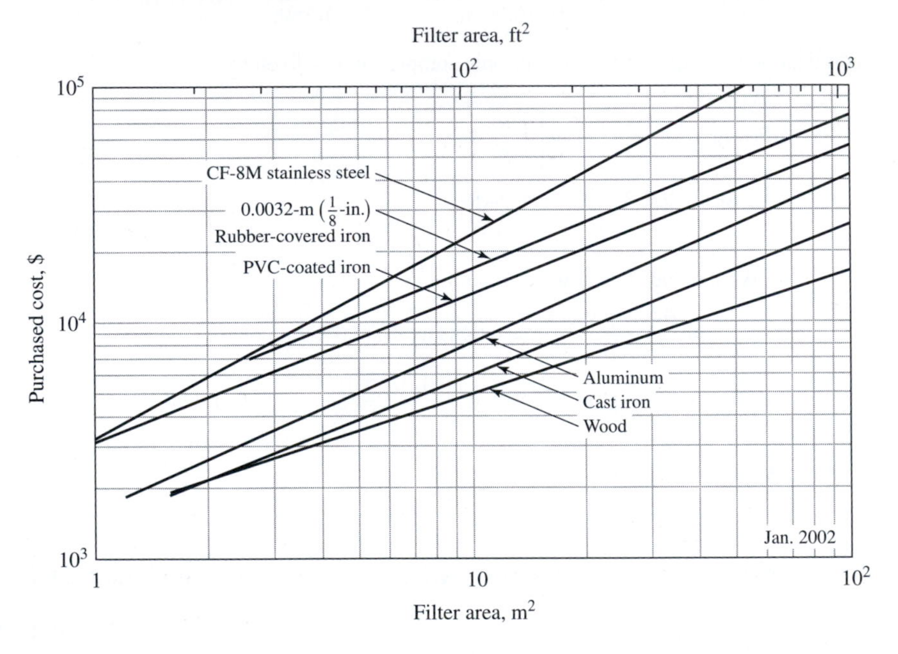

Figure 15-40
Purchased cost of plate-and-frame filters. Order-of-magnitude capital cost estimating data.

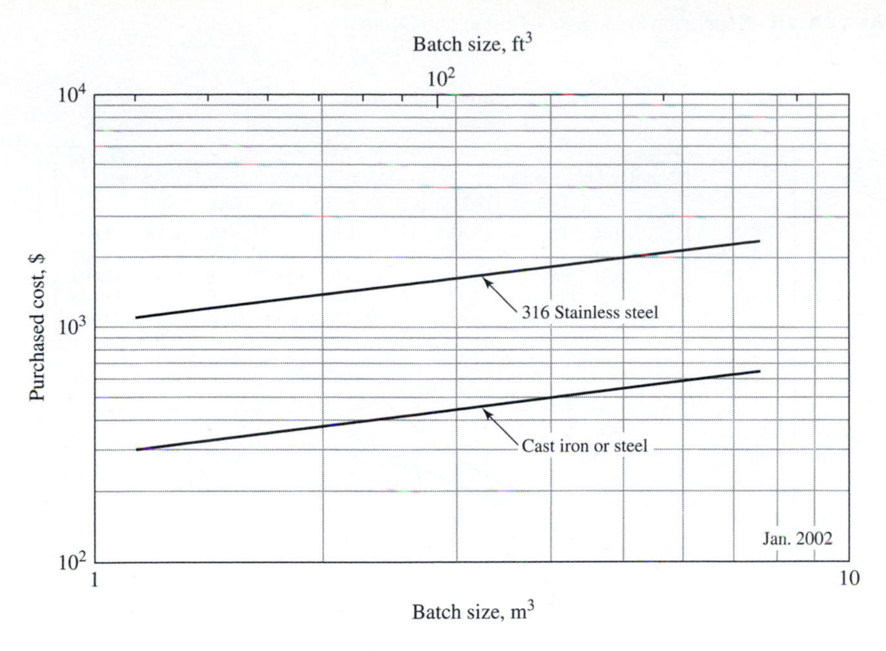

Figure 15-41
Purchased cost of cartridge-type filters

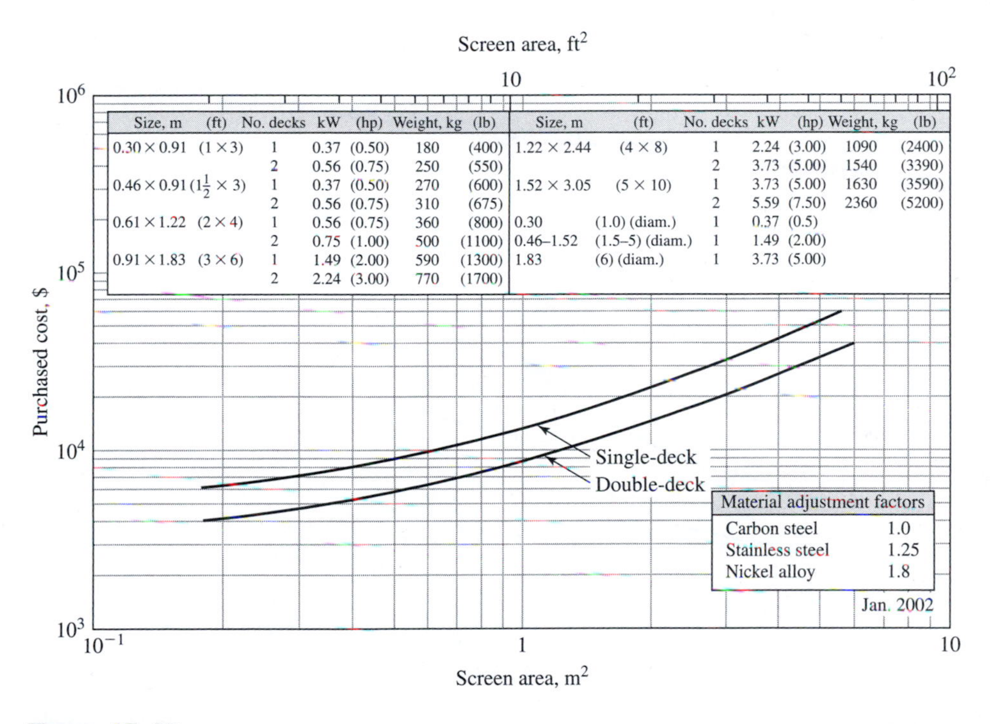

Size, m	(ft)	No. decks	kW	(hp)	Weight, kg	(lb)
0.30 × 0.91	(1 × 3)	1	0.37	(0.50)	180	(400)
		2	0.56	(0.75)	250	(550)
0.46 × 0.91	($1\frac{1}{2}$ × 3)	1	0.37	(0.50)	270	(600)
		2	0.56	(0.75)	310	(675)
0.61 × 1.22	(2 × 4)	1	0.56	(0.75)	360	(800)
		2	0.75	(1.00)	500	(1100)
0.91 × 1.83	(3 × 6)	1	1.49	(2.00)	590	(1300)
		2	2.24	(3.00)	770	(1700)
1.22 × 2.44	(4 × 8)	1	2.24	(3.00)	1090	(2400)
		2	3.73	(5.00)	1540	(3390)
1.52 × 3.05	(5 × 10)	1	3.73	(5.00)	1630	(3590)
		2	5.59	(7.50)	2360	(5200)
0.30	(1.0) (diam.)	1	0.37	(0.5)		
0.46–1.52	(1.5–5) (diam.)	1	1.49	(2.00)		
1.83	(6) (diam.)	1	3.73	(5.00)		

Material adjustment factors	
Carbon steel	1.0
Stainless steel	1.25
Nickel alloy	1.8

Figure 15-42
Purchased cost of vibrating screens with motor

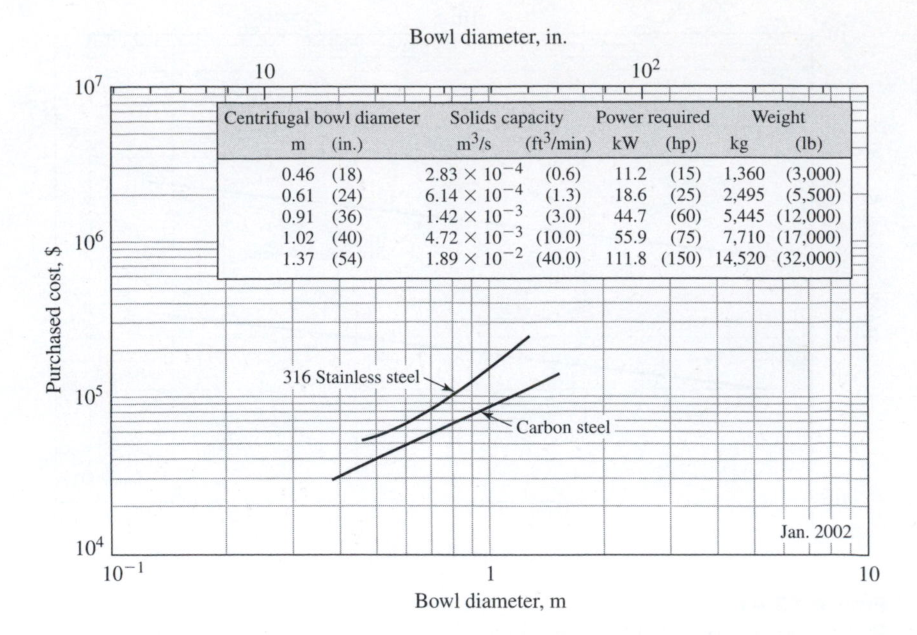

Centrifugal bowl diameter		Solids capacity		Power required		Weight	
m	(in.)	m^3/s	(ft^3/min)	kW	(hp)	kg	(lb)
0.46	(18)	2.83×10^{-4}	(0.6)	11.2	(15)	1,360	(3,000)
0.61	(24)	6.14×10^{-4}	(1.3)	18.6	(25)	2,495	(5,500)
0.91	(36)	1.42×10^{-3}	(3.0)	44.7	(60)	5,445	(12,000)
1.02	(40)	4.72×10^{-3}	(10.0)	55.9	(75)	7,710	(17,000)
1.37	(54)	1.89×10^{-2}	(40.0)	111.8	(150)	14,520	(32,000)

Figure 15-43
Purchased cost of centrifugal filters, continuous solid bowl. Price does not include motor and drive.

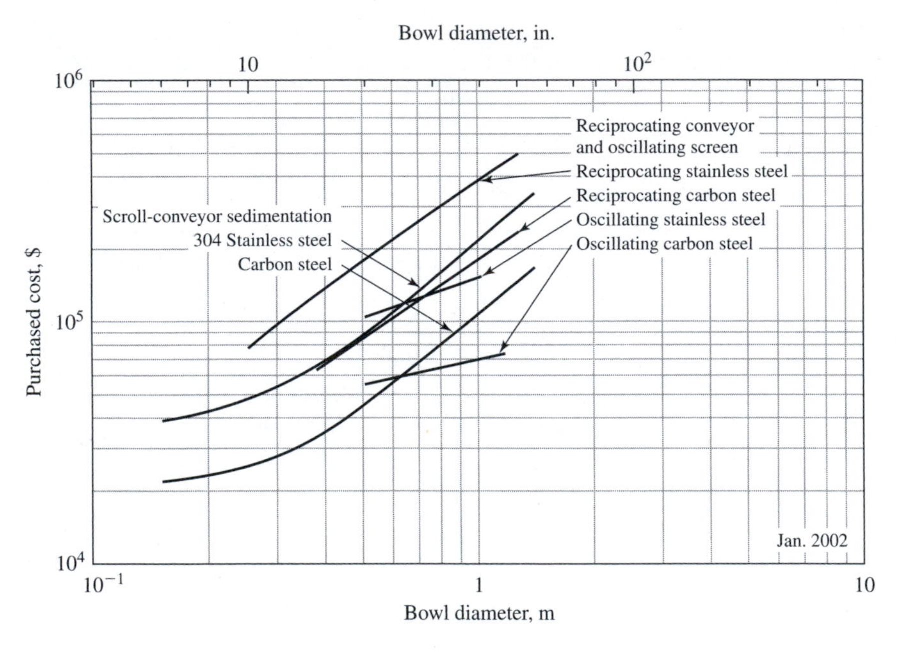

Figure 15-44
Purchased cost of centrifugal separators

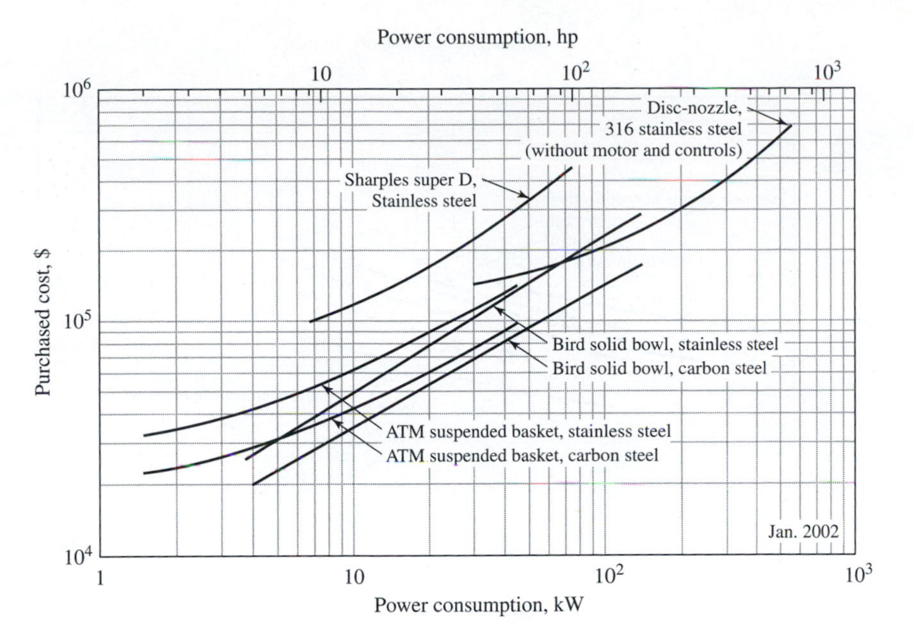

Figure 15-45
Purchased cost of centrifuges

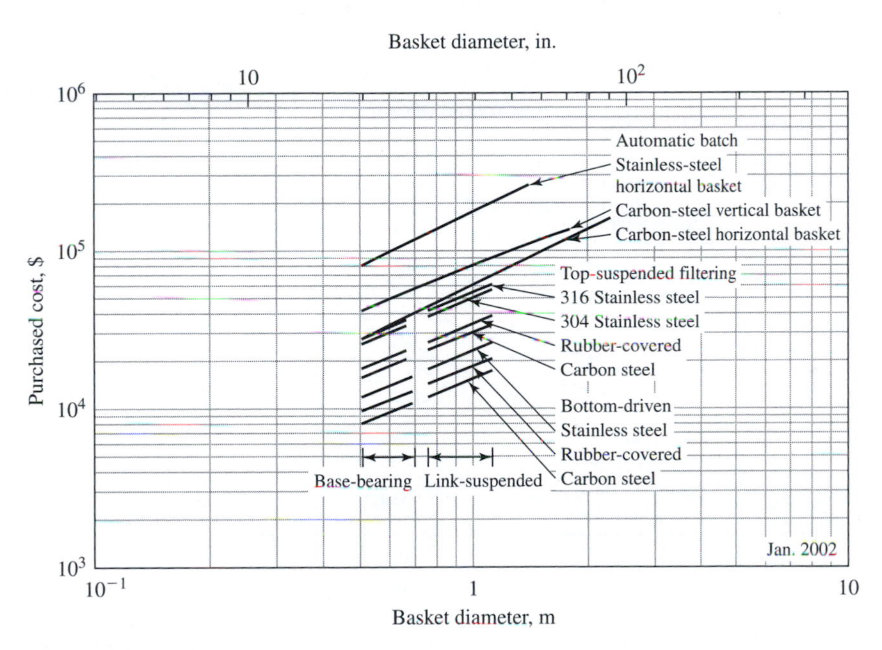

Figure 15-46
Purchased cost of centrifugal separators

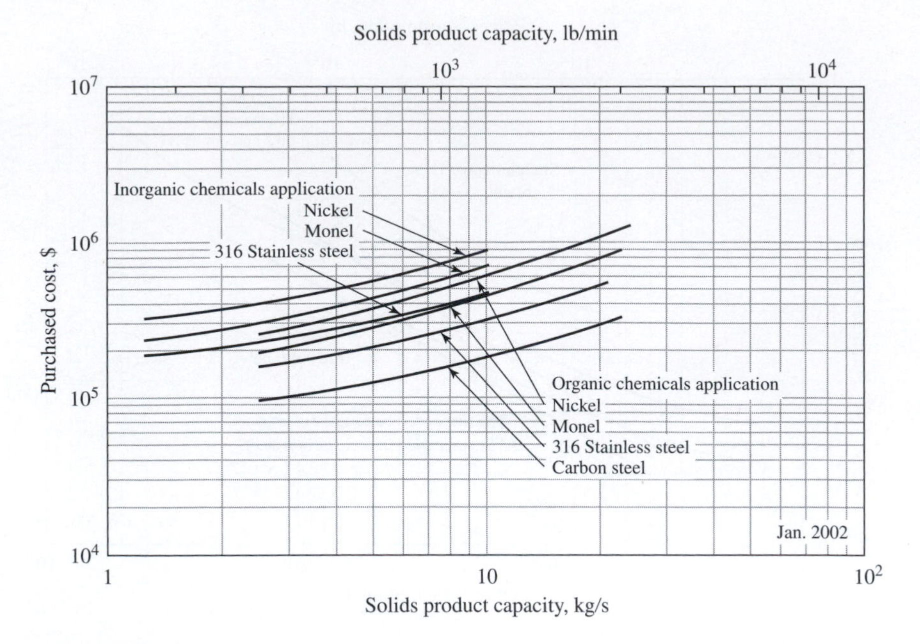

Figure 15-47
Purchased cost of centrifuges: solid bowl, screen bowl, and pusher types

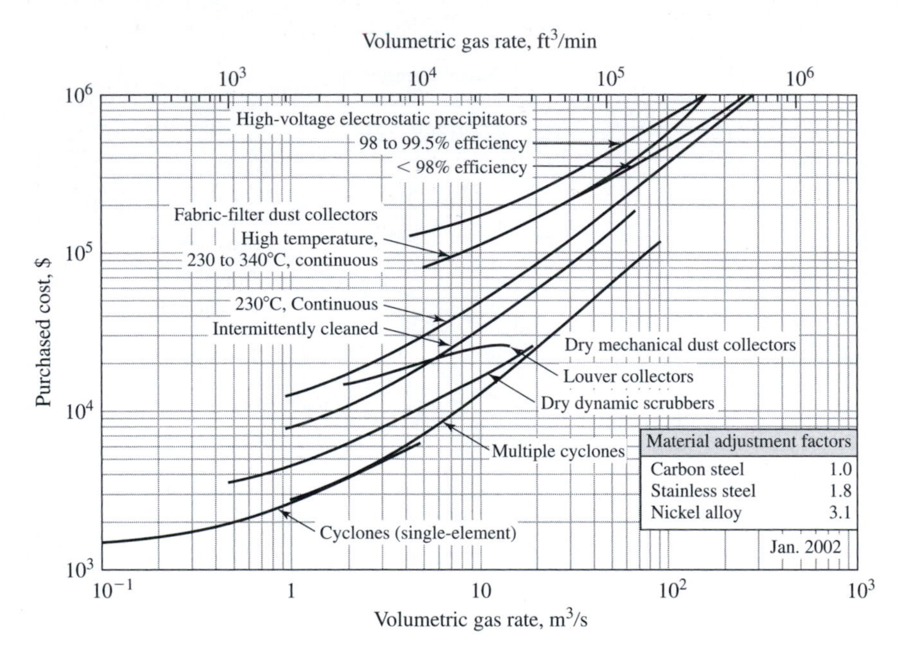

Figure 15-48
Purchased cost of dry, mechanical dust collectors, high-voltage electrostatic precipitators, and fabric-filter dust collectors

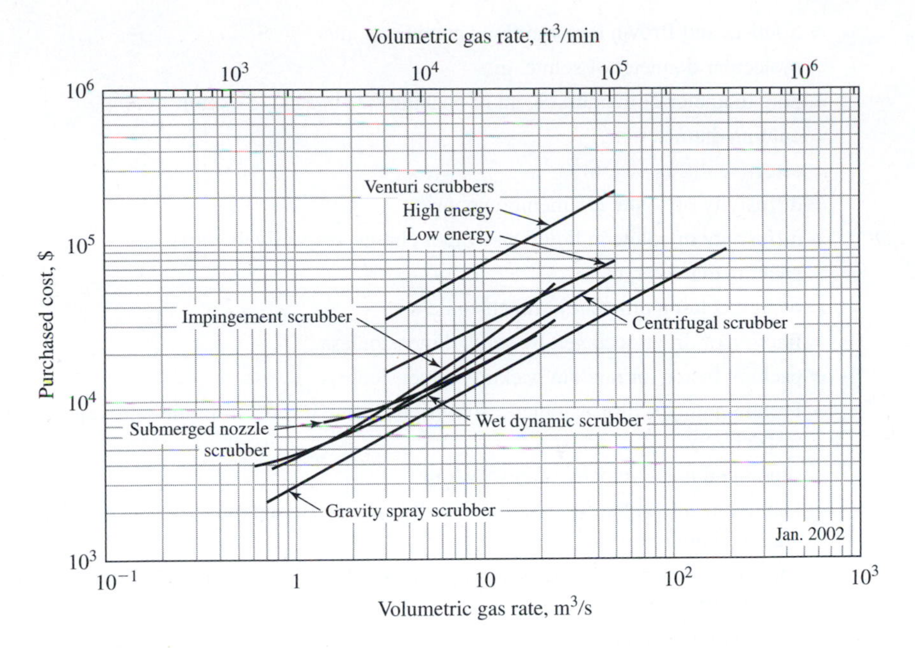

Figure 15-49
Purchased cost of wet dust collectors

NOMENCLATURE

a	= interfacial mass-transfer area per column volume, m^2/m^3
a_e	= interfacial contact area between liquid and vapor phase per unit volume of contactor, m^2/m^3
a_p	= surface area of packing per unit of packed-tower volume, m^2/m^3
a_v	= pore surface area per unit volume of membrane, m^2/m^3
a_w	= area of wetted packing, m^3
A_c	= total cross-sectional area of distillation column, m^2; cross-sectional area of dryer, m^2
A_d	= downcomer area of distillation column, m^2
A_D	= total filtration area available, m^2
A_i	= absorption factor for component i as defined in Eq. (15-32), dimensionless
A_M	= total membrane area, m^2
A_n	= net cross-sectional area used in a distillation column, m^2
A_S	= solids cross-sectional area in a dryer, m^2
c	= solute concentration, kg mol/m^3; subscripts F and P refer to solute concentrations at feed and permeate surfaces, respectively
c'_F	= concentration of adsorbate in feed, kg mol/m^3
C	= constant in Eq. (15-85), s^2/m^3
C'	= proportionality constant in Eq. (15-103), dimensionless
Ca_L	= capillary number defined by Eq. (15-23a), dimensionless

C_{sb} = Souders and Brown factor at flood conditions, m/s
d_m = molecular diameter of solute, m
d_p = pore diameter of membrane, m; or packing size in Fig. 15-7, min
D = column diameter, m
D_e = effective diffusivity, m^2/s
D_i = diffusivity of solute i in membrane, m^2/s
D_L, D_V = diffusivity of solute in liquid and vapor phases, respectively, m^2/s
D_p = packing diameter, m
E_o = overall column efficiency, dimensionless
f_S = fraction of dryer occupied by solid, dimensionless
F_p = packing factor for random packing, dimensionless
Fr = Froude number, dimensionless; subscripts L and V refer to liquid and vapor values, respectively
G = superficial mass velocity of gas, $kg/s \cdot m^2$
G_{max} = maximum allowable superficial velocity of gas (based on cross-sectional area of empty column or pipe), $kg/s \cdot m^2$
h = enthalpy of fluid, kJ/kg; subscripts L and V refer to liquid and vapor components of fluid, respectively
H_c = actual column height, m
H_i = Henry's law constant for component i, dimensionless
H_s = plate spacing height, m
ΔH = additional column height for operational purposes, m
HETP = height equivalent to a theoretical plate, m
HETS = height equivalent to a theoretical stage, m
HTU = height of a mass-transfer unit, m; subscripts E and R refer to mass transfer in extract and raffinate phases, respectively
k = individual mass transfer coefficient, kg $mol/s \cdot m^2$; subscript a refers to adsorption; subscript d to desorption; subscript e to the extract phase; subscript r to the raffinate phase; subscript L to the liquid phase, m/s; and subscript V to the vapor phase, m/s
$k_{d'}$ = drying rate constant, $m/K \cdot s$
K = equilibrium ratio for vapor/liquid equilibria, dimensionless; adsorption equilibrium constant defined by the ratio k_a/k_d, dimensionless
K_L = overall mass-transfer coefficient based on liquid phase, kg $mol/s \cdot m^2$
K_i = factor that accounts for the solubility of solute in a membrane, dimensionless
K_r = restrictive factor defined by Eq. (15-53a), dimensionless
K_V = overall mass-transfer coefficient based on gas phase, kg $mol/s \cdot m^2$
K' = partition ratio, defined as ratio of weight fraction of solute in extract phase to weight fraction of solute in raffinate phase, dimensionless
l = thickness of filtrate cake, m
l_F = thickness of the fictitious filtrate cake, m
l'_F = thickness of the fictitious filtrate cake required in a rotary vacuum filter, m
L = liquid flow rate, kg mol/s

L' = number of mols of liquid in a batch still at any time, kg mol; subscripts 1 and 2 refer to initial and final mols in column, respectively
dL = differential amount vaporized at any time from a batch distillation column, kg mol
L_B = total bed length for an adsorber, m
LES = length of equilibrium section in an adsorber, m
LUB = length of unused section in an adsorber, m
m = slope of equilibrium line defined by Eq. (15-40), dimensionless
$\dot{m}$ = mass flow rate, kg/s; subscripts L, V, or S designate mass flow rate for liquid, vapor, or solid, respectively
$\dot{m}'_V$ = volumetric flow rate of vapor, m^3/s
$\dot{m}''_V$ = molar flow rate of vapor, kg mol/s
M = molecular weight, kg/kg mol; subscripts L and V designate molecular weights of liquid and vapor components, respectively
n = number of individual components in a fluid, dimensionless; subscripts F and P refer to number of mols in feed and permeate for a membrane unit, respectively
N = number of theoretical stages or plates, dimensionless
N_{act} = number of actual stages or plates, dimensionless
N_B = number of equilibrium stages below feed stage on a separation column, dimensionless
N_D = number of equilibrium stages above feed stage on a separation column, dimensionless
N_i = molar transmembrane flux of component i subjected to a pressure driving force, $kg\ mol/s{\cdot}m^2$
N_{min} = minimum number of stages or plates, dimensionless
NTU = number of transfer units as defined by Eq. (15-13), dimensionless
p = total pressure, kPa; subscripts F, P, and R designate pressures at feed, permeate side, and retentate side of a membrane, respectively
$\bar{p}$ = partial pressure of a component in a mixture, kPa; subscripts F and P designate partial pressure of a component in membrane pores at feed and permeate surfaces
p_i^s = vapor pressure of solute, kPa
P_{M_i} = permeance of component i, m^3 (STP)$\cdot m/m^2{\cdot}s{\cdot}Pa$ or cm^3 (STP) $cm/cm^2{\cdot}s{\cdot}Pa$
$\bar{P}_{M_i}$ = permeability of component i, $kg\ mol{\cdot}m/m^2{\cdot}s{\cdot}Pa$
$\bar{q}$ = mols of saturated liquid on feed tray per mol of feed, dimensionless
q' = equilibrium loading or amount adsorbed per unit mass of adsorbent, cm^3/g or m^3/kg
q'_{max} = maximum amount adsorbed per unit mass of adsorbate, cm^3/g or m^3/kg
q_F = loading of adsorbent during each cycle, $kg\ mol/m^3$
q_S = heat transferred to solids in a dryer, kJ/s
R = operating reflux ratio, dimensionless
R_F = resistance of filter medium to passage of liquid
R'_F = resistance of filter medium to passage of air
R_K = resistance of filter cake to passage of liquid
R'_K = resistance of filter cake to passage of air
R_{min} = minimum reflux ratio, dimensionless
Re = Reynolds number, dimensionless; subscripts L and V refer to liquid and vapor values, respectively

s = compressibility exponent of filter cake, defined by Eq. (15-89), dimensionless
s' = constant, defined by Eq. (15-106*a*), dimensionless
Sc = Schmidt number, dimensionless; subscripts L and V refer to liquid and vapor values, respectively
S_i = stripping factor for component i as defined in Eq. (15-33), dimensionless
t_M = thickness of membrane, m
t_s = thickness of solid adsorbent bed, m
T_G = dry-bulb temperature of a gas, °C or K
T_{wb} = wet-bulb temperature of a gas, °C or K
$\Delta T_{\text{log mean}}$ = log mean temperature driving force, K
u = internal energy of fluid, kJ/kg
U = overall heat-transfer coefficient based on area of heat transfer, kJ/s·m^2·K
U' = overall heat-transfer coefficient based on volume of a dryer, kJ/s·m^3·K
v = volume, m^3; subscript a refers to the total volume of air drawn through a filter cake over a fixed time period; subscript av designates the total volume of air drawn through a filter cake per filter revolution; subscript d refers to dryer volume; subscript F refers to fictitious volume of filtrate per unit of filtering area required; subscript R refers to volume of filtrate obtained per filter revolution
V = vapor flow rate, kg mol/s
V_L = velocity of liquid phase, m/s
V_n = actual vapor velocity, m/s
V_{nf} = net vapor velocity at flood conditions, m/s
V_S = average flow velocity of solids in a dryer, m/s
V_V = velocity of vapor phase, m/s
w = mass of dry cake solids per unit volume of filtrate, kg/m^3
We = Weber number, dimensionless
x = mol fraction of a component in liquid phase, dimensionless; subscripts B, D, and F refer to mol fraction of a component in liquid phase at bottom, distillate, and feed locations in a column, respectively; subscripts R and P refer to retentate and permeate, respectively; subscripts LK and HK designate the mol fraction of the light and heavy key components in the liquid phase, respectively
x_i^* = mol fraction of component i in solute under equilibrium conditions, dimensionless
X = weight ratio of solute to feed solvent in raffinate phase, or weight ratio of liquid to dry solid in a dryer, dimensionless; subscripts F, R, and S refer to weight ratios of feed, raffinate, and solvent, respectively
y = mol fraction of a component in the vapor phase under equilibrium conditions, dimensionless
Y = weight ratio of solute to extraction solvent in the extract phase, dimensionless; subscript S refers to weight ratio of solvent
Z = height of a column or total height of mass-transfer zone, m

Greek Symbols

α = average relative volatility between two components, dimensionless; or specific cake resistance filtration defined by C/ρ_c, s^2/kg; or membrane separation factor

$\alpha_{i,j}$ = relative volatility between components i and j, dimensionless; or actual membrane separation factor between components i and j [Eq. (15-60)], dimensionless

$\alpha^*_{i,j}$ = separation factor between components i and j as defined by Eq. (15-57), dimensionless

$\alpha_{LK/HK}$ = relative volatility between light and heavy key components, dimensionless

α' = constant defined by Eq. (15-89)

β' = specific air suction cake resistance defined by C'/ρ_c, s^2/kg

β'' = constant defined by Eq. (15-106*a*), dimensionless

β^o = separation factor in liquid/liquid extractors defined by Eq. (15-41), dimensionless

$\gamma^{\infty}_{i,L}$ = activity coefficient of component i in liquid at infinite dilution, dimensionless

ε = extraction factor, dimensionless; or porosity of a membrane as defined by Eq. (15-51*b*), dimensionless

θ = time, s; subscript r designates residence time

Θ = relative volatility constant evaluated from Eq. (15-2), dimensionless; or stage cut in membrane separation defined as n_P/n_F, dimensionless

λ = ratio of slope of equilibrium line to slope of operating line, dimensionless

μ = viscosity of a fluid, Pa·s or cP; subscripts L and V refer to the viscosities of liquid and vapor phases, respectively

ρ = density of fluid or material, kg/m^3; subscript b refers to density of adsorbent bed; subscript c designates cake density, defined as mass of dry cake solids per unit volume of wet filter cake; subscripts L, V, and S refer to densities of liquid, vapor, and solid phases, respectively

σ = surface tension, dyn/cm; subscript c designates critical surface tension of various column packings

τ = tortuosity factor in membranes, dimensionless

Φ_i = driving force of component i across a membrane, Pa or kPa

ψ_a = fraction of filtration surface available for air suction in a rotary vacuum filter, dimensionless

ψ_f = fraction of total surface immersed in the slurry of a rotary vacuum filter, dimensionless

ω = fraction of adsorbent surface covered with adsorbed molecules, dimensionless

PROBLEMS

15-1 In a depropanizer described by King,[†] a six-component feed at 96°C and 2170 kPa has the following composition:

Component	Feed, mol %	*K* value at 96°C
Methane	26.0	15.0
Ethane	9.0	3.8
Propane	25.0	1.55
n-Butane	17.0	0.80
n-Pentane	11.0	0.38
n-Hexane	12.0	0.19

[†] C. J. King, *Separation Processes,* 2d ed., McGraw-Hill, New York, 1980.

The feed is to be separated in a sieve tray distillation column with a recovery of 98.4 percent of the propane in the distillate product and 98.2 percent of the n-butane in the bottom product. The feed quality is 66 percent vapor. The column is equipped with a partial condenser. What are the minimum number of stages and minimum reflux required for the separation? If a reflux ratio of 1.5 is selected, how many theoretical stages are required, and where is the feed location? Obtain answers to these questions with both a hand calculation and the use of an appropriate simulation program.

15-2 What reflux ratio was used in a C_2 splitter to obtain a 99.9 mol percent ethylene product using a 4.63-m-diameter column provided with 94 sieve trays? The partially vaporized (27 mol percent) feed enters the column at −26°C and 2000 kPa with the following hourly flow rate:

Component	kg/h
Methane	2.93
Acetylene	0.24
Ethylene	62,121.3
Ethane	11,974.5
Propylene	2.10
Propane	0.005

Also, determine the appropriate feed location in the column and the temperature profile throughout the column. To obtain this type of information will require the use of an appropriate computer simulation program.

15-3 A valve tray tower has been designed to separate a mixture of 60 mol percent benzene and 40 mol percent toluene into an overhead product containing 96 mol percent benzene and a bottom product containing 25 mol percent benzene. Calculations have shown that 6.1 theoretical stages will be required to obtain the desired separation conducted essentially at atmospheric pressure. The temperature at the top of the column is 82.8°C while at the bottom of the column it is 100.5°C. Assuming the reboiler acts as one theoretical stage, estimate the number of actual trays required. To simplify the calculation, assume that mixtures of benzene and toluene may be considered as ideal. At an average temperature of 91.6°C, the vapor pressure of pure benzene is 1070 mmHg, and the vapor pressure of pure toluene is 429 mmHg.

15-4 A benzene-toluene mixture consists of 40 mols benzene and 60 mols toluene. It is desired to reduce the residual benzene concentration in a batch distillation column down to 10 percent. What is the mol fraction of benzene in the receiver after the batch separation has been completed? Assume that the relative volatility of the benzene-toluene mixture remains constant at 2.90 throughout the separation process, vapor rate is considered to be constant, and column holdup is negligible.

15-5 Adsorption has been selected as the separation process to remove a hydrocarbon from a gas mixture by countercurrent scrubbing with a lean oil. The absorption column, packed with 0.0254-m metal Pall rings, must handle a gas rate of 900 kg/h and a liquid rate of 2700 kg/h. A gas velocity equal to 70 percent of the maximum allowable velocity at the given liquid and gas rates will be used. Densities of the gas and liquid are 1.20 and 881 kg/m^3, respectively. The viscosity of the oil is 20 cP. Under these operating conditions, estimate the required column diameter and the pressure drop through the column in pascals per meter of packed height.

15-6 A random-packed column is to be used for contacting 1365 kg/h of air with 1820 kg/h of water. The tower is packed with 0.0254-m ceramic Raschig rings. Assuming column operation is in the preloading range at 1 atm and 21°C, estimate the optimum diameter of the packed column to provide a minimum annual cost for the fixed charges and blower operating charges. In the analysis, use the following data:

Operating time = 8000 h/yr

Purchased cost of shell = \$1413/m^2 of packed volume

Purchased cost of packing = \$777/m^3

Annual fixed charges = 0.2 (cost of installed unit)

Cost of installed unit (including distributor plates, supports, and required auxiliaries) = 2(purchased cost of the shell and packing)

Cost of electric power = \$0.08/kWh

15-7 Select possible liquid/liquid extraction solvents for separating the following mixtures: (a) water and acetic acid, (b) water and acetone, (c) water and aniline, and (d) water and ethyl alcohol. For each case, indicate which of the two components will be the solute.

15-8 A countercurrent extractor is to be used in extracting a 60 mol percent methyl cyclohexane (MCH) and a 40 mol percent heptane mixture with pure aniline at 25°C. The raffinate product is to contain 1 mol percent MCH, and the extract product 98 mol percent MCH, both on a solvent-free basis. How many ideal stages will be required for this extraction process if a reflux ratio of 7.7 is to be used? For each 100 kg of feed, how much solvent is required and how much is removed in the solvent separator? Extraction data on this system are given by Varteressian and Fenske.†

15-9 For the conditions of Example 15-6, compute exit compositions for a spiral-wound membrane system that approximates crossflow. What membrane surface area will be required for a stage cut of 0.4?

15-10 Activated carbon particles with a mean diameter of 0.0017 m, surface area of 1100 m^2/kg, bulk density of 509 kg/m^3, particle density of 777 kg/m^3, and skeletal density of 2178 kg/m^3 are used in an adsorption bed to remove methane from a 90/10 mol mixture of hydrogen/methane. Operating conditions for the adsorption process are 25°C and 1013 kPa. Adsorption equilibria data for this adsorbent have been correlated with

$$n = \frac{2 \times 10^{-3} K_a p_a}{1 + K_a p_a} \qquad \text{mol/g adsorbent}$$

where n is the quantity adsorbed per unit mass of adsorbent and $K_a = 0.346$ atm^{-1}. Estimate the rate coefficient and evaluate the controlling factor for a superficial velocity of 0.3 m/s.

15-11 A counterflow tunnel dryer is to be used to provide 227 kg/h product with a 1 percent moisture content. The wet feed entering the dryer at 15°C contains 1.5 kg of water/kg of dry product. The dry bulk density is 560 kg/m^3. The specific heat of the dry material is 1.25 kJ/kg·K. Tests show that the critical moisture content of the material is about 0.4 kg of water/kg of dry product. The inlet air to the dryer has a temperature of 149°C, and the dried product leaves the dryer at 143°C. Since fresh air will be combined with recirculated air, the entering air will enter with a moisture content of 0.03 kg of water/kg dry air. The air will leave the dryer at 60°C. The maximum mass velocity of air that can be used with the solids being dried is 2.71 kg/s·m^2. Estimate the number of transfer units that will be required for the constant-rate section and the falling-rate section of the dryer. What is the length of the dryer if the length of a transfer unit, in meters, under these conditions is obtained from the empirical relation

$$\text{Length of transfer unit} = 72.77 t_{ts} G^{0.2}$$

where t_{ts} is the spacing between drying trays (assumed in this case to be 0.0381 m) and G the mass velocity of the air in kg/s·m^2.

†K. A. Varteressian and M. R. Fenske, *Ind. Eng. Chem.*, **29:** 270 (1937).

15-12 Determine the volumetric rate of steam generated in a falling-film evaporator that utilizes a vertical pipe to permit a uniformly distributed film of water to flow down the inside of the pipe. The latter has a length of 3 m and an inside diameter of 0.05 m. The inlet flow is 0.01 kg/s, and the saturation temperature can be assumed to be 100°C. Can the interfacial shear stress between the opposing flows of the water and the steam be neglected?

15-13 A triple-effect evaporator with parallel feed to the three stages is used to provide distilled water. The feedwater is available at 10 atm pressure and a temperature of 180°C. Hot, saturated steam at 20 atm and 215°C is available at 1 kg/s as the heating source. The operating pressures in the three stages should be selected to provide essentially equal temperature-difference values between the condensing steam and the evaporating water. Assume that the steam provided to each stage is condensed and may be added to the distilled water product. To minimize buildup of impurities in each stage, assume that 10 percent of the feedwater is continuously discarded. What is the flow rate of distilled water produced? If the feedwater at 10 atm had been flashed down to the final product pressure in the third stage of the evaporator, what additional amount of distilled water product would have been produced?

15-14 A mixture containing equal mass fractions of hexane, pentane, and butane is to be condensed in a shell-and-tube exchanger at an average pressure of 400 kPa. The total mass flow of the vapor is 200 kg/s. The mixture enters the condenser at 360 K and leaves at 340 K. Condensation takes place on the tube side of the exchanger. Water serves as the coolant entering the shell side at 290 K with a flow rate of 500 kg/s. Hewitt et al.† have estimated that the overall heat-transfer coefficient from the coolant to the condensate film interface can be assumed to be 800 $W/m^2 \cdot K$, and the heat-transfer coefficient for the vapor at the entrance to the tubes is 150 $W/m^2 \cdot K$. The specific heat of the vapor is 2 kJ/kg·K, and that for the coolant is 4.2 kJ/kg·K. Estimate the surface area required by the condenser to provide the required condensation, first by an approximate hand calculation using the Silver-Bell-Ghaly method and then by verifying this result with an appropriate computer simulation program.

15-15 A rotary filter operating at 2 r/min with a total filtering area of 0.75 m^2 has been found to deliver 0.3 m^3 of filtrate per minute with a pressure drop of 140 kPa across the filtrate. The fraction of filtering area submerged in the slurry is 0.2. Because of capacity limitations, another rotary filter is to be designed to handle the same slurry mixture. This unit will produce 3 m^3 of filtrate per minute and will operate with a pressure drop of 100 kPa while revolving at a speed of 1.5 r/min. If the fraction of filtering area submerged in the slurry remains at 0.2, estimate the total filtering area required for the new unit. What is the purchased cost of this new rotary filter? It may be assumed that no solids pass through the filter cloth, the cake is incompressible, and the resistance of the filtering medium is negligible in both rotary filters.

15-16 A slurry containing 1 kg of filterable solids per 10 kg of liquid is being filtered with a plate-and-frame filter press having a total filtering area of 25 m^2. This unit provides 5000 kg of filtrate during the first 2 h of filtration, starting with a clean unit and maintaining a constant pressure drop of 67 kPa. The resistance of the filter medium is negligible. The time required for washing and filtrate removal is 3 h per cycle. The pressure drop in this filter cannot exceed 67 kPa. The unit is always operated with a constant pressure drop.

The filter press is to be replaced by a rotary vacuum-drum filter with negligible filter medium resistance. This rotary filter can deliver the filtrate at a rate of 500 kg/h when the drum speed is 0.3 r/min. Assuming the fraction submerged and the pressure drop are unchanged, what drum speed is necessary to produce the amount of filtrate delivered in 24 h from the rotary filter match to the amount of filtrate obtained per 24 h from the plate-and-frame filter?

---

†G. F. Hewitt, G. L. Shires, and T. R. Bott, *Process Heat Transfer,* CRC Press, Boca Raton, FL, 1994.

APPENDIX

A

The International System (SI) of Units

As the International System (*Système International,* or SI) units become accepted in the United States, there will be a long transition period when both the U.S. Customary System (USCS) and SI will be in use simultaneously. The design engineer, in particular, will need to be able to think and work in both systems because of the wide variety of persons involved in design considerations. Accordingly, this text has used a mixture of the two systems.

The current International System units (SI) is a metric system of measurement that has been adopted internationally by the General Conference of Weights and Measures and is described in an International Standard (ISO 1000)[†] and in numerous other publications.[‡]

[†]International Standard, *SI Units and Recommendations for the Use of Their Multiples and of Certain Other Units,* ISO 1000-1973(E), American National Standards Institute, 1430 Broadway, New York, NY 10018.

[‡]The basic English document for SI is the National Bureau of Standards Special Publication 330, which can be obtained from the Superintendent of Documents, U.S. Government Printing Office, Washington, DC 20402, as document SD Catalog No. C13.10:330/3. This is the authorized English translation of the official document of the international body. For guidance in U.S. usage, the most widely recognized document in use is the ASTM Standard for Metric Practice available from the American Society for Testing and Materials, 1916 Race St., Philadelphia, PA 19103.

SI BASE UNITS ON WHICH THE ENTIRE SYSTEM IS FOUNDED

Name, Symbol	*Definition*
meter,[†] m—length	The meter is the length equal to 1 650 763.73 wavelengths in vacuum of the radiation corresponding to the transition between the levels $2p_{10}$ and $5d_5$ of the krypton-86 atom.
kilogram, kg—mass	The kilogram is a unit of mass (not force). A prototype of the kilogram made of platinum-iridium is kept at the International Bureau of Weights and Measures in Sèvres, France. (The kilogram is the only base unit having a prefix and defined by an artifact.)
second, s—time	The second is the duration of 9 192 631 770 periods of the radiation corresponding to the transition between the two hyperfine levels of the ground state of the cesium-133 atom.
ampere, A—electric current	The ampere is that constant current which, if maintained in two straight parallel conductors of infinite length and of negligible circular cross section, and placed 1 m apart in a vacuum, would produce between these conductors a force equal to 2×10^{-7} m·kg/s^2 (newton) per meter of length.
kelvin, K—temperature	The kelvin is the fraction $1/273.16$ of the thermodynamic temperature of the triple point of water. The kelvin is a unit of thermodynamic temperature T. The word (or symbol) *degree* is not used with kelvin. The Celsius (formerly centigrade) temperature is also used. Celsius temperature (symbol t) is defined by the equation $t = T - T_0$, where $T_0 = 273.15$ K. One degree Celsius (°C) is thus equal to 1 K. The term *centigrade* should not be used because of possible confusion with the French unit of angular measurement, the grade.

[†]The spelling *metre* (and *litre*) is commonly accepted internationally and is recommended by the ASTM. However, the spelling *meter* (and *liter*) is widely used in the United States and is the spelling used in this text.

mole, mol—amount of substance	The mol is the amount of substance of a system which contains as many elementary entities as there are atoms in 0.012 kg of carbon 12. When the mol is used, the elementary entities must be specified and may be atoms, molecules, ions, electrons, other particles, or specified groups of such particles.
candela, cd—luminous intensity	The candela is the luminous intensity, in the perpendicular direction, of a surface of $1/600\,000$ m^2 of a blackbody at the temperature of freezing platinum (2045 K) under a pressure of 101 325 kg·m/m^2·s^2.

SUPPLEMENTARY UNITS

Name, Symbol	*Definition*
radian, rad—plane, angle	The radian is the plane angle between two radii of a circle which cuts off, on the circumference, an arc equal in length to the radius.
steradian, sr—solid angle	The steradian is the solid angle which, having its vertex in the center of a sphere, cuts off an area of the surface of the sphere equal to that of a square with sides of length equal to the radius of the sphere.

DERIVED UNITS

Derived units are algebraic combinations of the seven base units or two supplementary units with some of the combinations being assigned special names and symbols. Examples are shown in Table A-1.

The SI system has a series of approved prefixes and symbols for decimal multiples, as shown in Table A-2.

The rules for conversion of units to SI units and the rules of their usage in the written form have been carefully outlined in a previous edition† and accordingly are not included in Appendix A. Table A-3 provides a detailed listing of factors used in converting U.S. Customary System units to SI units.

Time and space can be saved by the use of abbreviations. Unless the abbreviation is standard, the meaning should be explained the first time it is used. Examples of accepted abbreviations are shown in Table A-4.

†See M. S. Peters and K. D. Timmerhaus, *Plant Design and Economics for Chemical Engineers,* 4th ed., McGraw-Hill, New York, 1991, pp. 785–790.

Table A-1 Common derived units with special names and symbols acceptable in SI

Name	Symbol	Quantity	Expression in terms of SI base units	Expression in terms of other units
becquerel	Bq	Radioactivity	s^{-1}	
coulomb	C	Quantity of electricity of electric charge	$A \cdot s$	
farad	F	Electric capacitance	$m^{-2} \cdot kg^{-1} \cdot s^4 \cdot A^2$	C/V
gray	Gy	Absorbed radiation	$m^2 \cdot s^{-2}$	J/kg
henry	H	Electric inductance	$m^2 \cdot kg \cdot s^{-2} \cdot A^{-2}$	Wb/A
hertz	Hz	Frequency	s^{-1}	
joule	J	Energy, work, or quantity of heat	$m^2 \cdot kg \cdot s^{-2}$	$N \cdot m$
lumen	lm	Luminous flux	$cd \cdot sr$	
lux	lx	Illuminance	$m^{-2} \cdot cd \cdot sr$	lm/m^2
newton	N	Force	$m \cdot kg \cdot s^{-2}$	$J \cdot m^{-1}$
ohm	Ω	Electric resistance	$m^2 \cdot kg \cdot s^{-3} \cdot A^{-2}$	V/A
pascal	Pa	Pressure of stress	$m^{-1} \cdot kg \cdot s^{-2}$	N/m^2
siemens	S	Electric conductance	$m^{-2} \cdot kg^{-1} \cdot s^3 \cdot A^2$	A/V
tesla	T	Magnetic flux, density	$kg \cdot s^{-2} \cdot A^{-1}$	Wb/m^2
volt	V	Electric potential, potential difference, or electromotive force	$m^2 \cdot kg \cdot s^{-3} \cdot A^{-1}$	W/A
watt	W	Power or radiant flux	$m^2 \cdot kg \cdot s^{-3}$	J/s
weber	Wb	Magnetic flux	$m^2 \cdot kg \cdot s^{-2} \cdot A^{-1}$	$V \cdot s$

Table A-2 SI unit prefixes

Multiplication factor	Prefix	Symbol	Meaning (in United States)
10^{18}	exa	E	One quintillion times†
10^{15}	peta	P	One quadrillion times†
10^{12}	tera	T	One trillion times†
10^{9}	giga	G	One billion times†
10^{6}	mega	M	One million times
10^{3}	kilo	k	One thousand times
10^{2}	hecto	h	One hundred times
10	deka	da	Ten times
10^{-1}	deci	d	One-tenth of
10^{-2}	centi	c	One-hundredth of
10^{-3}	milli	m	One-thousandth of
10^{-6}	micro	μ	One-millionth of
10^{-9}	nano	n	One-billionth of†
10^{-12}	pico	p	One-trillionth of†
10^{-15}	femto	f	One-quadrillionth of†
10^{-18}	atto	a	One-quintillionth of†

†These terms should be avoided in technical writing because the denominations above 1 million are different in most other countries.

Table A-3 Common factors for converting U.S. Customary System units to SI units—alphabetical listing†

To convert from	To	Multiply by
abampere	ampere (A)	1.000 000*E+01
abcoulomb	coulomb (C)	1.000 000*E+01
abfarad	farad (F)	1.000 000*E+09
abhenry	henry (H)	1.000 000*E−09
abmho	siemens (S)	1.000 000*E+09
abohm	ohm (Ω)	1.000 000*E−09
abvolt	volt (V)	1.000 000*E−08
acre-foot (U.S. survey)‡	meter^3 (m^3)	1.233 489 E+03
acre (U.S. survey)‡	meter^2 (m^2)	4.046 873 E+03
ampere-hour	coulomb (C)	3.600 000*E+03
are	meter^2 (m^2)	1.000 000*E+02
ångstrom	meter (m)	1.000 000*E−10
astronomical unit	meter (m)	1.495 979 E+11
atmosphere (standard)	pascal (Pa)	1.013 250*E+05
atmosphere (technical = 1 kgf/cm^2)	pascal (Pa)	9.806 650*E+04
bar	pascal (Pa)	1.000 000*E+05
barn	meter^2 (m^2)	1.000 000*E−28
barrel (for petroleum, 42 gal)	meter^3 (m^3)	1.589 873 E−01
board foot	meter^3 (m^3)	2.359 737 E−03
British thermal unit (International Table)§	joule (J)	1.055 056 E+03
British thermal unit (mean)§	joule (J)	1.055 87 E+03
British thermal unit (thermochemical)§	joule (J)	1.054 350 E+03
British thermal unit (39°F)	joule (J)	1.059 67 E+03
British thermal unit (59°F)	joule (J)	1.054 80 E+03
British thermal unit (60°F)	joule (J)	1.054 68 E+03
Btu (International Table)·ft/h·ft^2·°F (*k*, thermal conductivity)	watt per meter-kelvin (W/m·K)	1.730 735 E+00
Btu (thermochemical)·ft/h·ft^2·°F (*k*, thermal conductivity)	watt per meter-kelvin (W/m·K)	1.729 577 E+00
Btu (International Table)·in./h·ft^2·°F (*k*, thermal conductivity)	watt per meter-kelvin (W/m·K)	1.442 279 E−01
Btu (thermochemical)·in./h·ft^2·°F (*k*, thermal conductivity)	watt per meter-kelvin (W/m·K)	1.441 314 E−01
Btu (International Table)·in./s·ft^2·°F (*k*, thermal conductivity)	watt per meter-kelvin (W/m·K)	5.192 204 E+02
Btu (thermochemical)·in./s·ft^2·°F (*k*, thermal conductivity)	watt per meter-kelvin (W/m·K)	5.188 732 E+02
Btu (International Table)/h	watt (W)	2.930 711 E−01
Btu (thermochemical)/h	watt (W)	2.928 751 E−01
Btu (thermochemical)/min	watt (W)	1.757 250 E+01
Btu (thermochemical)/s	watt (W)	1.054 350 E+03
Btu (International Table)/ft^2	joule per meter^2 (J/m^2)	1.135 653 E+04
Btu (thermochemical)/ft^2	joule per meter^2 (J/m^2)	1.134 893 E+04
Btu (International Table)/ft^2·h	watt per meter^2 (W/m^2)	3.154 591 E+00
Btu (thermochemical)/ft^2·h	watt per meter^2 (W/m^2)	3.152 481 E+00
Btu (thermochemical)/ft^2·min	watt per meter^2 (W/m^2)	1.891 489 E+02

(*Continued*)

Table A-3 ***Continued***

To convert from	To	Multiply by
Btu (thermochemical)/ft^2·s	watt per meter^2 (W/m^2)	1.134 893 E+04
Btu (thermochemical)/in.^{2}·s	watt per meter^2 (W/m^2)	1.634 246 E+06
Btu (International Table)/h·ft^2·°F (*C*, thermal conductance)	watt per meter^2-kelvin(W/m^2·K)	5.678 263 E+00
Btu (thermochemical)/h·ft^2·°F (*C*, thermal conductance)	watt per meter^2-kelvin (W/m^2·K)	5.674 466 E+00
Btu (International Table)/s·ft^2·°F	watt per meter^2-kelvin (W/m^2·K)	2.044 175 E+04
Btu (thermochemical)/s·ft^2·°F	watt per meter^2-kelvin (W/m^2·K)	2.042 808 E+04
Btu (International Table)/lb	joule per kilogram (J/kg)	2.326 000*E+03
Btu (thermochemical)/lb	joule per kilogram (J/kg)	2.324 444 E+03
Btu (International Table)/lb·°F (C_p, heat capacity)	joule per kilogram-kelvin (J/kg·K)	4.186 800*E+03
Btu (thermochemical)/lb·°F (C_p, heat capacity)	joule per kilogram-kelvin (J/kg·K)	4.184 000 E+03
bushel (U.S.)	meter^3 (m^3)	3.523 907 E−02
caliber (inch)	meter (m)	2.540 000*E−02
calorie (International Table)	joule (J)	4.186 80 *E+00
calorie (mean)	joule (J)	4.190 02 E+00
calorie (thermochemical)	joule (J)	4.184 000*E+00
calorie (15°C)	joule (J)	4.185 80 E+00
calorie (20°C)	joule (J)	4.181 90 E+00
calorie (kilogram, International Table)§	joule (J)	4.186 800*E+03
calorie (kilogram, mean)§	joule (J)	4.190 02 E+03
calorie (kilogram, thermochemical)§	joule (J)	4.184 000*E+03
cal (thermochemical)/cm^2	joule per meter^2 (J/m^2)	4.184 000*E+04
cal (International Table)/g	joule per kilogram (J/kg)	4.186 800*E+03
cal (thermochemical)/g	joule per kilogram (J/kg)	4.184 000*E+03
cal (International Table)/g·°C	joule per kilogram-kelvin (J/kg·K)	4.186 800*E+03
cal (thermochemical)/g·°C	joule per kilogram-kelvin (J/kg·K)	4.184 000*E+03
cal (thermochemical)/min	watt (W)	6.973 333 E−02
cal (thermochemical)/s	watt (W)	4.184 000*E+00
cal (thermochemical)/cm^2·min	watt per meter^2 (W/m^2)	6.973 333 E+02
cal (thermochemical)/cm^2·s	watt per meter^2 (W/m^2)	4.184 000*E+04
cal (thermochemical)/cm·s·°C	watt per meter-kelvin (W/m·K)	4.184 000*E+02
carat (metric)	kilogram (kg)	2.000 000*E−04
centimeter of mercury (0°C)	pascal (Pa)	1.333 22 E+03
centimeter of water (4°C)	pascal (Pa)	9.806 38 E+01
centipoise	pascal-second (Pa·s)	1.000 000*E−03
centistokes	meter^2 per second (m^2/s)	1.000 000*E−06
circular mil	meter^2 (m^2)	5.067 075 E−10
clo	kelvin-meter^2 per watt (K·m^2/W)	2.003 712 E−01
cup	meter^3 (m^3)	2.365 882 E−04
curie	becquerel (Bq)	3.700 000*E+10
day (mean solar)	second (s)	8.640 000 E+04
day (sidereal)	second (s)	8.616 409 E+04
degree (angle)	radian (rad)	1.745 329 E−02
degree Celsius	kelvin (K)	$T_K = t_{°C} + 273.15$
degree Fahrenheit	degree Celsius	$t_{°C} = (t_{°F} - 32)/1.8$
degree Fahrenheit	kelvin (K)	$T_K = (t_{°F} + 459.67)/1.8$
degree Rankine	kelvin (K)	$T_K = t_{°R}/1.8$

Table A-3 Continued

To convert from	To	Multiply by
°F·h·ft^2/Btu (International Table) (*R*, thermal resistance)	kelvin-meter^2 per watt (K·m^2/W)	1.761 102 E−01
°F·h·ft^2/Btu (thermochemical) (*R*, thermal resistance)	kelvin-meter^2 per watt (K·m^2/W)	1.762 280 E−01
denier	kilogram per meter (kg/m)	1.111 111 E−07
dyne	newton (N)	1.000 000*E−05
dyne·cm	newton-meter (N·m)	1.000 000*E−07
dyne/cm^2	pascal (Pa)	1.000 000*E−01
electronvolt	joule (J)	1.602 19 E−19
EMU of capacitance	farad (F)	1.000 000*E+09
EMU of current	ampere (A)	1.000 000*E+01
EMU of electric potential	volt (V)	1.000 000*E−08
EMU of inductance	henry (H)	1.000 000*E−09
EMU of resistance	ohm (Ω)	1.000 000*E−09
ESU of capacitance	farad (F)	1.112 650 E−12
ESU of current	ampere (A)	3.335 6 E−10
ESU of electric potential	volt (V)	2.997 9 E+02
ESU of inductance	henry (H)	8.987 554 E+11
ESU of resistance	ohm (Ω)	8.987 554 E+11
erg	joule (J)	1.000 000*E−07
erg/cm^2·s	watt per meter^2 (W/m^2)	1.000 000*E−03
erg/s	watt (W)	1.000 000*E−07
faraday (based on carbon-12)	coulomb (C)	9.648 70 E+04
faraday (chemical)	coulomb (C)	9.649 57 E+04
faraday (physical)	coulomb (C)	9.652 19 E+04
fathom	meter (m)	1.828 8 E+00
fermi (femtometer)	meter (m)	1.000 000*E−15
fluid ounce (U.S.)	meter^3 (m^3)	2.957 353 E−05
foot	meter (m)	3.048 000*E−01
foot (U.S. survey)‡	meter (m)	3.048 006 E−01
foot of water (39.2°F)	pascal (Pa)	2.988 98 E+03
ft^2	meter^2 (m^2)	9.290 304*E−02
ft^2/h (thermal diffusivity)	meter^2 per second (m^2/s)	2.580 640*E−05
ft^2/s	meter^2 per second (m^2/s)	9.290 304*E−02
ft^3 (volume; section modulus)	meter^3 (m^3)	2.831 685 E−02
ft^3/min	meter^3 per second (m^3/s)	4.719 474 E−04
ft^3/s	meter^3 per second (m^3/s)	2.831 685 E−02
ft^4 (moment of section)	meter^4 (m^4)	8.630 975 E−03
ft/h	meter per second (m/s)	8.466 667 E−05
ft/min	meter per second (m/s)	5.080 000*E−03
ft/s	meter per second (m/s)	3.048 000*E−01
ft/s^2	meter per second^2 (m/s^2)	3.048 000*E−01
footcandle	lux (lx)	1.076 391 E+01
footlambert	candela per meter^2 (cd/m^2)	3.426 259 E+00
ft·lbf	joule (J)	1.355 818 E+00
ft·lbf/h	watt (W)	3.766 161 E−04
ft·lbf/min	watt (W)	2.259 697 E−02
ft·lbf/s	watt (W)	1.355 818 E+00

(*Continued*)

Table A-3 ***Continued***

To convert from	To	Multiply by
ft·poundal	joule (J)	4.214 011 E−02
free fall, standard (g)	meter per second^2 (m/s^2)	9.806 650*E+00
gal	meter per second^2 (m/s^2)	1.000 000*E−02
gallon (Canadian liquid)	meter^3 (m^3)	4.546 090 E−03
gallon (U.K. liquid)	meter^3 (m^3)	4.546 092 E−03
gallon (U.S. dry)	meter^3 (m^3)	4.404 884 E−03
gallon (U.S. liquid)	meter^3 (m^3)	3.785 412 E−03
gal (U.S. liquid)/day	meter^3 per second (m^3/s)	4.381 264 E−08
gal (U.S. liquid)/min	meter^3 per second (m^3/s)	6.309 020 E−05
gal (U.S. liquid)/hp·h (SFC, specific fuel consumption)	kilogram per joule (kg/J)	1.410 089 E−09
gamma	tesla (T)	1.000 000*E−09
gauss	tesla (T)	1.000 000*E−04
gilbert	ampere	7.957 747 E−01
gill (U.K.)	meter^3 (m^3)	1.420 654 E−04
gill (U.S.)	meter^3 (m^3)	1.182 941 E−04
grad	degree (angular)	9.000 000*E−01
grad	radian (rad)	1.570 796 E−02
grain (1/7000 lb avoirdupois)	kilogram (kg)	6.479 891*E−05
grain (lb avoirdupois/7000)/gal (U.S. liquid)	kilogram per meter^3 (kg/m^3)	1.711 806 E−02
gram	kilogram (kg)	1.000 000*E−03
g/cm^3	kilogram per meter^3 (kg/m^3)	1.000 000*E+03
gram-force/cm^2	pascal (Pa)	9.806 650*E+01
hectare	meter^2 (m^2)	1.000 000*E+04
horsepower (550 ft·lbf/s)	watt (W)	7.456 999 E+02
horsepower (boiler)	watt (W)	9.809 50 E+03
horsepower (electric)	watt (W)	7.460 000*E+02
horsepower (metric)	watt (W)	7.354 99 E+02
horsepower (water)	watt (W)	7.460 43 E+02
horsepower (U.K.)	watt (W)	7.457 0 E+02
hour (mean solar)	second (s)	3.600 000 E+03
hour (sidereal)	second (s)	3.590 170 E+03
hundredweight (long)	kilogram (kg)	5.080 235 E+01
hundredweight (short)	kilogram (kg)	4.535 924 E+01
inch	meter (m)	2.540 000*E−02
inch of mercury (32°F)	pascal (Pa)	3.386 38 E+03
inch of mercury (60°F)	pascal (Pa)	3.376 85 E+03
inch of water (39.2°F)	pascal (Pa)	2.490 82 E+02
inch of water (60°F)	pascal (Pa)	2.488 4 E+02
in.^{2}	meter^2 (m^2)	6.451 600*E−04
in.^{3} (volume; section modulus)	meter^3 (m^3)	1.638 706 E−05
in.^{3}/min	meter^3 per second (m^3/s)	2.731 177 E−07
in.^{4} (moment of section)	meter^4 (m^4)	4.162 314 E−07
in./s	meter per second (m/s)	2.540 000*E−02
in./s^2	meter per second^2 (m/s^2)	2.540 000*E−02
kayser	1 per meter (1/m)	1.000 000*E+02
kelvin	degree Celsius	$t_{°C} = T_K - 273.15$
kilocalorie (International Table)	joule (J)	4.186 800*E+03
kilocalorie (mean)	joule (J)	4.190 02 E+03

Table A-3 *Continued*

To convert from	To	Multiply by
kilocalorie (thermochemical)	joule (J)	4.184 000*E+03
kilocalorie (thermochemical)/min	watt (W)	6.973 333 E+01
kilocalorie (thermochemical)/s	watt (W)	4.184 000*E+03
kilogram-force (kgf)	newton (N)	9.806 650*E+00
kgf·m	newton-meter (N·m)	9.806 650*E+00
kgf·s^2/m (mass)	kilogram (kg)	9.806 650*E+00
kgf/cm^2	pascal (Pa)	9.806 650*E+04
kgf/m^2	pascal (Pa)	9.806 650*E+00
kgf/mm^2	pascal (Pa)	9.806 650*E+06
km/h	meter per second (m/s)	2.777 778 E−01
kilopond	newton (N)	9.806 650*E+00
kWh	joule (J)	3.600 000*E+06
kip (1000 lbf)	newton (N)	4.448 222 E+03
kip/in.^{2} (ksi)	pascal (Pa)	6.894 757 E+06
knot (international)	meter per second (m/s)	5.144 444 E−01
lambert	candela per meter^2 (cd/m^2)	$1/\pi$ *E+04
lambert	candela per meter^2 (cd/m^2)	3.183 099 E+03
langley	joule per meter^2 (J/m^2)	4.184 000*E+04
league	meter (m)	[see footnote‡]
light-year	meter (m)	9.460 55 E+15
liter	meter^3 (m^3)	1.000 000*E−03
maxwell	weber (Wb)	1.000 000*E−08
mho	siemens (S)	1.000 000*E+00
microinch	meter (m)	2.540 000*E−08
micrometer (micron)	meter (m)	1.000 000*E−06
mil	meter (m)	2.540 000*E−05
mile (international)	meter (m)	1.609 344*E+03
mile (statute)	meter (m)	1.609 3 E+03
mile (U.S. survey)‡	meter (m)	1.609 347 E+03
mile (international nautical)	meter (m)	1.852 000*E+03
mile (U.K. nautical)	meter (m)	1.853 184*E+03
mile (U.S. nautical)	meter (m)	1.852 000*E+03
mi^2 (international)	meter^2 (m^2)	2.589 988 E+06
mi^2 (U.S. survey)‡	meter^2 (m^2)	2.589 998 E+06
mi/h (international)	meter per second (m/s)	4.470 400*E−01
mi/h (international)	kilometer per hour (km/h)	1.609 344*E+00
mi/min (international)	meter per second (m/s)	2.682 240*E+01
mi/s (international)	meter per second (m/s)	1.609 344*E+03
millibar	pascal (Pa)	1.000 000*E+02
millimeter of mercury (0°C)	pascal (Pa)	1.333 22 E+02
minute (angle)	radian (rad)	2.908 882 E−04
minute (mean solar)	second (s)	6.000 000 E+01
minute (sidereal)	second (s)	5.983 617 E+01
month (mean calendar)	second (s)	2.628 000 E+06
oersted	ampere per meter (A/m)	7.957 747 E+01
ohm-centimeter	ohm-meter (Ω·m)	1.000 000*E−02
ohm-circular mil per foot	ohm-millimeter^2 per meter (Ω·mm^2/m)	1.662 426 E−03
ounce (avoirdupois)	kilogram (kg)	2.834 952 E−02

(Continued)

Table A-3 ***Continued***

To convert from	To	Multiply by
ounce (troy or apothecary)	kilogram (kg)	3.110 348 E−02
ounce (U.K. fluid)	meter^3 (m^3)	2.841 307 E−05
ounce (U.S. fluid)	meter^3 (m^3)	2.957 353 E−05
ounce-force (ozf)	newton (N)	2.780 139 E−01
ozf·in.	newton-meter (N·m)	7.061 552 E−03
oz (avoirdupois)/gal (U.K. liquid)	kilogram per meter^3 (kg/m^3)	6.236 021 E+00
oz (avoirdupois)/gal (U.S. liquid)	kilogram per meter^3 (kg/m^3)	7.489 152 E+00
oz (avoirdupois)/in.^{3}	kilogram per meter^3 (kg/m^3)	1.729 994 E+03
oz (avoirdupois)/ft^2	kilogram per meter^2 (kg/m^2)	3.051 517 E−01
oz (avoirdupois)/yd^2	kilogram per meter^2 (kg/m^2)	3.390 575 E−02
parsec	meter (m)	3.085 678 E+16
peck (U.S.)	meter^3 (m^3)	8.809 768 E−03
pennyweight	kilogram (kg)	1.555 174 E−03
perm (0°C)	kilogram per pascal-second-meter^2 (kg/Pa·s·m^2)	5.721 35 E−11
perm (23°C)	kilogram per pascal-second-meter^2 (kg/Pa·s·m^2)	5.745 25 E−11
perm·in. (0°C)	kilogram per pascal-second-meter (kg/Pa·s·m)	1.453 22 E−12
perm·in. (23°C)	kilogram per pascal-second-meter (kg/Pa·s·m)	1.459 29 E−12
phot	lumen per meter^2 (lm/m^2)	1.000 000*E+04
pica (printer's)	meter (m)	4.217 518 E−03
pint (U.S. dry)	meter^3 (m^3)	5.506 105 E−04
pint (U.S. liquid)	meter^3 (m^3)	4.731 765 E−04
point (printer's)	meter (m)	3.514 598*E−04
poise (absolute viscosity)	pascal-second (Pa·s)	1.000 000*E−01
pound (lb avoirdupois)	kilogram (kg)	4.535 924 E−01
pound (troy or apothecary)	kilogram (kg)	3.732 417 E−01
lb·ft^2 (moment of inertia)	kilogram-meter^2 (kg·m^2)	4.214 011 E−02
lb·in.^{2} (moment of inertia)	kilogram-meter^2 (kg·m^2)	2.926 397 E−04
lb/ft·h	pascal-second (Pa·s)	4.133 789 E−04
lb/ft·s	pascal-second (Pa·s)	1.488 164 E+00
lb/ft^2	kilogram per meter^2 (kg/m^2)	4.882 428 E+00
lb/ft^3	kilogram per meter^3 (kg/m^3)	1.601 846 E+01
lb/gal (U.K. liquid)	kilogram per meter^3 (kg/m^3)	9.977 633 E+01
lb/gal (U.S. liquid)	kilogram per meter^3 (kg/m^3)	1.198 264 E+02
lb/h	kilogram per second (kg/s)	1.259 979 E−04
lb/hp·h (SFC, specific fuel consumption)	kilogram per joule (kg/J)	1.689 659 E−07
lb/in.^{3}	kilogram per meter^3 (kg/m^3)	2.767 990 E+04
lb/min	kilogram per second (kg/s)	7.559 873 E−03
lb/s	kilgoram per second (kg/s)	4.535 924 E−01
lb/yd^3	kilogram per meter^3 (kg/m^3)	5.932 764 E−01
poundal	newton (N)	1.382 550 E−01
poundal/ft^2	pascal (Pa)	1.488 164 E+00
poundal·s/ft^2	pascal-second (Pa·s)	1.488 164 E+00
pound-force (lbf)	newton (N)	4.448 222 E+00
lbf·ft	newton-meter (N·m)	1.355 818 E+00
lbf·ft/in.	newton-meter per meter (N·m/m)	5.337 866 E+01

Table A-3 ***Continued***

To convert from	To	Multiply by
lbf·in.	newton-meter (N·m)	1.129 848 E−01
lbf·in./in.	newton-meter per meter (N·m/m)	4.448 222 E+00
lbf·s/ft^2	pascal-second (Pa·s)	4.788 026 E+01
lbf/ft	newton per meter (N/m)	1.459 390 E+01
lbf/ft^2	pascal (Pa)	4.788 026 E+01
lbf/in.	newton per meter (N/m)	1.751 268 E+02
lbf/in.^{2} (psi)	pascal (Pa)	6.894 757 E+03
lbf/lb [thrust/weight (mass) ratio]	newton per kilogram (N/kg)	9.806 650 E+00
quart (U.S. dry)	meter^3 (m^3)	1.101 221 E−03
quart (U.S. liquid)	meter^3 (m^3)	9.463 529 E−04
rad (radiation dose absorbed)	gray (Gy)	1.000 000*E−02
rhe	1 per pascal-second (1/Pa·s)	1.000 000*E+01
rod	meter (m)	[see footnote‡]
roentgen	coulomb per kilogram (C/kg)	2.58 E−04
second (angle)	radian (rad)	4.848 137 E−06
second (sidereal)	second (s)	9.972 696 E−01
section	meter^2 (m^2)	[see footnote‡]
shake	second (s)	1.000 000*E−08
slug	kilogram (kg)	1.459 390 E+01
slug/ft·s	pascal-second (Pa·s)	4.788 026 E+01
slug/ft^3	kilogram per meter^3 (kg/m^3)	5.153 788 E+01
statampere	ampere (A)	3.335 640 E−10
statcoulomb	coulomb (C)	3.335 640 E−10
statfarad	farad (F)	1.112 650 E−12
stathenry	henry (H)	8.987 554 E+11
statmho	siemens (S)	1.112 650 E−12
statohm	ohm (Ω)	8.987 554 E+11
statvolt	volt (V)	2.997 925 E+02
stere	meter^3 (m^3)	1.000 000*E+00
stilb	candela per meter^2 (cd/m^2)	1.000 000*E+04
stokes (kinematic viscosity)	meter^2 per second (m^2/s)	1.000 000*E−04
tablespoon	meter^3 (m^3)	1.478 676 E−05
teaspoon	meter^3 (m^3)	4.928 922 E−06
tex	kilogram per meter (kg/m)	1.000 000*E−06
therm	joule (J)	1.055 056 E+08
ton (assay)	kilogram (kg)	2.916 667 E−02
ton (long, 2240 lb)	kilogram (kg)	1.016 047 E+03
ton (metric)	kilogram (kg)	1.000 000*E+03
ton (nuclear equivalent of TNT)	joule (J)	4.184 E+09
ton (refrigeration)	watt (W)	3.516 800 E+03
ton (register)	meter^3 (m^3)	2.831 685 E+00
ton (short, 2000 lb)	kilogram (kg)	9.071 847 E+02
ton (long)/yd^3	kilogram per meter^3 (kg/m^3)	1.328 939 E+03
ton (short)/h	kilogram per second (kg/s)	2.519 958 E−01
ton-force (2000 lbf)	newton (N)	8.896 444 E+03
tonne (metric ton)	kilogram (kg)	1.000 000*E+03
torr (mmHg, 0°C)	pascal (Pa)	1.333 22 E+02
township	meter^2 (m^2)	[see footnote‡]
unit pole	weber (Wb)	1.256 637 E−07

(*Continued*)

Table A-3 ***Continued***

To convert from	To	Multiply by
W·h	joule (J)	3.600 000*E+03
W·s	joule (J)	1.000 000*E+00
W/cm^2	watt per meter^2 (W/m^2)	1.000 000*E+04
W/in.^{2}	watt per meter^2 (W/m^2)	1.550 003 E+03
yard	meter (m)	9.144 000*E−01
yd^2	meter^2 (m^2)	8.361 274 E−01
yd^3	meter^3 (m^3)	7.645 549 E−01
yd^3/min	meter^3 per second (m^3/s)	1.274 258 E−02
year (calendar)	second (s)	3.153 600 E+07
year (sidereal)	second (s)	3.155 815 E+07
year (tropical)	second (s)	3.155 693 E+07

[†]Adapted from ASTM Standard for Metric Practice E 380-76. The conversion factors are listed in standard form for computer readout as a number greater than 1 or less than 10 with six or fewer decimal points. The number is followed by the letter E (for exponent), a plus or minus symbol, and two digits which indicate the power of 10 by which the number must be multiplied. An asterisk (*) after the sixth decimal place indicates that the conversion factor is exact and that all subsequent digits are zero. All other conversion factors have been rounded to the figures given. Where fewer than six decimal places are shown, greater precision is not warranted.

For example, 1.013 250*E+05 is exactly 1.013 250 $\times$ 10^5, or 101 325.0. Also 1.589 873 E−01 has the last digit rounded to 3 and is 1.589 873 $\times$ 10^{-1} or 0.158 987 3.

[‡]Since 1893, the U.S. basis of length measurement has been derived from metric standards. In 1959, a small refinement was made in the definition of the yard to resolve discrepancies both in this country and abroad which changed its length from 3600/3937 m to 0.9144 m exactly. This resulted in the new value being shorter by 2 parts in 1 million. At the same time, it was decided that any data in feet derived from and published as a result of geodetic surveys within the United States would remain with the old standard (1 ft = 1200/3937 m) until further decision. This foot is named the *U.S. survey foot.* As a result, all U.S. land measurements in U.S. Customary System units will relate to the meter by the old standard. All the conversion factors in this table for units referenced to this footnote are based on the U.S. survey foot rather than on the international foot.

Conversion factors for the land measures given below may be determined from the following relationships:

$$1 \text{ league} = 3 \text{ miles (exactly)}$$

$$1 \text{ rod} = 16\tfrac{1}{2} \text{ feet (exactly)}$$

$$1 \text{ section} = 1 \text{ square mile (exactly)}$$

$$1 \text{ township} = 36 \text{ square miles (exactly)}$$

[§]By definition, 1 calorie (cal) (International Table) is exactly 4.186 8 absolute joules which converts to 1.055 056 $\times$ 10^{-3} J for 1 Btu (International Table). Also, by definition, 1 cal (thermochemical) is exactly 4.184 absolute joules which converts to 1.054 350 $\times$ 10^{-3} J for 1 Btu (thermochemical). A *mean* calorie is one-hundredth of the heat required to raise the temperature of 1 g of water at 1 atm pressure from 0 to 100°C and equals 4.190 02 absolute joules. In all cases, the relationship between calorie and British thermal unit is established by 1 cal/(g·°C) = 1 Btu/(lb·°F). A *mean* Btu, therefore, is $\frac{1}{180}$th of the heat required to raise the temperature of 1 lb of water at 1 atm pressure from 32 to 212°F and equals 1.055 87 $\times$ 10^3 J. When values are given as Btu or calories, the type of unit (International Table, thermochemical, mean, or temperature of determination) should be given. In all cases for this table, conversions involving joules are based on the absolute joule.

*Exact equivalence.

Table A-4 Accepted abbreviations

Term	Abbreviation
American Chemical Society	ACS
American Institute of Chemical Engineers	AIChE
American Iron and Steel Institute	AISI
American Petroleum Institute	API
American Society of Mechanical Engineers	ASME
American Society for Testing and Materials	ASTM
American wire gauge	AWG
Atmosphere	atm
Average	avg
Barrel	bbl
Baumé	Bé
Biochemical oxygen demand	BOD
Boiling point	bp
Bottoms	btms
British thermal unit	Btu
Calorie	cal
Capacity	cap.
Catalytic	cat.
Centipoise	cP
Centistoke	cS
Chemically pure	CP
Concentrate	conc
Critical	crit
Cubic	cu
Cubic centimeter	cc
Cubic foot	cu ft or ft^3
Cubic foot per minute	cfm or ft^3/min
Cubic foot per second	cfs or ft^3/s
Cubic inch	cu in. or $in.^3$
Degree	deg or °
Diameter	diam
Dilute	dil
Distill or distillate	dist
Efficiency	eff
Electromotive force	emf
Equivalent	equiv
Evaporate	evap
Experiment	expt
Experimental	exptl
Extract	ext
Feet per minute	fpm or ft/min
Figure	fig.
Foot	ft
Foot-pound	ft·lb
Gallon	gal
Gallons per minute	gpm or gal/min
Height equivalent to a theoretical plate	HETP
Height of a transfer unit	HTU
Horsepower	hp
Hour	h
Hundredweight (100 lb)	cwt
Inch	in.
Inside diameter	ID or i.d.
Kilogram	kg
Kilometer	km
Kilovolt	kV
Kilowatt	kW
Liquid	liq
Logarithm (base 10)	log
Logarithm (base *e*)	ln
Maximum	max
Melting point	mp
Meter	m
Micrometer	μm
Mile	mi
Miles per hour	mph
Minute	min
Molecular	mol
Outside diameter	OD or o.d.
Overhead	ovhd
Page	p.
Pages	pp.
Parts per million	ppm
Pounds per cubic foot	lb/cu ft or lb/ft^3
Pounds per square foot	lb/sq ft or lb/ft^2
Pounds per square inch	psi or lb/$in.^2$
Pounds per square inch absolute	psia
Pounds per square inch gauge	psig
Refractive index	RI or *n*
Revolutions per minute	rpm or r/min
Second	s
Society of Automotive Engineers	SAE
Soluble	sol
Solution	soln
Specific gravity	sp gr
Specific heat	sp ht
Square	sq
Square foot	sq ft or ft^2
Standard	std
Standard temperature and pressure	STP
Tank	tk
Technical	tech
Temperature	temp
Thousand	M
Tubular Exchangers Manufacturers Association	TEMA
Volume	vol
Watt	W
Watthour	Wh
Weight	wt

APPENDIX

B

Auxiliary, Utility, and Instrumentation Cost Data[†]

CONTENTS

AUXILIARY COST DATA

INSTRUMENTATION

UTILITIES

[†]Costs reported are January 1, 2002, values except as noted.

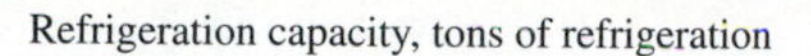

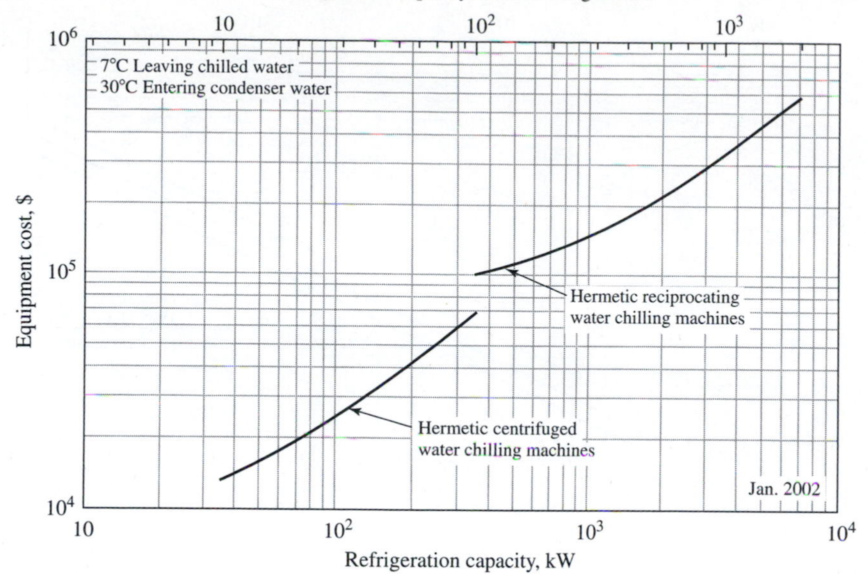

Figure B-1
Cost of air conditioning. Price includes compressor, motor, starter, controls, cooler, condenser, and refrigerant. It does not include cooling tower, pumps, foundations, ductwork, and installation costs.

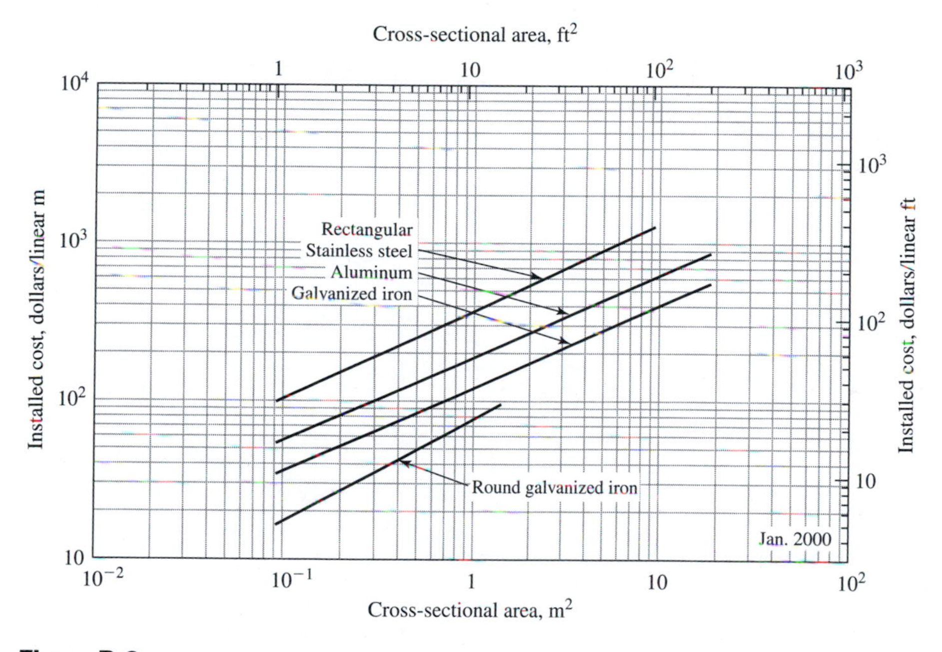

Figure B-2
Purchased cost of ductwork. Price is for shop-fabricated ductwork with hangers and supports installed.

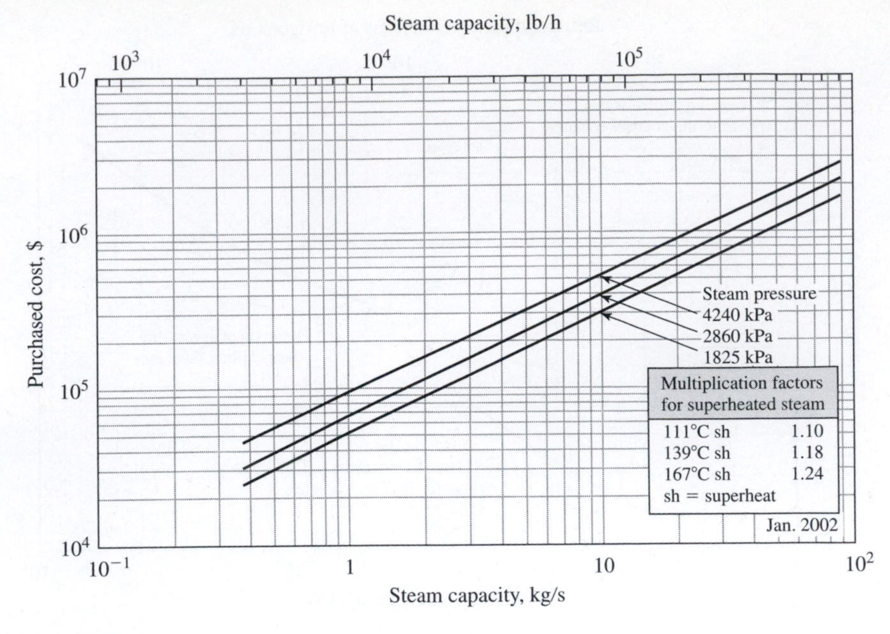

Figure B-3
Purchased cost of package boiler plants. Price includes complete boiler, feedwater deaerator, boiler feed pumps, chemical injection system, and shop assembly labor.

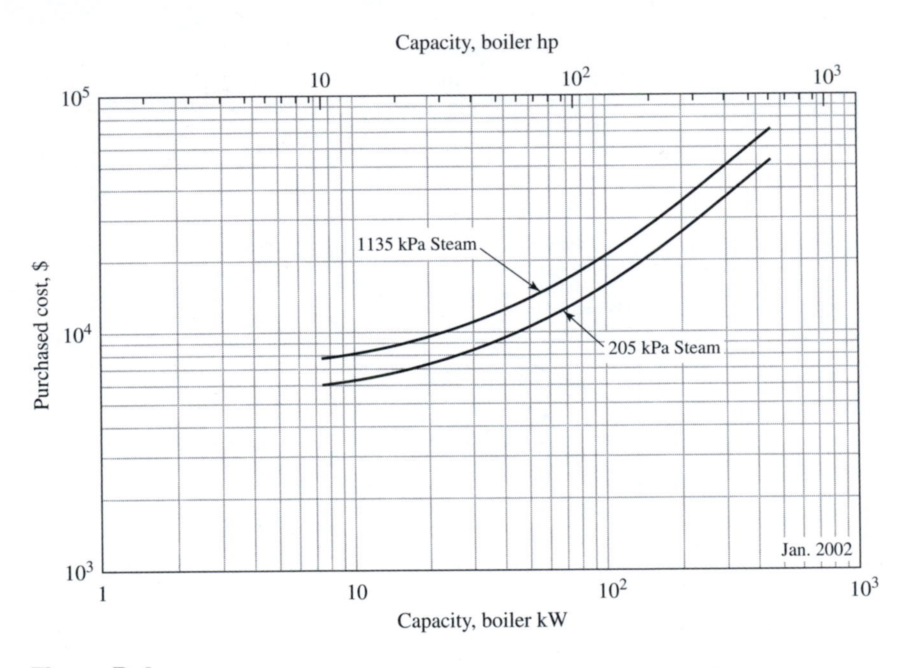

Figure B-4
Purchased cost of steam generators. Price is for packaged unit with steel tubes, gas-fired. Multiply by 1.02 for oil-fired boiler.

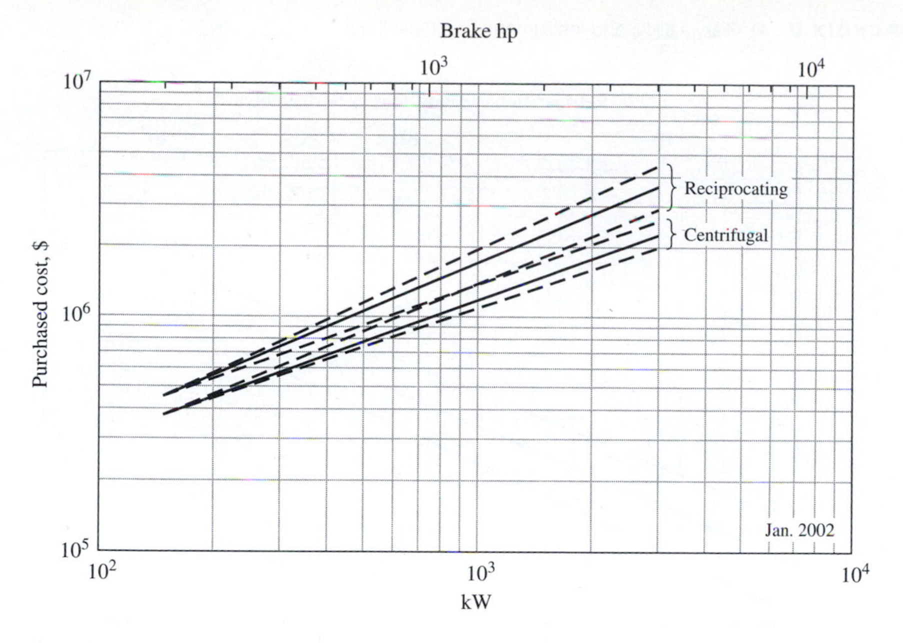

Figure B-5
Purchased cost of compressor plants. (Dotted lines indicate range of costs of the two types of plants.)

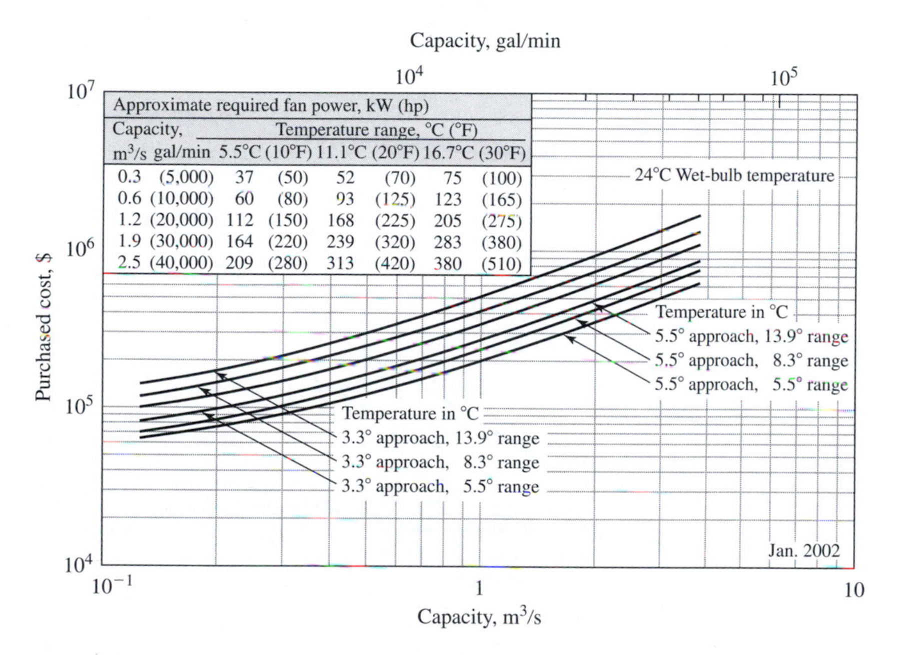

Figure B-6
Purchased cost of cooling towers. Prices are for conventional, wood-frame, induced-draft, cross-flow cooling towers. Price does not include external piping, power wiring, special foundation work, or field labor.

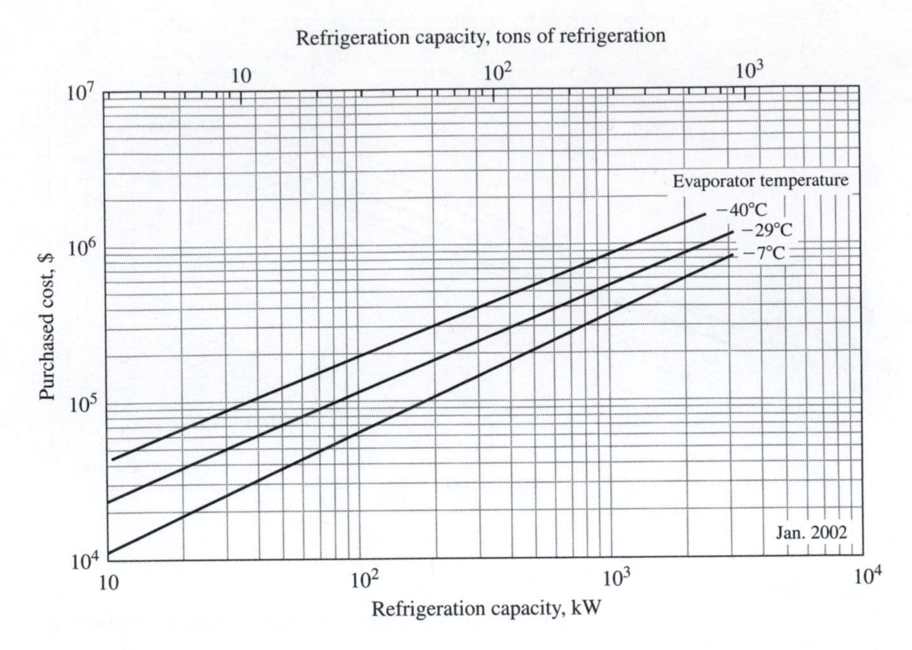

Figure B-7
Purchased cost of industrial refrigeration

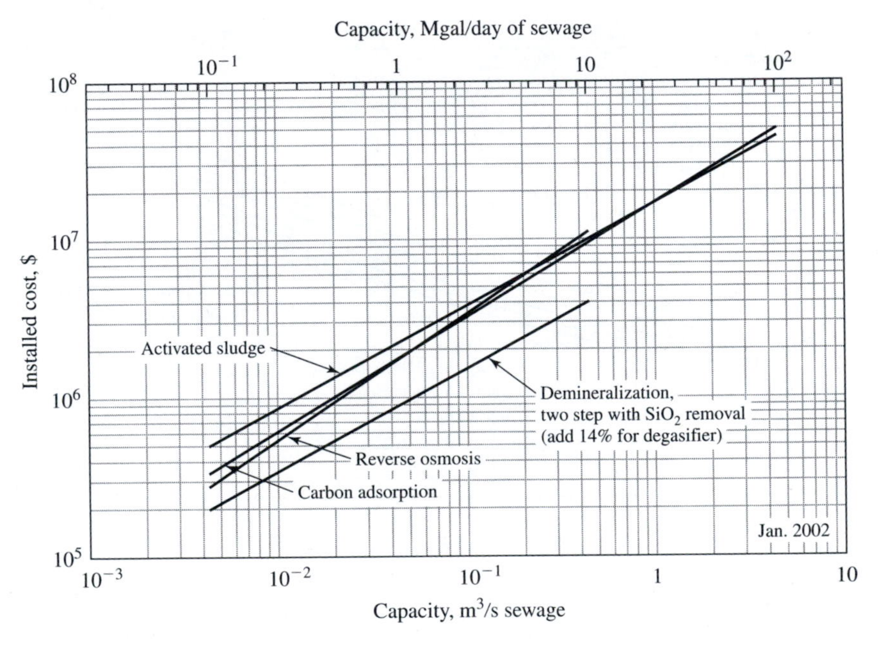

Figure B-8
Installed cost of wastewater-treatment plants

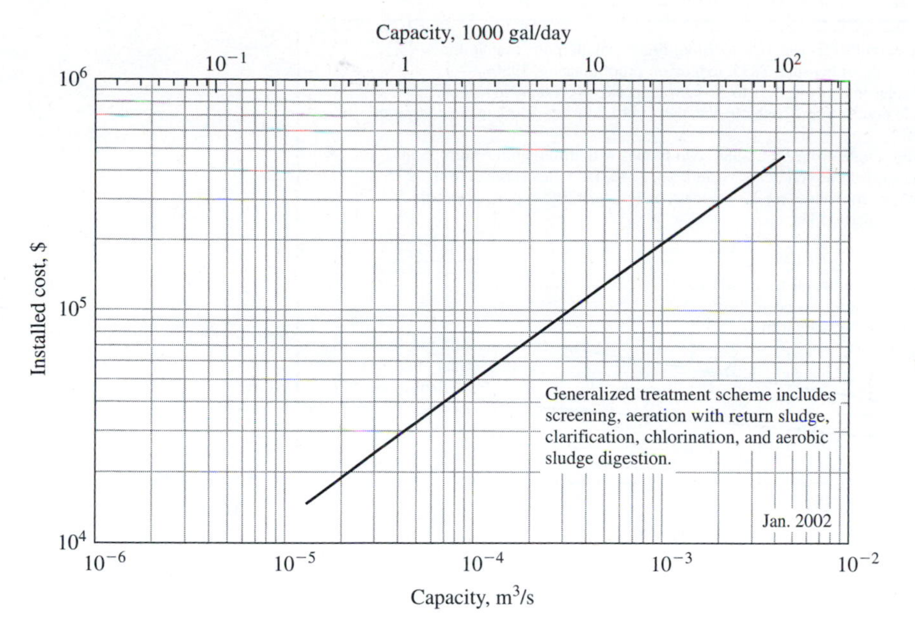

Figure B-9
Installed cost of small packaged wastewater treatment plants

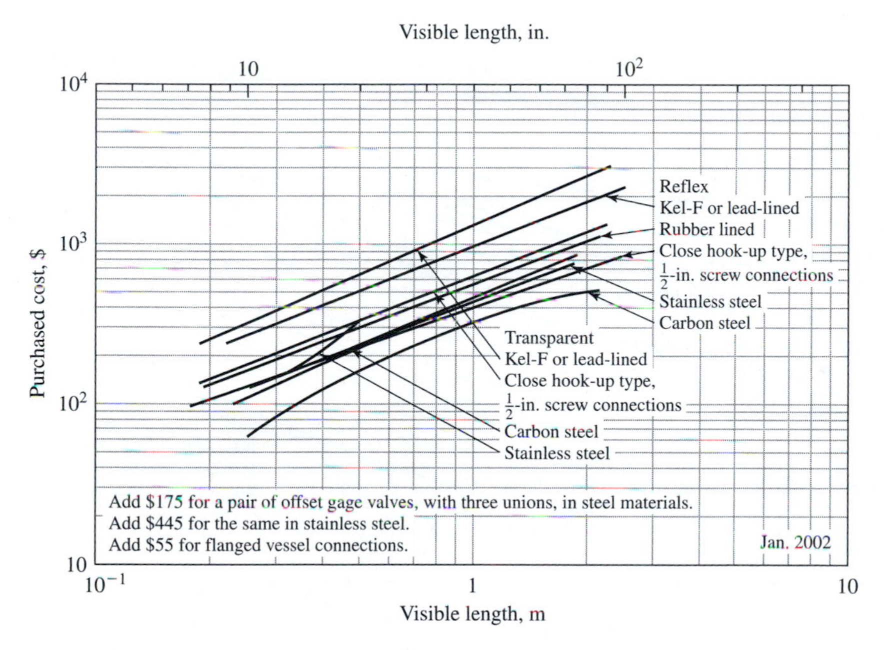

Figure B-10
Purchased cost of liquid-level gages, flat-glass type

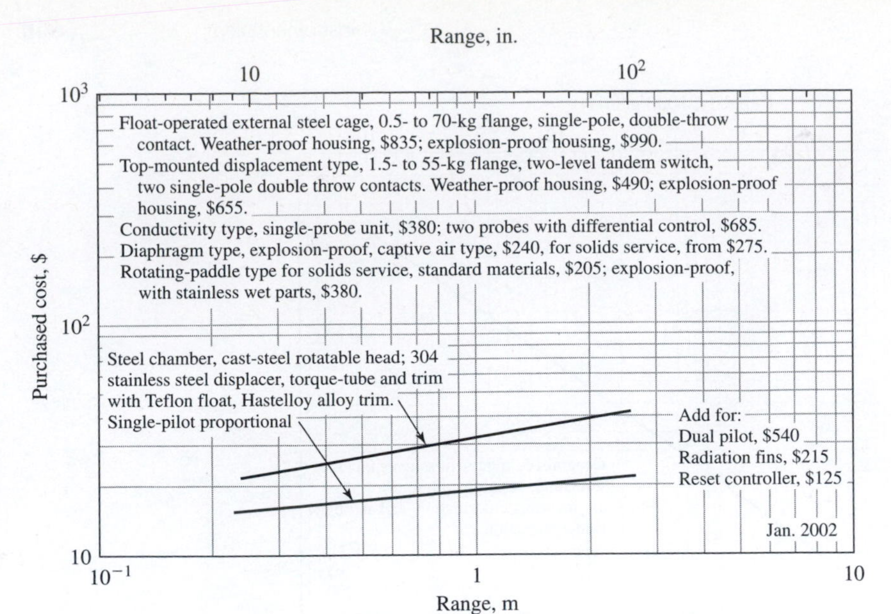

Figure B-11
Purchased cost of level controllers

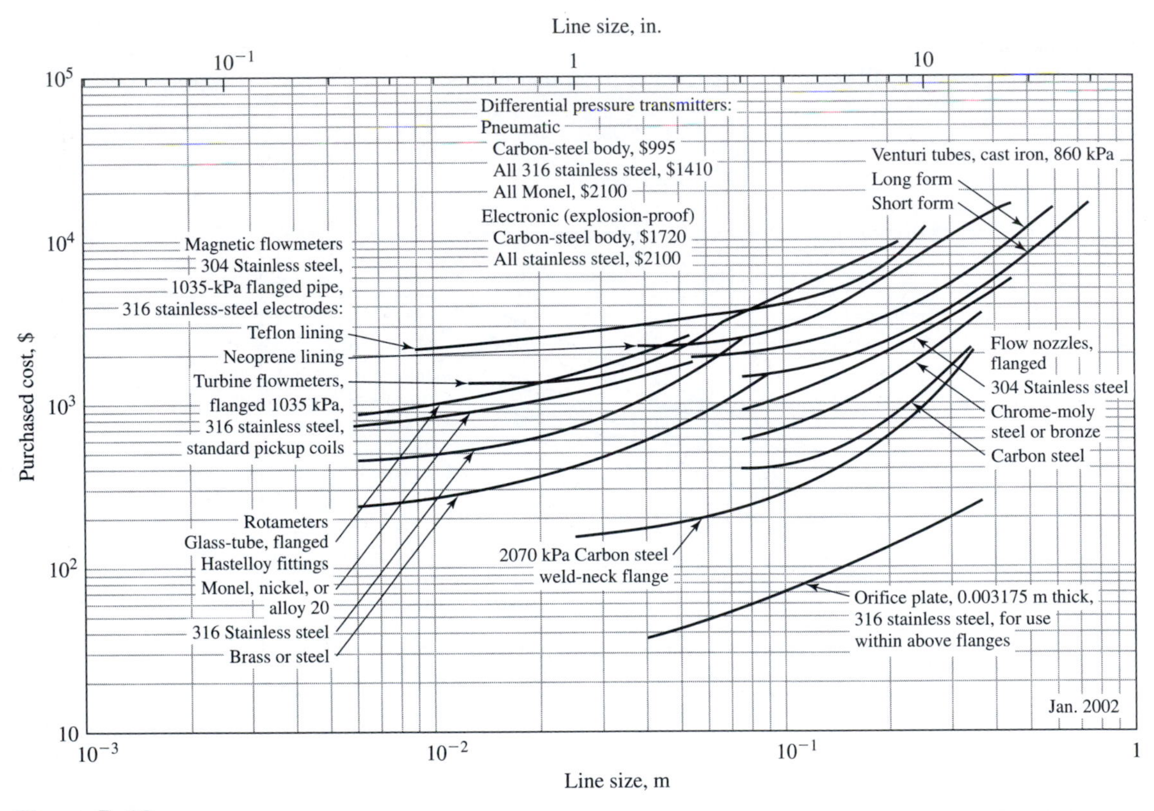

Figure B-12
Purchased cost of flow indicators

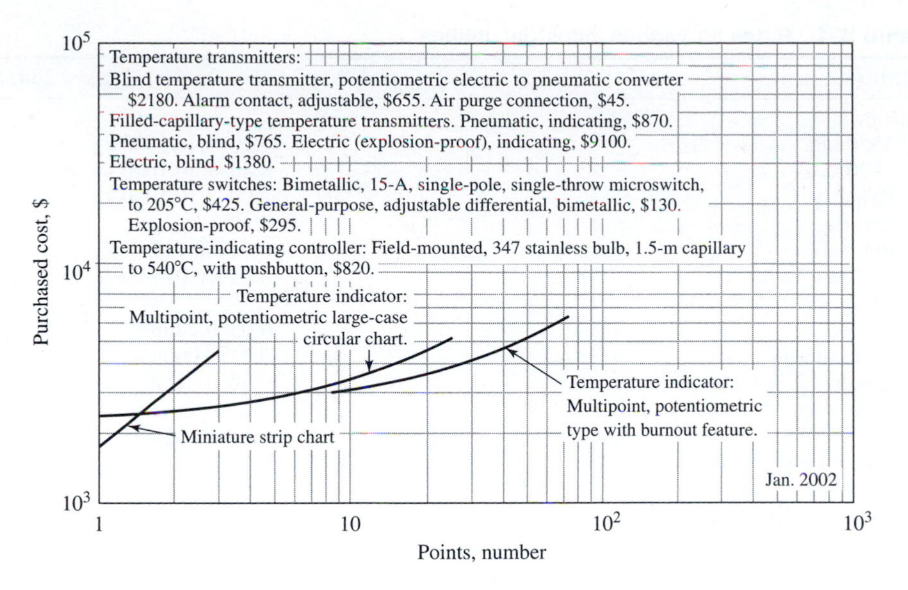

Figure B-13
Purchased cost of temperature recorders and indicators

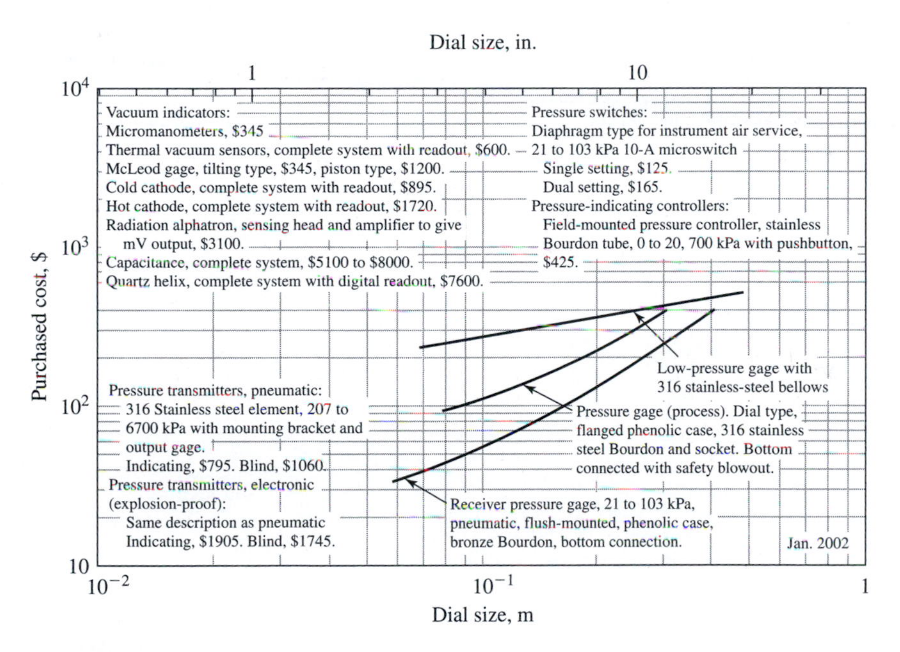

Figure B-14
Purchased cost of pressure indicators

Table B-1 Rates for various industrial utilities†

Utility	Cost range (January 2002), $
Steam:	
3550 kPa	7.70–9.40/1000 kg
790 kPa	4.40–7.50/1000 kg
Exhaust	2.00–3.50/1000 kg
Electricity:	
Purchased	0.040–0.13/kWh†
Self-generated	0.030–0.075/kWh
Cooling water:	
Well	0.05–0.22/m^3
River or salt	0.02–0.06/m^3
Tower	0.02–0.07/m^3
Process water:	
City	0.12–0.46/m^3
Filtered and softened	0.15–0.46/m^3
Distilled	0.60–1.10/m^3
Compressed air:	
Process air	0.25–0.70/100 m^3(SC)‡
Filtered and dried for instruments	0.45–1.30/100 m^3 (SC)‡
Natural gas:	
Interstate (major pipeline companies)	10.50–15.80/100 m^3 (SC)‡
Intrastate (new contracts)	10.50–16.00/100 m^3 (SC)‡
Intrastate (renegotiated or amended contracts)	11.50–19.00/100 m^3 (SC)‡
Manufactured gas	7.90–19.40/100 m^3 (SC)‡
Fuel oil	113.00–163.00/m^3 (18.00–26.00/bbl)
Coal	39.40–64.00/ton (metric)
Refrigeration (ammonia), to 1°C	0.64/kW refrigeration

†Highly dependent upon location.
‡For these cases, standard conditions (SC) are designated as a pressure of 101.3 kPa and a temperature of 15°C.

APPENDIX

C

Design Problems

CONTENTS

MAJOR PROBLEMS

MINOR PROBLEMS

PRACTICE SESSION PROBLEMS

MAJOR PROBLEMS†

Problem 1. Preliminary Design Details for an Ethylene Production Facility

The development of a conventional base-case design for the production of 500,000 t (metric)/yr of ethylene using propane as the feedstock is described in Chap. 3. The process flowsheets for this process are presented in Figs. 3-8 through 3-13. The flowsheets in sequence schematically show the details associated with the thermal cracking process, quenching, compression and acid gas removal, drying, deethanization and acetylene hydrogenation, cracked gas chilling and demethanization, MAPD hydrogenation, and product separation. Flowsheets for the propylene refrigeration, ethylene refrigeration, and steam process systems required in the preliminary design were not

†Time period of 30 days is recommended for individual student solution.

Table C-1

Utility	T_{in}, °C	T_{out}, °C	ΔT_{min}, °C	h, W/m^2·K
Cooling water	29	40	4.0	2215
MP steam (1650 kPa)	203	203	20.0	3180
LP steam (450 kPa)	148	148	10.0	3035
C_2H_4 refrig.	−101	−101	2.0	2525
C_2H_4 refrig.	−73	−73	2.0	2240
C_3H_6 refrig.	−46	−46	2.5	1590
C_3H_6 refrig.	−18	−18	2.5	1390
C_3H_6 refrig.	10	10	2.5	1135

included. For reasons of clarity, vessels, drums, and pumps were omitted from the six flowsheets presented.

Table 3-6 provides the design details for the thermal furnace while Table 3-7 presents the specifications and purchased cost of the base-case separation section shown in Fig. 3-13. The specifications and purchased cost data for the other process sections shown in Figs. 3-8 through 3-12 and in the process flowsheets for the propylene refrigeration, ethylene refrigeration, and the steam process systems were not included for brevity reasons.

To provide these design details requires the following information:

a. Develop flowsheets for the process sections that were not included.

b. Establish the location, duty, and area of all the heat exchangers, reboilers, and condensers required in the nine process flowsheets noted above.

c. Determine the capacity or size of those pieces of equipment in the nine process flowsheets that were not specified in the preliminary design.

Besides the assumptions provided in Table 3-3, thermal information on the utilities is given in Table C-1.

Problem 2. Formation of Styrene from Butadiene

In the early 1990s Dow developed a process to dimerize the butadiene in a crude C_4 stream (obtained from the ethylene process) to vinylcyclohexene using a proprietary copper-impregnated zeolite catalyst. The vinylcyclohexene is then oxidatively dehydrogenated to styrene by using a proprietary tin/antimony oxide catalyst. This approach utilizes the butadiene in the C_4 stream without resorting to an expensive extraction process to recover the butadiene.[†]

Develop a preliminary design to produce 4.5×10^8 kg/yr of styrene using the new Dow technology and determine the overall economics. The market price of the C_4 fraction listed in Table 3-10 was \$0.425 per kilogram (January 1, 2000). A typical crude C_4 has the composition in weight percent as shown in Table C-2.

Thus, butadiene has a value of 0.3(\$0.425), or approximately \$0.13 per kilogram. Assume that the C_4 stream after the butadiene removal may be returned to the ethylene

[†]U.S. Patent 5,329,057, July 12, 1994; *Chemtech,* 20 (May 1995).

Table C-2

Component	Weight percent
Butadiene	30
Isobutene	30
1-Butene	20
Cis-2-butene	7
Trans-2-butene	7
n-Butane	4
Isobutane	2

production facility at no cost and that the styrene can be sold for $0.66 per kilogram. The plant design that is to be developed should be safe to operate and comply with regulations. Every effort should be made to recover and recycle process materials in the most economic manner possible, and energy consumption should be kept to a minimum with the use of heat integration.

Problem 3. Production of Hydrogen for Use in Reformulated Gasoline

The required reduction of benzene, olefin, aromatics, and sulfur in gasoline and the reduction of aromatics and sulfur in diesel fuel mandated by the Clean Air Act Amendments will increase the hydrogen consumption in many refineries. Hydrogen is currently produced by either steam reforming of methane or partial oxidation of methane with high-purity oxygen and steam. A new autocatalytic reactor using air, methane, and steam has been proposed by Becker.†

In the proposed process, methane, steam, and air are each preheated to 600 to 700°C and fed into the catalytic reactor containing a bed of refractory nickel catalyst. Initial combustion occurs at a temperature of 1000°C as compared with 1400°C for the Shell-Texaco process. The combusted gas mixture passes through heat exchange and a heat recovery boiler before entering a multistage CO shift converter. The gas then passes through a CO_2 wash tower followed by a cryogenic separation unit, where the hydrogen is separated from the nitrogen, argon, and methane.

Prepare a preliminary cost estimate for a plant to produce 1.4 standard m^3/day of 99.0 percent hydrogen compressed to 3000 kPa. (Standard conditions are given as 15°C and 101.3 kPa.) The following design data should be used in this evaluation:

Kinetic Data Rate of methane reacting in kg mol/kg catalyst per hour is given by

$$-r = 1.96 \times 10^7 e^{-44,200/RT} \frac{\bar{p}_{CH_4}}{(1 + 4\bar{p}_{H_2})^2}$$

where $\bar{p}_{CH_4}$ is the partial pressure of methane and $\bar{p}_{H_2}$ the partial pressure of hydrogen, both in units of bar. The reaction on the catalyst is limited by the cracking reaction of

†Acknowledgment is given to E. R. Becker, Environex, for providing the design data for this problem.

methane, and the product gases containing CH_4, CO, CO_2, H_2, and H_2O exist in near equilibrium conditions.

Catalyst The refractory nickel catalyst is a spherical pellet, 0.005 m in diameter. The catalyst bed has a void fraction of 0.48 and a bulk density of 1200 kg/m^3. The catalyst is replaced annually.

Cost Data

Methane (100%), 3000 kPa	\$7.00/100 std m^3
Steam, 3000 kPa saturated	\$13.20/1000 kg
Power	\$0.07/kWh
Cooling water at 32°C	\$3.96/100 m^3
Catalyst cost	\$10.00/kg

Economic Guidelines

Project life of 10 yr

Annual effective interest rate of 9%/yr

Minimum acceptable rate of return of 15%/yr after income tax

Income tax rate 35%/yr of gross profit

Problem 4. Pentaerythritol Production

You are a design engineer with ABC Chemical Company, located at Wilmington, Delaware. Charles B. King, supervisor of the Process Development Division, has asked you to prepare a preliminary design of a new plant that will produce pentaerythritol.

The following memo has been given to you by Mr. King:

Dear Sir:

We are considering construction of a pentaerythritol plant at Louisiana, Missouri. At present, we have an anhydrous ammonia plant located at Louisiana, Missouri, and we own sufficient land at this site to permit a large expansion.

As yet, we have not decided on the final plant capacity, but we are now considering construction of a small plant with a capacity of 60 tons of technical-grade pentaerythritol per month. Any work you do on the design of this plant should be based on this capacity.

Sufficient water, power, and steam facilities for the new plant are now available at the proposed plant site. Some standby equipment is located at our existing plant in Louisiana, and we may be able to use some of this in the pentaerythritol plant. However, you are to assume that all equipment must be purchased new.

Please prepare a preliminary design for the proposed plant. This design will be surveyed by the Process Development Division and used as a basis for further decisions on the proposed plant. Make your report as complete as possible, including a detailed description of your recommended process, specifications and cost estimates for the different pieces of equipment, total capital investment, and estimated return on the investment, assuming we can sell all our product at the prevailing market price. We shall also be interested in receiving an outline of the type and amount of labor required, the operating procedure, and analytical procedures necessary. Present what you consider to be the best design. Although our legal department will conduct a patent survey to determine if any infringements are involved, it would be helpful for you to indicate if you know that any part of your design might involve patent infringements. Enclosed you will find information on our utilities situation, amortization policy, labor standards, and other data.

C. B. King, Supervisor
Process Development Division

Enclosure: Information for use in design of proposed pentaerythritol plant

Utilities: Water available for pumping at 65°F; cost, \$0.14 per 1000 gal. Steam available at 100 psig: cost, \$1.50 per 1000 lb. Electricity cost, \$0.07 per kWh.

Storage: No extra storage space is now available.

Transportation facilities: A railroad spur is now available at the plant site.

Labor: Chief operators, \$25.00 per hour; all helpers and other labor, \$20.00 per hour.

Depreciation: Straight line over 10 years.

Return on investment: For design calculations, we require at least a 20 percent return before income taxes on any unnecessary investment.

Income taxes: The income tax load for our company amounts to 35 percent of all profits.

Raw materials costs: All raw materials must be purchased at prevailing market prices.

Heats of reaction: With sodium hydroxide as the catalyst, 64,000 Btu is released per pound-mol of acetaldehyde reacted. With calcium hydroxide as the catalyst, 60,000 Btu is released per pound-mol of acetaldehyde reacted.

Heat-transfer coefficients: Steam to reactor liquid, 200 $Btu/h \cdot ft^2 \cdot °F$. Water to reactor liquid, 150 $Btu/h \cdot ft^2 \cdot °F$.

Problem 5. Formaldehyde Production Facility

Illinois Chemical Process Corporation
Plant Development Division
Urbana, Illinois

Gentlemen:

We are considering the production of formaldehyde, and we would like you to submit a complete preliminary design for a 70 ton/day formaldehyde plant (based on 37 weight percent formaldehyde) to us.

We have adequate land available for the construction of the plant at Centralia, Illinois, and sufficient water and steam are available for a plant of the desired capacity. Please make your report as complete as possible, including a detailed description of your recommended process, specifications and cost estimates for the different pieces of equipment, total capital investment, and estimated return on the investment, assuming we can sell all our production at the prevailing market price. We shall also be interested in receiving an outline of the amount and type of labor required, the operating procedure, and analytical procedures necessary.

Enclosed you will find information concerning our utilities situation, amortization policy, labor standards, product specification, etc.

Very truly yours,
A. B. Blank
Technical Superintendent
Centralia Chemical Company
Centralia, Illinois

ABB:jf
Enclosure

Utilities: Water is available for pumping at 70°F; cost, \$0.14 per 1000 gal. Steam is available at 200 psig; cost, \$1.50 per 1000 lb. Electricity cost is \$0.07 per kWh.

Storage: No extra storage space is now available.

Transportation facilities: A railroad spur is now available at the plant site.

Land: The plant site is on land we own which is now of no use to us. Therefore, do not include the cost of land in your analysis.

Labor: Chief operators, \$25.00 per hour; all helpers and other labor, \$20.00 per hour.

Depreciation: Straight line over 10 years.

Return on investment: For design comparison, we require an after tax return of at least 15 percent per year on any unnecessary investment.

Product specification: All formaldehyde will be sold as formalin containing 37.2 weight percent formaldehyde and 8 weight percent methanol. All sales may be considered to be by tank car.

MINOR PROBLEMS†

Problem 1. Optimum Temperature for Sulfur Dioxide Reactor

The head of your design group has asked you to prepare a report dealing with the use of a special catalyst for oxidizing SO_2 to SO_3. This catalyst has shown excellent activity at low temperatures, and it is possible that it may permit you to get good SO_3 yields by using only one standard-size reactor instead of the conventional two.

Your report will be circulated among the other members of your design group and will be discussed at the group meetings. This report is to be submitted to the head of your design group.

Some general remarks concerning your report follow:

1. The report should include the following:
   a. Letter of transmittal (the letter to the head of your group indicating that you are submitting the report and giving any essential results if applicable)
   b. Title page
   c. Table of contents
   d. Summary (a concise presentation of the results)
   e. Body of report
      (1) Introduction (a brief discussion to explain what the report is about and the reason for the report; no results should be included here)
      (2) Discussion (outline of method of attack on the problem; do not include any detailed calculations; this should bring out technical matters not important enough to be included in the summary; indicate assumptions and reasons; include any literature survey results of importance; indicate possible sources of error, etc.)

†Time period of 14 days is recommended for individual student solution.

(3) Final recommended conditions (or design if applicable) with appropriate data (a drawing is not necessary in this case, although one could be included if desired)

f. Appendix

(1) Sample calculations (clearly presented and explained)
(2) Table of nomenclature (if necessary)
(3) Bibliography (if necessary)
(4) Data employed

2. The outline as presented above can be changed if desired (e.g., a section on conclusions and recommendations might be included).

3. The report can be made more effective by appropriate subheadings under the major divisions.

The Problem A new reactor has recently been purchased as part of a contact sulfuric acid unit. This reactor is used for oxidizing SO_2 to SO_3, employing a vanadium oxide catalyst.

Using the following information and data, determine the temperature at which the reactor should be operated to give the maximum conversion of SO_2 to SO_3, and indicate the value (as percent) of this maximum conversion. Ten thousand pounds of SO_2 enters the reactor per day.

Air is used for the SO_2 oxidation, and it has been decided to use 300 percent excess air. Preheaters will permit the air and SO_2 to be heated to any desired temperature, and cooling coils in the reactor will maintain a constant temperature in the reactor. The reactor temperature and the entering air-and-SO_2 temperature will be the same. The operating pressure may be assumed to be 1 atm.

The inside dimensions of the reactor are 5 by 5 by 8 ft. One-half of the inside reactor volume is occupied by the catalyst.

Your laboratory has tested a special type of vanadium oxide catalyst, and on the laboratory's recommendation, you have decided to use it in the reactor. This catalyst has a void fraction of 60 percent (i.e., free space in catalyst/total, volume of catalyst = 0.6).

The reaction $2SO_2 + O_2 \rightarrow 2SO_3$ proceeds at a negligible rate except in the presence of a catalyst.

Your laboratory has run careful tests on the catalyst. The results indicate that the reaction is not third-order but is a complex function of the concentrations. Your laboratory reports that the reaction rate is proportional to the SO_2 concentration, inversely proportional to the square root of the SO_3 concentration, and independent of the oxygen concentration.[†]

This may be expressed as

$$\frac{dx}{d\theta} = k\frac{a - x}{x^{1/2}}$$

[†]Modern tests indicate that this may not be the case with some vanadium oxide reactors. However, the information given above should be used for the solution of this problem.

Table C-3

$k \times 10^4$, (lb mol/ft^3)$^{1/2}$/s†	Temperature, °C
14	350
30	400
60	450
100	500
210	550

†Applicable only to the conditions of this problem.

Table C-4

ΔF^0, cal/g mol	Temperature, °C
−9120	350
−8000	400
−7000	450
−5900	500
−4900	550

where

a = SO_2 originally present as lb mol/ft^3

x = lb mol of SO_2 converted in θ s of catalyst contact time per cubic foot of initial gas

k = specific reaction rate constant, (lb mol/ft^3)$^{1/2}$/s; this may be assumed to be constant at each temperature up to equilibrium conditions

The laboratory has obtained the data given in Table C-3 for the reaction rate constant. These data are applicable to your catalyst and your type of reactor.

The data of standard-state free-energy values at different temperatures given in Table C-4 were obtained from the literature. These data apply to the reaction

$$SO_2 + \frac{1}{2}O_2 \longrightarrow SO_3$$

The fugacities of the gases involved may be assumed to be equal to the partial pressures.

Problem 2. Heat Exchanger Design

To: Assistant Process Engineer
From: Dr. A. B. Green, Chief Design Engineer
Mountain View Chemical Company
Boulder, Colorado

We are in the process of designing a catalytic cracker for our petroleum division. As part of this work, will you please submit a design for a single-pass heat exchanger based on the information given below?

It is estimated that 200,000 gal/h of oil A must be heated from 100 to 230°F. The heating agent will be saturated steam at 50 psig.

The engineering department has indicated that the exchanger will cost $60.00 per square foot of inside heating area. This cost includes all installation.

You can neglect any resistance due to the tube walls or steam film; thus all the heat flow resistance will be in the oil film.

The cost of power is $0.07 per kilowatthour, and the efficiency of the pump and motor installation is estimated to be 60 percent.

Do not consider the cost of steam or exchanger insulation in your analysis.

The oil will flow inside the tubes in the heat exchanger. Following are data on oil A which have been obtained from *Perry's Chemical Engineers' Handbook*:

Average viscosity of oil A between 100 and 230°F = 6 cS

Average density of oil A between 100 and 230°F = 0.85 g/cm^3

Average specific heat of oil A between 100 and 230°F = 0.48 Btu/lb·°F

Average thermal conductivity of oil A between 100 and 230°F = 0.08 Btu/h·ft^2·°F/ft

We recommend that the Reynolds number be kept above 5000 in this type of exchanger. The tubes in the exchanger will be constructed from standard steel tubing. Tube sizes are available in $\frac{1}{2}$-in.-diameter steps. Tubing wall is 16 BWG.

We are particularly interested in the diameter of the tubes we should use, the length of the tubes, and the total cost of the installed unit. In case the exchanger length is unreasonable for one unit, what would you recommend?

Assume the unit will operate continuously for 300 days/yr. Fixed charges are 16 percent.

Remember that our company demands a minimum 20 percent return after income tax on all extra fixed-capital investments.

You may base your calculations on a total of 100 tubes in the exchanger.

Please submit this information as a complete formal report. Include a short section outlining what further calculations would have been necessary if the specification of 100 tubes in a one-pass exchanger had not been given.

Following are recommended assumptions:

You may assume that the Fanning friction factor can be represented by

$$f = \frac{0.04}{Re^{0.16}}$$

You may assume that the oil-film heat-transfer coefficient is constant over the entire length of the exchanger.

For simplification, assume that the oil-film heat-transfer coefficient may be represented by (standard heat-transfer nomenclature)

$$h = 0.023\frac{k}{D}Re^{0.8}Pr^{0.4}$$

where all variable values are at the average value between 100 and 230°F.

Problem 3. Design of Sulfur Dioxide Absorber

You are a member of a group of design engineers designing equipment for the recovery of SO_2 from stack gases.

The group leader has asked you to determine the optimum size of the SO_2 absorption tower. Specifically, you are asked to determine the height and cross-sectional area of the optimum absorption tower and to present your recommendations in the form of a formal report.

Your group has held several meetings to discuss the proposed overall design. Following is a list of conditions, assumptions, and data on which the group has decided to base the design:

100,000 ft^3/min of gas at 300°F and 1 atm are to be treated.

The entering gas contains 0.3 percent by volume SO_2 and 11.0 percent CO_2, with the balance being N_2, O_2, and H_2O.

The average molecular weight of the entering gas is 29.4.

The mol percent of SO_2 in the exit gas is to be 0.01 percent.

The entering and exit pressures of the absorption column may be assumed to be 1 atm for purposes of calculating the SO_2 pressures.

The zinc oxide process will be used for recovering the SO_2. In this process, a solution of H_2O, $NaHSO_3$, and Na_2SO_3 is circulated through the absorption tower to absorb the SO_2. This mixture is then treated with ZnO, and the $ZnSO_3$ formed is filtered off, dried, and calcined to yield practically pure SO_2. The ZnO from the calciner is reused, and the sulfite-bisulfite liquor from the filter is recycled.

The absorption tower will contain nonstaggered grids of the following dimensions:

Clearance = 1.5 in.

Height = 4 in.

Thickness = $\frac{1}{4}$ in.

Free cross-sectional area = 85.8%

Active absorption area per cubic foot of volume $a = 13.7$ ft^2/ft^3

The average density of the gas at the tower entrance can be assumed to be 0.054 lb/ft^3. The sulfite-bisulfite liquid has a density of 70 lb/ft^3 and can be considered as having a zero equilibrium SO_2 vapor pressure at both the inlet and outlet of the tower.

The sulfite-bisulfite liquid must be supplied at a rate of 675 lb/h·ft^2 of column cross-sectional area.

The optimum design can be assumed to be that corresponding to a minimum total power cost for circulating the liquid and forcing the gas through the tower. You may assume that this optimum corresponds to the optimum that would be obtained if fixed charges were also considered.

The following simplified equations are applicable for grids of the dimensions to be used:

$$K_g = 0.00222\,G_0^{0.8}$$

$$\frac{\Delta h_w}{L} = 0.23 \times 10^{-7} G_0^{1.8}$$

where

K_g = molar absorption coefficient, lb mol of component absorbed/ $h \cdot ft^2 \cdot atm_{\text{log mean}}$

G_0 = superficial mass velocity of gas in tower, $lb/h \cdot ft^2$

Δh_w = pressure drop through tower, in. of water

L = height of tower, ft

The liquid is put into the absorption tower by means of a nozzle at the top of the tower. The pressure just before the nozzle is 35 psig. Assume the pump for the liquid must supply power to lift the liquid to the top of the tower and compress the liquid to 35 psig. Use a 10 percent safety factor on the above pumping power requirements to take care of the friction in the lines and other minor losses.

The gas blower has an overall efficiency of 55 percent.

The pump has an overall efficiency of 65 percent.

Problem 4. Utilization of Liquid Methane Refrigeration for Liquefaction of Nitrogen and Oxygen

Management of a natural gas transmission company is considering using liquid methane as a heat sink in producing 210 tons/day of liquid nitrogen and 64 tons/day of liquid oxygen for a neighboring customer. Accordingly, one of the company's engineers has outlined a scheme for doing this. The process description is as follows.

Air is compressed from atmospheric conditions to 20 atm and then purified. The clean dry gas is then chilled by countercurrent heat exchange with liquid methane boiloff. The partially liquefied airstream serves as the reboiler stream for a dual-pressure air separation column. Before entering the 5-atm high-pressure lower section of the dual-pressure column, the high-pressure airstream is bled through a J-T valve. The bottoms product of the lower column is enriched to approximately 40 mol percent oxygen and is the feed for the 1-atm low-pressure upper column. The condensing vapors in the high-pressure column serve as a reboiler for the low-pressure column. High-purity N_2 is withdrawn from the top of the high-pressure column, reduced in pressure, and used as reflux for the low-pressure column. High-purity liquid oxygen is withdrawn as bottoms from the low-pressure column while high-purity nitrogen vapor is withdrawn from the top of this same column. The latter stream is warmed to room temperature, compressed to 20 atm, and then cooled in this same heat exchanger down to a low temperature. This high-pressure pure nitrogen stream is then condensed by countercurrent exchange with the liquid methane. After liquefaction, it is expanded to atmospheric pressure, and the resulting vapor is recycled and combined with the overhead stream from the low-pressure column to initiate the liquefaction cycle all over again.

As a chemical engineer, you have been asked to analyze the process and make recommendations. Start by making as complete a flowsheet and material balance as possible, assuming an 85 percent operating factor. Outline the types of equipment necessary for the process. Determine approximate duties of heat exchangers and sizes of major pieces of equipment. Instrument the plant as completely as possible, outlining special problems that might be encountered. List all the additional information which would be needed to completely finish the design evaluation.

Problem 5. Production of High-Purity Anhydrous Ammonia

A chemical company is considering the production of 450 t (metric)/day of high-purity anhydrous ammonia. The method selected is a high-pressure steam-methane reforming process. The process description is as follows:

The steam-methane reforming process produces ammonia by steam reforming natural or refinery gas under pressure, followed by a carbon monoxide shift, purification of raw synthesis gas, and ammonia synthesis. In the process, saturated and unsaturated hydrocarbons are decomposed by steam according to the basic equation

$$CH_4 + H_2O \longrightarrow CO + 3H_2$$

Feed streams high in olefins or sulfur require pretreatment. The primary reformer converts about 70 percent of a natural gas feed into raw synthesis gas in the presence of steam using a nickel catalyst. In the secondary reformer, air is introduced to supply the nitrogen required for ammonia. The heat of combustion of the partially reformed gas supplies the energy to reform the remainder of the gas after reacting with the oxygen in the air. High-pressure reforming conserves 30 to 40 percent in compressor horsepower over what is required in the low-pressure synthesis.

Next, the mixture is quenched and sent to the shift converter. Here CO is converted to CO_2 and H_2. If additional heat is still available after satisfying the water requirement for the shift reaction, a waste-heat boiler may be installed. Shift reactor effluent, after heat recovery, is cooled and compressed, then goes to the gas purification section. CO_2 is removed from the synthesis gas in a regenerative MEA (monoethanolamine) or other standard recovery system. After CO_2 removal, CO traces left in the gas stream are removed by methanation. The resulting pure synthesis gas passes to the oil separator, is mixed with a recycle stream, cooled with ammonia refrigeration, and goes to the secondary separator where anhydrous ammonia (contained in the recycle stream) is removed. Synthesis gas is then passed through heat exchange and charged to the catalytic ammonia converter. Product gas from the converter is cooled and exchanged against converter feed gas. Anhydrous liquid ammonia then separates out in the primary separator and, after further cooling, goes to the anhydrous ammonia product flash drum. The feed to the reforming section is normally in excess of 300 psig. The pressure, however, is not fixed and may be varied to provide optimum design for specific local conditions. Temperatures in the primary and secondary reformers are 760 to 980°C, while shift reaction temperatures are 370 to 450°C in the first stage and 230 to 290°C in the second stage. Ammonia synthesis is normally performed at 20.8 MPa. Temperature in the quench-type ammonia converter is accurately and flexibly controlled inside the catalyst mass to allow a catalyst basket temperature gradient giving a maximum yield of ammonia per pass, regardless of the production rate.

Analyze this process and make as complete a flowsheet and material balance as possible, assuming a 90 percent operating factor. Outline the types of equipment necessary for the process. Determine approximate duties of heat exchangers and sizes of the major pieces of equipment. What additional information will be needed to completely finish the preliminary design evaluation?

PRACTICE SESSION PROBLEMS[†]

Problem 1. Cost for Hydrogen Recovery by Activated-Carbon Adsorption Process[‡]

What is the cost, in cents per 1000 ft^3 (at STP) of 95 mol percent H_2, for recovering hydrogen of that purity from 10 million ft^3/day (at STP) of gaseous feed if the following conditions apply?

Feed composition by volume = 72.5 percent H_2 and 27.5 percent CH_4.

A hydrogen recovery method based on selective adsorption by activated carbon will be used.

For the carbon adsorption, three separate beds will be needed so that one bed can be in continuous use while the other two are being desorbed or reactivated. Base the recovery on a single pass and an absorption cycle of 1 h per bed.

A total of 0.00838 lb mol of the material is adsorbed per hour per pound of activated carbon.

The composition of the adsorbed phase is 96.8 mol percent CH_4, the balance being hydrogen.

The cost of activated carbon is \$3.65 per pound.

The annual amount of additional carbon necessary is 15 percent of the initial charge.

The total plant cost (equipment, piping, instrumentation, etc.) equals \$4,750,000.

The capital investment equals the total plant cost plus process materials (process materials are considered as auxiliaries; i.e., process materials are only the initial charge of activated carbon).

[†]The following problems are recommended for solution by students working in groups of two or three during a 3-h practice session. Many of these problems have been adapted from old Annual AIChE Student Contest Problems, as shown by footnotes for the individual problems. The original problems and winning solutions are available from the New York AIChE headquarters. Since these problems were originally in U.S. Customary System units, these practice session problems are also presented in the same units to minimize errors in unit conversions.

A useful procedure for using these practice session problems is to have students examine the original AIChE Student Contest Problem and Solution before the class period and then have someone (student or teacher) present a discussion on the problem during the first half-hour or hour of the practice session. Then the students can break up into groups and carry out the solution to the problem assigned during the last 2 h of the practice session. It is desirable for the students to have an opportunity to examine the correct solution to the problem immediately after they turn in their solution at the end of the 3-h session.

[‡]Adapted from 1947 AIChE Student Contest Problem.

The capital investment must be completely paid off (no scrap value) in 5 years.

The operating cost per year, including labor, fuel, water, feed, regeneration, fixed charges minus depreciation, and repairs and maintenance, equals $2,250,000 (operating costs not included in this value are the cost of the additional makeup carbon necessary and depreciation).

The plant operates 350 days/yr.

Problem 2. Adsorption Tower Design for Hydrogen Purification by Activated Carbon†

Your plant is producing 10 million ft^3/day (measured at STP) of a gas containing 72.5 volume percent H_2 and 27.5 volume percent CH_4. It is proposed to pass this gas through activated carbon to obtain a product gas containing 95 volume percent H_2. The activated carbon shows a preferential selective adsorption for the CH_4. An adsorption-desorption-regeneration cycle using three fixed beds will be used. One bed will be regenerated and purged over an 8-h period. During this period, the other two beds will be on alternate 1-h adsorption and 1-h desorption cycles to permit continuous operation. A single pass of the gas will be used.

Each individual bed may be designed to include a number of carbon-packed towers in parallel. The diameters of the individual towers must all be the same, and the diameter may be 6, 9, 12, or 15 ft.

Determine the following:

a. The number of individual units (or towers) in each bed to give the minimum total cost for the towers.

b. The height and diameter of the towers for the conditions in part (a).

Data and information previously obtained for the chosen conditions of the process (i.e., adsorption at 400 psia, desorption at 20 psia, and an average adsorption temperature of 110°F) are as follows:

A total of 0.0063 lb mol/h of the material is adsorbed per pound of activated carbon.

The composition of the adsorbed phase is 96.8 mol percent CH_4, and the balance is H_2.

The carbon has a bulk density of 0.30 g/cm^3.

The gas velocity in the adsorbers should not exceed 1 ft/s based on the cross-sectional area of the empty vessel. This applies to all the cycles including the adsorption, regeneration, and purge.

The feed gases enter the adsorbers at 400 psia and 80°F. The product gases leave the adsorbers at almost 400 psia and a temperature of 140°F. Regeneration includes 30 ft^3 (STP) of flue gas per pound of carbon and 80 standard ft^3 of purging air per pound

†Adapted from 1947 AIChE Student Contest Problem.

Table C-5 Cost data

Column diameter, ft	6	9	12	15
Cost per foot of length, $	1170	2140	3280	5840
Cost per tower for skirt or support, $	1950	3240	4540	5200

of carbon. The flue gas is at 600°F (its maximum temperature) and 5 psig, and the air is at 90°F and 5 psig. The air may reach a maximum temperature of 600°F at the start of the purging.

For each head (either top or bottom), add an equivalent cost of 5-ft additional length per vessel. Cost data are given in Table C-5.

Problem 3. Design of Rotary Filter for Sulfur Dioxide Recovery System†

As a member of a design group working on the design of a recovery system for SO_2, you have been asked to estimate the area necessary for a rotary vacuum filter to handle a zinc sulfite filtration. You are also to determine the power requirement of the motor necessary for the vacuum pump. Do not include any safety factors in your results.

The following conditions have been set by your group:

A slurry containing 20 lb of liquid per pound of dry $ZnSO_3 \cdot 2.5H_2O$ is to be filtered on a continuous rotary filter to give a cake containing 0.20 lb of H_2O per pound of dry hydrate.

One hundred pounds of $ZnSO_3 \cdot 2.5H_2O$ in the slurry mixture will be delivered to the filter per minute.

A drum speed of 0.33 r/min will be used, and a vacuum of 10.2 in. mercury below atmospheric pressure will be used.

The fraction of the drum area submerged will be assumed to be 0.25.

The fraction of the drum area available for air suction will be assumed to be 0.10.

The temperature of the slurry is 110°F, and the air into the vacuum pump can be considered to be at 110°F. For air at 110°F, $C_p/C_v = 1.4$. The temperature and pressure of the air surrounding the filter are 70°F and 1 atm, respectively.

The specific cake resistance and the specific air suction cake resistance can be assumed to be independent of pressure drop, drum speed, temperature, fraction of drum submerged, fraction of drum available for suction, and slurry concentration. However, to eliminate possible errors due to this assumption, a laboratory test should be run at conditions approximating the planned design. The results of these tests can be used as a basis for the design (i.e., cake and filtrate compositions and densities can be assumed to be the same for the design as those found in the laboratory).

The vacuum pump and motor have an efficiency (isentropic) of 85 percent.

†See Chap. 15 of this text for design basis.

Table C-6

Total area of filter	4.15 ft^2
Fraction of area submerged	0.20
Fraction of area available for air suction	0.10
Vacuum	9 in.Hg below atmospheric pressure
Slurry concentration	12 lb liquid/lb dry $ZnSO_3 \cdot 2.5H_2O$
Temperature	110°F
Drum speed	0.40 r/min
Density of wet cake leaving filtering zone	100 lb/ft^3
Pounds of liquid per pound of dry $ZnSO_3 \cdot 2.5\ H_2O$ in wet cake leaving filtering zone	0.6
Density of filtrate	68.8 lb/ft^3
Viscosity of filtrate	0.6 cP
Pounds of water per pound of dry $ZnSO_3 \cdot 2.5H_2O$ in final cake	0.20
Time interval	5.0 min
Volume of filtrate	0.95 ft^3
Volume of air at SC	8.5 ft^3

It can be assumed that all the $ZnSO_3 \cdot 2.5H_2O$ is filtered off and none remains in the filtrate.

The resistance of the filter medium can be assumed to be negligible.

Laboratory data compiled at the request of the SO_2 recovery design group (results of filtration of zinc sulfite slurry on an Oliver rotary vacuum filter) are shown in Table C-6.

Problem 4. Return on Investment for Chlorine Recovery System†

The off-gas from a chloral production unit contains 15 volume percent Cl_2, 75 volume percent HCl, and 10 volume percent $EtCl_2$. This gas is produced at a rate of 150 ft^3/min based on 70°F and 2 psig. It has been proposed to recover part of the Cl_2 by absorption and reaction in a partially chlorinated alcohol (PCA). The off-gas is to pass continuously through a packed absorption tower countercurrent to the PCA, where Cl_2 is absorbed and partially reacts with the PCA. The gas leaving the top of the tower passes through an alcohol condenser and then to an existing HCl recovery unit.

The reaction between absorbed Cl_2 and the PCA is slow, and only part of the absorbed Cl_2 reacts in the tower. Part of this PCA from the bottom of the tower is sent to a retention system where the reaction is given time to approach equilibrium. The rest of the PCA is sent to the chloral production unit. Ethyl alcohol is added to the PCA going to the retention system. Then the PCA is recycled from the retention system to the top of the absorption column, and ethyl alcohol is added at a rate sufficient to keep the recycle rate and recycle concentration constant.

Your preliminary calculations have set the optimum conditions of operations, and all costs have been determined except the cost for the absorption tower. Using the

†Adapted from 1949 AIChE Student Contest Problem.

following data and information, determine the yearly percent return on the capital expenditure:

The gases leaving the top of the tower contain 2 volume percent Cl_2 (on the basis of no PCA vapors present in the gas).
The PCA entering the top of the tower contains 0.01 mol percent free Cl_2.
The PCA leaving the bottom of the tower contains 0.21 mol percent free Cl_2.
The recycle rate is 200 gpm (entering the top of the absorption tower).
The gas rate at the top of the column is 27.2 lb mol/h.
The PCA entering the top of the tower has a density of 68.5 lb/ft^3 and an average molecular weight of 70.
The column is operated at a temperature of 35°C and a pressure of 1 atm.
The gases enter the bottom of the tower at 70°F and 2 psig.
No Cl_2 is absorbed in the alcohol condenser.
The column is packed with 1-in. porcelain Raschig rings, and a porcelain tower is used.

Laboratory tests with a small column packed with Raschig rings have been conducted. These tests were carried out at 35°C using PCA and off-gas having the same concentrations as those in your proposed design. These data have been scaled up to apply to 1-in. Raschig rings and are applicable to your column. The results are given in Table C-7.

The absorption of Cl_2 in the PCA follows Henry's law, and the following relation may be used for determining the number of transfer units:

$$N_t = \frac{y_1 - y_2}{\Delta y_m}$$

where

the log mean driving force $\Delta y_m = \dfrac{\Delta y_1 - \Delta y_2}{\ln(\Delta y_1/\Delta y_2)}$

$$\Delta y_1 = y_1 - y_1^*$$

$$\Delta y_2 = y_2 - y_2^*$$

Equilibrium data for Cl_2 and the PCA at 35°C and 1 atm are given in Table C-8. These data apply over the entire length of the column.

Table C-7

m_2G_2/L_2, based on conditions at dilute end of column	Height of a transfer unit (log-mean method), ft
0.178	3.6
0.571	4.8
1.04	6.2
1.53	7.6

Table C-8

Free Cl_2 in PCA, x, mol %	Free Cl_2 in gas (based on PCA vapor-free gas), y, mol %
0.1	2.0
0.2	4.0
0.3	6.0

The maximum allowable velocity (based on conditions of the gas at the inlet to the tower) is 1.5 ft/s. The column will operate at 60 percent of the maximum allowable velocity.

Necessary cost data are given in the following:

1. Tower:

Diameter, in.	12	18	24	30	36	42
Porcelain, $/ft of length	450	650	880	1170	1490	1850
Top or bottom heads, porcelain, each, $	390	550	750	970	1230	1520

2. Tower packing. One-inch Raschig rings, porcelain, at $24.00 per cubic foot.
3. Capital expenditure minus cost of absorption tower is $80,000.
4. Net annual savings (taking alcohol loss and all other costs, such as interest, rent, taxes, insurance, depreciation, maintenance, and other overhead expenses, into consideration) is $40,000.

Note: This $40,000 has been determined by developing an accurate estimate of the absorption tower cost and can be taken as the actual net savings.

Nomenclature for Problem 4

G = molar gas flow, lb mol/h
L = molar liquid flow, lb mol/h
m = slope of equilibrium curve, y/x
N_t = number of transfer units
y = mol fraction of chlorine in gas
y^* = equilibrium mol fraction of chlorine in gas

Subscripts

1 refers to bottom of tower.
2 refers to top of tower.

Problem 5. Economic Analysis of Chlorine Recovery System†

The data shown in Table C-9 have been obtained for a chlorine recovery system in which all values have been determined at the optimum operating conditions of retention time and recycle rates.

†Adapted from 1949 AIChE Student Contest Problem.

Table C-9

Mol % Cl_2 in exit gas	Total capital expenditure, $	Total annual operating costs (fixed costs, production costs, overhead, etc.), $	Gross annual savings by using process, $
0.2	368,000	336,000	456,000
1.0	304,000	316,000	444,000
2.0	272,000	296,000	420,000
5.0	212,000	268,000	332,000
10.0	136,000	236,000	188,000

Determine the following:

a. The percent Cl_2 in exit gas at breakeven point.

b. The maximum net annual savings and percent Cl_2 in the exit gas where it occurs.

c. The maximum percent return on the capital expenditure and percent of Cl_2 in exit gas where it occurs.

d. Which investment would you recommend and why?

Problem 6. Optimum Thickness of Insulation

Insulation is to be purchased for 3 miles of 10-in.-OD pipe carrying saturated steam at 250°F. The average air temperature for the year for the surroundings is 45°F.

It is estimated that the life period of the installation will be 20 years with negligible scrap value. The sum of fixed charges excluding depreciation is 10 percent, and maintenance is estimated to be 2 percent annually of the fixed-capital investment.

One company has submitted a bid that includes installation at a cost of $0.25 $D^{1.3}$ per lineal foot, where D is the outside diameter of the lagging in inches. Using the following data and equations, what thickness of insulation should be used for this job in order to give a 50 percent return on the full investment?

The line will be in continuous service 365 days/yr. The steam is valued at $1.30 per 1000 lb. The thermal conductivity of the lagging is 0.04 Btu/h·ft²·°F/ft. Steam-film and pipe resistance may be neglected.

Heat losses by conduction and convection from the surface may be calculated from

$$h_c = 0.42\left(\frac{\Delta T_s}{D'_o}\right)^{0.25}$$

where

ΔT_s = temperature difference between surface of lagging and air, °F

D'_o = OD, in.

h_c = Btu/h·ft²·°F

An average value of $h_r = 1.2$ Btu/h·ft^2·°F may be used to determine heat losses by radiation. This is an adjusted value such that total heat loss per hour may be calculated by the equation

$$\dot{q} = (h_c + h_r)A\,\Delta T_s$$

A mathematical setup with all necessary numbers and a description of the method for final solution will be satisfactory.

Problem 7. Capacity of Plant for Producing Acetone from Isopropanol†

Acetone is produced by the dehydrogenation of isopropanol according to the following reaction:

$$(H_3C)_2CH(OH) \xrightarrow{\text{catalyst}} CH_3{-}\overset{\overset{\displaystyle O}{\|}}{C}{-}CH_3 + H_2$$

The reaction can be assumed to be irreversible.

The catalyst used for the process decreases in activity as the amount of isopropanol fed increases. This effect on the reaction rate is expressed in the following:

$$k = \frac{0.000254\,NT}{V}\left(2.46 \ln \frac{1}{1-\alpha} - \alpha\right)$$

where

α = fraction of isopropanol converted to acetone and side products defined as (mols isopropanol converted)/(mols isopropanol supplied)

k = reaction rate constant, s^{-1}

N = lb mol of isopropanol fed to converter per hour

T = absolute temperature, °R

V = catalyst volume, ft^3

The feed rate of isopropanol is maintained constant throughout the entire process. The fresh catalyst has an activity such that $k = 0.30$ s^{-1}. After 10,000 lb of isopropanol per cubic foot of catalyst has been fed, $k = 0.15$ s^{-1}.

A plot of log k versus total isopropanol fed in pound-mols is a straight line.

The maximum production of acetone is 76.1 lb mol/h. This maximum production can be considered as that occurring at zero time (i.e., when $k = 0.30$ s^{-1}).

A constant temperature of 572°F and a constant pressure of 1 atm are maintained throughout the entire process.

The catalyst volume $V = 250$ ft^3.

$$\text{Efficiency} = \frac{\text{mols acetone produced}}{\text{mols isopropanol consumed}}(100) = 98\% \text{ at } 572°\text{F}$$

†Adapted from 1948 AIChE Student Contest Problem.

The catalyst can be restored to its original activity with a 48-h reactivation.

The catalyst will be discarded after the last operating period each year.

The unit will operate 350 days/yr (this includes reactivation shutdowns). The other 15 days are used for repairs and replacing the old catalyst.

If nine catalyst reactivations are used per year, what will the production of acetone be in pounds per year? Assume all acetone produced is recovered.

Outline your method of solution. The actual mathematical calculations are not necessary.

Problem 8. Equipment Design for the Production of Acetone from Isopropanol†

You are designing a plant for the production of acetone from isopropanol. Acetone is produced according to the following reaction:

$$(H_3C)_2CH(OH) \xrightarrow{\text{catalyst}} CH_3-\overset{\overset{\large O}{\|}}{C}-CH_3 + H_2$$

The reaction can be assumed to be irreversible.

The catalyst used in this process decreases in activity as the process proceeds. This effect on the reaction rate can be expressed as

$$k = \frac{0.000254}{V} NT \left(2.46 \ln \frac{1}{1-\alpha} - \alpha\right)$$

where

α = fraction of isopropanol converted to acetone and side products defined as (mols isopropanol converted)/(mols isopropanol supplied)

k = reaction rate constant, s^{-1}

N = lb mol of isopropanol fed to converter per hour

T = absolute temperature, °R

V = catalyst volume, ft^3

The catalyst must be regenerated periodically throughout the operation. The fresh catalyst has an activity such that $k = 0.30\ s^{-1}$.

The products from the reactor (unconverted isopropanol, acetone, side products, and some water from the impure isopropanol feed) are sent to a continuous distillation column where purified acetone is removed.

A distillation column, calandria, and condenser are now available in your plant, and you are to determine if these can be used for the purification step.

Using the following data, determine whether the column, calandria, and condenser are usable. If these are not satisfactory, determine the purchase cost of the necessary equipment. Do not buy any new equipment unless it is needed. The present equipment may be used in another part of the plant at a later date.

†Adapted from 1948 AIChE Student Contest Problem.

Feed rate of isopropanol is kept constant at 97 lb mol/h of isopropanol during the entire operation. A constant temperature of 572°F and a constant pressure of 1 atm are maintained throughout the entire operation.

The catalyst volume is 250 ft^3.

$$\text{Efficiency of conversion} = \frac{\text{mols acetone produced}}{\text{mols isopropanol consumed}}(100) = 98\% \text{ at } 572°\text{F}$$

The present distillation tower contains 50 actual plates. Your calculations have indicated that a reflux ratio of 1 : 1 will allow 98 percent of the acetone to be removed, assuming a 40 percent plate efficiency. The product material may be assumed as 100 percent acetone. These conditions are satisfactory, and you have made calculations at these conditions, giving the results shown in Table C-10.

The overall heat-transfer coefficient in the calandria is 250 Btu/h·ft^2·°F.

Saturated steam is available at 50 psig. Neglect any sensible heat transfer from steam condensate to boiling liquid. The temperature-difference driving force ΔT in the calandria may be assumed to be 90°F.

Your calculations have indicated that the present condenser on the column is satisfactory and may be used for this process.

The necessary data on the present equipment are given in Table C-11. Table C-12 shows the installed-cost data for any new equipment that must be purchased.

Estimate the steel distillation column on the basis of $160.00 per square foot of tray area. Column diameters should be even multiples of 6 in.

Table C-10

Per 1000 lb/h of acetone distilled and removed as product	
Column cross-sectional area, ft^2	6.4 (based on 12-in. plate spacing and 2-in. liquid depth)
Condenser area, ft^2	220
Steam, lb/h	1100

Table C-11

Number of plates in column	50
Column diameter	72 in.
Plate spacing	12 in.
Liquid depth on each plate	2 in.
Calandria heat-transfer area	104 ft^2
Calandria shell working pressure	60 psig

Table C-12 Steel heat exchangers

Surface, ft^2	Cost, $/$ft^2$
100	130
200	94
300	78
500	65

Problem 9. Quick-Estimate Design of Debutanizer[†]

You are the chief design engineer at a large petroleum refinery. The head projects engineer has asked you to make a preliminary design estimate for a proposed debutanizer. You are to present the following preliminary information to a group meeting within 3 h.

a. Number of plates for proposed debutanizer column
b. Diameter of proposed column
c. Outside tube area required for heating coils in reboiler (coils to contain saturated steam at 250 psia; average heat of vaporization of hydrocarbons in reboiler may be taken as 5000 cal/g mol)

The following information has been supplied by the project's engineer:

Charge stock from catalytic cracker to debutanizer totals 5620 BPSD (barrels per service-day). (See Table C-13.)
Debutanizer to operate at 165 psia.
Two fractions are to be obtained, OVHD and BTMS.
OVHD is to contain 98.5 percent of the butanes and lighter components with a contamination of 1.5 mol percent pentanes and pentenes.

The debutanizer must fractionate between nC_4 and iC_5.

Table C-13 Debutanizer charge stock from catalytic cracker

°API (60°F) = 100.3
Viscosity = 42 SSU at 120°F

Component	Mol percent	Molecular weight
C_2''	0.1	28
C_2	1.2	30
H_2S	2.1	34
C_3''	16.3	42
C_3	6.9	44
iC_4''	6.5	56
nC_4''	14.3	56
iC_4	10.8	58
nC_4	3.9	58
C_5''	11.9	70
iC_5	9.7	72
nC_5	2.3	72
C_6	11.8	86
C_7	2.1	96
C_8	0.1	112
	100.00	

[†]Adapted from L. J. Coulthurst, *Chem. Eng. Progr.*, **44:** 257 (1948).

A search through your debutanizer-design card file gives the information shown in Cards A to C.

Card A — Debutanizer: 30 trays

Feed 574 BPSD		Reflux 635 BPSD	OVHD 354 BPSD		BTMS 220 BPSD	
5897 lb/h		5630 lb/h	3114 lb/h		2783 lb/h	
84.6 mol/h		100.6 mol/h	56.0 mol/h		28.6 mol/h	
Mol %		Composition same as OVHD	Mol %		Mol %	
C_3''	12.4	Reflux ratio = 1.8 : 1	C_3''	18.8	C_4's	1.4
C_3	6.1		C_3	9.2	C_5, 400°F	98.6
C_4's	47.9		C_4's	71.6		100.0
C_5, 400°F	33.6		C_5, 400°F	0.4		
	100.0			100.0		
°API = 68.8		°API = 110.7	°API = 100.7		°API = 39.7	
Basis feed:		119 mol %	66.2 mol %		33.8 mol %	

Card B — Values of K at 165 psia

Listings derived principally from the data of Scheibel and Jenny, *Ind. Eng. Chem.*, **37:** 80 (1945).

Hydrocarbon	260°F	280°F	285°F	290°F	300°F
n-Butane	1.75	2.00	2.06	2.13	2.25
n-Pentene	1.15	1.30	1.34	1.39	1.47
Isopentane	1.03	1.19	1.23	1.28	1.37
n-Pentane	0.90	1.06	1.11	1.16	1.25
n-Hexane	0.46	0.58	0.61	0.64	0.70
n-Heptane	0.24	0.31	0.32	0.35	0.39
n-Octane	0.13	0.17	0.18	0.19	0.21

Card C — Recommended overall heat-transfer coefficients (based on outside coil heating area)

	Debutanizer service	Transfer rate, Btu/h·ft^2·°F
Condenser	Butane to water	100–110
Reboiler	Gasoline to steam	120–140
Reboiler	Gasoline to hot oil	50–75
Exchanger	Gasoline to gasoline	80
Preheater	Gasoline to steam	110–120
Cooler	Gasoline to water	75–90

Problem 10. Economic Analysis of Formaldehyde-Pentaerythritol Plant†

You are a member of a firm doing consulting work on chemical engineering design. G. I. Treyz Chemical Company of Cooks Falls, New York, has asked your firm to determine the advisability of adding a pentaerythritol (PE) production unit to its present

†Adapted from a real-life situation for one of the coauthors of this text.

formaldehyde plant. You have been sent to Cooks Falls to analyze the situation and obtain the necessary details.

A conference with G. Victor Treyz, owner of the plant, supplies you with the following information:

The present formaldehyde plant cost \$1,140,000 several years ago and is in satisfactory operating condition. It produces formaldehyde by the oxidation of methanol. The yearly fixed charges (interest, rent, taxes, insurance, and depreciation) at the plant amount to 15 percent of the initial investment.

The plant capacity is 100,000 lb of formalin (37.2 percent HCHO, 8 percent CH_3OH as inhibitor, and 54.8 percent H_2O by weight) per day.

Miscellaneous costs (salaries, labor, office expenses, laboratory supplies, maintenance, repairs, communication, sales, silver catalyst replacement, etc.) amount to \$360,000 per year.

The overall efficiency of conversion of methanol to formaldehyde is 80 percent; that is, 0.8 lb of CH_3OH is converted to HCHO per pound of CH_3OH provided.

The total cost of utilities (fuel, electricity, steam, water, etc.) equals 5 percent of the total cost of producing the formaldehyde.

Mr. Treyz has included all the smaller costs, such as insurance benefits, in the overhead cost so that the total cost of the present operation is the sum of fixed charges, overhead cost, utilities, and methanol.

The proposed PE plant is to produce 6000 lb of final pentaerythritol per day using the inhibited formalin produced at the plant. The basic reaction involving lime, acetaldehyde, and formaldehyde can be assumed to go to completion. Only 70 percent of the PE produced in the reaction is obtained as the final product. Any costs due to the presence of excess $Ca(OH)_2$ can be neglected. The calcium formate formed must be discarded.

The initial installed cost of the proposed plant is \$500,000. It can be assumed that the yearly cost of the new operation minus raw materials costs will be 40 percent of the initial installed cost. This 40 percent includes fixed charges, overhead, utilities, and all other expenses except raw materials.

Both plants operate 350 days/yr.

Mr. Treyz feels he can sell 6000 lb of PE per day at the present market price. He can also sell 100,000 lb of formalin per day, but he is willing to make the new investment if it will give him better than a 30 percent yearly return on the PE initial plant investment. Ignore income tax effects and working capital.

Following is a list of prices supplied by Mr. Treyz. All these prices are f.o.b. Cooks Falls, New York, and they are to be used for the cost estimate.

Methanol, carload lots	\$0.60/gal = \$0.091/lb
Formaldehyde (or Formalin), 37.2% HCHO, 8% CH_3OH	\$0.10/lb
Acetaldehyde	\$0.46/lb
Lime as pure CaO	\$40.00/ton
Pentaerythritol	\$0.72/lb

From the preceding information, determine the following:

a. The present profit per year on the formaldehyde unit.

b. The total profit per year if the PE unit were in operation.

c. Yearly percent return on the PE initial plant investment.

d. Should the Treyz Company make the investment?

Problem 11. Operating Time for Catalytic Polymerization Reactor to Reach Minimum Allowable Conversion†

A catalytic polymerization plant is to operate continuously with a special catalyst with properties and results as given below. Under the following specified operating conditions, how many days could each reactor remain on stream from the time of fresh (catalyst age zero) catalyst charge until the percent of propylene conversion through the catalyst bed drops to 93.75 percent? What will be the pressure drop across the catalyst bed at that time? How often should a new recharged unit come on stream if all units are to operate on identical staggered schedules (i.e., how many days will each unit be down for dumping and recharging)?

Total feed stream is 15,000 bbl/day and is 40 percent by volume propylene and 60 percent by volume propane.

Five reactors are available, each holding 20,000 lb of catalyst.

All reactors will operate on identical staggered schedules.

The unit will be operated with the same flow rate in four of the reactors while the fifth one is down for dumping and catalyst recharging, with this dumping and recharging requiring at least 5 days but no more than 8 days.

The temperature in the reactors will be maintained constant at 430°F. At this temperature 0.715 bbl of polymer is obtained for every barrel of propylene converted.

$$\text{Catalyst age factor } A = \frac{\text{gallons of polymer produced since catalyst recharging}}{\text{pounds of catalyst charged}}$$

$$A = \frac{\left[\int_0^D F(\theta)\,d\theta\right] (\text{conversion factors to give gallons of polymer})_{\text{arith.avg.}}}{\text{pounds of catalyst}}$$

θ = time, days

D = time in days at which $\Delta p/F^2$ is to be calculated

$F = F(\theta)$ = reactor total feed rate, 10^3 bbl/day

Δp = pressure drop across reactors, psi

At the operating pressure and temperature of 430°F, the following data apply for the special catalyst used:

In a plot of log $\Delta p/F^2$ versus catalyst age, A is linear with $\Delta p/F^2 = 0.2$ at $A = 0$ while $\Delta p/F^2 = 100$ at $A = 84.90$. (The resultant equation is $\log \Delta p/F^2 = 0.03179A - 0.699$.)

The conversion of propylene to polymer at the time of fresh catalyst charge (catalyst age zero) is 97.66 percent.

The data shown in Table C-14 apply for the catalyst and indicated temperature (linear interpolation is satisfactory).

†Adapted from 1974 AIChE Student Contest Problem.

Table C-14

A	$\frac{\text{\% conversion of propylene at catalyst age } A}{\text{\% conversion of propylene at catalyst age zero}}$
0	1.00
10	0.995
20	0.99
30	0.98
40	0.97
50	0.96
60	0.935
70	0.91
80	0.87
90	0.82
100	0.76

Problem 12. Sizing and Costing of Multicomponent Distillation Column for Biphenyl Recovery Unit†

The feed to a multicomponent distillation column is shown in Table C-15 (in order of decreasing volatility).

A carbon steel, bubble-cap distillation column with 10 trays operated at an average pressure of 400 mmHg, an average temperature of 280°C, and a reflux ratio of 8 : 1 will give an overhead product containing 98 percent of the entering biphenyl and a negligible amount of methyl biphenyl and heavier components. Based on the following data and assumptions, estimate the cost of the distillation tower, including installation and auxiliaries (*not* including reboiler and condenser) at present:

Pressure drop in the column can be neglected, and average temperature and pressure can be used for calculations.

Ideal gas laws apply to the gas mixture under these conditions.

Liquid feed is at its boiling point.

Table C-15

Component	Feed rate, lb mol/h
Toluene	0.488
Naphthalene	1.599
Biphenyl	16.835
O-Methyl biphenyl	0.208
P-Methyl biphenyl	0.333
M-Methyl biphenyl	0.121
Diphenylenemethane (Fluorene)	1.029
Phenanathrene	0.769
M-Terphenyl	1.919
	23.301

†Adapted from 1975 AIChE Student Contest Problem.

Column operates adiabatically, and constant molal overflow assumption is acceptable.

The average molecular weight of the gas can be taken as that of biphenyl or $C_6H_5 \cdot C_6H_5 = 154.2$.

The average specific gravity of the liquid in the column is 0.72.

The tray spacing is 24 in. with a 2-in. slot liquid seal.

Assume the surface tension of the liquid is 20 dyne/cm.

Size and cost the column, using an 85 percent safety factor on the maximum allowable vapor velocity.

Problem 13. Design of Reactor for Coal Conversion to Nonpolluting Fuel Oil (Plus Partial Solution)†

A plant is being designed to produce low-sulfur oil from coal under the conditions outlined below. A major concern in the design is to minimize the volume of the reactor, and you are to carry out some preliminary studies for the reactor system. Specifically, you are to determine the total volume of the reactor if it is operated isothermally at 800°F for the case of a single, ideal, plug-flow reactor operation and for the case of a single, back-mix (continuous stirred-tank reactor) reactor system with the conditions and assumptions as outlined in the following.

Operating Conditions Plant is to produce 50,000 bbl/day (based on 60°F) of low-sulfur oil (0.4 weight percent sulfur) from coal. Table C-16 gives the specifications for the coal feed and the product oil.

Coal in the slurry is 35 percent by weight with the balance being recycled oil of the same composition as the product oil.

A nickel-molybdenum on alumina catalyst in the form of $\frac{1}{8}$-in. spheres is used with a desulfurization activity A_s of 1.25 and a bulk density of 42.0 lb/ft^3.

The following assumptions apply for the reactor system:

Pressure of 2500 psia and negligible pressure drop across the reactor.

25,000 ft^3 of gas at SC (SC = 60°F and 1 atm) flows to the reactor per barrel of slurry feed (based on 60°F).

The gas to the reactor contains 85 percent hydrogen; the other 15 percent is methane with negligible H_2S content.

Yield of product is 4.2 bbl of product oil (at 60°F) per ton of coal (as received).

Average molecular weight of fuel oil is 301.

No hydrogen, methane, or hydrogen sulfide is dissolved in the slurry.

Necessary heating and cooling units are available so reactors can be assumed to operate isothermally at 800°F.

Partial pressure of hydrogen for plug-flow reactor can be assumed as constant at the arithmetic average of entrance and exit pressures.

†Adapted from 1976 AIChE Student Contest Problem.

Table C-16

Coal feed		Oil product	
Bulk density, lb/ft^3 = 45.0		4.4° API = density of 64.97 lb/ft^3 at 60°F Density = 44.8 lb/ft^3 at 800°F	
†Proximate analysis, weight percent		Boiling distribution: true boiling point cut, weight percent	
Moisture	1.5	C_5, 400°F	8.1
Ash	10.3	400–650°F	32.1
Volatile matter	35.5	650–975°F	22.1
Fixed carbon	52.7	975°F+	37.7
Total	100.0	Total	100.0
†Ultimate analysis, weight percent		Ultimate analysis, weight percent	
Carbon	70.2	Carbon	90.2
Hydrogen	4.6	Hydrogen	8.5
Nitrogen	1.0	Nitrogen	0.8
Sulfur	3.6	Sulfur	0.4
Oxygen	10.5	Oxygen	—
Ash	10.1	Ash	0.1
Total	100.0	Total	100.0

†See *Perry's Chemical Engineers' Handbook* for discussion of these and other methods of analysis (6th ed., p. 9-4). Proximate and ultimate analyses in this case were carried out with air-dried coal samples; so the oxygen and hydrogen in the "moisture" reported in the proximate analysis are included in the ultimate analysis.

Assume the term $1 + K_{HS}\bar{p}_{HS}$ in the rate equation stays constant for the plug-flow reactor at the arithmetic average of the entering and exit values.

Assume negligible volume change during the reaction so that $C_s = C_{s_o}(1 - X_s)$.

Of the fuel oil passing through the reactor, 15 weight percent is vaporized in the reactor section, and this can be doubled to 30 percent on a molal basis considering different volatilities of the components.

The carbon in the coal that is lost to the gas stream is converted to CH_4, C_2H_6, C_3H_8, and C_4H_{10} in equal-volume amounts so that the average carbon/hydrogen ratio of the resultant gas is 0.35714 on a mol basis considering hydrogen as being 1.008 lb/lb mol of hydrogen.

All the nitrogen in the coal that is lost is converted to gaseous NH_3.

All the sulfur in the coal that reacts goes to H_2S.

Reactor sizing will be based on the rate equation for the desulfurization reaction

$$-r_s = k_s A_s \frac{C_s^2 \bar{p}_H}{C_{S_o}(1 + K_{HS}\bar{p}_{HS})}$$

where

$-r_s$ = rate of sulfur removal, lb mol/h·lb catalyst

k_s = reaction rate constant, ft^3/h·lb cat·psia

$k_s = \exp(14.76 - 55{,}000/RT) = 7.405 \times 10^{-4}$ at 800°F

K_{HS} = adsorption constant for H_2S inhibition (psia)$^{-1}$

$= 0.10 \exp(1200/RT) = 0.162$ at 800°F

$R = 1.987$ Btu/lb mol·°R
T = temperature, °R
A_s = desulfurization activity
C_s = sulfur concentration in slurry, lb mol/ft^3
$\bar{p}_{\rm H}$ = hydrogen partial pressure, psia
$\bar{p}_{\rm HS}$ = hydrogen sulfide partial pressure, psia

The reactor performance equations are as follows:

For plug-flow

$$\frac{W}{Q} = C_{S_o} \int_{X_{S_i}}^{X_{S_f}} \frac{dX_s}{-r_s}$$

For back-mix (CSTR)

$$\frac{W}{Q} = C_{S_o} \frac{X_{S_f} - X_{S_i}}{-r_s}$$

where

W = catalyst charge, lb
X_s = fractional conversion of sulfur
C_{s_o} = concentration of sulfur in slurry feed to reactor, lb mol/ft^3
Q = volumetric feed rate of slurry, ft^3/h
i = inlet value
f = final value

Suggestions Base material balances on 1 ton (2000 lb) of coal as received. Integrate rate expression for plug flow analytically (not graphically or by approximations). See information as provided for initial part of solution presenting necessary material balances for the conditions given for this problem, and understand what was done.

Problem 13. Partial Solution

Figure C-1 shows the first part of the solution to Problem 13 dealing with the design of a reactor for coal conversion to nonpolluting fuel oil.

Material Balances Choose as basis 1 ton of coal as received (2000 lb) to produce 4.2 bbl of product oil at 60°F

$$\text{Fuel produced} = 4.2\ \text{bbl}\left(42\frac{\text{gal}}{\text{bbl}}\right)\left(\frac{1}{7.48}\frac{\text{ft}^3}{\text{gal}}\right)\left(64.97\frac{\text{lb}}{\text{ft}^3}\right) = 1532\frac{\text{lb prod. oil}}{\text{ton coal}}$$

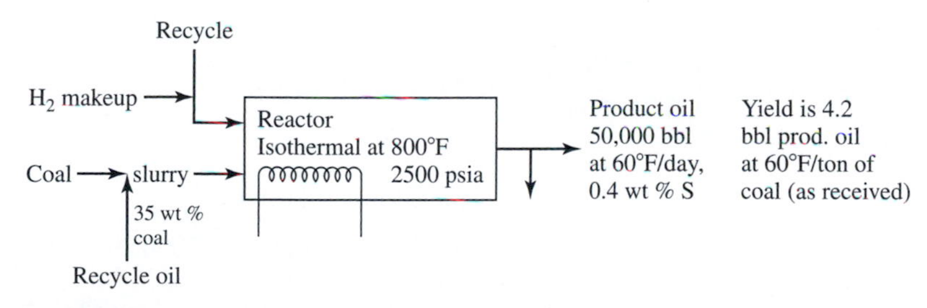

Figure C-1

Table C-17 Material balance, overall—2000 lb coal

	Feed		Prod. oil		Difference,	
Component	wt %	lb	wt %	lb	lb	
C	70.2	1404	90.2	1381.95	22.05	
H	4.6	92	8.5	130.23	−38.23	gain in H
N	1.0	20	0.8	12.26	7.74	
S	3.6	72	0.4	6.13	68.87	
O	10.5	210	—	—	—	
Ash	10.1	202	0.1	1.53	200.47	
Total	100	2000	100	1532.1		

Hydrogen Material Balance 22.05 lb of C in coal is burned to CH_4, C_2H_6, C_3H_8, or C_4H_{10} as 0.35714 mol C per mol H.

$$\text{Hydrogen used to burn C} = \left(\frac{22.05}{12.011}\text{mol C}\right)\left(\frac{1}{0.35714}\frac{\text{mol H}}{\text{mol C}}\right)\left(1.008\frac{\text{lb H}}{\text{mol H}}\right)$$
$$= 5.19 \text{ lb H}$$

$$\text{Hydrogen used to make NH}_3 = \left(\frac{7.74}{14.007}\text{mol N}\right)\left(3\frac{\text{mol H}}{\text{mol N}}\right)\left(1.008\frac{\text{lb H}}{\text{mol H}}\right)$$
$$= 1.67 \text{ lb H}$$

$$\text{Hydrogen used to make H}_2\text{S} = \left(\frac{65.87}{32.064}\text{mol S}\right)\left(2\frac{\text{mol H}}{\text{mol S}}\right)\left(1.008\frac{\text{lb H}}{\text{mol H}}\right)$$
$$= 4.14 \text{ lb H}$$

$$\text{Hydrogen used to make H}_2\text{O} = \left(\frac{210}{16}\text{mol O}\right)\left(2\frac{\text{mol H}}{\text{mol O}}\right)\left(1.008\frac{\text{lb H}}{\text{mol H}}\right)$$
$$= 26.46 \text{ lb H}$$

$$\text{Hydrogen gain} = 38.23 \text{ lb}$$
$$\text{Total H used} = 75.69 \text{ lb H}$$
$$\frac{75.65}{2.016} = 37.544 \text{ lb mol H}_2$$

Material Balance, at Reactor Inlet, for Slurry Concentration Basis—2000 lb Coal. For slurry, 35 wt % is coal and 65 wt % is oil.

$$(2000 \text{ lb coal})\left(\frac{0.65 \text{ lb oil}}{0.35 \text{ lb coal}}\right) = 3714 \text{ lb of recycle oil}$$
$$= \frac{3714 \text{ lb}}{44.8 \text{ lb/ft}^3 \text{at } 800^\circ\text{F}}$$
$$= 82.9 \text{ ft}^3 \text{ oil at } 800^\circ\text{F}/2000 \text{ lb coal}$$

$$\text{Volume of coal} = 2000 \text{ lb}/(45 \text{ lb/ft}^3) = 44.44 \text{ ft}^3$$

Total volume of slurry to reactor

$$= 82.9 + 44.44 = 127.34 \text{ ft}^3/2000 \text{ lb of coal fed}$$

Sulfur content of fuel oil in slurry

$$= (3714)(0.004) = 14.856 \text{ lb S}$$

Sulfur content of coal in slurry

$$= (2000)(0.036) = 72.00 \text{ lb S}$$

$$\text{Total} = 86.86 = \frac{86.86}{32.066} = 2.709 \text{ lb mol S}$$

Concentration of sulfur entering reactor in slurry

$$= C_{s_o} = \frac{2.709}{127.34} = 0.0213 \text{ lb mol/ft}^3$$

At Reactor Outlet

$$\text{Oil} = \overset{\text{Recycl.}}{3714} + \overset{\text{Prod.}}{1532} = 5246 \text{ lb} = \frac{5246 \text{ lb}}{44.8 \text{ lb/ft}^3} = 117.1 \text{ft}^3 \text{at } 800^\circ\text{F}$$

$$\text{Sulfur in outlet oil} = (5246)(0.004) = 20.98 \text{ lb} = \frac{20.98}{32.066} = 0.65444 \text{ lb mol}$$

Concentration of sulfur in oil leaving reactor

$$= C_{s_f} = \frac{0.65444}{117.1} = 0.00559 \text{ lb mol/ft}^3$$

$$X_{S_f} = 1 - \frac{C_{S_f}}{C_{S_o}} = 1 - \frac{0.00559}{0.0213} = 0.738 \cong 0.74$$

$$X_{S_f} = \frac{C_{S_o} - C_{S_f}}{C_{S_o}} \qquad \text{assuming constant fluid volumetric flow rate}$$

Material Balance for Gas at Entrance and Exit of Reactor Basis—2000 lb Coal. 25,000 SCF of gas/bbl of slurry at 60°C is given as the condition

$$\text{Bbl of slurry at } 60^\circ\text{F/2000 lb coal} = \left(\overset{\text{ft}^3 \text{ coal}}{\frac{2000}{45}} + \overset{\text{ft}^3 \text{ oil at } 60^\circ\text{F}}{\frac{3714}{64.97}}\right)\left(7.48\frac{\text{gal}}{\text{ft}^3}\right)\left(\frac{1 \text{ bbl}}{42 \text{ gal}}\right)$$

$$= 18.1 \text{ bbl/2000 lb coal}$$

$$\text{Gas to reactor} = (25{,}000)(18.1) = 452{,}450 \text{ SCF of gas/2000 lb coal}$$

Gas is 85% H_2; so

$$(0.85)(452{,}450) = 384{,}583 \text{ SCF } H_2/2000 \text{ lb coal.}$$

$$\frac{359(520)}{492} = 380 \text{ ft}^3\text{/mol at SC of 60°F and 1 atm}$$

$$\frac{384{,}583}{380} = 1014 \text{ lb mol } H_2/2000 \text{ lb coal}$$

Gas is 15% CH_4; so

$$452{,}450\left(\frac{0.15}{380}\right) = 179 \text{ lb mol } CH_4/2000 \text{ lb coal}$$

$$\text{Total} = 1193 \text{ lb mol } (H_2 + CH_4)/2000 \text{ lb coal}$$

$$p_H \text{ at entrance to reactor} = \frac{1014}{1193}(2500) = 2125 \text{ psia}$$

$$p_H \text{ at entrance} = 0 \text{ (given)}$$

Material Balance, Gas at Exit of Reactor Basis—2000 lb Coal. Amount of fuel oil entering reactor is 3714 lb, or

$$\frac{3714}{\text{avg mol wt } 301} = 12.34 \text{ lb mol/2000 lb coal}$$

Fuel oil is 15 wt % or 30 mol % vaporized; so $12.34(0.30) = 3.7$ mols of fuel oil are vaporized in reactor, leaving 8.64 mols of fuel oil in liquid and 3.7 mols of fuel oil in gas at reactor exit.

Assume no mols of H_2 or CH_4 or NH_3 or H_2S are dissolved in the liquid.

Total mols of H_2 in gas at exit	$= 1014 - 37.544$ (used in reactor)	$=$	976.5	lb mols
Total mols of H_2O in gas at exit	$= \frac{26.46}{2.016}$	$=$	13.1	lb mols
Total mols of NH_3 in gas at exit	$= \frac{7.74}{14.007}$	$=$	0.55	lb mols
Total mols of H_2S in gas at exit	$= \frac{4.14}{2.016}$	$=$	2.06	lb mols
Total mols of CH_4 in gas at exit	$=$	$=$	179.0	lb mols
Total mols of fuel oil in gas at exit	$=$		3.7	lb mols
Total exit gas, mols	$=$		1174.91	lb mols/2000 lb coal

Total entering gas, mols = 1193 lb mols/2000 lb coal

$$p_H \text{ at reactor exit} = \frac{976.5}{1174.91}(2500) \cong 2075 \text{ psia}$$

$$p_{HS} \text{ at reactor exit} = \frac{2.06}{1174.91}(2500) \cong 4.4 \text{ psia}$$

This completes the major work on material balances needed for solving this problem.

Now proceed, using these results and other information given in the problem, to complete the reactor design analysis requested.

Problem 14. Material Balance for Alkylation Plant Evaluation†

The simplified diagram of a catalytic alkylation unit is shown in Figure C-2. In the reactor, butylene and isobutane react to form C_8 "alkylate" according to the following reaction:

$$C_4H_8 + C_4H_{10} \longrightarrow C_8H_{18}$$

The unit is to produce product alkylate at a rate of 1700 m^3/day (10,693 bbl/day).

The yield is 1.72 m^3 alkylate per m^3 butylene consumed; 1.10 m^3 of isobutane are consumed per 1.0 m^3 butylene consumed.

The reactor effluent is to contain 75 volume % isobutane.

It may be assumed that the recycle is pure isobutane and that propane, alkylate, and *n*-butane are completely recovered as pure products in the columns. Propane and *n*-butane do not react.

Under these conditions,

a. How much of each feed stream is required in m^3/day and in bbl/day?

b. How much isobutane must be recycled in m^3/day and in bbl/day?

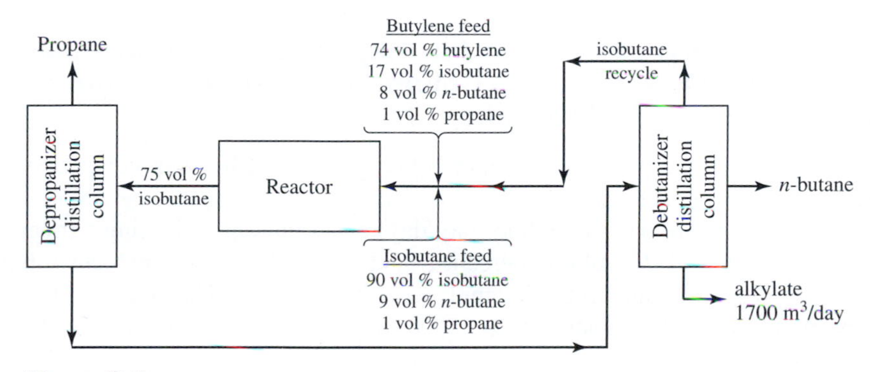

Figure C-2

†Adapted from 1977 AIChE Student Contest Problem.

Problem 15. Cost for Producing Butadiene Sulfone†

A design has been completed for the production of butadiene sulfone from 1,3-butadiene and sulfur dioxide. Given the following results from the completed design and the basic company requirements as listed, determine what the selling price of the butadiene sulfone product should be in dollars per pound based on:

a. The case where the company demands a 20 percent continuous nominal interest rate of return after taxes on any investment (i.e., discount rate r is 20 percent).

b. The case where the company demands a 20 percent finite effective end-of-year interest rate of return after taxes on any investment (i.e., discount rate i is 20 percent).

The total capital investment for the complete plant is \$300,000. Production is 4500 t (metric)/yr of butadiene sulfone.

A patent royalty charge of 5 percent of the annual sales value before taxes must be paid.

Working capital is 10 percent of the fixed capital investment or \$27,000.

Special start-up costs for the first year *only* are 10 percent of the fixed capital investment.

The plant operates 330 days/yr (90 percent on-stream factor).

The income tax rate for the company is 48 percent of the gross profits.

The straight-line method is used for calculating depreciation cost.

Calculation of costs per year gives a total of \$758,000 that includes all costs except those for royalty, income taxes, and start-up in the first year of operation. The \$758,000 includes the annual depreciation cost of \$27,000.

The annual costs and the annual income are considered, by company policy, as end-of-year lump-sum transactions.

Life period of plant is 10 years. Scrap value of plant at end of life is zero.

Neglect interest during construction period and value-of-land effects.

Note: Rate-of-return calculations for this company must be based on discounted cash flow procedures to account for the time value of money.

Problem 16. Cost of Reboiler for Alkylation Unit Heat Pump Fractionator if Fractionation Column Operation Is Assumed to Be at Minimum Reflux Ratio‡

An alkylation unit heat pump fractionator is being designed for continuous operation to meet the following conditions:

Feed is 30,000 bbl/day (47,500 kg mol/day) of a liquid at its boiling temperature with a composition of 24 volume percent *i*-butane and 76 volume percent *n*-butane (23.3 mol percent *i*-butane and 76.7 mol percent *n*-butane). The product stream as *i*-butane-rich overhead product is to be 5000 bbl/day (17,000 lb mol/day) of which

†Adapted from 1970 AIChE Student Contest Problem.

‡Adapted from 1980 AIChE Student Contest Problem.

400 bbl/day is n-butane with the rest being i-butane (composition is 91.7 mol percent i-butane and 8.3 mol percent n-butane).

The temperature of the materials at the bottom of the tower is 90°F.

At 90°F, the equilibrium vapor pressure of i-butane is 62 psia, and that for n-butane is 44 psia.

Raoult's law and Dalton's law hold for the mixtures involved.

Relative volatility is constant for all compositions of the mixture at the value obtained at the bottom of the tower (90°F temperature).

An alkylation unit provides 30×10^6 Btu/h of heat for reboiling in the fractionation column.

The overall heat-transfer coefficient U in the reboiler is 120 Btu/(h·ft^2 outside area·°F).

The ΔT (constant) in the reboiler for use in the equation $\dot{q} = UA\,\Delta T$ is 20°F.

The heat of vaporization for the mixture in the reboiler can be taken as 152 Btu/lb.

The cost for the reboiler is \$25/ft^2 of outside heat-transfer area.

Neglect pressure drop in the column.

Assume ideal liquids and ideal gases.

McCabe-Thiele assumptions apply for the fractionation column.

Physical properties	i-Butane	n-Butane
Molecular formula	C_4H_{10}	C_4H_{10}
Molecular weight	58.12	58.12
Density, lb/gal (at 60°F)	4.69	4.87

(One barrel is 42 gal measured at 60°F.)

What would be the cost of the reboiler if the column is assumed to operate at the minimum reflux ratio? What would be the reboiler cost if operation were at 1.2 minimum reflux ratio?

Problem 17. Incremental Investment Comparison for Two Conversions for a Dicyanobutene Reactor System†

As one step in the process of making nylon, dichlorobutene (DCB) is reacted with sodium cyanide to form dicyanobutene (DNB). A CSTR is used to carry out the reaction, and special controls and materials of construction are required.

You are asked, on the basis of the simplifying assumptions given in the following, to make a preliminary estimate of the incremental return on investment when conditions for two conversions are compared. At a later time, using details as given in the 1981 AIChE student contest problem, it may be possible to extend the scope of this problem to include all effects such as losses and major effects of working capital, but these items will not be considered at this time.

†Adapted from 1981 AIChE Student Contest Problem.

Following are the conditions for the problem: Upon going from 80 to 85 percent conversions, where the rate of reaction in kilograms of DCB converted per minute per kilogram of DCB charged is measured as 0.0169 at 80 percent conversion and 0.0131 at 85 percent conversion, what is the incremental return on the extra capital investment required under the following conditions? A single continuous stirred tank (back-mix) reactor is to be used in each case. The reaction is

$$C_4H_6\,Cl_2 + 2NaCN \xrightarrow{\text{catalyst}} C_4H_6(CN)_2$$

Component	*Molecular Weight*
NaCN	49
DCB	125
DNB	106

A total of 90×10^6 kg of DNB is to be produced per year (8000 h). The catalyst is $NaCu(CN)_2$ with a molecular weight of 138.6, and 0.038 kg of copper with an atomic weight of 63.6 is needed per kilogram of DCB charged to the system. The catalyst solution is as follows:

	Wt %
$NaCu(CN)_2$	16.5
NaCN	17.3
H_2O	76.2

Liquid density $= 1.15 \times 10^3$ kg/m^3

The DCB charge is 100 percent DCB with a liquid density of 1.16×10^3 kg/m^3. None of the cyanide in the catalyst is consumed in the reaction, and there is no catalyst loss. Therefore, consider the catalyst initial cost as a component of the total fixed-capital investment based on the amount of catalyst needed for a 2-h operating period.

The sodium cyanide solution added for the reaction is as follows (enough is always charged for 100 percent DCB conversion, no matter what the actual conversion is):

	Wt %
NaCN	26.0
Na_2CO_3	1.0
NH_3	0.3
NaOH	0.2
H_2O	72.5

Liquid density $= 1.13 \times 10^3$ kg/m^3

The hydrogen cyanide solution added for pH control is only about 1 percent by weight of the NaCN stream, and the HCN stream effects can be neglected for this calculation.

Reactor purchased costs are given in Table C-18 with straight-line interpolation or extrapolation to 30,000 gal acceptable.

The equipment fixed-capital investment for the entire system, taking all costs for heat exchanger equipment, pumps, piping, installation, etc., into account, is equal to the initial cost of the catalyst solution plus 4.5 times the purchased cost of the reactor. Assume none of the unreacted materials can be recovered.

Table C-18

Working capacity, gal	Cost, $
5,000	108,000
10,000	204,000
12,500	230,000
15,000	260,000
24,000	390,000

The density of the average reaction mixture is 1.14×10^3 kg/m^3.
The cost of raw materials is as follows:

DCB solution = \$0.62/kg

Catalyst solution = \$0.30/kg

NaCN solution = \$0.082/kg

In making the investment comparison, ignore working capital and consider only the fixed-capital investment with an annual charge for depreciation of 8 percent of the fixed-capital investment.

Assume all operating costs except those for raw materials and depreciation are constant for the two conversion cases under consideration. Do not consider the effects of income taxes.

Problem 18. Recycle Compressor Power Cost for Methanation Unit of SNG Plant†

In the design of a substitute natural gas (SNG) plant for producing SNG from coal, you have been asked to determine the power cost, in dollars per year, for compressing the recycle gas for the bulk methanation portion of the plant with a single bulk methanator using electric motor-driven compressors of conventional centrifugal type with an adiabatic efficiency of 75 percent and a maximum compression ratio per stage of 3.5. Following are the special conditions for your design in this preliminary evaluation:

The feed is a mixture of CH_4, CO, CO_2, H_2, and N_2 with pollutant H_2S having been removed, and only a small amount of CO_2 is present. The gas will be fed to a single bulk methanator at a pressure of 360 psia and temperature of 450°F. The feed rate for the initial gas brought into the system (before recycle gas is added) is 390×10^6 standard ft^3/day, where a standard cubic foot is defined as at 60°F and 1 atm. The critical methanation reaction is $CO + 3H_2 \rightarrow CH_4 + H_2O$ with an exothermic heat of reaction of −95,404 Btu/lb mol CO. Ignore any heat effects due to the CO_2 methanation reaction or overall heat losses.

The initial entering gas (before recycle is added) contains 63.4 mol percent hydrogen, and 90 percent of this entering hydrogen reacts by the methanation reaction given in the preceding with a final methanator exit gas temperature of 900°F. There are no

†Adapted from 1982 AIChE Student Contest Problem.

other reactions. The molecular weight of the gas mixture entering the methanator has been calculated to be 15.5 and for the exit gas is 16.5.

The average heat capacity of the gas in the reactor can be taken as 0.47 Btu/(lb·°F), and this average applies over a temperature range of 450 to 900°F. Enough gas must be recycled through the compressor to the single-bulk-methanator entry-gas stream so that the temperature rise from 450 to 900°F is maintained.

The inside diameter of the methanator to be used has been calculated to be 24.8 ft, and the methanator height of the catalyst-packed bed is 12.4 ft.

The brake power efficiency of the electric motor, defined as

$$\frac{\text{power delivered to the turbine}}{\text{power provided as electric energy}}(100)$$

is 80 percent. The value of k can be taken as 1.35, where k is the ratio of the heat capacity of the gas at constant pressure to the heat capacity at constant volume.

The gas is cooled before it enters the compressor to a temperature such that the exit gas from the compressor will be at 450°F and 360 psia, assuming ideal adiabatic compression. To obtain the compressor entering-gas flow rate, assume that $p_1 v_1^k = p_2 v_2^k$. Neglect pressure drop due to the cooler.

The following equation, obtained from *Perry's Chemical Engineers' Handbook,* can be used to determine the pressure drop in the methanator:

$$\Delta p = \frac{2 f G^2 L (1-\varepsilon)^{3-h}}{D_p \rho \phi_S^{3-h} \varepsilon^3}$$

where

Δp = pressure drop, lb/ft^2

f = friction factor, assumed to be 1.0 for this case

G = fluid superficial mass velocity based on empty methanator cross-sectional area, lb/s·ft^2

L = actual depth of methanator bed in ft plus 3 ft added to account for Δp due to nozzles, distributor, and supports

ε = voidage (fractional free volume of packing), taken as 0.40 for this case

h = exponent (function of Reynolds number) and assumed to be 2.0 for this case

D_p = average catalyst particle diameter, equal to 0.0238 ft for this case

ρ = fluid density, lb/ft^3, based on average temperature and entering pressure (0.464 lb/ft^3 for this case)

ϕ_S = shape factor for solid catalyst particles (1.0 for this case)

Under the given conditions, determine:

a. The power cost for driving the recycle gas compressor in dollars per year if the purchase cost for electricity is \$0.06/kWh and the plant operates 330 days/yr.

b. The net present worth of the initial fixed-capital costs plus the discounted operating cost of the recycle gas compressor over a 20-year life period when the discount rate is 11 percent per year with interest compounding annually and expenses paid annually at the end of year. Use of the series present-worth factor is applicable for this case with $i = 0.11$ since inflation effects cancel out by assuming that the purchase cost of power will also inflate. Capital cost for the related installation can be taken as $5,000,000, which is the initial fixed-capital investment for the related equipment to be used in determining the net present worth.

Problem 19. Process Design for Wood Pulp Production Plant†

In a process for producing bleached wood pulp from southern pine trees, chipped wood is charged to a digester where it is heated to 346°F and cooked with a water-alkali mixture. After cooking, the mixture is passed to a blow tank where the pressure is reduced to atmospheric, and part of the water evaporates. The mixture is then passed through a series of washers and filters (brown stock washers) where the pulp is removed and the remaining liquid (black liquor) is sent to a multiple-effect evaporator for concentration, followed by chemical treatment to prepare the liquid for recycling to the digester.

For the following conditions, you are to determine (a) the flow rate in tons per day and the weight percent of dissolved solids for the black liquor leaving the brown stock washers system passing to the evaporators and (b) the rate of final bleached dry pulp production in tons per day.

Cycle time per batch is 185.75 min with eight digesters in continuous use, so there are 62.02 digester charges per 24-h day.

Density of the solid wood before chipping is 40.6 lb/ft^3, and it contains 52 percent H_2O with the remaining 48 percent being cellulose, lignin, and various carbohydrates. After chipping, 1 ft^3 of the chips contains 12.5 lb of dry wood and 13.5 lb of water. (This corresponds to a void fraction of 0.36.)

The volume of each digester is 4150 ft^3 with the chipped wood being added to fill the digester to its full capacity.

The cooking liquor added to the digester takes up the void space and contains 10.25 weight percent of alkali as dissolved solids with the balance being water.

The liquid in the digester has a ratio of 3.5 lb of total liquid including the water in the wood per pound of dry wood.

In the digestion step, 45 weight percent of the dry wood is left as unbleached pulp, which goes on through the process and comes out of the brown stock washers as pulp product for further treatment. The final yield of bleached pulp is 90 percent of the unbleached pulp yield. The other 55 percent of the dry wood in the digester charge (except for a small amount of turpentine which is removed from the digester and has a negligible effect on the total amounts) is dissolved in the cooking liquor and passes through the system to the evaporators as dissolved solids. It is later burned in a recovery furnace for its heat value.

In the pressure reduction from the digester to the blow tank, 14 weight percent of the stream leaving the digester is lost as steam released in the blow step.

†Adapted from 1983 AIChE Student Contest Problem.

In the brown stock washers, water is used to wash the final pulp in the last filter, and the dilution factor D used is 4, where the pounds of wash water used per pound of dry pulp in the wet filter product is D plus the amount of water in the wet filter product per pound of dry pulp.

The pulp product from the last washer contains 82 weight percent water and 18 weight percent unbleached pulp. This is sent to the final bleaching unit where a 90 percent conversion of the unbleached pulp to bleached pulp product occurs.

Note that a total material balance around the brown stock washers system shows that the entering stream from the blow unit plus the wash water added at the final filter must equal the weight of the wet pulp product plus the black liquor product. This reduces to a material balance result of the tons per day of black liquor to the evaporators being equal to the tons per day of the stream from the blow unit to the brown stock washers plus $(D - 1)$(tons/day of dry pulp in the wet pulp product).

Problem 20. Paraffin Removal by Extractive Distillation with Dimethylphthalate

Figure C-3 represents a flowsheet prepared by a junior design engineer for part of an aromatics plant which provides paraffin removal by extractive distillation with dimethylphthalate (DMP). Check this design and make recommendations concerning the design conclusions of the junior design engineer who worked on the project. If

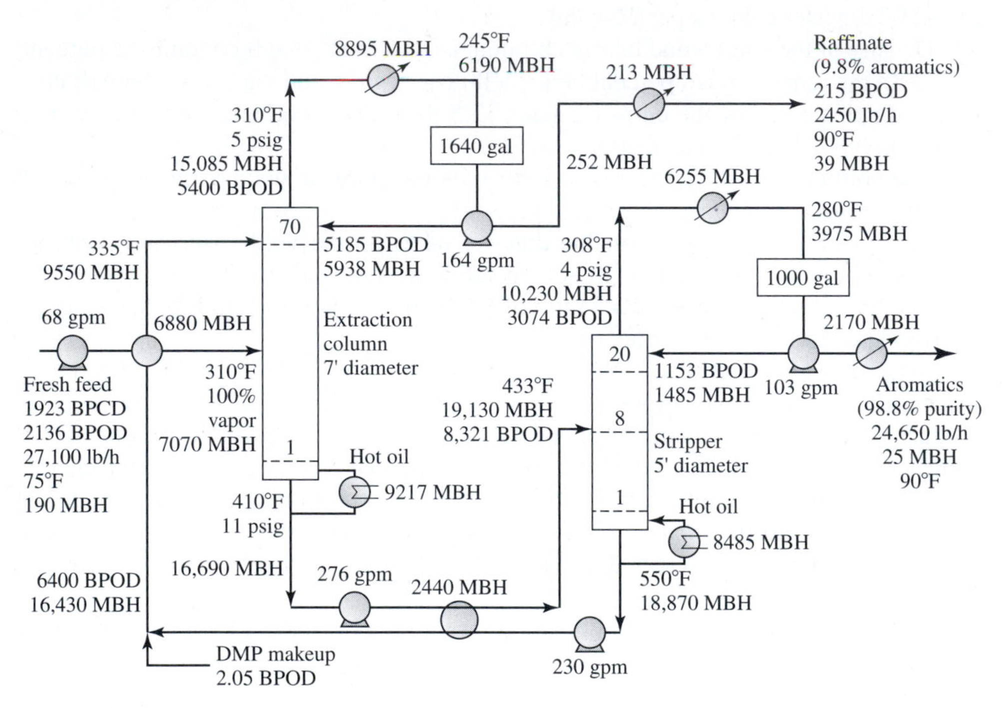

Figure C-3
Paraffin removal by extractive distillation with dimethylphthalate

there is an inconsistency in the results, indicate where the error is, what must be done to correct the error, and what would be the magnitude of the error if it were not corrected. The basis for the design is as follows:

1. The operating factor for the process is 90 percent.
2. The tray efficiency of the columns is 60 percent.
3. The fresh feed is of the composition shown in Table C-19.
4. The addition of DMP enhances the relative volatility of the paraffins and naphthenes an average of 25 percent above their normal values.
5. $(L/D)_{min}$ of the extraction column is 20.
6. $(L/D)_{min}$ of the stripper is 0.5.
7. $(L/D)_{act} = 1.2\ (L/D)_{min}$.
8. Dimethylphthalate properties with a formula of $C_6H_4(CO_2CH_3)_2$ are given in Table C-20.
9. Since the DMP makeup is so small compared to the other streams, it may be neglected in the overall heat and material balance.
10. Pump sizes may be assumed correct even though each one includes a fixed safety factor.

Table C-19

Component		BPCD	Volatility relative to o-xylene
C_8 paraffin	C_8-Pn	11	
C_8 napthene	C_8-N	24	1.52
C_9 paraffin	C_9-Pn	68	
C_9 napthene	C_9-N	9	
Ethylbenzene	EB	112	1.22
Para-xylene	Px	358	1.155
218°F paraffin	281°F-Pn	49	1.15
Meta-xylene	Mx	739	1.135
Ortho-xylene	Ox	500	1.0
292°F paraffin	292°F-Pn	34	0.90
Heavies	300°F-C_9	19	0.70
		1923	

Table C-20

Temperature, °F	Vapor pressure, psia	Liquid enthalpy, Btu/lb	Vapor enthalpy, Btu/lb
60	—	0	—
100	—	12	—
200	—	42	—
300	0.2	73	—
400	1.7	110	251
500	8.2	148	281
550	17.0	169	293
600	—	191	306

Duties of condensers and reboilers are given in MBH (million Btu per hour). BPCD is the designation for barrels per calendar day, and BPOD is that for barrels per operating day. Note that the number of plates in each column is indicated by the number at the top of the column. Significant temperatures are given wherever necessary.

Problem 21. Cooling and Condensing a Reactor Effluent

The stream from the reactor in the vinyl chloride process consists of 12.60 kg/s of vinyl chloride, 13.42 kg/s of 1,2-dicloroethane, and 7.35 kg/s of hydrogen chloride at 500°C and 2635 kPa. This stream is to be cooled and condensed to 6°C, and the pressure is to be reduced to 1215 kPa. A three-step process is suggested to achieve this task as follows: (1) Cool the effluent in a heat exchanger at 2635 kPa to the dew point temperature; (2) provide adiabatic expansion across a valve to 1215 kPa; and (3) further cool and condense in another heat exchanger at 1215 kPa to 6°C.

Determine the heat duties in each of the two heat exchangers, and provide cooling curves for each, assuming that the pressure drop in each heat exchanger is being neglected.

Since there are other design arrangements that could have provided the cooling and condensing of the reactor effluent, is the arrangement suggested initially the optimal choice? Show at least one other process arrangement that could have been used, and provide an analysis comparing that arrangement with the suggested arrangement.

Problem 22. Cooling of a Hot Flue Gas

Hot flue gas leaves a waste heat boiler at 149°C and a dew point of 37.7°C and needs to be cooled to 32°C. The flow rate of the flue gas is 2.52 kg/s. Cooling water is available at 26.8°C but cannot be heated beyond 49°C. A pressure drop no larger than 5 kPa is allowed.

What type of cooling unit is recommended? What are the dimensions or size of the unit and the purchase cost of the unit on January 2002?

Problem 23. Process Options for Separating a Gas Mixture

In a secondary oil recovery scheme, carbon dioxide is sometimes injected into an oil well to stimulate the recovery process. The composition of the gas on a volume basis coming out of the well after the injection is given in Table C-21.

Devise three alternative schemes to create a methane-rich fuel, recycle CO_2, and a liquefied natural gas that is a H_2S- and CO_2-free, C_2^+ hydrocarbon stream. In these alternative schemes develop suitable process flowsheets for the separation process required. Provide an approximate material balance based on a flow rate of 25 m^3/s at 40°C and a pressure of 2 MPa. Identify the key equipment required. Note any potential problems that may have an overall effect on the solution.

Table C-21

Gas	Volume composition
CO_2	90.25
N_2	0.16
C_1	4.76
C_2	1.75
C_3	1.52
i-C_4	0.22
n-C_4	0.58
i-C_5	0.21
n-C_5	0.22
C_6	0.16
C_7^+	0.12
H_2S	0.05

In the analysis note that CO_2 and ethane form an azeotrope at 65 mol percent CO_2. This azeotrope can be accommodated in an azeotropic distillation arrangement with the addition of a solvent-agent C_4^+ hydrocarbon stream.

Problem 24. Ultrafiltration with a Tubular Membrane

Ultrafiltration tests with a 0.015-m tubular membrane at a Reynolds number of 25,000 provided a permeate flux of 40 $L/m^2 \cdot h$ and 75 percent rejection for a 5% polymer solution. The polymer has an average molecular weight of 30,000, and the estimated diffusivity is 5×10^{-9} m/s. Neglect the effect of molecular diffusion in the pores.

a. Predict the fraction of the polymer solution rejected for a flux of 20 $L/m^2 \cdot h$. Predict the maximum rejection.

b. Estimate the fraction rejected for the low-molecular-weight fraction of the polymer with a molecular weight of approximately 10,000.

c. If the selective layer is 0.2 μm, does molecular diffusion have a significant effect on the rejection for case (a)? Discuss your conclusions.

Problem 25. Heat Integration of the Ethylene Separation Section

The separation section for the base-case ethylene preliminary design is shown in Fig. 3-13. The heating and cooling requirements in this section are provided by 12 heat exchange units. These units with their heat exchange areas are enumerated in Table 3-7. Since there is considerable overlap in the temperature intervals observed among many of these heat exchange units, there may be an opportunity to achieve some energy savings with the application of pinch technology. Information for this heat integration process is given in Table C-22.

Note that several of the streams that are to be cooled require a propylene refrigerant and cannot be included in the heat integration. With the remaining heat exchanges,

Table C-22 Stream thermal data on ethylene separation section heat exchange units

Equipment code	Stream type[†]	Temperature interval, °C		Heat capacity rate, kJ/s·°C	Enthalpy change rate, kJ/s
		In	Out		
E-601	Hot	−29.48	−30.12	24,440	−15,642
E-602	Cold	−6.99	−6.92	277,725	19,440
E-603	Hot	45.21	45.01	22,082	−4,416
E-604	Cold	114.51	133.35	416	8,063
E-605[‡]	Cold	44.5	100	23	1,248
E-605[‡]	Hot	113.6	57.93	23	−1,248
E-606	Hot	127.73	100	23	−638
E-607	Hot	57.93	43.3	23	−328
E-608	Hot	39.08	39.07	307,513	−30,751
E-609	Cold	49.23	49.24	307,513	29,433
E-610	Hot	40.49	40.44	28,643	−1,432
E-611	Cold	108.28	117.88	94	890
E-612	Hot	86.04	35	9	−507

[†]Hot refers to a stream that is to be cooled, and cold refers to a stream that is to be heated.
[‡]Refers to the cold and hot streams of E-605 (see Fig. 3-13).

determine the best heat integration network that can be developed if only the ethylene separation section is considered. (Heat integration of the base-case ethylene process involves 44 heat exchange units that can be included in the heat integration process but will require the aid of suitable software to accomplish the integration task.)

Problem 26. Recovery of Germanium from Optical Fiber Manufacturing Effluents[†]

The manufacturing process for making optical fibers involves high-temperature oxidation of silicon tetrachloride ($SiCl_4$) and germanium tetrachloride ($GeCl_4$) to form glass particles (SiO_2 and GeO_2) which are incorporated into a glass preform rod. This rod is subsequently drawn in a furnace to produce optical fiber. Germanium tetrachloride is added to increase the refractive index of the glass core in the optical fiber preform. It is known from experimental studies that the oxidation of $GeCl_4$ to GeO_2 proceeds to only 25 percent completion whereas oxidation of $SiCl_4$ is nearly complete. In addition, particle deposition is only 50 percent efficient, resulting in further losses of germanium. Due to this loss and the high cost of germanium, a need exists for developing a process to recover germanium from optical fiber manufacturing effluents.

Presently effluent gases from each preform production unit are drawn into a small packed-bed scrubber. The scrubbing liquid is an aqueous NaOH solution. A single fan unit draws the effluents into the scrubbers. Due to operating requirements, it is not possible to make a tight seal between the effluent stream outlet and the inlet to the scrubbing system; hence, the effluent stream gets diluted with a large amount of room air as it enters the scrubber.

[†]Adapted from 1990 AIChE Student Contest Problem.

Within the scrubbers, the $GeCl_4$ is removed from the gas stream by absorption and converted to soluble particles according to the following reaction:

$$GeCl_4 + 5OH^- \longrightarrow HGeO_3{}^- + 4Cl^- + 2H_2O \quad (1)$$

A preliminary design has been prepared for a new scrubber system that uses an existing fan, but replaces the individual scrubbers with a single scrubber. Your task is to find the optimum mol fraction of $GeCl_4$ in the scrubber effluent, using the information given below.

Process Data

Feed gas flow rate to the scrubber is 500 ft^3/min or 633 g mol/min.

Mol fraction of $GeCl_4$ in feed gas to scrubber is 0.001475.

Feed gas temperature and pressure are 25°C and 101 kPa. These values may be assumed to be constant throughout the scrubber.

Use 1 molar sodium hydroxide solution in deionized water as the scrubbing solution.

The pH of the scrubbing solution must not decrease below 12.

A scrubber diameter of 0.762 m (2.5 ft) has been selected, and $\frac{1}{2}$-in. Raschig rings are to be used as the packing.

Reaction (1) is rapid compared to mass transfer, so the $GeCl_4$ mol fraction in the liquid phase can be assumed to be zero. Hence, the number of gas-phase transfer units is given by

$$N_g = \ln \frac{y_1}{y_2}$$

where y_1 and y_2 are the gas-phase mol fractions of $GeCl_4$ in and out of the scrubber, respectively.

The height of a gas-phase transfer unit H_g is estimated to be 0.4 m.

Economic Data

The scrubber will operate 24 h/day, 335 days/yr.

It is estimated that each kilogram of $GeCl_4$ recovered in the scrubber saves \$100 on the purchase of $GeCl_4$ (this value includes the purchase price of $GeCl_4$ and the cost of recovering $GeCl_4$ from the Ge absorbed in the scrubber).

The cost of raw materials, utilities, and all other elements of the total product cost (except depreciation, which is included below) may be neglected, because of the high value of the $GeCl_4$ relative to these costs.

The purchased cost for the scrubber and packing is given by

$$C = 18{,}590 + 7160Z$$

where C is the purchased-equipment cost in dollars and Z is the height of the packing in the scrubber in meters.

The objective function to maximize is the first-year net present worth (NPW) of the scrubbing process:

$$\text{NPW, \$/yr} = (\text{savings due to recovery of } GeCl_4 \text{ over and above that recovered by present process}) (1 - \Phi) - 1.41C$$

where Φ is the income tax rate of 0.35 and C the purchased-equipment cost. The factor 1.41 includes a Lang factor of 4 to obtain the fixed-capital investment, depreciation (straight-line for 6 years), income tax rate, a 10 percent allowance for auxiliary equipment, and a capital recovery factor of 0.3784 (for discrete income and discrete compounding at 30 percent per year discount rate).

APPENDIX

D

Tables of Physical Properties and Constants

CONTENTS

Table D-1 General engineering conversion factors and constants†

Length

1 inch	2.54 centimeters
1 foot	30.48 centimeters
1 yard	91.44 centimeters
1 meter	100.00 centimeters
1 meter	39.37 inches
1 micrometer	10^{-6} meter
1 mile	5280 feet
1 kilometer	0.6214 mile

Mass

1 pound‡	16.0 ounces
1 pound‡	453.6 grams
1 pound‡	7000 grains
1 ton (short)	2000 pounds‡
1 kilogram	1000 grams
1 kilogram	2.205 pounds‡

Volume

1 cubic inch	16.39 cubic centimeters
1 liter	61.03 cubic inches
1 liter	1.057 quarts
1 cubic foot	1728 cubic inches
1 cubic foot	7.481 U.S. gallons
1 U.S. gallon	4.0 quarts
1 U.S. gallon	3.785 liters
1 U.S. bushel	1.244 cubic feet

Density

1 gram per cubic centimeter	62.43 pounds per cubic foot
1 gram per cubic centimeter	8.345 pounds per U.S. gallon

1 gram mol of an ideal gas at 0°C and 760 mmHg is equivalent to 22.414 liters

1 pound mol of an ideal gas at 0°C and 760 mmHg is equivalent to 359.0 cubic feet

Density of dry air at 0°C and 760 mmHg	1.293 grams per liter = 0.0807 pound per cubic foot
Density of mercury	13.6 grams per cubic centimeter (at −2°C)

Pressure

1 pound per square inch	2.04 inches of mercury
1 pound per square inch	51.71 millimeters of mercury
1 pound per square inch	2.31 feet of water
1 atmosphere	760 millimeters of mercury
1 atmosphere	2116.2 pounds per square foot
1 atmosphere	33.93 feet of water
1 atmosphere	29.92 inches of mercury
1 atmosphere	14.7 pounds per square inch

Temperature scales

Degrees Fahrenheit (°F)	1.8°C + 32
Degrees Celsius (°C)	(°F − 32)/1.8
Kelvins (K)	°C + 273.15
Degrees Rankine (°R)	°F + 459.7

Power

1 kilowatt	737.56 foot-pounds force per second
1 kilowatt	56.87 Btu per minute
1 kilowatt	1.341 horsepower
1 horsepower	550 foot-pounds force per second
1 horsepower	0.707 Btu per second
1 horsepower	745.7 watts

Table D-1 *Continued*

Heat, energy, and work equivalents

	cal	Btu	ft·lb	kWh
cal	1	3.97×10^{-3}	3.086	1.162×10^{-6}
Btu	252	1	778.16	2.930×10^{-4}
ft·lb	0.3241	1.285×10^{-3}	1	3.766×10^{-7}
kWh	860,565	3412.8	2.655×10^{6}	1
hp·h	641,615	2545.0	1.980×10^{6}	0.7455
J	0.239	9.478×10^{-4}	0.7376	2.773×10^{-7}
liter·atm	24.218	9.604×10^{-2}	74.73	2.815×10^{-5}

	hp·h	J	liter·atm
cal	1.558×10^{-6}	4.1840§	4.129×10^{-2}
Btu	3.930×10^{-4}	1055	10.41
ft·lb	5.0505×10^{-7}	1.356	1.338×10^{-2}
kWh	1.341	3.60×10^{6}	35,534.3
hp·h	1	2.685×10^{6}	26,494
J	3.725×10^{-7}	1	9.869×10^{-3}
liter·atm	3.774×10^{-5}	101.33	1

Constants

e	2.7183
π	3.1416

Gas law constants

R	8.3144 J/g mol·K
R	1.987 cal/g mol·K
R	82.06 cm^3·atm/g mol·K
R	10.73 (lb/in.^{2})·ft^3/lb mol·°R
R	0.730 atm·ft^3/lb mol·°R
R	1545.0 (lb/ft^2)·ft^3/lb mol·°R
g_c	32.17 ft·lbm/s^2·lbf

Analysis of air

By weight: oxygen, 23.2%; nitrogen, 76.8%; by volume: 21.0% oxygen; 79.0% nitrogen
Average molecular weight of air on above basis = 28.84 (usually rounded to 29)
True molecular weight of dry air (including argon) = 28.96

Viscosity

1 centipoise	0.001 Pa·s
1 centipoise	0.01 g/s·cm
1 centipoise	0.000672 lb/s·ft
1 centipoise	2.42 lb/h·ft

†See Table A-3 for SI conversion factors and more exact conversion factors.
‡Avoirdupois.
§Thermochemical equivalent.

Table D-2 Viscosities of gases

Coordinates for use with Fig. D-1

No.	Gas	*X*	*Y*	No.	Gas	*X*	*Y*
1	Acetic acid	7.7	14.3	29	Freon 113	11.3	14.0
2	Acetone	8.9	13.0	30	Helium	10.9	20.5
3	Acetylene	9.8	14.9	31	Hexane	8.6	11.8
4	Air	11.0	20.0	32	Hydrogen	11.2	12.4
5	Ammonia	8.4	16.0	33	$3H_2 + 1N_2$	11.2	17.2
6	Argon	10.5	22.4	34	Hydrogen bromide	8.8	20.9
7	Benzene	8.5	13.2	35	Hydrogen chloride	8.8	18.7
8	Bromine	8.9	19.2	36	Hydrogen cyanide	9.8	14.9
9	Butene	9.2	13.7	37	Hydrogen iodide	9.0	21.3
10	Butylene	8.9	13.0	38	Hydrogen sulfide	8.6	18.0
11	Carbon dioxide	9.5	18.7	39	Iodine	9.0	18.4
12	Carbon disulfide	8.0	16.0	40	Mercury	5.3	22.9
13	Carbon monoxide	11.0	20.0	41	Methane	9.9	15.5
14	Chlorine	9.0	18.4	42	Methyl alcohol	8.5	15.6
15	Chloroform	8.9	15.7	43	Nitric oxide	10.9	20.5
16	Cyanogen	9.2	15.2	44	Nitrogen	10.6	20.0
17	Cyclohexane	9.2	12.0	45	Nitrosyl chloride	8.0	17.6
18	Ethane	9.1	14.5	46	Nitrous oxide	8.8	19.0
19	Ethyl acetate	8.5	13.2	47	Oxygen	11.0	21.3
20	Ethyl alcohol	9.2	14.2	48	Pentane	7.0	12.8
21	Ethyl chloride	8.6	15.6	49	Propane	9.7	12.9
22	Ethyl ether	8.9	13.0	50	Propyl alcohol	8.4	13.4
23	Ethylene	9.5	15.1	51	Propylene	9.0	13.8
24	Fluorine	7.3	23.8	52	Sulfur dioxide	9.6	17.0
25	Freon 11	10.6	15.1	53	Toluene	8.6	12.4
26	Freon 12	11.1	16.0	54	2,3,3-Trimethylbutane	9.5	10.5
27	Freon 21	10.8	15.3	55	Water	8.0	16.0
28	Freon 22	10.1	17.0	56	Xenon	9.3	23.0

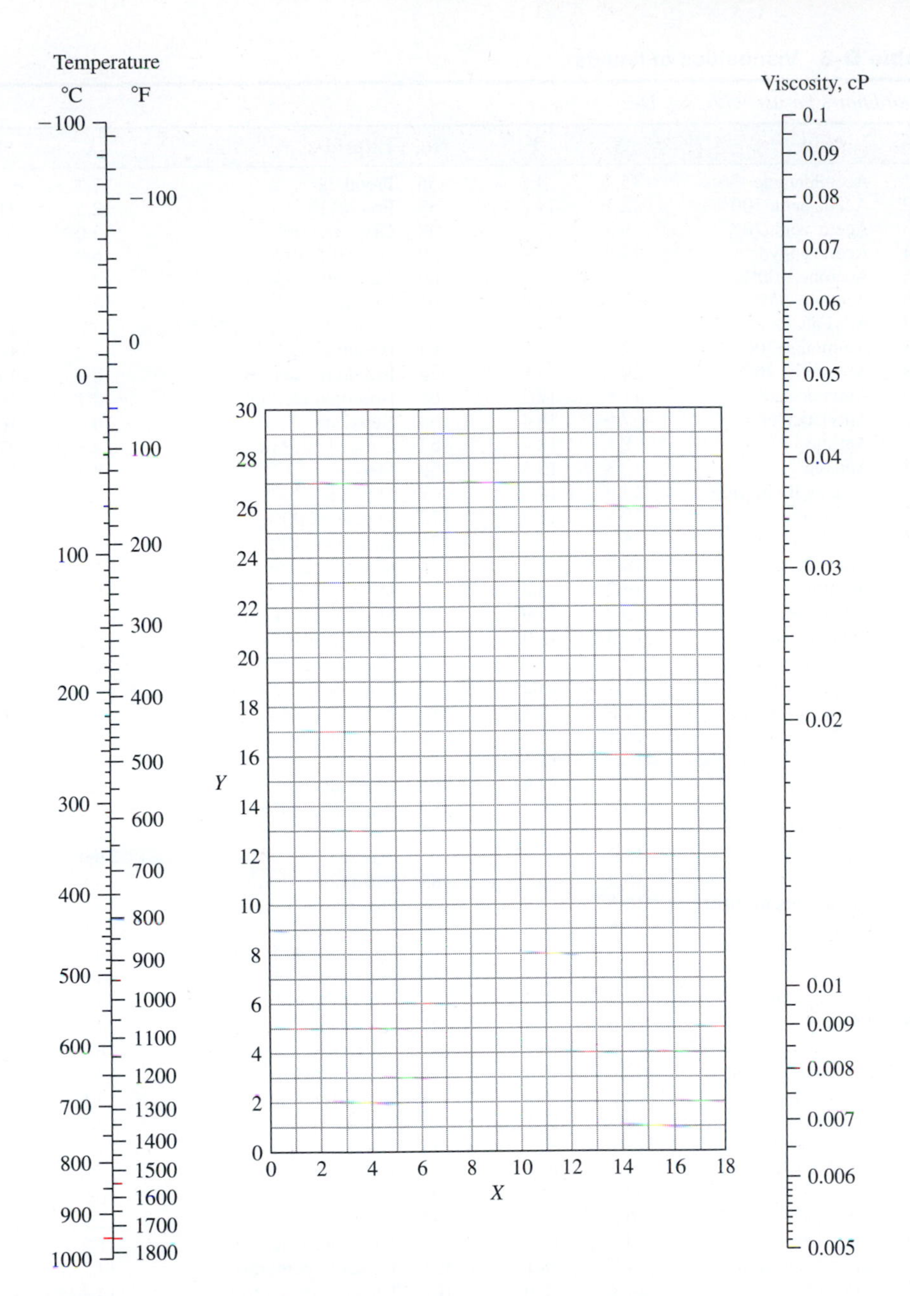

Figure D-1
Viscosities of gases at 1 atm. (For coordinates see Table D-2.)

Table D-3 Viscosities of liquids

Coordinates for use with Fig. D-2

No.	Liquid	X	Y	No.	Liquid	X	Y
1	Acetaldehyde	15.2	4.8	56	Freon 22	17.2	4.7
2	Acetic acid, 100%	12.1	14.2	57	Freon 113	12.5	11.4
3	Acetic acid, 70%	9.5	17.0	58	Glycerol, 100%	2.0	30.0
4	Acetic anhydride	12.7	12.8	59	Glycerol, 50%	6.9	19.6
5	Acetone, 100%	14.5	7.2	60	Heptene	14.1	8.4
6	Acetone, 35%	7.9	15.0	61	Hexane	14.7	7.0
7	Allyl alcohol	10.2	14.3	62	Hydrochloric acid, 31.5%	13.0	16.6
8	Ammonia, 100%	12.6	2.0	63	Isobutyl alcohol	7.1	18.0
9	Ammonia, 26%	10.1	13.9	64	Isobutyric acid	12.2	14.4
10	Amyl acetate	11.8	12.5	65	Isopropyl alcohol	8.2	16.0
11	Amyl alcohol	7.5	18.4	66	Kerosene	10.2	16.9
12	Aniline	8.1	18.7	67	Linseed oil, raw	7.5	27.2
13	Anisole	12.3	13.5	68	Mercury	18.4	16.4
14	Arsenic trichloride	13.9	14.5	69	Methanol, 100%	12.4	10.5
15	Benzene	12.5	10.9	70	Methanol, 90%	12.3	11.8
16	Brine, $CaCl_2$, 25%	6.6	15.9	71	Methanol, 40%	7.8	15.5
17	Brine, NaCl, 25%	10.2	16.6	72	Methyl acetate	14.2	8.2
18	Bromine	14.2	13.2	73	Methyl chloride	15.0	3.8
19	Bromotoluene	20.0	15.9	74	Methyl ethyl ketone	13.9	8.6
20	Butyl acetate	12.3	11.0	75	Naphthalene	7.9	18.1
21	Butyl alcohol	8.6	17.2	76	Nitric acid, 95%	12.8	13.8
22	Butyric acid	12.1	15.3	77	Nitric acid, 60%	10.8	17.0
23	Carbon dioxide	11.6	0.3	78	Nitrobenzene	10.6	16.2
24	Carbon disulfide	16.1	7.5	79	Nitrotoluene	11.0	17.0
25	Carbon tetrachloride	12.7	13.1	80	Octane	13.7	10.0
26	Chlorobenzene	12.3	12.4	81	Octyl alcohol	6.6	21.1
27	Chloroform	14.4	10.2	82	Pentachloroethane	10.9	17.3
28	Chlorosulfonic acid	11.2	18.1	83	Pentane	14.9	5.2
29	Chlorotoluene, ortho	13.0	13.3	84	Phenol	6.9	20.8
30	Chlorotoluene, meta	13.3	12.5	85	Phosphorus tribromide	13.8	16.7
31	Chlorotoluene, para	13.3	12.5	86	Phosphorus trichloride	16.2	10.9
32	Cresol, meta	2.5	20.8	87	Propionic acid	12.8	13.8
33	Cyclohexanol	2.9	24.3	88	Propyl alcohol	9.1	16.5
34	Dibromoethane	12.7	15.8	89	Propyl bromide	14.5	9.6
35	Dichloroethane	13.2	12.2	90	Propyl chloride	14.4	7.5
36	Dichloromethane	14.6	8.9	91	Propyl iodide	14.1	11.6
37	Diethyl oxalate	11.0	16.4	92	Sodium	16.4	13.9
38	Dimethyl oxalate	12.3	15.8	93	Sodium hydroxide, 50%	3.2	25.8
39	Diphenyl	12.0	18.3	94	Stannic chloride	13.5	12.8
40	Dipropyl oxalate	10.3	17.7	95	Sulfur dioxide	15.2	7.1
41	Ethyl acetate	13.7	9.1	96	Sulfuric acid, 110%	7.2	27.4
42	Ethyl alcohol, 100%	10.5	13.8	97	Sulfuric acid, 98%	7.0	24.8
43	Ethyl alcohol, 95%	9.8	14.3	98	Sulfuric acid, 60%	10.2	21.3
44	Ethyl alcohol, 40%	6.5	16.6	99	Sulfuryl chloride	15.2	12.4
45	Ethyl benzene	13.2	11.5	100	Tetrachloroethane	11.9	15.7
46	Ethyl bromide	14.5	8.1	101	Tetrachloroethylene	14.2	12.7
47	Ethyl chloride	14.8	6.0	102	Titanium tetrachloride	14.4	12.3
48	Ethyl ether	14.5	5.3	103	Toluene	13.7	10.4
49	Ethyl formate	14.2	8.4	104	Trichloroethylene	14.8	10.5
50	Ethyl iodide	14.7	10.3	105	Turpentine	11.5	14.9
51	Ethylene glycol	6.0	23.6	106	Vinyl acetate	14.0	8.8
52	Formic acid	10.7	15.8	107	Water	10.2	13.0
53	Freon 11	14.4	9.0	108	Xylene, ortho	13.5	12.1
54	Freon 12	16.8	5.6	109	Xylene, meta	13.9	10.6
55	Freon 21	15.7	7.5	110	Xylene, para	13.9	10.9

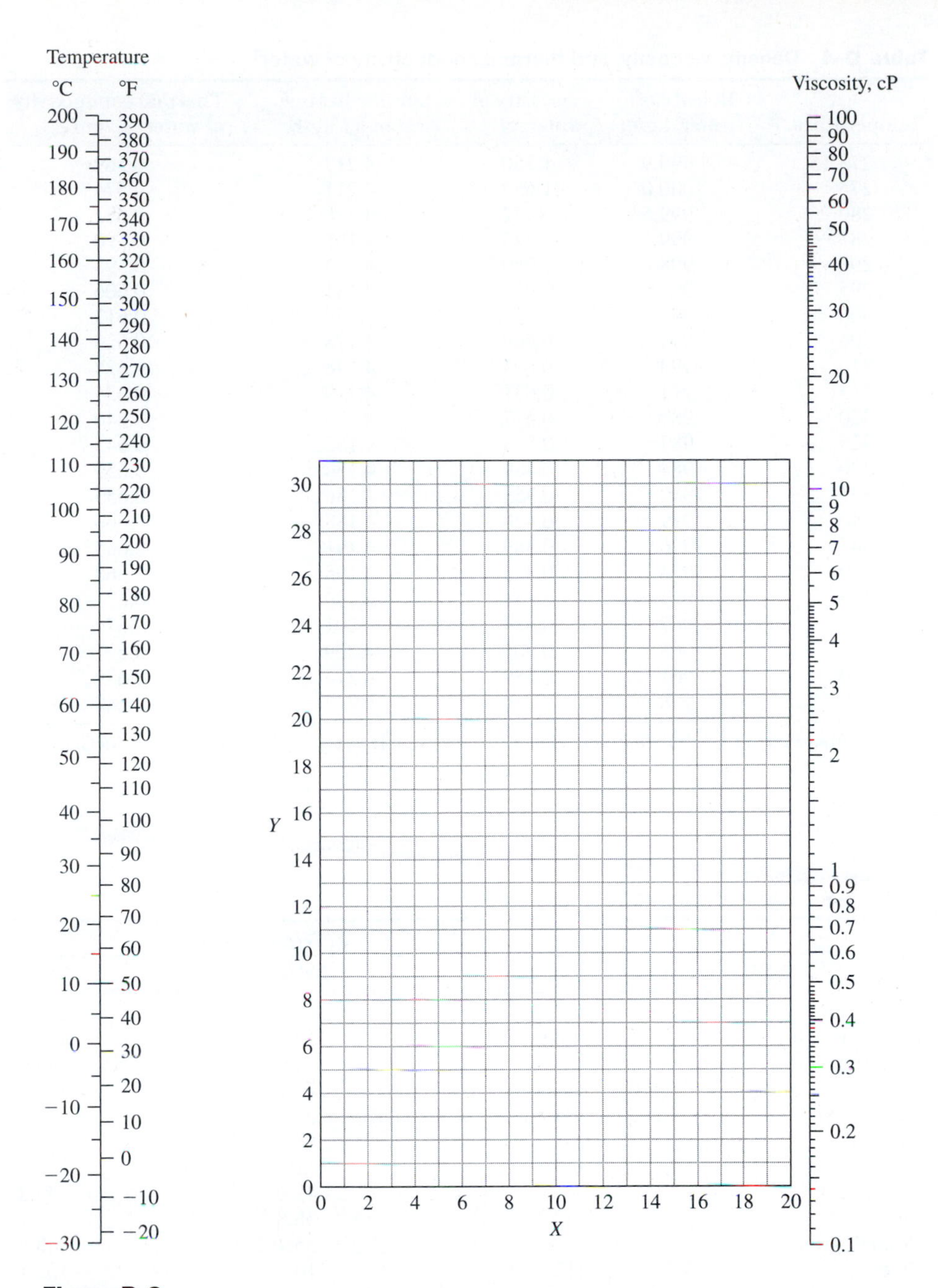

Figure D-2
Viscosities of liquids at 1 atm. (For coordinates see Table D-3.)

Table D-4 Density, viscosity, and thermal conductivity of water†

Temperature, K	Density of water, kg/m^3	Viscosity of water, cP	Specific heat of water, kJ/kg·K	Thermal conductivity of water, W/m·K
273.15	999.9	1.750	4.217	0.569
275	1000.0	1.652	4.211	0.574
280	999.5	1.422	4.198	0.582
285	999	1.225	4.189	0.590
290	998	1.080	4.184	0.598
295	997	0.959	4.181	0.606
300	996	0.855	4.179	0.613
305	995	0.769	4.178	0.620
310	993	0.695	4.178	0.628
315	991	0.631	4.179	0.634
320	989	0.577	4.180	0.640
325	987	0.528	4.182	0.645
330	984	0.489	4.184	0.650
335	982	0.453	4.186	0.655
340	979	0.420	4.188	0.660
345	976	0.389	4.191	0.665
350	974	0.365	4.195	0.668
355	971	0.343	4.199	0.671
360	967	0.324	4.203	0.674
365	963	0.306	4.209	0.677
370	961	0.289	4.214	0.679
373.15	958	0.279	4.217	0.680

†Adapted from *Perry's Chemical Engineers' Handbook,* 7th ed., J. H. Perry and D. W. Green, eds., McGraw-Hill, New York, 1997.

Table D-5 Thermal conductivity of metals

Metal	k, W/m·K		
	0°C	100°C	300°C
Aluminum	202.5	205.9	230.1
Brass (70–30)	96.9	103.8	114.2
Cast iron	55.4	51.9	45.0
Copper	387.7	377.3	366.9
Lead	34.6	32.9	31.1
Nickel	62.3	58.8	55.4
Silver	418.8	411.9	—
Steel (mild)	—	45.0	43.3
Tin	62.3	58.8	—
Wrought iron	—	55.4	48.5
Zinc	112.5	110.8	102.1

Table D-6 Thermal conductivity of nonmetallic solids

Material	Temperature, °C	k, W/m·K
Asbestos-cement boards	20	0.74
Bricks:		
Building	20	0.69
Fire clay	200	1.00
	1000	1.64
Sil-O-Cel	204	0.073
Calcium carbonate (natural)	30	2.25
Calcium sulfate (building plaster)	25	0.43
Celluloid	30	0.21
Concrete (stone)	...	0.93
Corkboard	30	0.043
Felt (wool)	30	0.032
Glass (window)	...	0.5–1.05
Rubber (hard)	0	0.15
Wood (across grain):		
Maple	50	0.19
Oak	15	0.21
Pine	15	0.15

Table D-7 Thermal conductivity of liquids

Liquid	Temperature, °C	k, W/m·K
Acetic acid:		
100%	20	0.171
50%	20	0.346
Acetone	30	0.176
	75	0.164
Benzene	30	0.159
	60	0.151
Ethyl alcohol:		
100%	20	0.182
	50	0.151
40%	20	0.388
Ethylene glycol	0	0.265
Glycerol:		
100%	20	0.284
	100	0.284
40%	20	0.448
n-Heptane	30	0.140
Kerosene	20	0.149
Methyl alcohol:		
100%	20	0.215
	50	0.197
40%	20	0.405
n-Octane	30	0.144
Sodium chloride brine, 25%	30	0.571
Sulfuric acid:		
90%	30	0.363
30%	30	0.519
Toluene	30	0.149
Water	0	0.554
	93	0.678

Table D-8 Thermal conductivity of gases

Gas	Temperature, °C	k, W/m·K
Air	0	0.0242
	100	0.0317
	200	0.0391
Ammonia	0	0.0222
	50	0.0272
Carbon dioxide	0	0.0147
	100	0.0230
Chlorine	0	0.0074
Hydrogen	0	0.1731
	100	0.2233
Methane	0	0.0303
	50	0.0372
Nitrogen	0	0.0242
	100	0.0311
Oxygen	0	0.0245
	100	0.0320
Sulfur dioxide	0	0.0086
	100	0.0119
Water vapor	93	0.0275
	315	0.0443

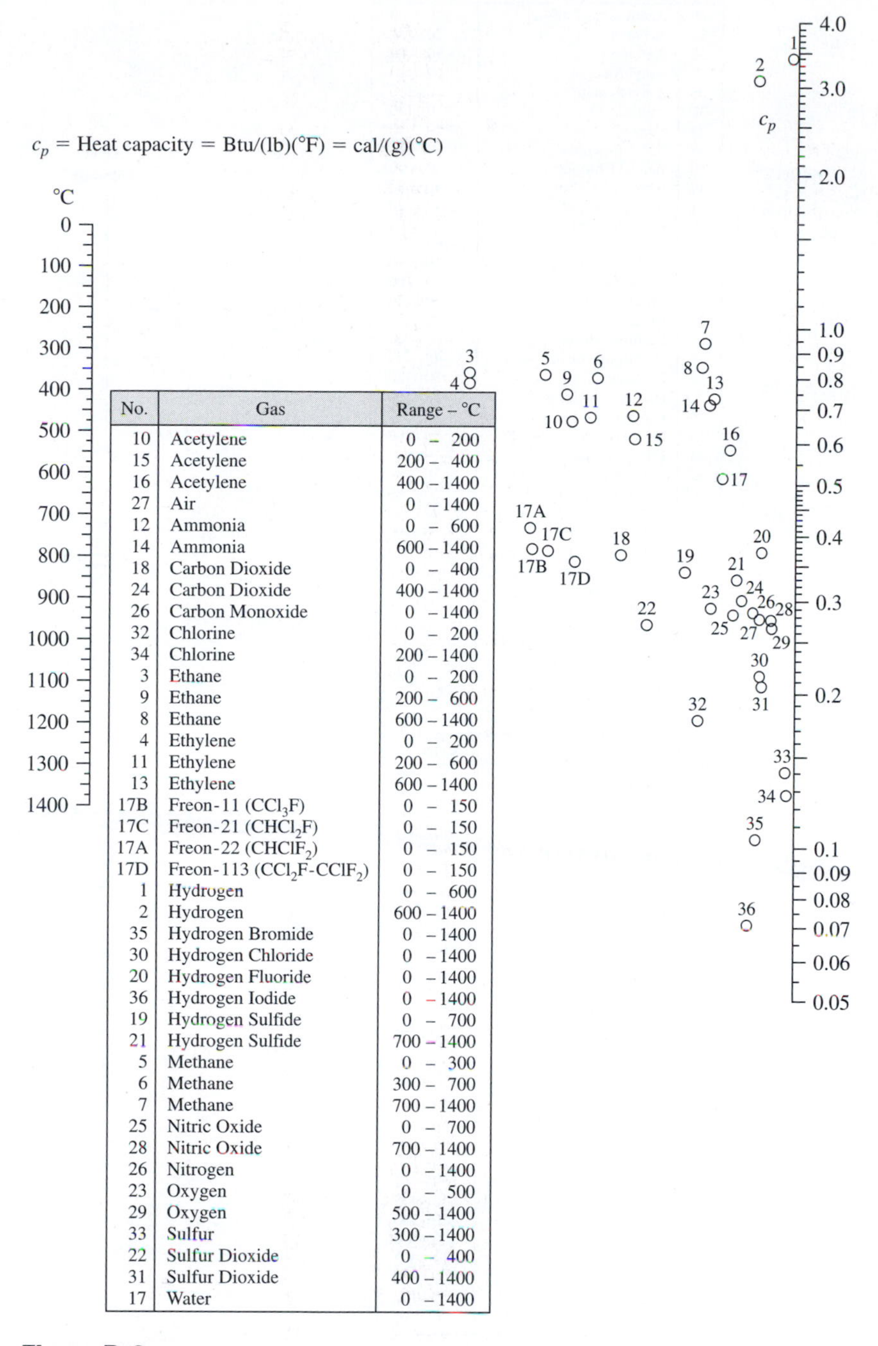

No.	Gas	Range – °C
10	Acetylene	0 – 200
15	Acetylene	200 – 400
16	Acetylene	400 – 1400
27	Air	0 – 1400
12	Ammonia	0 – 600
14	Ammonia	600 – 1400
18	Carbon Dioxide	0 – 400
24	Carbon Dioxide	400 – 1400
26	Carbon Monoxide	0 – 1400
32	Chlorine	0 – 200
34	Chlorine	200 – 1400
3	Ethane	0 – 200
9	Ethane	200 – 600
8	Ethane	600 – 1400
4	Ethylene	0 – 200
11	Ethylene	200 – 600
13	Ethylene	600 – 1400
17B	Freon-11 (CCl_3F)	0 – 150
17C	Freon-21 ($CHCl_2F$)	0 – 150
17A	Freon-22 ($CHClF_2$)	0 – 150
17D	Freon-113 (CCl_2F-$CClF_2$)	0 – 150
1	Hydrogen	0 – 600
2	Hydrogen	600 – 1400
35	Hydrogen Bromide	0 – 1400
30	Hydrogen Chloride	0 – 1400
20	Hydrogen Fluoride	0 – 1400
36	Hydrogen Iodide	0 – 1400
19	Hydrogen Sulfide	0 – 700
21	Hydrogen Sulfide	700 – 1400
5	Methane	0 – 300
6	Methane	300 – 700
7	Methane	700 – 1400
25	Nitric Oxide	0 – 700
28	Nitric Oxide	700 – 1400
26	Nitrogen	0 – 1400
23	Oxygen	0 – 500
29	Oxygen	500 – 1400
33	Sulfur	300 – 1400
22	Sulfur Dioxide	0 – 400
31	Sulfur Dioxide	400 – 1400
17	Water	0 – 1400

Figure D-3
Heat capacities of gases at 1 atm pressure. To obtain heat capacity in kJ/kg·K, multiply by 4.1868.

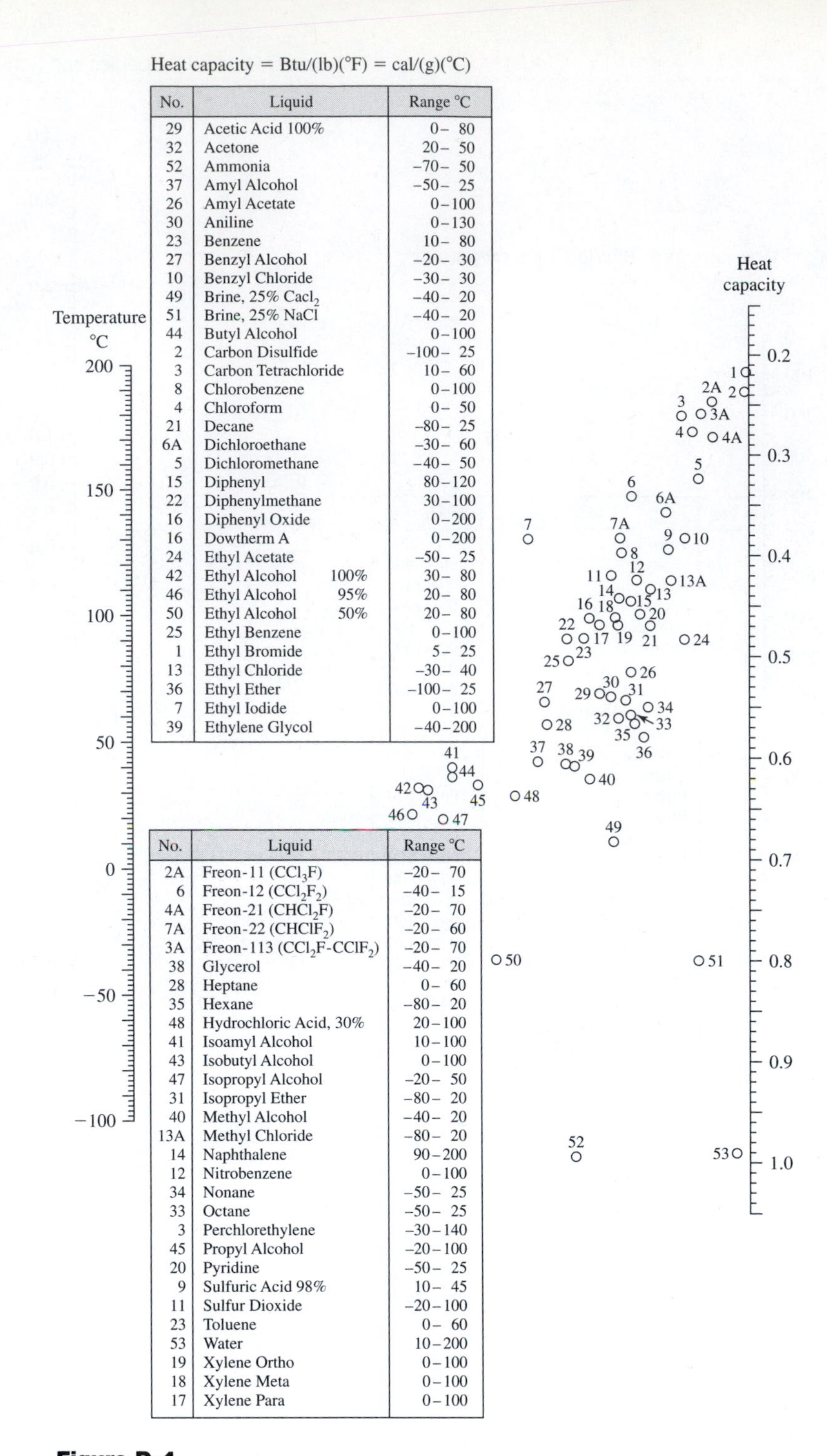

No.	Liquid	Range °C
29	Acetic Acid 100%	0– 80
32	Acetone	20– 50
52	Ammonia	−70– 50
37	Amyl Alcohol	−50– 25
26	Amyl Acetate	0–100
30	Aniline	0–130
23	Benzene	10– 80
27	Benzyl Alcohol	−20– 30
10	Benzyl Chloride	−30– 30
49	Brine, 25% $Cacl_2$	−40– 20
51	Brine, 25% NaCl	−40– 20
44	Butyl Alcohol	0–100
2	Carbon Disulfide	−100– 25
3	Carbon Tetrachloride	10– 60
8	Chlorobenzene	0–100
4	Chloroform	0– 50
21	Decane	−80– 25
6A	Dichloroethane	−30– 60
5	Dichloromethane	−40– 50
15	Diphenyl	80–120
22	Diphenylmethane	30–100
16	Diphenyl Oxide	0–200
16	Dowtherm A	0–200
24	Ethyl Acetate	−50– 25
42	Ethyl Alcohol 100%	30– 80
46	Ethyl Alcohol 95%	20– 80
50	Ethyl Alcohol 50%	20– 80
25	Ethyl Benzene	0–100
1	Ethyl Bromide	5– 25
13	Ethyl Chloride	−30– 40
36	Ethyl Ether	−100– 25
7	Ethyl Iodide	0–100
39	Ethylene Glycol	−40–200

No.	Liquid	Range °C
2A	Freon-11 (CCl_3F)	−20– 70
6	Freon-12 (CCl_2F_2)	−40– 15
4A	Freon-21 ($CHCl_2F$)	−20– 70
7A	Freon-22 ($CHClF_2$)	−20– 60
3A	Freon-113 (CCl_2F-$CClF_2$)	−20– 70
38	Glycerol	−40– 20
28	Heptane	0– 60
35	Hexane	−80– 20
48	Hydrochloric Acid, 30%	20–100
41	Isoamyl Alcohol	10–100
43	Isobutyl Alcohol	0–100
47	Isopropyl Alcohol	−20– 50
31	Isopropyl Ether	−80– 20
40	Methyl Alcohol	−40– 20
13A	Methyl Chloride	−80– 20
14	Naphthalene	90–200
12	Nitrobenzene	0–100
34	Nonane	−50– 25
33	Octane	−50– 25
3	Perchlorethylene	−30–140
45	Propyl Alcohol	−20–100
20	Pyridine	−50– 25
9	Sulfuric Acid 98%	10– 45
11	Sulfur Dioxide	−20–100
23	Toluene	0– 60
53	Water	10–200
19	Xylene Ortho	0–100
18	Xylene Meta	0–100
17	Xylene Para	0–100

Figure D-4
Heat capacities of liquids. To obtain heat capacity in kJ/kg·K, multiply by 4.1868.

Table D-9 Specific gravities of liquids

Pure liquid	Formula	Temperature, °C	Specific gravity
Acetaldehyde	CH_3CHO	18	0.783
Acetic acid	CH_3CO_2H	0	1.067
		30	1.038
Acetone	CH_3COCH_3	20	0.792
Benzene	C_6H_6	20	0.879
n-Butyl alcohol	$C_2H_5CH_2CH_2OH$	20	0.810
Carbon tetrachloride	CCl_4	20	1.595
Ethyl alcohol	CH_3CH_2OH	10	0.798
		30	0.791
Ethyl ether	$(CH_3CH_2)_2O$	25	0.708
Ethylene glycol	$CH_2OH{\cdot}CH_2OH$	19	1.113
Glycerol	$CH_2OH{\cdot}CHOH{\cdot}CH_2OH$	15	1.264
		30	1.255
Isobutyl alcohol	$(CH_3)_2CHCH_2OH$	18	0.805
Isopropyl alcohol	$(CH_3)_2CHOH$	0	0.802
		30	0.777
Methyl alcohol	CH_3OH	0	0.810
		20	0.792
Nitric acid	HNO_3	10	1.531
		30	1.495
Phenol	C_6H_5OH	25	1.071
n-Propyl alcohol	$CH_3CH_2CH_2OH$	20	0.804
Sulfuric acid	H_2SO_4	10	1.841
		30	1.821
Water	H_2O	4	1.000
		100	0.958

Note: The values presented in this table are based on the density of water at 4°C and a total pressure of 1 atm.

$$\text{Specific gravity} = \frac{\text{density of material at indicated temperature}}{\text{density of liquid water at 4°C}}$$

Density of liquid water at 4°C = 1000 kg/m^3.

Table D-10 Specific gravities of solids

Substance	Specific gravity
Aluminum, hard-drawn	2.55–2.80
Brass, cast-rolled	8.4–8.7
Copper, cast-rolled	8.8–8.95
Glass, common	2.4–2.8
Gold, cast-hammered	19.25–19.35
Iron:	
Gray cast	7.03–7.13
Wrought	7.6–7.9
Lead	11.34
Nickel	8.9
Platinum, cast-hammered	21.5
Silver, cast-hammered	10.4–10.6
Steel, cold-drawn	7.83
Tin, cast-hammered	7.2–7.5
White oak timber, air-dried	0.77
White pine timber, air-dried	0.43
Zinc, cast-rolled	6.9–7.2

Note: The specific gravities as indicated in this table apply at ordinary atmospheric temperatures.The values are based on the density of water at 4°C.

$$\text{Specific gravity} = \frac{\text{density of material}}{\text{density of liquid water at 4°C}}$$

Density of liquid water at 4°C. = 1000 kg/m^3.

Table D-11 Properties of saturated steam[†]

Temperature, K	Abs. pressure, kPa	Volume, m³/kg Liquid	Volume, m³/kg Vapor	Heat of vaporization, kJ/kg	Specific heat, kJ/kg·K	Viscosity, Pa·s × 10⁶	Thermal conductivity, W/m·K × 10³
273.15	0.611	1.000	206.3	2502	1.854	8.02	18.2
275	0.697	1.000	181.7	2497	1.855	8.09	18.3
280	0.990	1.000	130.4	2485	1.858	8.29	18.6
285	1.387	1.000	99.4	2473	1.861	8.49	18.9
290	1.917	1.001	69.7	2461	1.864	8.69	19.3
295	2.617	1.002	51.94	2449	1.868	8.89	19.5
300	3.531	1.003	39.13	2438	1.872	9.09	19.6
305	4.712	1.005	29.74	2426	1.877	9.29	20.1
310	6.221	1.007	22.93	2414	1.882	9.49	20.4
315	8.132	1.009	17.82	2402	1.888	9.69	20.7
320	10.53	1.012	13.98	2390	1.895	9.89	21.0
325	13.51	1.013	11.06	2378	1.903	10.09	21.3
330	17.19	1.016	8.82	2366	1.911	10.29	21.7
335	21.67	1.018	7.09	2354	1.920	10.49	22.0
340	27.13	1.021	5.74	2342	1.930	10.69	22.2
345	33.72	1.024	4.683	2329	1.941	10.89	22.6
350	41.63	1.027	3.846	2317	1.954	11.09	23.0
355	51.00	1.030	3.180	2304	1.968	11.29	23.3
360	62.09	1.034	2.645	2291	1.983	11.49	23.7
365	75.14	1.038	2.212	2278	1.999	11.69	24.1
370	90.40	1.041	1.861	2265	2.017	11.89	24.5
373.15	101.33	1.044	1.679	2257	2.029	12.02	24.8
375	108.15	1.045	1.574	2252	2.036	12.09	24.9
380	128.69	1.049	1.337	2239	2.057	12.29	25.4
385	152.33	1.053	1.142	2225	2.080	12.49	25.8
390	179.4	1.058	0.980	2212	2.104	12.69	26.3
400	245.5	1.067	0.731	2183	2.158	13.05	27.2
410	330.2	1.077	0.553	2153	2.221	13.42	28.2
420	437.0	1.088	0.425	2123	2.291	13.79	29.8
430	569.9	1.099	0.331	2091	2.369	14.14	30.4
440	733.3	1.110	0.261	2059	2.46	14.50	31.7
450	931.9	1.123	0.208	2024	2.56	14.85	33.1
460	1171	1.137	0.167	1989	2.68	15.19	34.6
470	1455	1.152	0.136	1951	2.79	15.54	36.3
480	1790	1.167	0.111	1912	2.94	15.88	38.1
490	2183	1.184	0.0922	1870	3.10	16.23	40.1
500	2640	1.203	0.0766	1825	3.27	16.59	42.3
510	3166	1.222	0.0631	1779	3.47	16.95	44.7
520	3770	1.244	0.0525	1730	3.70	17.33	47.5
530	4458	1.268	0.0445	1679	3.96	17.72	50.6
540	5238	1.294	0.0375	1622	4.27	18.1	54.0
550	6119	1.323	0.0317	1564	4.64	18.6	58.3
560	7108	1.355	0.0269	1499	5.09	19.1	63.7
570	8216	1.392	0.0228	1429	5.67	19.7	70.7
580	9451	1.433	0.0193	1353	6.40	20.4	76.7
590	10,830	1.482	0.0163	1274	7.35	21.5	84.1
600	12,350	1.541	0.0137	1176	8.75	22.7	92.9
610	13,730	1.612	0.0115	1068	11.1	24.1	103
620	15,910	1.705	0.0094	941	15.4	25.9	114
625	16,910	1.778	0.0085	858	18.3	27.0	121
630	17,970	1.856	0.0075	781	22.1	28.0	130
635	19,090	1.935	0.0066	683	27.6	30.0	141
640	20,270	2.075	0.0057	560	42	32.0	155
645	21,520	2.351	0.0045	361	—	37.0	178
647.3[‡]	31,700	3.170	0.0032	0	∞	45.0	238

[†]Adapted from F. P. Incropera and D. P. DeWitt, *Fundamentals of Heat and Mass Transfer,* 4th ed., J. Wiley, New York, 1981.
[‡]Critical temperature.

Table D-12 Heat exchanger and condenser tube data†

Tube OD, in.	BWG	Wall thickness, in.	ID, in.	Flow area per tube, in.^{2}	Surface per linear ft, ft^2 Outside	Surface per linear ft, ft^2 Inside	Weight per lin ft, lb steel
½	12	0.109	0.282	0.0625	0.1309	0.0748	0.493
	14	0.083	0.334	0.0876	0.1309	0.0874	0.403
	16	0.065	0.370	0.1076	0.1309	0.0969	0.329
	18	0.049	0.402	0.127	0.1309	0.1052	0.258
	20	0.035	0.430	0.145	0.1309	0.1125	0.190
¾	10	0.134	0.482	0.182	0.1963	0.1263	0.965
	11	0.120	0.510	0.204	0.1963	0.1335	0.884
	12	0.109	0.532	0.223	0.1963	0.1393	0.817
	13	0.095	0.560	0.247	0.1963	0.1466	0.727
	14	0.083	0.584	0.268	0.1963	0.1529	0.647
	15	0.072	0.606	0.289	0.1963	0.1587	0.571
	16	0.065	0.620	0.302	0.1963	0.1623	0.520
	17	0.058	0.634	0.314	0.1963	0.1660	0.469
	18	0.049	0.652	0.334	0.1963	0.1707	0.401
1	8	0.165	0.670	0.335	0.2618	0.1754	1.61
	9	0.148	0.704	0.389	0.2618	0.1843	1.47
	10	0.134	0.732	0.421	0.2618	0.1916	1.36
	11	0.120	0.760	0.455	0.2618	0.1990	1.23
	12	0.109	0.782	0.479	0.2618	0.2048	1.14
	13	0.095	0.810	0.515	0.2618	0.2121	1.00
	14	0.083	0.834	0.546	0.2618	0.2183	0.890
	15	0.072	0.856	0.576	0.2618	0.2241	0.781
	16	0.065	0.870	0.594	0.2618	0.2277	0.710
	17	0.058	0.884	0.613	0.2618	0.2314	0.639
	18	0.049	0.902	0.639	0.2618	0.2361	0.545
1¼	8	0.165	0.920	0.665	0.3271	0.2409	2.09
	9	0.148	0.954	0.714	0.3271	0.2498	1.91
	10	0.134	0.982	0.757	0.3271	0.2572	1.75
	11	0.120	1.01	0.800	0.3271	0.2644	1.58
	12	0.109	1.03	0.836	0.3271	0.2701	1.45
	13	0.095	1.06	0.884	0.3271	0.2775	1.28
	14	0.083	1.08	0.923	0.3271	0.2839	1.13
	15	0.072	1.11	0.960	0.3271	0.2896	0.991
	16	0.065	1.12	0.985	0.3271	0.2932	0.900
	17	0.058	1.13	1.01	0.3271	0.2969	0.808
	18	0.049	1.15	1.04	0.3271	0.3015	0.688
1½	8	0.165	1.17	1.075	0.3925	0.3063	2.57
	9	0.148	1.20	1.14	0.3925	0.3152	2.34
	10	0.134	1.23	1.19	0.3925	0.3225	2.14
	11	0.120	1.26	1.25	0.3925	0.3299	1.98
	12	0.109	1.28	1.29	0.3925	0.3356	1.77
	13	0.095	1.31	1.35	0.3925	0.3430	1.56
	14	0.083	1.33	1.40	0.3925	0.3492	1.37
	15	0.072	1.36	1.44	0.3925	0.3555	1.20
	16	0.065	1.37	1.47	0.3925	0.3587	1.09
	17	0.058	1.38	1.50	0.3925	0.3623	0.978
	18	0.049	1.40	1.54	0.3925	0.3670	0.831

†The data provided in this table are in the USCS units used by the heat exchanger manufacturers in the United States.

Table D-13 Steel pipe dimensions[†]

Nominal pipe size, in.	OD, in.	Schedule no.	ID, in.	Flow area per pipe, in.^{2}	Surface per linear ft, ft^2		Weight per lin ft, lb steel
					Outside	Inside	
⅛	0.405	40[‡]	0.269	0.058	0.106	0.070	0.25
		80[§]	0.215	0.036	0.106	0.056	0.32
¼	0.540	40	0.364	0.104	0.141	0.095	0.43
		80	0.302	0.072	0.141	0.079	0.54
⅜	0.675	40	0.493	0.192	0.177	0.129	0.57
		80	0.423	0.141	0.177	0.111	0.74
½	0.840	40	0.622	0.304	0.220	0.163	0.85
		80	0.546	0.235	0.220	0.143	1.09
¾	1.05	40	0.824	0.534	0.275	0.216	1.13
		80	0.742	0.432	0.275	0.194	1.48
1	1.32	40	1.049	0.864	0.344	0.274	1.68
		80	0.957	0.718	0.344	0.250	2.17
1¼	1.66	40	1.380	1.50	0.435	0.362	2.28
		80	1.278	1.28	0.435	0.335	3.00
1½	1.90	40	1.610	2.04	0.498	0.422	2.72
		80	1.500	1.76	0.498	0.393	3.64
2	2.38	40	2.067	3.35	0.622	0.542	3.66
		80	1.939	2.95	0.622	0.508	5.03
2½	2.88	40	2.469	4.79	0.753	0.627	5.80
		80	2.323	4.23	0.753	0.609	7.67
3	3.50	40	3.068	7.38	0.917	0.804	7.58
		80	2.900	6.61	0.917	0.760	10.3
4	4.50	40	4.026	12.7	1.178	1.055	10.8
		80	3.826	11.5	1.178	1.002	15.0
6	6.625	40	6.065	28.9	1.734	1.590	19.0
		80	5.761	26.1	1.734	1.510	28.6
8	8.625	40	7.981	50.0	2.258	2.090	28.6
		80	7.625	45.7	2.258	2.000	43.4
10	10.75	40	10.02	78.8	2.814	2.62	40.5
		60	9.75	74.6	2.814	2.55	54.8
12	12.75	30	12.09	115	3.338	3.17	43.8
16	16.0	30	15.25	183	4.189	4.00	62.6
20	20.0	20	19.25	291	5.236	5.05	78.6
24	24.0	20	23.25	425	6.283	6.09	94.7

[†]The data provided in this table are in the USCS units used by the pipe manufacturers in the United States.
[‡]Schedule 40 designates former "standard" pipe.
[§]Schedule 80 designates former "extra-strong" pipe.

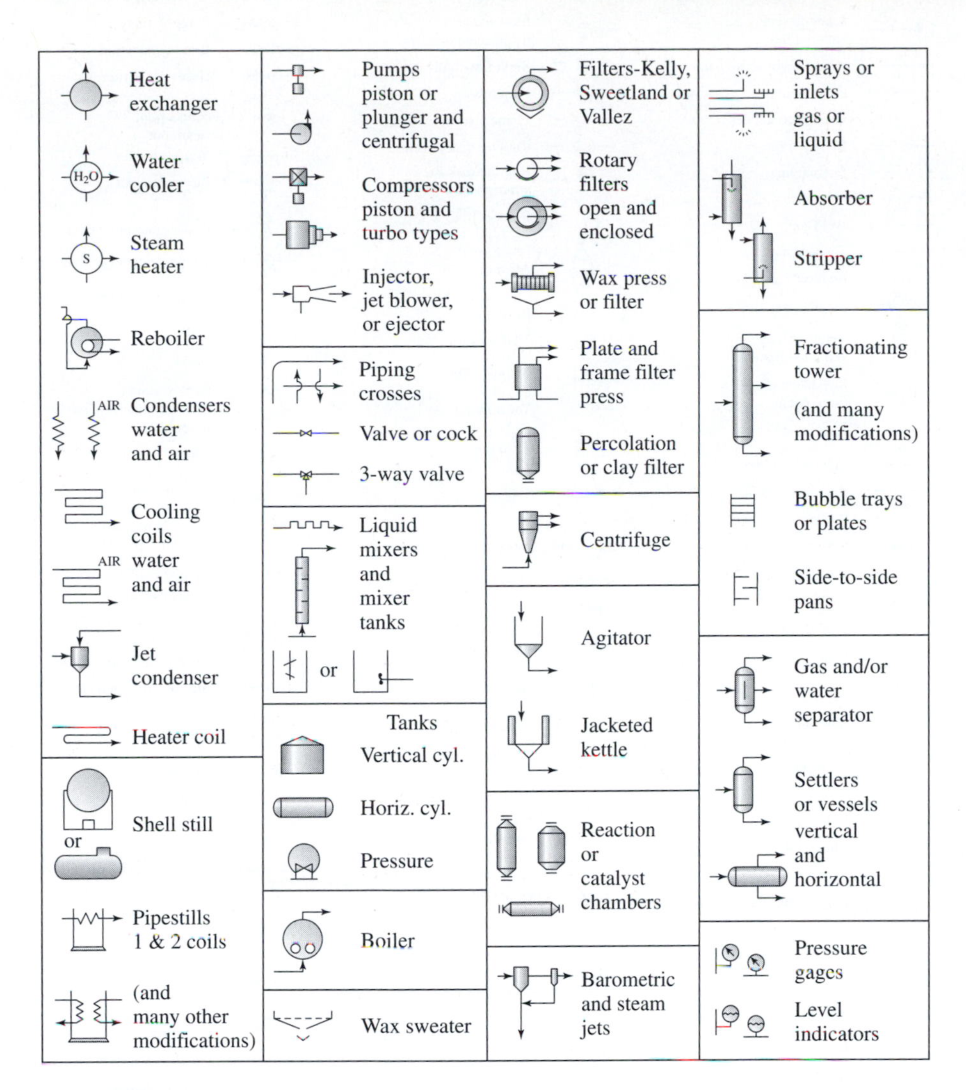

Figure D-5
Equipment symbols

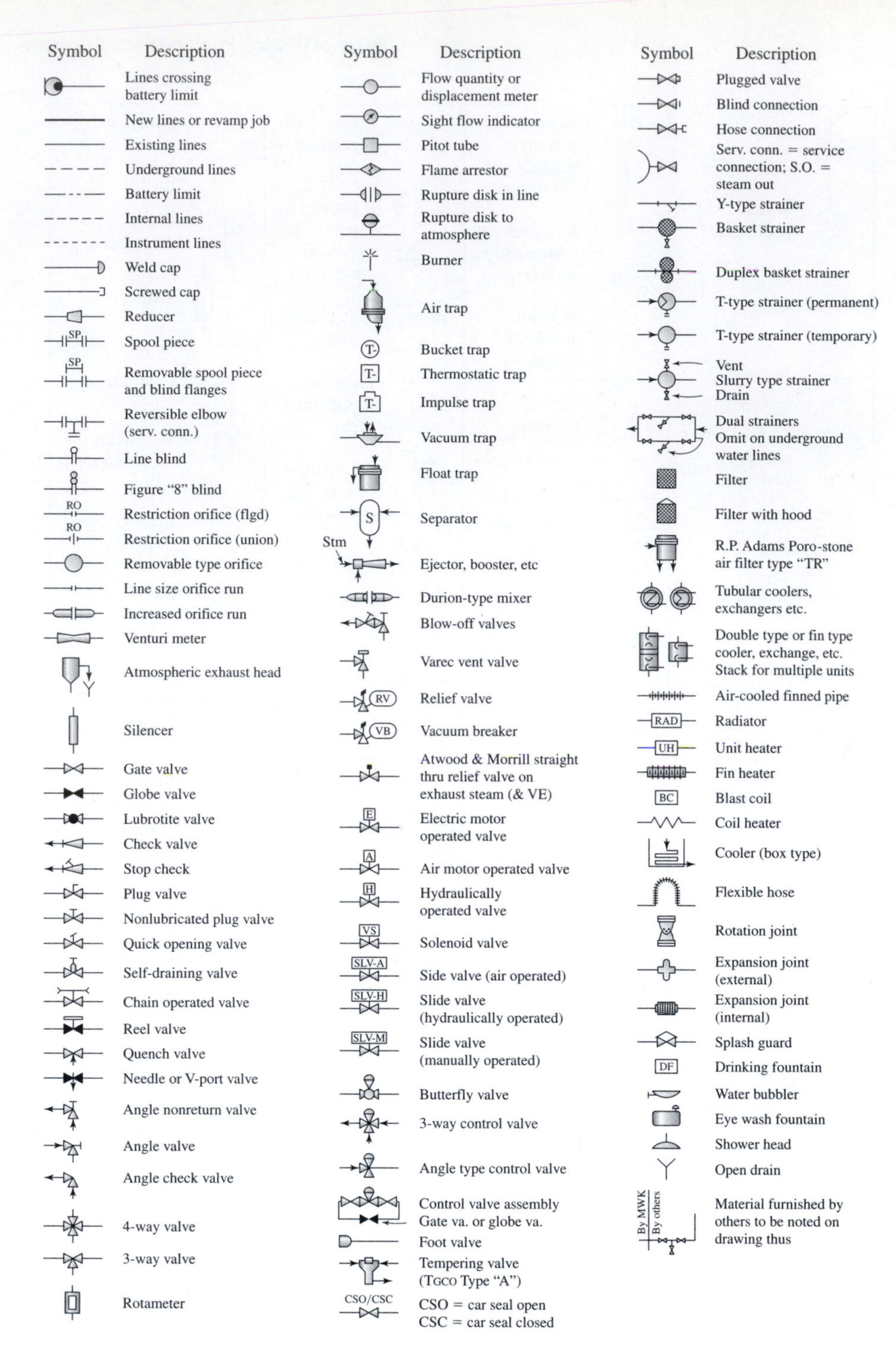

Figure D-6

Flowsheet symbols, particularly for detailed equipment flowsheets. *(Courtesy of the M. W. Kellogg Co.)*

Table D-14 International atomic weights

Substance	Symbol	Atomic weight	At. no.	Substance	Symbol	Atomic weight	At. no.
Actinium	Ac	227.03	89	Mercury	Hg	200.59	80
Aluminum	Al	26.98	13	Molybdenum	Mo	95.94	42
Americium	Am	243	95	Neodymium	Nd	144.24	60
Antimony	Sb	121.75	51	Neon	Ne	20.179	10
Argon	A	39.948	18	Neptunium	Np	237	93
Arsenic	As	74.92	33	Nickel	Ni	58.70	28
Astatine	At	210	85	Niobium			
Barium	Ba	137.33	56	(Columbian)	Nb	92.91	41
Berkelium	Bk	247	97	Nitrogen	N	14.007	7
Beryllium	Be	9.012	4	Nobelium	No	259	102
Bismuth	Bi	208.98	83	Osmium	Os	190.2	76
Boron	B	10.81	5	Oxygen	O	16	8
Bromine	Br	79.904	35	Palladium	Pd	106.4	46
Cadmium	Cd	112.41	48	Phosphorus	P	30.975	15
Calcium	Ca	40.08	20	Platinum	Pt	195.09	78
Californium	Cf	251	98	Plutonium	Pu	244	94
Carbon	C	12.011	6	Polonium	Po	209	84
Cerium	Ce	140.12	58	Potassium	K	39.098	19
Cesium	Cs	132.91	55	Praseodymium	Pr	140.91	59
Chlorine	Cl	35.457	17	Promethium	Pm	145	61
Chromium	Cr	52.00	24	Protactinium	Pa	231	91
Cobalt	Co	58.93	27	Radium	Ra	226.03	88
Copper	Cu	63.55	29	Radon	Rn	222	86
Curium	Cm	247	96	Rhenium	Re	186.2	75
Dysprosium	Dy	162.50	66	Rhodium	Rh	102.91	45
Einsteinium	Es	254	99	Rubidium	Rb	85.47	37
Erbium	Er	167.26	68	Ruthenium	Ru	101.1	44
Europium	Eu	152.0	63	Samarium	Sm	150.4	62
Fermium	Fm	257	100	Scandium	Sc	44.96	21
Fluorine	F	19.00	9	Selenium	Se	78.96	34
Francium	Fr	223	87	Silicon	Si	28.09	14
Gadolinium	Gd	157.25	64	Silver	Ag	107.868	47
Gallium	Ga	69.72	31	Sodium	Na	22.990	11
Germanium	Ge	72.59	32	Strontium	Sr	87.62	38
Gold	Au	197.0	79	Sulfur	S	32.06	16
Hafnium	Hf	178.49	72	Tantalum	Ta	180.95	73
Helium	He	4.003	2	Technetium	Tc	97	43
Holmium	Ho	164.93	67	Tellurium	Te	127.60	52
Hydrogen	H	1.008	1	Terbium	Tb	158.93	65
Indium	In	114.82	49	Thallium	Tl	204.37	81
Iodine	I	126.90	53	Thorium	Th	232.04	90
Iridium	Ir	192.2	77	Thulium	Tm	168.93	69
Iron	Fe	55.85	26	Tin	Sn	118.69	50
Krypton	Kr	83.80	36	Titanium	Ti	47.90	22
Lanthanum	La	138.91	57	Tungsten (Wolfram)	W	183.85	74
Lawrencium	Lr	260	103	Uranium	U	238.03	92
Lead	Pb	207.2	82	Vanadium	V	50.94	23
Lithium	Li	6.941	3	Xenon	Xe	131.30	54
Lutetium	Lu	174.97	71	Ytterbium	Yb	173.04	70
Magnesium	Mg	24.31	12	Yttrium	Y	88.91	39
Manganese	Mn	54.94	25	Zinc	Zn	65.38	30
Mendelevium	Md	258	101	Zirconium	Zr	91.22	40

APPENDIX

E

Heuristics for Process Equipment Design

The use of heuristics or rules of thumb in process design is a valuable tool not only in selecting appropriate types of equipment to perform a specific function but also in setting reasonable values for many of the variables associated with the process operation. A number of the more useful heuristics developed through past experience of chemical engineers are listed in alphabetic order for some of the more widely used process equipment.[†]

COMPRESSORS, FANS, AND BLOWERS

Fans operate near atmospheric pressure with pressure drops generally less than 15 kPa. Efficiencies vary from 60 to 80 percent.

Pressure drops in blowers normally do not exceed 300 kPa. Efficiencies range from 70 to 85 percent.

Compressors are used for high-pressure operation, from 200 kPa to 400 MPa. Staged compression is usually employed when the compression ratio is greater than 4 to avoid excessive temperatures. In multistage units the compression ratio should be about the same in each stage.

Exit temperatures for compression of diatomic gases should not exceed 200°C.

Efficiencies of reciprocating compressors are about 70 percent at a compression ratio of 1.5, 80 percent at a compression ratio of 2, and 80 to 85 percent at a compression ratio of 3 to 5.

[†]Modified from material presented by J. M. Douglas, *Conceptual Design of Chemical Processes,* McGraw-Hill, New York, 1988, G. D. Ulrich, *A Guide to Chemical Engineering Process Design and Economics,* J. Wiley, New York, 1984, S. M. Walas, *Chemical Process Equipment—Selection and Design,* Butterworth-Heinemann, Newton, MA, 1988, H. Z. Kister, *Distillation Design,* McGraw-Hill, New York, 1992, and D. R. Woods, *Process Design and Engineering Practice,* Prentice-Hall, Upper Saddle River, NJ, 1995.

Efficiencies of large centrifugal compressors, 3 to 50 m^3/s at suction, are 75 to 80 percent. Rotary compressors have efficiencies of 70 percent, except liquid-liner types which have efficiencies around 50 percent.

COOLING TOWERS

Water in contact with air under adiabatic conditions cools to the wet-bulb temperatures. In commercial units saturation of the air to 90 percent is feasible.

Countercurrent induced-draft towers are the most common cooling towers since they are able to cool water to within 1 to 2°C of the wet-bulb temperature. However, the cooling tower size increases rapidly as the difference between the exit water and ambient wet-bulb temperatures approaches this minimum.

Pressure drop in the cooling tower in standard practice does not exceed 0.5 kPa. Water circulation rates generally are 2.5 to 10 $(m^3/h)/m^2$, and the air rates are 6350 to 8800 $(kg/h)/m^2$.

Evaporation losses are 1 percent for every 5°C of cooling range. Windage or draft losses of mechanical draft cooling towers are 0.1 to 0.3 percent. Blowdown of 2.5 to 3.0 percent of the circulation is necessary to prevent excessive salt buildup.

DISTILLATION AND GAS ABSORPTION

Distillation usually is the most economical method of separating liquids, superior to extraction, adsorption, crystallization, or others.

Tower operating pressure is determined most often by the temperature of the available condensing medium, 30 to 48°C if cooled by water or by the maximum allowable reboiler temperature.

Sequencing of columns for separating multicomponent mixtures is based on these heuristics: (1) Perform the easiest separation first, that is, the one least demanding of trays and reflux, and leave the most difficult to the last. (2) When neither relative volatility nor feed concentration vary widely, remove the components one by one as overhead products. (3) When the adjacent ordered components in the feed vary widely in relative volatility, sequence the splits in the order of decreasing volatility. (4) When the concentrations in the feed vary widely but the relative volatilities do not, remove the components in the order of decreasing concentration in the feed.

Economically, the optimum reflux ratio is about 1.2 to 1.25 times the minimum reflux ratio, and the optimum number of trays is close to twice the minimum number of trays. To allow for uncertainty it is advisable to increase the number of trays by an additional 10 percent above the calculated number.

Tray spacings for reasons of accessibility are generally fixed at 0.6 to 0.86 m.

Peak efficiency of trays is at values of the vapor factor $F_s = V\rho^{1/2}$ in the range of 1.2 to 1.5 m/s $(kg/m^3)^{1/2}$. This range of F_s establishes the diameter of the column. Roughly, linear velocities are 0.6 m/s at moderate pressures and 1.8 m/s in vacuum.

Tray efficiencies for separating light hydrocarbons and aqueous solutions are 60 to 90 percent; for gas absorption and stripping the efficiencies are 10 to 20 percent.

Pressure drop per tray is on the order of 0.75 kPa.

The optimum value of the Kremser-Brown absorption factor $A = K(V/L)$ is in the range of 1.25 to 2.0.

Packed towers with a random and structured packing are particularly well suited to columns with diameters less than 1 m and where low pressure drop is desirable. The ratio of column diameter to packing size should be at least 15. For gas rates of 0.25 m^3/s use 0.0254-m packing; for gas rates of 1 m^3/s or more use 0.051-m packing. Packed towers should operate near 70 percent of the flooding rate.

Reflux drums usually are horizontal with a liquid holdup of 300 s when half-full. Reflux pumps are generally oversized by 25 percent.

For towers 1 m in diameter, add 1 m at the top of the tower for vapor disengagement and 2 m at the bottom of the tower for liquid level and reboiler return. Limit height of towers to around 50 m because of wind load and foundation stability.

DRIVERS AND POWER RECOVERY EQUIPMENT

For drivers under 75 kW, electric motors are used almost exclusively.

Efficiency of motors is 85 to 95 percent, for steam turbines it is 42 to 78 percent, and for gas engines and turbines it is 28 to 38 percent.

Steam turbines are competitive above 75 kW. They are speed-controllable. Frequently they are employed as spares in case of power failure.

Gas expanders for power recovery may be justified at capacities of several hundred kilowatts; otherwise any needed pressure reduction in the process is effected with throttling valves.

DRYING OF SOLIDS

Drying times range from a few seconds in spray dryers to 1 h or less in rotary dryers and up to several hours or even several days in tunnel shelf or belt dryers.

Rotary cylindrical dryers operate with superficial air velocities of 1.5 to 3.0 m/s, sometimes up to 10 m/s when the material is coarse. Residence times are 5 to 90 min. Holdup of solid is 7 to 8 percent. An 85 percent free cross-section is generally assumed for design purposes.

Drum dryers for pastes and slurries operate with contact times of 3 to 12 s, with evaporation rates of 15 to 30 $kg/m^2 \cdot h$.

Pneumatic conveying dryers normally take particles 1 to 3 mm in diameter but up to 10 mm when the moisture is mostly on the surface. Air velocities are 10 to 30 m/s. Single-pass residence times are 0.5 to 3.0 s, but with normal recycling the average residence time is brought up to 60 s.

Fluidized-bed dryers work best on particles of a few tenths of a millimeter in diameter, but particles up to 4 mm in diameter have been processed. Gas velocities of twice the minimum fluidization velocity are recommended.

For spray dryers, surface moisture is removed in about 5 s, and most drying is completed in less than 60 s. Parallel flow of air and stock is most common.

EVAPORATORS

Long tube vertical evaporators with either natural or forced circulation are most popular. In forced circulation, the linear velocities in the tubes are 4.5 to 6.0 m/s.

Elevation of boiling point by dissolved solids results in differences of 1.7 to 5.5°C between solution and saturated vapor.

When the boiling point rise is appreciable, the economic number of effects in series with forward feed is 4 to 6. When the boiling point rise is small, minimum cost is obtained with 8 to 10 effects in series.

In backward feed, the more concentrated solution is heated with the highest-temperature steam so that the heating surface is lessened, but the solution must be pumped between stages.

The steam economy of an N-stage system is approximately $0.8N$ kg evaporation/kg of outside steam.

EXTRACTION (LIQUID-LIQUID)

The dispersed phase should be the one that has the higher volumetric rate except in equipment subject to back-mixing, where it should be the one with the smaller volumetric rate. It should be the phase that least wets the material of construction. Since the holdup of the continuous phase usually is greater, that phase should involve the less expensive or less hazardous material.

Mixer-settler arrangements are limited to at most five stages. Mixing is accomplished with rotating impellers or circulating pumps. Settlers are designed on the assumption that droplet sizes are about 150 μm in diameter. In open vessels, residence times of 30 to 60 min or superficial velocities of 0.15 to 0.5 m/min are provided in settlers. Extraction stage efficiencies commonly are taken as 80 percent.

Spray towers generally only provide a single stage for the extraction process.

Packed towers are used when 5 to 10 stages are required. Dispersed phase loadings should not exceed 1 $(m^3/min)/m^2$. Packed towers are not too satisfactory when the interfacial tension is greater than 10 dyne/cm.

Sieve tray towers use tray spacings of 0.15 to 0.6 m. Tray efficiencies are in the range of 20 to 30 percent.

Pulsed packed or sieve tray towers may operate at frequencies of 90 cycles/min and amplitudes of 6 to 25 mm. Interfacial tensions as high as 30 to 40 dyne/cm have no adverse effect.

Reciprocating tray towers operate at 100 to 150 strokes per minute with a plate spacing of 0.025 to 0.15 m. Power requirements of this tower are much lower than those for pulsed towers.

Rotating disk contactors or other rotary agitated towers realize HETS in the range of 0.1 to 0.5 m, providing high extraction efficiencies.

FILTRATION

Processes are classified by their rate of cake buildup in a laboratory vacuum leaf filter: rapid, 0.1 to 10.0 cm/s; medium, 0.1 to 10.0 cm/min; slow, 0.1 to 10.0 cm/h.

Rapid filtering is accomplished with belts, top-feed drums, or pusher-type centrifuges; medium-rate filtering is accomplished with vacuum drums, disks, or peeler-type centrifuges; slow-filtering slurries are handled in pressure filters or sedimenting centrifuges.

Continuous filtration should not be attempted if 0.3-cm cake thickness cannot be formed in less than 5 min.

Clarification with negligible cake buildup is accomplished with cartridges, precoat drums, or sand filters.

Laboratory tests are advisable when the filtering surface is expected to be more than a few square meters, when cake washing is critical, when cake drying may be a problem, or when precoating may be needed.

HEAT EXCHANGERS

In a shell-and-tube exchanger, assume countercurrent flow with the tube side for corrosive, fouling, scaling, and high-pressure fluids and the shell side for viscous and condensing fluids. For other flow geometries, the log-mean temperature difference correction factor F should not be less than 0.85.

For a shell-and-tube exchanger with tubes of 0.019-m OD, 0.0254-m triangular spacing, and 4.87-m length, a shell with a 0.305-m diameter accommodates about 9 m^2 of surface area; a 0.61-m shell diameter, about 37 m^2; and a 0.91-m shell diameter, about 102 m^2.

Minimum temperature approach is 10 to 25°C with ambient coolants and 5°C or less with refrigerants. Cooling water inlet temperature is typically about 30°C, and outlet temperature about 45°C.

Pressure drops are 10 kPa for boiling conditions and 20 to 60 kPa for other services.

Heat-transfer coefficients for estimating purposes in $W/m^2{\cdot}K$ are as follows: water to liquid, 850; liquid to liquid, 285; condensers, 850; liquid to gas, 30; gas to gas, 30; reboiler, 1100. Maximum flux in reboilers is approximately 31,500 W/m^2.

Double-pipe heat exchangers are competitive at duties requiring 10 to 20 m^2.

Stainless steel plate-and-frame exchangers are 25 to 50 percent cheaper than comparable stainless steel shell-and-tube units.

Compact plate and fin exchangers provide about 3 to 4 times more heat-transfer area per unit volume than that obtained with shell-and-tube exchangers.

Air coolers provide areas of about 15 to 20 m^2/m^2 of bare surface. About 5 to 12 kW of fan power is required to remove 1000 kW of energy with a temperature approach of 30°C or more.

Thermal efficiency of fired heaters is 70 to 75 percent. Radiation rate is about 38,000 W/m^2; convection rate is about 12,000 W/m^2.

INSULATION

Up to 350°C, 85 percent magnesia is the preferred insulation. Diatomaceous earth is used up to 870 to 1050°C, and ceramic refractories are used at higher temperatures.

Cryogenic equipment use porous powders down to $-196°C$ and multilayered insulation at lower temperatures.

Optimum thickness varies with temperature: 0.013 m at 90°C, 0.0254 m at 200°C, and 0.031 m at 315°C. Under windy conditions, insulation thickness should be increased by 10 to 20 percent.

MIXING AND AGITATION

Intensities of agitation with impellers in baffled tanks are measured by power input and impeller-tip speeds. Blending requires 0.04 to 0.1 kW/m^3 and a tip speed of less than 0.04 m/s. Agitation of liquid-liquid mixtures requires approximately 1 kW/m^3 and a tip speed of 0.08 to 0.1 m/s.

Typical proportions for a stirred tank relative to the tank diameter D are as follows: liquid level $\simeq D$; turbine impeller diameter $\simeq D/3$; impeller level above bottom $\simeq D/3$, impeller blade width $\simeq D/15$; four vertical baffles with width $\simeq D/10$.

Gas bubbles sparged into the bottom of a tank will result in mild agitation at a superficial velocity of 0.3 m/s and severe agitation at 1.2 m/s.

Power to drive a mixture of a gas and a liquid can be 25 to 50 percent less than the power to drive the liquid alone.

In-line blenders are adequate when contact time of a second or two is sufficient, with power inputs of 0.02 to 0.04 kW/m^3.

PUMPS

Depending upon the type of pump and the flow conditions, a specified normal pump suction head must be maintained to avoid damage to the pump. The normal range is 1 to 6 m of liquid.

Centrifugal pumps are for flows of 10^{-3} to 0.3 m^3/s and a 150-m maximum head; multistage for 1.25×10^{-3} to 0.7 m^3/s and a 1600-m maximum head. Efficiency is 45 percent at 6×10^{-3} m^3/s, 70 percent at 0.03 m^3/s, and 80 percent at 0.6 m^3/s.

Axial pumps are for flows of 1.25×10^{-3} to 6 m^3/s, a 12-m head, with a 65 to 85 percent efficiency.

Rotary pumps are for flows of 6×10^{-5} to 0.3 m^3/s, a 15,000-m head, with a 50 to 80 percent efficiency.

Reciprocating pumps are for flows of 6×10^{-4} to 0.6 m^3/s, a 300,000-m maximum head, with a 70 percent efficiency at 7.5 kW, 85 percent efficiency at 37.5 kW, and 90 percent efficiency at 375 kW.

REACTORS

The rate of reaction must be established in the laboratory, and the residence time or space velocity and product distribution generally must be established in a pilot plant.

Dimensions of catalyst particles normally are 0.1 mm in fluidized beds, 1 mm in slurry beds, and 2 to 5 mm in fixed beds.

The optimum proportions of stirred tank reactors are with liquid level equal to the tank diameter, but at high pressures slimmer proportions are economical.

Power input to a homogeneous reaction stirred tank is 0.1 to 0.3 kW/m^3, but 3 times this amount when heat is to be transferred.

Ideal CSTR (continuous stirred tank reactor) behavior is approached when the mean residence time is 5 to 10 times the length of time needed to achieve homogeneity, which is accomplished with 500 to 2000 revolutions of a properly designed stirrer.

Batch reactions are conducted in stirred tanks for small daily production rates or when the reaction times are long or when some condition such as feed rate or temperature must be programmed.

Relatively slow reactions of liquids and slurries are conducted in continuous stirred tanks. A battery of four or five in series is most economical.

Tubular flow reactors are suited to high production rates at short residence times (seconds or minutes) and when substantial heat transfer is needed. Embedded tubes or shell-and-tube construction then is used.

For conversions under about 95 percent of equilibrium, the performance of a five-stage CSTR battery approaches plug flow.

VESSELS (DRUMS)

Drums are relatively small vessels to provide surge capacity or separation of entrained phases. Liquid drums are horizontal. Gas/liquid separators are vertical.

Optimum length/diameter ratio for drums is 3, but a range of 2.5 to 5 is common.

Holdup time is 5 min half-full for distillation reflux drums and 5 to 10 min for a product feeding another tower. In drums feeding a furnace, 30 min half-full is allowed.

Knockout drums located ahead of compressors to remove liquid droplets should hold no less than 10 times the liquid volume passing through per minute.

Liquid/liquid separators are designed for settling velocity of 0.05 to 0.07 m/min.

Gas velocity in gas/liquid separators is given by $V = k(\rho_L/\rho_V - 1)^{1/2}$ m/s, with $k = 0.1$ with a mesh deentrainer and $k = 0.03$ without a mesh deentrainer. (Entrainment removal of 99 percent is achieved with mesh pads of 0.1 to 0.3-m thickness.) For vertical pads, the value of k is reduced by a factor of $\frac{2}{3}$.

VESSELS (PRESSURE)

Design temperature is set 30°C above the operating temperature between −30 and 350°C; higher safety margins are used outside the given temperature range.

The design pressure is 10 percent or 70 to 175 kPa over the maximum operating pressure, whichever is greater. The maximum operating pressure is taken as 175 kPa above the design pressure.

For vacuum operation, design pressures are 200 kPa outside and full vacuum inside.

Corrosion allowance is 9 mm for known corrosive conditions, 3.8 mm for noncorrosive streams, and 1.5 mm for steam drums and air receivers.

Allowable working stresses are one-fourth of the ultimate strength of the material. Maximum allowable stress depends strongly on the operating temperature.

VESSELS (STORAGE TANKS)

For less than 4 m^3, use vertical tanks on legs; between 4 and 40 m^3, use horizontal tanks on concrete supports; beyond 40 m^3, use vertical tanks on concrete foundations.

Storage capacity of 30 days often is specified for raw materials and products, but it depends on connecting transportation equipment schedules. Capacities of storage tanks should be at least 1.5 times the capacity of this transportation equipment schedule.

Liquids subject to evaporation losses may be stored in tanks with floating or expansion roofs for conservation.

APPENDIX

Software Useful for Design

Software availability guide[†‡]

Design use	Designation	Provider[§]
Flowsheet creation and optimization		
Overall		
	NETOPT	Simsci
	SuperTarget	Linnhoff
	SYNPHONY	Epcon
Heat exchange		
	Hyprotech.HX-Net	Hyprotech
Heat integration		
	Aspen Pinch	Aspen Tech
	HEXTRAN	Simsci
Separation trains		
	Aspen Split	Aspen Tech
	Hyprotech.DISTIL	Hyprotech
Process simulators		
Overall		
Steady-state		
	Aspen Plus	AspenTech
	BioPro/SuprePro	Intelligen
	CHEMCAD	Chemstations
	Hyprotech.Process (Hysys)	Hyprotech
	PRO/II	Simsci
	PROSIM	BR&E
	ProSimPlus	Prosim
Dynamic		
	Aspen Dynamics	Aspen Tech
	gPROMS	BR&E
	Hyprotech.Plant (Hysys)	Hyprotech

Continued

Design use		Designation	Provider[§]
	Specific equipment simulation		
		Adsorption	
		ADSIM	Aspen Tech
		Heat exchangers	
		AeroTran	Aspen Tech
		CC-THERM	Chemstations
		Hetran	Aspen Tech
		Reactors	
		REACT	ChemEng
		Chemical Kinetics Simulator	Almaden
Process economic evaluation			
		Aspen IPE	Aspen Tech
		BioPro/SuprePro Designer	Intelligen
		CHEMCAD	Chemstations
		CostPlus	Epcon
		Hyprotech.Economix	Hyprotech
Process optimization			
		Aspen Optimizer	Aspen Tech
Material-transfer and piping			
		CHEMPRO	Epcon
		Hyprotech.Pipesys	Hyprotech
		Hyprotech.Pipesim	Hyprotech
		INPLANT	Simsci
		SINET	Epcon

[†]This is a listing compiled for software available in the beginning of 2002 and reflects only *some* of the software available at this time.

[‡]AIChE maintains an extensive online listing of available software at:
http://www.cepmagazine.org/features/software/

[§]Contact information for software developer provided below. Bold type indicates name of software developer for use in the software availability table.

Company	Telephone number	Internet address
Almaden Research Center	408-927-2420	www.almaden.ibm.com/st/msim/
Aspen Technologies	617-949-1000	www.aspentech.com
AEA Technology Engineering Services, Inc.	888-827-2356	www.software.aeat.com/cfx/
Bryan **R**esearch **&** **E**ngineering, Inc.	409-776-5220	www.bre.com
ChemEng Software and Services	44 (0)1308 862 880	
Chemstations Inc.	800-243-6223	www.chemstations.net
Epcon International	800-367-3585	www.epcon.com
Hyprotech	403-520 6000	www.hyprotech.com
Intelligen, Inc.	908-654-0088	www.intelligen.com
Linnhoff March	44 (0)1606 815100	www.linnhoffmarch.com
ProSim	33 5 62 88 24 30	www.prosim.net
Process **S**ystems **E**nterprises, Ltd	44 (0) 20 8563 0888	http://www.psenterprise.com
Simsci	714-579-0412	www.simsci.com

REFERENCE INDEX

SUBJECT INDEX

A

B

C

F

G

H

I

J

K

L

M

N

O

R